HEAVY DUTY EQUIPMENT TECHNOLOGY

A SYSTEMS APPROACH

Scott A. Heard

PEARSON

Boston Columbus Indianapolis New York San Francisco
Amsterdam Cape Town Dubai London Madrid Milan Munich Paris Montréal Toronto
Delhi Mexico City São Paulo Sydney Hong Kong Seoul Singapore Taipei Tokyo

Editor-in-Chief: Andrew Gilfillan
Product Manager: Anthony Webster
Program Manager: Holly Shufeldt
Project Manager: Rex Davidson
Editorial Assistant: Nancy Kesterson
Team Lead Project Manager: Bryan Pirrmann
Team Lead Program Manager: Laura Weaver
Director of Marketing: David Gesell
Senior Product Marketing Manager: Darcy Betts
Field Marketing Manager: Thomas Hayward
Procurement Specialist: Deidra M. Skahill
Creative Director: Andrea Nix

Art Director: Diane Y. Ernsberger
Cover Designer: Cenveo
Cover Image: (from top left to bottom right) Aomnet7/Shutterstock, Mr.Teerasak Khemngern/Shutterstock, Alexey Stiop/Shutterstock, Maria Jeffs/Shutterstock, Budimir Jevtic/Shutterstock
Full-Service Project Management: Lindsay Bethony, Lumina Datamatics, Inc.
Composition: Lumina Datamatics, Inc.
Printer/Binder: R.R. Donnelley&Sons
Cover Printer: Phoenix Color
Text Font: Helvetica Neue LT Std 55 Roman 9/14

Unless otherwise indicated herein, any third-party trademarks that may appear in this work are the property of their respective owners and any references to third-party trademarks, logos or other trade dress are for demonstrative or descriptive purposes only. Such references are not intended to imply any sponsorship, endorsement, authorization, or promotion of Pearson's products by the owners of such marks, or any relationship between the owner and Pearson Education, Inc. or its affiliates, authors, licensees or distributors.

Copyright © 2016 by Pearson Education, Inc. or its affiliates. All Rights Reserved. Printed in the United States of America. This publication is protected by copyright, and permission should be obtained from the publisher prior to any prohibited reproduction, storage in a retrieval system, or transmission in any form or by any means, electronic, mechanical, photocopying, recording, or otherwise. For information regarding permissions, request forms and the appropriate contacts within the Pearson Education Global Rights & Permissions department, please visit www.pearsoned.com/permissions/.

Library of Congress Cataloging-in-Publication Data

Heard, Scott A.
  Heavy duty equipment technology : a systems approach / Scott A. Heard. -- 1st edition.
    pages cm
  ISBN 978-0-13-237362-3 -- ISBN 0-13-237362-9
   1. Hydraulic machinery--Textbooks. 2. Hydraulic machinery--Maintenance and repair--Textbooks. 3. Machinery, Kinematic of--Textbooks. 4. Hydraulic engineering--Textbooks. I. Title.
  TJ840.H398 2015
  621.8'042--dc23

2015021207

10 9 8 7 6 5 4 3 2 1

ISBN 10: 0-13-237362-9
ISBN 13: 978-0-13-237362-3

Brief Contents

chapter 1	Heavy Duty Equipment Repair and Safety	1
chapter 2	Fasteners, Seals, Gaskets, Bearings, Personal Tools, Shop Equipment, and Field Service Repair	30
chapter 3	Machine Servicing	71
chapter 4	Hydraulic Systems—The Basics	101
chapter 5	Direct Current Electrical Systems	137
chapter 6	Prime Movers	175
chapter 7	Diesel Engine Starting Systems	214
chapter 8	Charging Systems	239
chapter 9	Electronics and Multiplexing Systems	259
chapter 10	Clutches, Torque Converters, Torque Dividers, and Retarders	301
chapter 11	Gearing Fundamentals and Manual Transmissions	325
chapter 12	Powershift Transmissions	348
chapter 13	Electric Drive Systems	389
chapter 14	Drivelines, Transfer Cases, Auxiliary Drives	412
chapter 15	Drive Axles, Power Dividers, Final Drives	429
chapter 16	Advanced Hydraulic Systems	459
chapter 17	Hydrostatic Drive Systems	505
chapter 18	Heavy-Duty Hydraulic Frictional Brake Systems	536
chapter 19	Wheeled Machine Steering Systems	559
chapter 20	Track-Type Machine Steering Systems	585
chapter 21	Machine Frames and Suspension Systems	604
chapter 22	Wheels, Tires, Hubs, and Pneumatic Systems	623
chapter 23	Undercarriage Systems	645
chapter 24	Operator's Station	664
	Glossary	693
	Index	697

Contents

chapter 1   **Heavy Duty Equipment Repair and Safety**   1
Introduction   2
Material Safety Data Sheet   5
Personal Protective Equipment   6
What Is Heavy Equipment?   8
What Keeps Changing?   18
How Much Does It Cost to Purchase?   18
How Much Does It Cost to Run?   19
What Is DownTime?   19
How Long Should It Last?   20
What Happens Then?   21
What Is an HDET?   23
What Does an HDET Do?   23
What Kind of Person Would Make an Excellent HDET?   24
Where Some Personality Traits Might Lead You To   25
What Does an Employer Look for in an HDET?   25
Where Do HDET's Work?   26
What Are the Benefits of Being an HDET?   26
What About After?   26
Taking Care of Our Environment   27
Summary   29

chapter 2   **Fasteners, Seals, Gaskets, Bearings, Personal Tools, Shop Equipment, and Field Service Repair**   30
Introduction   31
Need to Know   31
Friction Type Bearings   41
Summary   69

chapter 3   **Machine Servicing**   71
Introduction   72
Need to Know   73
Fluid Storage   89
Summary   99

chapter 4   **Hydraulic Systems—The Basics**   101
Introduction   102
Why Use Hydraulic Systems?   103
Need to Know   106
Force, Pressure, and Area   108
Force versus Pressure   109

Flow   110
Basic Hydraulic System Components   112
Check Valves   125
Orifices   125
Accumulator   126
Hydraulic Schematic Symbols   127
Open Center Hydraulic System   128
Hydraulic System Maintenance   130
Hydraulic System Troubleshooting   130
The Real Deal   131
Hydraulic System Repairs   132
Nice to Know   133
Summary   135

chapter 5 **Direct Current Electrical Systems   137**
Introduction   138
Introduction to Basic Direct Current Electrical Systems   139
Need to Know   140
What Is an Electron?   140
Ohm's Law   145
Watt's Law   145
Magnetism and Electromagnetism   145
DC Electrical Circuits   146
Circuit Types   147
Practical Uses for Ohm's Law   151
Wiring Schematics and Symbols   151
Basic Circuit Components   153
Circuit Protection   157
Electrical Testing Equipment   164
Summary   173

chapter 6 **Prime Movers   175**
Introduction   176
What Is a Prime Mover?   176
Need to Know   177
The Four-Stroke Cycle   180
Diesel Engine Accessory Systems   191
Getting the Most Life Out of a Diesel Engine   204
Nice to Know   205
Reconditioning   208
Summary   212

chapter 7 **Diesel Engine Starting Systems   214**
Introduction   215
Need to Know   217
Electric Starter Motor Assembly   217
Air Starting Systems   229

Hydraulic Starting Systems   231
Starting Control Circuit Interlock Systems   232
Starting Circuit Relays   232
Starting System Troubleshooting   233
Summary   238

chapter 8   **Charging Systems   239**
Introduction   240
What the Charging System Does   240
Alternator Operating Principles   241
Current Rectification   246
Charging System Preventive Maintenance Practices   254
Summary   257

chapter 9   **Electronics and Multiplexing Systems   259**
Introduction   260
Need to Know   260
Electrical versus Electronics   261
Semiconductors   262
Diode   262
Transistors   266
Resistors   269
Capactitors   271
Digital versus Analog   273
Signal Conditioning   273
Integrated Circuits   274
Central Processing Units   274
Logic Gates   275
Computer Memory   275
ECMs   277
What Makes Electronic Components Fail?   278
The Electronic Information Cycle   279
Electronic System Wiring   280
ECM Inputs and Outputs   280
ECM Outputs   286
CAB Displays   288
Software   289
Multiplexing   289
Controller Area Network Bus System   289
Electronic Systems for Diagnostics   293
Fault Codes   293
Electronic Service Tools   295
Live Data   295
Snapshots   295
Wireless Machine Communication   295
Calibrating   296
Electronic Service Information   296

ECM Replacement 296
Wiring Diagnostics–Schematics, Meter Usage 297
Nice to Know 297
Summary 299

chapter 10 Clutches, Torque Converters, Torque Dividers, and Retarders 301

Introduction 302
Clutches, Torque Converters, and Hydraulic Retarders 302
Torque Converters 311
Torque Dividers 319
Hydraulic Retarders 321
Summary 323

chapter 11 Gearing Fundamentals and Manual Transmissions 325

Introduction 326
Gear History 326
Manual Transmissions 335
Summary 346

chapter 12 Powershift Transmissions 348

Introduction 349
What Is a Powershift Transmission? 349
Need to Know 352
Hydraulically Actuated Clutches 353
Powershift Transmission Hydraulic System 354
Countershaft Powershift Transmissions 355
Full Powershift Countershaft Transmission 367
Gear Selection 368
Hydraulic System Operation–Power Shift Transmission 369
Planetary Powershift Transmissions 377
Powershift Transmission Maintenance 382
Powershift Transmission Troubleshooting 383
Transmission Removal 385
Powershift Transmission Reconditioning 386
Summary 387

chapter 13 Electric Drive Systems 389

Introduction 390
Electric Drive History 392
Generators 398
Conductors 401
Electric Drive Cooling System 401
Electronic Controls 403
Electric Motors 405
High-Voltage System Grounding and Bonding 407
Braking Resistor 407
Maintenance 408

Insulation Testing and Maintenance 408
Operator Complaints 409
Electric Drive Diagnostics and Repairs 409
Summary 410

chapter 14 **Drivelines, Transfer Cases, Auxiliary Drives 412**
Introduction 413
Machine Driveline 413
Transfer Cases 422
Auxiliary Drives 423
Operator Complaints Related to Drivelines 424
Driveline Maintenance 424
Drive Shaft/U-Joint Repair 425
Transfer Case/Auxiliary Drive Repair 426
Summary 428

chapter 15 **Drive Axles, Power Dividers, Final Drives 429**
Introduction 430
Environmental Concerns 430
Need to Know 430
Drive Axle Types 430
Differential 441
Power Dividers 453
Final Drives 455
Summary 458

chapter 16 **Advanced Hydraulic Systems 459**
Introduction 460
Need to Know 461
Variable Displacement Pumps 464
Load-Sensing Pressure-Compensated Systems 470
Pilot Controls 475
Main Control Valves 476
Other Hydraulic Components 482
Excavator Hydraulic Systems 485
Excavator Main Control Valves 490
Electrohydraulic Systems 494
Advanced Hydraulic System Diagnostics 498
Hydraulic System Testing 498
Hydraulic System Problems 501
Component Replacement Procedures (Safe Start-Up) 501
Causes of Component Failures 502
Summary 503

chapter 17 **Hydrostatic Drive Systems 505**
Safety First 506
Introduction 506
Hydrostatic Drive Arrangements 507

Hydrostatic System Overview   510
Dual Path Hydrostatic System   510
Hydrostatic Circuit Operation   512
Hydrostatic System Components   516
Hydrostatic System Valves   526
Other Hydrostatic System Components   527
Operator Controls   530
Hydrostatic Electronic Control Systems   530
Diagnostics   531
Hydrostatic System Adjustments and Repairs   533
Summary   534

chapter 18   **Heavy-Duty Hydraulic Frictional Brake Systems**   536
Introduction   537
Need to Know   538
Frictional Brake Hydraulic Systems   543
Friction-Type Foundation Brakes   544
Hydraulic Brake Application Systems   548
Nice to Know   557
Summary   557

chapter 19   **Wheeled Machine Steering Systems**   559
Introduction   560
Need to Know   560
Mechanical Steering Systems   561
Hydraulic Steering System Operation   573
Summary   583

chapter 20   **Track-Type Machine Steering Systems**   585
Introduction   586
Need to Know   586
Track Machine Mechanical Steering Systems   587
Summary   603

chapter 21   **Machine Frames and Suspension Systems**   604
Introduction   605
Need to Know   605
Track Machine Suspension Systems   617
Summary   621

chapter 22   **Wheels, Tires, Hubs, and Pneumatic Systems**   623
Introduction   624
Need to Know   624
Tires   627
Wheel Hubs   633
Pneumatic Systems   637
Summary   644

chapter 23   **Undercarriage Systems   645**
Introduction   646
Need to Know   646
Undercarriage Configurations   647
Track Frame   648
Recoil Spring   651
Sprocket   651
Chain Assembly   653
Master Link   654
Track Pads   656
Undercarriage Wear   658
Measuring Undercarriage   659
Rubber Tracks   661
Diagnosing Undercarriage Problems   661
Summary   663

chapter 24   **Operator's Station   664**
Introduction   665
Need to Know   665
Machine Controls   665
Operator Displays/Machine Gauges   666
Operator Protection—ROPS/FOPS/Canopies   672
HVAC Systems   675
Servicing and Repairing HVAC Systems   686
Summary   691

Glossary   693

Index   697

Dedication

A very special thank you to my wife, Penny, and sons, Max and Vaughan, for their support and understanding throughout the many hours I spent working on this project over the last six years. Seeing this book to completion has been a huge challenge and I have sacrificed a lot of quality family time during this period. For your patience and understanding, I will be forever grateful.

Love, Scott

Preface

THE STORY BEHIND THE BOOK

The purpose of this text is to provide students enrolled in college-level Heavy Duty Equipment Technician programs with a resource that is both easy-to-follow yet detailed enough to provide the knowledge to master the operating fundamentals of most heavy equipment machine systems.

By sharing my experience and knowledge with students I explain in simple terms how the various systems of heavy equipment machines work and how to repair and maintain them.

I have had an interest in heavy equipment from a young age when I loved watching machines work on a variety of jobs such as digging basements for houses, building and maintaining roads, and loading gravel trucks. The sounds and smells of diesel-powered machines have fascinated me for many years. This fascination turned into a curiosity of wanting to know how they worked.

I decided to turn this interest in heavy equipment into a career and enrolled in the Heavy Equipment Technician program at Sir Sandford Fleming College in Lindsay, Ontario. After graduation I went on to complete an apprenticeship and became a Red Seal licensed Heavy Duty Equipment Technician (HDET) over 20 years ago. While attending college I discovered there was a need for a quality up-to-date textbook that would cover most of the subject areas taught for the program. When I attended college there was only one textbook available that only covered some of the systems we were learning about; in addition, it was poorly written and contained outdated information.

Since then, an additional textbook has entered the market that covers some of the areas that most college-level heavy equipment programs include; however, there are some important topics that it overlooks. For example, engines, basic electrical systems, and system diagnostics are not included.

My experience as a student gave me a perspective of what it was like to rely on less-than-ideal textbooks as references.

A few years ago I accepted the position of professor at Centennial College in Toronto. While there, I taught over 30 different courses in all subject areas of heavy equipment and truck and coach technician programs. I had to use over five different textbooks to cover these subject areas. This experience made the need for a single comprehensive textbook apparent.

I feel this book will provide a quality textbook for most heavy equipment subject areas. The intent of this book is to help the reader identify the components that make up each system, understand the operating fundamentals of systems, and understand how to maintain, diagnose, and repair common problems that can occur with these systems.

This book should be a valuable reference for instructors in Heavy Equipment Technician programs to support their course curriculum.

STRENGTHS OF THIS VOLUME

There are existing textbooks that try to cover some of the systems found on heavy equipment machines but most fall short in several areas. One of the books uses the input of several authors, which leads to differences in writing styles between chapters. This book is written by one author, which makes a more consistent chapter-to-chapter read.

Good illustrations are also key to explaining how systems work. This book features several hundred illustrations and photographs to help the reader visualize what is being explained in the text.

There is a chapter dedicated to diesel engines and related features unique to diesel engines used for heavy equipment. There is also a chapter covering operator stations. Operator stations have become more complex over the last few years and this chapter presents the reader with information on how HVAC and gauge displays work.

Because I have access to new machines and their service information, the book covers the most current technology concerning heavy equipment machines. I have also recently attended several equipment manufacturers' new model training courses to stay current with the latest machine technology and have incorporated this up-to-date technology in the book.

In addition I have included many personal experiences that I have encountered during my days as an HDET, as they relate to the subject matter being discussed.

ORGANIZATION OF THE BOOK

This book is organized in a logical manner starting with the first chapter that explains what an HDET does and how to work safely around heavy equipment. Safety is a big part of this book and each chapter starts with a discussion about the safety concerns related to its content.

Most chapters are arranged to first give the reader some theory-based knowledge and then moves into the specifics of how the system works. The chapter finishes with how problems with the system are identified (troubleshooting) and how they are repaired and serviced.

Shop activities are included in most chapters as a suggested way for students to apply what they have learned throughout the chapter.

Progress checks and end-of-chapter questions reinforce the main points of each chapter.

Basic hydraulic and electrical chapters are at the beginning of the book because the principles they cover apply to many other systems.

A total of nine chapters are related to drivetrain systems with electronic controls included where applicable.

I hope you find reading this book as fulfilling as I did writing it.

Scott Heard

ACKNOWLEDGMENTS

Deven Wilson- Deere & Company
David Bussan- Deere & Company
Paul Hoggarth- Nortrax
Susan Hudson- Caterpillar Inc.
Preeti Chockalingham- Caterpillar Inc.
Carmen Moore
Jason Hiles
Tim Allan- Centennial College
Rhonda Laursen- Toromont Caterpillar
Adam Shebib- Toromont Caterpillar
James Halderman- Pearson Author
Maxim Heard- MH Design
TDI Tech Development

Brandon Richards- KTI
Camilla Gustafson- Volvo Construction Equipment
Dan Wallace- DD Consulting

TECHNICAL AND CONTENT REVIEWERS

John Bright- Hibbing Community College
Craig Defendorf- University of Alaska, Anchorage
James Martin- Tarrant County College, South Campus
Shannon McCarty- Ashland Community and Technical College
Larry Thompson- Texas State Technical College
John Yinger- Ozarks Technical Community College

SUPPLEMENTS

ONLINE INSTRUCTOR RESOURCES

- Image Bank. This is a download of all the images in the text.
- Instructor's Resource Manual.
- Power Points.
- Download Instructor Resources from the Instructor Resource Center

To access supplementary materials online, instructors need to request an instructor access code. Go to www.pearsonhighered.com/irc to register for an instructor access code. Within 48 hours of registering, you will receive a confirming e-mail including an instructor access code. Once you have received your code, locate your text in the online catalog and click on the Instructor Resources button on the left side of the catalog product page. Select a supplement, and a login page will appear. Once you have logged in, you can access instructor material for all Pearson textbooks. If you have any difficulties accessing the site or downloading a supplement, please contact Customer Service at http://247pearsoned.custhelp.com/.

chapter 1
HEAVY DUTY EQUIPMENT REPAIR AND SAFETY

LEARNING OBJECTIVES

After reading this chapter, the student should be able to:

1. Describe what a heavy equipment machine is.
2. Identify the different types of industries where heavy equipment may be found working in.
3. Describe the main types of propulsion systems used for heavy equipment.
4. Describe the characteristics of a good heavy equipment technician.
5. Explain how to work safely.
6. Name the major manufacturers of heavy equipment.
7. Identify different personal protective equipment.
8. Describe what a heavy-duty equipment technician would be expected to do on a daily basis.
9. Explain some of the factors that a machine model could refer to.

KEY TERMS

Car body 11
Drivetrain 10
Ejector 11
Hydrostatic 12
Lock out 5
Packer 13
Payload 9
Powershift 10
Scraper 4
Stick 11
Tag out 5

INTRODUCTION

SAFETY FIRST Safe work practices cannot be emphasized enough when working on or near heavy equipment.

Being a heavy-duty equipment technician (HDET) can be a dangerous occupation. There will be many tasks that need to be performed that will have differing amounts of risk to your health and well-being. As long as the HDET recognizes the risks being taken, takes steps to prevent a mishap, and always uses the appropriate safety equipment, the technician will be able to go home after work in the same physical condition as when he or she started.

Working safely as a heavy equipment technician means a "safe attitude" must be developed. This could come naturally for some people, but for other people who may be accident prone, they will need to try harder to recognize the signs that lead to an unsafe situation.

This chapter will discuss the safe attitude and how to work productively while being safe. Each of the following chapters will feature specific safety discussions for the chapter's subject.

Later in this chapter, different types of heavy equipment will be discussed as well as what an HDET does and what makes a good HDET.

BECOMING A SAFE HDET Heavy equipment machines are designed to do the work of many human workers and, as a result, are much stronger than human skin and bones. Even the smallest skid steer that is capable of lifting just 700 lb can break, cut, or crush any part of the human body if it gets a chance. You *must* recognize the forces a machine is designed to create to respect it and prevent yourself from being injured or killed. ● **SEE FIGURE 1–1** to get an idea of the damage a heavy equipment machine can do.

There are thousands of HDETs who work on heavy equipment every day and won't get injured. There are also many who

FIGURE 1–1 An example of the damage a heavy equipment machine can do.

>  **WEB ACTIVITY**
>
> Caterpillar has an excellent website dedicated to safety-related information when operating, servicing, repairing, or transporting their machines. Visit http://safety.cat.com to find a classic video called "Shake Hands with Danger." Seeing this video should give you a strong visual image that will stay with for the rest of your career. There are many other videos and safety information on this site that if viewed will be time well spent.

will get minor cuts, scrapes, and bruises that will heal completely. Unfortunately, there will be a few who sustain serious injuries on the job, and in rare cases, some will die.

Two simple practices will ensure that you will always make it home at the end of the day in the same physical condition as when you started the day: minimizing risks and proper use of PPE (personal protective equipment).

MINIMIZING RISKS. There are many potential dangers you may encounter when working on a piece of heavy equipment, and before you can minimize the risk to you from it, you must know what it is. When you are first starting out in the trade, you should be made aware of the many dangers involved with working on heavy equipment. You won't know all the potential hazards involved with repairing or servicing a machine. It is the responsibility of your employer to make sure that you have a basic knowledge of how to work safely and to recognize what a potential hazard is.

Many potential hazards are visually obvious and can be dealt with easily with a little common sense. For example, if a cover needs to be removed to access a part of the machine that needs attention, you should be able to judge whether the cover can be lifted by yourself or if you need the assistance of someone else or maybe a crane.

There are many potential hazards that are not as obvious as the previous example, and this is where you must develop a sixth sense to be able to predict what may happen when you do something. For example, a common task that an HDET will perform is to repair a hydraulic system leak. Taking a hose off may seem like a fairly low-risk task. However, there could be serious consequences that result from removing a hose that has a lot of hydraulic pressure trapped in it. It is sometimes impossible to know exactly what potential hazard may be involved in the task you are performing, but if your sixth sense tells you something isn't right, you should rethink what you are about to do.

If you are unsure about something that you are about to do, then simply *stop*. This is when you need to access the machine's service information manual or ask someone who has performed the same task to find out about the potential hazards in what you are about to do. Based on the information you receive, you must take steps to minimize the risk of injury before proceeding.

Heavy equipment can have many sources of potential energy that can cause harm to you if allowed to. Potential energy can be in many forms and, if released unintentionally, can be deadly. Some of these sources will be very apparent and others won't. A few examples are:

1. any part of the machine that is movable but is not resting on the ground. All equipment should be lowered to the ground before parking the machine (blades, buckets, etc.). If the equipment is to be worked on while raised in the air, it must be properly suspended. Most service manuals will illustrate how to do this and may give weights and recommend tooling to do this.
2. electrical storage devices such as batteries and capacitors. You will see lots of warnings about both these devices, so be aware of the potential hazards.
3. tires on heavy equipment machines will contain enormous potential energy and need to be respected.
4. hydraulic accumulators are similar to a battery or capacitor and are a device that can store a lot of hydraulic energy.
5. any part of a hydraulic system can contain enormous pressure that could be released in a deadly manner if an HDET unknowingly opens the system.
6. springs that are under tension can be hazardous.
7. any part of a machine that has been deformed from its original shape. If a part of a machine has been damaged, it could be under a great deal of pressure that is waiting to be released.
8. a running machine. It has many potential hazards such as the pinch points, moving belts, fans and shafts, and hot surfaces.

Any reckless or careless action could leave you seriously injured. There may be pressure from someone (a supervisor, customer, coworker, etc.) to make you hurry up to finish the job as soon as possible and get the machine back to work quickly. This is when you have to weigh the benefits of working safely, getting the machine running, being productive, and performing the task correctly. Working safely should always take priority in this equation. If you cut corners to get the machine back to work quickly, you may end up injured to the degree that you miss work or worse then you won't be doing anything productive.

SAFETY TIP

A former coworker once had a finger broken while changing a cutting edge (the blade part of the machine) on a **scraper** that had partially fallen off.

The machine had continued to work with a loose section of cutting edge and wedged some material under one end of it. This cutting edge section was approximately 3 ft long, 10 in. wide, and ¾ of an inch thick. When the remaining bolts were undone, a huge release of energy made the cutting edge jump and the HDET had his hand in the way. The result was a trip to the hospital and a hand cast that kept the technician from working for a few weeks.

It is inevitable that if you are going to become a heavy equipment technician, you will get hurt because of the interaction with a piece of heavy equipment. You will be climbing on it, up it, and under it and sometimes with greasy hands on slippery surfaces, with a handful of tools, and with a number of distractions. Minor injuries (cuts, scrapes, and bruises) are part of the job but can be minimized with the proper use of PPE.

It is your actions that will matter most. Ultimately, it is the work practices you choose to use that will keep you safe and will make sure that you get home at the end of the workday in the same condition that you left in.

This doesn't mean you should be aware of only what *you* are doing.

If you are working in a shop, you should also be aware of what others working around you are doing. You could use all the best practices and be the safest HDET possible, but if someone working close by has an accident that causes harm to you, your own safe practices have meant nothing to you.

When working on machines, you should notice many warning labels that are placed on the machine. You will also see many caution and warning sections in a machine's operation and service information manuals. You must understand what these labels mean and take precautions based on what the label is trying to tell you. ● SEE FIGURE 1–2 for an example of warning labels.

You can minimize risk of injury by using safe work practices. These include:

1. using the proper steps, handrails, and catwalks when climbing, descending, or entering machines and always using three-point contact (two hands and one foot or two feet and one hand).

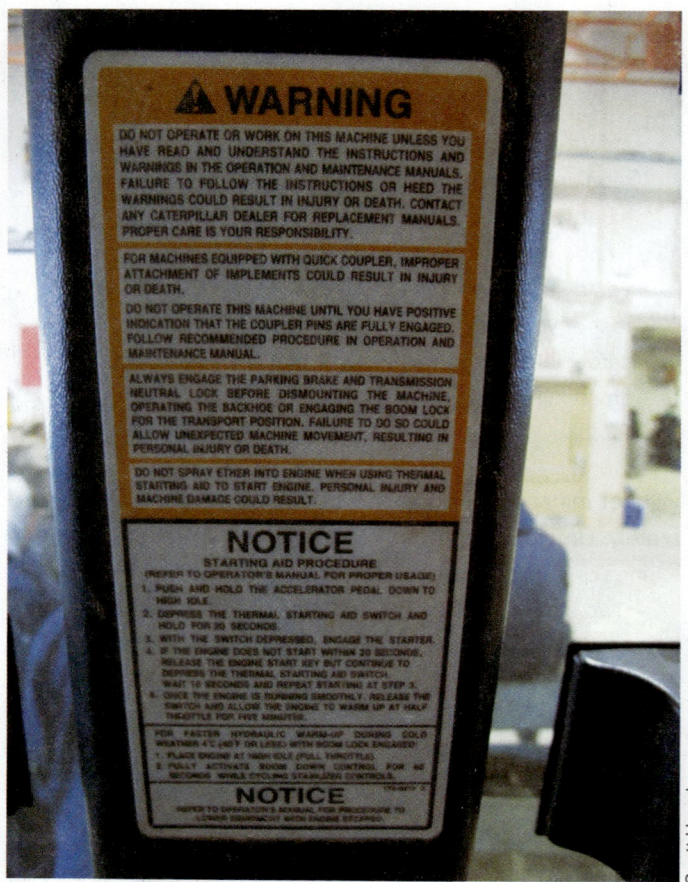

FIGURE 1–2 An example of warning labels you may find on equipment.

2. staying clear of exposed moving parts.
3. avoiding exposure to high-pressure fluids.
4. being aware of any noxious engine exhaust that could accumulate in your work area.
5. preventing fires, having a fire extinguisher close by, and knowing how to use it.
6. preventing battery explosion.
7. handling chemicals safely, and if uncertain about safe handling or use of any chemical products, contacting the chemical supplier for a material safety data sheet (MSDS). The MSDS describes physical and health hazards, safe use procedures, and emergency response techniques for chemical substances. Follow MSDS recommendations to handle chemical products safely.
8. preparing for an emergency by keeping a proper first aid kit close by or at least knowing where the nearest one is.
9. not starting or running a machine from anywhere but the operator's seat.
10. lowering all work equipment and engaging all parking brake devices when leaving a machine.

> **SAFETY TIP**
>
> When I was starting out in the trade and working as an apprentice for a major equipment manufacturer's dealer, I was assigned to help a licensed journeyman remove a transmission from a track loader. This job meant the machine's cab had to be tilted ahead to be able to lift the transmission out. The shop crane was needed and was attached to the top of the cab. I had been given a heads up about working with this particular technician and told that he is easily frustrated and likes to take short cuts to get the job done quickly. When I noticed what he had attached the crane to the cab with, I understood why I needed to be aware of what he was doing. He used two ⅜ in. eye bolts to lift the cab when the recommended lifting device is a ½ in. lifting link. The eye bolts are okay for a straight lift but are not designed for any side pull. As the cab came up, it pivots on two pins so the initial straight pull soon turns into a side pull the higher the cab gets. Luckily, I had cleared out from under the cab by the time the eye bolts failed and the cab came crashing down. This was a close call and a good education of the potential hazard created by improper use of tooling and poor work practices.
>
> The senior technician whom I was working with had a serious elbow injury from a previous incident that was no doubt caused by a lack of safe work practices.
>
> Remember that the actions of others working with you and near you can also be dangerous to yourself.

11. avoiding all work site hazards (overhead wires, underground wires, gas lines, people/equipment nearby) if running a machine.
12. taking no riders on a moving machine unless it was designed for it.
13. being sure that the area is clear around you when running a machine and having a spotter if you aren't sure.
14. keeping your workplace tidy to avoid tripping hazards.
15. making use of proper hoisting, lifting, and blocking tool usage and techniques.
16. understanding how to operate a machine safely before running it and knowing all its safety features before starting it.
17. removing all jewelry before starting work.
18. always sounding the machine's horn before starting the machine's engine and before moving the machine.
19. chocking wheels when servicing or repairing equipment.
20. using steering and loader arm safety brackets when servicing or repairing machine and are working in potential pinch point areas.
21. reading and understanding all safety warnings.
22. before attempting to lift something without a crane or assistance, knowing how much weight you are about to lift and the proper technique to lift safely. Lift with your knees and not your back, and keep your back straight. There are online calculators that you can access to help determine the safe amount of weight that can be lifted depending on how you are lifting it.
23. not allowing anyone under the influence of drugs or alcohol near heavy equipment.
24. wearing high-visibility work clothes.
25. not approaching a working machine until you confirm the operator has made eye contact with you and signals that it's okay to approach.
26. staying clear of the working radius of a machine.
27. using proper fall arrest apparatus if working higher than 8 ft above ground level.
28. not using a machine's bucket as a work platform.
29. if operating a machine, checking all machine safety features are working properly and using them.
30. when working on a machine, using proper **lock out** and **tag out** procedures. All HDETs should be aware of their companies' or the equipment owners' lock out and tag out procedure and follow it to the letter. Lock out and tag out procedures will ensure that a machine does not get started or operated unless the person who tagged the machine gives the approval to do so. Besides a good safety practice, this is a good way to prevent damage to a machine's components if it is started without fluids or filters installed.

MATERIAL SAFETY DATA SHEET

There are many different materials that you may come into contact with when working on a machine and you should be aware of any hazardous ones. This is possible with the MSDS system. Any fluid or other substance that is used either to keep a machine running or to repair a machine should have an MSDS for it. This sheet will provide you with critical information that you will need to take any precautions or use specific PPE to protect yourself from harm.

WEB ACTIVITY

Find the website of a supplier of PPE and look at a variety of different types of glasses, boots, gloves, and so on. Try to imagine what an HDET may need to use and what are some specifications that you should look for when purchasing PPE.

Safety is an attitude. If you think accidents will always happen to someone else, you are in for a rude awakening. If you think there is a good chance that you can get hurt and take steps to prevent that from happening, then it is likely you won't get hurt.

The best safety tool you have is between your ears. You should develop a sixth sense after a while and know when something doesn't seem right. You should be able to predict if what you are about to do will result in an unsafe incident, and if you feel something isn't quite right, you should stop and rethink the situation. This will sometimes go against your desire to get the machine back up and running, but your safety should take priority over the urgency to complete the job quickly. Safe work practices will many times come down to using common sense.

If you are working off-site at a customer's workplace, there are many additional hazards you need to be aware of. Every work site will have its own unique set of safety challenges, and this may mean taking a safety orientation course for that particular work site before you will be allowed to work on the site. If there is heavy equipment traffic, you need to know what the rules of the work site are. Things such as who has the right of way, is there any overhead wires or conveyors that could be a problem, or is there hazardous material on site are examples of specific safety concerns you would want to be aware of.

All these potential safety hazards can be minimized with proper work practices followed and the proper use of PPE. Always refer to the machine's service manual to make yourself clear as to exactly what you are doing, and if you are unsure or don't feel comfortable doing something, *don't do it!*

SAFETY TIP

While working at a limestone quarry, I was called to the rock drill on site that was having problems. It was a cold winter morning and the complaint was the drill function wouldn't work. After arriving and talking to the operator, I started my basic checks with the drill running. Because I wasn't familiar with the machine, I wanted to read the manual a little to try to get an idea what might cause this problem. I had just gone back to the service truck to have a look at the manual only to look up to see the operator with his head under the suspended drill steel looking up into the hollow steel to see if it was plugged. This is a machine that uses massive amounts of air flow and pressure to make the drill bit act like an impact hammer and blow the waste material up and out of the hole, which could be over 100 ft deep. Just as I was about to yell over and tell him to get away, the ice jam that was causing the drill not to work let go. The force blew the operator several feet away toward a 70 ft drop-off. Luckily for him, he didn't go over the edge and get the full force of the air pressure. He did get a very quick lesson knowing what 300 psi and 400 cfm of air flow feels like!

This is an example of someone who should have made himself aware of the hidden hazard waiting to be unleashed before he put himself in a dangerous situation.

PERSONAL PROTECTIVE EQUIPMENT

The term *PPE* is used to describe all the safety equipment used to protect an HDET from injury or harm. You should at some point in your career automatically feel uncomfortable doing certain tasks without using the proper PPE. For example, if you pick up a hammer, the second thing you should pick up are safety glasses (if you aren't already wearing them), and it shouldn't feel right swinging a hammer without safety glasses on. Different companies will have different policies for employee use of PPE, and the minimum standard set may not always be enough protection for an HDET. Don't be afraid to go beyond the minimum requirements for PPE if necessary. Some examples of PPE and when they should be used are the following:

Safety glasses: Your eyes are an exposed and sensitive part of your body. There are many tasks related to working on heavy equipment that can cause both temporary and permanent damage to eyes. Most machines that you work on will be dirty or leaking fluids, so if you are working under a component, you should expect to get something in your eyes if you aren't wearing eye protection. Flying paint chips are particularly dangerous to the eyes, so if there is any activity (hammering,

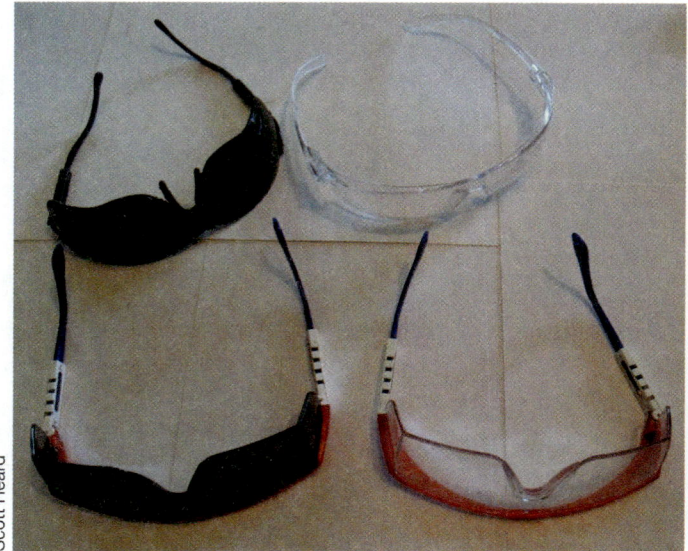

FIGURE 1–3 A variety of safety glasses.

FIGURE 1–4 A pair of work boots doing their job.

prying) that may cause a piece of paint from a machine to be propelled toward you, you should have eye protection on.

If it isn't a company policy to wear eye protection, you should make it your own to ensure that you get the maximum use out of the two eyes you were given. Any safety glasses that meet ANSI Z87.1-2003 and CSA Z94.3-2002 specifications will be sufficient, but if they are comfortable to wear, clean, and unscratched, you will likely be using them more. A good investment is a quality pair of safety glasses and a good decision is to use the right glasses for the job you are doing.
● **SEE FIGURE 1–3** for a variety of safety glasses.

A recent survey found that 60% of workers who suffered eye injuries weren't wearing any eye protection, and of those who did, 40% were wearing the wrong type. There are many different types of safety glasses. Make sure that you use the ones that will give you maximum protection for the job you are doing.

It is ironic that many people don't think twice to wear sunglasses under bright light conditions or just to look good but then hesitate to wear safety glasses when there is an obvious risk of eye injury.

Safety boots: Foot protection is critical when working on heavy equipment and should also be a minimum mandatory piece of safety gear. Full ankle–covering boots, and not shoes, should be worn to fully protect your feet. Good-quality boots are also a good investment because they will last longer and be more comfortable to wear. An easy identification of approved footwear is the Green Patch symbol, which equates to the following regulatory standards: CSA Standard CAN/CSA-Z195-M92 or ANSI Standard Z41-1991.

Good oil-resistant tread on the soles are also important to maintain a solid footing. ● **SEE FIGURE 1–4** for a pair of work boots doing their job.

Head protection: Some workplaces will make it mandatory that HDETs wear full hard hats at all times. Full hard hats can be cumbersome when working on heavy equipment because of the tight spaces you will sometimes have to get into. Other companies will allow HDETs to wear bump caps on the job. This will give protection from minor bumps and prevent minor cuts and scrapes. Bump caps are a good compromise between no protection and a small inconvenience that will prevent injuries and they should be worn as much as possible. Hard hat standards are set out in regulations such as CSA-Z94.1–05 and ANSI Z-89.1–2009. The "type" of hard hat determines the type of blow to the head it will protect the wearer from (top only or top and side) and the "class" determines how much electrical conductivity the hard hat can allow.

Hand protection: This is worn to protect an HDET's hands from being scraped and cut as well as having toxins enter the body through the skin. The use of latex or nitrile gloves has become very popular and would be a good habit to get into. Some equipment dealers have made a policy of mandatory use of hand protection at all times, and this has resulted in a reduction of lost time accidents.

Hearing protection: Some tools used to repair and service heavy equipment such as air tools as well as running machines can exceed safe ear safety thresholds and require the HDET to use hearing protection.

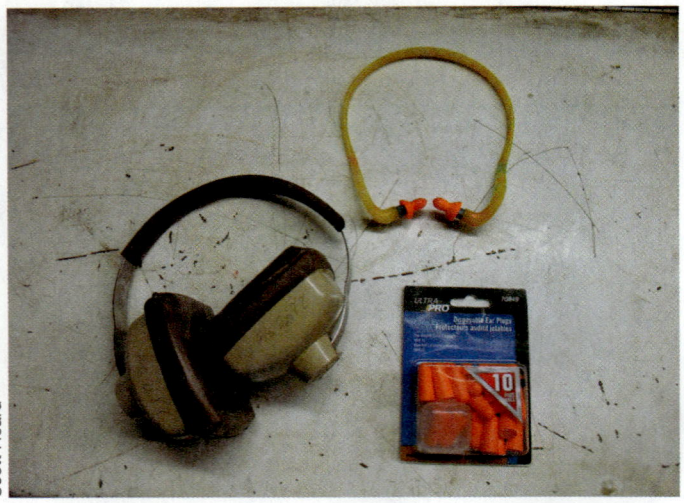

FIGURE 1–5 A variety of hearing protection devices.

Hearing protection can be in the form of ear plugs that are inserted into the ear or muffs that go over the ear or both. Caution must be exercised when using hearing protection because you will have a reduced sense to warn you of possible danger. ● **SEE FIGURE 1–5** for a variety of hearing protection devices.

Respiratory protection: Most heavy equipment will have some kind of dust on it after it has been on the work site for a period of time. When working on a machine, you will be disturbing this dust, and if you suspect that it may be harmful to you or if you are unsure what the dust is, you should be using respiratory protection. There are different levels of protection from simple dust masks to fully sealed respirators. Breathing protection is wise part of PPE.

Back support: There are many components of a machine that are heavy and, therefore if not handled properly, could result in a back injury. Proper lifting techniques should be practiced (lift with your knees and not your back) and hoisting devices should be used when necessary. There are back support apparatus that will help to prevent back injury and should be used if there are excessive or repetitive weights being lifted manually.

Coveralls/safety vest: Coveralls are part of an HDET's work attire and should be considered part of PPE. They will cover most exposed skin and prevent many minor scrapes, burns, and cuts. Many coveralls worn today will incorporate a high-visibility feature such as a reflective cross on the back and two reflective stripes on the front. This will ensure that the HDET is seen when working near a running machine.

Extreme weather protection: If you are required to work in extreme hot or cold conditions, you should be dressed for the conditions and limit your exposure by layering up in the cold and wearing as little as safely possible in the heat. Keeping covered from the sun is also a good idea.

Progress Check

1. This would *not* be a safe practice if a HDET would:
   a. approach a working machine from behind
   b. use safety glasses at all times
   c. wear ankle-covering work boots
   d. wear high-visibility coveralls
2. MSDS stands for:
   a. Material Standards Department of Safety
   b. Mega Safe Digging Society
   c. material safety data sheet
   d. multi-safe data standards
3. If someone is practicing three-point contact, he or she will be:
   a. using a special type of fastener
   b. in contact with a machine with two hands and one foot or two feet and one hand
   c. in contact with a machine with a hand, a foot, and a safety harness
   d. in a machine's seat with a three-point seat belt on
4. Two good ways an HDET will prevent personal injury is to:
   a. use safety glasses and two-point contact
   b. use work boots and issue a work plan
   c. do everything that a senior HDET does and work quickly
   d. minimize risks and use PPE properly
5. This would *not* be a normal part of PPE that an HDET would use:
   a. a cooling vest for extremely hot conditions
   b. green patch work boots
   c. hearing protection for working near running machines
   d. eye protection when grinding

WHAT IS HEAVY EQUIPMENT?

Before we talk about what an HDET is and does, let's talk about the machines that an HDET will be maintaining and repairing.

Today's heavy equipment has evolved from agricultural machines, which evolved from the first self-powered machines developed to replace horsepower (the four-legged variety).

The first mass-produced steam-powered thrashers and tractors with wheels appeared around the first of last century and could do the work of several men and animals. Soon

FIGURE 1–6 One of the first track-type machines.

after these self-powered wheeled machines were created came a track drive system that was invented to allow these heavy machines to work on soft ground without getting stuck.
● **SEE FIGURE 1–6** for an early heavy equipment machine with tracks. This led to developing farmland that was once thought to be too wet to access. As soon as a plow was hooked on to the back of one of these machines, man realized the potential this new-fangled apparatus had and the sky has been the limit as far as what a mobile machine can be designed to do. If someone can imagine how a mobile machine can do a job better, chances are there is an engineer or technician working on a solution to make it happen.

Today, it is possible for a small group of skilled workers to operate all the heavy equipment needed to run a large-scale mining operation from a remote location hundreds of miles away. There is the ability not just to run the equipment but to have it communicate any problems and have an HDET start the diagnostic and troubleshooting process from a remote location as well. It is truly amazing to think of how far heavy equipment development has progressed in just over 100 years.

From the beginning when all a machine had to do was pull around some cultivating tools, it has become more specialized to accomplish different jobs. Heavy equipment machines are used in general groups of industries that require specific types of machines to get the job done. Some machines are more versatile and will be used in more than one area.

In general, heavy-duty equipment is any mobile machine that is designed to lift heavy objects, dig, push, cut, carry, compact material, or drill into it. Material can mean earth, rock, wood, water, compost, garbage, or any man-made substance or object.

The major manufacturers of heavy equipment are the following:

Caterpillar: http://www.cat.com
John Deere: http://www.deere.com/en_US/industry/construction/construction.page?
Komatsu: http://www.komatsuamerica.com/
Volvo: http://www.volvo.com/constructionequipment/na
Bobcat: http://www.bobcat.com
JCB: http://www.jcb.com/
Liebherr: http://www.liebherr.com/en-GB/default_lh.wfw
Terex: http://www.terex.com/main.php
Atlas Copco: http://www.atlascopco.com
Dynapac: http://www.dynapac.com/
Hitachi: http://www.hitachi-c-m.com/global/products/index.html
Case: http://www.casece.com/
Doosan: http://www.doosanequipment.com/
JLG: http://www.jlg.com/en-US/Products.html
Kawasaki: http://www.kawasakiloaders.com/
Kobelco: http://www.kobelcoamerica.com/Pages/homepage.aspx
Kubota: http://www.kubota.com
Link-Belt: http://www.linkbelt.com/
Manitowoc: http://www.manitowoccranes.com/
P&H: http://www.phmining.com/en/PHMining.htm
http://www.phmining.com/PHMining
Pettibone: http://www.gopettibone.com/
Herrenknecht: http://www.herrenknecht.com/products.html

MACHINE MODEL DESIGNATION Heavy equipment machines are categorized by their type and are given model numbers by their manufacturers. When machine manufacturers put a model number on a machine, it will usually relate to one specification of that machine. Sometimes it will be the machine's horsepower (hp), its empty operating weight (tons or tonnes), its maximum **payload** capacity (tons or tonnes), its volume capacity (cubic yards or meters), or its lifting capacity. As models are updated or modified, a letter after the model number will change. Machines are not categorized by the year they are made like automobiles. These are loose guidelines that different machine manufacturers will change as they see fit and some machine model numbers won't relate to anything in particular.

An example of how a model number relates to a machine characteristic would be the Cat 621 scraper. The volume of material that it can haul is roughly 21 yd^3 of material, which is the 21 in 621. The first series of these machines were designated 621. After the first major redesign of this machine, Cat added the letter B to the model number (621B). However, the next 621 redesign was the 621E and the next one after that was the 621F.

INDUSTRY GROUPS

There are several different general types of industries where you will find heavy equipment working in. However, some types of machines could be adapted for performing almost any type of heavy work. You will notice some types of machines will be used in several industry groups and some machines such as excavators can be modified to perform specialized work. Here are some general industry groups and some examples of different types of machines that are used for that industry:

- **Demolition:** excavators, skid steers, dozers, articulated trucks
- **Forestry:** skidders, harvesters, forwarders, crawler dozers, feller bunchers
- **Light construction and landscaping:** backhoe loaders, skid steers, mini excavators, small dozers, small compactors, small wheel loaders, telehandlers, aerial work platforms, small horizontal drills, trenchers
- **Landfill:** compactors, dozers, scrapers, wheel loaders
- **Mining:** front shovels, excavators, dozers, wheel loaders, rock drills, portable crushers, off-highway trucks, draglines, scooptrams
- **Pipeline:** pipelayers, excavators, dozers, graders, rock drills
- **Heavy construction:** graders, dozers, excavators, articulated trucks, cranes, track loaders, tunnel boring machines, large horizontal drills, rock drills
- **Rail equipment:** cranes, specialized rail maintenance equipment (tie driller, anchor applicator), heavy-duty forklifts
- **Road building and maintenance:** graders, dozers, excavators, pavers, grinders, compactors, articulated trucks
- **Material handling:** telehandlers, heavy-duty forklifts/material handlers
- **Lifting/shoring:** track-type cranes, rubber tired cranes, pile driving cranes
- **Aggregate/recycling:** excavators, screening plants, tub grinders, wheel loaders, dozers, trucks

MACHINE TYPES

TRACK-TYPE TRACTOR. This is a Caterpillar term that goes back to the first track machines that used tracks. Some other common terms are *bull dozer*, *dozer*, or *crawler tractor*. Dozers can have engines that produce between 75 and 1000 hp. They can have mechanical drivetrains or be hydrostatically driven. They will have different types of blades attached to the front for pushing material or other machines and are sometimes equipped with rippers mounted at the back of the

TECH TIP

Machines are made to be mobile by their drivetrains. The **drivetrain** starts at the prime mover (usually a diesel engine) and ends at the machine's tires or tracks. There are three general types of drivetrains that any type of heavy equipment machine will have.

Hydrostatic drive machines: They will use hydraulic pumps driven by the prime mover to send fluid to hydraulic motors that will turn the machines' tracks or tires to make the machines travel.

Mechanical drivetrain machines: They either will use a clutch and manual shift transmission (not very common anymore) or will use a torque converter and **powershift** transmission that will mechanically send power through other driveline components to drive the tires or tracks of the machines.

Electric drive machines: They either will use a remote electric power source provided the machine doesn't have to travel too fast or far or will have a diesel engine drive a generator, which will be used to power electric drive motors that will drive the machines' tracks or tires.

Machines with tires that are built on a one-piece frame are sometimes called straight-framed machines and are steered by a steering axle or two or more steered tires. Machines that are based on two frames that hinge side to side are called articulated machines. They are steered by a hydraulic system that shifts the hinge point to the left or right.

Most track machines are straight framed and are steered by changing the speed between the two tracks. Some wheeled machines will steer this way also and are called skid steers.

FIGURE 1–7 A large dozer equipped with a ripper.

FIGURE 1–9 A twin-engine scraper.

FIGURE 1–8 A large excavator.

machine that are meant to loosen material so it can be moved. ● **SEE FIGURE 1–7** for a very large dozer with a ripper.

EXCAVATORS. These are a type of machine that is designed to dig holes and trenches or just move material. Some other terms are *hi hoe* or *hoe*. An excavator is usually on tracks, but some are mounted on wheels. The tracks or wheels are driven by the hydraulic system. The upper structure or **car body** is able to swing around indefinitely and has a counterweight on the back of it. At the front of the car body, the boom is mounted and moved with hydraulic cylinders. The boom has a **stick** attached to it and the stick has a bucket attached to it. Excavators are classified by their weight and can have different boom, stick, and bucket arrangements as well as having other tools in place of the bucket. ● **SEE FIGURE 1–8** for a large excavator.

SCRAPERS. Commonly called an earth mover or more formally a tractor-scraper, these are articulated four-wheeled machines that pull a bowl that collects material (earth) when the cutting edge "scrapes" the ground. After the bowl is full, an apron swings down in front of the bowl to hold the material in. The scraper will then carry the material to another location and an **ejector** pushes the material out of the bowl after the apron is lifted. Scrapers can be single- or twin-engine machines. Single-engine scrapers will need to be pushed by a dozer to get loaded and will get stuck fairly easily in less than ideal traction conditions. Twin-engine scrapers can sometimes self-load and have the capability to hook together with two other scrapers to assist each other with loading. Once loaded, scrapers can travel quickly (up to 40 mph) and, depending on the model, can carry up to 50 yd^3 of material. ● **SEE FIGURE 1–9** for a twin-engine scraper.

GRADERS. A grader has four mechanically driven wheels at the back and two wheels that turn and lean at the front. It has a moldboard or blade in the middle of the machine that can move in many different directions to grade the surface it is travelling over. Most graders are also articulated and some have front wheels that are hydrostatically driven. Graders are classified by the width of their moldboard and the widest is 24 ft long. They can also have rippers and snow wings attached to make them more versatile. ● **SEE FIGURE 1–10** for a grader working.

CRANES. Cranes can be mounted on highway worthy trucks, be made to travel off-road on large tires, or be mounted on tracks. They can have fixed metal frame booms (lattice booms) or hydraulically extendible multisection booms. They can lift from a few hundred pounds to several thousand tons and can lift several hundred feet high and away from the crane. Cranes are rated by their maximum lifting capacity. They can also have different attachments to do different jobs

FIGURE 1–10 A grader working.

FIGURE 1–11 A crawler crane.

FIGURE 1–12 A large off-highway truck getting loaded.

FIGURE 1–13 An articulated truck getting loaded.

such as pile driving or hole boring. ● **SEE FIGURE 1–11** for a crawler crane.

OFF-HIGHWAY TRUCKS. These trucks are mostly configured with a dual-wheeled rear axle and two steering wheels at the front of the truck. They start at 35 ton payload (weight they can haul) and can haul over 400 tons at speeds of just under 50 mph. Trucks under 250 tons will have a mechanical drivetrain, but once trucks get over 250 tons payload, they can have either a mechanical or an electric drivetrain. Caterpillar's largest truck has a payload of 400 tons and its engine is rated at over 3700 hp! Trucks are loaded by a wheel loader or shovel, which should be matched to take four buckets or passes to load one truck. Their dump box is sometimes called a body. ● **SEE FIGURE 1–12** for a large off-highway truck getting loaded.

ARTICULATED TRUCKS. These trucks usually have six wheels, and all wheels can drive if requested. They are articulated behind the front wheels to steer and can haul up to 40 tons at 35 mph. They can travel through very adverse conditions because of their mechanical all-wheel drive system. They can be loaded with an excavator wheel loader. Some articulated trucks will have an ejector for dumping material instead of the more common dump box. ● **SEE FIGURE 1–13** for an articulated truck getting loaded.

SKID STEER LOADERS. A common brand name that is often used for these machines is Bobcat. The skid steer is like a Swiss army knife because it can remove its bucket and over 100 different attachments can be installed in its place. They have a horsepower range of 20–100, use a **hydrostatic** drive system to make them very maneuverable, and can be equipped with either tires or tracks. ● **SEE FIGURE 1–14** for a track-type skid steer.

TELEHANDLERS. These machines can be seen mostly around housing projects. They have a telescopic boom that is usually

FIGURE 1-14 A track-type skid steer machine.

FIGURE 1-15 Telehandler in action.

FIGURE 1-16 Pavement grinder in the shop.

equipped with a set of forks but, like the skid steer, can also be equipped with other attachments. They can be hydrostatically driven or have a mechanical drivetrain. They are classified by their maximum lifting capacity, and some can lift 15,000 lb to a height of 55 ft. Most of these machines will also have four wheel steering (crab steering) as an option. ● **SEE FIGURE 1-15** for a telehandler.

PAVING EQUIPMENT. If you have driven over an existing highway that is being worked on and felt or seen a series of grooves in the pavement, they were made by a grinder or cold planer. Large grinders have a drum that has up to 200 carbide teeth on it that spins at 100 rpm to remove up to 12 in. of pavement at a time. These machines travel on three or four small hydrostatically driven tracks while the drum is mechanically driven. They are sometimes called cold planers. The machine that then lays new asphalt is called a paver. It will travel slowly on rubber-padded tracks and put down a consistent layer of fresh asphalt. ● **SEE FIGURE 1-16** for a pavement grinder in the shop.

COMPACTORS. These machines come in various configurations, but their main purpose is to compact material as they move over it. Compactors (**packers**) can have pneumatic tires, smooth steel drums, lots of dull protrusions on the drums (sheepsfoot), or sharper bumps on four drums for chopping and packing garbage (chopper wheels). Most packers with drums vibrate to increase their effectiveness. They can be small walk behind units or large ones weighing up to 60 tons with over 500 hp engines. They can be hydrostatically or mechanically driven. Some will have small dozer blades as well.
● **SEE FIGURE 1-17** for a smooth drum compactor.

TRACTOR LOADER BACKHOE. Commonly called backhoes, these are very versatile machines that have an excavator arrangement (boom, stick, and bucket) at the back and a loader arrangement at the front (boom and bucket). There are two legs (stabilizers)

HEAVY DUTY EQUIPMENT REPAIR AND SAFETY 13

FIGURE 1–17 Smooth drum compactor.

FIGURE 1–19 Forwarder in action.

FIGURE 1–20 A forestry harvester.

FIGURE 1–18 Tractor loader backhoe.

FIGURE 1–21 A typical skidder.

that drop down when the excavator is being used, and the stick is usually extendable to be able to dig deeper holes. They mostly have a mechanical four-wheel drive power train and run from 50 to 120 hp. Some backhoes will also have quick attach buckets to enable other attachments to be connected to the machine. ● **SEE FIGURE 1–18** for a tractor loader backhoe.

FORWARDERS. These machines are used in forestry applications to transport cut logs through rough and steep terrain to other locations. They are six- or eight-wheel drive, articulated steering machines that have a hydraulic boom with a grapple on the end to lift the logs onto the back of itself for transport. ● **SEE FIGURE 1–19** for a forwarder.

HARVESTERS. These machines are used to cut down trees, and some are equipped with length cutting attachments. They have a grapple to grab the tree and a hydraulically driven saw that cuts the tree off. They can be mounted on tracks or rubber tires. Some track machines can work on much steeper hills because the upper structure can stay level while the tracks follow the contour of the ground. ● **SEE FIGURE 1–20** for a harvester in action.

SKIDDERS. These machines will pick up one end of several logs and skid them to a different location. They can have either a cable on a winch or a hydraulic grapple to do this. They are articulated steering machines with a mechanical drivetrain and large rubber tires. ● **SEE FIGURE 1–21** for a typical skidder.

14 CHAPTER 1

FIGURE 1–22 A large pipelayer hoisting a large pipe.

FIGURE 1–23 A large wheel loader.

PIPELAYERS. When pipelines are constructed whether in remote or urban areas, the pipelayer will be used to lift and manipulate the welded-together sections of pipe. Most times there are three or more pipelayers working together lifting the same section of pipe. They are based on a dozer that has its blade removed, a counterweight on one side, and a boom on the opposite side. They are rated as to their maximum lifting capacity and range from 20 to 100 tons. Smaller pipelayers may have rubber tracks on one side if they are to be working on pavement so they don't damage the pavement.
● **SEE FIGURE 1–22** for a large pipelayer.

WHEEL LOADERS. These machines are used in many different industries to move and load material. They are articulated steering machines, and all but the smallest (½ yd) are mechanical drive with some of the largest ones being electric drive (Letourneau). They are rated by their bucket capacity, and the largest wheel loader has a capacity of 53 yd^3 and 160,000 lb with a 2300 hp engine! Some wheel loaders will also have quick attach couplers that will allow the bucket to be removed and other attachments to be installed. Some large wheel loaders are equipped with tire handling equipment.
● **SEE FIGURE 1–23** for a large wheel loader.

WHEEL/TRACK DOZERS. These machines are based on articulated wheel loaders that have their boom and bucket replaced with a blade frame and blade. They are mostly used for haul road maintenance and light material dozing. John Deere has recently introduced a hi-speed dozer that has four tracks.
● **SEE FIGURE 1–24** for the John Deere hi-speed dozer.

SCOOPTRAMS. Underground mining machines need to be low and compact because of the tight quarters they work in. Scooptrams are used to load trucks (mucking) from the face

FIGURE 1–24 John Deere's hi-speed dozer.

of a fresh blast in an underground mine. They are low-profile machines that feature articulated steering and mainly differ from conventional wheel loaders by the location of the cab. It is located at one side of the machine just above ground level. They can have buckets up to 10 yd^3 and engines of over 400 hp.
● **SEE FIGURE 1–25** for a scooptram.

TRACK LOADERS. Track loaders or crawler loaders are a versatile machine that can dig, move material, and backfill trenches. Older track loaders were mechanically driven but are now all hydrostatically driven. This gives them much faster cycle times and higher production. They can have buckets as big as 4.5 yd^3 and engines up to 300 hp. Sometimes they are equipped with forks that fit on top of the bucket so the machine can be used as a heavy-duty off-road forklift. ● **SEE FIGURE 1–26** for a track loader.

DRILLING MACHINES. Mining operations need drilling machines to drill holes in the material they are mining so the blasting crew can come along after and load the holes with explosives and

HEAVY DUTY EQUIPMENT REPAIR AND SAFETY 15

FIGURE 1–25 A scooptram.

FIGURE 1–26 Track loader.

blast apart natural formations. The loosened material can then be transported to a crushing and processing facility. There are a few variations on mining drills. Some are vertical drills, meaning they can drill holes straight down or on a slight angle. They

FIGURE 1–27 An underground jumbo drill.

can drill holes up to 12 in. in diameter and several hundred feet deep. Other drilling machines can drill multiple holes at once into a rock face or overhead and are mostly used in underground operations.

Horizontal drills are used to run cables or pipe underground. Their bits can be directed in any direction, and they can drill holes several hundred feet long. They are rated by the amount of torque they can produce and the amount of pullback force they can generate. Very large horizontal drills can generate over 90,000 lb/ft of torque and 1,000,000 lb of pullback force and need a 1600 hp engine to drive it. ● **SEE FIGURE 1–27** for an underground jumbo drill.

TRENCHER. A trencher is similar to a large chainsaw on tracks that is designed to dig a trench and move the dug material to the side of the trench. The hydraulically driven chain will have some type of digging teeth or bits on it that will be forced into the ground as it is rotated. As the material is removed, an auger will move the material to the side.

GROOMERS. Ski hills around the world rely on a piece of heavy equipment that is called a groomer. A groomer is on a set of wide tracks to make it stable on steep slopes, and it will pull some different tools to groom the snow the way the operator wants it. It can be anywhere from 100 to 400 hp and can have some very complicated electronically controlled hydraulic systems on it.

CONTAINER HANDLERS AND HEAVY-DUTY FORKLIFTS. Shipping containers that move goods between continents need to be moved from trucks or trains to ships and back again. These containers are also stored in yards and loaded at factories. To move and stack these containers, there are machines that have specialized booms and clamps that will pick up loaded containers and then are able to stack them several containers

FIGURE 1–28 A stacker or container handler.

FIGURE 1–30 A rubber-tired excavator.

FIGURE 1–29 Mining shovel loading a truck.

FIGURE 1–31 Machine attachment.

high. Because of this they are sometimes called stackers.
- **SEE FIGURE 1–28** for a stacker.

MINING SHOVELS. They are similar to excavators in that they have a boom, stick, and bucket; however, the bucket is pushed away from the machine to fill it. The bucket can also open once it is over the box of the truck it is loading. They are usually used for loading large mining trucks that are over 150 ton capacity and will fill them with three to four full buckets.
- **SEE FIGURE 1–29** for a mining shovel loading a truck.

RUBBER-TIRED EXCAVATOR. The upper structure of a rubber-tired excavator is similar to a track-type excavator. The lower section of this type of machine can vary from being an off-road type of drivetrain to a more highway truck-type arrangement. The advantage for this machine over its track-type cousin is that it can travel between jobs without the need for a float. One type of rubber-tired excavator is called a Gradall and has a unique boom and bucket arrangement.
- **SEE FIGURE 1–30** for a rubber-tired excavator.

ATTACHMENTS. An attachment is any type of tool that can be added to a machine to make it perform a different function than it was intended to. Attachments are usually hydraulically driven and have a set of quick couplers to drive them and attach them to the machine. Some attachments will have their own power source. ● **SEE FIGURE 1–31** for an attachment of a machine.

Progress Check

6. A machine model may indicate these three characteristics of the machine:
   a. empty weight, length, horsepower
   b. loaded weight, bucket capacity, horsepower
   c. width, height, fuel tank capacity
   d. weight, engine torque, boom length

HEAVY DUTY EQUIPMENT REPAIR AND SAFETY

WEB ACTIVITY

Go to the website of a heavy equipment manufacturer and look at the different models of machines it produces. Look at the model numbers and see if they relate to a characteristic of the machine. Also see what kind of drivetrain it uses, how much horsepower the engine has, and what level of emissions the engine is compliant with.

7. A scraper will carry material in its:
   a. box
   b. bucket
   c. hopper
   d. bowl

8. These are the three typical types of drivetrain arrangements found on heavy equipment:
   a. mechanical, hydraulic, pneumatic
   b. mechanical, solar, electric
   c. hydrostatic, mechanical, electric
   d. hydrostatic, hydraulic, pneumatic

9. The main difference between an excavator and a mining shovel is:
   a. a mining shovel doesn't have tracks
   b. mining shovels are always electrically driven
   c. an excavator has a short boom
   d. an excavator pulls its bucket to fill it

10. A scooptram is a machine that:
    a. is used for open pit mining
    b. hauls material out of an underground mine
    c. loads trucks in an underground mine
    d. will dig tunnels in an underground mine

WHAT KEEPS CHANGING?

Like automobiles, new machines are constantly evolving, and there are several factors that drive this evolution today. Although machine evolution may not appear to be as dramatic as what occurs in the automotive world, there have been some very significant developments in machine design and technology over the last few years. These changes will have an effect on today's HDETs and are the reason why the HDETs need to constantly be involved in upgrading their skills and knowledge:

1. **Emission regulation:** This has been the driving force of the biggest change in diesel engine technology in the last 50 years. It will continue to force machine evolution in the future as well.

2. **Operator comfort:** The biggest factor in whether a machine is as productive as possible is the person sitting in the operator's seat. If that person can be more comfortable and not be fatigued easily, then in theory the machine will be more productive. Machine cabs and controls have changed dramatically over the years to improve operator comfort.

3. **Fuel efficiency:** The biggest operating expense for heavy equipment is fuel. Engineers are always trying to make machines more fuel efficient.

4. **Machine efficiency:** As technology advances, engineers are constantly looking at ways to make machines more efficient. Lately, this has been accomplished with a greater use of electronic controls for the engine, drivetrain, and hydraulic systems.

5. **Machine diagnostic systems:** As machines have their electronic systems expanded to incorporate more sensors and electronic control modules (computers), the diagnostic system used to allow the HDET to troubleshoot problems will change.

HOW MUCH DOES IT COST TO PURCHASE?

Heavy equipment can range from being relatively cheap to very expensive to purchase, operate, maintain, and repair. Some skid steers can be bought for as little as $20,000 while a large mining shovel with an 80 yd^3 bucket can be as much as $20 million. These prices all relate to the volume of work the machine is designed to do. If the small skid steer was to fill the 80 yd^3 bucket of the mining shovel, it would take a very long time. However, it was designed to do small finesse work without creating a lot of extra damage and not move massive amounts of material. The mining shovel could dig a swimming pool in less than one bucket, but it would make quite a mess doing it and would cost several thousands of dollars just to move the machine to the jobsite.

New machines are purchased at a dealership similar to a car dealer, while used equipment can be bought privately, through a dealership or at an auction. ● **SEE FIGURE 1–32** for a heavy equipment dealership.

Medium-size equipment can easily cost $500,000 to purchase new. In spite of the seemingly high purchase price of equipment, most machines can easily pay for themselves in a

FIGURE 1–32 Dealership.

few years if they are operated and maintained properly and the work they perform is profitable.

HOW MUCH DOES IT COST TO RUN?

A machine's operating costs are also related to the size of the machine. As the machine's purchase price goes up, so does its operating expense. The small skid steer may only need 5 gallons of fuel to run all day while some very large machine could use that much fuel in five minutes. ● **SEE FIGURE 1–33** for a machine getting fuelled up for the day.

FIGURE 1–33 A machine getting fuelled up. Fuel is typically the highest operating cost and could range from $25/day to well over $1000/day.

Machine maintenance costs also rise as machines get bigger. A complete service on the small skid steer would require a few filters and maybe 20 gallons of fluids. This may cost less than $100. A very large wheel loader would need over 700 gallons of fluids and a skid full of filters. The cost for just the fluids and filters could be $15,000–20,000. While the operating costs (fuel, oil, filters) of a machine are of little concern to an HDET, the repair costs and downtime should be.

Most replacement parts and components such as hydraulic pumps are also relative to the size of machine they are for. For example, the small skid steer's hydraulic pump that will displace 10 gallons per minute is likely $350. This compares to the large wheel loader's one of four 500 gallons per minute pumps that could be well over $30,000 each.

While operating and replacement parts costs can be huge, the most important part of a machine being out of service because of regular service or an unplanned breakdown is the cost of downtime.

WHAT IS DOWNTIME?

No matter what a machine costs to run per hour, it should produce more in terms of dollars-worth of work per hour than it costs. When a machine is out of service because of a repair, service, or adjustment, it is producing $0 worth of work just as a car sitting at a stoplight is getting 0 miles per gallon. The unscheduled time that a machine is nonproductive is called downtime.

All heavy equipment machines are designed to perform a certain amount of work within a certain amount of time. For example, a 30 ton excavator should be able to fill its bucket at a maximum depth of say 25 ft, lift that bucket out of the trench, dump it on a pile, and return the bucket to the bottom of the trench within a specific amount of time. This is called a digging cycle, and if the operator of that machine feels that the machine isn't able to complete the cycle fast enough, then an HDET will be called on to find the problem and rectify it. In this case, the machine is underperforming and not working to its potential. Whether it's a wheel loader moving material, a grader grading a road, an excavator digging a trench, or a paver paving a road, that machine must perform a minimum amount of work per hour, or it will be costing someone money and not making him or her a profit even though it may still be producing.

However, if a machine is not producing at all when it normally should be, this is called downtime. Downtime could be a result of a few different things.

1. All machines need service. If the servicing is done outside of normal operational hours, then it would *not* be

heavy duty equipment repair and safety 19

FIGURE 1–34 Machine in a repair shop.

FIGURE 1–35 Machine being delivered to a jobsite.

considered downtime, but if it is done when the machine should be working, then it would be downtime. Most regular servicing is scheduled to be done after normal working hours to minimize downtime.

2. If a machine is in a shop for a major rebuild during its off-season, this would not be downtime. However, some industries don't have an off-season, so the rebuild time would be considered downtime if the machine would normally be expected to work during that time.
● **SEE FIGURE 1–34** for a machine in a repair shop.

3. If a machine has a coolant leak as a result of a poor shop repair, blows a hydraulic hose, or has a flat tire when it is expected to be working and productive, this would be considered downtime. Downtime or unplanned nonproductive time leads to planned production that is lost and has to be made up for. Not only are the repairs added expenses but making up the lost production is. This will mean paying the operator and any other work crew involved with the job extra, and if the machine is down long enough, it could mean getting a replacement machine. If it happens during the company's peak production time, it will likely have to rent a replacement machine because all other machines in the fleet will be working.

A small machine might have a rental replacement cost of $50/hr, medium size $200/hr, large size $500/hr, and very large size more than $1000/hr. There are also costs associated to having a replacement machine brought in (floated) to the jobsite. ● **SEE FIGURE 1–35** for a replacement machine being delivered to a jobsite. The out-of-service or down machine could be holding up a crucial highway job or some other time-sensitive job that has penalties of thousands of dollars per hour on top of all the other expenses. An extreme case of a machine being down could result in an extra cost of $100,000 or more.

HOW LONG SHOULD IT LAST?

Heavy equipment is designed to withstand severe working conditions and climates. However, all things mechanical will eventually fail after an expected period of time, or if the machine's design parameters are exceeded, there will be a premature failure. If factory service requirements are adhered to and the machines are operated within their design limits, most machines can be expected to operate 15,000–20,000 hours without any major repairs. This is a very general rule of thumb, and there are lots of factors that will decrease this ideal longevity. There are also practices that could extend machine life as well.

Unlike automobiles that have their service requirements and life expectancy measured in miles or kilometers because of their purpose, most heavy equipment use hours as a gauge to measure machine life. Some equipment made for hauling material will have an odometer and an hour meter, but it is usually the hour meter that is followed.

A typical year of work for a machine is 2000 hours. This is based on an 8-hour day, 40-hour week, and 50 weeks per year. Again this is a general rule of thumb. There are machines that are used for two shifts a week that could accumulate over 4000 hours and others that are used for special jobs may only see 700 hours or less.

If we use 2000 hours/year as a guide, then a 15,000–20,000-hour life span would equate to 7–10 years.

CASE STUDY

A sewer and water main contractor has a job to perform that involves installing a new 60 in. high-pressure water main under a major city intersection. The pipe comes in 20 ft sections that weigh 8 tons each. The only way of doing this job is to shut the intersection down, remove the pavement, open cut (excavate) the material, install the pipe, cover it, and repave the road. The catch is that the governing body for the road will only allow the intersection to be shut down starting at 8 P.M. and it must be reopened by 6 A.M. the following morning. If the road is not reopened on time, there is a penalty of $5,000/hr that it is closed.

The job is mainly performed with a large excavator and a crane (for lifting the pipe sections) with some supporting equipment such as a medium-size wheel loader, two sheep's foot compactors, portable light towers, and some dump trucks.

The job starts on time with all machines and manpower in place. The intersection is excavated by midnight and the pipe starts to get installed at 12:30 A.M. Just as the third length of pipe is being installed, the crane develops a hydraulic leak at one of the hoses going to its main hoist hydraulic motor.

Luckily, the contractor had an HDET on-site and he was quickly summoned to get the leak fixed so the crane could keep installing pipe. The problem was a simple fix of replacing the hose and topping up the hydraulic tank. The next problem was where would the HDET get a hose at 2 A.M.?

The task of getting parts for a machine that is down can uncover many problems at the best of times, but this gets magnified if the parts are needed after normal business hours. Most major equipment dealers have a 24-hour emergency parts person on call, but there is still the question of will they have the part in stock and how far away from the machine is the dealer store.

In this case, the crane dealer did not offer 24-hour parts service, so the HDET had to be resourceful and locate a hydraulic hose supplier who would open up and hopefully be able to make a hose that matched.

The HDET wasn't able to get a hose shop to make a hose until 5 A.M., and because the hose was an uncommon size, the HDET had to be creative and find some fittings that could be used to make the hose work. By the time he got the hose and fittings on the machine and the hydraulic tank refilled, it was 7:30 A.M. The crane was back to work at 7:45 A.M. and the job was completed at 9:30 A.M.

This ultimately cost the contractor a $17,500 penalty, plus 3.5 hours of overtime for the operators and work crew as well as inconvenience caused to many commuters who had to find an alternate route around the closed intersection.

The situation could have been worse if the HDET wasn't able to find hose shop that would open early and if he wasn't able to make the hose work. The value of a good HDET cannot be overstated, and this should be recognized by employers.

Some machine downtime is inevitable and should be expected by the equipment owner. However, a good HDET will help to keep machine downtime to a minimum with many positive results arising from this.

The equipment owner will be making money and keeping production on schedule, the equipment operator will be doing what he or she is paid to do, and the value of a good HDET and his or her support network should be recognized, which in turn will give the HDET job security.

WHAT HAPPENS THEN?

Unlike old automobiles that end up in the crusher and recycled to make refrigerators and soup cans, heavy equipment machines will get reconditioned a few times before they get scrapped. This is possible because a large heavy machine needs a large frame, and it is very rare that this frame is ever pushed beyond the point of being able to be repaired.

There comes a point in the life of a machine where its owner has to decide whether it's worth fixing or not and whether it can be relied upon to be productive on a regular basis if it is repaired or reconditioned. This is where the value of a good HDET should be recognized. A new machine is a big investment and a lot of that investment is for a machine that can be relied upon to be productive. If a machine can be repaired properly, there is no reason why it can't be put back into service and given a second, third, or even fourth life.

CASE STUDY

One quarry that I worked at had bought a 100 ton rock truck with around 30,000 hours on it at an auction to use for spare parts. The plan was to dismantle it and use the major components for the other similar trucks in the fleet when the need arose. When it was looked at a little closer and found to be in fairly good condition, it was decided to put some money into the truck and put it into regular service. A few years later the truck was still running with over 50,000 hours on it with just regular repairs and maintenance required. Then one day it showed up at the shop with a strange noise under the cab. After a cleaning and inspection, there was a 3 ft long crack that had opened up in the main section of the frame that is roughly 5 ft tall. The easy thing to do would have been to dismantle the truck and cut up the frame. The downside of this would be purchasing a replacement. At that time a new truck was a little over $1,000,000 to purchase. After removing all surrounding components to reveal the crack and allowing the welder to have good access to it, the welder felt confident that a good repair was possible, and the decision was made to repair the crack. The welder drew the frame back into place, welded the crack, and reinforced it to make it better than new! The truck was then reassembled, put back into service, and eventually made it past the 60,000-hour mark with no major repairs.
● **SEE FIGURE 1–36** to see the machine's hour meter.

FIGURE 1–36 Rock truck hour meter.

WEB ACTIVITY

Search the Internet to find a video that shows the process of performing a certified rebuild on a machine. Take note of the "before" and "after" state of the machine.

I recently read an article about a fleet of mining trucks that were purchased new for a large scale mine and have been in a rigorous maintenance and repair before failure program since they were new. There are some trucks that have surpassed the 100,000-hour mark.

Most major equipment dealers will have a certified rebuild program. This is a program where the customer brings a tired machine with a lot of hours into the dealers shop. A group of HDETs will strip the machine to the bare frame and totally rebuild it with new or rebuilt components. The frame will get checked for cracks and repaired if necessary; the machine will be given all of the latest updates (improvements) for that model and a fresh coat of paint. After this process, the machine is given back to the owner with a new machine warranty.

There are also retrofit programs where a higher Tier engine could be installed in a machine to make it emission compliant or new electronic systems installed to update a machine.

If it's decided that a machine is not going to be repaired and put back into service, it will most likely be sent to an auction where there are always buyers looking for a deal. They are willing to take a risk that the machine is a good investment even if it may need some major repairs. Here again the knowledge and skill of an HDET must be relied upon to get this machine back into reliable working condition.

Eventually, a machine will be taken out of service because:

1. it is too hard or impossible to get parts for. Some owners will have parts made if they can't buy them from a dealer.
2. it is not fuel efficient. This is a big part of operating expenses, and if a machine is costing too much in fuel to keep running, it could be parked in favor of a newer more fuel-efficient machine.
3. it is not emission compliant. This is something new that is happening in California where some Tier 1 machines can no longer be run.
4. it is not productive enough. New machines are more productive and an older one may just not be able to keep up any longer.
5. operators don't want to run it. Older machines are likely not too comfortable to run for an 8–12-hour day when

When the economy slows down, new machine sales will likely slow down, which could lead to dealer cut backs, but as long as there is a minimum of activity in most industry groups, there will always be a need for quality HDETs to maintain, repair, and rebuild existing fleets of machines.

WHAT IS AN HDET?

HDET is a relatively new term. Past titles for someone who repaired and serviced heavy equipment were diesel mechanic, heavy-duty mechanic, heavy equipment mechanic, and some derogatory titles that aren't worth repeating. Lately, skill trades in general have been given an elevated status. Some have coined the term *gold collar* in reference to skilled trade occupations because of the higher than average income that skilled workers receive. Up until recently there was a push from the elementary and high school levels to get graduates into degree programs at universities and colleges. The trades were thought to be something to fall back on if a career from a postsecondary education didn't work out. Eventually, many shop or technical programs at the high school level were phased out for various reasons.

This trend not to recognize skill trades as an excellent career choice is starting to reverse, and it has been driven mostly by the lack of new people getting into the trades to replace older skilled workers who are retiring.

For young people or anyone looking to skilled trades as a career, there are many choices available. For people who are mechanically inclined, becoming an HDET is a great choice. Being mechanically inclined means being curious about how things work and being interested in repairing and servicing heavy equipment. But there is more that is required to be an HDET than just being mechanically inclined.

The term *mechanic* implies you will only work on something mechanical. This relates to gears, chains, shafts, nuts and bolts, and so on. An HDET must be mechanically inclined, but they must also be electrically, electronically, and fluid power inclined. Because heavy equipment uses many different systems besides mechanical, the HDET needs a much wider breadth of knowledge to understand how a machine works and how to keep it running properly.

WHAT DOES AN HDET DO?

You may already have a preconceived notion of what a typical HDET does in his or her daily responsibilities. If you assumed he or she performs repairs and maintenance to heavy

FIGURE 1–37 A machine that has been taken out of service.

compared to new ones, and if a company wants to keep good operators, it will need to keep updating its machines.

6. it becomes unsafe. This could be a judgment call or because of a change in safety regulations.

● **SEE FIGURE 1–37** for a machine that has been taken out of service.

THE ROLE OF AN HDET. Until the day after that a piece of heavy equipment is dismantled for scrap metal, there will be a need for an HDET to maintain and repair it to keep it productive. An HDET will likely be needed to dismantle the machine at the end of its life.

Even an equipment owner who buys a brand-new machine will need to rely on an HDET to maintain a new machine with regular servicing and adjustments. New machines will still break down on occasion as well, and this will require the services of a dealer HDET.

FIGURE 1–38 An HDET servicing a machine.

equipment, you would be correct. But what does this actually mean? In the chapters that follow, you will learn about the many different systems needed to make a machine work, and these are the systems that an HDET will be expected to understand. You will initially need to know how they work in general, and as you gain experience, you will be expected to service, diagnose, and repair the specific systems that are on a machine.

Servicing a machine involves changing fluids and filters, taking fluid samples, lubricating wear points, checking for proper operation of machine systems and safety features, and recording all this information along the way. ● **FIGURE 1–38** shows an HDET servicing a machine. Maintenance of heavy equipment is the key to keeping it productive as long as possible, and it involves way more than just changing oil. A good maintenance program will lead to a good predictive repair program. This will allow the machine owner to plan repairs around his or her production schedule and not have unplanned repairs force production schedule changes. More and more companies that rely on heavy equipment are buying into preventive maintenance, but there are still many operations that work by keeping their fingers crossed and using the "fix it when it breaks" method.

When it comes to repairing machines, you can expect to perform apparently simple tasks like repairing leaks or making minor adjustments and also be expected to rebuild hydraulic components that may need to have as much cleanliness as an operating room. This is what's so great about being an HDET. There is an endless array of tasks that you will be challenged with, and they are all doable with enough time and support. You will also likely need to order parts for the machine you are repairing or servicing.

You may at first be overwhelmed by all the different systems you will be expected to know, and each machine will have its own variation on a particular system. This is a normal feeling, and you shouldn't ever be expected to know everything. When you are confronted with a new problem, you need to take the basic concepts of how mechanical, electrical, and fluid power works and apply them to the problem. By using this basic knowledge and having some good service information available, there is nothing you shouldn't be able to solve. There will always be machine problems that will present a challenge, but with the confidence you gain as you move through your career, you will come to enjoy these challenges.

You could be part of a major machine rebuild that takes a group of HDETs several months to complete, part of a team of HDETs who are required to assemble a large piece of mining equipment on-site, or working by yourself on a mini-excavator that won't start. There is an almost endless amount of possibilities of where an HDET career could lead you. Your personality will likely eventually lead you to where you should be working.

WHAT KIND OF PERSON WOULD MAKE AN EXCELLENT HDET?

Anyone could become an HDET if he or she so desire, but it can be a physically and mentally demanding occupation.

Some general physical and mental characteristics that would help a person make a great HDET would be if he or she:

1. is mechanically inclined. You should be comfortable with basic mechanical concepts and be able to learn new mechanical systems and concepts fairly easily. A mechanically inclined person will enjoy working with his or her hands and with tools and will not mind getting dirty.

2. is physically fit. There will be lots of lifting, bending, climbing, and getting into awkward positions, so an above-average fitness level will help. If you enjoy sports outside of work, this will coincide with becoming an excellent HDET.

3. enjoys learning new things all the time. The heavy equipment world is constantly evolving and you need to

embrace change and have a desire to always learn new things.

4. is a good communicator. You have to effectively communicate with customers, fellow HDETs, parts people, and supervisors. Part of communicating is also listening, and the price of miscommunication can be very high.

5. is creative. Being an HDET is not always an exact science and many times you need to be creative or think outside the box to overcome problems.

6. is an efficient problem solver. Most of what an HDET does is problem solving, and if you can repair a broken machine in an efficient manner, you will be a valuable asset to your employer.

7. is strong with electrical and hydraulic concepts. These two areas can be challenging because they rely on the HDET to be able to understand things that aren't as easily obvious as with something mechanical. Most machines will have moderately to extremely complicated electrical and hydraulic systems, and an HDET needs to excel in these areas.

8. is a safety conscious worker. Getting hurt on the job isn't good for you or your employer. There are many hazards that you need to be aware of to keep you and those around you safe and productive. This mentality should also carry over to your home life.

9. is naturally curious. If you wonder how things work and like to figure out what makes something function, then you will enjoy working with heavy equipment.

10. enjoys challenges. Being an HDET is not easy and you will be faced with challenges on a daily basis. A good HDET will thrive on being challenged.

FIGURE 1–39 Typical repair shop.

minimal supervision. He or she will have to assume a great deal of responsibility by being the "go to" person.

If a person likes to work with people close by and be in a fairly controlled environment, then he or she should be part of a shop team where he or she will do some diagnostic work but mostly remove and install components that have been diagnosed to be faulty. He or she could also be doing component rebuilds, which means following a set of step-by-step instructions closely. ● SEE FIGURE 1–39 for a typical heavy equipment repair shop.

If a person doesn't like to be creative and likes to know what he or she is doing for the day when he or she starts, then should be part of a rebuild facility where they will recondition components that have been exchanged with good ones. There is a closely controlled environment that usually has very limited interaction with customers.

WHERE SOME PERSONALITY TRAITS MIGHT LEAD YOU TO

If a person likes to be challenged, is an independent thinker, and likes to be creative, then he or she should be in field service where he or she will be presented with a lot of diagnostic troubleshooting tasks and many kinds of repairs. Part of the attraction of a field service position is not knowing what you are going to be doing next or where you might be for the next job. The field service person should also like to interact with customers (operators, owners, foremen) and like to work with

WHAT DOES AN EMPLOYER LOOK FOR IN AN HDET?

If an employer is looking to hire an HDET they will want someone who is reliable, is punctual, is trainable, enjoys learning, and is responsible for their actions. Efficiency is something desirable that an employer would look for in an HDET. This means the HDET will always make the best use of his or her time. An example would be if an HDET is working on a large machine that requires a lot of climbing to get the job done, then an efficient HDET will think ahead and try to predict what tools he

HEAVY DUTY EQUIPMENT REPAIR AND SAFETY

or she may need so he or she won't have to waste time going back and forth to his or her toolbox.

WHERE DO HDET'S WORK?

An HDET could work anywhere in the world where heavy equipment is working. As you travel around and take notice of your surroundings, you are not likely to take too long to see a piece of heavy equipment. If a machine breaks down where it's working, then an HDET will be getting it running again. This would be a field service HDET. If a machine is brought to a repair shop, whether it's a dealer, independent shop, or a contractor that has its own HDETs, then these are shop HDETs, and they will almost always stay in the shop. If a machine has components that are removed and sent to a facility for reconditioning, then the facility will have HDETs who work and stay in their shop. If an HDET is running a truck that is equipped to perform machine maintenance on the work site, then a lube technician could go from site to site doing this.

WHAT ARE THE BENEFITS OF BEING AN HDET?

If you become an HDET, you will get a great deal of satisfaction in knowing that you have been given a malfunctioning machine or component and you can give it back to the customer in working order. There is a lot of other benefits as well like good pay, but the feeling of accomplishment you will get when you see a previously dead machine come to life, make noise and load a truck, lift a concrete pipe, or pave a road is hard to duplicate.

If you become a good HDET, then you will very likely never be out of work. In the 25 years since I started my career as an HDET, I was only out of work for about a month and have had many job opportunities come my way. I have made an above-average wage with good benefits, drove a company vehicle for a number of years, and have attended many training courses that were paid by my employer.

WHAT ABOUT AFTER?

Some years down the road after you have been an HDET for a while and are maybe ready for the next big challenge or just ready for a change, there will be lots of opportunities for you. There are many positions that you will be able to put your skill and knowledge toward.

Some jobs that could be the next step are:

Shop foreman: A shop foreman or lead hand will oversee several HDETs and be expected to be someone to fall back on when difficulties are encountered.

Field service supervisor: A field service supervisor will ideally be a former field service HDET and will take calls from customers who have machines down or who need a mobile HDET for other reasons. He or she will then assign the incoming workload to the field service HDETs working for him or her.

Technical advisor: It is inevitable that at some point every HDET will come across a problem that they he or she can't overcome. He or she will need someone to turn to for help and that would be a technical advisor. A technical advisor will typically have worked for an equipment dealer for a number of years and accumulated a great deal of knowledge and experience that they can pass on to help other HDETs.

Technical trainer/educator: All HDETs will at some point need to upgrade themselves by going to a training course. Training courses are typically run by former senior HDETs who have taken a new challenge and become a technical trainer. This is a demanding job because the students in the course are expecting the trainer to know everything about the topic at hand. While this is ideal, it is not always possible. A similar occupation is an educator in a technical/trade school where the school will hire and train HDETs to teach students the basics of how machines work.

● **SEE FIGURE 1–40** for an instructor teaching a class of future HDETs.

Sales representative: Some HDETs who are looking for a change find that moving into a sales position is a good

FIGURE 1–40 Instructor teaching future HDETs.

fit. Equipment dealers and suppliers need a sales force to sell new equipment, parts, and service.

TAKING CARE OF OUR ENVIRONMENT

Being an HDET means you can have a big influence on the environment. Machines that are running properly will use a minimum amount of fuel, emit the minimum amount of exhaust emissions, and generally have as little impact on our environment as possible. An HDET will ensure a machine is working as efficiently as possible.

When repairing or servicing a machine, fluids and components need to be replaced, and the removed components need to be handled responsibly. For example, if fuel, oil, and coolant filters or batteries used with any machine are not disposed of properly, they can create serious environmental problems. Some good work practices are:

1. Never pour waste fluids onto the ground, down a drain, or into any water source. ● SEE FIGURE 1–41 for a proper storage container of waste oil.
2. Air conditioning refrigerants can damage the atmosphere. Government regulations may require using a certified service center to recover and recycle used refrigerants.
3. If uncertain about the safe disposal of waste, contact your local environmental or recycling center or your dealer for more information.
4. Diesel engines that are not running properly will have a major negative impact on the air we breathe.
5. Treat the environment as though it is something special that you want to leave to your children.

FIGURE 1–41 Waste oil storage container.

Progress Check

11. This factor would *not* be one influence that initiates changes in heavy equipment machine design:
    a. fuel efficiency
    b. weight reduction
    c. operator comfort
    d. emission reduction legislation

12. The term *downtime* in relation to heavy equipment machines means:
    a. how long a machine can stay underground
    b. the time a machine can download information to a computer
    c. the time a machine spends with a flat tire
    d. the time a machine is out of service for unscheduled repairs or service

13. An HDET would *not* normally be required to do this:
    a. track the operating expenses for a machine
    b. order replacement parts for a machine
    c. adjust operating systems to improve machine performance
    d. try to improve the designed performance parameters of a machine

14. This would be the highest cost for operating expenses for a machine owner:
    a. tires
    b. fuel
    c. tracks
    d. depreciation

15. Typically, a machine would accumulate this many hours of operating time per year:
    a. 1000
    b. 2000
    c. 4000
    d. 10,000

SHOP ACTIVITY

Find the following types of machines and identify the main components on them:

- **Excavator:** boom, stick, bucket, car body, counterweight, main pumps
- **Track-type tractor:** track frame, sprocket, blade lift cylinders, ROPS, final drive
- **Wheel loader:** loader frame, hinge point, ROPS, counterweight, rear axle, transmission
- **Tractor loader backhoe:** oscillating axle, backhoe boom, stabilizers, transmission
- **Compactor:** hinge point, ROPS

CASE STUDY

A case study that proves the value of a safe and efficient HDET:

An HDET named Brian is called to a mining operation where a large rock truck has developed a brake fluid leak. It is a hot summer afternoon and the machine has been working for most of the day. Brian checks in with the quarry foreman Dave first to see where the truck is and if it's okay to go into the working mine where the truck is. Brian gets permission to work on the truck while Dave announces on the operations radio that there will be a service truck entering the quarry and for the other operators to be aware. Brian turns on the flashing beacon on his service truck and is then escorted by Dave to where the disabled truck is.

The technician may have saved time and went directly to the truck, but he would have risked being run over by other trucks that would maybe not have seen him because they were unaware of him coming to the dead truck. He may also have not bothered to turn on his beacon and become a speed bump.

Brian asks for pylons to be set up around the truck as a precaution and he also places wheel chocks behind the truck's wheels. He then completes an initial walk-around to check for any potential hazards. He could have not bothered doing an initial inspection of the area and machine and missed a loose rock hanging off the side of the truck that could have easily been dislodged and caused a serious injury.

Brian tags the machine as disabled with a "DO NOT START" tag on the steering wheel and places his lock on the battery master switch to lock out the machine. If he hadn't done this, there is a possibility of someone coming along and trying to start or move the truck while he is working on it. By this time Dave is asking how long before the truck is fixed and Brian tells him these are necessary safety steps that he needs to take to stay safe on this job. He then asks Dave to tell the other operators to slow down when they go by him and Dave complies.

Brian then starts his diagnostic procedure. He notices the fluid leak coming from inside the top of the frame rail. This means the truck's box will have to be raised to identify exactly where the fluid leak is coming from. Brian raises the truck's box and installs the box lock pins. He could have forgotten about this because he may have been able to look over the frame rails and not get under the raised box but he wanted to feel safe.

Brian eventually had to climb over the frame rail to find the leak. After finding the leaking hose, he then loosened the cap on the brake fluid reservoir to stop any more fluid leak and to take any pressure off the failed hose that he will have to replace. This also was a step he could have skipped, but this is a step that prevented hot oil from burning his skin when the hose was removed.

Brian made sure that he was wearing all appropriate PPE for this job: coveralls with high-visibility marks on them, safety glasses, boots, and gloves.

After removing the failed brake hose, Brian capped off the tubes that were leaking oil to prevent any further leakage onto the ground. He then left the disabled truck and had Dave announce on the quarry radio that he was leaving the quarry. He also gave Dave an estimated time that he would return with a new hose to get the truck running again.

Brian wanted to stop for a coffee at his favorite coffee shop on the way to the hose shop, but he knew Dave was anxious to get the truck running so he put it off for now. He had a couple of options for getting a new hose and went with the store that was a little more expensive but did good work and would put a rush on if Brian asked them to.

It wasn't too long and Brian was on his way back to the quarry with a new hose and some oil to top up the brake reservoir. He now knew the tools he would need to do the job and quickly installed the hose and had Dave give him a hand topping up the brake oil. The reservoir was near the cab of the truck, which meant lifting a couple pails of oil 10 ft up the truck's ladder. He could have done this himself, but it was a lot safer having someone help.

The truck was started and Brian then operated it to check for leaks. Dave wanted the operator to take it back to work, but Brian needed to make sure that the brakes were working properly first so he ran the truck for a few minutes. Brian tested the brake system and felt confident that everything was working as it should. This last check was also time well spent because of the possible consequences that could have resulted had Brian let the operator drive the truck with faulty brakes.

SUMMARY

1. Becoming a safe HDET is a must because serious injury or death can occur if unsafe work practices aren't used. This includes developing a safe attitude.
2. Heavy equipment can cause a great deal of damage if it is allowed to get out of control. There are also many sources of stored potential energy on a machine such as hydraulic, electrical, mechanical, and pneumatic.
3. Warning labels must be understood and are placed on machines and in service literature to warn HDET of hazards and dangers.
4. Some examples of safe work practices include lowering all equipment, using wheel chocks, locking articulated steer machines from turning, releasing pressure slowly from hydraulic systems, and using proper lock out and tag out procedures.
5. PPE such as approved safety glasses and footwear should be worn to prevent injuries. Minimum levels of PPE usage start with government and company policy for your jurisdiction and workplace.
6. Today's heavy equipment evolved from steam-powered agricultural tractors.
7. Heavy equipment can be found in a wide cross-section of industry and construction activities such as heavy construction, material handling, forestry, and pipeline construction.
8. There are many different manufacturers of heavy equipment. Some specialize in producing a few types of machines while major players such as Caterpillar, John Deere, Volvo, and Komatsu produce a wide variety of types and sizes of machines.
9. Some types of machines such as scrapers are designed for very specific applications while others such as excavators can be used for many different types of jobs.
10. Machines can be classified by the type of drivetrain they use such as mechanical, hydraulic, or electric.
11. Their model numbers can sometimes indicate their weight or load capacity.
12. The largest machines will be used in open-pit mining operations.
13. Low-emission regulation, fuel efficiency improvement, and operator comfort are three big factors that lead to design changes in heavy equipment.
14. Downtime because of unplanned repairs or maintenance is unproductive and costly.
15. Most machines produced today should last between 15,000 and 20,000 hours without major repairs if the machine is properly maintained and operated.
16. HDETs play a key role in helping to achieve maximum life out of machines by ensuring that proper maintenance and repairs are done. An HDET is a valuable part of any organization that operates heavy equipment.
17. Machines can be reconditioned many times before they are taken out of service.
18. Some characteristics of a person who would make an excellent HDET include being mechanically inclined, having a love of learning, enjoys problem solving, enjoys challenges, and being creative.
19. Most HDETs are rarely out of work because of the constant need for people with a specialized skill set required to look, repair, and maintain heavy equipment.
20. HDETs have a responsibility to work to help preserve the earth's natural environment.

FASTENERS, SEALS, GASKETS, BEARINGS, PERSONAL TOOLS, SHOP EQUIPMENT, AND FIELD SERVICE REPAIR

LEARNING OBJECTIVES

After reading this chapter, the student should be able to:

1. Identify different fasteners by size and type.
2. Explain how to properly torque threaded fasteners.
3. Explain the purpose of a washer.
4. Identify different types of seals.
5. Explain how to properly size a seal.
6. Explain what makes bearings fail.
7. Describe the various tools an HDET should own.
8. Describe common shop tools.
9. Explain how to safely set up and start an oxy-acetylene torch set.
10. Explain how to safely set up and operate a stick welder.
11. Explain how to safely set up and operate an MIG welder.

KEY TERMS

Lip-type seal 38
Oxy-acetylene 60
Tapered roller bearing 42
Tensile strength 34
Thread crest 33
Thread pitch 33
Torque multiplier 53

INTRODUCTION

SAFETY FIRST Many minor accidents, lost time accidents, and even some deaths of HDETs are related to improper usage, selection, and maintenance of tools.

Simple hand tools need to be maintained and used correctly to ensure that no injuries occur when an HDET is servicing or repairing a machine. Many scrapes, cuts, and bruises have resulted from hand tools such as wrenches, ratchets, and pry bars failing because of having too much force applied. Other times HDET injuries are caused by not using the correct body position when applying force to a hand tool. You should brace yourself when applying force to a hand tool and be ready for a sudden movement if the tool slips or if the fastener unloads quickly. You should always keep in mind that as more force is applied to a hand tool, there is a greater chance of the tool slipping and the potential for injury increases in direct relationship to that. When using hand tools, try to pull the tool toward you, and if that isn't possible, push the tool with an open palm. Try to anticipate what your hands or other body parts will contact if the tool slips. ● **SEE FIGURE 2–1** for the proper body positioning when using a torque wrench. Using gloves will prevent cuts and scrapes when pulling on wrenches and ratchets.

Damaged or poorly maintained hand tools are also a common source of injuries. Screwdriver tips that are damaged, ratchets that aren't lubricated, chisels that are mushroomed, and hammer heads that are loose are all examples of hand tools that can be major factors in the cause of injuries.

Power tools need extra respect when being used as they can create higher forces and, therefore, cause more severe injuries. Air-powered ratchets and impact guns use air pressure to tighten and loosen fasteners and must be used with care. Electric power tools such as grinders and drills can cause electrical shock injuries if they are damaged or used in wet conditions and can cause cuts and abrasions if used improperly. Safety glasses should always be worn when using power tools.

Tools such as hydraulic presses and wrenches that use hydraulic pressure need to be used with caution to prevent serious injury. These injuries can result from oil injection or by reaction of the forces created by the tool. Pressing parts together or apart can create forces of over 100 tons, and the HDET should use extreme caution when dealing with these extreme forces. Improper use and assembly of press tooling can lead to flying parts and injury or death.

Heating and cooling tools and processes have obvious safety concerns related to their use. Mild to severe burns and frostbite can be avoided by proper usage of torches, welding equipment, and heating and cooling equipment as well as using proper PPE while handling these tools or equipment.

This chapter will discuss different fasteners, seals, bearings, and gaskets that you will find on a heavy equipment machine. Personal hand and power tools as well as some special tools that you will likely use will be discussed. Shop equipment, welding equipment setup, and operation will also be discussed. The chapter will finish with a look at field service repair and maintenance work.

NEED TO KNOW

FASTENERS A heavy equipment machine is made up of many different component assemblies that are securely fastened to the machine's main frame. Examples of component assemblies are engines, transmissions, axles, cabs, hydraulic pumps, and steering systems. Each main component will be made up of many subcomponents or individual parts, and these also need to be fastened together properly so the machine stays in service. ● **SEE FIGURE 2–2** for the fasteners that hold an engine to a machine's main frame.

Fasteners are a very important part of a heavy equipment machine and an HDET needs to understand the differences between various types of fasteners, how to identify them, proper inspection procedures, and proper installation and torqueing procedures.

When installed and tightened properly, a fastener will create a specific clamping force that is designed to hold two or more parts or components together.

FIGURE 2–1 Proper body position when using a torque wrench.

FIGURE 2–2 The fasteners that hold an engine to a machine's main frame.

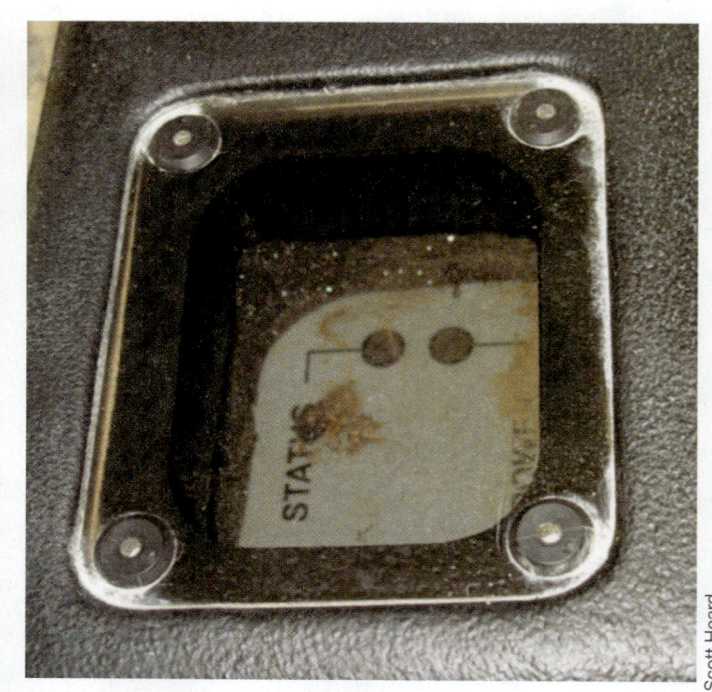

FIGURE 2–3 Rivets holding two parts together.

A common cause of machine downtime is one or more fasteners that don't perform as they should. Many very expensive component failures have been a result of issues directly related to fasteners. Sometimes it can be a manufacturing flaw, but more likely the result of the fastener failure is:

1. overstressed from being over tighten
2. left loose
3. vibrated loose
4. incorrect application
5. overstressed from excessive external forces

The most common type of fastener is the threaded fasteners, and they are sized in either metric or standard (imperial) dimensions. Threaded fasteners include bolts (sometimes called cap screws), screws, nuts, and studs. One type of non-threaded fastener is called a rivet. Rivets are a onetime-use fastener mostly used for light fastening duties. ● **SEE FIGURE 2–3** for rivets holding two parts together. Rivets come in a variety of diameters and lengths and are installed with a rivet gun. Rivets are removed by drilling or grinding the rivet head off.

FIGURE 2–4 Normal fastener rotation.

Almost all threaded fasteners found on heavy equipment will tighten when turned in a clockwise direction if you looked at the top of it. This is called right-hand rotation because the top of bolt will turn to the right when tightened. ● **SEE FIGURE 2–4** for normal threaded fastener rotation.

There are rare instances when left-hand threaded fasteners are used, and this is usually when a fastener is used on a rotating part. If the part's normal direction of rotation may make

32 CHAPTER 2

TECH TIP

An easy way to remember proper fastener rotation direction for fasteners that tighten when turned clockwise (right-hand thread) is "righty tighty, lefty loosey."

An easy way to think of this is to picture looking down on top of a fastener with a mark at 12 o'clock. As it is tightened, the mark will rotate to the right from its starting point.

a right-hand thread fastener loosen, then a left-hand threaded fastener should be used.

The most common threaded fasteners are bolts, studs, and screws. These will mate with nuts or components that have female threads. The most common style of nut is a six-sided or a hex nut. There are several variations of this common nut for special applications. A couple of examples are acorn nut, jam nut, and self-locking nut.

A threaded fastener's threads are based on the principle of an inclined plane. This principle relates to a bolt's function that moves it in an axial direction as it is rotated. The bolt's threads match up with the thread of either a nut or a threaded hole in a part. The degree of angle of the inclined plane relates to the **thread pitch**. This is measured in threads per inch with standard fasteners and in millimeters between **thread crest** for metric. ● SEE FIGURE 2–5 for the description of a typical hex head male threaded fastener.

Both metric and standard fasteners are loosely grouped as either coarse or fine. Coarse fasteners have their treads spaced farther apart. This will give the fastener a more robust thread that is easier to start and harder to damage. The majority of fasteners used on heavy equipment are coarse thread. Fine thread fasteners can be torqued to a slightly higher value (roughly 10%), tend to stay tight better, and are used where a more precise clamping force is required.

Bolt length is measured from the bottom of the bolt head to the bottom of the threaded portion of the fastener. Threaded fastener diameter sizing is based on the fasteners outside threaded diameter, which is called its major diameter and its length.

Bolts and screws can have full-length threads, which is typical for shorter lengths or just be partially threaded. The non-threaded portion of the bolt links is called the bolt shoulder. Bolts are sometimes called cap screws.

Cap screw heads are most commonly six sided or hex shaped, some are 12 pointed, or they may have a socket head that is recessed with a hexagon shape or Torx shape.

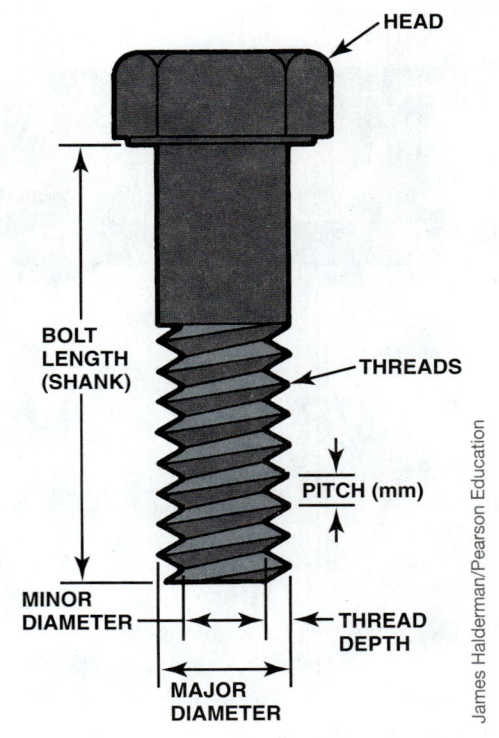

FIGURE 2–5 Typical threaded fastener description (Halderman, *Automotive Technology*, 5th ed., Fig. 8–1).

Bolt and nut fastener diameters start at $\frac{1}{4}$ in. (standard) or M8 (metric) and can go up to 2 in. or 50 mm and even larger. ● SEE TABLE 2–1 for threaded fastener sizing. Screw diameter sizing that you will likely see used on heavy equipment starts at #6 (standard) and goes to #12 or M4 (metric) to M7.

Threaded screws are used for fastening smaller components such as light covers or dashes and electrical components together.

There are many variations of regular bolts that are used for special applications. A couple of examples are plow bolts for cutting edges and pusher bolts for disassembling components.

Screw heads will be made to accept one of several types of tools so the fastener can be tightened or loosened. Typical styles are slotted, Philips, Robertson, Torx, and Allen.

There are some simple tools that are used to identify thread pitch (thread pitch gauge) and fastener size that should be part of every HDET's tool collection.

There are several additional ways to classify threaded fasteners such as:

Tensile strength: This relates to the maximum stretch that a fastener will withstand before becoming deformed. If a fastener becomes deformed because of excessive loading, it should not be used again.

FASTENERS, SEALS, GASKETS, BEARINGS, PERSONAL TOOLS, SHOP EQUIPMENT, AND FIELD SERVICE REPAIR

3/8"x 1"- UNC (16 tpi- threads per inch)

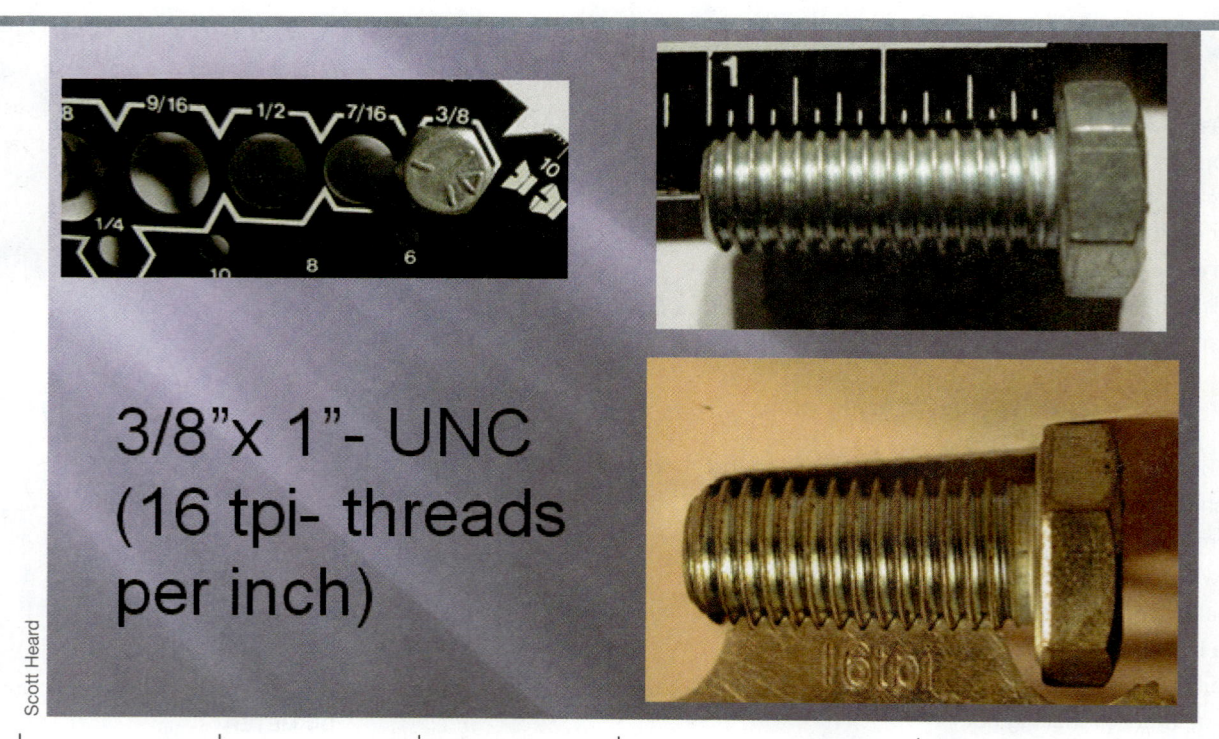

	Full Size Body Diameter		Width Across Flats		Width Across Corners		Head Height			Radius of Fillet		Thread Length for Bolt Lengths	
			Basic and									6 in. and Shorter	Over 6 in.
Nominal Size	Max	Min	Max	Min	Max	Min	Basic	Max	Min	Max	Min	Nom.	Nom.
	A		B		C		H			R		L	
1/4	0.260	0.237	0.438	0.425	0.505	0.484	0.172	0.188	0.150	0.030	0.010	0.750	1.000
5/16	0.324	0.298	0.500	0.484	0.577	0.552	0.219	0.235	0.195	0.030	0.010	0.875	1.125
3/8	0.388	0.360	0.562	0.544	0.650	0.620	0.250	0.268	0.226	0.030	0.010	1.000	1.250
7/16	0.452	0.421	0.625	0.603	0.722	0.687	0.297	0.316	0.272	0.030	0.010	1.125	1.375
1/2	0.515	0.482	0.750	0.725	0.866	0.826	0.344	0.364	0.302	0.030	0.010	1.250	1.500
5/8	0.642	0.605	0.938	0.906	1.083	1.033	0.422	0.444	0.378	0.060	0.020	1.500	1.750
3/4	0.768	0.729	1.125	1.088	1.299	1.240	0.500	0.524	0.455	0.060	0.020	1.750	2.000
7/8	0.895	0.852	1.312	1.269	1.516	1.447	0.578	0.604	0.531	0.060	0.020	2.000	2.250
1	1.022	0.976	1.500	1.450	1.732	1.653	0.672	0.700	0.591	0.090	0.030	2.250	2.500
1 1/8	1.149	1.098	1.688	1.631	1.949	1.859	0.750	0.780	0.658	0.090	0.030	2.500	2.750
1 1/4	1.277	1.223	1.875	1.812	2.165	2.066	0.844	0.876	0.749	0.090	0.030	2.750	3.000
1 3/8	1.404	1.345	2.062	1.994	2.382	2.273	0.906	0.940	0.810	0.090	0.030	3.000	3.250
1 1/2	1.531	1.470	2.250	2.175	2.598	2.480	1.000	1.036	0.902	0.090	0.030	3.250	3.500
1 3/4	1.785	1.716	2.625	2.538	3.031	2.893	1.156	1.196	1.054	0.120	0.040	3.750	4.000
2	2.039	1.964	3.000	2.900	3.464	3.306	1.344	1.388	1.175	0.120	0.040	4.250	4.500
2 1/4	2.305	2.214	3.375	3.262	3.897	3.719	1.500	1.548	1.327	0.190	0.060	2.750	5.000
2 1/2	2.559	2.461	3.750	3.625	4.330	4.133	1.656	1.708	1.479	0.190	0.060	5.250	5.500
2 3/4	2.827	2.711	4.125	3.988	4.763	4.546	1.813	1.869	1.632	0.190	0.060	5.750	6.000
3	3.081	2.961	4.500	4.350	5.196	4.959	2.000	2.060	1.815	0.190	0.060	6.250	6.500
3 1/4	3.335	3.210	4.875	4.712	5.629	5.372	2.188	2.251	1.936	0.190	0.060	6.750	7.000
3 1/2	3.589	3.461	5.250	5.075	6.062	5.786	2.313	2.380	2.057	0.190	0.060	7.250	7.500
3 3/4	3.858	3.726	5.625	5.437	6.495	6.198	2.500	2.572	2.241	0.190	0.060	7.750	8.000
4	4.111	3.975	6.000	5.800	6.928	6.612	2.688	2.764	2.424	0.190	0.060	8.250	8.800

TABLE 2–1

Threaded fastener sizing.

Standard fasteners are graded by their tensile strengths in grades of bolts with grade 1 being the lowest and grade 12 being the highest. ● SEE TABLE 2–2 to understand how different grades relate to tensile strength. Standard fastener grades are identified by the number of lines on the bolt head while metric fasteners will have a number stamped on their head. Typical standard fasteners found on heavy equipment will be grade 8, have a tensile strength of 150,000 psi, and have six lines on their head. A typical metric

SAE BOLT DESIGNATIONS

SAE GRADE NO.	SIZE RANGE	TENSILE STRENGTH PSI	MATERIAL	HEAD MARKING
1	$\frac{1}{4}$ through $1\frac{1}{2}$	60,000	Low or medium carbon steel	
2	$\frac{1}{4}$ through $\frac{3}{4}$ $\frac{7}{8}$ through $1\frac{1}{2}$	74,000 60,000		
5	$\frac{1}{4}$ through 1 $1\frac{1}{8}$ through $1\frac{1}{2}$	120,000 105,000	Medium carbon steel, quenched & tempered	
5.2	$\frac{1}{4}$ through 1	120,000	Low carbon marten-site steel*, quenched & tempered	
7	$\frac{1}{4}$ through $1\frac{1}{2}$	133,000	Medium carbon alloy steel, quenched & tempered	
8	$\frac{1}{4}$ through $1\frac{1}{2}$	150,000	Medium carbon alloy steel, quenched & tempered	
8.2	$\frac{1}{4}$ through 1	150,000	Low carbon Marten-site steel*, quenched & tempered	

TABLE 2–2

Bolt grades and their tensile strength (Halderman, *Automotive Technology* 5th ed., Chart 8–2).

*Martensite steel is steel that has been cooled rapidly, thereby increasing its hardness. It is named after a German metallurgist, Adolf Martens.

fastener will have 10.9 stamped on its head, and this equates to a tensile strength of 150,000 psi. The 10.9 rating relates to the tensile strength being measured in MPa.

The type of material they're made from: The material they are made from usually relates to their tensile strength. High tensile strength bolts will be made from fine grade steel. Occasionally, stainless steel or brass could be used for fastener material.

The type of coating applied to the bolt's surface: Different coatings are usually applied to threaded fasteners to minimize corrosion and or friction. Corroded fasteners will be difficult to remove and will eventually fail prematurely. Phosphate and oil coatings will provide better clamping loads while zinc plating will resist corrosion.

Threaded nuts are also classified by threaded diameter, by thread pitch, by material they're made from, by material they're coated with, if they are designed to be anti-loosening, by their height, and by the outside diameter of opposing flats. Nuts are almost always six-point hexagonally shaped unless they are for special purposes. ● **SEE FIGURE 2–6** for a variety of threaded nuts.

Some ways to prevent nuts from loosening are nylon inserts that create friction on the bolt, distorted nuts (sometimes called Stover nuts), and nuts with distorted lips that are simply called lock nuts.

Threaded fasteners create a clamping force when tightened properly. ● **SEE TABLE 2–3** for the clamping forces created by different sizes of fasteners. To create this clamping force, the fastener needs to be tightened or torqued to a specific amount and with specific conditions. The torque amount can be accurately measured with a torque wrench and or tightening sequence. The fastener's thread conditions can affect the clamping force also. The threads should not be damaged, and if a threaded fastener is being installed into a blind hole or a sealed hole, you must be certain that the blind hole is cleaned out so that the fastener will not bottom out in the hole. If the fastener's threads are damaged or it bottoms out, it will not be creating sufficient clamping force. The threads may

FIGURE 2–6 A variety of threaded nuts and a threaded fastener sizing gauge.

FASTENERS, SEALS, GASKETS, BEARINGS, PERSONAL TOOLS, SHOP EQUIPMENT, AND FIELD SERVICE REPAIR

ASTM A354 GRADE BD/SAE GRADE 8

BOLT SIZE (in.)	TPI	PROOF LOAD (lbs)	CLAMP LOAD (lbs)	TIGHTENING TORQUE (ft lbs) LUBRICATED	PLAIN
1/4	20	3,800	2,850	6	12
5/16	18	6,300	4,725	12	25
3/8	16	9,300	6,975	22	44
7/16	14	12,750	9,563	35	70
1/2	13	17,050	12,788	53	107
9/16	12	21,850	16,388	77	154
5/8	11	27,100	20,325	106	212
3/4	10	40,100	30,075	188	376
7/8	9	55,450	41,588	303	606
1	8	72,700	54,525	454	909

TABLE 2-3

Clamping forces created by different sizes of fasteners.

need to be clean and dry or lubricated with oil or other types of lubrication. Final clamping force of the threaded fastener will be greatly influenced as to what condition the threaded fastener is in. The machine's service information will advise you of this, and if it doesn't mention any special conditions for tightening fasteners, assume clean and dry threads and standard torque values.

Proper torqueing of fasteners is usually done by using a torque wrench and tightening the fastener until the torque wrench clicks or indicates that the proper torque has been reached. The second method of torqueing threaded fasteners is called torque turn. This requires an initial torque plus a certain amount of degrees of additional rotation to achieve the desired clamping force. Torque wrench usage will be covered later in this chapter.

Some threaded fasteners will be described as torque to yield and should be discarded after one use because they will be deformed or stretched when torque to specification. Threaded fasteners used in corrosive conditions will likely require some type of anti-seizing compound applied to them, but before using this type of material, machine's service information should be checked.

ANTI-LOOSENING DEVICES/METHODS. Fasteners will tend to loosen; to prevent this, there are liquids that can be applied to fasteners that create a bond between the threads of a set of mating threads. There are several types of locking washers, and sometimes a manufacturer requires a component to be deformed by staking it with a center punch to prevent loosening.

THREADED FASTENER INSPECTION. Threaded fasteners should be inspected before reuse. This can be a simple quick visual check to look for cleanliness of threads, thread damage, straightness, and head damage. For fasteners that have a more important function such as holding a cylinder head to a block, a more thorough inspection should take place and include checking for stretched bolts. This is sometimes done with a special gauge that acts as a template. Many manufacturers will instruct the technician to discard cylinder head bolts, no matter what their condition or type.

THREAD REPAIR. Damaged threads should be repaired to make sure that their intended clamping force can be achieved again. Depending on how severe the damage is there are different repair methods, tools, and supplies that can be used. If a bolt or nut has major thread damage, it should be replaced.

For light damage, a thread chaser or thread file can restore the threads. You need to make sure to use the correct pitch for the damaged thread. A tap or die can also be used.

For major thread damage of a component with a female thread, there are different threaded inserts that can replace the original threads. The component is drilled out oversize, threaded with a special tap, and a thread insert is installed with a special tool. If the repair is done properly, the threads should withstand the same amount of clamping force as the original thread.

MAKING THREADS. Occasionally, you may need to make threads. To make threads on a rod, a die is used. To make threads in a component, you would first select the proper drill size for the tap and then tap new threads in the hole. Drilling and tapping fluid should be used to make cutting new threads with a tap easier.

WASHERS. When threaded fasteners are used, washers are quite often used in combination to spread the clamping force out on the component being fastened.

There are several types of washers but the most common would be a flat washer that is used for the previously stated purposes. Lock washers are sometimes used to lock fasteners in place once they are torqued; however, their use is becoming less common. There are several different types of lock washers but the most common being a split ring made from spring steel.

Washers can be made from many different materials and could be coated with different materials to prevent corrosion.

RETAINING/LOCATING DEVICES. Many parts of a machine such as rods, shafts, bearings, and gears need to be held into place

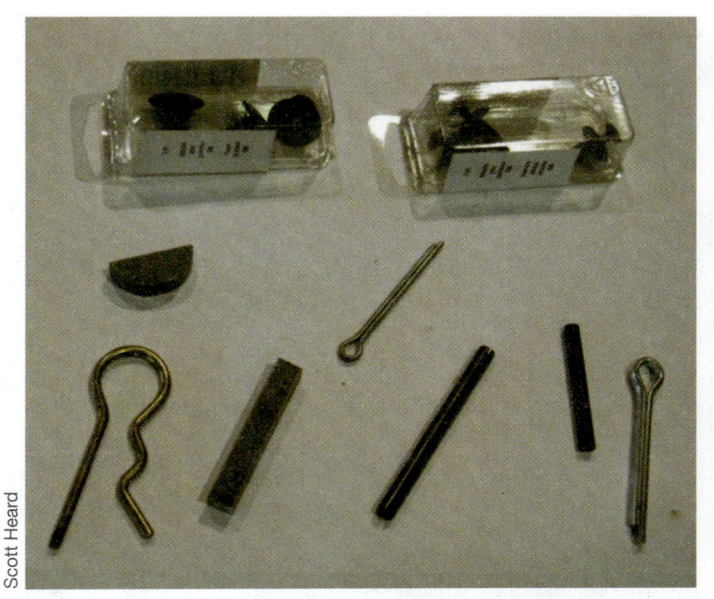

FIGURE 2–7 A variety of different types of retainers.

FIGURE 2–8 A leaking seal.

or located in a specific position. There are several different devices to do this.

Some examples are cotter pins, roll pins (also called split pins), dowels, circlips, and snap rings. ● **SEE FIGURE 2–7** for a variety of different retaining devices.

Care needs to be taken when installing these retainers so as to not damage them by overstretching or deforming them. It's a good practice to discard these fasteners upon removal and install new ones at reassembly.

To lock something onto a shaft such as a gear or pulley and prevent rotation between these two parts, a key can be used. These can be woodruff keys that are semicircular or square stock keys. Keys fit into machined slots on both parts to lock the two parts together.

SEALS. A typical heavy equipment machine will use many different types of seals. Seals are necessary to keep fluids inside compartments, separated from another fluid, or to keep dirt out of compartments. There are many different types and styles of seals, and they are made of many different types of materials.

A very common repair is to fix a leak because of a failed seal so it's important to understand the different types of seals and what they are made of. ● **SEE FIGURE 2–8** for a leaking seal.

Seals can be grouped into two main categories: static, the seal works between two stationary components, and dynamic, the seal works between two components that move in relation to each other. Static-type seals can be broken down into different types of seals such as O-rings, D-rings, backup rings, and flat rings. Dynamic seals can be metal-to-metal piston ring-type seals, taper-faced metal-to-metal seals, lip-type

> **WEB ACTIVITY**
>
> Search "fluid seals" and see how many different manufacturers there are and check out some different types of seals. What's the biggest seal you can find? Where could you find the biggest seal on a heavy equipment machine?

seals, lip-type metal-supported seals, and spring-loaded lip-type metal-supported seals.

O-RING, D-RING, SQUARE-RING SEALS. These are flexible seals made from an elastomeric compound made of a base polymer that is mainly a natural or synthetic rubber. They will usually stay circular in shape in use but can sometimes fit into an odd-shaped groove or over a noncircular-shaped component. Their cross-section shape determines if they are called an O-ring, D-ring, or square-ring. For the remainder of this section, all these seals will be called O-rings because they are the most common.

They are sized by their overall inside and outside diameter and cross-sectional diameter. Some different types of O-ring material are silicone, fluorocarbon, neoprene, silicone, and nitrile. Seal material is based on the type of fluid it will be exposed to and the fluid's normal operating temperature. This

FIGURE 2–9 A variety of different seals.

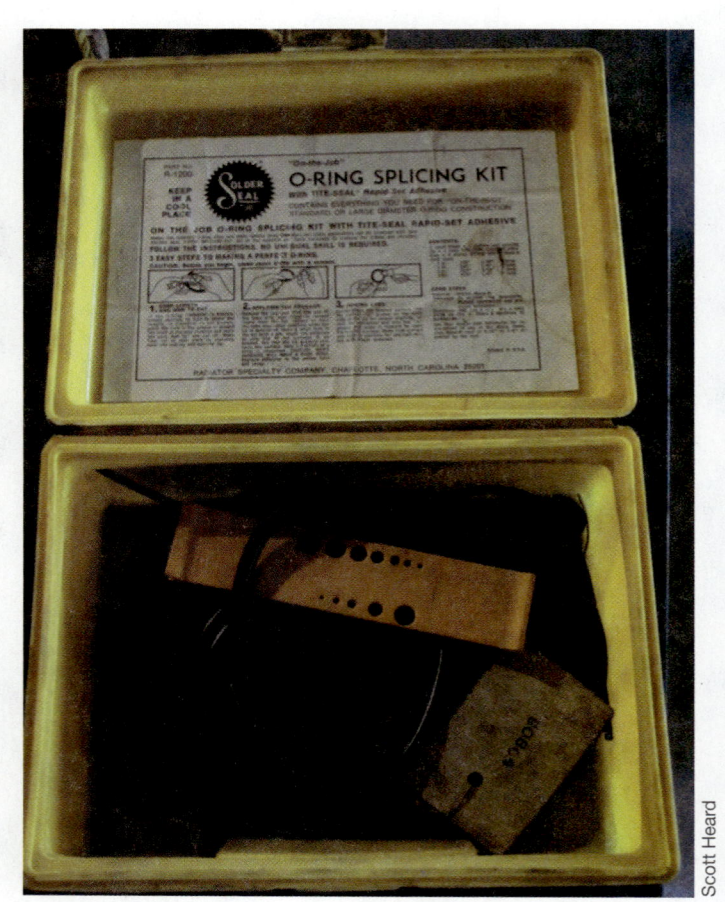

FIGURE 2–10 An O-ring repair kit.

means the seal material has to be compatible with the fluid and withstand the temperatures it will be subjected to. If an O-ring is used for some noncompatible fluid or out of its temperature range, it can't be expected to seal properly. If an O-ring is overheated, it will become hard, crack, and allow a fluid leak. Heat is the leading cause of O-ring failure.

If an O-ring is used to seal two components, it will fit into a groove on one component and should get compressed by about 1/3 of its diameter when the second component is fastened to the first to provide a positive seal. O-ring seals that are used with fluid fittings can be used to fit over top of the fitting and seal against a beveled shoulder in a component when the fitting is threaded into it. O-rings could also seal the fitting on its face when they fit into a groove in the fitting.

Some O-ring-type seals that are used around the circumference of a shaft or cylinder rod will require a second ring that is called a backup ring. This will be a harder nylon material that is placed directly beside the O-ring to provide support for it in the O-ring groove around the outside circumference of a component. ● **SEE FIGURE 2–9** for a variety of different seals.

Any time components are removed for repair or service, it is common practice to replace all O-ring-type seals. This is a cheap way to ensure that there won't be any seal-related leaks.

The groove that the O-rings fit into must be clean and corrosion free.

O-ring-type seals can be custom-made from bulk lengths. O-ring-making kits will come with measuring and cutting tools and special adhesives to join the two ends of the O-ring. ● **SEE FIGURE 2–10** for an O-ring repair kit.

O-ring seals should be lubricated when installed, and sometimes grease is used to hold the seal in place upon installation. You should also be sure that they aren't twisted. A good set of seal picks will help with removing and installing O-ring seals.

LIP-TYPE SEALS. Lip-type seals are designed to have one or more sharp elastomeric sealing surfaces creating a seal between the seal lip and the mating circumferential surface of a component or shaft. Some lip-type seals will rotate on a stationary shaft and others will be stationary with the shaft rotating inside the seal. Lip-type seals are usually dynamic seals, meaning that they will seal between two moving components or one stationary and one moving component. These seals can be made from different types of material such as hard urethane, flexible rubber, or a combination of either and a steel backing support. Lip-type seals could have oil pressure acting on the lip to help apply more force to the lip and thereby create

FIGURE 2–11 A variety of lip-type seals.

a more effective seal. ● **SEE FIGURE 2–11** for a variety of lip-type seals.

Some lip-type seals will use spring pressure behind the lip to assist with creating a tighter seal. You should always check this type of seal after installation to make sure that the spring is still in place.

Lip-type seal installation can be tricky and care needs to be taken to not damage the seal on installation. Depending on the type of lip-type seals, there could be several different installation procedures.

Urethane or elastomeric lip-type seals that are installed on the inside of hydraulic cylinder head will need to be carefully folded without creasing the seal. This will effectively decrease the inside diameter of the seal allowing it to fit inside the cylinder bore and then expand into the seal groove. There are special tools to do this properly.

Metal-supported lip-type seals will usually be installed with an interference fit into a recess in the component. The metal backing or support for this type of seal will usually have a soft vinyl like coating on it, which will fill any inconsistencies between the metal backing and the component it is installed into. The ideal way to install a metal-supported lip-type seal is with a seal installer tool that will center itself on the seal and apply pressure only to the outside diameter of the metal support. This tool will likely have a solid part in the center that allows the technician to drive the seal in with a hammer. Care needs to be taken when installing this type of seal to install it evenly and to stop driving when the seal has bottomed out.

Removing lip-type seals could involve using heel bars, pry bars, or drilling holes in the metal support and inserting screws that are pried on or pulled out with a slide hammer and a coarse threaded adapter.

Grooves or recesses for lip-type seals should be clean and dry before seal installation. The seal lip should be lubricated with before component assembly.

Some lip-type seals will have their lips ride on a wear sleeve. The wear sleeve is a thin piece of hardened steel that is designed to wear over time and should be replaced when the seal is replaced. This prevents the rotating motion from wearing on the actual components. Wear sleeves are installed by gently heating them and using a special tool to slide them over the components they are protecting or pressed on with special tools. When the sleeve cools, it will shrink onto the component for a tight fit. Engine crankshafts will commonly use wear sleeves.

Some lip-type seals and wear sleeves come as an assembly and must not be separated. They will require a special installation tool and are mainly used for rear crankshaft seals on engines.

METAL-TO-METAL FACE SEALS. These are sometimes called Duo-cone seals, and they feature two hard special alloy metal rings that are precisely hardened and lap fitted. The hard smooth surfaces run against each other and create a seal on their faces. The fluid that the seal is containing will provide some lubrication and cooling for the seal.

They will have a large elastomeric toric O-ring seal behind them. The O-rings will keep pressure on the metal seals after component assembly as well as complete a seal between the outer circumference of the metal seal and the component they are mounted in. One metal seal is stationary while the other rotates against it. This type of seal is most commonly used for wheel and final drive seals. Duo-cone seals are very effective at keeping oil in and dirt out. ● **SEE FIGURE 2–12** for a Duo-cone type of seal.

Duo-cone seals can be reused, provided they don't exceed recommended wear specifications and don't show other signs of failure such as pitting or cracks. Wear is measured with Duo-cone seals by seeing how wide the contact pattern is between the two metal parts. Duo-cone toric O-ring seals must be completely dry and clean during installation; however, the metal faces must be lubricated before component assembly. Special installation tools are recommended to be used for installing these seals.

PISTON RING-TYPE SEALS. This style of seal is typically found in turbochargers and used to seal the turbo shaft. They are also found in torque converters and transmissions. They are a simple hardened metal ring that creates a dynamic seal between two rotating parts and are very similar to piston rings. Care must be taken when installing this type of seal to not overstretch or scratch them.

GASKETS. Gaskets can be very simple in design and made from one type of material. They may need to seal almost no pressure, or they can be very complex multi-material structures that could be required to seal several thousand psi of pressure such as when used for sealing between an engine block and its cylinder head. Gaskets can be made from materials such as paper, cork, plastic blends, steel, rubber blends, steel rubber laminations, and composite fiber blends. They should be designed to seal two stationary parts and the pressures that are built up inside one or more compartments. Although the

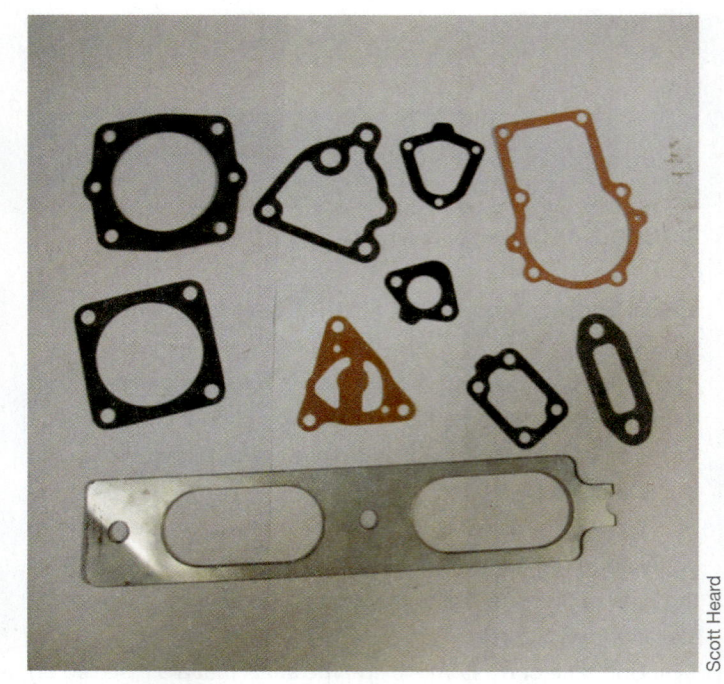

FIGURE 2–13 A variety of gaskets.

components that a gasket is sealing are stationary, the gasket must allow for slight relative movement created by temperature or pressure changes. They should also allow for slight surface imperfections. ● **SEE FIGURE 2–13** for a variety of gaskets.

Most gaskets should be installed on clean dry surfaces. This will require a thorough cleaning procedure of old gasket material. Appropriate care and PPE usage should be practiced when removing old gaskets.

Care must be taken when cleaning old gasket material off of metal surfaces. If you are too aggressive with your cleaning procedure, it is possible to remove metal from the component being cleaned. Some cleaning tools you may use for this are scrapers, knives, emery cloth, sandpaper, fiber pads on pneumatic die grinders.

BEARINGS. Bearings play an important part in keeping heavy equipment machines working. They are designed to greatly reduce friction between moving parts by keeping them separated and allowing relative motion. If two parts are allowed to contact each other while there is relative motion, then the increase in friction will create heat. If enough heat is created, the parts will start to weld themselves together, similar to inertia welding. Almost all bearings are made from metal, but some light-duty bearings could be nonmetallic.

Any part of the machine that moves or rotates will have some type of bearing to allow movement and rotation freely between the part and one or more other parts without damage occurring. Bearings can allow partial rotation, low-speed full rotation, and high-speed full rotation, and some are used for

FIGURE 2–12 A Duo-cone-type of dynamic seal.

> **WEB ACTIVITY**
>
> Go to http://www.timken.com/AntiFriction/player.html to learn more about bearings. This is a great interactive webpage that has lots of information about the different types of bearings and how to install and maintain them.

controlling axial movement. The size and type of bearing used mostly depends on the amount of load and speed during normal operation between the two parts the bearing is separating.

FRICTION TYPE BEARINGS

PLAIN BEARINGS

FRICTION BEARINGS (PLAIN BUSHINGS). This type of bearing can be used for slow linear movement like an extendable backhoe stick, partial slow rotation of components like bucket linkages, or very fast rotating components like engine camshafts.

Plain spherical bearings provide mostly radial support but will also limit axial movement. This type of bearing is commonly found at the ends of hydraulic cylinders where their outer diameter is an interference fit in the cylinder rod and barrel bore while the bearing's inside diameter accepts a smooth pin. The pin secures the cylinder to the machine and the bearing allows movement between components. This type of bearing is made from hardened steel and is usually lubricated with high pressure grease.

Plain bearings used in engines are multilayered thin metal circular strips that can be either one or two piece semicircular design. One piece circular plain bearings will be an interference fit into a bore in the engine and rely on pressurized oil to support the rotating component that turns inside of it. An example of this type of bearing is a camshaft bearing.
- **SEE FIGURE 2–14** for a cam bearing.

Plain bearings are usually round and flat on both inside and outside diameters with one or more fluid application holes drilled through. There could also be diagonal grooves on the bearing surface to allow a more even grease film. There needs to be a specific amount of clearance between the bearing and the part it is supporting so the fluid can separate the components. ● **SEE FIGURE 2–15** for a plain bearing.

Occasionally, plain bearings will be mounted on rotating shafts but they are typically installed with an interference that's in the bore of a component. For bearing installation, this may require cooling the bearing or pressing the bearing into place with special tools. Once again bearing and component

FIGURE 2–14 Cam bearing.

FIGURE 2–15 A plain bearing.

cleanliness is critical to proper installation. Bearing removal could include using a puller, carefully cutting the bearing, or welding on the bearing surface and allowing the welds to cool. This last process will shrink the bearing.

ANTI-FRICTION

BALL BEARINGS. Ball bearings are typically used to allow independent rotation of a component and a shaft and could be used for low- or high-speed applications. They are designed to withstand mainly radial loads and limited axial loads. They are for relatively low loads because the load is concentrated on a few small contact points. Ball bearings are typically an assembly with an outer race, inner race, balls, and a bearing cage. Bearing cages are meant to keep the rolling element of the bearing separated and evenly spaced apart. They can be made of light, soft metal, or hard-machined metal depending on the bearing design.

FASTENERS, SEALS, GASKETS, BEARINGS, PERSONAL TOOLS, SHOP EQUIPMENT, AND FIELD SERVICE REPAIR

TECH TIP

Lack of lubrication and neglect by a machine operator led to a major bearing failure on a machine at a worksite where I was formerly employed.

The main production loader at this particular quarry had an automatic greaser that would occasionally malfunction. This was easy to detect visually because of a lack of grease at the grease points. There would also be audible clues (squeaking) to tell the operator that something wasn't being lubricated properly.

One day a squeaking noise was reported by the operator to the foreman but he was told to keep working for the rest of the shift. The squeak was coming from the loader boom upper pin. The pin on this machine is about 40 in. long and 8 in. in diameter. The pin is stationary in the machine and the loader boom has two plain bearings pressed into it that are about 12 in. wide each. Grease normally keeps the pin and bearings separated.

A couple of hours later the joint became so dry that the 8 in. diameter pin seized in the boom bearings and broke into two pieces. The operator was lucky that the boom didn't come through the cab when the pin failed.

The resulting repair took several days because the boom had to be removed, the pin and bearings had to be melted out, and new parts had to be installed. The repair probably cost in excess of $25,000 in parts and labor, not to mention machine downtime.

This was an expensive lesson in the importance of a properly lubricated bearing. ● SEE FIGURE 2–16 for the pin that failed.

FIGURE 2–16 Wheel loader boom pin that failed.

Ball bearing races will usually be an interference fit for either the outer diameter of the outer race or the inner diameter of the inner race. This means installation will require a gentle heating or cooling process and a proper puller tool is used to install the bearing. Low-speed ball bearings could be lubricated by grease while medium- to high-speed bearings will be lubricated with oil. Ball bearing assemblies could also incorporate seals to keep both contamination out and lubrication in. ● SEE FIGURE 2–17 for a ball bearing.

ROLLER BEARINGS. Roller bearings will be used in applications where a shaft exerts high radial loads with very minimal axial loads (10% of axial load). Three types of roller bearings are spherical, cylindrical, and needle. Spherical-type bearings have barrel-shaped rollers, will handle some axial load, and can tolerate misalignment of the two separated components. Cylindrical- or needle-type bearings have straight-shaped rollers. A typical use for a cylindrical roller bearing is to support the planet pinion gears in final drives and powershift transmissions. Needle bearings are a type of roller bearing that uses many small diameter rollers. Needle bearings can be found in U-joint caps.

Similar to ball bearings, roller bearings will have an inner race, outer race, rollers, and roller cage. Roller bearings could come preassembled or have a separate outer race. Roller bearings can be used in partial and full rotation and low- to high-speed applications. ● SEE FIGURE 2–18 for a roller bearing.

Removing a roller bearing could require puller tools, a hammer and punch, cutting tools, or heating tools. Roller bearing installation could require gentle heating, cooling, or puller tools. Care must be taken to only push on the races when installing.

Depending on their application and speed, roller bearings could be lubricated by grease or oil.

TAPERED ROLLER BEARINGS. Tapered roller bearings have rollers that are cone shaped and the angle of the cone shape matches the angle of the race surfaces. These bearings have high axial and radial load handling characteristics. Tapered roller bearings are commonly used for wheel bearings and in differentials for the pinion gear shafts and crown gear assemblies.

Tapered roller bearings have the ability to be preloaded, which ensures that the rollers are in full contact of the races all the way around the bearing. They will come as two separate

FIGURE 2–17 A ball bearing.

FIGURE 2–18 A roller bearing.

pieces commonly called a cup and cone. The cup is the tapered outer race and the cone is an assembly of the tapered inner race with the rollers and a bearing cage. ● **SEE FIGURE 2–19** for a tapered roller bearing.

Removing tapered roller bearings requires similar procedures to roller and ball bearing removals; however, bearing installation will require an adjustment procedure.

Tapered roller bearing preload adjustment could involve adding or removing shims or a fastener torqueing procedure. The goal of this procedure is to ensure that every roller is supporting an equal part of the load. To confirm this, a specific rolling resistance must be created, which equates to proper bearing preload. It is important throughout the bearing adjustment procedure to lubricate the rollers and to rotate the component to allow proper bearing seating to get an accurate reading.

FIGURE 2–19 A tapered roller bearing.

FASTENERS, SEALS, GASKETS, BEARINGS, PERSONAL TOOLS, SHOP EQUIPMENT, AND FIELD SERVICE REPAIR

STEP #1. HAND SPIN WHEEL

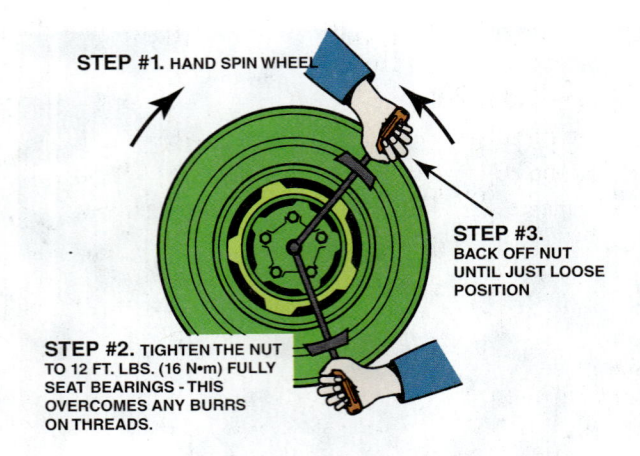

STEP #3. BACK OFF NUT UNTIL JUST LOOSE POSITION

STEP #2. TIGHTEN THE NUT TO 12 FT. LBS. (16 N•m) FULLY SEAT BEARINGS - THIS OVERCOMES ANY BURRS ON THREADS.

STEP #5. LOOSEN NUT UNTIL EITHER HOLE IN THE SPINDLE LINES UP WITH A SLOT IN THE NUT – THEN INSERT COTTER PIN.

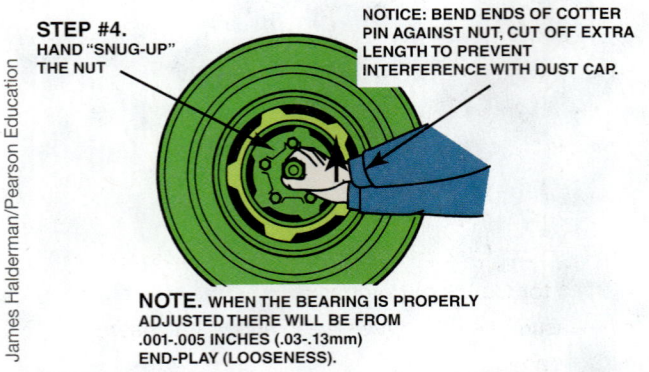

STEP #4. HAND "SNUG-UP" THE NUT

NOTICE: BEND ENDS OF COTTER PIN AGAINST NUT, CUT OFF EXTRA LENGTH TO PREVENT INTERFERENCE WITH DUST CAP.

NOTE. WHEN THE BEARING IS PROPERLY ADJUSTED THERE WILL BE FROM .001-.005 INCHES (.03-.13mm) END-PLAY (LOOSENESS).

FIGURE 2–20 A tapered roller bearing adjustment procedure.

James Halderman/Pearson Education

Preload measurement is usually accomplished by using a fish scale to measure the pounds or kilograms of force required to rotate the supported component. ● SEE FIGURE 2–20 for an example of a tapered roller bearing adjustment procedure.

BEARING FAILURES. Bearings rarely fail as a result of a design or manufacture defect. Machine design engineers select bearings for use based on normal operating loads and conditions, regular maintenance being done, and an expected life span. If the machine is operated, maintained, and repaired according to the manufacturer's recommendations, the machine's bearings should last several thousand hours. They usually fail as a result of one of the following causes:

- **Overloading:** For example, if a rock truck is constantly overloaded, this extra weight will overload the machine's wheel bearings and shorten their life.
- **Overspeeding:** For example, if a machine gets stuck and one wheel receives all the drive torque from the differential, it will be overspeeding one wheel bearing and could cause it to fail.

WEB ACTIVITY

Go to http://www.emersonbearing.com/technical-toolbox.html to see a variety of different bearing failure types, possible causes, and what to do to prevent it happening again.

TECH TIP

When reconditioning components with bearings, the bearings should be thoroughly cleaned with a water-free cleaning solvent like Varsol and inspected. Look for imperfections such as pitting, scoring, brinelling, corrosion, chipping, smearing, discoloration, and flaking of the surface. *Do not* blow-dry bearings with full shop air pressure as this may overspeed the roller elements and cause damage.

Reusing nonfriction-type bearings is possible if they pass a thorough visual inspection and feel smooth when rotated by hand.

Do not unpack new bearings until you are ready to install them.

New bearings should always be lubricated with the oil they will normally run in after they are installed. Ideally, you should wear latex gloves when handling new bearings as the moisture from your hands will start corroding the bearing.

If new bearings need to be heated to be installed, *do not* overheat them. Use a temperature crayon to identify when the correct temperature has been reached. Bearings should never be heated directly with a torch. Some methods include hot oil, oven, and induction heater. To cool a bearing for installation, you could use liquid nitrogen, use dry ice, or put the bearing in a freezer overnight.

- **Contamination:** Bearings need a specific type of clean lubrication and will fail if the lubrication is compromised because of contamination from dirt, water, metal, coolant, air, fuel, or other fluids.
- **Misalignment:** If the bearing isn't designed to handle misalignment, failure will occur when there are excessive and misdirected loads created because of this.
- **Lack of lubrication:** Low oil levels or lack of grease will cause shortened bearing life.

44 CHAPTER 2

FIGURE 2–21 Failed bearing roller.

Improper installation: Bearings that are misaligned, overheated, or damaged from improper installation (e.g., not pushing on races) will fail prematurely.
● **SEE FIGURE 2–21** for failed bearings.

Progress Check

1. Which one of the following actions would *not* be normally taken when installing a threaded fastener:
   a. torque to proper specification
   b. use lubrication only if service information requests
   c. make sure blind holes are clean and dry
   d. use anti-loosening liquid on its thread
2. Threaded fastener tensile strength refers to:
   a. how the threads are formed
   b. how much torque it takes to deform the fastener
   c. coating strength
   d. amount of torque needed to tighten it properly
3. Fluid compatibility when referring to seals means:
   a. the seal is made of the same base material as the fluid
   b. the seal is made to withstand fluid contamination
   c. the seal is made to withstand extreme pressures
   d. the seal material will not degrade when exposed to the normal system fluid
4. If a threaded fastener is described to be left hand, this means:
   a. it will be used only on the left side of the machine
   b. it will tighten when the top of it is turned to the left
   c. its threads lean to the left
   d. it is designed for left-handed people
5. This would be the largest screw size:
   a. #4
   b. #6
   c. #8
   d. #10
6. This is a typical use for a Duo-cone-type seal:
   a. sealing hydraulic hoses
   b. sealing final drive hubs
   c. sealing turbocharger shafts
   d. sealing bucket pins
7. What must all gaskets allow for when creating a seal between two parts?
   a. temperature changes that create part movement
   b. negative pressure spikes
   c. highly corrosive fluids
   d. fastener detorqueing
8. Occasionally, an HDET will weld on the race of a plain bearing. This will:
   a. be part of a bearing removal process
   b. take up excessive clearance
   c. make sure it doesn't move
   d. fill in the extra lubrication grooves
9. Some bearings will have a cage as part of their assembly that will:
   a. help distribute oil evenly
   b. keep dirt away from the rollers
   c. help preload the bearing
   d. keep the rollers or balls separated
10. When installing some bearings, a fish scale is used to:
    a. help torque the nut properly
    b. measure the rolling resistance after it has been preloaded
    c. weigh the balls or rollers before they are installed
    d. weigh the torque wrench to make sure it's calibrated

PERSONAL TOOLS The types of basic hand tools an HDET needs have not changed much in recent years. However, an increasing amount of specialty tools may be required when working on today's machines. New lighter and stronger materials are being used to manufacture tools and new manufacturing processes will also continue to improve tool quality and longevity.

An HDET will be required to have a minimum selection of hand and power tools to be able to effectively service, diagnose, and repair heavy-duty machines. This minimum selection will vary according to several factors such as whether the HDET is:

- Self-employed
- Working for a contractor
- Working for a government institution
- Working for an equipment dealer
- Working in a component rebuild facility
- Working as a field service technician
- Working on one or more types of equipment
- Working on one or more makes of equipment

FASTENERS, SEALS, GASKETS, BEARINGS, PERSONAL TOOLS, SHOP EQUIPMENT, AND FIELD SERVICE REPAIR

- Working on one or more sizes of equipment
- Working on a lube truck

We will look at the tools that a typical HDET working in a heavy equipment general repair shop would have in his or her toolbox. All sized tools will need to be duplicated in both standard or imperial and metric sizes.

Tools are a very costly investment and should be purchased with quality, comfort, and value kept in mind. Some companies will provide an annual tool allowance to HDETs to offset the cost of buying tools. HDET apprentices will be expected to start into the trade with a minimum amount of basic tools and then will typically acquire more tools as they are needed and their budget allows.

Most tools will carry a lifetime warranty that provides free replacement from the manufacturer as long as the tool hasn't been abused. Generally speaking, higher priced tools are higher quality and will have a better guarantee. Tool manufacturer and dealer support should also be a consideration when purchasing tools. If a tool fails through no fault of your own, you need it replaced as soon and with as little hassle as possible. Some tool retailers will be easier to deal with than others, and you are best to ask other HDETs what their experience has been like dealing with tool retailers.

Some tool manufacturers will have their sales reps travel around in cube vans and service trucks providing convenient sales and service directly to HDETs.

PROPER TOOL USAGE. Hand tools are designed to last a long time as long as they are used properly and used for the job they are designed for. For example, a screwdriver is not a pry bar and should never be used as one. Instead of trying to save time and not bother getting the proper tool for the job, you will likely damage the component you're working on and the tool you're using. Hand tools should be maintained to get maximum life out of them and properly repaired if not in good working condition. For example, punches and chisels should be dressed and not have mushroom heads. Ratchets that slip should be repaired or replaced.

Proper tool maintenance will also result in a decreased chance of injury. If an improper tool is used, chances are it will slip or break just when you may be applying a lot of force on it, resulting in cuts, scrapes, and bruises.

Proper tool usage also includes applying no more torque or force to the tool than it is designed for. Avoid double wrenching and using pipes on ratchets for added leverage.

The following is a list of the common tools needed to perform most general service and repair procedures on heavy equipment starting with hand tools.

WRENCHES. Combination wrenches start from $\frac{1}{4}$ to $1\frac{1}{2}$ in. and 6 to 36 mm. They are open on one end and closed (box

---

TECH TIP

Keeping tools clean and organized is a good investment in time. You should always make time to put all tools away either at the end of a job or at the end of the day. There will be lots of time wasted looking for tools otherwise. An organized toolbox is a sign of a good HDET. Try to logically set up your tools, so the most commonly used tools are easy to access.

Try to avoid lending tools unless you can absolutely trust the person borrowing them to return them as soon as they are done with them and in the same condition they were in when they were given out.

This practice also goes for shop equipment as well. Putting shop tools that are shared back to where they should be will make a more efficient shop. Any tools that are broken should be repaired or replaced as soon as possible.

Keeping a tidy workplace is also a time well spent. A good shop should designate a little time each day to allow tools to be put back, old parts to be thrown out, and floors to be swept.

FIGURE 2–22 A set of combination wrenches.

ended) on the other. A box end wrench will completely fit over a fastener to get maximum engagement with it while an open end wrench will only grab the fastener at two flats or sides. The box end wrench will minimize the chance of wrench slippage. Most box end wrenches come in 12-point configuration. This means that it could fit onto a typical 6-point (see fasteners) bolt or nut at 12 different rotational locations around the fastener. To lessen the chance of slippage, a 6-point box end wrench would be the best choice. ● **SEE FIGURE 2–22** for a

FIGURE 2–24 A set of hex wrenches.

FIGURE 2–25 A set of sockets, ratchets, and extensions.

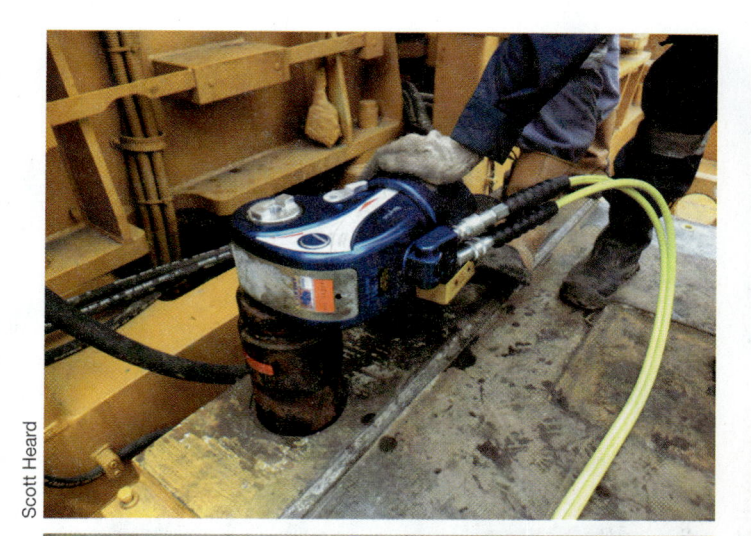

FIGURE 2–23 A hydraulic wrench.

set of combination wrenches. Larger wrenches are sometimes required depending on how big the fasteners are.

There are many other types of wrenches that may be needed by the HDET such as stubby, extra-long, ratcheting, crescent, brake line, offset, hammer, hex, double box end, and double open end. Ratcheting wrenches have become more popular in recent years and are great for quick fastener removal after the fastener has been broken loose.

Hydraulic wrenches can be used for high torque fasteners. ● **SEE FIGURE 2–23** for a hydraulic wrench.

Hex or torx wrenches (sometimes called Allen keys) are used to remove and install hex and torx socket-type fasteners. Hex wrenches will come in both metric and standard sizes.
● **SEE FIGURE 2–24** for a set of hex wrenches.

SOCKET/RATCHET/ACCESSORIES. Socket sets needed are $\frac{1}{4}$, $\frac{3}{8}$, $\frac{1}{2}$, and $\frac{3}{4}$ in. drive with both standard length and deep sockets in all but $\frac{3}{4}$ in. drive sets. Deep sockets allow the socket to fit over a fastener that is threaded past a nut. Chrome sockets are for hand use only while impact sockets are used with impact guns and are stronger but bulkier than chrome sockets. Swivel impact sockets are a great investment that allow an impact gun to be used on an angle other than 90° to the fastener. Various lengths of ratchets and drive extensions will be required as well as flex head ratchets that provide more options when working in tight spaces. One inch drive sets will likely be required for larger fasteners and are usually provided by the employer. Some very large machines could require even larger $1\frac{1}{2}$ in. or spline drive sockets for some very large fasteners. ● **SEE FIGURE 2–25** for a set of sockets, ratchets, and extensions. Many different adapters are available that can give sockets and ratchets of differing drive sizes interchangeability.

FASTENERS, SEALS, GASKETS, BEARINGS, PERSONAL TOOLS, SHOP EQUIPMENT, AND FIELD SERVICE REPAIR

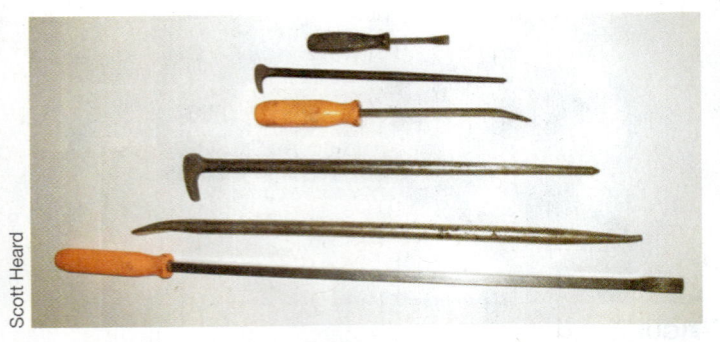

FIGURE 2–26 A set of pry bars.

FIGURE 2–27 A variety of squeezing, holding, and turning tools.

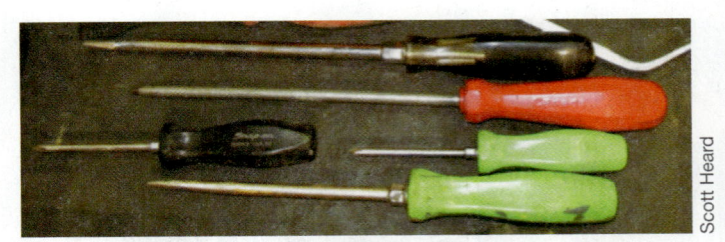

FIGURE 2–28 A variety of screwdrivers.

FIGURE 2–29 A variety of hammers.

For example, an adapter could be used on a $\frac{3}{8}$ in. ratchet to drive a $\frac{1}{2}$ in. drive socket. Swivel adapters also come in all drive sizes to allow the ratchet to be used offset from the socket.

PRY BARS. A good selection of pry bars, lining bars, and heel bars will be used on a regular basis to separate, move, and align different components. Pry bars will allow the technician to use the mechanical advantage of a lever to multiply the force applied to the tool to perform high-force tasks. ● **SEE FIGURE 2–26** for a set of pry bars.

PLIERS, LOCKING PLIERS, ADJUSTABLE WRENCHES, PIPE WRENCHES, CHAIN WRENCHES. These hand tools will be used by an HDET to grab, squeeze, manipulate, and turn various parts and components while working on heavy equipment. ● **SEE FIGURE 2–27** for a variety of squeezing, holding, and turning tools. Pipe wrenches and chain wrenches can be used to manipulate larger round components. Care must be taken to not use these wrenches on delicate surfaces. Pipe wrenches come in sizes of 8 to 48 in. in length. Chain wrenches can also hold unusually shaped components.

SCREWDRIVERS. Heavy equipment machines use a wide variety of fasteners. To install and remove many light-duty fasteners, a technician will quite often need a screwdriver. Screwdrivers come in many different lengths and tip styles. Some examples of styles of tips are straight or slotted, Phillips or star, Robertson, Torx, and Hex. As mentioned earlier, screwdrivers should *not* be used for pry bars. ● **SEE FIGURE 2–28** for a variety of screwdrivers.

HAMMERS. A variety of different types of hammers will be used by an HDET on a regular basis. Hammers such as ball peen, soft blow, and large sledgehammers will all be used for a variety of tasks. Hammers will come in different weights to allow the HDET to select the proper hammer for task at hand. Usually, if five to six good strikes with a hammer isn't moving whatever you are trying to move, then you need a bigger hammer. ● **SEE FIGURE 2–29** for a variety of hammers.

CHISELS AND PUNCHES. There are many different types of chisels and punches needed to service and repair heavy equipment, and a good-quality complete set of punches and chisels will be necessary. Brass punches will be needed to move parts without damaging them. Punches and chisels will come in a wide range of sizes and should be maintained (dressed) to minimize the chance of chipping when struck

FIGURE 2–30 A selection of chisels and punches.

by a hammer. ● **SEE FIGURE 2–30** for a selection of chisels and punches.

DRILLING TOOLS. Drilling holes in components may be necessary on occasion, and a good set of drill bits will make this task easier. Drill bits should be sharpened when they start to need excessive force applied to the drill machine or don't cut through material evenly. Left-handed drill bits are a great investment and will many times remove broken bolts themselves without using other types of extractors.

BROKEN FASTENER REMOVERS. Sooner or later an HDET will come across a broken bolt or stud that needs to be removed. There are several different types of broken fastener removers. They are a hardened tool that will get jammed into a hole that is drilled into the broken fastener. The trick is to not break the removal tool when trying to remove the broken bolt or stud because this can turn into a bigger repair than repairing the broken fastener.

CUTTING AND FILING TOOLS. Sometimes machine components, parts, and other materials may need to be cut or filed. Some hand tools that could be used to do this are hacksaws, metal snips, wire cutters, utility knives, files, and heavy-duty scissors. Side cutters are used to cut wire and plastic ties.
● **SEE FIGURE 2–31** for a variety of cutting and filing tools.

SNAP RING PLIERS. To install and remove snap rings, a special plier like tool is used. These tools will have different sizes and types of ends and will be angled to allow easier tool access to the snap ring. For large diameter snap rings, the tool may have a ratcheting feature that will hold the snap ring open.
● **SEE FIGURE 2–32** for a variety of snap ring pliers.

PULLERS AND DRIVERS. HDETs should have one or two light-duty general purpose pullers in their toolbox. Pullers will be used on occasion to pull apart and push together

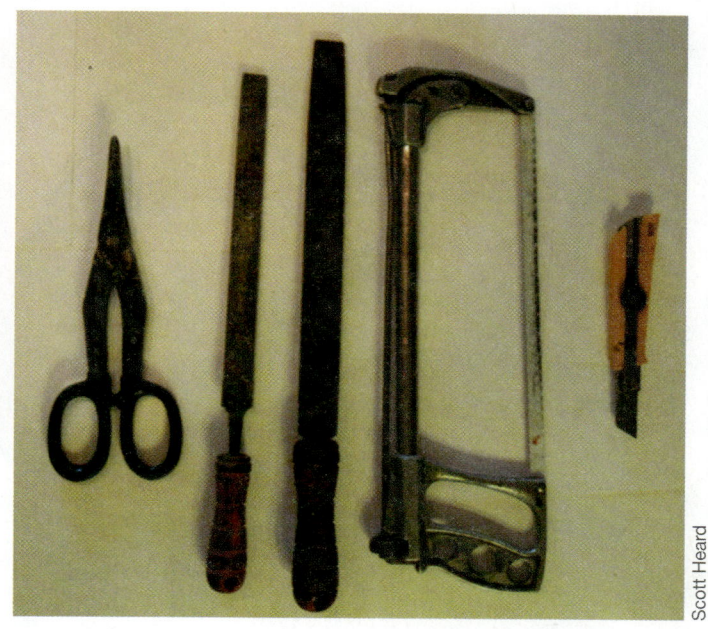

FIGURE 2–31 A variety of cutting and filing tools.

FIGURE 2–32 A variety of different snap ring pliers.

components such as gears, bearings, pulleys, and shafts. A common type is a two- or three-leg puller that uses the legs to pull from a gear off a shaft when a threaded and pointed bolt is turned in against the shaft. Other types of pullers are H-bar, slide hammer, and hydraulic. Pullers can have different adapters and arrangements attached to allow them to do different things. Heavy-duty pullers are usually supplied by the employer. Light-duty drivers can be used to install small seals and bearings. ● **SEE FIGURE 2–33** for a light-duty puller.

TORQUE WRENCHES. Fasteners are required to have a specific amount of torque applied to them to create a clamping force. This will require the use of a torque wrench to ensure that the fastener is not under-torqued or over-torqued. There are different types of torque wrenches: click, beam, electronic,

FASTENERS, SEALS, GASKETS, BEARINGS, PERSONAL TOOLS, SHOP EQUIPMENT, AND FIELD SERVICE REPAIR

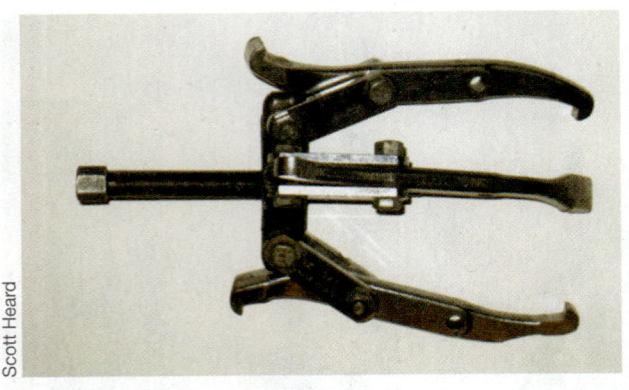

FIGURE 2–33 A light-duty puller.

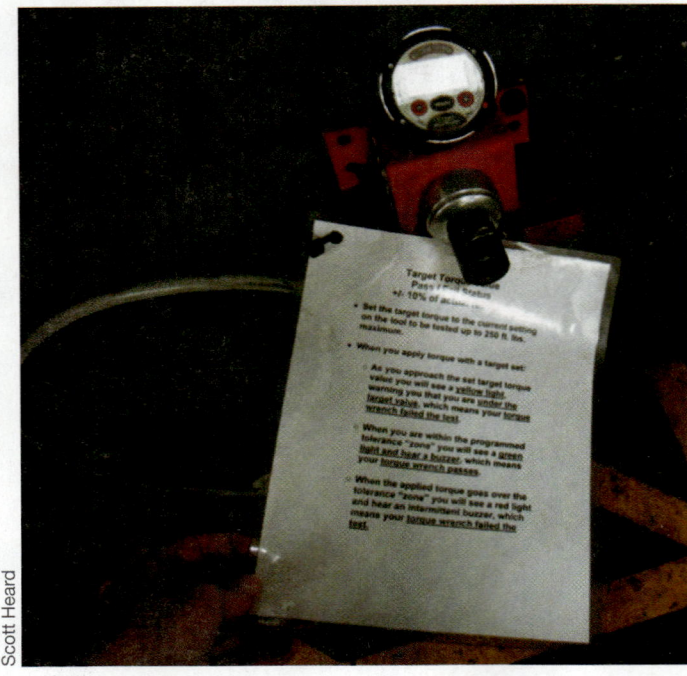

FIGURE 2–34 Two types of torque wrenches and a torque wrench calibrator.

FIGURE 2–35 A selection of cleaning tools.

and rotary dial. ● **SEE FIGURE 2–34** for different styles of torque wrenches. Torque wrenches will also come in different drive sizes including $\frac{1}{4}$, $\frac{3}{8}$, $\frac{1}{2}$, $\frac{3}{4}$, and 1 in. drive. To use a torque wrench properly, you should pull only where the hand grip is and with a slow steady pull. Avoid the use of extensions with torque wrenches as the extension will take away some of the torque intended for the fastener. Do not use a torque wrench as a power bar. Torque wrenches should be calibrated according to the method and interval suggested by its manufacturer.

CLEANING TOOLS. Cleaning parts or components will be a common part of what an HDET does. Some common tools for doing this are wire brushes, scrapers, razor blades, nylon brushes, seal picks, mineral spirit sprayer, and emery cloth holder block.
● **SEE FIGURE 2–35** for a selection of cleaning tools.

FASTENER REPAIR TOOLS. Frequently, an HDET will come across some damaged threads for fasteners. To repair these threads, the HDET should have thread chasers, taps, and dies for common sizes of metric and standard fasteners up to $\frac{5}{8}$ in. and 17 mm.
● **SEE FIGURE 2–36** for a selection of fastener repair tools.

ELECTRICAL TEST AND REPAIR TOOLS. An HDET will need to diagnose and repair electrical systems on machines. This will require tools such as a multimeter, soldering iron, heat gun, terminal extractor tools, terminal crimpers, clamp-on ammeter, and wire strippers. Battery maintenance and cable repair tools should be part of common tools owned by an HDET. A good-quality multimeter is money well spent. See Chapter 5 for details on multimeters. ● **SEE FIGURE 2–37** for a variety of electrical test and repair tools.

INFRARED TEMPERATURE GUN. Accurate temperature measurement of different machine components will be part of many diagnostic procedures. An infrared temperature gun should be in every HDET's toolbox because it is not very expensive.
● **SEE FIGURE 2–38** for an infrared temperature gun.

CLAMPING TOOLS. Many repair jobs will require an extra set of hands. If those hands aren't available, you will need to improvise with some type of clamping device such as a C-clamp, spring clamp, or quick clamp.

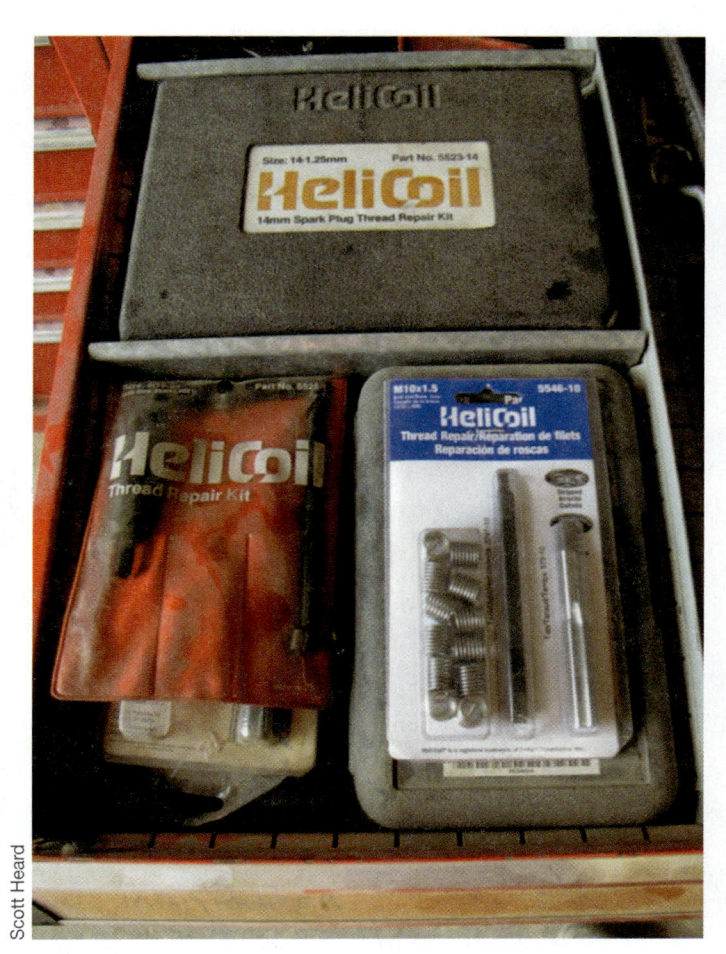

FIGURE 2–36 A selection of thread repair tools.

TUBE CUTTING AND FLARING TOOLS. Some machines may use automotive-type steel brake lines. When these lines need repairing, you will need tools to cut the line to length, install a compression fitting, and flare the line. Small pipe cutters and a flaring tool kit are needed for this.

MEASURING TOOLS. Occasionally, an HDET will need to measure components for different service or repair procedures. Some basic rough measuring tools that should be in your toolbox would include a heavy-duty tape measure 25 ft long and a 12 in. steel ruler that can serve as a small straight edge. For precision measuring, a 0–8 in. dial-type vernier caliper and a set of feeler gauges are a minimum. ● **SEE FIGURE 2–39** for a variety of measuring tools. Feeler gauges are used to measure the clearance between two components, and they are simply thin strips of metal that are accurately sized in steps of one thousandth of an inch or in one hundredth of millimeter increments. An 8 in. caliper (dial type or digital) is a very versatile and accurate measuring tool because it can measure inside and outside diameters as well as depth and do so to within one thousandth of an inch.

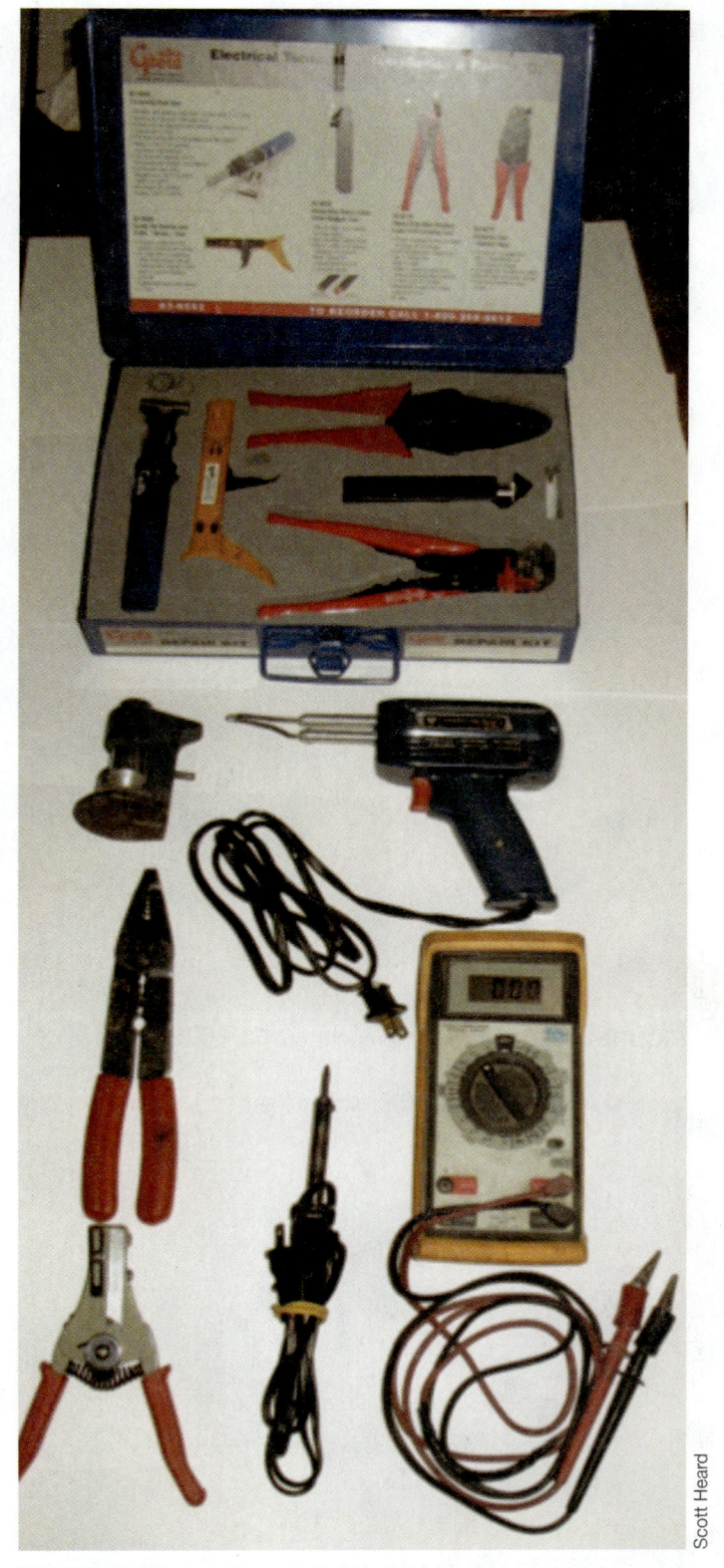

FIGURE 2–37 A variety of electrical test and repair tools.

SEAL PICKS. A lot of time will be spent repairing leaks on machines, and part of doing that is replacing seals. A good-quality set of seal picks will be used frequently. ● **SEE FIGURE 2–40** for a good-quality set of seal picks.

FASTENERS, SEALS, GASKETS, BEARINGS, PERSONAL TOOLS, SHOP EQUIPMENT, AND FIELD SERVICE REPAIR

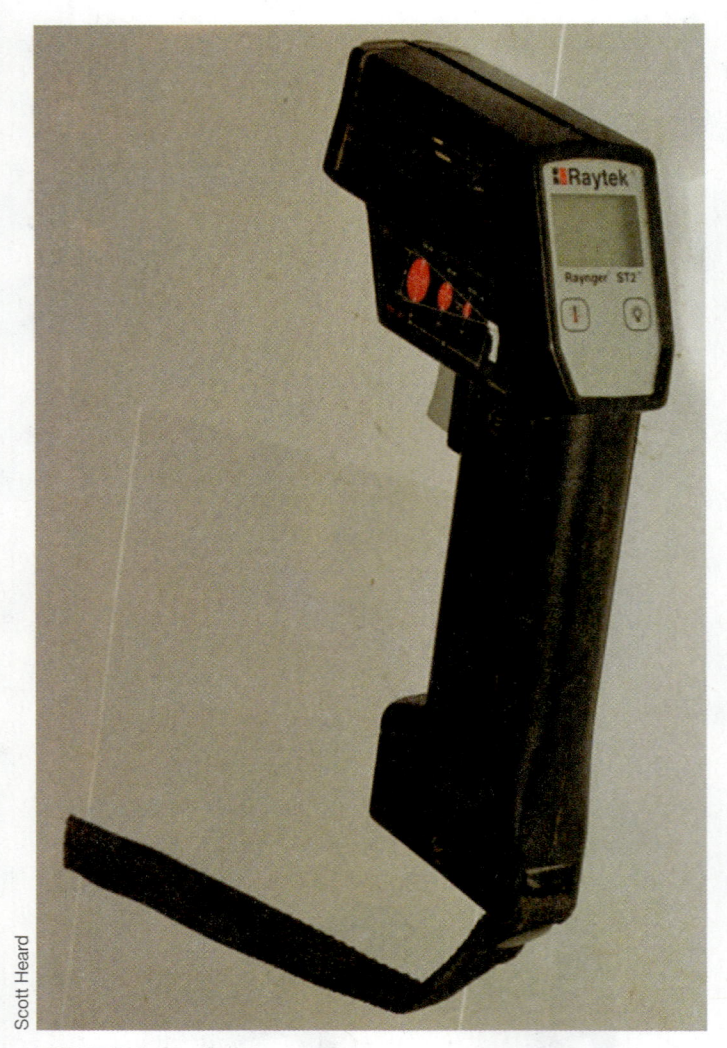

FIGURE 2–38 An infrared temperature gun.

FIGURE 2–39 Basic measuring tools.

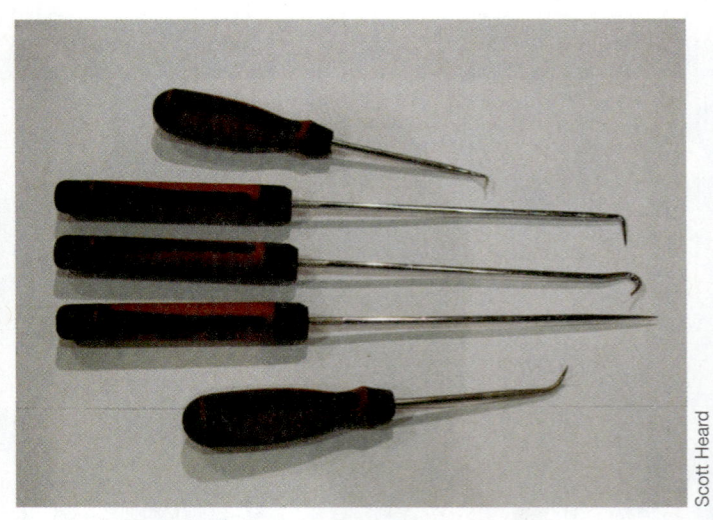

FIGURE 2–40 A good-quality set of seal picks.

ELECTRIC AND PNEUMATIC POWER TOOLS
Buying good-quality power tools with a good warranty will be a good investment. All pneumatic power tools should be oiled with air tool oil on a regular basis. Corded electric tools should have their cords inspected regularly. A minimum selection of power tools to have as an HDET is:

- **Pneumatic impact guns:** $\frac{3}{8}$ and $\frac{1}{2}$ in. drive
- **Pneumatic ratchet:** $\frac{3}{8}$ in. drive
- **Pneumatic hammer/chisel:** with a variety of bits
- **Pneumatic die grinder:** with a variety of bits
- **Pneumatic blow gun:** for parts cleaning (should have a vented nozzle to limit pressure)
- **Pneumatic drill:** preferably with $\frac{1}{2}$ in. chuck
- **Electric grinder:** 4–5 in. disc capacity
- **Electric drill:** $\frac{1}{2}$ in. chuck
- **Cordless electric impact gun:** handy for small fasteners

● **SEE FIGURE 2–41** for a selection of power tools.

SPECIAL TOOLS
Many specialized tools are needed to service and repair heavy equipment machines. If an HDET is going to get into component reconditioning such as engine rebuilding, transmission rebuilding, hydraulic pump, or valve rebuilding, then he or she will definitely need special tools. Special tools are also needed to perform many diagnostic procedures. They will normally be purchased by the HDET's employer. If the HDET is self-employed, then the technician will have to purchase any special tools as required. A few examples of special tools are as follows.

DOWEL PULLERS/STUD REMOVERS. Many component reconditioning jobs will require dowels and studs to be removed. There are special tool kits to do this for various sized dowels and studs.

SURFACE RECONDITIONING TOOLS. For inside round surfaces that are rough or rusty, hone stones and flexible hones are attached to a drill to renew the surface. Hydraulic cylinders and

FIGURE 2–41 A selection of power tools used by an HDET.

engine liners are two uses for hones. For flat surfaces, a lapping table or surface is used with an abrasive cloth to smooth out imperfections. Hydraulic pump components can be reconditioned this way.

HYDRAULIC FLOW METER. Hydraulic flow meters are a valuable troubleshooting tool for machine hydraulic systems. There are a few different types of flow meters, and there will be a wide variety of fittings needed to hook a flow meter up to a machine. Flow meters are usually installed to check pump flow during a diagnostic procedure, but other uses could include measuring and adjusting oil flow to an attachment. Extreme caution needs to be taken any time a flow meter is used because of the potential for hot high-pressure fluid leaks. Hydraulic system testing tools are shown in Chapters 4 and 16.

PRESSURE GAUGES. A variety of pressure gauges should be part of any shop's special tools. Pressure gauges will be used to measure system pressures for many different machine systems. They will range from negative pressures to 10,000 psi.

They should also read in the units of measure appropriate for the machine. This could be imperial or metric. A selection of test fittings, hoses, and adapters will also be needed. Care must be taken to ensure that all diagnostic fittings are capable of withstanding the maximum system pressure they are connected to.

Gauges can be mechanical or electronic. Electronic gauges will use pressure transducers to convert pressure into an electrical signal that is displayed on a readout device. Electrical gauges are safer to use because the display device doesn't contain high-pressure hot fluid like a mechanical gauge.

To connect gauges to a machine may require installing test fittings and adapters.

TACHOMETER. This special tool will be used for diagnostic procedures that relate to speed measurement. Typically, engine speeds are measured, but drivetrain speeds such as shafts and wheels may also need to be measured. Some tachometers are handheld while others will have a pickup unit that is wired to a separate display device (see Chapter 6).

HYDRAULIC PULLERS. Hydraulic pullers will be used for many disassembly and assembly procedures but mostly for drivetrain and steering/suspension components. When two parts are assembled with an interference fit, they will likely need to be pressed apart. An interference fit occurs when the outside diameter of a part is slightly larger (a few thousandths of an inch) than the inside diameter of the part it is mated to. The disassembly of these parts may require several tons of force.

A hydraulic pump unit that could be electric or air driven supplies the oil flow and pressure (sometimes over to 10,000 psi) supplies oil to hydraulic rams that vary in diameter and length for different pulling jobs. The cylinders are then attached to different adapters that are connected to the parts to be pulled. Care must be taken when pulling and pressing parts because of the extreme forces involved. You should always try to make sure that the part will have limited travel when it pops loose because there will likely be a sudden release of energy. ● **SEE FIGURE 2–42** for a hydraulic puller and pump unit.

Hydraulic pullers can also be hand actuated with either a hand pump supplying oil flow or a threaded cylinder that is part of the puller.

TORQUE MULTIPLIER. Some large fasteners will need to be tightened with a great deal of torque. To increase the amount of torque that can be created by hand with a ratchet or torque wrench, a torque multiplier is used. A torque multiplier uses

FASTENERS, SEALS, GASKETS, BEARINGS, PERSONAL TOOLS, SHOP EQUIPMENT, AND FIELD SERVICE REPAIR

FIGURE 2–42 A hydraulic puller and pump unit.

a set of gears to increase the torque input and transfer it to the output drive. The torque multiplier must be held stationary while it is being used. Torque multiplication can vary from 4:1 to over 20:1. Caution needs to be taken when using a torque multiplier because of the extreme forces that can be generated. ● SEE FIGURE 2–43 for a torque multiplier.

MACHINE-SPECIFIC DIAGNOSTIC, RECONDITIONING, AND ADJUSTING TOOLS. There are many different diagnostic tools that may be only useful for certain models or families of machines from one manufacturer. There are many different specialized tools that will only be used for a specific recondition or adjustment procedure on one model of machine. These tools can only be purchased from the equipment manufacturer's dealer and will be purchased by the machine owner or dealer. ● SEE FIGURE 2–44 for a special tool group for adjusting a mechanical fuel system.

TECH TIP

One job I was required to perform was to change the front suspension strut on a 100 ton rock truck. This required lifting and blocking the front of the truck and removing the brake caliper, tire and wheel assembly, and steering arm. Before the strut was removed, I tried to remove the steering knuckle that was press-fitted to the bottom of the suspension cylinder rod.

I hooked up a 100 ton press and started applying pressure. When the gauge indicated 70 tons, I made sure that I was well protected if it came loose. After applying 100 tons of force and it still didn't move, I released the pressure. I then applied some penetrating fluid and moderate heat and again applied 100 tons of force. There was still no movement.

After being on the truck for several years of pounding and with a buildup of rust, it was obvious there needed to be more force applied.

The whole assembly was then sent out to a machine shop that had a 300 ton press. It took everything their press had to break the knuckle free, and when it did let go, it shook the entire building!

MEASURING TOOLS. Other measuring tools that may be used occasionally would be inside micrometers, outside micrometers, depth micrometers, calipers, telescoping gauges, and bore gauges. These tools can accurately measure to dimensions of ten thousandths of an inch or hundredths of a millimeter or smaller if required.

Outside micrometers may be used for a variety of measuring functions but typically are used to measure shaft diameters. They typically come in measuring ranges of 1 in. increments. There should be a master gauge with each micrometer that is

FIGURE 2–43 A torque multiplier.

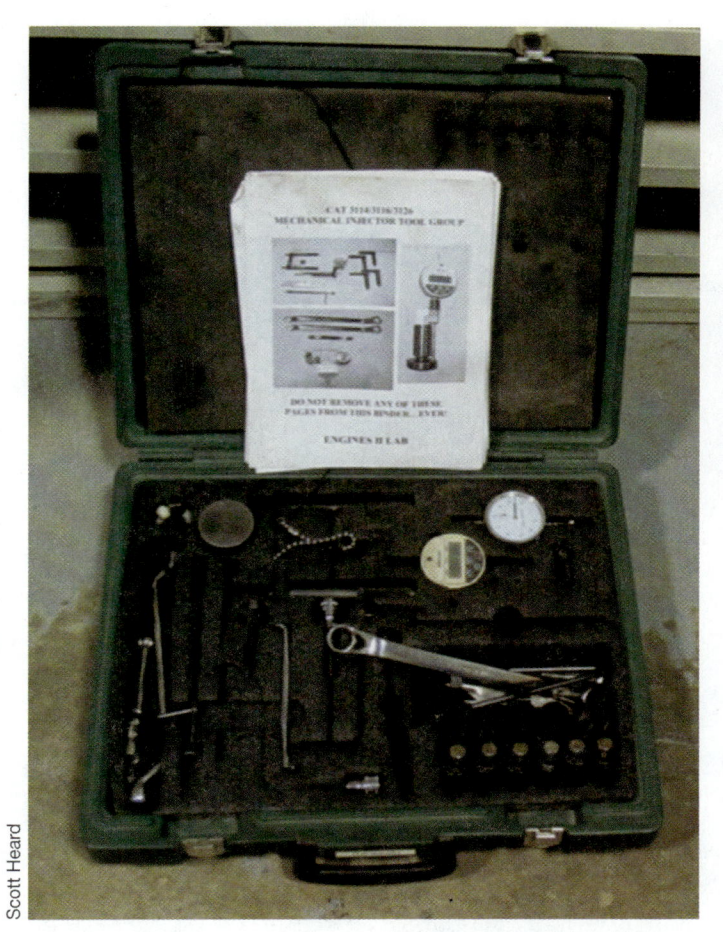

FIGURE 2–44 A special tool group for adjusting a mechanical fuel system.

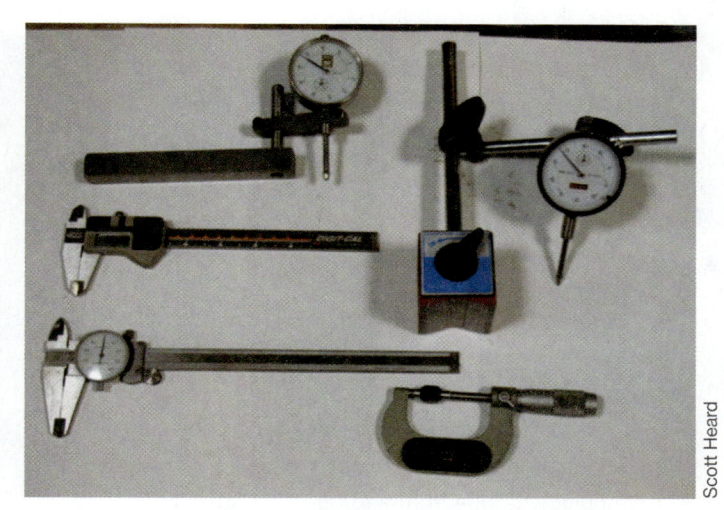

FIGURE 2–45 A variety of special measuring tools.

used to zero the micrometer before each use. An adjustment can be made if the micrometer doesn't zero on the master gauge.

An outside micrometer is used to measure outside diameters of components by having a finely threaded spindle lightly touch one side of the components while the anvil of the micrometer is held against the other side. For a standard micrometer, the thread pitch for the spindle makes it travel 25 thousandths of an inch for one revolution. The micrometer thimble will be marked in 25 equal increments that, when lined up with the sleeve, will indicate to the nearest thousandths of an inch the measured component outside diameter. A metric micrometer will have its increments indicating hundredths of a millimeter and one revolution will be 0.5 mm.

Inside micrometers are used to measure inside diameter dimensions of components. This measuring device functions similarly to the outside micrometer but will fit inside the component. There is a ball at each end of it that lightly fits inside the component being measured.

Telescoping gauges are another tool used to measure inside diameters and are simply spring-loaded gauges that fit inside the component being measured and are locked at the maximum inside diameter of the components. They are then removed while still being locked and measured with an outside micrometer.

Depth micrometers will be used to measure the difference between the heights of two surfaces. Once again the threaded measuring parts of the micrometer use a threaded spindle, but a depth micrometer has a flat base that is used for a reference point to compare the surface the base is on to another surface.

Dial indicators can be used for different measuring jobs depending on what they are mounted to. An example where they are commonly used is to measure shaft endplay where they are mounted to a magnetic base to hold the body of the dial indicator stationary. The indicator's pointer is then rested on the end of the shaft, and when the shaft is moved, the movement is measured with dial travel. ● **SEE FIGURE 2–45** for a variety of special measuring tools.

These special measuring tools are typically provided by the HDET's employer.

Progress Check

11. A box end wrench will:
    a. be used for removing screws
    b. reduce the chance of wrench slippage
    c. increase the torque applied to the fastener
    d. always have 12 points
12. The term *Robertson* is associated with:
    a. pry bars
    b. sockets
    c. punches
    d. screwdrivers

FASTENERS, SEALS, GASKETS, BEARINGS, PERSONAL TOOLS, SHOP EQUIPMENT, AND FIELD SERVICE REPAIR

13. The proper method to operate a torque wrench is to:
    a. use two hands to pull it
    b. pull the grip only with a slow steady pull
    c. pull the grip with a hard fast pull
    d. pull until it clicks and go an extra 30°
14. Maintaining punches and chisels means they should be:
    a. polished regularly
    b. lubricated with chisel grease
    c. dressed to eliminate mushroomed heads
    d. painted or rustproofed
15. A chain wrench is used to:
    a. turn or hold large or irregular shaped objects
    b. tighten and loosen chains
    c. multiply torque from a torque wrench
    d. take the place of a ratchet for small fasteners

SHOP EQUIPMENT Equipment repair shops will be equipped with a wide variety of shop equipment and specialized tooling. Generally speaking, the better-equipped shops will be equipment dealers. However, some large contractors and independent repair shops will also be very well equipped to service and repair heavy equipment. Some examples of shop equipment are as follows.

JACKING EQUIPMENT. Many repair procedures performed in shops will require machines to be jacked up off the shop floor. For smaller machines such as skid steers, medium-duty 3–5 ton hydraulic floor jacks could be used. For larger machines, 10–20 ton hydraulic floor jacks could be used, and for large heavy equipment, specialized jacking tools are available from equipment manufacturers and specialty tool companies. ● **SEE FIGURE 2–46** for a heavy-duty floor jack.

TIRE HANDLING AND INFLATING EQUIPMENT. Rubber-tired machines that are being repaired in a shop will quite often need their tires removed and installed as part of a repair process or just for tire service. For machines larger than a skid steer, this will require some special tooling. If an overhead crane is able to be used for tire handling, a set of tire slings should be used for safe handling of tires. It should be noted here that safe handling of tires must start with deflation and removal of valve cores. If it's not possible to use an overhead crane for tire handling, then other specialized types of tire handling equipment must be used. This could include customized forklifts and truck-mounted tired handling equipment or even heavy equipment machines with tire handling attachments. ● **SEE FIGURE 2–47** for a tire handling machine.

PRESSING EQUIPMENT. Almost all shops will have at least one hydraulic press. Hydraulic presses will range from 5 to several

FIGURE 2–46 A heavy-duty floor jack.

FIGURE 2–47 A tire handling machine.

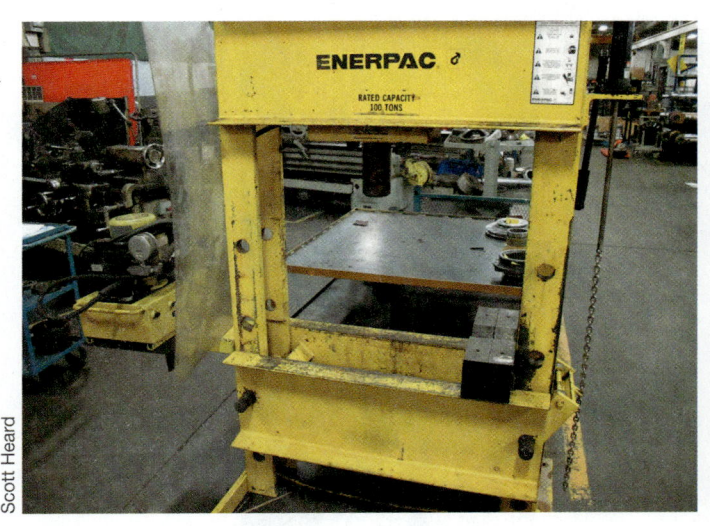

FIGURE 2–48 A hydraulic shop press.

hundred tons in capacity. They are used for many disassembly and assembly procedures and for many different components. Caution must be taken when using presses because of the enormous forces created and the possibility of flying parts. Heavy mesh screens should be put in place when a pressing procedure is started and whatever it takes to prevent parts from flying should be done. ● **SEE FIGURE 2–48** for a hydraulic press in a shop.

Portable presses use a hand power pump unit and will have a variety of different attachments. An arbor press is also hand powered and multiplies human force through a rack and pinion gear set. Always be aware of the potential for uncontrolled movement of tools or components when using any kind of press.

HOISTING/LIFTING EQUIPMENT. Most shops will have equipment to perform heavy lifting such as overhead cranes, wall- or post-mounted cranes, mobile engine hoists, and chain falls. HDETs should be properly trained before they use any hoisting or lifting equipment. Some equipment could have lift capacities to over 100 tons. All HDETs should be trained on the proper use of overhead cranes by qualified personnel.

Many accidents in shops have been a result of improper use of lifting equipment. Overhead cranes should never be used for side pulling. The lifting hook should always be centered over the lifting point for a single point lift.

All lifting and hoisting equipment should undergo regular inspections by qualified personnel. ● **SEE FIGURE 2–49** for an overhead shop crane.

HOISTING, LIFTING, AND RIGGING EQUIPMENT. There will be a variety of hoisting, lifting, and rigging equipment such as multileg chains, load adjusters, chain falls, nylon slings, lifting

FIGURE 2–49 An overhead shop crane.

links, eye bolts, cable slings, and custom-made lifting brackets. The HDET should first of all be sure of how much weight is being attempted to lift and then decide on the proper rigging needed and mentally plan the lift. Most service information will include component weights, but care must be taken to be sure that the weight given is for the exact same component and nothing additional. Always be prepared for unexpected load movement and keep yourself and others clear of the danger zones.

Attending a course on proper lifting and rigging techniques and equipment would be a good idea if you are unsure about performing safe lifts. Crosby has a wealth of great information on its website and offers training courses as well.

CHAINS. Heavy equipment repair shops should have a good selection of chains available for a technician to perform a variety of lifting and rigging tasks that will be part of repairing and servicing heavy equipment machines. Knowledge of chain strength and proper usage of chain is critical when the safety of you and your coworkers is at stake.

FASTENERS, SEALS, GASKETS, BEARINGS, PERSONAL TOOLS, SHOP EQUIPMENT, AND FIELD SERVICE REPAIR

> **WEB ACTIVITY**
>
> Go to www.theCrosbygroup.com to find lots of valuable information on hoisting and rigging hardware. There are also downloadable apps that can be installed on smartphones to help you perform safe lifts.

Chain strength is determined by three factors: The first factor being the tensile strength of the material that the chain is made of, the second being the cross-sectional size of the links, and the third being the condition of the chain. Chains are graded for their tensile strength, and link strength will increase as link cross-section size increases.

The old saying of "a chain only being as strong as its weakest link" is very true. Chains should be inspected on a regular basis for any damage and stretched links.

Chain hooks are usually one of two types. Grab hooks will be able to grab one link and change the length of the chain. Slip hooks can be used to go back over the chain to make a choke assembly. A slip hook should have a properly functioning safety latch as part of it.

Multileg chain assemblies are used to lift components or machines at more than one lifting point. These chain assemblies are constructed starting with a ring that has chains attached to it with link assemblies. The ring will also have grab hooks to allow leg lengths to be changed for lifting adjustments. ● **SEE FIGURE 2–50** for a chain assembly.

CABLE ASSEMBLIES. Some shops will use cable assemblies or slings for lifting procedures. These assemblies will come premade and should be inspected regularly for damage such as stretching, pinches, and fraying. Care should be taken to not allow cables to pull against sharp objects. ● **SEE FIGURE 2–51** for a cable assembly. Cable assembly capacity is mainly based on cable diameter and sizes range from $\frac{3}{8}$ to 2 in. and greater.

NYLON SLINGS OR STRAPS
These lifting devices are popular because they rarely damage the component they are lifting; however, they can be easily damaged if pulled against a sharp object. They also need to be inspected regularly like cable slings. Lifting capacity of nylon slings is listed on a tag that is sewn to the sling, and many widths and lengths are available. ● **SEE FIGURE 2–52** for a selection of nylon slings.

LOAD ADJUSTERS. Many components being lifted will not be uniformly shaped and will not have an obvious center of

FIGURE 2–50 A chain assembly.

FIGURE 2–51 A cable assembly for lifting.

FIGURE 2–52 Nylon lifting slings.

FIGURE 2–53 A load leveler.

FIGURE 2–54 An example of blocking equipment.

gravity. Other components will need to be lifted at one set angle or a changing angle. This will mean that the load will need to be adjusted to lift properly. Some examples where this may be necessary are engines, transmissions, and hydraulic pumps. Load leveling devices can change the center of gravity when lifting, allowing the load to change its angle. One device is a load leveler that will use a threaded rod to shift the center of gravity. Another way is to use a "come-along," which is a hand-cranked chain and ratchet device to shorten the distance between a lifting point and the lifting hook. ● **SEE FIGURE 2–53** for a load leveler.

BLOCKING EQUIPMENT. Once a machine or component is lifted or hoisted, it must be suspended securely so that it isn't resting on something that may be damaged and so can't easily be knocked over. A variety of blocking equipment such as hardwood blocks, jack stands, heavy-duty stands, or custom-made stands will be used for this. Wheel chocks will ensure that a rubber-tired machine can't roll away, and many shops will make it mandatory that any machine in a shop or being serviced anywhere has its wheels chocked. ● **SEE FIGURE 2–54** for an example of blocking equipment.

FASTENERS, SEALS, GASKETS, BEARINGS, PERSONAL TOOLS, SHOP EQUIPMENT, AND FIELD SERVICE REPAIR

> **TECH TIP**
>
> If you are blocking up something made of metal with a metal stand and the mating surfaces are fairly flat and smooth, it's a good idea to put a thin piece of softwood or rubber between the stand and component to create some friction, should something bump against the component being blocked up. This will help to prevent the component/machine from falling off the blocking or stands.

> **SAFETY TIP**
>
> A recent news story reported that a high school student was killed while cutting open a used 45 gallon drum. The fumes inside the drum ignited and caused fatal injuries to the student. An innocent shop project turned into a tragedy. This is a sobering reminder of the dangers of using cutting equipment.
>
> Be sure to *never* apply excessive heat to sealed containers or containers that may have flammable materials in them.

HEATING, CUTTING, AND COOLING EQUIPMENT. Many repair procedures will require components to be heated, cut, or cooled. The following section will outline some equipment that could be used for this. An HDET needs to use extra caution when using equipment to cut, heat, or cool. With extreme temperature changes, there could be pressure changes that can cause serious injuries and death. There is always the chance of starting an uncontrolled fire when heating and cutting so knowledge of fire extinguisher locations and how to operate them is crucial.

Appropriate PPE should always be worn when using this equipment. Typical minimum is leather gloves, long sleeves and pants (coveralls ideally), ankle-covering work boots (done up to the top and tops covered), safety glasses, face shield, or shaded goggles.

OXY-FUEL TORCHES
Almost all shops will have at least one set of **oxy-acetylene** or oxy-fuel torches. Acetylene is the most common fuel used; however, some shops may use propane or natural gas. With different tips, they can be used for metal cutting, gas welding, or heating.

A typical setup will have one bottle of oxygen and one bottle of acetylene secured to a cart with wheels so it can be moved around the shop. The bottles will have pressure regulators

FIGURE 2–55 A typical oxy-acetylene setup.

threaded to them to reduce the bottle pressure to usable levels. When they are refilled, oxygen bottles will have around 2,200 psi of pressure and acetylene bottles around 250 psi. There will be a hose assembly connecting the regulator to flash back arrestors and a torch handle. Oxygen hoses are identified by green coloring and acetylene hoses are red. Acetylene connections use left-hand threads that are identified by a grooved line on the hose's nut.

The torch handle will have shutoff valves to allow the HDET to regulate and stop the flow of gas to the tip that is attached to the handle.

A cutting tip attachment will have another valve that will increase the flow of oxygen when the valve handle is depressed. ● **SEE FIGURE 2–55** for a typical oxy-acetylene setup.

There are different sizes of cutting tips available for different cutting jobs with the sizing related to the hole size where the gas exits the tip. A gouging tip will be curved to help when starting to cut a hole in the middle of a piece of metal.

Tip cleaners will be used to keep the tip orifices clear of slag.

For all operations that use an oxy-fuel setup, torch setup procedures are the same as follows (assume all components are disconnected):

1. crack both bottle valves slightly open and then close. This will purge any impurities from the bottle valve and not allow them into the regulator.

2. install regulators onto bottles
3. install anti-flashback devices at regulators (these can also be installed at handle)
4. install hoses at regulators
5. install torch handle to hoses
6. install appropriate size and type of tip to be used (cutting, heating, welding)

For lighting a cutting tip–equipped oxy-acetylene, torch setup follow these steps:

1. make sure that torch handle valves are closed and regulator pressure adjusters are backed out
2. open bottle valves slowly (easier on regulators) and no more than ¾ of a turn (easier to turn off in a hurry if needed)
3. adjust acetylene pressure (never more than 15 psi) to the torch manufacturer's recommended setting (usually 5–8 psi)
4. adjust oxygen pressure to the torch manufacturer's recommended setting (usually 15–30 psi)
5. open torch handle valves slightly to purge lines and then close valves and recheck pressure settings
6. open the torch handle oxygen valve (cutting tip valve should still be closed)
7. open the acetylene valve on the handle slightly and light with sparker
8. Open the cutting tip oxygen valve and adjust both valves until a neutral flame is achieved
9. make sure that full-regulated oxygen pressure should come out of the center orifice of the tip when the lever is depressed on the cutting tip

Once the flame is adjusted properly, the material to be cut is heated until it starts to melt. The oxygen lever is then depressed. This will increase the temperature of the metal and blow away the molten metal.

Cutting metal is easier if started at an edge, and travel speed and angle are dictated by the thickness of material being cut. Thinner material ($\frac{1}{8}$ in. or less) requires a steeper lead angle, and for material that is $\frac{1}{2}$ in. thick and up, the torch should be close to 90° to the work piece.

Oxy-acetylene torches can be used for heating, and depending on how much heat is needed, there are different tips available. For light heating jobs, a welding tip could be used; for medium heating jobs, a cutting tip could be used; and for extra heat, there are large multi-flame tips commonly called rosebuds. Rosebud tips will consume a lot of fuel and oxygen,

SAFETY TIP

- Do not light a torch with anything other than a friction lighter or sparker.
- Store fuel and oxygen bottles separately.
- Do not apply heat to any fuel or oxygen cylinders.
- Make sure that there is some type of flash back arrestor between the bottles and the torch tip.
- Wear eye and ear protection. Light intensity of the cutting flame requires the use of a welding helmet with lens shade of 5–6.
- Always remove fumes and vapors produced by the process from the area. Extra care should be taken when working with metal covered with paint, coatings, or cladding. Individual tolerances and safety regulations may also warrant the use of breathing protection.
- Always know where sparks are landing. Always remove combustible and flammable materials from cutting/welding area.
- Never cut piping or vessels that contain flammable materials.
- Never cut a vessel that has held flammable materials. All vessels that have held flammable materials contain vapors from that material. If cutting must be done, proper procedures must be followed.
- Never use any petroleum oil or grease to lubricate any part of oxy-fuel system. Pure oxygen lowers the kindling temperature of any petroleum product that will cause a fire.
- Never use pure oxygen to blow off cloths and other combustibles. Pure oxygen will lower the materials kindling temperature and cause the material to ignite easier.
- Never operate acetylene at pressures greater than 15 psi (103 kPa) and never exceed the withdrawal limitations of the acetylene cylinder.
- Make sure that hot parts are identified so other people are aware if you need to leave them unattended.

and there is a greater risk of flash back when using this type of tip. ● **SEE FIGURE 2–56** for a pair of rosebud heating tips.

PLASMA CUTTER Many shops will have a plasma cutter to be used for a variety of cutting jobs. Plasma cutters use an electric arc and a shielding gas to create a plasma "flame" that

FIGURE 2–56 A pair of rosebud heating tips.

FIGURE 2–57 A plasma cutter.

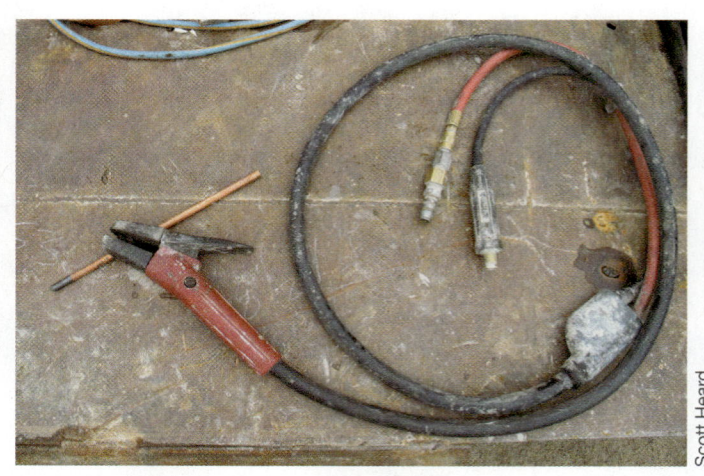

FIGURE 2–58 Arcair cutting equipment.

harnesses the energy of the fourth state of matter. They are fast and very easy to use, and with a little practice, an HDET can plasma cut with great precision. They are very nice for cutting light gauge metal because they don't build a lot of heat up in the metal, which usually would distort it. ● **SEE FIGURE 2–57** for a plasma cutter.

ARCAIR CUTTING EQUIPMENT For deep high-volume cutting, there is an attachment that can be added to an arc welding machine that is called an arc/air arrangement. Some typical uses for this tool are cutting off welded-on bucket cutting edges and tooth adapters or removing liners from haul truck boxes.

It uses special carbon electrodes and has a compressed air line going to it as well. The material to be cut will have the welder's ground clamp attached to it. The welding machine is started after the equipment is attached to the positive cable and an air line is attached to the adapter. When an electrode starts an electric arc and the handle is squeezed to allow air through the adapter, the cutting process is started. This equipment will create a lot of sparks, noise, and fumes so it should be done in a partitioned and well-ventilated area.
● **SEE FIGURE 2–58** for Arcair cutting equipment.

HEATING TOOLS Some repair procedures will require parts to be heated for assembly or disassembly. For light heating jobs, an electric heat gun may be sufficient (heating shrink tube). For more heat, other tools such as a propane torch, oxy-fuel torch, oil bath heaters, toaster ovens, or an induction heater could be used. ● **SEE FIGURE 2–59** for different heating equipment.

It will be important to heat parts to a specific temperature and not overheat them as the parts' metallurgical structure will start to change and the parts will likely fail prematurely. To prevent this, you should use a heat crayon that will indicate when the part is at a specific temperature. The crayon is rubbed on the part, and when it melts, you know it is at the right temperature. Heat crayons come in different heat ranges.
● **SEE FIGURE 2–60** for a heat crayon.

COOLING EQUIPMENT Some repair procedures require parts to be cooled. Many bearing installations will be easier if the bearing or part that it is installed on is cooled. This could be as simple as using a food freezer or getting dry ice or liquid nitrogen to super cool parts. Special caution and PPE needs to be taken and worn when using super cooling equipment to prevent frostbite.

FIGURE 2–59 A variety of heating equipment.

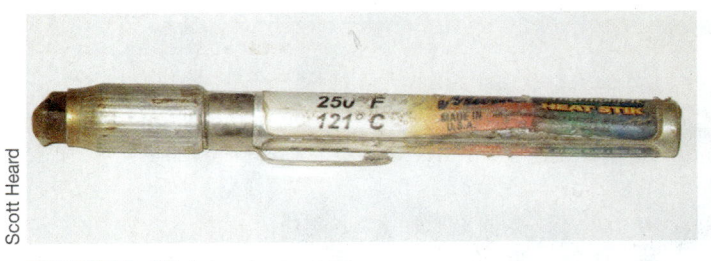

FIGURE 2–60 A heat crayon.

UNDERCARRIAGE TOOLING. When undercarriage on a track machine is measured and repaired, there will be special tools and equipment needed. This includes undercarriage measuring tools (mechanical or ultrasonic), track pin press, idler, and roller handling equipment. ● **SEE FIGURE 2–61** for a track pin press.

LARGE AIR TOOLS. Most heavy equipment shops will have large air compressors to provide large volumes of air to air tools. These tools include $\frac{3}{4}$ and 1 in. drive impact guns and bead blasting equipment. Shop air compressors need to be maintained to ensure that they are reliable. ● **SEE FIGURE 2–62** for large air-powered tools.

HOSE MAKING EQUIPMENT
Some shops will have their own hose making equipment. These are hydraulic presses with special dies and adapters that are used to press apart old hoses with reusable fittings and press together new fittings onto bulk hose lengths. See a picture of a hose press in Chapter 4.

DRILL PRESSES AND GRINDERS Most shops will have at least one drill press and either a bench grinder or one on a pedestal. Drill presses should be used with sharp bits and bit lubrication to make drilling easier. Grinders should have their stones dressed to keep the stone flat. Their work piece rest should also be adjusted close to the stone as the stone wears.

CLEANING EQUIPMENT. Heavy equipment that has been working on a job site will be dirty when it gets to the shop. Larger shops will have a wash bay with water cannons and pressure washers to wash whole machines and large components. Other cleaning equipment could include hot tanks, soda baths, bead blasting cabinets, and parts' cleaners that use mineral spirits. ● **SEE FIGURE 2–63** for parts' cleaning equipment.

FLUID EVACUATION, FILTERING, AND REFILLING EQUIPMENT. Larger shops will be set up to do a lot of machine servicing, which will include a lot of fluid changes. To speed this process up, the shop will likely have a vacuum pumping system that will suck the fluid out of machines and into a holding tank. This fluid is then taken off-site by a recycling company.

FASTENERS, SEALS, GASKETS, BEARINGS, PERSONAL TOOLS, SHOP EQUIPMENT, AND FIELD SERVICE REPAIR

FIGURE 2–61 Some examples of track pin presses.

FIGURE 2–62 Large air-powered tools.

New fluids are then pumped in from bulk tanks to speed up refilling. Retractable hose reels will be used and could have digital readouts that can be set to shut off once a desired amount of fluid has been pumped.

Filter carts will add an extra degree of filtration to new fluid as it is pumped into a machine. A filter cart is a unit on wheels and has a pump that draws fluid into it and pushes it through one or more filters. Some filters are equipped with a particle counter that will show the cleanliness of the fluid as it leaves the cart. Once it leaves the cart, it is pumped into the machine. See pictures of these systems in Chapter 3.

FIGURE 2–63 Parts cleaning equipment.

WELDING EQUIPMENT. Part of an HDET's duties will be to perform welding repairs or welding fabrication tasks.

Welding machines will be used for a variety of different functions such as repairing broken metal components, removing bearing races, fabricating new components, fabricating tools, and adding antiwear hard facing to buckets and blades.

Once again, depending on whom the HDET works for, the level of welding skill required will change. For example, some large contractors or companies could have welding technicians who will handle all welding jobs. Other companies could bring in a welding contractor for big welding jobs and leave the smaller jobs to the HDET. Smaller companies could rely on the HDET for all welding jobs.

Regardless of whom an HDET works for, a good understanding of welding equipment and basic welding technique is important.

Welding is as much an art as it is a skill and to be good at any art takes practice. If an HDET only welds on rare occasion, he or she cannot be expected to produce weld quality comparable to a full-time welder. The most common equipment and basic equipment setup will be covered here. To get good at welding takes practice and much trial and error.

The main focus when welding should be to end up with a structurally sound weld that will be as strong as the parts that are welded together. The welding procedure should not damage other parts in the process either.

Weld penetration refers to how deep the welding bead has gone below the surface of the work piece. Good penetration will create a strong weld. Shallow penetration may look nice but will easily fail. One of the keys to good welding is heat management. This means using the welding equipment to create enough heat to combine the work piece with the filler rod or wire but not too much heat that may burn holes in the work piece or damage machine components.

This section is merely an introduction to the different welding processes and equipment terminology. There are many excellent books written just on welding if you need to explore the subject further.

OXY-FUEL WELDING Welding with an oxy-fuel setup uses the same equipment from the bottles to the torch handle as the cutting torch setup. The most common fuel used is acetylene. A welding (sometimes called brazing) tip is installed on the handle. Welding tips come in various sizes and should be changed according to how much heat is desired and the thickness of the work piece.

The process involves heating the two parts to be welded until they become molten and then adding a filler rod to the puddle. There are different types of filler rod for different types of work piece metal.

The trick to gas welding is to work on the puddle and add filler rod as needed and then allow the weld to cool. There will be a lot of heat spreading through the parts you are welding, and this should be kept in mind to prevent other damage.

Oxy-fuel equipment setup is as follows:

1. same as cutting equipment setup from bottles to torch handle
2. select welding tip that is appropriately sized for welding job
3. lighting sequence is same as cutting except there is only two torch handle valves to adjust the flame

Key points to good oxy-fuel welding:

1. with neutral flame, heat parts until molten and add filler rod slowly while working puddle with flame
2. take care to not overheat parts and allow time for puddle to solidify.

● **SEE FIGURE 2–64** for an oxy-fuel process.

STICK WELDING OR SHIELDED METAL ARC WELDING (SMAW) Stick welding machines are likely present in every heavy equipment repair shop because they are relatively inexpensive, reliable, and easy to operate. The stick welding machine is a transformer that will take high-voltage (220 or 440) AC and transform it to provide low-voltage/high-current AC to

FIGURE 2–64 An oxy-fuel process.

weld with or convert into DC positive or negative current. The amount of current is adjustable and will produce more or less heat through the electrode and into the work piece. Welding machines could be powered by shop voltage of 120–600VAC or be driven by a gas or diesel engine.

A stick welding machine creates an electrical short circuit between the material being welding and the welding rod, and this will create a very high temperature. This high temperature melts the work pieces and the electrode together. Electrodes (rods) are a wire rod that is coated with a material that becomes slag on the weld bead when the weld cools. The coating on the rod shields the weld bead from impurities and makes the bead cool at a controlled rate. Electrodes are available with many different wire material types, sizes, and coating types. They are identified with a standardized number and letter system that indicates their tensile strength and coating type. Electrode sizes range from $\frac{1}{16}$ to $\frac{3}{8}$ in. An example of a typical general purpose rod commonly used is E7018 $\frac{1}{8}$ in. in diameter.

Rods are consumed as they melt and need to be replaced regularly. This is one drawback of stick welding. The other is the slag that forms on the weld bead that has to be completely cleaned off before another bead can be laid on top of it. Cleaning slag requires a chipping hammer and wire brush.

With different machine settings and electrode types, stick welding machines can weld any type of metal and in any position. This of course depends on the skill level of the HDET running the machine. If possible, weld in the flat or horizontal position. Vertical and overhead welding takes more skill and practice.

Because a lot of electric current is passed through the work piece if SMAW is being done on a machine, the battery ground cable and all ECMs (machine onboard computers) should be disconnected to prevent any possibility of damage being caused to the machine's electrical system.

Stick welding equipment setup and operation is as follows:

1. clean and dry the weld area
2. select the appropriate rod for the job
3. connect the welding cables to the welding machine for the type of welding—DC positive, DC negative, or AC
4. adjust the amperage setting on the machine (higher amps for thicker rod)
5. connect the ground clamp to the work piece
6. put on all appropriate PPE (cover all exposed skin and use proper shade in welding helmet)
7. install rod into electrode holder
8. tap or scratch electrode against work piece to start the arc and maintain the proper rod angle, movement, and travel speed as the rod gets consumed
9. when electrode is almost gone quickly, pull electrode holder away to stop the arc
10. let the bead cool and clean off the slag thoroughly

Key points to achieving a good stick weld:

1. choose the correct rod that best suits the weld position and work piece material
2. choose the correct machine settings for a starting point and adjust if necessary
3. make sure that the work piece is clean and dry
4. hold the electrode at the correct angle for the position of weld and maintain the angle
5. move the rod at the correct speed (and weave pattern if necessary) to achieve good penetration

● **SEE FIGURE 2–65** for an arc welding machine.

FIGURE 2–65 An arc welding machine.

MIG (METAL INERT GAS) WELDING OR GMAW (GAS METAL ARC WELDING)

Another common type of welding process is called MIG welding. The two main differences between MIG and stick welding are MIG uses a continuous feed wire instead of individual electrodes and a shielding gas replaces the electrode coating to protect the bead when it is forming. The transforming part of the machine is similar, but there needs to be a wire feed and gas supply mechanism. The filler wire comes in spools, and like electrodes, it comes in different diameters and compositions. The machine has a hose/lead coming out of it that feeds the wire, electrical current, and gas to a gun. The gun has a trigger that when pulled will feed wire and start the gas flowing.

The inert gas used is usually argon, but other gas mixes can be used. A bottle will be mounted on or near the welding machine and a flow regulator will be used to adjust the proper flow of gas to the work piece.

Unlike stick welding that leaves a crusty protective coating (slag) over the weld bead, the MIG welding process just leaves the newly formed weld bead that can be welded on top of immediately if necessary.

The main advantages of MIG welding are that it's fast and relatively easy.

MIG welding equipment setup and operation is as follows:

1. install the proper wire spool in the feed drive
2. attach the regulator to the inert gas bottle
3. turn on the power supply
4. pull the trigger and adjust the gas flow and wire feed speed
5. attach the ground clamp as close to the work piece as possible
6. make sure that all appropriate PPE is worn
7. hold gun close to work piece and pull trigger
8. hold gun proper with a loose grip and travel at proper speed to achieve good penetration
9. deposit weld bead as long as necessary
10. release trigger and pull gun back to stop

Key points to achieving a good MIG weld bead:

1. use correct wire size and type
2. use correct gas type and gas flow rate
3. adjust machine voltage and wire feed speed
4. hold gun at correct angle and travel at correct speed

● **SEE FIGURE 2–66** for an MIG welding machine.

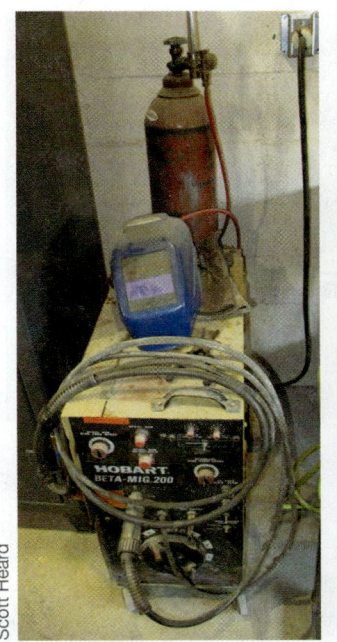

FIGURE 2–66 Two examples of MIG welding machines.

FIELD SERVICE HDET

It will quite often be very difficult and expensive for a machine owner to bring a machine into a repair shop for repairs or regular maintenance. With larger machines, it is simply not worthwhile transporting a machine to a repair shop for anything less than extensive major repairs.

Therefore, many machines will never see the inside of repair shop unless they are in for a major reconditioning procedure. Mobile lube trucks are able to perform services on the machines that aren't able to get to a service shop.

If a machine breaks down on a job site, this is where the field service HDET comes into the picture. A field service technician will have a service truck equipped to do everything from repairing leaks to performing some very major repairs.

Some field service vehicles are as simple as a well-equipped van or pickup truck while others are fully equipped with air compressors, cranes, torches, welders, generators, complete set of hand tools, power tools, a hose press, many special tools, and a stock of commonly used parts. A field service truck will have a laptop loaded with machine service information and diagnostic interface hardware and software for the HDET to diagnose machines. There will also be room to carry large components to and from the job site.

● **SEE FIGURE 2–67** for an example of a fully equipped field service truck.

FASTENERS, SEALS, GASKETS, BEARINGS, PERSONAL TOOLS, SHOP EQUIPMENT, AND FIELD SERVICE REPAIR

FIGURE 2–67 Three examples of field service trucks.

CASE STUDY

Being a service field service technician is a very demanding job. Not only is there unfriendly weather elements to deal with, there is a great deal of pressure placed on you as soon as you arrive at a disabled machine to get it working as soon as possible. A good field service technician is a very valuable team member to any heavy equipment service/repair organization.

One of the more challenging jobs I had as a field service technician started off with a call to check a scraper that had a wheel leak. By the time I got to the machine, there was no operator or supervisor around, and the machine was sitting in the middle of a muddy field. Just getting to the machine was an adventure.

Once at the machine, an initial inspection showed that there was a leak at the inside of the right front wheel, but it was caused by the entire wheel assembly starting to fall off.

This meant the machine had to be fixed where it sat. The first call was to the large tire repair company we dealt with. A tire service technician arrived with a tire handling truck. He blocked the machine up and removed the tire. This allowed me to remove the axle shaft and final drive, which revealed why the wheel was falling off. The fasteners holding the bearing retainer on the hub had backed off and allowed the wheel assembly to work its way off the spindle.

I then removed the hub and brake assembly, which revealed damage to the spindle. The spindle was then removed and taken to a machine shop for repair. New bearings and seals were ordered, and all other parts were inspected for damage. Some slight damage was found on the brake shoes but was deemed not bad enough to require new parts.

The repaired spindle was picked up a few days later and installed. The remaining components were installed, and the hub was installed with the proper torqueing procedure done on its bearings.

After completing the hub and final drive reassembly, the tire guy put the wheel and tire back on and let the machine off its blocks. The brakes were adjusted and the oil level was topped up; the machine was started and drove to check brake operation. The machine was then put back to work.

This machine was out of service for a week, which meant a replacement machine had to be floated to the job and several thousand dollars were spent on new parts plus the cost of my labor and service truck time as well as the machine shop repair bill.

This repair was a result of some fasteners that didn't do their job and an example of the kind of situations that field service technicians may find themselves in.

Progress Check

16. What is used for an identification mark on acetylene hose assembly nuts?
    a. capital "a" stamped in it
    b. letters "ac" stamped in it
    c. the word "fuel" stamped in it
    d. a grooved line
17. What should be put in place for safety reasons when using a shop press?
    a. a screen
    b. a tarp
    c. a heavy sheet of steel
    d. a clear piece of plastic
18. Besides the cross-sectional size of chain links and their condition, what else determines how strong they are?
    a. color code
    b. shape
    c. material's tensile strength
    d. coating on them
19. Heat crayons will be used to:
    a. mark an area to be heated on a component
    b. let you know when a component reaches a specific temperature during a heating process
    c. help clean an area to be soldered
    d. let you know when a component has been overheated during normal operation
20. This is a major disadvantage of stick welding when compared to MIG welding:
    a. it can only be used for flat position welding
    b. the slag must be cleaned off after each weld bead
    c. the wire spool needs to be changed frequently
    d. the gas pressure needs to be adjusted frequently

SHOP ACTIVITIES

- **Threaded fasteners:** Find five different threaded fasteners and use measuring tools and a thread pitch gauge to identify their size, grade, and thread pitch.
- **Seals:** Find five different seals and identify them by their type and size.
- **Bearings:** Identify three different types of bearings. Inspect them and record any defects found.
- **Hand tools:** Go to the main toolbox in the shop and inspect all hand tools. Record any missing or damaged tools.
- **Hydraulic pullers:** Look at the information tag on a hydraulic cylinder and record the maximum force it can create. What pressure is needed to create this force?

SUMMARY

1. Improper tool usage and body positioning can lead to injury.
2. There are many types of fasteners used on heavy equipment machines. Fastener problems are quite often the root cause of failures such as leaks and failed components.
3. The most common type are threaded fasteners such as bolts and nuts.
4. Threaded fasteners are sized by their thread section diameter and length measured from bottom of head to and of threaded section.
5. Standard bolt threads are classified as threads per inch while metric bolts are classified as distance between threads in millimeters.
6. The strength of a bolt relates to the tensile strength of the material used to make the bolt and is measured in psi

(pounds per square inch) or MPa (megapascals). Standard bolts are classed by grade with grade 3 being a weak light-duty bolt while a grade 8 bolt would be for a heavy-duty application. Metric bolts are numerically classified with a factor of 10 times their MPa strength.

7. When threaded fasteners are tightened, they create a clamping load on the components they are holding together because the fastener is stretched and acts like a rubber band.
8. Applying a specific amount of torque to a fastener will create a specific amount of stretch and clamping load.
9. A torque to yield fastener will be stretched to the point of deformation and should not be reused.
10. Threaded fasteners should always be inspected for damage before installation and thread lubricants should only be used when specified.
11. There are many different types of locating/retaining fasteners such as split pins, woodruff keys, and rivets.
12. Seals come in many different varieties and are used between two components to keep fluids or gases from escaping.
13. Static seals are used between two stationary components, and a common type is called an O-ring because of its cross-section shape. They are usually made from an elastomeric type of material and fit into a groove in one component. It is good practice to always replace all seals when performing service or repairs.
14. Dynamic seals are used when two components move in relation to each other, and there need to be a seal between them to contain liquid or gas. A common example is a lip-type seal for a rotating shaft. Metal-to-metal face dynamic seals are commonly used for wheel seals.
15. Gaskets are used to create a seal between two components that are fastened together and can be made from paper, cork, or other heat-resistant materials. An example would be a cylinder head gasket.
16. Bearings are used to allow relative motion between two components. Friction-type bearings need a lubricant (grease or oil) to separate the bearing from a component while antifriction bearings have a rolling member that keeps to races separated.
17. Antifriction bearings can be ball, roller, needle, or tapered roller type and will have inner and outer races that allow the rolling member to roll on. Cages will keep the rolling members separated.
18. Tapered roller bearings need to be adjusted during installation to create a preload.
19. Bearing failures usually result from things such as lack of lubrication, misaligned components, improper installation, or overspeeding.
20. Bearings can be reused but a close inspection should check for imperfections such as pitting, scoring, rust, or corrosion. Any imperfections noted require bearing replacement.
21. All HDETs will work with a wide variety of tools. High-quality tools should be purchased and cared for with pride. Proper tool usage includes using the right tool for the job and not overstressing tools.
22. There are many different types of wrenches with the most common being combination wrenches with one open end and one box end on a common shaft. Ratchet wrenches will speed up fastener handling.
23. Hydraulic wrenches can be used to apply very high torque to large fasteners.
24. Socket sets are used with ratchets to loosen and tighten threaded fasteners such as bolts and nuts. Ratchets can be $\frac{1}{4}$, $\frac{3}{8}$, $\frac{1}{2}$, $\frac{3}{4}$, 1, or $1\frac{1}{2}$ in. drive.
25. Other hand tools required by HDETs include pry bars, pliers, screwdrivers, hammers, chisels and punches, cutting and filing tools, and cleaning tools.
26. Measuring tools needed by HDETs include dial calipers, outside micrometers, feeler gauges, and dial indicators.
27. Electrical testing and repair tools needed include multimeters, clamp-on ammeter, wire strippers, and soldering iron.
28. Common power tools used by HDETs include pneumatic or battery-powered impact guns, pneumatic or battery-powered drills, and grinders.
29. Special tools sometimes used by HDETs include hydraulic pullers, fuel system adjustment tooling, and torque multipliers.
30. Shop tools and equipment found in heavy equipment repair shop include hydraulic presses, hydraulic track presses, jacking/blocking equipment, hoisting/lifting/rigging equipment, and large tire handling equipment.
31. *All* tools and equipment used in the repair and servicing of heavy equipment *must* be inspected for damage prior to use. If any tool or piece of equipment is found to be defective, it must be taken out of service either permanently or until it can be repaired to new condition. Many injuries have occurred as a result of using tooling that is damaged and fails during use.
32. Heating, cooling, and cutting procedures are common when working on machines. This includes using torches, liquid nitrogen, and other specialized equipment. Proper PPE usage must be followed to prevent injuries when heating, cooling, or cutting.
33. Oxy-fuel torches can use either acetylene or propane for fuel. The pressure regulators must be adjusted to provide proper pressure to the torch handle from the pressurized cylinders. Torch handles have valves to allow the final flame adjustment for the type of heating or cutting taking place, and the proper start-up and shutdown sequence must be followed.
34. Repairs using welding equipment are common tasks for HDETs. SMAW (stick) and GMAW (MIG) equipment are the common types of welding machines used.
35. Stick weld quality results depend on operator experience (rod angle and speed), type of rod, machine settings, and weld position. Slag deposits must be thoroughly cleaned off as weld beads are laid down.
36. MIG weld quality results depend on operator experience (gun angle and speed), type of wire, type and flow of gas, machine settings, and weld position.
37. Field service HDETs will have fully equipped service trucks and travel to machines to service and repair them on-site.

chapter 3
MACHINE SERVICING

LEARNING OBJECTIVES

After reading this chapter, the student should be able to:

1. Describe how machine service intervals are determined.
2. Describe the properties of the different fluids used in a machine.
3. Explain how to properly change machine fluids.
4. Describe how to take accurate fluid samples.
5. Describe how to interpret fluid sample reports.
6. Describe the key properties of engine coolant.
7. Explain the importance of machine fluid cleanliness.
8. Describe fluid viscosity.
9. Explain what should happen to waste fluids.

KEY TERMS

Cartridge filter 94
Fluid sampling 87
ISO viscosity rating 86
SAE viscosity rating 86
Service intervals 76
Spin-on filter 94
Viscosity 86
Waste oil container 85

INTRODUCTION

SAFETY FIRST Machine servicing can become a routine part of being an HDET, but there are specific concerns you should be aware of to ensure that safety is also part of the routine. Some examples are:

Disable the machine: Make sure that the machine is secured from anyone else attempting to start it or move parts of the machine while you are working on it. Perform lock out and tag out procedures to prevent unplanned machine movement. The minimum requirements for this is to disconnect machine battery power either by shutting off and locking out the main battery switch or by removing the main ground cable from the battery or battery pack (two or more batteries that are hooked together). Other steps include locking out the machine's implement controls, ensuring that the machine won't roll by using wheel chocks or checking the parking brake operation and placing warning signs on the machine to alert anyone else that the machine is being serviced. Some machines with electronic controls will allow the technician to disable the machine electronically and keep it in a service mode until the technician enters a password to allow normal operation. ● SEE FIGURE 3–1 for a machine that is locked out and ready for a service.

Stored fluid power energy: There could be trapped pressure in many fluid compartments that can cause serious injury or death if it is allowed to be released in an uncontrolled manner. As servicing a machine usually means changing fluids, the technician must be aware of potential hazards in pressurized fluid. Depressurizing procedures should be referenced in the machine's service literature. This could be as simple as loosening the hydraulic tank cap but could also mean depleting stored pressure in an accumulator.

Hot fluids and components: Servicing machines will include handling hot fluids and being close to hot components. Proper PPE usage and common sense will prevent burns.

Belly pans and heavy guards: Draining fluids may involve removing heavy belly pans, skid plates, or other heavy guarding. They could be several hundred pounds and could be awkward to handle. Care must be taken to ensure that these heavy components are handled safely when they are being removed and installed. If a skid pan has hinges, make sure that they are intact before removing its fasteners. Otherwise, have a jack below or a chain or sling from above to lower it safely.

FIGURE 3–1 A machine that is ready to be serviced safely.

Machine component mechanical locks: A machine can have one or more locking devices to prevent unwanted machine component movement. Some examples are boom locks, hood locks, steering locks, cab locks, and box locks. If servicing a machine requires working under a raised boom, cab, or hood, you must ensure that these components are securely held in place. Many injuries and deaths have occurred because a technician didn't bother to use these locking devices properly. ● SEE FIGURE 3–2 for a locking device properly installed.

Slips and falls: Servicing a machine will involve handling oils, greases, and other fluids. Fluid spills should be cleaned up as soon as possible before they become a hazard. Wearing work boots that have treads that are in good condition will help prevent slips and falls.

Fall arrest: If you are servicing a machine that requires you to be working higher than 3 m or 10 ft from the next safe surface, you must wear an appropriate fall arrest device.

FIGURE 3–2 A machine's boom lock properly installed.

Ladders/steps: If ladders or steps are needed to service a machine, they must be in good repair and secure.

Climbing/descending: Always face the machine when climbing and descending machine steps and ladders and use three-point contact (at least two hands and one foot or two feet and one hand are in contact with the machine). Make sure that all steps will support you before trusting them fully. Use caution when carrying supplies and tools up and down machines. Do not jump down from a machine even if it's only a couple of feet. Many twisted ankles and other medical problems have resulted from a technician not using all the machine's steps when getting off of it.

Moving parts: Use extreme caution around any moving machine part or component. Servicing equipment may require you to be close to moving parts for a variety of reasons. You may need to check oil levels when the machine is running, for example. Be aware of any moving part of a machine that may pose a danger such as drive shafts or belts.

Hazardous materials: You may encounter hazardous materials when servicing a machine. Take appropriate steps to protect yourself against materials such as engine coolant, used oil, solvents, and brake dust. Protective gloves should be worn when handling hazardous materials at minimum and possibly respiratory protection as well. If you are unsure as to what you are handling, stop and educate yourself or ask someone who knows and then take the necessary steps to protect yourself.

NEED TO KNOW

MACHINE SERVICING When starting out as an HDET, you will likely spend a fair bit of time doing machine services. This may seem like a mundane task to some, but it's one of the best ways to get familiar with heavy equipment machines if you are new to the trade or new to any particular machine. This may also be a good opportunity to see how the machine sounds, feels, and operates normally so that if you are called out to the machine to troubleshoot a problem, you will be able to think back and remember what it should run like.

Machine services are the key to a solid preventive maintenance program that leads to a predictive repair program and are, therefore, one of the most important ways of getting maximum up time from the machine. Predictive repairs are more cost effective than unplanned repairs. Many expensive repairs can be traced back to poor maintenance procedures, extending service intervals, or not doing some parts of a service.

Some machines will have grease fittings that are hard to get to or even hard to find. For example, wheel loaders will have an oscillating rear axle that needs regular greasing for its bushings. The grease fittings to do this are not always visible or easy to access. Not greasing these bushings will eventually turn into a very expensive repair that could see the machine in the shop for over a week, and some parts may need machining if they are worn past the bushing and into the housing.

The first part of performing any service is to get yourself familiar with the machine. Whether you are new to the trade or have been an HDET for 20 years but have never seen the model of machine that you are about to service, you should spend a few minutes reading the operation and maintenance guide of the specific machine you are about to work on. This small manual should be within reaching distance from the operator's seat. ● **SEE FIGURE 3–3** for an Operation and Maintenance Manual of a machine.

If it is missing, it should be replaced as there is important safety and operational information in it that needs to be available for anyone operating or working on that machine. There could be safety features unique to this machine that you need to be familiar with. If it's possible to view the latest electronic version of this manual, this would be ideal as it will have the most current information available.

Part of a machine service should include checking that all safety features are working properly before putting the machine back to work. This is not just a good practice, but

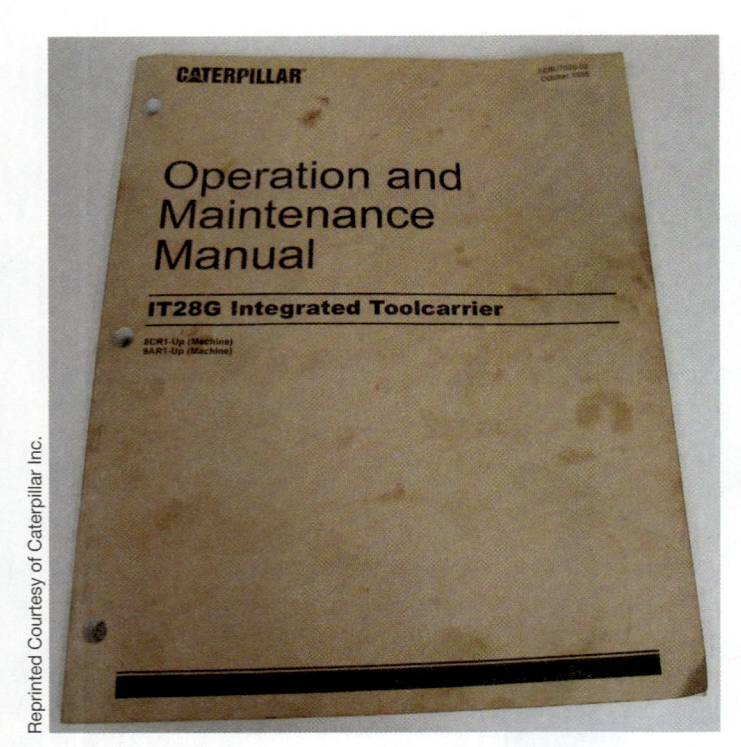

FIGURE 3–3 Operation and maintenance manual.

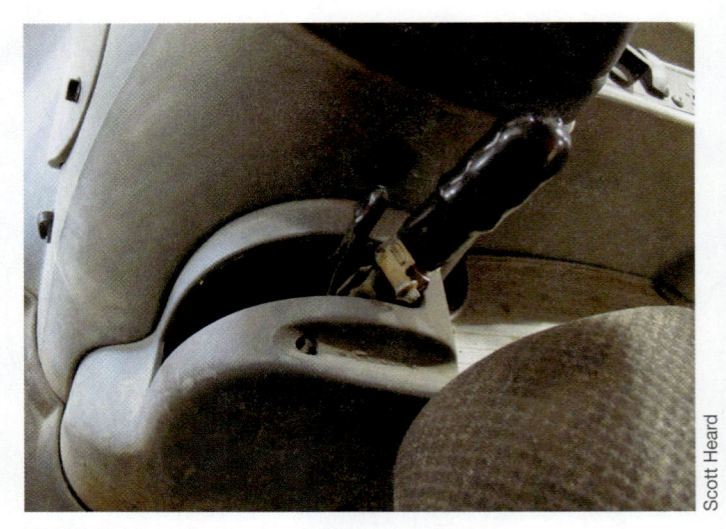

FIGURE 3–4 Machine parking brake control.

MACHINE SAFETY SYSTEM CHECKS

STARTING SYSTEM INTERLOCK. Depending on the type of machine you are working on, there will be certain machine conditions to be met before the operator can start it. The transmission controls have to be in neutral, hydraulic controls locked out, and so on; the operator have to be in the seat with the seat belt on, parking brake may need to be applied, and so on.

For an example of testing the interlock system of a machine that has a powershift transmission, you should try to start it with the transmission control in forward or reverse. Depending on the machine, it may or may not allow the engine to start, but it should not engage the transmission until the transmission control is cycled back through neutral.

PARKING BRAKE OPERATION. Because of the huge variety in parking brake systems, there is no way to make a general statement on how to check their operation. Some machines will need to be held at a certain grade and others will need the parking brake to hold against the machines' drivetrain. However, this is a key part of any machine's service procedure.

Always refer to the manual for the proper procedure to ensure that the parking brake holds the machine. ● **SEE FIGURE 3–4** for a machine's parking brake control.

SERVICE BRAKE OPERATION. The same statement applies here as for the parking brake operation. Some maintenance procedures will include a measurement of the service brake friction material.

WINDSHIELDS/WINDOWS. They should be free of cracks, pitting, and distortion. The wipers/washers should also be checked for proper working order.

BACKUP/TRAVEL ALARM. If the machine is equipped with one of these, it should actuate when the machine reverses or travels and be loud enough to be heard over the noise of the machine.

> **TECH TIP**
>
> If any of the machines safety features are not working properly, it is the responsibility of the HDET to make this known to the machine owner, operator, or both. This is also a good time to suggest the best way to repair any faulty machine systems or components.
>
> This is similar to what car dealerships do to create business for themselves. When a car comes in for a regular service, there is likely an inspection done at the same time and a well-trained technician will likely find something that needs attention.
>
> This is also a good time to find out if there are any recalls, service bulletins, or updates that the machine should have. This includes software updates for the machine's electronic system.

if there happened to be an accident that resulted from one of the machine's safety features not working properly and you were the last person to work on the machine, you may have some explaining to do. Some machine safety-related features are listed here along with a general procedure to test them. As always, check each machine's specific manual to find the proper inspecting/testing procedure for the machine you are working on.

MACHINE IMPLEMENT CONTROLS. You should check to ensure that they operate smoothly, return to neutral freely, and actuate the implements as they are supposed to when the machine is running. Also if there is an implement control lock out device, you should ensure that it is working properly—no implements are moved when the lock out is on and levers/controls are moved.

MACHINE STEERING SYSTEM. Whatever types of steering system the machine uses, its proper operation should be checked before the machine is released back to the operator. If the machine has a steering wheel, it should turn freely and the machine should respond to steering inputs appropriately. Any other type of steering controls should also be checked for proper operation. If the machine has a secondary steering system, its operation should be checked at service time. See Chapter 19 on wheeled steering for an explanation of this system.

SEAT CONDITION AND ADJUSTMENTS. The operator's seat may be equipped with a variety of adjustments. If there are adjustments as part of the seat, they should allow the operator to sit comfortably and reach all controls while allowing proper vision from the operator station.

SEAT BELT. Most machines will be equipped with at least a lap belt and some are now coming with a three-point belt system. These safety devices should be inspected for damage and adjustability. They should also have their installation date checked and be replaced if regulations require this. Most seat belts will be required to be replaced within three years of the date of installation or within five years of the date of manufacture. Seat belts should have a date of installation or date of manufacture tag attached to them.

STEERING OR BOOM LOCKS. Many machines will come with locking devices that are required to be used when the machine is being serviced or repaired. These could be locks that prevent cylinder movement to prevent machine steering or boom movement or other unwanted movement. If a machine came with these locks and they are missing, the technician must not release the machine without replacing them. ● **SEE FIGURE 3–5** for a steering lock of an articulated machine.

DOOR AND WINDOW LATCHES AND HANDLES. The machine's door and window latches need to work properly to allow the operator to leave the operator station in case of emergency.

FIRE SUPPRESSION EQUIPMENT. Some machines will be equipped with standalone fire extinguishers or a total machine fire suppression system. Fire suppression systems must be

FIGURE 3–5 Steering lock for an articulated machine.

inspected on a regular basis. This may need to be done by a technician with specific training for this equipment.

CAB ESCAPE TOOLS OR MECHANISMS. Some machines may come with special tools or devices that allow the operator to exit the cab through a window by either smashing or removing it. This equipment should be inspected on a regular basis and be replaced if missing.

ALL WORK LIGHTS, REFLECTIVE DEVICES, WARNING LIGHTS, AND CAB LIGHTS. Most machines will have a number of lights and reflective devices that should work properly. Outside work lights and flashing lights should be inspected for proper operation to allow the machine to be safely operated in less than normal lighting conditions and to be seen by others as well. Operator warning lights need to be inspected and repaired if not working properly.

ALL GUARDING THAT COVERS MOVING PARTS OR HYDRAULIC LINES. All machines will have some type of protective guarding to protect either operators or technicians from moving parts or fluid leaks causing injuries. If these guards are damaged or missing, they must be repaired or replaced.

ROPS (ROLLOVER PROTECTION SYSTEM) AND FOPS (FALLING OBJECT PROTECTIVE SYSTEM). Most machines produced today will have protective devices that are designed to protect the operator from machine rollover or something falling on the machine. These devices need regular inspection to ensure their integrity.

STEPS AND HANDHOLDS. Almost all machines will have some form of handrail or handhold and many will have steps and ladders as well. These should be inspected for damage and need to be repaired if broken or replaced if missing.

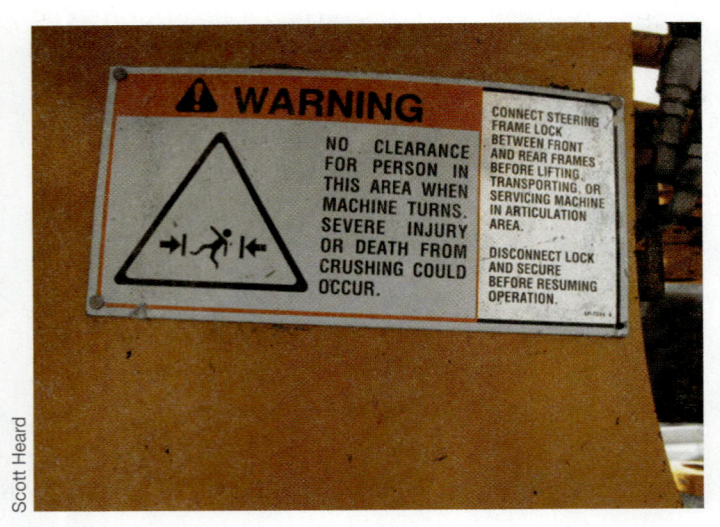

FIGURE 3–6 A typical warning sign.

ALL SAFETY SIGNS/DECALS/PLATES. There will be many cautions, warnings, and safety alerts that should be visible to the machine operator or technician. If they are dirty, damaged, or missing, this should be corrected by cleaning or replacing them. ● SEE FIGURE 3–6 for a warning decal.

ENGINE GOVERNOR CONTROL. The machine's engine rpm control device should be checked for smooth operation and return to low idle freely. The engine shutoff should also work properly.

ALL MIRRORS. Most machines will have at least one mirror that allows the operator to see behind and around the machine. These mirrors should be checked for clarity, cleanliness, and adjustment.

BACKUP AND VISION CAMERAS. Like mirrors, cameras that are mounted on a machine should be checked for cleanliness, focus, aim, and proper operation.

CIRCLE CHECKS Visual inspections are sometimes called walk-around inspections or circle checks. Operators should do circle checks each day before starting and operating the machine, but an HDET should also always do a check at the start of a service and continue a visual inspection as the service progresses. If any problems are found at the start of a service, it may allow time to order parts and repair small problems before the machine is put back into service. With experience, your eyes will be drawn to unusual sights when looking at a machine. When you see shiny metal where there should be paint, it should tell you something has been rubbing against this component that probably shouldn't be and, if left to continue, will eventually cause a problem beyond paint missing. There may also be loose fasteners that should be retorqued or perhaps a weld has broken.

Visual checklists are sometimes available online from the machine manufacturer's website. Caterpillar has a variety of machine checklists on its website: http://safety.cat.com.
● SEE FIGURE 3–7 for a machine inspection checklist of an articulated truck.

Checklists provide a reminder to the HDET or operator of all the points that should be checked during a visual inspection. They also provide the equipment owner or manager with written proof that the inspection has been done. They should also provide space for comments that can be recorded if future repairs are needed because of something noticed during an inspection. For example, if a hose has been rubbing on something but doesn't need replacing immediately, it can be planned for a later date when the parts and machine are available. ● SEE FIGURE 3–8 for a hose that has had a hole worn through it and should have been noticed during a visual inspection.

If little problems are spotted during the visual inspection and corrected, they will be much cheaper to fix and result in less downtime than if repairs are put off to a later time.

Circle checks should be started and stopped where the entrance to the cab is. The first time around the machine the inspection should be from the ground to eye level. This will involve looking under the machine for leaks and damage. The second time around will be from eye level up to the top of the machine.

As you become more experienced, you will be able to notice problem areas to focus on when doing machine visual inspections. The problem areas will change with different types of machines, the conditions they work in, and how old the machine is. Many small problems will turn into bigger problems if they aren't first noticed and then corrected.

Machines will become dirty after they work on the job and may become caked with mud, dirt, dust, or other materials such as coal dust, limestone, or asphalt that are part of the machines' working environment. It can be difficult to do a proper inspection on a dirty machine, but before washing the machine, it should still be given a visual inspection anyway. If the machine has been leaking, you should still be able to trace the leak to the general area and maybe even pinpoint the source on a very dirty machine. The leak will likely keep an area clean and provide a path to follow to find the leak's source. If the machine is pressure-washed, it may be difficult to find a slow leak as the clues given by the washed-off dirt will be gone.

WHEN TO SERVICE? Machine **service intervals** are mainly determined by machine working hour accumulation and calendar time intervals. All machines will have an hour

76 CHAPTER 3

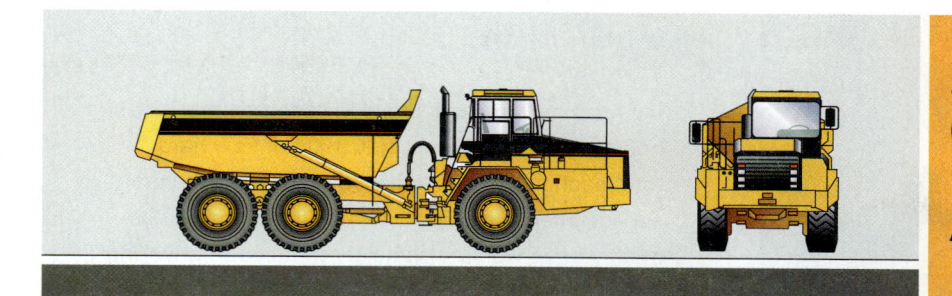

Safety & Maintenance Checklist:

Articulated Trucks

SAFETY.CAT.COM™

Operator/Inspector_____ Date_____ Time_____
Serial Number_____ Machine Hours_____

What are you inspecting?	✓	What are you looking for?	✓	Evaluator Comments

For more information, please refer to the Operation and Maintenance Manual or any other applicable manuals and instructions for this product. If you have questions, please contact your local Caterpillar dealer.

FROM THE GROUND

What are you inspecting?	What are you looking for?		
Overall Machine	Loose Or Missing Nuts & Bolts, Loose Guards, Cleanliness		
Lights	Broken lamps, lenses, operation		
Grab Irons, Steps, Handholds	Condition And Cleanliness		
Brakes, Suspension	Leaks, Damage, Wear		
Tires, rims, hub, stem caps	Inflation, Leaks, Damage, Wear		
Underneath Machine	Leaks, Damage, People Working		
Steering hydraulic system	Leaks, worn hoses, damaged lines		
Body	Damage, wear, distortion		
Fuel tank, hydraulic tank	Level, Leaks		
Transmission Oil level	Fluid level		
Suspension Cylinders	Leaks, Measure cylinder height		
Frame & Hoist Cylinders	Cracks, Leaks, Damage, Wear		
All Axles	Leaks, Damage, Wear		
Welds, castings; mounting bolts and pads	Cracks; damaged bolts or pads		

ENGINE COMPARTMENT

Engine Oil	Fluid Level		
Engine Coolant	Fluid Level		
All Hoses	Cracks, Wear Spots, Leaks		
All Belts	Tightness, Wear, Cracks		
Overall Engine Compartment	Trash Or Dirt Buildup, Leaks		
Radiator	Debris, Damage, Leaks		

ON THE MACHINE, OUTSIDE THE CAB

Battery Compartment	Connections, Fluid level		
Fire Extinguisher	Charge, Damage		
Windshield Wipers & Washers	Wear, Damage, Fluid Level		
Air Filter	Restriction Indicator		

INSIDE THE CAB

Gauges, lights, switches	Damage, operation		
Seat	Adjustment, Brake Travel		
Seat Belt, Buckle & Mounting	Damage, Wear, Adjustment		
Horn, backup alarm, lights	Proper Function		
Windows and Mirrors	Condition, clean, adjust		
ROPS	Damage		
Overall Cab Interior	Cleanliness		

HTTP://SAFETY.CAT.COM/CHECKLISTS

V0611.2

CAT, CATERPILLAR, their respective logos, "Caterpillar Yellow" and the POWER EDGE trade dress, as well as corporate and product identity used herein, are trademarks of Caterpillar and may not be used without permission. © 2011 Caterpillar All Rights Reserved

CATERPILLAR®

FIGURE 3–7 Machine inspection checklist for an articulated truck (Reprinted Courtesy of Caterpillar Inc.).

FIGURE 3–8 A hose that has worn a hole through it.

FIGURE 3–9 Machine hour meter.

meter. The hour meter will sometimes run when the electrical system is energized, but most will track hours only when the engine is running. ● **SEE FIGURE 3–9** for a machine's hour meter.

An alternative way to schedule services is to track machine fuel consumption. See the following Tech Tip. For every service interval reached, all other services required at a shorter service interval than the one being done are also required to be done. In other words, if the machine is due for a 500-hour service, then the 250- and 50-hour services should be done as well. There are certain conditions that can change service intervals, and this will be noted in the maintenance manual. The main condition that could shorten a service interval is if the machine works in an extreme environment.

TECH TIP

If a machine service interval is based on machine's hours of use, is this any indication of how hard that machine has worked? No, it isn't. If a standard service interval is 250 hours and if a machine has mostly been working with light material or travelling empty, then will it need to be serviced as often as the same machine that has worked to its maximum capability every hour of that 250 hours? The answer again is no.

A true indication of how hard a machine works is to track its fuel consumption. Most machine manufacturers will have information available to tell when a machine should be serviced based on the amount of fuel it has consumed. The harder a machine works, the more fuel it uses, and therefore, the sooner it should be serviced. This is similar to your car or pickup manual that will always say if you are towing a trailer or driving in dusty conditions, you should change the oil sooner.

This is called fuel consumption–based servicing and could easily be monitored with many newer machines that are electronically capable of measuring and recording fuel consumption.

Equipment owners will be trying to extend service intervals as long as possible, and this may be one way to do that.

Other ways are to monitor oil/fluid condition with fluid sampling or to add an aftermarket oil condition monitoring system.

Extending service intervals could save the equipment owner thousands of dollars in fluids filters and reduced downtime. However, the machine manufacturer should be consulted first, and there is always a possibility of reducing component life when service intervals are extended.

MACHINE SERVICE INTERVALS The following are typical machine service interval examples. Always check each machine's maintenance manual for its specific service interval and the items/components to be serviced.

There will be maintenance points that are to be done only when required. Some examples are filling the windshield washer reservoir, cleaning the radiator fins, and replacing the ether bottle. The following are examples of hourly or calendar time–based service intervals:

Daily or every 10 hours: This should be the responsibility of the operator and will include fluid level checks, visual checks, and possible lubrication of grease points. Some examples are inspect the seat belt, test the backup alarm, check engine oil level, check hydraulic fluid level, check exhaust fluid level, and check transmission oil level.

Weekly or every 50 hours: This service may be done by the operator or HDET and will involve a few other checks that require less frequent attention. Some examples are check tire pressure, track tension, clean cab air filter, and drain fuel tank of water. Some grease fittings could also be required to be serviced.

Monthly or 250 hours (the initial 250 hours will likely require all filters to be changed): This service used to be the normal engine oil change interval and a lot of older machines will still need this.

For many years, the common length of time that engine oil was recommended to stay in the engine was 250 hours or monthly. Today, because diesel engines run so clean the interval is commonly 500 hours. Engine oil change intervals can be lengthened by monitoring oil condition by sending samples to a laboratory for testing. More information on extending oil change intervals follows later.

This service and all others following will be done by an HDET. Newer machines will need some more checks and lubrication, and some companies will have all machine fluid samples taken at this interval. Some examples of other 250-hour service points are brake system test, engine oil sample, and belt inspection. Again some grease fittings will need servicing.

Three months or 500 hours: This service usually involves all oil filters being changed. Most engine oil will be changed at 500 hours now. The standard used to be 250 hours but because of improvements to oils, filters, and how clean the engine's run there is no need to change it that often. This also creates less waste. All oils should be sampled at this interval if they weren't at 250 hours. All service points from previous service intervals are done at this point.

Six months or 1000 hours: The main point for this service is usually a transmission oil change if the machine has a powershift transmission. A typical safety check at this interval is to check the ROPS. All service

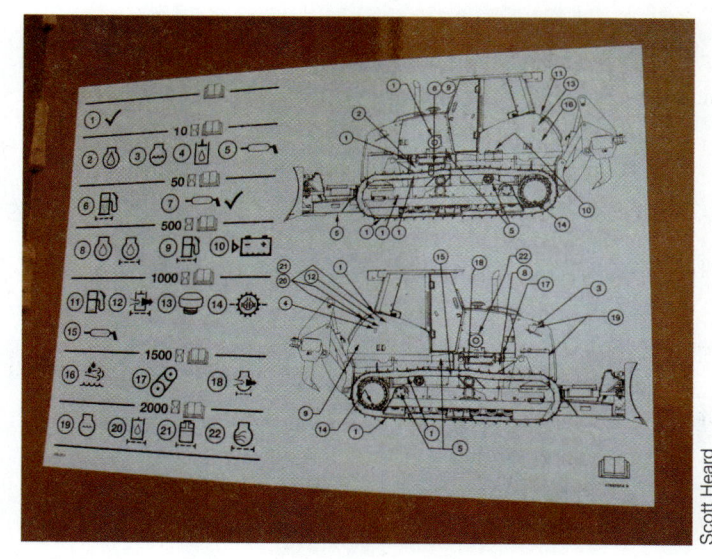

FIGURE 3–10 Service interval chart.

points from previous service intervals are done at this point.

One year or 2000 hours: The main points for this interval are usually differential and hydraulic oil changes (if conventional fluids are used). Some more in-depth checks may include fuel injection system clearances, engine valve clearance adjustment, accumulator pressure checks, and brake wear checks. This is usually when all machine fluid and filters get changed. All service points from previous service intervals are done at this point.

Three years or 6000 hours: If the machine uses long-life engine coolant, it will likely need an additive added. Some long-life hydraulic and drivetrain fluids can be used for up to 6000 hours if certain conditions are met. Machine seat belts should be replaced every 3 years after they are installed or every 5 years after their manufacturer. All service points from previous service intervals are done at this point. Some hydraulic oils will allow a 6000-hour change interval if certain conditions are followed. Exhaust filter change or cleaning could be required.

Six years or 12,000 hours: If long-life engine coolant is installed, it should be replaced. All service points from previous service intervals are done at this point.
● **SEE FIGURE 3–10** for a typical service interval chart.

Progress Check

1. Machine service intervals are mostly based on:
   a. how dirty the filters look
   b. machine running hours and calendar time
   c. the amount of oil the engine burns
   d. miles or kilometers of machine travel

2. When checking the safety features of a machine, which of the following statements is correct:
   a. parking brake operation is the same on all machines
   b. steering systems don't need to be checked regularly
   c. cab door operation isn't considered a safety feature
   d. work lights and reflectors that are on a machine should be working and visible

3. When performing a visual inspection or circle check, this is *not* one part of the procedure:
   a. taking oil samples
   b. looking for fluid leaks
   c. looking for missing guards
   d. looking for damaged steps

4. The service interval for which machine fluid has recently been changed from 250 to 500 hours in many machines:
   a. brake fluid
   b. engine oil
   c. transmission fluid
   d. hydraulic fluid

5. The service interval when all machine fluids and filters are most likely to be changed is:
   a. 250 hours
   b. 500 hours
   c. 1000 hours
   d. 2000 hours

COMMON MACHINE SERVICE PROCEDURES

ENGINE AIR FILTERS/FILTRATION DEVICES. Diesel engines consume massive amounts of air, and considering that the machines they are working in will likely be working in a dusty environment, their air filtration system needs to be maintained on a regular basis. A machine will likely have an air pre-cleaner device and an air filter assembly with a replaceable element in it. (See Chapter 6 for more information.) Pre-cleaners will eliminate larger dust and dirt particles and should be inspected, emptied, and cleaned when required.

Air filter element change intervals should *only* be based on specified engine's maximum air intake restriction levels or a maximum calendar time. Once the restriction level is exceeded or usually a 12-month period maximum has elapsed, then it is necessary to change the filter or filter elements.

Care must be taken when changing air filters so that dirt and dust aren't allowed into the engine's intake. It is also very important to make sure that the new filter fits and seals properly in its housing and the filter cover is installed securely. Any unfiltered air that enters an engine will carry dust with it and lead to a shortened life of the engine.

ENGINE OIL CHANGE. A proper engine oil change requires the oil to be warmed up before it is drained. This ensures that any contamination that may have settled to the bottom of the oil pan will likely be drained out with the old oil as it will be mixed up with warmed-up oil. When the oil is warmed up to at least 125°F, set the engine rpm to low idle and obtain a live oil sample if the engine is equipped with a sample valve (see oil sampling later in this chapter). Stop the engine and make sure that you have the proper filter and enough of the right kind of oil before proceeding.

Some machines will have a simple drain plug that is removed to drain the oil or have a valve that is opened to allow the oil to be metered as slow or as fast as needed. Drain plugs are sometimes magnetic so any metallic particles in the oil are collected to it and can be inspected at service time. If the oil pan just has a plain plug, make sure that you have enough drain container capacity before you remove the plug. You don't want to be trying to put the plug back in while oil is still draining to prevent an overflow. It's also a good idea to have a wand-type magnet handy in case you drop the plug in the drain container.

Drain plugs should be cleaned off before reinstalling, and they may also have some type of washer or O-ring seal that should be replaced. Care must be taken when installing drain plugs that they aren't overtightened. This could lead to damaged threads and a leaking oil pan.

After the drain plug is installed and the oil filter has been replaced, new engine oil is added and brought up to at least the add mark on the dipstick. The dipstick should always be wiped clean before inserting back into the engine. Once the engine oil is topped up, the HDET should crank the engine with the fuel injection system shut off if possible until oil pressure can be read. This will fill the oil filter with clean oil and pressurize the engine's bearings before the engine starts. The engine can then be started and ran for a few minutes. It should then be stopped and allowed to rest for 5–10 minutes for the oil to drain back to the pan. After this the oil level can be checked and topped up to its full level. Take care to not overfill the oil level. ● **SEE FIGURE 3–11** for an HDET checking the oil level of a diesel engine.

FIGURE 3–11 HDET checking oil level.

Some fleets will be equipped with evacuation valves on the engine oil pan only, on some of the other fluid reservoirs, or on all fluid reservoirs. These evacuation valves allow the machine's oil to be drawn out by a vacuum source and into a waste oil container. This is faster and cleaner than draining oil by removing drain plugs. The new oil can also be pumped back in through the evacuation valve, which also saves time. The valves should have protective caps that need to be in place when the machine is running to prevent oil contamination. Some shops are equipped with new oil pumping systems that can have handles that will allow you to set the exact amount of fluid you want to be pumped. The handle will shut off as soon as the quantity of fluid that you requested has been pumped. This is a nice feature that lets you do something else while the fluid is pumping.

ENGINE OIL FILTERS. These will be changed every time the oil needs to be changed. They could be spin-on–type filters that are one piece and have a metal case as part of them and could need a filter wrench to loosen them. They could also be a cartridge type that only requires the actual filter element to be changed while the outside metal housing is part of the filter assembly and is cleaned and reused. The cartridge-type filter housing should have its seal replaced each time the filter is changed. To increase filter capacity large displacement engines will use multiple filters.

The practice of filling oil filters with clean oil is always debatable. It should be avoided because the possibility of introducing unfiltered oil and contamination from funnels and containers into the engine is greater. If the filter is installed empty and it is possible to crank the engine over for 10–15 seconds before it starts, this should be done. This will prime the lubrication system and reduce wear upon initial start-up after the new oil is installed.

With any spin-on filter, it is always a good practice to write the machine hours and date of service on the filter with a permanent marker. This will be a way to confirm the service hours on the filter since the last service.

TRANSMISSION OIL CHANGE. Some machines will combine their transmission oil with hydraulic or differential oil. It is also possible with larger machines that there may be a separate tank for transmission oil as well as the transmission housing being a reservoir. Be aware of machines with more than one drain plug for a particular housing so you drain all the oil that is necessary.

As with performing an engine oil change, the transmission should also be warmed up before draining the oil. The fluid should be sampled when running if there is a sampling valve. There may be a magnetic drain plug that should be inspected

TECH TIP

Spin-on oil filters should be installed properly so they stay tight but are easy to remove at the next service. After the old filter is removed, clean the filter base, lubricate the filter seal with clean engine oil, and thread onto the base until the seal makes contact with the base. The filter should then be tightened an extra amount. This is usually half to three quarters of a turn and no more. Many filters will have this information on the side of them. ● **SEE FIGURE 3–12** for a spin-on–type oil filter's tightening information.

Larger filters will require the use of a strap-type wrench to loosen and tighten them.

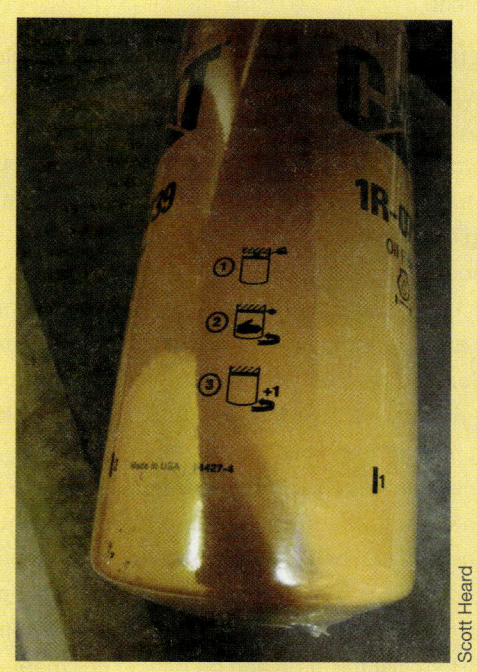

FIGURE 3–12 A spin-on oil filter's tightening information.

TECH TIP

I once had to remove a filter base and then put it in a vice to remove the filter with a hammer and chisel. This occurred after someone had installed it during a previous service and figured if a little tight was good, then a whole lot tight must be better! Resist the urge to overtighten filters and heed the instructions for tightening properly.

MACHINE SERVICING **81**

and cleaned off after it is removed. Ensure that there is enough capacity in the drain container for the amount of oil that is being drained before the drain plug is removed.

Transmission oil level is usually checked with the engine running and warm so you may have to refill the oil to a cold or stopped engine; mark before starting the engine and then warm up the oil and finalize the level.

TRANSMISSION FILTER CHANGE. Transmission filters are usually changed every 500 hours. There is likely a suction screen that will need to be removed and cleaned as well. Transmission filters could be spin-on or cartridge type, and there may be multiple filters depending on the quantity of oil in the system. Some cartridge filters will have a drain plug in the bottom of them to allow the housing to be drained before removing the filter element.

HYDRAULIC FLUID CHANGE. Hydraulic fluid should also be warmed up and then sampled before it is drained. By warming the fluid, you will be pressurizing the tank. An important point to remember here is to release hydraulic tank pressure before attempting to drain the tank. This could be as simple as slowly loosening the filler cap or the tank may have a pressure release valve. Most hydraulic tanks will have a valve to control the fluid drain rate. Sometimes an air-bleeding procedure needs to be done after changing the reservoir fluid to ensure that the pumps don't get air pushed into their inlet. Always consult the machine manual before starting a machine immediately after changing its hydraulic fluid.

Some machines will provide the capability to pump the refill oil in through the machines' filters. Otherwise unless the oil's cleanliness can be verified, it should be ran through an external filter before it is pumped into the tank. It will be tempting to remove the screen in the filler neck of the tank to fill the tank faster. Try to resist this urge and be patient. In Chapter 4, I have recalled an expensive mistake that occurred when someone didn't want to wait for the oil to flow through the screen.

HYDRAULIC FILTER CHANGE. There are several different types of hydraulic filters (pressure, inlet, return, case drain) that may need to be changed. Some hydraulic filters are located in the reservoir and extra care should be taken when changing these to not drop anything in the tank. Some filters will be part of an assembly that will have a screen around the filter.

FUEL SYSTEM FILTERS/WATER SEPARATORS. Fuel filters are usually changed every 500 hours. They are either spin-on type, cartridge type, or a glass or metal square-shaped fuel filter. They can incorporate a water separator, or this can be a separate item. Water separators may need to be serviced more frequently than the filters, and this will depend on the water

FIGURE 3–13 Typical fuel water separator.

content of the machine's fuel. ● **SEE FIGURE 3–13** for a typical fuel water separator.

Most machines will have a primary filter that could be a simple screen or spin-on filter/water separator and one or more secondary fuel filters. Secondary filters should not be filled up before they are installed because of the potential expensive damage that could occur if dirty fuel is allowed to enter the high-pressure fuel system. Some fuel filters will actually have a valve that will make it impossible to fill them before being installed.

Fuel filter changes can lead to much frustration if the fuel system is not bled properly. Machines will have different methods for bleeding air from the fuel system. There could be manually or electrically operated priming pumps. It's okay to fill the primary fuel filter before installing it because the fuel leaving it will still be filtered by the secondary filter.

If air is not bled properly from the low-pressure fuel system and the engine is cranked over, the machine could be down for a long time.

FUEL TANK DRAIN. An often overlooked maintenance point is to drain off any accumulated water and other contamination that has settled to the bottom of the fuel tank. If allowed to build up, it will eventually make its way through the fuel system. This is sometimes made easy if the machine has a drain

SAFETY TIP

All HDETs should practice safe and healthy work habits when servicing machines. This includes cleaning up oil spills when they occur to prevent slip and fall hazards, wearing gloves that will protect against hot oil and oil contaminants from entering the body, and releasing any pressure in reservoirs before draining oil. Hands should always be washed before eating. All normal PPE should be worn when servicing machines (safety glasses, safety boots, coveralls, and gloves). Extra PPE could be required if the machine has been working in hazardous materials.

FIGURE 3–14 Fill and drain plugs on a final drive.

tap at the bottom of the tank. Machines with water separators may not have a tank drain.

AXLE OR FINAL DRIVE OIL CHANGE. The oil used in these compartments is usually a higher viscosity and therefore should be warmed up as much as possible before draining to get all contamination out. The warmer the oil, the faster it drains as well. Beware of built-up pressure when removing the drain plugs for these compartments. There may be a breather for it, but it will likely be plugged, which will cause the oil to quickly rush out and likely be all over you if you aren't careful. Close attention should be paid to where the proper oil level should be when filling final drives and axles. Most axles will need to be filled until the oil starts to run out the fill plug opening. However, some could have a dipstick or a sight glass instead. ● **SEE FIGURE 3–14** for the fill and drain plugs on a final drive.

AXLE FILTERS. Some larger machines will use filters to keep their axle fluid clean, and these will likely be spin-on type.

COOLANT ADDITIVE TOP UP OR COOLANT CHANGE. Although coolant change intervals have gotten longer over the years, this is still an important part of machine servicing. Before the coolant change interval, there will likely need to be an additive top-up service. This could mean simply pouring into the coolant reservoir a specific quantity of additive or a spin-on filter/additive may need to be changed.

There are a lot of engine components that are exposed to the engine coolant, and if it isn't serviced regularly and properly, there could be some expensive repairs that result.

As a cooling system operates under pressure, the pressure needs to be released before draining the coolant. There will likely be warnings related to safely relieving cooling system pressure, and these should be closely adhered to.

It will be impossible to drain all coolant from the machine but as much as possible should be. Refilling a system will take patience as there is a chance of creating air pockets in the engine if it is filled to fast. Some machines require special tooling to fill the system from the bottom of the rad.

COOLANT CONDITION OR FILTER CHANGE. Some engines will need to have a coolant filter changed regularly. There will be one or two valves that need to be closed before the filter can be removed. ● **SEE FIGURE 3–15** for a diesel engine coolant filter.

Coolant should be tested for freeze and boil-over protection. This can be done with a hydrometer or refractometer. Be aware that hydrometers for EG- and PG-type coolants are different (see coolants later in this chapter).

BRAKE FLUID CHANGE. Some machines may have a dedicated fluid for the brake system. This fluid should also be changed on a regular basis according to the maintenance guide. This will then require a brake-bleeding procedure to eliminate any or all air that

FIGURE 3–15 A diesel engine coolant filter.

MACHINE SERVICING 83

may have entered the brake fluid system. (See Chapter 18 for more information on hydraulic brake systems.)

CAB FILTER CHANGE. Most machines today will have an HVAC (heating, ventilation, and air-conditioning) system that will keep the cab pressurized to keep out dust and keep the temperature at a comfortable level. This system relies mainly on air flow and will not work properly with dirty filters. These filters are often overlooked when a machine is serviced, and if left dirty, they will likely lead to other problems.

DPF (DIESEL PARTICULATE FILTER) CLEANING. These filters will require removal and cleaning by specialized equipment. This is a newer maintenance procedure that may need to have new procedures and equipment used to complete the service. See Chapter 7 on prime movers for more information on DPFs.

BATTERY MAINTENANCE. Good battery condition is critical to having a dependable machine. If the machine has a conventional battery (see Chapter 5), it should have its electrolyte level checked. All batteries should be clean, dry, and secure in their compartment.

MACHINE LUBRICATION. Most machines will have at least one location or grease point that needs regular lubrication. The majority of machine grease points require daily lubrication, which should be done by the operator. There will likely be one or two grease fittings that will be hard to access and will likely get overlooked by the operator. When you are servicing a machine, you should look at every grease point and see if it has been getting grease and if not investigate why. If a grease fitting is hard to get to, try running a hose from it and put the fitting in an easier location to access.

Greasing machines will require the use of a manual handheld grease gun, an electric or air-powered grease gun, or a shop air-powered grease gun. Caution should be taken when operating a grease gun because very high pressures (5000 psi and higher) can be created quite easily.

Because of the high pressures created when greasing, care must be taken to not overgrease sometimes. For example, if a belt-driven fan has a set of bearings that need grease, do not keep pumping grease until you see it come out. You will likely blow the seal out by doing this. Try to keep in mind how the grease will act against the seal for whatever you are greasing. If the seal lip faces in toward the grease, then it's possible to blow the seal out with pressure. If the lip points out, then the grease will flow past the seal. Most machine grease points will be arranged with the seal out, and when greasing, the time to stop pumping is when you see grease move past the seal.

FIGURE 3–16 Automatic greaser pump unit.

Many machines will now have automatic greasing systems installed on them. These systems will need regular attention themselves to ensure that they keep the machine lubricated. Auto greasers will have a reservoir of grease that needs to have a minimum level of grease of the proper viscosity in it. The pump assembly will send grease out through a series of valves and lines to the grease points that would normally get greased with a grease gun.

Auto greasers should be inspected for proper operation and for any damage to the grease lines or tubes. Some systems will shut down if one line becomes plugged. Some systems will be controlled electrically and can be set to run on time intervals while others will cycle every time the machine goes into reverse. ● **SEE FIGURE 3–16** for an automatic greaser.

USED FLUID DISPOSAL/RECYCLING

When machines are serviced, there will be used fluids that need to be dealt with. There could be as little as 1 or 2 gallons or several hundred. Any amount of waste fluid needs to be dealt with responsibly and properly. Please consider the negative effects to our environment if you are tempted to dispose of waste fluids in an improper way. You will also face huge fines or jail time if caught disposing fluids illegally.

> **WEB ACTIVITY**
>
> Search "used oil recycling" to see how many different companies are in the business of collecting and processing waste fluids into usable fluids. See if there is one near you.

FIGURE 3–17 A used oil tank.

Fluids can be collected in a **waste oil container** at a shop or in a mobile lube truck. There are several companies that will come to shops with tanker trucks, remove the used oil, and take it to a facility where it can be recycled. Mineral-based oil can be recycled and sold as recycled oil or mixed with virgin oil. ● SEE FIGURE 3–17 for a used oil tank.

Used coolant can also be recycled and will be collected in a separate tank.

MACHINE SERVICING DOCUMENTATION Another important part of a preventive maintenance program is the practice of accurate record keeping. Usually, there will be a machine manager responsible for tracking when machines are due for service and keeping records of what services have been done. The information source for these records is the HDET. Service reports should be made for each service and details of the inspection and service need to be clearly recorded. This can be a simple checklist with space for comments or an electronic form. Any problems found during a routine service should be acted on, if not at the same time as the service is being done, then at the next best time. This can be initiated by a communication process that may generate a work order for the problem.

These records are also a valuable tool for the machine owner if the machine is ever put up for sale. The owner can show the potential purchaser how the machine has been maintained and, along with fluid sample reports and repair history, can give a complete machine history.

Progress Check

6. Air filter elements should only be changed based on calendar time and by:
   a. putting a light inside to check for cleanliness
   b. tapping it against a tire to see how much dust comes out
   c. measuring exhaust color
   d. checking air inlet restriction
7. Care should be taken when removing plugs from warm axles because:
   a. the fluid will be highly corrosive
   b. the fluid will likely be pressurized
   c. the fluid will likely cause third-degree burns
   d. there will be metal slivers on them
8. A typical transmission oil filter change will occur at:
   a. 100 hours
   b. 250 hours
   c. 500 hours
   d. 2000 hours
9. Most secondary fuel filters should not be prefilled before installation because:
   a. this will make the system hard to bleed
   b. the primary filter will receive unfiltered fuel
   c. the injectors will run dry
   d. there will be unfiltered fuel going to the high-pressure fuel system
10. What is one possible result of overgreasing a component?
    a. there won't be enough clearance between moving parts
    b. a lip-type seal could get blown out
    c. the grease will get too thin from too much pressure
    d. there will be cross-contamination of fluids

MACHINE FLUIDS Fluids are used in machines for many purposes. Most fluids will be needed for lubricating system components. The fluid will create a film between parts that are moving in relation to each other, and this film prevents metal-to-metal contact of the components. Transmission fluid, for example, will keep shafts floating in a film of oil. This is called hydrodynamic suspension. Other fluids like engine coolant are used for cooling components, while others are needed for both lubrication and cooling. Hydraulic and brake systems use fluid to transfer energy, cooling, sealing, and lubrication.

It is very important to use the proper viscosity, type, and quality fluid for each system. All systems on a machine are designed to be used with a fluid made to a certain standard. Most machine manufacturers will have their own branded fluids that have been made to their standards, and they will strongly recommend them to be used in their machines. They may also give minimum standards that an aftermarket fluid must meet if the owner wants to use another kind of fluid. If a machine is still under warranty, it is wise to use whatever fluids the machine manufacturer recommends and keep records that can prove this fact in case there is a repair dispute. Warranty claims can be denied based on the use of wrong fluids.

VISCOSITY. Fluid **viscosity** refers to the fluid's resistance to flow. When it comes to recommending proper fluid viscosities, machine manufacturers will provide a viscosity chart with the machine's maintenance guide. Fluid viscosity requirements will change with ambient air temperature (outside air temperature in the immediate vicinity of the machine). Generally speaking, as the ambient temperature where the machine is working warms up from winter to summer, then thicker fluids should be used and the opposite is true from summer to winter. If the fluid is to stay in the compartment for 500 hours, you must try to roughly predict what the highest and lowest temperatures the machine will be operating in and use the recommended viscosity for that temperature range.

Fluid viscosity is measured by standards set by two organizations. They are SAE (Society of Automotive Engineers) and ISO (International Organization for Standardization). Both organizations have developed test methods that measure how fast a fluid flows through a fixed orifice when it is at a certain temperature. The faster it flows, the lower its viscosity is or the thinner it is. It will be given a lower number that relates to the lesser amount of time it took to flow through the orifice. The opposite is true for a thicker fluid, and the longer it takes, the higher its viscosity number. To compare two fluids that you may be familiar with, think of water as a very low-viscosity fluid and shampoo as a very high-viscosity fluid. Typical SAE viscosity numbers for machine fluids go from SAE 0W to SAE 60. An SAE number that has a W following it has been tested at a lower temperature and the W stands for winter. ISO viscosity numbers go from ISO VG (viscosity grade) 22 to ISO VG 100. ● **SEE FIGURE 3–18** for a viscosity chart for diesel engine oil.

Gear oil has a different numbering system that starts at SAE 75W and goes to SAE 140.

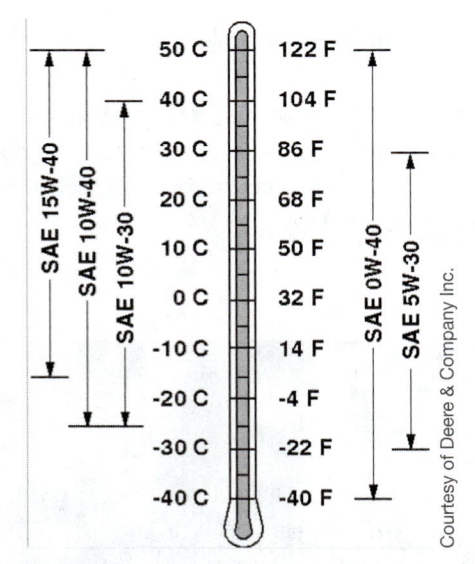

FIGURE 3–18 Viscosity chart for diesel engine oil.

Some fluids are called multi-viscosity such as 10W-30 engine oil. They will act like a lower viscosity fluid when cold, which means they flow better but will resist thinning out when they warm up. This is done with additives called viscosity improvers. If a fluid thins out too much, it won't provide the proper oil film between rotating components. Some typical viscosity numbers are:

Diesel engine oil: SAE 10W-30 or 15W-40
Hydraulic fluid: SAE 10W or ISO 32
Powershift transmission fluid: SAE 30
Axle fluid: SAE 30
Final drive fluid: SAE 50
Brake fluid: 10W

ADDITIVES. Machine manufacturers will generally recommend that no aftermarket additives need to be added to any oils for a machine. There are some exceptions to this such as an axle that has friction material in it for an anti-spin device.

There are countless fluid additives available that their manufacturers claim will improve performance, increase fuel economy, stop leaks, and lower emissions. Resist the temptation to add a fluid additive with the hope to gain anything besides a lighter wallet and use the right fluid that meets or exceeds manufacturer's specifications.

FLUID CLEANLINESS Machine fluids need to be at a specific level of cleanliness. Fluid can look clean to the naked eye but may not be anywhere near as clean as it should be for the system it is in. Solid contaminate particles in fluids are measured in microns, and 1 micron is equal to 1 millionth of a meter or

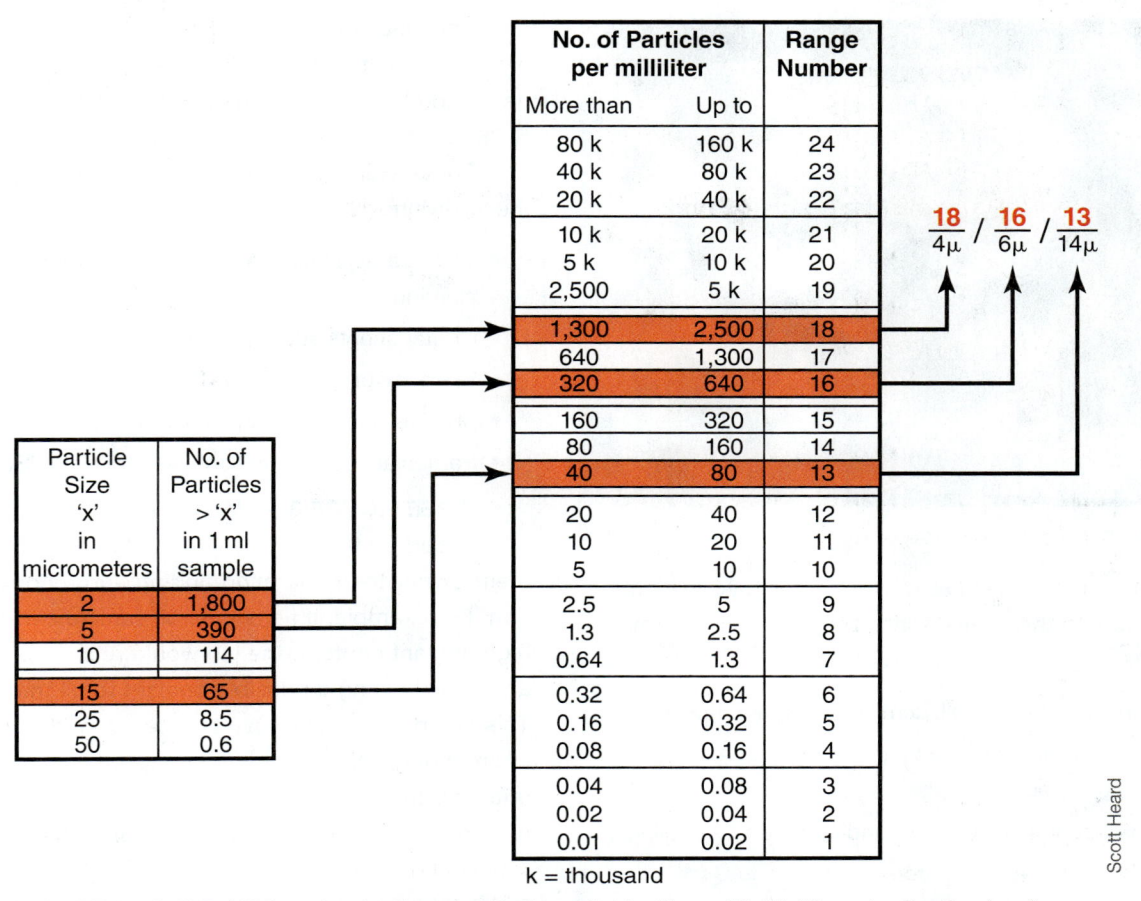

Particle Size 'x' in micrometers	No. of Particles > 'x' in 1 ml sample
2	1,800
5	390
10	114
15	65
25	8.5
50	0.6

No. of Particles per milliliter		Range Number
More than	Up to	
80 k	160 k	24
40 k	80 k	23
20 k	40 k	22
10 k	20 k	21
5 k	10 k	20
2,500	5 k	19
1,300	2,500	18
640	1,300	17
320	640	16
160	320	15
80	160	14
40	80	13
20	40	12
10	20	11
5	10	10
2.5	5	9
1.3	2.5	8
0.64	1.3	7
0.32	0.64	6
0.16	0.32	5
0.08	0.16	4
0.04	0.08	3
0.02	0.04	2
0.01	0.02	1

$\frac{18}{4\mu} / \frac{16}{6\mu} / \frac{13}{14\mu}$

k = thousand

FIGURE 3–19 ISO 4406 contamination chart using a fluid with an 18-16-13 contamination level.

39 millionths of an inch. To put this into perspective, the smallest particle that can be seen by the naked eye is 40 microns. Some amount of contamination is acceptable, and this is determined by the engineers that design the components that come into contact with the fluid. Some high-pressure fuel systems will require fuel to be filtered to remove contamination as small as 2 microns! The fluid cleanliness requirements are mostly dependent on the clearances between moving parts and system pressures.

There is an ISO standard that precisely measures how clean a fluid is. To measure fluid cleanliness, a fluid sample is run through a particle counter. A particle counter is a precise measuring device that uses either laser or LED technology to measure the size and quantity of particles in the fluid. The most widely accepted way of describing how clean a fluid is the ISO 4406 standard. This standard measures the quantity of contaminants in three different size categories (4, 6, and 14 microns). The number of particles counted for each size is then compared to a chart that has quantity range numbers. The numbers increase as the number of particles increase.
● **SEE FIGURE 3–19** for the ISO 4406 contamination chart with an example of a fluid that meets an 18-16-13 cleanliness code.

FLUID SAMPLING Part of a good maintenance program is **fluid sampling**. It takes a little extra time and is an extra expense, but it is a very good investment in the long run. As mentioned previously, the ideal way to take an oil sample is to use a sampling valve when the machine is running and the fluid is warm (120°F minimum) and circulating. Many machines will have these valves installed at the factory and are either a quick coupler for low-pressure compartments or a threaded fitting for high-pressure compartments. A hose and fittings connect the oil sample bottle to the sample valve. The sample bottle is roughly 100 ml and should only be filled three quarters full. If a sample is taken with the fluid circulating, this is called a live sample and is the most accurate sample of what is in the compartment. Sampling ports can be added to any pressurized fluid system on a machine. ● **SEE FIGURE 3–20** for a machine's oil sampling ports.

The next best way is to draw a sample out of the compartment is with a hand-actuated vacuum pump. The oil sample bottle is threaded onto the pump and is removed and capped when filled up. It is important when taking fluid samples to be as clean as possible and not contaminate the sample.

MACHINE SERVICING

FIGURE 3–20 Oil sample ports.

It is important to send as much accurate information with the oil sample to the analysis lab. The minimum information should be:

- Machine's make, model, serial number, and hours
- Fluid type, brand, viscosity, and number of hours it has been in the machine
- Compartment it came out of and method of sampling

Once the laboratory receives the sample, it is put through a series of tests and compared against a sample of brand-new fluid of the same properties (brand, type, viscosity, etc.).
● **SEE FIGURE 3–21** for an oil sample lab test machine.

Some examples of what a fluid sample lab tests for are the following:

- Wear particles such as iron, lead, tin, chromium, copper, aluminum
- Contaminants such as water, fuel, coolant, dirt (silicon)
- Viscosity of fluid—to see if it has changed
- pH level—to see if it is acidic
- Particulate contamination level—given in ISO code

● **SEE FIGURE 3–22** for an oil sample report.

Based on the test results, the lab may alert the equipment owner to an alarming finding and recommend an action. For example, if an engine oil sample came back with a high coolant content, the lab would recommend resampling and investigating a possible coolant to oil transfer problem. This could potentially save thousands of dollars if detected soon enough. If the sample analysis report shows nothing unusual, then the owner simply puts the report in the machines service/repair file. As mentioned before, trends can

FIGURE 3–21 Oil sample lab test machine.

88   CHAPTER 3

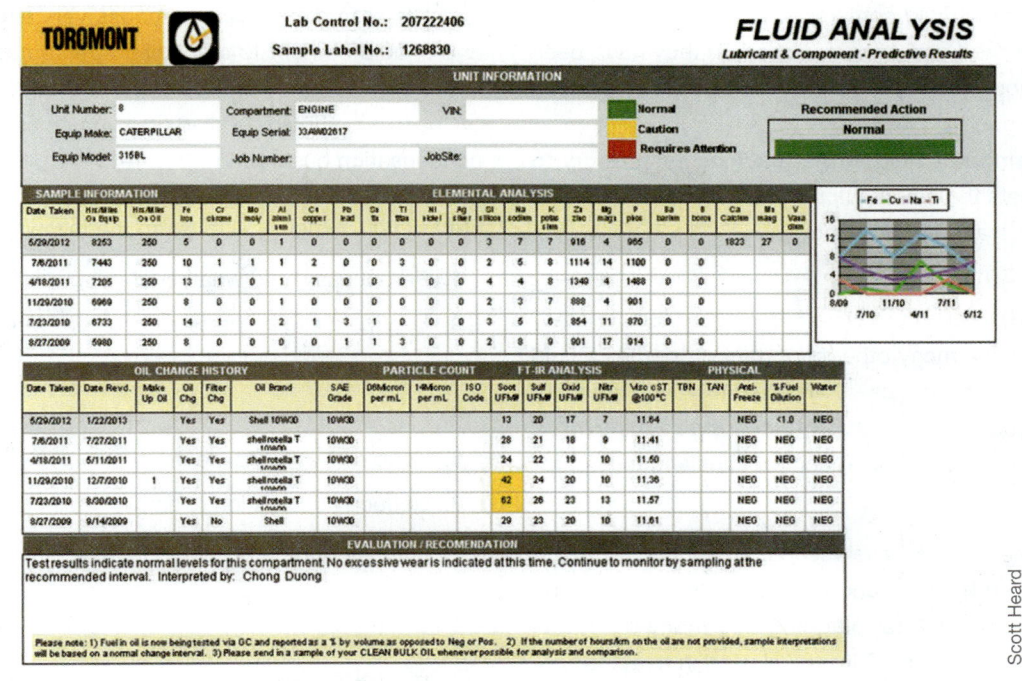

FIGURE 3–22 An oil sample report.

FIGURE 3–23 A portable particle counter.

TECH TIP

Fluid samples for machines are like giving a blood sample to your doctor. Fluid sampling is mainly done to monitor what is happening inside machine systems, but there are other benefits as well. If a machine has had regular fluid sample taken from when it was new, there will be a nice history of the machine's servicing information.

Sometimes trends of wear particles are seen, and predictive repairs can be planned based on these trends.

Fluid samples can be taken and analyzed for any oil or coolant compartment, and fluid sampling is another key to a good preventive maintenance program.

be followed if one or more elements are seen to be higher than normal.

Some dealers are equipped with portable particle counters that can read fluid particle contamination levels right on the machine. ● **SEE FIGURE 3–23** for a portable particle counting machine.

A particle counter will tell you if the fluid is contaminated with solid particles, but it can't identify what the particles are made of.

FLUID STORAGE

Proper fluid storage should be a major concern for any heavy equipment owner. Sealed containers are a must to prevent dirt, water, and air from entering the oil and causing damage. Ideally, even brand-new oils and fluids should be filtered before

MACHINE SERVICING **89**

going into a machine. Fluid containers such as pails and barrels need to be stored and sealed properly if they have been opened. They should be kept in a dry, cool, and low-humidity location if possible.

Before pouring any fluid into a machine system, try to ensure that it meets the manufacturer's recommended cleanliness standard for that compartment.

MACHINE FLUID PROPERTIES
Besides viscosity grading of fluids, there are many other important properties a fluid should have before it can be used. Always check with the manufacturer maintenance information to see the required fluid properties for any system. Other machine fluids and their properties are as follows.

DIESEL FUEL. The type of diesel fuel recommended for a machine is determined by the manufacturer. Improper fuel used in a machine can cause filter plugging and many other fuel system component problems. Diesel fuel used in heavy equipment can fall under two classifications. No.1 diesel has a lower viscosity and used in the winter while No.2 diesel is slightly higher in viscosity and used in the summer. Diesel fuel suppliers will adjust the viscosity of fuels as seasonal temperatures change.

Sulfur content in diesel fuel is very important with today's low-emission engines. Most engines designed for North American use now require the use of ULSD (ultra-low-sulfur diesel) with a sulfur content of less than 0.0015% (15 ppm). Some fuels used in the past had sulfur contents over 1%.

Biodiesel has gained popularity lately and has been approved by most manufacturers to use if it meets recognized standards. The organic part of biodiesel is limited to no more than a few percent (5–20%) of the diesel fuel content. Biodiesel is labeled B5 or B20 to identify the organic content. ● **SEE FIGURE 3–24** for how biodiesel is identified.

FIGURE 3–24 Biodiesel identification.

Water in diesel fuel can cause very expensive problems and should be kept to a specified percentage minimum. This is done with water separators and fuel tank drains. Fuel tanks on machines should be filled up as much as possible to reduce condensation buildup.

ENGINE OIL. Engine oil must do the following 10 things:

1. Permit easy cold weather starting
2. Lubricate engine parts and prevent wear
3. Reduce friction in an operating engine
4. Protect against corrosion and rust
5. Keep engine parts clean
6. Reduce combustion chamber deposits
7. Fight soot buildup
8. Cool engine parts
9. Seal combustion chamber
10. Be non-foaming

The ideal engine oil for any diesel engine will use a base oil combined with the proper additives to address these 10 points.

Using the oil recommended by the engine manufacturer in a machine is crucial to get the maximum service life from engine components that are in contact with the oil as well as being able to maximize oil change intervals.

Some machines will come from the factory with break-in oil that will need to be changed at 250 hours. This may also be required to be used after an engine is reconditioned.

Proper engine oil type is recommended by two characteristics. Viscosity is the first and the second is its rating according to API (American Petroleum Institute) or another engine oil property rating organization. Most major engine manufacturers will create their own oil standard.

API has set engine oil standards for many years. They started in the mid-1900s by setting minimum standards that engine oil should meet. They looked for properties such as anti-oxidation, anti-foaming, and anticorrosion to help engine manufacturers get the most life and efficiency out of their engines. The first designation for gas engines was SA (S stands for spark ignition) and the first for diesel was CA (C standing for compression ignition). As engine technology moved forward, it put different demands on the engine oil and every few years SAE would create a new minimum standard. As the new standards came out, the letter designation would change. Today's diesel engines will need from a CG-4 to a CJ-4 spec oil. ● **SEE FIGURE 3–25** for the label on a container of diesel engine oil.

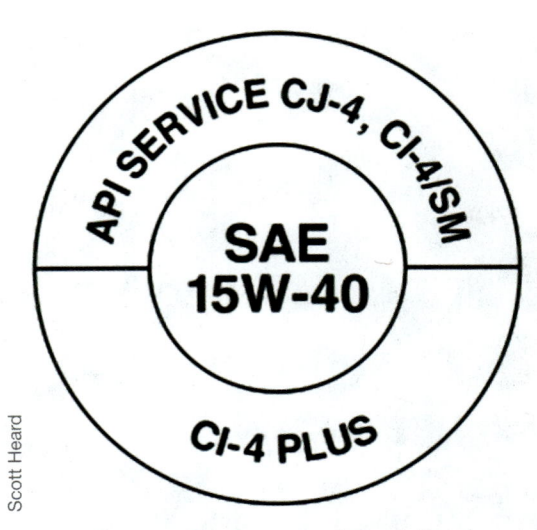

FIGURE 3–25 The label on a container of diesel engine oil.

FIGURE 3–26 Diesel engine coolant.

CASE STUDY

A very expensive lesson was learned once from refilling an engine with the wrong specification engine oil. A limestone quarry had for many years used rock trucks that had two-stroke engines in them. These engines required a straight-grade viscosity engine oil to be used year-round. When the quarry owners decided to upgrade their trucks, the new trucks they bought had four-stroke Caterpillar engines in them. These engines required a multigrade viscosity oil that vary between 10W-30 and 15W-40 depending on ambient temperature. One of the quarry mechanics who was doing a service on one of the new rock trucks made a mistake and filled up the new engine with straight 30 viscosity. This happened to be during the cold winter months, and one particularly very cold morning, shortly after the oil change, the truck was started and had a major engine failure because of the wrong viscosity oil. The results of this wrong engine oil usage was a $100,000 parts bill, several thousand dollars in labor, as well as a truck that was out of commission for several weeks.

Using the proper oil for today's low-emission engines is critical to ensuring their longevity and staying within the low-emission regulations.

Synthetic engine oil can be used as long as it meets the recommended standards and has the proper viscosity. The main advantage of synthetic oil is its cold weather low viscosity or ease of flowability in low temperatures and resistance to oxidation at high temperatures.

All new oil is backward compatible, meaning the newest specification can be used for older engines as well.

ENGINE COOLANT. Most diesel engine coolant is made up of three things. They are glycol, water, and additives. Water alone should never be used in a cooling system unless in case of emergency. A minimum of 30% glycol/70% water should be used, and a more common mix is 50/50. Distilled or deionized water should be used if needed to top up, but the safest way is to add a premixed coolant solution, which is usually 50/50.

Coolant additives will protect the metal surfaces that the coolant is exposed to by preventing foaming, rust, corrosion, liner pitting, and scale buildup. Additives will get depleted as the engine accumulates hours and will need to be topped up at certain service intervals.

Glycol in the coolant helps prevent freezing and boiling of the coolant. A 50/50 concentration will give a freeze protection down to –34°F and a boil protection of up to 223°F. Overconcentration of glycol (70% and higher) will start to decrease these properties. There are two main types of coolant used, and they are ethylene glycol and propylene glycol based.

Many machines come with long-life coolant commonly called ELC (extended life coolant), which will be able to stay in the machine from 6000 to 12,000 hours with only additive top-ups. ● **SEE FIGURE 3–26** for an example of a diesel engine coolant.

Engine coolant can also be sampled and analyzed. This is also a good practice especially if long-life coolant is expected to be in the system for five or more years.

HYDRAULIC FLUID. It has been said that the most important component in a hydraulic system is the hydraulic fluid. Hydraulic fluid not only has to clean, cool, lubricate, and seal; its main function is to transfer energy. Hydraulic fluid in a high-pressure system will be compressed (roughly 1% for every 2000 psi) and relaxed constantly. This will put an extra strain on the fluid.

The most common type of hydraulic fluid is petroleum based and is usually called mineral oil, hydraulic oil, or just oil. It is the most economical and easiest to find. Some machine manufacturers will allow or recommend that diesel engine oil can be used in the hydraulic system because many of the requirements that an engine oil must meet apply to hydraulic oil.

You will quite often see the letters AW associated with hydraulic oil. This stands for antiwear, and the oil will have a higher zinc and phosphorus content. Other additives in the oil will include detergents, rust inhibitors, anti-oxidants, and anti-foaming agents.

To make a hydraulic fluid have a more stable viscosity over a wide range of temperatures, it will have viscosity improvers added.

Some machines can use water- or glycol-based hydraulic fluids. These fluids are used for their fire-resistant properties. There are also biodegradable ester synthetic-based hydraulic fluids that are used for their environmentally friendly properties. These alternative fluids are much more expensive than mineral oil–based hydraulic fluids and will need special consideration for service intervals and filter compatibility. For example, bio hydraulic oil is much more sensitive to water content and has to be changed if water content exceeds 0.10%. ● **SEE FIGURE 3–27** for a container of biodegradable hydraulic fluid.

POWERSHIFT TRANSMISSION FLUID. Powershift transmissions have friction discs in them that require specific additives to allow them to function properly and last. Friction discs that run in fluid need an oil with friction modifiers added to it. Almost all powershift transmissions will have a torque converter driving it. A torque converter uses the fluid as a power transfer medium that relies on hydrodynamic principles. The fluid in a powershift transmission also needs to float rotating shafts, transfer heat, and provide anticorrosion, anti-oxidation, and antiwear functions. One powershift fluid specification that is required to be met by many manufacturers is Caterpillar TO-4.

AXLE AND FINAL DRIVE FLUID. If the axle or final drive doesn't have friction materials in it that will be exposed to the lubricating fluid, it will likely require an oil classed as a gear oil. A common oil used for axles and final drives is classified as GL-5.

FIGURE 3–27 Biodegradable hydraulic fluid (SH photo).

BRAKE FLUID. Machines that have an automotive type of brake system (master cylinder and wheel cylinders or calipers) will require a glycol-based brake fluid that will be in either DOT 3 or DOT 5 spec. Brake fluid must be able to withstand extremely high temperatures and be able to absorb moisture.

Machines that use petroleum-based brake fluid may use oil from its hydraulic system or have its own brake oil reservoir. It will likely be a low-viscosity fluid like 10W.

GREASE. Grease is used to lubricate moving parts that don't have their own lubrication system. It is a lubricant that is usually used for slower moving parts that are exposed to high forces. It needs to be replenished regularly because it will get squeezed out from between the parts it is lubricating and also dry out.

Grease needs to do the following:

1. Reduce friction and wear
2. Provide corrosion protection
3. Protect bearings from water and contaminants
4. Resist leakage, dripping, and throw off
5. Resist change in structure and consistency during usage
6. Be compatible with seals

Base materials for grease are mostly either petroleum, mineral, or synthetic with calcium, aluminum, or lithium as the thickener. Additives are then combined with the base material to provide all these qualities. Grease additives may include the following:

- Oxidation inhibitors—prolong the life of a grease
- EP agents—guard against scoring and galling
- Anticorrosion agents—protect metal against attack from water
- Antiwear agents—prevent abrasion and metal-to-metal contact

Grease specifications are mostly created by the NLGI (National Lubricating Grease Institute). The required grease for machine lubrication can vary widely, but the most common type of grease used for lubricating the slow-moving high-force parts such as excavator boom, stick, and bucket pins is EP (extreme pressure) 2. Grease viscosity ranges for grease are 000 (thinnest for extreme cold) to 6 that is solid. The most common grease viscosities are 1 (like soft margarine) and 2 (like soft peanut butter). Grease used in automatic greasers will be thinner (0 or 00) because it has to flow through hoses, tubes, and valves. ● SEE FIGURE 3–28 for a variety of greases.

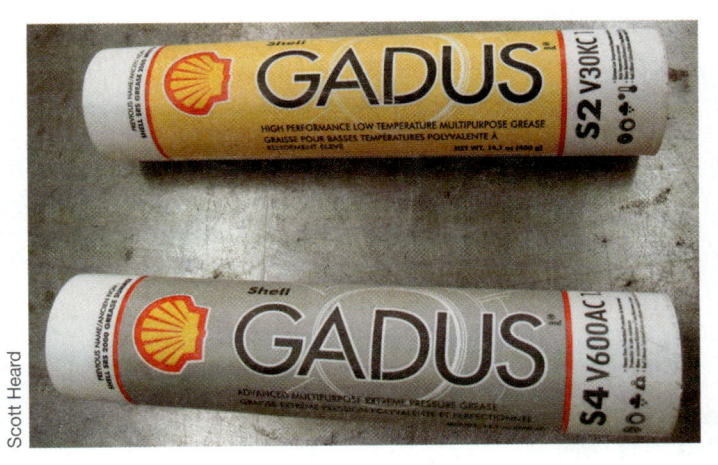

FIGURE 3–28 A variety of greases.

> **WEB ACTIVITY**
>
> Search the Internet and find three manufacturers of grease that could be used for greasing heavy equipment. List the manufacturers and the brand names of the grease.
>
> Search the Internet and find three manufacturers of greasing equipment. Describe three different types of greasing tools/equipment you find.

MACHINE FILTERS Filters can be found on just about all machine fluid systems: both liquid and gas. Machine filters need to keep contamination out of fluids to maintain fluid quality and keep components lasting as long as they are designed to. The fluid is pushed through the filter media (paper or synthetic material) where the contamination is trapped and disposed of when the filter is changed or cleaned. Tier 4 engines will likely feature exhaust filters that trap particulates so they don't enter the atmosphere.

The filters for a machine are designed to hold a normal amount of contamination from when they are installed to the next service interval when they are required to be changed. A filter will not do its job when it is left on the machine to long, there is an abnormal amount of contamination, or it is the wrong filter for the application it's being used for.

Many liquid filter bases will have either a restriction indicator to show how plugged the filter element is or a bypass switch to alert the operator when the filter bypass has opened or both.

LIQUID FILTERS. Machine liquid filters are needed to collect solid contamination from the system fluid it is filtering. They should be designed to let the full flow of the system through without causing an excessive pressure drop. As the filter gets older and collects more and more contamination, the pressure drop will increase across it. Most filter bases and some filters will incorporate a filter bypass valve that will open when the filter becomes plugged and can't allow the proper oil flow through it. Many filter bypass mechanisms used now will incorporate a switch that is used to alert the operator with a warning light or set a fault code when the filter gets plugged.

Solid liquid contamination comes from many sources, and it can be broken down into five categories:

1. Built-in: This is any type of loose material that is left in the system at the time of manufacture such as metal grinding, hardware, Teflon tape, and O-rings.

2. Ingressed: This is any type of contamination that is introduced into the system from the environment the machine is working in. This could be through the reservoir vent or a leaking cylinder rod. If a system has a breather to allow ventilation to the outside atmosphere, then moisture and contamination can be brought in. Any time there is a fluid leak there is a chance that contamination could enter the system.

3. New fluids: Fluids that are used to replace a drained fluid at service time or top up a low level can have contamination in them. Many machine manufacturers request that hydraulic fluid should always be filtered before it is added to the machine's tank.

MACHINE SERVICING 93

4. **Induced:** When a system is opened up by an HDET, there is always a possibility that a contaminant could get in. Systems should only be open as long as needed to do a service, repair, or top up a fluid level.

5. **In operation:** This type of contamination comes from normal wear of components inside the system.

Contamination could be many things and small levels of contamination in any fluid are acceptable. Some types of contamination are the following:

1. **Dirt:** This could be a chemical makeup of many things depending on where the machine is working, but the main component of dirt is silicon. Depending on the size of dirt particles, there could be very little damage being done, or it could be acting like a grinder on engine components.
2. **Metallic:** Steel, brass, aluminum, or copper are some types of metal contaminants.
3. **Water:** Water mixed with any petroleum-based fluid is bad news.
4. **Air:** In fluid is never a good thing.
5. **Biological organisms:** This is common in fuel tanks that have biodiesel in them.
6. Acids, sludges, and oxidation.
7. Other types such as rubber and paint.

Filters can be located in the pump's inlet (usually a screen), in the pump's outlet (pressure filter), or before the fluid returns to tank (return filter). ● **SEE FIGURE 3–29** where filters could be located in a system.

Filters can be designed to remove almost anything in a fluid system; however, most filters will just remove a certain size of particle. Particles are measured by micron size, which is 1/1,000,000th of a meter. The smallest particle a human can see with the naked eye is 40 microns. The smallest particle size you would commonly see being filtered would be 4 microns for some hydraulic or fuel systems.

Filter efficiency ratings are calculated by a scientific test that introduces a uniform micron size of test dust into the filter. The ratio of particles that are captured to those that pass through the filter is an accurate measurement of the filter's efficiency. This is called the beta ratio. Liquid filters are of two main types:

1. **Spin-on filters:** They can be replaced as a unit each time a filter change is required. The filter assembly is a metal housing with a threaded base for mounting the filter media, filter media end plates, and a spring to keep the media assembly in place inside the housing. ● **SEE FIGURE 3–30** for cutaway examples of spin-on filters.

2. **Cartridge filters:** They differ from spin-on filters in that the metal housing is reused and the filter media is the only part that is replaced. Many manufacturers will call the filter part of a cartridge-type filter an element. When a cartridge filter element is changed, the O-ring that seals the housing should be changed as well. ● **SEE FIGURE 3–31** for a cartridge filter assembly.

The canister must be cleaned out before it is installed.

Filtration of contamination is always a balance between effectiveness or efficiency and restriction. The better a filter stops contamination, the more it will restrict fluid flow through it.

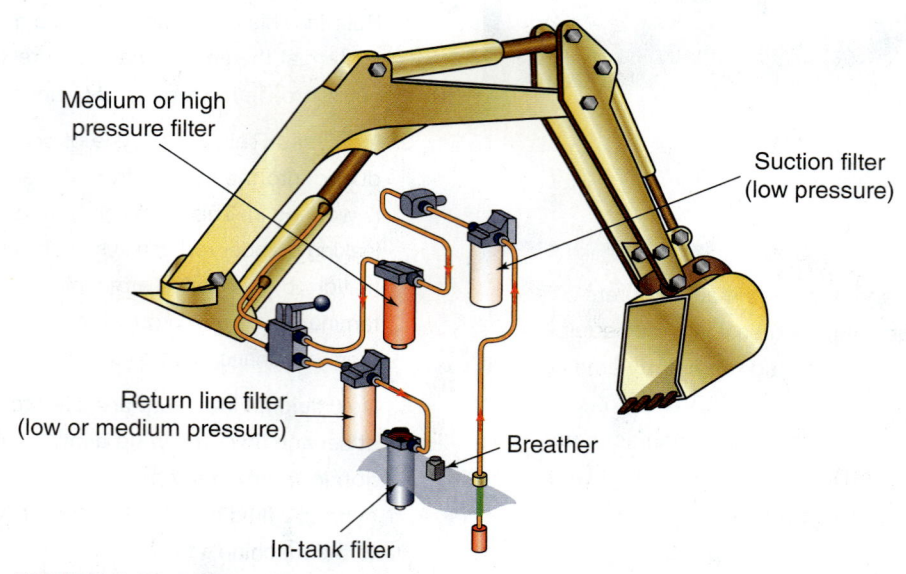

FIGURE 3–29 Different locations for filters (Donaldson illustration).

FIGURE 3–30 Spin-on filter cutaways.

FIGURE 3–31 A cartridge filter assembly.

FIGURE 3–32 An air filter assembly.

All filters can be recycled. This doesn't mean they can be cleaned and reused, but because there will be a fair amount of metal that is part of either a spin-on filter or a cartridge-type element if old filters are sent to a recycling facility, most of the filter can be made into something else eventually.

AIR FILTERS. There are several systems that need clean air and, therefore, need to have air filters that require maintenance. Most liquid reservoirs on machines will have breathers that allow the transfer of atmospheric air in and out of the reservoir. This air could be a major source of contamination if it wasn't filtered. Breathers will vary from a coarse mesh element in a throwaway metal or plastic housing to a finer foam filter that is washable or replaceable in a reusable housing.

The main air filtration system needed for a machine will be the diesel engine air filter used to clean the massive amounts of air that is used to support combustion of fuel inside the engine.

Most diesel engines used on heavy equipment will use a dual element filter system. One housing will contain an inner and outer filter. The outer filter is intended to clean all the air going into the engine and the inner is present there mainly as a backup. The inner is sometimes called a safety filter. Other terms for the filters are primary and secondary.
● **SEE FIGURE 3–32** for an air filter assembly.

MACHINE SERVICING 95

Any contamination that is allowed to enter the engine will shorten its life. Air filter cleanliness is measured in inches of water restriction. Smaller, lower horsepower engines that are non-turbocharged will have a maximum allowable filter restriction of 30 inches of water restriction while turbocharged engines should have no more than 25 inches of water restriction. These values are general rules of thumb and specific engine service information should be referenced when in doubt.

EXHAUST FILTERS. Underground mining machines have used exhaust scrubbers or filters for many years, but with new emission reduction regulations taking effect, almost all machines will eventually have a DPF on them.

These filters will require periodic removal and cleaning based on hourly interval or contamination from an engine failure.

SERVICING IN THE FIELD
Machine servicing isn't just restricted to a shop activity. Any machine that is larger than 30–40 tons will be expensive to move from the job site to a shop for service. Float moves can easily be several hundred dollars and for large machines could be several thousand dollars. To service medium to large equipment, it becomes more economical both in terms of float costs and having the machine leave the job site to have a field service HDET or lube technician and a lube truck come to the machine. If a truck is set up to do machine services, it will have several oil containers for different new fluids and one or more waste oil container. It will have hose reels for pumping oil and a vacuum pump to suck up the waste oil. There will also be a supply of filters, oil sample bottles, rags, oil absorbent, and other servicing-related supplies. The truck will also have a filter drain rack and barrel for the old filters. It will also likely be equipped with basic tools at a minimum so the HDET can perform minor repairs as the service is being done. Some lube trucks will also deliver fuel and could be equipped as full field service repair trucks with cranes, air compressor, torches, and welding machine.

Large equipment fleets will have their own lube truck, and many dealers will have one or more lube trucks working out of each branch location. ● **SEE FIGURE 3–33** for a fully equipped lube truck.

FIGURE 3–33 A fully equipped lube truck.

Progress Check

11. Fluid viscosity is determined by:
    a. boiling the fluid and measuring the thickness
    b. pouring the heated fluid through a fixed orifice
    c. freezing the fluid and measuring the thickness
    d. vaporizing the fluid and measuring the density

12. The smallest particle a human eye can detect is:
    a. one millionth of an inch
    b. one millionth of a centimeter
    c. 10 microns
    d. 40 microns

13. API ratings for diesel engine oil start with a "C." The letter C stands for:
    a. compression
    b. combustion
    c. centigrade
    d. combination

14. "Built-in" contamination comes from:
    a. normal component wear
    b. dirt introduced when the system is serviced
    c. cross-contamination from other systems
    d. contamination left in the system from manufacturing processes

15. This is one type of contamination an engine oil sample would *not* detect:
    a. fuel
    b. transmission oil
    c. coolant
    d. water

16. Tech A says No. 1 diesel fuel has a higher sulfur content than No. 2 diesel fuel. Tech B says a machine's fuel tank needs to have water drained from it regularly. Who is correct?
    a. Tech A only
    b. Tech B only
    c. Tech A and B are both correct
    d. Tech A and B are both wrong

17. Tech A says fuel consumption can be tracked and used to determine service intervals. Tech B says machine running hours are the only way to determine service intervals. Who is correct?
    a. Tech A only
    b. Tech B only
    c. Tech A and B are both correct
    d. Tech A and B are both wrong
18. Tech A says grease that is labeled EP means it can withstand extreme pressure. Tech B says automatic greasers will use a lower viscosity grease. Who is correct?
    a. Tech A only
    b. Tech B only
    c. Tech A and B are both correct
    d. Tech A and B are both wrong

SHOP ACTIVITY

Go to a machine and then find and describe the following:
- All fluid drains
- The correct type and viscosity fluid for each machine system
- All machine system filters
- Any fluid sample ports

CASE STUDY

Let's look at a complete 2000-hour machine service procedure for a large wheel loader.

Max is an HDET and is informed that the main production loader for the limestone quarry he works at is due for a complete service in two days. He gets a list of required fluids, filters, oil sample bottles, and any other supplies required to complete this service. ● **SEE FIGURE 3–34** for a wheel loader being serviced.

He then looks up the machine manual in the OEM service information and prints off the applicable sections. He takes a few minutes to go through all the procedures that should be performed.

On the day of the service, Max gathers all the filters, fluid sample bottles, and any other supplies he will need and checks to see if there are enough of the fluids needed for this service. He also checks the waste oil container capacity to be sure that there is enough room for all the machine's fluids.

Max first goes to the machine and performs an initial visual inspection. This includes checking all fluid levels to see if the levels look normal. Up to this point it has been the operator's responsibility to check machine fluid levels that are easy to check by either sight glass or dipstick. This would include engine oil, transmission, hydraulic, and coolant levels.

FIGURE 3–34 A wheel loader being serviced.

While doing a circle check on this wheel loader, Max checks the tires for damage; wheel fasteners and rims for looseness and damage; bucket teeth and adapters for damage and excessive wear; loader frame and bucket linkage for damage; hydraulic hoses and cylinders for damage and leaks; steps and handholds for damage; lights and wiring for damage; any guarding or covers that are damaged or missing; windows and wipers for damage; drivetrain, engine, and hydraulic oil for leaks; engine belts for damage; exhaust system for leaks; and the charge of the fire suppression system.

Max then enters the cab to check the hour meter and the electronic display for any fault codes and also looks for any warning lights or gauges that are not functioning. He also performs a quick visual check in the cab looking for damaged machine controls, cracked or broken windows, and a functioning seat belt. The next step performed by Max is to start the engine and listen for unusual noises while looking at the exhaust condition. Throughout this inspection, abnormal findings are recorded on an inspection sheet and recommendations are made as to what further action should be taken.

Once the machine is started, Max will start to warm it up by increasing its rpm and operating the boom and bucket functions while moving the machine back and forth and steering as well. During this warm-up procedure, Max will be paying close attention to how the machine operates in response to his inputs. He will be looking and listening for unusual noises, jerky movements, and vibrations. The powershift transmission is used to load the engine by performing stall tests, and this will also warm up the drivetrain oil (see Chapter 12 to find out more about stall testing). The hydraulic system could also be used to load the engine by performing hydraulic stalls. Depending on the type of hydraulic system loader has, a hydraulic stall can be performed by bottoming out one of the machine's cylinders and continuing to try and move the cylinder. This will keep the hydraulic system at its relief valve setting, which creates a load for the engine as well as warming up the hydraulic fluid.

In a perfect world, this warm-up procedure would be long enough to bring all machine fluids to normal operating temperature. Most times it will not be practical to do this. However, at minimum the engine coolant should be warmed up to at least open the engine thermostat. If fluids are brought up to normal operating temperature before they are changed, this will ensure that the maximum amount of contamination is drained out with the old fluid.

Now that the fluids are warm in the main systems (engine, hydraulic, and transmission), Max can take live oil samples from these fluids with the machine running at low idle. He is careful in cleaning any dirt away from the valves before taking a sample. An engine coolant sample is taken by a vacuum pump, that is, to coolant from the radiator top tank.

Depending on the ambient temperature, it may be hard to get axle oil warmed up. Travel the machine as much as possible to at least circulate the oil. Max pulls the machine into the shop and gets someone to help line up one final drive drain plug to the right position for draining.

At this point, Max has secured the machine with wheel chocks and locked out and tagged out the machine to be sure that no one else will attempt to run it while he is servicing it.

He will now take oil samples from the axles and final drives with a vacuum pump, and as he does this, he writes on the sample bottle where the oil is from.

Max will now start to drain fluids starting with the final drive that is lined up. This machine is equipped with an oil evacuation system for the engine oil, hydraulic fluid, and transmission oil. Max hooks the suction hose to the engine drain and turns on the vacuum pump. He then changes the engine oil filters and drivetrain filters. After he releases the hydraulic tank pressure, he starts to change the hydraulic filters. By the time he has changed all these filters, the engine oil has been drained, and he can refill it with the correct oil from a pump system. He sets the quantity on the pump handle so it will shut off at the exact amount needed and then switches the suction hose to the transmission housing. While the transmission is getting drained, Max changes the engine air filters and cab filters. All filters have the date and machine hours written on them for future reference.

The machine's automatic greasing can be topped at this point by using the air-powered shop grease keg pump. The machine's batteries and cables are inspected now. Max looks for proper electrolyte levels, corrosion, and secure batteries and cables. The ROPS is also inspected for damage and loose fasteners.

Once the transmission oil is drained, there are two suction screens that need to be removed, inspected for abnormal wear particles, cleaned, and installed with new O-rings. The transmission is now filled with the proper oil type and quantity. At this point, Max starts the machine and checks the engine and transmission levels and is able to move the machine to line up the next closest final drive drain plug. The machine is once again locked out and Max starts to drain the second final drive. The suction hose is now switched to the hydraulic tank and started and the axle oil is drained. Max keeps a close eye on the drain container for the axle oil because it quickly fills and needs to be switched to an empty one.

At this point in the service, Max changes the fuel filters and drains the bottom of the fuel tank until clean fuel is seen. After installing new filters, the fuel system is prime to purge any air.

The hydraulic tank is refilled by pumping the oil through an external filter to ensure its cleanliness. Max goes through the air-bleeding procedure as described in the manual to prevent air from reaching the machine's pumps. Max can now start the machine to line up the last two final drive drain plugs. After these final drives have been drained, the axles and final drives are refilled until oil just starts to come out the fill plug openings. The plugs are installed and torqued to spec.

If the engine coolant needed to be changed at this service, then it would be drained into an appropriate container and marked as to its contents. The cooling system would get refilled slowly to allow air to escape and would also be bled of air (see Chapter 6 to learn about air bleeding).

Once Max rechecks all fluid levels, he starts the machine and checks for leaks. As he did a great job and found no leaks, he can take the machine out of the shop and run it enough to warm it up. One last check is made for leaks, that all covers have been reinstalled and that all machine safety features are working properly. After this last step, the machine is ready to go back to work.

Max will now clean the shop up and cut open an engine, hydraulic, and transmission filter to look for abnormal wear particles. If none are found, he will dispose of all old filters after they have drained into the filter recycling bin. A company will come by and pick up the old filters for recycling. Max then fills out the company's servicing documentation forms so there is a written record of the service, orders new filters to replace the ones that were used, and checks the new oil levels in the shop.

SUMMARY

1. Machine servicing is an important part of normal HDET routine and includes machine inspections, changing fluids and filters, and taking fluid samples.
2. Safety concerns related to machine servicing include making sure that the machine is locked out and tagged out, using wheel chocks and boom/steering locks, and releasing system pressure.
3. Service information can be found in the operation and maintenance guide that should be in the machine's cab, or it can be referenced electronically through the machine manufacturer's service information system.
4. While servicing machines, there are many safety-related system checks and inspections that should be performed. A few examples would include steps and handrails, seat belts, and hydraulic lock out systems.
5. Circle checks are part of all service procedures and would include the previous examples as well as a few examples such as checking fluid levels, checking for leaks, and looking for damaged components. Machine manufacturers will provide inspection checklists for each machine.
6. Machine service intervals are determined by running hours accumulated on the machine and calendar time. A typical engine oil and filter service would occur at the 250 hours or every 6 month interval, whichever occurs first. Fuel consumption tracking can also be used for determining service intervals.
7. For each interval, all earlier service points should be covered. For example, if a machine was getting a yearly/2000 hour service, all monthly and daily checks/services should be done as well.
8. Oil changes are best done when the oil has been warmed up and circulated first. This will drain as much contamination out as possible.
9. Install spin-on oil filters as per instructions to avoid overtightening.
10. Machine lubrication is an important part of the operator's job, but there may be components that the HDET greases during services only. All grease fittings should be checked for damage or plugging.
11. Always dispose used fluids and filters responsibly. There are many fluid and filter recycling companies that specialize in this.
12. Documenting service procedures is a critical part of machine ownership.
13. Most fluids in machines are needed to lubricate and cool the components they are in contact with.
14. Fluid properties such as type and viscosity must be correct for the compartment and operating temperatures they are exposed to.
15. Fluid viscosity refers to how fast a fluid can flow through a certain size orifice at a certain temperature. Viscosity standards are set by SAE and ISO.

16. Fluid cleanliness levels are measured against an ISO standard that assesses how many particles of a certain size are in a certain volume of fluid. For example, an ISO cleanliness code rating of 18-16-13 for particles sizes of 4, 6, and 14 microns means there are an acceptable amount of particles of that size in the sample size of fluid.

17. Fluid sampling is an important part of servicing machines. Small samples of fluid can be taken (much like a blood sample) and are then sent to a lab to be analyzed. A sample report is returned to the machine owner and gives a good indication of what is happening inside the fluid compartment. Some elements that are tested for are iron, copper, coolant, and silicon.

18. When replacing fluids during a service, the fluid properties must match the recommended properties as outlined in the machine's service information. This is particularly important if the machine is still under warranty. For example, a hydraulic fluid may need to have a certain level of anti-foam additive to be acceptable for use in the hydraulic system.

19. Filters are needed on machines to collect contamination and hold it until the next service when the filter is removed and either replaced or cleaned. They can keep liquids or gases clean such as engine oil or engine intake air and are designed to collect minimum sizes of particles that are measured in microns (one millionth of an inch). All filters have a maximum capacity of contamination that they can hold and are designed to easily hold the normal expected wear particles for each service interval.

20. Spin-on filters are an assembly that have their own threaded end cap, metal shell, and filter element. The assembly is replaced during a service, and if sent to a recycler, the materials in a filter assembly are separated and recycled.

21. A cartridge-type filter is one that has a shell that is reusable with a replaceable element.

22. Most hydraulic and transmission filter housings will incorporate a filter bypass that allows fluid flow even if the filter becomes plugged and will have a filter bypass sensor that will turn on a light to tell the operator that the filter is plugged.

23. Engine air filter assemblies usually have two elements with a secondary or safety element that is meant to provide a backup if the primary filter fails.

24. Mobile lube trucks can be equipped to perform a complete service on the work site.

chapter 4
HYDRAULIC SYSTEMS— THE BASICS

LEARNING OBJECTIVES

After reading this chapter, the student should be able to:

1. Calculate force, pressure, and area problems.
2. Identify safety concerns related to hydraulic systems.
3. Explain the fundamental differences between flow and pressure.
4. Explain Pascal's law.
5. Describe how a fixed positive displacement hydraulic pump works.
6. Explain how pump displacement is measured.
7. Explain pump efficiency.
8. Describe the advantages of a hydraulic system over a mechanical system.
9. Identify and explain the operation of the main system components.
10. Calculate hydraulic system problems based on force, pressure, and area.
11. Describe a directional control valve.
12. Explain hose and tube sizing.
13. Describe basic hydraulic system troubleshooting procedures.

KEY TERMS

Actuator 116
Aeration 112
Area 107
Atmospheric 135
Bar 110
Dash size 119
Directional control valve 115
Fixed displacement 113
Flow 110
Gear pump 113
Hose 119
Levers 104
Micron 123
Motor 118
Pascal's law 106
Pressure 109
Pump 106
Reservoir 114
Seal 120
Torque 118
Tube 120
Viscosity 121
Volume 110

INTRODUCTION

SAFETY FIRST Minor injuries such as burns and serious injuries or even death have resulted from technicians not being aware of potential dangers inherent to hydraulic systems. Hydraulic systems use hot, high-pressure fluids to transfer large amounts of energy, and personal harm can occur quickly if a technician isn't proactive in preventing an accident.

Incidents can happen as a result of improper testing methods or use of improper test equipment such as using underrated hoses or gauges. Use of inadequate replacement parts such as hoses, seals, or tubes can result in the rupture of these components, and not only the escape of hot, high-pressure fluid but the sudden uncontrolled movement of a machine could have disastrous results. A little common sense and proper use of PPE will go a long way to keeping you safe when working on hydraulic systems. You should also always refer to the machine's specific manufacturer's service information before performing any servicing or repairs to hydraulic systems.

These safety concerns also apply to hydraulic tools such as presses, which can operate at extreme pressures. You will see lots of warnings and cautions on machines and in service information so don't overlook them and become another statistic. Here are a few types of specific safety concerns that you should be aware of and some tips that help you avoid injuring yourself or others.

OIL INJECTION As you will learn in this chapter, heavy equipment hydraulic systems can operate at high pressures that are typically around 3000–5000 psi and can exceed 10,000 psi. This high pressure can be used to perform many high-force machine operations, but if that pressure is allowed to escape and your skin or other body parts are exposed, some very serious consequences can occur. Skin can be punctured easily by high-pressure fluid, and if hydraulic fluid gets into your bloodstream, there is a high chance of blood poisoning and death could result.

One former coworker of mine had one finger seriously disfigured by simply looking for a leak and not realizing he had put his finger over a high-pressure spray of fluid. He now has a permanent reminder of the dangers of high-pressure leaks. I also recently heard of a technician who was using a grease gun to try and unblock a plugged grease fitting had a high-pressure stream of grease piercing his skin when the grease gun hose failed the technician. He has had a series of operations to remove the grease and remains unable to return to work.

If you are looking for the source of a hydraulic leak on a machine and see oil dripping, this should raise a red flag for you. Although the oil looks like it is dripping harmlessly, it may in fact be spraying out under high pressure and waiting to do harm if you put your hand in its path. Be very cautious and use a piece of cardboard or wood to identify where the leak is.
● **SEE FIGURE 4–1** for how to safely look for a hydraulic leak.

An oil injection accident is very serious and can easily result in an amputated limb or worse. To see the result of an oil injection injury, visit the Fluid Power Safety Institute at http://www.fluidpowersafety.com/fpsi_alert-19.html.

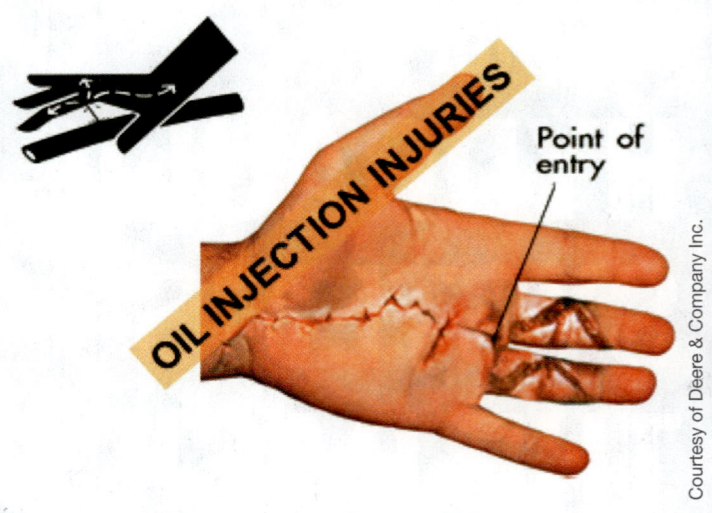

FIGURE 4–1 An oil injection injury.

TRAPPED PRESSURE Pressure can be trapped inside many hydraulic components even after the machine is shut off. If you need to replace a hose, tube, or other component at any time, the pressure inside that component must be reduced to safe levels. This may not be easy to do, and you should always refer to the appropriate service manual for proper procedures. Releasing pressure or bleeding off pressure can be very dangerous if done improperly. Always keep in mind if you are releasing trapped hydraulic pressure, there's a good chance that some part of the hydraulic system is going to move. *Always* refer to the machine manufacturer's service information to learn how to release pressure safely.

CRUSHING HAZARDS A crushing hazard is present on a machine where there is the potential of moving parts to squeeze or crush you or your body parts. A good example would be the pivot point or articulation joint of an articulated steering wheel loader. If a machine has hydraulic functions that can create crushing hazards, you must be very wary that you do not put yourself in a vulnerable position when working on the hydraulic system. Make sure any component that could move and cause a crushing injury will be mechanically held in place. Steering locks and boom cylinder supports are two examples of mechanical locks that must be installed if you're going to be working near crushing hazards.

FIGURE 4–2 An example of a warning you may see on a machine.

FIGURE 4–3 A warning for the possibility of a fire.

If you are testing or adjusting a running piece of equipment, your senses should be on high alert. If someone else is running the machine you are working on, you must keep clear communications with that person to let him or her know what you want done. This is sometimes difficult as a running machine can be very noisy but the person operating the machine must be absolutely sure what is required. If you are unsure of the operator's capabilities, stop and find someone who you feel confident with. Even the smallest mistake when working on a running machine could be deadly. ● **SEE FIGURE 4–2** for an example of a safety warning you may see in a machine.

BURNS Hydraulic systems can generate high amounts of heat, and this should be another safety concern for the technician. Burns can occur at just over 100°F. You need to make sure that you take precautions to avoid getting burnt if you are servicing or repairing a hydraulic system that has just recently been operating. You should consider using a heat gun to ensure safe component temperature before proceeding. At times it will be unavoidable to work on a machine with hot hydraulic fluid, and if this is the case, you should use the proper PPE to keep yourself protected.

SLIPS AND FALLS Hydraulic fluid is slippery by nature, and if that fluid is leaked onto a surface that is walked on or a grab handle that is needed, a slip hazard is created. This could be a problem not only with a machine operator slipping and falling, but the technician is also at risk from fluid that has spilled or sprayed as a result of servicing or repairing the hydraulic system. I've seen hydraulic leaks in machine cabs bad enough to have created slip hazards on the cab floor, truly an accident waiting to happen. Make sure that all spilled fluid is cleaned up and leaks that create slip hazards are repaired.

FIRE HAZARDS Hydraulic fluid is mostly mineral-based oil, which is flammable. There have been more than one machine catch fire and burn up because of a hydraulic leak that has either been sprayed onto a turbocharger or been ignited by welding sparks or a torch. You need to be careful when welding or using a torch on a machine that has a hydraulic leak. ● **SEE FIGURE 4–3** for a fire hazard warning you may see on a machine.

ENVIRONMENTAL CONCERNS A hydraulic system should be a sealed system, and if it is intact, there are minimal environmental concerns. However, if your actions as a technician cause a fluid leak by opening a system, the fluid leak *must* be contained. Drain buckets or trays, fluid absorbent cloth and socks, or floor dry should be used to contain leaks. See some examples of these types of materials at http://www.americanproducts1.com/. The result of not containing fluid leaks can be very harmful to the environment and very costly with examples of fines in the $20,000 per square foot reported.

WHY USE HYDRAULIC SYSTEMS?

A hydraulic system that is found on a heavy equipment machine can be loosely defined as any system that is driven with mechanical energy and transfers that energy into a fluid medium,

HYDRAULIC SYSTEMS—THE BASICS

which then transfers it back into mechanical energy to perform work. In other words, energy is transferred through fluid (the medium).

Hydraulic systems can be used for many different functions on heavy equipment machinery such as moving buckets, steering machines, and lifting loads. To be a competent HDET, it is mandatory that you understand thoroughly the fundamentals that are covered in this chapter.

This chapter is placed near the front of this book because of its importance. The principles of hydraulics apply to many other systems used to make a piece of heavy equipment work. You will see in following chapters that engines use variations of simple hydraulic systems (lubrication, fuel, coolant systems); transmissions use hydraulic systems to actuate clutches; brake systems use hydraulic systems; and to understand electrical basics, the fundamentals of pressure and flow learned here will be a great benefit.

Hydraulic systems have been used on heavy equipment for many years now. They were first introduced on machines by equipment manufacturers to replace a variety of mechanical systems that used devices such as clutches, shafts, gears, chains, and cables.

Mechanical systems are very efficient at transferring power with friction losses accounting for the majority of their inefficiency. Hydraulic systems aren't quite as efficient because of the friction and heat that fluid flow and internal leakage create. A general rule is that hydraulic systems are about 85% efficient, which means that 15% of the amount of energy going into a hydraulic system is lost mainly through heat.

However, hydraulic systems have many advantages over mechanical systems. Some of them are as follows:

Flexibility: The output of a hydraulic system is a cylinder or motor and their location on a machine is only limited to whether there can be fluid plumbed to it through hoses, tubes, or swivel joints. Mechanical drive is limited to components such as chains, cables, shafts, and gears that have to physically engage with each other, which put limitations on the flexibility of the machine's design.

Component cleanliness: Most of the hydraulic system itself is sealed from the environment and runs in a fluid (usually oil) that promotes long life of components. A lot of mechanical drive components are exposed to the environment that shortens component life and makes regular lubrication and maintenance mandatory.

Component protection: Hydraulic systems will have overload protection built into them to prevent failures caused by abuse. Mechanical drive system abuse is limited to the operator being able to judge when a component is being overloaded. If an operator of a mechanical system on a machine isn't careful, the machine will either stall or something will break.

Speed and direction: The speed and direction of the hydraulic cylinder or motor can be easily and smoothly changed with valves and not completely related to engine rpm unlike a mechanical system that needs clutches and gears to change speed and direction. This requires a certain level of skill and finesse.

Multifunctional: Most hydraulic systems allow more than one component or function to be used at a time; this may not be the case with a mechanical system.

The following is a comparison of mechanical and hydraulic machines: ● **FIGURE 4–4** shows an old excavator that used mechanical drive and gravity to move the machine's bucket in response to operator commands. Its diesel engine drives a master clutch, which in turn drives a series of gears. These gears had a clutch and brake attached to their output that if engaged or held made a winch turn or stop. The winch turns a drum loaded with cable, and the cable is fed through a series of pulleys that eventually attaches to a component such as the bucket or stick. Also a lot of the machine's movement relied on help from gravity that sometimes worked against what the operator wanted the machine to do. Controlling these clutches requires a great deal of effort and skill on the part of the operator as the clutches are mechanically controlled by a set of **levers** and foot pedals that are attached to linkages connected to the clutches. There is a lot of maintenance and adjustments required to keep this machine working properly.

In ● **FIGURE 4–5**, you could imagine the resulting operator fatigue created by pushing and pulling this combination

FIGURE 4–4 Cable excavator.

FIGURE 4–5 An old machine with long levers and pedals.

FIGURE 4–6 A newer excavator with shorter joysticks.

of levers and foot pedals over a typical 8–12-hour workday. Each lever and foot pedal would have a travel of about 5 in. and would not be very easy to move.

For this cable excavator, there would be cables that may fail and need to be replaced, winches to maintain, clutches to adjust, pulleys to lubricate, and levers and linkages to maintain and adjust.

Mechanical Excavator Powerflow Components	Hydraulic Excavator Powerflow Components
Diesel engine, master clutch, gears, shafts, bearings, winch, cable, pulley	Diesel engine, tank, pump, valve, hoses, cylinders, motors

Hydraulic excavators use fluid power to manipulate the main parts of the machine. Pumps, cylinders and motors use hydraulic fluid to transfer energy. The operator also only needs to move two levers a maximum of 3 in. in any direction to control the bucket movement.

The decrease in operator effort needed to control this machine means the operator is a lot less fatigued at the end of a workday.

Although there is an increase in reliability by reducing the amount of moving parts, there is still a need to properly maintain hydraulic systems to get the maximum design life from the various components. Many machines today have gone a step further and feature electronic joysticks that require even less effort. You are able to pull a joystick with one finger easily, which in turn could control a hydraulic system that is lifting 30 tons of material or more! Try to imagine what's next: voice-activated machines?

MACHINE HYDRAULIC SYSTEM EVOLUTION The first fluid used to transfer work was water. Water power was used to drive many different devices in the mid-1800s such as sawmills and gristmills. If settlers of the day needed to process wheat into flour or logs into lumber, they would take their raw material to a facility located beside a river where the power of flowing water was captured and transferred into mechanical drives that would cut wood and grind flour. ● **SEE FIGURE 4–7** for an operational flour/saw mill.

Using this source of flowing water to power mechanical devices is a good example of how a fluid with energy was harnessed to perform work. This type of hydraulic system is

FIGURE 4–7 Operational water-powered grist mill.

> **WEB ACTIVITY**
>
> See a water-powered saw mill at https://www.youtube.com/watch?v=TNrYS1ar9Zg.

HYDRAULIC SYSTEMS—THE BASICS 105

considered to be a hydrodynamic system. This means energy is transferred by a lot of fluid movement or flow in loosely sealed components with not much pressure being created. In Chapter 10, this type of hydrodynamic system is discussed under torque converters.

The early hydrodynamic type of hydraulic system was still a long way from the hydraulic systems first used on heavy equipment.

For a manufacturer to incorporate a hydraulic system on a piece of heavy equipment, it had to:

1. Be driven by a mobile prime mover (diesel engine) so the machine could be operated wherever there was diesel fuel available
2. Create and transfer enough force to perform the mechanical task required (e.g., lift a bulldozer blade or turn a winch)
3. Be easily controlled to allow the machine's operator to manipulate the machine

This type of system is considered to be a hydrostatic system. This means energy is transferred by pushing fluid through a tightly sealed system, which in turn can take advantage of fluid pressure.

The fluid used in most hydraulic systems today is mineral-based oil and will be discussed later.

Progress Check

1. A hydraulic system that transfers energy through the motion of fluid in a loosely sealed device is called a:
    a. hydrostatic system
    b. hydromechanical system
    c. hydropneumatic system
    d. hydrodynamic system
2. Oil injection is:
    a. a way to get more power out of a hydraulic system
    b. a serious safety concern you need to be aware of with any hydraulic system
    c. what happens in a hydraulic cylinder
    d. easy to treat with disinfectant
3. Any piece of equipment supported by a hydraulic system should be:
    a. safe to work under if there are no leaks
    b. safe to work under if the oil is cold
    c. secured mechanically before work is done on it
    d. safe to work under
4. These are all advantages that a hydraulic system has over a mechanical system *except*:
    a. flexibility
    b. less chance of damage from overload
    c. components are less expensive
    d. easier to use more than one function at a time
5. Hydraulic systems work on the principle of:
    a. fast moving fluid in loosely sealed components
    b. using fluid to transfer energy in tightly sealed components
    c. using compressed fluid to store energy
    d. using an air/oil mixture to transfer energy

NEED TO KNOW

As stated previously, a hydraulic system is one that converts the mechanical rotary output energy from a prime mover (most commonly, a diesel engine) into fluid power (flow and pressure), which is then converted back into mechanical movement. All systems need to have a **pump** to create fluid flow, which is then controlled by one or more valves that direct the fluid to an actuator (cylinder or motor). The actuator then converts this fluid movement back into mechanical movement. The example of the hydraulic excavator in the introduction uses a diesel engine to turn a pump that moves fluid, which is directed by control valves actuated by the operator. This fluid can then be sent to several cylinders or motors to mechanically move one or more components of the machine.

The movement of fluid in any hydraulic system is created by a difference in pressure. It starts at the pump inlet where a low pressure is created and moves fluid from the tank into the pump. The pump then pushes the fluid through the system at a higher pressure as it returns back to tank to a lower pressure. This is a key point to understand: fluid can only flow if there is a pressure differential.

In the mid-1600s, Blaise Pascal was experimenting with a hydraulic press and came up with a theory that is now called **Pascal's law**.

Pascal's law states "pressure created by an external force acting on a fluid inside a sealed container applies the same amount of pressure to any and all surfaces equally and at right angles to the inside of that container." This fact or law simply means that any force applied to a liquid will be transferred by that liquid to other parts of the system that it is exposed to.
● **SEE FIGURE 4–8** for how pressure is transferred through a liquid.

A liquid is the ideal medium to use in a hydraulic system as it takes the shape of whatever container it is poured into, and it is also practically incompressible. For argument's sake, we will consider hydraulic fluid to be noncompressible, but in

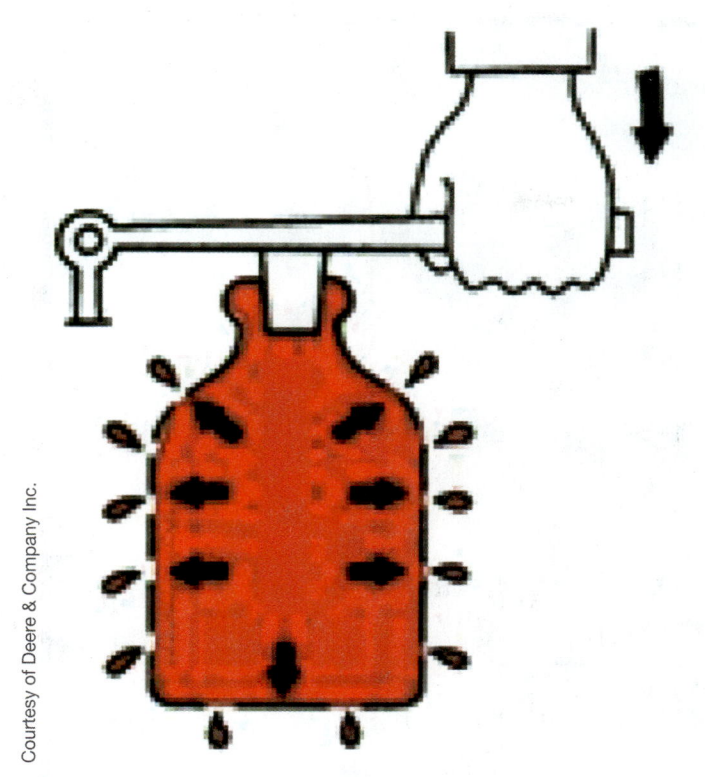

FIGURE 4–8 How pressure is transferred through a liquid.

fact it will compress roughly 0.5% volume for every 1000 psi applied to it. This is not the case with a gas such as air that is easily compressed.

This is also one reason why air in a hydraulic system will result in poor system performance. If you've experienced driving a car that has air in its hydraulic brake system, you know what the term *spongy* means. When the brake pedal is pushed, it feels like stepping on a sponge because you need to compress the air before you are able to move the brake fluid. Although air (pneumatic) systems can transfer lots of force, the amount of force needed to drive an excavator bucket into the ground can only be created practically by a hydraulic system that uses a liquid. There is also a time delay in a pneumatic system as the air is compressed to build pressure. If you have driven a vehicle with air brakes, you would have noticed a slight delay from the time you push the brake pedal to when the vehicle starts to slow due to the time it takes for the pressure to rise while the air is being compressed. There is very little delay with the actuation of a hydraulic system, which is another desirable feature.

> **Prove It Yourself**
>
> You can prove liquids can't be compressed: Take a simple squirt gun and fill it with water. If you now try to squirt the water and there is air in the gun, you could probably block the nozzle and still be able to squeeze the trigger a little bit. You are compressing the air in the gun, not the water. Once all the air is purged from the gun and the nozzle is blocked again, you would now find the trigger won't move unless there is a leak somewhere else in the squirt gun. You have now proved a liquid is incompressible.

> **Math Time**
>
> Some simple math skills are required here to fully understand the basic principles of a hydraulic system. If you need to brush up on your math skills, there are a variety of online sites that you can visit and improve. One example is at http://www.math.com/.
>
> **Area** calculations need to be applied to find the surface area of circles. Two formulas that can be used to calculate the area of a circle are as follows:
>
> $A = \pi r^2$ (where π is a constant value of 3.14 and r is the radius or distance from the center of the circle to the outside of the circle)
>
> or
>
> $A = d^2 \times .7854$ (where d is the diameter of the circle or the distance from one side to the other through the center of the circle)
>
> Hydraulic cylinders have round bores in which the round piston travels. It's the surface area of the piston that needs to be considered to understand how fluid pressure applied to it results in a specific amount of force. ● SEE FIGURE 4–9 for an example of circle dimensions you might find for a cylinder.
>
> To find out the area of the 3 in. diameter circle, let's apply the formulas:
>
> $A = 3.14 \times 1.5$ in. $\times 1.5$ in.
> $= 7$ in.^{2}
>
> $A = 3$ in. $\times 3$ in. $\times .7854$
> $= 7$ in.^{2}

HYDRAULIC SYSTEMS—THE BASICS

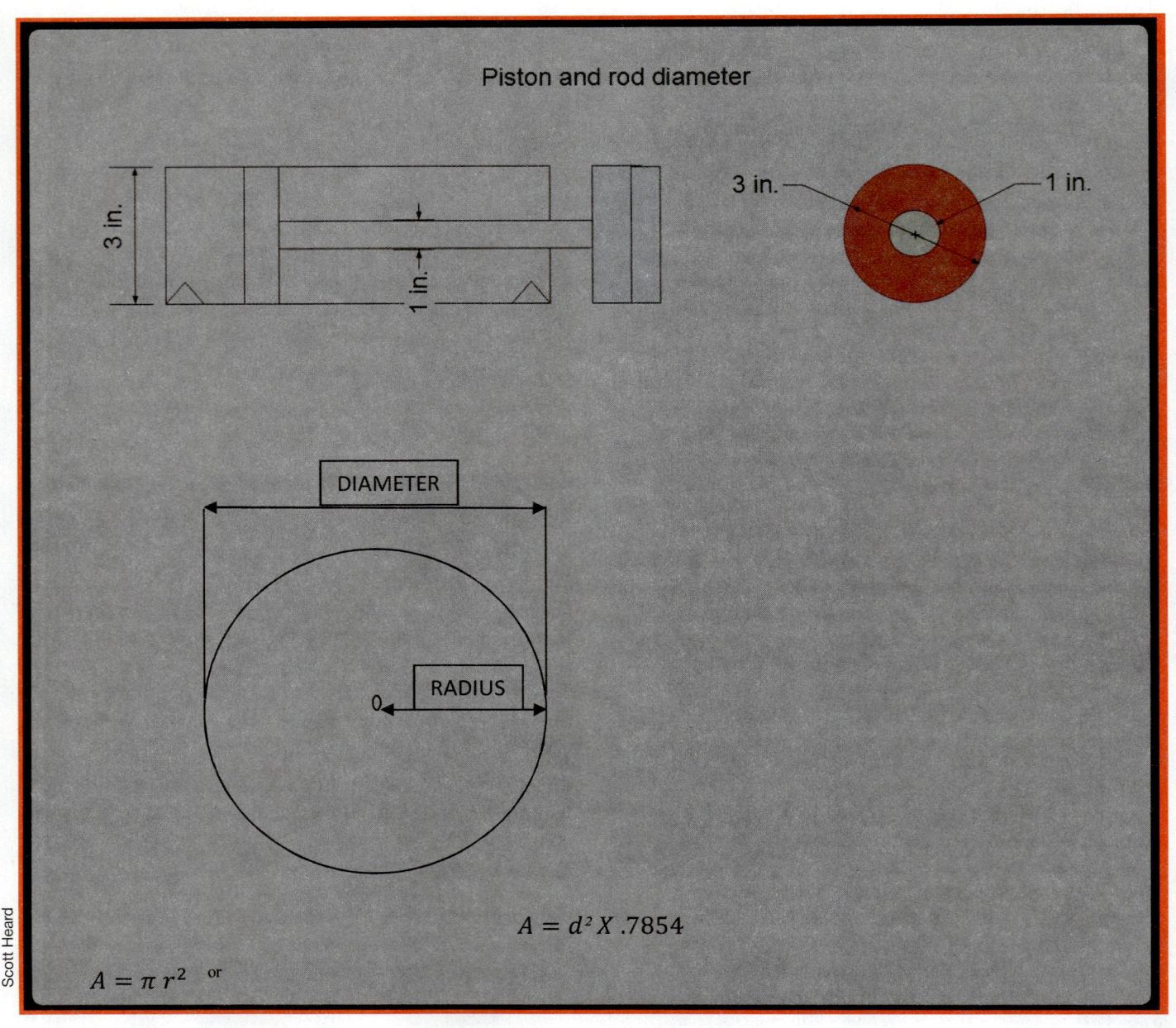

FIGURE 4–9 Circle dimensions for a cylinder.

$A = \pi r^2$ or $A = d^2 \times .7854$

FORCE, PRESSURE, AND AREA

To understand how energy is transferred through a hydraulic system, you need to understand the relationship between force, pressure, and area.

If we apply Pascal's law to a simple hydraulic bottle jack, we can see how a great amount of weight can be lifted with a relatively small amount of input force. Pascal's law states that $F = P \times A$ (F is force, P is pressure, and A is area). ● **SEE FIGURE 4–10** for how pressure acts in a bottle jack.

? Helpful Hint

To remember this important formula, the letters F, P, and A can represent other words such as Foul = Pizza and Anchovies! Use your imagination to help you remember this important equation.

When the jack handle has a force applied to it, a force is applied to the pump. The force is transferred to the oil in the pump section of the jack. Applying Pascal's law, you can see

108 CHAPTER 4

Bottle Jack Force Formula

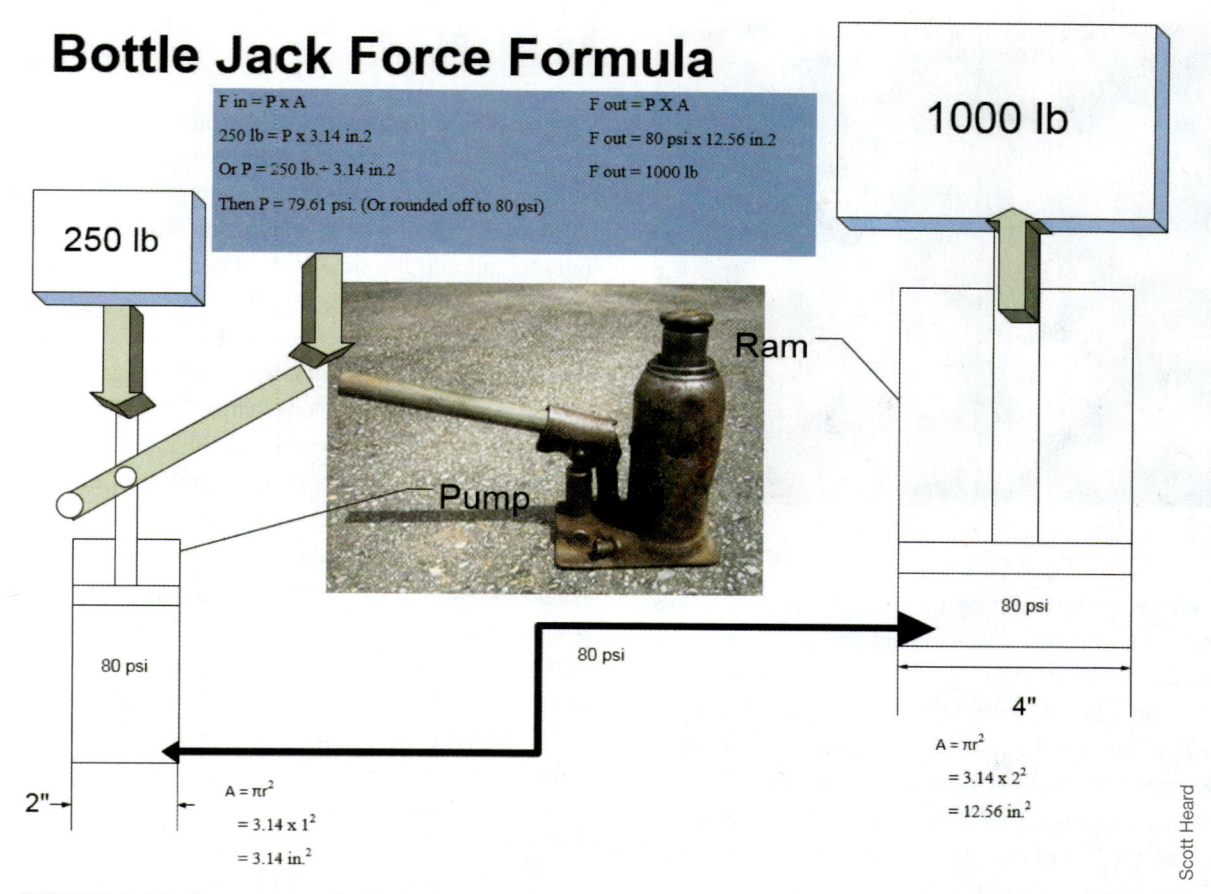

F in = P x A

250 lb = P x 3.14 in.2

Or P = 250 lb. ÷ 3.14 in.2

Then P = 79.61 psi. (Or rounded off to 80 psi)

F out = P X A

F out = 80 psi x 12.56 in.2

F out = 1000 lb

250 lb

1000 lb

Ram

Pump

80 psi

80 psi

80 psi

2"

4"

$A = \pi r^2$
$= 3.14 \times 1^2$
$= 3.14$ in.^{2}

$A = \pi r^2$
$= 3.14 \times 2^2$
$= 12.56$ in.^{2}

Scott Heard

FIGURE 4–10 Pressure in a bottle jack.

how the fluid transfers this force to the bottom of the ram (cylinder). The fluid acts on the bottom surface area and pushes the ram up. The amount of pressure developed by the pump depends on the load that is on top of the ram. In this example, there is a load of 1000 lb that is on top of the ram. To move the ram up, there needs to be slightly more force acting on the ram. To get it to move, the pump needs to be pushed with enough force to apply 80 psi fluid pressure to the bottom of the ram.

FORCE VERSUS PRESSURE

What is the difference between force and pressure? Force is the amount of linear push or pull applied to an object. If a large truck was stuck in a muddy field and a bulldozer with a winch hooked its cable to it, it might take 60,000 lb of pulling force to free the truck. If that same truck was pushed with a bulldozer, the dozer would need to create 60,000 lb of pushing force to free the truck. The force applied to the truck is linear in this case and is the same value whether it is pushing or pulling.

Pressure is an amount of force applied to a specific area. If a woman who weighs 100 lb has a pair of high heels on and balances on one heel that has a surface area of 1 in.^{2}, she would be applying 100 psi pressure to the floor. If she changed shoes and the heel had only a ½ in.^{2} surface area, she would be applying 200 psi pressure to the floor. If she switched to shoes that had 2 in.^{2} surface area, she would apply 50 psi pressure to the floor.

Pressure in terms of fluid power applies the same principles of force exerted on a certain area, and this pressure can be measured with a gauge.

If a pressure gauge was installed somewhere before the nozzle at the end of a garden hose and the hose was turned on at the house with the nozzle closed, you would read the maximum pressure for that water system (let's say, its 60 psi). This is because there is a complete restriction or resistance to the flow of water through the hose. If you then started to open the nozzle and watched the gauge, you would see the pressure decrease as the flow of water increased. As the resistance through the nozzle decreases and allows more water flow, the pressure falls in direct proportion.

HYDRAULIC SYSTEMS—THE BASICS

FIGURE 4–11 A variety of pressure gauges.

From this you could conclude that pressure is created in a hydraulic system by the resistance to fluid flow.

To get a better idea of pressure values, consider this: If you were standing on a beach beside the ocean with the air temperature at 70°F and have a pressure gauge in one hand that had a scale called absolute, it would read 14.7 psia (pounds per square inch absolute). This is the weight of the column of air that reaches the top of our atmosphere acting on 1 in.² of the beach. This gauge will read 0 only in a complete vacuum.

If you were standing at the same spot on the beach and have a gauge with the pressure scale that considers atmospheric pressure in the other hand, it would read 0 psig (pounds per square inch gauge). This is the most commonly used type of gauge.

Some pressures you may have read with a psig gauge would be a bicycle tire at 20–50 psi, a car tire at 30–40 psi, a household water pressure system at 50 psi, a vehicle's engine oil pressure at 40 psi, and a properly charged fire extinguisher at 175 psi. If a gauge reads only in psi, it's assumed to be a psig gauge.

Don't be confused as pressure measurements are the same whether they are measuring pressure of a gas or liquid. It is still the amount of force applied to a specific surface area.
● SEE FIGURE 4–11 for a variety of pressure gauges.

Occasionally, it may be necessary to read less than atmospheric pressure, and a gauge for this will usually be calibrated in units of inches of mercury (in Hg) or inches of water (in H$_2$O). This is a more accurate way of reading negative pressure (sometimes called vacuum). There are very few occasions where there will be a less than atmospheric pressure inside a mobile hydraulic system. One example might be a negative pressure reading between the pump inlet and the tank.

FLOW

Fluid **flow** in a hydraulic system can perform work if it is directed to an actuator (cylinder or motor). In real terms, an excavator can only dig a hole if fluid moves the rods in and out of its cylinders. The cylinder rods move because there is fluid moving and the flow of fluid is created by the system's pump.

To understand the concept of the rate of fluid flow you must consider units of volume first and then relate them to time. **Volume** is measured in gallons (either US Standard or imperial) or liters (metric). Flow units are gallons per minute (gpm, usually US gallons) or liters per minute (lpm). ● **SEE FIGURE 4–12** for how volume of fluid is calculated and relates to movement.

To sum up, a hydraulic force increase is gained by applying force from a certain size of surface area to a fluid and having the pressure created in that fluid transferred to a larger surface area. The volume of fluid directed to a hydraulic cylinder's piston relates to how far the piston will move. The amount of fluid volume is provided by the pump. Therefore, the rate of pump flow determines how fast a piston moves. The pressure in a system is created by the load or resistance to the movement of the load.

These are very important basic principles to know in the understanding of basic hydraulics. You must be confident that you know the relationship between pressure and force as well as flow and volume.

> **TECH TIP**
>
> **Units of Measure**
>
> The units of measure we have mentioned so far are imperial or US Standard units. This system is commonly used to measure values such as temperature (°F), flow (gpm), and pressure (psi) in North America and Great Britain. Hydraulic systems can be measured using the metric system, which is common in most other parts of the world. If you work on a machine that is manufactured in Europe or Asia, the service information will most likely be given in metric units. Temperature is measured in Celsius (°C), force in Newtons (N), area in centimeters squared (cm²), pressure in kilopascals (kPa), kilograms per centimeter squared (kg/cm²), or **bar**, and flow in lpm. Search the internet for a hydraulic unit conversion tool.

110 CHAPTER 4

Bottle Jack Volume/Distance Formula

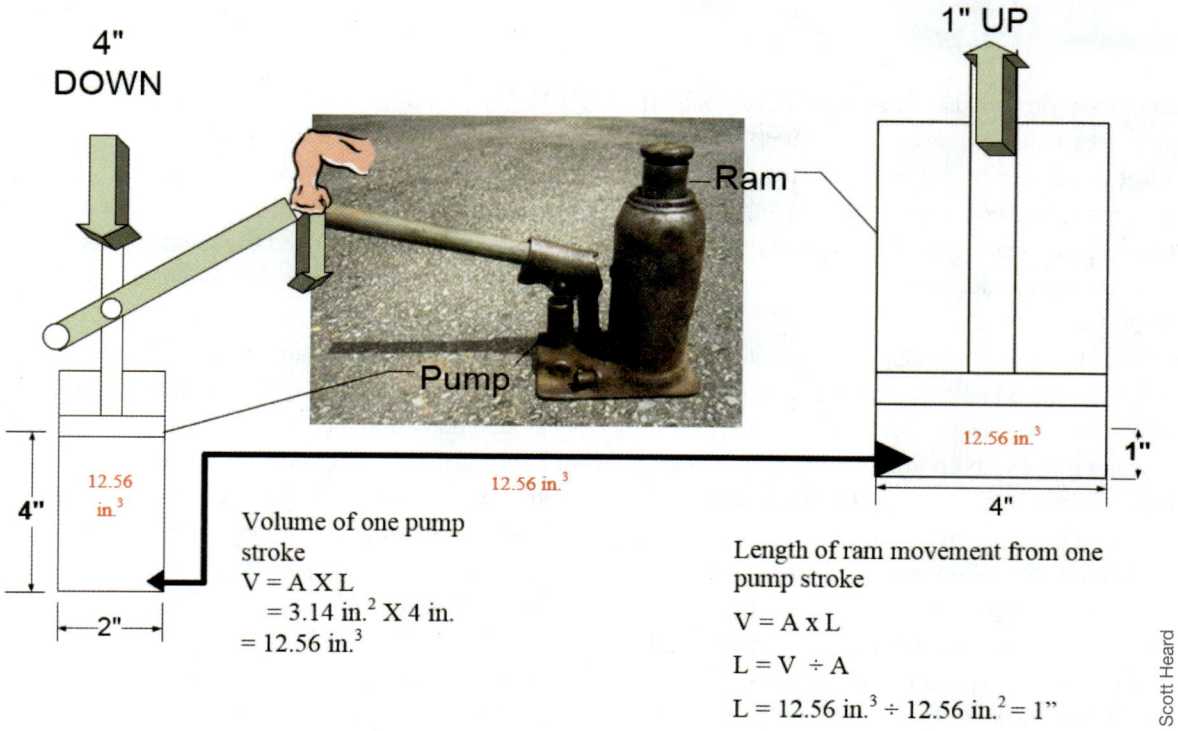

4" DOWN

1" UP

Ram

Pump

12.56 in.³

Volume of one pump stroke
V = A X L
= 3.14 in.² X 4 in.
= 12.56 in.³

Length of ram movement from one pump stroke
V = A x L
L = V ÷ A
L = 12.56 in.³ ÷ 12.56 in.² = 1"

Scott Heard

FIGURE 4–12 How volume of fluid is calculated and relates to movement.

Here's a practical example of where these basic principles apply. If the operator of an excavator complains to you that the bucket is moving slow, would you check for the proper pressure or flow? If you said flow, you are correct. The rate of flow in a hydraulic system only dictates how fast actuators move and has nothing to do with how much force is created by the cylinder or motor.

If the same operator complains that the bucket is not digging into the ground with the same amount of force it used to, would you check the pressure or flow? If you said pressure, you are right. The amount of fluid pressure an actuator has applied to it directly relates to the amount of force the actuator can create.

Remember that pressure is created in a hydraulic system by the resistance to fluid flow. The resistance can be caused by the fluid flowing through hoses, tubes, or valves, by intentional restrictions called orifices, or by the load on the circuit actuator.

Progress Check

6. Most heavy equipment machines have a hydraulic system that is driven by this prime mover:
   a. electric motor
   b. hydraulic motor
   c. gasoline engine
   d. diesel engine
7. If the area of two cylinder's pistons is different and the pressure applied to them is the same, the result is:
   a. the bigger piston will move faster
   b. the bigger piston will have more output force
   c. the smaller piston will move slower
   d. the smaller piston will have more output force
8. If the volume of two hydraulic cylinders is different and the same amount of flow is sent to them, the result is:
   a. the smaller cylinder will move faster
   b. the bigger cylinder will move faster
   c. the smaller cylinder will have more force
   d. the bigger cylinder will have more force
9. The metric unit of measurement for pressure is:
   a. psi
   b. gpm
   c. kpi
   d. kPa
10. Pressure × area equals:
    a. volume
    b. force
    c. speed
    d. Newtons

HYDRAULIC SYSTEMS—THE BASICS

BASIC HYDRAULIC SYSTEM COMPONENTS

To make a complete mobile hydraulic system, there needs to be the following components: prime mover, pump, reservoir, pressure control valve, directional control valve, actuators, conductors, fluids, and fluid conditioners. ● SEE FIGURE 4–13 for a wheel loader and its main hydraulic system components.

Let's look at these individual components.

PRIME MOVER Most heavy equipment hydraulic systems use a diesel engine as the prime mover or power source. Some specialized equipment or circuits could use an electric or air motor as the prime mover. The prime mover's function is to drive the pump. Typically, this is done by turning a pump shaft that is either splined or keyed to a gear that is driven by the diesel engine. It could also be done by a belt and pulley or a drive shaft and U-joints. Depending on the drive ratio between the pump and the prime mover, the pump could be driven faster, slower, or at the same speed that the prime mover is rotating. If a pump is directly driven by a diesel engine, the pump turns when the engine rotates. Occasionally, a pump could be driven through a clutch so it could be started and stopped on demand. Hydraulic systems must be designed to not overload the prime mover. If a hydraulic system is not adjusted correctly it could overload the machines prime mover. ● FIGURE 4–14 shows a typical pump being driven by a prime mover.

PUMP The pump is literally the heart of any hydraulic system. Just as the heart beating in your chest pumps blood, the hydraulic system pump's function is to move oil or create flow. Remember what causes pressure in a hydraulic system? If you said the pump, you need to repeat the next sentence five times. Pumps only create flow; pressure is caused by resistance to flow.

All pumps have an inlet and an outlet port. Most pumps have a bigger inlet than outlet to ensure that the pump is never starved or always has more than enough inlet fluid. Also if the pump's inlet port is bigger, the fluid velocity (speed of oil through the inlet) will be slower so there is less chance of turbulence and **aeration**. To ensure maximum pump life, it is mandatory that the pump receives an unrestricted supply of cool clean fluid with no excessive air mixed in.

As the pump starts to rotate, a pressure drop is created at the pump's inlet and fluid is forced into the pump's inlet port from the tank or reservoir because of the lower pressure created at the pump inlet. This fluid gets pushed out of the reservoir to the pump inlet and the pump now pushes it out the pump outlet, through the system, and back to the tank where it

FIGURE 4–13 Wheel loader and its main hydraulic system components.

FIGURE 4–14 A hydraulic pump driven by a prime mover.

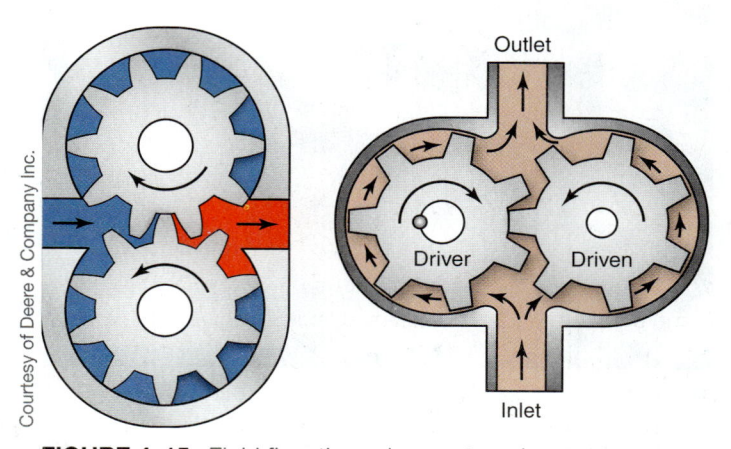

FIGURE 4–15 Fluid flow through an external gear-type pump.

is ready to enter the pump again. For every gallon of oil that is pushed out, there is another gallon behind it.

These pumps are considered to be positive displacement pumps. If for some reason the outlet of a positive displacement pump is blocked off and because the hydraulic fluid is non-compressible, this will cause a pump failure such as having its drive shaft break or the pump housing will split.

There are several types of hydraulic pumps, but we will just talk about external gear-type pumps in this chapter. Some other common types of pumps used for heavy equipment hydraulic systems are internal gear, vane, and piston. ● **SEE FIGURE 4–15** to help understand how the fluid flows through an external gear-type pump.

External gear-type pumps are very common in hydraulic systems because of their simplicity and tendency to last a long time if proper maintenance is performed regularly. They are considered to be a **fixed displacement** type of pump, meaning that for every revolution that the input shaft makes, it displaces a fixed volume of fluid. These pumps are usually used with open center-type control valves that allow pump flow to continue through it and back to the tank.

A fixed displacement pump can only vary its flow rate by having the rpm of the prime mover vary that is driving it. In other words, the faster the pump turns, the more oil it moves. All fixed displacement pumps are rated by their displacement per revolution in cubic inches or cubic centimeters. They could also be rated by the volume they pump during certain amount of time at a certain speed. This is usually in gallons per minute, but it could be liters per minute as well. A variable displacement type of pump can vary its output volume regardless of the speed of the pump's drive shaft. Variable displacement pumps will be discussed in Chapter 16.

The term *external* **gear pump** simply means the two rotating shafts inside the pump have gears with teeth on their outside diameter that rotate inside the pump housing. One shaft is driven by the prime mover and is usually attached to a drive gear by splines or a keyway. As the drive gear turns, the other gear is driven by it as the two gears mesh together. The oil flow in this pump starts at its inlet and then continues around the outside of both gears. Between each pair of adjacent gear teeth and the inside of the pump housing, a small volume of oil is carried around the inside of the pump. The oil then leaves through the pump outlet as more oil comes along behind it.

The clearances inside the pump are small enough that the fluid creates a seal between the gear teeth and the pump housing. The ends of the gears are usually sealed to the pump housing by brass wear plates that have a film of oil between them and the gears. The rotating gears should never contact the pump housing or the endplates.

If the pump becomes worn and clearances increase, internal leakage occurs and the pump's efficiency decreases. Pump efficiency is a calculation based on how much fluid the pump inlet receives and how much fluid leaves the pump's outlet. The difference is how much internal leakage the pump has and is measured at a set fluid temperature, pump rpm, and with the outlet fluid restricted to create a certain pressure. As the outlet pressure increases, leakage increases and pump efficiency will decrease. Pump efficiency is usually in the 85–90% range for a new unit. ● **SEE FIGURE 4–16** to help understand the components that make up an external gear-type pump.

Gear-type pump housings can be made of aluminum, cast iron, or steel. The shafts usually ride in plain bushings that are pressure lubricated to keep the shafts floating in hydraulic fluid. The pump housing sections will have seals between them, and sometimes the housing section that mounts the pump to its prime mover will have a bleed hole to indicate if the shaft seal is leaking. Some pump housings contain more than one set of gears that are divided by section housings. These are called

FIGURE 4–16 Exploded view of an external gear-type pump.

FIGURE 4–17 A multi-section external gear-type pump.

multi-section pumps and allow the prime mover to drive up to four different pumps with one drive because the pumps use the same housing. These multi-section pumps could use a common inlet port. ● SEE FIGURE 4–17 for a typical multi-section gear pump.

RESERVOIRS Hydraulic **reservoirs** are more commonly known as hydraulic tanks. Their main purpose is to hold a supply of hydraulic fluid for the system to use as needed. Most tanks will become pressurized to provide a positive flow to the pump inlet but some are open to atmospheric pressure. Tank position in relation to the pump is important. If the tank is located above the pump, gravity will assist pump inlet flow by the fluid weight creating head pressure. However, if the tank is below the pump, the tank must be pressurized to help get the fluid to the pump inlet. Tanks can be pressurized naturally by the expanding oil after it heats up or can have air pressure added from an external source. ● SEE FIGURE 4–18 for a typical hydraulic tank.

Tank pressure is limited by a relief valve, and this valve will also relieve any negative pressure created in the tank after the fluid cools down and contracts. Hydraulic tanks provide fluid cooling by convection through tank walls and allow air and large contamination to separate from the fluid before it is circulated back through the system. You can actually calculate the amount of heat loss through the walls of the

FIGURE 4–18 A typical hydraulic tank.

reservoir with a simple formula. Search the internet for hydraulic formulas and see if you can find one for tank heat loss. Contamination separation occurs because the tank will usually have one or more baffles in it to slow the return fluid down so the fluid doesn't immediately go straight back to the pump inlet. As the return fluid slows, contamination will settle out of the fluid and collect at the bottom.

Tanks can be made from steel or plastic and can be part of a machine's frame. They will have a minimum of two ports: one for pump inlet and the other for return oil from the system. Most tanks will have more than two ports as there are sometimes more than one pump being fed from the same tank and more than one circuit returning oil to it.

There will be some kind of fluid level checking device (dipstick, sight glass, electrical/electronic sensor) to allow the operator to monitor fluid level as well as a capped opening at the top to allow refilling. All tanks will have a means to drain the hydraulic fluid when necessary and some will have access covers that can be removed to allow a thorough cleaning if needed.

Hydraulic tanks may also contain a pump inlet strainer to prevent large contamination from entering the pump.

CASE STUDY

I had an experience where an operator complained of a strange noise coming from a machine, and after some investigation, I found a 2 lb piece of cast steel hose fitting/nozzle right at the pump inlet. During a service, someone had dropped it into the tank, and eventually it was drawn up through the pump inlet hose to the pump inlet that is about 5 ft above the bottom of the tank. The pump started to chew it up, requiring the machine to be down for several days as I rebuilt the hydraulic pump. This was a case where a pump inlet strainer would have saved a costly repair.

PRESSURE CONTROL VALVE

Any type of hydraulic system needs to have its maximum system pressure limited to prevent component failures, and this is performed by a pressure control valve or more commonly called a pressure relief valve or simply relief valve. If there was no pressure control valve and the machine operator kept trying to direct fluid into an actuator after it came to the end of its stroke, something would fail. This would happen because there would be a maximum restriction to fluid flow, and as fluid is not compressible, something would have to give. If the system uses a fixed and positive displacement type of pump to move fluid and the fluid has nowhere to go, then component failures such as blown seals, ruptured hoses, and tubes would occur or components such as the pump, valves, or actuators would fail.

The pressure control valve simply limits maximum system pressure to a safely designed value, and when this setting is reached, the valve opens and allows the fluid to flow back to the tank. A simple system might be limited to 3000 psi. Some hydraulic presses have a relief valve setting as high as 10,000 psi, and most excavators will have a main relief pressure set at around 5000 psi.

A simple pressure control valve consists of a metal body with three ports, a movable piston, and a spring. The pressure setting the valve is set at is controlled by the amount of spring pressure applied to the movable piston. When the fluid pressure overcomes the spring pressure, the piston uncovers a port and pump flow will go to tank. This spring pressure is usually adjustable by adding or removing metal shims behind the spring or turning a threaded adjuster in or out.

A more accurate and more common type of pressure control valve is the pilot operated type. This type uses an additional piston and spring. The pilot piston now controls movement of the main piston by trapping oil behind it. The valves' setting is controlled by the pilot piston's spring. When a relief valve just starts to open, this is called its cracking pressure. A simple relief valve will have a fairly wide spread between the cracking pressure and the fully open pressure. A pilot-controlled relief valve will have a much narrower gap, and this is what makes it more accurate. Pressure control valves are normally very reliable but can occasionally stick open or closed if there is contamination in the fluid or on rare occasions when springs fails. ● **FIGURE 4–19** illustrates a typical main system pilot-type pressure control valve. When you are adjusting relief valves, never go over the specified maximum system pressure.

DIRECTIONAL CONTROL VALVE

A **directional control valve** (DCV) allows the operator to direct pump flow to one or more actuators (cylinder or motor). The most common type of DCV is a spool-type valve. There is a cylindrical spool inside the valve body that is shifted in a bore to open and close ports.

The oil flow for a single circuit starts with pump oil that the DCV can send to one side of the cylinder's piston and direct the oil from the other side of the piston back to tank.

If there is more than one double-acting cylinder in the system, a multi-circuit DCV will send oil to either end of those cylinders to extend or retract their rods. If there is a rotary actuator or motor as part of the system, the DCV will allow the operator to change rotation of the motor by sending oil to either of two ports on the motor.

A monoblock spool-type DCV consists of a single cast iron body with at least four external ports that allow the connection of tubes or hoses. The ports will connect to the pump outlet, tank inlet, circuit A port, and circuit B port. If the monoblock DCV is used for more than one circuit, then each spool section will have its own A and B ports. A and B ports will deliver pump oil and receive return oil to and from the actuator through hoses and or tubes. Inside the valve body, the spool is controlled by the operator to redirect pump flow out to A or B port depending on which way the spool is moved. If pump oil is directed out of A port, then B port will receive the return oil flow from the actuator and the spool valve directs it out of the tank port of the valve. This common type of DCV is a 3 position 4 way valve (spool can be moved to 3 positions and there are 4 ports allowing oil flow in and out of the valve).

Multi-section DCVs are made up of individual sections for each circuit and then stacked together and held with long bolts or rods. Some multi-section DCVs can have 10 or more sections that each has an A and a B port. The assembly will have an inlet section with the pump port and an outlet section with the tank port. ● **SEE FIGURE 4–20** for a typical DCV and ● **FIGURE 4–21** for the same valve in schematic form.

DCVs will most likely also house other valves such as main relief valves, anti-cavitation valves, and port relief valves.

AT CRACKING PRESSURE

AT FULL-FLOW PRESSURE

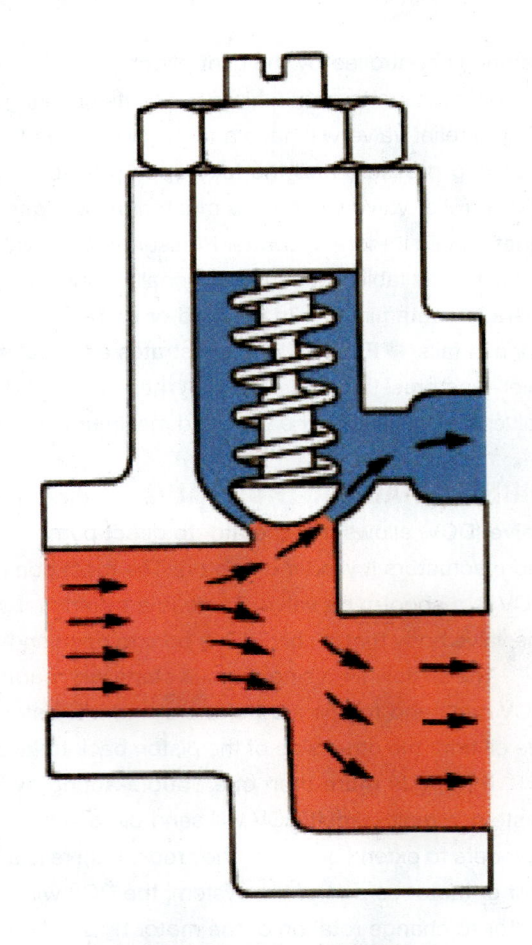

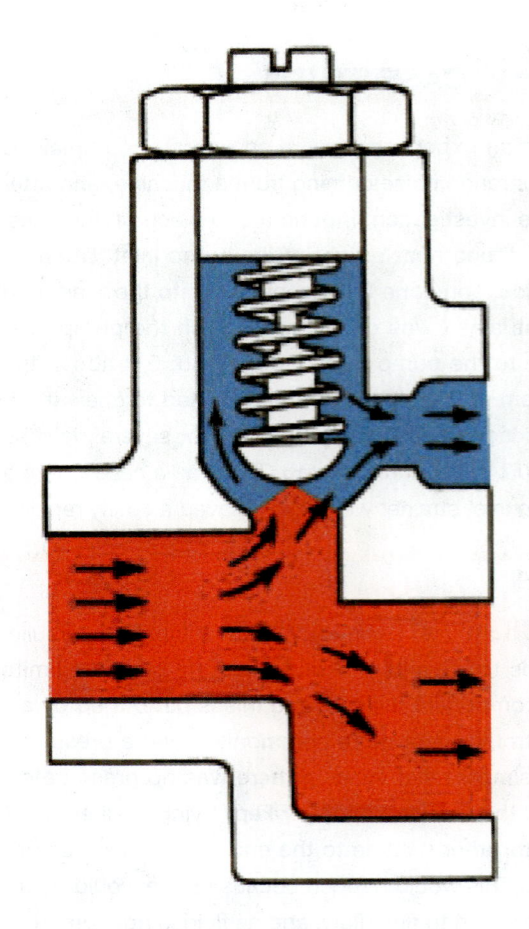

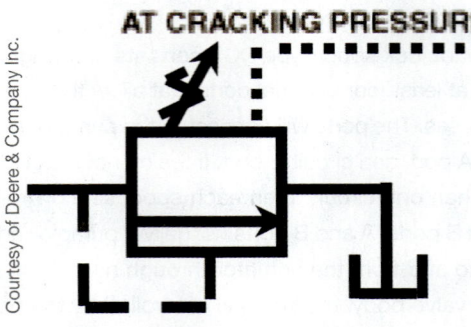

Courtesy of Deere & Company Inc.

FIGURE 4–19 A pressure relief valve.

Scott Heard

FIGURE 4–20 A cutaway view of a typical 2 spool 3 position, 4 way directional control valve.

These valves may or may not be needed depending on the type of circuit the DCV is part of.

DCV spools can be controlled several different ways. The simplest way is a direct mechanical link to either a hand-controlled lever or a foot pedal physically moving the spool in the bore. The spools can also be moved by lower hydraulic pressure (pilot pressure) working on the end of the spool, electrically or pneumatically.

ACTUATORS Linear Actuators (Cylinders): **Actuators** perform the opposite function of pumps. They convert the hydraulic energy created by the pump back into mechanical energy to

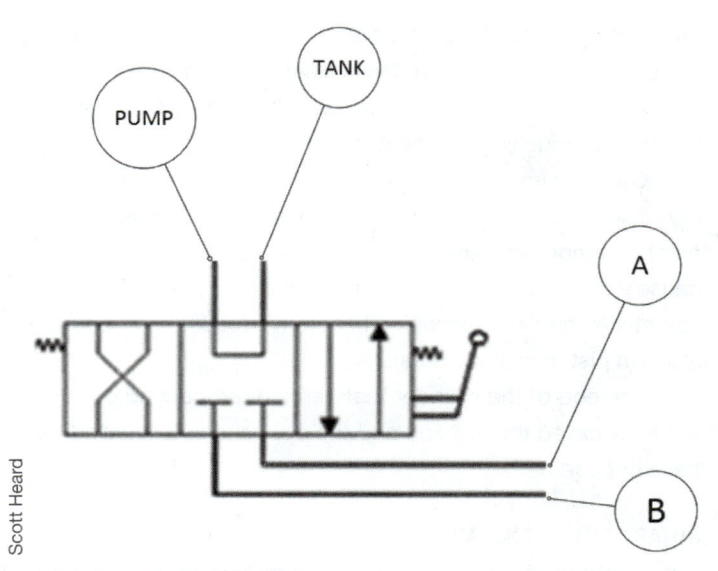

FIGURE 4–21 Directional control valve in schematic form.

perform the work the machine's operator has requested. They are the "business end" of the hydraulic system. Hydraulic cylinders create movement in a straight line and are called linear actuators because they create a linear push or pull force. The speed of a linear actuator will change according to the rate of oil flow that is fed to it. ● SEE FIGURE 4–22 for a cutaway view of a typical cylinder and what its parts are called.

A linear actuator or cylinder is a sealed barrel with a movable piston inside of it. Attached to the piston is a rod that moves with the piston and slides through the center of a removable cap called the rod end. The opposite end of the rod is attached to a device on the machine that needs to be moved such as the ripper frame (● SEE FIGURE 4–23). Cylinders can be either double acting or single acting. Single-acting cylinders will use pump flow to move the rod one way and gravity will return it as the oil is drained out toward the reservoir. An example of an application for a single-acting cylinder would be one that is used for a dump truck hoist.

Double-acting cylinders are more common on heavy equipment. They have fluid flow into one side of the barrel and act on the piston to create movement. The opposite side of the piston will push fluid out of the barrel and toward the reservoir through the DCV. To move the piston and rod in the opposite direction, oil is moved into the opposite end of the barrel where it acts on the opposite side of the piston. The surface area of the piston that the oil acts on is called the effective area.

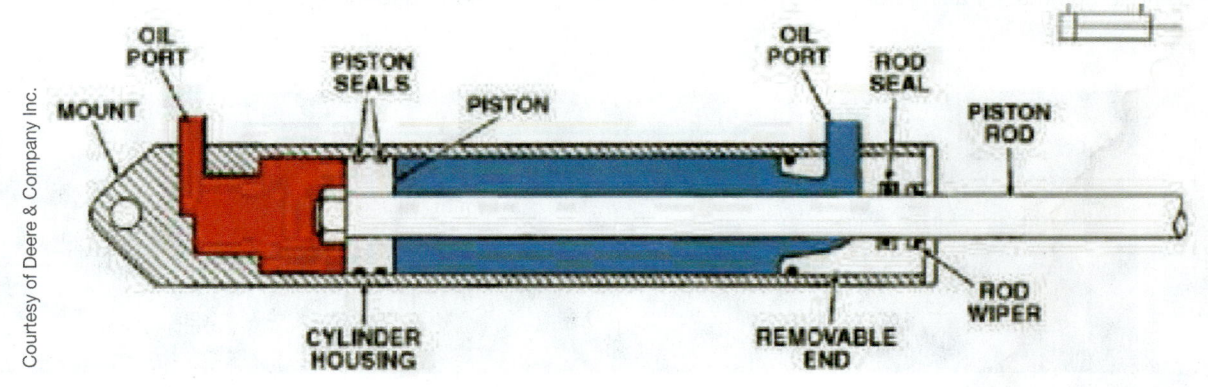

FIGURE 4–22 A cutaway view of a cylinder.

HYDRAULIC SYSTEMS—THE BASICS   **117**

FIGURE 4–23 Hydraulic cylinders used to move a ripper.

Because a double-acting cylinder has the rod take up volume inside the cylinder and an effective area is lost on one end of the piston, it will retract faster and have less force compared to when it extends. Double-acting cylinders are sometimes referred to as unequal displacement cylinders or differential cylinders.

Double-rod cylinders have two rods connected to a single piston. The rods move together with the piston out of opposite ends of the cylinder. These cylinders are equal displacement cylinders and the rod speed doesn't change regardless of the direction the piston moves. They are mostly used for steering cylinders on backhoe loaders. ● SEE FIGURE 4–24 for a double-rod cylinder.

The force output of a cylinder can be calculated by using the $F = P \times A$ formula. If the surface area of the piston is known and the pressure applied to it is known, then by multiplying these values the force output is found.

Cylinders are attached to a machine in a variety of ways such as through pins, bolts, and ball sockets.

Pistons are attached to the rods in a variety of ways such as by using large nuts, by using bolts, or by simply threading on. Pistons will have one or more seals on their outside diameter to create a seal between the inside of the barrel and the piston and will usually have a fiber wear sleeve to keep the metal piston from touching the metal barrel.

Cylinder end caps are attached to the barrels in a variety of ways such as through bolts, key wire, or threads. The most common problem with cylinders is seal failures. This can cause either an internal or an external leak. An internal leak can sometimes be detected with a heat gun because oil bypassing a blown piston seal will create heat.

The end of the cylinder that has the rod coming out of is naturally called the rod end and the opposite end is called the base or head end.

ROTARY ACTUATORS (MOTORS). To get a rotational action output from a fluid power system, a rotary actuator is needed. A rotary actuator is also called a hydraulic **motor**. Some common uses of a hydraulic motor would be to provide swing drive for an excavator, turn a drill steel for a rock drill, or turn a cooling fan. Oil flow from the pump is sent to one port of the motor and the pressure drop across the motor causes shaft rotation or **torque** output. As the fluid rotates the shafts, it leaves the motor through the second main port. To reverse rotation of the motor, fluid is sent to the motor in the opposite direction.

Rotary actuators are very similar in construction to pumps, and the three main types are gear, vane, and piston. Visually a rotary actuator can usually be differentiated from a pump by a third port called a case drain port that allows normal internal fluid leakage to return to tank.

Rotary actuators will vary their output shaft speed as the oil flow to them varies. Hydraulic motors are classified by their type and by the amount of fluid required to make the output shaft complete one rotation. This is typically measured in cubic inches per revolution (cir). The amount of torque coming out of a rotary actuator will vary with the amount of pressure applied to the motor's movable components. Once again the pressure created is dependent on the load on the motor's shaft. The

FIGURE 4–24 Double-rod cylinder.

FIGURE 4–25 A cutaway view of a rotary actuator.

formula for calculating output torque from a hydraulic motor is $T = psi \times cir/2\pi$. ● **SEE FIGURE 4–25** for a cutaway of a rotary actuator.

CONDUCTORS AND FITTINGS Fluid conductors are needed to move the hydraulic fluid from one component to another. These conductors are more commonly called hoses, tubes or lines, and fittings.

Hoses are flexible and are mostly used when two components need to move in relation to each other. They are measured by their inside diameter and can vary in size from $\frac{1}{8}$ of an inch to 4 in. They can be made to withstand very high pressure or moderate vacuum. Hydraulic hose is usually reinforced with steel strands, alternating with layers of rubber. There are many variations of hydraulic hose for different applications. Go to http://www.parker.com or http://www.gates.com to download and view their catalogs to learn more about hydraulic hoses and fittings. Hydraulic hoses are attached to components in many different ways. On each end of a hose is either a permanent crimp on fitting or removable fitting that provides a mating seal to whatever component the hose is attached to. As the technician will spend a great deal of time replacing hydraulic hoses, it is important to become familiar with common hose sizing and connector types. Hoses may eventually fail from repeated flexing, but if they rub against another component, are overheated, or are overpressurized, they will fail prematurely.

> **CASE STUDY**
>
> A conductor that fails under pressure can be very dangerous. I once heard a hose fail on a large excavator when I was on the same job site repairing another machine. I was at least 500 yards away, and it sounded like a shotgun going off. Hoses that have failed near an operator's cab have sometimes caused serious injury or death to the operator and others have injured bystanders.
>
> This example should highlight the need to replace any worn or damaged conductor or fitting as soon as possible.

The safest way to know you are replacing a failed hose with the proper one is to go by the manufacturer's part number. Never attempt to mix different styles or types of hoses, tubes, or fittings.

Hose sizing is always determined by hose inside diameter. Hoses that are sized by fractions of an inch are then converted to a dash size. One **dash size** equals $\frac{1}{16}$ of an inch. For example, a hose that has a $\frac{1}{2}$ in. inside diameter is considered to be a dash 8 size and a hose that has a 2 in. inside diameter is considered to be a dash 32 size. There are several different guidelines for hydraulic hose construction such as SAE, ISO, and EN standards. Hydraulic hoses are usually rated to burst at no less than four times their normal working pressure. For example, a hose that normally would only see 3000 psi maximum is designed to not burst until 12,000 psi. See the following table for sizing information.

Never replace a hose with one that is rated for less than system working pressure or undersized as an extra restriction will be created.

Hose Inside Diameter (in inches)	Dash Size
1/16	1
1/8	2
1/4	4
3/8	6
1/2	8
3/4	12
1	16
1 1/4	20
1 1/2	24

● **SEE FIGURE 4–26** for a hose that needs to be replaced.

When you are replacing hoses, always install all clamps that were removed and secure the hose to prevent it from moving against some other component. ● **SEE FIGURE 4–27** for a securely fastened hose.

Hydraulic tubing is made of high ductile annealed steel and is always measured by the outside diameter. Steel **tube** inside diameter is therefore based on its pressure rating. For example, if two pieces of 1 in. hydraulic tubing were compared, one being rated for 1000 psi and the other for 5000 psi, the higher pressure tube would have a significantly smaller inside diameter. This simply means the tube's wall thickness increases as its pressure capacity increases. Hydraulic tubing has a burst pressure three to five times higher than normal working pressure.

Hydraulic connectors or fittings come in many shapes, varieties, styles, and sizes. Connectors are used to connect components and hoses or tubes. The type of seal created by a connector is either by flexible seal to metal or by metal to metal. Some common styles of connectors are JIC (Joint Industry Committee), SAE, NPT (National Pipe Tapered), and ORFS (O-Ring Face seal). ● **SEE FIGURE 4–28** for a variety of different styles of connectors. To see many types of connectors, you can go to http://www.gates.com/catalogs-and-resources/catalogs.

Hydraulic **seals** are a critical part of hydraulic systems because they keep the hydraulic fluid contained inside of components, hoses, and tubes whenever they are joined together. Seals are made of many different types of materials, and they

FIGURE 4–26 A hydraulic hose that should be replaced.

FIGURE 4–27 A securely fastened hose.

FIGURE 4–28 A variety of hydraulic connectors.

> **WEB ACTIVITY**
>
> See http://www.parker.com/portal/site/PARKER/ to download a catalog of many different types of hydraulic seals.

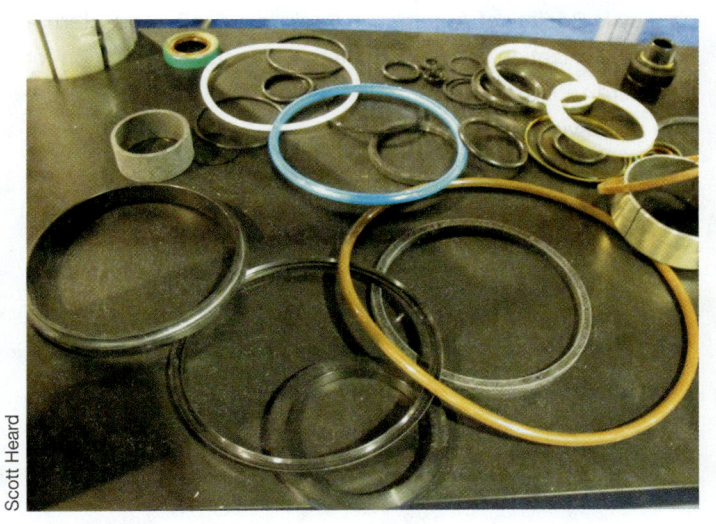

FIGURE 4–29 A variety of hydraulic seals.

are sized by their diameters and cross-section. Most seals have a round cross-section and are called O-rings; however, there are variations on this as well, such as D-rings and square rings. Machine manufacturers install seals that are designed to be compatible with the particular hydraulic fluid used for the machine and must be able to withstand maximum system pressure and temperatures.

Seals that have been overheated or are incompatible with the type of hydraulic fluid used in the system will fail and leak. Whenever changing any component that uses a seal, it's always a good practice to replace the accompanying seals. New seals should protrude about $1/3$ of their diameter before component installation. When the component is installed, the seal will compress and contain the oil as it is intended to. ● **SEE FIGURE 4–29** for a variety of seals used for sealing hydraulic systems.

HYDRAULIC FLUIDS Manufacturers of machines 30–40 years ago were not very concerned with the type of hydraulic fluid used in their hydraulic systems. System pressures were relatively low and internal moving parts had relatively loose tolerances; therefore, hydraulic fluid quality was not a big concern. Over time, system temperatures and pressures have increased, hydraulic systems became more complex, and machine owners wanted longer life from their hydraulic systems. Manufacturers started to become more concerned with the type of hydraulic fluid used in their system. Manufacturers typically recommended engine oil be used in lower pressure hydraulic systems, and this may still be the case for some machines. However, most manufacturers now use a specific hydraulic fluid as factory fill and require it as replacement fluid when a change is required to keep the machine's warranty valid.

It could be argued that hydraulic fluid is the most important component in any hydraulic system. Here's what the hydraulic fluid needs to do:

1. The most obvious purpose for a hydraulic fluid is to transfer the energy created by the pump to the actuator. As mentioned earlier, fluids are not compressible; however, in actual fact, hydraulic fluids will compress approximately 1% at 2000 psi. This is a small enough factor to not worry about when understanding basic concepts. But for higher pressure systems, the fluid will be compressed and expanded enough to have an effect on fluid longevity.
2. It must lubricate all moving parts inside the hydraulic system. This could include items such as the pump, valves, and actuator components.
3. The hydraulic fluid needs to be able to transfer heat from hot system components to cooler parts of the system or quite often to a cooler.
4. The hydraulic fluid needs to carry any contamination inside the system to the filter where it is collected.
5. The hydraulic fluid needs to seal clearances between moving parts. Lower pressure systems usually have bigger tolerances and this is not so critical, but high pressure systems have very close tolerances and fluid quality is very important to maintain this seal.

The most common type of hydraulic fluid used is mineral-based oil. However, with environmental concerns playing a big part in most industries, many biodegradable nonmineral-based fluids are now being used. ● **SEE FIGURE 4–30** for a pail of biodegradable hydraulic fluid.

Some machine applications require fire-resistant fluid, which is either water or glycol based. These base fluids will then have additive packages blended into them according to what the hydraulic system designer requires.

With all hydraulic fluids, it's the additive package that will wear out in hydraulic fluid and not the base fluid.

Hydraulic fluid viscosity is a very important consideration when either topping up or refilling a hydraulic reservoir. **Viscosity** is a measurement of how easily hydraulic fluid flows.

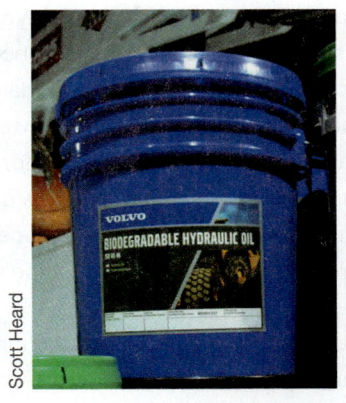

FIGURE 4–30 Biodegradable hydraulic fluid.

It is measured scientifically by pouring a fixed amount of fluid through a fixed orifice and measuring how fast it will flow through the orifice and into a container. Two common units that are used to measure this flow are SUS (Saybolt universal seconds) and cSt (centistokes). The longer the fluid takes to flow, the thicker it is and it will be given a higher viscosity number. There are two organizations that define the viscosity ratings according to their own specific tests. The North American agency is Society of Automotive Engineers (SAE). The international organization is called ISO (International Organization for Standardization). ● **SEE FIGURE 4–31** for a viscosity test device and a viscosity chart.

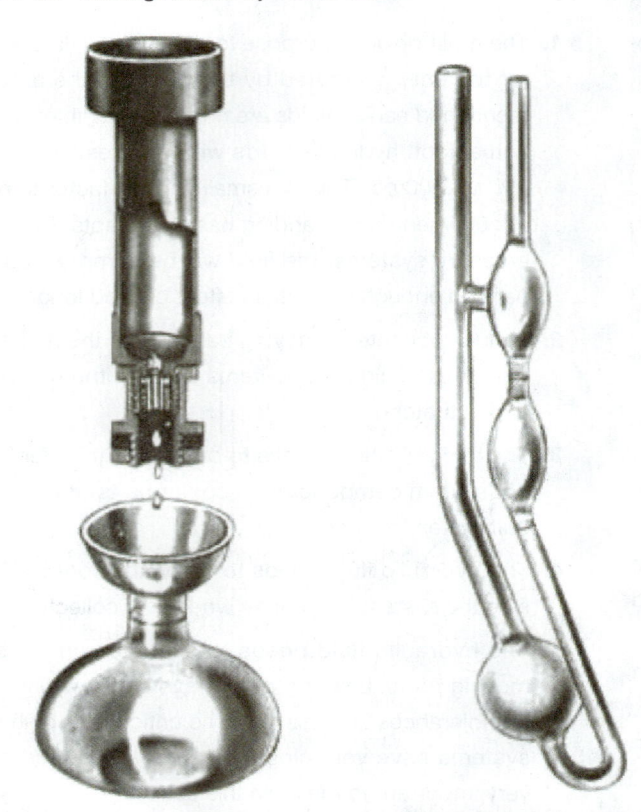

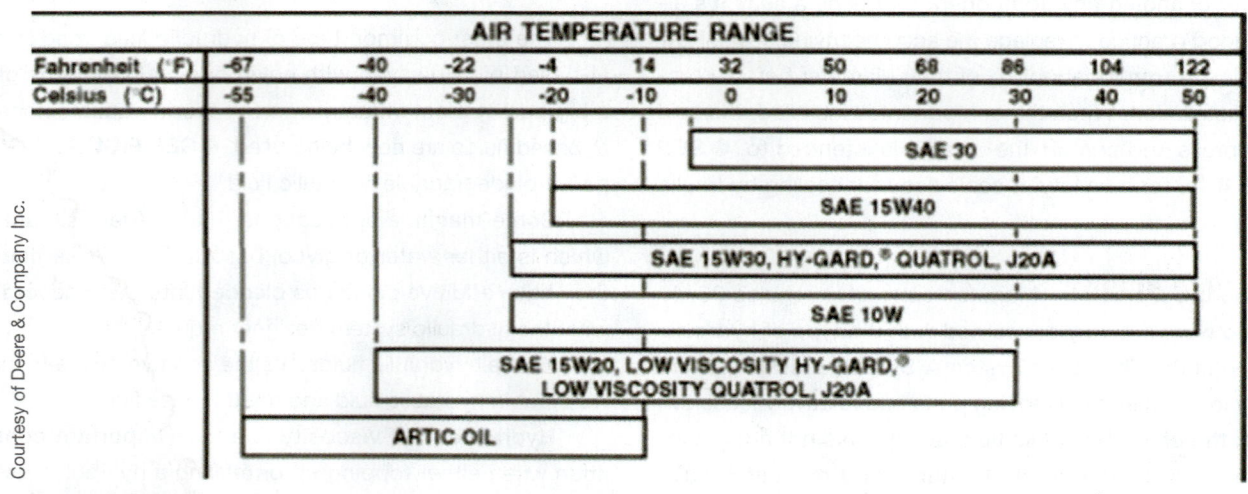

AIR TEMPERATURE RANGE											
Fahrenheit (°F)	-67	-40	-22	-4	14	32	50	68	86	104	122
Celsius (°C)	-55	-40	-30	-20	-10	0	10	20	30	40	50

- SAE 30
- SAE 15W40
- SAE 15W30, HY-GARD, QUATROL, J20A
- SAE 10W
- SAE 15W20, LOW VISCOSITY HY-GARD, LOW VISCOSITY QUATROL, J20A
- ARTIC OIL

FIGURE 4–31 A viscosity test device and a viscosity chart.

Typical viscosity of factory fill hydraulic fluid for machines used in North America will be SAE 10 or ISO 32. Hydraulic fluid viscosity recommendations are based on the expected ambient air temperature the machine will be working while that fluid is in the system.

FLUID CONDITIONERS Fluid conditioning is required to keep the hydraulic fluid clean and at the proper operating temperature. It can be broken into three categories: filtration, cooling, and heating.

FILTERS. Filtration is by far the most important factor in keeping a hydraulic system operating for as long as it was designed to. The three most important words in hydraulic system maintenance are clean, clean, and clean. Fluid filtration starts at the filler opening for the system reservoir. There is usually a mesh screen to prevent large particles from getting into the fluid. Unfortunately, this also slows down the flow of oil that is getting poured into the tank so a lot of the time this screen is removed to speed up tank refill or top up. Try to avoid this if possible as topping up or refilling a hydraulic tank is the most likely source of system contamination. Another type of filtration device that is often neglected is the tank breather.

The next part of fluid filtration (if used) is the pump inlet strainer or sometimes called the suction strainer. This strainer, like the filler opening strainer, is also a coarse mesh because it is only meant to catch any large particles before they get to the pump inlet. If the mesh is too fine, there would be an excessive restriction at the pump inlet and the pump would starve for fluid.

The main part of system fluid filtration takes place usually when the fluid is returning to tank, and not surprisingly, this is called a return filter. Sometimes the fluid is filtered after it leaves the pump, and this is called a pressure filter.

There are two main types of filters: spin on and cartridge type. Spin-on filters are a one-piece assembly that consists of a cylindrical-shaped steel housing containing a filter element that threads onto a filter base. When serviced, the spin-on filter is completely replaced. The difference between these two is that the cartridge-type filter only has the filter element replaced and the steel housing is reused. The base of these filters will have a valve that allows fluid flow through the filter unit if the filter becomes plugged. This valve is called a bypass valve. Sometimes, the bypass valve will have an indicator to tell the operator that the filter is plugged. This could be a visual device or electronic sensor that will display a message or warning light on the machine's dash. Fluid typically flows down through the outside of the filter elements and up through the center of the base. ● **SEE FIGURE 4–32** for a spin-on hydraulic filter.

FIGURE 4–32 A spin-on hydraulic filter.

> **WEB ACTIVITY**
>
> To learn more about filtration, you can view and download information at http://www.parker.com /portal/site/PARKER/ or see the fluid and filtration "Technical Reference Guide" at http://www.donaldson .com/en/ih/support/000721.pdf.

Filter elements can be made of paper or synthetic fiber, and they're usually reinforced with metal screen. Filters are expected to clean contamination that is much smaller than the human eye can see. This contamination is measured in microns. A **micron** is a very small measurement based on a meter. A meter is slightly longer than 3 ft or roughly 39 in. long. If that meter was divided into 1 million equal divisions, one of those parts would be 1 μ in size. To put that into perspective, the smallest particle visible to the human eye is 40 μ in size. A typical hydraulic fluid filter element will filter particles as small as 5 μ.

Fluid cleanliness is measured according to an ISO standard that relates to how many particles of a certain size are in certain volume of fluid. The ISO 4406 document outlines testing procedures and will classify contamination into two or three categories. A smaller number means less particles and a cleaner fluid. See an example of a cleanliness code in ● **FIGURE 4–33** to get a better idea how fluid cleanliness is categorized.

For example, an ISO code of 17/15/12 indicates there are 641–1300 particles = 4 μm/ml, 161–320 particles = 6 μm/ml, and 21–40 particles = 14 μm/ml of fluid. You will find most

HYDRAULIC SYSTEMS—THE BASICS **123**

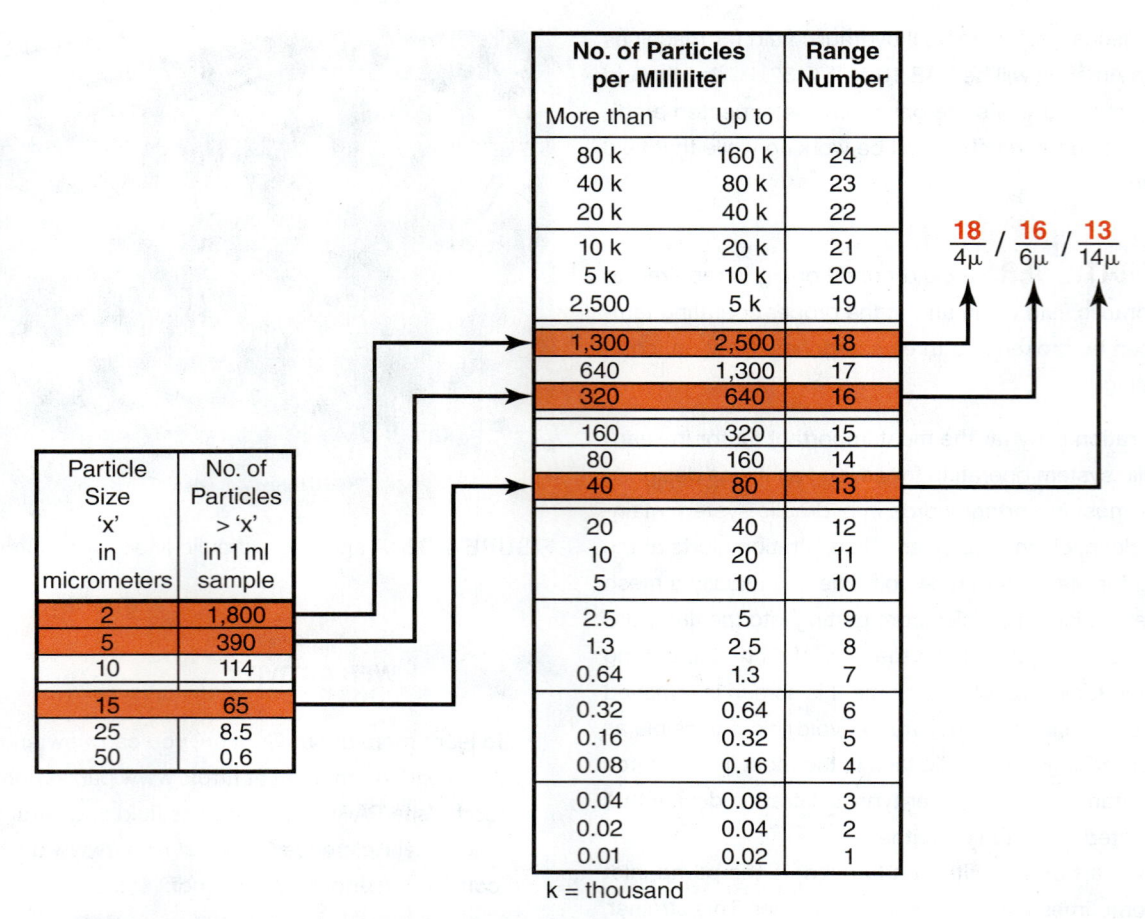

| No. of Particles per Milliliter || Range Number |
More than	Up to	
80 k	160 k	24
40 k	80 k	23
20 k	40 k	22
10 k	20 k	21
5 k	10 k	20
2,500	5 k	19
1,300	2,500	18
640	1,300	17
320	640	16
160	320	15
80	160	14
40	80	13
20	40	12
10	20	11
5	10	10
2.5	5	9
1.3	2.5	8
0.64	1.3	7
0.32	0.64	6
0.16	0.32	5
0.08	0.16	4
0.04	0.08	3
0.02	0.04	2
0.01	0.02	1

k = thousand

$\dfrac{18}{4\mu} / \dfrac{16}{6\mu} / \dfrac{13}{14\mu}$

Particle Size 'x' in micrometers	No. of Particles > 'x' in 1 ml sample
2	1,800
5	390
10	114
15	65
25	8.5
50	0.6

FIGURE 4–33 An example of fluid cleanliness ISO code.

FIGURE 4–34 A filter cart and particle counter.

machine manufacturers will recommend that even brand-new fluid gets filtered before adding it to a machine. If possible, a filter cart should be used to filter any oil added to a machine, and its cleanliness can be measured with an inline particle counter.
● **SEE FIGURE 4–34** for a filter cart and particle counter.

COOLERS. A typical maximum operating temperature for the fluid in a machine's hydraulic system is 150°F. Most systems are designed to run at this level and are most efficient at this temperature. If the hydraulic system runs hotter than this, there is a risk of component failure and hydraulic fluid break down.

Some experts say keeping hydraulic fluid from overheating is just as or more important than keeping it clean. Hydraulic fluid breaks down quickly when overheated, and its viscosity changes. The importance of keeping the fluid cool is often overlooked. To maintain the ideal operating temperature, most hydraulic systems will use a hydraulic fluid cooler. These will usually be located near the engine radiator and will be cooled by air as it gets moved by the engine coolant fan. However, some machines will use a separate hydraulic cooling fan to move air past the cooler. In some cases, a thermostat similar to an engine coolant thermostat is used to regulate hydraulic fluid temperature.

HEATERS. In some extreme northern climates, some hydraulic systems will use a fluid heater to prevent component damage because of high fluid viscosity and possible pump starvation. These will likely be an electric element style similar to an engine coolant heater.

CHECK VALVES

Check valves play an important role in most hydraulic systems and can be considered to be a DCV. They are a simple device that allows fluid to flow freely in one direction and stops the flow in the opposite direction. A good example of a check valve is the load check that is used in a circuit to block fluid flow from draining out of a loaded cylinder before enough pressure can be built up to move it. Typically, it will be in the boom-up circuit for a wheel loader.

In this case, the load check is placed between the directional spool valve and the port leading to the cylinder. ● SEE FIGURE 4–35 for a load check valve.

Another use for check valves is to prevent cavitation in a circuit. If the valve is placed in the circuit so that one side of it is exposed to tank oil and the other to a part of a circuit under certain circumstances, the valve will open and allow oil into the circuit from the tank.

ORIFICES

Orifices can be used for two purposes in a hydraulic system:

1. **Pressure differential:** As mentioned previously, pressure is created by any resistance to flow so when a pressure differential is desired, an orifice is placed in the path of oil flow to build up pressure on the upstream side of the

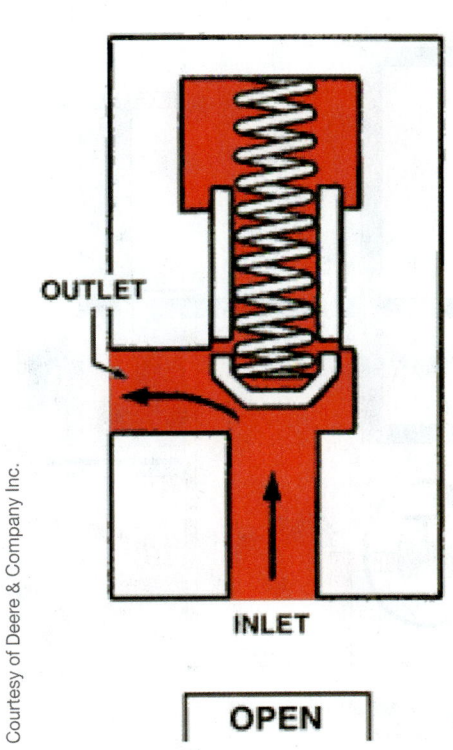

FIGURE 4–35 A check valve.

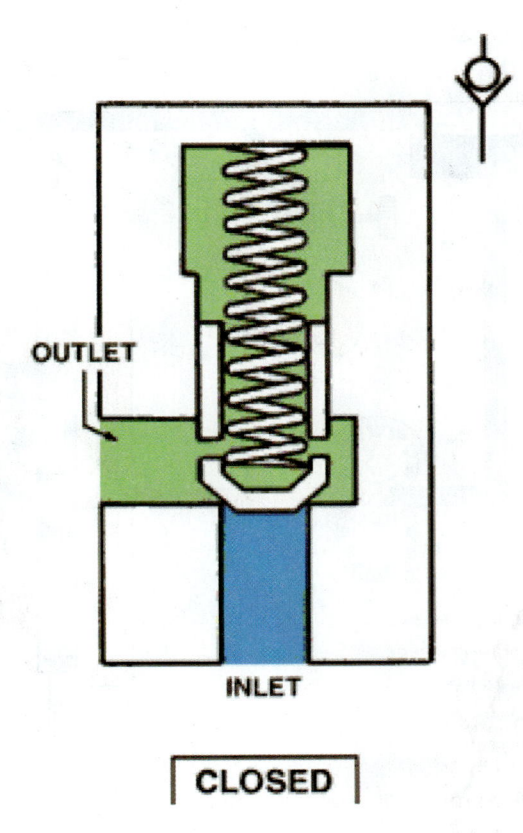

HYDRAULIC SYSTEMS—THE BASICS 125

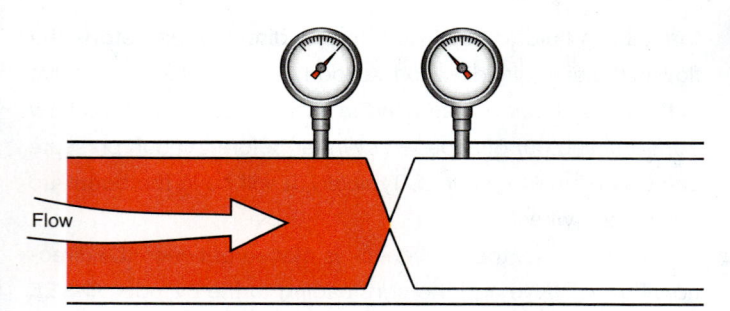

FIGURE 4–36 An orifice or restriction.

orifice. The downstream side will have a lower pressure as long as oil flows across the orifice.

2. Flow control: If there is a need to decrease flow to a certain part of a circuit, an orifice can be used to do this.

● **SEE FIGURE 4–36** and study how an orifice could affect both pressure and flow in the circuit.

Most orifices are fixed and can't be varied while others are adjustable or variable. For example, if a hydraulic attachment is added to a machine many times, it will need a variable orifice incorporated to its circuit to control the circuit flow to the attachment. ● **SEE FIGURE 4–37** for a variable orifice. Variable orifices are usually used when a hydraulic attachment is added to a machine, and the flow to operate it needs to be adjusted.

ACCUMULATOR

An accumulator can be used as

1. hydraulic energy storage device
2. protecting systems against pressure spikes
3. providing a hydraulic spring in some systems

There are three different types of accumulators: diaphragm, bladder, and piston type. The piston type is the easiest to understand because you just need to imagine a cylinder with a sealed movable piston inside. The cylinder is capped on both ends with one cap having a service fitting for nitrogen gas and the opposite cap having a fitting for hydraulic fluid.

An accumulator needs a precharge of nitrogen gas that is applied with a charging kit and nitrogen cylinder. Hydraulic system fluid flows into the opposite end of the cylinder and will move the piston toward the gas side, which compresses the gas. Once the nitrogen gas is compressed, and the oil on the opposite side is trapped in the accumulator, there is now a potential energy available to be used when required.

Accumulators can come in many different sizes and pressure ratings. Bladder-type accumulators will have a rubber bladder keeping the fluid and nitrogen separate. Diaphragm-type accumulators will have a diaphragm separating the fluid and nitrogen.

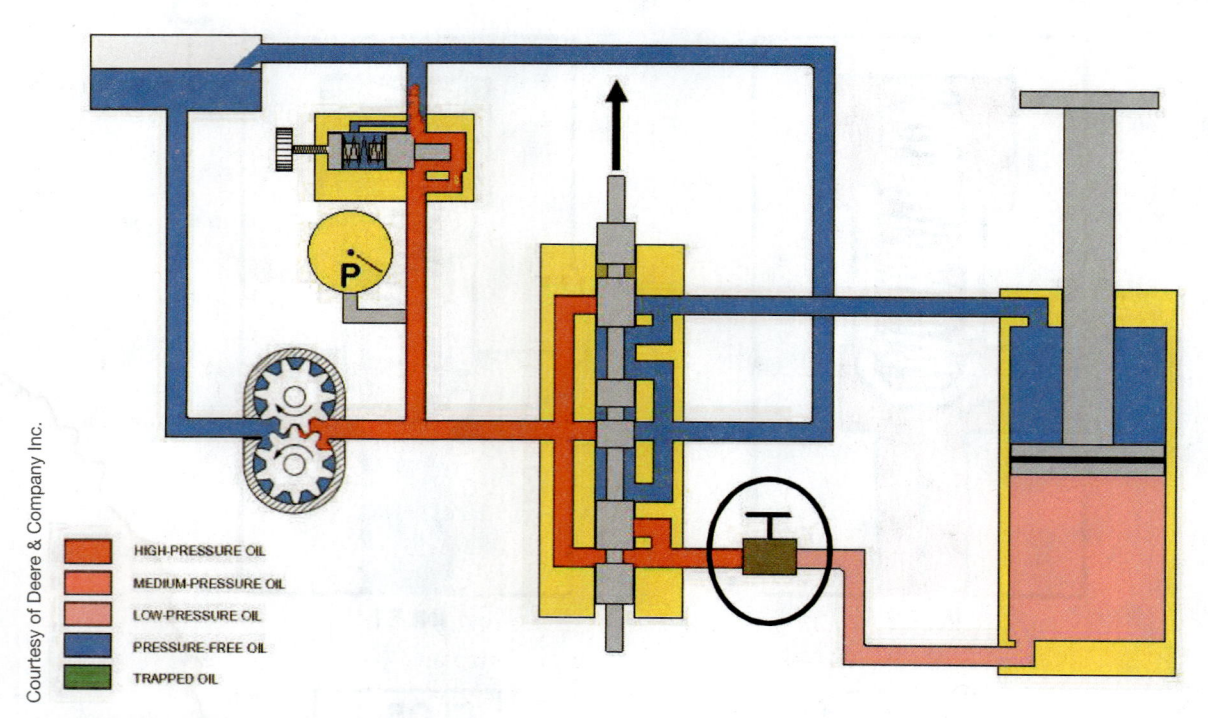

HIGH-PRESSURE OIL
MEDIUM-PRESSURE OIL
LOW-PRESSURE OIL
PRESSURE-FREE OIL
TRAPPED OIL

FIGURE 4–37 A variable orifice.

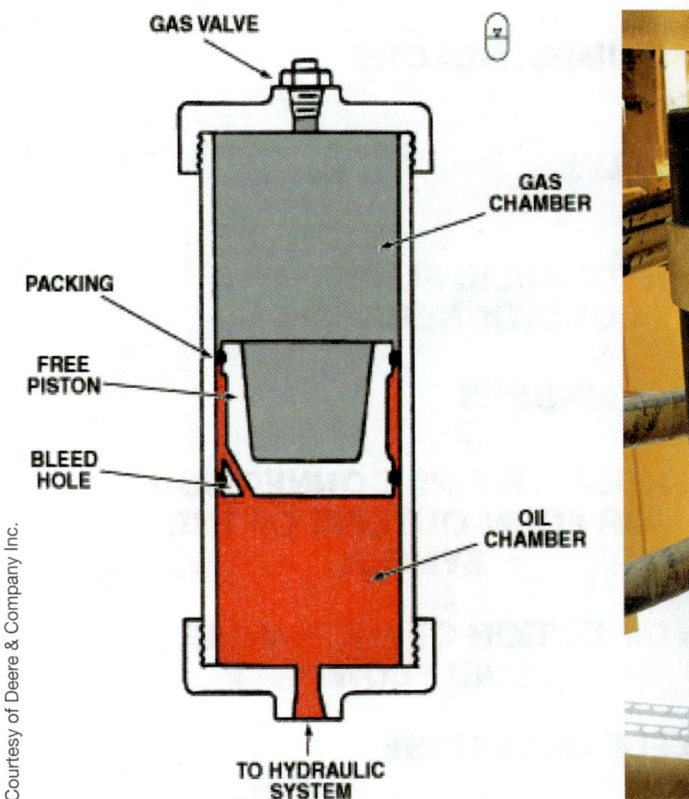

FIGURE 4–38 A piston-type accumulator.

Accumulators can be used for storing low pressure fluid, for a brake system, for a ride control system, for a suspension system, and for many others. ● **SEE FIGURE 4–38** for a piston-type accumulator.

HYDRAULIC SCHEMATIC SYMBOLS

To understand how the hydraulic components work together to make a hydraulic system, you will need to understand a hydraulic schematic. A schematic is a group of symbols that represent the various components of a system. There could be some variation with some symbols between different manufacturers, but they should be close enough that you can still understand what they represent.

Some shapes are the following:

1. Circles represent rotary motion (pumps or motors).
2. Squares represent pressure control valves (pressure relief, reducing, etc.).
3. Rectangles represent DCVs.
4. Diamonds represent fluid conditioners.
5. Pill shape represents an accumulator. See Figure 4-39 for a few examples of hydraulic symbols.

Progress Check

11. When a hydraulic pump is termed *fixed displacement*, it:
    a. will always be used with a closed center directional control valve
    b. can vary its gpm at a fixed rpm
    c. has a fixed rpm
    d. has a fixed output for each revolution
12. A hydraulic reservoir will have return oil slowed down so the fluid is cooled and:
    a. can let large contamination settle out
    b. can have more time to be pressurized
    c. will have more time to depressurize
    d. allow more time for the fluid heater to keep it warm
13. This kind of pressure relief valve has a closer range between cracking and being fully open:
    a. pressure reducing
    b. pilot operated
    c. simple relief
    d. flow reducing

HYDRAULIC SYSTEMS—THE BASICS 127

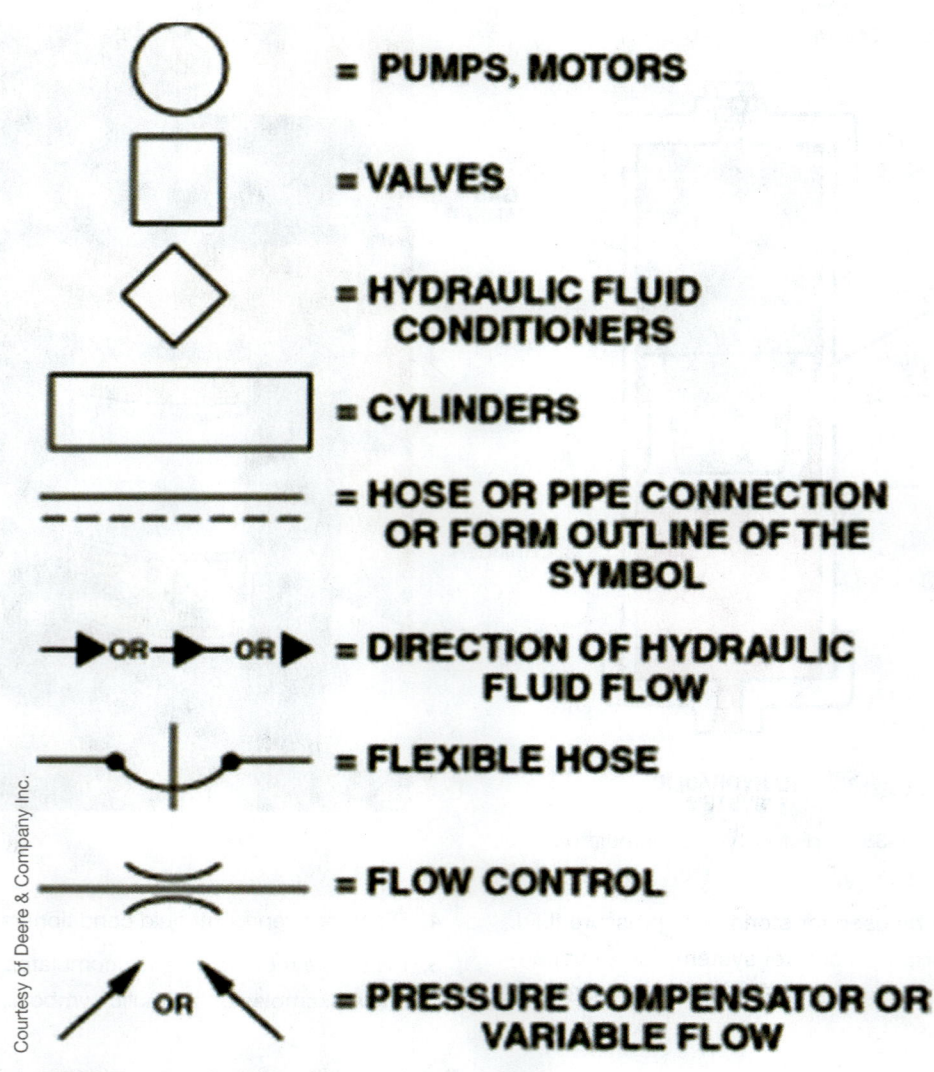

○	= PUMPS, MOTORS
□	= VALVES
◇	= HYDRAULIC FLUID CONDITIONERS
▭	= CYLINDERS
———	= HOSE OR PIPE CONNECTION OR FORM OUTLINE OF THE SYMBOL
▶ OR ▷ OR ▷	= DIRECTION OF HYDRAULIC FLUID FLOW
⌒	= FLEXIBLE HOSE
⌣	= FLOW CONTROL
↗ OR ↑	= PRESSURE COMPENSATOR OR VARIABLE FLOW

FIGURE 4–39 Common hydraulic schematic symbols.

14. A double-rod cylinder is considered to:
    a. be an unequal displacement cylinder
    b. have a faster retract than extend
    c. be an equal displacement cylinder
    d. be needed when two rods are needed to travel the same way

15. If it is a ¾ in. hose, this means it has:
    a. a dash size of $3/16$
    b. an ID dash size of –12
    c. an OD of ¾ in.
    d. a wall thickness of ¾ in.

OPEN CENTER HYDRAULIC SYSTEM

Now that we have identified the components of a hydraulic system, let's see how they work together in a typical two-circuit open center system. If a fixed displacement pump is used, it is usually matched with an open center type of DCV. This allows pump flow to travel through the DCV and return to tank if the pump is turning and none of the spool valves are moved.

● SEE FIGURE 4–40 for a pictorial schematic of a backhoe loader's front loader boom and bucket hydraulic system and ● FIGURE 4–41 for the same system in true schematic form with the bucket section's spool moved to send oil to the bucket cylinder.

If the valve spools are in neutral, the fluid will go through the cooler and filter and then return to tank. If the operator moves a spool to direct flow to an actuator, some of the pump flow is now diverted to the actuator and the return fluid is returned to tank through the DCV. Also included in the DCV's body are the following:

1. circuit relief valves with anti-cavitation function to protect the cylinders from external shocks

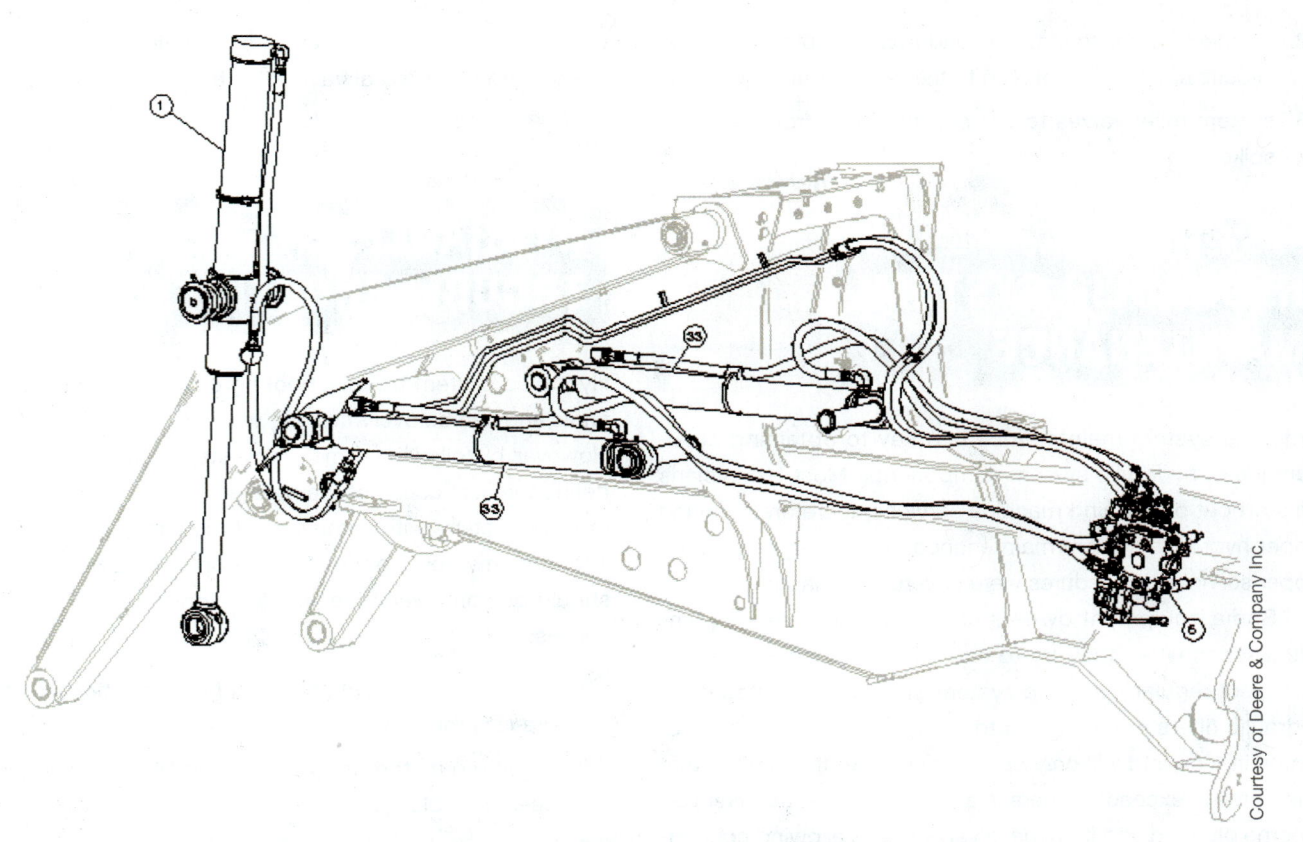

FIGURE 4–40 Pictorial schematic of a backhoe loader's front lift and tilt hydraulic system.

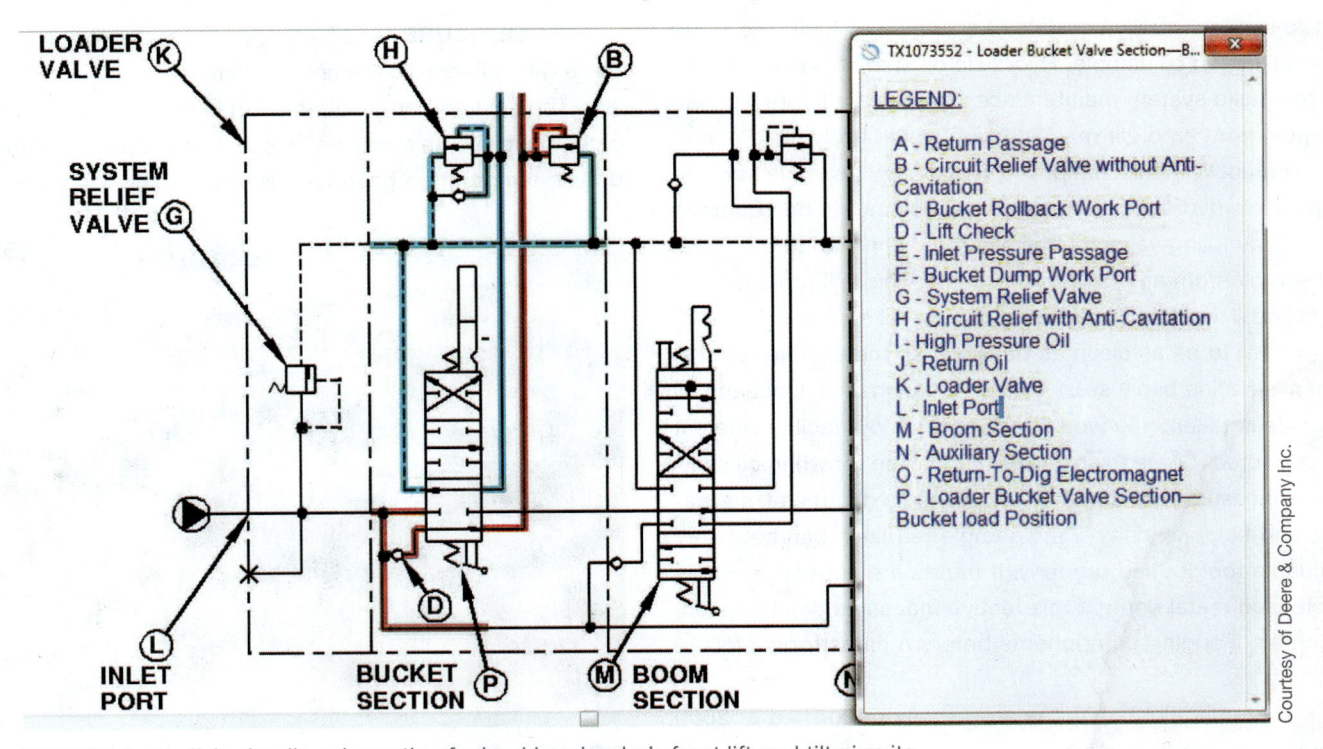

FIGURE 4–41 A hydraulic schematic of a backhoe loader's front lift and tilt circuits.

LEGEND:

A - Return Passage
B - Circuit Relief Valve without Anti-Cavitation
C - Bucket Rollback Work Port
D - Lift Check
E - Inlet Pressure Passage
F - Bucket Dump Work Port
G - System Relief Valve
H - Circuit Relief with Anti-Cavitation
I - High Pressure Oil
J - Return Oil
K - Loader Valve
L - Inlet Port
M - Boom Section
N - Auxiliary Section
O - Return-To-Dig Electromagnet
P - Loader Bucket Valve Section—Bucket load Position

HYDRAULIC SYSTEMS—THE BASICS 129

2. lift check valves to stop the load from dropping when that circuits spool is first moved to the raise position
3. system relief valves to protect the pump from pressure spikes

HYDRAULIC SYSTEM MAINTENANCE

Hydraulic system maintenance is a key to obtaining maximum life of hydraulic system components. Most technicians will start out performing machine services; therefore, learning proper hydraulic system maintenance now is very important. Proper servicing procedures were covered in Chapter 3.

Some equipment owners have had some very expensive lessons when they try to extend maintenance intervals or neglect regular hydraulic system maintenance altogether. Hydraulic filters are designed to hold a maximum amount of contamination and will only be effective up to that limit. If that time frame is exceeded, there is a possibility that the filter can become plugged and its bypass valve opens allowing contaminated oil to circulate through the system. Likewise, hydraulic fluid is designed to last a specific amount of machine hours and should be changed when recommended.

A typical hydraulic system maintenance interval will be performed at between 500 and 1000 hours or six months. This basic service usually consists of changing all hydraulic filters and taking an oil sample. Hydraulic oil sampling is a key part in hydraulic system maintenance. To obtain a hydraulic fluid sample from an older machine, you would likely need a vacuum pump. With this pump, you would draw a sample from the machine's hydraulic tank. Newer machines will be equipped with a sampling valve that allows the technician to obtain an oil sample from an operating machine. This will provide what is called a live sample. When obtaining oil samples, it is very important to be as clean as possible, so the sample result is not misleading because of added contamination. Once the oil sample is taken, you would then send it to a facility where it is measured. Some things that are checked for when oil sample is measured are contaminants such as dirt, metal, air, water, and fluid viscosity. Fluid sample readings can be a very valuable tool if used properly. If trends are noticed, such as increased metal content, preventive measures can be taken, such as changing components before a catastrophic failure occurs.

Hydraulic fluid changes are usually performed at 2000 hours or yearly. These maintenance intervals can vary greatly between different manufacturers, machine types, and operating conditions, so always refer to manufacturer's service information.

HYDRAULIC SYSTEM TROUBLESHOOTING

Hydraulic system troubleshooting can seem complicated at times and can be overwhelming when you are first starting out. However, by following some simple steps you should be able to define the root cause of the problem. A good troubleshooting procedure starts with gathering information. The first source of information would be the machine operator/owner, and there should be some very specific questions asked. These are as follows:

1. What is the problem (this must be clearly defined as a hydraulic problem)?
2. When did it start (is this a new problem or has it always been a problem)?
3. Under what conditions does it happen (a certain engine speed, operating temperature, load)?
4. Are there unusual noises, vibrations, or smells (what is the operator sensing)?
5. Has any work been done on the machine lately?

● **SEE FIGURE 4–42** for a field service call to repair a machine with a hydraulic system problem.

The next source of information is the machine itself. For older machines, this may be as simple as seeing what the hydraulic temperature gauge reads and visually inspecting the

FIGURE 4–42 Field service diagnostic call.

system for obvious problems such as leaks, kinked lines, bent, or scored rods. On the newest machines, you will likely be able to hook up your laptop and gather a great deal of stored information such as operator input, engine operating information, and hydraulic system operation values. This information could then be placed into a file that could be attached to a work order or service report.

If you aren't familiar with the hydraulic system, this would be a good time to do a little reading and get to know the system. You can't fix a problem if you don't know how the system is designed to work. In other words, you need to know what is normal and what is abnormal.

While gathering information, you need to verify the operator's complaint. Sometimes operator information can be misleading or just inaccurate, so you need to operate the machine yourself to make sure that the stated problem exists.

Complaints related to hydraulic systems usually fall into two categories. Either a hydraulic function is to slow or it doesn't have enough power. Remember slow hydraulic functions are related to lack of flow and weak hydraulic functions relate to low pressure.

Some other complaints could be hydraulic fluid overheating, unusual noises/vibrations, fluid leaks, or unresponsive functions.

Bucket open	3.0 seconds
Bucket closed	3.5 seconds
Stick open	3.0 seconds
Stick closed	4.5 seconds
Stick open	4.5 seconds
Boom up	3.5 seconds
Boom down	2.5 seconds

FIGURE 4–43 A typical machine cycle time chart.

THE REAL DEAL

Let's go through an excavator operator complaint of a slow bucket function. Remember the first step is to gather information. Let's assume we asked the operator the typical questions and his response was the bucket is slow opening. It started to slow down last week and has slowly got worse. It doesn't seem to matter if the machine is hot or cold and happens whether the bucket is full or empty. There are no unusual noises, vibrations, or smells, and nobody has worked on the machine for at least four months.

Any troubleshooting procedure should be started by checking basic system integrity or "simple stuff first." This would include checking hydraulic fluid level, looking for fluid leaks, checking proper machine lubrication, and checking proper engine operation. This would lead you into verifying the complaint by operating the machine.

The complaint is about speed (slow bucket), so you should already be thinking about some kind of flow problem. As this complaint involves one circuit, you should verify that all other circuits are working properly. To do this, you would operate all functions one at a time and then in combination with each other. If you're unsure about whether these functions are operating properly, you would refer to the machine manufacturer's service information. Within this information, operational checks called cycle times should be found. ● **SEE FIGURE 4–43** for a typical cycle time specification chart.

Cycle times relate to how fast a linear actuator will cycle from fully retracted to fully extended, or vice versa. Rotary actuators will also have cycle times specifications listed. To properly check the cycle time, certain conditions must be met. These would include a certain level for hydraulic fluid temperature, a certain engine rpm (usually high idle), and a specific position that the machine should be tested in. Any time a complaint is to be verified that relates to cycle times, the specified procedure must be adhered to.

If you performed the cycle time checks for this machine and found other functions to be slow as well as the bucket, you would now need to change strategies. However, let's assume this machine was tested and found to have only a slow bucket open function. The specification is 3.0 seconds and you verify it to be 4.8 seconds.

We'll also assume all basic checks were performed and found to be okay. Remember that troubleshooting is a process of elimination so by confirming that there is only a bucket function problem we have eliminated any other problems and can now focus on the bucket circuit. Remember that "simple stuff first" still applies. Technicians sometimes have a tendency to jump to conclusions. For example, if this particular model of machine has had a history of the bucket cylinder piston coming loose and causing a slow bucket open condition, you may conclude this is the problem and proceed to remove the bucket cylinder. By playing the odds, you may be right, but if that cylinder is removed, taken to a shop, disassembled, and found to be okay; now what? This could be a very expensive mistake. By going against the odds and following a logical troubleshooting procedure, you would check out the operator controls first. We'll assume this is an older machine with mechanical linkage

HYDRAULIC SYSTEMS—THE BASICS 131

from the levers to control valves. Upon further inspection, you find the nut that holds the ball joint for the bucket open linkage has backed off. After tightening this nut and then operating the machine, you confirm the cycle time for bucket open is now within specification. The backed off nut caused the DCV to not move as far as it should have, and therefore, there wasn't enough fluid flow going to the bucket cylinder to open it, which resulted in a slow bucket problem and complaint. You won't always find a simple fix like this for troubleshooting hydraulic problems, but you need to look for the simple fixes first.

HYDRAULIC SYSTEM REPAIRS

Hydraulic system repairs can vary from simple adjustments to extensive component removal and disassembly. If the machine has a catastrophic hydraulic component failure, there is a possibility that every part of the hydraulic system would need to be removed, cleaned, or replaced.

The majority of repairs performed on hydraulic systems are related to leaks. Leaks can be caused by defective seals, defective hoses, defective fittings, defective tubes, and other defective components such as pumps, valves, and actuators. Remembering safety first with hydraulic systems, you must always make sure that any pressure contained in the component you are repairing is relieved. This could be a very simple procedure such as moving all control levers through all positions or a very complicated procedure where accumulators need to be discharged and gauges installed to confirm pressure relief. Do not take any shortcuts with this part of the repair.

You also need to consider what part of the machine could move while performing hydraulic repairs. If hydraulic pressure is sealed in a circuit, it could be holding a loader boom in the air; for example, if you were to just loosen the hose for that part of the circuit and were in the wrong position when doing so, the result could be fatal. Once you have verified the system is safe to work on, you can now proceed with the repair.

Let's assume you need to change a hose going to a cylinder and have verified there is no pressure present in the hose. Wearing the appropriate PPE, make sure that you have enough capacity in your drain buckets to capture any oil that will leak out. Once the hose is removed, you should cap the open fittings with plastic caps. This will prevent any system contamination. If you need to leave the machine, you must tag the machine as not operational and perform a lock out procedure according to your company's policy. The ideal replacement part would be the exact part number purchased from the machine manufacturer's dealer. If this is not possible, have the hose matched to original specification at a reputable hose supplier. Anything short of this, you will be asking for problems. To install a new hose properly, new seals should be used and proper torque specifications adhered to for installing the fasteners that were removed. After the hose is installed, the fluid level must be checked and corrected.

Some machines must have their actuators in a certain position to confirm proper level and may require the engine stopped or running. The machine should then be started by you, and then with the engine running slowly, check to confirm proper operation. If it works okay, run the machine normally for a few minutes while putting maximum pressure on the new hose and then let the operator put the machine back to work. These extra few minutes of confirming your repair can save a lot of potential problems.

Using the example that we just went through in troubleshooting a hydraulic system, let's assume that the slow bucket open complaint was also related to a slow boom problem and the root cause was suspected to be a pump failure. This troubleshooting procedure was performed by a field service technician for your company and you work in his shop. ● **SEE FIGURE 4–44** for an excavator working. You are issued a work order to replace the appropriate pump. Let's assume you have talked to the field service technician, and he removed the case drain filter from that pump and found it to be loaded with metal particles. If a machine is suspected of having a failing hydraulic pump, it should not be run any longer than absolutely necessary. After positioning the machine to most easily replace the pump, you refer to the manufacturer's recommendations for this procedure.

After replacing the pump, you now need to perform complete hydraulic system maintenance on this machine to reduce the chance of a repeat failure from leftover contamination. Most manufacturers will also recommend a kidney-loop procedure. This is similar to performing a dialysis procedure on the

FIGURE 4–44 An excavator working.

machine, whereby an external filter is used to filter the machine hydraulic fluid. This external filter is usually called a filter cart. Then fluid sampling should be performed, and the machine should not be put back to work until the fluid is completely clean according to manufacturer's specifications. An accurate service report should now be completed and filed to be kept with this machine's records.

NICE TO KNOW

At the start of this chapter, we talked about harnessing the power of a flowing river to power mechanical devices to cut wood. We will now look at a modern hydraulic device used to split wood. The hydraulic log splitter is a machine used to take a cut length (usually 16 in.) of tree and split it into smaller pieces so it can be easily handled. The split wood is then usually placed into a wood stove where it burns to create heat energy to warm the inside of a house or shop. There are many variations of log splitters, but we will take a look at the most common type that uses a gas-powered internal combustion engine to rotate a hydraulic pump and a hydraulic cylinder to push a steel wedge through the piece of log. ● **SEE FIGURE 4–45** for a typical log splitter. You may be wondering what a log splitter has to do with heavy equipment. A log splitter is a simple single circuit fluid power device that is easy to understand, and you can easily see its components. The principles that apply to its hydraulic system apply to any hydraulic system. A prime mover drives a pump that creates fluid power and that fluid power is used to perform work.

The prime mover is a gasoline-powered internal combustion engine. It drives the hydraulic pump, and in this case, the type of pump used is an external gear pump. The only purpose of this pump is to create oil flow. This is a very important point to be clear on; remember that any pump used in any hydraulic system is only used to create flow. If you're thinking pumps create pressure, stop and reread the last sentence. Pressure is created by the resistance to oil flow. A simple way to think of this is by imagining pumping up a flat tire on your bicycle with a hand pump. It's very easy at the start because there is no resistance to the pump's flow but gets harder as the pressure inside the tire gets higher. The pump now has to overcome the added resistance that the stiffening tire is producing. The tire is creating a resistance to the pump's flow.

Looking at the picture of the log splitter, you can see the pump used is a simple external gear-type pump. You can tell this because of the oval shape of the housing and by imagining the two external gear shafts inside of the housing. As we learned before, this type of pump is a positive displacement pump, meaning as soon as that gasoline engine starts up, the pump rotates and pushes oil.

The DCV is an open center type. This control valve would most likely have a pressure relief valve built into it. The other common component that you see is the hydraulic filter. Two components you don't see are the linear actuator or cylinder, and the hydraulic reservoir. The cylinder is attached to the wedge that travels back and forth, and the hydraulic reservoir is mounted between the two tires.

Now, let's do a little math and see if we can figure out how much force this log splitter will produce.

Let's assume the piston inside the cylinder has a 3 in. diameter and the pressure relief valve is set at 2000 psi. What is the maximum force that can be produced by the cylinder when the rod is extended?

First, we need to calculate the piston area to determine the amount of surface area the pump flow is acting on.

$$A = d^2 \times .7854$$
$$= 3 \text{ in.} \times 3 \text{ in.} \times .7854$$
$$= 7 \text{ in.}^2 \text{ (rounded from 7.0686 in.}^2)$$

Now we can find the force output.

$$F = P \times A$$
$$= 2000 \text{ psi} \times 7 \text{ in.}^2$$
$$= 14{,}000 \text{ lb of force on extend}$$

What would be an easy way to increase this force?

We could either increase the size of the piston (not so easy) or increase the relief valve setting. Let's say we increased it to 2500 psi. Let's make a new calculation and see how much force increase we get.

$$F = 2500 \text{ psi} \times 7 \text{ in.}^2$$
$$= 17{,}500 \text{ lb of force on the extend}$$

FIGURE 4–45 A typical log splitter.

HYDRAULIC SYSTEMS—THE BASICS 133

Increasing the relief valve setting has a direct effect on the output force. When testing a piece of heavy equipment for a low hydraulic power complaint, you can see why it might be important to confirm the system relief valve setting. Now let's see how much force is created when the cylinder is retracted. Do you think there'll be a big difference?

We now need to consider the rod diameter of the cylinder. Let's assume the rod is 1 in. in diameter. Because this rod is getting retracted into the cylinder barrel, it's taking up volume and also reducing the effective area on the piston. Let's calculate the area of the rod.

$$\text{Rod area} = d^2 \times .7854$$
$$= 1 \text{ in.} \times 1 \text{ in.} \times .7854$$
$$= .7854 \text{ in.}^2$$
$$\text{Piston area} - \text{Rod area} = \text{Effective area}$$
$$7 \text{ in.}^2 - .7854 \text{ in.}^2 = 6.2 \text{ in.}^2$$

Now we can calculate the force on retract.

$$\text{Retract force} = P \times A$$
$$= 2500 \text{ psi} \times 6.2 \text{ in.}^2$$
$$= 15{,}500 \text{ lb}$$

So you can see there is a lot less maximum force created by this cylinder when it is retracted.

This is the case for any double-acting cylinder because of the difference in effective surface area of the rod side of the piston. The larger the rod in relation to the piston, the less force that can be created on the rod retract cycle.

Now that we know how strong the log splitter is, let's think about how fast it will be. First, we need to know how much oil is required to move the cylinder full stroke both ways. Then we can figure out the pump volume needed to give a certain cycle time. To do this, we need to know the stroke length of the cylinder. We can see by ● **FIGURE 4–46**, the

FIGURE 4–46 Log splitter cylinder.

stoke length is 30 in. To move the piston out, we can calculate the volume by:

$$\text{Rod extend volume} = \text{Piston area} \times \text{Stroke length}$$
$$= 7 \text{ in.}^2 \times 30 \text{ in.}$$
$$= 210 \text{ in.}^3$$
$$\text{Or in US gallons} = 210 \text{ in.}^3 \div 231 \text{ in.}^3/\text{gal}$$
$$= 0.9 \text{ US gal}$$
$$\text{Rod retract volume} = \text{Piston effective area} \times \text{Stroke length}$$
$$= 6.2 \text{ in.}^2 \times 30 \text{ in.}$$
$$= 192 \text{ in.}^3$$
$$\text{Or in US gallons} = 192 \text{ in.}^3 \div 231 \text{ in.}^3/\text{gal}$$
$$= .83 \text{ US gal}$$

Because the rod is a lot smaller in diameter than the piston and therefore doesn't take up much room in the cylinder, the amount of fluid to move the rod in is very close to what it takes to extend the rod.

To know how fast the piston rod will travel, we now need to consider the pump flow rating. Let's assume a rating of 10 gpm at an engine speed of 3000 rpm. We will be calculating the cylinder's cycle times.

C.T. (cycle time) rod extend

$$= \frac{\text{Cylinder volume in gallons}}{\text{Pump flow in gallons per minute}} \times 60$$

(multiplying by 60 gives the answer in seconds)

$$\text{C.T. rod extend} = \frac{0.9 \text{ gal}}{10 \text{ gpm}} \times 60$$

C.T. rod extend = 5.4 seconds

$$\text{C.T. rod retract} = \frac{\text{Cylinder volume in gallons}}{\text{Pump flow in gallons per minute}} \times 60$$

$$\text{C.T. rod retract} = \frac{0.83 \text{ gal}}{10 \text{ gpm}} \times 60$$

C.T. rod retract = 4.98 seconds

If you were to build a log splitter, you would want to know how much horsepower would be required to drive the pump. There is an easy way to calculate this. The most volume the pump would put out is 10 gpm and the highest system pressure would be 2500 psi. Therefore, the horsepower needed to keep the pump turning at maximum flow and pressure would be:

$$\text{Hydraulic horsepower} = \text{psi} \times \text{gpm} \times .0007$$
$$= 17.5 \text{ hp}$$

Progress Check

16. Cycle times are measured as part of a diagnostic check to see if:
    a. the circuit has the required flow through it
    b. all the functions work at the same speed

c. the circuits will have enough power
d. the controls are backward

17. The first part of a good troubleshooting procedure is:
    a. checking main relief pressure
    b. checking cycle times
    c. gathering information
    d. checking oil temperature

18. A port relief valve will:
    a. protect the circuit from cavitation
    b. provide extra cooling for that part of the circuit
    c. protect one part of the circuit from external shock loads
    d. allow return fluid to go back into the supply

19. A filter cart will:
    a. provide an external filter after a system is contaminated
    b. make moving filters around easier
    c. be attached to the machine when an extra filter is needed
    d. crush old filters for recycling them

20. Hydraulic fluid sampling:
    a. should only be done when a component has failed
    b. should be part of a good preventive maintenance program
    c. is most accurate when taken from the tank drain
    d. is not very cost effective

SHOP ACTIVITY

- Go to a machine and choose one hydraulic circuit. Make a schematic to represent the circuit with the common symbols from Figure 4–37.
- Go to a machine and choose an actuator. Closely estimate the cylinder's dimensions (piston diameter, rod diameter, piston stroke length). Using a relief pressure of 3000 psi, calculate maximum possible force output for extend and retract and cycle times for extend and retract.
- Go to a machine and describe the procedure you would use with a tool and material list to complete a total hydraulic system service for this machine.
- Disassemble a linear actuator and write a detailed report on its dimensions and condition.
- Disassemble a gear pump and write a detailed report on its dimensions and condition.

SUMMARY

1. Safety concerns related to hydraulic systems include burns, high-pressure fluid injection injuries, sudden pressure release, and slips and falls.
2. A hydraulic system can be defined as a system that transfers energy through a fluid medium.
3. Hydraulic systems have several advantages over mechanical systems such as greater flexibility for locating components, components stay cleaner as they are running in oil, components are protected from damage by relief valves, speed and direction of actuators is easily controlled, and actuating more than one actuator is easy.
4. Hydrodynamic fluid power systems use the energy of high volumes of moving fluid to transfer energy while hydrostatic systems mostly rely on fluid pressure to transfer energy.
5. Fluids are almost incompressible (1% per 1000 psi applied).
6. Some math skills are needed to help understand fluid power systems: area calculation of circles ($A = d^2 \times .7854$ or $A = \pi r^2$) and volume calculations ($V = A \times L$).
7. Pascal's law states that "pressure created by an external force acting on a fluid inside a sealed container applies the same amount of pressure to any and all surfaces equally and at right angles to the inside of that container."
8. The formula $F = P \times A$ is the foundation for understanding energy transfer in a hydraulic system.
9. Force applied to a liquid is transferred through the liquid.
10. Force is a linear push or pull and pressure is the result of force applied to an area.
11. **Atmospheric** pressure can be measured in psi and is the weight of a 1 in.^{2} column of air. At sea level and an ambient air temperature of 70°F, atmospheric pressure is 14.7 psi.
12. A pressure gauge that reads psia does not account for atmospheric pressure while a gauge that indicates psig combines atmospheric pressure with the pressure it is reading.
13. Fluid flow in a hydraulic system can create work if directed to actuators (cylinders or motors) and starts with the system's pump. It is measured in units of volume per unit of time: gallons per minute or liters per minute.
14. Pressure in a hydraulic system is directly proportional to the fluid's resistance to flow and is mostly created by the

load on the actuator. It is measured in psi (standard units), MPa (megapascals), or kg/cm² (metric).
15. All hydraulic systems must have the following minimum components: prime mover, pump, reservoir (tank), relief valve, directional control valve, filter, and fluid conductors (hoses and tubes).
16. The prime mover for heavy equipment machines is a diesel engine although some backup systems will use an electric motor.
17. Pumps create flow, and fluid power systems use positive displacement pumps. Simple systems will use external gear-type pumps and are fixed displacement. To change the flow rate for this type of pump, the rpm needs to change.
18. Pump displacement is measured in cubic centimeters per revolution or cubic inches per revolution.
19. External gear-type pumps have two shafts. One is driven by the prime mover and the second shaft is driven by the first.
20. Reservoirs for hydraulic systems hold a supply of hydraulic fluid and should allow returning oil from the system to slow down and let large contamination settle out before it heads to the pump inlet.
21. Pressure control valves are normally closed spool-type valves that are held seated with springs. They will limit system or circuit pressures and can be adjusted with screws or shims.
22. DCVs are controlled by the operator to direct pump flow to an actuator and allow the retuning oil from an actuator to return to the reservoir. Most machine systems will have more than one circuit and DCVs will have multiple spools. They can have a one-piece housing (monoblock) or multi-section valves that are stacked together.
23. Actuators can create linear movement (cylinders) or rotary movement (motors). They receive oil from the DCV and oil that leaves the actuators returns to the tank through the DCV.
24. Cylinders can be single acting, which are not common, or double acting. A double-acting cylinder has a piston with a rod attached inside a barrel. Seals separate oil on both sides of the piston and seal the rod to the cap of the barrel.
25. Double-rod cylinders have a rod on both sides of the piston.
26. Motors have oil flow through them and the result is shaft that turns. Its speed is determined by the flow rate of oil to the motor.
27. Hoses can be sized by dash size that relates to the inside diameter of the hose. One dash size equals $1/16$ in.
28. Seals are used when hoses and tubes join together or components are connected together.
29. Hydraulic fluids are usually mineral oil based but can be biodegradable organic based fluid if needed. Proper fluid viscosity is critical to ensuring proper system operation.
30. Hydraulic filters are needed to remove contamination from system fluid. Sometimes screens are placed near pump inlets but must not restrict flow. The most common place for filters is the return to tank line.
31. Many hydraulic systems will need coolers to keep hydraulic fluid temperatures from exceeding safe temperatures (usually 150°F).
32. Check valves and orifices are commonly used in systems to control fluid flow.
33. Accumulators can be used to store hydraulic energy and use nitrogen gas on one side of a piston or bladder to compress.
34. Hydraulic schematics are like a road map for hydraulic systems and use symbols to represent different components.
35. Open center hydraulic systems use DCVs that allow pump flow to flow through them and return to tank.
36. Hydraulic system maintenance recommendations should be followed closely and include regular fluid and filter changes along with fluid sampling.
37. Hydraulic system troubleshooting procedures should start with simple checks such as fluid levels and temperatures. Flow and pressure principles should be kept in mind when looking for causes of hydraulic system problems.
38. Common checks for system performance are cycle time checks and pressure and flow measurements.

chapter 5
DIRECT CURRENT ELECTRICAL SYSTEMS

LEARNING OBJECTIVES

After reading this chapter, the student should be able to:

1. Describe what an electron is.
2. Explain what makes an electron move.
3. Describe what happens when an electron moves.
4. Explain what electrical pressure is.
5. Explain what current flow is.
6. Explain what electrical resistance is.
7. Describe what an electrical load is.
8. Describe basic electromagnetic principles.
9. Identify the difference between a series and a parallel circuit.
10. Explain what Ohm's law is.
11. Perform circuit calculations using Ohm's law.
12. Identify the components that make a simple series circuit.
13. Explain and identify common circuit faults.
14. Describe troubleshooting procedures.
15. Recognize ISO electrical symbols.
16. Describe what a machine wiring diagram or schematic is.
17. Explain the physical qualities of insulators, conductors, and semiconductors.
18. Explain the properties of magnetism and electromagnetism.
19. Define volts, current, resistance, and power and the units of measure for each.
20. Explain how to use a digital multimeter.
21. Identify and define properties of the three basic electrical circuit types.
22. Identify and explain an open, shorted, grounded, or high-resistance circuit.
23. Using Ohm's law, calculate for any of its variables when two are known.
24. Apply Ohm's law to a circuit to calculate voltage, current, and resistance.
25. Identify hazards encountered with lead-acid storage batteries.
26. Identify different types of batteries.
27. Explain battery construction, sizing, and capacity.
28. List safety precautions and procedures for boosting batteries.
29. Explain connections in multiple battery circuits.
30. Explain open circuit voltage and resistance measurements using a voltmeter.
31. Explain battery maintenance and testing methods.
32. Describe how to load test a battery.

KEY TERMS

ampere 139
connector 160
electrolyte 155
electron 140
ground 170
load 163
ohm 145
open 149
relay 162
resistance 144
short 170
switch 160
voltage 143

INTRODUCTION

SAFETY FIRST Although most machine electrical systems use relatively low voltage (12V or 24V) as opposed to 120V used for household systems, there are still some serious safety issues you need to keep in mind when servicing or repairing basic electrical systems. Some machines, such as the one shown in ● **FIGURE 5–1**, use high voltage for drive motors, and these systems will need several hundred volts to function. Typically, the high-voltage systems for driving these machines will be serviced by industrial electricians and an HDET will usually not be expected to work on these systems without taking training specific to high voltage. You should be familiar with which components and wiring are related to the high-voltage systems on these machines if you do work on them.

There are also dangers that could be present with newer fuel injection systems that use high voltage. These are usually clearly labeled as high voltage, but in case the labels are dirty or missing, you should be extra cautious when working around electronic fuel injectors and their wiring.

Many machine safety features are electrically controlled and can be integrated into any or all other systems on a machine. Features such as not allowing an engine to crank or run under certain conditions or not allowing the hydraulic system to operate unless certain conditions are met could be disabled by mistake if you are working on the related electrical system. Therefore, it is mandatory that you understand these safety features and test that they work properly before handing the machine back to the operator.

You may also find it necessary to perform various tests on a running machine and may have to put some of your body parts near moving belts and pulleys. Use extreme caution when working near any moving parts and keep all long hair and loose clothing tucked away. Jewelry should be removed when working near electrical systems as the metal can cause a short to occur and a severe burn could result. Pay attention to all safety warnings. ● **SEE FIGURE 5–2** for a typical warning you may see.

Batteries are very dangerous components of the electrical system. When most batteries are charging, dangerous hydrogen gas is produced and can be ignited very easily. If you have ever witnessed a battery explode, you will gain a new respect for the potential energy that is stored in a machine's battery. I personally was very close to one that blew up in a coworker's face and since have been very leery when near a battery. It was a very scary experience that I have never forgotten and luckily my coworker was wearing safety glasses that saved him from permanent eye damage. ● **FIGURE 5–3** shows the warning related to battery safety.

To see the results of a battery explosion, go to http://www.rayvaughan.com/battery_safety.htm.

FIGURE 5–2 Safety warning for high voltage.

FIGURE 5–1 Wheel loader that uses high-voltage electric drive system.

FIGURE 5–3 Safety warning for batteries.

INTRODUCTION TO BASIC DIRECT CURRENT ELECTRICAL SYSTEMS

Today's heavy equipment machines can have very complex direct current (DC) electrical systems. Even small machines like skid steer loaders, as seen in ● **FIGURE 5–4**, can have a number of electronic control modules (ECMs) integrated into their electrical system, and some machines will have more than 10 ECMs interconnected. A typical machine from 30 to 40 years ago, like the one in ● **FIGURE 5–5**, would have a complete electrical schematic for the machine's electrical system that would fit on this single page. It would just require a starting system to start the engine, charging system to charge the battery, and maybe some simple electrical circuits such as lights, wipers, and a heater fan if it had a cab. Machine manufacturers

FIGURE 5–4 Skid steer loader.

FIGURE 5–5 Old bulldozer with a simple electrical system.

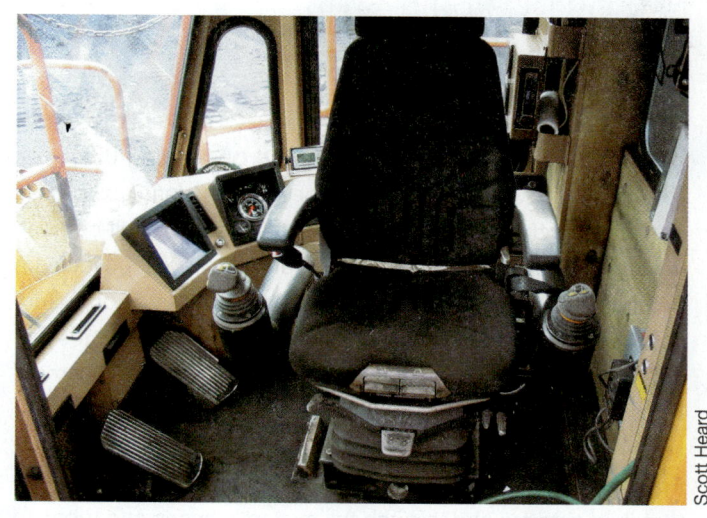

FIGURE 5–6 A wheel loader that is completely "fly by wire."

have made electrical systems more complex as a result of several factors. Operator safety and comfort, machine reliability, fuel efficiency, reduced emissions, system data storage, and easy troubleshooting are some of the reasons why an electrical schematic for today's machines can take up many pages. ● **SEE FIGURE 5–6** for a total fly-by-wire operator's station. Although the complexity of electrical systems has increased immensely in the last 10 years, the need for a technician to know basic DC electrical fundamentals has not changed.

It's a pretty safe bet that tomorrow's machines will have a more complex electrical system since advanced features such as blind spot cameras, GPS telemetry, tire pressure monitoring systems, and laser leveling systems are becoming fairly common now. In the future, you will likely see more electric hybrid propulsion system used (see Chapter 13) and more wireless communication between different parts of the machine. Use of fiber optics is also being tested on machines.

This chapter will help you understand DC electrical circuit principles, what makes DC circuits work, and how to diagnose and repair them.

It is imperative the principles and practical testing methods of voltage (electrical pressure), **amperage** (electron flow), and resistance be learned and mastered upon the completion of this chapter for you to be a confident and competent technician.

> 🚗 **WEB ACTIVITY**
>
> Search the Internet to find a machine that uses an electric hybrid system or see the link to Komatsu (http://www.komatsuamerica.com/equipment/excavators/25001-70000lbs/hb215lc-1) for an excavator with an electrical hybrid system.

NEED TO KNOW

This chapter is placed right after the basic hydraulic systems chapter for a reason. If you happen to be reading this book in sequence, you would have learned about a basic hydraulic system, and the principles that apply to it should still be fresh in your mind. If you haven't read Chapter 4 recently, it would be helpful if you read through the "Need to Know" section before proceeding with this chapter. There are many similarities between hydraulic and electrical systems. The benefit of learning a hydraulic system first is that you can visualize what's going on inside the system by seeing the result of hydraulic fluid moving. For instance, you know there is a hydraulic tank full of fluid and that fluid gets pushed through the system's hoses and tubes to transfer energy and perform work. You can see this work being done when a cylinder moves in or out, and if a cylinder is attached to a bucket, then the bucket moves. A similar kind of action also occurs in the electrical system, but the big problem is, it's hard to visualize anything moving through solid wires.

There are already certain things you likely know about a machine's electrical system. You probably know the machine's battery is a storage device with a huge amount of potential energy. Can you imagine trying to crank a large displacement diesel engine over by hand on a cold winter morning? It's very hard to do by hand even slowly when the engine is warm; now imagine how much energy the battery must store to turn the engine fast enough for it to start (usually around 200 rpm). So you can think of a battery as a tank full of electrons waiting to go to work when the operator completes a circuit for them to travel through. You may also know the job of the machine's alternator is to restore any lost potential energy in the battery just like a hydraulic pump charging a hydraulic accumulator after its pressure drops.

You can see and hear electrical devices such as the wiper motor moving the windshield wipers, the backup alarm beeping when the machine reverses, and lights flashing on the roof of a machine. These are all things you know happen, but you may have a hard time imagining how or why they happen and what to do when they stop working. After reading this chapter, you should have a clear understanding of what makes these things work and how to fix them when they don't work.

When trying to understand electricity, there are certain things that you must assume happen in the electrical system because you can't see them, like the fact that there are extremely small particles moving at extremely high speed inside the wires and components that make up a machine's electrical system. Instead of hydraulic fluids flowing through hoses and tubes, you now have to imagine **electrons** flowing through solid copper wires and transferring energy. If you can make this leap and assume this is what is happening, then understanding electricity will be much easier. ● SEE TABLE 5–1 for a comparison of electrical and hydraulic components.

ELECTRICAL COMPONENTS	HYDRAULIC COMPONENTS
Battery	Accumulator
Alternator	Pump
Wires	Hoses
Fuse	Relief valve
Motor	Motor
Switch	Directional control valve

TABLE 5-1

A comparison of electrical and hydraulic components.

Electricity can be created in several different ways, but for now we will consider two sources of electricity for a machine's electrical system. The first is the chemical device called the battery that supplies electrical pressure or voltage for starting the machine's engine and when there are high electrical system loads. Batteries will be discussed in depth later on. The other source is the machine's alternator that creates electron flow through magnetic induction to power the machine's electrical circuits and to charge up the battery (electron storage tank). Alternators and charging system operation will be discussed in depth in Chapter 8. Alternators and batteries are necessary to get electrons moving, which in turn energizes electrical circuits.

WHAT IS AN ELECTRON?

An **electron** is a very small part of a very small particle called an atom. Atoms are the building blocks for all matter in the universe. The diameter of an atom is so small that it would require some 200 million of them side by side to form a centimeter-long line, which means most atoms are .1–.5 nanometers (1 billionth of a meter) in diameter.

There are different theories as to how atoms are structured. As an HDET, you don't need to be overly concerned with atomic theories, but you should try to understand what electrons are and what makes them move.

The easiest way to think of an atom is to imagine it is similar in structure to our solar system where there is a center mass like our sun and particles rotating around it like the planets. These particles remain in a balanced orbit because of their natural electrical charges. The center mass is the nucleus, which contains two other very small particles called the neutron and

140   CHAPTER 5

proton. The electrons are much smaller than the atom itself and scientists estimate them to be 200,000 times smaller than the nucleus.

Protons have a positive charge and neutrons have no charge. The particles orbiting around the nucleus are called electrons and they have a negative charge. The electrons are arranged in rings around the nucleus that are also called valence shells. Because the charges of the proton and electron are opposite, they attract each other, and this keeps the atom balanced. All atoms have an equal number of protons and electrons. Remembering that the electron has a negative charge is a key to understanding how electricity works.

You are probably wondering how these electrons get moved if an atom is balanced and stable. Different atoms have different amounts of protons and electrons, and as the number of electrons in each new outer ring increases, there is a limit to the maximum number of electrons in these rings. Once a new outer ring is started, there will only be one electron in it. This single electron is called a free electron, and when this free electron gets moved to another atom by an opposite or positive charge, the result is electric current flowing. Any atom with one electron in its outer shell makes an ideal conductor of electricity because this single electron is easily moved to another atom that has a free electron. Some elements with one electron in their outer ring are gold, silver, and copper.
● **SEE FIGURE 5–7** to understand the rings and free electron of a copper atom. The free electron in copper's outer ring makes it an ideal conductor and it's much more economical than gold and silver. Not only is the size of an electron hard to comprehend but the speed at which it moves is also. Electrons will move at the speed of light!

You could think of the free electrons as small drops of hydraulic fluid that could easily be moved to transfer energy if they were influenced by another force and a conductor is the hollow part of the hose they travel through.

Elements with the maximum number of electrons in their outer shell are very stable and are called insulators. Insulators do not conduct or allow electricity to flow because their outer shell is full and the electrons are hard to move. ● **SEE FIGURE 5–8** for an atom with a full outer shell. Materials that have atoms with full outer shells such as rubber, plastic, and glass are stable, electrically speaking. This makes them good insulators, and this is why they are used for covering the copper core of wires.

Some other atoms have four electrons in their outer shell and are called semiconductors because they are neither good conductors nor good insulators. Semiconductors can be swayed to be either a conductor or an insulator and will be discussed in Chapter 9.

CONVENTIONAL OR ELECTRON THEORY?
Just as there are different theories as to how electrons, protons, and neutrons combine to make up atoms, there are different theories as to which direction electrons flow in a DC circuit. In reality, electrons flow out of the negative battery post because they are attracted to the unlike charge at the positive post. This is the electron theory. The theory that you need to remember is the conventional theory that states electrons flow from the positive post out through the circuit and back to the negative post. Because they travel so fast, it doesn't really matter which way they flow; however, the key is to remember electrons always flow from a higher to a lower pressure because of an electromotive force. Once again this compares to a hydraulic system in that hydraulic fluid always flows from a higher to a lower pressure. When discussing electrical systems or looking at machine electrical schematics, electron flow is considered to be conventional or positive to negative. ● **SEE FIGURE 5–9** for conventional current flow.

WHAT MAKES ELECTRONS MOVE?
If a machine operator wants to turn the lights on in a machine, the operator reaches for the light switch on the control panel and pushes the

FIGURE 5–7 A copper atom.

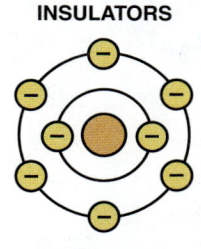

FIGURE 5–8 An atom with full valence shell.

DIRECT CURRENT ELECTRICAL SYSTEMS 141

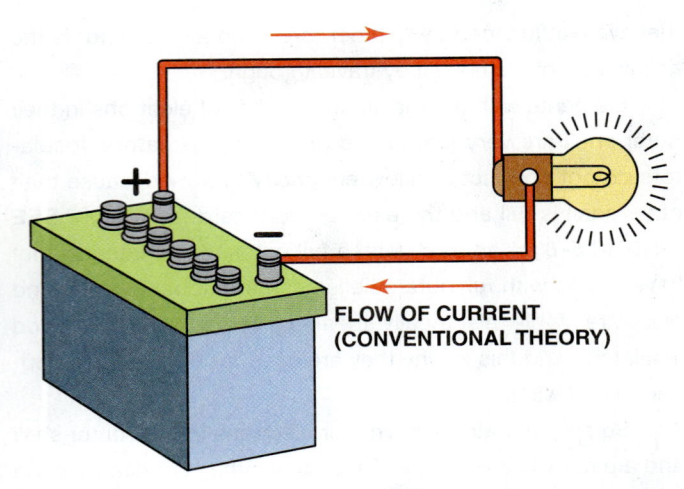

FIGURE 5–9 A conventional current flow.

button or turns the switch on. In a simple DC electrical circuit, what the operator has just done is created a path for electrons to flow from the battery positive post to the switch, through the light circuit's wiring, to the lights, and back in to the battery through the negative post. To compare this to a hydraulic system, it would be like moving a control valve to allow fluid movement that starts at the tank and gets pushed by the pump to the control valve (or switch) through the hoses and tubes to a cylinder where the fluid can move the piston and the load and return back to the tank. ● SEE FIGURE 5–10 for a comparison of electrical and hydraulic circuits.

By closing a switch or completing a circuit, the operator has created a path for the electrons to flow through. It is the potential charge difference between the terminals of a properly charged battery that creates the electrical pressure to make electrons move toward protons. This is the force that causes electrons to flow through wires to the load (in this case, the machine lights). Once the electrons travel through the load,

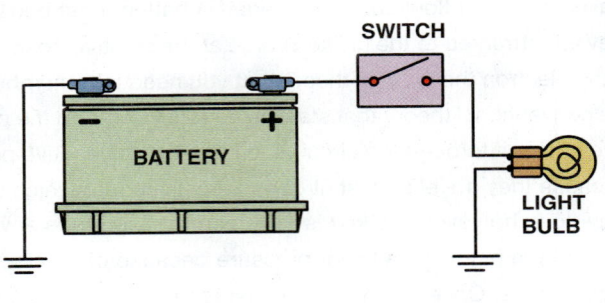

FIGURE 5–11 How a voltmeter reads voltage.

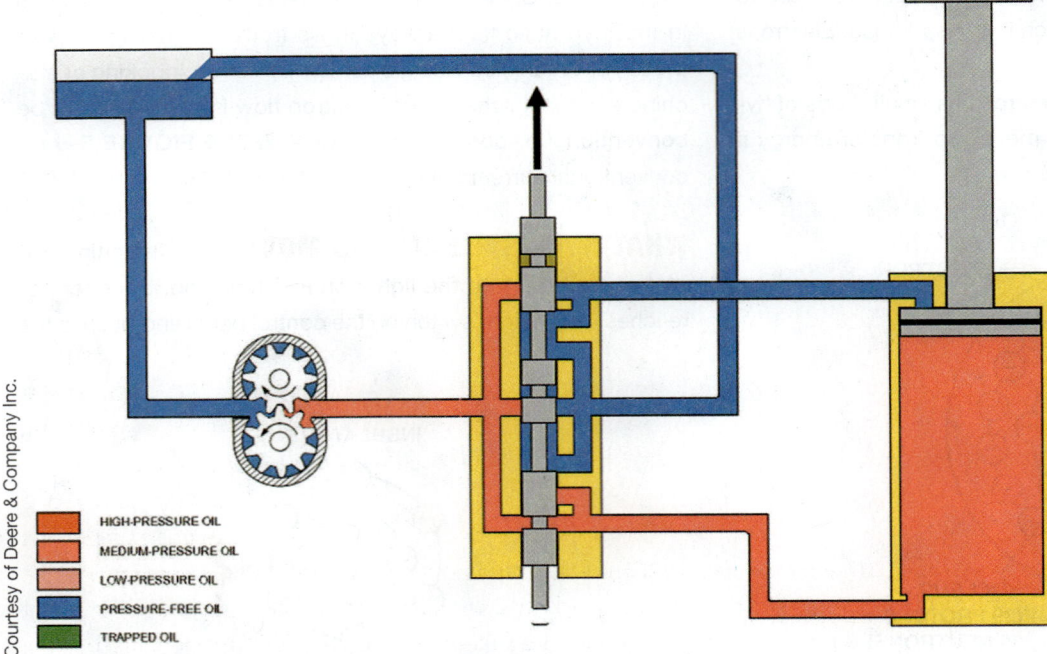

FIGURE 5–10 Compare oil flow and electron flow.

- HIGH-PRESSURE OIL
- MEDIUM-PRESSURE OIL
- LOW-PRESSURE OIL
- PRESSURE-FREE OIL
- TRAPPED OIL

142 CHAPTER 5

FIGURE 5–12 A 24V light bulb off and on.

they can now go back to the battery in through the negative post or the point of lower pressure.

As mentioned earlier, a properly charged battery is like an electron tank that has been pressurized. The electrical pressure in the battery is created by the machine's alternator or the electron pump. The term for electrical pressure is *voltage* and is represented by the letter "V." Remember, electrons will always flow from a higher potential pressure or voltage to a lower voltage, and this is the force that makes electrons move. As battery voltage level or pressure drops, fewer electrons will be pushed through a circuit, and if it keeps dropping, the result may be lights dimming, windshield wipers slowing, or an engine cranking slower.

Voltage is also referred to as electromotive force (EMF). This is why it is also represented by the letter "E" sometimes.

There are two different DC electrical systems used for heavy equipment machines, and they are the 12V and 24V systems. The main difference is simply the 24V system uses double the electrical pressure to move electrons. It would be similar to increasing the pressure relief valve setting from 1200 to 2400 psi for a hydraulic system.

Voltage measurements are taken with a voltmeter that compares the electrical pressure reading between any two points of a circuit with two leads that are connected to the meter. This would be like taking two hydraulic pressure readings at two different points in a hydraulic circuit and reading the difference. Some circuits will only require very small amounts of voltage, which are measured in millivolts (mV) or 1 thousandth of a volt, and this is also read with a voltmeter set to a different scale. ● **SEE FIGURE 5–11** to understand how a voltmeter reads voltage.

HOW DO ELECTRONS PERFORM WORK?
Electrons can perform work by creating magnetic fields that can cause movement or by creating heat. When current flows through a conductor, a magnetic field is created around the conductor, and this magnetism is what creates rotational movement in electric motors and linear movement in relays and solenoids. Electrons can also transfer energy from a battery to light a light bulb or to warm a glow plug. The lights get bright because the electrons are forced through a very small wire inside the light bulb. This wire gets very hot because when the electrons are forced through the light bulb element they create friction and the hot wire becomes very bright. ● **SEE FIGURE 5–12** for how a light bulb changes when current flows through it.

HOW IS ELECTRON FLOW MEASURED?
Electron flow is measured in units called amperes or amps. Later when we talk about using Ohm's law to calculate electrical problems, you might see amperes represented in the formula as "A" for amps or "I" for intensity. To imagine the amount of electrons moving through a simple lighting circuit for a machine is very difficult once again because of their size and the speed at which they move. A typical lighting circuit on a machine with a 24V DC system could have a current flow of 10 amps. One ampere is the amount of electrons that cross any point in a circuit in one second. The actual number is 6.28×10^{18} or more than 6 billion, billion electrons; therefore, if the 10 amp circuit was close to its current limit, there would be more than 60 billion, billion electrons moving through the circuit's wires every second. Likewise, if a starter circuit has 800 amps flowing through it, then there are more than 4,800 billion, billion electrons flowing through the starter cables each second.

It's hard to imagine or relate anything to numbers that big. What you do know is it takes a lot more amperage to crank an engine over than to light a light bulb just as it would take a lot more hydraulic fluid flow to move a cylinder with a 6 in. diameter than one with a 1 in. diameter.

The instrument used to measure current flow is called an ammeter, and it can be compared to a hydraulic flow meter that measures hydraulic fluid flow in gallons per minute. ● **SEE FIGURE 5–13** to understand how an ammeter works.

Very small amounts of current flow will sometimes need to be measured in a circuit. This current is measured in milliamps (mA) or a thousandth of an amp and is also measured with an ammeter but set to a different scale. A conventional ammeter must be hooked directly into the circuit to measure current flow. This is called hooking it up in series.

DIRECT CURRENT ELECTRICAL SYSTEMS

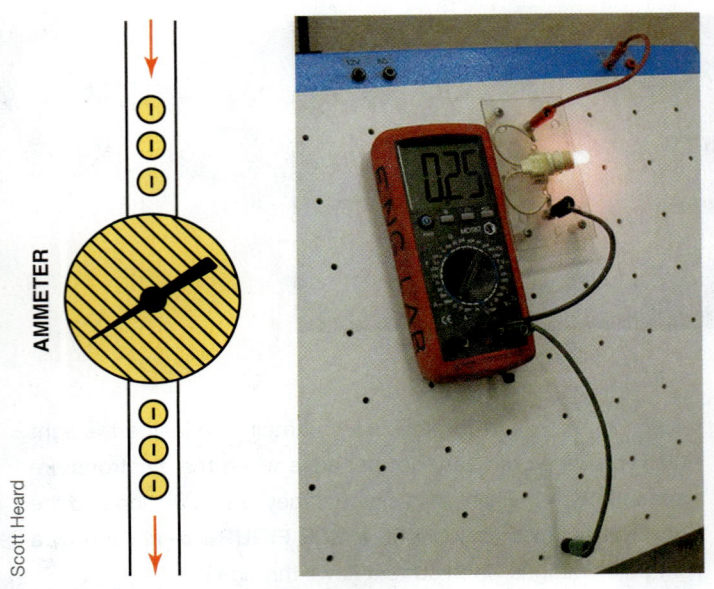

FIGURE 5–13 How an ammeter works.

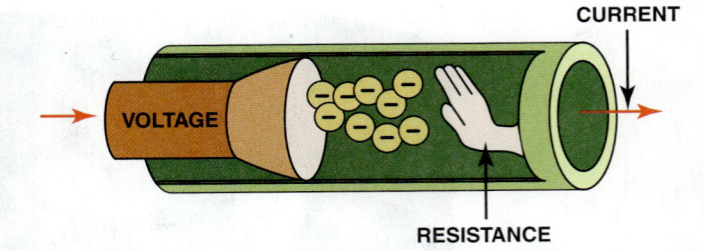

FIGURE 5–14 The effect of resistance in an electrical circuit.

WHAT IS ELECTRICAL RESISTANCE?

Electrical **resistance** is anything in an electrical circuit that slows down current flow. This could be the inherent resistance in any component such as a light bulb, a fan motor, or a solenoid, any wire or connector in the circuit, or any unwanted resistance such as corroded or dirty terminals. Resistance is caused by the friction created as the electrons move through conductors and components. Once again this is similar to a hydraulic system in that everything that the hydraulic fluid flows through creates resistance of varying degrees.

Resistance in a wire is increased as the wire's length increases, diameter decreases, or temperature increases. The wire's dimensions are a very important consideration when replacing or adding a conductor to an electrical system so excessive resistance isn't created. This is similar to a hydraulic hose being replaced with one that is too small. The result would be excessive restriction or resistance to the oil flow.

Resistances can be intentionally placed in circuits to control current flow, and they get a similar effect to orifices that are used to control fluid flow in hydraulic systems. Resistances create a pressure or voltage drop just as orifices do. ● SEE FIGURE 5–14 for the effect of resistance in an electrical circuit.

Electrical resistance is measured in units called ohms and represented by the Greek letter Ω. This unit got its name from Georg Ohm.

Small amounts of resistance can be measured in milliohms (mΩ) or one thousandth of an ohm and very large amounts of resistance can be measured in kiloohms (kΩ) (1 kΩ = 1000 Ω) or megaohms (MΩ) (1 MΩ = 1,000,000 Ω). These measurements are taken with an ohmmeter. For example, a typical glow plug will have 1–3 Ω of resistance.

Progress Check

1. What part of the atom moves through a conductor?
    a. electron
    b. nucleus
    c. proton
    d. neutron
2. The electrical force that moves electrons is called:
    a. amperage
    b. voltage
    c. insulator
    d. resistance
3. Why is copper used for wiring in a machine's electrical system?
    a. bends easy
    b. doesn't corrode
    c. good insulator
    d. good conductor
4. Electrical pressure is called:
    a. amperage
    b. voltage
    c. current
    d. resistance
5. The part of the atom that is negatively charged is called:
    a. negatron
    b. neutron
    c. positron
    d. electron
6. Electron theory states that electrons will flow from:
    a. positive to negative
    b. negative to positive
    c. neutral to positive
    d. ground to neutral
7. If the outer shell of an atom has one free electron, it is an ideal:
    a. conductor
    b. semiconductor
    c. semi-insulator
    d. insulator

8. Resistance will be measured with a(n):
   a. ammeter
   b. flow meter
   c. ohmmeter
   d. voltmeter

OHM'S LAW

To mathematically relate the three elements of electricity to each other, a German scientist in the early 1800s named Georg Ohm created the formula that states one volt of pressure is required to force one amp of current flow through one **ohm** of resistance. This became the mathematical formula to state the relationship between volts, amps, and ohms that is called Ohm's law.

If two of the three values are known, Ohm's law can be used to calculate the third or unknown value.

The formula is stated as:

$E = I \times R$ (where E is electromotive force or voltage; I is intensity or current; R is resistance)

The same formula can also be written as:

$I = E \div R$ or
$R = E \div I$

An example of using Ohm's law to find the amount of current that should be flowing through a circuit that has a 24V source and a resistance of 3Ω would be:

$I = E \div R$
$= 24V \div 3\Omega$
$= 8$ amps

This means a circuit with a 24V battery connected to a 3Ω resistance will have 8 amps of current flowing through it. To confirm your answer, simply plug all the numbers in to the equation and see if they equate.

We will look at other Ohm's law problems when the different types of circuits are looked at.

WATT'S LAW

Electrical power is a measurement of the energy consumption or the energy needed to perform work in an electrical circuit. It is calculated by multiplying the circuit supply voltage by the amount of amperage flowing through the circuit ($P = I \times E$). The result is a measurement using the unit of watts. This value can be directly converted to horsepower by dividing. This means that if a circuit consumes 746 watts of e is also using 1 horsepower of energy to keep it going. Looking at a simple light bulb circuit that uses a 12V battery and has 10 amps flowing through it when the switch is closed, we can calculate the circuit is consuming 120 watts of power or roughly 1/6 of a horsepower. If another 12V circuit had 62 amps flowing through it, there would be 744 watts of power or almost 1 horsepower used for that circuit. To power that same circuit with a 24V battery, there would only need to be 31 amps flowing through it even though it consumes the same power. If a 24V starter was under a heavy load to turn over a large cold diesel engine, it may need 400 amps of current to make it turn. This would equate to slightly less than 13 hp to turn over that engine ($24 \times 400/746$).

MAGNETISM AND ELECTROMAGNETISM

The properties of magnetism and electromagnetism play an important role in any machine's electrical system. It is important that you understand what electromagnetism is, how it can be created, and how it is manipulated to perform work.

You have probably at some point experimented or played with magnets and realized that any magnet has two poles, north and south. If two magnets have opposite poles exposed to each other, they will attract to each other, and if they have their like poles exposed to each other, they will try to push away from one another. This action is caused by the magnets' fields (lines of flux). The fields have been proven to leave the north pole and enter the south pole outside the magnet and go from south to north inside the magnet. ● **SEE FIGURE 5–15** for how the lines of flux move from north to south.

The strength of a magnetic field is measured by the intensity of the lines of force that surround the magnet, and these lines are called flux lines. The stronger a magnet is, the denser its flux lines are and the farther out they reach. Magnetism can occur in either of three forms:

1. Natural magnets or "lodestone" have magnetite, a natural material that has magnetic properties.

2. Man-made magnets are some combination of iron or steel that has been exposed to strong magnetic fields and retain magnetism.

3. Electromagnets are created when a conductor (wire) has current flowing through it. This wire will have a magnetic

DIRECT CURRENT ELECTRICAL SYSTEMS

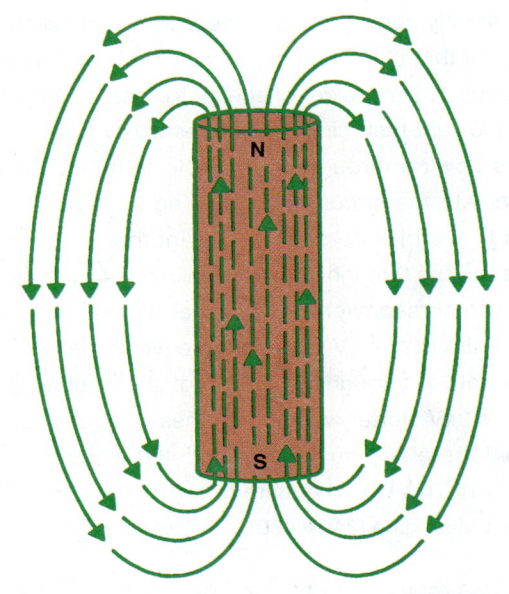

FIGURE 5–15 See how flux lines move from north to south.

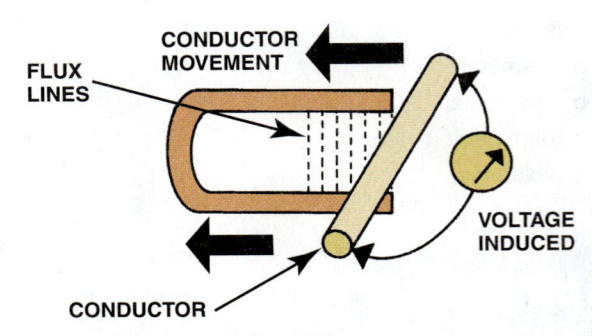

FIGURE 5–16 Electron movement in a conductor caused by crossing flux lines at 90°.

field around it, and as more current flows, the magnetic field increases. To make this magnetic field strong enough to do practical work, it is necessary to loop the wire into several coils. To make it even stronger, an iron core is placed in the center of the coil of wire. Magnetic flux lines have the strange property of going through conductive material easier than nonconductive material. In other words, because iron is a better conductor than air, to encourage the flux lines to travel and make a stronger magnet, an iron core is used for an electromagnet. This type of electromagnetism makes electric motors such as a starter motor work. To make the magnetic field stronger, there needs to be either more current flowing or more turns of wire. Electromagnets are used in many devices that are part of a machine's electrical system such as relays, solenoids, and motors.

Magnetism can also cause electron movement in a conductor. If a wire is passed through a magnetic field at 90°, this will cause electron movement in the wire. This is called induction. ● **SEE FIGURE 5–16** for how this occurs.

As with increasing electromagnets strength, once again if the wire is wound into a coil, there will be an increase in current flow. Also if the wire is moved faster or the magnetic field is increased, more current will flow. This is the basis for how a machine's alternator can turn mechanical rotation in to current flow and is called induction (this will be discussed in depth in Chapter 11). Some speed sensors will also generate a voltage signal through magnetic induction, which is then translated into a speed signal.

Progress Check

9. According to Ohm's law, in a 12V circuit, if resistance goes up:
   a. voltage goes down
   b. current goes down
   c. power goes up
   d. current starts to reverse
10. When calculating electrical power consumption, you need to know:
    a. ohms and watts
    b. watts and volts
    c. volts and ohms
    d. amps and volts
11. If a wire is passed through a magnetic field at right angles:
    a. current flow is induced in the wire
    b. the wire is drawn to the north pole
    c. the wire is drawn to the south pole
    d. the wire heats up
12. Electrical power is measured in:
    a. amps
    b. watts
    c. volts
    d. horsepower
13. To make the strongest electromagnet, you would make:
    a. a loop of wire
    b. several loops of wire and reverse the current flow
    c. several loops of wire around an iron core
    d. several loops of wire around a piece of plastic

DC ELECTRICAL CIRCUITS

A machine's complete electrical system can be broken down into subsystems such as the diesel engine starting system, the charging system, various electrical accessories systems, and electronic control and monitoring system if so equipped.

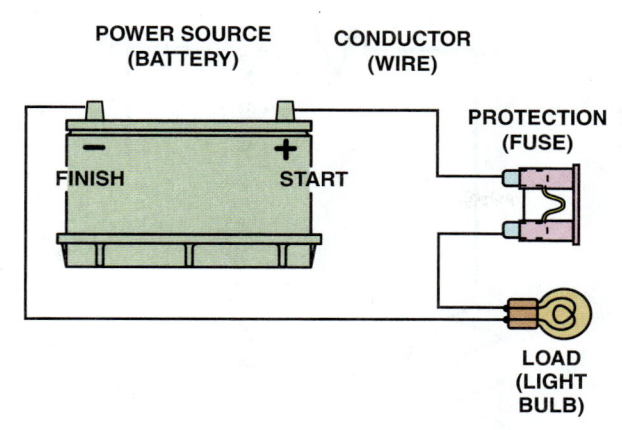

FIGURE 5–17 A simple electrical circuit.

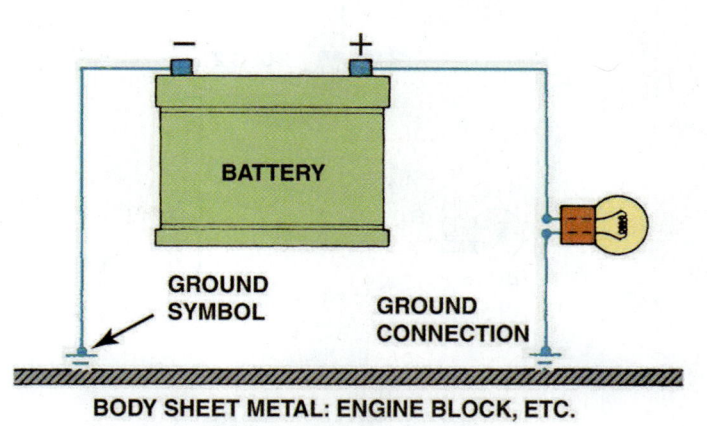

FIGURE 5–18 A circuit with a common ground.

Each of these subsystems can have one or many circuits in them. A circuit is a path for electricity to flow through and perform a function such as light a light bulb, move windshield wipers, make a gauge move, or control a transmission solenoid.

A simple DC circuit must include the following minimum components:

1. voltage source (battery)
2. conductors (wires)
3. resistance (load)

To be more realistic, there should be at least two other components, and they are:

1. switch to control the circuit
2. circuit protection, usually a fuse

To make a complete circuit, there must be a way for electrons to flow from a high-voltage level to a low-voltage level. ● **SEE FIGURE 5–17** for the schematic view of a simple electrical circuit.

DC CIRCUIT POSITIVE SIDE DC electrical circuits can be thought of as having two sides: positive and negative. The positive side is sometimes called the hot side or insulated side and usually consists of everything that starts at the positive battery terminal and goes to the positive side of all the loads. It is almost always the part of the circuit that is fused and switched.

DC CIRCUIT NEGATIVE SIDE The negative side of the circuit is quite often called the ground side or non-insulated side and goes from the opposite end of the load to the negative battery terminal. Most circuits will use the machine frame as part of the ground side and is called the common ground. Using a common ground eliminates a lot of wires because if the negative post of the battery is connected to the frame, then to complete any circuit the ground for that circuit just has to be connected to the machine's frame. A circuit can still be controlled by switching the ground, but this isn't very common. Many electrical problems can be traced to faulty circuit grounds. ● **SEE FIGURE 5–18** for a circuit with a common ground.

CIRCUIT TYPES

SERIES CIRCUIT There are three types of circuits that are used for a machine's electrical systems. The simplest being a series circuit. A DC series circuit is a circuit where electrons have only one path to circulate through from the voltage source (battery) to the load or loads and back to the battery. ● **SEE FIGURE 5–19** for a simple series circuit. The simplest series circuit would be a battery with one length of wire connecting the positive terminal post to one light bulb terminal and another length of wire connecting the negative terminal post to the light bulb's other terminal. There is one complete path for the electrons to be pushed through the light bulb to make it light up.

A series circuit could have more than one resistance or load as long as it is connected in series or one after the other. Total circuit resistance for a series circuit is equal to the sum of all circuit resistances. A series circuit with one load will use or drop all source voltage or pressure at that load. In other words, as there is only one path for the current to travel through once the circuit is complete, all voltage from the source is used to push the electrons through the load. To put this in practical terms, if a 12V battery was connected to a single light bulb, you should be able to measure 12V across or in parallel to the

DIRECT CURRENT ELECTRICAL SYSTEMS 147

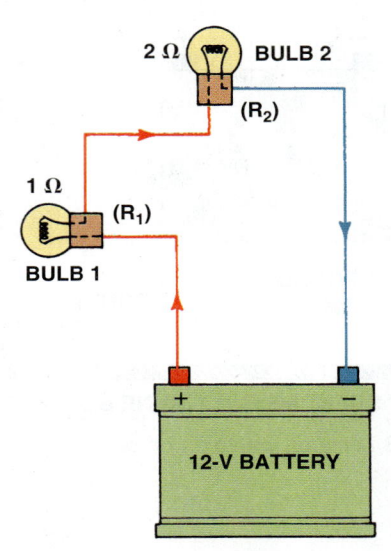

FIGURE 5–19 A simple series circuit.

FIGURE 5–20 Performing a voltage drop measurement.

light bulb connections. ● SEE FIGURE 5–20 for how to measure the voltage drop across the only load in a simple series circuit.

If the series circuit has more than one load, the voltage drop at each load will be in proportion to the load's resistance and the source voltage will be reduced as it pushes electrons through each load. The higher the load's resistance, the more voltage is required to push electrons through it. If the loads are equal in resistance, then the voltage drops will be equal and the sum of these voltage drops will equal the source voltage.
● SEE FIGURE 5–21 for a 12V series circuit with three equal resistances.

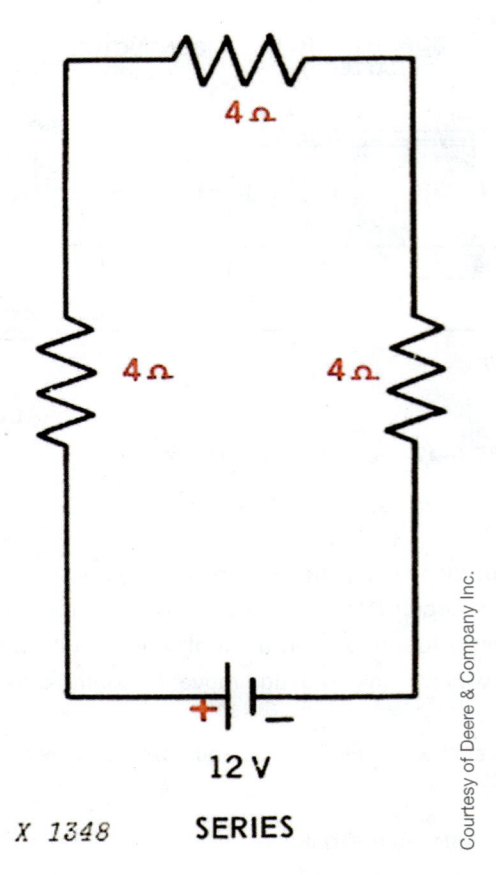

FIGURE 5–21 A 12V DC circuit with three equal series resistances.

If the series circuit has more than one resistance that are of different value, then it will take different amounts of voltage to push the electrons through the varying resistances. Once again the sum of all voltage drops should equal the source voltage.

If any part of a series circuit is **open** or not complete, there will be no electron flow and whatever load or resistance the circuit is designed to energize will not work. Once the circuit is complete, the current flow will be the same at any point in the circuit.

If there is excessive resistance at any point in the circuit, the designed or required amount of current will not flow through the circuit and the load may not operate properly. This will affect all loads in the circuit if there is more than one. In practical terms, the circuit's light could be dimmer or the wipers will work slower than they're supposed to.

OHM'S LAW FOR SERIES CIRCUITS ● SEE FIGURE

5–22 to understand the relationship of volts, ohms, and amps

148 CHAPTER 5

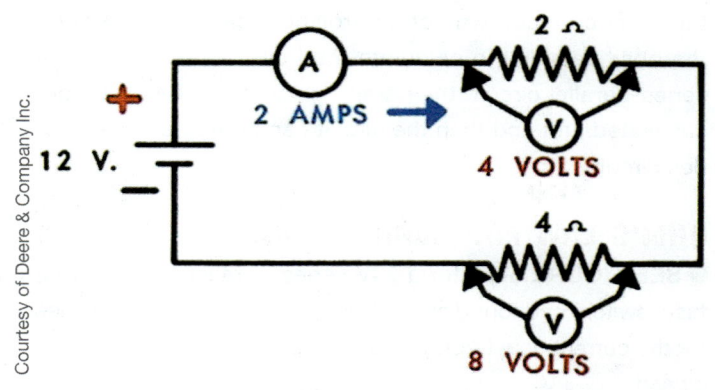

FIGURE 5–22 Understanding Ohm's law in a simple series circuit.

VOLTS	AMPS	OHMS
12		10
	6	4
24	1.8	
12		.12

TABLE 5–2

Fill in the blanks using Ohm's law.

in a series circuit. This is how a series circuit with two resistances of different values will look. If you imagine the second resistance is a light and the first one is a switch with dirty contacts, what do you think happens with the brightness of the light? Calculate the missing values in ● **TABLE 5–2** using Ohm's law.

PARALLEL CIRCUIT A parallel circuit has more than one path for current to flow through. The same minimum number of components is required for a parallel circuit as a series circuit (battery, wires, fuse, switch, load) except there must be at least two loads or resistances. ● **SEE FIGURE 5–23** for a light circuit that is a parallel circuit.

Parallel electrical circuits will provide more than one path for current flow. An easy way to think about this is imagine a river flowing and splitting into two parts around an island. One part of the river has a nice smooth sandy bottom and the other part of the river is full of rocks, logs, weeds, and other obstructions. After the river divides into two different parts, it joins back up into one river. As you can probably imagine, there'll be more water flowing through the smooth bottom section of the river, as opposed to the one with all the obstructions. If you measured the water flow before the river split and after the river split, you would find the same total amount of flow.

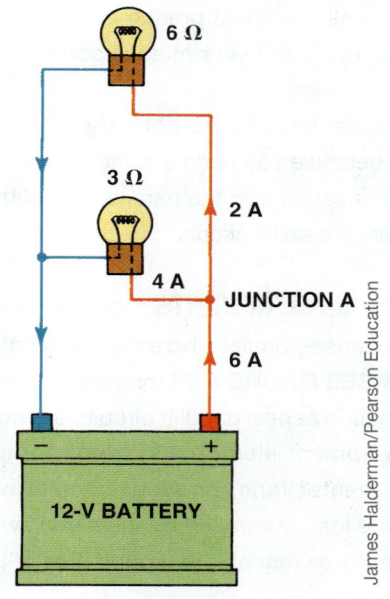

FIGURE 5–23 A parallel circuit with two loads.

This concept of water flow taking the path of least resistance is directly in relation to what occurs in a parallel electrical circuit. Parallel circuits can have more than two branches or legs and current flow through all legs will always be greater in the branch with the least resistance.

A machine's electrical system will typically use parallel circuits for lights, where the left and right light will still be part of the same circuit only and will be controlled together. Normally, machine parallel circuits will have equal resistances. So how would you think current flow is affected when split between two equal resistances? Yes, current flow should be equal between the two equal resistances. The other key point to remember with parallel circuits is the total resistance of a parallel circuit is always less than the single smallest resistance. The point may seem a little confusing, but because there is now more than one path for current flow, it is easier for the electrons to be pushed through the circuit.

OHM'S LAW FOR PARALLEL CIRCUITS To calculate total circuit resistance (R_t) in a parallel circuit, you assign each resistance to a number (R_1, R_2, R_3, etc.) and use the following formula:

$$1/R_t = 1/R_1 + 1/R_2 + 1/R_3 + \cdots$$

Remember the total circuit resistance should be less than the smallest individual resistance.

Once the total circuit resistance is found, you can find the total circuit current flow with $I = E/R$. Now to find the current in each branch, if the resistances are equal, you can divide the total circuit current flow by the number of resistances. Otherwise, simply apply Ohm's law to each branch using the source voltage. When all individual branch currents are found, they can be added up, and if your math is correct, it should equal the total circuit current.

True parallel circuits are not very common for electrical systems because you need a switch to control the circuit, which will be in series with the parallel part of the circuit. This makes it a series-parallel circuit.

SERIES-PARALLEL CIRCUITS The last type of circuit is a combination series-parallel where there are both types of circuits used. ● **SEE FIGURE 5–24** for a series-parallel circuit.

Combination series-parallel circuits are commonly used in heavy equipment electrical systems for lighting, horn, glow plugs, heater fans, or wiper circuits where there is more than one load controlled by the same switch. It is possible to have the switch on the ground side of the circuit, but the most common way of controlling a circuit is to switch the positive side. To calculate total circuit resistance of a series-parallel circuit, the parallel portions of the circuit are calculated first and then the circuit can be treated like a series circuit.

OHM'S LAW FOR SERIES PARALLEL CIRCUITS

● **SEE FIGURE 5–25** for a 24V series-parallel circuit with a fuse, switch, and four glow plugs in parallel. Look at the values for the current flow through the glow plugs when the switch is closed.

Now look at ● **FIGURE 5–26** for the same circuit, which shows the switch has a dirty contact that will add resistance (1 ohm) when it is closed. Calculate the new current values according to the added resistances. What has changed and how would this affect glow plug operation?

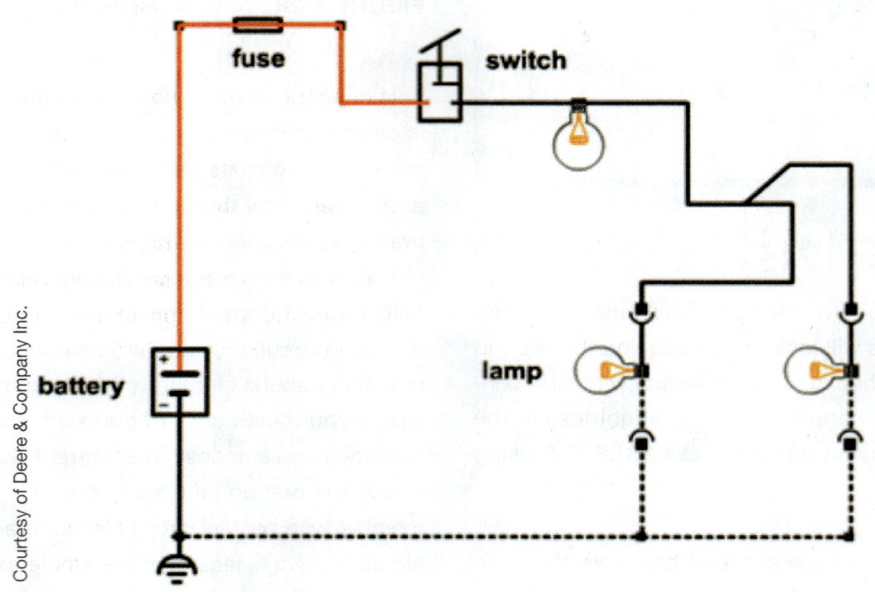

FIGURE 5–24 A series-parallel circuit.

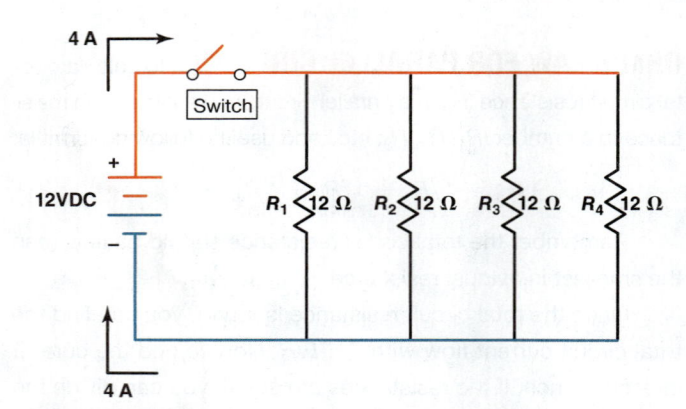

FIGURE 5–25 A complete series-parallel circuit (Halderman, *Automotive Technology*, 5th ed., Fig. 41–21).

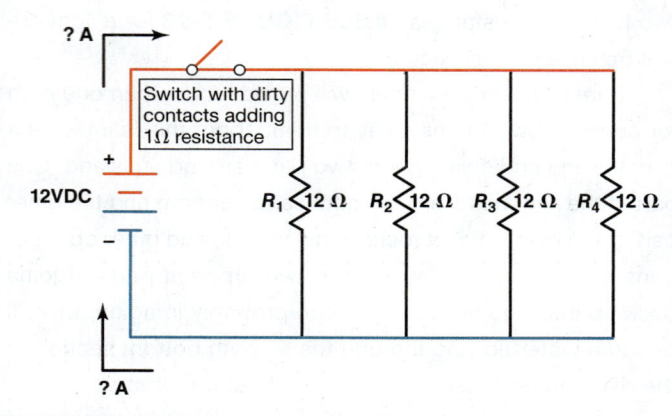

FIGURE 5–26 Glow plug circuit with a dirty switch. (Halderman, *Automotive Technology*, 5th ed., Fig. 41–22).

PRACTICAL USES FOR OHM'S LAW

When some technicians hear Ohm's law, their eyes will roll back and they will start wondering what the point is. It's true you probably won't be performing calculations on a regular basis through the normal duties of servicing and repairing heavy equipment, but by practicing these calculations you would gain a better understanding of the relationship between volts, amps, and resistance. Because you know the source voltage is either 12V or 24V, you can easily predict what happens with the other two parts of the equation. If resistance increases, current decreases, and if resistance decreases, current increases. For example, a circuit with excessive resistance such as dirty switch contacts will reduce the amount of current flow through it.

WIRING SCHEMATICS AND SYMBOLS

A wiring schematic or diagram is like a road map for a machine's electrical system. Just as a road map tells the names of roads and shows where they start and end, the schematic will tell you what the manufacturer has named all the machine's wires as well as wire junctions or connectors and where the wires come from and go to. The wiring schematic is a necessity when trying to troubleshoot an electrical problem because without it would be like driving in a foreign country with no map or compass. You may have a good sense of direction and eventually get to where you want go, but the journey would be a lot easier and stress free with a good map. This is very similar when you are looking for the cause of an electrical problem. Your job is made much easier with an electrical schematic to guide you along the way.

Some manufacturers will have colored schematics for their machine's electrical system and include a great deal of extra information such as electrical component values and part numbers for components. However, a lot of machines will only have black and white schematics that provide minimal extra information. Not so long ago all schematics were printed on either large sheets or multipage sheets of paper. If a machine's electrical system is shown on many pages, it will be broken down into individual systems or circuits. It is gettir to see paper schematics now as they are mostly av electronically, meaning they have to be viewed on desktop computer.

It is important to use the exact schematic of the machine you are working on as manufacturers will change wiring systems on a regular basis and you could be misled using the wrong schematic.

Wiring schematics will have many different symbols on them that represent the many different components that are part of the machine's electrical schematic. ● **SEE FIGURE 5–27** for some common electrical symbols.

There will be some variations between machine manufacturers, but in general they should be universal.

Progress Check

14. The battery's negative terminal will be part of this side of the circuit:
    a. hot side
    b. ground side
    c. cold side
    d. high side
15. Total circuit resistance in a parallel circuit will always be:
    a. higher than the highest resistance
    b. higher than the smallest resistance
    c. less than the highest resistance
    d. less than the lowest single resistance
16. If resistance in a series circuit goes up, then current flow in the circuit:
    a. goes down
    b. goes up
    c. always doubles
    d. stays the same but reverses
17. If a 12V battery is part of a complete series circuit that has a load of 10 ohms, the current flow is:
    a. 120 watts
    b. 120 amps
    c. 1.2 amps
    d. 1.2 milliamps
18. If one resistance in a four-branch series circuit has a lower value than the others, this branch will:
    a. need more voltage to energize it
    b. have more current flowing through it than the others
    c. have less current flowing through it than the others
    d. consume less power than the others

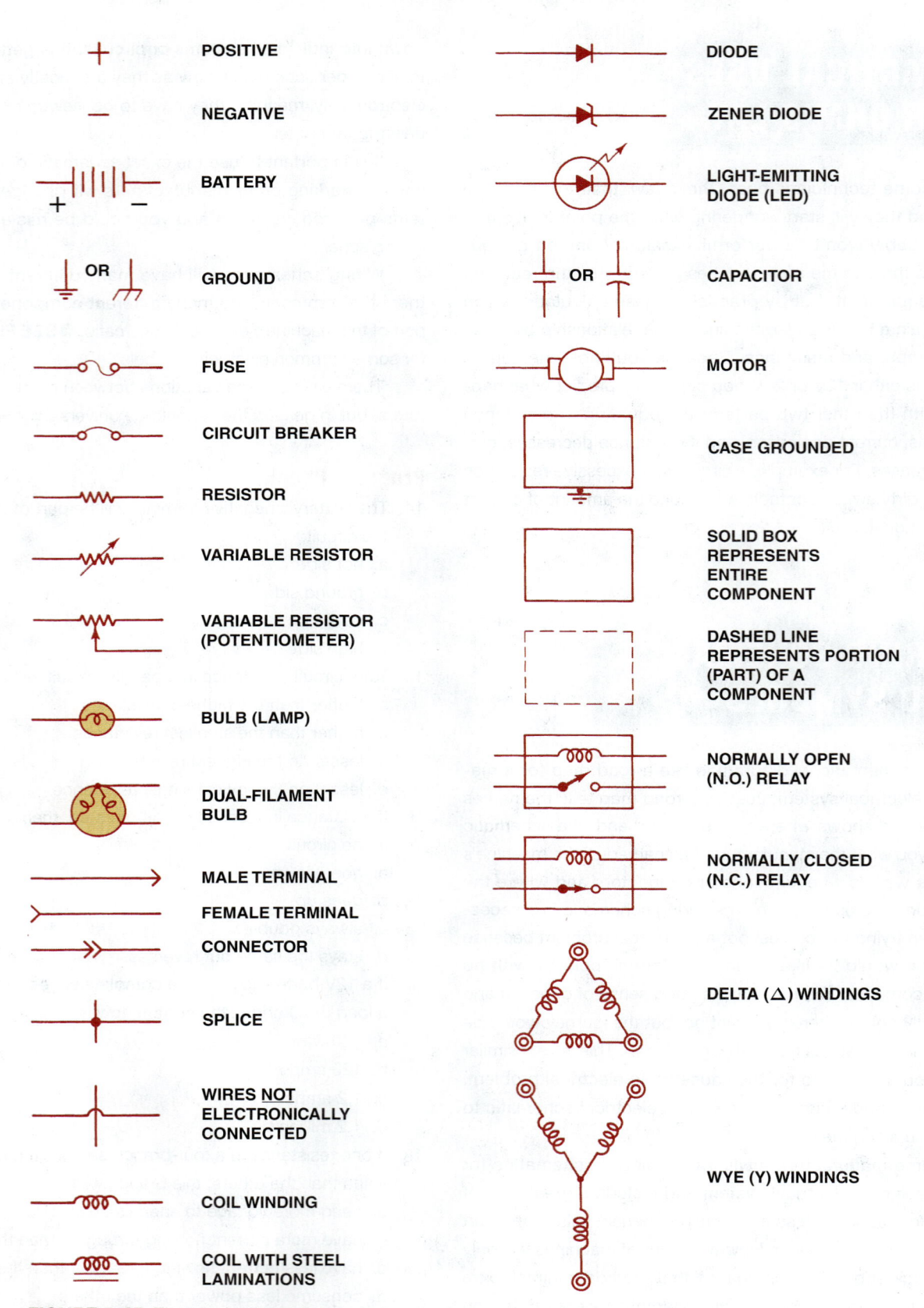

Symbol	Name	Symbol	Name
+	POSITIVE	▶︎▎	DIODE
–	NEGATIVE	▶︎▎	ZENER DIODE
⊢|⊢|⊢|⊢	BATTERY	(LED symbol)	LIGHT-EMITTING DIODE (LED)
⏚ OR ⏚	GROUND	⊣⊢ OR ⊣(	CAPACITOR
—◠—	FUSE	—(M)—	MOTOR
—◠ ◠—	CIRCUIT BREAKER	(box with ground)	CASE GROUNDED
—/\/\/—	RESISTOR	(solid box)	SOLID BOX REPRESENTS ENTIRE COMPONENT
—/\/\/— (arrow)	VARIABLE RESISTOR	(dashed box)	DASHED LINE REPRESENTS PORTION (PART) OF A COMPONENT
—/\/\/— (wiper)	VARIABLE RESISTOR (POTENTIOMETER)	(relay NO)	NORMALLY OPEN (N.O.) RELAY
—⊙—	BULB (LAMP)	(relay NC)	NORMALLY CLOSED (N.C.) RELAY
—⊙—	DUAL-FILAMENT BULB	(triangle windings)	DELTA (Δ) WINDINGS
—▶	MALE TERMINAL	(Y windings)	WYE (Y) WINDINGS
—⟩	FEMALE TERMINAL		
—⟫	CONNECTOR		
—•—	SPLICE		
(crossover)	WIRES NOT ELECTRONICALLY CONNECTED		
—∽∽∽—	COIL WINDING		
—∽∽∽— (with lines)	COIL WITH STEEL LAMINATIONS		

FIGURE 5–27 Electrical schematic symbols.

BASIC CIRCUIT COMPONENTS

The simplest circuit you may find on a machine would consist of a power source (battery), two conductors (wires), a load (light, motor, solenoid), a circuit protection device (fuse), and a circuit controller (switch). Let's have a look at these different electrical components.

THE BATTERY SYSTEM The battery system is a necessity for an electrical system. Because heavy equipment could have either a 12V or 24V system and all mobile equipment batteries are 12V, this means to power a 24V system there needs to be at least two batteries connected together. You may find four or more batteries connected together on some machines. This is why the term *battery system* is used. A battery system is needed to:

1. Provide electrical energy to power electrical devices if the machine's engine is not running.
2. Provide electrical energy to the starter to crank the engine.
3. Provide electrical energy when the engine is running if the alternator can't satisfy electrical demand.
4. Act like an accumulator to store and stabilize voltage.

The use of 6V batteries was fairly common for heavy equipment electrical systems 30–40 years ago, but at least 99% of the machines today will use only 12V batteries. However, if more than one battery is used, it will depend on how they are connected as to how much voltage comes out of the battery system. Two 12V batteries hooked in parallel (positive to positive and negative to negative) will still produce 12V while two 12V batteries connected in series will produce 24V.

● **SEE FIGURE 5–28** for how two 12V batteries are connected to make it either a 12V or a 24V system. If you are replacing a set of batteries, make a sketch before removing any cables. Always disconnect and connect ground cables first and last.

THE LEAD-ACID BATTERY A battery is a device that converts chemical energy to electrical energy and stores electrical energy in chemical form. It has two dissimilar plates (lead) that react with an **electrolyte** (acid) solution, which will produce electron flow when the battery's terminals are connected to a complete circuit. When fully charged, the battery will provide the

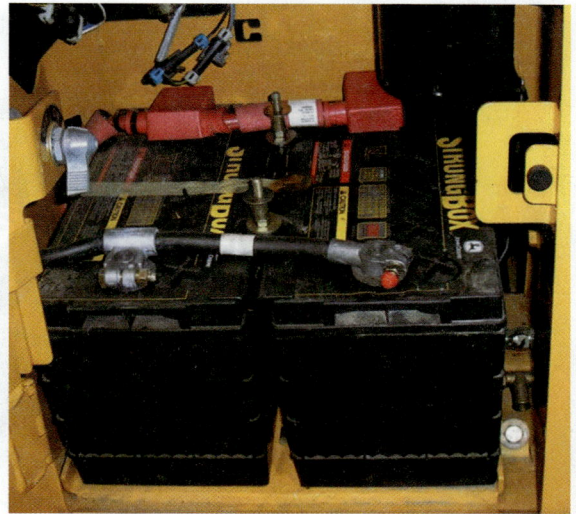

FIGURE 5–28 Two different battery connections.

potential force or pressure needed for the electrical system to operate properly. The electrical pressure difference between the battery's terminals gives the machine's electrical system the ability to operate electrical circuits.

Batteries are in a state of either discharging (even when the machine is at rest) or charging when the machine is running and the alternator is producing current.

We will first look at how a battery is constructed. A fully charged 12V battery is really six 2.1V batteries or cells connected in series and housed in one case that should produce 12.6V. ● **SEE FIGURE 5–29** for how a battery is constructed. The three main components inside the battery that make it work are the electrolyte, the negative plates, and the positive plates. The positive and negative plates are stacked alternately and separated by porous insulators. The most common type of battery is termed the *lead-acid battery* because its plates are

DIRECT CURRENT ELECTRICAL SYSTEMS 153

FIGURE 5–29 Cutaway view of a battery.

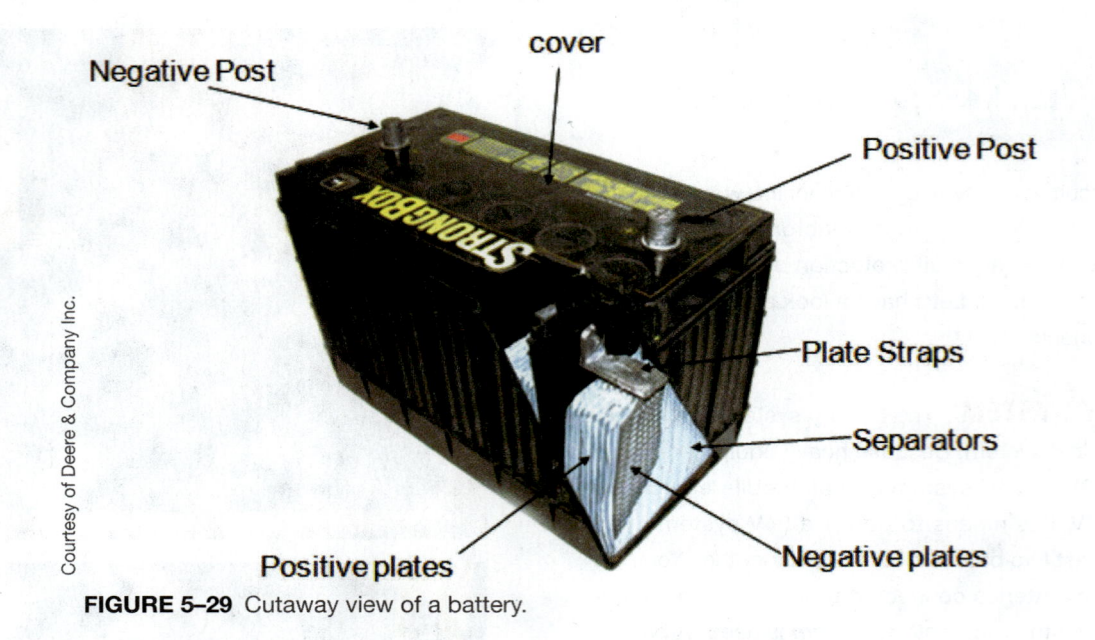

made of two different types of lead and the electrolyte is a type of acid. Let's see what makes up the three main components of a conventional lead-acid battery:

Electrolyte: A solution of 64% water (H20) and 36% sulfuric acid (H2S04) is present when the battery is fully charged. This mixture will have a specific gravity of 1.270, meaning it is 27% heavier than water. As the battery discharges, the electrolyte will move more toward a water solution or have a lighter specific gravity. This is why a discharged battery that has been subjected to freezing temperatures should be suspected to be frozen.

All batteries will self-discharge at a rate of .001 specific gravity points per 24-hour period at a constant 29°C (84°F). The discharge rate increases as temperature increases and decreases as temperature decreases. If the machine is not used for a period of time, the batteries must be maintained or stored in a cool place. This chemical process of discharging is reversed when the battery is charged and the electrolyte is returned to its fully charged specific gravity. When a battery is in the charging state, the chemical reaction will cause

> **+ SAFETY TIP**
>
> If a battery is suspected of being frozen, *do not* boost or charge it until it is warmed up. If a battery's electrolyte freezes, it will likely distort the plates, which could cause an explosion if charged or boosted.

> **+ SAFETY TIP**
>
> Electrolyte or battery acid is a very dangerous substance. If you get any on your skin, you should immediately flush the affected area with water. PPE should always be worn. Baking soda will neutralize battery electrolyte. Every shop and service truck should have a box on hand.

hydrogen gas to be produced. This gas is very dangerous and will smell like rotten eggs. The gas also evaporates water, which will need to be replaced periodically in a conventional battery.

Negative plates: Sponge lead (Pb) that is gray in color is applied to a frame to make a plate. The plate hangs in the electrolyte solution.

Positive plates: Lead dioxides (PbO2) make up the positive plates and are brown in color.

To increase the capacity or current output of the battery, the size and number of plates are increased. The lead plates are what gives a battery most of its weight.

Each cell's negative plates are connected together and positive plates are connected together. All six cells' negative plates are then connected to the negative terminal or post and, likewise, the positive plates are connected to the positive post. To make batteries require less maintenance, different blends of lead materials are used to make the plates. This causes less gassing and reduces water loss. This allows a battery to be constructed without the caps that a conventional battery would be needed to allow venting and access to the cell for water top up. Batteries made like this are called low-maintenance or maintenance-free batteries.

There are several types of batteries used for heavy equipment. The conventional type is identified by the removable caps over each cell that allows the electrolyte to be tested and topped up. Because of the design of this style of battery, there will be a loss of electrolyte due to gassing that occurs when the battery charges. This style of battery is becoming less and less popular because of the higher maintenance required.

Another style is the low-maintenance battery that has removable caps, but because of different materials used for its plates that reduce the gassing during the charging process, it rarely requires maintenance.

The most popular style of battery is the maintenance-free battery that is sealed for life. There is no way of checking or testing its electrolyte, and because of its different plate material, it produces very little gas and, therefore, doesn't lose electrolyte.

Slowly gaining popularity are gel cells and glass mat batteries. These are a more expensive version of maintenance-free batteries that feature a gel type of electrolyte but are supposed to be very durable. These batteries, however, come at a price premium and have special requirements when charging or boosting is required.

Battery ratings: Batteries are mostly rated with two systems. The first is cranking performance where a battery will be tested at two temperatures to see how many amps it can produce for a certain amount of time at a certain temperature. This rating is in either CA (cranking amps) or CCA (cold cranking amps).

The second rating is called reserve capacity (RC) and is used to predict how long a machine can run if the alternator stops working. This rating is in AH (amp hours).

Battery maintenance: Battery maintenance is often overlooked, and a lack of maintenance for the battery system is often the cause of a dead machine. Simple things such as checking electrolyte level, keeping the battery clean and secure, and checking and cleaning connections will help get full life from a battery. Distilled water should only be used when topping up battery electrolyte. Regular battery inspections should be a part of any good maintenance program. If a large fleet of machines is able to get an extra season out of the battery in each machine because of a little extra maintenance, that could add up to thousands of dollars saved without including the cost of machine downtime. There is also a huge environmental benefit to having batteries last as long as possible.

Because of the constant chemical to electrical change (self-discharge, discharge, or charge) and the harsh environment it must withstand in a machine (vibration, dirt, temperature extremes), the battery has a limited life. Proper care (cleaning, adding water, and charging) will extend the life of the battery.

Cleaning of batteries could be as simple as washing the top of the battery with clean water. If dirt is allowed to accumulate and gets moist, this provides a path for electron flow between the two terminals. Keep caps in place when cleaning and charging. Terminals should be covered or protected to prevent corrosion, which will eventually cause excessive resistance. Batteries need to be secured against vibration, which will damage plate material. Machines will need to have some type of battery hold-down device, and this should be inspected regularly. ● SEE FIGURE 5–30 for a properly secured battery. Battery cable ends need to be kept tight and the cables should be secure so they don't rub against part of the machine and expose the wire. Machines have caught fire because of a battery or cable being loose and shorting out.

Battery testing: Four tools that may be needed to test a battery are a voltmeter, a hydrometer, a load tester, and an electronic battery analyzer. A voltmeter is

FIGURE 5–30 A properly secured and maintained battery pack.

DIRECT CURRENT ELECTRICAL SYSTEMS

TABLE 5-3

ITEM	MEASUREMENT	SPECIFICATION
Stabilized OCV 12.5 or more	Percent charged	100
Stabilized OCV 12.4	Percent charged	75
Stabilized OCV 12.2	Percent charged	50
Stabilized OCV 12.0	Percent charged	25
Stabilized OCV 11.7 or less	Percent charged	0

Battery OCV versus state of charge.

Scott Heard

used to check open circuit voltage (OCV) of a battery. Batteries should be maintained at an OCV of 12.6 or greater. To determine the stabilized OCV, the surface charge on the battery needs to be removed if the machine has been run in the last 10 hours. To remove the surface charge, turn on three or four work lights (or any light load) and leave them on for three to five minutes. If the machine has not been run for the last 10 hours, this step is not necessary. The battery OCV can now be checked at the battery by putting the negative lead of a multimeter on the negative terminal leaving the battery system and the positive lead on the positive terminal leaving the battery system. ● SEE TABLE 5–3 for how OCV compares to percentage charged.

A hydrometer is used to check specific gravity of a conventional battery's electrolyte. Special care and the proper PPE must be used when drawing electrolyte from a battery.

A load tester is used to simulate placing a heavy load on a battery. The OCV is then checked during and after the load is applied. The battery should have an OCV of at least 12.4 before doing this test. The proper method is to apply a load equal to half the battery's CCA for 15 seconds and read the OCV at the end of the load test. If the battery voltage drops below 9.6, the battery should be recharged and retested. If it fails again, it should be replaced.

Electronic testers are now becoming very popular and are said to be more accurate and less harmful to the battery. Visit at http://www.midtronics.com to see electronic battery testers.

Battery charging: If a battery is found to have a lower than 75% charge, it should be charged before being put back into service and the reason for its lower charge should be investigated.

Battery charging should only be done in a well-ventilated area. Most battery chargers used now are automatic, meaning they will decrease charging rates as the battery gets closer to 100% charged. Always follow the instructions for the particular charger you are using, and remember if you smell rotten eggs, *do not* create any spark or flame as you risk creating an explosion. The best practice to follow when charging a battery is "long and low." This means the longer the time taken and the lower the amperage, the easier it is on the battery.

Boosting batteries: If a machine is dead or won't crank the engine over and you don't have a battery charger, you will probably need to boost the dead machine with a running machine. Taking your time when doing this will ensure not only it is done safely but it is also easier on both electrical systems. The ideal practice once the boosting cables are hooked up is to give the dead machine's battery lots of time to be charged instead of just hooking up cables and immediately trying to crank the dead machine.

When hooking up cables, the key is to make the last ground connection away from the dead machine's battery. After the dead machine is running, this *must* be the first connection to unhook. Because you have just charged up the dead battery, there is likely hydrogen gas present, and when the first cable is unhooked, you are opening a circuit that will create a spark. If this spark is away from the battery, there is little chance of explosion.

Most manufacturers (check with the machine's service information to be sure) will want the cables hooked up as follows:

1. positive booster cable to dead machine positive battery system output cable
2. positive booster cable to live machine positive battery system output cable
3. negative booster cable to live machine negative battery system output cable
4. negative booster cable to dead machine ground away from battery
5. For unhook sequence, follow these steps in reverse.

Progress Check

19. A conventional lead-acid battery will have:
    a. 2 cells
    b. 6 cells

c. 12 cells
d. 24 cells

20. When charging a machine's battery, this is the best advice:
   a. high and hard
   b. long and low
   c. soft and low
   d. fast and short

21. If a conventional lead-acid battery is found to have low electrolyte levels, you should add:
   a. sulfuric acid
   b. tap water
   c. depends if its maintenance free
   d. distilled water

22. The battery rating called reserve capacity relates to:
   a. cold starting capacity
   b. how long the machine could run without the alternator working
   c. the maximum voltage the battery can produce
   d. the maximum rpm the battery can crank the engine over

23. If you tested for a voltage drop between the battery positive post and the positive cable, this would be an acceptable finding:
   a. 12.6V DC
   b. 2.1V DC
   c. 1.2V DC
   d. .1V DC

CIRCUIT PROTECTION

An important part of any electrical circuit is its circuit protection. If a circuit fault called a short occurs and allows an excessive amount of electrons to flow, there could be enough heat generated in the circuit to start a fire. If a circuit protection device is in the circuit and working properly, this will never happen. To make a comparison to a hydraulic system, again you could consider an electrical circuit protector to be similar to a pressure relief valve that protects hydraulic components from damage.

There are two common types of circuit protection devices: fuses and circuit breakers. A third type that is less common but could still be found is a fusible link. One other new type of circuit protection starting to make its way into machine's electrical systems is an electronic virtual fuse that uses electronic components to sense excessive current flow and open the circuit to prevent damage. This system will be discussed in Chapter 10.

FUSES Fuses are usually placed in the positive side of the circuit as close to the battery as possible. Most machines will have a fuse box located in the cab out of the elements, but this is not always the case. The fuse box will contain a series of fuses that are dedicated to individual circuits. Fuses are designed to fail and will open if there is too much current flowing through them. They do this by getting too hot from the excessive current and melting. All circuits are designed to have a maximum amount of current flowing through them, and if this is exceeded, then the fuse will open and protect the circuit components from damage. Fuses that are open need to be replaced with ones of the same amperage rating. Never replace a blown fuse with one of a higher value. If you replace a blown fuse and it blows again, you need to find where the problem is that is causing the excessive current flow.

There are many different types of fuses. ● **SEE FIGURE 5–31** for some examples.

Fuses can come in many different amperage ratings from .5A to 100A but common sizes are 10A, 15A, 20A, and 30A. Fuses are color coded to make identification easy. ● **TABLE 5–4** shows how the fuses are colored for their ratings.

CIRCUIT BREAKERS Circuit breakers perform the same function as fuses but are made to be reset and not replaced.

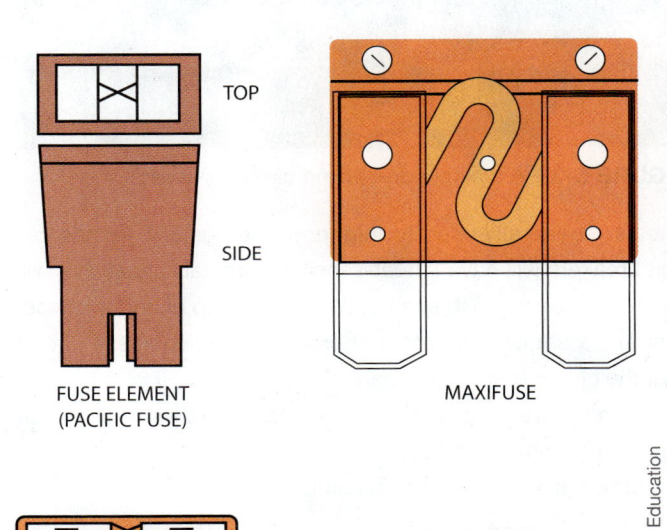

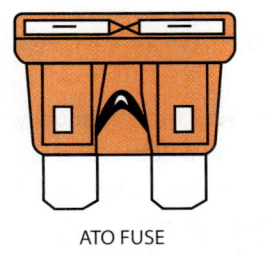

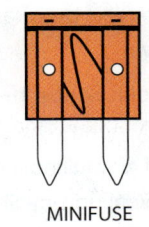

FIGURE 5–31 A variety of fuses.

| FUSE (BLADE-TYPE) COLOR CODES ||
AMPERAGE RATING	COLOR
1	Black
3	Violet
4	Pink
5	Tan
7-1/2	Brown
10	Red
15	Light Blue
20	Yellow
25	Natural (White)
30	Light Green

TABLE 5-4

Color identification of fuses.

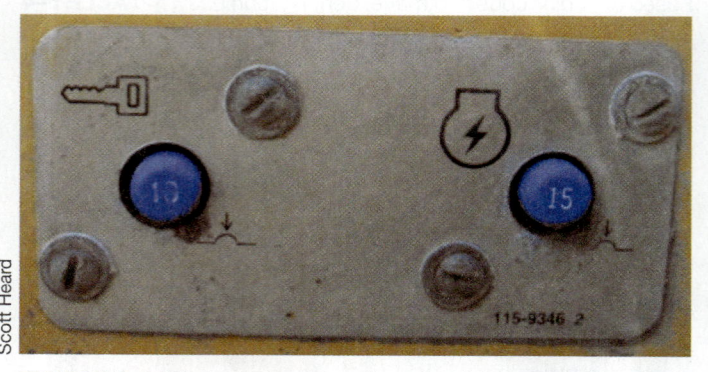

FIGURE 5-32 A typical non-cycling circuit breaker.

They are generally used for higher amperage circuits. All circuit breakers will have a heat-sensitive bimetallic strip that will change shape when it gets hot. When the strip changes shape, it opens a set of contacts that creates an open in the circuit that the circuit breaker is a part of.

Circuit breakers also come in many different ratings but are usually from 25A to 100A in capacity.

There are two types of circuit breakers:

1. Cycling-type breakers will reset themselves after they cool down.
2. Non-cycling-type breakers need to be reset manually after they open. This is done by pushing a button or a tab inside of a small hole on the breaker. ● **SEE FIGURE 5-32** for a typical non-cycling type of circuit breaker.

FUSIBLE LINKS A fusible link is put in to a circuit, and like a fuse, it is a weak link that is designed to fail if excessive heat causes it to melt. It is basically a wire that is four sizes smaller than the wires used in the rest of the circuit. Fusible links are usually identified with a tag.

CONDUCTORS The term *conductor* is used for any material that will allow easy electron movement through it. Conductor is usually the term substituted for wire when referring to machines. Copper wires are ideal conductors and the most economical choice to be used for heavy equipment wiring and the critical parts of electrical components such as solenoid and motor windings. A good insulator is the opposite of a conductor in that it is very stable and doesn't allow electron flow. To insulate copper wiring, so there aren't stray electrons going where they shouldn't, the wire is insulated with a cover. The ideal materials used for insulation are very stable elements such as plastic and rubber that have the maximum number of electrons in their outer shell. PVC is the most common and economical insulator for wires.

Wiring is sized for a circuit to allow electrons to flow through it unobstructed or with very little resistance. If you were to repair a wire and mistakenly used a smaller size wire, you are creating an added resistance. The flow of electrons would be restricted, excessive heat may be created, and the device that receives the electrons to perform a function would not operate as designed. For example, if there was a light bulb that should light, it would be dimmer than it should be.

Wires are sized by gauge numbers set by a standard called the American Wire Gauge (AWG). As the gauge number increases, the wire diameter decreases.

Wires are also sized in metric dimensions that relate to the copper cores cross-section as measured in mm^2.

● **SEE TABLE 5-5** for a comparison of AWG and metric wire sizing.

GAUGE/METRIC CONVERSION TABLE		
METRIC GAUGE		ENGLISH GAUGE
0.22 MM	SMALL Diameter Wire	24 GA
0.35 MM		22 GA
0.5 MM		20 GA
0.8 MM	LARGE Diameter Wire	18 GA
1.0 MM		16 GA
2.0 MM		14 GA
5.0 MM		10 GA

TABLE 5-5

Comparison of AWG and metric wire sizes.

Wire resistance varies with copper core diameter, length, condition, and heat. Multi-strand wires are mostly used because of their flexibility.

Wire insulation can have any color, and some machine manufacturers will use different color combinations to identify the circuit a wire is used for. There are also many different letter and number combinations printed on wires to identify them. This wire identification system can be used to help troubleshoot electrical problems when you are using a wiring schematic.

WIRING HARNESS AND CONNECTORS A machine's main wiring harness will have a group of wires bundled together and will be covered with a plastic sheath or cloth-like material that is secured to a frame rail of a machine. Most machine circuits will branch off from the main part of the harness. Machine manufacturers will usually do this to make the most economical use of conductors. Wiring harnesses should be secured to the machine at least every 12–18 in. From the main harness, there will be subsystem harnesses branch off that could be used for transmissions, engines, or other options or attachments such as extra lights, HVAC units, wipers, or any other extra accessories. These subsystem harnesses will sometimes start at a bulkhead. This is a multi-wire connector that will be secured and go through to a cab wall or another part of the machine's sheet metal. ● **SEE FIGURE 5–33** for a bulkhead.

Connectors are necessary to join wires to each other or to components. They come in many different styles and are identified by their manufacturer. ● **SEE FIGURE 5–34** for several different styles of connectors.

Some examples are Weatherpack, Deutsch Transportation (DT), Sure-Seal, Metri-Pack, and Packard.

Connectors will be crimped or soldered or crimped and soldered to the wire. (Some manufacturers will not recommend that wires be soldered.) Most crimping procedures will require a specific type of crimper to do the job properly, and you should follow the specific instructions for the crimpers you are using. One part of the connector will be the male or pin and the other part will be the female or socket. There will be some kind of seal to prevent dirt and moisture from getting into the mating area and starting corrosion. Special extractor tools are needed to remove wire ends from connector plugs. ● **SEE FIGURE 5–35** for the extractor tools needed to remove pins and sockets from connectors. When the connectors are mated together, they will allow current flow. Connectors could be for a single wire or grouped together with 70 or more wires being secured to the male and female junctions.

SWITCHES All the circuits used in a machine's electrical system will at some point need to be opened and closed, and

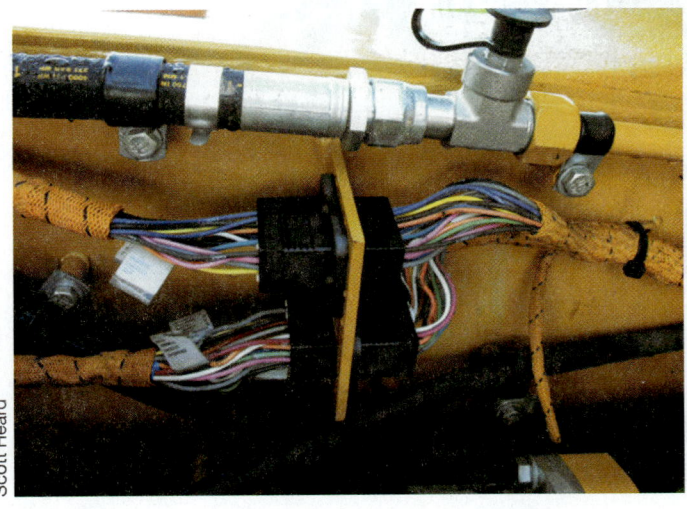

FIGURE 5–33 Machine wiring harness bulkhead.

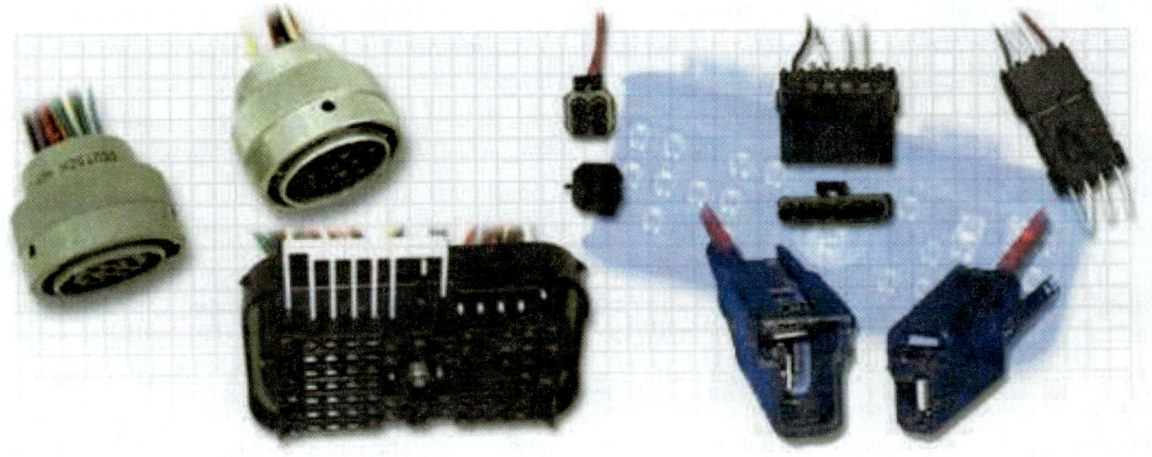

FIGURE 5–34 Several different styles of wiring connectors.

JDG 362 EXTRACTOR/INSERTER (BLUE)

JDG 363 EXTRACTOR/INSERTER (RED)

JDG 785 EXTRACTOR/INSERTER (WHITE)

JDG 361 EXTRACTOR/INSERTER (YELLOW)

JDG 777 EXTRACTION TOOL (GREEN)

JDG 939 EXTRACTION TOOL

Pull Type

FIGURE 5–35 A variety of connector pin extractor tools.

Courtesy of Deere & Company Inc.

this means they will need to have a **switch** included in the circuit. Most circuits will use a switch on the power side, but some will use a ground side switch. Switches can also be used with resistors to vary motor speeds or dim lights.

There are many different types of switches, and they're usually classified as to how many positions they have, how many poles or terminals they have, their type of contact terminals, and other details like how they are operated. Some different types of switch control could be pushing, pulling, toggle, rotary, pedal switches, and limit switches. Switches can also be controlled by temperature, pressure, or position (by magnetic field or mechanically).

A simple switch would be a single pole, single throw, where the switch has two terminals, and when actuated it either opens or closes contacts to provide a path for the electrons to flow through or complete the circuit. ● **SEE FIGURE 5–36** for some common switches and their schematic representation.

Another type of common switch would be engine starter switch. These switches usually have at least three positions: off, on or ignition, and start, and sometimes an accessory position is used. The start position is considered a momentary position, because when the key is released from the start position, spring returns it to the on position.

Switches can also have many different types of terminals to connect them to the wiring harness. Conventional switches are a mechanical device and, therefore, will eventually wear out, but they are usually trouble-free if they are located out of the elements. However, if exposed to harsh conditions, they can corrode and cause problems. Mechanical switches are being replaced by electronic switches.

160 CHAPTER 5

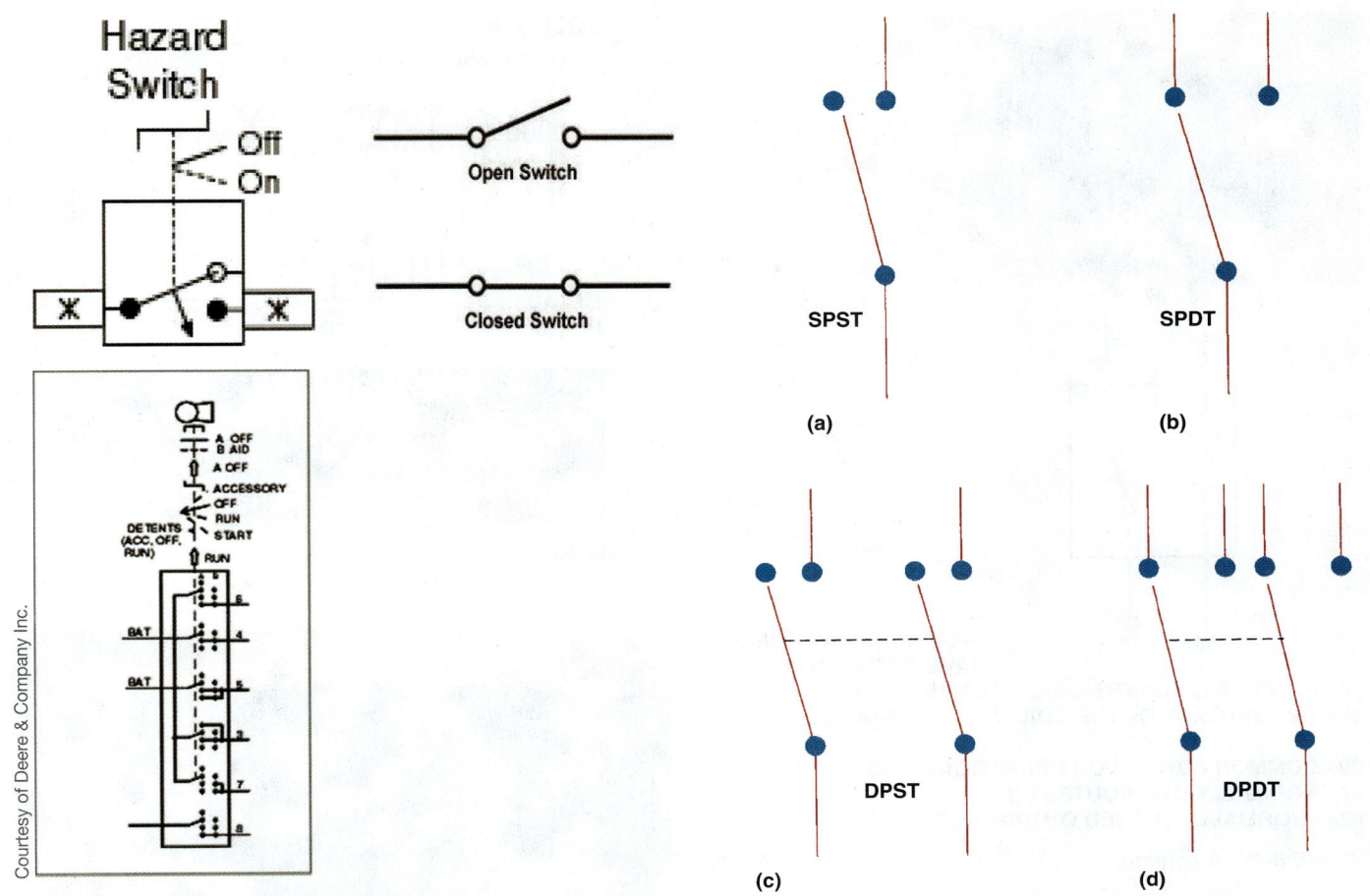

FIGURE 5–36 Some common switches.

RELAYS Relays are common components for heavy equipment electrical systems. Sometimes called magnetic switches, **relays** perform the job of controlling a high-current circuit with a low-current switching circuit. They do this by using an electromagnet to open and close heavier contacts that control the high-current circuit. Most new machines will use many relays to control many different circuits. There are lots of different types of relays and are made in a wide range of current capacity.

One common relay has five terminals to connect wires to the terminals. ● SEE FIGURE 5–37 for a common relay. Two terminals are used for the control circuit, and they are numbered 85 and 86. One of these will be connected to ground and the other will be switched to power through another switch. When there is a ground and power fed to terminals 85 and 86, an electromagnet is energized inside the relay that will pull an armature in to close the contacts for the high-current switch. Terminal number 30 is a supply for the high-current side and 87 and 87a are the two terminals for the high-current side output. Terminal 87a is the output for the normally closed part of the relay. This terminal will be energized until the relay is energized. Terminal 87 is the normally open terminal and is most commonly used. When the control side of the relay is energized (86 and 85) and closes the contacts, terminal 87 will send current out to the high-current circuit the relay is part of.

This is the most common type of relay, but there are several variations on this that could include only three or four terminals and they could be numbered differently. However, their purpose is the same.

SOLENOIDS Solenoids are an electromagnetic device used to convert current flow into a linear mechanical movement. They can be used for controlling oil flow in transmission and hydraulic control circuits, moving starter pinion drives into the flywheel, fuel system control, and any other strong short linear action that needs to be done. A solenoid is a coil of wire surrounding an iron core. The iron core is the movable part or plunger that moves when the coil winding is energized. The iron core either pulls in or pushes out depending on the way the current flows through the windings. Solenoids will usually have to overcome spring pressure when they are energized. The spring will reverse the motion created by the magnetic field. This type of solenoid will only travel full distance both ways without stopping in between.

DIRECT CURRENT ELECTRICAL SYSTEMS 161

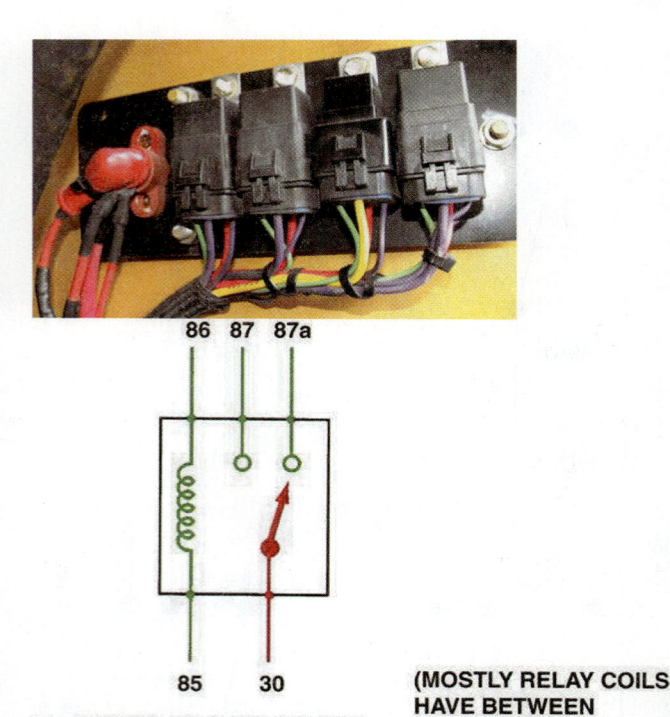

86 - POWER SIDE OF THE COIL
85 - GROUND SIDE OF THE COIL
(MOSTLY RELAY COILS HAVE BETWEEN 50–150 OHMS OF RESISTANCE)

30 - COMMON POWER FOR RELAY CONTACTS
87 - NORMALLY OPEN OUTPUT (N.O.)
87a - NORMALLY CLOSED OUTPUT (N.C.)

FIGURE 5–37 A common relay.

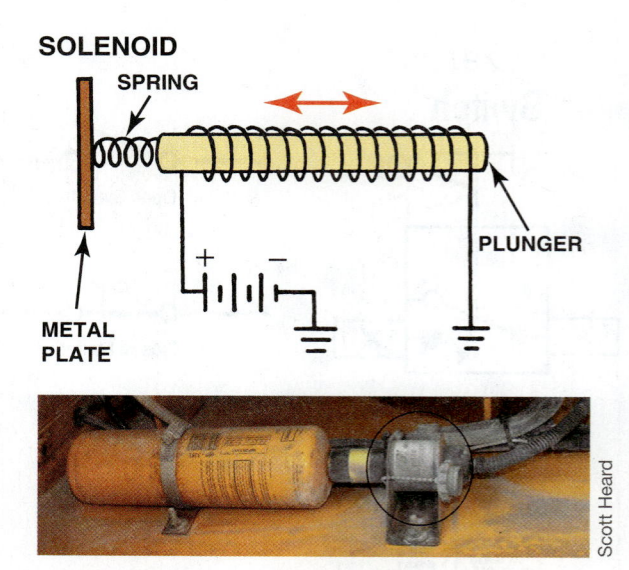

FIGURE 5–38 A simplified solenoid.

FIGURE 5–39 A variety of different lights for machines.

A solenoid could have two coils of wire in the same housing that is called the pull in coil and the hold in coil. The pull in coil will be lower resistance to allow a higher current through it. This will create a stronger magnetic field to overcome the spring pressure. Once the plunger or armature is pulled in, the hold in coil can easily hold the armature in. The hold in coil doesn't need to create as much magnetic force; therefore, it can have a higher resistance. The hold in coil will not develop a lot of heat, and this will make the solenoid last longer. These solenoids are typically used to move starter pinions into the ring gear of an engine. ● **SEE FIGURE 5–38** for a simplified look at a solenoid and a solenoid used for an ether injection system.

Another type of solenoid is a proportional type. These solenoids are usually used in relation to a hydraulic system where a pressure needs to be varied. These will receive a PWM (pulse width modulated, see Chapter 9) signal and will move in proportion to the signal.

ELECTRICAL LOADS There are many different loads that electrical circuit's power. The **load** is the device that the circuit powers when current flows through it. Some different loads are:

LIGHTS. Incandescent lights will use a very fine wire inside a vacuum-sealed glass container to create light. The wire creates so much resistance that it will glow white hot, and it is designed to withstand this heat for thousands of hours. Halogen lights produce an intense hot and white light. They must be handled with care when being replaced because even the oil from your skin will drastically reduce their life span.

LED lights are a semiconductor device that create light when energized and can produce many different colors based on their application. They draw very little current and are becoming more popular as work lights on machines. HID (high-intensity discharge) lights are also a very efficient type of light that are becoming more popular on machines. ● **SEE FIGURE 5–39** for a variety of different lights found on a machine.

MOTORS. Electric DC motors operate when current is fed through field coils that are mounted in the outside housing and through coils in the armature (center shaft). The magnetic

162   CHAPTER 5

FIGURE 5–40 An electric motor.

fields created by these two components oppose each other and rotation of the armature occurs. Current is fed through carbon brushes that slide on a commutator, which is part of the armature that allows the armature to rotate and still be energized. The armature will be connected to whatever device needs to be rotated, such as a starter motor pinion, wiper motor arm, or fan motor cage. DC motors will be discussed in more detail in Chapter 7. DC motors can be used to drive backup steering or brake pumps, windshield washer pumps, and heater motors.
● SEE FIGURE 5–40 for an electric motor that drives an engine prelube pump.

ENGINE PREHEATING DEVICES. Glow plugs and inlet heaters do similar things and operate on a similar basis. They offer low resistance and, therefore, have a lot of amperage going through them to ground. When their circuit is opened and this high current flow, it creates a lot of heat, and this will warm the engine intake air. Inlet heaters are often compared to a toaster grid. They are designed to withstand high amounts of heat. ● SEE FIGURE 5–41 for an engine inlet heater.

HORNS/BACKUP ALARMS. Electric horns and backup alarms will use current to move a diaphragm back and forth at high speed to make sounds. This is the same principle that is used for audio speakers.

FIGURE 5–41 An engine inlet heater.

Progress Check
24. A fuse will create this in a circuit if it senses too much current:
    a. high resistance
    b. short
    c. ground
    d. open

DIRECT CURRENT ELECTRICAL SYSTEMS

uit breakers can reset themselves if they:
  a. are the non-cycling type
  b. open softly
  c. are the cycling type
  d. are the non-fusible type

26. This is the largest wire size:
  a. 18
  b. 6
  c. 0
  d. 000

27. A five-pin relay has numbered terminals 30, 85, 86, 87, and 87a. The normally open terminal is:
  a. 30
  b. 86
  c. 87
  d. 87a

28. A solenoid is an electromagnetic device that:
  a. creates rotary motion
  b. controls a high-current circuit
  c. creates linear motion
  d. senses motion

ELECTRICAL TESTING EQUIPMENT

MULTIMETER A multimeter is a combination electrical measuring tool. Older units showed readings with an analog scale and sweeping needle and on rare occasions are still a preferred tool to the newer and more commonly used digital multimeter. ● SEE FIGURE 5–42 for an analog multimeter. For the past 20 years or so, the digital multimeter has been the preferred electrical system diagnostic tool. ● SEE FIGURE 5–43 for a digital multimeter. The multimeter is a combination of voltmeter, ammeter, and ohmmeter. Some multimeters will also measure temperature and speed with the proper attachments connected to them. A wise investment for any HDET is the purchase of a high-quality multimeter with a good warranty.

VOLTMETER. To measure voltage (electrical pressure), you will use the voltmeter function of the multimeter. The meter can read both AC and DC voltages. For most machine voltage readings, the meter's rotary dial will be turned to DC voltage, which is usually indicated with two straight lines. Two wire leads are connected to it and these leads are placed at any two points in the circuit to measure the voltage differential between the two points. Voltage can be read in volts or millivolts. There are 1000 millivolts in 1 volt.

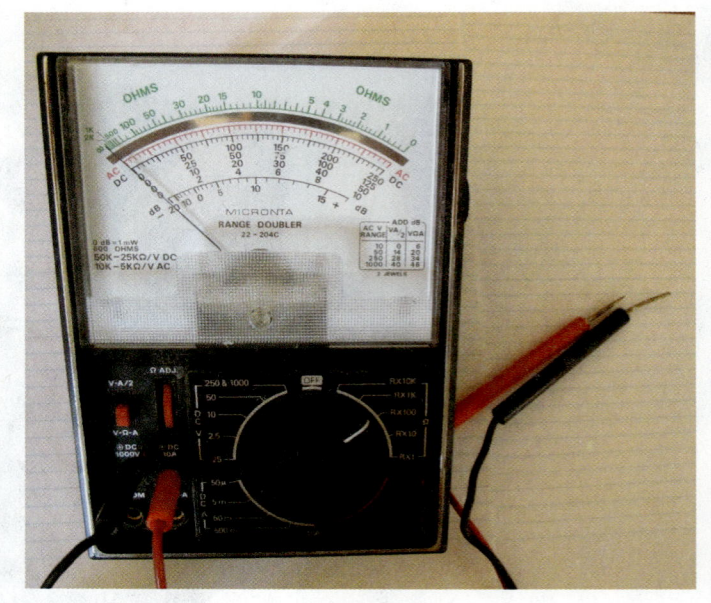

FIGURE 5–42 An analog multimeter.

If the positive or red lead is placed on the positive side of a circuit and the negative or black lead is placed on the ground side of the circuit, you are most likely checking maximum system potential voltage. This is called connecting the meter in parallel to the circuit. If the meter is "auto-ranging," it will automatically display the actual voltage reading. If the leads are backward, you will see a negative sign before the numerical value. If the meter is not auto-ranging, you will have to set the meters dial to the appropriate range setting, and this should be the closest value higher than you expect to see. This kind of voltage testing can determine if there is sufficient system or circuit voltage for proper circuit operation. Normal voltage readings for 12V systems should be between 11.5V and 13.5V, depending on battery state of charge and how many loads are working. If the circuit is not closed, you would be checking OCV. ● SEE FIGURE 5–44 for a voltmeter connected to a circuit.

The voltmeter is also used to perform voltage drop testing. This is really a test for excessive resistance in a live circuit. If the meter's positive lead is placed somewhere on the positive side of a live circuit and the negative lead is placed anywhere past the positive lead toward ground, the meter will read how much voltage is consumed or dropped between the two leads.

If the leads are on either side of a wire, connector, fuse, or switch, the voltmeter should read no more than .3V for a 12V system or .6V for a 24V system. Any reading higher than this will indicate excessive resistance (likely corrosion) because it is taking more pressure to push the electrons through the corrosion. This is a very valuable test because it is performed on a live circuit and can be used to check for too much resistance

164 CHAPTER 5

FIGURE 5–43 A digital multimeter.

anywhere in the circuit. This could also be used to troubleshoot for a bad ground.

AMMETER. There are two ways you can measure the current (electron flow) for an electrical circuit.

1. With the multimeter set on the amp setting, you need to make the meter part of the circuit you're testing or placed in series in the circuit. To do this, the circuit is opened at any point and each lead of the meter is connected with the opened part of the circuit. The circuit is then energized and all the current flowing through the circuit will now flow through the meter, which can be measured. Most meters will only read low amperage circuits usually up to 15 amps. ● **SEE FIGURE 5–45** for a multimeter checking amperage in a horn circuit.

2. The other way to measure current flow and electrical circuit is with a clamp-on meter. ● **SEE FIGURE 5–46** for an all-in-one clamp-on ammeter. A clamp-on ammeter could be attached to regular multimeter or could be a stand-alone unit. The clamp part of the meter simply surrounds the wire that has current flowing through it and measures the amount of magnetism in the wire and converts this into an amperage value. This is called an inductive pickup.

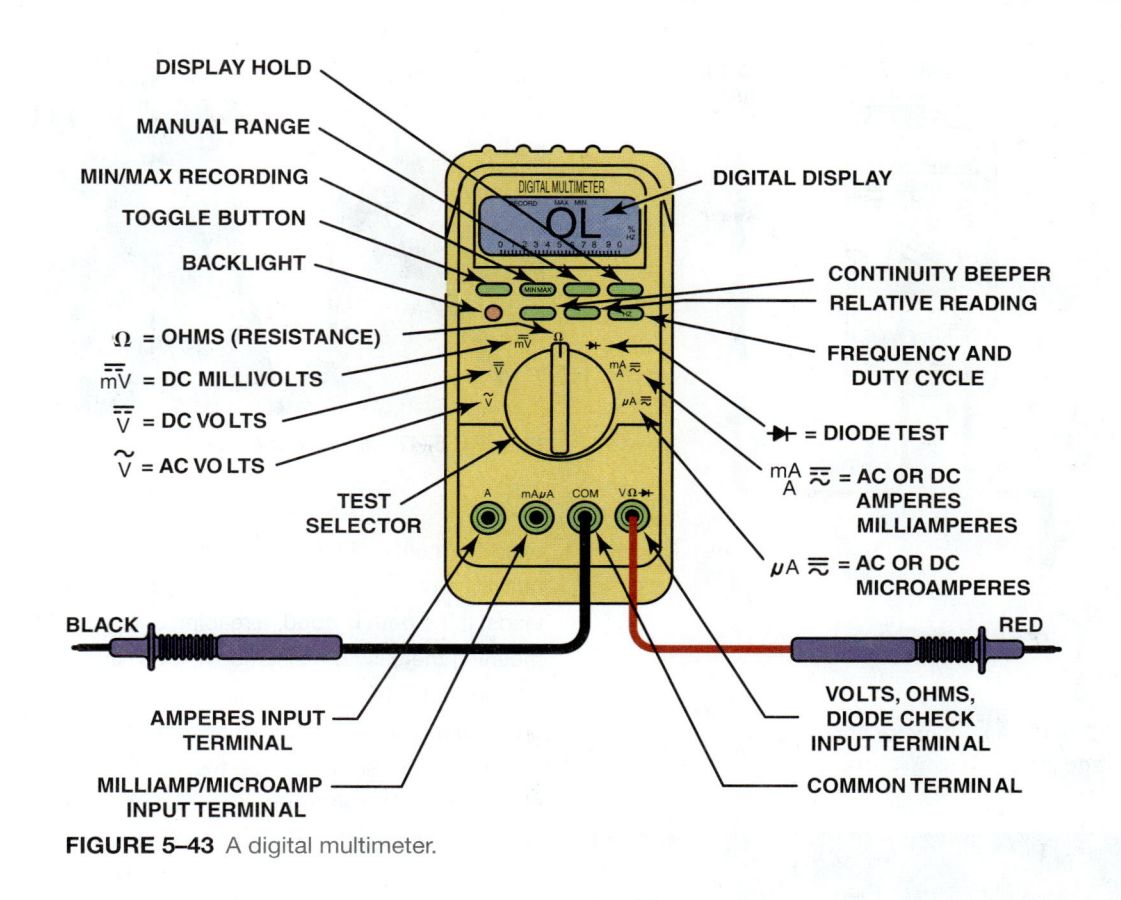

FIGURE 5–44 A voltmeter connected to a circuit.

DIRECT CURRENT ELECTRICAL SYSTEMS **165**

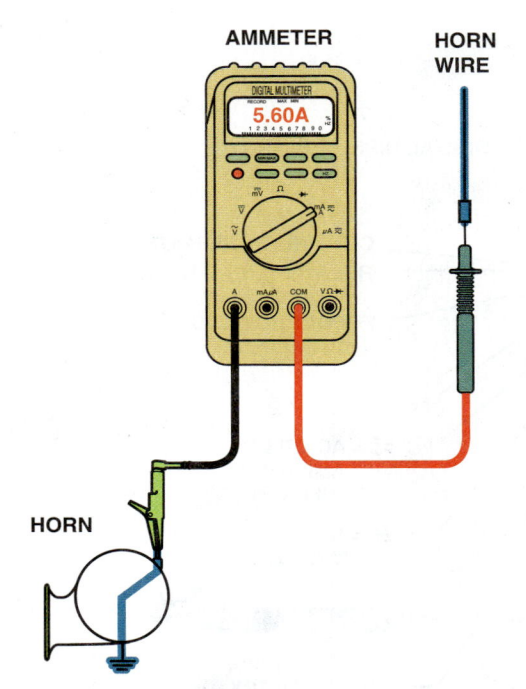

FIGURE 5–45 A multimeter checking amperage in a horn circuit.

FIGURE 5–46 Clamp-on ammeter.

If used as part of a regular multimeter, the rotary dial will need to be set to mV. Clamp-on ammeters can read very small or very large amounts of current flow and can be very useful when troubleshooting.

OHMMETER. To measure resistance in a part of a circuit or component, you need to set the multimeter to an ohms setting, de-energize the circuit or isolate the component, and place the two leads of the meter across the part of the circuit or the component you want to measure. The most common time to use an ohmmeter would be to check the resistance value of a coil

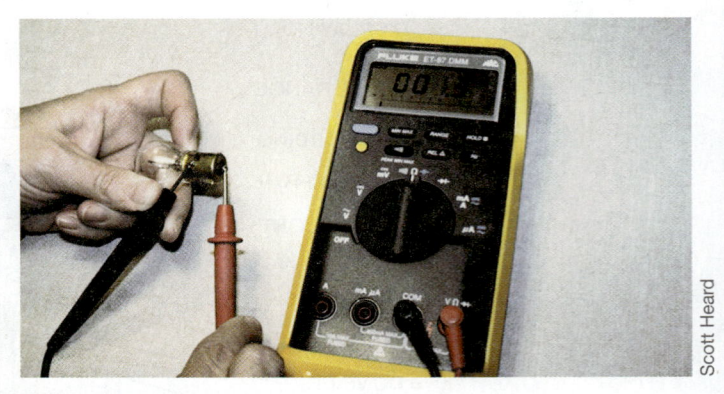

FIGURE 5–47 Ohmmeter in use.

used for a solenoid. An ohmmeter is also used to check continuity of a circuit or if there is a complete path between the two leads. If the path is good, a reading of 0 ohms will be displayed, meaning there is no resistance. If the path is open, a reading of OL (out of limits) will be displayed. ● **SEE FIGURE 5–47** for an ohmmeter being used to test an incandescent light bulb.

Incandescent bulb-type test lights should not be used for electrical troubleshooting because of potential damage that can be done to electronic components.

ONBOARD DIAGNOSTICS Almost all machines sold today will have some type of fault code display system. This could be helpful in finding electrical circuit problems. These systems will be covered in Chapter 10.

LAPTOP COMPUTER Today's machines will almost always have at least one ECM onboard and there will likely be a connector to hook up a laptop to read live data from the machine. This could be valuable for electrical system troubleshooting by being able to read things such as voltage output and switch status.

ELECTRICAL SYSTEM FAULT DIAGNOSTICS When diagnosing a complaint that an electrical system is not operating properly, the same basic guidelines apply as when you are troubleshooting any other system. Verify the complaint, learn the system, gather information, list possible causes of problems, start with easiest causes first, eliminate possible causes until problem is found, repair problem, and test system to verify repair.

Here are a few examples of common complaints related to electrical systems:

1. The wipers only work on high speed.
2. One tail light is out.
3. The heater fan doesn't always work when turned on.
4. The glow plug light doesn't come on.

5. The fuel gauge won't go above half.
6. The wiper is slow.

Most complaints can be put into three categories.

1. The circuit doesn't work (tail light is out)
2. The circuit works sometimes (intermittent)
3. The circuit doesn't work properly (wiper is slow)

The first step in troubleshooting is verifying the complaint. You need to know how the system should normally work to be able to do this. This may require you to read the operator's guide or some service manual information. Make sure that the information applies to the machine you are working on or you could be heading in the wrong direction right from the start. You may also want to refer to the machine's schematic at this time as well to understand the system. Once you understand how the system works, you can then confirm whether the operator's complaint is valid.

If the stated problem exists, then you need to check the "simple stuff first." After checking for stored or active fault codes, you would start a visual inspection. If a fault code is found that relates to the operator complaint, you would look up the troubleshooting procedure related to that fault code.

● **SEE FIGURE 5–48** for a fault code displayed.

The visual inspection includes:

1. Look for bare wires that could ground a component or short across to another component.
2. Look for missing or worn conduit. This could indicate a wire problem.
3. Look for loose or broken connectors and wires.
4. Inspect batteries for:

- corroded terminals
- loose terminals or battery posts
- dirty condition
- damp condition
- cracked case
- proper electrolyte level

5. Check alternator belt tension.
6. If your visual inspection does not indicate the possible malfunction, but your inspection does indicate that the machine can be run, turn the key switch to the IGN position. Try out the accessory circuits, indicator lights, and gauge lights. How does each of these components work? Look for sparks or smoke, which might indicate shorts.
7. Start machine. Check all gauges for good operation and check to see if system is charging or discharging.
8. In general, look for anything unusual.

These simple steps are often overlooked and many times will reveal a problem or at least give you a starting point for further troubleshooting of electrical circuit faults.

ELECTRICAL CIRCUIT FAULTS Once you have identified one or more circuits that are malfunctioning, this is a good time to refer to the electrical schematic. Looking at a bundle of wires behind an instrument panel can be intimidating at first glance, but by looking at a schematic for a few minutes, you should be able to identify the circuit or circuits you are looking for and find some wire numbers and colors to look for on the machine. Looking at a machine's schematic can also be intimidating if you are not familiar with it. This is a good time to find a quiet place where you can study the wiring diagram and make some notes. It's also a good idea to make yourself a simple wiring diagram of the circuit you are looking for. This way you can decide if the problem is part of a series circuit or a parallel circuit. If the problem is in one branch of a parallel circuit, only that branch will be affected.

There are three parts of a simple circuit where the fault can occur:

1. between the battery and the controlling switch
2. between the switch and the load
3. between the load and the ground

The load component can also be defective, and this should be kept in mind when looking for the cause of an electrical problem. If the circuit wiring checks out fine, then the load component must be at fault.

FIGURE 5–48 A fault code.

DIRECT CURRENT ELECTRICAL SYSTEMS **167**

There are four types of electrical circuit faults that you need to be able to solve to correct a machine electrical problem. They are as follows:

1. excessive resistance
2. open
3. grounded
4. shorted

You should have a good understanding of the differences between the four faults and the troubleshooting procedures used to find these faults. Let's look at these faults and how to perform a diagnostic procedure to find them.

EXCESSIVE RESISTANCE As mentioned earlier, all components in an electrical circuit will cause resistance to current flow. Usually, this resistance is relatively small and is accounted for when the circuit is designed.

Excessive resistance in any part of a series circuit will result in improper load operation such as a dim light or lights, slow wipers, quiet horn, or slow heater fan. If the excessive resistance occurs in one branch of a parallel circuit, it will only affect that branch. If excessive resistance is very high, the current flow could be completely blocked, and this would be similar to an open circuit.

Some causes of excessive resistance are the following:

1. loose, corroded, dirty, or oily terminals
2. wire size too small
3. strands broken inside the wire
4. poor ground connection to frame

Many of the previous causes would be found with a good visual inspection.

This troubleshooting procedure involves voltage drop testing. This is where practical use of Ohm's law comes in. The circuit must be energized to find the source of this problem. You may think an ohmmeter would be the choice to locate an excessive resistance problem, but because an ohmmeter puts out very little voltage (usually around 1 volt), the circuit would check out OK.

Check the voltage supply to the switch and compare to battery OCV. There should be no more than a .3V drop or change in voltage across any connector or switch in a 12V circuit. This is a maximum of .6V in a 24V circuit. For this example, let's say the battery OCV is 12.6 and the voltage at the switch is 12.55. You can verify this voltage drop by putting the two voltmeter leads on either side of the switch. You would read .05V in this case. This means the switch contacts are clean and not the problem. You would then go from the output of the switch to the load, or if there are joins in the harness between these components, you could check along the way. When checking at the positive side of the load, you find a voltage drop of .7V. This indicates excessive resistance, and by moving back toward the switch, you can locate the exact point of high resistance.

This looks easy on paper and should make sense. When trying to do this on a large piece of equipment, it could be a tough job because of the distances between components and where they are located. This may require the help of an assistant and some lengthened leads for your meter.

Now that you have located the fault, you would repair it and verify that your repair has solved the problem. ● **SEE FIGURE 5–49** for how to test for a voltage drop.

OPEN An open circuit fault simply means a part of the circuit that is normally complete has opened. This is just like opening a switch and results in the load not working (light stays off, wipers don't move, etc.). An open can occur on the positive side or ground side of the circuit, or in the load component itself. ● **SEE FIGURE 5–50** for a simple circuit with an open.

Some causes of open circuits are the following:

1. broken wire
2. disconnected component terminal or connector
3. pins inside a connector not making contact
4. blown fuse or open circuit breaker
5. failed switch or component
6. disconnected ground wire

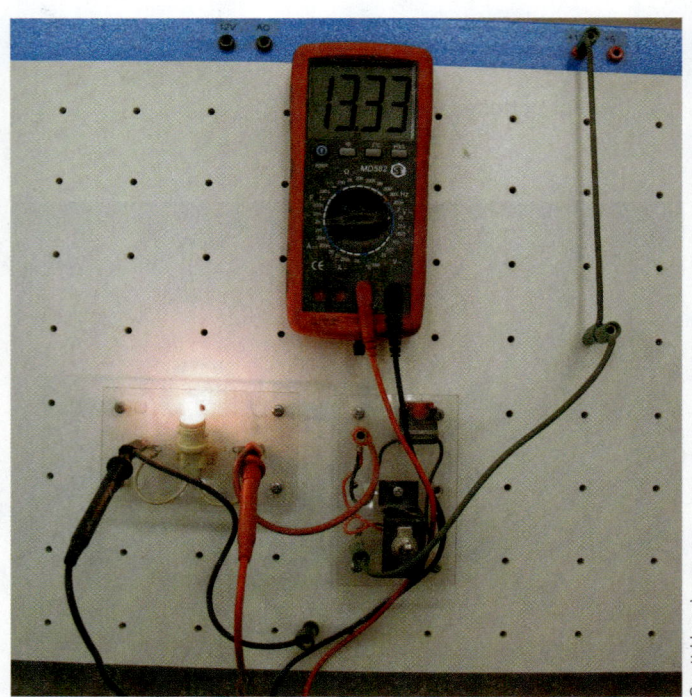

FIGURE 5–49 How to test for a voltage drop in a circuit.

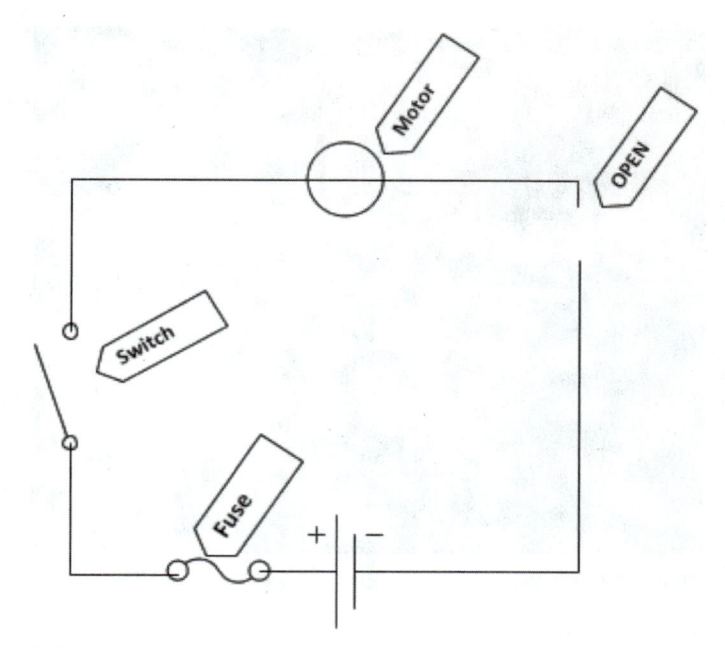

FIGURE 5–50 A simple circuit with an open.

To locate an open circuit, a good place to start is to check the fuse because a blown fuse is an open (even though the cause of the blown fuse is likely another fault). Assuming the circuit fuse is good, you would then start checking along the circuit for loss of voltage. This would start at the switch input, and if battery voltage was found there, you would proceed along to the load. In this example, the open is between the switch output and the load. If the open was past the load on the ground side, you would place the positive lead on a known good voltage source and move the negative lead along the ground side until the open was found. You then make a repair and verify this solved problem.

GROUNDED CIRCUIT A grounded circuit (also called **short to ground**) is created when a part of the positive side of the circuit contacts the ground before the load. When the load is bypassed, there is no resistance to current flow, and if this occurs past the fuse, the fuse should open to prevent overheating of the circuit. If the **ground** occurs before the circuit protection, the result could be a fire. If the fault occurred between the battery and the fuse, the problem should be obvious as there will be a burnt wire that needs to be replaced. A ground is usually caused by a part of a wire having its insulation worn through which exposes the copper to part of the ground side of the machine. If a wire or harness is moving or another part of the machine is moving against the wire or harness, the insulation will be worn through. ● **SEE FIGURE 5–51** for a simple circuit with a short-to-ground fault.

We will assume the fuse has been blown and you have replaced the fuse once, which has failed as well.

To isolate the location of a grounded circuit, set the multimeter to ohms and perform a continuity test. This test will confirm the part of the positive side of the circuit that is connected to the ground side. You will place one of the leads to a good ground and work along the circuit past the fuse isolating any part of the circuit by disconnecting connectors or opening the switch as you go. The part of the circuit that is connected to ground will be indicated by a 0.00 Ohms reading on the multimeter. Repair the ground fault and check circuit operation.

SHORT CIRCUIT The type of fault called a short occurs when the normal path for current flow is diverted to another load.

A shorted circuit causes components in separate circuits to operate when a switch in either circuit is turned on. An example of this could be the wipers operating when a light switch is turned on.

Like a grounded circuit, a short could be the result of wires rubbing together from a harness being moved by vibration or a component moving it. Shorted circuits can also result in a blown fuse if a switch tries to energize a load it isn't designed for. A shorted circuit with a blown fuse will need to be checked with an ohmmeter and check for continuity with other circuits. Remember to de-energize all circuits that you are testing with an ohmmeter.

A process of elimination should reveal the location of the short as you work through the circuit. ● **SEE FIGURE 5–52** for a simple circuit with a short circuit fault. Once the fault is found, make the appropriate repair and check all related circuits.

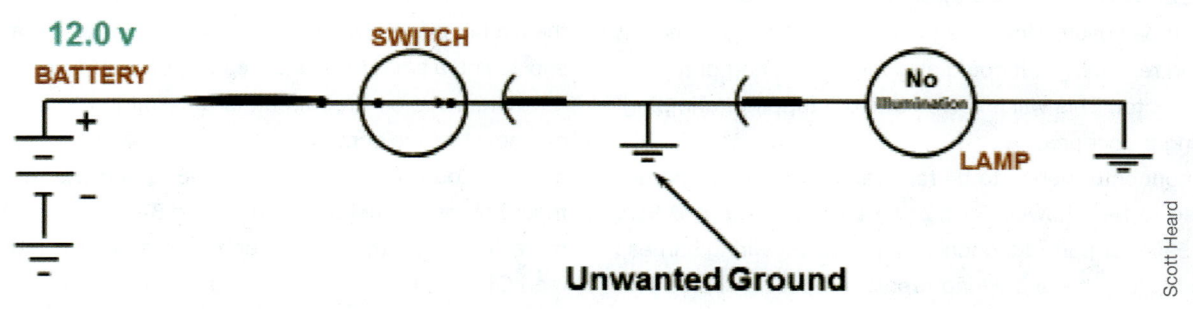

FIGURE 5–51 A simple circuit with a short-to-ground fault.

DIRECT CURRENT ELECTRICAL SYSTEMS 169

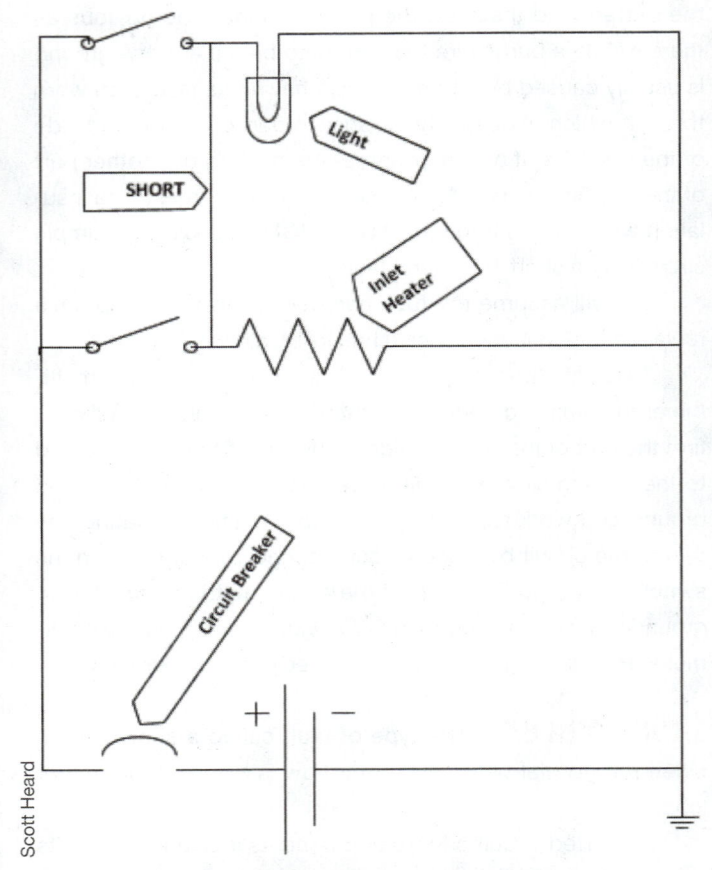

FIGURE 5–52 A simple circuit with a short circuit fault.

WIRE REPAIR To repair circuit wiring after you locate a fault, you need to have certain tools and supplies. These include the following:

1. wire cutters/strippers
2. a sharp knife
3. connector crimpers
4. specific connector tools (crimpers, extractors)
5. wire joiners
6. shrink tube
7. soldering supplies

● **SEE FIGURE 5–53** for these tools and supplies.

Wire repair usually involves cutting out a damaged section of wire and replacing it. It could also mean repairing or replacing a connector. This will involve cutting, stripping, crimping, and sealing procedures.

If a connector needs to be repaired, the specific instructions need to be followed for that type of connector. ● **SEE FIGURE 5–54** for part of a connector repair. If a wiring harness is opened to complete a wiring repair, it must be sealed and secured after the repair is complete.

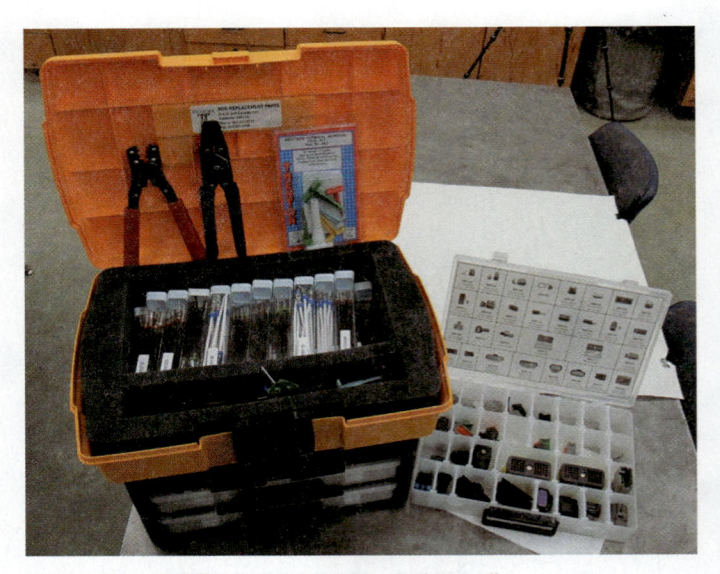

FIGURE 5–53 Wiring repair tools and supplies.

COMPONENT TESTING

SWITCHES. Switches are simple devices and the test procedure for them is also simple. With the multimeter set to ohms, disconnect the switch and test whether the contacts are opening or closing when they should. However, just because there appears to be continuity, don't assume the switch is not part of a circuit problem. To ensure switch operation is good, put the switch back into the circuits and perform a voltage drop test across the switch. Even a small amount of corrosion at the contact points could be enough to not allow full current through the switch. Any switch tested for voltage drop should never create a larger voltage drop than .1V. Switches are typically replaced and thrown away as they are usually sealed units. ● **SEE FIGURE 5–55** for a switch being tested.

RELAYS. Because relays are used so often, you should become very familiar with how they operate and know how to go about troubleshooting one. They are a very simple device yet seem confusing to many technicians. You should practice testing relays to become familiar with multimeter usage. The first part of the relay to be tested is to control circuit. To test this, simply place your own meter on the ohms setting, and with the relay removed from the circuit, place your meter leads on terminals 85 and 86. You should find a small resistance reading of .5 ohms or less.

Once the control circuit is determined to be intact, you can now test the contacts between 30, 87, and 87a. Place your meter leads on terminals 30 and 87a and the meter should read 0 ohms. To test between 30 and 87, you would place your meter leads on those two terminals and your meter should read OL to indicate an open circuit as this is the normally open contact.

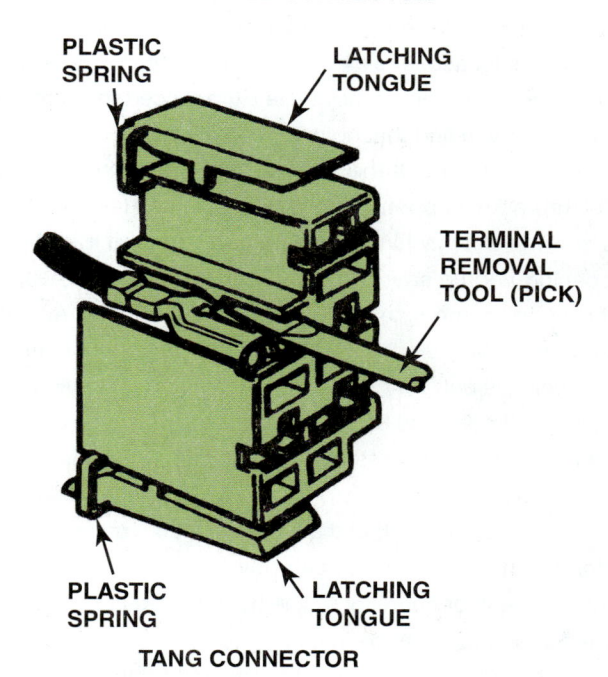

FIGURE 5–54 A part of a connector repair.

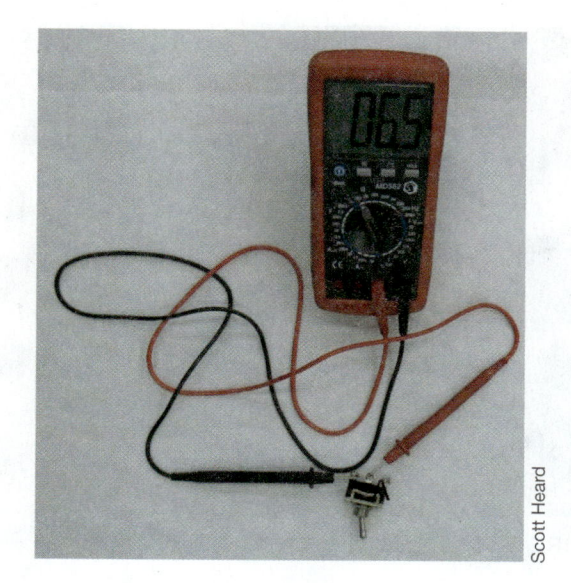

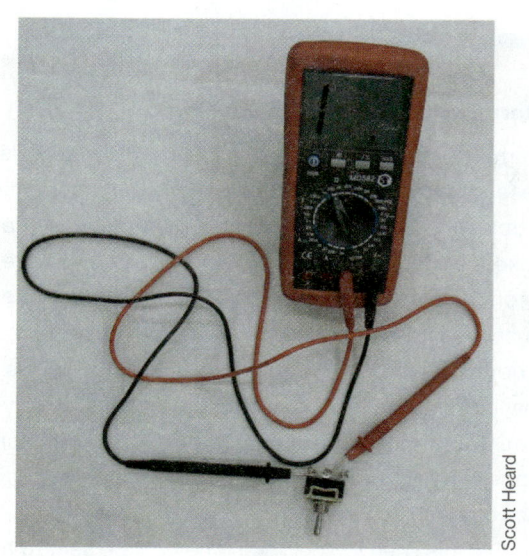

FIGURE 5–55 A switch being tested.

The next part of relay testing is to see if the relay is working and closing the contacts between 30 and 87. You'll need to connect a power source of the same voltage that the relay is rated for to terminal 85 or 86. Then connect a ground lead to the opposite control terminal (85 or 86). As soon as the second lead is connected, you should hear a click, which indicates that the magnet has just closed the contacts. With the contacts closed, you would now place your ohmmeter leads across 30 and 87. You should read 0 ohms, which means the contacts are close. This test procedure should indicate that the relay is operable. However, it may still have problems, because there is no real load across the high-current contacts. The most accurate way of testing the high-current side integrity is to perform voltage drop testing with the relay in the circuit. ● SEE FIGURE 5–56 for the relay test procedure.

SOLENOID. Testing a solenoid is a fairly simple procedure that mainly sees you testing the resistance value of the coil winding. You should be able to find this value in the machine's service information system under specifications. Another way of testing a solenoid would be to perform a voltage drop across it when it is energized.

DIRECT CURRENT ELECTRICAL SYSTEMS 171

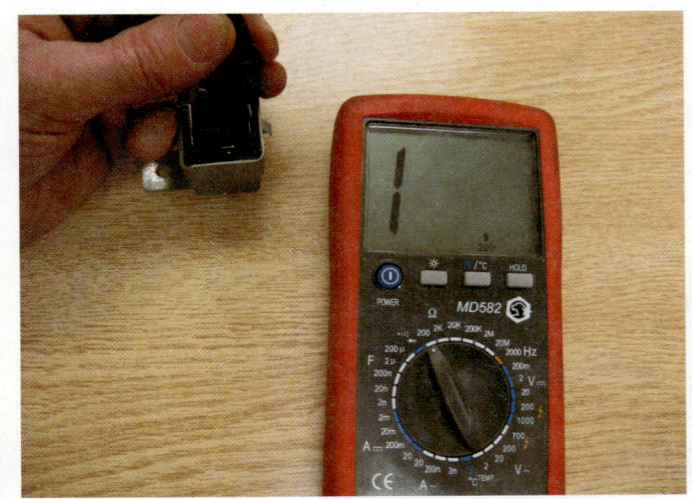

FIGURE 5–56 The relay test procedure.

🔧 CASE STUDY

Intermittent Problem

Intermittent means "occurring at irregular intervals," and these kind of electrical problems are the most difficult to troubleshoot because you can't fix something that's not broken. If the problem doesn't occur when you are working on the machine, you are taking educated guesses at its cause. To find the root cause of these problems, you need to be there when the problem is occurring. This is where clear communication with the operator is crucial so you can understand the exact conditions that make this problem happen. This is also where electronic fault code logging capabilities with newer machines comes in very handy.

For this scenario, let's assume you are sent to a newer wheel loader with an electrical problem, and when you get to the machine, you ask the operator what the problem is. He tells you that when it rains or is very humid, the horn starts beeping. Unfortunately, the weather is nice and dry the day you are there so the problem is not occurring at that time. You continue anyway with your basic visual checks. When checking for fault codes, you find several related to the communication system for the machine. You record these code numbers and clear them and perform a thorough visual inspection especially near the horn wiring. Before leaving the machine, you ask the operator to record the conditions that are present when the problem occurs next time.

After leaving the machine, you do some research on the horn circuit and find it is connected with the payload system. When the operator beeps the horn to signal the truck he has just loaded to drive on, the signal to the horn resets the scale system to tell the ECM that another truck is going to be loaded. You record all wire numbers and harness locations related to the horn circuit. You ask your fellow technicians if they have experienced the same problem and you find that no one has.

You receive a call the following week for the same problem and it happens to be a very damp day. When you arrive at the machine, you ask the operator if it is still happening and he says it happened three or four times, first thing in the morning and hasn't happened since. After more questioning, he tells you it seems to happen when the machine is bouncing over rough terrain. You check for logged fault codes and find the same communication error ones that were cleared last time. You look up the troubleshooting procedure for these and try to see what would relate to the conditions that the problem occurs under. The switch for the horn is in the cab obviously and is not affected by weather conditions. The horn harness has been inspected and found to be intact.

The horn switch is an input to the ECM because it is part of the payload system and the ECM then sends a 24V signal out to the horns. You start to focus on the electronic system and find the compartment where the ECM is located is exposed to the elements and is very wet inside. You then start to wiggle wires, and only after misting water on the connector that goes into the ECM and while wiggling the harness, does the horn start to beep.

After inspecting the connector, you find it is cracked, and this has allowed water and dirt to get into the connection and has allowed a short to occur, which under the right conditions causes the horn to beep. You proceed with a connector repair and test the circuit while simulating rain and vibration and the horn only beeps when the button is pushed by the operator. You also check for fault codes and no new ones are found.

This is an example of an electrical problem that I came across and you may find yourself troubleshooting similar problems someday.

By following a step-by-step procedure, you will be able to find and repair any electrical problem.

Progress Check

29. The multimeter will have an ammeter setting and if set to check amps:
    a. the red lead is moved and both leads are connected in parallel with the load
    b. the black lead is moved and both leads are connected in series with the load
    c. the red lead is left out and the black lead becomes an inductive pick up
    d. the red lead is moved and the leads are connected in series with the load

30. This fault will cause slow wipers or dim lights:
    a. open
    b. short
    c. high resistance
    d. ground

31. This fault will always cause the circuit protection to fail:
    a. open
    b. short
    c. high resistance
    d. ground

32. When performing voltage drop testing, you are testing for this fault:
    a. open
    b. short
    c. high resistance
    d. ground

33. When testing a relay, you hear a click. This means:
    a. its fuse just opened
    b. it's working fine and that's all you need to hear
    c. its control circuit has just closed the contacts
    d. the high-current side has too much resistance

SHOP ACTIVITY

Go to a machine and identify one circuit. For that circuit:
- Operate circuit to see if it is working properly.
- Perform voltage drop testing at each component and wiring junction.
- Isolate the load and perform resistance test.
- Operate circuit and perform current measurement.
- Draw a schematic for the circuit using common symbols and record your findings from steps 1–4 on the schematic.
- Describe the procedure to find open, short, ground, and high resistance for this circuit.

SUMMARY

1. Some machines could have high-voltage systems that have the potential to injure or kill you. Pay attention to all high-voltage warnings on machines and in service information.
2. Older machines had very basic electrical systems with maybe two or three circuits for an electric starter, charging system, and lights. Newer machines have evolved to include a multitude of electrical circuits but basic electrical circuit operating principles have not changed.
3. Electrical principles align with hydraulic system principles. Electrons must flow to actuate an electrical function (turn on a light), and to make electrons flow, there must be a difference in electrical pressure in a circuit.
4. Electrons are the negatively charged particles of atoms and rotate around an atom's nucleus in a series of rings.
5. An atom that has an outer ring with one electron in it will make an ideal conductor because the electron can be moved from one atom to another. In an electrical circuit, a conductor is a wire, which is similar to a hose or tube in a hydraulic circuit.
6. Copper is an ideal conductor because it has one electron in its outer shell.
7. The opposite element to a conductor, electrically speaking, is an insulator. It has a full shell of electrons in its outer shell and, therefore, is very stable. Insulator materials are used for covering copper wires (rubber, plastic).
8. There are two main theories stating in which direction electrons move in DC electrical circuits. Electron theory

says negative to positive while conventional theory says positive to negative.

9. Direct current (DC) circuits and systems are used for the 12V and 24V systems found on almost all heavy equipment. DC means the electrons flow in one direction.
10. Electrons always flow from a higher voltage to a lower voltage.
11. Voltage (V) is the electrical pressure needed in a complete circuit to make electrons flow and can be referred to as electromotive force (EMF). It is measured in volts with a voltmeter. The machine's battery is the main source of electrical pressure.
12. Current (A) is the volume of electron flow that occurs in an operating circuit and is measured in amperes or amps with an ammeter.
13. Resistance in an electrical circuit opposes current flow and is measured in ohms (Ω) with an ohmmeter.
14. Electrical units of measure (V, A, and Ω) can be expressed in larger or smaller units by factors of 10 using prefixes. Common examples are milli (one thousandth) and mega (one millionth).
15. The mathematic relationship between volts, amps, and ohms is expressed by the formula: $E = I \times R$ or Volts = Amps $\times$ Resistance. This is Ohm's law and can be used to calculate one of the values when the other two are known.
16. Watt's law states that electrical power is equal to Volts $\times$ Amps and is measured in watts; 746 watts is equal to 1 horsepower.
17. Magnetism plays an important role in electrical circuits. As current flows through a conductor, a magnetic field is created around it, and this is how an electromagnetic field can be created.
18. A complete circuit provides a path for electrons to flow in a loop that starts and ends at the battery. It normally includes a switch, conductors, circuit protection, voltage source, and a load (resistance).
19. DC circuits can be split into two halves. The positive side goes from the positive terminal of the battery to the positive side of the load and the opposite side of the load is connected to the ground side that goes to the negative terminal of the battery.
20. Three circuit types are series, parallel, and series-parallel.
21. A series circuit means there is only one path for current to flow.
22. A parallel circuit has more than one path for current to flow through and a series-parallel circuit is a combination of both.
23. Series circuits can have one or more loads and the source voltage will decrease proportionately across each resistance in the circuit. Series resistances accumulate to give a total circuit resistance that is equal to the sum of all separate resistances. Use the formula: $R_t = R_1 + R_2 + R_3 + \cdots$.
24. Excessive or unwanted resistance anywhere in a circuit will negatively affect the circuit operation. This could be the corrosion on switch terminals.
25. Parallel circuit total resistance is equal to less than the lowest single resistance and can be calculated with the following formula: $1/R_t = 1/R_1 + 1/R_2 + 1/R_3 + \cdots$.
26. Electrical schematics are like road maps for the electrical system on a machine. There are symbols to represent different electrical components and lines represent the wires to show how they are connected.
27. Batteries are chemical devices that can keep a potential voltage stored. Almost all machine batteries are called 12V batteries, and there are really six batteries connected in series. They are termed *lead-acid batteries* and contain two types of lead-covered plates with liquid acidic electrolyte surrounding them. The specific gravity of the electrolyte changes as the battery charge level changes. Batteries can produce dangerous levels of hydrogen gas when charging.
28. A fully charged battery will have a voltage of 12.6.
29. Most batteries are low maintenance or maintenance free, which means the electrolyte levels stay consistent.
30. Boosting dead batteries must be done carefully to avoid an explosion. The important point is to make the last connection with booster cables away from the battery to a good frame ground.
31. Circuit protection is part of all electrical circuits and will open the circuit if excessive current flows through it. Circuit breakers and fuses are the most common types.
32. Wire size is measured by gauge and a bigger gauge size indicates a smaller wire diameter. It can also be measured in mm^2. Most low-amperage circuit wiring is 14 gauge while battery cable is 000.
33. There are many different types of wiring connectors that are used to join one or more wires or a cluster of wires together. Special tools are needed to repair most connectors.
34. There are many different types of switches used to open and close circuits.
35. Relays are an electromagnetic switching device that allows a low-current circuit to control a high-current circuit.
36. Solenoids create linear movement when energized by having a coil of wire energized, which then creates a magnetic field.
37. Some different types of loads found on machines are resistive (incandescent lights, inlet heaters), motors (electromagnetic), and semiconductors (LED lights).
38. All HDETs should know how to use a multimeter and apply Ohm's law to find the source of electrical problems.
39. Multimeters have at least the following functions: voltmeter, ammeter, and ohmmeter. They can also have other functions such as diode tester, temperature probe, and graphing meter.
40. Clamp-on ammeters can measure current flow by surrounding a wire and converting the magnetic field strength into a current value.
41. Electrical circuit troubleshooting starts with verifying the complaint and then eliminating possible causes starting with the simplest first.
42. Thorough visual inspections can reveal many problems such as bare or loose wires or corroded terminals.
43. Four types of electrical faults are open circuit, short to power, short to ground, or excessive resistance.
44. Different testing procedures are used for the different types of faults.
45. Wire repairs could include crimping, soldering, heat shrink, and connector repair.
46. Component testing can detect faulty electrical components.

chapter 6
PRIME MOVERS

LEARNING OBJECTIVES

After reading this chapter, the student should be able to:

1. Describe the operational fundamentals of diesel engines.
2. Identify the major components of a diesel engine.
3. Interpret the power, torque, and fuel consumption ratings of diesel engines.
4. Describe the various properties of diesel engine exhaust.
5. Interpret the current and future emission regulations.
6. Describe a low-power diagnostic procedure.

KEY TERMS

Camshaft 188
Cavitation 199
Crankshaft 187
Horsepower 180
Long block 182
Piston 177
Prime mover 176
Reciprocating 212
Torque 178

INTRODUCTION

FIRST When working near prime movers, there are several safety concerns you should be aware of to keep yourself from being injured or killed.

Because diesel engines create a great deal of heat, you should be aware of the many hot components and surfaces you could come into contact with. This is especially important around exhaust and cooling system components and in particular any newer reduced emission engines that need huge amounts of heat in their exhaust treatment systems. Some exhaust components can reach over 1000°F.

Pressurized engine components can also be a cause of personal injury if not respected. Cooling system, lubrication system, and fuel system injuries can contain pressurized liquids that can also cause harm. Pressure levels can range from 10 psi in cooling systems to today's engines that use common rail fuel systems that can create over 30,000 psi.

Prime movers create rotation, and they usually have some external rotating parts that pose lots of hazards that the HDET needs to pay attention to. Things such as belts, pulleys, cooling fans, and drive shafts will all cause serious injury if allowed to.

Heavy-duty diesel engines will likely have some very heavy components that could pose crushing and pinching hazards when repairs or service is being done. This should be done with the proper lifting and blocking procedures and equipment used.

Diesel engines need a variety of fluids to operate and some of these fluids can be harmful to your health if not handled properly.

● **SEE FIGURE 6–1** for some examples of warning labels you may see on or near a machine's prime mover.

If an electric generator is driven by the prime mover, it could create very high voltage (up to 600 volts), and this should be given a great deal of respect. Do not touch anything you are not familiar with.

LOOK AFTER OUR ENVIRONMENT Used engine oil and coolant should be disposed of properly and should be recycled whenever possible. Ground water sources, plants, and animals can all be seriously affected by contamination from engine fluids. There are severe fines that will be assessed if anyone is found to be dumping used or new engine fluids illegally.

WHAT IS A PRIME MOVER?

The term *prime mover* refers to the primary power source that creates a rotational torque output for a piece of heavy equipment to be mobile and operate its implements. In simpler terms, the prime mover is what powers the machine. Most prime movers used for heavy equipment will have their output directly drive one or more of the following:

1. a wet or dry clutch
2. a torque converter
3. an electric generator
4. one or more hydraulic pumps
5. a drive shaft

When the first self-powered or mobile heavy equipment was developed around the turn of the last century, the prime mover was steam power. This was considered an external combustion heat engine because the heat source (wood or coal) was located outside of the engine's rotating components.
● **SEE FIGURE 6–2** for a functional steam shovel.

The steam-powered equipment of the day was fueled by coal because it was a cheap and plentiful source of energy. As you can see in the picture, emissions were not a concern. As time moved on and technology progressed, engineers found they could only get a limited amount of power out of a steam engine so they started to develop the internal combustion

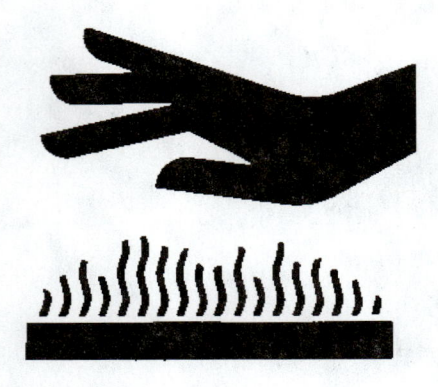

FIGURE 6–1 Some typical warning signs you may see near the prime mover.

FIGURE 6–2 An operational steam shovel.

engine. Rudolph Diesel came up with the compression ignition engine that was originally fuelled with peanut oil but later burned refined crude oil.

Almost all machines sold today use a diesel engine as their prime mover; however, there are other types of internal combustion engines (gasoline, natural gas, bio-fueled, and propane reciprocating engines) as well as electric motors used. Prime mover power ratings used for heavy equipment can range from just under 20 hp to several thousand horsepower and torque ratings of 20 ft-lb to several thousand foot-pounds. Hybrid systems are just now being implemented as prime movers, and these will use a combination of internal combustion and stored electric or hydraulic power. Some underground mining equipment will use a diesel engine to move the machine to a work location and then use an electric motor to drive its hydraulic system. Some very large mining shovels will only use electricity to power all functions of the machine.

This chapter will focus mainly on diesel engines. There are some very good text books already written on diesel engines and their fuel systems, so this chapter will mostly look at the external systems that may be unique to diesel engines that are used in heavy equipment. External systems would include cooling, low-pressure fuel, intake, and exhaust systems. However, we will start off with a brief discussion about how a four-stroke-cycle compression ignition engine works.

Major improvements have been made in recent years to improve the diesel engine in terms of fuel efficiency, longevity, and emission reductions. These improvements have mostly been the result of new fuel injection technology and electronic control systems. The engine's electronic control and monitoring systems will be covered in Chapter 10.

NEED TO KNOW

DIESEL ENGINES Internal combustion engines produce a rotating torque as a result of a thermal event (controlled combustion) that occurs inside a sealed combustion chamber. The combustion chamber is cylindrical in shape and contains a movable metal plug inside it called a **piston**. You could compare the piston in the cylinder to a bullet in a gun barrel. The controlled thermal reaction on the top of the piston results in a pressure increase that will force the piston down inside the combustion chamber similar to the bullet being pushed down the gun barrel when the gunpowder explodes behind it. The bullet leaves the barrel but the piston is pushed back up the cylinder because it is attached to a crankshaft. The piston can repeat the process over and over while the bullet leaves the barrel to never return. Once the piston is "reloaded" or pushed back up the cylinder, it can transfer the force from the next thermal reaction to the crankshaft again. The crankshaft will then rotate the engine's flywheel and the flywheel can now transfer this rotary motion to whatever is attached to it (clutch, torque converter, hydraulic pump, etc.).

There are several variations of the internal combustion reciprocating heat engine. Most North American cars, light-duty pickup trucks, and light-duty equipment use gasoline-fueled spark-ignited piston engines. This engine design is favorable to light-duty applications because of its light weight and low production cost in comparison to the diesel engine.

Back in the day when agricultural and heavy equipment manufacturers decided they needed a prime mover that was durable, had lots of torque at low speed, and was fuel efficient, they looked to the diesel engine to satisfy these needs. Diesel engine design and function lends itself perfectly for use in heavy equipment.

The diesel engine is synonymous with compression ignition engines. Compression ignition means that the combustion process is started as a result of air that has been compressed in the cylinder. This combustion process is what makes the diesel engine so different from the spark-ignited engine. When air is compressed by the piston, it is also heated up, and when it is heated to a high enough temperature, it will support combustion. This is called the autoignition temperature, and diesel fuel has a minimum autoignition temperature of 380–500°F (200–250°C).

A diesel engine needs a high-pressure fuel system to inject fuel into the combustion chamber at precisely the right moment when the air is hot enough to ignite the fuel. Because the basis of compression ignition relies on squeezing air enough to create high temperatures to ignite fuel, the engine design needs to be different. It differs from spark-ignited engines that have a compression ratio of between 7:1 and 11:1 and has compression ratios from 15:1 to 23:1. Compression ratio refers to the volume of a cylinder above the piston and below the cylinder head when the piston is at its lowest point of travel versus the volume when the piston is at its highest point of travel. To increase compression ratios, the piston must travel farther up in the cylinder, and the only way to do this is to have a longer throw on the crankshaft. ● SEE FIGURE 6–3

for a look at the main internal components of a diesel engine. The crankshaft throw is the distance from the centerline of the crankshaft to the centerline of the connecting rod journal on the crankshaft. The connecting rod transfers motion and torque between the piston and crankshaft. A longer crankshaft throw will create more torque at the crankshaft as the piston is driven down on the power stroke. As the piston is being driven down on the power stroke, the connecting rod big end (attached to the crankshaft) will swing out to one side of the engine. The farther out the center of the crankshaft rod journal is from the crankshaft centerline when they are at right angles, the more torque is developed. **Torque** is force times distance so if force or distance is increased, then torque output is increased. Ideally, maximum force on the top of the piston should be developed when the piston is just starting to travel down on the power stroke.

It is this by-product of the need for a diesel engine to have high compression ratios that gives it more torque than a gasoline engine. Another benefit to having a longer stroke is that the force on top of the piston has a longer time to act on the crankshaft, and this translates to the diesel engine having

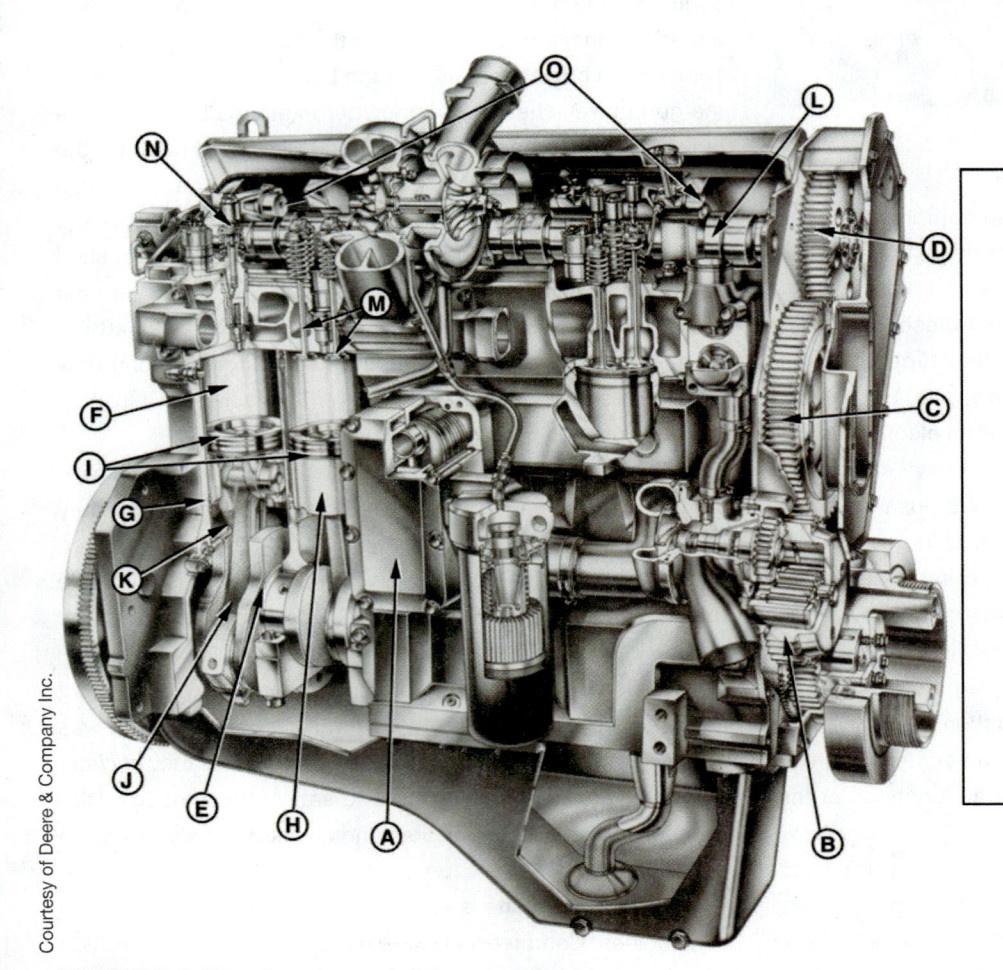

a-	Oil cooler
b-	Oil pump
c-	Idler gear
d-	Cam gear
e-	Crankshaft
f-	Cylinder
g-	Liner seals
h-	Piston
i-	Piston rings
j-	Connecting rod
k-	Cooling jet
l-	Camshaft
m-	Valves
n-	Follower
o-	Follower shaft

FIGURE 6–3 Diesel engine main internal components.

(Continued)

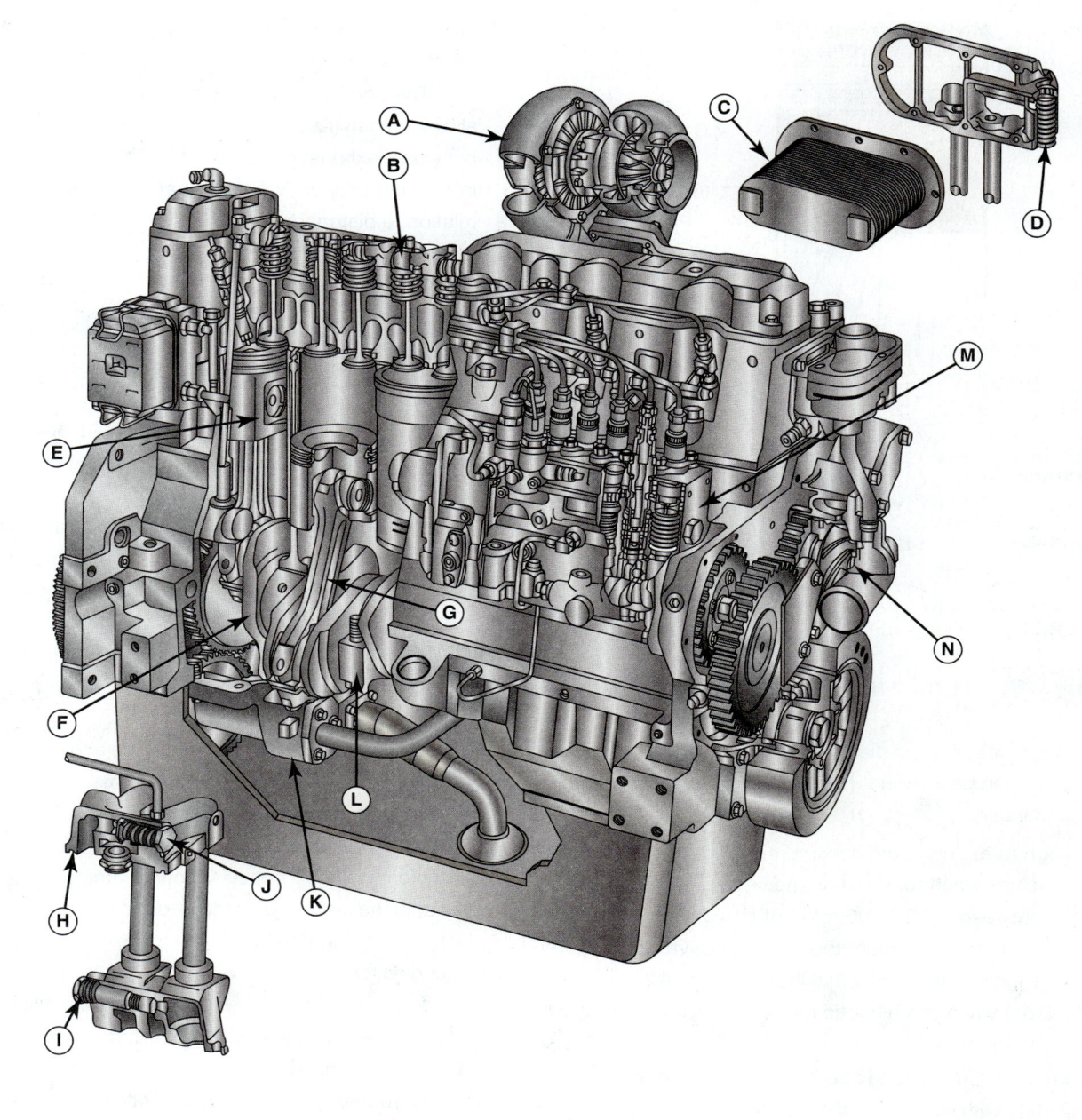

A—Turbocharger
B—Rocker Arm Shaft
C—Oil Cooler
D—Oil Cooler Bypass Valve
E—Piston
F—Crankshaft
G—Connecting Rod
H—Oil Filter Base
I—Oil Pressure Regulating Valve
J—Oil Filter Bypass Valve
K—Oil Pump Assembly
L—Main Bearing Cap
M—Fuel Injection Pump
M—Coolant Pump

Courtesy of Deere & Company Inc.

FIGURE 6–3 Diesel engine main internal components

excellent low-speed, high-torque characteristics. ● **SEE FIGURE 6–4** for how a greater torque is produced with a longer crankshaft throw.

Piston stroke refers to the total length of travel in the combustion chamber a piston makes from the highest (TDC—top dead center) to the lowest (BDC—bottom dead center) point of travel, or vice versa. A four-stroke engine, therefore, will require the piston to make four complete strokes to fully complete the combustion process.

All diesel engines used for heavy equipment will be multi-cylinder, and the cylinders will fire in a specific order. This is called the firing order and will always start with the number one piston. The firing order is created to minimize the counteracting forces on the crankshaft, which in turn will make the engine run as smooth as possible. The number one piston will usually be the one closest to the front of the engine and the opposite end from the flywheel. For an inline six-cylinder engine, the firing order will always be 1-5-3-6-2-4. This firing order will

PRIME MOVERS **179**

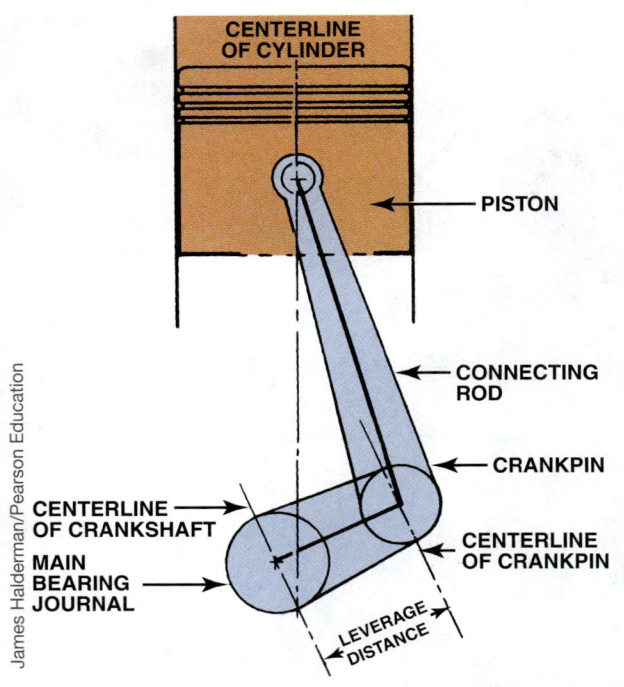

FIGURE 6–4 Longer crankshaft throw equals more torque.

need to be referenced when adjusting valves or timing the fuel injection system.

The combustion process couldn't be completed without allowing fresh clean air into the combustion chamber and forcing waste exhaust gases to leave. This is done with the use of intake and exhaust valves that are timed to open and close in relation to piston travel.

THE FOUR-STROKE CYCLE

A four-stroke diesel engine will have its piston travel timed to its valve train to ensure that valve opening and closing occurs at precisely the right time. Most engines will have these timed events fixed. Some engines will have mechanisms that provide variable timing for either the exhaust or the intake valve, or both. ● SEE FIGURE 6–5 for the four-stroke cycle.

TECH TIP

When describing a diesel engine's power output and operating parameters, there are a few terms you need to understand:

High idle: This is the maximum speed a diesel engine is governed to. For most heavy equipment, this is between 1900–2200 rpm but could be as high as 3000 rpm. Compare this to automotive spark-ignited engines that have a maximum speed of around 5000 rpm while some exotic racing engines have speeds of up to 20,000 rpm!

Low idle: The engine speed at the lowest governor setting with no load on the engine.

Rated speed: The speed at which the engine makes its maximum horsepower. For example, if its high idle is 2000 rpm, its rated speed is usually around 1800 rpm.

Peak torque: The speed at which the engine makes its maximum amount of torque because cylinder pressure is at its highest. Another example is if the same engine has a high idle of 2000 rpm, its peak torque speed could be around 1600 rpm.

Lug condition: If the load on the engine keeps increasing, the engine rpm keeps decreasing, and if it drops below peak torque speed, the engine is now in a lug condition.

Full load speed: This is the speed the engine can maintain when it is under its maximum machine load (torque converter stall and hydraulic stall).

BSFC: Brake specific fuel consumption is a measure of the engine's fuel consumption at rated speed.

Torque: The amount of work produced at the engine's flywheel. If work equals force times distance and if an engine had a flywheel with a 1 ft radius and could create a force of 500 lb when measured at the outside diameter, it would be said to be producing 500 ft-lb of torque.

Torque rise: This is the percentage of torque output rise from torque output at high idle speed to its peak torque output.

Brake horsepower: The power measured at the flywheel by an engine dynamometer. **Horsepower** is the rate of work done by an engine. One horsepower is produced when an engine puts out 550 ft-lb of torque per second. Horsepower = Rpm × Torque/5252. The metric equivalent is the kilowatt and can be found by multiplying horsepower by 0.746.

Torque converter stall speed: The speed an engine can maintain if it is driving a torque converter that is at stall.

Hydraulic stall speed: The speed the engine is brought down to when the hydraulic system is at main relief pressure.

180   CHAPTER 6

4 Stroke Cycle

Intake   **Compression**   **Power**   **Exhaust**

FIGURE 6–5 The four-stroke cycle.

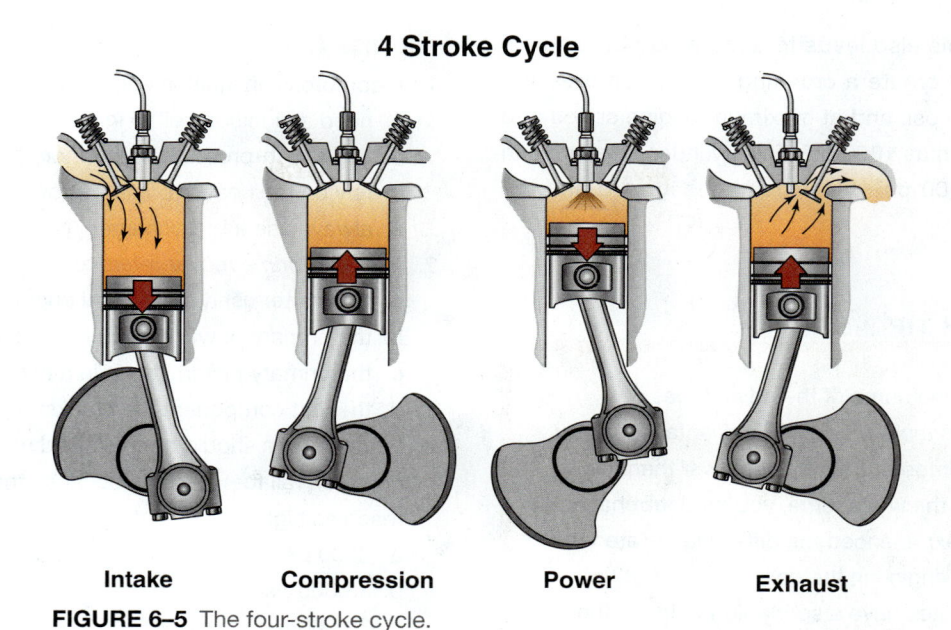

Let's go through the four strokes that a piston makes starting on the intake stroke. The intake valve is opened when the piston is still moving up on the exhaust stroke and the exhaust valve is still open. This is called valve overlap, and it allows the remaining exhaust gases to be flushed out with the intake air coming into the cylinder. As the piston moves down, it creates a void that is filled with filtered air that flows through the air cleaner. As the piston starts to move up from BDC, the valve closes and the cylinder is sealed. The air is now compressed and heated up. Near the top of piston travel, fuel is injected into the hot air and starts a thermal expansion event. This rapid increase in pressure in the cylinder forces the piston down from TDC to BDC to create the power stroke. The power stroke transfers force from the top of the piston to the connecting rod, which turns the crankshaft where it can be used to drive whatever is attached to the flywheel end of the crankshaft (clutch, torque converter, hydraulic pump). As the piston moves farther down on the power stroke, the exhaust valve opens and the exhaust gases start to leave the combustion chamber. At BDC, the piston reverses, and as it moves up the exhaust valve, is kept open until the piston goes past TDC slightly. These four stokes occur over 720° of crankshaft rotation.

You have probably realized that a four-stroke engine needs to perform a lot of work to get the combustion process complete. There are three other strokes that the piston needs to complete before any work is produced on the fourth stroke. There are also other power robbing engine support systems that need to be operated such as:

1. intake and exhaust valve mechanisms
2. fuel systems
3. lubrication systems
4. cooling systems
5. electrical systems

Most of the power gain that is realized on the power stroke is consumed by the energy it takes to complete the other three strokes and by keeping all the other systems functioning. Heat loss also accounts for a big part of the energy loss. Although diesel engine efficiency has risen dramatically lately with the advances in electronic controls, most engines are still lucky to get higher than the 45% efficiency mark. This means that for every gallon of fuel burned, just 45% of the energy value of that fuel is actually converted to torque output. This is still much better in comparison to a gasoline engine that has an efficiency rating of 35%, although this too continues to improve.

All diesel engines used in current production heavy equipment use the four-stroke design. Up until about 20 years ago, some machines used two-stroke diesel engines, which were simpler but not very fuel efficient or environmentally friendly when it came to exhaust emissions. In theory, a two-stroke engine should have twice the output as a four-stroke engine, but this is not the case. You may still come across these engines in some machines, but they are slowly being fazed out of use.

As mentioned before, one of the benefits of a diesel engine is its durability. The main reason for this is due to the fact that it can develop maximum torque at a relatively low speed. Because of the low speed, there is much less friction created between the engine's moving parts, which equates to less wear.

The diesel engine is also built to withstand high compression and combustion pressures that lead to it being

PRIME MOVERS

more robust, and this also leads to a more durable engine. A diesel engine will create a cranking speed compression pressure of 300–500 psi, and at maximum engine speed and load, it will be as high as 1800 psi. This cylinder pressure can increase to over 3000 psi when the fuel is ignited and the piston is near TDC.

TECH TIP

We have discussed some of the advantages of a diesel engine but there are also disadvantages. If you live where the temperature may get lower than the freezing mark in the winter time, you no doubt have heard about or experienced the difficulties related to starting a diesel engine in the cold. Why is that?

Gasoline engines have a spark plug to ignite the fuel/air mixture while diesel engines rely on high cylinder temperatures to ignite the fuel sprayed into the cylinder, and cold morning start-ups don't promote this process. In fact, almost everything good about diesel engines go against cold start-ups. They usually use higher viscosity oil for their lubrication so this oil creates higher friction when it is cold and thick and, therefore, slows down cold engine cranking, which reduces compression pressure and temperature. Internal engine components such as pistons, rings, and liners that are cold not only absorb heat but also don't seal it, which also leads to lower compression temperatures. Also the fact that diesel engines have higher compression ratios means the starting system (starter, batteries, wiring) has to be much stronger and in top condition to overcome this higher cranking resistance. Any deficiency with the starting system will equate to an engine that is hard to start.

In very cold temperatures of –40°F and lower, diesel fuel can start to solidify as well. What can be done to counteract this? Starting aids such as glow plugs and inlet heaters will help to raise the temperature inside the combustion chamber. Coolant and oil heaters will help raise the temperature of the engine itself. Extra battery capacity will also help a diesel engine start better in the cold because if the engine turns faster, it will create more heat in the combustion chamber.

Diesel fuel heaters could also be added to the fuel system to reduce the chance of the fuel solidifying or waxing up.

Progress Check

1. A compression ignition engine will:
   a. need an ignition coil to ignite the air fuel mixture
   b. use a high-pressure carburetor
   c. rely on the heat of compression to ignite fuel
   d. always use a turbocharger
2. The term *prime mover* refers to:
   a. the starter used on a diesel engine
   b. the primary power source for a heavy-duty machine
   c. the primary piston stroke in a two-stroke engine
   d. the first component the crankshaft turns
3. Extra caution should be exercised when working near common rail fuel systems because the fuel pressure could reach as high as:
   a. 3000 psi
   b. 13,000 psi
   c. 30,000 psi
   d. 300,000 psi
4. The next stroke after the power stroke in a four-stroke cycle is the _____ stroke:
   a. intake
   b. exhaust
   c. injection
   d. compression
5. This would *not* be considered a cold starting aid for a diesel engine:
   a. turbocharger
   b. glow plug
   c. air inlet heater
   d. coolant heater

LONG BLOCK COMPONENTS

The term *long block* refers to the main mechanical components that are needed to complete the combustion process. If you were looking at a long block, you would see the engine block with an oil pan on the bottom and the cylinder head(s) with the valve cover(s) on top. The long block does not include intake or exhaust manifolds, cooling, or fuel systems. A long block assembly does include the following:

1. pistons and rings
2. connecting rods
3. cylinder liners or sleeves
4. cylinder head
5. intake and exhaust valves
6. crankshaft and bearings
7. valve drivetrain
8. cylinder block

In contrast to a long block assembly, a short block assembly is strictly the main components of the lower end of a diesel engine (block, crankshaft, connecting rods, pistons).

Let's take a look at the components that make up a long block assembly.

PISTONS. Pistons are designed to create a pumping action inside the engine's cylinder and are used to move air in, exhaust gas out, compress air, and transfer the forces of expanding gas on the power stroke to the connecting rod. Diesel engines need pistons capable of withstanding very high compression pressures (350–700 psi) and high combustion pressures of 3000 psi plus. The piston also needs to withstand great amounts of heat that can exceed 3000°F. The shape of the top of the piston will have a big effect on the engine's efficiency and output characteristics. There are many variations on piston top shape and design from being completely flat to raised to dished out and anything in between. Some pistons will have pockets machined out to give clearance for the valves when the piston moves up to TDC. The newest style of heavy-duty piston is termed the Monotherm made by Mahle, and it's a one-piece forged steel piston that has a unique skirt shape that reduces weight yet maintains strength. The Monotherm piston has a "Mexican hat" design on its top.

Pistons are attached to the connecting rod with a piston pin. This pin allows the piston to pivot on the rod as the piston moves up and down and the connecting rod swings back and forth. Piston pins need to withstand a great deal of force on a concentrated area and are usually held in place with snap rings that fit into grooves in the piston and center the pin in the piston. ● SEE FIGURE 6–6 for a variety of different piston designs.

Pistons have grooves machined in the top outside diameter of them to hold a variety of piston rings in place. Sometimes there will be ring groove inserts made from a different material than the rest of the piston. Because of the high heat that pistons have to deal with, they will usually need to be cooled with engine oil. This is sprayed underneath the piston and the

FIGURE 6–6 A variety of different piston designs.

PRIME MOVERS **183**

heat is transferred from the piston to the oil. The oil is sprayed from cooling jets that must be directed at a specific spot on the piston.

PISTON RINGS. There are two types of piston rings used: compression rings and oil control rings. Compression rings must create a seal between the piston and cylinder wall. The cylinder head and gasket will seal the top of the combustion chamber. These rings are crucial to the diesel engine's performance because of the need for a good seal to raise pressures and temperatures on the compression stroke and to keep combustion pressure sealed. This will keep all the force created on the power stroke pushing on top of the piston. If this seal is less than ideal, then engine performance will suffer and blowby is the result. Blowby occurs when combustion pressure goes past the piston rings, and into the crankcase, this is wasted energy and will result in low-power complaints from the machine operator. Each piston will typically have three compression rings spaced closely apart near the top of the piston crown. Piston compression rings are flexible and will fit into a groove in the piston. The rings have springlike properties that help keep them out tight against the cylinder. Compression and combustion pressure will also get behind the compression rings to help the sealing effect. Any contamination that reaches the combustion chamber will lead to excessive wear of the rings or liners, which will start to diminish the engine's designed output. Compression pressures can be measured with special tooling and a gauge, and a less than specified pressure could indicate worn compression rings.

The lowest ring on the piston is an oil control ring. This ring is used to distribute and collect a thin film of oil on the cylinder liner to help lubricate and seal the compression rings. These rings need assistance in being held out against the liner so that they can be effective in controlling oil so they will usually have some kind of spring behind them. ●**SEE FIGURE 6–7** for a variety of piston ring designs.

CONNECTING RODS. The piston is attached to the crankshaft with a connecting rod that has a cap that can be separated. The cap is fastened together with either bolts or studs and nuts once the cap is installed over the crankshaft journal. There will be a two-piece bearing that fits between the connecting rod and the crankshaft journal. This is called the big end of the rod. The other end of the rod is called the small end and has a piston pin go through it and a bore in the piston to keep the two parts attached to each other. The piston pin will fit into a plain bushing in the connecting rod. The newest style of connecting rod is called a fractured cap. When the rod is manufactured, it is laser etched and fractured at the big end to give the rod a natural

FIGURE 6–7 A variety of different piston rings.

mating surface. This will make a more secure connection when the cap is fastened back together. Other more traditional ways of making cap to rod mating surfaces involve a machining process with dowels or keyways to help secure the connection. The goal of engine manufacturers with all moving components is to keep them strong while reducing weight. Weight reduction equals a reduction in parasitic losses as it takes energy to keep moving parts in motion. ●**SEE FIGURE 6–8** for a fractured cap connecting rod from a heavy-duty diesel engine.

CYLINDER LINERS. Light-duty (150 hp and less) diesel engines, like gasoline engines, will not likely use liners, but they will have the engine block precisely machined to provide cylindrical holes for the pistons to travel up and down. Most heavy-duty engines, however, will use a wet sleeve type of cylinder liner. This is a replaceable sleeve that fits in the engine block and will be exposed to engine coolant on most of its outside diameter. This allows the engine to be reconditioned more economically and quicker because the sleeves alone can be reconditioned or replaced and the entire block doesn't have to be reconditioned.

The inside surface of the liner that is exposed to piston travel will have a crosshatch pattern ground into it when manufactured or reconditioned. This crosshatch pattern will serve to retain a small amount of engine oil when the oil control ring deposits it there on the piston upstroke. The oil film will help to transfer heat to the liner where it then be transferred to the engine coolant. Cylinder liner crosshatch inspection is a good indication of how well the engine has been maintained. A high hour or poorly maintained engine will have this crosshatch pattern worn off and look shiny. This will lead to low-power and high-oil consumption. ●**SEE FIGURE 6–9** for a cylinder crosshatching.

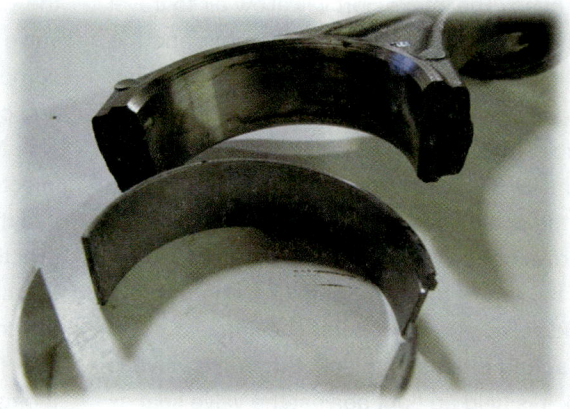

FIGURE 6–8 A heavy-duty fractured cap connecting rod.

FIGURE 6–9 Cylinder liner crosshatch pattern.

Liner scoring is a result of contamination that has entered the combustion chamber or an overheated engine.

A wet sleeve will have seals on its outside diameter top and bottom to keep the engine coolant from leaking into the oil pan or above the liner and into the combustion chamber. The engine coolant that is contacting the liner will remove excessive heat from the liner so it won't distort.

Most liners will have a flange around the top that is used to seat the liner in the engine block. There is a crucial dimension related to this called liner protrusion, and this is measured from the top of the block deck to the top of the liner flange. There is usually a protrusion specification (the liner is slightly higher than the block) of .003–.005 in., and this is needed to provide the proper amount of compression to the cylinder head gasket when the cylinder head is installed on the block. The proper amount of gasket compression will seal the combustion chamber at the top of the cylinder. If the protrusion is not high enough, say .001 in. or less, then the head gasket will likely fail and repairs are needed. Some engine manufacturers will describe a repair process that may include machining the block and placing shims under the liner flange to bring the liner up high enough to provide the proper liner protrusion.

Some diesel engines will use a dry-type cylinder liner that is completely surrounded by the metal of the engine block. The engine block will have coolant circulating through it to provide cooling to the combustion chamber. Dry sleeves will usually have to be pressed in and pulled out with special tools.

CYLINDER HEAD. A diesel engine's cylinder head needs to perform a variety of functions, but when related to the combustion process, it must provide a seal on top of the combustion chamber. There will always be some kind of gasket between the cylinder head and the cylinder sleeve/block. The gasket is needed to allow for differences in expansion rates between components as temperatures change. It also completes the seal at the top of the combustion chamber. Diesel engines produced today will have the intake and exhaust valves located in the cylinder head. There could be two, three, or four valves per cylinder. The valves open and close as required and provide an opening between the combustion chamber and the intake and exhaust

PRIME MOVERS **185**

ports in the cylinder head. The intake port will allow air to travel from the intake manifold to the combustion chamber and the exhaust port will allow exhaust gas to travel to the exhaust manifold from the combustion chamber. The cylinder head will likely have replaceable valve seats that can be changed when reconditioning the head. This is faster and more economical than reconditioning the entire head if there were no valve seats.

The intake and exhaust valves will be held closed by spring pressure. At the top of the valve stem are one or more grooves that will hold valve spring keepers. The keepers will in turn hold the valve spring retainer in place. When the valve is installed in the head, it is held up tight against the valve seat as the spring is trying to expand between the top of the cylinder head and the bottom of the spring retainer. Almost all diesel engine cylinder heads will use replaceable valve stem guides and some will use oil seals on top of the guide to stop oil from seeping into the combustion chamber past the valve stem. There will be a difference in angles between the valve face and the valve seat. This is needed to create a tight seal when the valve is closed. Any irregularities related to the valve seal will compromise the power output for that cylinder as compression and combustion pressure is lost.

Most diesel engines produced today are designed as direct injection. This means the fuel injector is spraying fuel directly into the combustion chamber. Many older engines were indirect injection, which meant the cylinder head housed a precombustion chamber that had the fuel injector spray fuel into it and start the combustion process outside the main combustion chamber.

Most cylinder heads will have coolant and oil passages in them and some will also have fuel passages as well depending on the type of fuel system used. Cylinder heads need to be securely attached to the block because of the high pressures generated during combustion. They will be attached by a series of bolts or studs and nuts. Cylinder heads can be single cylinder or multi-cylinder with some covering up to eight cylinders.
● **SEE FIGURE 6–10** for a typical two-valve cylinder head.

FIGURE 6–10 A typical two-valve cylinder head.

TECH TIP

The HDET of today will not likely routinely recondition or perform major repairs on diesel engines. This is a very general statement but has become a trend over the years for various reasons. Of course there are exceptions to this, and it all depends on the company you will work for or if you become self-employed.

As with all major components used on a piece of heavy equipment, the engine itself if maintained properly will last a long time (15,000–20,000 hours is common) without needing any major internal repairs. The engine's support systems (cooling, fuel, electrical) will require the most attention. Engine rebuilding is a specialized sub-trade and requires unique skills and tools. If an engine is diagnosed to have a major internal problem (worn liners, broken piston ring, bent connecting rod), the engine will most likely be sent to a dedicated engine repair facility for repair or the HDET will install an exchange unit. Some major engine manufacturers will provide only exchange components (cylinder head assemblies, piston/liner kits, etc.) supplied from a central rebuild facility. This provides a better control over component quality but removes the need for an HDET to have engine reconditioning skills. If a company does get into the engine rebuilding part of the industry, it will usually train certain HDETs for that part of the trade and keep them in that part of the business.

INTAKE AND EXHAUST VALVES. Intake and exhaust valves are a poppet type of valve that are normally closed and mechanically opened. They have a stem that fits inside a valve guide, which may have a seal on top of it. Some exhaust valves are hollow and filled with sodium to help absorb combustion heat. The valves are made of materials that will withstand the high pressures and temperatures of the combustion chamber.

CRANKSHAFT AND BEARINGS. The **crankshaft** is the heaviest moving component in the diesel engine and for a good reason. It must be strong enough to withstand many different and extreme forces acting on it at once. The engine's pistons create a reciprocating motion and the connecting rods transfer this motion to the crankshaft where it is sent out the engine as rotating torque. In a multi-cylinder engine, the crankshaft must also help smooth out the pulsing action occurring between the various cylinders that are on different strokes of their cycle as the firing order progresses. ● **SEE FIGURE 6–11** for a four-cylinder engine crankshaft. To keep this heavy-duty piece of cast iron rotating, it must rotate on a film of pressurized oil. This oil starts at the oil pump and gets sent to the engine's crankshaft bearings through passages in the engine block. Crankshaft main bearings get covered in a thin film of oil that allows the crankshaft to effectively float in oil to keep it from contacting the metal of the bearings in the engine block. This is called hydrodynamic suspension. Crankshaft main bearings are a split-shell design (two pieces) to allow the main bearing cap to be removed for crankshaft removal and installation. The upper half of the bearing will have a hole through it to allow the oil from the engine block to get through the bearing. It will also have a groove leading from the hole around the bearing face to allow the oil to create a solid film. Split-shell bearings will have a locating lug that must align with the block and cap to ensure that the bearing is installed properly. Because of the high cylinder pressures and high torque developed by a diesel

FIGURE 6–11 A four-cylinder crankshaft.

FIGURE 6–12 A set of heavy-duty diesel engine main and connecting rod bearings.

engine, there is a very high load on the engine oil that is trying to keep the crankshaft from contacting the bearings.

Some light-duty diesel engines will use a tunnel block design where crankshaft is preassembled with its bearings and installed into the block from one end.

There is also connecting rod bearings that keep the bottom of the connecting rod suspended in a film of oil. This oil is fed through the crankshaft in passages called rifle drillings from the main bearing to the connecting rod journal. ● **SEE FIGURE 6–12** for a set of main and connecting rod bearings. Engine bearings, whether they are crankshaft, connecting rod, or camshaft, are all made with some similarities. They have a hard steel backing and layered with softer tin-like material that is sometimes called babbitt. This is to allow small particles of contamination to embed in the bearing and not damage the much more expensive hardened and polished component rotating inside them (crankshaft, camshaft, or gear). Inspection and interpretation of used bearings is a good way to understand what has been happening inside the engine. Go to Mahleclevite's website at http://catalog.mahleclevite.com/bearing/ for a great visual reference to use as a guideline for bearing inspection.

Crankshaft bearing surfaces are ground smooth, hardened, and polished to create a very smooth surface for the engine oil to work with. There will be a specific amount of clearance between the crankshaft and the bearing, and this is measured when the engine is reconditioned.

Another bearing needed to keep the crankshaft in place is called the thrust bearing or washer. The thrust bearing controls any end-to-end movement that may be created. This is typically limited to .005 in.

Some manufacturers will incorporate the thrust washer with one of the main bearings. Crankshafts need to be balanced so they don't self-destruct as a result of the many opposing forces acting on them. This is done with weights that can be part of the forging, or some larger crankshafts will have additional counterweights bolted on to oppose or counteract these forces.

The crankshaft will most likely have a flywheel attached to one end of it. This will provide a few functions. It will assist in smoothing out firing pulses between the different cylinders, it will provide a way for the engine starting system to crank the engine to start it, and it will be the output for whatever the engine is trying to rotate.

Some four- or eight-cylinder engines will use balance shafts to counteract vibrations in the engine. These are gear-driven eccentric weights that must be timed when the engine is assembled.

VALVE DRIVETRAIN. The valve drivetrain (valve train) is necessary to coordinate the timing of intake and exhaust valve opening to the position of the piston in its travel. Its components are timing gears, **camshaft**(s), and valve actuators. The valve drivetrain will turn the rotating motion of the cam into reciprocating motion of the valves. The type of valve actuator depends on whether the engine is overhead cam or cam-in-block. Cam-in-block design used to be the norm, and this design, as the name implies, has the camshaft mounted in the block, cam followers or lifters moving according to the cam profile, pushrods moved by the followers, and a rocker arm transferring this motion to the valve stems. ● **SEE FIGURE 6–13** for a cam-in-block design.

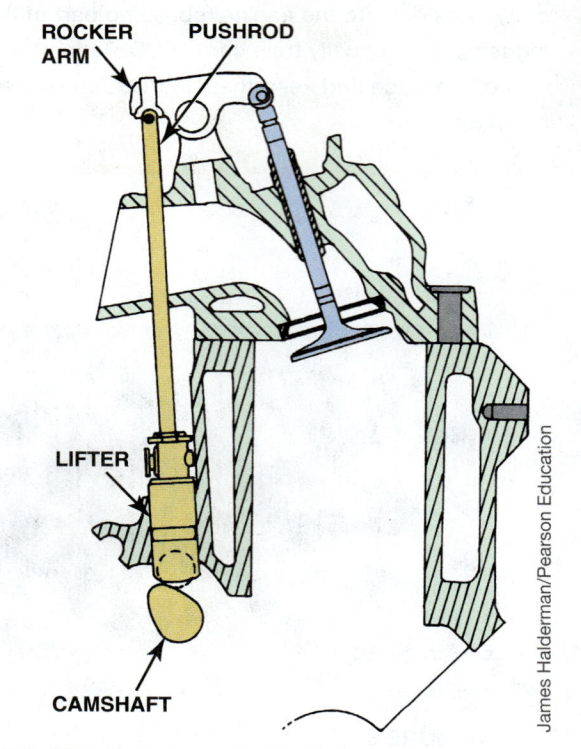

FIGURE 6–13 A typical cam-in-block design.

188 CHAPTER 6

FIGURE 6–14 A diesel engine using an overhead cam arrangement.

FIGURE 6–15 A typical heavy-duty diesel engine block.

An overhead cam design engine has, as its name implies, the cam or cams mounted in or on the cylinder head(s). The cam shafts will have their cam profiles followed by a rocker arm that will transfer motion to the valve stem or by buckets that will directly act on the valve stem. Some engine designs incorporate variable valve timing into the valve drivetrain to make the engine more efficient according to differing loads and conditions. ● SEE FIGURE 6–14 for a diesel engine with an overhead cam arrangement.

If an engine is designed with a four-valve-per-cylinder head, it will use a valve bridge to actuate pairs of intake or exhaust valves together.

CYLINDER BLOCK. The cylinder block is the foundation for the diesel engine. There are several different configurations of engine blocks that are possible; however, there are typically only two that are used for heavy equipment. The most common would be an inline cylinder block, meaning that all cylinders are in line and the most common size would be a six cylinder. The other style would be a V-style block where there are two banks of cylinders that form a V if you looked at the engine's flywheel. V blocks could have from 6 to 24 cylinders.

The cylinder block for a diesel engine has to be much stronger than a gasoline engine. More strength used to equal more weight, but advancements in materials technology have allowed designers to make blocks stronger and lighter over the years. Weight is generally not a big concern for heavy equipment as the engine is sometimes even used as a counterweight.

The engine block must withstand the high pressures and forces created by the high combustion pressures and the high torque output of the diesel engine. ● SEE FIGURE 6–15 for a heavy-duty diesel engine block.

FLYWHEEL AND VIBRATION DAMPER. There are typically two devices mounted to the ends of the crankshaft that may be considered part of the long block assembly. At the front of the engine is a type of vibration damper, and as the name implies, its purpose is to smooth out any vibration pulses that may occur. These dampers could be a simple outer steel ring bonded with rubber to a driving flange that is bolted or pressed on to the front of the crankshaft. Another type of damper is the viscous damper that has a metal ring that floats in a silicon fluid inside a sealed housing that is also attached to the front of the crankshaft. Many engines will use these dampers as timing indicators if they are keyed to the crankshaft.

The opposite end of the crankshaft will have a flywheel bolted to it. The flywheel will have a ring gear pressed on to it that allows the starter motor drive gear to engage with its teeth and rotate the crankshaft to get it started. The flywheel also helps smooth out ignition pulses by creating an inertia effect in between power strokes. The flywheel and flywheel housing will have a series of threaded holes to allow other components to be attached to them such as clutches, torque converters, transmissions, and hydraulic pumps. The flywheel housing will also have threaded holes for the rear engine mounts to be located on.

Progress Check

6. The term *Monotherm* refers to:
   a. the temperature a piston reaches when under full load
   b. a new style of lightweight steel piston
   c. a new style of super strong connecting rod
   d. one temperature that diesel fuel ignites at

7. The number of valves per cylinder a cylinder head will *not* have is:
   a. 2
   b. 3
   c. 4
   d. 6

8. *Blowby* is a term used to describe:
   a. how much exhaust leaks by the exhaust valve
   b. how much fuel leaks past the injector
   c. the combustion pressure that leaks into the crankcase
   d. the compression pressure that leaks into the cooling system

9. This will help the compression rings create a tight seal:
   a. compression and combustion pressure
   b. spring pressure
   c. oil pressure
   d. fuel pressure

10. This component is sometimes used to counteract engine vibrations:
    a. balance shaft
    b. camshaft
    c. damper
    d. both a and c

TECH TIP

Emission Regulations

Much of today's diesel engine design technology is driven by various government-mandated regulations regarding what leaves the exhaust pipe. Starting in 1996 there has been a steady tightening of diesel engine exhaust emission regulations. On-highway diesel vehicle regulations have led the way but off-highway regulations are following close behind. ● SEE FIGURE 6–16 for how emission regulation has changed.

Emission reduction has been focused on particulate matter, hydrocarbons, and nitrous oxides (NO_x).

Diesel engine designers have had a difficult time trying to keep ahead of these regulations especially since most of the technology they incorporate to reduce emissions make the engine less efficient and place higher loads on the engine's cooling system.

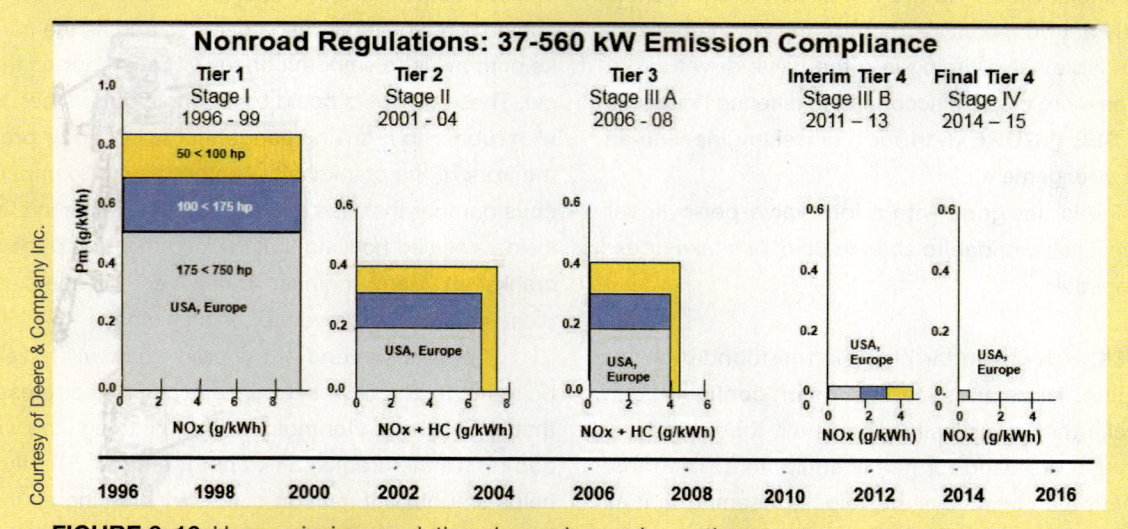

FIGURE 6–16 How emission regulations have changed over the years.

DIESEL ENGINE ACCESSORY SYSTEMS

The following systems are necessary to keep a diesel engine running:

1. intake and exhaust system
2. cooling system
3. lubrication system
4. fuel supply system

INTAKE AND EXHAUST SYSTEMS You could think of a diesel engine in simplified terms as an air compressor with a fuel system attached to it. The intake and exhaust system allows the engine to ingest an unlimited amount of clean air and to get rid of spent exhaust gases efficiently and quietly as well. ● **SEE FIGURE 6–17** for a typical intake and exhaust system.

By the nature of their lean burn design, diesel engines use massive amounts of air to operate properly. A diesel engine at low idle with no load could have an air/fuel ratio of 160:1. This fact plus the fact that most heavy equipment works in less than ideal conditions means that the intake system must be very effective at removing dust and contamination before the air enters the engine. It only takes a small amount of dust to be ingested into the cylinders to create permanent damage to cylinder liners, piston rings, valves, and valve seats.

Intake systems start with pre-cleaning devices that can be as simple as a screen. There are also many different types of components that will pre-clean intake air before it gets to the main filter system. These pre-cleaners will create a vortex when the air is drawn in toward the intake manifold or turbocharger. This vortex will force all heavy particles such as sand or heavy dust to the outside where they can be collected and drawn out through the exhaust system or gathered in a temporary container where they are cleaned out manually.

The more effective a pre-cleaning system is, the longer between air filter changes. The most common type of air filters for diesel engines are folded paper. There are still some machines that use oil bath–type filters, but these are not seen very much these days. Pleated paper–type filters like the one in ● **FIGURE 6–18** are the most common way of ensuring only clean air reaches expensive internal engine components.

Some smaller machines have gone to different styles of filters that aren't the typical round paper design. Most heavy equipment will use a two-stage filtration system that will have an inner and outer filter. The inner filter, sometimes called a safety filter, is only meant for a backup if the outer filter is damaged and lets contaminants through. When required, a typical service will need the outer filter to be changed and the inner filter only being changed for every third outer. Of course this depends on machines operating conditions and manufacturers recommendations. Air filter service should mainly be based on inlet restriction. This is the measurement in inches of water of negative pressure that is created between the filtration system and the outlet of the filter housing.

As the filters collect more contamination and become plugged, there will be a higher restriction value. This can be indicated by a simple restriction gauge that uses a spring to hold a movable disc in place. Higher restriction will move the disc farther and the distance it moves translates into higher inches of water. This is also indicated as a change in color to make it simple for an operator to see when the machine's air filter needs servicing. There could also be electronic sensors that monitor air filter restriction, and these will either turn a warning light on or log a fault code or both. ● **SEE FIGURE 6–19** for an air filter restriction gauge.

Naturally aspirated (non-turbocharged) engines should have their filters changed at 25 in. of water restriction and turbocharged engines should have their filters changed at 30 in. of water restriction. This is a general rule of thumb and is a good guideline to go by.

Replaceable filters should be left in place as long as possible to within the machine's maintenance guide recommendations for calendar time (usually one year maximum) or until the maximum restriction is reached. There will be a good chance you will allow more contamination into the intake by just removing the filter to inspect it. Visual inspection is *not* a reliable indication of air filter condition and should only be used for diagnosing engine problems. Many air filters are changed unnecessarily because they look dirty.

It is possible to wash air filters and there are fleet operations that find this cost effective; however, it only takes one failed engine resulting from a failed washed air filter to eat up all money saved by not replacing filters. If you do consider this is a risk worth taking, make sure that the filter washing company will stand behind its product if there is an issue.

> **WEB ACTIVITY**
>
> A great source of intake and exhaust information can be found at http://www.donaldson.com/en/engine/shoptalk/index.html.

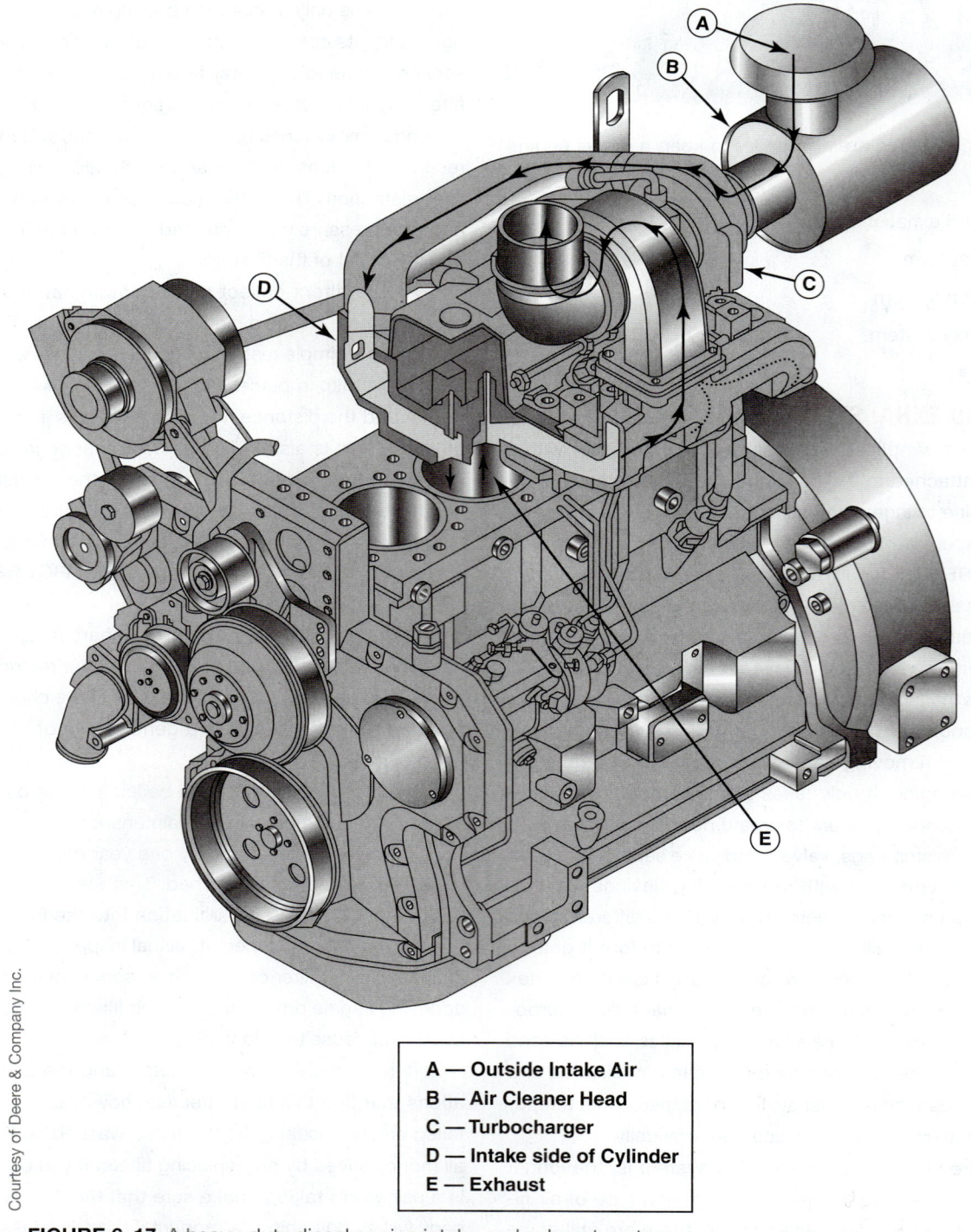

A — Outside Intake Air
B — Air Cleaner Head
C — Turbocharger
D — Intake side of Cylinder
E — Exhaust

FIGURE 6–17 A heavy-duty diesel engine intake and exhaust system.

Intake air piping for a diesel engine can consist of a simple straight pipe from the outlet side of the air filter assembly to a very complex network of piping that could include twin turbochargers, air cooled aftercoolers, coolant cooled aftercoolers, and EGR coolers. ● **SEE FIGURE 6–20** for a complex air inlet system.

Naturally aspirated engines will have piping (hoses and or tubes) that goes from the air filter housing outlet to the intake manifold. The intake manifold then directs the clean air into the individual intake ports on the cylinder head(s). The term *naturally aspirated* means as the piston moves down in the cylinder on the intake stroke, it will create a pressure drop naturally. The

192 CHAPTER 6

FIGURE 6–18 A typical two-stage air filter.

FIGURE 6–20 A complex air inlet system on a modern diesel engine.

FIGURE 6–19 Air filter restriction gauge.

engine then is relying on atmospheric pressure to push air into the lower pressure void. If the machine happens to be working in higher altitudes or the air filter is plugged, then it will likely be lacking the proper amount of air in the cylinder for complete combustion. This will result in black smoke out the exhaust pipe.

Except for small light-duty diesel engines, all of today's heavy equipment will use one or two turbochargers to create a positive pressure in the intake manifold. This will ensure that the engine will always get more than enough air for complete combustion.

Turbochargers use waste heat and pressure from the engine's exhaust to spin a finned wheel called a turbine. This turbine then turns a shaft that turns another finned wheel called the impeller. As exhaust temperature and pressure rise when the engine speeds up or comes under load, the turbine speed increases and so does the impeller speed. As the impeller speeds up, it provides more boost pressure in the intake manifold. Boost pressure can exceed 40–50 psi in some engines, and turbochargers can rotate well over 100,000 rpm. Because of the shaft speed and heat surrounding the turbo, it needs a good supply of oil from the lubrication system, and it is usually fed directly from the main oil gallery.

One problem that is created when the intake air is pressurized is that it is heated up. If you have ever felt the outlet of an air compressor, you will have realized it gets very hot. The problem with hot air is that it is less dense and, therefore, contains less oxygen. Less oxygen content means there can be less fuel ignited. To cool this charged or pressurized air, it is sent through an aftercooler, sometimes also called an intercooler. The most common way to cool it is to send it through a heat exchanger that is mounted in front of the engine coolant radiator where the moving air from the engine cooling fan will remove the heat. This can reduce charge air temperatures in the range of 150–250°F. Some engines will use an engine coolant cooled heat exchanger and some will use both.

Turbocharger technology has changed dramatically in the past 10 years, and today most reduced emission heavy-duty diesel engines are using variable geometry turbochargers. Prior to this, fixed geometry turbos used wastegates that simply dumped exhaust gas past the turbine to limit the turbo boost pressure. This wastegate opening is based on how high the boost pressure reaches. Wastegates could be simple mechanical devices controlled by a diaphragm that senses boost pressure or controlled electronically.

Fixed geometry turbos would only be most efficient when the engine was under full load. This resulted in turbo lag and a general less than ideal output in any other engine operating condition than full load. Variable geometry turbos can change their output to match engine operating conditions. There are a few different ways to vary turbo output. Some manufacturers will rotate a series of vanes in the turbine housing and others will slide a collar in the housing sideways. Either method simply makes the turbo more effective for the amount of exhaust pressure it is exposed to. The output of a variable turbo will be precisely regulated by an electronic control module (ECM) that may be mounted on the turbo. Some turbos will have a speed sensor that will monitor their shaft speed and send this information to the ECM.

Some turbochargers will have their center housing and ECM cooled with engine coolant. ● SEE FIGURE 6–21 for a turbocharger with variable geometry technology.

Exhaust systems used to be very simple back in the day, and up until 15 years ago, some small diesel engine exhaust could be as simple as an exhaust manifold that directed exhaust gases from separate cylinders to a central outlet and from there maybe to a pipe or some kind of sound deadening device.

Exhaust systems, like intake systems, also vary greatly and have most recently gotten more complex as emission regulations have become more stringent. The simplest today being a naturally aspirated engine with a pipe from the exhaust manifold connected to a sound deadening device (muffler) to another pipe to direct the exhaust away from the machine.

With the coming Tier 4 emission regulations, diesel engine manufacturers have needed to change exhaust systems from systems that just needed to quiet and direct exhaust to systems that manage and control the chemical makeup of what goes out the pipe.

Most engines that fall under Tier 3 emission regulations will use exhaust gas recirculation (EGR) technology. It is a system that redirects a portion of the engine's exhaust gas through a cooler and back into the intake manifold. This reduces combustion temperatures by reducing the amount of oxygen in the cylinder, which in turn reduces NO_x formation. EGR systems are variable with differing types of metering valves and, depending on the manufacturer, used to adjust the amount of intake air that comes from the exhaust. These valves are electronically controlled, and an ECM will regulate the amount of exhaust that is sent back into the exhaust based on current operating conditions. ● SEE FIGURE 6–22 for a modern diesel exhaust system with EGR.

Machines with low-emission engines may have devices such as diesel particulate filters (DPFs), diesel oxidation

> **WEB ACTIVITY**
>
> To see information on the latest emission regulations and an animation on a Tier 4 interim engine, go to http://www.deere.com/en_US/rg/index.html.

FIGURE 6–21 Turbocharger with variable geometry technology.

FIGURE 6–22 A modern diesel engine exhaust system with EGR.

FIGURE 6–23 A modern exhaust system with a DPF, DOC, and SCR.

FIGURE 6–24 Variable valve timing mechanism.

catalysts (DOCs), and selective catalytic reduction (SCR) systems being added as part of the exhaust systems.

A DPF is a device designed to trap almost all particulate matter (this is what makes black smoke) in it, and then when the DPF gets hot enough, the collected soot is burned off. This is done two ways. First, if the engine is producing high enough exhaust temperatures, it will be done automatically, and this is called "passive" regeneration. If the exhaust gas temperature doesn't get high enough, then the engine must perform an "active" regeneration. This is an ECM-controlled process that artificially creates high enough temperatures to burn out the collected soot or regenerate the DPF. Occasionally, the DPF must be removed from the machine for a more thorough cleaning process. The DPF will have a number of sensors mounted on it to monitor pressures and temperatures and send this information to an ECM.

A DOC is a device that creates a chemical reaction with the exhaust as it passes through to reduce hydrocarbons and carbon monoxide. DPFs and DOCs have been used in the underground mining industry for many years to reduce exhaust emissions in an environment where air quality is critical.
● **SEE FIGURE 6-23** for an exhaust system with a DPF, DOC, and SCR.

SCR is a system that treats the exhaust after it leaves the DOC to reduce NO_x. A liquid urea/water mixture is injected into the exhaust and this requires a controller, sensors, and injection system.

Crankcase vapors used to be collected and vented to the atmosphere, but on many machines now, these are collected and sent back into the intake manifold.

Exhaust brakes can be part of heavy equipment machines that travel at higher speeds such as rock trucks. These are

> **TECH TIP**
>
> It is important to note here that any repairs done to a diesel engine that is Tier 2 or newer must *not* result in increased exhaust emissions. If HDETs are caught modifying or tampering with any emission-related component, they face fine or imprisonment. This means that if the engine came from the factory with a certain type of component, that same type of unmodified component must be used as the replacement part. Examples include fuel injectors, pistons, turbochargers, mufflers, and pre-cleaners.

systems that turn the engine into a load, which is then transmitted through the driveline as a braking force. The fuel system stops injecting fuel first and then the exhaust valves are opened slightly just as the piston nears TDC on compression. There is a solenoid-controlled, oil-actuated mechanism incorporated into the exhaust valve actuators. This in effect makes a variable valve timing system and a common name for this system is "Jake" brake.

Engine exhaust brakes are used to lessen the wear and heat generated by the machine's driveline brake systems. A diesel engine exhaust brake could also be a simple mechanical device that merely blocks off exhaust flow coming out of the engine. ● **SEE FIGURE 6–24** for a variable valve timing mechanism.

COOLING SYSTEMS In simple terms, the diesel engine cooling system is a heat exchanging system. Most diesel engines manufactured today use a circulating pressurized liquid to control the temperature of the engine. ●SEE FIGURE 6-25 for a typical diesel engine liquid cooling system.

Deutz is one manufacturer that offers diesel engines that are air cooled. These engines rely on a cooling fan that moves air past engine components that have many cooling fins attached to them so the heat is dissipated to the moving air. ●SEE FIGURE 6-26 for an air cooled engine.

When the engine is at normal operating conditions, the liquid cooling system will absorb heat energy from around the combustion chamber and shed most of this heat at the radiator. Some heat is dispersed through the engine components themselves to the surrounding air. Engine cooling systems are necessary to maintain a consistent operating temperature for all engine components. This operating temperature is usually around the 180–195°F range. A consistent operating temperature will help the engine be at its most efficient and least polluting state. This is because the engine's components are designed to fit together properly at this regulated temperature.

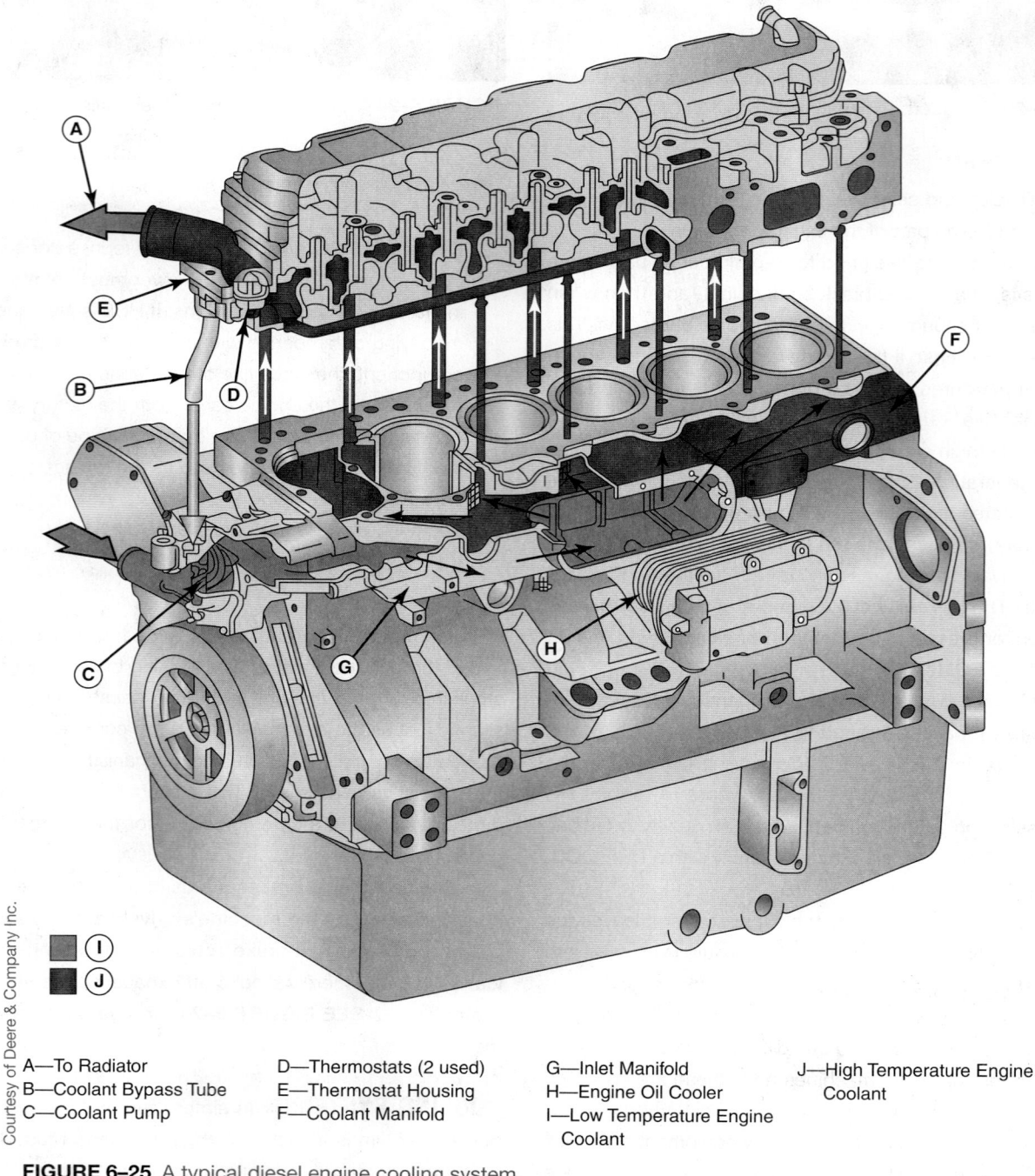

A—To Radiator
B—Coolant Bypass Tube
C—Coolant Pump
D—Thermostats (2 used)
E—Thermostat Housing
F—Coolant Manifold
G—Inlet Manifold
H—Engine Oil Cooler
I—Low Temperature Engine Coolant
J—High Temperature Engine Coolant

FIGURE 6-25 A typical diesel engine cooling system.

196 CHAPTER 6

FIGURE 6–26 An air cooled diesel engine.

FIGURE 6–27 A centrifugal-type water pump.

The cooling system should also assist in bringing the engine up to operating temperature as quickly as possible. This is especially critical with today's strict emission regulations.

Another job of the cooling system is to regulate temperatures of other system fluids such as drivetrain, brakes, and hydraulic systems as well as the engine oil. A more recent requirement of the cooling system is to cool exhaust gas for EGR systems. The EGR cooler is required to drop the exhaust gas temperatures from 1100 to 400°F.

The main components of a typical cooling system are water pump, radiator, thermostat, pressure regulator, cooling fan hoses, coolant, coolers, coolant heaters and cab heaters.

WATER PUMP. Liquid cooling systems need a water pump to circulate the coolant around the system. These water pumps are a non-positive displacement centrifugal-type pump. ● **SEE FIGURE 6–27** for a typical water pump.

The water pump draws engine coolant in to its inlet from the bottom of the radiator. The coolant leaves the water pump and will circulate through the engine components that have coolant passages.

RADIATOR. The radiator is simply a heat exchanger that will dissipate the heat that the coolant has collected when it circulates around hot engine components. There are different styles of radiators, but the most common type will have a tank on top that allows engine coolant to flow down through many smaller tubes that have fine fins attached to them. These fins will then transfer the heat in the coolant to the air that is passed by the fins from the cooling fan. The tubes and fins are copper or aluminum because of their excellent heat transfer properties. The radiator tanks can be steel, aluminum, or plastic. Air flow and clean fins are critical to enabling an efficient heat transfer, and if either the fins get dirty or the air flow slows down, then an overheated engine is likely to result. ● **SEE FIGURE 6–28** for a comparison of a clean radiator to one that needs cleaning. Other types of radiators will circulate the coolant from side to side or up and down, but their method of getting rid of heat is still the same.

THERMOSTAT. The regulator will open when the engine reaches operating temperature. The thermostat senses the coolant temperature at the top of the cylinder head, and when the coolant gets hot enough, the thermostat opens against spring pressure and the coolant will then circulate through the radiator.

PRESSURE REGULATOR Pressure-regulating devices are an important part of a cooling system because they raise the boiling point of the coolant. For every 1 psi, the coolant is pressurized and the boiling point is raised by 3°F. This is important because if the coolant is allowed to boil and change into a vapor, there will not be a good heat transfer through it. As the coolant passes by hot components, it must be in the liquid state to effectively transfer heat. A typical pressure-regulating valve will limit the cooling system pressure to 15 psi.

These valves are held closed by a spring and will open if the pressure exceeds the spring value. They can be incorporated in the radiator cap or in a separate reservoir tank cap.

COOLING FAN. The engine cooling fan will be an important part of the cooling system because to get the heat transfer to from the coolant to the atmosphere, there needs to be a good airflow past the radiator. This is usually done by a belt-driven fan, but more often the fan is driven by a hydraulic motor. Having a hydraulically driven fan allows the machine designers to move the radiator away from the typical location in front of the engine. This could provide better visibility

PRIME MOVERS 197

FIGURE 6–28 A clean radiator versus one that needs to be cleaned.

FIGURE 6–29 Comparison of the belt driven fan and the hydraulically driven fan.

for the operator and easier access for servicing and repairs. Hydraulically driven fans are capable of variable speeds that will save fuel and allow faster engine warm-ups than directly driven fans. Many machines today will also feature a hydraulically driven cooling fan that is capable of reversing the fan direction. This feature will give the fan a self-cleaning function that will dislodge most built-up dust and dirt that has collected on the radiator fins.

● **SEE FIGURE 6–29** for a comparison of a belt drive cooling fan and a hydraulically driven fan. Some engine cooling fans are engaged with an air or electric clutch that saves fuel. There are also aftermarket systems that will vary the pitch of the fan to save fuel and reverse air flow.

HOSES. Engine coolant needs to be transferred between components with as little restriction as possible and without being able to escape. For a number of years, highway truck operators have faced stiff fines for running a truck with any kind of leak. This was driven by the fact that most engine coolant is a toxic substance and, therefore, harmful to the environment. It is quite possible heavy equipment could face the same scrutiny in the future, but any leaks that do occur should be repaired as soon as possible to protect the environment and especially wildlife that may think coolant is a sweet treat. Hoses used to be mainly made of rubber, but silicone is being used more all the time because of its longer life. Hose clamps have also changed to prevent leaks. Clamps that use small Belleville washers to maintain a constant pressure are being used more frequently.

● **SEE FIGURE 6–30** for a hose clamp that is leaking.

COOLANT. Engine coolant has changed greatly over the last 15 years, all in the name of extending the interval that it needs

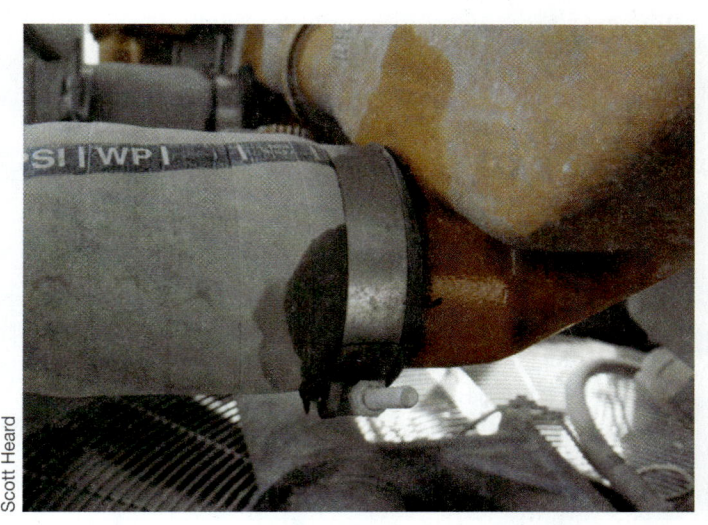

FIGURE 6–30 A hose clamp that is leaking.

to be changed at and to make components that are exposed to it last longer as well. Ethylene glycol-based coolant was once the standard and was commonly colored green. It was usually recommended to be changed once a year. This created two problems. If the schedule was adhered to, there would be thousands of gallons of antifreeze that had to be disposed of or reprocessed yearly. This no doubt led to a great expense and a great burden to the environment. The second problem was if the coolant wasn't changed on a yearly basis, there would be premature failures of the cooling system components because the coolant lost a lot of its protective additives.

Today's long-life diesel engine coolant is usually able to stay in the machine for a minimum of three years and sometimes up to six years. Some antifreezes are organic based and, therefore, much more environmentally friendly. Some cooling systems feature a coolant filter that will remove contaminants and maintain the additives in the coolant. Other coolant will just need an additive recharge once a year that is a simple process of pouring in a quantity of additives to the reservoir. One of the keys to getting the most life out of the machine's cooling system components is to always check the maintenance guide to see what the recommended coolant is for that machine and follow the regular service intervals.

Poor-quality coolant could also result in liner **cavitation**. This occurs when the vibrations created by the combustion pressures inside the cylinder cause small bubbles to form on the outside of the cylinder liner. If the bubbles collapse rapidly or implode and are allowed to directly contact the liner, they will eat away the surface of the liner. If this happens long enough, the liners could be pitted all the way through and allow coolant into the combustion chamber.

COOLERS. As heavy equipment will use several other fluids to operate, there will usually be a need to cool those fluids.

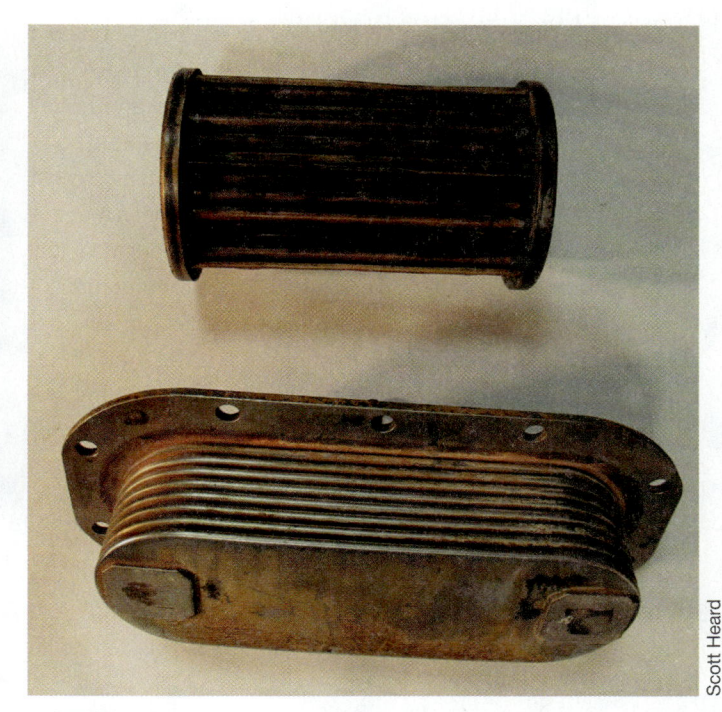

FIGURE 6–31 Two different types of fluid coolers you could find on a machine (plate and bundle).

Fluids can be cooled by air flowing past a heat exchanger or the diesel engine cooling system can be used to transfer heat from other systems such as braking, transmission, and hydraulic when their fluid temperatures become hotter than engine coolant temperature. These coolers can also assist with warming up other system fluids if the engine warms up first, or vice versa. Coolers that use engine coolant can be either of two styles. The most common being the bundle type, whereby engine coolant flow going into the cooler is divided into several small copper tubes and out into a return manifold where it is sent back into the cooling system. The fluid being cooled is sent into one end of the cooler to flow around the outside of the individual tubes where heat is easily transferred to the engine coolant and it will exit the cooler at the other end. A bundle type of cooler is identified by its round shape. It will have two hoses or tubes connected to the system fluid that requires cooling.

The other style of cooler is called the plate type. It differs from the bundle style by having -flat hollow plates, not round tubes, that carry the fluid being cooled through them. Plate-type coolers will be mounted in a coolant manifold where the coolant is freely circulating past the plates and will have two hoses or tubes connected to the outside of it to allow flow of the fluid being cooled through the cooler. ● **SEE FIGURE 6–31** for the two types of fluid coolers used.

COOLANT HEATERS. Engine coolant heaters are needed to assist with cold starts for machines working in cold climates. One type is commonly called a block heater and is a simple heating element usually installed in a cooling jacket in the block, but it could be an external mount with its own pump and hoses connecting it to the cooling system. This type of heater is powered by 120V AC or 240V AC, which means the machine must be near a source of hydro. If the machine is not near a hydro source, another type of coolant heater uses the machine's fuel supply and battery as its power source. One manufacturer of these heaters is Proheat, and they can be installed on any machine that can spare about 2 cubic feet of space. They are an ECM-controlled unit that can be programmed to start anytime in a seven-day workweek. They use an electric pump to circulate coolant through a diesel-fired heat exchanger (similar to a household oil furnace) that has a circular pipe going around it. The unit will also have a DC electric pump that will pump fuel to the heat exchanger. When started up, it will ignite the heat exchanger and start pump coolant through it to the engine. It will shut off when the engine is at about 120°F. This will make the engine start much easier.

CAB HEATER. Another function of the cooling system is to provide heat for the operator's station. There will be a supply and return hose going to the cab heater. The heater is just a small heat exchanger (heater core) that will have a fan to circulate heat from the heater core to the cab.

LUBRICATION SYSTEMS
Diesel engine lubrication systems are a crucial support system to ensure the longevity of the engine. All moving parts in the engine must be separated by a good supply of clean oil of the proper viscosity and properties to prevent any metal-to-metal contact. If bare metal is allowed to move against a moving or stationary metal component, the road to failure has begun. ● **SEE FIGURE 6–32** for a typical lubrication system.

The main components of the lubrication system are oil pan, oil pump, pressure regulating valve, oil passages/hoses/tubes, oil filters, oil coolers, spray nozzles, pre-lube systems, and oil.

OIL PAN. This is where the engine oil is stored and returns to and can be thought of as the reservoir. It will hold enough oil to allow the oil to have large contaminants settle out of it before it is drawn into the oil pump. It will also have some kind of measuring device, usually a dipstick to allow the operator to check the oil level before starting the machine and sometimes even after its running. There could also be an electronic sensor that could be used by the monitoring system on the machine that would set a fault code if the oil was too low. In the future, you may see oil condition–monitoring sensors as well. Oil pans on small engines are usually stamped metal (one-piece) units, whereas medium and large engines will use cast aluminum oil pans that could be one-piece or multipiece units.

Oil pans are secured to the bottom of the engine block with a series of fasteners around their perimeter and will have a one-piece or multipiece gasket to seal between the two components. All oil pans will have a drain plug that is removed at service time and must not be over torqued when installing. Some oil pans on large diesel engines will incorporate a valve that allows a controlled flow of oil when opened. This is particularly helpful if you run out of drain pans while doing an oil change. A popular feature that large fleets of equipment owners like is a quick evacuation system that allows a quick coupler to be attached to a remote fitting that is connected to the bottom of the drain pan. When servicing a machine equipped with this system, the technician will connect a vacuum source to the quick coupler and the oil is drawn out of the oil pan.

OIL PUMP. The oil pump creates flow from the oil in the pan to the oil filters, coolers, galleries, passages, and any other components that need lubrication such as turbochargers, air compressors, and fuel injection pumps. The oil pump is usually driven by a crankshaft gear and could be either a gerotor or an external gear-type pump. It will get its supply of oil through a suction screen that prevents any large contamination from entering the pump. The suction screen will be near the bottom of the oil pan. ● **SEE FIGURE 6–33** for the two different types of oil pumps used for diesel engines.

OIL FILTRATION. The oil filtration system used for diesel engines can vary from a simple single spin-on- or cartridge-type full-flow filter to a series of filters and a combination of full-flow and bypass-type filters. There could also be a centrifugal-type filter.

OTHER LUBRICATION SYSTEM COMPONENTS
The oil will then be fed to a variety of components that differ from manufacturer to manufacturer and model to model. A typical flow will go from the oil pump outlet to the oil filter first. There will be a pressure control valve at this point as well to limit maximum oil pressure. This should only open when the engine is started on a cold morning or if the oil viscosity is too thick. Engine oil pressure is usually limited to 75 psi. Some engines will have an oil temperature regulator that will prevent oil flow to the cooler until the oil temperature raises enough to open the valve and start the cooling process. A typical operating pressure for a heavy-duty diesel engine is somewhere around 30–50 psi; however, this could vary greatly between different engines, oil temperature, oil viscosity, and engine rpm.

Some engines will have a secondary oil pump called a scavenge pump. This is used to ensure that the main oil pump doesn't starve for oil when the machine may be working on extreme angles. The scavenge pump will have its inlet pickup tube draw oil from the opposite end of the oil pan that the main oil pump picks its oil up from. It will then discharge this oil in to the main sump area of the oil pan. The scavenge pump could be a separate section of the main pump or could be a completely separate pump driven from a different source.

Engine Lubrication System Serial Number (200,000—)

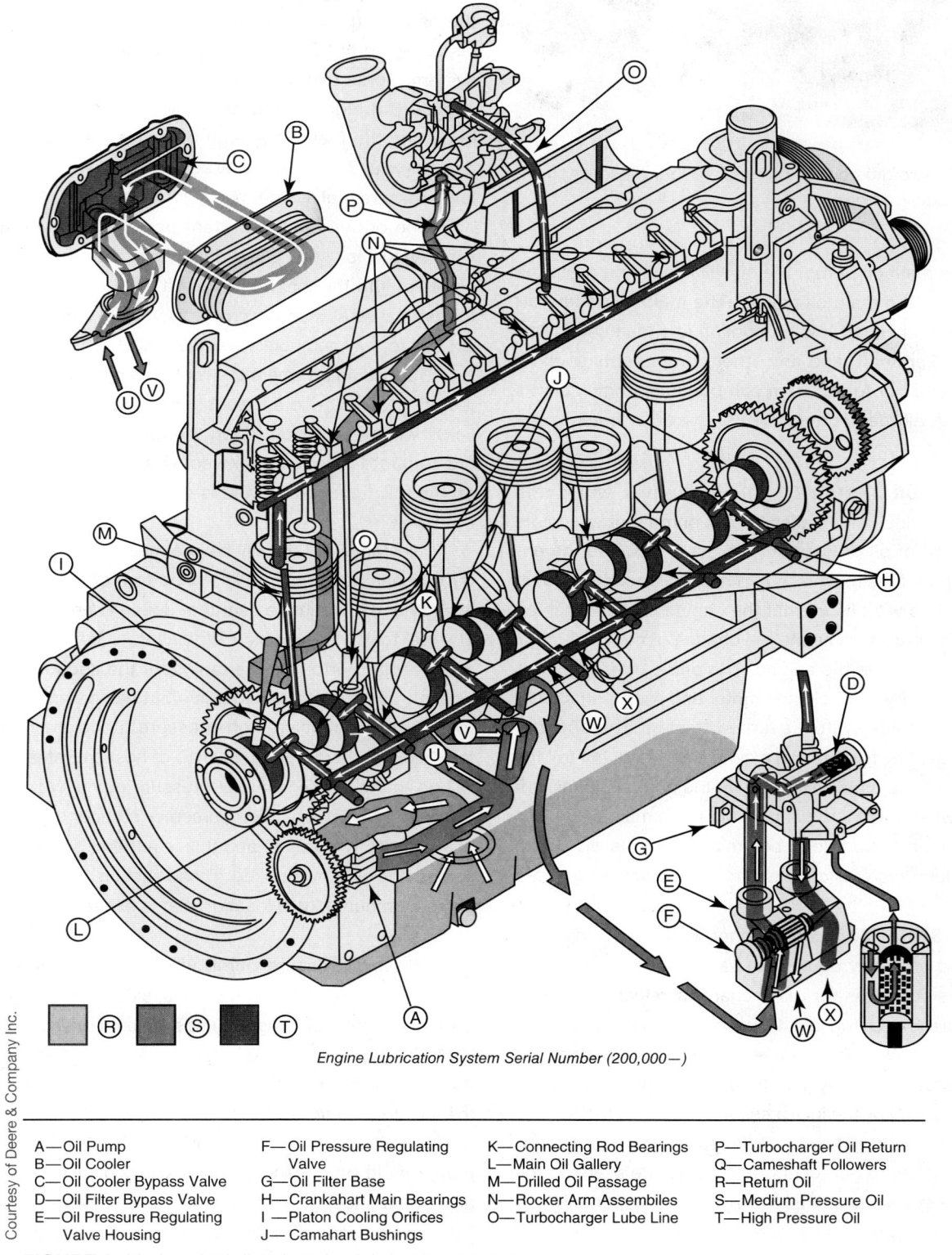

A—Oil Pump
B—Oil Cooler
C—Oil Cooler Bypass Valve
D—Oil Filter Bypass Valve
E—Oil Pressure Regulating Valve Housing
F—Oil Pressure Regulating Valve
G—Oil Filter Base
H—Crankahart Main Bearings
I—Platon Cooling Orifices
J—Camahart Bushings
K—Connecting Rod Bearings
L—Main Oil Gallery
M—Drilled Oil Passage
N—Rocker Arm Assemblies
O—Turbocharger Lube Line
P—Turbocharger Oil Return
Q—Cameshaft Followers
R—Return Oil
S—Medium Pressure Oil
T—High Pressure Oil

FIGURE 6-32 A typical diesel engine lubrication system.

FIGURE 6–33 Two different oil pumps used for diesel engines (gerotor and external gear).

From the main oil pump, the oil will likely then go to the turbocharger or the main oil gallery. The main oil gallery directs it to the crankshaft, piston cooling jets, camshaft, timing gears, rocker shaft, and any other components such as an air compressor or fuel system components. Engine oil will then leak past all these components and drain back to the oil pan.

PRE-LUBRICATION SYSTEMS. Many engines will come equipped with a supplemental engine oil pump that will create a small amount of oil pressure in the system before the engine is started and the main oil pump has a chance to build up oil pressure. This will ensure that the engine is never starved for oil and will virtually eliminate the chance of metal-to-metal contact or dry starts in the engine. The pump used for these systems is driven by a DC electric motor and will automatically turn on when the key switch is turned to start. The operator simply holds the key to start and a simple ECM will control the sequence of the pre-lube system. Once the pump is running, it will shut off when there is 5 psi built up in the main oil gallery. There will then be a short delay before the starter is allowed to engage. Pre-lube systems can be retrofitted to any engine.

FUEL SUPPLY SYSTEMS. As mentioned earlier, fuel injection systems used on heavy equipment vary immensely between manufacturers as well as the many changes in technology that have taken place over the last 10 years. Several books could be dedicated to fuel injection alone so we will just discuss the low-pressure fuel supply system here. Many low-power problems can be caused by the low-pressure side of the fuel system and these quite often get overlooked, so a good understanding of what it takes to get fuel from the machine's tank to the high-pressure system is critical to becoming a capable HDET. Remember the high-pressure side of the fuel system will *not* work properly if there isn't a good steady supply of clean fuel with no air in it. ● **SEE FIGURE 6–34** for a typical low-pressure fuel supply system.

The supply starts at the fuel tank or tanks of the machine that could be mounted anywhere.

Newer Caterpillar scrapers have the tank mounted at the back of the bowl so the fuel has to travel a significant distance (up to 30 ft) and height to get to the high-pressure system. Some machines will have the tank outlet above the engine (Cat high track dozers), which will help a great deal in getting the fuel to the high pressure fuel system. Gravity can be your friend or work against you when it comes to low-pressure fuel systems.

One of the most important parts of any fuel tank is its vent. As fuel is used up in the tank over a workday, there must be air let into the tank to replace it. If the vent gets plugged, this will cause all kinds of problems, mainly the machine will be starved for fuel and could eventually stall.

Depending on where the vent is located and how it is protected, it could need a lot of attention to keep it clean. Some vents will be incorporated in the fuel tank cap while others are separate. The fuel tank vent is an often overlooked item on the maintenance list and is often the cause of low-power problems.

Fuel tanks will usually have a means to drain off any water that has accumulated in the bottom of the tank, and this is another maintenance item that doesn't see much attention. The fuel tank will likely have a pipe inside that will pick up fuel close to the bottom of the tank so if the accumulated water doesn't get drained off, that is what will be end up heading toward the expensive high-pressure fuel system. From the tank pickup, there will be a series of hoses or steel tubes that will go to the inlet of the low-pressure pump. Most machines will have at least a suction screen or strainer somewhere between these two points and this is another service point that sometimes gets overlooked. These suction screens can sometimes be mounted in a location that is either not easily seen or accessible.

The low-pressure pump will be one of three different styles. There are diaphragm pumps, external gear pumps, and electric pumps. Diaphragm pumps are usually driven off a camshaft that will pulse a diaphragm back and forth to create fuel flow. The pump will have two check valves to make sure that the fuel flows in the right direction and could also have a manual operation feature for priming purposes. The external gear pump could be driven off any auxiliary gear drive and could also use a check valve and relief valve. ● **SEE FIGURE 6–35** for a low-pressure fuel pump. Pressure is limited to between

A—Fuel Return Line B—Fuel Tank Cap

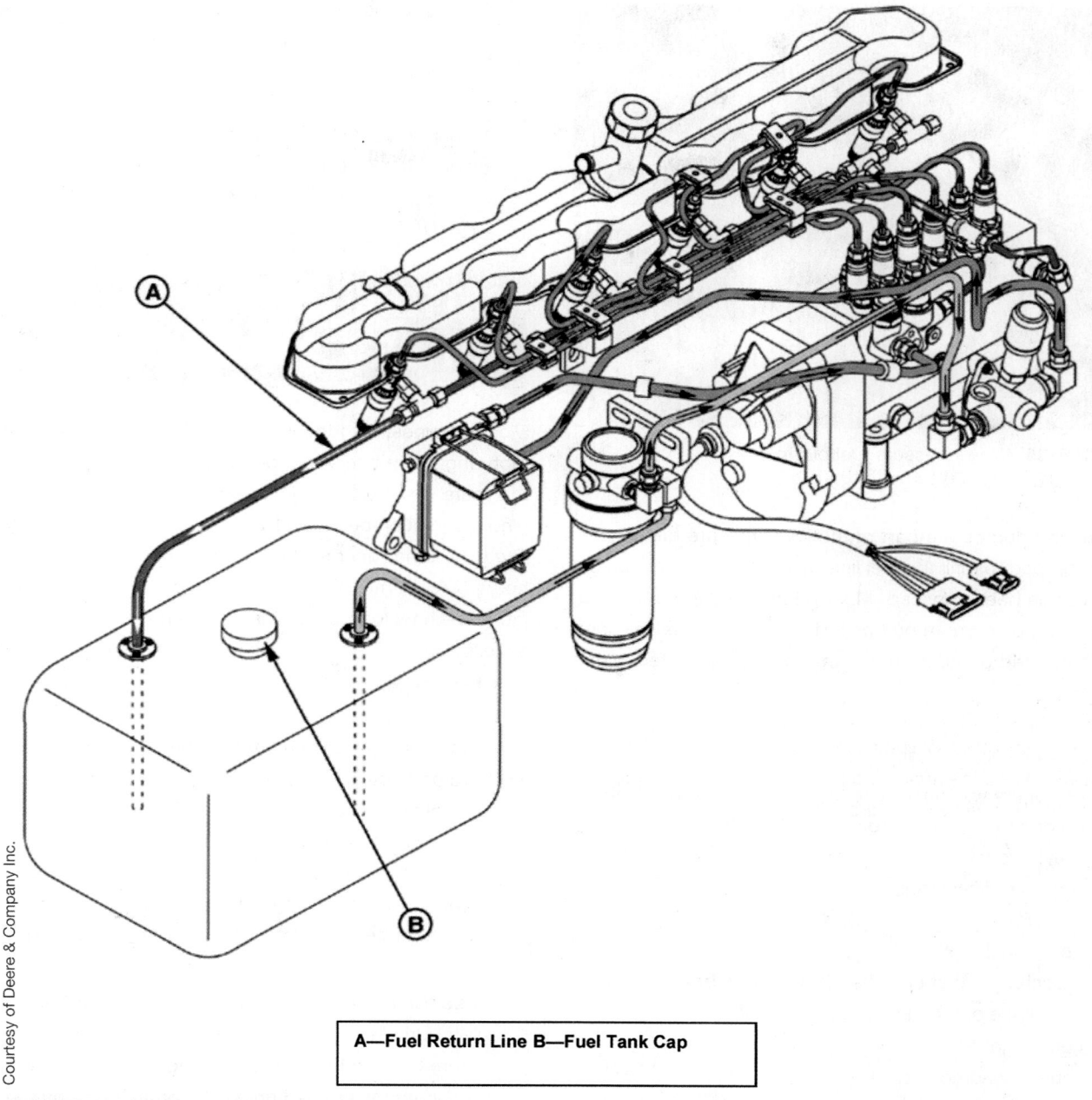

FIGURE 6–34 A typical low-pressure fuel system.

20 and 50 psi by a simple spring loaded relief valve. If the fuel tank is located where there is a chance the fuel could drain back out of the high-pressure fuel system, a check valve will likely be located somewhere in the pump outlet line to prevent this. The pump's output volume is much greater than what the fuel injection system requires under full load. This excess fuel is used to provide a cooling function to other fuel system components and will sometimes be used to cool other components like an engine ECM.

After leaving the low-pressure pump, the fuel will be filtered in one or more types of filters. The most common type of filter is the spin-on type that is usually replaced at 500-hour intervals. Some engines will use a cartridge-type filter that just needs the filter element to be replaced. There may also be a water separator that could be part of a filter or a separate item. Even a small amount of water entering the high-pressure fuel system can cause severe and expensive damage so regular service of the water separator is critical.

The fuel is sent to the filter base and then is forced through the filter media and on to the high-pressure fuel system. Any excess fuel that is not required by the high-pressure fuel system is then returned back to the fuel tank.

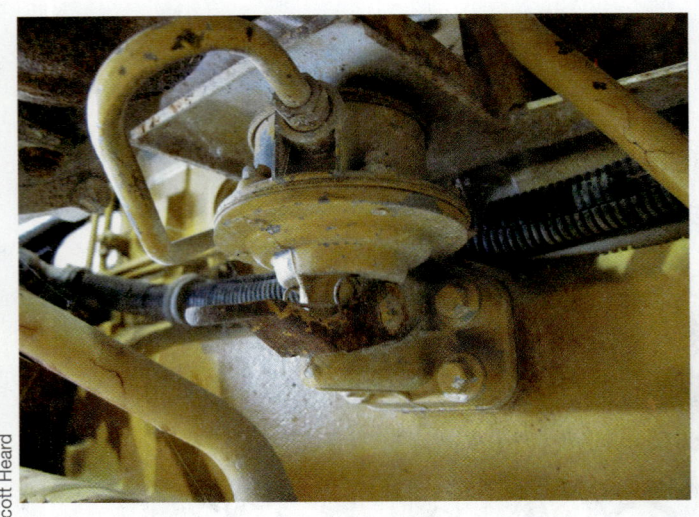

FIGURE 6–35 A low-pressure fuel pump.

Priming pumps are part of the low-pressure fuel system and are needed to refill the fuel lines if the machine has run out of fuel or has been serviced. Most priming pumps are manually operated diaphragm pumps but more machines are coming with electric priming pumps that can be operated with a switch.

Progress Check

11. Pre-cleaners will mostly use this effect to remove larger types of contamination from the intake air:
    a. gravitational
    b. pressure differential
    c. vacuum
    d. centrifugal
12. A turbocharger that can change its output based on different engine operating conditions is called a:
    a. centrifugal type
    b. variable geometry type
    c. variable impeller geometry type
    d. varying turbine type
13. The valve that regulates coolant flow to the radiator is called a:
    a. thermopile
    b. rad regulator
    c. pressure relief valve
    d. thermostat
14. The two types of engine oil pumps used for diesel engines are:
    a. external gear and diaphragm
    b. external gear and gerotor
    c. internal gear and piston
    d. internal gear and gerotor
15. During normal engine operation if a DPF does not get hot enough to burn off the collected soot inside of it, it will need to:
    a. perform an active regeneration
    b. perform a passive regeneration
    c. perform a reverse flow regeneration
    d. be removed for thorough cleaning

GETTING THE MOST LIFE OUT OF A DIESEL ENGINE

Diesel engines, with the improvement of materials and manufacturing, have become more reliable and longer lasting over the years. Twenty years ago, it was common to plan for an engine recondition between 10,000 and 12,000 hours of service. Today, it is not uncommon to run a diesel engine for 20,000 hours before it needs any serious attention. Some factors that influence how long before an engine needs major repairs are as follows:

Maintenance: The easiest way to maximize longevity from an engine is to perform all regular maintenance according to the manufacturer's guidelines. This includes oil and filter changes, air filter changes, and fuel filter changes. A fluid-monitoring program will tell the technician or machine owner if the maintenance program needs to be adjusted and will give notice to upcoming failures so they can be caught in time before causing major damage to engine components.

Use the recommended fluids: Engines are designed to run with a specific oil or coolant and this should be adhered to as closely as possible. Fluid additives should be avoided unless recommended by the engine manufacturer. This also includes using the proper fuel.

Use the recommended filters: The engine manufacturer will recommend specific filters be used and this should be adhered to whenever possible.

Operating conditions: The engine should be operated within the expected parameters for which the machine was designed to maximize its life expectancy. If the machine is constantly working at full load, then a shortened life span should be expected.

Minor repairs: If minor engine repairs are attended to when they first show up, then no other related failure should occur. If they are put off, then they will likely lead to bigger problems. Think of an air inlet pipe that has a broken bracket. If the repair isn't made soon enough, the pipe may wear through and the engine could be destroyed in a matter of hours.

NICE TO KNOW

DIAGNOSING, TESTING, ADJUSTING, AND REPAIRING DIESEL ENGINES Some engine problems will be obvious. If a connecting rod is hanging out the side of the block, you have a pretty good idea what the repair process is going to be. ● **SEE FIGURE 6–36** for an engine with a major failure. Some options for an engine with a major failure may be remove the engine and send it out for repair, get an exchange short block assembly, and swap over the external components or replace the entire engine. For light-duty engines, the last option might be viable; however, repair and exchange are the two most likely options. You would then install the new/rebuilt/exchange engine and make sure that it works as specified by running the machine for a short period. At some point in the repair process, a technician should have done a failure analysis to determine the root cause of the failure.

The above example is an extreme case where the repair process leaves very few options. For most other engine problems, the HDET will be required to understand how the engine or engine system works, how to diagnose the problem, how to repair the problem properly so it doesn't recur, and how to test the engine to know the problem is solved.

Even simple things such as finding and fixing fluid leaks can be a complex problem-solving exercise that will test the limits of an HDET's knowledge, skill set, and perseverance.

There is no way to go through even a fraction of the specific problems that could be found on a diesel engine, but this section will try to give the reader some intuition and general knowledge to help guide you through some situations.

One part of diagnostics that has changed in the last few years is that one of the first checks to perform is to check for electronic fault codes. Even for something as obvious as a connecting rod out the block, there could be a fault code logged that could be a clue as to the root cause of the failure. Fault codes will be explained further in Chapter 9.

The following are a few common engine problems the HDET will need to diagnose.

FLUID LEAKS Diesel engines have a few different fluids running through them so if there is a complaint about a leak, the first step is to identify it and make sure that it is an engine fluid leak. ● **SEE FIGURE 6–37** for an engine with a fluid leak. Some fluids are obvious like fuel and coolant. There may be other system fluids near the engine so if it isn't obvious what the fluid is, you may want to level off all fluids to their proper level and run the machine for a while to see which fluid level drops. It could be that the leak has been caused by someone overfilling a fluid.

If the leak is near the engine fan, it is most likely being blown around making it difficult to find. You must be very careful, and you may even find stopping the fan temporarily will be the only safe and accurate way to find the leak. Keep in mind that fluids always run downhill! This may seem obvious, but if a leak starts out in one spot and runs along a hose or tube to another location, you need to trace it back to its origin.

Sometimes it will be necessary to recreate conditions that make the leak occur. Certain temperature, pressures, or engine

FIGURE 6–36 An engine with a major failure.

FIGURE 6–37 An engine with a fluid leak.

PRIME MOVERS 205

> **HANDS-ON ACTIVITY**
>
> The next time you are near a parked machine that has been working on the job, have a look around it, and there's a good chance you will see a fluid leak. Start to look around and see where the leak is originating from. You may be surprised.
>
> Of course you need to keep your personal safety in mind.

> **HANDS-ON ACTIVITY**
>
> Go to a machine that is running, listen for any unusual noises and carefully feel for any unusual vibrations. Try this at different engine speeds and loads. You need to train your senses to pick up these unusual sounds and vibrations. But you first need to know what is a normal sound or vibration, and this comes with experience and paying attention to what the machine is telling you.

loads can trigger leaks. The HDET will need to get as much information as possible from the operator and try to make the leak happen. It's always a good idea to clean the area you suspect has the leak after an initial inspection. If dirt has collected around the leak area, the cleanest spot near the dirt is the source of the leak. This is because the leaking fluid will wash away the dirt.

All leaks are a result of a failed gasket, O-ring, or component. If the failed part is on the outside of the engine, this should be fairly obvious. If the failed part is on the inside of the engine, it will take a little more work to find it. For example, if there is engine coolant in the engine oil, there could be several places for this to occur. Some faulty components causing this are liner seals, oil cooler, turbocharger, air compressor, head gasket, or other locations that may have engine oil and coolant close together. The troubleshooting procedure would likely involve pressurizing the cooling system and trying to expose and isolate the suspect component. This may involve taking the oil pan or valve cover off. Once the failed part is identified, you need to determine the cause of its failure before you just replace the failed component. Otherwise, the failure will recur at some point. For example, perhaps a cooling system gasket has failed. It's not very likely the gasket itself was the cause of the failure. Maybe the system pressure was too high, there was a fastener not torqued properly, or the gasket wasn't installed properly. If anyone of these were the case, then just replacing the gasket would be a temporary fix.

ENGINE NOISES To determine the cause of an engine noise, you must first know if the noise is abnormal and then determine if the engine is the cause of it. This could be a problem if you have never heard this model of engine run before or heard the same engine in a different machine. See if the noise is in synch with the engine rpm or if it is reliant on machine travel speed or implement related. Once you have confirmed the noise is from the engine and know what conditions the noise is present under, then you must then try to isolate it to a specific part of the engine if possible. The condition(s) it takes to make the noise happen will be good clues as to the source of it. It takes a trained ear to identify where noises are coming from; there are listening tools (e.g., a stethoscope) to help this process, and don't be afraid to get a second opinion from a fellow tech. Heavy clunking or knocking noises (thud, thud, thud) will likely be related to the bottom end (crankshaft, connecting rods, pistons). Heavy rapping noises may be fuel system related (timing is out for one or more cylinders). Lighter metallic noises (tick, tick, tick) are probably related to the valve drivetrain.

ENGINE VIBRATIONS Engine vibrations can come from a defective engine mount, any rotating engine part that is out of balance, or a misfiring cylinder causing an imbalance. These vibrations also need to be isolated to the engine. If it is possible stop all drivetrain rotation and accessories the engine is driving one at a time to eliminate these as a cause. Some obvious external sources of engine vibrations are vibration dampers and pulleys and their fasteners. Internally, there are many possible sources, and the vibration should be narrowed down to whether it's the same frequency as the pistons or valves. Some possibilities are any internal engine counterbalance device, crankshaft bent, crankshaft weight, or a misfiring cylinder.

EXCESSIVE VISUAL EXHAUST EMISSIONS Up to about 30 years ago, it was acceptable to see a steady stream of black smoke coming out of the exhaust pipe of some machines. Prior to electronically governed engines, it was quite normal to see puffs of black smoke being discharged from the exhaust pipe.

● **SEE FIGURE 6–38** for an excessive amount of exhaust smoke emission. This would happen as the engine came under load the governor would overfuel the engine momentarily. If someone changed the governor settings and modified the fuel and aneroid adjustments, there would be a steady stream of black smoke. Some owners and operators thought this meant the machine was making more power that way but in fact the black smoke is a result of incomplete combustion.

FIGURE 6–38 An excessive amount of exhaust smoke emission.

Today's Tier 3 and 4 diesel engines that are used in most heavy equipment will not emit any smoke when at normal operating temperature and at any load. This means that any smoke coming from the exhaust pipe should be cause for concern. By reading the color of the exhaust, you can relate this to a probable cause. If the exhaust color is:

Blue: Blue exhaust from an engine that is up to operating temperature is an indication of engine oil getting into the combustion chamber and being burnt.

Oil level and viscosity should be checked first because an over-full level or too thin of a viscosity could lead to the engine oil getting burnt. A plugged crankcase ventilation tube will cause a buildup of crankcase pressure, and this can lead to oil being forced into the cylinder where it gets burnt. All engines will use or consume a quantity of oil between services. John Deere suggests excessive engine oil consumption is anything exceeding I gallon of oil for 400 gallons of fuel (400:1). If you are unsure about an engine's oil consumption, it will need to be monitored closely after an oil change for at least 100 hours. Some causes of oil entering the combustion chamber are piston rings worn/stuck, piston rings broken/weak, ring gap not staggered, liner wear/glazing, valve guide/stem wear, valve stem seal(s) worn or missing, and turbocharger seal worn.

White: White smoke will indicate either coolant is entering the combustion chamber or a sign that raw fuel is coming out the exhaust. If coolant leaking into a cylinder is suspected, you will likely be able to smell it as well (think of how a bakery smells when something sweet is baking). Raw fuel out the exhaust is a sign that there is no combustion occurring in one or more cylinders.

ABNORMAL OIL PRESSURE Engine oil pressure should stay within a specified range according to the engine manufacturer. This will typically be between 15 and 70 psi. If the oil pressure falls outside of the normal range, there will likely be a warning alarm come on and the HDET will have to find the cause for this. You should start with simple things such as checking the oil level and viscosity and looking for leaks and then check the warning system for proper operation. Ask the operator if there are any unusual noises, vibration, or loss of power. To confirm the oil pressure, you should connect a mechanical gauge directly into the main oil gallery to see exactly what it is. If the gauge confirms there is a low oil pressure, the next step may be to remove the oil filter, cut it open, and look for metal particles. If some are found, this may indicate a failing part that has caused excessive clearance and low oil pressure. It could also be from a faulty oil pressure relief valve.

FLUID TRANSFERS Occasionally, engine fluids may transfer from their normal location to mix with another fluid. For example, there may be coolant in the engine oil, fuel in the oil, or oil in the coolant. Some possible failed components that could cause this are engine oil cooler, head gasket, fuel injector seal, and air compressor. The only way this can happen is when one fluid is under greater pressure than the other and is allowed to mix together.

COOLANT TEMPERATURE OUT OF NORMAL RANGE Engine coolant can be overheated or overcooled. Overheating could be caused by poor coolant flow, poor air flow past the radiator, externally plugged radiator, combustion gases entering the coolant, or low level of coolant. It could also be caused by overloading the machine. An engine can also be overcooled, and this could be caused by a faulty thermostat or the cooling fan rotating too fast.

EXCESSIVE BLOWBY Engine blowby can be difficult to measure and determine if it is excessive. Most manufacturers will have rough guidelines as to what is excessive, and it sometimes comes down to a judgment call. HDETs could make a good decision if they performed a cylinder leak down test or compression test. These tests require special tools and will assess internal cylinder sealing integrity.

CYLINDER MISFIRES A rough running engine is said to have a misfire, and this can be caused by a faulty fuel injection component, a faulty valve, or a faulty piston ring or liner.

TUNE-UPS Diesel engine tune-ups need to be performed on a fairly frequent basis. There could be a requirement to change fuel nozzles and set the valves. Generally speaking, modern diesel engines require less maintenance and usually involve

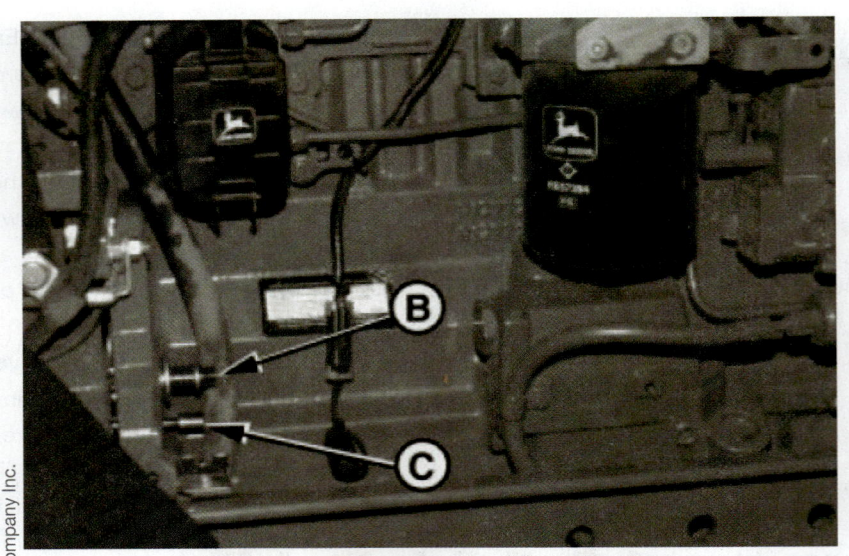

Flywheel Turning/Locking Tools

A—Timing Hole Plug
B—JDG820 Flywheel Turning Tool
C—JDE81-4 Timing Pin

FIGURE 6–39 A timing bolt location in the flywheel housing of an engine.

setting only valve clearances and any mechanical fuel injector settings if applicable. The following is a valve set procedure for an inline six-cylinder engine:

1. A valve set procedure starts with getting the engine's number one cylinder at TDC on the compression stroke. Some engines will have a timing bolt or pin that will lock the engine in place when the correct position is found while other engines may have an indicator on the vibration damper that must be lined up with a pointer. ● SEE FIGURE 6–39 for the timing pin location in the flywheel housing of an engine.

2. Adjust valve clearance on No. 1, 3, and 5 exhaust valves and No. 1, 2, and 4 intake valves.

3. Turn crankshaft 360°. Lock No. 6 piston at TDC compression stroke (C).

4. Adjust valve clearance on No. 2, 4, and 6 exhaust valve and No. 3, 5, and 6 intake valves.

● SEE FIGURE 6–40 for an HDET adjusting valve clearance.

RECONDITIONING

A diesel engine is designed to operate for a certain number of hours under certain conditions and have the proper level of preventive maintenance done along the way. If all these

FIGURE 6–40 A setting valve clearance.

conditions are ideal, then it's likely that the engine will last 15,000–20,000 hours. If a machine averages 2000 hours per year, then this could equate to up to 10 years of service before it will need reconditioning. However, if any of these ideals are not met, the engine could need reconditioning or repair much sooner.

Reconditioning is a repair procedure that will bring the engine back to new standards. It could be done in the machine if necessary; this is called an in-frame rebuild. This is only practical if the engine is medium-duty or larger and the crankshaft does not need to be removed. The engine will get stripped down to its block and rebuilt with new or exchange components.

However, ideally it will be removed and sent to a facility with trained technicians and proper tooling. The engine will first get cleaned and then disassembled carefully with the technician taking note of any unusual wear patterns. The crankshaft and camshaft will get measured and inspected and be sent out to a machine shop for repairs if necessary. The block will get inspected and checked for cracks only if a problem is suspected. All engine support system components will get reconditioned or replaced.

The rebuild process will go in this order:

1. The cleaned and inspected crankshaft and camshaft will get installed with new bearings.
2. New or reconditioned liners, pistons, and connecting rods with new bearings will get installed and liner protrusion will be measured.
3. The cylinder head will get reconditioned with new valve guides and seats and inspected for any other defects and then installed with a new gasket.
4. The valve train will be inspected, installed, timed, and adjusted.
5. The high-pressure fuel system will be tested or reconditioned and installed.
6. The oil and water pumps will be replaced or rebuilt.
7. The intake and exhaust manifolds will be inspected and installed with new gaskets or O-rings.

FIGURE 6–41 An engine being tested on a dynamometer.

8. All covers and the oil pan will be installed with new gaskets and O-rings.
9. The engine will be filled with engine oil and all final adjustments will be made.
10. The engine will be connected to a dynamometer and will be run through a series of load tests to ensure that there are no leaks and it is operating at its specified horsepower and torque output.

● **SEE FIGURE 6–41** for an engine being tested on a dynamometer.

CASE STUDY

This section will take you through a diagnostic procedure for a low-power complaint.

Low-power complaints are one of the most common complaints, but some other engine-related complaints could be for vibrations, noises, excessive smoking, leaks, and not starting.

An HDET is sent to a job site where a medium-size wheel loader is having an engine problem. The operator tells the technician the machine won't fill the bucket like it used to and he thinks it's an engine problem.

The HDET now starts the diagnostic process called troubleshooting. This is when technicians put their past experience, general knowledge about a machine, and a logical step-by-step process to work to find the root cause of this problem. It can be more of a process of elimination or finding what is working right than it is about what isn't working right.

First, the technician gathers as much information as possible from the operator. Using the low-power scenario, the following are some questions and responses:

Q: When did you first notice the problem?
A: It seems to have got worse in the last week.
Q: Has the machine been worked on lately?
A: No. Just a flat tire last week.
Q: Has the machine been serviced regularly?
A: Yes, it has been.
Q: Does the problem happen all the time or only under certain conditions?
A: It seems okay in the morning and slowly gets worse as the day goes on.
Q: Are there any other operational problems that you have noticed with this machine?
A: No, everything else seems okay.

Q: Are all the fluid levels normal?
A: Yes.
Q: Does it use any more or less fuel than normal?
A: It's hard to say but it seems pretty normal.

When getting information from an operator, try to separate facts from opinions. The main thing is to have them try to give specific information about what the exact problem is and under what conditions it occurs.

Now that the technician has some preliminary information, it would be time to start verifying the complaint. This is why knowing the exact complaint is important. If the complaint is not clearly stated or is given by a secondhand source, the technician may waste valuable time looking for a problem that doesn't exist.

The complaint that the machine won't fill the bucket could be a result of several things. It could be a hydraulic problem, a torque converter problem, or an engine problem. To verify this complaint, the technician checks to see if there is a low-power problem happening now. The operator mentioned earlier that the problem seems to get worse as the day goes on. The technician decides to check the machine to see if it is working okay in the morning. To check for a low-power problem, the technician performs a stall test and measures engine speeds. This test should identify whether it is a hydraulic, drivetrain, or engine problem. To perform this test, the technician will first check low and high engine speeds with no load and compare them to specification. This is done with a the machine's transmission in neutral, hydraulics in neutral, and the governor set to minimum and maximum speeds with the right pedal in the cab. If a machine has a tachometer, the engine speed is read from it, and if it doesn't, the engine speed must be read by a special tool called a photo tachometer.

While performing the stall tests, the HDET is also listening for any unusual noises, feeling for abnormal vibrations, and looking for unusual smoke colors and quantities from the exhaust pipe.

Let's say the speeds at no load were within spec (high idle 2200 ± 50 rpm and low idle 750 ± 50 rpm). The next test is to perform a torque converter stall test. This is when the transmission is put into its highest gear (third or fourth for a wheel loader), and with the service brakes applied, the governor is set to maximum rpm. A typical engine speed for this would be 1950 ± 50 rpm. Let's say this is on the low side but still within spec. If this speed was too high, it would indicate a torque converter problem and may explain why the machine has a hard time filling the bucket. If it was too low, it would indicate low engine power.

The next test is a hydraulic stall test. This is done by stalling a hydraulic function with the engine at maximum speed and checking the engine speed. Some newer machines may not have a hydraulic system that will be able to load the engine easily by this test. A typical engine speed for a hydraulic stall is 2050 ± 50 rpm. Let's say this test was within spec. The last test is a double stall where both a torque converter stall and a hydraulic stall are done at the same time. A typical speed for this test is 1850 ± 50 rpm. Let's say this test gave a speed of 1750 rpm or indicated an engine low-power problem.

The next step now that a problem has been verified is to run the machine and see if there is an operational problem that is noticeable. Again this test is done in the morning so it may not be an accurate indication of what the operator is complaining about. The machine seems to fill the bucket alright so the technician decides to return later in the day.

When the HDET returns, the double stall speed test shows a speed of only 1450 rpm. There has been a very noticeable loss of power and the bucket is very hard to fill. The complaint has now been verified.

Simple checks are done first that go in order of the following:

1. If the machine has an electronic fault recording system, it should be checked for any fault codes that may have been logged. The HDET would look for fuel system or intake and exhaust system-related faults.
2. Visual checks of the air inlet system for any loose clamps, wear points, and damaged components. Any restriction of intake air or loss of boost pressure will lead to a low-power problem.
3. Check governor control mechanism.
4. Check the fuel supply system for any obvious problems such as loose connections, bent tubes, pinched hoses, or valves half closed. The fuel tank vent should also be checked for cleanliness.

At this point, the HDET notices the fuel tank vent is covered in mud and then asks when the machine is fuelled up each day. ● **SEE FIGURE 6–42** for a plugged fuel

FIGURE 6–42 A plugged fuel tank breather.

tank breather. The operator responds that it gets fuelled in the morning before the machine is started. When the HDET attempts to remove the fuel tank fill cap, it seems to be stuck to the tank, and when pulled on hard enough, it lets go with a big whooshing sound of air moving. There had been a vacuum created in the tank as the fuel level dropped throughout the day because of the plugged tank breather.

If the HDET had overlooked this simple check, he may have moved on to more complex testing procedures that involved installing gauges and using special tools to look for a problem that wasn't there.

This is an example of why you should never overlook simple visual checks first when troubleshooting.

Progress Check

16. When looking for a fluid leak, it would likely be:
    a. near where the drip is seen
    b. near the cleanest wet spot
    c. near the dirtiest damp spot
    d. near the lowest wet spot
17. When checking for an engine vibration, you should:
    a. drive the machine at top speed
    b. run all the implements at once
    c. check it at low idle because it is quieter
    d. run the engine at whatever speed and load that creates the vibration
18. If a tune-up is required for a modern diesel engine, the HDET would most likely:
    a. change all the filters
    b. check and adjust valve clearance
    c. change the fuel injectors
    d. check and adjust the timing
19. A reconditioned engine will:
    a. always have new pistons installed
    b. only get components replaced if they don't meet minimum specifications
    c. have gaskets reused if they look okay
    d. never last as long as a new engine
20. This is one question you would *not* ask an operator when investigating a low-power complaint:
    a. how much do you make an hour?
    b. when did the problem start?
    c. has there been any work done on the machine lately?
    d. what conditions does this problem occur under?

SHOP ACTIVITY

- Go to a machine with a diesel engine and record the following: does it have a turbocharger, does it have an engine ECM, is it overhead cam, what type of low-pressure fuel pump does it have, what drives the fan, where is the thermostat located, is there any other coolers on the engine, and does the exhaust system have any aftertreatment devices?
- List the engine oil viscosity recommendations for an ambient temperature of −20°C and 80°F.
- What type of pre-cleaner does the engine have?

PRIME MOVERS 211

Summary

1. There are several safety concerns related to diesel engines that HDETs need to be aware of, so appropriate PPE can be used or actions taken to prevent injury or death. Some examples are high-pressure fuel systems, rotating parts, and hot components.

2. The term *prime mover* refers to the device that provides the rotational torque to drive all other systems on a machine. For mobile heavy equipment, it is primarily a diesel engine (compression ignition), but in some cases, it could be an electric motor or a spark-ignited engine.

3. When heavy equipment machines were first invented, steam engines heated with coal were the prime movers. These were external combustion engines.

4. Internal combustion engines create a thermal event inside a sealed chamber and harness the force of expanding gas to provide a rotational torque that drives other devices such as torque converters and pumps.

5. Low-emission regulations have changed diesel engine technology immensely over the last decade. Off-road engines sold in North America are categorized by different Tier ratings based on their horsepower and when they were produced. Tier 3 engines were introduced around 2007, Tier 4 interim were introduced around 2010, and Tier 4 final started appearing in 2013.

6. Diesel engines rely on compression of intake air to heat it up enough that when diesel fuel is injected, it will ignite.

7. Diesel engines use reciprocating pistons inside a cylinder to act as a pump to make the engine breath and to transfer the pressure of expanding gas acting on top of them to transfer that energy into a crankshaft. Pistons travel up and down inside a cylinder, and each full length travel distance is a stroke. Bottom of travel is BDC (bottom dead center) and top of travel is TDC (top dead center).

8. Some diesel engines use a two-stroke principle, but these engines are almost obsolete now. All engines produced for heavy equipment today need four piston strokes to operate.

9. Pistons transfer their **reciprocating** motion to connecting rods that can rotate on a journal on a rotating crankshaft.

10. The top of the combustion chamber is sealed with the cylinder head. It will house at least two valves (one intake and one exhaust) per cylinder that open and close to allow air in and exhaust out at the right time. A special head gasket seals combustion pressures in the combustion chamber. The head also has coolant and oil circulating through it.

11. To create high temperatures inside the combustion chamber, high compression ratios are needed. This is a comparison of the cylinder volume with the piston at BDC to when it's at TDC. Diesel engine ratios are usually between 15:1 and 24:1, and this gives a compression pressure of between 300 and 500 psi.

12. The four strokes that complete one cycle for a compression ignition engine are intake (piston moves down with intake valve open), compression (piston moves up with valves closed), power (fuel is injected and combustion forces piston down), and exhaust (piston driven up with exhaust valve open). The four strokes take place over 720° of crankshaft travel.

13. Today's diesel engines have efficiency ratings of between 40% and 45%.

14. Most diesel engines used in heavy equipment will have an upper rpm limit of approximately 2000, which equates to a lower piston speed and contributes to engine longevity.

15. There are many different types of pistons made from different types of materials. Highest horsepower engines use steel pistons today. They have grooves around the top to accept piston rings. Rings create a seal and control oil, usually top two are compression and one oil control.

16. Except for very low horsepower diesel engines, most will have cylinder liners. These are removable sleeves that fit into bores in the engine and have coolant surround most of their outer circumference. The inside surface should have a crosshatch pattern with a specific roughness to retain oil.

17. The diesel engine crankshaft needs to be heavy and strong to withstand many opposing forces applied to it. It will rotate in a film of oil that is created between it and a set of friction bearings.

18. Valve trains open and close the valves at precisely the right time in relation to piston travel. Two different configurations are cam in block and overhead cam.

19. HDETs will perform valve sets regularly, which will ensure that the proper clearance is maintained in the engine valve train.

20. The cylinder block is the backbone of the engine and will have oil and coolant circulating through it.

21. The crankshaft will need a vibration damper on it to even out combustion pulses. A flywheel will be fastened to the rear of the engine and provide a means for the starter to crank the engine, help to stabilize rpm fluctuations, and provide a means to drive drivetrain or hydraulic components.

22. Diesel engine intake and exhaust systems are designed to allow the engine to breathe in clean air in unrestricted quantities and expel spent gases quickly and quietly.

23. Air filtration systems for off-road machines have a tough job because conditions are often extremely dusty. Pre-cleaners can remove larger particles usually through centrifugal effect but need to be checked for plugging. Air filters are usually two-stage assemblies with an inner filter as a backup.

24. Non-turbocharged engines are considered to be naturally aspirated while turbocharged engines provide a positive pressure to the engines intake manifold. Many low-emission engines feature two turbochargers and many will have one that has variable geometry technology. Most modern turbocharged engines will have an intercooler to cool down intake air after it has been pressurized.

25. One common way to reduce emissions is to include an EGR (exhaust gas recirculation) system. These systems meter some exhaust gas through a cooler and back into the intake.
26. Modern low-emission diesel engine exhaust systems have changed drastically over the last decade to include one or more of the following: diesel particulate filters (DPFs), diesel oxidation catalysts (DOCs), and selective catalytic reduction (SCR).
27. The majority of diesel engines use a liquid coolant-type cooling system. One manufacturer produces an air cooled diesel engine.
28. Cooling systems need to get an engine to operating temperature (190°F is typical) as quick as possible and maintain it no matter what load on the engine is or the ambient temperatures. Thermostats regulate the flow of coolant to the radiator based on coolant temperature.
29. Cooling systems need to be pressurized to raise the boiling point of the coolant. Coolant is mostly glycol based and now most types are long life. Water pumps can be belt or gear driven and are centrifugal-type pumps.
30. Radiators are heat exchangers that transfer heat from the coolant to the atmosphere when air is pushed past them. Dirty radiators will prevent heat transfer.
31. Cooling fans are driven by belts and pulleys or hydraulic motors. Hydraulic motors can reverse the fan direction to help clean out accumulated dust and dirt.
32. Coolant can also be used to cool other machine fluids like transmission oil. Oil coolers are heat exchangers that transfer heat into engine coolant from other systems.
33. Engine lubrication systems use the engine oil and a pump to supply pressurized oil to the main moving parts of the engine. The pumps can be external gear or gerotor style. Engine oil type and viscosity specifications must be closely adhered to.
34. Engine fuel supply systems are divided into two parts: the low-pressure supply side and the high-pressure delivery side. The high-pressure system's purpose is to inject the right amount of fuel at the right time and in a finely atomized manner. There are many variations of high-pressure fuel systems and all are electronically controlled high-pressure common rail-type now. They operate at well over 20,000 psi and can deliver multiple injections per four-stroke cycle.
35. The low-pressure system starts with the fuel tank. The tank breather should be maintained to allow air to enter the tank as the fuel level drops. Low-pressure pumps get fuel from the tank through a water separator/screen and push it through a secondary filter before it gets to the high-pressure fuel system. Priming pumps will allow the HDET to bleed air out of the fuel system after servicing the filters or if the machine runs out of fuel.
36. Engine services at the right interval are critical to keeping an engine running as long as possible without needing major repairs. Manufacturers will specify the types of fluids the engine should have in it, and sticking with OEM filters is always wise.
37. Depending on the application and operating conditions, an engine should last 15,000–20,000 hours before needing reconditioning. Facilities that are equipped to completely rebuild engines should also be equipped with dynamometers so the engine can be tested before it is put back into service.
38. Engine diagnostic procedures start with verifying the complaint and then performing simple checks. Some common engine-related problems are leaks, noise, vibration, low power, overheating, excessive smoke, low oil pressure, and excessive blowby.

chapter 7
DIESEL ENGINE STARTING SYSTEMS

LEARNING OBJECTIVES

After reading this chapter, the student should be able to:

1. Identify all main components of a diesel engine starting system.
2. Describe the similarities and differences between air, hydraulic, and electric starting systems.
3. Identify all main components of an electric starter motor assembly.
4. Describe how electrical current flows through an electric starter motor.
5. Explain the purpose of starting systems interlocks.
6. Identify the main components of a pneumatic starting system.
7. Identify the main components of a hydraulic starting system.
8. Describe a step-by-step diagnostic procedure for a slow cranking problem.
9. Describe a step-by-step diagnostic procedure for a no crank problem.
10. Explain how to test for excessive voltage drop in a starter circuit.

KEY TERMS

Armature 218
Brushes 218
Commutator 221
Disconnect switch 235
Field coil 218
Hold in 218
Pull in 218
Starter interlock 232
Starter relay 223

INTRODUCTION

SAFETY FIRST Some specific safety concerns related to diesel engine starting systems are as follows:

- Battery explosion risk
- Burns from high current flow through battery cables
- Strain injuries from lifting heavy starters and batteries (some are over 65 lb)
- Burns from battery electrolyte
- Fire hazards from sparks and hot wires
- Unexpected cranking or starting of engine
- Injuries from sudden release of stored energy (electrical, hydraulic, air pressure)

THE IMPORTANCE OF STARTING SYSTEMS A functional machine needs a running engine, and if the engine doesn't crank, it doesn't start. A properly operating and reliable starting system is a must for keeping a machine productive. For many years, diesel engines have mostly used electric motors to crank them over to start the combustion process. For some applications, an air or hydraulic motor will create the torque needed to turn the engine over.

Many years ago, diesel engines were sometimes started with a smaller gas engine called a pup engine. ● SEE FIGURE 7–1 for a pup engine on an older diesel engine. Another way to get a diesel engine started was to start it on gasoline and then switch it over to run on diesel fuel. This was a complex solution to a simple task because the engine had to have a way to vary its compression ratio, and it needed a spark ignition system and a carburetor. As 12V electrical systems became more popular and electric motor design improved, electric starters were able to get the job done. Many large diesel engines will use a 24V starting system for even greater cranking power. ● SEE FIGURE 7–2 for a typical arrangement of a heavy-duty electric starter on a diesel engine.

A diesel engine needs to rotate between 150 and 250 rpm to start. The purpose of the starting system is to provide the torque needed to achieve the necessary minimum cranking speed. As the starter motor starts to rotate the flywheel, the crankshaft is turned, which then starts piston movement. For a small four-cylinder engine, there doesn't need to be a great deal of torque generated by a starter. But as engines get more cylinders and bigger pistons, a huge amount of torque will be needed to get the required cranking speed. Some heavy-duty 24V starters will create over 200 ft-lb of torque. This torque then gets multiplied by the gear reduction factor between the starter motor pinion gear and ring gear on the engine's flywheel. This is usually around 20:1. ● SEE FIGURE 7–3 for how a starter assembly pinion engages with the flywheel ring gear.

FIGURE 7–2 A typical arrangement of a heavy-duty electric starter on a diesel engine.

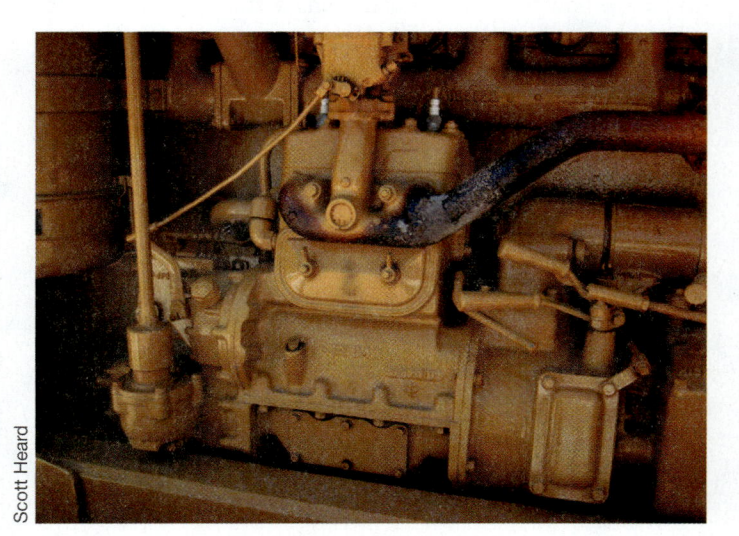

FIGURE 7–1 A pup engine starter motor.

FIGURE 7–3 A starter assembly engaging with a flywheel ring gear.

DIESEL ENGINE STARTING SYSTEMS   215

FIGURE 7–4 A double starter assembly arrangement.

Some larger engines will need two or more starters to do this. Some starters for large diesel engines will create over 15 kW or 20 hp! ● **SEE FIGURE 7–4** for a double starter arrangement.

When a starter motor starts to turn the engine over, its pistons start to travel up in the cylinders on compression stroke. There needs to be between 350 and 600 psi of pressure created on top of the piston. This is the main resistance that the starter has to overcome. This pressure is what is needed to create the necessary heat in the cylinder so that when fuel is injected it will ignite. If the starting system can't crank the engine fast enough, then the compression pressure and heat won't be high enough to ignite the fuel. If the pistons are moving too slowly, there will be time for the compression to leak by the piston rings. Also the rings won't get pushed against the cylinder, which again allows compression pressure to leak into the crankcase. When this happens, the engine won't start or it starts with incomplete combustion. Incomplete combustion equals excessive emissions. This is another reason to have a properly operating starting system.

The faster a starter can crank a diesel engine, the faster it starts and the quicker it runs clean.

This engine cranking task is much more difficult in colder temperatures especially if the engine is directly driving other machine components such as hydraulic pumps, a torque converter, or a PTO (power take-off) drive shaft. Cold engine oil adds to the load on the starter, and this load may increase by three to four times what it would normally be in warmer weather. Engine oil that is the wrong viscosity (too thick) for the temperature will greatly increase the engine's rolling resistance. Adding to this problem is the fact that a battery is less efficient in cold temperatures.

When engineers design a cranking system, they must take into account cold weather cranking conditions and will quite often offer a cold weather starting option. This would likely include one or more of the following: bigger or more batteries, higher output starter, larger battery cables, battery blankets, oil heaters, diesel fired coolant heater, electric immersion coolant heater (block heater), and one or more starting aids like an ether injection system or an inlet heater.

One more recent difficulty added to starting systems is a result of electronic controls on some engines. Some ECMs may need to see a minimum number of engine revolutions at a minimum speed before it will energize the fuel system. This equates to longer cranking times and more strain on the cranking system. Some electronic engines will crank for five seconds or longer even when the engine is warm before the ECM starts to inject fuel and the engine starts.

It's important that a machine's starting system works properly and you should be aware of how the main components of a system work. This will give you the knowledge needed to make a proper diagnosis when you get a complaint of an engine cranking slowly or not at all.

If an engine doesn't start, then a machine isn't working, and instead of making money, it's costing money. The better you know how to diagnose and repair a starting system problem, the more valuable you will be as an HDET. There are lots of technicians who are good at changing starters whether the starter is faulty or not. Many times the cause of a starting complaint is something other than the starter.

If a starter is used properly, it will last for well over 10,000 starts. The biggest factor in reducing the life of an electric starter is overheating from over-cranking. *Never* run the starter for more than a 30-second stretch, and if it does run that long, then wait at least two minutes between cranks to allow the starter to cool.

For engines up to 500 hp, electric starting systems will be used for 99% of the applications. For any size engine, air and hydraulic starting systems are an option; however, they will likely only be used for special applications and usually for engines over 500 hp.

TECH TIP

This chapter focuses on diesel engine starting systems because at this time they are the most popular type of engine used for heavy equipment. However, there may be some other types of engines used in the future such as natural gas–powered engines.

Natural gas engines are used for many stationary power applications, and many are similar to diesel engines but with lower compression ratios, different fuel system, and a spark ignition system. Because of the lower compression ratio, they will put a lower demand on the starting system.

The job of any starter motor assembly is to take a stored energy (electric, air, or hydraulic) and convert it into mechanical rotation to crank the engine fast enough to begin the engine's ignition sequence.

NEED TO KNOW

The most common type of starting system uses electrical energy; however, compressed air and hydraulic energy can be used as well.

The following are the main comparable components of the three main types of starting systems:

ELECTRIC STARTING SYSTEM	AIR STARTING SYSTEM	HYDRAULIC STARTING SYSTEM
Electric starter motor assembly	Air motor starter assembly	Hydraulic motor starter assembly
Battery cables	Air lines	Hydraulic hoses
Starter relay	Relay valve	Directional control valve
Starter interlock system	Starter interlock system	Starter interlock system
Battery(ies) or capacitor	Air tank	Hydraulic accumulator
Starter switch	Starter switch or valve	Starter switch or valve
Wiring harness	Wiring harness (optional)	Wiring harness (optional)

We'll first examine the different electric starter motor designs, next discuss air and hydraulic starter motors, and then look at the control circuit for starters.

ELECTRIC STARTER MOTOR ASSEMBLY

An electric starter will take stored electrical energy from a battery (or sometimes a capacitor) and convert it into torque at the starter's pinion gear. The pinion then engages with the ring gear that is part of the engine's flywheel and turns the flywheel that rotates the engine's crankshaft. ● SEE FIGURE 7–5 for a cutaway of a starter and its main parts.

There are two main types of electric starter motor assemblies:

- **Direct drive (pinion is driven directly by the armature):** A direct drive electric starter has a motor that is designed to generate high torque at low speed and operate at high speed with low torque (the motor will sometimes exceed 5000 rpm) for a short length of time. It will use a solenoid actuated shift lever to push out the pinion to engage it with the ring gear before or just as the armature (rotating shaft in the motor assembly) starts turning.

- **Gear reduction (higher speed motor output to a gear reduction and then to pinion):** A gear reduction starter (planetary or pinion reduction) is designed to use a smaller higher speed electric motor to produce higher cranking torque with the same or less electrical power consumed. The heaviest and bulkiest part of a direct drive starter is the motor so by reducing motor size and weight the engineers have saved space and weight. Some direct drive starters are twice the weight as a comparable output gear reduction starter. Although this isn't a big concern for a large machine, you will be thankful for the lighter weight whenever it comes time to change the starter.

Gear reduction starter motor assemblies can have their motor offset from the output shaft or use planetary gears and have the motor shaft in-line with the output shaft.

DIRECT DRIVE STARTER COMPONENTS

- Starter housing: Center section that holds the pole shoes and field coils in place.

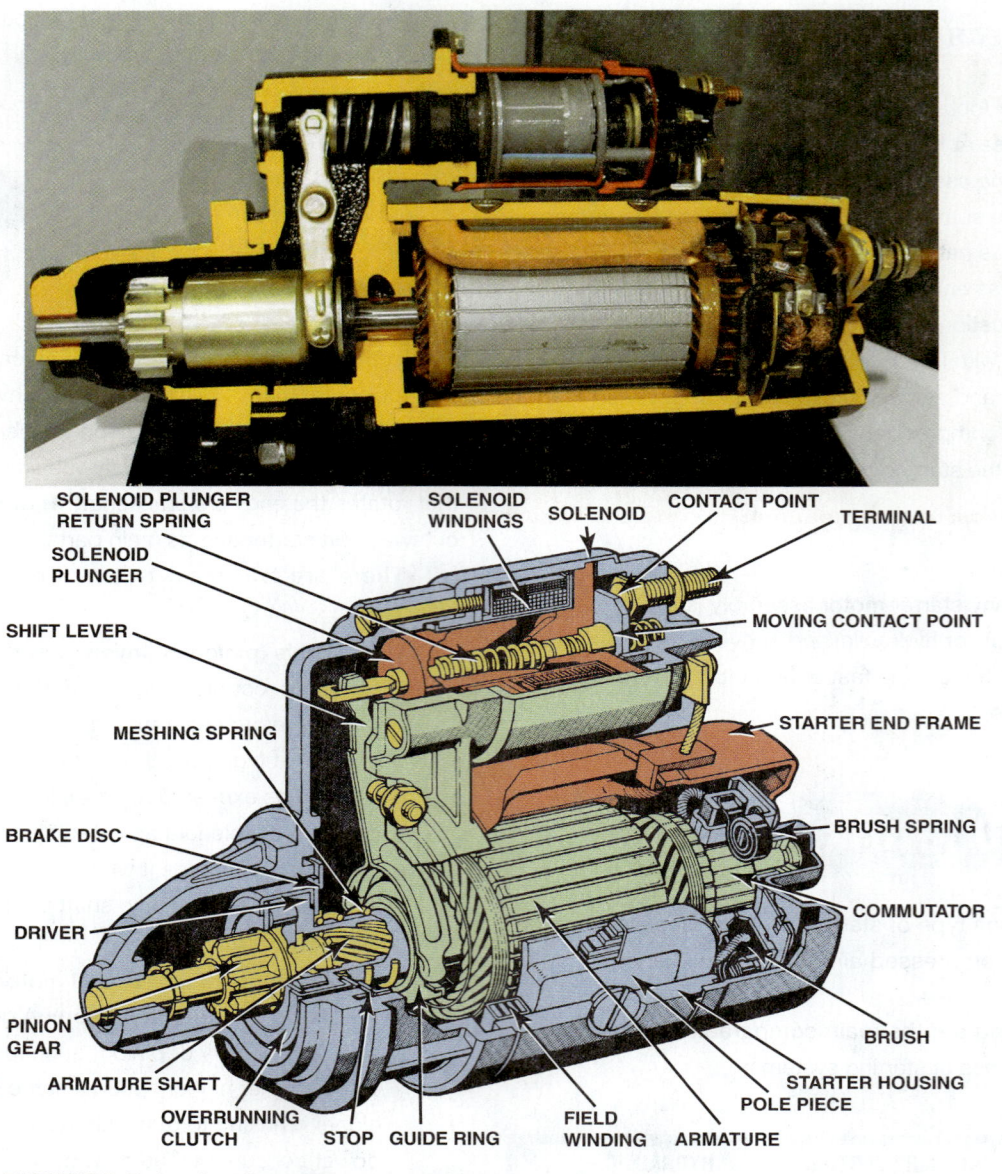

FIGURE 7-5 An electric starter cutaway.

- **Nose piece**: The drive end of the starter where the pinion gear is located. Holds the shift lever in place and supports the armature shaft with a bushing.
- **End cap**: Opposite end of the starter from the nose piece. Supports brush holder assembly and the other armature shaft bushing.
- **Armature**: The rotating part of the motor that has several windings that have each of their ends loop to a commutator bar. It will have splines to drive the starter drive.
- **Brushes**: Contact the commutator bars and transfer electrical current to the armature.
- **Brush holders**: Spring loaded to keep the brushes in contact with the armature.
- **Field windings**: Heavy copper windings that create a strong magnetic field when current flows through them.
- **Pole shoes**: Iron cores for the field windings that help to increase magnetism.
- **Solenoid**: Has two windings (**pull-in** and **hold-in**) that get energized by the starter control circuit and magnetically move a plunger. The plunger is connected to a heavy contact disc that is a switch. The switch will send current from the battery terminal to the field windings. The plunger could also be connected to a shift lever that will move the pinion.
- **Pinion gear**: The starter output that engages with the flywheel and cranks the engine.

- Overrunning clutch: Drives the pinion from the armature shaft but will not allow the armature to be driven by the ring gear.
- Shift lever: Used to push the pinion out to engage with the ring gear.

GEAR REDUCTION STARTER COMPONENTS

- Motor components (armature, brushes, brush holder, field windings, pole shoes) are the same as a direct drive starter.
- Gear reduction: The armature shaft will have a gear output that will drive an intermediate gear that drives the pinion gear shaft.
- Solenoid: The solenoid performs the same electrical functions as the direct drive starter but may directly push the pinion gear out.
- Overrunning clutch: Same as direct drive.

● **SEE FIGURE 7–6** for a gear reduction starter.

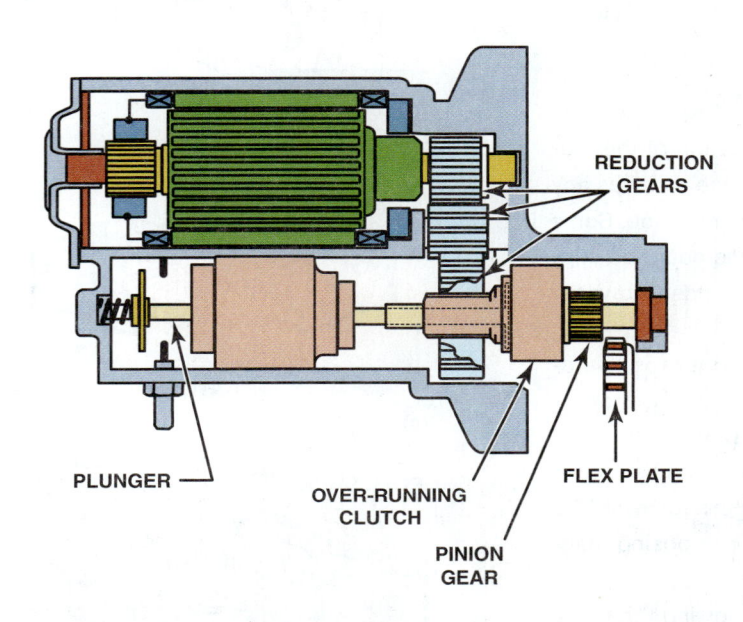

FIGURE 7–6 A gear reduction starter.

DIESEL ENGINE STARTING SYSTEMS 219

FIGURE 7–7 Current flow causing opposing magnetic fields.

ELECTRIC MOTOR OPERATION
The action of the motor section of the electric starter assembly is the same as any brush-type motor used on a heavy equipment machine. Some other applications of motors that use brushes could be supplementary steering or brake pump motors, windshield wiper motors, and hood lift motors.

The basic principle behind any electric motor operation is the arranging of opposing magnetic fields so that rotation is created. This is the same force that will try to keep two like poles of permanent magnets apart. If you have ever played with magnets, you will be familiar with this force. ● SEE FIGURE 7–7 for how current flow can cause opposing magnetic fields.

As discussed in Chapter 5, if current is flowing through a conductor, then there will be a magnetic field created around the conductor. If you want to build a stronger magnetic field, then loop an insulated wire into coils, put an iron core through the middle of the coil, and increase the current flow through the wire.

If opposing or attracting magnetic fields can be arranged between two components, then relative motion occurs. This of course can only happen if the magnetic forces are strong enough to overcome the resistance that it is opposing. Opposing magnetic fields are also used to make solenoid plungers move.

We'll start to look at a simple motor with one pair of field windings and an armature with one loop of wire. In a simple DC electric starter motor, one magnetic field is created in the stationary motor housing and is generated between a pair of field windings/pole shoe electromagnets. ● SEE FIGURE 7–8 for the field windings/pole shoes in a starter motor housing. The pole shoes are the iron core of the electromagnet, and when the field windings are energized, a strong magnetic field is created. One field winding will act like a north pole and the field winding opposite will act like a south pole. This magnetic

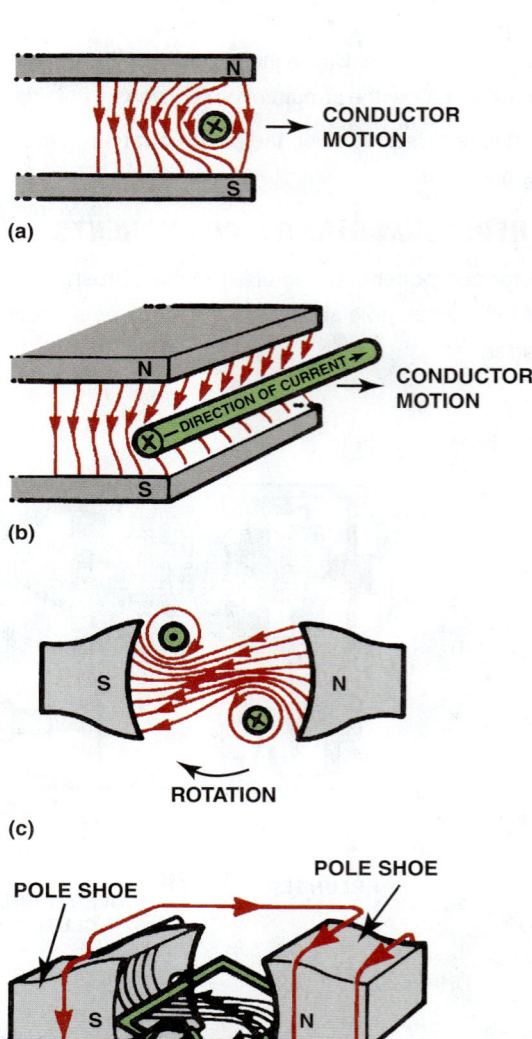

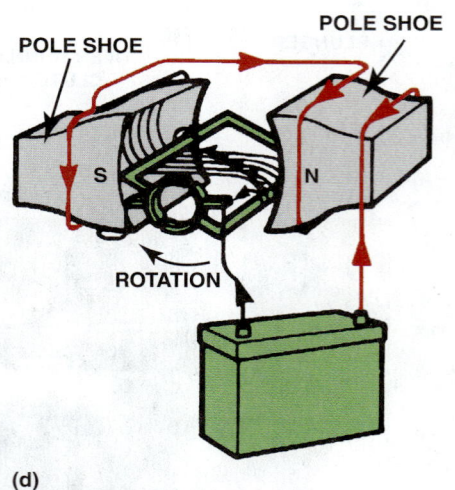

FIGURE 7–8 The field coils/pole shoes in a starter motor housing.

field is like the field occurring between the ends of a horseshoe magnet. The field windings are heavy flat copper and appear to be bare wire but are coated with a thin varnish to keep the loops insulated from itself and from the pole shoe. If this insulation fails, there will be a short circuit fault.

The other magnetic field in the simple starter motor is created in a loop of wire (winding) that is the armature. The armature operates inside the magnetic field that has been created by the field windings/pole shoes. When the armature winding is

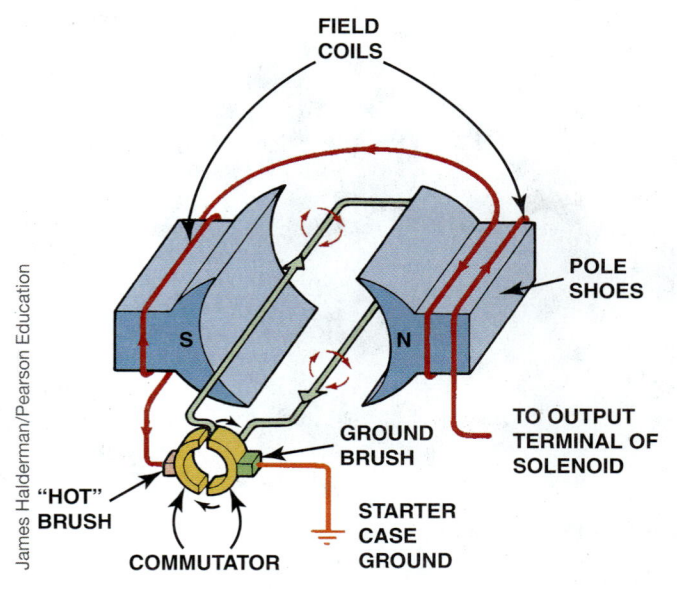

FIGURE 7–9 How do the armature winding interact with the field poles.

WEB ACTIVITY

Search the Internet for homemade electric motors and you will see a variety of simple designs. Why not see if you can make your own motor. Hopefully curiosity will get the better of you. This will also help you to understand how magnetism and electricity work together to make mechanical movement.

It's not hard to make an electric motor. ● SEE FIGURE 7–10 for a simple homemade motor. The field magnets are permanent magnets but could be made stronger with field coils.

FIGURE 7–10 A simple homemade electric motor.

energized, its surrounding magnetic field interacts with the pole shoe/field windings magnetic field. The armature shaft is supported on two bearings in each end of the motor assembly to allow rotation. When the two opposing magnetic fields create an unbalance, the armature will rotate to try to become magnetically balanced. If the magnetic fields are timed correctly to oppose each other at the proper location, then the armature will rotate. In other words, as the loop of armature winding is energized, it will have a surrounding magnetic field on each of the two longest sections of the winding that run parallel to the armature shaft. These sections of the winding will react to the field coils' magnetic fields and the reacting force will try to push the armature winding away. As the armature rotates, the winding will have its direction of current flow reverse as soon as the ends of the winding swap places with the brushes feeding it current. This allows the same loop of wire to be continuously pushed away from the field coil magnetic field. ● SEE FIGURE 7–9 for how the armature winding interacts with the field poles.

The armature shaft is the rotating output of any electric motor. The armature windings are embedded into grooves that run along the length of the armature core. The armature core serves the same purpose as the field windings/pole shoes—that is, to increase the magnetic field surrounding the energized windings. The windings are insulated from each other and from the core by a thin coat of varnish. Shorts and grounds may occur if this varnish coating fails and exposes the bare copper. ● SEE FIGURE 7–11 for a typical armature and how the loops of wire are embedded into the core. Each end of an armature winding is connected to one **commutator** bar. The commutator bars are insulated from each other but are exposed on their outer surface to allow connection to the motor brushes. You should also note that the windings are offset to the commutator bars that they connect to. This is to place the winding at the correct location relative to the field coil magnetic field so that maximum torque is created on the armature by the opposing magnetic fields.

You've probably noticed by looking at the photo of the armature that there is more than one loop of wire and more than one set of commutator bars. To provide continuous and steady rotation of an electric motor, there needs to be a constant opposing magnetic field. As more windings are added to the armature, it's possible to do this. As one loop of wire's

DIESEL ENGINE STARTING SYSTEMS

FIGURE 7–11 A typical armature and how the loops of wire are embedded into the core.

> **WEB ACTIVITY**
>
> Search the Internet for commutation animation and see if you can find an animation of an electric motor working. This should help you understand the interaction between the armature windings, commutator bars, and brushes.

FIGURE 7–12 A set of brushes riding on an armature commutator.

FIGURE 7–13 A set of brush leads.

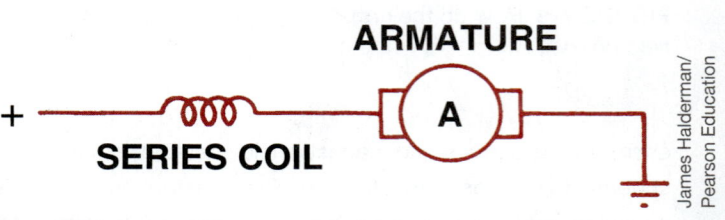

FIGURE 7–14 Current flow through a series wound motor.

commutator bars comes into contact with a set of brushes to get energized, the rotation it creates moves another set out of contact.

To get current flowing to an armature winding, there must be a connection that can allow relative motion in order for the armature to be able to rotate. The motor's brushes and the armature's commutator bars create this connection. ● SEE FIGURE 7–12 for a brush riding on an armature commutator.

The motors brushes are held in contact with the commutator bars by spring pressure. The brushes will eventually wear down and not make proper contact. They are made of a hard conductive combination of carbon and copper material and will constantly transfer current to and from the armature commutator. This current is the full current flow that the starting motor will see whenever it is energized by the starter solenoid. This could be as much as 3000 amps when the motor is first engaged. ● SEE FIGURE 7–13 for a starter motor set of brush leads. The brush holder assembly is held in place by the end housing of the starter. The brushes have a flexible lead that attaches to a terminal on the brush holder assembly. One brush will supply voltage to the armature from the field coil while the brush opposite will be connected to ground. Most heavy-duty starters will have two or three sets of brushes to energize more armature windings at the same time. This will in theory double or triple the motors output.

When a series wound electric motor is first energized and the armature is stationary, the motor will make its maximum torque. This is because there should be very little resistance in the motor components that have current flowing through them. ● SEE FIGURE 7–14 for the current flow through a series wound starter motor assembly. As it is shown, if you think of current flowing in the conventional direction, it starts at the

222 CHAPTER 7

battery positive, goes to the first set of field windings, and then goes through the rest. A series wound starter could have two, four, or six field coils. This current flow path will change if the starter is a shunt wound or compound (combination shunt and series). Different field winding arrangements will give a starter motor different characteristics such as good low speed torque or good speed limiting.

Heavy-duty starters will sometimes have an external ground cable to ensure a good connection to ground. Light-duty starters will be internally grounded. This means they rely on their mounting fasteners to make a good connection to the flywheel housing, which in turn should get a good connection through the engine block to battery ground. This, however, is sometimes a source of problems and some repairs include putting an external ground cable on to the starter housing.

In theory, if energized, an electric motor should keep increasing in speed until it self-destructs. A self-limiting phenomenon called counter-electromotive force (CEMF) will not allow an electric motor to keep increasing in speed. Another outcome between magnetism and conductors happens when a conductor is moved through a magnetic field. Current flow is generated because of something called induction. This is the basic principle behind generators and alternators. As the armature windings move past the field coils, there is a voltage induced in the armature. This voltage opposes the voltage that is flowing into the winding from the brushes. The two opposing voltages reduce the overall size of the magnetic field around the armature windings. The higher the speed that the armature spins, the greater the CEMF produced. Therefore, the motor self-governs its speed as the speed increases.

Remember that it was stated earlier that the motor makes its maximum torque at 1 rpm and the torque reduces as the speed increases. This is exactly how you would want a starter motor working. In other words, when the engine is stationary, it will take the most power to start it cranking, and then as the cranking rpm increases, it should take less torque and this is how the starter motor delivers its torque and speed. The CEMF reduces current flow by opposing the current flowing into the armature and this in turn reduces torque as speed increases. This is a proportionate reaction. ● **SEE FIGURE 7–15** for the graph showing the speed, torque, and horsepower output of a heavy-duty electric starter.

STARTER MOTOR SOLENOID All heavy-duty electric starter motors will use a solenoid assembly to do two things. They will pull a plunger in to move the drive pinion out and will close a heavy-duty contact switch. There are typically four terminals on a heavy-duty starter solenoid. The two smaller ones are the "S" terminal, which is energized by a key switch or **starter relay**, and the "G" terminal, which supplies the ground to the solenoid. The two larger terminals are "B+" for the main battery positive cable and a motor terminal that sends current from the solenoid to the motor. Sometimes the ground wire isn't needed as the solenoid may be grounded internally.

A starter solenoid will have two sets of wire windings in it. The pull-in and the hold-in windings get energized to pull

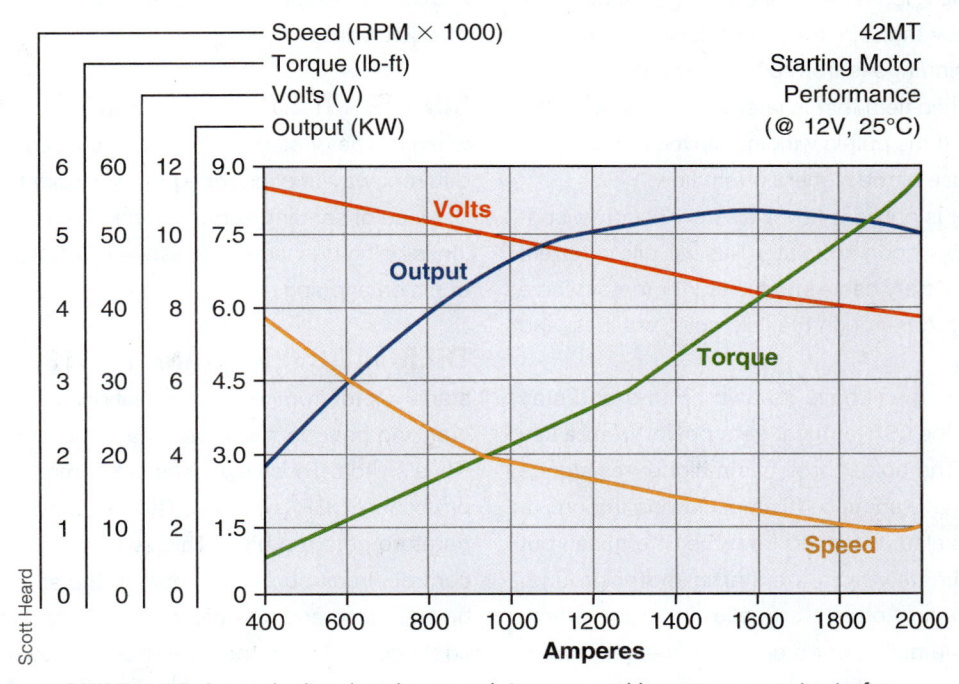

FIGURE 7–15 A graph showing the speed, torque, and horsepower output of a heavy-duty electric starter.

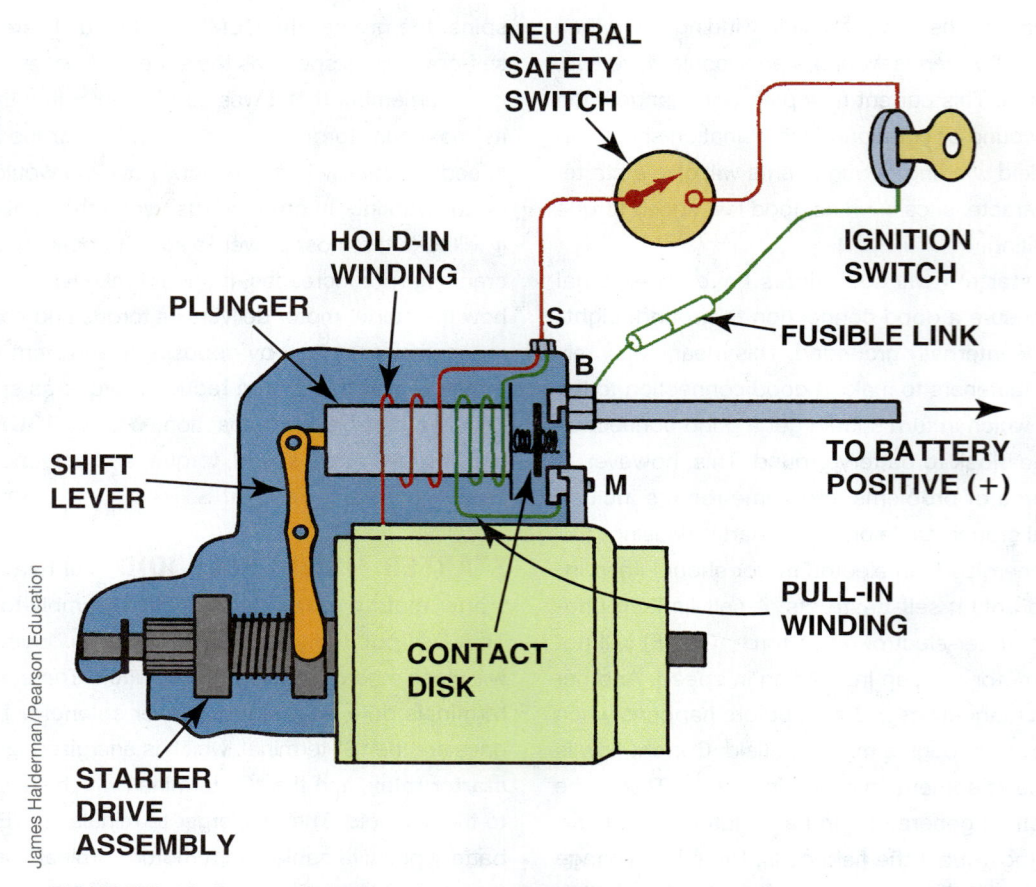

FIGURE 7–16 Current flow through the hold-in and pull-in windings in a typical solenoid.

the solenoid plunger in against return spring pressure. The pull-in winding will use a lower-resistance, bigger gauge wire to let more current flow through it. Current flow through typical solenoid pull-in windings is from 30 to 60 amps. This will create a strong magnetic field that is needed to overcome the initial spring pressure. If the pull-in winding stayed energized, it would overheat because of the higher current flow.

When the plunger is pulled in, it will also connect two contacts with a heavy copper contact plate. This will allow current to flow from the main battery cable into the starter motor where it starts heading to ground through the field coils, brushes, and armature.

The hold-in winding will have its own ground and stay energized as long as the "S" terminal gets power from a key switch or start relay. The hold-in coil has a higher resistance and will typically only have about 5–10 amps flowing through it.

The pull-in coil is also energized from the S terminal, but it gets its ground ultimately from the starter motor ground. Once the solenoid contact plate starts allowing current flow to the starter motor, the pull-in coil's ground turns into a positive part of the circuit. As the pull-in coil has a positive voltage on each end of its winding, there is no potential difference and, therefore, current stops flowing through the coil. ● **SEE FIGURE 7–16** for how current flows through the two windings of a solenoid.

SOFT START STARTERS You may hear the term *soft start* when a type of starter is being discussed. The term refers to a different way of connecting the solenoid to the motor with the purpose of sending some current flow to the motor as the pinion is being pushed out. This will help to ease the engagement of pinion and ring gear.

THERMAL OVER-CRANK PROTECTION To protect a starter motor from excessive heat damage caused by extended cranking periods, the motor may incorporate a thermally sensitive switch. These switches are commonly called over-crank protection (OCP) switches. The switch will open when the temperature gets too high. This switch will be part of the starter control circuit so when it opens, the starter solenoid will not be energized and, therefore, the starter cannot operate until it cools down. When the starter cools down, the switch closes and starter operation goes back to normal. ● **SEE FIGURE 7–17** for a starter with a thermal OCP switch.

FIGURE 7–17 A starter with an OCP switch.

STARTER DRIVES
The business part of the starter is the pinion gear, which is driven by the starter drive. For a direct drive starter, the drive is turned directly from the armature shaft by splines. Gear reduction starters will have one or more gears between the armature shaft and the starter drive. The drive pinion will have between 7 and 11 teeth that must be of the same pitch as the ring gear teeth they are engaging with. A typical ratio between the starter's pinion and the flywheel ring gear is 15:1. This would equate to a pinion speed of 2250 rpm if the engine is turning at 150 rpm.

The drive mechanism must provide a means of not allowing the starter's armature to be driven by the engine's flywheel. If we take the ratio of 15:1 and calculate potential pinion speed if the engine is running at low idle of 700 rpm, we would see the pinion would spin at 10,500 rpm. This would surely damage a starter motor that is designed to operate at a maximum speed of 5000 rpm. To prevent this from happening, an overrunning or one-way clutch will be between the pinion gear and the armature shaft.
- **FIGURE 7–18** shows an overrunning clutch for a starter drive.

An overrunning clutch works like a ratchet in that it provides a positive drive action in one direction but allows slippage in the reverse direction. Overrunning clutches can get worn and allow slippage in both directions.

The drive mechanism also engages the pinion before full motor torque is starting to turn the pinion. This will reduce damage to the pinion and the ring gear. To engage the pinion with the ring gear, a mechanism will shift the pinion out along the pinion drive shaft. Most drive shafts will have a helix or very coarse thread shape machined on them with which the pinion drive engages. Its purpose is to slightly rotate the pinion as it is pushed into the ring gear so that it will engage easier. The pinion gear teeth will usually have a taper machined into the leading edge of the tooth that will assist with engagement. These features are to prevent the pinion from trying to be a milling machine and take material off the ring gear teeth. If you hear someone mention that a ring gear has been milled, you now know what this means.

Direct drive and planetary drive starters will use a shift lever to move the pinion drive out toward the ring gear. The shift lever is actuated by the solenoid plunger and pivots on a cross pin. The bottom part of the shift lever has a fork with wear buttons that ride in a slotted part or the drive. This allows the drive to rotate and still be allowed to be pushed out by the shift lever.
- **SEE FIGURE 7–19** for how a shift lever and drive assembly work together.

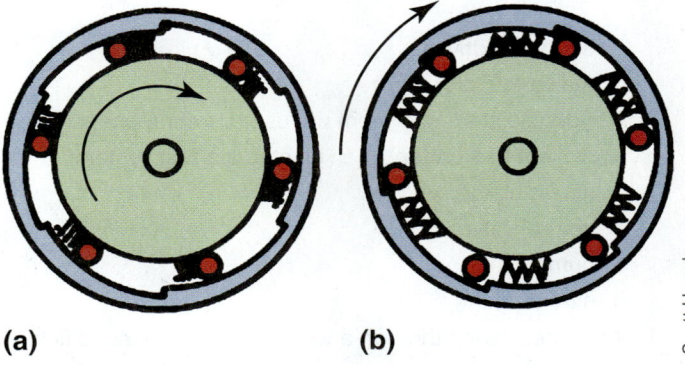

(a) **(b)**

FIGURE 7–18 An overrunning clutch for a starter drive. (A) A starter shaft driving pinion gear. (B) A pinion gear is allowed to rotate separately from starter shaft.

Progress Check
1. If an electric starter is run for 15 seconds, it should be allowed to cool down for how many seconds before running again:
   a. 15
   b. 30
   c. 90
   d. 120
2. An electric direct drive starter will drive the pinion gear with:
   a. the solenoid plunger
   b. the armature shaft

DIESEL ENGINE STARTING SYSTEMS

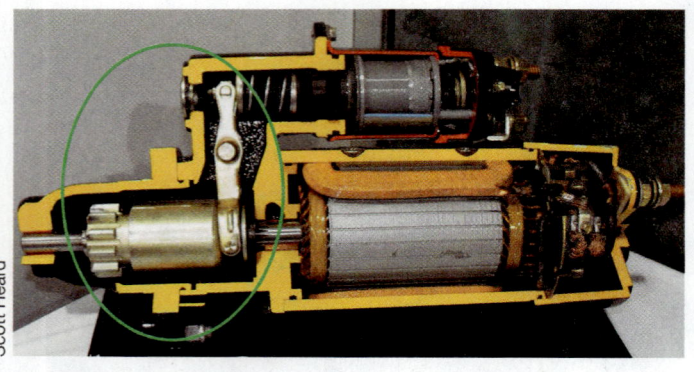

FIGURE 7–19 A shift lever and drive assembly.

   c. the brush drive
   d. the vane motor shaft
3. A hydraulic starting system will use this component as the equivalent to a battery in an electric starting system:
   a. accumulator
   b. control valve
   c. pump
   d. nitrogen tank
4. If current flowing through a wire creates a magnetic field around the wire, then to increase the strength of the magnetic field, you would:
   a. increase the length of the wire
   b. increase the diameter of the wire
   c. decrease the diameter of the wire
   d. increase the amount of current flowing through the wire
5. An electric drive starter motor will make maximum torque:
   a. before the solenoid plunger moves
   b. at maximum speed
   c. when the engine is running
   d. just as it starts to rotate
6. The S terminal on an electric starter solenoid is for:
   a. receiving current from a relay or switch
   b. sending current to the motor
   c. special applications and not normally used
   d. sensing when the starter is turning
7. The solenoids hold-in coil will have less current flowing through it than the pull-in coil because:
   a. it gets energized after the pull-in coil
   b. it only gets half system voltage
   c. it has a separate resistor
   d. its wire is a smaller gauge
8. Some starters will have an OCP switch. This stands for:
   a. only crank positive
   b. over-center pinion
   c. overdrive clockwise pinion
   d. over-crank protection
9. If the ratio of starter pinion teeth to ring gear teeth was 20:1 and the engine was turning at 180 rpm, then the pinion must be turning at:
   a. 900 rpm
   b. 2000 rpm
   c. 3600 rpm
   d. 4800 rpm
10. The starter's overrunning clutch will allow:
    a. the starter pinion to spin with the flywheel but not drive the starter motor
    b. the armature to increase output torque
    c. the pinion to be pushed back away from the ring gear
    d. the armature to increase speed output

ELECTRIC STARTER MOTOR REMOVAL, DISASSEMBLY, COMPONENT TESTING, AND INSTALLATION

One of the last things to be done when diagnosing a starting system problem is to remove the starter. However, if a proper troubleshooting procedure has been followed and there appears to be a problem with the starter motor, then it must be removed. Electric starting system troubleshooting procedures are discussed later in this chapter.

The following are key points to remember when replacing a starter:

- Always disconnect the battery cables from the battery first.
- Clean the area around the starter because the flywheel housing may be wet (be part of a drivetrain fluid system) and any dirt that drops into the housing will now be circulating through drivetrain components.
- Always mark the wires as they are removed.
- Starters can be awkward and heavy to remove so call for help or attach a lifting device before the last bolt is taken out.
- Put the old and new starter side by side to make sure that you have the right one (all terminals are the same) and the nose piece is oriented correctly.
- Make sure that you have the right mounting gasket if required.
- Support the main battery stud with a wrench when tightening the battery cable terminal nut.
- Clean any dirty terminal before installing and put protective covers back over terminals.

The ring gear should always be inspected when a starter is removed to look for excessive wear. Although the ring gear will likely be in its normal resting position, the engine should be rotated by a helper while the whole ring gear is inspected.

BENCH TESTING AN ELECTRIC STARTER

A starter bench test should be done after any starter is removed. This will require the starter to be put in a vice to hold it securely. First, check the drive mechanism to see if it turns free in one direction and locks in the other. The gear should also be inspected for excessive wear. A power source is then connected to the starter B+ terminal and a ground to the starter housing and the S terminal is energized. The pinion should be pushed out firmly. When the pinion is turning, a tachometer can be used to measure no load speed and an ammeter can measure current draw. A typical no load current draw is 75–150 amps and a maximum speed could be as high as 5000 rpm depending on the starter. These two measurements can then be compared to specifications. If the measurements are out of spec, then consider the following causes:

Diagnose No-Load Test

Fails to Operate—Low Current Draw (approximately 25 amps)

- Open series field circuit
- Open armature windings
- Defective brush contact with commutator

Fails to Operate—High Current Draw

- Grounded terminal or fields
- Seized bearings

Low Speed—Low Current Draw

- High internal resistance
- Defective brush contact with commutator

Low Speed—High Current Draw

- Excessive friction
- Shorted armature
- Grounded armature or fields

High Speed—Low Current Draw

- Open shunt field circuit

High Speed—High Current Draw

- Shorted series field coils

If the measurements are within spec and the drive appears good, then you should think twice before installing a new starter.

Once a starter is removed and found to be defective, it will most likely be replaced with a new or exchange unit. However, sometimes a situation may call for a starter to be reconditioned by the HDET who removes it.

The following is a list of starter components that should be tested and the kind of test done. You should always consult the starter manufacturer's service information for the particular starter you are working on.

- Field coils: Check for opens and shorts.
- Armature windings: Check for opens and shorts (growler tool is used).
- Armature commutator: Check for shorts, insulation depth, and bar appearance. If necessary, clean up with crocus cloth or fine sandpaper (do not use emery cloth).
- Armature condition: Check for shaft straightness and bearing mount surface.
- Solenoid: Check for opens and shorts, energize, and check that it pulls in plunger.
- Drive shift mechanism: Check for positive movement both ways.
- Brushes: Measure length.
- Brush holder: Check for brush to base ground.
- Brush springs: Check tension.

After reconditioning the starter, perform a no-load test to confirm that the starter is okay to install on the machine.

Starter motor installation requires all fasteners to be torqued to specification and wires to be secured to prevent insulation wear through.

BATTERY CABLES—SIZING, REPAIR

Due to the fact that the starting circuit is the highest current drawing part of a machine's electrical system, it is critical that the battery cables and connections that run between the battery system and the starter are sized right and are in good condition. Any less than ideal sizing or connection condition will create excessive resistance, which will translate into a voltage drop and slow cranking. Battery cables are purposely large to give very little resistance to current flow. Usually, this resistance is less than 1 milliohm. ● **SEE FIGURE 7–20** for a chart that indicates minimum cable size for a given length and system voltage.

Ideally batteries should be located as close as possible to the machine's starter. This is not always the case though and some machines will have cables over 20 ft long. These cables will be subject to vibration and damaged from other machine parts. Part of an HDET's duties will at some point include repairing or replacing battery cable assemblies. Care should be taken relating to the following points:

- Proper cable sizing (diameter) to avoid unnecessary resistance. It is better to oversize cables if you are unsure.

DIESEL ENGINE STARTING SYSTEMS **227**

Determining Starting Circuit Cable System Cable Size

System Voltage and Type	Maximum Circuit Voltage Drop per 100 Amps	Total Cranking Circuit Length in Inches				
		100	200	300	400	500
12-Volt High-Output Heavy-Duty (4.5–7.8 kW)	w/Single Path** to-and-from Starter 0.075 Volts	00	000	In Parallel: 0000 or 2–0 / 2–00 / 2–000 / 2–0000	2–0000 Positive Cables / 2–0000 Negative Cables	
	w/Dual Path* to-and-from Starter 0.075 Volts (0.150 per LEG)	00		000	In Parallel: 0000 or 2–0 / 2–00 / 2–000	
12-Volt Super Heavy-Duty (7.9–8.5 kW)	Single Path** 0.060 Volts	00	000	In Parallel: 0000 or 2–0 / 2–00 / 2–000 / 2–0000	2–0000 Positive Cables / 2–0000 Negative Cables	
	Dual Path** 0.120 per LEG	00		000	0000 / 2–00 / 2–000 / 2–0000 In Parallel	
24-Volt Heavy-Duty (7.5–9.0 kW)	0.20 Volts	6 / 4 / 2	1 / 0	00	000	In Parallel: 0000 or 2–0 / 2–0

* See Fig. 1: Dual Path Circuit
** See Fig. 2: Single Path Circuit

Note: Cable Should meet or exceed SAE J1127.

FIGURE 7–20 Minimum cable size for length and system voltage.

- Proper terminal to cable installation to ensure longevity and corrosion protection.
- Proper cable security. Battery cables should be secured every 18–24 in. to prevent insulation being worn off from other parts rubbing on it. ● SEE FIGURE 7–21 for how a battery cable should be secured.
- Seal all connections to prevent corrosion. There are many types of electrical connection sealants. Be sure to use one after installing battery cable terminals or connections.

Some tools and equipment required to make battery cable repairs are as follows:

- Heavy-duty cable cutter
- Heavy-duty insulation stripper
- Heavy-duty terminal crimper
- Propane torch
- Solder
- Shrink tube
- Terminals

FIGURE 7–21 A properly secured battery cable.

STARTER CONTROL CIRCUIT The starting control circuit can be a very simple series circuit with a key or push button switch in the cab that is fed power from a fuse. It will then send power to the starter solenoid when the key or push button switch is closed. ● SEE FIGURE 7–22 for a simple starter control circuit. Most starting circuits will use a relay between the control switch and the starter solenoid. This will be the case on all but the smallest machines. The relay will be used to handle higher current flow that needs to go to the starter solenoids on larger starter motors. Some very large starters could even have two relays between the switch and the solenoid.

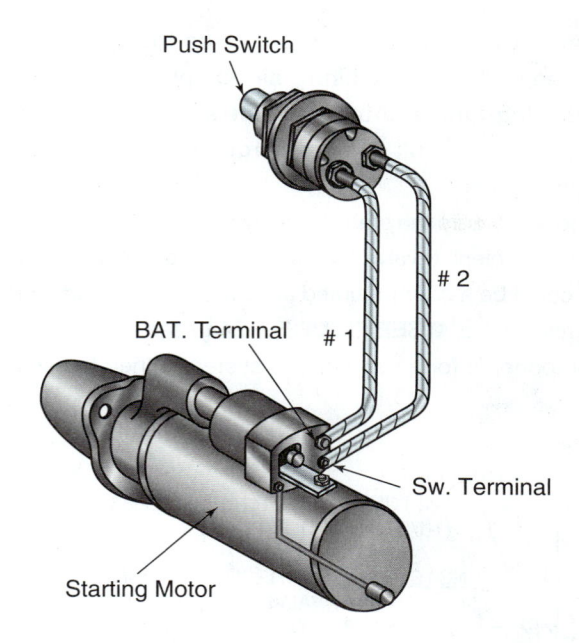

FIGURE 7–22 A simple starter control circuit.

- If the starter has been running longer than 30 seconds at a time, it will stop the starter cranking the engine and not allow it to engage for 60 seconds to allow it to cool down.
- Fault codes can be generated such as when the control circuit sends a signal to the starter relay, but there is no engine rpm sensed at the engine speed sensor.

The starter control circuit could tie into other machine electrical circuits such as hydraulic, drivetrain, and lighting. Some examples of what this interaction might do are as follows:

- The hydraulic pump could be electrically de-stroked to decrease the load on the starter.
- The work light circuit could be de-energized to provide more current for the starter.
- The transmission neutralizer circuit could be actuated to ensure that the transmission doesn't try to drive the machine.
- The HVAC circuit could be de-energized to provide more current flow for the starter.

FIGURE 7–23 A keypad used to start a machine.

Control circuits can also be very complicated and include several switches, relays, sensors, and ECMs. Many machines today that use electronic controls will use keypad buttons to control the starter control circuit. This could incorporate a security system where the operator has to key in a password before the machine will start. ● **SEE FIGURE 7–23** for a keypad used to start a machine.

Other starting circuit features that could be part of an electronic system are as follows:

- If the ECM sees that the engine is turning more than 300 rpm, it won't allow the starter to engage.

AIR STARTING SYSTEMS

Different engine applications could call for an alternative starting system to the electrical one just discussed. The environment the machine is working in could be flammable and require a spark-proof machine or the cost of replacing batteries in extremely cold environments is seen to be excessive. One alternative is to use a dedicated air supply to spin an air-powered starter motor assembly.

There are some advantages to having an air driven starter. They are much lighter and, therefore, have a higher power to weight ratio than a comparable output electric starter. There is no chance of an air starter overheating from over-cranking. Because of their simple design, there is very little that goes wrong with them. The most problematic area that can cause trouble with an air starter assembly is excessive moisture in the air system that can freeze in cold weather.

One disadvantage is how fast the air supply is depleted when the starter is engaged. Most starting tanks will empty within 20 seconds. If the air tank does deplete before the engine starts, this means charging the tank with an external air source from a shop air line, other machine, or service truck.

An air starter will generate high cranking speed and torque so that under normal conditions the engine should start before the starter air tank runs out.

DIESEL ENGINE STARTING SYSTEMS

There are two main types of air starter motors. One is a vane type that uses sliding vanes in a rotor to convert air flow into mechanical movement. The other type is called turbine, and its rotation is created by air flow pushing on the blades of one or more turbine wheels.

If you look back to the chart comparing air, hydraulic, and electrical starting systems, the main differences are the energy supply, type of motor, air lines, and system control.

The machine will most likely have an air compressor to provide air for other pneumatic systems and to keep the starter air tank charged up. Once the engine starts, it is then up to the machine's air compressor to recharge the starting tank and the machine's other supply tanks. The air starting tank will be charged to between 110 and 150 psi.

To send air to the starter, a relay valve will be controlled by an electric solenoid valve that is activated by the key switch or there could be a floor-mounted air relay valve to send air to the main relay valve. ● **SEE FIGURE 7–24** to see the arrangement of components for an air starting system. When the solenoid

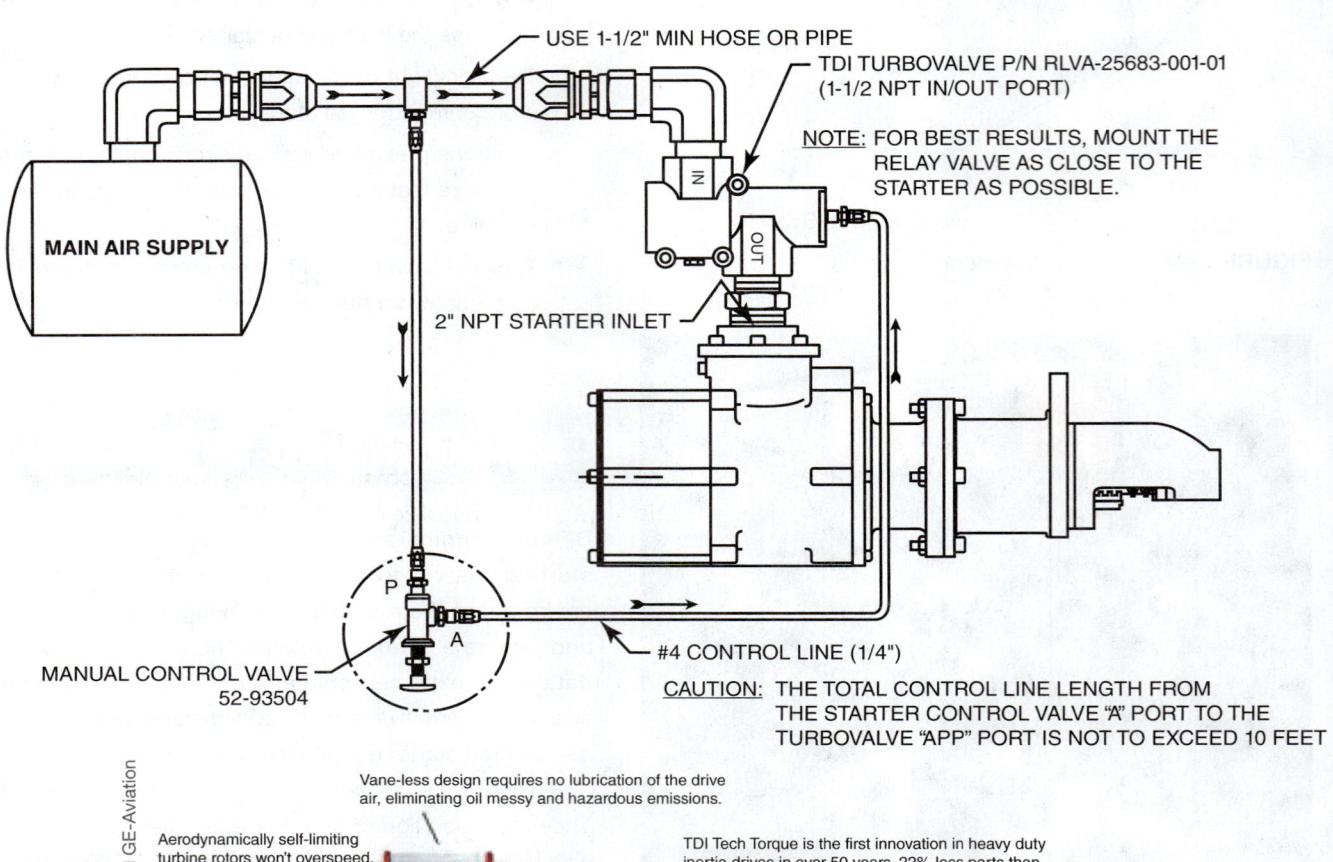

FIGURE 7–24 Components of an air starting system.

valve is energized, it will send air to the relay valve that will open to allow tank air into the starter motor. There are two main types of starter motors: vane and turbine. The motors create shaft rotation that usually has its speed reduced and torque increased through a gear reduction. The torque is then sent out through a drive pinion to engage with the flywheel. Vane-type motors will need lubrication and will usually have a small amount of diesel fuel drawn into the motor inlet during starter engagement.

It is important to have clean dry air entering air starters and their control circuit. Problems with moist air are magnified in the winter with relay valves freezing and sticking. Air leaks and air restrictions are the only other concern with air starter systems. The motors will last a long time, and if they are found to be worn out, repair kits can be installed to renew the starter assembly.

HYDRAULIC STARTING SYSTEMS

Another nonelectric starting system is one that uses hydraulic fluid to rotate a hydraulic starter motor. The motor will then rotate a drive gear in the same manner as typical electric starters. Hydraulic start systems have an accumulator that keeps hydraulic fluid stored under pressure until needed. A control valve is actuated to send pressurized fluid to the motor to get the motor turning. The motor is a fixed displacement axial piston unit, and its shaft drives the pinion gear directly.

SEE FIGURE 7–25 for a hydraulic starting system. The control valve could be floor mounted, cable operated, or controlled electrically through an LCD screen touch pad called a human–machine interface (HMI).

The accumulator for this system has a pre-charge of 1500 psi of nitrogen, and when the oil is pumped into it, the pressure builds to 3000 psi.

This system will have a backup hand pump that could be used to charge the accumulator.

If the system doesn't operate, then perform a good visual inspection. Then check the accumulator pre-charge pressure and the oil pressure after the accumulator has been charged. If these pressures are good, then look for restrictions or leaks past the accumulator toward the control valve. Make sure that the valve is moving as it should, and if there is still a problem, you may have to install pressure gauges throughout the system to see if there is oil pressure getting past the control valve.

As with any fluid power system, cleanliness is crucial so check for fluid contamination. For information on accumulator service and repair, refer to Chapter 16.

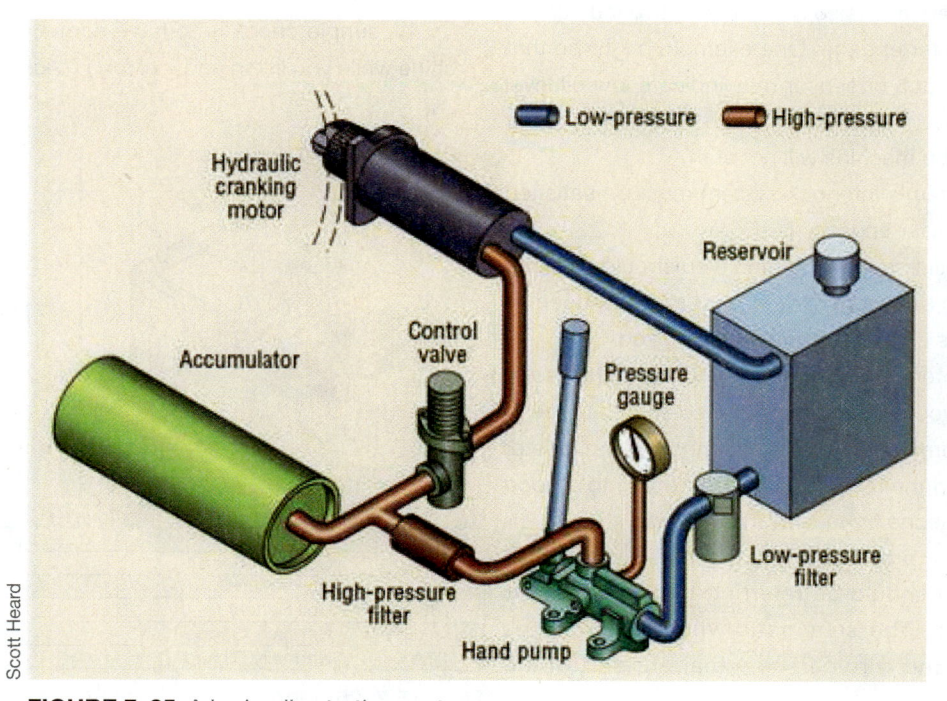

FIGURE 7–25 A hydraulic starting system.

STARTING CONTROL CIRCUIT INTERLOCK SYSTEMS

Most machines will have some kind of safety-related or damage-preventing system that will not allow the engine to either crank over or start under certain circumstances. There could be one or more conditions that have to be met before the operator's input to try and start the machine will result in a signal going to the starter solenoid. This could be done with a simple two-position switch or with several switches and one or more ECMs. The interlock system will likely be part of the starter control ground side but could be on the positive side. It will complete that part of the circuit by closing switches or changing sensor values.

Here are a few examples of machine conditions that may need to be satisfied:

- Seat belt fastened
- Safety bar lowered
- Door closed
- Operator in seat
- Gear selector in neutral
- Hydraulic controls in neutral
- Parking brake engaged
- Password entered

All these conditions can be verified with switches, sensors, or a keypad. There may also be certain active fault codes that won't let the starter engage. One example might be that the engine oil level switch or sensor may indicate low oil level and be setting an active fault code. Until the oil is topped up or the sensor repaired, the machine will not start.

The starting system's interlock system must be satisfied before it will let the starter engage.

If you are unfamiliar with a machine, you should read the operator's manual to see what switches or sensors need to change state to let the starter solenoid be energized.

To know if the interlock system is letting the starter solenoid get energized, you could confirm the solenoid is receiving power to the S terminal. To do this, connect a multimeter positive lead to it with the meter's other lead going to a good ground and have someone try to start the machine.

To troubleshoot a problem with a machine's **starter interlock** system, you should be referring to the machine's electrical schematic. The schematic will tell you all the sensors, switches, and wire numbers that are part of the system.

STARTING CIRCUIT RELAYS

Relays were looked at in Chapter 5 but a quick review here is beneficial. A relay will be part of two different circuits. One part will be a control circuit that gets energized by the operator's starter control circuit. This is the low current flow part of the relay and is simply a coil of light gauge wire that will create a magnetic field when energized. The second part of the relay is the high current section that will have a constant power feed that will get sent out of the relay depending on the state of the relay's control circuit.

When the control coil is energized, the magnetic field created will close the switch or contacts on the high current side.

The most common relay will have terminal numbers that indicate the part of the relay they are connected to. ● FIGURE 7–26 shows the terminal numbers for this common relay.

Some large starters will need a heavy-duty relay like the one shown in ● FIGURE 7–27.

Some relays are internally grounded, meaning the control coil has one end of it connected to the housing of the relay. This means the relay body must be grounded.

Some starters will have a relay mounted directly on them, and these are usually called integrated magnetic switches (IMS).

Starter relays are sometimes overlooked as a problem part of the starting circuit. They are fairly easy to troubleshoot once you understand how they work. Sometimes the hardest part is finding them!

A simple check is to have someone try to start the machine while you listen and feel for a "click." This will tell you that

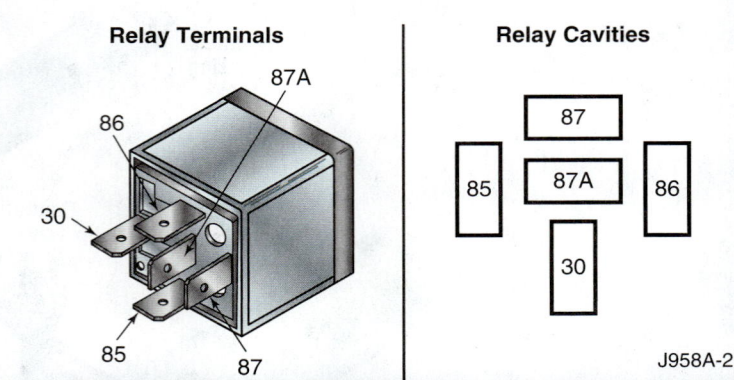

Number	Identification
30	Common Feed
85	Coil Ground
86	Coil Battery
87	Normally Open
87A	Normally Closed

FIGURE 7–26 Terminal numbers for a common relay.

FIGURE 7-27 A heavy-duty starter relay.

at least the relay is getting energized and the control coil has a good ground. If you hear a click, then check the voltage at the high current supply. If this is good, then check the high current output for voltage while someone tries to start the machine again. You could check for an excessive voltage drop across the high current terminals of the relay. If this checks okay, then the starter solenoid should be getting voltage if there are no wiring issues.

STARTING SYSTEM TROUBLESHOOTING

The basic steps of troubleshooting should be followed here and as always start with the following.

VERIFY THE COMPLAINT The two most common problems you will hear about with a starting system are no cranking and slow cranking. To some operators, not starting means not cranking, and in any case, where there's a starting complaint, you need to see for yourself what is happening. If the problem is happening now, then get the operator to show you. If the problem is intermittent, then ask when and under what conditions the problem is occurring. You can try to recreate it or you may have to return at another time.

Always check to see if the problem isn't simply part of the starting interlock system or that there is extra resistance to cranking the engine being caused by other factors (wrong engine oil viscosity, hydraulic system load, drivetrain load).

Once you have verified there is a starting system problem, then you can move on to diagnose it.

TECH TIP

For any type of starting system, when diagnosing a no crank problem and you try to engage the starter, you should listen and feel for the starter solenoid actuating. If you hear and feel a heavy clunk, this should tell you the control circuit is good, the solenoid is working, and the starter may be trying to turn over a seized engine. To confirm this, closely watch the crankshaft vibration damper when the solenoid is actuated. This will require some extra help. If the damper moves slightly, there's a good chance the engine has some serious issues.

You should also turn on some work lights to see if they dim while trying to crank the engine. If they dim, this should tell you the starter motor is drawing current.

CHECK THE SIMPLE STUFF FIRST There are many checks you can make before getting any tools out. A good HDET sensory check (look, listen, smell, feel) may detect the problem. Visually look for loose or corroded cable and wire connections, battery defects (cracked case, missing vent cap, dirty top, loose/broken/missing hold down), loose starter mounting fasteners, loose switches or relays, and loose or broken control switches. With electrical systems your nose may tell you that the starter has been overheated or some wiring has melted. Carefully feel for hot wires and components. Air and hydraulic systems should be carefully checked for leaks. Never use your bare hands to check for air or hydraulic leaks. Use a piece of cardboard or wood. Check to see if there are other electrical system problems by turning on lights, wipers, or fans.

CHECK THE BATTERY SYSTEMS The battery condition and charge state should always be confirmed when looking into any starting problem. After a thorough visual inspection (checking electrolyte level if possible), check open circuit voltage (OCV) of the battery systems. This should be a minimum of 12.4V for a 12V system and 24.8 for a 24V system. See Chapter 5 for more details on battery testing. If OCV is good, perform a load test or electronic battery condition test. If the battery condition indicates it needs a charge, then hook up a charger and start the charging process (see Chapter 5 for details).

DIESEL ENGINE STARTING SYSTEMS

Once a good battery condition has been confirmed and a visual inspection of the rest of the system hasn't uncovered any defects, then you can move on to determining whether the problem is in the starter control circuit or the main power circuit. This will be done with available voltage checks.

Checking voltage at the starter solenoid is a good place to start. Have an assistant try to start the engine while you are checking for voltage at the S terminal on the starter solenoid. If there is battery voltage at the S terminal, then you need to focus on the starter assembly and battery cables. If there is less than battery voltage, you will need to focus on the starter control circuit.

Once you have decided which part of the system is defective, you should be checking for available voltage and good ground through the system. Voltage should be very close to battery voltage, but if you find voltage and it is less than it should be, then you should start performing voltage drop testing throughout the control circuit. This will be done with the circuit energized. You are looking for excessive voltage loss that will indicate high resistance. The likely places for this are connection points such as wiring harness connectors and switch or relay terminal. ● SEE FIGURE 7–28 for how to check for voltage drops through the starter control circuit.

The control side is all about sending power to the S terminal but relays will need a good ground to operate. Some likely problems in the control side are the following:

- Faulty key switch
- Faulty interlock switches/sensors
- Faulty relay(s)
- Loose, corroded, broken, or dirty wires
- Open circuit protection device

If you find the problem is past the S terminal, then keep in mind that the starter motor needs voltage from the solenoid and a good ground back to the battery. A quick way to determine if battery cables, the master switch, or any connections

> **SAFETY TIP**
>
> At this point in the diagnosis it will be tempting to "jump the starter." This means using a jumper wire to energize the S terminal from the B+ terminal on the solenoid. This is how many people have been injured or killed over the years. Try to avoid this at all cost. If you insist on doing this, you must ensure that the fuel system won't inject fuel and that the drivetrain and hydraulic system are neutralized if the engine is cranked over by jumping it. There's also the possibility of causing electronic component damage when jumping circuits.

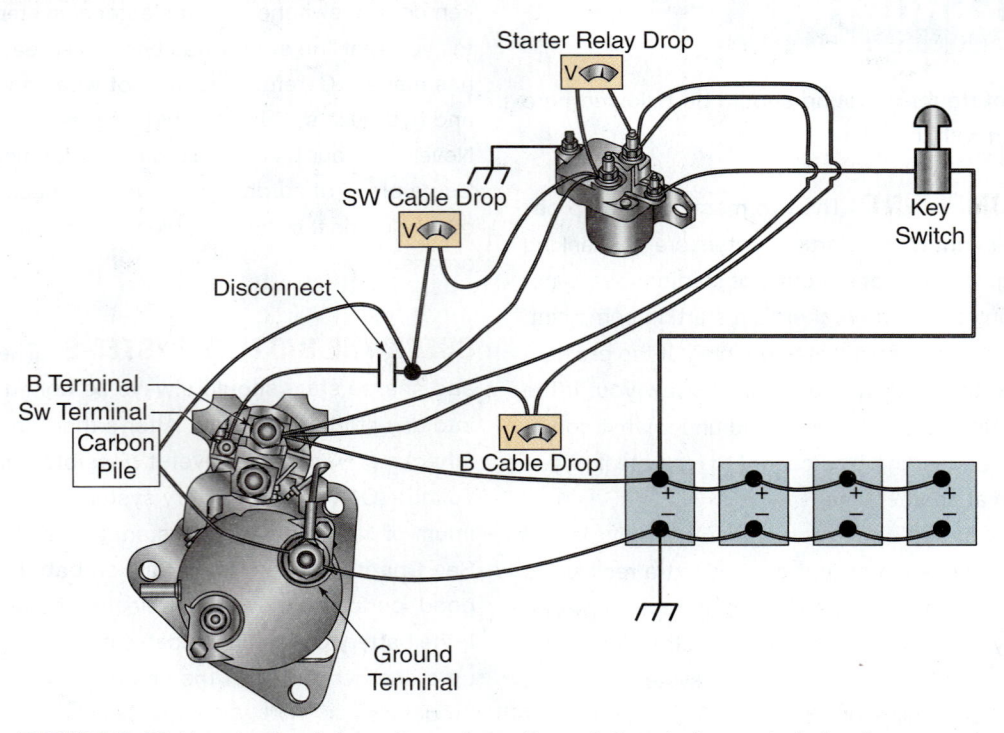

FIGURE 7–28 How to check for voltage drops through the starter control circuit.

234 CHAPTER 7

TECH TIP

Most heavy equipment machines will have a battery **disconnect switch** (sometimes called master switch). It is usually on the ground side but is occasionally on the positive side of the system. These heavy-duty switches take a lot of current flow and are often overlooked when diagnosing a starting problem. A simple voltage drop test across the switch terminals will detect a problem. There should be no more than a 0.2V drop for a 24V system and no more than a 0.1V drop for a 12V system.

are a problem is to bypass all this with a good set of booster cables. Take care not to cause sparks near the battery when doing this. You will be going from the battery system's positive to the starter B+ terminal and from the battery system's negative to the starter ground post or starter housing. If this temporary battery cable arrangement gets the engine cranking, then you will have a better idea where the problem lies. You could then take one cable off at a time and try it again. Remember to disconnect at the starter first and be careful with the live cables.

If all cables, connections, and the master switch check out fine and the starter still won't engage when the S terminal is energized, then you should take the starter off and bench test it.

Some problems that you may find with the starter assembly and battery cables are as follows:

- Loose, broken, corroded, dirty battery cable connections
- Faulty starter drive
- Loose starter mounting fasteners
- Worn ring gear teeth
- Faulty motor components

For a no crank problem with either an air or a hydraulic starting system, the same general diagnostic procedure applies. Determine if the problem is in the starter control or the starter motor energy supply circuit. Pressure checks may need to be done with gauges to see how far pressure is getting.

CHECK FOR SLOW CRANKING

To verify a slow cranking problem, you may need to measure engine speed with a tachometer or it may be obvious enough that you don't need to. You should be able to find a spec for minimum cranking speed at a certain ambient temperature.

If the cranking speed measures slow, then you must determine if the problem is because of too much resistance against the crankshaft.

Don't forget the simple stuff first and battery testing before moving into instrument testing.

Because the starter is getting engaged, you can assume the control circuit is working properly. The main test you will perform is a voltage drop test. This will be the best way to confirm wire and connection condition because you will perform it with the wire's conducting current.

● **FIGURE 7–29** shows where to connect your voltmeter to perform a voltage drop test. This test is to see how much resistance there is in the battery cables and their connection points. A carbon pile-type load tester is used to create current flow. This is preferable to using the starter for a couple of reasons. First, safety wise there is always the possibility of injury if you are near rotating

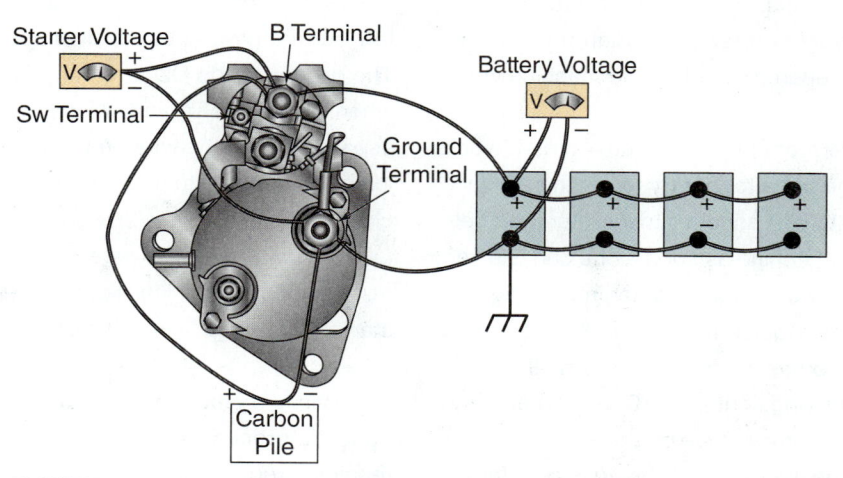

FIGURE 7–29 Where to connect a voltmeter to perform voltage drop testing on the high current part of the starting system.

DIESEL ENGINE STARTING SYSTEMS 235

parts if the engine is cranking. Second, the carbon pile is an adjustable steady load that will give a more accurate result when using the starter for putting a load on the high current circuit.

● SEE FIGURE 7-30 for the maximum allowable voltage drops for different locations in a 12V and 24V system.

If you find one part of the circuit to have an excessive voltage drop, you should be able to isolate the part of the circuit to see exactly where the high resistance is.

Once the problem is found and you make the necessary repairs, then try to start the engine and see if the cranking speed is normal.

LOCATION	12V	24V
Negative battery post to the starting motor negative	0.5 volts	1.0 volts
Drop across the disconnect switch terminals	0.5 volts	1.0 volts
Positive battery post to the positive terminal of the starting motor	0.5 volts	1.0 volts
Solenoid "Bat" to the solenoid "Mtr" terminal	0.2 volts	0.4 volts
Across starter relay load terminals	0.3 volts	0.6 volts

FIGURE 7-30

Maximum allowable voltage drops for different parts of the circuit.

CASE STUDY

Fred was sent out to a machine for a starting complaint issue. When he arrived at the machine, it was working so he asked the operator what the problem was. The operator told Fred that lately he has had to work the key on and off several times to get the machine to start. Once the starter engages, it starts fine. Fred asks the operator if he can try it, and when Fred tries to start the machine several times, it starts without hesitation. Fred asks the operator when the problem occurs, and the operator says it can happen anytime he tries to start the machine. Fred performs a quick visual inspection of the machine's batteries, wiring, relay, and starter and finds everything looking normal. He then takes a few minutes to look at the electrical system service information and schematics related to the start circuits. He finds information on the starter interlock system and reads through the troubleshooting tree to be prepared for when the machine won't start again.

Fred has the operator call him the next time it won't start and asks that the operator not try to get the machine going before he gets to it. A couple of days later Fred gets a call and gets to the machine as soon as he can. He verifies that the machine won't start and begin his diagnostic procedure. He has a quick look around for anything obvious and checks to see if any other electrical circuits are not operating by trying lights, HVAC fan, wipers, and horn. This will also take any surface charge off the battery. Everything seems to be working as it should except for the starter. He checks battery OCV and finds it to be 12.5V, which indicates a good charge.

Fred then asks the operator to get ready to start the engine. Fred attaches the positive lead of his voltmeter to the S terminal and the negative to a good frame ground and gives the operator the signal to start the machine. The display on the meter reads 0V. Fred then begins to work back through the circuit to see where the problem is.

He finds the starter relay and checks for available voltage at the high current terminal and reads 12.3V. He again asks the operator to try start the engine and finds that there is 12.3V at the relay low current in terminal and doesn't hear it click.

He then checks for continuity to ground from the relay low current ground terminal with the meter set to ohms. The display reads OL (over limit) but changes to 0 ohms when the wire is jiggled. When Fred pulls on the wire, it separates from the wire end terminal. When a temporary jumper is installed to give the relay a good ground, the machine starts fine.

Fred makes a new wire, installs it, and seals the terminals with an electrical sealer. When the operator tries to start the machine several times, the starter engages each time.

This is an example of a starting system problem diagnostic procedure that shows a typical diagnostic and repair procedure.

Progress Check

11. These are two values measured during a starter bench test:
    a. no load torque and speed
    b. no load speed and voltage
    c. no load speed and current draw
    d. full load speed and current draw

12. This is the largest battery cable size:
    a. 0
    b. 00
    c. 1
    d. 2

13. The purpose of the starter interlock system is to:
    a. stop the engine from cranking until one or more safety-related conditions are met
    b. lock the starting pinion out for a minimum of 15 seconds
    c. lock the flywheel in place for security reasons
    d. prevent the machine from starting if the operator has been drinking

14. A machine that has a starter relay will use it to:
    a. shift the drive pinion out
    b. send current to the S terminal on the starter solenoid
    c. increase current flow to the starter motor
    d. switch the G terminal at the starter solenoid

15. This is the first step in troubleshooting a starting system complaint:
    a. check battery voltage
    b. listen for the solenoid clicking
    c. check starter current draw
    d. verify the complaint

16. This testing tool will verify a slow cranking problem:
    a. voltmeter
    b. tachometer
    c. ammeter
    d. ohmmeter

17. The faster a starter motor turns:
    a. the less CEMF that is generated
    b. the more torque that is generated
    c. the less current that is drawn from the battery
    d. the more current that is drawn from the battery

18. A starter's brushes will transfer current between:
    a. the solenoid and the motor
    b. the individual field coils
    c. the field coils and the armature windings
    d. the hold-in and pull-in coils

19. If a voltage drop test is done and the result is outside of specification, this means:
    a. the battery needs to be charged
    b. there is a bad ground
    c. there is a short
    d. there is excessive resistance in the circuit

20. When diagnosing a no crank problem and you find sufficient voltage being supplied to the S terminal but the starter doesn't turn, you can safely say:
    a. the control circuit is operating fine
    b. the brushes are worn out
    c. the armature is grounded
    d. the battery terminals are loose

SHOP ACTIVITY

- Go to a machine with an electric starting system and list the main components of the system.
- Draw an electrical schematic showing the main components of the starting system.
- Describe a step-by-step procedure to replace the starter.
- Inspect the battery cables leading to the starter and record any defects.
- Are there any safety interlock devices to prevent the starter from engaging? If yes, test these devices to ensure that they are working properly.

Summary

1. Some safety concerns related to starting systems are battery explosion risks, burns from overheated wires, strains from lifting heavy components, and injuries related to sudden release of stored energy.
2. Diesel engine starting systems need to produce torque that will turn the engine's crankshaft fast enough to start the combustion process. This is usually between 150 and 250 rpm.
3. Some older engines used a "pup" engine to crank the diesel engine. They were a small gasoline engine that drove the flywheel pinion through a transmission.
4. Most starting systems use DC electric motors to drive a pinion gear to rotate the engine's flywheel.
5. They can be 12V or 24V motors, and all starter motor output torque gets increased because of the gear ratio between the pinion and the flywheel. Usually, this is 20:1.
6. For larger diesel engines, they will need more powerful starter motors to crank them. Some engines will use two starter motors.
7. Other factors that necessitate larger starter motors are cold engine oil, engine driven accessories, and electronically controlled engines.
8. Overheating electric starter motors is the leading cause of failure. *Do not* engage a starter for longer than 15 seconds straight. Wait two minutes to allow the starter to cool.
9. Electric starter motors are the most common type used on heavy equipment. Other types are hydraulic and pneumatic.
10. Electric starter assemblies convert electric DC into rotational torque output at the starter pinion gear. The gear engages with the engine's flywheel ring gear.
11. Two types of electric starter motors are direct drive and gear reduction.
12. DC brush-type motors use current flow to create opposing magnetic fields, which in turn rotate the motor shaft.
13. The motor housing has field coils and pole shoes that create one field. Arranged in pairs on opposite sides of housing.
14. Armature rotates in center of housing and has many separate loops of heavy copper wire laid into grooves of laminated steel core.
15. Armature is fed current through brushes to commutator bars.
16. Each loop on the armature creates the second magnetic field when current flows through it.
17. Starter motors are designed to make maximum torque at low rpm and are self-limiting speed wise by CEMF (counter-electromotive force).
18. Starter solenoids will move the starter pinion into the flywheel and close a set of contacts to send current to the starter motor.
19. Solenoids have two windings: pull in and hold in.
20. Some starters feature an OCP (over-crank protection) switch that disengages the starter if it overheats.
21. Most starter pinions incorporate an overrunning clutch that prevents the engine's flywheel from driving the starter pinion.
22. Starter replacement requires cleaning area, marking wires, carefully removing wires, using a lifting device if necessary, cleaning gasket area, and torqueing all fasteners to spec.
23. Bench testing starters requires a power source and connecting starter. Operate starter and measure pinion rpm and current draw.
24. Many problems can be traced to bench test results that aren't in specification.
25. Further diagnostics can reveal starter component problems. Reconditioning starters is mostly left to specialty shops.
26. Starter/battery cable sizing is critical to proper operation of starting system. When replacing existing or installing new cable, *do not* undersize cable. Make sure that cables are secure.
27. Starter control circuit will allow operator to send current to starter solenoid.
28. Pneumatic starting systems use compressed air as an energy source and an air-powered motor to provide pinion drive torque. Provides plenty of torque but will deplete air supply quickly and can freeze up.
29. Hydraulic starting systems use pressurized hydraulic fluid that is stored in an accumulator to rotate a hydraulic motor. The motor output then turns a pinion gear.
30. Starting circuit interlock systems prevent engine cranking unless certain conditions are met such as seat belt is latched, doors are closed, and transmission is in neutral. They will keep the solenoid circuit open.
31. Relays are a common part of starter control circuits.
32. Starting system troubleshooting starts with verifying the complaint. If there is an interlock circuit, make sure that it is not open.
33. Voltage drop testing is done with current flowing through circuits and will determine if there is excessive resistance.

CHARGING SYSTEMS

LEARNING OBJECTIVES

After reading this chapter, the student should be able to:

1. Explain how an AC generator creates electron flow.
2. Describe the main components of an alternator.
3. Explain how a voltage regulator works.
4. Describe how to perform a voltage drop test on alternator wiring.
5. Explain how to perform a load test on an alternator.
6. Describe how to perform a quick test of the charging system.
7. Describe what full fielding an alternator is.
8. Explain what the set point of an alternator is.

KEY TERMS

Brushes 244
Brushless 244
Cooling fan 243
Corrosion 249
Rectification 246
Rectifier bridge 247
Regulator 241
Rotor 243
Set point 252
Stator 244
Voltage drop 257
Zener diode 249

INTRODUCTION

SAFETY FIRST Some specific safety concerns related to charging systems include the following:

- Battery explosion risk
- Burns from high current flow through battery cables
- Strain injuries from lifting heavy alternators (some are over 65 lb) into awkward places
- Injuries from body parts and loose clothing getting caught in belts
- Sparks and hot wires from battery cables and wires shorting to ground

WHAT THE CHARGING SYSTEM DOES

A machine charging system is needed to maintain sufficient battery charge so the batteries can energize the engine starting system. For an electric starting system, this will be the biggest electrical load. The charging system also provides current flow to keep the single or multiple battery systems' charge high enough to operate the machine's electrical circuits when the engine is running.

The main component of the charging system is the alternating current generator commonly called the alternator. An alternator is an energy conversion device that will take an input of mechanical rotation and convert it into an output of electrical energy. It first changes torque into AC (alternating current) and then into DC (direct current). This will be the type of current needed to charge the machine's battery or battery pack and to operate the various electrical circuits. Almost all alternators used for heavy equipment machines will be belt driven from a crankshaft-mounted pulley.

Some electric drive machines will use an inverter to charge the machine's batteries and run electrical circuits. The inverter will change some of the machine's main generator's AC output into DC. This eliminates the need for a conventional alternator.

Alternators can be classified into two main voltage groups, that being 12V and 24V, and are sized or rated by the maximum amount of amperage they can produce. Amperage ratings for 12V alternators can range from 30 to 300 amps and 24V alternators will range from 15 to 150 amps. There are also two different design types called brush and brushless. Some alternators will be described by how many wires that are connected to it. They are also classed by the type of voltage regulator they have, whether they are self-excited or not and the type of mounting design. ● **SEE FIGURE 8–1** for a typical alternator arrangement on a machine.

FIGURE 8–1 A typical alternator arrangement on a machine.

An alternator that pivots to adjust the belt is said to be hinge mounted as opposed to the type that is fastened solid to the engine. This is called pad mounted.

The alternator output will almost always go through a heavy 6–10 gauge wire to the machine's starter B+ terminal. This will allow the alternator to be connected to the battery positive cable and the machine's main power feed. Another common destination for the alternator output wire to go to is if the machine has a breaker for the alternator, the alternator

TECH TIP

A good comparison can be made between a machine's charging system and a hydraulic system. If the battery is like an accumulator, then as its pressure drops there needs to be a pump to recharge the accumulator. Instead of pumping oil like a hydraulic pump, the charging system's alternator will pump electrons back into the battery to be converted into chemical energy storage. The alternator will also pump electrons to maintain system pressure and keep the electrical loads working just as the hydraulic pump keeps oil flowing to keep hydraulic actuators working.

TECH TIP

The term *battery pack* refers to the fact that a machine may have more than one battery to provide enough cranking power to start the engine. These batteries could be connected in series, parallel, or series-parallel.
● **SEE FIGURE 8–2** for a 24V multiple battery system.

FIGURE 8–2 A 24V battery pack.

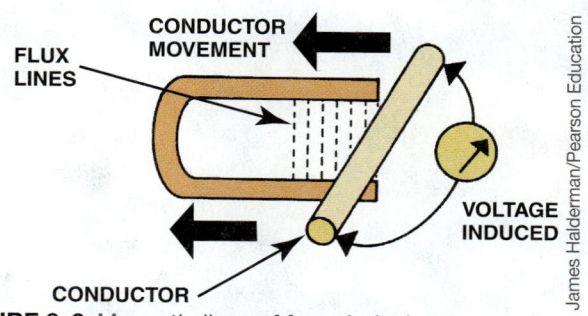

FIGURE 8–3 Magnetic lines of force inducing current flow in a conductor.

Ideally, the lines of force are cut at a 90° angle by the conductor. ● **SEE FIGURE 8–3** for how this works.

The magnetic lines of force in the alternator are created by the rotor. It spins inside stationary windings called the stator. The rotor creates several alternating north and south magnetic poles that are passed by a series of windings that make up the stator. The stator windings are the conductors that have electricity induced in them. The stator is where the output of the alternator starts.

The electricity produced in the stator is AC that needs to be converted or rectified into DC to make it usable for the machine's electrical system. This rectification is accomplished with a series of semiconductor devices named diodes. See Chapter 9 for a better explanation of semiconductors and diodes.

The alternator's output is controlled by a voltage regulator that also contains electronic semiconductors. The **regulator** will turn on and off current flow to a coil of wire that is part of the rotor. The rotor coil current flow will make the rotor an electromagnet with several sets of north and south poles. When the current is turned off to the rotor's coil, there is no output from the alternator.

ALTERNATOR DRIVE The alternator drive mechanism for a typical machine is a belt and pulley system that is almost always driven off a crankshaft pulley at the front of the engine. Three variations of pulleys are possible:

1. Single V
2. Double V
3. Multi-groove or multi-rib

● **SEE FIGURE 8–4** for the different alternator drive belt styles.

Belt tension must be maintained to keep the alternator pulley turning. If an alternator is producing 150 amps at 24V, it will need almost 5 hp transferred into its drive shaft (24 × 150 = 3600 watts ÷ 746 = 4.82 hp).

will feed the breaker and then on the other side of the breaker there will be a wire to go to the B+ terminal of the starter to connect to the battery pack.

ALTERNATOR OPERATING PRINCIPLES

The simple way to think about how an alternator creates electricity is to compare it to the relationship between a conductor or wire and a magnetic field and relative motion between the two.

As mentioned in Chapter 5 to create electron flow from mechanical motion, you simply have to move a wire in and out of a magnetic field or move the magnetic field over a wire.

CHARGING SYSTEMS

FIGURE 8–5 The keyway for an alternator pulley.

belt. Belt tension is maintained by a mechanical adjusting mechanism that will pivot the alternator nearer to or farther away from the driving pulley or by a spring loaded idler pulley. Proper belt tension is an important maintenance check and will be discussed later.

Some belts will only drive the alternator while others will be used to drive other components such as the engine cooling fan, water pump, or, not as common, an air compressor. Light-duty diesel engines will likely have a serpentine belt that will drive several components.

DRIVE PULLEYS Alternator output increases with rotational speed. As a machine's diesel operates typically between 650 and 2100 rpm, a speed differential between the alternator pulley and engine speed is needed to spin the alternator fast enough. Most alternators are limited to 6000 rpm, which means the alternator drive ratio is precisely chosen to produce high output at idle yet stay below maximum speeds. Also factored is alternator brush and bearing wear that increases with speed and the fact that amperage output curves tends to flatten out with increasing speed.

For large bore diesel engines found in machines, the standard pulley ratio is approximately 2.7:1—every engine crankshaft rpm produces 2.7 alternator rotor shaft revolutions. In recent years, 3 or even 3.1:1 is becoming common. At 2000 rpm from the crankshaft, the alternator will turn 6000 rpm. Some slow rpm diesel engines may use as high as a 5:1 ratio in comparison to automotive engines using as low as a 2:1 ratio. Most pulleys are locked to the alternator shaft with a key and held in place against a shoulder with a washer and nut. Special tools may be needed to hold the shaft and pulley when removing the nut and to pull the pulley off the shaft. ● SEE FIGURE 8–5 for a keyway of an alternator pulley.

FIGURE 8–4 Different alternator drive belt styles.

The belt needs to be properly tensioned so it doesn't allow slippage if too loose and doesn't put excess strain on the alternator bearings or other components driven by the

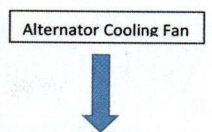

FIGURE 8–6 The cooling fan of an alternator.

ALTERNATOR COOLING
Most alternators rely on air flow to cool internal components. If equipped with a **cooling fan**, the alternator fan must rotate in a direction that will push air through the unit. Typically, an alternator is located close to the engine's cooling fan, and this will help cool the alternator as well. Cooling fans can be inside or outside of the alternator housing.

When operating in environments where a spark from an alternator's brush could trigger an explosion or cause a fire, the alternators are sealed and heat is radiated through the housing. Some very large alternators can even be oil cooled. ● SEE FIGURE 8–6 for the cooling fan of an alternator.

ALTERNATOR COMPONENTS
The following section lists and describes the main components that make up a typical alternator. ● SEE FIGURE 8–7 for a cutaway of an alternator and its main components.

ALTERNATOR HOUSING
Most alternators are vented to allow cooling air to pass through it and cool the components inside the alternator. The housing is usually a two-piece aluminum construction but larger alternators can be three piece. The half that has the shaft running through it is called the drive end (DE) and the opposite end is called the slip ring end (SRE). The SRE will house the brushes for a brush-type alternator and the voltage regulator if it is the internal type. The housings will be held together with long screws or long small diameter bolts.

Some applications require explosion-proof or spark-free alternators. Their housings will be totally sealed. The components inside this type of alternator must be designed to handle extra heat that builds up.

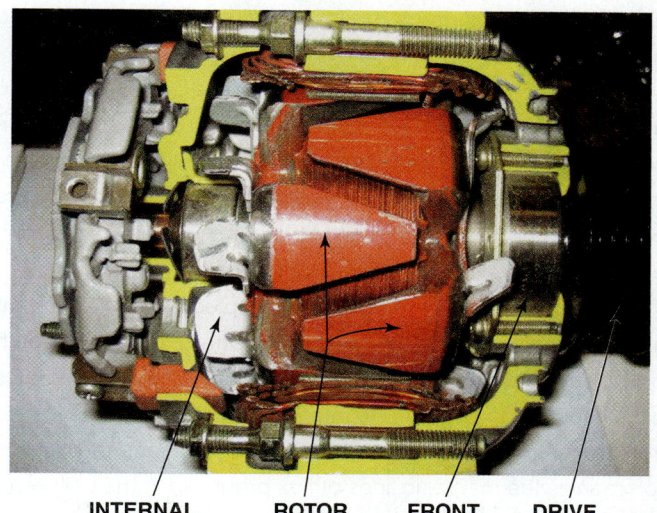

INTERNAL COOLING FAN | ROTOR POLES | FRONT BEARING | DRIVE PULLEY

FIGURE 8–7 A cutaway of an alternator and its main components.

BRUSH-TYPE ROTOR
Rotating magnetic fields induce the flow of electrical current in an alternator. The magnetic fields are formed by the rotor, which is driven by the machine's engine and is the only moving part of an alternator.

The **rotor** consists of a shaft that is suspended on two bearings. One end of the shaft will ride in a bearing that is pressed into the back half of the alternator housing. The other end will run through a bearing that is pressed into the front housing. The front half of the shaft will have a pulley attached to it to drive the rotor. The shaft has a soft iron core pressed onto it, which has a coil of 500–600 turns of insulated wire

CHARGING SYSTEMS 243

wrapped around it. Each end of the wire coil is connected to one of two conductive slip rings on the rotor shaft.

Graphite **brushes** feed current into the coil by contacting the slip rings. Lightweight springs help the brushes maintain contact with the slip rings. The wire coil and slip rings are electrically insulated from the rotor shaft. Energizing the rotor's wire coil with a normal current flow of 2–9 amps produces a strong electromagnetic field.

Over the coil of wire there are two pieces of soft iron that are arranged into claws or pole pieces. One is to create a south pole and the other is to create a north pole. The pole pieces have two purposes. One is to intensify the electromagnetic field (remember iron has a lower reluctance than air) and the other is to organize the magnetic lines of flux produced in the coil by arranging them into north and south poles on alternating claws of the rotor. When the coil is energized, each of the claws or pole pieces will be either a north or a south magnetic pole, which will create an alternating sequence of north and south poles when the rotor is rotated. ● SEE FIGURE 8–8 for a brush-type rotor.

An alternator has between 12 and 16 pole pieces altogether but 14 is more common for heavy-duty alternators. Passing current through the coil magnetizes the rotor claws, thereby forming alternating poles of magnetism north–south–north–south on the rotor.

The output of the alternator is determined by a couple of the alternator's physical features. One is the size and number of windings in the stator that are cut by the magnetic lines of force. The second is the strength of the magnetic field of the rotor. Increasing or decreasing the current flow through the rotor winding will change the magnetic field strength. Usually, the maximum practical amperage is 5 amps or less and voltage is no more than half of the alternator's total output voltage. Controlling the voltage and current flow to the rotor and, therefore, the strength of its magnetic field is the job of the voltage regulator. ● FIGURE 8–9 shows how alternate magnetic poles are formed around the rotor by the magnetic field in the rotor's electromagnetic coil.

BRUSHLESS ROTOR

A brushless alternator operates on the same principles as any other alternator. The main difference is the rotor design. Instead of having the coil of wire rotate with the rotor assembly, a **brushless** alternator will have a stationary rotor coil with the pole pieces rotating over top of it. There is a very small clearance between the coil and pole pieces. As the coil is stationary, it can be wired directly to the voltage regulator. This will achieve the same effect of creating rotating magnetic poles when the pole pieces are rotated over the energized coil but will eliminate the need for brushes and the repair time for replacing worn-out brushes. ● SEE FIGURE 8–10 for a brushless alternator rotor.

FIGURE 8–8 Brush-type rotor.

SELF-EXCITING ALTERNATORS

Some alternators are classed as self-exciting and require residual magnetism left in the rotor iron core to "kick-start" the alternator before current flow can begin to be generated. Residual magnetism is a small amount of magnetism left in the rotor when the alternator is manufactured or after the machine it is in shuts off. Without any current passing through the rotor coil, the residual magnetism will be enough to create an electromagnet out of the rotor and it will begin to induce current in the stator windings, which in turn supplies current to the voltage regulator. Normal alternator operation using a regulator will resume once current is supplied to the regulator.

A self-exciting alternator is distinguished by a single positive battery cable connected to the alternator. No external battery current is supplied to switch on the voltage regulator or excite the field inside the rotor.

STATOR

The **stator** is the main stationary generating component of the alternator. If an alternator has an open housing, you should be able to see the stator. It is made of three separate lengths of insulated wire that are wrapped in a series of loops around a slotted metal frame. The loops are exposed to the rotor's multiple magnetic north–south poles passing them as the rotor spins in the alternator. Alternating current is then produced in the stator windings.

The three sets of windings are offset with each loop 120° apart from the next winding's loop. This will stagger the AC voltage pulses from each winding to the next. This arrangement

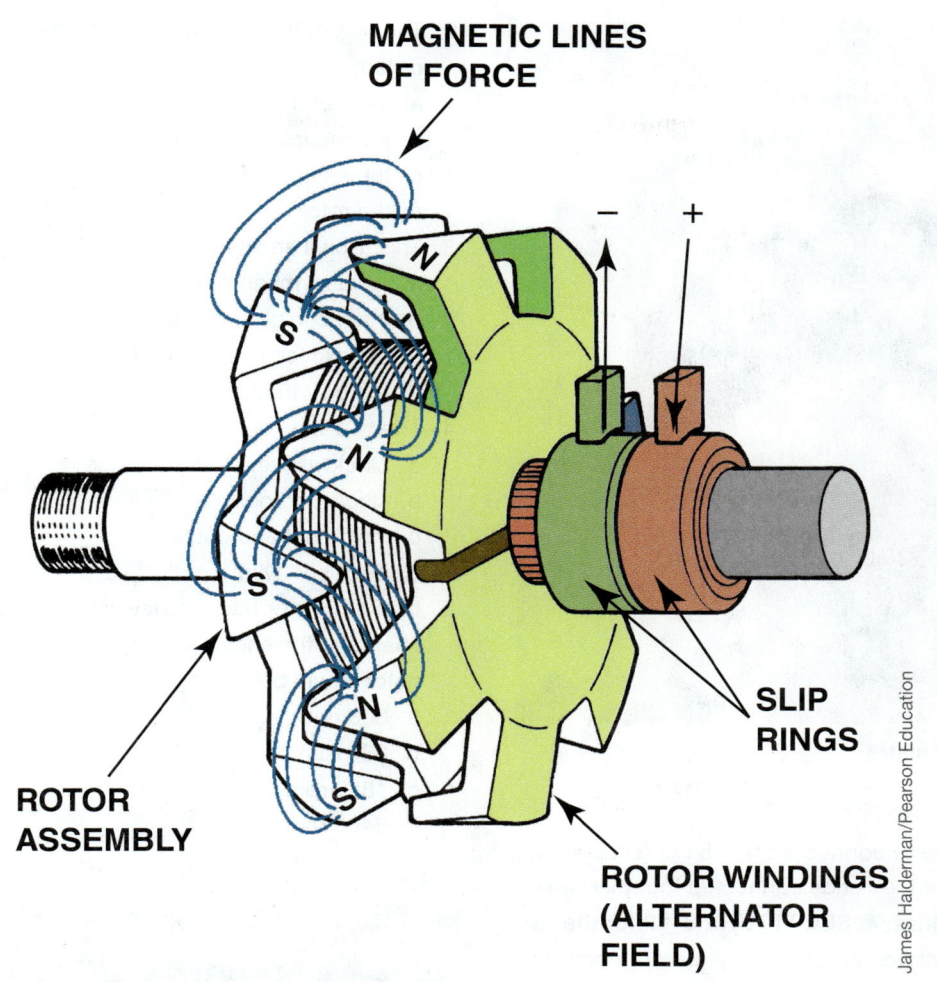

FIGURE 8–9 Alternate magnetic poles formed around the rotor by the magnetic field in the rotor's electromagnetic coil.

produces a more even flow of current from the alternator, and this is called three-phase voltage.

A larger stator having more loops, more turns of wire in each loop, and thicker wire will have higher maximum output amperage than one with fewer loops, less wire, and thinner wire. Some stators will have double loops instead of using larger gauge wire. ● SEE FIGURE 8–11 for a typical stator.

STATOR CONNECTIONS The ends of each of the three windings of the stator are connected together into two different possible configurations—wye and delta connected. Both types of windings produce three-phase AC, but voltage and amperage output characteristics differ. The ends of the stator windings are connected to diodes that change the AC voltage to DC. This is called rectification.

Windings connected into a wye-type configuration, which, as the name suggests, resembles the letter "Y," have four connection points. Three ends of each of the windings are connected together to a point called the neutral junction.

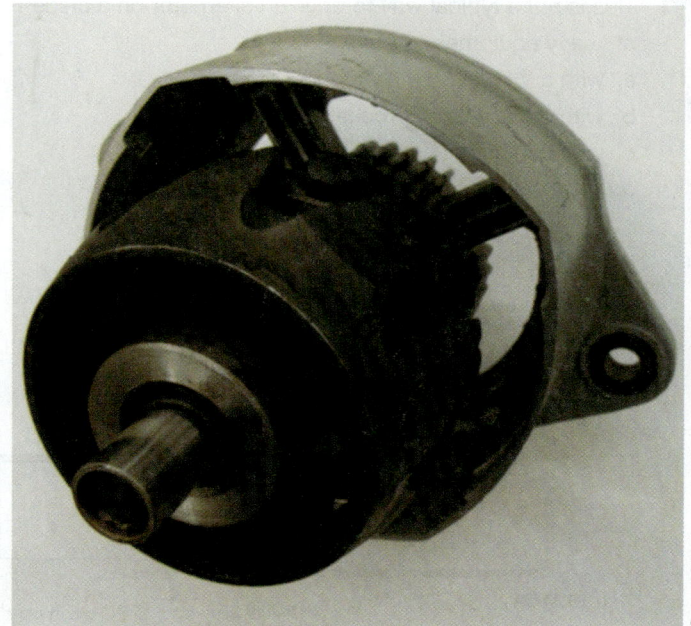

FIGURE 8–10 A brushless alternator rotor.

CHARGING SYSTEMS 245

FIGURE 8–11 A typical stator.

Alternators with delta connected windings (shaped like the symbol "delta") have their ends connected together, which puts the windings in series. ● SEE FIGURE 8–12 for the two types of stator connections and how they are connected to the rectifier.

Progress Check

1. This is *not* a typical way to drive an alternator on a piece of heavy equipment:
   a. with a V belt
   b. with a chain and sprocket
   c. with a multi-rib belt
   d. with multiple V belts

2. The part of the alternator that creates magnetic fields is called:
   a. the stator
   b. the magnafluxticator
   c. the rectifier
   d. the rotor

3. This allows an alternator to "self-start":
   a. residual magnetism
   b. magnetic glow plugs
   c. the exciter wire
   d. the B+ terminal

4. The two ways a stator could be wound are:
   a. clockwise and anticlockwise
   b. single and double
   c. inside out and outside in
   d. delta and wye

5. If an alternator had a pulley with a 3:1 ratio to the crankshaft and the engine was turning 1800 rpm, then the alternator would be turning at:
   a. 300 rpm
   b. 600 rpm
   c. 1800 rpm
   d. 5400 rpm

CURRENT RECTIFICATION

Alternators produce AC, which is acceptable for operating many household or industrial electrical devices. However, AC cannot charge a battery or operate most machines' electrical circuits. Converting the AC to usable DC is referred to as **rectification**.

AC is produced in the stator due to the influence of the rotor's magnetic fields. Alternate north–south poles passing over windings will alternately push and pull electrons. Moving the electrons in two different directions gives stator current flow its AC characteristic. The speed at which the lines of force cut the conductors,

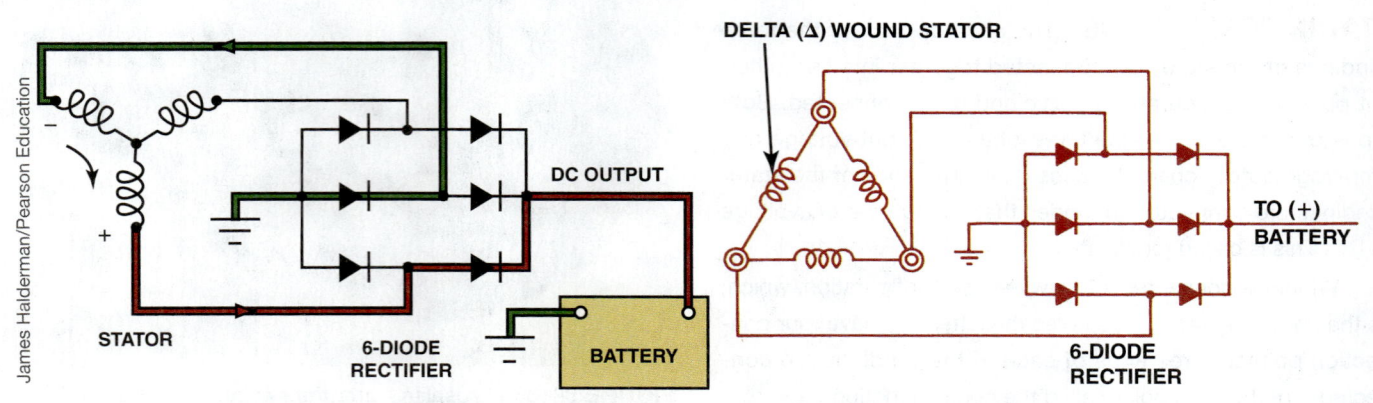

FIGURE 8–12 The two types of stator connections and how they are connected to the rectifier.

246 CHAPTER 8

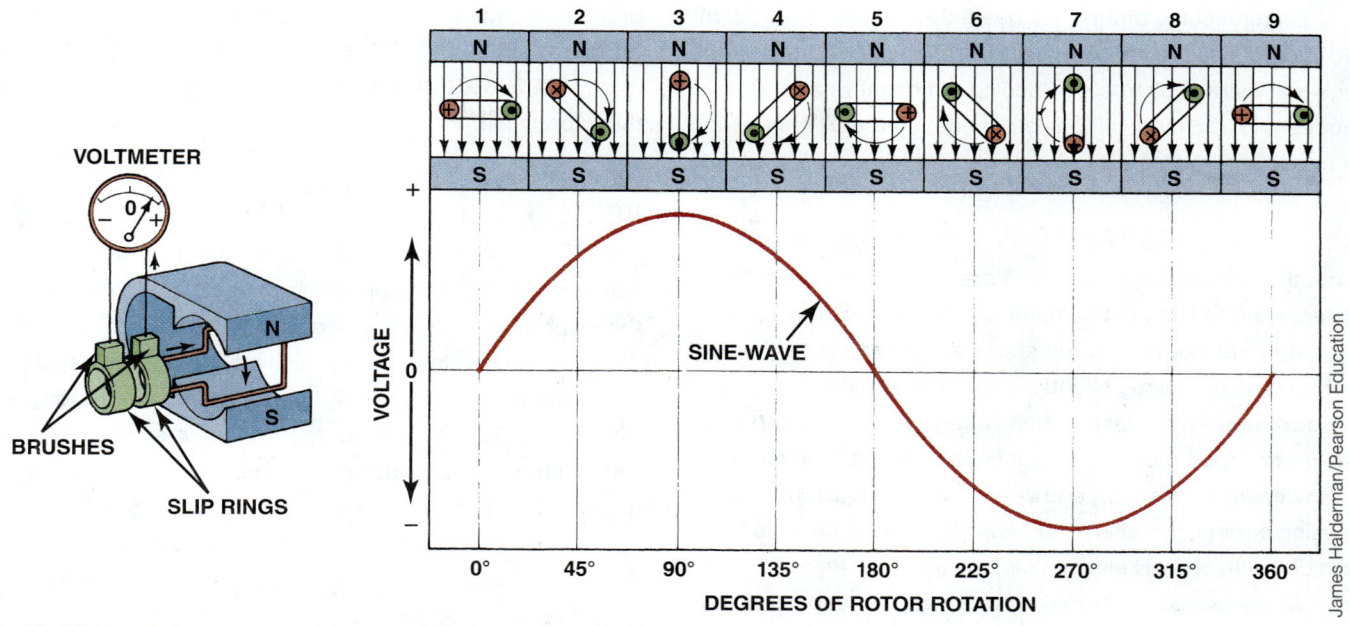

FIGURE 8–13 An example of how an AC sine wave is produced.

the angle at which the magnetic field cuts the stator conductors, the number of conductors, and the wire gauge will determine the amount of amperage induced in the stator. ● **SEE FIGURE 8–13** for an example of how an AC sine wave is produced.

Diodes are connected to each of the three stator wire ends of either delta or wye wound stators. Each stator winding will produce one of three phases of AC. A minimum of six diodes is required to completely rectify all three phases of AC into DC. The silicon diodes making up the rectifier behave like a one-way electrical check valve. The two diodes connected to each winding will only allow a positive voltage potential to leave the output of the rectifier. If only a single diode was used at the end of the windings, only half the AC sine wave is rectified. Two diodes enable full-wave rectification. ● **SEE FIGURE 8–14** for how AC is rectified to DC.

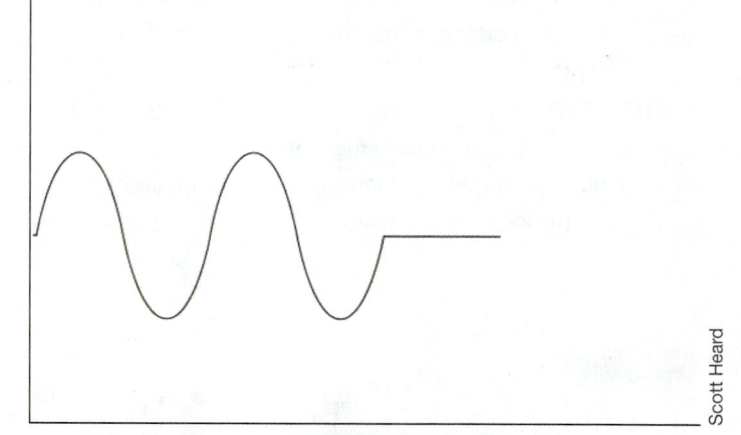

FIGURE 8–14 How AC current is rectified to DC.

RECTIFIER BRIDGE Rectifying diodes are connected together into a single unit called the **rectifier bridge**. Positive diodes deliver current to the battery terminal of the alternator and negative diodes allow current flow from ground. The metal shells of the diodes are commonly pressed into a metal plate or heat sink to secure them and keep them from overheating.

Positive diodes prevent the discharge of the battery through the alternator when the engine is not running. If the battery is connected incorrectly, a tremendous amount of current can flow through the diodes and stator winding as well as the battery cable to the alternator. For this reason, circuit protection of the alternator is sometimes used. This may be in the form of a fusible link, a maxi fuse, or a breaker. ● **SEE FIGURE 8–15** for a rectifier bridge assembly.

FIGURE 8–15 A rectifier bridge assembly.

CHARGING SYSTEMS   **247**

Larger rectifier bridges are needed when alternator output increases. These rectifier bridges are substantially larger for the higher amperage alternator. Spacing out negative and positive diodes and adding cooling fins provide more efficient cooling of bridge assemblies.

SMOOTHING CAPACITORS Capacitors are used to smooth alternator AC ripple and prevent electromagnetic interference (EMI). In the alternator, one is connected across the output to act like an electric shock absorber. When the output voltage increases slightly, the capacitor will charge and absorb the new increase. When voltage drops, the capacitor will drain, topping up the output voltage and is then ready for a new charge. This happens very quickly with each pulse. A missing or defective alternator capacitor can cause radio noise and EMI with chassis electronic modules.

When checking for parasitic draws, the capacitor may give a false indication of current draw as it charges for a few seconds after the battery is disconnected. If batteries are disconnected for even a short time, they will spark when connected while the capacitor charges.

DIODE TRIO Some alternators will have their rotor field voltage supplied with current from the stator. This current needs to be rectified first and this is done with a diode trio. These three diodes will perform single-wave rectification of each phase of the alternator's windings. Single-phase rectification means only a maximum of half the alternator's output can excite the rotor. ● SEE FIGURE 8–16 for how a diode trio is connected to the rotor circuit.

VOLTAGE REGULATORS Every alternator will have a voltage regulator as part of its assembly that will control the voltage output of it and ultimately control the system voltage of the machine's electrical system.

Voltage regulators are classified as either external or internal; however, the majority of alternators used today have internal regulators—that is, the regulator is inside the alternator housing. Regulators can also be categorized by the circuit-type connections used to supply current to the rotor, which in turn is used to induce "field excitation." Knowing the type of field excitation circuit used is helpful when developing

TECH TIP

Rectification versus Inversion
Conversion of current from AC to DC is called rectification. Converting DC to AC requires "inversion" or inverting the unchanging polarity of DC into an AC sine wave. Inverters are used to take onboard 12 volt DC and change it into 120 volt AC to operate some accessories, tools, or lighting. To change current, inverters take DC voltage and pass it through one or more pairs of switching transistors. Switching the transistors on and off at 60 times per second while supplying alternate sides of a primary winding transformer duplicates AC. By alternating the flow of DC into the transformer, AC is produced at the secondary winding transformer. The quality and complexity of the inverter controls how closely the output resembles a true AC sine wave.

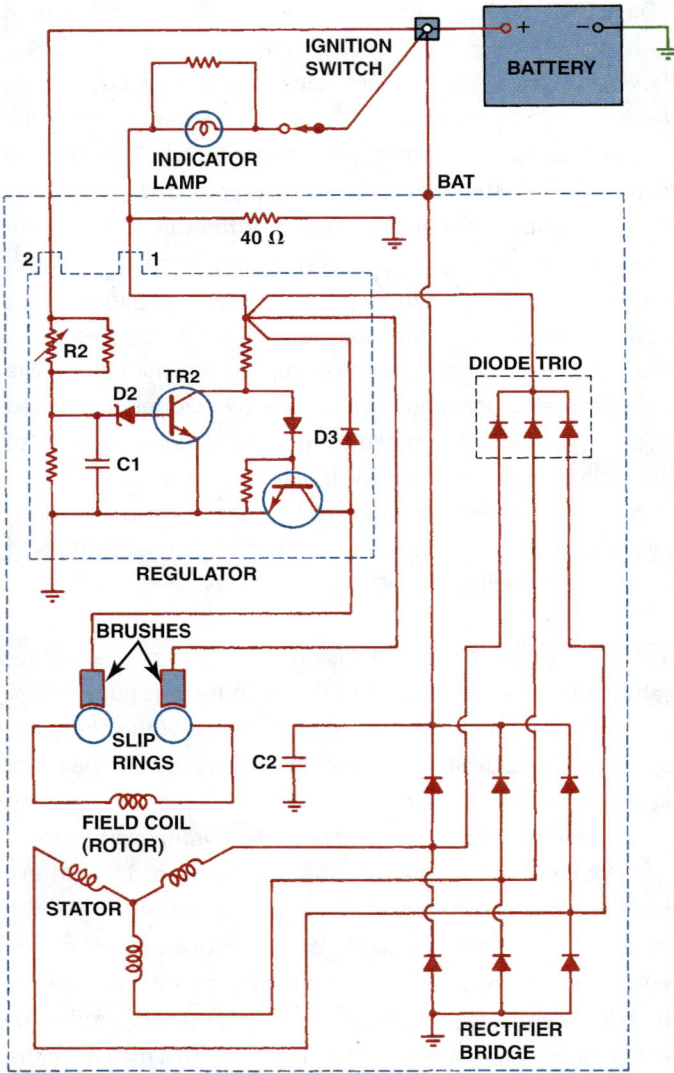

FIGURE 8–16 How a diode trio is connected to the rotor coil circuit.

248 CHAPTER 8

diagnostic strategies for testing alternators. The two types are as follows:

1. "A"-type regulators regulate the field current by controlling the ground side of the field circuit. One rotor brush is connected to the alternator output or B+, and the other is switched to ground through the regulator.

2. "B"-type regulators control the battery positive supply to the rotor. One brush is connected directly to ground, and the regulator varies battery positive voltage supplied to the other brush. If the electronic regulator fails or develops a resistive ground due to **corrosion**, it commonly causes the alternator to overcharge because system voltage is sensed through the ground and battery positive.

The voltage regulator output goes to the rotor field coil. For a brushless-type alternator, this is straightforward. For an alternator with brushes, the output will need to go through a set of brushes to energize the rotor coil.

Most regulators are electronic or are simply an electronic switching device composed principally of transistors, resistors, and diodes to form a completely static unit containing no moving parts. (A more complete explanation of these devices will follow in Chapter 9.)

The transistors are used to switch the alternator field current on and off and are controlled by the resistors and the Zener diode. ● SEE FIGURE 8–17 for a voltage regulator with the cover off.

A **Zener diode** is a special diode that will break down and permit a reverse flow of current when the voltage reaches a certain value, without damaging the semiconductor material. This diode is the trigger that senses voltage and turns the transistor on or off to limit charging system voltage.

Another diode found in many alternators is the field discharge diode. The field discharge diode provides an alternator current path to protect the transistors from induced high voltage from the alternator field windings. The sudden stopping of field current by the transistor and subsequent collapsing of the magnetic field causes an induced voltage in the rotor windings.

Another electronic component in regulators is a thermistor. The RT thermistor is a temperature-compensating resistor. Its resistance varies with temperature and controls the operating point of the Zener diode so that a high system voltage is produced in cold weather, and a lower system voltage in warm weather.

REGULATOR OPERATION The following is a summary of how a typical alternator voltage regulator works. ● SEE FIGURE 8–18 for the schematic of this voltage regulator. The main electronic components of it are resistors, capacitors, diodes, and transistors.

When the voltage appearing at the output terminal of the alternator rises to a predetermined value (14.2 volts), the voltage "turns on" the Zener diode. Transistor T-1 is turned on, which turns off T-2 and shuts off the current applied to the rotor coil. This effectively stops alternator output.

When the system voltage drops below the predetermined value, the Zener diode stops conducting, T-1 turns off, and T-2 turns on. When transistor T-2 is switched on, field current again is supplied to the alternator, which energizes the rotor coil to "turn on" the alternator.

The operation of transistor T-2 is in effect like a switch, turning the alternator field current on and off as the electrical supply varies due to the varying electrical load. This action occurs many times a second, so fast that it cannot be detected in the alternator output with a normal voltmeter.

FIGURE 8–17 A voltage regulator with cover off.

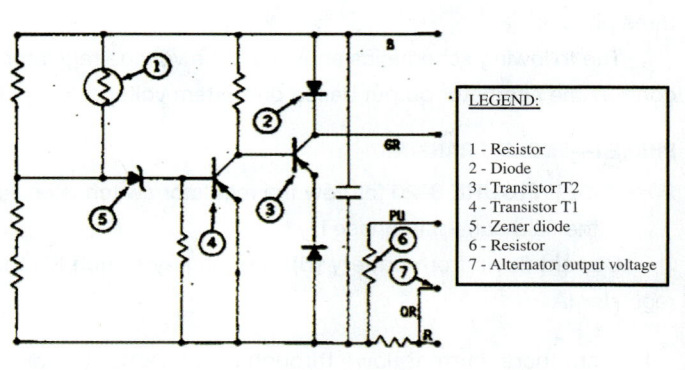

FIGURE 8–18 A schematic of this voltage regulator.

LEGEND:
1 - Resistor
2 - Diode
3 - Transistor T2
4 - Transistor T1
5 - Zener diode
6 - Resistor
7 - Alternator output voltage

CHARGING SYSTEMS

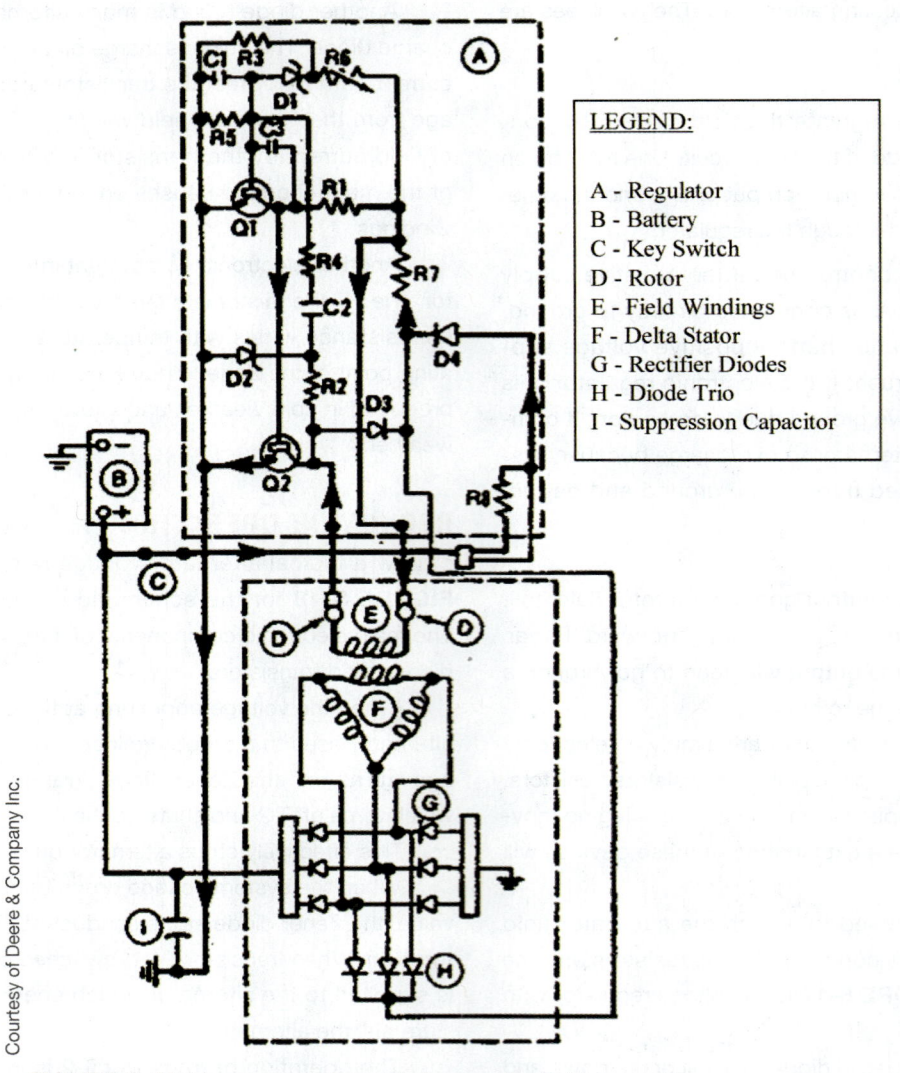

FIGURE 8–19 How a regulator switches on the field current in phase I.

LEGEND:

A - Regulator
B - Battery
C - Key Switch
D - Rotor
E - Field Windings
F - Delta Stator
G - Rectifier Diodes
H - Diode Trio
I - Suppression Capacitor

The solid-state regulator is either mounted on rear of alternator or mounted internally. It controls output by controlling current through field. In operation, regulator has the following three phases.

The following schematics show how the voltage regulator controls the alternator output based on system voltage:

PHASE I—ENGINE CRANKING

- SEE FIGURE 8–20 for how the regulator switches on the field current in phase I.

Current flows from battery (B) through key switch (C) to regulator (A).

1. From there, current flows through resistors R8, R7, and R1 to transistor Q2, turning it on.

2. Transistor Q2 then provides a path so current can flow through field (E) to ground, enabling alternator to generate electricity.

3. Zener diode D1 prevents flow of current to transistor Q1. A Zener diode is a special type of diode that will not permit current to pass until voltage reaches a certain preset level. If voltage exceeds that level, current can pass through the Zener diode.

PHASE II—GENERATING ELECTRICITY

- SEE FIGURE 8–21 for how the regulator switches on the field current in phase II.

The key switch (C) and rectifier bridge diodes (G) all have equal voltage. Therefore, no current flows from key switch to regulator.

250 CHAPTER 8

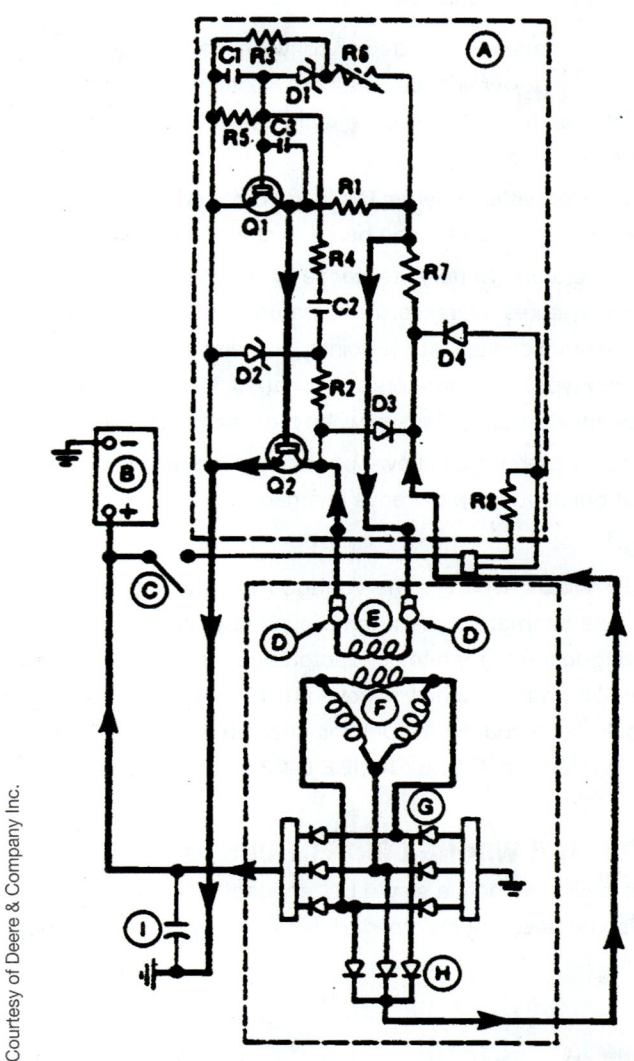

FIGURE 8–20 How a regulator switches on the field current in phase II.

1. Current, now coming from rectifier bridge diodes, flows through resistors R7 and R1 to turn on transistor Q2.
2. Transistor Q2 still provides path to ground through field (E), enabling alternator to generate electricity.
3. As the field is rotating, it does generate electricity. AC is induced in the stator windings (F). The rectifier bridge converts it to DC, providing current to run electrical accessories and charge the battery.
4. Output voltage still has not reached critical voltage of the Zener diode D1.

PHASE III—SHUTOFF

● **SEE FIGURE 8–22** for how the regulator switches off field current.

Output voltage reaches critical voltage of Zener diode D1.

1. Current can now pass through Zener diode D1 to turn on control transistor Q1.
2. Current from transistor Q1 cuts off voltage to transistor Q2, turning it off. There is now no path to ground for current through field (E).
3. Current through field is shut off instantly, and alternator stops generating electricity.
4. With transistor Q2 off, system voltage starts to drop in unrestricted fashion until it falls below the Zener diode D1 critical voltage. When this occurs, Zener diode D1 and transistor Q1 switch off and Q2 turns on again.
5. Phases II and III are repeated many times per second to maintain voltage at proper level.

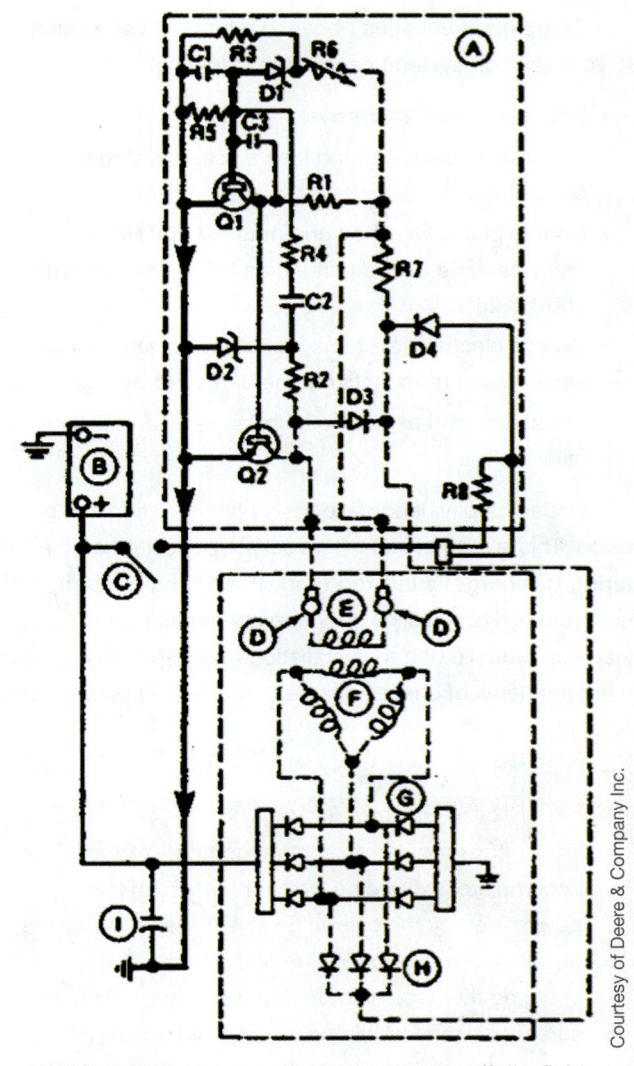

FIGURE 8–21 How a regulator switches off the field current.

CHARGING SYSTEMS 251

VOLTAGE REGULATION SET POINT Alternators must be capable of precisely controlling the DC output to the proper level. They need to charge the batteries to their fully charged level yet not too high and cause damage to the battery's or the vehicle's electrical system. Voltage regulation for 12 volt systems will establish a maximum charging voltage, known as the **set point**. Charging voltage set point averages between 13.5 and 14.6 volts. This is 1.5–2.0 volts above 12.6 volt open circuit voltage for a typical 12 volt battery. Twenty-four volt systems use 27–29 volts for a typical set point.

Once the alternator begins to produce current, the difficulty becomes to maintain the output voltage constant regardless of the load placed on the electrical system.

The Zener diode in the voltage regulator controls the charging set point.

There are some models of alternators that allow their set point to be changed manually to compensate for different conditions.

Charging at voltages above 15 volts (12 volt system) and 31 volts (24 volt system) causes the following:

- Batteries to gas excessively
- Batteries to overheat and lose electrolyte through electrolysis
- Battery plates will shed grid material, buckle, and generally become heat damaged as the temperature rises above 125°F
- Vehicle electrical systems, control modules, and other electronic components can be damaged by high voltage
- Premature and extensive bulb failure and LED light failure

Undercharging leads to battery plate sulfation and grid corrosion. This is a condition where sulfate deposited on the plates during discharge is left too long. If left long enough, sulfate turns to a hard crystalline structure and cannot be driven off by charging. Multiple battery installations are especially vulnerable to the problems of uneven charge rates causing plate sulfation.

> **TECH TIP**
>
> The voltage regulator set point compares to an accumulator's charging cutout pressure that is determined by the charging valve. When the cutout pressure is reached, the pump flow will stop charging the accumulator. When the hydraulic pressure drops to a certain point, then the pump will start to charge the accumulator again.

Factors affecting the set point include the following:

- Type of batteries: Flooded or conventional batteries charge at lower voltages than no maintenance or AGM batteries. AGM batteries are more easily damaged by overcharging.
- States of battery charge: Discharged batteries have low resistance to current compared to charged batteries.
- Temperature: Battery resistance to charging increases as temperatures decrease. Temperature sensors in voltage regulators can adjust set points. To warm up the battery, some types of alternators charge at 16.5 volts for the first few minutes after start-up when the weather is cold.
- Engine speed cycle: Low-speed operation requires higher set points to keep batteries charged.

BRUSH ASSEMBLY The voltage regulator output for a brush-type alternator is sent through a set of brushes. These are spring-loaded graphite conductors that ride on the rotor's copper slip rings. Brushes will eventually wear out or the slip ring could get a coating of contamination on them to not allow a good connection. ● **SEE FIGURE 8–22** for a brush assembly.

ALTERNATOR WIRING CONNECTIONS Self-exciting alternators will use only a single battery cable. A ground cable may also be used but no other additional wires connect to the

> **TECH TIP**
>
> The vehicle's electrical system can be severely damaged by high-voltage spikes if batteries are disconnected accidentally or intentionally while the alternator is charging. As the rotor's magnetic fields do not disappear immediately and the battery is unable to absorb current, output voltage can suddenly rise to levels that damage sensitive electronic devices. Load dumping is a feature alternators can have, which temporarily suppresses these high-voltage spikes. This usually involves using specialized diodes in the rectifier bridge, which become resistive rather than conductive at a specific voltage level. The diodes called transient voltage suppression (TVS) diodes will temporarily resist high voltage and automatically reset when the overvoltage goes away. The best practice is never to disconnect batteries when the engine is running.

FIGURE 8–22 A brush assembly.

FIGURE 8–23 An alternator with a heavy output wire and a ground wire.

alternator. The following are some other wires and their identification that you may see on an alternator.

REMOTE VOLTAGE SENSE (S). A sensing wire connected directly to battery voltage is compared with the regulator's set point and determines the alternator's output. Without this wire, the difference between the battery positive and ground voltage are used to sense system voltage. This small gauge wire has voltage present at all times. A remote sensor wire will provide a more accurate voltage reading at the batteries. Otherwise, a voltage drop between the alternator and the battery would produce an overcharge condition.

IGNITION EXCITE (I). This small gauge wire has voltage present only when the ignition switch is in the run position. Current through this wire switches the voltage regulator on and excites the alternator. If equipped with a dash warning light, current will pass from the switch and to the light. When current flows only one way from the ignition switch to the regulator and the alternator is not charging, the light stays on. Charging voltage appears at the "I" terminal after the alternator begins to charge, which provides battery positive at both sides of the light and turns off the charging system warning.

When the key is turned on, current flows through the warning bulb to regulator terminal 1 or ignition (I). Also connected to this regulator pin is the output from the diode trio. When alternator voltage equals and begins to slightly exceed battery voltage, no potential difference exists between the two ends of the warning bulb filament and the light goes out. If alternator voltage becomes excessive—exceeding battery voltage significantly, the light will re-illuminate. If the alternator does not charge, the bulb stays on.

> **WEB ACTIVITY**
>
> Search the Internet for heavy-duty alternator manufacturers and see what the maximum amperage rating is for one of their 24 volt alternators.
>
> Look at the specs of an alternator to see what the maximum rpm is for it. How long is their warranty and do they offer exchange units?

BATTERY POSITIVE CABLE (B+). This large gauge wire, 6 AWG or larger, has voltage present at all times. It likely goes to the starter B+ terminal but could go to a circuit protection device first.

NEGATIVE CABLE. On some alternators, particularly high output ones, a large gauge wire, 4 AWG or larger, is connected to battery or chassis ground. This prevents the engine block from conducting high amperage the alternator may produce and minimizes voltage loss.

RELAY (R) TERMINAL. This terminal is connected directly to one phase of the stator winding. If connected to the stator, the AC can supply input to a tachometer or hour meter or operate a frequency sensitive starter lock out relay, which disables the cranking circuit when the engine is running. The frequency of the output from the "R" terminal translates to engine speed so if an engine ECM (electronic control module) read this output as 300 rpm or higher, it would assume the engine is running and not allow the starter circuit to be energized.
● **SEE FIGURE 8–23** for an alternator with a heavy output wire and a ground wire.

CHARGING SYSTEMS 253

Progress Check

6. A rectifier bridge assembly will only use these semiconductor components:
   a. diodes
   b. transistors
   c. Zener diodes
   d. capacitors
7. This electronic component will trigger the regulator to turn the alternator on and off:
   a. transistor
   b. thermistor
   c. diode
   d. Zener diode
8. If an alternator has a diode trio, it will:
   a. supply the voltage regulator with rectified current flow from the stator
   b. supply the rotor with a PWM signal
   c. boost the output of the alternator for peak demands
   d. feed the stator with rectified voltage
9. An alternator with an "S" terminal will need it to:
   a. start the alternator charging
   b. stop the alternator charging
   c. sense battery voltage directly
   d. stop the alternator from draining the battery when the engine isn't running
10. If an alternator has an "I" terminal, it will be used to:
    a. send power to the ignition circuit
    b. indicate to the ECM how much current the alternator is producing
    c. turn on the voltage regulator and excite the alternator
    d. turn on the rotor directly

CHARGING SYSTEM PREVENTIVE MAINTENANCE PRACTICES

The charging system requires the following procedures as part of regular machine maintenance or as needed:

1. Cleaning cable terminals, wiring, and alternator connection points or corrosion. Alternator surfaces should be cleaned until they are free of accumulations of dirt, grease, and dust. Air passages need to be unobstructed to allow air to easily pass through. All connection points must be clean and free from corrosion because voltage is sensed between ground and battery positive. Anticorrosion sealer should be applied to all exposed connections.

FIGURE 8–24 Belt tension gauge.

2. Mounting brackets should be inspected for loose bolts and to allow correct belt alignment. Broken and loose mounting may indicate damage from engine torsional vibration. If other accessory drive system components are functioning correctly, a sturdier model of alternator may be required.
3. Condition of pulleys, idlers, belts, and belt tension. A loose belt will slip and cause undercharging. Tensioners must be correctly aligned operating perpendicular to the belt. Multi-grooved belts should be checked for cracks on both sides. If they extend completely across the belt, the belt should be replaced. The back side of the belt should not be worn and glazed. Belt tension should be confirmed with a tension gauge. ● SEE FIGURE 8–24 for a belt tension gauge.

Care must be taken when adjusting belt tension and prying on the alternator not to damage the housing or other engine components. Always replace multi-belt arrangements as sets.

Idler pulleys should be checked for bearing play.

CHARGING SYSTEM DIAGNOSIS
Charging system problems will show up to an operator in different forms. There could be starting complaints (likely slow cranking) that originate as a charging system problem, there could be dim lights that are caused by a subpar charging system, or there could be a machine fire if the charging system has a major fault.

How do you know if the charging system is working properly? Charging system malfunctions are often identified by battery condition.

1. Overcharged batteries indicate one or all of the following:
   - Excessive resistance at voltage sensing lead contact at alternator or electrical system
   - Open voltage sensing circuit

- Defective voltage regulator
- Improperly adjusted voltage regulator
- One shorted battery in a battery pack

2. Undercharged batteries indicate one or all of the following:

- Loose drive belts
- Corroded, broken, burnt, or loose wiring connections
- Undersize battery cables
- One or more defective batteries
- Batteries too far from sensing lead contact
- Missing sensing lead contact
- Defective voltage regulator
- Improperly adjusted voltage regulator
- Defective rectifier bridge–shorted or open diodes

An undercharged battery in a multiple battery system of two or more batteries that becomes sulfated is due to uneven rates of discharge and charging. The slightest resistance in battery cables or longer distances from the alternator could cause batteries to charge at different rates. The alternator will sense the voltage of the highest battery and lower the charging rate leaving one or more batteries in a continuous state of discharge. Prolonged discharge will increase a battery's resistance leaving it undercharged even further.

Other machine electrical functions that are operating erratically or not at all could be an indication of charging system problems.

As with other system problems, initial diagnosis should include a visual inspection. You would start by getting information from the operator and then check the drive belt and pulley for wear, alignment, cracks, and glazing. Then check the alternator wiring for corrosion, loose wires/connections, blown fuses, open circuit breakers, and worn insulation.

You should have a look at the machine to see if there have been any extra lights or other electrical loads added. If so, then a higher output alternator may be needed.

The alternator's part number and model should be checked to see if the right one is on the machine.

The machine's battery condition should be checked before any alternator or wiring tests are done. See Chapter 5 to review battery checking and testing procedures. If the battery or batteries are less than 12.4 OCV, they should be charged and checked for defects.

There are some simple tests that will indicate charging system condition. They are the following:

1. Put the positive "+" lead of a multimeter on the "B+" terminal of the alternator. Put the negative "−" lead on the ground terminal or on the frame of the alternator. Put a suitable ammeter around the positive output wire of the alternator.

2. Turn off all electrical accessories. Turn off the fuel to the engine. Crank the engine for 30 seconds. Wait for two minutes to cool the starting motor. If the electrical system appears to operate correctly, crank the engine again for 30 seconds.

NOTE: Cranking the engine for 30 seconds partially discharges the batteries to do a charging test.

3. Start the engine and run the engine at full throttle.

4. Check the output current of the alternator. The initial charging current should be equal to the minimum full load current or greater than the minimum full load current. Refer to the alternator specifications for the correct minimum full load current. To check the alternator output, a clamp-on ammeter should be placed around the alternator B+ wire close to the alternator.

Refer to Table 8–1 for test results and follow-up actions. All cable and wiring connections should be checked before replacing the alternator.

5. After approximately 10 minutes of operating the engine at high idle the output voltage of the alternator should be 14.0 ± 0.5 volts for a 12 volt system and 28.0 ± 1 volts for a 24 volt system (● **SEE TABLE 8–1**).

6. After 10 minutes of engine operation, the charging current should decrease to approximately 10 amps. The actual length of time for the decrease to 10 amps depends on the following conditions:

- The battery charge
- The ambient temperature
- The speed of the engine

If an alternator is found to be defective, it is usually replaced with a new or rebuilt one. The drive pulley and fan are usually switched over from the old alternator. Care should be taken when doing this so no damage is done to the alternator, the fan, or the pulley. ● **SEE FIGURE 8–25** for how to use special tools to change an alternator pulley.

The replacement alternator should have all wiring connections checked and cleaned and the drive belt should be inspected and tensioned properly. The batteries should be fully charged before starting the engine after a replacement alternator has been installed.

CHARGING SYSTEMS **255**

Current at start-up	The voltage is below specifications after 10 minutes.	The voltage is within specifications after 10 minutes.	The voltage is above specifications after 10 minutes.
Less than the specifications	Check the exciter circuit of the ignition switch and circuit. If okay, replace alternator.	Turn on all accessories. If the voltage decreases below the specifications, replace the alternator.	—
Decreases after matching specifications	Replace the alternator.	The alternator and the battery match the specifications. Turn on all accessories to verify that the voltage stays within specs.	Replace the alternator.
Consistently exceeds specifications	Test the battery. Test the alternator again.	The alternator operates within the specifications. Test the battery(ies).	Replace the alternator. Inspect the battery(ies) for damage.

TABLE 8–1

Fault Conditions and Possible Causes

FIGURE 8–25 How to use special tools to change an alternator pulley.

Once the replacement alternator is installed, the system should be checked for proper operation by performing the simple test described earlier.

ALTERNATOR BENCH TESTING Not many repair facilities are set up to repair alternators, and like other components, it is not usually worth the time for an HDET to spend rebuilding one.

If a shop does have a bench to test alternators, it can be used to verify that the alternator that was removed does have a problem or rebuilt units can be tested to make sure that they're okay to put on a machine. Bench testing will put a load on the alternator and its output voltage and current flow can be measured and sometimes graphed and printed out.

The following table is an example of alternator outputs for several different alternators:

ALTERNATOR A ALTERNATOR B	VOLTS	RATED AMPS	OUTPUT TEST [OUTPUT TEST @ 3000–4000 ROTOR RPM] AMPS (HOT)	FIELD CURRENT [FIELD CURRENT @ 20–27°C (70–80°F); @ 10V FOR 12V SYSTEMS AND 20V FOR 24V SYSTEMS] AMPS	WINDING RESISTANCE [WINDING RESISTANCE @ 20–27°C (70–80°F)] OHMS
A	12	65	54–62	3.2–3.7	4.0–4.7
B	24	42	37–42	1.8–2.3	12.0–13.0

> **TECH TIP**
>
> **What makes alternators fail?**
>
> Alternators will generally last for a long time, and outside of belt tension, they are maintenance free. There are some things that will decrease the life span of an alternator though. They are the following:
>
> **Heat:** Excessive heat will damage diodes and could fail wire insulation.
>
> **Vibrations:** Any unbalance of the alternator shaft will cause an unbalance and cause the alternator components like bearings to fail early.
>
> **Dust, oil, grease:** If an alternator has brushes, the contact surface at the slip rings needs to be clean. Oil leaks close by should be repaired quickly.
>
> **Excessive loads:** If an alternator is constantly operating at its maximum output, it will not last as long as it should. A higher capacity alternator may have to be installed if extra loads have been added to a machine.

ALTERNATOR CABLE VOLTAGE DROP TEST To test the positive cable for excessive resistance, voltage between the alternator and the batteries is measured. With the engine running at 1500 rpm and the alternator loaded to 75% of its output capacity, voltage is measured at the alternator and batteries. If the difference is greater than .5 volts in a 12 or 24 volt circuit, all positive and ground wire cable connections should be checked. Acceptable cable **voltage drop** readings are less than .5 volts on both positive and negative sides of both 12 and 24 volt systems.

Progress Check

11. The following condition would *not* lead you to look at the charging system as the cause of the problem:
    a. slow cranking
    b. dim lights
    c. batteries are wet
    d. the horn doesn't work
12. If an alternator's drive belt was over-tightened, the following could happen:
    a. alternator bearing failure
    b. battery overcharging
    c. belt slippage
    d. fan blades bent
13. The maximum total voltage drop between the alternator and the battery pack for a 24 volt system is:
    a. 24V DC
    b. 12V DC
    c. .5V DC
    d. .5V AC
14. If an alternator is "full fielded," this should:
    a. restore residual magnetism
    b. make the stator create its maximum magnetism
    c. make the alternator produce its maximum output
    d. make the rotor become a single magnet
15. The maximum voltage you should read from a 24 volt alternator is:
    a. 13.5V DC
    b. 24V DC
    c. 27V DC
    d. 29V DC

> **SHOP ACTIVITY**
>
> - Go to a machine in your shop and perform a visual inspection of the charging system. Look for belt condition, alignment and tension, wire damage and corrosion, and battery condition.
> - See if there is a tag on the alternator and see what information is on it.
> - Perform a quick test of the charging system to see if it is working as it should.

SUMMARY

1. Some safety concerns related to charging systems are rotating belts and pulleys, battery explosion, and burns from hot wires or sparks.
2. All machines produced in the last few decades have a charging system that includes an AC generator, commonly called an alternator.
3. Alternators are belt driven and convert rotary motion into DC flow that keeps the machine's battery voltage level high enough to properly operate all machine electrical systems.
4. Alternators are rated by their voltage output (12 or 24V DC) and amperage output (25–300 amps). There are also different mounting and shaft styles, brush type, or brushless and if they are self-exciting.
5. Alternator output wire is usually a heavy wire (6–10 gauges) and either leads to the B+ terminal on the starter

CHARGING SYSTEMS   257

or goes to a breaker first. The current can then flow to the battery pack.

6. Alternator current flow starts with the principle of moving a conductor (wire) through a magnetic field.
7. The two parts of the alternator that produce current flow are the rotating rotor and the stationary stator. The rotor creates several sets of alternating magnetic fields and the stator is three lengths of wire arranged in many loops. The stator has AC voltage induced in it and sends the current to the rectifier where it is converted to DC.
8. Alternator drive belts can be V-belt or multi-rib style and proper tension must be maintained. Drive pulley diameter is determined to give the alternator proper rpm through normal engine rpm range.
9. Brush-type rotor has coil of wire inside alternating pole fingers. Each end of wire is connected to slip rings. Brushes ride on slip rings to supply current to coil to create a set of rotating magnetic fields.
10. Brushless-type rotor has a stationary coil with pole pieces rotating over top of it.
11. Self-exciting rotors rely on residual magnetism in rotor iron pole pieces to initiate induction in stator.
12. Stator is three wires configured in one of two ways. Wye or delta connections indicate how the wires are connected. Loops are wound around a laminated core.
13. Stator output is in AC form.
14. A set of diodes called a rectifier bridge will convert the AC to DC.
15. Some alternators use a diode trio to supply the voltage regulator with DC.
16. A voltage regulator will control alternator output voltage levels and ultimately the system voltage level. It uses several electronic components such as transistors, diodes, and resistors to cycle the rotor field coil circuit on and off. This occurs many times per second.
17. The voltage set point is the voltage value that the regulator tries to maintain when the alternator is turning. It will be slightly higher than OCV. For example, a 12V system is 13.5–14.6V DC and a 24V system is 27–29V DC.
18. Alternator brushes are a graphite material that ride on the rotor slip rings and transfer current into the rotor.
19. Depending on the type of alternator, it could have from one to five wires connected to it. A heavy gauge wire will be used for the output.
20. Charging system maintenance includes check wire connections, clean and tighten if necessary, check mounting hardware and brackets, check belts and pulleys for damage or excessive wear, and check belt tension.
21. Charging system quick check includes put red lead of multimeter on alternator output and black to a good ground; place a clamp-on ammeter around output wire; crank engine for 30 seconds to lower battery OCV; start engine and run at high idle, turn on several heavy loads, and watch for voltage increase and amperage output; and compare readings to specifications.
22. Perform voltage drop tests on alternator wiring to ensure that all connections and wire are in good condition.
23. When diagnosed to be defective, alternators are replaced with new or exchange units. Take care to ensure that wire connections are right and secure.
24. Pulleys are usually switched from old to new alternator. Use patience and proper tools.
25. Tension belt to specification.

chapter 9
ELECTRONICS AND MULTIPLEXING SYSTEMS

LEARNING OBJECTIVES

After reading this chapter, the student should be able to:

1. Explain how a semiconductor material is created and functions.
2. Describe what an electronic device is.
3. Explain how a diode works.
4. Explain how a bipolar transistor works.
5. Explain how an electronic control module (ECM) works.
6. Describe how the different types of memory is used in an ECM.
7. Explain what inputs and outputs to an ECM are.
8. Explain the term *multiplexing*.
9. Describe how a typical electronic multiplexing network functions.
10. Describe how electromagnetic interference is counteracted.
11. Explain how a technician performs electronic troubleshooting.
12. Explain how a payload system works.
13. Describe how the different types of sensors work.
14. Explain what the different outputs from an ECM are.

KEY TERMS

Biased 263
Bipolar 266
Diode 263
Field-effect transistor 266
Inputs 281
Outputs 286
Semiconductor 262
Silicon 262
Transistor 266

INTRODUCTION

SAFETY FIRST Most electronic devices found on machines are generally low voltage and low current, and there is little inherent danger when working with them. There is, however, some electronic fuel injectors that have well over 50V sent to their solenoids from the engine ECM. You may see a warning to be aware of high voltage when working around certain fuel systems. ● SEE FIGURE 9–1 for an example of a high-voltage warning.

Many electronic devices will control machine systems that if not working properly can have serious negative health and safety consequences. It's imperative that any time a system's electronic control or monitoring system has been serviced or repaired that the machine is checked for proper operation before the machine is put back into service.

NEED TO KNOW

Chapter 5 dealt with basic electrical fundamentals such as what electrons are, what makes them move through circuits, and how their movement performs work in a machine's electrical system. After reading Chapter 5 and answering the Progress Check questions correctly, you should have a good grasp of basic DC electrical circuits, what makes them work, why they sometimes don't work, and how to fix them when they don't. You should now be ready to learn about more advanced electrical systems that have electronic components as part of them. You may have been able to disregard these advanced circuits a few years ago when machines were much simpler electrically but not today.

Thirty years ago, most machines had a simple electrical system that consisted of a starting and charging system and maybe some light circuits, a wiper circuit, and a heater fan circuit. Electronics were first introduced on machines to enhance the machine's system monitoring capabilities. The original monitoring system consisted of a few mechanical gauges and warning lights. Electronic monitoring and warning systems were added to let the operator know if there was a problem such as low engine oil pressure or high engine coolant temperature without having to keep looking at gauges. This led to a more productive and reliable machine. From there the use of electronics has increased dramatically, and there are several reasons for this, such as:

1. **Operator comfort and efficiency:** Machine controls that are electronic are much easier to move and can sometimes be customized for an individual operator. This will make the operator less fatigued and more efficient and will in turn make the machine more efficient.

2. **Machine reliability:** Machine systems can be monitored with electronics closely, and if abnormal temperatures, speeds, or pressures are detected, the electronic system will alert the operator and can possibly override the operator to change the way the machine is working to try and correct the abnormality. Electronics have even allowed machine problems to be communicated to someone at a central location like a dealership or maintenance shop many miles away where an initial troubleshooting procedure can start.

3. **Safety concerns:** There are electronic systems on machines dedicated to reducing accidents such as cameras that display blind spots to the operator, stability control to help prevent loss of control, and interlock systems that won't allow a machine to start and operate unless certain safety-related conditions are met.

4. **Diagnostics:** HDETs are able to use the electronic system to access fault codes and view live system data such as pressures and temperatures without installing extra gauges. The machine's service information will be in electronic form as well. This will allow the most current information to be available to the HDET. ● SEE FIGURE 9–2 for a computer connected to a machine.

5. **Fuel economy:** Electronic systems have greatly reduced machine fuel consumption by making engines, drivetrain, and hydraulic systems more efficient.

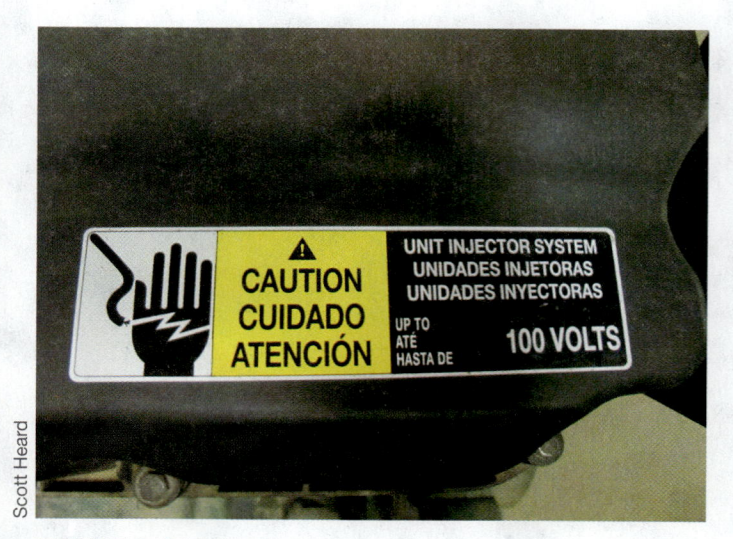

FIGURE 9–1 High-voltage warning.

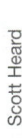

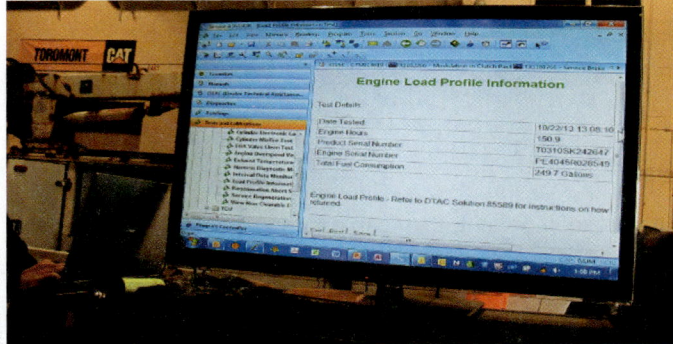

FIGURE 9–2 Computer connected to a machine.

FIGURE 9–3 Clean exhaust.

6. **Engine emission reduction:** Environmental concerns have led to adding more electronic controls and sensors to diesel engines to monitor and control engine operation so that harmful exhaust emissions can be reduced. Engines that run cleaner with the help of electronics and this extends the oil change interval, which is good for the environment in many ways. ● **SEE FIGURE 9–3** for the clean exhaust of an electronically controlled diesel engine when under full load.

7. **Data logging functions:** Machine system data such as pressures, speeds, and temperatures can be recorded and reviewed to see how productive a machine and an operator is. This information can also be used to diagnose problems with a machine.

Some machines will have 10 or more separate ECMs that can communicate with each other and be able to control all machine systems along with many sensors, input devices, and actuators. The continuing addition of more electronic components and systems to machines is inevitable as machines evolve. An HDET needs to have a strong understanding of how these components and systems work to be able to keep a machine running longer, repair it faster, and help it be more efficient.

If you learn how electronic components and systems work and become proficient at using electronic systems to your benefit, you will be a valuable HDET.

ELECTRICAL VERSUS ELECTRONICS

As a general rule, the difference between something that is either electrical or electronic is that an electronic device will have no moving parts and has one or more parts of it made from a semiconductor type of material. An electronic device is considered to be in a solid state because there are no moving parts in it. An electrical device can have moving parts such as a switch, create motion through magnetism such as a solenoid or motor, or create heat or light by current flow and resistance.

A comparison would be a conventional switch with contacts and moving parts would be considered an electrical device while a transistor that can also be used as a switch is a solid-state electronic device because it has no moving parts. ● **SEE FIGURE 9–4** for an electrical switch and a transistor.

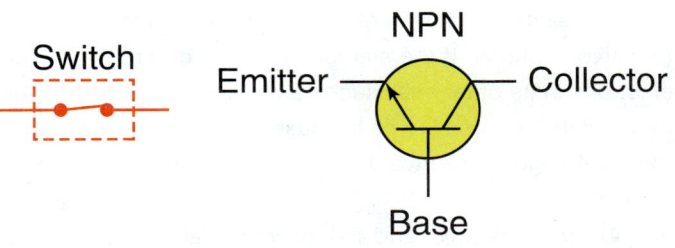

FIGURE 9–4 Switch and transistor.

ELECTRONICS AND MULTIPLEXING SYSTEMS 261

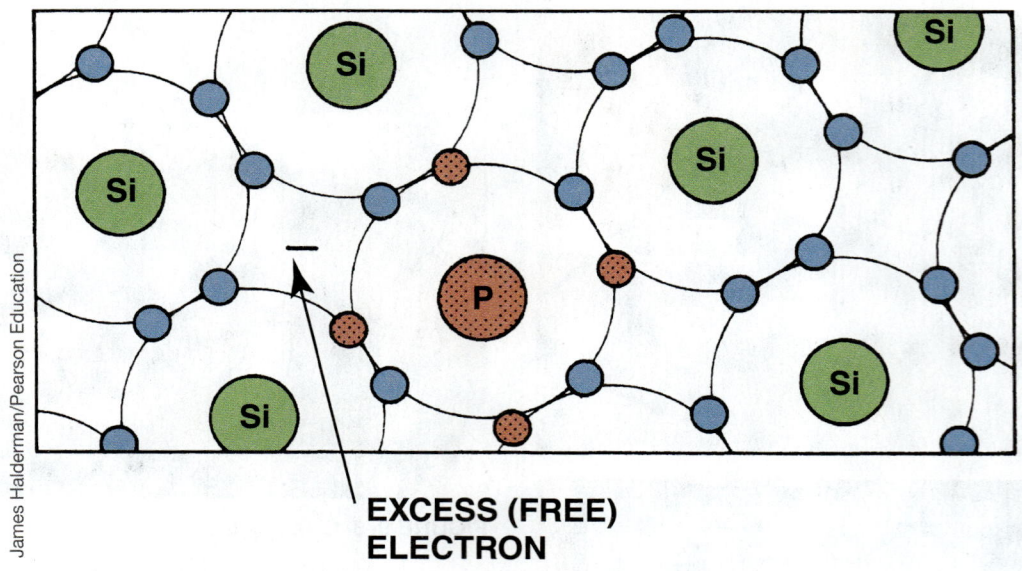

FIGURE 9–5 An N-type material.

SEMICONDUCTORS

Chapter 5 discussed insulators and conductors and which material's atomic structure makes an ideal insulator or conductor. To review, an ideal conductor has one free electron in its outer shell that is easy to move when voltage is applied to it while an ideal insulator has a full shell of electrons that resists movement. A **semiconductor** can be either an insulator or a conductor. An ideal semiconductor element has four electrons in its outer shell. The most common elements used to make a semiconductor are **silicon** and germanium. In their pure natural state, these elements aren't much good for electrical circuits. However, if they are slightly changed or doped with another material to make them impure, then they can be either a conductor or an insulator under certain conditions. If the base element is doped with phosphorous, arsenic, or antimony that all have five electrons in their outer shell, this new material is called N type (negative) because it has an excess of negatively charged electrons. ● **SEE FIGURE 9–5** for an N-type material.

The same base element can be doped to make a P-type (positive) material. If the element used for doping has three electrons in its outer shell such as boron or indium, then the new material is called P type because the shortage of electrons makes it slightly positive. ● **SEE FIGURE 9–6** for a P-type material.

On their own N- and P-type materials are not much typically used electrically. However, when they're combined into one component, they can be very effective for controlling current and voltage. When they are part of a circuit, the attraction of the power source's negative and positive charges on the N- and P-type materials will make them either conductors or insulators. The simplest solid-state device that uses semiconductor material is a diode.

DIODE

A comparison can be made between a hydraulic component and the electronic diode. A diode is like a check valve in that it allows current to flow one way but not the other. It is made of the two types of doped materials previously mentioned that are joined together. Depending on how a diode is oriented

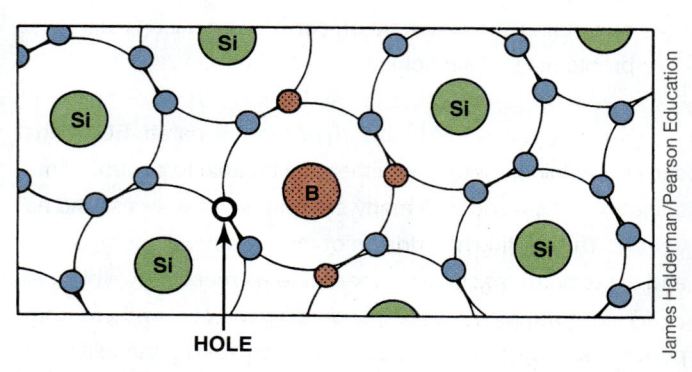

FIGURE 9–6 How a P-type material is made.

262   CHAPTER 9

FIGURE 9–7 Diode.

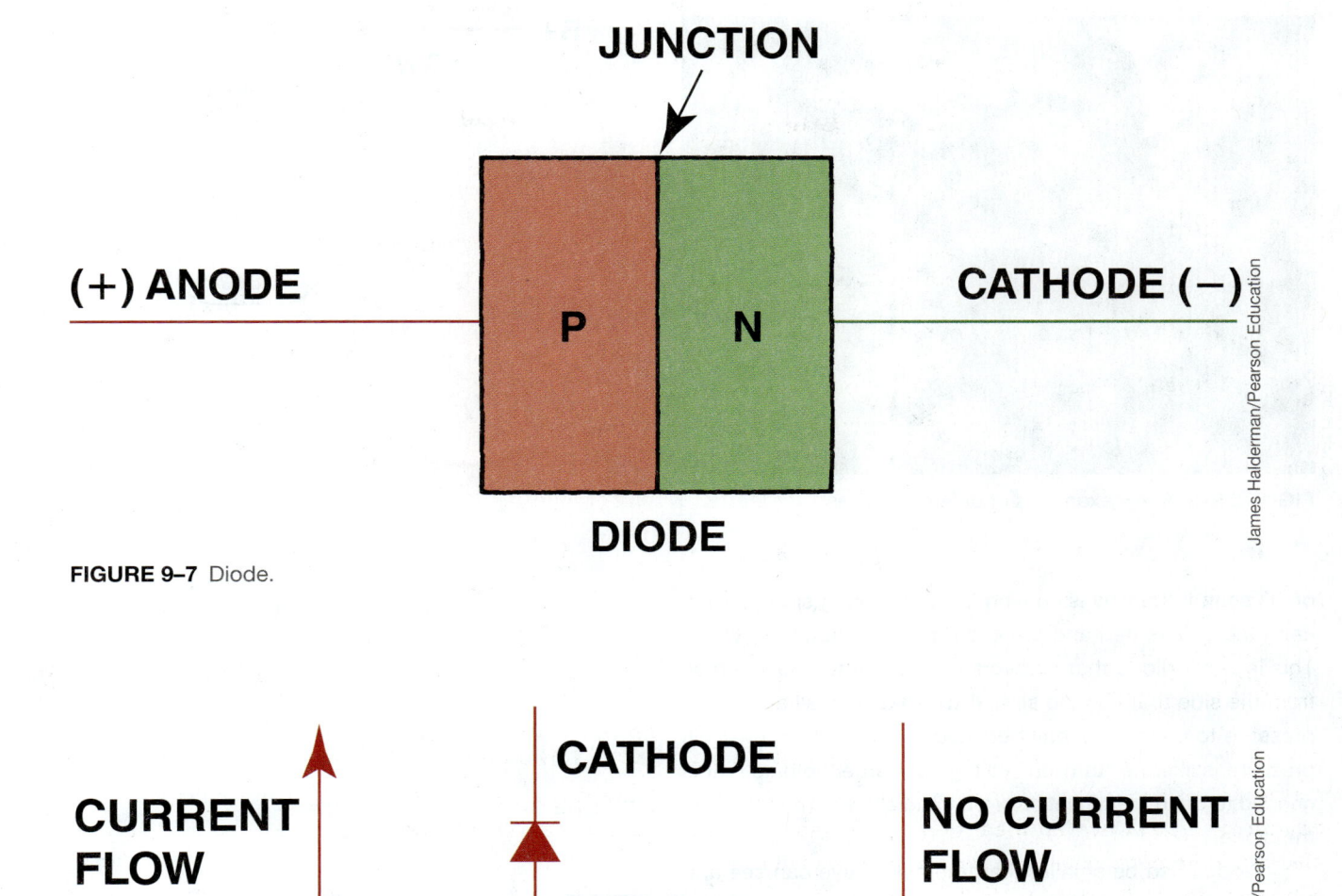

FIGURE 9–8 The symbol for a diode.

in the circuit will determine if it will allow current flow or not.
● **SEE FIGURE 9–7** for a diode.

A **diode** has a positive terminal called the anode and its negative terminal is called the cathode. A larger diode will be marked to identify the cathode with a solid stripe around the circumference of it toward the cathode end. The area in the middle of the diode is called the depletion region, and depending on what happens in this part of the diode will determine whether the diode will allow current to flow through it or whether it will block it. If the P and N materials are attracted to the ends of the diode by an external power source, then the depletion region is left void. The diode will not allow current flow and is said to be reverse **biased**. This is when the anode is connected to the negative side of the power source. ● **SEE FIGURE 9–8** for the symbol of a diode and how it can allow current flow one way and block it the other way.

If the diode is in a circuit and the anode is connected to the positive side of the circuit, then the depletion region will allow current flow through. This is when the diode is connected in the forward-biased orientation.

To forward bias (allow current flow) a diode, there has to be a minimum amount of voltage present to arrange the depletion region. This is called the barrier voltage. Doped germanium has a barrier voltage of about .3V DC. Doped silicon has a barrier voltage of about .6V DC.

You can think of the diode in forward bias as having its gate open to allow current flow or being like a check valve with its ball off its seat. To compare a diode to a check valve, think of the check valve having a light spring holding the ball

ELECTRONICS AND MULTIPLEXING SYSTEMS 263

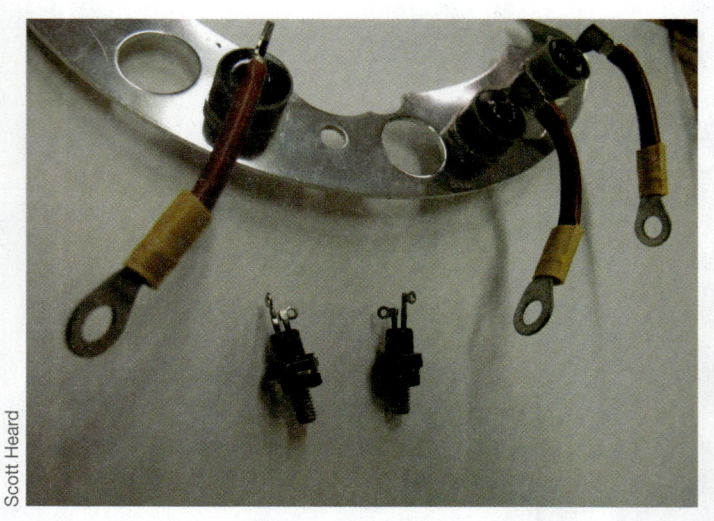

FIGURE 9–9 A few examples of different diodes.

on its seat. If the flow is coming from the spring side, it helps keep the ball seated and there is no flow through the valve. This is like a diode that is reverse biased. If the flow comes from the side that has the seat, it will take a small amount of pressure to unseat the ball because of the light spring. This pressure is like the "turn-on" voltage or barrier voltage that is needed to open the depletion region and allow current through the diode.

Diodes can be smaller than the naked eye can see and only be rated for microamps, or they can be as big as a pill bottle and able to handle over 100 amps. ● SEE FIGURE 9–9 for a few examples of different diodes.

Any diode is designed to handle a maximum amount of current flow, and if this is exceeded, the diode will either fail and become an open that won't allow current flow in either direction or be shorted to allow current flow both ways undeterred.

To test a diode it should be isolated from the circuit and a multimeter with a diode setting is needed. At this setting, the meter will produce the necessary amount of voltage (.3–.7V DC) to make the diode conduct current in the forward bias. The diode will be checked for either an open or a short. With the meter's positive lead connected to the anode and its negative lead connected to the cathode, a good diode will read .3 or .7 depending on its type, and when the leads are reversed, the display should read OL. A diode could be checked in a live circuit for the proper voltage drop across it.

Diodes that are used to limit current flow in a DC circuit to one direction may seem redundant, for since a DC circuit should only flow current in one direction anyways, but there are instances where undesired reverse flow is possible. This happens when a coil that is energized to create an electromagnet such as one

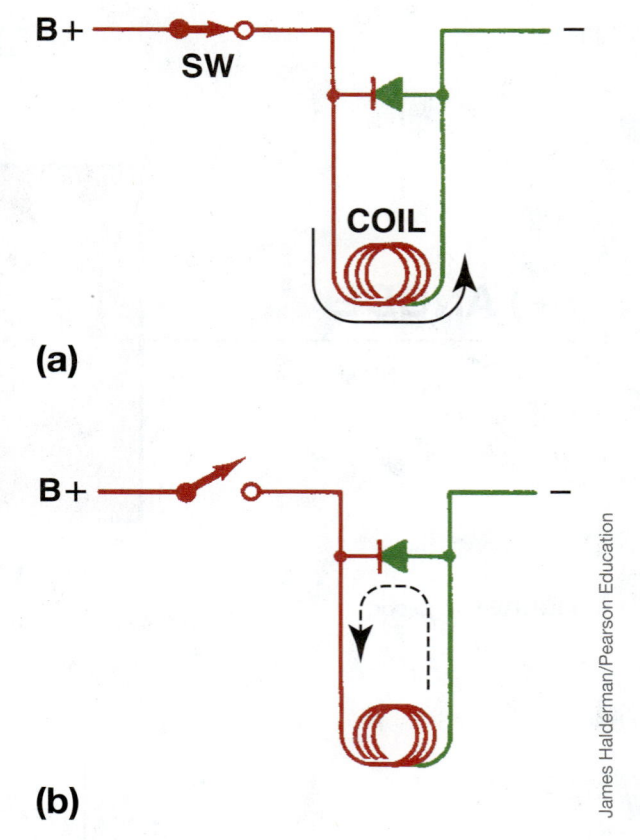

FIGURE 9–10 A clamping diode.

used in a solenoid or relay is de-energized. There will be a high voltage induced back into the circuit. This voltage can be damaging to other components in the circuit if not controlled. If a diode is connected in parallel to the coil in a reverse bias fashion, it will allow the coil to function normally when energized, but when current stops flowing and the collapsing magnetic lines of force induce voltage, the diode directs the current flow back into the coil where it loops between the coil and the diode until it drops to below .7V DC. At that point, the diode stops allowing current flow, but the voltage has lowered enough that no damage can be done. When a diode is used for this purpose, it is called a clamping diode or de-spiking diode. ● SEE FIGURE 9–10 for how a diode is used as a clamping diode.

Many machine circuits that use a diode will have them located in the fuse block. They plug in like a fuse so they can easily be removed, tested, and replaced if needed. They can also be located anywhere in a machine's wiring harness. ● SEE FIGURE 9–11 for a diode located near an air-conditioning compressor clutch.

RECTIFYING DIODE Another important use for diodes on heavy equipment is when they are used to convert AC power into DC. If you read through Chapter 8, you would have learned

FIGURE 9–11 A diode located near an AC compressor.

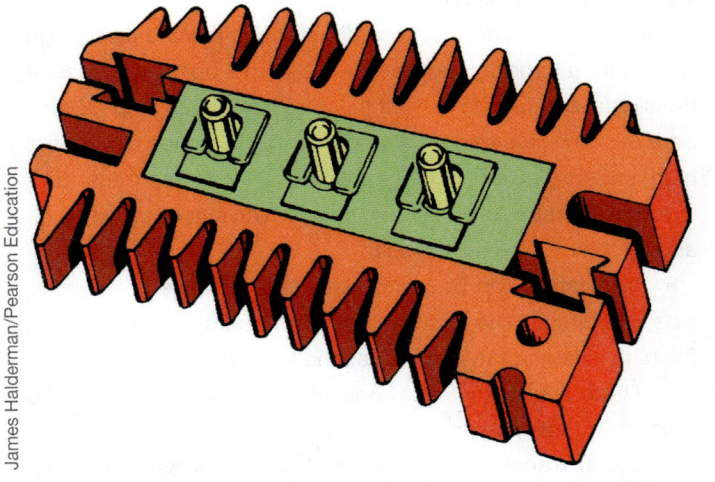

FIGURE 9–12 A set of diodes used for rectifying AC to DC.

REVERSE CURRENT

ANODE　　　　　CATHODE

ZENER DIODE SYMBOL
FIGURE 9–13 The symbol for a Zener diode.

FIGURE 9–14 The symbol for an LED.

that an AC generator uses a set of diodes to rectify or change the AC that is created in the stator into usable DC. When diodes are used to rectify, they effectively take the negative part of the oscillating AC sine wave and change it to a positive relatively flat DC voltage. If only a single diode was used to do this, only half the AC wave would be used. However, if a set of four diodes are used (this is called a diode bridge), full-wave rectification occurs with the complete AC wave made positive or made into usable DC. ● **SEE FIGURE 9–12** for a set of diodes used for full-wave rectification.

ZENER DIODE Another type of diode that is used mainly in an alternator is called a Zener diode. Its material is heavily doped to give it a special feature. It acts like a normal diode by allowing current flow in the forward direction and blocking it in the reverse direction, but if a certain amount of voltage (called the Zener point) is applied to it in reverse direction, it will allow current flow in the reverse bias mode with no damage to the diode. In an alternator's voltage regulator, a Zener diode will sense system voltage and ultimately turn the alternator on or off based on the Zener point. ● **SEE FIGURE 9–13** for the symbol of a Zener diode.

LIGHT-EMITTING DIODES Light-emitting diode (LED) lighting is becoming quite common in the automotive world and recently with heavy equipment as well. LEDs are similar to a normal diode in their construction except they will produce photons or light energy when forward biased in a circuit. They are made with gallium-based materials that are given different doping additives to make their N and P materials give them different colors when they are energized. The action of a diode giving off light is called electroluminescence. This different material will require them to have a much higher turn-on voltage, from 1.5 to 2.2V DC to light them up. LEDs can be very small and used for warning indicator lights or bigger ones are used as a set for brake or running lights. They are much more efficient than incandescent lights, will last longer, and are faster acting. ● **SEE FIGURE 9–14** for the symbol of an LED.

Most diodes will only be able to handle 20–30 mA of current, and for this reason, they must always be wired into a circuit that has a resistor in series with it.

ELECTRONICS AND MULTIPLEXING SYSTEMS

FIGURE 9–15 An LED work light assembly.

> **TECH TIP**
>
> LED lights are generally perceived as having very low power consumption. When compared with an incandescent light, the LED light is very efficient. However, at the moment, HID Xenon lights and fluorescent tubes are still more efficient in terms of the lumen/watt output (lumens is a measure of brightness). In real-life conditions, an HID will have a maximum output of 85 lm/W compared with an LED's maximum of 50 lm/W, making it roughly 30% more efficient. For example, an LED light with the same light output as an HID light would consume 60W compared with 40W.

Exterior work LED light assemblies for machines are packaged in a housing that also includes resistors, heat sinks, mounting hardware, and the primary optics, that is, a lens made of silicon that directs the light out of the component. To get the right light pattern from an LED light, there are secondary optics formed from a reflector or a lens, or a combination of both. ● SEE FIGURE 9–15 for an LED work light assembly.

PHOTODIODES Photodiodes are a semiconductor device that will allow current flow through them when exposed to light. They can be used to automatically turn on headlights or dim dash lights. ● SEE FIGURE 9–16 for the symbol of a photodiode.

TRANSISTORS

A **transistor** is an electronic solid-state device that can act like an electrical relay and switch a high-current circuit on or off at the command of a low-current controlling circuit. It can also act like an amplifier that can vary the high-current circuit output by varying the low-current control circuit. Keep in mind that a high current for a transistor is typically 1–3 amps, but some types of transistors can handle current flows as high as 25 amps. Compare this to relays that can handle up to 100 amps. To control high amperage circuits, electronic circuits can use transistors to switch the control circuit of a relay, which in turn switches on and off the high-current circuit.

Transistors can be switches for the positive or negative side of a circuit. The positive side is sometimes referred to as the high side and the negative as the low side.

There are many variations of transistors, but the two most commonly used in heavy equipment electronics are the bipolar transistor and the **field-effect transistor** (FET), which are sometimes called unipolar transistors. The most common type of FET is the MOSFET (metal oxide semiconductor FET).

A **bipolar** transistor is similar to two diodes that are combined back to back with the two similar parts of the diodes joined in the center as one. Therefore, it will have two PN junctions or depletion regions that give it its unique characteristics. The transistor is really a combination of three doped elements

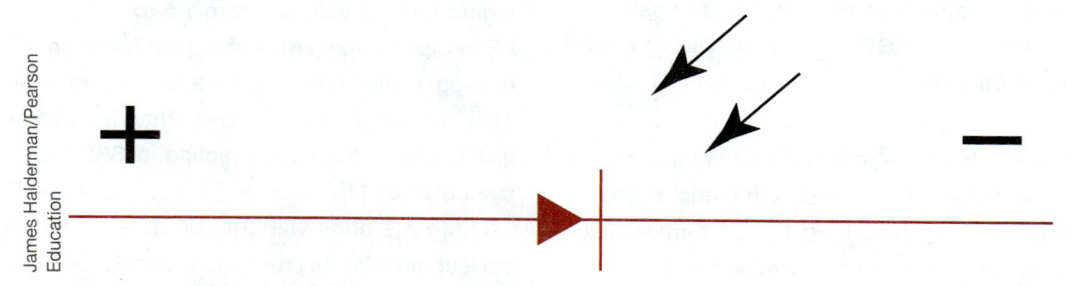

FIGURE 9–16 The symbol for a photodiode.

FIGURE 9–17 Transistor symbols.

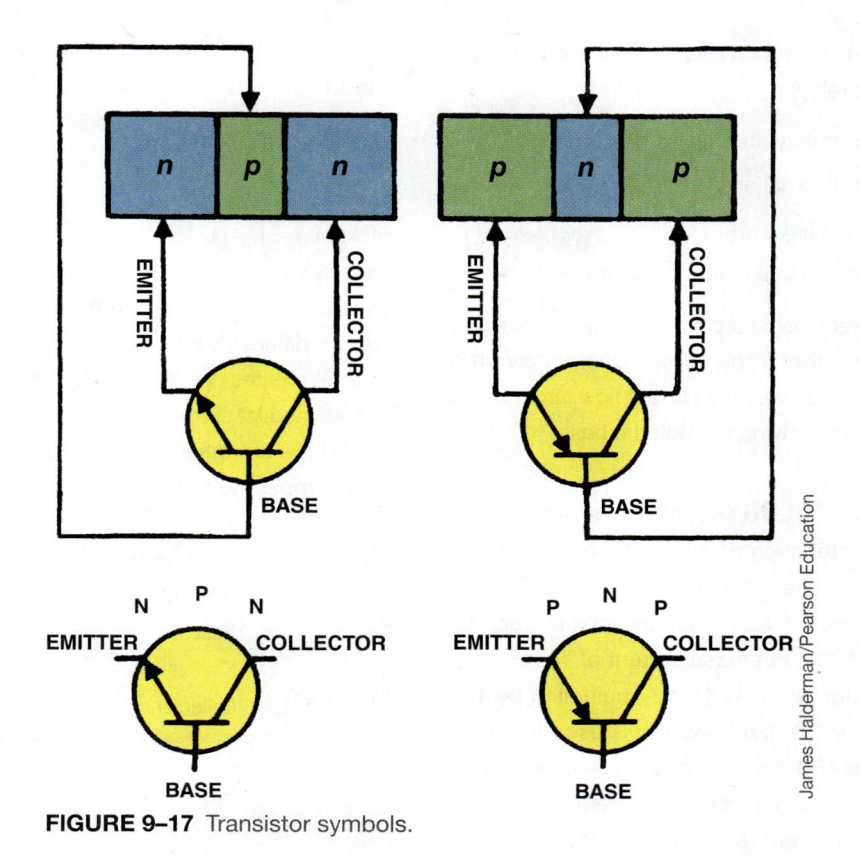

that are stacked together like a sandwich. Two N type and one P type make an NPN transistor or two P type and one N type make a PNP transistor. These three elements are called the collector, base, and emitter and will have their own terminals to allow the transistor to be connected to circuits. The symbol for both types is shown in ● FIGURE 9–17.

The arrow for the transistor symbol is always on the emitter side and always points to the N material when viewing it in terms of conventional flow. The arrow could also be thought of as pointing in the direction of conventional current flow.

Here's what the different parts of the bipolar transistor do:

Base: The base will control what happens with the rest of the transistor. If it gets a voltage signal that is high enough to turn on the transistor (just like the barrier voltage of .3–.7V DC for a diode depending on the material), then current will flow from the base to the emitter and from the collector to the emitter. This signal is sometimes called the threshold voltage. The base current will be part of a circuit that will always be restricted by a resistor to limit its current flow.

Collector: It will supply the transistor with current from a power source or be connected to ground depending on the transistor configuration. Once the transistor is turned on, it will send current to the emitter.

Emitter: This is the high-current outlet for this transistor. It can also be connected to the positive or negative side of a circuit. It will receive current flow from the base and the collector although the majority is from the collector.

NPN transistors are more commonly used in heavy equipment electronics. These will switch the low side of a circuit or complete a ground path.

TRANSISTOR OPERATION When you are trying to understand how a transistor functions in a specific circuit, there are two facts you must remember. First, an NPN transistor is turned *on* by applying voltage to the base leg and it is turned *off* by removing voltage from the base leg. Second, the current through the base circuit is always much smaller than the current across the collector circuit. Changing the base current a little results in a big change in the collector current. This is like opening a main directional control valve with a pilot control valve and having the two flows combine. The current that flows through the emitter circuit is always the largest current because the base current gets added to the collector current. A transistor working this way is acting as an amplifier.

ELECTRONICS AND MULTIPLEXING SYSTEMS

The transistor (solid state relay) has the following advantages over a mechanical relay:

- The solid-state relay can switch faster.
- The solid-state relay is smaller.
- The solid-state relay will not wear out.
- There is no high-voltage coil induction concerns.

Transistors are different from mechanical relays. A mechanical relay acts as a switch that turns current completely on or completely off. A transistor can vary the current flow according to the amount of current that is flowing through the base.

TRANSISTOR TERMINOLOGY There are many terms that describe the characteristics of a specific transistor. For example, transistor current gain describes how much greater the collector circuit current is than the base circuit current. If a transistor has a gain of 100 and a base current of 10 mA, then the current in the collector circuit is 10 mA multiplied by 100, which equals 1000 mA, or 1 A. If a transistor is used for amplification and if the gain is 100, the collector current will always be 100 times higher than the base current. There is of course a limit to the amount of current flow any transistor can handle without damage. Once the transistor base voltage is high enough that no more current can flow through the collector–emitter section, the transistor is said to be saturated.

Transistors have many other ratings that are similar to those for diodes. Transistors can be rated according to the following conditions:

- How fast the transistor can turn on and off.
- How much heat the transistor can handle.
- How much current leaks through a transistor when the transistor is supposed to be turned off.

DARLINGTON PAIR A Darlington pair consists of two transistors wired together similar to two relays being connected. If one relay was energized and the high current it switched went to turn on a second relay that was able to handle a lot more current, the base of the low-current transistor is turned on and the current flowing out of its emitter then goes on to the base of the second transistor. The second one is rated for a much higher current flow and will be able to send that out of its emitter.

DRIVERS Transistors, like diodes, can be microscopic and number in the thousands inside of a machine's ECMs. It would be very unusual to see a transistor used anywhere else on a machine other than inside an electronic device such as an ECM. If they are used to send high current out of an ECM, they will be called drivers.

A driver is a transistor that converts a small digital signal into a higher-current output that can perform work such as turn on a relay coil, energize a solenoid, or send out a pulse width modulated (PWM) signal.

MOSFET TRANSISTOR Most computers now use MOSFET transistors for switching functions because they are simple and inexpensive to manufacture and don't require resistors. They operate differently than bipolar transistors. When the gate is energized, it creates an electric field that allows a conducting channel to open and let current flow between the source and drain.

There are different terms for the parts of this transistor as they compare to the bipolar transistor as shown here:

BIPOLAR	FET
Base	Gate
Emitter	Source
Collector	Drain

TECH TIP

Electrostatic discharge (ESD) is something to avoid when working with electronics. This is the charge that builds up on your skin and discharges to give you a lift when you are just about to touch a door knob. It can reach several thousand volts.

One of the few downfalls of electronic components is their sensitivity to voltage spikes and how easily they can be damaged. In some cases, it will only take an application of as little as 30 volts to destroy an electronic device.

All heavy equipment manufacturers recommend that HDETs wear some type of grounding device when working with a machine's wiring as there is a possibility of a static shock damaging an electronic component. This could mean wearing a grounding bracelet that goes around the wrist of the HDET and attaches to a good ground point on the machine or standing on an antistatic mat if performing bench work on anything with sensitive electronics. This may seem like an inconvenience but is a lot less of a trouble and expense than replacing an ECM that has failed because of a static shock.

Most sensitive electronic devices will come in special antistatic packaging and shouldn't be unpacked until they are about to be installed.

MOSFET

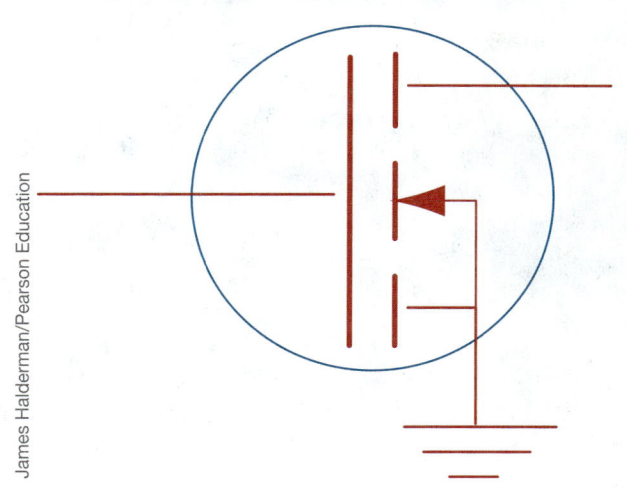

FIGURE 9–18 A MOSFET transistor and its symbol.

MOSFET transistors are highly sensitive to static electricity and can be easily damaged if exposed to high-voltage spikes or excessive current. ● **SEE FIGURE 9–18** for a MOSFET transistor and its symbol.

TRANSISTOR TESTING Since a transistor is much like a diode, it will be tested much the same way. With your digital multimeter (DMM) on the diode setting, you will be arranging the two leads on the three terminals and checking for opens or shorts. ● **SEE FIGURE 9–19** for how a DMM is used to test a transistor.

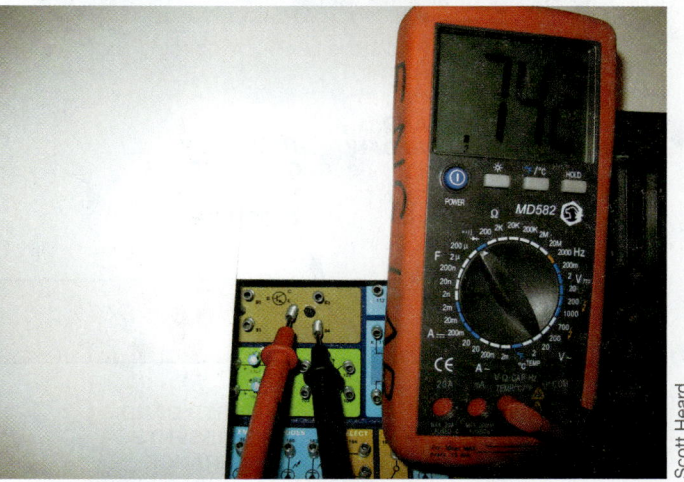

FIGURE 9–19 How to use a DMM to test a transistor.

TECH TIP

At some point while reading this chapter, you will likely ask yourself: Why do I need to know all this electronic stuff when I can just read fault codes and go from there? HDETs typically have a curious nature of wanting to know how things work. However, if you just can't see the point, try to make the comparison of knowing what makes an engine work. As long as a machine's engine is working fine, who cares what's inside it? But when it doesn't work right, you need to know what's inside it to figure out why it isn't working right. Even though most of what you learn about with electronics can't be seen, heard, or felt by you, you still should understand what makes them work. Sooner or later you'll run into some part of an electronic system that isn't doing what it should, and instead of wondering what happens inside that box with all the wires coming out, you should be comfortable in knowing there's no magic going on.

RESISTORS

Electrical resistance, to varying degrees, is inherent in all components and conductors that electrical current flows through. If the resistance is kept below a designed level, it won't create unwanted problems and the circuit that it's in will work fine.

Some circuits need a specific amount of resistance to operate properly, and in some cases extra resistance is added intentionally in the form of a resistor. Resistors are used in many different kinds of circuits for a variety of different functions.

Resistors are used to limit current flow and create a voltage drop in a circuit. A resistor can be directly compared to an orifice in a hydraulic system because an orifice will restrict oil flow and create a pressure drop in a hydraulic circuit.

ELECTRONICS AND MULTIPLEXING SYSTEMS **269**

FIGURE 9–20 Resistors used for a circuit board.

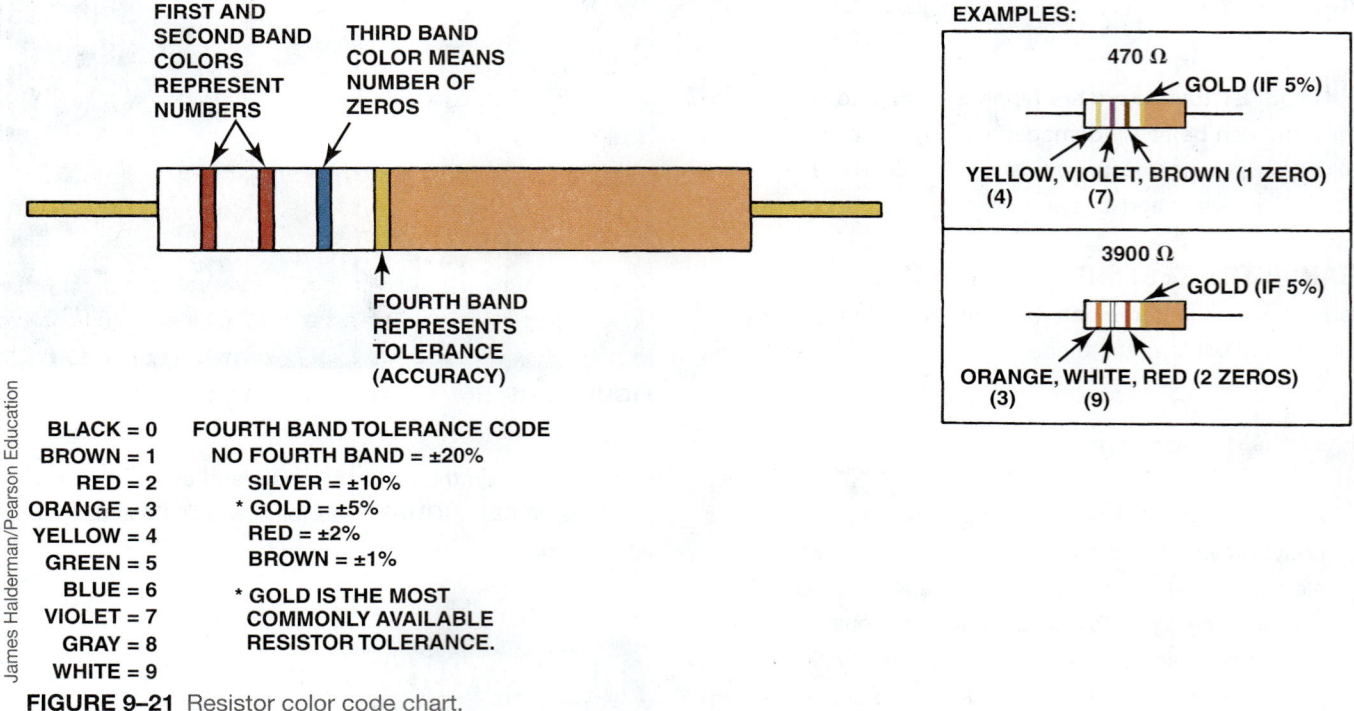

FIRST AND SECOND BAND COLORS REPRESENT NUMBERS

THIRD BAND COLOR MEANS NUMBER OF ZEROS

FOURTH BAND REPRESENTS TOLERANCE (ACCURACY)

BLACK = 0
BROWN = 1
RED = 2
ORANGE = 3
YELLOW = 4
GREEN = 5
BLUE = 6
VIOLET = 7
GRAY = 8
WHITE = 9

FOURTH BAND TOLERANCE CODE
NO FOURTH BAND = ±20%
SILVER = ±10%
* GOLD = ±5%
RED = ±2%
BROWN = ±1%

* GOLD IS THE MOST COMMONLY AVAILABLE RESISTOR TOLERANCE.

EXAMPLES:

470 Ω
GOLD (IF 5%)
YELLOW, VIOLET, BROWN (1 ZERO)
(4) (7)

3900 Ω
GOLD (IF 5%)
ORANGE, WHITE, RED (2 ZEROS)
(3) (9)

FIGURE 9–21 Resistor color code chart.

On a smaller scale, resistors are used in electronic devices to also reduce current flow and create voltage drops. ● **SEE FIGURE 9–20** for resistors used on a circuit board.

The type of simple resistor that is shown in Figure 9–20 has two terminals and a colored insulating shell and is filled with a mostly carbon material that provides the actual resistance. Like a hydraulic orifice, it will also get warm, but this heat will simply be dissipated by its shell to the surrounding air. Resistors used for electronics will be rated for 2 watts or less. This means if they have too much current flowing through them and exceed 2 watts, they will fail. These small resistors have a color code scheme that indicates the amount of resistance created by them. ● **SEE FIGURE 9–21** for the color code chart of resistors.

These types of carbon resistors are considered to be fixed value. This means their resistance value doesn't change. An example of a common use for a fixed resistor is to control voltage spikes near coils when the coil is de-energized. ● **SEE FIGURE 9–22** for how a resistor is used to de-spike a coil.

Other types of resistors are stepped resistors and are able to have their value change. Stepped resistors can be used to control the speed of a motor. They could be part of a switching circuit that will not add any resistance for high

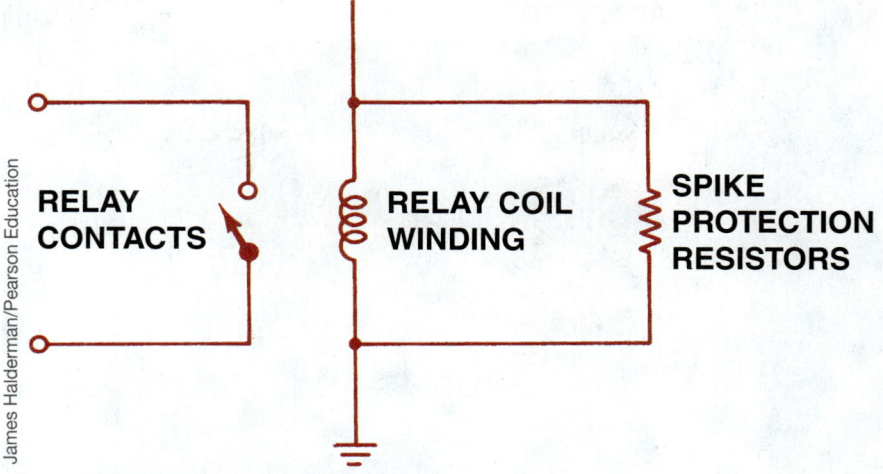

FIGURE 9–22 A resistor that is used to de-spike a coil.

Variable Resistor

FIGURE 9–23 A stepped resistor.

speed but will add more resistance in the circuit to the motor for lower speeds. This resistance is added in defined amounts and in series so that for each lower speed the switch's added resistance accumulates. ● **SEE FIGURE 9–23** for a stepped resistor.

CAPACITORS

A capacitor is a simple electrical device that can absorb an electrical charge, store it, and then release it back into the circuit it is part of. It can store a potential charge on the surface of its conducting material when the capacitor is connected to a higher-voltage source. Capacitors are also sometimes called condensers.

You can make a direct comparison between a capacitor in an electrical system to an accumulator in a hydraulic system. Accumulators can also store hydraulic energy for later use and act as a hydraulic shock absorber to protect against pressure spikes. Capacitors are used in timer circuits, for voltage spike suppression, and for electrical circuit noise limiting. There are different styles and sizes of capacitors, but they all work on the same principle.

A common type of capacitor is constructed from two conducting sheets or plates, one positive and the other negative, that are separated by an insulating material called a dielectric. The insulating material can be paper, plastic, mica, glass, ceramic, air, or vacuum. The better the dielectric material is at insulating, the longer the capacitor will keep its charge. These three sheets are layered and rolled into a cylinder that serves as a protective cover. One of the sheets will have a wire lead attached to it and the other sheet is connected to the housing. The housing then becomes the ground for the capacitor.

The plates can be an aluminum foil or a thin film of metal applied to opposite sides of a solid dielectric. ● **SEE FIGURE 9–24** for an illustration of a capacitor.

The operation of a capacitor is relatively simple. It is connected to a live circuit, and if the circuit's power source, a battery for instance, has a higher charge than the capacitor, then one of the plates will get an excess of electrons from the negative terminal and the other plate will end up with a deficit of electrons.

ELECTRONICS AND MULTIPLEXING SYSTEMS 271

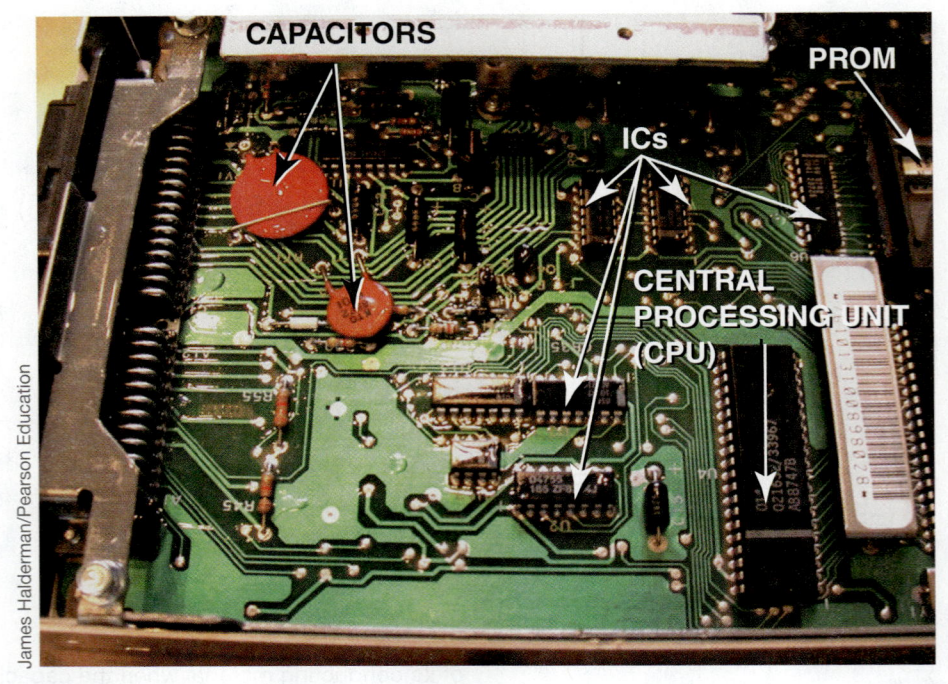

FIGURE 9–24 An illustration of a capacitor.

This is the charging cycle, and it happens until the plates are at the same voltage or charge differential as the power source. If the capacitor was removed from the circuit or the circuit opened, the charge will remain in the capacitor. The length of time it remains charged depends on how good the dielectric is at insulating. There is an electrostatic field between the two plates when the capacitor is charged that keeps the charge potential. The size of the surface area of the plates will determine how much charge potential the capacitor can hold.

If the capacitor is connected back into a circuit that has a lower potential charge than it does, the capacitor will discharge its stored charge back into the circuit. If the voltage potential changes in a circuit, the capacitor will either charge or discharge and this will smooth out any voltage fluctuations in the circuit. This characteristic will make it work well as a voltage spike suppressor. If allowed, capacitors will quickly discharge the majority of their charge and then slowly dissipate the rest.

When a capacitor is used as a timer, it will work with a resistor in series to store a charge and slowly send the charge back into the circuit after the normal voltage supply has been shut off. The unit of measure for a capacitor is the farad. Most capacitors used in electronic circuits are much less than 1 farad and are quite often as small as 1 microfarad or even 1 picofarad. Ultracapacitors can store enough energy to crank a diesel engine over and some are used for energy storage with hybrid electric machine drives.

Progress Check

1. In electronic terms, a driver is:
   a. the base path for a transistor
   b. the N material of a diode
   c. the transistor input to an ECM
   d. another term for a transistor output from an ECM
2. A semiconductor material will have:
   a. four electrons in its outer shell
   b. eight electrons in its outer shell
   c. one electron in its outer shell
   d. no electrons in its outer shell
3. The term *solid state* refers to:
   a. the state of an electrical switch that is closed
   b. an electronic device with no moving parts
   c. a type of transistor that is made for sub-zero temperatures
   d. the type of capacitor with no moving parts
4. The part of the transistor that controls that transistor's output is called:
   a. the N material
   b. the P material
   c. the B material
   d. the base
5. LEDs are based on this type of material:
   a. silicon
   b. germanium
   c. disodium
   d. gallium

DIGITAL VERSUS ANALOG

When trying to understand how a computer and its electronic sub-components function, you need to understand that computers rely on a digital language to transfer information. A comparison of digital to analog with something that you are familiar with should help.

Measurement of time can be by a clock with hands that are continuously moving past stationary numbers. This could be considered an analog measurement because the hands that are tracking time are constantly moving. A clock that displays time with only numbers and changes in defined steps is considered to be digital. When a digital clock tracks time, it is like taking snapshots of time and sequencing them in order.

Electrically speaking, if you were to watch an AC sine wave on an oscilloscope, you would see an analog waveform that is constantly changing in value from positive to negative because the scope is tracking the moving voltage value onto the screen. ● SEE FIGURE 9–25 for how an AC sine wave looks like on an oscilloscope.

A digital comparison would be to watch a DC voltage that is being switched on and off. If this happened at regular intervals, you would see a square or rectangular wave depending on the on/off intervals. A switching DC voltage is a good example of what a digital signal looks like. Let's assume a 12V DC battery is supplying voltage to a turn signal light in your car that has a flasher unit turning the light on and off. If an oscilloscope is measuring this voltage, there are two distinct values of 12V DC and 0V DC. This would be seen as a square wave or a distinct on/off signal or a high and low signal. ● SEE FIGURE 9–26 for how a DC square wave looks like on an oscilloscope.

This is an illustration of the type of signal a computer likes to see. The on/off signal compares to a 0 and a 1 or high and low. Computers are designed to work with digital values and the only form of information a computer processor recognizes is either a 0 or a 1. All information that goes to a computer to be processed (read, sorted, stored) must be in this form. This is called the binary number system.

FIGURE 9–25 An AC sine wave on an oscilloscope.

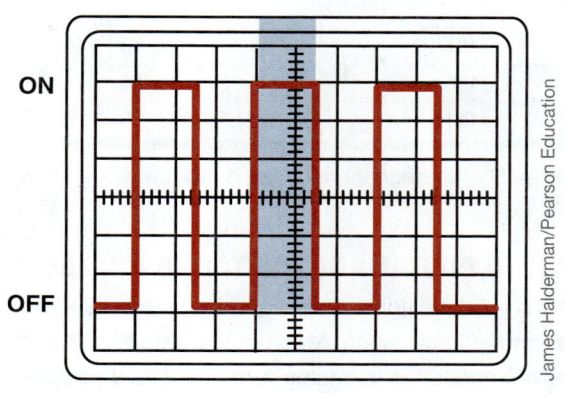

FIGURE 9–26 A DC square wave on an oscilloscope.

Unlike the decimal number system that we are familiar with that uses digits from 0 to 9 to describe any numerical value, the binary number system only uses 0s and 1s. All alpha and numeric information can be transmitted in binary form. ● SEE FIGURE 9–27 for how the decimal numbers 1–10 are written in binary form.

In computer terms, each 0 or 1 is a bit and a series of 8 bits is a byte. One thousand bytes is a kilobyte (KB), 1 million bytes is a megabyte (MB), and 1 billion bytes is a gigabyte (GB). These terms should be familiar to you if you have looked at any computer or electronic equipment specifications because their memory storage capacity is measured in GB.

WEB ACTIVITY

There are many sources of information that can easily be found on the Internet that explains how the binary number system works. If you are curious to know how information is converted into the binary system, search "binary system," do a little research, and learn how it works.

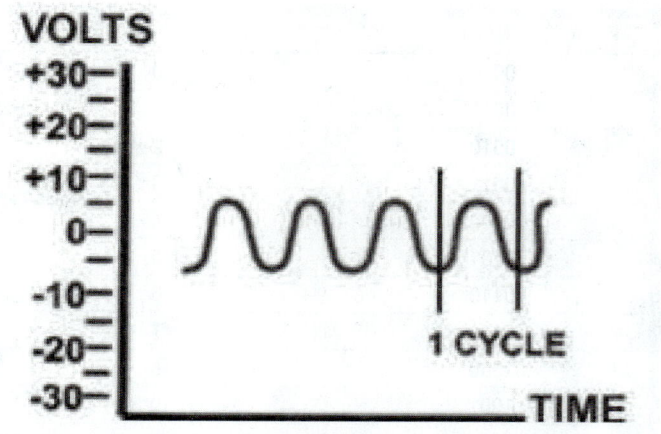

SIGNAL CONDITIONING

Information coming in and going out of a computer will likely need to be modified. The term *signal conditioning* means that raw data is changed into computer friendly information. This could include signal amplification, filtering, and digital conversion.

BINARY	DECIMAL
0000	0
0001	1
0010	2
0011	3
0100	4
0101	5
0110	6
0111	7
1000	8
1001	9
1010	10
1011	11
1100	12
1101	13
1110	14
1111	15

FIGURE 9–27 The decimal numbers 1–10 are written in binary form.

ANALOG TO DIGITAL AND DIGITAL TO ANALOG CONVERTERS

As previously mentioned, computers know only digital forms of information. Much of the data that is sent to a machine's computer is in analog form such as a variable voltage signal from a sensor. Before the computer can process the info, it must be converted into digital form. The variable analog information is broken down into distinct time slices called samples and then each slice is quantified into values that are in digital forms of combinations of 0s and 1s. The smaller the samples or the thinner the slices, the more accurate the conversion is.

For example, a sensor that is measuring a machine's liquid temperature may be sending this temperature value as an analog voltage value. The computer needs to understand this signal so that it can send a signal to a gauge to give the operator an accurate indication of what the temperature is. Let's say the sensor signal voltage could range from 0 to 5V DC and the computer is able to read 1 byte at a time. If 1 byte has 256 different possibilities, then each sample could be divided into 256 equal voltage values or each division of each sample could be a step of .02V DC. This would allow the computer to accurately interpret the input voltage from the sensor and convert this into an accurate temperature. The computer will then send this data to a gauge.

If the gauge is controlled by a variable voltage, then the reverse situation applies. When the computer sends out digital information to an analog device, the computer will need to convert the temperature's digital equivalent to a variable voltage. This can be done with diodes and transistors.

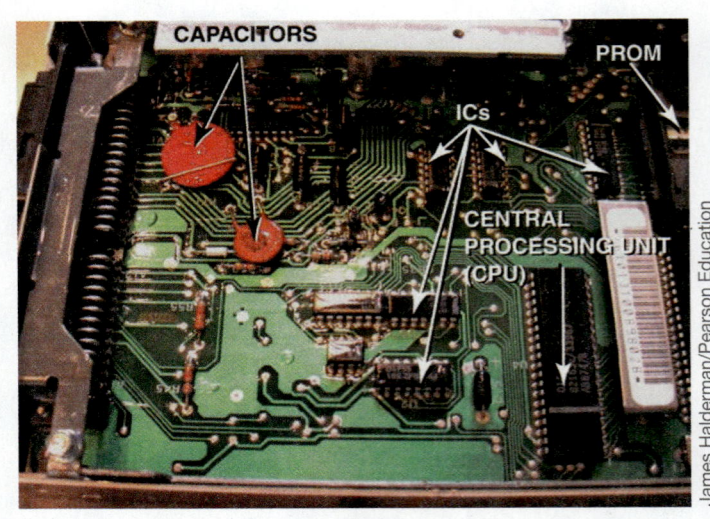

FIGURE 9–28 An integrated circuit that is part of an ECM.

Computers don't necessarily have to see absolute 0s and 1s to process information. For example, the analog to digital converter can be set up to convert any low range of values, say from 0 to 1.5V DC as the low value that gets converted to a 0, while anything over 1.5V DC is converted to a 1. These parameters are totally flexible and set in the computer's operating memory.

INTEGRATED CIRCUITS

An integrated circuit could be any combination of resistors, diodes, and transistors that are combined and connected (integrated) onto one common platform like a printed circuit board. An integrated circuit will be much more reliable and faster than having separate components in separate units. With the amazing size reduction of electronics in the last few years, the size of machine ECMs has been dramatically reduced while their performance and capability has increased tremendously. ● **SEE FIGURE 9–28** for an integrated circuit that is part of an ECM.

CENTRAL PROCESSING UNITS

Sometimes called the microprocessor, the central processing unit (CPU) is where all the data processing takes place in an ECM.

CPUs will work in a timed rhythm that is called its clock speed that is measured in Hz or cycles per second. Many

> **TECH TIP**
>
> **ECU or ECM or?**
>
> Different manufacturers will have different names for the electronic control devices they use. Whatever the term, they all more or less perform the same function of processing information from inputs and sending out command signals to outputs. The following are common terms:
>
> ECU—electronic control unit
> ECM—electronic control module
> ECU—engine control unit
> PCM—power train control module
> CCU—chassis control unit
>
> From here on, the term *ECM* will be used when talking about a machine's electronic control unit.

CPUs operate at speeds of 32 MHz or 32,000,000 cycles per second. Another computer speed measurement is baud rate. This is how many bits per second a computer can process. This can be 500,000 or higher.

It's mind boggling to try and think how fast a computer can process vast amounts of information. As CPUs have got smaller, they are able to do more, perform faster, and function with less power requirements.

Processors work according to how the programs stored in the ECM's memory tell them to. The programs will use computer logic to make the processor logically process information.

LOGIC GATES

The digital information that flows through an ECM is processed mainly with the use of logic gates. The way the gates are programmed to process information is what makes it seem as though a computer can "think" logically.

A logic gate is a set of transistors that are set up to act like switches and give an output signal based on one or two input signals. These signals are in digital form, meaning they are a 1 or 0 (on or off). Logic gates are the foundation of the information processing function of a computer. The different ways they work are summarized in a truth table. Truth tables are charts that show how the inputs (0 or 1) to a logic gate result in an output of a 1 or 0. Logic gates will likely work together with other gates and electronic components to make a computer "think" and make decisions based on incoming or stored information. Remember that there could be thousands of these gates able to fit on the space inside this "0" and some computer chips will contain over a billion gates!

● **SEE FIGURE 9–29** for the symbols of the basic logic gates and their truth tables.

The following are five basic types of logic gates:

AND gate: Both of the two inputs for this gate have to be a 1 for the output to be a 1.

OR gate: Either one or the other of these gates' two inputs can be a 1 to give the output a 1.

NOT gate: This gate is sometimes called an inverter gate and has just one input that is always reversed for its output.

NAND gate: This is similar to the AND gate except its output will be opposite to that of the AND gate.

NOR gate: Either one or both of the inputs is 1 but its output is 0.

Computer logic is based on instructions it receives from a programmed memory. An example of how logic is used to control something is when a machine has a starter interlock system that prevents the engine from cranking unless certain conditions are met. These conditions are monitored by sensors and their output will go to a set of logic gates in the CPU. Imagine an operator wanting to start a skid steer, and unless the door is closed AND the seat switch is closed AND the seat bar is down AND the correct password is entered, the machine's ECM will not allow the starter to be energized. This process can be done very quickly with a set of logic gates in the CPU.

COMPUTER MEMORY

A computer needs memory to be able to operate and process data it receives. Computer memory is millions of "cells" that are set to 0 or 1. Examples of types of memory cells include a flip-flop, a charged capacitor, magnetic particles aligned, or a single spot burned onto a disc.

There are two types of memory in a computer: volatile and nonvolatile. Nonvolatile or permanent memory will be retained no matter what. Here are three examples of permanent memory:

ROM (read-only memory): The computer can only read the contents of this memory and can't change it (write

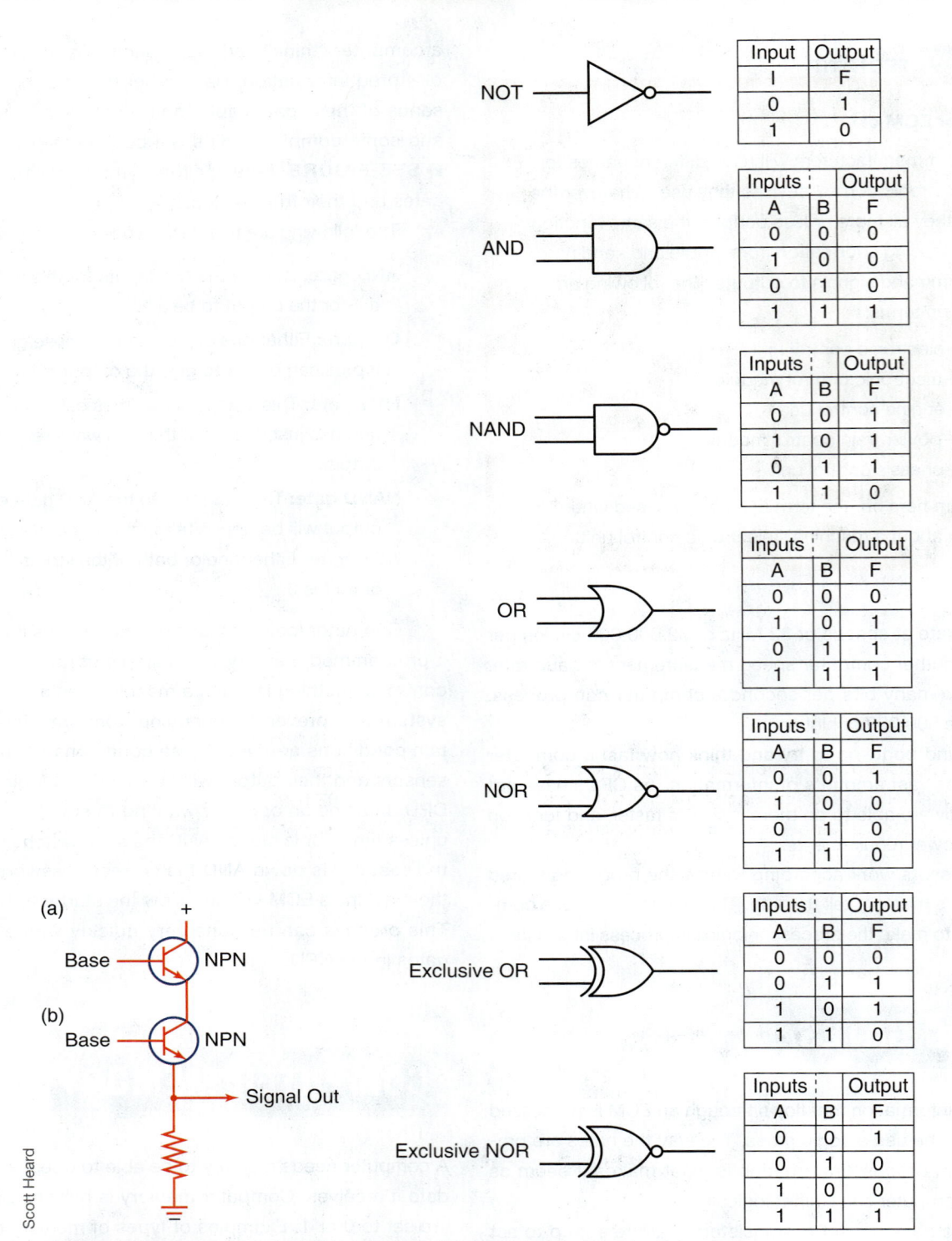

FIGURE 9–29 The symbols for the basic logic gates and their truth tables.

NOT	Input	Output
	I	F
	0	1
	1	0

AND	Inputs A	Inputs B	Output F
	0	0	0
	1	0	0
	0	1	0
	1	1	1

NAND	Inputs A	Inputs B	Output F
	0	0	1
	1	0	1
	0	1	1
	1	1	0

OR	Inputs A	Inputs B	Output F
	0	0	0
	1	0	1
	0	1	1
	1	1	1

NOR	Inputs A	Inputs B	Output F
	0	0	1
	1	0	0
	0	1	0
	1	1	0

Exclusive OR	Inputs A	Inputs B	Output F
	0	0	0
	0	1	1
	1	0	1
	1	1	0

Exclusive NOR	Inputs A	Inputs B	Output F
	0	0	1
	0	1	0
	1	0	0
	1	1	1

to it). Think of this memory as the permanent memory. Part of the ROM is built into the computer's main processor and is usually programmed when the processor is manufactured.

PROM (programmable ROM): Another type of ROM called PROM is outside the main processor in a separate integrated circuit chip and can be altered or programmed after the computer is first used.

EEPROM (electronically erasable PROM): This type of memory can be changed by external programming but is retained when power is removed. Heavy equipment machine ECMs will have their personality software stored in EEPROM. For example, if a common engine that is used in more than one type of machine has its ECM programmed at the factory, the specific data for that machine will be stored in the EEPROM. This would include things such as fuel injection timing, variable turbo control information, and coolant temperature limits. To change this memory, a process called flashing can be performed anytime with the right software and password.

Re-flashing an ECM's memory will wipe out its EEPROM that contains all the instructions for the ECM to be able to control the system it is intended to. The memory is then reloaded with new instructions. This is typically done by an HDET who connects a laptop to the machine and then downloads a file of the latest software update from the manufacturer. The new file is then installed after the old software is removed.

Volatile memory is lost when the power is removed from the computer.

RAM (random-access memory): It is a type of temporary memory that the computer can both read from and write to anytime. This memory is usually erased when it is powered down but it can be wired to direct battery power or have its own small battery. This is also called keep alive memory (KAM). Writing to memory means that information is being stored in the memory.

There is also another type of RAM called nonvolatile RAM that will retain stored data even when power is shut off.

> **TECH TIP**
>
> If you are ever requested to re-flash a machine's memory, make sure that you have the right software, the right hardware, a good Internet connection, enough time to do it, and the factory's helpline in case something goes wrong.
>
> I was once witness to a re-flash procedure that went wrong during a dealer training course and left a machine dead for several days before the problem got sorted out.

ECMs

Most machines produced today will have at least one ECM installed on it and possibly more than a dozen. The ECM will gather information from a variety of inputs. The ECM's processor will process the information, compare it to stored data and send out signals directly, or condition them before they go to actuators or displays or store information. ECMs can be located directly on major components such as engines and transmissions or centrally located together somewhere like in a machine cab. They can control an entire machine, a major component such as an engine or a transmission, or small subcomponents such as a turbocharger or an engine's selective catalytic reduction (SCR) system. Some ECMs are even integrated into joysticks and display panels.

Because electronic devices create heat when they operate, this heat needs to be taken away, or it will build up and destroy components in the ECM. Some engine ECMs are cooled with diesel fuel or engine coolant that circulates through a heat exchanger inside the ECM while others rely on air movement over heat exchanger fins to dissipate excessive heat.

A typical wiring connection to an ECM is to have two 70 pin connectors securely sealed and fastened to its housing.
● **SEE FIGURE 9–30** for a machine's ECM mounted to an engine.

Sometimes an extra ground wire is attached to the ECM housing. Part of the machine's service information will identify all the connector pins for the ECM.

ECMs are usually sealed at the factory to prevent dust and moisture contamination as well as tampering. They contain a wide assortment of electronic devices such as

FIGURE 9–30 A machine ECM mounted to an engine.

ELECTRONICS AND MULTIPLEXING SYSTEMS

FIGURE 9–31 The inside of a machine's ECM.

integrated circuits, CPUs, memory chips, drivers, A/D converters, and circuit boards. These are part of the hardware of the electronic system. ● **FIGURE 9–31** shows the inside of a machine's ECM.

An HDET can communicate with and diagnose an ECM a couple of different ways. Sometimes a machine's display in its cab will provide access to the machine's ECM or an HDET can connect a laptop to the machine that is loaded with service software. This will be discussed more later on.

ECMs are very rarely the root cause of a machine problem, but if they do need to be replaced, it is possible to sometimes buy an exchange unit. The newly installed ECM will almost always need to have software installed based on the machine's serial number. This is usually done only by the machine manufacturer's dealer and a dealer password is required.

WHAT MAKES ELECTRONIC COMPONENTS FAIL?

Many of the conditions that electronics don't like are what must be tolerated on a daily basis when they are part of a machine's electrical system. The reason electronic components fail is because of one or more of the following:

- dust exposure
- extreme temperatures
- moisture
- vibrations
- shock loads
- corrosive atmosphere
- possible voltage spikes

As time has progressed, electronics manufacturers have produced very reliable components that will withstand the rigors of being part of a heavy equipment machine.

ECMs themselves very rarely fail unless they are exposed to moisture or excessive voltage spikes. The first problem is quite often caused by someone pressure-washing too close to wiring connectors. The second problem is likely caused by someone touching an exposed terminal or wire and causing a voltage spike by an ESD (electrostatic discharge), by an incorrect battery boost, or by an improper welding machine connection on a machine.

HDETs should always ground themselves to the machine's frame to prevent ESD, always follow proper battery boost procedures, and if welding on a machine check to see what the manufacturer recommends. Some manufacturers will say to put the ground cable as close as possible to the weld

TECH TIP

One machine I worked on had all the machine's ECMs (seven of them) located in a partially sealed compartment outside of the cab. This machine seemed to have gremlins in the electronic system that would cause some very strange and intermittent problems. Some days that were excessively damp saw the machine's horn blowing and wipers working randomly on their own with a multitude of fault codes logged. Many of these problems were eventually traced back to being caused by where the ECMs were located. The compartment was open just enough that it could collect dust and condense moisture on the bottom of its lid. This combination of wet corrosive limestone mud from the machine's working conditions caused havoc at the ECM connectors and a series of modifications to the housing eventually helped to minimize the problem.

Another machine had major electronic problems and was out of service for two days while two experienced HDETs tried to trace the problem down. The machine's operator had requested that the cab get pressure-washed one weekend. This may have been okay but the machine's main ECM was located on the floor of the cab and its connectors were caked in wet mud over a couple of nights. When the operator returned Monday, the problem started with the machine not starting and got worse from there. Eventually, two ECMs had to be replaced to get the machine back to work.

(always good practice), disconnect the batteries, and might even suggest disconnecting all machine ECMs from the wiring harness.

THE ELECTRONIC INFORMATION CYCLE

A machine's electrical/electronic system is simple in structure when its main component arrangements are looked at. There are one or more ECMs that get information sent to them from a variety of input sources. ● SEE FIGURE 9–32 for an ECM and its inputs and outputs.

The information then gets processed, analyzed, and compared to stored data. The ECM then sends out signals to one or more outputs. The output signals will be in response to the inputs received or based on instructions from the ECM's programmed memory.

The type of signal an ECM sends and receives is varied and depends on the type of component it comes from or goes to. The different types of signals can be:

Variable voltage: This is an analog type of signal, meaning it is a changeable voltage and can change voltage level any time or stay the same. Many sensors will send a variable voltage signal to the ECM. This type of signal will have to be converted to digital form for it to be processed.

Variable resistance: Some sensors will vary their resistance based on what they are exposed to (pressure or temperature mostly).

On or off, positive or ground, high or low: This type of signal is ideal for an ECM because it can easily be translated into 0s and 1s. Hall-effect sensors and switches create this type of signal.

Variable frequency: If a speed sensor generates an AC voltage, its frequency will change as the speed of the component being measured changes. Because it varies and is not an on/off signal, it will be considered an analog signal.

PWM: These signals are in digital format. They are a pulsed voltage that turns on and off from 0 to 5V DC. The frequency of the pulse stays the same but the length of on time can change within each cycle. This change in length of on time is called the duty cycle.

● SEE FIGURE 9–33 for how a PWM signal varies.

Sensors can input PWM signals and solenoids can receive PWM signals. If a coil receives a PWM signal, it will be like getting an average voltage that is below the maximum voltage based on the amount of on time or the duty cycle.

One manufacturer that uses a lot of PWM signaling will pulse the signal at either 500 Hz or 5000 Hz. PWM signals can come from position, level, pressure, and temperature sensors or can be an ECM output to a proportional solenoid.

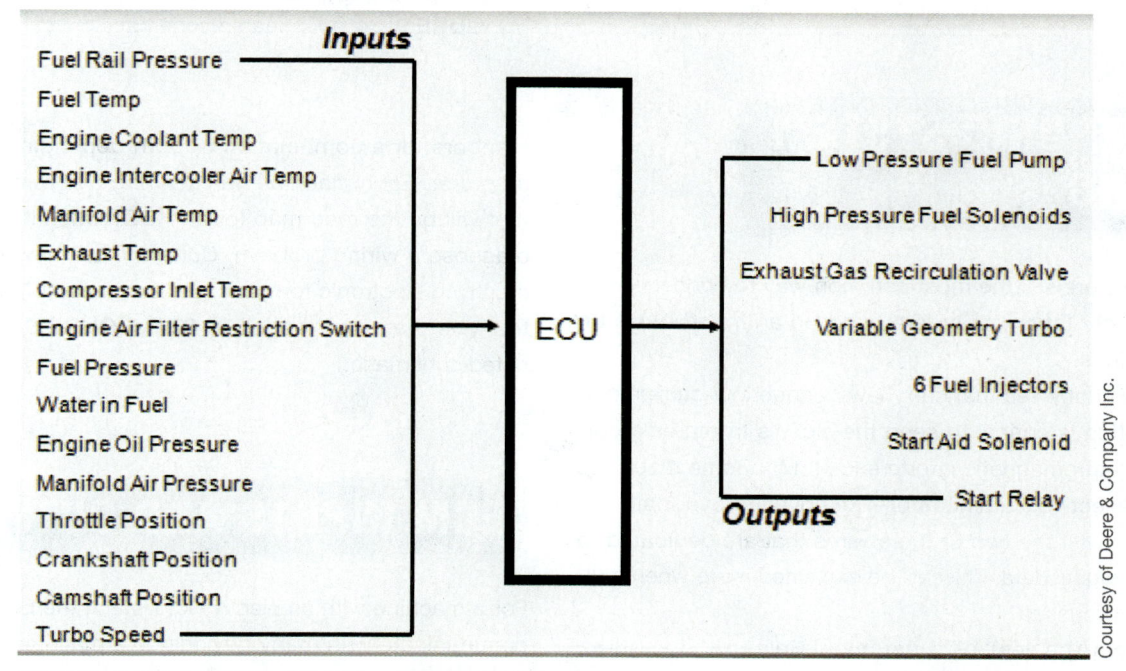

FIGURE 9–32 An ECM with its inputs and outputs.

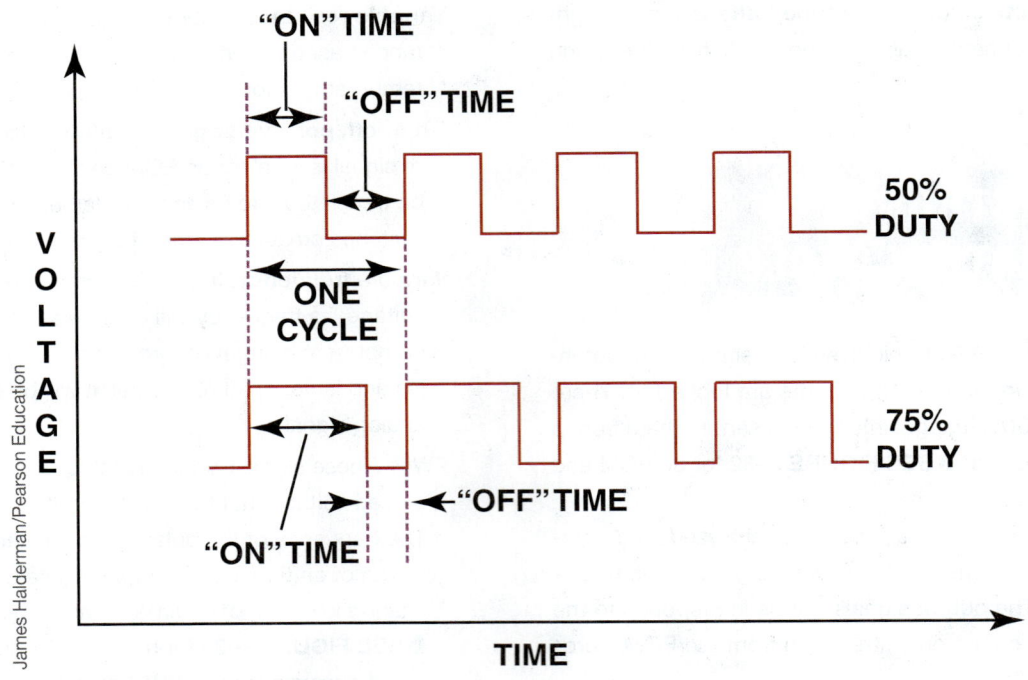

FIGURE 9–33 How a PWM signal varies.

Reference voltage: Many sensors need a voltage sent to them that will then get changed and sent back to the ECM as a signal voltage. The voltage sent out from the ECM is called a reference voltage. A common value for a reference voltage is 5V DC, but some manufacturers use an 8V DC reference voltage.

These different types of signals are part of the electronic information cycle.

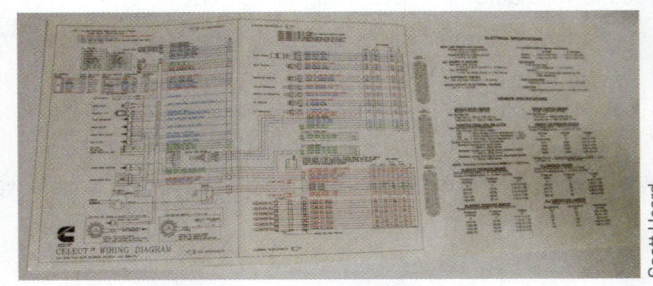

FIGURE 9–34 A laminated schematic.

ELECTRONIC SYSTEM WIRING

Copper wires are still the most common way to connect electronic devices. Fiber optics are used on a limited basis on some machines.

Shielding of wires that carry low-voltage/low-current digital information is critical to keep the signals from being corrupted by electromagnetic interference (EMI). Some machines will use a reflective aluminum foil–like covering over data link wires or will twist the two or three wires that are dedicated to transferring digital data. This will be explained more when multiplexing is discussed.

Just like the rest of a machine's electrical system, the electronic system wiring will be identified with colors, numbers, or a combination of both depending on the manufacturer. Each machine has a wiring diagram or schematic that will be the road map for an HDET when it comes time to diagnose a wiring problem. Current machine schematics are mostly in electronic form but can be printed off and laminated to make them last longer. ● **SEE FIGURE 9–34** for a laminated schematic.

ECM INPUTS AND OUTPUTS

For a machine with an electronic system, the ECM(s) needs to communicate with many devices on a machine. The operator will need to tell an ECM a wide variety of commands related

280 CHAPTER 9

TECH TIP

You may have heard of the term *fly by wire* as it relates to modern aircraft. This means the pilot's controls are only electronic commands that go to one or more computers. The computers then make decisions that are based on the information they have stored in their memory and then control several actuators to make the plane do what the pilot wants it to. Does it sound familiar?

Many new machines are also fly by wire. This means that all the operator controls are devices that create electronic messages that are sent to an ECM. This includes everything from turning on lights, to starting the engine, to propelling the machine, to controlling its hydraulic system. The ECM then decides what is that best way to make the machine do what the operator wants it to. Some old school operators say this takes away a lot of the feel that an operator used to get when moving joysticks and drivetrain controls. Many newer generation operators meanwhile are used to the feel of fly by wire joysticks from playing video games and won't know any different.

Many machines with electronic joysticks have the ability to change the sensitivity of joysticks to allow for different operator preferences. Many machine joysticks will have much more than the standard two (X and Y) axis outputs (ahead, back, left, right). Some can have three axes (twist clockwise and counterclockwise) along with six or more switches for outputs. ● SEE FIGURE 9–35 for an electronic joystick with multiple outputs.

FIGURE 9–35 An electronic joystick with multiple outputs.

---

to starting and operating the machine (inputs), there are many sensors on a machine that need to send information to the ECM (inputs), and there are many devices that are controlled by an ECM (outputs). A modern fully electronic machine could have over a hundred different inputs and outputs to and from the machine's ECMs.

ECM INPUTS Inputs to an ECM can come from many different devices and can be mainly grouped into switches and sensors. The signals sent to the ECM as inputs may need to be conditioned before being processed or, if they are in a digital state, can go directly to the processor.

SWITCH INPUTS Switches provide input signals to the ECM that are a perfect digital-type signal because a switch can be in only one of two states—on or off. An ECM can read this as a 0 for off and a 1 for on.

Switches can turn on and off either power or ground. The contacts of a switch have two contact states: either open or closed.

- When switch contacts are open, no signal is provided to the corresponding input of the ECM.

- When switch contacts are closed, either a ground signal or a voltage signal is passed through the switch contacts to the corresponding input of the ECM.

SWITCH TYPES

ON/OFF TOGGLE, ROTARY, PUSH BUTTON, ROCKER. These are simple switches that are mechanically actuated and will have a set of contacts, each with a terminal that connects the switch to the wiring harness and then to the ECM.

These switches can send either a power or a ground signal back to the ECM. To test an on/off switch, the best way is to check its output with a voltmeter when it is installed in the machine if possible. Otherwise, a continuity test could be done with the switch removed. ● SEE FIGURE 9–36 for how to test a simple switch.

Some examples of circuits where simple on/off switches that are used as ECM inputs are lights, parking brake, transmission neutralizer disable, and safety interlock enable.

MULTI-POSITION SWITCHES. Another type of switch used for an input is the multi-position switch. It can have a power or

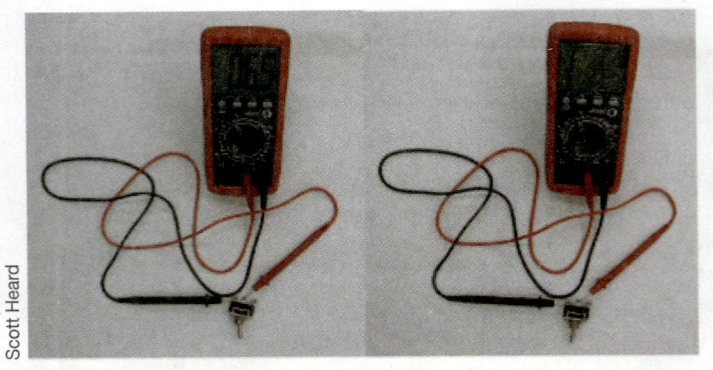

FIGURE 9–36 How to test a simple switch.

ground fed to it and depending on its position will send a signal out of one of its other contact terminals to the ECM.

Some inputs that are switches but have more than on/off positions are transmission shift lever switch (R, N, 1–8 speeds), box hoist control (hoist, lower, hold, float), retarder switch (off, on, automatic), throttle speed dial (1–10), heated seat/mirror (off, low, high), and key switch (off, ignition, accessory, glow plug). Although these switches have more than two positions, they are still considered digital inputs because the switch position is a defined location and the ECM can recognize the switch's input as a combination of 0s and 1s. ● SEE FIGURE 9–37 for a multi-position switch.

REED SWITCH. A reed switch is a magnetically actuated switch that has a set of movable contacts in a sealed capsule. The contacts are connected to external leads, and when the movable contacts are exposed to another part of the switch that is magnetized, the contacts close. They can be used for level switches to warn of a fluid level that is too low.
● SEE FIGURE 9–38 for how a reed switch works.

HALL-EFFECT SWITCHES. A Hall-effect switch uses magnetic fields to create an on/off signal that is sent to the ECM. The Hall effect refers to the principle that when a magnetic field is applied perpendicularly to a current-carrying conductor, a difference in potential appears on opposite sides of the conductor through which the electric current is flowing.

Based on this principle, Hall-effect switches detect changes in the magnetic field and provide a switched output voltage. They provide contactless switching and, consequently, can be deployed in robust packages that protect against environmental elements.

Because they provide increased reliability in harsh environments, Hall-effect sensors are used for on/off switch detection for doors, seat bars, and key ignition switches.
● SEE FIGURE 9–39 for a Hall-effect switch.

FIGURE 9–37 A multi-position switch.

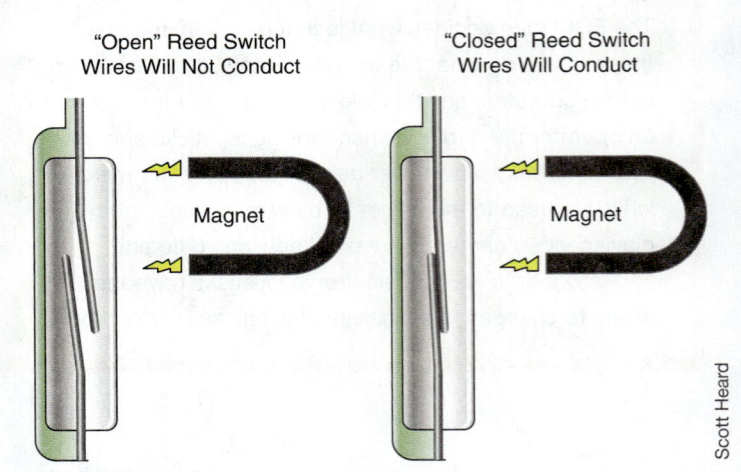

FIGURE 9–38 How a reed switch works.

FIGURE 9–39 A Hall-effect switch.

OTHER SWITCH USES Some examples of other switches used to send on/off signals to the ECM are pressure switches (filter bypass, travel alarm, engine oil pressure, and backup alarm), operator control panel switches, auxiliary

> **TECH TIP**
>
> Operator keypads and sealed switch modules provide a method for the operator to input commands directly into an ECM. Keypads allow an operator to enter their identity or other numerical values such as truck numbers or material density. They are also a way for an HDET to enter a Service Mode to check live values, fault codes, and change parameters.
>
> Sealed switch modules are a combination of touchpad switches and an ECM. The switches control many different machine components such as lights, wipers, ride control, and seat height. These switches are a direct input to the ECM in the module.
> - **SEE FIGURE 9–40** for a sealed switch module.
>
> **FIGURE 9–40** A sealed switch module.

is located near a spinning shaft. The shaft or component will have a set of teeth that pass by the sensor, and when this happens, an AC voltage is generated in the sensor's coil of wire. The changing reluctance of the teeth and gap between the teeth passing by will change the magnetic field of the sensor's permanent magnet and this will induce a voltage in coil. This happens because the reluctance is different between the teeth and the air in between the teeth.

As the shaft spins faster, the sensor will create more voltage and at a faster frequency. As the AC voltage sine wave oscillates faster, its cycles per second or Hertz increase and this will be seen by the ECM as a speed increase.

Sometimes these sensors are used for position sensing as well. If the toothed wheel is just for providing a signal and not for transferring torque, there will be one tooth missing to indicate a specific spot in the component's rotation. This will show up at the computer as an irregularity in the signal. A typical use for this is a camshaft or crankshaft speed and position sensor.

This type of sensor will always have two wires coming out of it, and it can easily be tested while operating with a multimeter that is set on AC voltage providing of course that it's safe to do so.

The sensor could also be tested for its resistance with an ohmmeter and with the sensor isolated from the circuit.
- **SEE FIGURE 9–41** for a variable reluctance sensor.

Some other common names for this type of sensor are pulse generator, AC generator, and speed sensor.

AC generators can be tested by using a voltmeter set on AC that is connected to the sensor's terminals and measuring its output as the tone wheel is passed by.

VARIABLE RESISTANCE SENSORS. Some examples of this type of sensors are as follows.

Thermistor temperature sensors. A thermistor-type temperature sensor will change its resistance as the temperature of whatever it is sensing changes. This change of resistance will likely be lower as the temperature goes higher. This type of thermistor is called a negative temperature coefficient (NTC) sensor. There are positive temperature coefficient (PTC) sensors that work the opposite way but they aren't as common.

Thermistors can be one, two or three wire units. A one wire thermistor will be create a variable series resistance to ground signal through its body. A two wire unit will have the second wire connected to ground instead of using the thermistors body. A three wire unit changes a reference voltage sent to it into a variable voltage based on the changing temperature the thermistor is exposed to. ● **SEE FIGURE 9–42** for a thermistor.

implement switch, and high coolant temperature and limit switches that change states when moved mechanically by a component.

SENSORS Sensors are needed to send a signal to an ECM that is other than an on/off type of signal. This could be any type of changing value such as position, temperature, speed, fluid level, or pressure.

Many types of sensors will have the electronic components needed to send a usable signal to the ECM directly inside the sensor housing.

VARIABLE RELUCTANCE SENSOR. This is a type of speed sensor that will create a variable frequency and voltage signal. It is a permanent magnet that is wrapped with a coil of wire and

ELECTRONICS AND MULTIPLEXING SYSTEMS

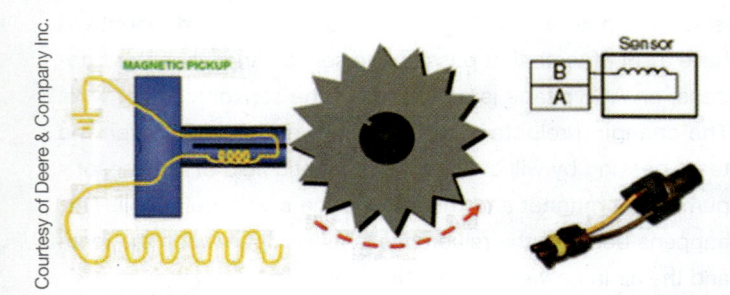

FIGURE 9–41 A variable reluctance sensor.

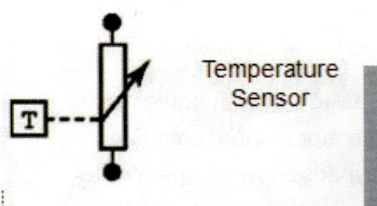

FIGURE 9–42 A thermistor.

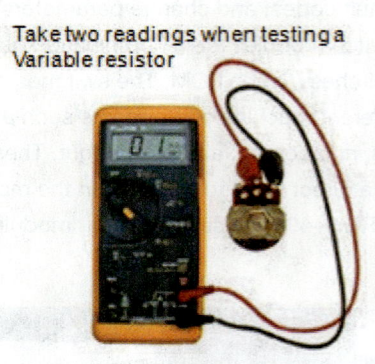

FIGURE 9–43 Testing a potentiometer.

Thermistors can be used to sense the temperature of air, fuel, oil, coolant, and exhaust fluid.

Testing a thermistor is done by measuring the resistance it is creating and comparing the finding to a chart that gives the sensor's resistance value for certain temperatures.

Potentiometers. Another use of a variable resistor is a device that will turn mechanical motion into a variable voltage signal. One type is called a potentiometer or pot. A potentiometer can be thought of as a voltage divider. It is a three-terminal resistor assembly that has one of its terminal's fed positive voltage that is usually less than full system voltage (typically 5 volts). This terminal feeds a resistor wire that is circular in shape and its other end goes to a second terminal that goes to ground. The third terminal will be connected to a rotating arm that has a wiper or movable contact at one end of it. The output of the third terminal will go to an ECM. As the wiper contact moves around the resistor wire, it will send a variable voltage out the third terminal that is based on the wiper position. The amount of voltage depends on how much resistance the supply voltage has to go through, which is determined by the position of the wiper. The computer can now interpret the amount of voltage sent to it as a position of the wiper. A typical use for a pot is for an accelerator position sensor. This is commonly called a throttle position sensor (TPS) and could be connected to a foot pedal or hand-controlled dial. A foot-actuated TPS will have a spring to return the foot pedal and give the operator a feel of resistance. This signal will be an input to the engine ECM to tell the computer how far the operator is pushing his foot down on the governor pedal. ECMs will be explained later in this chapter. A potentiometer is thought of as a voltage divider because part of the supply voltage goes to the signal wire and the rest goes to ground.

A potentiometer can be tested by measuring the resistance between the movable contact and the ground terminal as the movable wiper is rotated. ● **SEE FIGURE 9–43** for how to test a potentiometer. Alternatively a potentiometer can be tested by supplying power and ground to the sensor and measuring the signal voltage while moving the wiper.

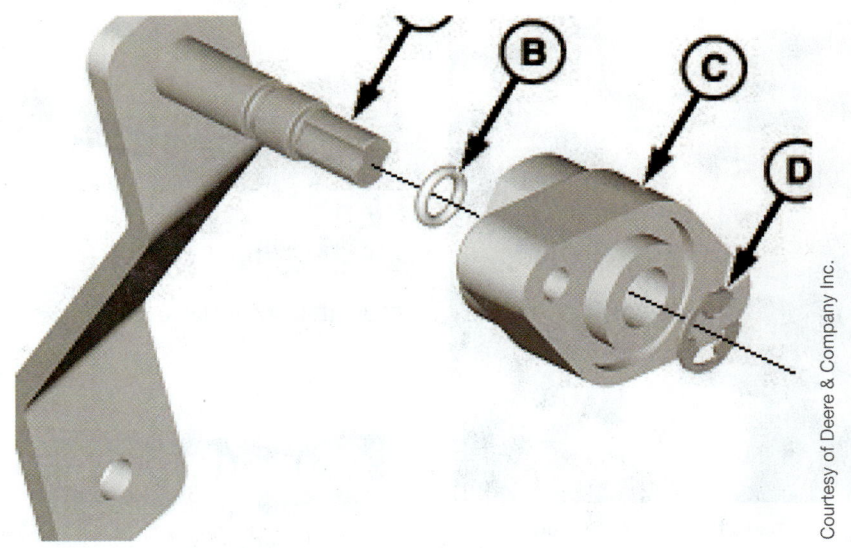

FIGURE 9–44 Position sensor that sends a PWM signal.

PWM TEMPERATURE, PRESSURE, AND LEVEL SENSORS
Some manufacturers prefer sensors that send a PWM signal to the ECM. This type of signal is ideal for an ECM because it is an on/off signal that varies in the time it is on or off.

These sensors will use a variety of technologies to measure the values that they are monitoring and then with the use of built-in electronics will convert the value into a PWM signal and send it to the ECM. They will be three-wire sensors with one being a reference voltage, one a ground, and the third a signal voltage. PWM sensors can be tested by using a multimeter with PMW capabilities and measuring the output signal. It will typically range between 10% and 90%.

PWM POSITION SENSOR
Position sensors will translate the position of a movable mechanical component on a machine into an electrical signal. There are different types of position sensors and one of them sends a PWM signal to the ECM based on the position of the component it is attached to. An example might be the throttle pedal of a machine or the boom of a loader. This sensor contains the electronics to convert a rotational or linear position into a PWM signal. One end of the component's travel would be a 10% duty cycle signal while the other end would be a 90% signal. ● **SEE FIGURE 9–44** for a position sensor that sends a PWM signal.

GROUND SPEED RADAR SENSOR
Some wheel loaders will have an optional automatic differential lock feature. The differential lock is solenoid controlled and one of the inputs that the power train ECM needs to decide if the lock should be engaged is the machine's actual ground speed. Travel speed measured with a radar speed sensor that uses radar technology and sends a digital signal to the ECM.

FIGURE 9–45 A variable capacitance type of sensor.

PIEZOELECTRIC PRESSURE SENSORS
Some semiconductor crystal materials are unique structures that can generate voltage when they are squeezed and move when they have voltage applied to them. These materials are termed piezoelectric. Piezoelectric sensors can be used as pressure sensors because they change their voltage output based on how hard they get squeezed. Internal electronics in these sensors will increase this voltage into a usable signal that is sent to the ECM.

VARIABLE CAPACITANCE PRESSURE SENSORS
This type of sensor is used to measure pressures such as air, fuel, or oil. This will be a three-wire sensor that has a reference voltage, a ground wire, and a signal wire. The medium that is to be measured will be exposed to a ceramic disc that has a flat wound coil connected to it. The pressure moves the coil in relation to a stationary steel disc. This action will change the capacitance of the sensor. With the help of some other electronics in the sensor, this varying capacitance will be converted to a varying voltage signal going back to the ECM.
● **SEE FIGURE 9–45** for a variable capacitance type of sensor.

ELECTRONICS AND MULTIPLEXING SYSTEMS   285

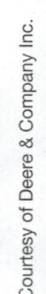

FIGURE 9–46 A thermocouple used to measure exhaust temperature.

FIGURE 9–47 A Hall-effect speed sensor.

THERMOCOUPLE These are sensors that are made from two dissimilar types of metals that will generate a voltage based on how hot they get. They are typically used for sensing exhaust temperature. ● SEE FIGURE 9–46 for a thermocouple used to measure exhaust temperature.

HALL-EFFECT SENSORS These are three-wire sensors that contain a permanent magnet and some electronics to detect relative motion and position. They are commonly used as crankshaft and camshaft speed and position sensors but can be used for other purposes.

The sensor will take a reference voltage and energize a semiconductor material. The output signal is created when a timing disc or tone wheel that is attached to a moving component (e.g., crankshaft) passes by the sensor. The timing disc will have windows that will alter the magnetic field of the permanent magnet in the sensor. This changing magnetic field will then create a change in voltage through the semiconductor. This changing voltage is then converted to a digital signal, which sent out the signal wire to the ECM.

These sensors provide two distinct levels of current depending on whether the magnetic field is applied or not. While current levels vary depending on the sensor manufacturer, they typically range from 5 to 8 mA for low-level (no magnetic field) and 11 to 14 mA for high-level (presence of magnetic field) signals. ● SEE FIGURE 9–47 for a Hall-effect speed sensor.

DATA LINK One other type of input an ECM may need is information from another ECM. If a machine has more than one ECM, they will most likely need to communicate with each other and this is called multiplexing (to be discussed more later). Shared information is in the form of messages that are sent in a series of 1s and 0s.

ECM OUTPUTS

The **outputs** from an ECM start at the CPU as decisions that are based on programming stored in the ECM's memory. They will start out as very small electrical signals in digital form (0s and 1s). These signals will then go to an amplifier to increase their strength or a converter where the digital signal gets changed to an analog one.

The output signals can create an action or the output signals can provide information to another ECM. The ECM will use different types of output signals to control the system components. The types of electrical signal are the following:

- High-side PWM proportional driver outputs
- On/off high-side driver outputs
- Low-side or sinking driver outputs
- Sensor power supply outputs (reference voltage)
- Digital information outputs

The ECM monitors the circuits of the output components. If the ECM determines that an abnormal condition exists in one of the circuits, the ECM will log a diagnostic code for the component.

Some output signals will go directly to actuators that perform mechanical work such as small solenoids or motors. Other signals will turn on relays that control a component that performs mechanical work. Some outputs are simple tasks such as turning on warning lights and buzzers and sending signals to gauges.

An output can be a positive voltage signal and come from a high-side driver or can be a negative signal and come from a low-side driver. Driver is another term for transistor.

Outputs can also be digital information that is sent to one or more other ECMs.

There are output actuators that will provide a feedback signal back to the ECM to let the ECM know that the signal

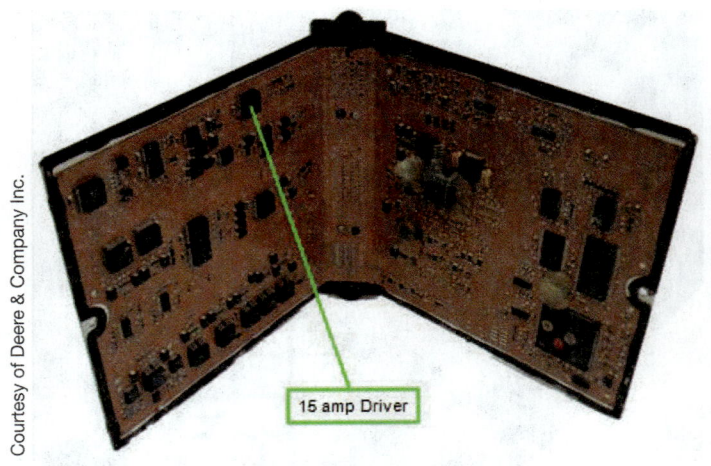

FIGURE 9–48 A driver inside an ECM.

it sent to that actuator has done what it was supposed to do.
- SEE FIGURE 9–48 for a driver inside an ECM.

HIGH-SIDE PWM PROPORTIONAL DRIVER OUTPUTS
An ECM output that is in PWM form on the high side can be a much faster responding and more accurate signal than a variable voltage signal. It is also easy for the ECM to create because it can be created from a digital signal. This digital signal will control a driver (transistor) and pulse it on and off to send out a PWM output in a duty cycle form.

PWM outputs can be sent to a variety of devices, but a common use for one is to accurately modulate oil flow by controlling a proportional solenoid for a hydraulic directional control valve.

An example of where a proportional solenoid is used would be when one is part of an electrohydraulic system that will be used to regulate pilot oil pressure on the end of a main control valve spool (directional control valve [DCV]) that is part of a hydraulic circuit. The ECM can send a PWM signal to the solenoid based on a request from an operator input (electronic joystick). The solenoid will only receive the duty cycle that the ECM determines will create enough oil pressure to move the main control valve far enough to correspond to what the operator wants. ● SEE FIGURE 9–49 for a proportional solenoid that is actuated with a PWM signal.

ON/OFF HIGH-SIDE DRIVER OUTPUTS
These outputs start as 0s and 1s from the ECM. A 1 signal will turn a transistor on and its output will be sent out one of the transistor's terminals. This could directly turn on a small solenoid, or motor. It could also complete the low-current power circuit for a relay that could energize the solenoid for an engine's starter.

LOW-SIDE OR SINKING DRIVER OUTPUTS
Another type of output from an ECM could be one that completes a

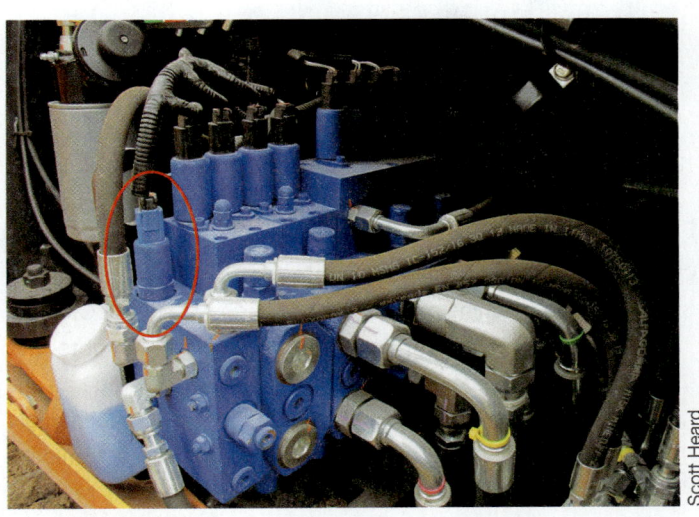

FIGURE 9–49 A proportional solenoid that is actuated with a PWM signal.

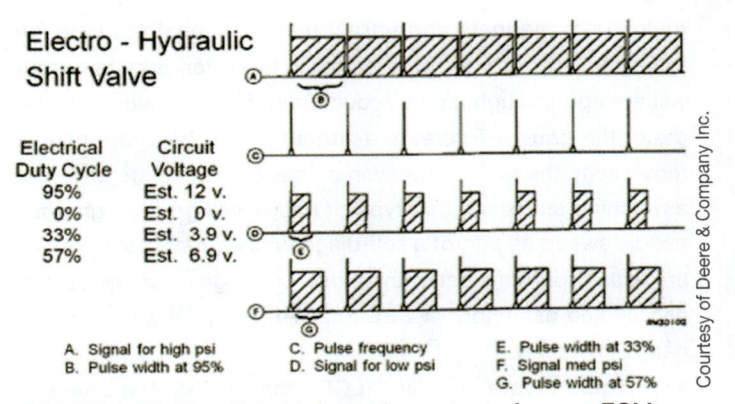

FIGURE 9–50 The digital information outputs from an ECM.

path to ground for a circuit. If a component has a steady power supply, the switching of a ground signal to it will energize the component. A couple of examples could be dash warning lights, solenoids, or relays.

SENSOR POWER SUPPLY OUTPUTS (REFERENCE VOLTAGE)
Many sensors will need a reference voltage sent to them to be able to operate and send a signal back to the ECM. This is most likely a 5V DC signal but could be 8V DC or higher.

DIGITAL INFORMATION OUTPUTS
Most machines today will have more than one ECM, and they will need to share information back and forth. To do this, an ECM will have one or more outputs for digital data that will connect to one or more wires called a data link. These wires are part of a networking system called a multiplexing network. ● SEE FIGURE 9–50 for the digital information outputs from an ECM.

ELECTRONICS AND MULTIPLEXING SYSTEMS 287

CAB DISPLAYS

Operators need to stay informed as to how their machine is performing to help them to stay within its designed limits. Typically, hydraulic, power train, and engine systems could have their fluid level, pressure, speed, and temperatures monitored. To keep the operator informed of the status of these systems, the machine cab can have a wide variety of ECM outputs to do this. The simplest will be LED warning lights that come on when a safe value is exceeded. There could be an audible alarm as well. Other visual monitors that get direct signals from an ECM will be gauges and display panels. Electronically controlled mechanical gauges will likely have their needles moved by a pair of electromagnets or by stepper motors. A stepper motor is a series of electromagnetic coils arranged in a circle that surround a rotor that is a set of two permanent magnets. The ECM will send voltage to the coils that will then make them magnets and attract the permanent magnets on the rotor. The output of the motor is the rotor, and its output will then go through a gear reduction before it rotates the needle of the gauge. To create a seamless and accurate needle movement, the motor rotates in as many as 2000 steps. When a machine starts up, this type of gauge will go through a full needle sweep as part of a self-diagnostic check. These gauges are usually part of a set of three or four gauges that will be part of a sealed assembly. ● SEE FIGURE 9–51 for an electronic gauge assembly.

Liquid crystal displays (LCD) can provide the operator with an alphanumeric way to read live data, warnings, and fault codes.

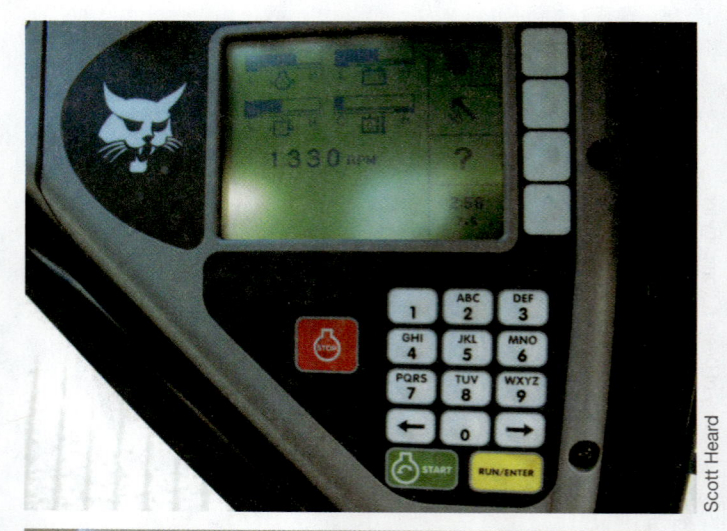

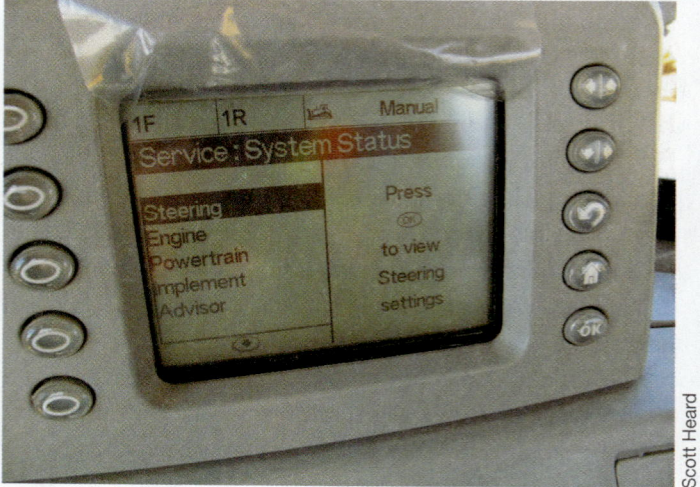

FIGURE 9–52 An LCD display dash.

Some machines will be equipped with a video display screen that can be customized to show one or more virtual gauges or a combination of gauges and alphanumeric messages. Some displays will even have a touch screen option. ● SEE FIGURE 9–52 for an LCD display dash.

Progress Check

6. The language that is only comprised of 0s and 1s is called:
    a. bipolar language
    b. binomial language
    c. binary language
    d. decimal language
7. The combination of 8 bits is called:
    a. an eight
    b. a byte
    c. a kilobyte
    d. a kilobit

FIGURE 9–51 An electronic gauge assembly.

8. Sampling refers to:
   a. taking a sample of electricity from a sensor
   b. dividing an analog signal into equal time divisions
   c. taking a sample of pressure from a hydraulic system
   d. taking a sample of temperature from a hydraulic system
9. The baud rate of a processor is:
   a. its processing speed measured in MHz
   b. its processing speed measured in bits/sec
   c. its processing speed measured in bits/Hz
   d. a measure of how fast it can store information
10. This type of memory is usually erased when a processor is powered down:
    a. EEPROM
    b. ROM
    c. KAM
    d. RAM

SOFTWARE

Most of what has been discussed so far in this chapter has dealt with hardware. The term *hardware* describes physical objects (parts or components) that make up a computer system. If you think about a desktop computer, it would be the major components such as the screen, keyboard, mouse, DVD player/writer, memory chips, hard drive, and all the electronic components that are part of these (diodes, transistors, capacitors). The programmed information that is in digital form and that is stored in a computer's memory that makes all these components work is called software. A machine's electronic system needs software for its electronic system to work. In fact, machines equipped with one or more ECMs would stop working if there was major software problems. Changing a machine's software programming for a machine that has a complex electronic system can dramatically change the operating characteristics of a machine. Typically, through one machine model's production run, that model will have changed or updated its software several times. Throughout a machine's life it is common that it will need to have the ECM software updated. This has traditionally been done by an authorized HDET who has gone to the machine and plugged a laptop into the machine to perform a software update. This is called re-flashing the machine's software and it involves rewriting the EEPROM. With GPS links available on most machines, now this can even be done remotely and automatically.

This updating can correct many operational problems the machine may be having, can change the fuel efficiency and power of the machine, and could also change the self-diagnostic software of the machine.

An electronically controlled machine's software will give it its operating characteristics. For example, a hydraulic excavator that is electronically controlled will have its hydraulic system's speed and power and the engine's power and torque defined by the instructions stored in the machine's ECMs.

For a machine with either a mechanical or a hydrostatic drivetrain that is electronically controlled, the drivetrain's speed and torque outputs will be electronically managed by the ECMs software.

MULTIPLEXING

As most machines made today will have more than one ECM on it, there needs to be a way for multiple ECMs to communicate with each other. The term *multiplexing* refers to the ability for two or more ECMs to share data over a network. The network is called a data bus. This will allow one sensor to be used as an input to more than one ECM and, if necessary one actuator to be controlled by more than one ECM, This will simplify the machines electrical system by reducing the number of wires and connectors used. With a reduction of wires comes a reduction of cost and complexity as well as a more reliable machine because of the reduced chances of wiring problems.

For multiple ECMs to be able to communicate effectively, they must speak the same language. A machine's electronic system needs to transfer a high volume and wide variety of data. Even though all ECMs use binary language, this doesn't mean that complex commands and information sharing can be effective without a higher level of structured communication.
● **SEE FIGURE 9–53** for a multiplex network.

CONTROLLER AREA NETWORK BUS SYSTEM

Most heavy equipment manufacturers use the controller area network (CAN) system as a method of structuring digital communication throughout the machine's electrical/electronic system. This includes all the sensor and actuator information previously mentioned as well as ECM-to-ECM communication.

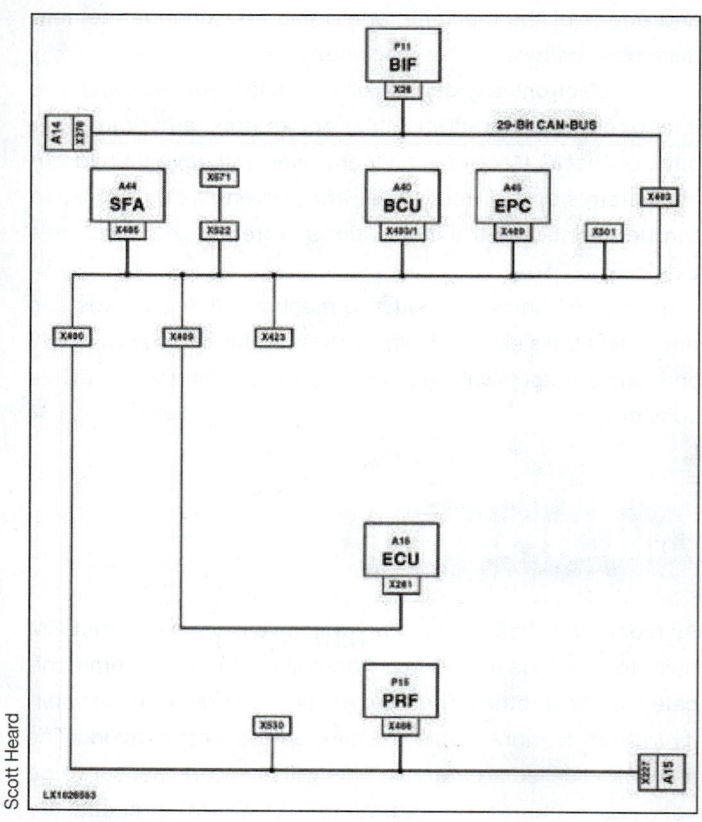

FIGURE 9–53 Multiplex network.

it wouldn't be long before they would become a chaotic mess of gridlock with no one getting anywhere.

CAN protocol is a base structure for how data is transferred over a network. CAN data travels in individual packets or frames that could consist of over 100 bits each with each frame split into seven different fields.

This is part of the standardization that CAN protocol creates. These data frames travel at up to 1,000,000 bits per second along the data bus. The CAN structure or protocol ensures that shared information is identified as to where it came from and what kind of information it is.

All information put on any part of the network can be received by all ECMs and the ECMs will then decide if the information is important to them. This will allow a sensor to be an input to whatever ECM is closest to it regardless of whether the ECM really needs the sensor's information. This will reduce the amount of wires on the machine.

Just about any type of electronic information such as component numbers, serial numbers, engine speed, throttle position, signals to dash displays, and engine shutdown commands can be communicated over the CAN bus system.

● **SEE FIGURE 9–54** for how information is structured into frames to be sent out on the network.

There is also a built-in diagnostic part of each frame that will send an error message back to the frame's source if the frame wasn't received properly.

Here's an example of how a CAN functions on a running machine: An engine speed sensor signal is sent to the engine ECM (this is the only ECM it is connected to). This signal came to the ECM through the sensor signal wires, the harness connector, and then a wire that goes to one of the large (usually

CAN was developed by Robert Bosch as a system to unify the way data bus systems should operate. A data bus system can be thought of as an information highway. The CAN protocol, therefore, can be compared to the laws that dictate how cars and trucks must be operated on streets and highways. If there were no rules to regulate vehicle movement on our roads,

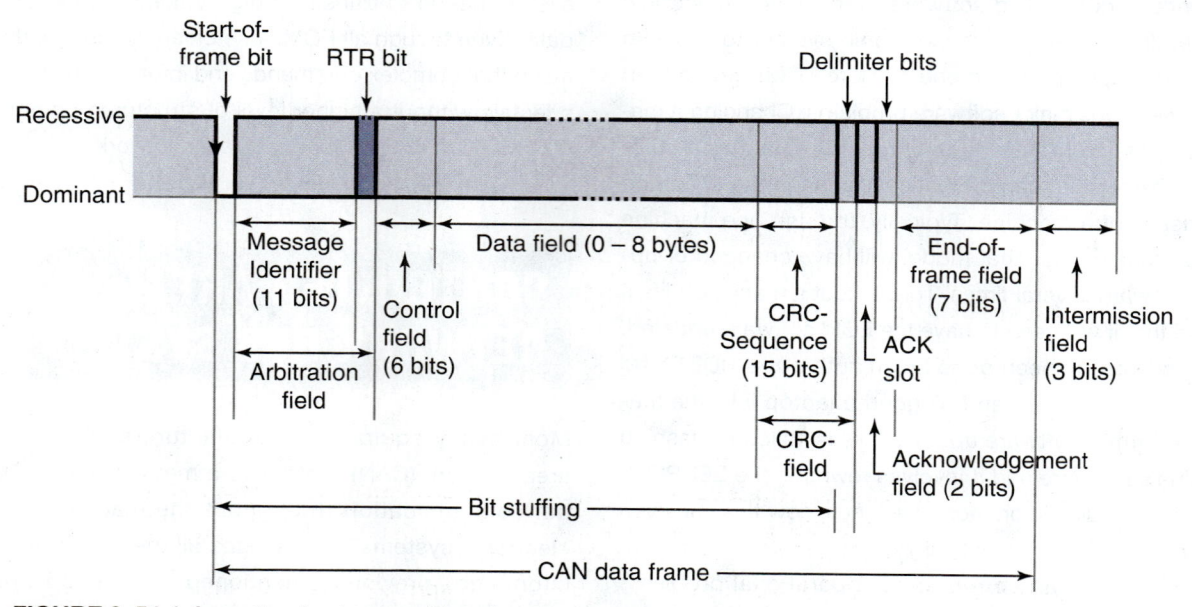

FIGURE 9–54 Information is structured into frames to be sent out on the network.

290 CHAPTER 9

70 pin) connectors on the side of the ECM. The signal now goes into the engine ECM. None of this is considered part of the CAN so far. The engine ECM processes the signal and sends out the signal as a series of samples on the CAN in the proper digitized frame format. If the machine has an ECM controlling the cab display (cab ECM) and there is a tachometer on the dash, it will take that speed sensor's now-digitized signal and convert it to a signal the tachometer can use to display the engine speed. There may be other ECMs that need to know how fast the engine is turning such as a transmission controller or hydraulic pump controller. These ECMs can also take the signal off the network and use it for their own purposes such as changing transmission gear ranges or reducing the pump flow. The engine ECM could also use the signal to increase the amount of fuel injected to try to keep the engine from lugging. If the rpm is too fast, there could be a warning light and buzzer activated by the cab display ECM and a fault code would be stored.

The CAN data link consists of two wires that are twisted at a rate of one twist every centimeter of length. This ensures that any chance of signals being corrupted by EMI (electromagnetic interference) is eliminated. Also, if an induced electrical spike occurred in the twisted pairs, the spike would have an equal and opposite effect on each wire so the signal would not be corrupted. This is because the voltage differential between the two wires compared to the idle voltage (2.5V DC volts) would remain consistent.

The two wires carry mirror images of the digital information that is being sent. The wires are called CAN low and CAN high. If no information is being transferred, then both wires have 2.5V DC on them. This occurs when the bus is idle. As information is sent out over the network, the CAN high wire will switch from 2.5 to 3.75V DC and the CAN low wire will switch from 2.5 to 1.25V DC. This switching occurs at the exact same time. It is the voltage differential between the two wires that makes the digital signals. If during a message transmission both wires are at 2.5V DC, this equates to a "0," and if the wires switch (CAN high to 3.5V DC and CAN low to 1.5V DC), this is seen as a "1." The 0s and 1s combine to form packets of information and can transfer at high speed along the CAN system to exchange a huge volume of data in a short amount of time. ● **SEE FIGURE 9–55** for how the CAN high and CAN low signals work and what live data looks like.

There is a main pair of twisted wires that are the backbone of the network and they will have sub-bus-twisted pairs of wires connected to them that run to each individual ECM.
● **SEE FIGURE 9–56** for the structure of an actual CAN.

These twisted pairs of wires that connect all ECMs on the network will transfer information at very high speeds (up to 1 MB/s or 1 million bits per second) both ways. It is possible to have over 100 ECMs connected to them and the maximum number is based on how fast the information can travel. The maximum distance that each ECM can be from the next one is also regulated. Individual ECMs are called nodes in CAN terms.

Some networks will use a twisted triple group of wires for transferring data. There will still be the two CAN wires and a third wire to assist in shielding the CAN wires.

At the farthest ends of the network backbone, the twisted pairs will have a terminating resistor joining them. This will be a 120 ohm resistor that will not allow the data on the network to bounce back and forth excessively, causing confusion on the network.

As the network wires are still just copper wires, they are susceptible to damage and problems like any other machine wires. You may have to diagnose a CAN problem at some point, so here are some tips.

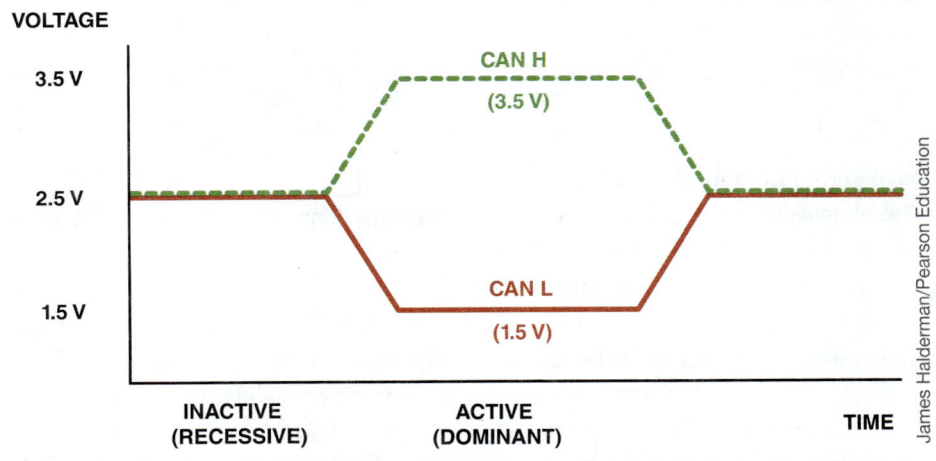

FIGURE 9–55 How the CAN high and CAN low signals work and what live data looks like.

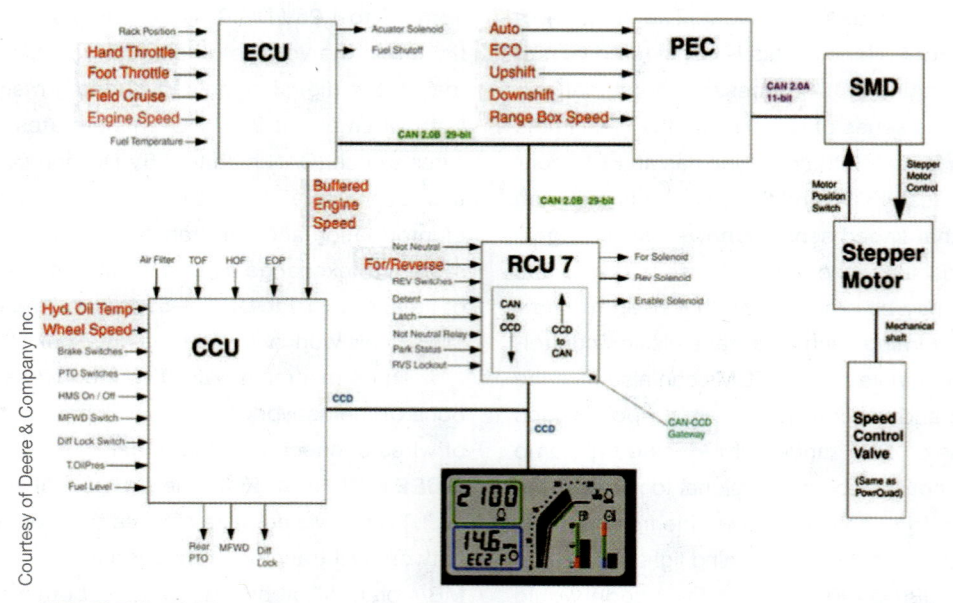

FIGURE 9–56 The structure of an actual CAN.

CAN NETWORK DIAGNOSTICS

As already mentioned, CAN uses two wires to enable communication between the nodes. There will be two points on the circuit where CAN high and CAN low are joined with two 120 ohm buffer resistors. As this means there are two resistors wired in parallel, the total resistance of the two will be around 60 ohms. Knowledge of this will prove extremely useful when diagnosing a fault with a CAN circuit. ● **SEE FIGURE 9–57** for how to test the terminating resistors.

One of the initial measurements that should be carried out is resistance between the CAN high and CAN low terminals where there is easy access. There will usually be easy access at the diagnostic plug. If the main bus lines were still intact, you should expect to have a resistance of around 60 ohms.

If there was an open on either the high or the low of the main bus line, you should find the resistance to be 120 ohms. If this was found to be the case, you would then use basic electrical diagnostic procedures to locate the open after isolating the ECMs.

Another relatively common problem with CAN is for the high and low lines to have shorted together. If this has happened, there would be a total network failure and all of the systems using the network would have operational problems. The process of measuring the resistances at the diagnostic connector plug will help identify the fault. The resistance across the two terminals if the CAN wires are shorted would be around 0 ohms.

Now that you have discovered a short between CAN high and low, you will need to locate the fault. What you should do now is disconnect the individual ECM sub-bus lines, one at

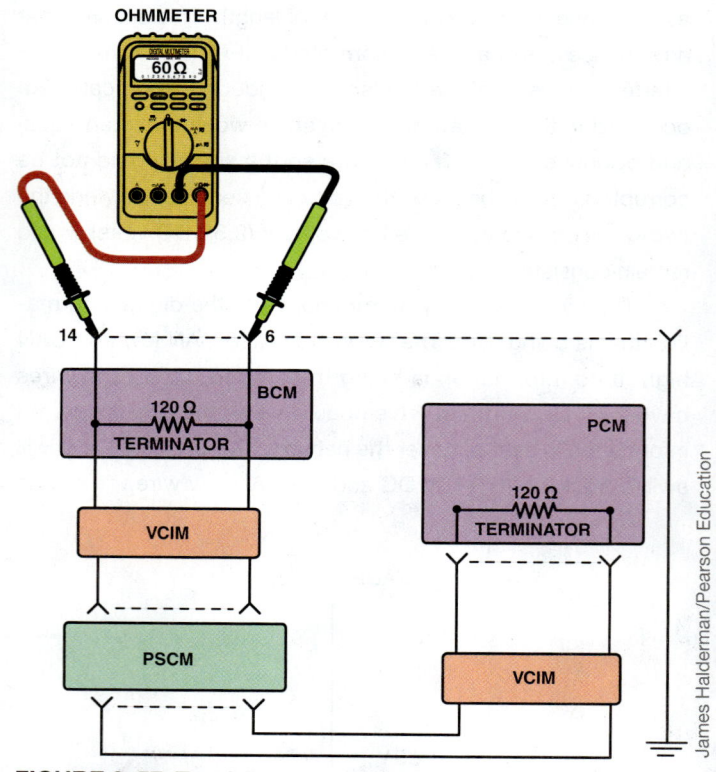

FIGURE 9–57 Terminating resistor testing.

a time, and monitor the resistance value on the screen of the ohmmeter. If, as you disconnect one of the sub-bus lines, the resistance of the circuit goes back up to 60 ohms, you know that the short is in the pair of sub-lines that have just been disconnected. If at no point does the resistance across the two terminals increase to 60 ohms, the short must lie in between

292 CHAPTER 9

the two main bus lines. It is now purely going to be a matter of basic electrical diagnosis to find the location of the specific short.

Diagnosis of an open circuit sub-bus line is slightly more complex. Through purely measuring resistance at the diagnostic plug, there will be no evidence of a fault. In this instance, it would be advisable to use a tester capable of communication with a CAN. The tester is normally plugged into the diagnostic plug, and one of its functions will be a bus check function (the terminology will vary between manufacturers). This function involves the tester carrying out a "roll call." This will involve all of the nodes that are able to hear the roll call replying back to the tester. At the end of the roll call, the tester will list all of the nodes that replied. Through referring to a wiring diagram, you should be able to tell which ECM is not replying. That means the problem must lie with either the sub-line connecting the ECM to the network or the ECM itself. The loom should be inspected and tested first using basic electrical test procedures, and if there is nothing wrong with the loom, then the fault can lie with only the ECM.

DOES EVERY MACHINE USE CAN?

Unfortunately, unlike the automotive industry, there is no mandatory standardized system that all heavy equipment manufacturers use. However, most major manufacturers use a CAN-based SAE information transfer protocol called J1939. This differs slightly from the CAN protocol by defining the baud rate and using only five of the seven fields. There is also a focus on parameter group numbers (PGN) that standardize how the source of information is described.

Some machines will have a combination of types of communication protocols that can interact through interfaces while others will have multiple CANs that are identified as CAN 1 and CAN 2.

TECH TIP

If the wires that make up the CAN bus system require repairing, you must thoroughly follow the manufacturer's instructions or the CAN signals will be corrupted. This will include the proper type of wire, solder, or connectors to use and how many times to twist the wire. If the wire repair is not done properly, there could be many problems related to poor communications over the bus and you could have created more problems.

ELECTRONIC SYSTEMS FOR DIAGNOSTICS

One of the biggest advantages of a machine with an electronic system is that it will help diagnose problems with the machine. This can be a problem with the electronic system itself or any other machine system that is being controlled or monitored.

The engineers who design a machine determine what system parameter values fall within a normal range. These are system values such as voltage, current, temperature, pressure, and fluid levels. The machine's ECMs have these normal values stored in their ROM and EEPROM, and when one of these monitored values falls outside the normal range, a fault code is logged at minimum and further action may take place such as turning on a warning light or an audible alarm. An extreme action would be for the ECM to shut the machine down as a result of a monitored value out of range. Extremely low engine oil pressure could prompt the ECM to shut the engine down, for example.

The use of fault codes can be a good starting point for an HDET, but the same principle of checking the "simple stuff first" still applies.

A lot of newer machines will have service menus that are accessible from the machine's in-cab display. These menus can allow HDETs to check fault codes, view live system values such as pressures and temperatures, and sometimes allow different components to have their operating parameters changed.

FAULT CODES

Fault codes indicate that there is something abnormal with the machine that the electronic monitoring system has detected. Keep in mind, however, if a machine's monitoring system doesn't show any codes, it doesn't necessarily mean the machine is working fine. When starting a troubleshooting procedure, it's a good habit to check for fault codes as part of the initial checks. The fault codes detected could give some clues as to the root cause of the problem you are looking for. There are still lots of things that could be abnormal, but they just aren't detected by the electronic monitoring system so don't lean on fault codes to tell you everything about a malfunctioning machine. Fault codes can also be called diagnostic trouble codes (DTCs), trouble codes, and error codes.

There are two main types of fault codes:

Active: If a fault code is active, this means it is present or happening right now. These types of codes can't be cleared until the problem is corrected.

Logged or stored: This type of code means that the fault has occurred at some point but it isn't active. These codes can be cleared but should be recorded first.

Fault codes can usually be read from a machine's cab display and always from a connected laptop. Along with the code number, the number of times the fault has occurred will be shown as well as the hour meter reading of the first time it occurred. The manufacturer will provide a troubleshooting procedure for every fault code.

Different machine manufacturers will have different code identification systems. Most codes will communicate three pieces of information: which ECM the code came from, what component or wiring circuit is faulty, and what type of fault it is. ● **SEE FIGURE 9–58** for a fault code that is read from a machine's ECM.

One manufacturer identifies the ECMs with a three-digit module identifier (MID) number, the component with a three- or four-digit component identifier (CID) number, and the type of fault with a two-digit failure mode indicator (FMI) number. An example would be MID 087 (chassis ECM), CID 3535 (front axle speed sensor), and FMI 05 (current below normal).

The FMI numbers are fairly standardized for all manufacturers. The following table is an example of how one manufacturer's FMI numbers correspond to different types of faults.

TECH TIP

Clearing fault codes is a common practice that an HDET will perform at various times. This could be done during a troubleshooting procedure or when a machine is being serviced. It's always a good practice to check for fault codes, record them, and then clear them because it is usually a simple process.

Clearing them could be done through the machine's dash display or through a connected laptop.

Remote monitoring of machines via satellite or cell technology can also provide reading and clearing fault code procedures.

FMI NO.	FAILURE DESCRIPTION
00	Data valid but above normal operating range
01	Data valid but below normal operating range
02	Data erratic, intermittent, or incorrect
03	Voltage above normal or shorted high
04	Voltage below normal or shorted low
05	Current below normal or open circuit
06	Current above normal or grounded circuit
07	Mechanical system not responding properly
08	Abnormal frequency, pulse, or period
09	Abnormal update
10	Abnormal rate of change
11	Failure mode not identifiable
12	Bad device or component
13	Out of calibration
14	Special instruction
15	N/A
16	Parameter not available
17	Module not responding
18	Sensor supply fault
19	Condition not met

FIGURE 9–58 Fault code that is read from on a machine's ECM.

Another manufacturer uses the J1939 standard for fault codes. It is a two-part code with the first part being the two- to four-digit suspect parameter number (SPN), which will identify the faulty component, and the second part is the two-digit FMI number.

FIGURE 9–59 The J1939 diagnostic connector port of a machine.

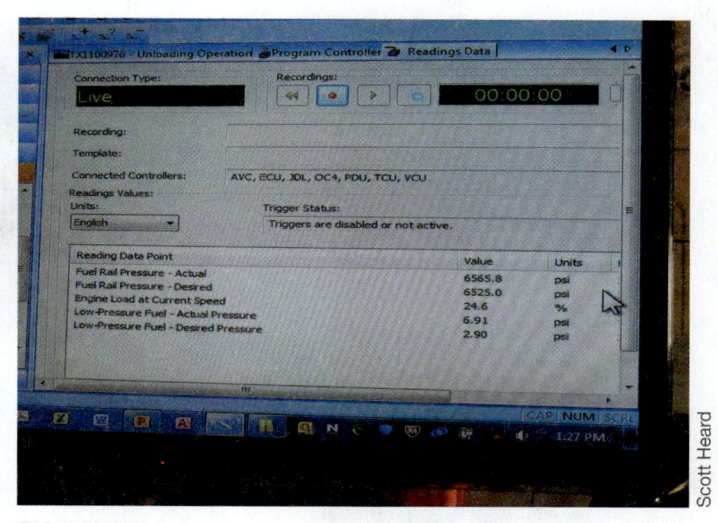

FIGURE 9–60 Screenshot of a live data graph from a John Deere backhoe.

ELECTRONIC SERVICE TOOLS

If more in-depth diagnostic procedures need to be performed, then the HDET will connect an electronic service tool (EST), usually a laptop computer, to the machine. There will be a diagnostic connector port on the machine and a manufacturer-specific interface is usually needed to allow communication between the machine and the EST. These could be a proprietary connector or a SAE J1939–type connector. ● **SEE FIGURE 9–59** for the J1939 diagnostic connector port of a machine.

Many of today's interface devices will provide wireless Bluetooth communication to the EST. With the proper software and sometimes a password, the HDET can perform many diagnostic procedures, view live data, download stored data, change operating parameters, perform calibrations, and disable or enable different machine functions. There will likely soon be, if there isn't already, smartphone applications to allow an HDET to perform monitoring and diagnostics with their smartphones.

LIVE DATA

When diagnosing a machine operational problem, it will be helpful to read live system information such as temperatures, pressures, and speeds. For machines with an electronic monitoring system and a display in the cab, this will be easy to do through a service menu. A connected laptop will also be able to view and record system data. With a laptop, there could be a graphing function that can turn the data into a graph with the capability of graphing multiple values with different colors for easy identification. For example, if you were diagnosing an overheating torque converter problem, you may want to graph torque converter temperature, engine coolant temperature, wheel speed, transmission output shaft speed, and engine speed. This could be all viewed on one screen. ● **SEE FIGURE 9–60** for a screenshot of a live data graph from a John Deere backhoe.

SNAPSHOTS

Another valuable diagnostic tool is to be able to take snapshots of data. This could be set up through a laptop to start with a trigger value of one machine data value that is chosen by the HDET to start the snapshot. Then a set of values for a chosen period of time before and after the trigger event starts is recorded. If the same example of an overheating torque converter problem is used, the trigger could be when the torque converter outlet temperature reaches 230°F. The same set of values could be recorded and saved in the machine's ECM to be viewed later. This data snapshot could also be converted into a file and exported to a computer.

WIRELESS MACHINE COMMUNICATION

Many machines are now being equipped with GPS communications links. This enables many of the previously mentioned diagnostic and data recording/viewing functions to be done

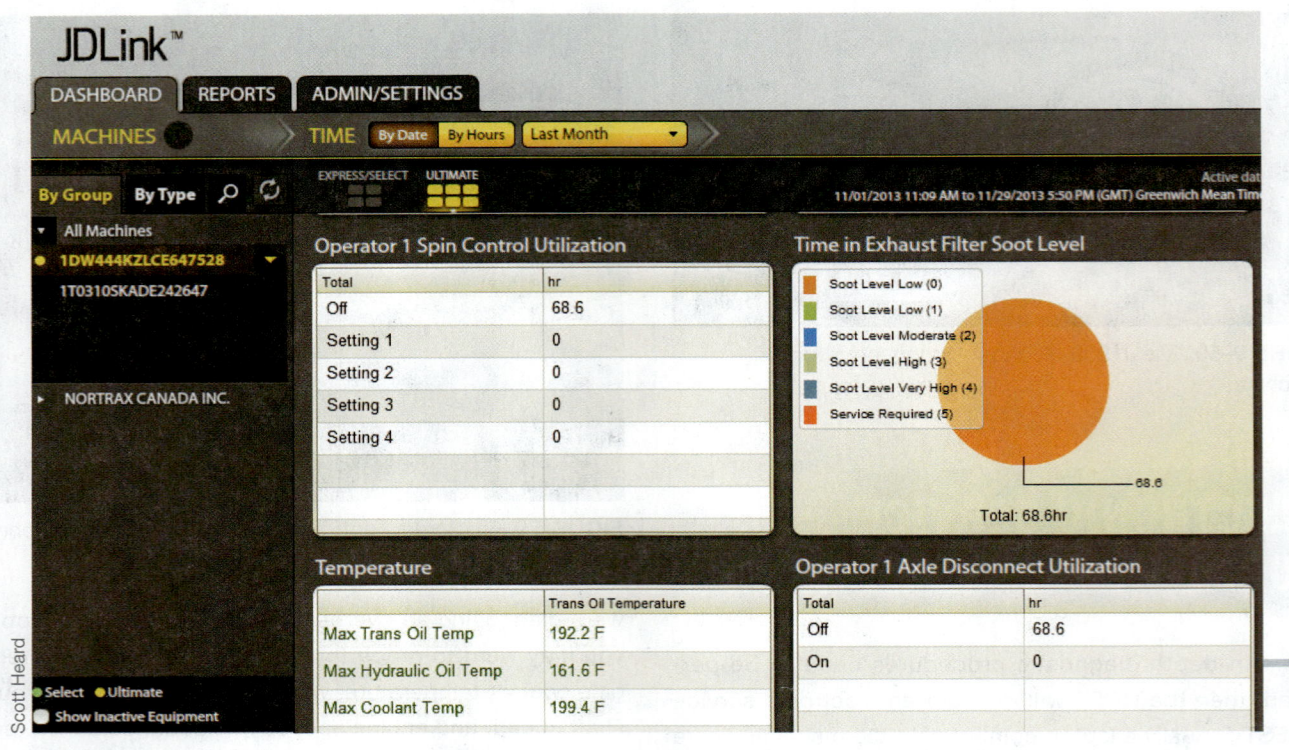

FIGURE 9–61 JD Link.

from a remote location. ● **SEE FIGURE 9–61** for a screenshot of John Deere's JD Link system.

CALIBRATING

Calibrating procedures can be performed from the machine display or from a connected ESC. A calibrating procedure matches an ECM input from either an operator input or a machine sensor to a physical dimension or speed. This will serve as a reference point or a zero point for the ECM to provide proper control of the machine component. It will require a series of actions by an HDET that must be followed in a specific sequence.

Calibrations could be required when a machine isn't working right or sometimes it may be necessary to calibrate a new sensor when one is installed. This will ensure that the sensor is sending the proper information to the ECM. A wheel loader's bucket position sensor will usually need to be calibrated when installed so that the ECM knows when the bucket is fully open, fully closed, and flat on the ground. Other calibration procedures could be part of other repairs and may include calibrating steering position sensors, pump control solenoids, and articulation sensors.

ELECTRONIC SERVICE INFORMATION

Before computers became popular and relatively inexpensive, a machine owner would need a paper manual as a reference for when the machine needed service or repair. As machines became more complicated, the manuals got bigger and bigger. Some machines had multiple volume manuals that would be 3–4 in. thick each. They were expensive (several hundred dollars), were bulky, usually had lots of greasy fingerprints on them, and may not apply to all machines of the same model.

It became much more practical to have machine manuals in electronic form. This started in CD form and then moved to DVDs. Most manufacturers have their service information available as web-based content. Any updates to service information can now be easily changed and the information can be printed if needed. ● **SEE FIGURE 9–62** for Caterpillar's version of electronic service information called SIS.

ECM REPLACEMENT

On rare occasions when performing a troubleshooting and repair procedure on a machine's electronic system, it will be

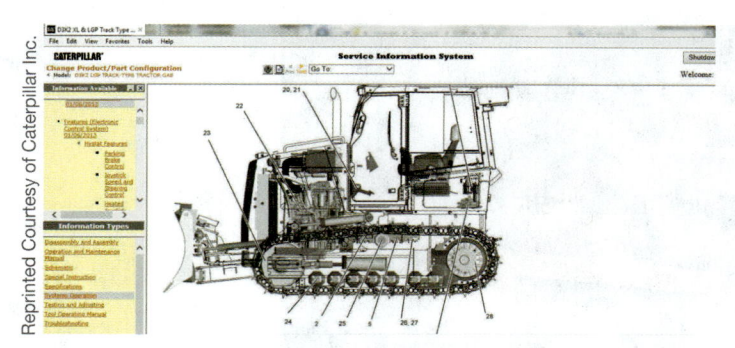

FIGURE 9–62 Caterpillar's version of electronic service information called SIS.

necessary to replace an ECM. There will be specific instructions in the machine's service information on how to do this properly. These instructions *must* be followed. The last part will likely involve reloading the software. You will need to make sure that you are installing the correct version and this may mean updating all the machine's ECMs to make them all compatible.

Proper ESD (electrostatic discharge) prevention procedures must be followed according to the manufacturer's service information to avoid damaging the new ECM.

WIRING DIAGNOSTICS—SCHEMATICS, METER USAGE

To make diagnosing an electronic problem easier, you should follow the recommended procedure in the service manual from the beginning without skipping steps. Referring to the correct electrical schematic for the machine will help a great deal as well as getting to know the system if you are unfamiliar with it. Having a good quality multimeter and knowing how to use it will be essential. As mentioned in previous chapters, don't overlook the simple things that may seem too obvious to be the cause of the problem.

NICE TO KNOW

OTHER ELECTRONIC FEATURES The addition of electronic systems to machines has enabled a wide array of extra features to be incorporated into or added alongside the existing electronic system. Some examples are the following:

Payload systems: By using pressure transducers to convert hydraulic pressure into weight, machines such as wheel loaders and trucks can have systems added to either the hydraulic system or the suspension system that will record and display the weight of a bucket of material or the box of a truck.

Camera or radar object detection systems: Machines will have many blind spots that the operator can't properly see can create safety hazards for workers near an operating machine. Many machines will have an optional warning system that will detect objects that the machine is getting close to and warn the operator before a chance of contact is close. Other systems will have cameras that show the blind spots on a cab display. ● **SEE FIGURE 9–63** for a camera used for blind spot viewing.

Machine security systems: The use of electronics that are part of the starting system can allow only people with a password to start and operate a machine. GPS security systems can disable a machine if it is outside of a certain geographical area. This is called electronic fencing or geofencing.

Grading/leveling/sloping systems: Any machine that can dig, level, and grade material such as excavators, dozers, and graders can have systems added that will be able to use laser or GPS guidance to help guide the operator in keeping a grade or slope or even take over the controls of the machine to automatically grade and slope. These systems increase the productivity of a machine and reduce manpower on the job site. ● **SEE FIGURE 9–64** for a machine grading system display.

Traction/skid control: The integration of electronic systems into drivetrain controls has allowed a machine to be programmed with antispin and antiskid features. This is easily done with speed sensors, electronically controlled brakes, and software.

Telemetry: Most machines have an option that allows an owner or HDET to remotely access the machine's vital information and perform some preliminary diagnostic troubleshooting if there is a problem. It could also be possible to re-flash an ECM with GPS and cell phone–based technology.

Progress Check

11. All the information that is stored in an ECM's permanent memory that gives instructions to the processor is called:
    a. software
    b. hardware
    c. firmware
    d. instruction ware

ELECTRONICS AND MULTIPLEXING SYSTEMS 297

FIGURE 9–63 A camera used for blind spot viewing.

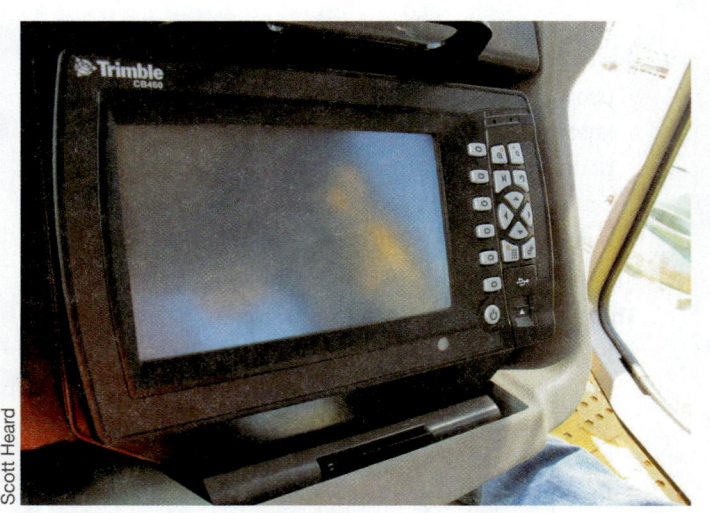

FIGURE 9–64 A machine grading system display.

12. The term for a network of computers on a machine that are able to communicate to each other is:
    a. multi-network
    b. community network
    c. controller area network
    d. modular communication

13. The CAN voltage at idle is:
    a. 3.75V DC
    b. 1.25V DC
    c. 2.5V DC
    d. 0V DC

14. If you accessed the fault code menu of a machine and wanted to know of any faults that are happening presently, you would look for:
    a. permanent faults
    b. short-term faults
    c. stored faults
    d. active faults

15. Sometimes sensors will have to be _____ for the ECM that it is sending a signal to so that the ECM will know its "zero point."
    a. calibrated
    b. re-flashed
    c. turned on
    d. metered up

> **SHOP ACTIVITY**
>
> Go to a machine, and if it has an electronic system:
> - Try to find an ECM on it.
> - Find a two-wire sensor and describe what you think it does.
> - Find a three-wire sensor and describe what you think it does.
> - See if there are any stored or active fault codes.
> - See if there are any sensors that can be recalibrated.
> - Try to find the twisted wires that are used for the data link.

298 CHAPTER 9

SUMMARY

1. Safety concerns related to electronic systems include: exposure to high voltage, machine system controls not functioning properly.
2. Electronics were first used on machines to monitor machine systems and warn operators if temperature and pressure values exceeded normal ranges.
3. Operator controls that are electronic are much easier to operate and reduce fatigue.
4. Machine reliability has increased due to enhanced machine monitoring.
5. Machines are safer because of electronic features like blind spot cameras.
6. System diagnostics have improved because of electronic features like fault codes and live data viewing.
7. Fuel economy has improved because of electronically controlled engines.
8. Engine emissions have dramatically declined because of electronic engines.
9. Data logging is possible due to the powerful computers found on machines.
10. Generally speaking an electronic device is one that has one or more components that have no moving parts.
11. The components that don't move are considered to be "solid state."
12. A comparison would be an electrical switch that has moving parts and a transistor with no moving parts. Both devices can do the same thing but the transistor is electronic.
13. As mentioned in Chapter 5 an ideal conductor has one free electron in its outer shell and an ideal insulator has a full outer shell.
14. Atoms that have 4 electrons in their outer shell are ideal semi-conductors and they can be conductors or insulators.
15. When they are "doped" with a small amount of another atom they can be N- or P- type material.
16. A diode is made up of two materials, N- and P- type semi-conductors. It is used in a circuit like an electrical check valve. It will allow current flow one way and stop it if it tries to reverse. Its positive terminal is called the Anode and the negative terminal is the Cathode. They can be smaller than the eye can see and only handle micro amps or large enough to handle hundreds of amps.
17. Diodes are very common in electronic devices and can be used for eliminating induced current in coils or rectifying AC output from alternators.
18. Zener diodes will allow reverse current flow and are commonly used in voltage regulators on alternators.
19. Light emitting diodes (LEDs) are made with materials that produce light when forward biased. They are becoming more popular for machine lighting and have been used in machine dash displays for many years.
20. Photodiodes are a type of diode that allows current flow through them when exposed to light. They can be used to control light circuits.
21. Transistors are an electronic device that have three terminals. The base terminal controls the current flow from the collector to the emitter. It can act like a relay or an amplifier.
22. Electro static discharge (ESD) can destroy electronic components. As a technician you should take precautions to avoid this.
23. Resistors are used in electronics to reduce current flow and create voltage drops.
24. Capacitors are used in electronic components to: store electrical energy, accept voltage spikes and can be used as a timer. They can be directly compared to an accumulator in a hydraulic system. Ultra capacitors can be used to stored large amounts of energy for starting systems and hybrid functions.
25. A digital electronic signal is one that is either on or off. Compare this to opening and closing a switch. Analog electrical signals can be converted to digital signals.
26. Computer logic gates are the basis for how a computer "thinks." They allow a computer to process information and are made of diodes and transistors.
27. Computers need memory to store permanent programmed information and temporary information that is lost when the ECM (electronic control module) powers down.
28. Almost all machines produced today will have at least one ECM. They receive electrical information (signals) from sensors and switches (inputs), compare it to stored information and send out electrical signals (outputs) to control machine systems or display information to the operator. This is the electronic information cycle.
29. An ECM can send and receive a variety of different types of signals such as: variable voltage, variable resistance, on/off, variable frequency, variable duty cycle (PWM).
30. Electronic systems will mainly use copper wires to transfer electrical signals. For main data transmission wires they will need to be shielded or twisted. This will prevent the signals from being corrupted. Machine electrical schematics will guide technicians around the electronic system like a road map.
31. Inputs to an ECM can be from a variety of switches, sensors or other ECMs. Sensor signals may need to get converted to digital information and this would happen in an analog to digital converter in the ECM.
32. ECM outputs could control dash displays, relays for higher current flow components or be information sent to another ECM.
33. Electronic cab displays for machine monitoring can range from LED warning lights to LCD screens or electronically controlled gauges.
34. ECMs need software programming that is customized to the particular machine the ECM is installed in. Software can be updated by "flashing" new programming into the ECM with a connected laptop or through a satellite link.

35. Machines with multiple ECM's will have an electronic communication network to allow the ECMs to communicate. This is called a CAN (controller area network) Bus system and sometimes referred to as a multiplex network.

36. All machine ECM's, sensors, switches, displays, and wiring can be considered to be part of the CAN Bus network.

37. The main part of the CAN Bus network is the data communication wires. These are a twisted pair or wires but can sometimes be three wires and are the "backbone" of the network. One wire will be CAN high and the other CAN low; this refers to the voltage levels of the signals being transmitted.

38. At both ends of the wires there will be a 120 Ohm resistor called a terminating resistor.

39. Diagnosing CAN Bus faults starts with checking resistance between the CAN high and low wires.

40. Electronic systems provide a means to diagnose the machine systems they control. Fault codes tell operators and technicians: the ECM that is reporting the fault, components that is faulty, and the type of fault that is occurring.

41. Active codes are present all the time, and logged codes have occurred at some time but aren't currently active.

42. Electronic Service Tools (EST) can be used to troubleshoot machine faults that can be detected by the machines electronic system. A laptop computer is connected through a communication adapter. Fault codes and live data can be read.

43. Many machines can now communicate remotely to technicians through satellite or cell phone link technology.

chapter 10
CLUTCHES, TORQUE CONVERTERS, TORQUE DIVIDERS, AND RETARDERS

LEARNING OBJECTIVES

After reading this chapter, the student should be able to:

1. Identify the main components of a mechanical clutch.
2. Explain how a mechanical clutch works.
3. Explain how to adjust a mechanical clutch.
4. Identify the main components of a torque converter.
5. Explain the operating principle of a torque converter.
6. Describe how a variable torque converter works.
7. Describe what a lock up torque converter does.
8. Identify the main components of a torque divider.
9. Explain how a torque divider differs from a torque converter.

KEY TERMS

Clutch disc 304
Cooler 309
Impeller 311
Lock up clutch 316
Pressure plate 303
Release bearing 307
Rotary flow 312
Rotor 321
Stator 311
Turbine 311
Vortex flow 312

INTRODUCTION

SAFETY FIRST Some specific safety concerns related to clutch, torque converters, torque dividers, and retarders include the following:

- Strain injuries from lowering and lifting clutches into place
- Burn injuries from working near hot torque converters, retarders, and related components
- Injuries from rotating parts
- Injuries from improperly securing machine while working on it

When working with dry-type clutches, try not to disturb friction disc dust any more than possible. There is a possibility of this fine dust causing serious breathing problems. Appropriate breathing protection should be used. Refer to the machine or clutch manufacturer's service information for more specific information.

CLUTCHES, TORQUE CONVERTERS, AND HYDRAULIC RETARDERS

A typical heavy equipment machine that uses a mechanical drivetrain will have a diesel engine for its prime mover. The engine's flywheel will transfer torque to either a clutch or a torque converter. The output torque of these devices will then get transferred into a transmission that will change the torque and speed between its input and output through a series of gears. The output of the transmission will then go on to drive either tracks or tires to propel the machine.

A friction-type clutch can be used to start and stop the power flow from the engine to an output such as a manual transmission, a powershift transmission, or a drive shaft. It operates by using friction to transfer torque between its input and output. There are other less common uses for mechanical clutches on heavy equipment. An example would be for starting and stopping power flow to winches on older cranes and pipe layers or for paving equipment that needs a large drum driven. Many stationary diesel engines will also use mechanical flywheel clutches to drive things such as crushers, screening plants, drill rigs, water pumps, and generators. Many graders use friction-type clutches to drive powershift transmissions.

A torque converter is also used to transfer torque and allow the engine to keep running even though the transmission is engaged. A torque converter can also multiply the torque output of an engine. It uses fluid to transfer torque between its input and output.

A torque divider is a combination of a torque converter and a set of gears that will transfer torque from the engine's flywheel to a transmission.

A hydraulic retarder will provide a braking action through the drivetrain of a machine. Like torque converters, they use fluid to transfer torque between their input and output; however, they decrease torque throughput. You would typically find a retarder on a machine that is capable of travelling at a fairly high speed (25–50 mph).

This chapter will mainly discuss mechanical flywheel clutches, torque converters, torque dividers, and hydraulic driveline retarders used with mobile machines.

FRICTION-TYPE FLYWHEEL CLUTCHES A mechanical flywheel clutch is used to allow an operator to start and stop the torque transfer from a running diesel engine's flywheel to an output. This output is typically a transmission that is a manual shift (see Chapter 11) or a powershift (see Chapter 12), but it could also be anything that needs rotational torque to drive it such as a drive shaft, pulley, or sprocket. ● SEE FIGURE 10–1 for how a clutch is used in a mechanical drivetrain.

Manual shift transmissions are rarely seen used in anything but very light equipment that is produced today. Most machines today use torque converters, hydraulic pumps, or electric generators to transfer power from the prime mover to the rest of the drivetrain.

Part of the clutch will be driven by the engine's flywheel and part of it will drive the transmission or other driveline component. These parts will be discussed shortly.

This type of clutch uses friction to transfer torque through it. It will allow some slippage between the clutch being fully released and not transferring any torque to when it is fully engaged and providing direct drive with zero slippage. You can compare what happens in a friction-type clutch to what happens with the brakes on a car as the brakes are applied and the car comes to a stop. The stationary friction material (shoes or pads) at the wheel ends press against a rotating part (rotor or drum) and the friction created will slow down and stop your car. The stationary frictional material tries to get the rotating part to match its speed of 0 rpm.

Instead of slowing and stopping your car, friction generated by a clutch will get a vehicle or machine moving. The stationary part of the clutch should match the speed of the rotating part of the clutch after it is engaged.

Think about a machine with a mechanical clutch and manual transmission that is sitting stationary with the engine running on a hard-level surface and the operator wants to get it moving. The transmission will be engaged in a low gear ratio and the clutch disengaged to start with. The operator will then slowly engage the clutch to start turning the transmission input shaft. With

302 CHAPTER 10

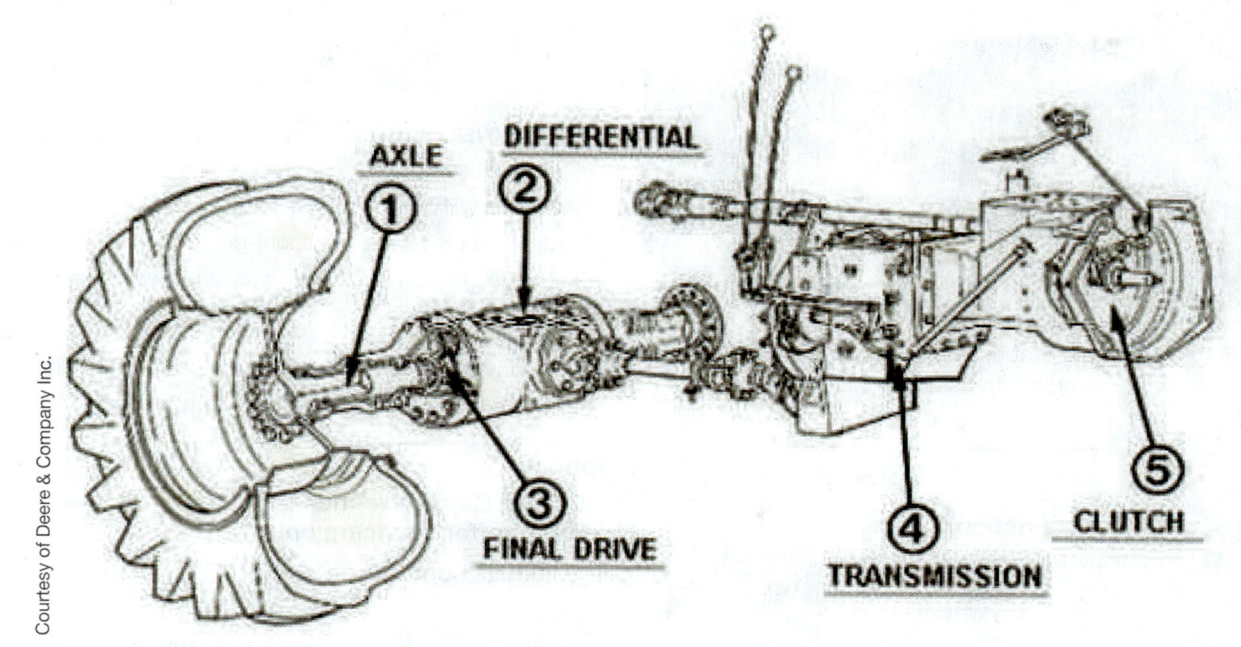

FIGURE 10–1 Clutch as part of a mechanical drivetrain.

little rolling resistance, the machine will start moving easily and the operator shouldn't have to slip the clutch much or increase the engine rpm. The friction of the clutch will easily overcome the resistance of the load and quickly turn the transmission input shaft at the same speed as the engine flywheel. This is called direct drive or 1:1. With little slippage, there will be very little heat created. Once these two components are turning at the same speed, they should stay that way until the operator releases the clutch to either change gears or stop the machine movement. If there is any speed difference between the flywheel and the transmission input shaft after the clutch is fully engaged, there is definitely a problem and this is termed a *slipping clutch*. A clutch that slips for any length of time should be detected by anyone with a sense of smell. The smell of burnt clutch lining won't soon be forgotten after the first time it's detected. If you have been to a tractor pull, you may have experienced this unpleasant odor.

However, when the machine is loaded and stopped in soft material or facing uphill with a high rolling resistance (or hooked onto a pulling sled), then the operator will likely have to slip the clutch and increase the engine rpm to keep it from stalling. This will put a much higher load on the friction material of the clutch and create more heat in the clutch because of the slippage. Too much clutch slippage and heat will decrease the life of a clutch. One way manufacturers have been able to decrease the negative effects of too much heat is to run the clutch in oil. The clutch housing will have some oil circulating through it that shouldn't get any higher than $\frac{1}{3}$ of the way up the clutch.

This may at first seem counterintuitive, but with the right clutch material, there will be no more slippage than a dry clutch and the oil will carry the heat away to a cooling circuit. The oil will also provide a cushion effect when the clutch is being engaged.

The mechanical clutch differs from a hydraulic clutch in that it is spring applied and mechanically released, whereas the hydraulic clutch is spring released and oil applied. Hydraulic clutches are discussed in Chapter 12.

There aren't many machines produced today that use mechanical clutches, but there is a very good chance that at some point you will be asked to change, adjust, or diagnose a clutch problem so it is important that you know how a mechanical clutch works and the terminology related to one.

Some clutches will directly drive the transmission's input shaft while others will drive their own clutch shaft that will likely drive a universal joint that could then go on to drive a drive shaft; the drive shaft could then drive a transmission or other component that needs rotational torque.

One of the most common uses for a mechanical clutch is on some graders that use a multi-disc oil-cooled clutch to drive a powershift transmission. ● **SEE FIGURE 10–2** for a grader that uses a mechanical flywheel clutch to drive its powershift transmission.

Some track drive machines will have what is called a direct drive option that will use a flywheel clutch and manual transmission. Other applications may be for paving equipment or cranes.

The main parts of a clutch are the clutch cover, **pressure plate**, clutch discs, springs, and release mechanism. Multi-disc clutches will also have intermediate plates. Other clutch-related parts are the flywheel, the clutch brake, the pilot bearing,

CLUTCHES, TORQUE CONVERTERS, TORQUE DIVIDERS, AND RETARDERS 303

FIGURE 10–2 A grader that used a mechanical flywheel clutch to drive its powershift transmission.

and sometimes a clutch shaft and clutch housing. ● **SEE FIGURE 10–3** for a heavy-duty clutch.

The main components of a clutch assembly are described here.

CLUTCH COVER. The clutch cover (sometimes called the backing plate) houses the other clutch components and is bolted to the flywheel. It will provide a base for the clutch springs to be able to push on the pressure plate. Light-duty clutch covers will be made of stamped metal while heavy-duty ones will be cast metal.

CLUTCH DISCS. The **clutch disc** (sometimes called friction plate) is the driven component of the clutch that has friction material riveted or bonded or both on to a metal disc. It is splined

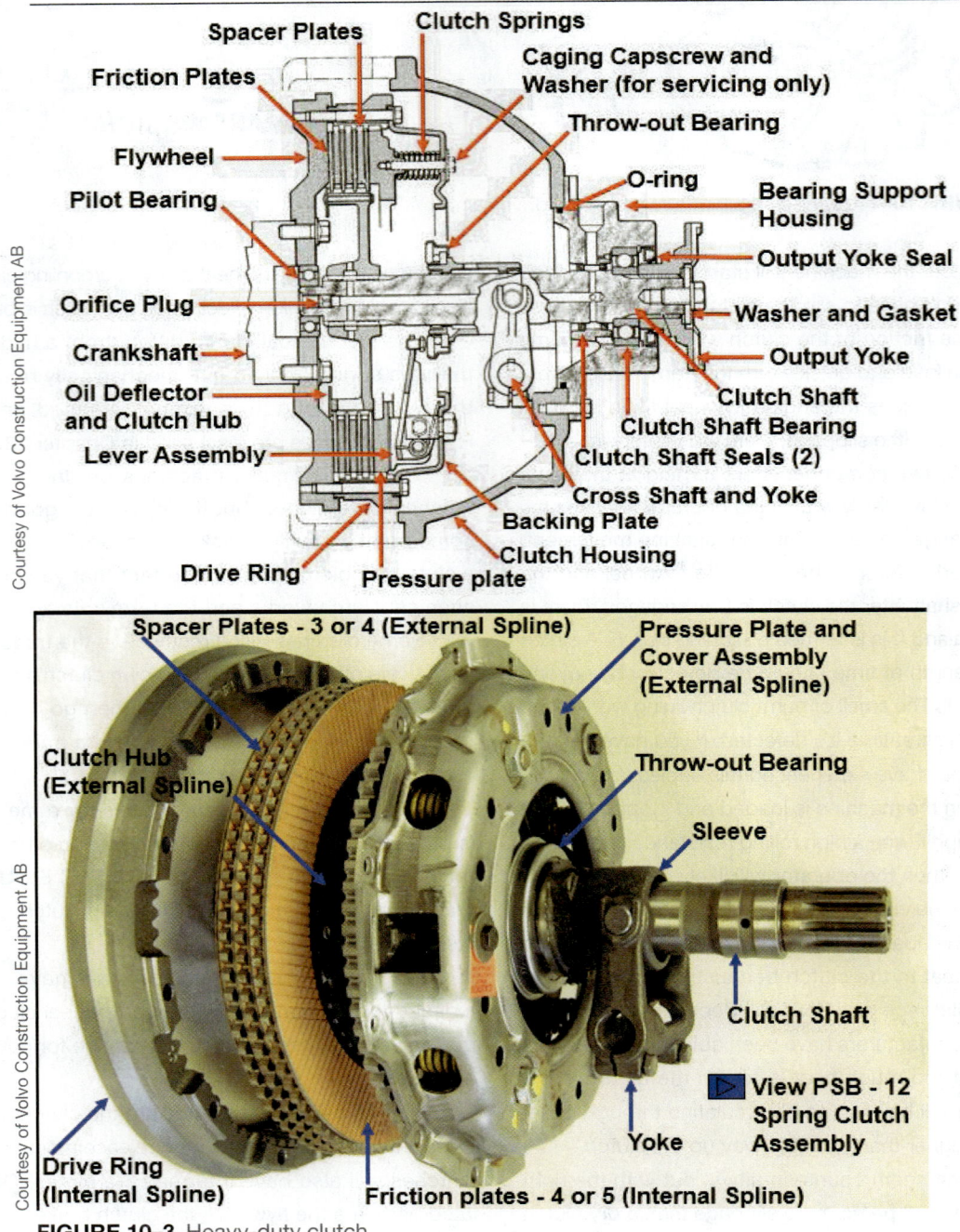

FIGURE 10–3 Heavy-duty clutch.

304 CHAPTER 10

to the transmission input shaft, and when it gets squeezed by spring pressure between the pressure plate and the face of the flywheel, it will transfer torque from the engine to the transmission. The clutch disc is designed to absorb lots of heat and varying amounts of shock load. To increase the torque capability of a clutch assembly, sometimes multiple discs are used. Some clutches could have four or more discs. This simply increases the surface area of clutch material and provides more friction capacity. All discs could be splined to the transmission input shaft or one could be and the second is splined to the first.

Clutch disc friction material can be made of several different types of material such as organic paper and cork or ceramic mixtures. Different types of friction material will have different coefficients of friction. This means how effective the material will grab its mating surface. The clutch manufacturer will also decide on the type of material based on other factors such as:

- maximum torque load the clutch will have to handle
- whether it is wet or dry
- the type of duty cycle the machine places on the clutch
- expected life of the clutch

Clutch discs can be a solid disc or have a center hub with dampening springs to help absorb shock loads. ● **SEE FIGURE 10–4** for a variety of different styles of clutch friction discs.

Ceramic discs will be star shaped with small friction pucks fastened to the metal base.

The clutch disc will be the main wear component of the clutch, and once the friction material wears past a minimum thickness the clutch disc must be replaced. Heavy-duty clutch manufacturers will provide specifications for clutch disc minimum thickness dimensions.

If a light-duty clutch is removed for any reason, it makes sense to always replace at least the friction discs and usually the complete assembly. As oil-cooled discs will wear very little, there may be wear indicators or grooves that will determine if a disc is reusable. Even though a disc is reusable, it usually makes sense to replace it any time a clutch is removed considering the work involved to remove a clutch.

PRESSURE PLATE. This is the drive part of the clutch that rotates with the flywheel and clutch cover.

The pressure plate is a movable spring loaded and machined disc that will apply pressure to the clutch disc when released toward the flywheel. It will be moved away from the flywheel when the operator wants to release the clutch. Most clutches will have some kind of adjustment mechanism to move the pressure plate as the clutch disc friction material wears and becomes thinner. The back of the pressure plate will have lugs cast into it that serve as attachment points for the clutch release mechanism and the clutch springs. ● **SEE FIGURE 10–5** for a typical pressure plate.

FIGURE 10–4 A variety of different styles of clutch friction discs.

FIGURE 10–5 A typical pressure plate.

CLUTCHES, TORQUE CONVERTERS, TORQUE DIVIDERS, AND RETARDERS

TECH TIP

Over-center-type clutches will use the same principle for transferring torque as all friction clutches previously mentioned, but they will have a different way of having their pressure plate held against the clutch disc. A set of levers and pivots that are part of the clutch assembly will push on the pressure plate with a great deal of force, enough to slightly compress the clutch disc and some of the linkage. When the lever is moved a little more in the same direction, the pressure will be slightly reduced, but one part of the linkage will have gone over center and will now be held there by a stop. It will be locked in place by the expanding force of the clutch disc and linkage. The linkage is said to be over center. This action engages the clutch with a mechanical lock instead of spring pressure. To release the clutch, the lever is pulled back, which moves the linkage back over center and pulls the pressure plate back. ● **SEE FIGURE 10–6** for an over-center clutch.

This action can be compared to pushing a heavy ball up a 100 ft long steep hill and 5 ft after it crests the hill there is a ledge the balls stops on it to keep it there. Now to let the ball down, you just have to push it back up 5 ft and let it go!

Can you think of any other mechanical devices that use this over-center principle?

FIGURE 10–6 An over-center clutch.

The pressure plate will have spring force originating from the clutch cover pushing it toward the flywheel to keep the discs clamped to the flywheel for positive clutch engagement. Light-duty clutches will have a diaphragm-type spring and heavy-duty clutches will either use coil springs or have an over-center mechanism that will lock the pressure plate in place up against the discs.

The surface of the pressure plate that mates with the clutch disc starts out with a finish on it that will be relatively rough by design. This is to give the clutch disc a good surface to grab. If a clutch is slipped excessively throughout its life, this finish will become more polished and the clutch will be less effective. This is called glazing. This finish can be restored through a grinding/sanding process.

Pressure plates should be checked for signs of overheating, cracks, and grooves. Light-duty clutches should always be replaced as an assembly.

INTERMEDIATE PLATES. These plates are used when a clutch has more than one clutch disc to provide a machined surface for the friction material to grab. They will have lugs around the outside of them to engage with slots in the flywheel. Like the pressure plate, the intermediate plates are driven by the engine flywheel and will have a rough surface finish. Intermediate plates are sometimes called separator plates and will alternate with clutch discs for multi-disc clutches. If a clutch has four discs, it will use three separator plates.

Intermediate plates need to be checked for warpage and dishing when a clutch is removed. The lugs should also be checked for notches or excessive wear.

CLUTCH RELEASE MECHANISM
The operators will most likely push on a clutch pedal in the machine's cab to release the clutch but a lever may be used alternatively. From the pedal there could be a mechanical linkage, cable assembly, or some hydraulic components transferring force down to the clutch cross-shaft. The cross-shaft will go through clutch cover either above or below the transmission input shaft where it moves a fork that will move a release bearing. The release bearing is commonly called the throw out bearing as well. This will indirectly move the pressure plate away from the flywheel. ● SEE FIGURE 10–7 for a typical release bearing.

The **release bearing** is needed to allow the nonrotating clutch fork to be able to pivot and either push or pull on the bearing to start to pull the pressure plate away from the flywheel. This is done either by direct action on the diaphragm-type spring fingers or by pushing on levers. The levers will in turn pull the pressure plate back against spring pressure. This action will allow slippage and stop the transfer of torque to the clutch discs that are between the pressure plate and flywheel.

FIGURE 10–7 A typical release bearing.

Mechanical linkage for clutch release is simply a series of rods and bell cranks to the clutch release lever, which is keyed or splined to the clutch cross-shaft.

A cable assembly will do the same as the mechanical linkage but will eliminate a few extra moving parts. There will be a sheathed cable attached to the clutch pedal on one end and the clutch release lever at the other.

If a clutch is released hydraulically, the clutch pedal will push a plunger into a master cylinder. This will force fluid out of the master cylinder through a hose to a slave cylinder. The slave cylinder will then have its rod move out of it that is attached to a lever that will rotate the cross-shaft. Some heavy-duty applications will have a hydraulic boost mechanism similar to a brake booster to decrease the operator effort and increase the release force when pushing on the clutch pedal. ● SEE FIGURE 10–8 for a hydraulic actuator of a mechanical clutch.

One of the maintenance points for many heavy-duty clutches will be to grease the throw out bearing. *Do not over-grease the throw out bearing of a dry clutch because of the possibility of getting grease on the clutch disc.*

FLYWHEEL
Although the flywheel is fastened to the engine's crankshaft, it is integrated to the operation of a friction-type clutch. The flywheel drives the clutch cover through a ring of fasteners around the outside circumference of the flywheel. The clutch cover in turn drives the pressure plate and any intermediate plates. The flywheel rearward machined face is the mating surface for the clutch disc. It can also drive the intermediate plates if so required.

TECH TIP

Some machines will use clutches to drive rotating equipment that has nothing to do with propelling the machine. An example would be to drive the drum on a pavement grinding machine. Some of these machines have their grinding drum belt driven, and to start and stop the belt drive, a clutch is used. This will likely be a heavy-duty clutch and its release mechanism could be controlled by an electric switch. The electric switch will actuate an air or hydraulic solenoid that will send air or oil to a cylinder. The cylinder will have its rod connected to the clutch release lever, and from there, the release mechanism is the same as a heavy-duty clutch.

There could also be clutches that use air to engage the clutch instead of spring pressure. Air pressure is fed into a tube with a swivel to allow rotation and fills a pneumatic tube to apply pressure to the pressure plate.

These systems are much easier for the operator to control.

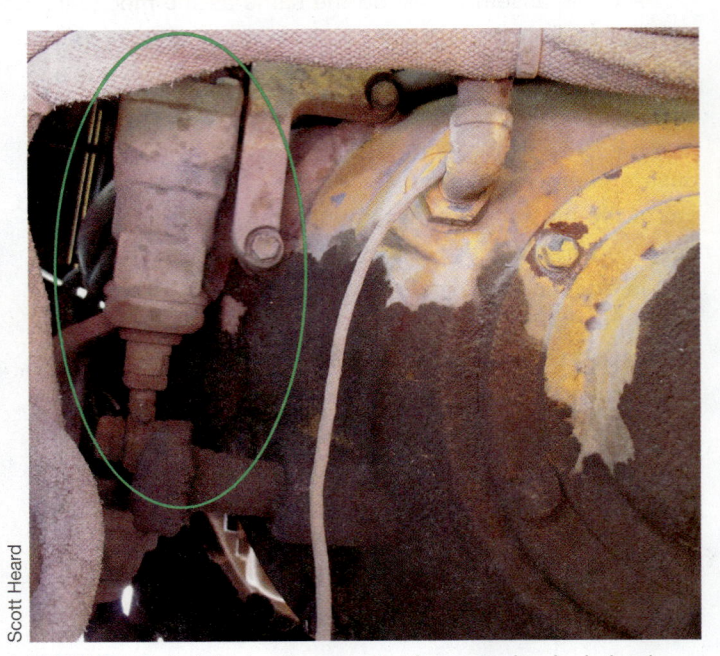

FIGURE 10–8 A hydraulic actuator for a mechanical clutch.

FIGURE 10–9 Two different types of flywheels.

There are two main types of flywheels: flat and pot type. A flat flywheel is just that and the clutch cover will bolt to it. Some flat flywheels will have a drive ring bolted to it that will make it effectively the same as a pot-type flywheel. Pot-type flywheels will have their face recessed and will have driving slots or teeth machine into their circumference. ● **SEE FIGURE 10–9** for the two different types of flywheels.

It is common practice to have a flywheel sent to a machine shop for reconditioning any time a clutch is removed. The machine shop will grind it flat and make the surface the proper roughness again. There will be a minimum thickness specification that the flywheel should *not* be allowed to go under, which should be checked during and after reconditioning. The minimum thickness will ensure that the pressure plate springs can apply the proper clamping force to the clutch discs.

CLUTCH BRAKE Some clutch assemblies will incorporate a clutch brake. This will be used to slow down and stop the transmission input shaft to assist with changing gears on the fly. It could be a simple flat disc with some friction material on its face or could be a band-type brake. It provides a braking effect between the transmission shaft and the transmission housing.

PILOT BEARING Most flywheel/clutch/transmission arrangements use a pilot bearing to support the engine end of the transmission shaft. In the center of the flywheel, there will be a bearing where the end of the transmission shaft or clutch shaft rests. It is a ball or roller bearing and should be replaced any time the clutch is removed. The outer race is an interference fit in the flywheel while the transmission shaft fits loosely in the pilot bearing's inner race.

CLUTCH COOLING SYSTEM
All clutches will generate heat when engaging and releasing because of the friction material sliding against an opposing metal surface (flywheel, pressure plate, or intermediate plate). If this heat becomes excessive, the friction material will start to fail and the metal contact surfaces will become discolored and can crack or warp. ● SEE FIGURE 10–10 for an example of overheated clutch components.

A dry-type clutch will dissipate heat to the air surrounding it. The cooling effect is limited to the air temperature in the clutch housing.

A wet clutch will transfer heat to the oil that it is exposed to. Oil-cooled clutches will have oil sent to their shaft through cross-drillings where it is allowed to spill out onto the clutch disc. This oil will likely originate from a lubrication circuit of the transmission and will absorb heat from the clutch discs, pressure plate, and flywheel. The oil will then flow to a **cooler** where the heat is dissipated. The cooler could be air or engine coolant cooled.

FIGURE 10–10 Overheated clutch components.

WEB ACTIVITY

Search for manufacturers of heavy-duty clutches and find one that offers multi-disc clutches.

Look for the specifications of one heavy-duty clutch and find the following information: torque capacity, disc spline diameter, number of springs, disc material type, flywheel housing bolt pattern, and release mechanism options.

CLUTCH MAINTENANCE
Because a clutch is designed to wear, it must be adjusted to maintain the proper amount of free play in the release mechanism. As the clutch discs wear, the clutch springs will slowly keep moving the pressure plate toward the flywheel. Because the release mechanism's purpose is to pull the pressure plate away from flywheel, this wear will change the clearance between the clutch release fork and the release bearing because the release bearing will have moved toward the fork. There is a specific dimension that must be kept between the fork and the bearing so there isn't pressure applied to the throw out bearing all the time. If there were no clearance, this would be similar to operators keeping his foot on their clutch pedal all the time. This will make the release bearing fail prematurely.

Free play is measured at the clutch pedal for a mechanically released clutch and at the clutch release lever for a hydraulically released clutch. It is measured by moving either the pedal or the lever from its relaxed position to a point where you start to feel resistance. This will be when the fork contacts the release bearing.

To adjust a clutch, you will likely need help from an assistant because it can only be done with the clutch released. With the clutch released, there will be a mechanical way to move the pressure plate closer to the flywheel that will restore the proper free play. Some clutches will have a large thread that will turn the whole pressure plate while others may have a way to move the release levers.

The release bearing will usually need to be greased occasionally. Be careful and *do not* over-grease the release bearing. Follow the service manual for grease types, intervals, and amounts.

If a clutch is controlled by mechanical linkage, there may be grease fittings on the pivot points of the linkage that need to be greased regularly.

DIAGNOSING CLUTCH PROBLEMS
Operator complaints that may lead to clutch repair or adjustment are the following:

- no drive
- jerky engagement
- won't disengage
- slipping
- strange smell
- fluid leak
- spongy pedal
- vibration

As with all operator complaints, the first step is to verify it, then check the simple things first, and then find out under what condition the problem is occurring. This may include different loads, engine rpm, transmission gear, and position of clutch pedal.

Here are some problems and possible causes related to mechanical clutches:

COMPLAINT	SYMPTOM	POSSIBLE ROOT CAUSE
Noisy	Chirping, squealing, grinding, or growling noise	• Insufficient pilot bearing or throw out bearing lubrication • Friction material gone off disc
Chatter	Clutch grabs or jerks during engagement	• Oil or grease contamination on disc friction material • Worn or distorted clutch components • Misalignment between chassis and driveline components • Loose or broken transmission or engine mounts
Vibrations	Cab, pedal, or other components shaking	• Part of friction material off of disc • Bent clutch or transmission input shaft
Slipping*	Clutch cannot fully transmit engine torque and slips under load (typically in higher gears)	• Normal wear • Oil or grease contamination (new clutch) • Incorrect release system adjustment • Improper flywheel machining (step/cup dimensions)
Not releasing	Clutch does not fully release; disc continues to turn the input shaft and transmission gears Gears grind when shifting Won't go into gear	• Excessive free play • Air in the hydraulic system • Binding or worn release system components • Slave or master cylinder leak • Bent or distorted clutch components • Pilot bearing damaged by misalignment • Input shaft corroded or damaged
Hard pedal	Clutch is difficult to actuate	• Release system components sticking, binding, and improperly lubricated • Worn pedal bushings • Worn or damaged fork

*Some slipping is normal and necessary for smooth engagement, but clutch should not slip once it is engaged.

To determine if clutch slips too much:

- Set parking brake, start engine, and leave at low idle
- Put transmission in high gear
- Slowly engage clutch, and if:
  - engine stalls immediately, clutch is OK
  - engine gradually slows and then stalls, clutch is slipping partially
  - engine continues to run when clutch is fully engaged, clutch is slipping excessively

CLUTCH REPAIR/REPLACEMENT If clutch problems can't be resolved with clutch adjustment, linkage adjustment, or lubrication of throw out bearing, then the clutch will need to be removed for repair, reconditioning, or replacement.

This may require the transmission or engine to be removed. If the clutch output is a drive shaft, then the clutch alone can be removed once the drive shaft is taken out of the way.

If the transmission is removed, this will likely require the back of the engine to be supported.

Care should be taken to disturb as little clutch dust as possible because there could be unhealthy particles in the dust. Breathing protection should be worn as a precaution.

Once the clutch is removed, a thorough inspection should be performed to analyze why it failed. Look for signs of overheating (discoloration) that may indicate abuse.

When a new clutch is installed, to save a lot of trouble with clutch and transmission alignment, there are special alignment tools that should be used when installing a new clutch. It will go through the clutch discs and into the pilot bearing.

The clutch cover should be installed with patience by going through the proper fastener tightening and torqueing procedure. This will slowly and evenly draw the clutch into the flywheel.

There may be a break in period for a new clutch where it will have some fast initial wear that will make a first clutch adjustment necessary soon after it's installed. Check with the clutch manufacturer for the proper initial adjustment interval.

Progress Check

1. The part of the clutch that sends torque out through (or output) is the:
   a. pressure plate
   b. clutch disc
   c. release mechanism
   d. flywheel
2. This is the highest wear part of the clutch:
   a. the pressure plate
   b. the clutch disc
   c. the release mechanism
   d. the flywheel
3. Another term for the throw out bearing is the:
   a. over-center bearing
   b. master bearing
   c. slave bearing
   d. release bearing
4. When a clutch is adjusted, this dimension should also be checked and corrected if necessary:
   a. pedal travel
   b. free play
   c. pressure plate travel
   d. pilot bearing clearance
5. This will *not* cause a clutch to slip:
   a. friction material worn off disc
   b. poor adjustment
   c. leaking slave cylinder
   d. broken springs

Other types of clutches are discussed next.

DOG CLUTCH Dog clutches are crude and simple mechanical drive transfer devices that aren't used very much anymore but could be seen on older equipment. Two rotating parts of the dog clutch have mating notches and lugs. One of the parts is mechanically forced into the other part to make both parts rotate together.

ONE-WAY CLUTCH OR SPRAG CLUTCH One-way clutches are used to allow free rotation between two components

> **TECH TIP**
>
> An old grader I had to work had an almost complete mechanical blade maneuvering system. It had a set of dog clutches that were engaged by the operator to drive shafts that would in turn mechanically lift and turn the blade. The grader had a gearbox in the cab above the steering wheel where the dog clutches were housed and levers would directly engage the clutches when needed. The controls were stiff and rough to engage. It's very unlikely that an operator who is used to today's machine controls would operate this old grader for more than an hour!

in one direction but will lock them together in the opposite direction. A set of spring loaded balls or rollers on ramps will get wedged between the ramp and a circular housing to lock the two components together.

Two uses for this type of clutch are for torque converter **stators** and for starter motor drive pinions.

EXPANDING SHOE CLUTCH This type of clutch could be mistaken for a drum and shoe brake system. The components are practically the same, but the purpose is to transfer drive not slow a vehicle down. The shoe assembly is able to rotate because of a hydraulic swivel that lets hydraulic fluid transfer to the clutch cylinder. Typically, the drum is the drive part of this clutch and the shoes are the driven. When the shoes expand, they will drive a shaft that they are attached to. This type of clutch is used on cranes and winches. ● **SEE FIGURE 10–11** for an expanding shoe type of clutch.

TORQUE CONVERTERS

A machine with a torque converter will be able to seamlessly transfer torque from its engine's flywheel to its transmission. It will allow the engine to idle with the transmission in gear. When the operators of the machine want to start torque transfer through the torque converter, they will simply speed up the engine.

Unlike a mechanical clutch that relies on friction, a torque converter will use the inertia of moving fluid to transfer torque. This principle is known as hydrodynamics, which means fluid movement. Fluid couplers have two components (**impeller** and **turbine**) that use a fluid medium to transfer torque between them.

FIGURE 10–11 An expanding shoe type of clutch.

Torque converters are considered to be a type of fluid coupler, but with the addition of a third component (stator), they also have the advantage of being able to multiply engine torque.

A torque converter has three main components to perform torque transfer. ● **SEE FIGURE 10–12** for the main components of a typical torque converter.

One is driven by the engine flywheel and is called the impeller although it is sometimes called a pump. As it spins at engine speed, it will throw oil with centrifugal force at the turbine. The turbine is the torque converter output and will drive the transmission's input shaft. The other main component is the stator. Its job is to redirect the oil that leaves turbine back to the impeller. While a simple fluid coupler will be able to transfer torque using just an impeller and turbine when a torque converter uses a stator, it will also be able to multiply the engine's torque. All of these components will have a series of evenly spaced curved fins or vanes to direct the oil flow.

The vanes of both the impeller and the turbine start at the outside circumference and follow the bowl shape toward the center. Instead of the fins being level all the way across, there is a smaller bowl-shaped hollow split guide ring that takes some space in the middle of both components. It is meant to help direct the oil as it moves through the spaces between the vanes.

A good way to visualize how a fluid coupler works is to imagine two fans facing each other and are very close together. If you were to turn one fan on, you can imagine that the other fan will start to turn. If you also put a shield around the fans, this would help to keep the air directed at the driven fan. This is what happens with the oil inside a fluid coupler. The driven fan could drive something with its shaft just from the air flow created by the first fan. Think of the fan blades as primitive vanes. ● **FIGURE 10–13** shows one fan driving the other.

Another way to visualize the moving oil is to imagine what happens to a fluid that is sitting in a stationary bowl. If the bowl had ridges going from its center to the outer edge of the bowl in a slight spiral shape and the bowl got spun, you could imagine that the fluid would get flung out to the outside of the bowl. This is done by centrifugal force. This is how the hydrodynamic force is created. This bowl is like the impeller.

If a second similar bowl was suspended and faced the impeller bowl, you should be able to picture the second bowl (turbine) being driven by the force of the oil from the force of it leaving the first one. You should also be able to imagine that as the impeller bowl is spun faster, there will be more energy in the oil that will also spin the turbine bowl quicker. This is also a simple fluid coupler. It will transfer torque from one component to another, but without a stator, it won't be very effective. The closer the two components are moving at the same speed, they will be getting closer to what is termed *rotary flow*. **Rotary flow** occurs when the oil is mostly moving in the same direction as the impeller. True rotary flow is hard to achieve if there is any load on the turbine. ● **SEE FIGURE 10–14** for a rotary flow.

The problem with a fluid coupler is that when the turbine starts to have a higher load placed on it and slows down, the fluid it returns to the impeller will try to slow the impeller down. This is a reduction in efficiency.

In a torque converter, if the load on the turbine stops it completely or stalls it while the impeller oil flow is trying to turn it, this is called **vortex flow**. The closer a fluid coupler gets to vortex flow, the more it needs a stator to redirect the oil back into the impeller.

A torque converter with a stator will be changing between rotary and vortex flow as the load on the turbine changes and as the impeller speed changes. This seamless speed change between the impeller and turbine is one of the biggest advantages to a torque converter.

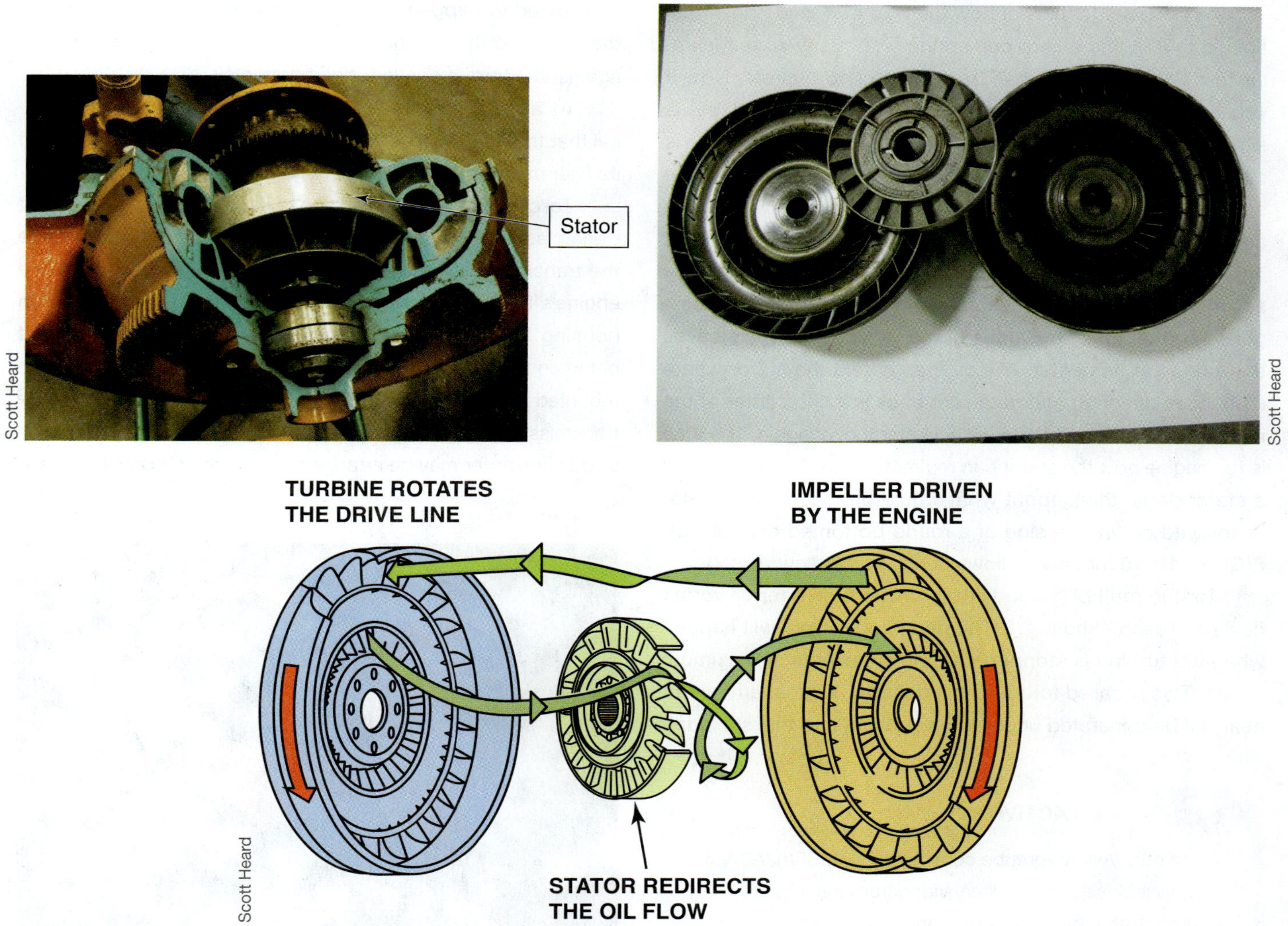

FIGURE 10–12 The main components of a typical torque converter.

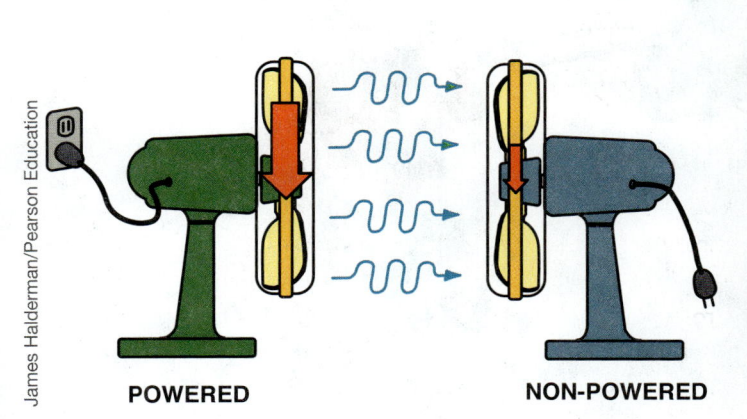

FIGURE 10–13 One fan driving another.

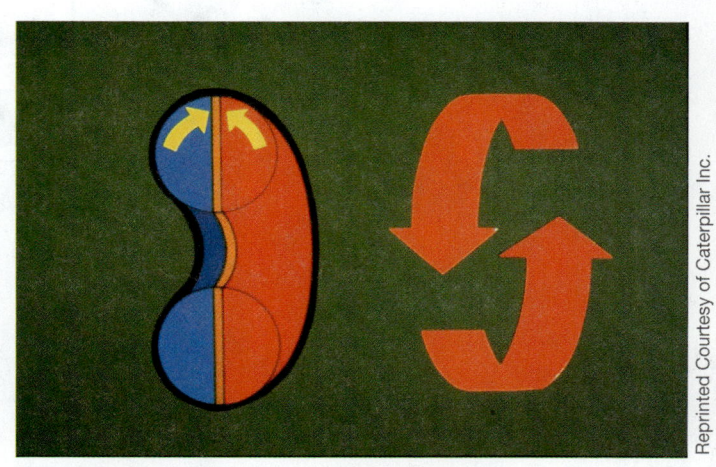

FIGURE 10–14 An illustration of rotary flow.

CLUTCHES, TORQUE CONVERTERS, TORQUE DIVIDERS, AND RETARDERS

A good way to think of how the fluid moves during vortex flow is to imagine a light coil spring with many wraps (similar to a Slinky toy) that has been made into a circle. Now if you imagine that the coils are tracing the path of the oil, this is similar to what happens during vortex flow. When the turbine is stalled and the impeller is spinning fast, it's like the coils of the spring are close together. Then as the turbine starts to move and the impeller slows down, it's like the coils get farther apart. The oil flow during vortex flow is continuous, but to clarify, let's say it starts at the outer circumference of the impeller vanes, goes into the opposing vanes of the turbine, and then leaves the turbine at its center. The oil then enters the stator where it changes direction and gets sent back into the center of the impeller where the continuous cycle starts over again. The trick is to imagine how the stator can redirect the oil. To picture what a stator does, think about what would happen to water that is sprayed down one side of a round bottomed cup. ● **SEE FIGURE 10–15** for how oil flows during vortex flow.

Torque multiplication with a torque converter at vortex flow can reach almost 4:1. The most vortex flow will happen when the turbine is stopped and the impeller is at maximum speed. This is called torque converter stall. A great amount of heat will be generated under this condition and this shouldn't be allowed to happen for long. The heat will be generated by the friction of the oil that is travelling at a high rate of speed against the torque converter components.

As a torque converter uses oil to transfer torque, it is critical that the proper viscosity and type of oil is used. This oil and its filter must be changed on a regular basis as well.

Torque converters will most likely drive a powershift transmission. The torque converter could be mounted inside the transmission housing with the whole unit attached to the engine's flywheel housing. It could also be in the transmission housing but be separate from the engine with a drive shaft between the two housings. The torque converter could also be mounted to the flywheel housing in its own housing and drive the transmission with a drive shaft. ● **SEE FIGURE 10–16** for how a torque converter may be arranged in a machine's drivetrain.

> **WEB ACTIVITY**
>
> Go to http://www.youtube.com/watch?v=IeCEmJA0WsI to view a classic U.S. military video from the 1950s that demonstrates how a torque converter works.

FIGURE 10–15 Oil flow during vortex flow.

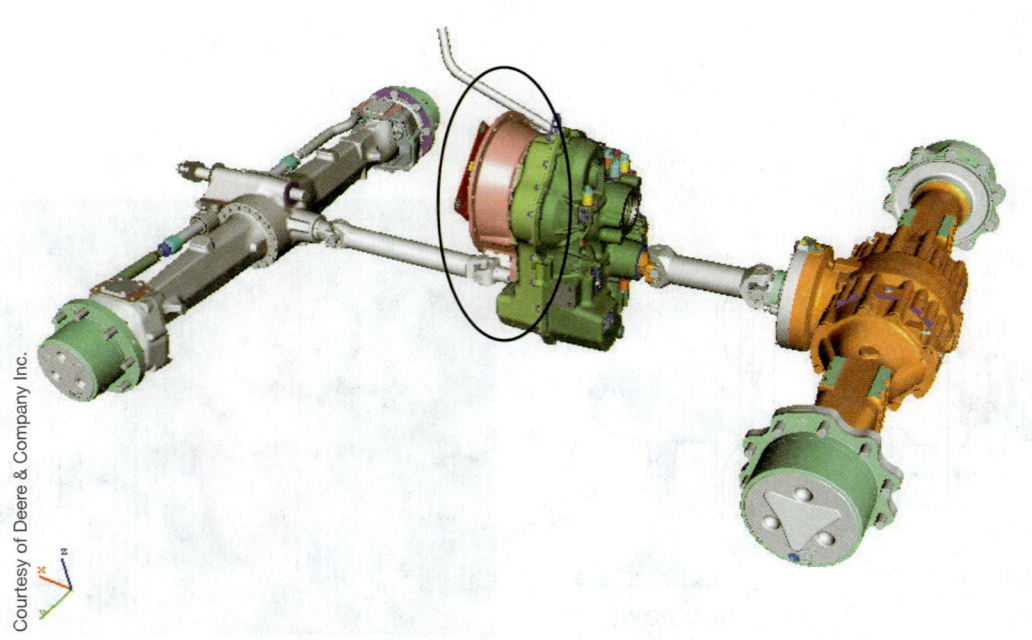

FIGURE 10–16 How a torque converter may be arranged in a machine's drivetrain.

FIGURE 10–17 A torque converter with pump driven by it.

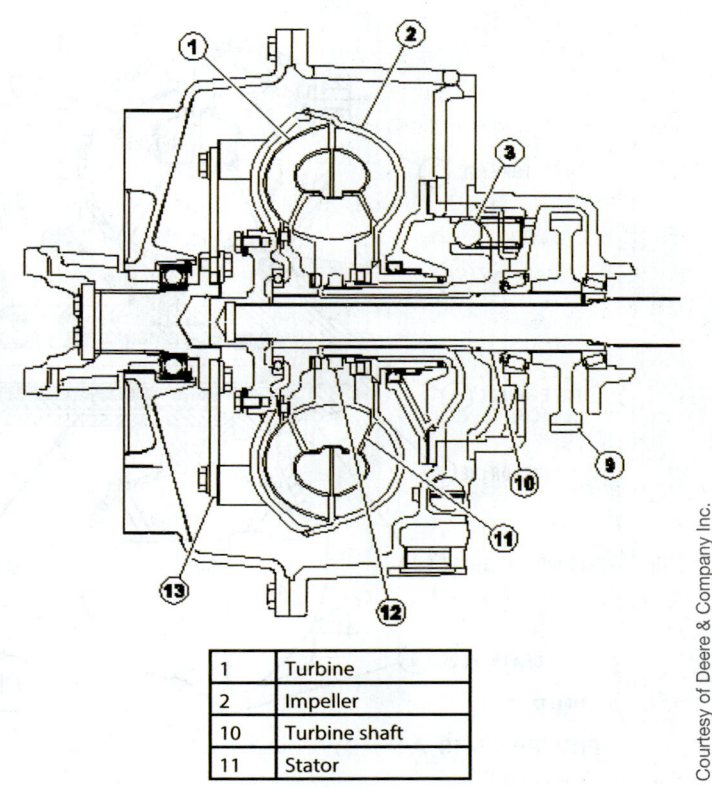

FIGURE 10–18 The main parts of a single-stage single-phase torque converter.

1	Turbine
2	Impeller
10	Turbine shaft
11	Stator

Many torque converter housings will also have pump drives coming off of them. There will be one or more gears driven by a gear that is driven by the impeller. Each gear will have a spline drive coming off of it to drive a pump or drive shaft for a PTO. Remember the impeller is connected to the engine flywheel so it can be used to drive other components. ● SEE FIGURE 10–17 for a torque converter with pump driven by it.

DIFFERENT CONFIGURATIONS OF TORQUE CONVERTERS
Torque converters can be configured to have different characteristics for different machine applications such as low travel speed and high torque multiplication, high travel speed and little torque multiplication, or a combination of both.

The simplest variation will be termed a *single-stage single-phase torque converter*. This means the impeller is one piece and the stator is stationary. A typical application for this is for a small slow speed machine like a light dozer. ● SEE FIGURE 10–18 for the main parts of a single-stage single-phase torque converter.

The stator will actually impede oil flow as the torque converter moves toward rotary flow but this isn't likely to happen too often with this type of machine. This machine will mostly need torque multiplication from the torque converter. For this application the torque converter has its own housing attached to the engine's flywheel housing and drives the transmission with a drive shaft.

The engine's flywheel will drive the impeller at the same rpm the engine is turning. The impeller is at the back of the torque converter housing and forms a shell around the outside of the turbine. It will be supported at the back of the housing with a bearing. The turbine is at the front of the housing and fits very closely to the impeller. Their outer circumference mating faces will only be a few thousands of an inch apart. This clearance is critical to the efficient operation of the torque converter because excessive clearance will equate to internal leakage as the impeller is sending oil to the turbine. When a torque converter is being reconditioned, this clearance dimension should be checked carefully.

The turbine rides on bearings in the housing and impeller and its output will be a splined shaft out the rear of the torque converter. The stator fits in between the turbine and impeller, and in this configuration, it will be splined to the housing to keep it stationary.

The next variation of a torque converter will be a single-phase dual-stage type. ● SEE FIGURE 10–19 for a torque converter with a freewheeling stator.

The impeller, turbine, and stator arrangement are the same as the single-stage single-phase type.

The term *dual phase* indicates that it has a stator that will freewheel in one direction. This will happen when there is a light load on the turbine and the torque converter is getting close to rotary flow.

CLUTCHES, TORQUE CONVERTERS, TORQUE DIVIDERS, AND RETARDERS   315

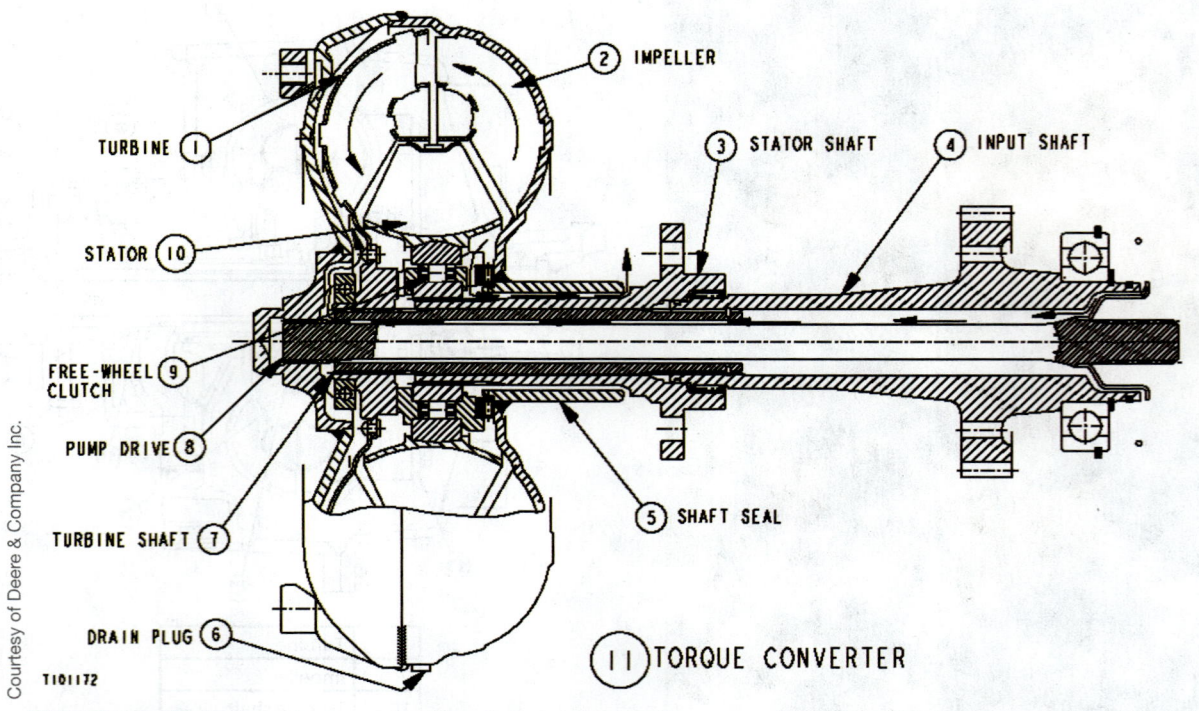

FIGURE 10–19 A torque converter with a freewheeling stator.

There will be a one-way clutch that locks the stator when it tries to rotate one way for rotary flow or higher load conditions but will unlock when the load decreases for a more rotary flow situation. This will reduce resistance caused during rotary flow if the stator was stationary. A typical application for this type is a backhoe loader where most of the torque multiplication comes from the transmission, and the machine needs to also travel relatively fast under light load when travelling on a road between jobs.

Other variations of torque converters are discussed next.

VARIABLE IMPELLERS
Torque converters that use variable impellers are sometimes called variable capacity torque converters and are commonly used for larger wheel loaders. The operator can vary the drive between the flywheel and the impeller or between the second section of a two-piece impeller and a modulated oil-applied clutch (see Chapter 12 to learn more about this type of clutch). This will partially drive the impeller or the second section of the impeller and, therefore, not send as much oil to the turbine. This will lower the torque output of the torque converter. Think of this as a machine having an impeller that can change its size and output. ● **SEE FIGURE 10–20** for a variable capacity torque converter.

The main reason for this is to be able to cut back the torque sent out to the rest of the drivetrain to reduce wheel spin and have more power available to operate the hydraulic system of the machine. Some machines will reduce the impeller clutch pressure every time there is a transmission shift to reduce shock loads on the drivetrain.

Another term for this feature is *reduced rimpull*. Older machines had a lever that controlled a spool valve that would vary the oil pressure to the clutch. Newer machines will do this electronically.

Some wheel loaders will incorporate the clutch control into the left pedal. The first part of brake pedal travel will be controlling the variable torque converter clutch and will decrease the torque flow more as the pedal is pushed farther. If the pedal is pushed far enough, it will apply the brakes. These machines are designed to be run with the engine at high idle all the time and travel speed is mostly varied by using the left pedal. With the engine at high idle, maximum hydraulic flow is available anytime.

LOCK UP TORQUE CONVERTERS
There will always be a certain amount of slippage with a torque converter. In other words, a torque converter will never be able to transfer 100% of the flywheel torque through it. Because of the necessary clearances between the moving parts of a torque converter, there will be internal leakage. This equates to a loss of efficiency.

To eliminate this problem, many torque converters will have a **lock up clutch** that will lock the impeller to the stator. The clutch is oil applied and spring released and could be single or multi-disc (see Chapter 12 to learn more about this type of clutch). ● **SEE FIGURE 10–21** for a lock up torque converter.

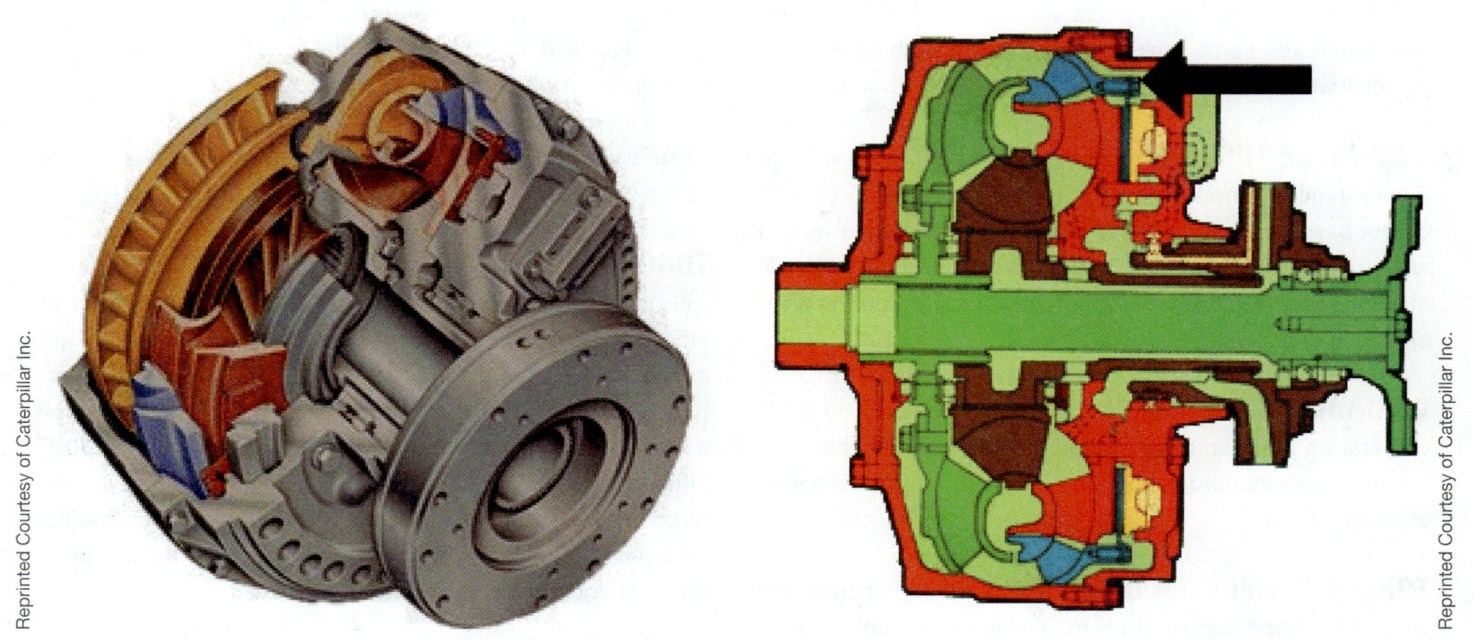

FIGURE 10–20 A variable capacity torque converter. The arrow points to the second impeller.

This will give a direct drive through the torque converter just like an engaged mechanical clutch. In machines such as rock trucks, articulated trucks, and scrapers, the clutch will lock the torque converter most of the time in all gears but first and reverse. As the machine accelerates and changes gears, the lock up clutch will momentarily unlock to help give the machine a smoother shift. In the lowest gear and reverse, the torque converter will remain unlocked until the load decreases enough as deemed by the ECM. On newer machines, the clutch will be controlled electronically and an ECM can do this automatically, or if the feature is turned off, the operator has control of the lock up. Wheel loaders may also use this system to make the machine more efficient when travelling. Some wheel loaders are designed to be load-and-carry machines, meaning they will fill their bucket and then travel a long distance to dump it.

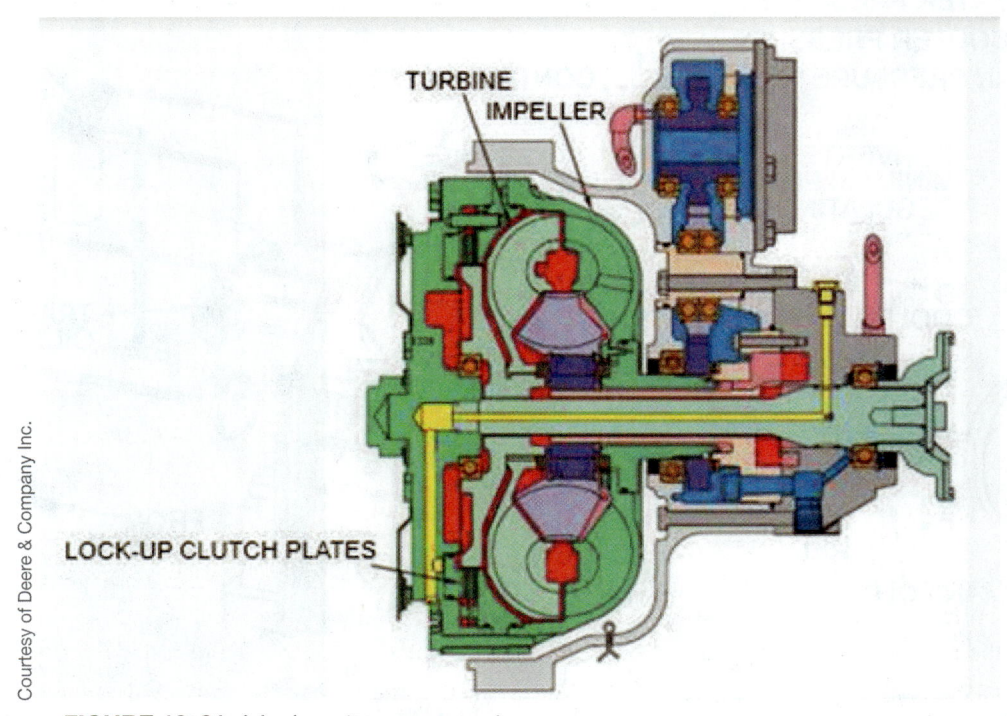

FIGURE 10–21 A lock up torque converter.

CLUTCHES, TORQUE CONVERTERS, TORQUE DIVIDERS, AND RETARDERS

Lock up torque converters will also be able to let the engine provide some braking action if the machine is going downhill whether the machine has an engine brake or not.

TWO-PIECE TURBINES Some torque converters will use a two-stage turbine to allow the torque converter to vary its output based on load changes. Like the variable impeller, this will also be controlled by a hydraulic clutch. Again this will make the torque converter more effective over a wider range of operating conditions.

VARIABLE VANE STATOR Some torque converters will use stators that can have the pitch of their vanes changed. This will make the torque converter more efficient under a variety of conditions.

TORQUE CONVERTER OIL SYSTEM Torque converters will need a good supply of clean, air-free, and relatively cool oil. This will usually come from the transmission oil system; however, some machines could have a separate oil system for the torque converter. Most heavy-duty torque converters will have their oil pressure regulated coming in and going out of the housing. A typical regulated inlet pressure is just under 100 psi and there will be a valve to regulate the oil pressure in the torque converter to a maximum of 130 psi. This will ensure that the torque converter housing is always filled with oil.
● **SEE FIGURE 10–22** for a pictorial schematic of a transmission oil circuit with the torque converter included.

If there is a lock up clutch, it will have to have a regulated oil supply as well. It will likely be regulated at the same pressure as the clutches in the transmission. This will be 250–350 psi.

TORQUE CONVERTER OIL COOLING SYSTEM There will be a lot of heat generated by the torque converter at times, and this oil will need to be cooled down so it doesn't oxidize and become ineffective. There will likely be temperature sensors at the torque converter outlet that will turn on a warning light or buzzer when the oil gets to around 230°F. This oil will typically be sent to a cooler when it leaves the torque converter. This could be an oil to air cooler on smaller machines but will likely be an oil to engine coolant type cooler on most equipment. ● **SEE FIGURE 10–23** for an oil cooling circuit.

TORQUE CONVERTER PROBLEMS The biggest operator complaint related to torque converters is there is a lack of drive power. This could be related to other systems on the machine, but once it is narrowed down to the torque converter, it could be caused by a few different things. The second most common problem is oil overheating. Again there could be a few

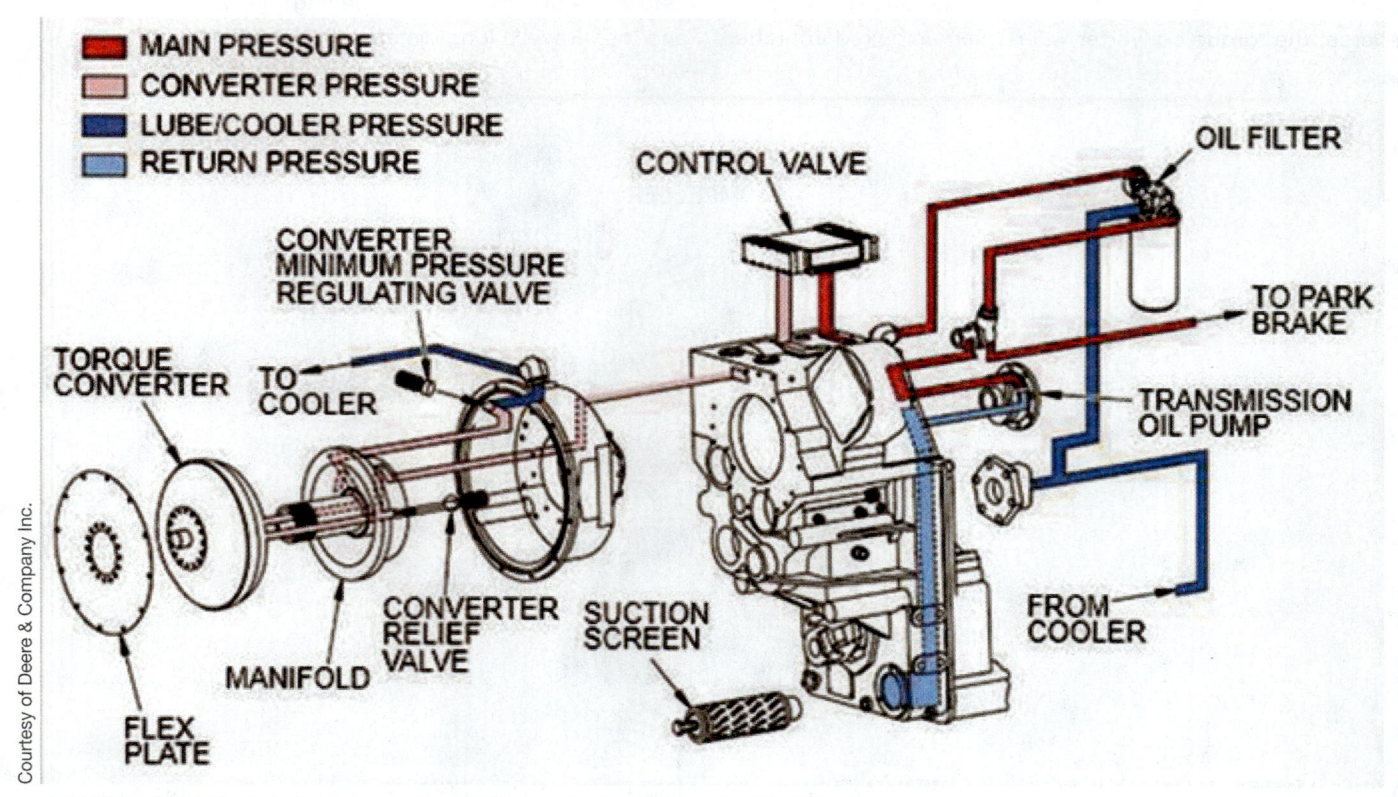

FIGURE 10–22 A pictorial schematic of a transmission oil circuit with the torque converter included.

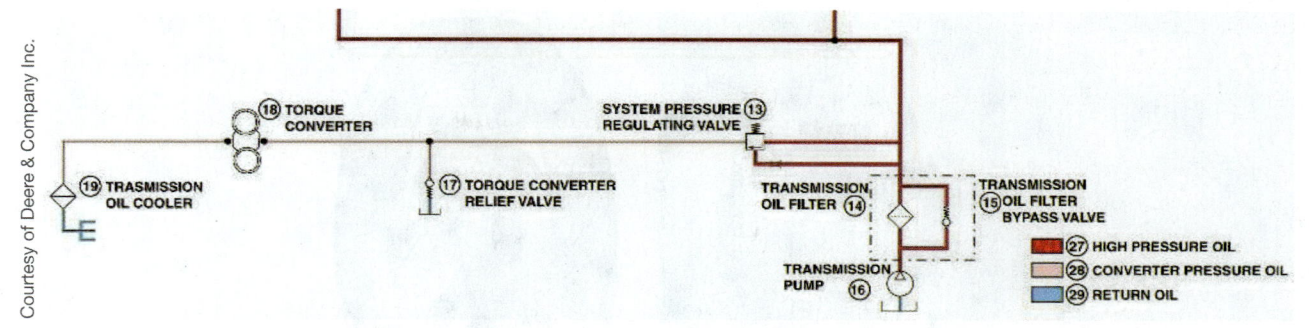

FIGURE 10–23 An oil cooling circuit.

different reasons for a torque converter to overheat, but one of the biggest causes is the operator working in too high of a gear range. This will make the torque converter try to compensate for the wrong gear ratio by staying in vortex flow too long, which will then overheat the oil.

The following are some operator complaints that may lead to finding a problem with a torque converter:

- no drive
- low drive power
- drivetrain overheating
- fluid leak
- vibration

After you have verified the complaint and performed a visual inspection, start to check for simple things and then work toward more compicated diagnostic procedures such as installing pressure guages. Newer machines with electronics integrated into the drivetrain could help by logging fault codes related to the torque converter or by checking logged data such as gear selection, engine rpm, and drivetrain oil temperature.

One fairly easy test that could indicate a possible torque converter problem is a torque converter stall test. Check the machine manufacturer's service information for the proper test procedure for the specific machine you are looking at.

Here is a general test procedure for performing a torque converter stall. Monitor oil temperature closely to avoid overheating the oil when performing this test:

1. Make sure that oil type, viscosity, and level is correct.
2. Get drivetrain oil to normal operating temperature (usually 160–180°F).
3. Chock wheels and make sure that no one enters immediate area surrounding the machine.
4. Put machine into highest gear possible.
5. Apply service or parking brakes.
6. Move engine governor control to maximum.

FIGURE 10–24 Dash of a loader in a stall condition.

7. Read engine rpm as soon as it stabilizes.
8. Compare to specification.

If engine rpm is too high, this could indicate a torque converter problem. ● SEE FIGURE 10–24 for the dash display of a loader in a stall condition.

You will notice the top left gauge has moved toward the yellow area to indicate the torque converter oil is getting hot.

TORQUE DIVIDERS

Many medium to large tracker-type dozers will include a torque divider as part of their drivetrain.

A torque divider is a combination of a torque converter and a set of planetary gears that provide torque multiplication two different ways. When there is a load on the machine that causes the torque divider to multiply torque, the torque converter section will supply 70% of the torque with the remainder supplied by the gear set. ● SEE FIGURE 10–25 for a torque divider.

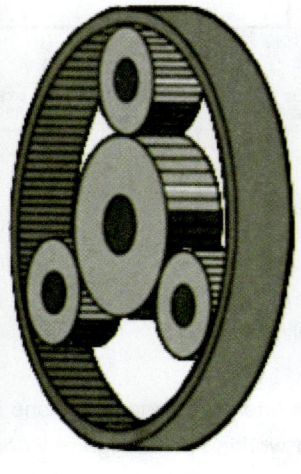

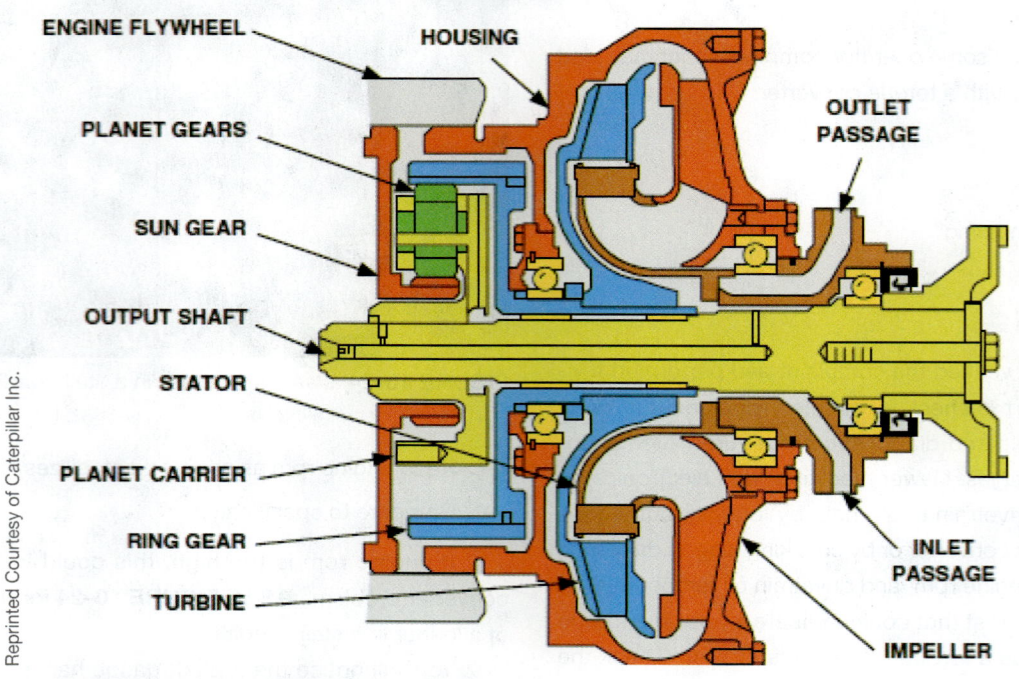

FIGURE 10–25 An illustration of a torque divider.

The torque converter is a simple single-stage single-phase unit, but it has parts of a planetary gear set connected to it (see Chapter 12 for a complete explanation of planetary gear sets). The flywheel-driven impeller is connected to the sun gear and the turbine is connected to the ring gear. The output shaft of the torque divider is driven by the planet carrier.

When the torque converter is in rotary flow during a low load condition, there will be very little relative movement between the impeller and turbine, and therefore, the planetary gear set will also be spinning as a unit with no torque multiplication.

When a load starts to slow down the output shaft, this will slow the planet carrier, and since the engine rpm hasn't changed, there is a speed difference between the sun gear and planet carrier. This will cause the planet gears to rotate, which then causes the turbine to slow down. As there is now a difference in speed between the impeller and turbine, the torque converter starts to move into vortex flow, which multiplies torque. Torque is also multiplied through the planetary gear set.

If the output shaft were to completely stop in a stall condition while the impeller was still moving, the turbine would actually be turning in reverse. The maximum torque output of a torque divider happens just before the turbine starts to reverse rotation.

320   CHAPTER 10

The higher torque multiplication characteristics of the torque divider make it ideal for heavy dozing applications.

HYDRAULIC RETARDERS

The term *retarder* can refer to a few different types of devices meant to slow the travel speed of a machine. The retarder that will be discussed here will be part of a drivetrain for a machine that is capable of a high travel speed of over 25 mph (haul trucks and scrapers). It is almost always incorporated into a powershift transmission housing and could be at the transmission input or output. However, some retarders are between the torque converter and flywheel and some are stand-alone units mounted somewhere along the driveline past the transmission. If a retarder is located ahead of the torque converter, then the torque converter must have a lock up clutch that is activated for the retarder to provide driveline braking. ● **SEE FIGURE 10–26** for a cutaway view of a retarder used on an articulated rock truck.

A hydraulic retarder can be thought of as the opposite of a torque converter and also works on hydrodynamic principles. Its purpose is to slow the machine down without the use of wearable service brakes and it does this by directing oil at a rotor that is connected to either the input or the output shaft of the machine's transmission. They are ideal for holding a machine at a steady speed when going down a long grade without using service brakes.

They will generate a great deal of heat because like any kind of braking system a retarder is simply an energy conversion device. It will convert the kinetic energy of a moving machine into heat. This heat will be absorbed into oil, which then must be cooled down. A properly functioning oil cooling system is a must for a machine with a retarder.

Retarders are fairly simple in design and only have one moving part, which is called the rotor. The **rotor** will have vanes or fins on both sides and will rotate inside a housing with a stator on one or both sides. The stators are stationary and will also have fins. These fins are much simpler in design than torque converters because their purpose is to discourage oil flow.

When the operator or ECM wants the retarder to slow the machine down, a valve will be activated that will send oil to the retarder housing. The housing is normally empty so as oil enters the housing, the stator fins will try to keep the oil stationary. As the oil contacts the moving rotor, the rotor will be slowed down. As the rotor is part of the driveline, the machine will be slowed down. The horsepower equivalent of a retarder on a 300 hp machine could be as high as 800 hp. To achieve a higher retarding effect, the valve has to simply allow more oil into the housing. To see a schematic of the oil circuit for a retarder, ● **SEE FIGURE 10–27**.

The oil will be heated up as it is impacted by the rotor and will be sent to a cooler when it leaves the housing. The cooler is most likely coolant cooled and transfers heat into the engine's cooling system.

The control valve could be air or electrically activated. The operator could control the system with a lever on the steering column or a foot pedal. ● **SEE FIGURE 10–28** for a foot pedal control of a retarder.

It could also be automatically controlled by an ECM.

RETARDER PROBLEMS The most common complaint that could lead you to a retarder problem is that it isn't slowing the machine as good as it should. Make sure that the operator is using the system properly. Most newer machines will prevent improper operation.

After the complaint has been verified and if the machine has an electronically controlled retarder, you should check for fault codes. Oil condition should be checked. Foamy oil will not provide good retarder function.

The next common problem with retarders is that the oil is overheating. If this is the case, after you confirm the system is being operated correctly, then a few checks with a heat gun may find that the oil isn't reaching the cooler, the oil isn't circulating, or the cooler is plugged either internally or externally. An infrared thermometer will be useful for trying to find oil overheating problems.

RETARDER REPAIR OR REPLACEMENT Since the retarder is a simple device, it will be rare that something goes wrong with it. If it is part of a transmission or torque converter, then it should be inspected and reconditioned if necessary when these components are repaired or reconditioned.

FIGURE 10–26 A cutaway view of a retarder.

Transmission Retarder ON

1. Retarder Housing
2. Temperature Sensor
3. Flow Valve
4. Orifice
5. Transmission Cooler
6. Check Valve
7. Retarder Control Valve
8. Retarder Charge Pressure Test Port
9. Main Oil
10. Control Oil
11. Return/Suction Oil
12. Converter Oil
13. Cooler/Lube Oil

Courtesy of Deere & Company Inc.

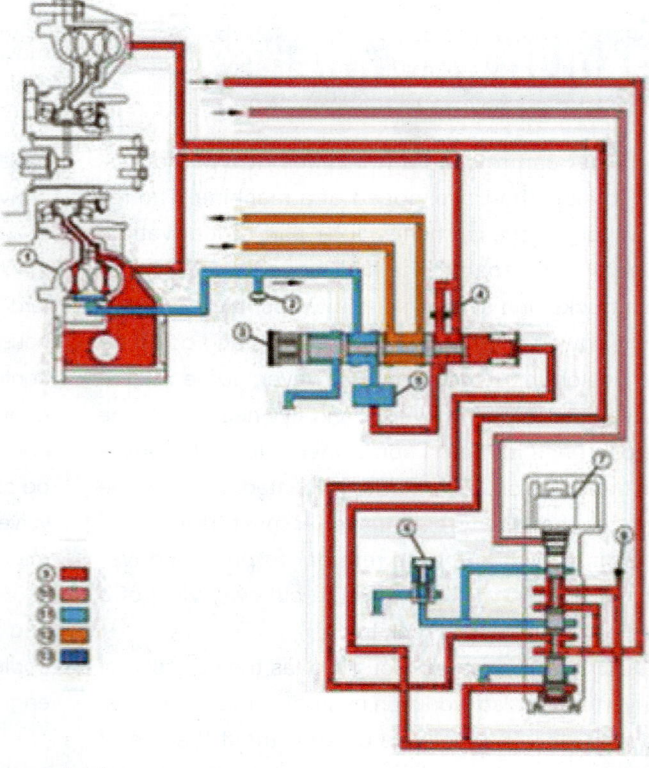

FIGURE 10–27 A schematic of the oil circuit for a retarder.

FIGURE 10–28 A foot pedal control for a retarder.

Progress Check

6. The driven member of a torque converter is the:
   a. lock up clutch
   b. stator
   c. impeller
   d. turbine

7. If a torque converter has a one-way clutch, it will work with the:
   a. lock up clutch
   b. stator
   c. impeller
   d. turbine

8. Torque multiplication is possible when a torque converter is in this mode:
   a. rotary flow
   b. vortex flow
   c. lock up
   d. dual phase

9. A torque divider will make its maximum torque when:
   a. the impeller almost stops
   b. the sun gear stops
   c. the turbine is about to turn backward
   d. the planet carrier is at maximum speed

10. A hydraulic retarder will use this moving component to slow a machine down:
    a. the planet carrier
    b. the turbine
    c. the rotor
    d. the stator

322 CHAPTER 10

SHOP ACTIVITY

- Find a machine with a torque converter. After performing a circle check, start the machine and move the machine to a safe location with no one else close by.
- Put the machine in its lowest gear. With the brakes released, does the machine start to travel with the engine at low idle?
- With the service brakes applied, will the machine start to move when the engine is revved up?
- With the transmission in a higher gear, will the machine start to move when the engine is revved up?
- With the machine in its highest gear and the service and parking brakes applied, what is the highest engine rpm achieved? What is this speed called?

SUMMARY

1. Safety concerns related to clutches, torque converters, torque dividers and retarders are: heavy component lifting injuries, burns from hot components, clutch material dust
2. A diesel engine's flywheel can drive a variety of different components such as: clutch, torque converter or torque divider
3. A friction type mechanical flywheel clutch allows the operator to transfer torque from one drive component to a driven component. Typically the diesel engine flywheel is the drive member and a transmission (manual or powershift) is the driven member
4. Friction between independent clutch components must be strong enough to allow torque transfer through it. A disengaged clutch doesn't allow torque transfer through it while an engage clutch provides 1:1 direct drive.
5. The main components of the friction type clutch are either drive: clutch cover, pressure plate or drive: clutch discs\s. There will also need to be a release bearing, actuator mechanism
6. To increase the torque capacity of a friction clutch designers will increase spring pressure and increase friction surface area. Extra clutch discs and intermediate plates can be added
7. The clutch cover mounts to the engine flywheel, and drives the pressure plate. Springs will try to move the pressure plate towards the flywheel and clamp one or more friction discs to the flywheel
8. Clutch release mechanisms can be linkage, cable or hydraulic master and slave cylinder arrangements
9. Over center type friction clutches use mechanical linkage instead of spring force to create a clamping force
10. Heat is the biggest enemy of clutches and is caused by slippage when clutches are between full engagement and release. Some clutches run in oil to keep them cool and are called wet clutches
11. Friction type clutches will need regular adjustments as the friction material wears away to maintain a specific clearance between the pressure plate, flywheel and clutch disc/s
12. Common operator complaints related to friction clutches are: noisy, won't engage, won't release
13. Clutch repair usually requires removal of complete assembly which usually requires engine or transmission removal. The flywheel face should also be reconditioned at this time
14. Other types of clutches are: dog clutch, one way clutch and expanding shoe clutch
15. Torque converters transfer torque seamlessly from an engine's flywheel to the machines transmission. Fluid is used as the medium to transfer torque through a torque converter by using the inertia of centrifugal force
16. The three main components of a simple torque converter are: the engine driven impeller, the stator and the turbine (output member). All of these components have a series of curved fins or vanes to direct the fluid
17. Oil is directed towards the turbine from the impeller and as oil exits the turbine it flows through the stator which redirects it back to the impeller
18. When the turbine has a light load on it the oil flow in the torque converter is said to be rotary flow. When the turbine has a large load on it the oil flow is said to be vortex flow
19. When the turbine is stationary and the impeller is turning the torque converter is said to be in stall
20. When vortex flow occurs there is a multiplication of torque through the torque converter. This can be as high as 4:1. Vortex flow also creates a lot of heat in the oil and if allowed to increase too high can damage
21. There are several different configurations of torque converters that can be used for different applications. Lock up torque converters have a clutch that can lock the impeller to the turbine for true 1:1 drive. Variable capacity torque converters have a second impeller with a clutch to provide a variable amount of fluid flow. Stators can have a one way clutch to allow them to freewheel in low load conditions. Some torque converters can also have two piece turbines and variable pitch stators

22. Torque converters need to have an oil supply system and valves to regulate the pressure inside the housing. The oil also needs to be cooled and the torque converter oil supply is usually from the power train oil system
23. Operator complaints related to torque converters can be: not enough power, overheating, noises or vibrations
24. Checking torque converter stall speeds can determine if a torque converter has excessive leakage
25. Torque converter repair will require removal, disassembly and inspection. Major components can be reused if deemed Ok but bearings seals and gaskets should always be replaced
26. A torque divider is a combination of a planetary gearset and a torque converter. It will be used for high torque low speed applications like bulldozers. The planetary gearset provides an extra torque multiplication when loads on the turbine become high
27. Hydraulic retarders can be part of a machines drivetrain and are used to slow down a travelling machine. They are found in machines that travel at high speed such as haul trucks and articulated trucks
28. They have two main components: a housing and a rotor and work opposite to a torque converter. The housing and rotor have fins and as oil is let into the housing it acts to slow the rotor down because the fins cause turbulence. The rotor is connected to other driveline components such as: transmission, torque converter or drive shaft
29. Hydraulic retarders are actuated by allowing oil into the housing by an operator controlled valve. Oil temperature rises quickly and oil will be sent directly to a cooler

FUNDAMENTALS AND MANUAL TRANSMISSIONS

LEARNING OBJECTIVES

After reading this chapter, the student should be able to:

1. Explain how torque is transferred through gears.
2. Calculate gear ratios.
3. Describe the different types of gears used in heavy equipment.
4. Compare levers to gears.
5. Describe how a manual countershaft transmission works.
6. Explain the different types of shifting mechanisms for manual transmissions.
7. Describe a troubleshooting procedure for an operator complaint.
8. Explain what an auxiliary transmission is.
9. Explain how a shifter mechanism works.

KEY TERMS

Countershaft 336
Gear reduction 329
Helical 333
Overdrive 329
Ratio 329
Reverse idler 337
Spur 333
Synchronizer 337
Underdrive 329
Yoke 337

INTRODUCTION

SAFETY FIRST When working with gears and manual transmissions, there are some specific safety concerns an HDET should be aware of. As always proper PPE needs to be worn. This is a minimum of safety boots and glasses.

Some repair procedures will require the use of pullers to remove gears with several tons of force. Be aware of the possibility of a sudden release of energy when using pullers. There should be a way to control the travel of the part you are trying to remove like using a large nut or plate on the end of the shaft.

When required, you may have to strike gears to remove them. Never do this on their teeth and always use something soft like a brass punch or plastic hammer. Gear teeth are brittle and flying pieces will penetrate skin and eyes easily.

There may be sharp edges on gear teeth and pinch points when working in close quarters. Wear gloves when handling gears.

Many gears and transmission subassemblies will be heavy. Know your safe lifting limit and how much something weighs before trying to lift it by yourself or with a lifting device.

You will likely need to remove snap rings when working with gears. Special caution should be taken when doing this. Gloves and glasses should be worn because they are made of spring steel and have a tendency to take flight when being removed or installed.

GEAR HISTORY

The concept of transferring torque through a set of gears has been around for centuries. In North America in the 1800s, wooden gears were used to transfer the drive from water turbines to drive shafts in water-powered saw and grist mills. In fact, there are still some in operation today! ● SEE **FIGURE 11–1** for a set of wooden teeth of a gear.

A simple gear can be any disc with a set of evenly spaced teeth around its outer edge circumference. A gearset will include two or more gears in mesh with the same tooth spacing (pitch): one gear supplies the drive (input) while the other is driven (output). Some gears simply transfer motion and are called idler gears.

Compound gears will have more than one set of teeth using a common hub.

GEARS IN HEAVY EQUIPMENT Gears are used for many purposes on a heavy equipment machine and there are lots of different features about gears that you need to know. A lot of

FIGURE 11–1 A set of wooden teeth for a gear.

service procedures such as checking fluid levels and changing fluids and filters can greatly affect the life of gears. When gears wear out or fail, there is likely a fairly extensive and expensive repair procedure that follows. Improper repairs to components with gears can lead to major repairs. Learning about gears will help you understand how many different components function on heavy equipment and emphasize the importance of proper maintenance and repair procedures.

After discussing gears, this chapter will move into looking at manual transmissions and how their gears work together to change speed and torque in a drivetrain.

A machine's diesel engine will provide an output of horsepower, which is a factor of speed and torque. The diesel engine itself will likely have one or more sets of gears in it (see Chapter 6 for more information on prime movers). The engine's output needs to be converted to make the machine work. This can be done electrically, hydraulically, or mechanically. Mechanical power transfer can be done with chains and

FIGURE 11–2 An excavator's Rotex gear.

TECH TIP

When gears mesh together and are under load, they will transfer that load to the bearing, shaft, and component they are mounted to. The higher the resistance to turning the output gear creates a higher load. As the load increases, the meshing gears will try to push away from each other. For an extreme example, think about a fully loaded 400 ton rock truck that is stopped on a steep grade going up. Now that's resistance! When the operator wants to get the truck going, he will put the governor pedal to the floor with the transmission in gear. Approximately and 10,500 ft/lb of torque will be transferred and multiplied many times through several sets of meshing gears and all that torque will be placed on just a few individual teeth! Those teeth will transfer the load to the gear body, shaft, bearings, and components. As long as all those components are designed, assembled, and maintained properly, the truck will get moving and keep moving. Any flaw in any of those gears and related parts will lead to a messy failure.

sprockets, cables and pulleys, friction or gears. As mentioned in Chapter 10, if a machine has a mechanical drivetrain, it will have either a clutch or a torque converter bolted to the engine's flywheel. After that there will likely be many combinations of gears to get the engine's power to the tracks or tires. Transmissions, power dividers, axles, and final drives all make use of gears to change speed, torque, and direction of power flow.

All machines will use a set of gears for something. Even a hydraulic excavator will use several gearsets to make it work. It will have timing gears in its engine, a set of gears to drive its tracks, and a large gear to make the upper structure rotate. This is sometimes called the Rotex gear or Rotex bearing because it is actually a combination of gear and bearing. ● SEE FIGURE 11–2 for an excavator's Rotex gear.

This is an example of an internal gear because the teeth are on the inside of the gear.

Using gears to transfer torque is a very efficient method when compared to other power transfer methods such as fluid power or electrical. With gears there is only a slight loss of efficiency through gear tooth contact friction; in fact, most gear combinations used on a machine will have a theoretical efficiency of 98–99.5%. While still minimal, there is a higher loss of efficiency through bearing and lubricant friction (2%) and is mainly the result of the lubricant's viscosity between the meshing teeth and any fluid the gears and bearings are running through.

Gears used in heavy equipment drivetrains are cast from low to medium carbon steel and alloy steel for strength and then machined and ground to the required finish before they are heat treated for hardness. Their designed strength depends on their application and intended life span. Their teeth must be designed to handle both the surface loading and the bending load for their application.

They are usually mounted on shafts that are supported by bearings. Bearings allow rotation of the shaft and support the load on the gear and shaft. Gears can be connected to shaft's various ways. They can be splined, keyed, bolted, and pressed on to a tapered surface or a combination of these methods. Some gears will float on a shaft, meaning they will just be supported and not transfer drive to the shaft.

Most meshing gears have their center axes parallel to each other. Crown and pinion gears sets in differentials will have intersecting axes where their axes are at 90° to each other and some rear engine scraper drive differentials will have a less than 90° angled gearset. ● SEE FIGURE 11–3 for two examples of gear alignment (parallel and 90°).

GEARS AS LEVERS Two gears that mesh together can be compared to the action that occurs with a lever. As a gear is driven and mounted on a shaft, the force applied to the teeth on the gear's outer circumference will cause the shaft to turn. This is like a constant prying motion applied to the driven gear's teeth by the drive gear.

If two gears are in mesh and the drive gear's radius is half that of the driven gear, the torque increase will be double when comparing the torque input to the output. This is because of the leverage effect between the gears.

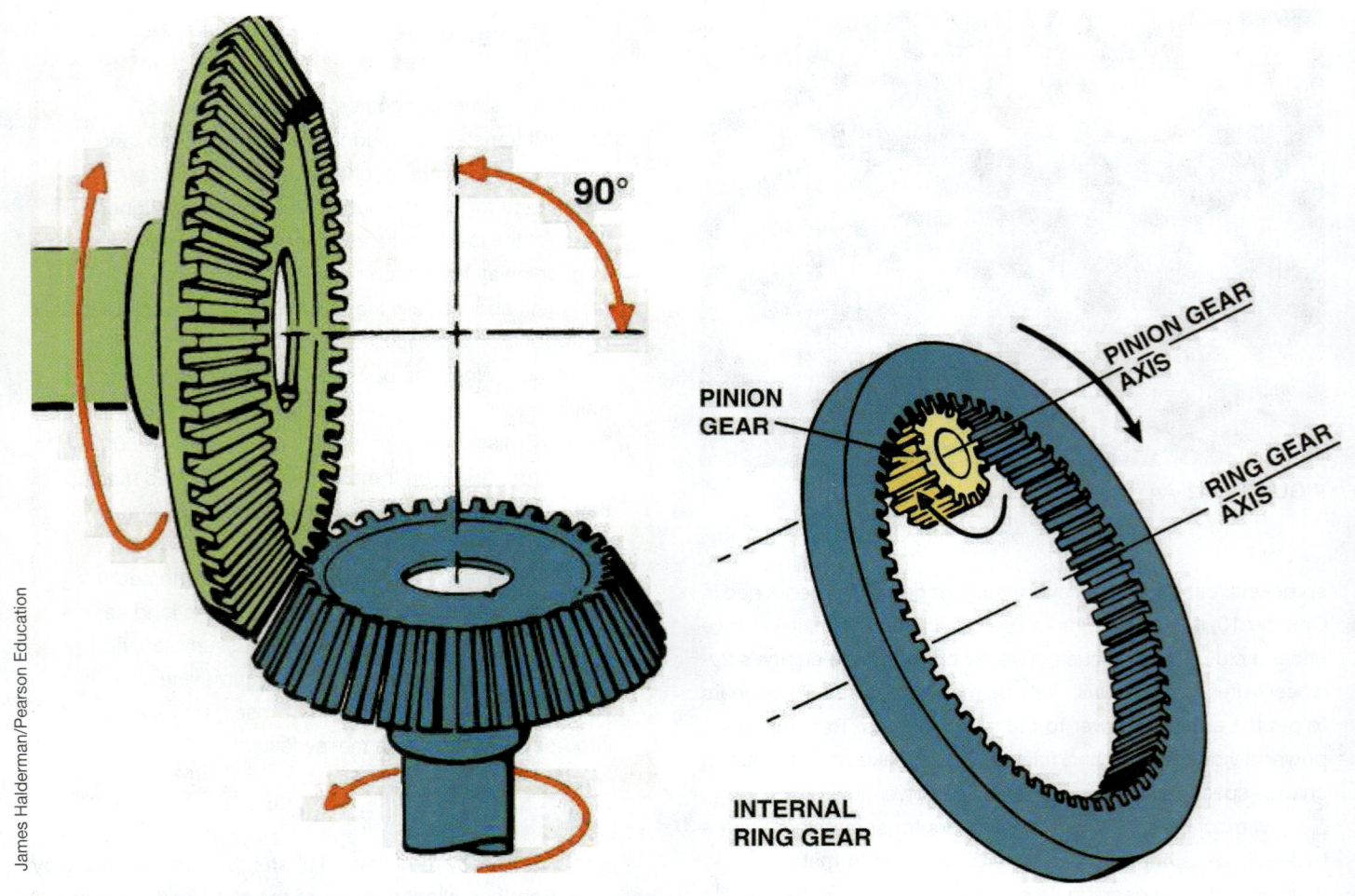

FIGURE 11–3 Two examples of gear alignment (parallel and 90°).

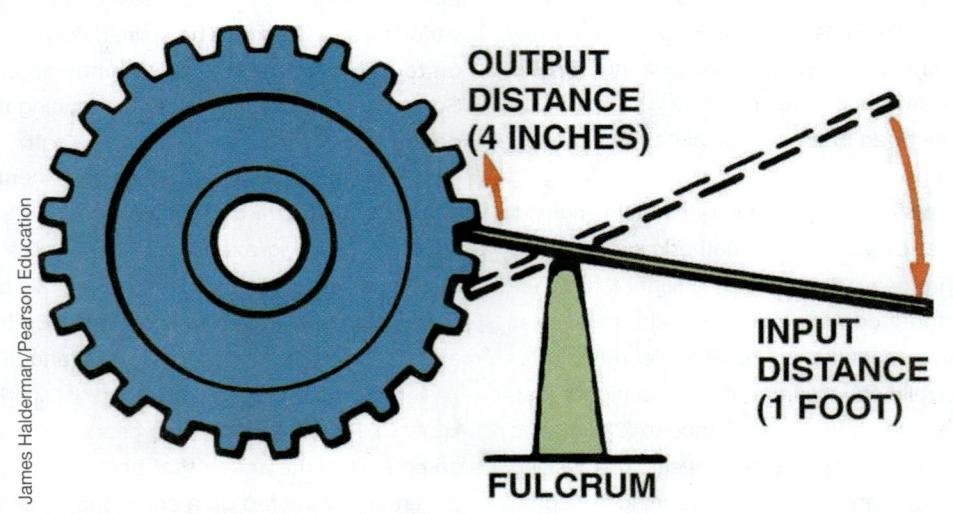

FIGURE 11–4 How gears and levers relate.

For example, if a (drive) gear with a 6 in. (½ ft) radius is turned with 100 ft-lb of torque and is meshed with a (driven) gear that has a 12 in. (1 ft) radius, the output of the driven gear will be 200 ft-lb. When the teeth mesh, there will be 200 lb of force applied to the larger gear (100 ft-lb ÷ ½ ft) at its radius. The 200 lb of force will give the output shaft a torque of 200 ft-lb (200 lb × 1 ft). You can now see how a mechanical advantage (or disadvantage) can be gained based on the radii of the two gears. ● **SEE FIGURE 11–4** for how gears and levers relate.

Gears are a mechanical means of moving rotation (although some gearsets can create linear movement) from a drive source to a driven output. Speeds between the two can be equal, or there can be a change in the drive-to-driven relationship—either faster or slower. As speed changes, so does the amount of torque.

GEAR RATIOS. The amount of torque increase or decrease between a drive gear and driven gear will be in direct relationship with their gear ratio. Any **ratio** is a comparison between two things. For example, if a mixture of epoxy and hardener called for a ratio of 3:1, you would squeeze out a quantity of hardener and then three-times that amount of epoxy. This is a comparison of volume between epoxy and hardener and, if measured correctly, would make the epoxy/hardener mix set properly.

A Gear ratio relates to the number of teeth on each gear in a pair, which in turn relates to their radius.

Gear ratios are expressed as the number of revolutions that the drive gear needs to turn to make the driven gear turn one complete revolution. To calculate a gear ratio, you would count the number of teeth on both gears and divide the driven by the drive. An example would be 40 teeth on the driven and 10 on the drive. This would give a 4:1 ratio. This really means it takes four revolutions of the drive gear to make the driven gear turn one revolution. This is considered to be an **underdrive** ratio or **gear reduction** because the drive speed is under the output speed. If one tooth was added to each gear, their relationship would change to a lower numerical ratio of 3.73 (41 ÷ 11). This would produce less torque at the output for the same input. These are popular rear axle ratios for pickup trucks.

If there was a need to have the output (driven) gear turn faster than the input, there needs to be more teeth on the drive gear than the driven. This is called an **overdrive** ratio because the drive speed is over the driven speed. For example, if the drive gear has 40 teeth and the driven has only 20, that would make a ratio of 0.5:1. This means it only takes one half of a turn for the drive gear to rotate the driven gear one revolution.

GEAR TORQUE VERSUS SPEED Gear teeth transfer torque from one gear to another using one or more teeth at a time. Torque is work and is most commonly measured in foot pounds (standard) or Newton meters (metric). For example, a shaft is directly driven by an engine that is creating 1000 ft-lb of torque and a gear is attached to the shaft that has 100 teeth with a radius of 1 ft. This means one of its teeth can transfer 1000 lb of force (1000 ft-lb divided by 1 ft) to a second gear. We'll say the engine is running at 1000 rpm. If the second gear being driven by the first gear is the same diameter with the same number of teeth, then it will have 1000 lb of force acting on one of its teeth. If the second gear is driving a shaft, that shaft will now have an output of 1000 ft-lb of torque, and as it has the same number of teeth, it will rotate at 1000 rpm. This is a direct drive ratio or 1:1 where the speed and torque values between drive and driven are the same. The gears would be turning in opposite directions. ● **SEE FIGURE 11–5** for a direct drive ratio.

If another gear had a radius of 0.5 ft, had 50 teeth, and is also being driven by the first gear, then its shaft will create half the amount of torque because the same 1000 lb of force is only 0.5 ft from the center of the shaft. The other factor that changes is speed. If the drive gear is still rotating at 1000 rpm, then the driven gear will turn at 2000 rpm. This is called an overdrive ratio. So as you can see, there is a trade-off of speed versus torque. If you think about speed and torque transfer through a set of gears and put speed and torque on opposite sides of a balancing scale, they would alternate moving up and down equally. ● **SEE FIGURE 11–6** for an overdrive ratio.

If the second gear was twice the radius as the drive gear and had 200 teeth, then it would rotate at 500 rpm or half the speed but its shaft would have twice the torque that the drive gear has or 2000 ft-lb. This is because the same 1000 lb of

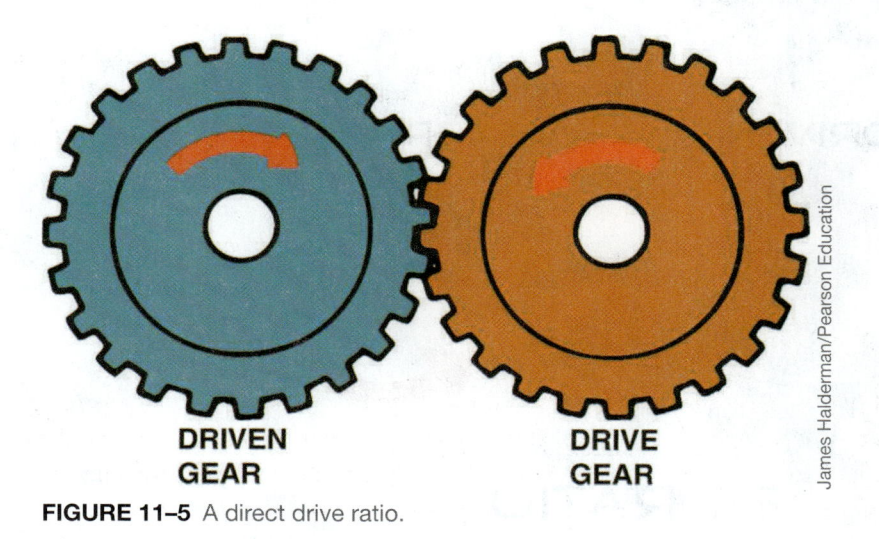

FIGURE 11–5 A direct drive ratio.

GEARING FUNDAMENTALS AND MANUAL TRANSMISSIONS

24 TEETH

8 TEETH

DRIVEN GEAR

DRIVE GEAR

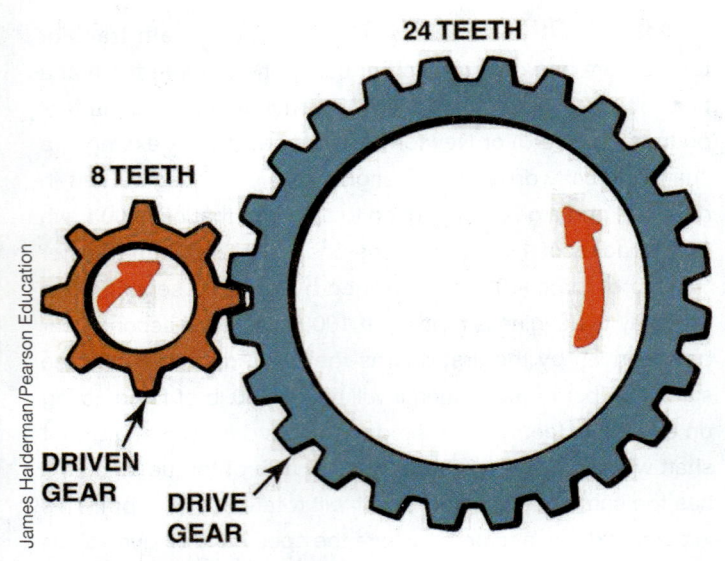

FIGURE 11–6 An overdrive ratio.

the same. This third gear is called an idler gear and will have no effect on speed or torque. ● **SEE FIGURE 11–8** for how an idler gear affects rotation.

At some point in the career of HDETs, they will be inspecting gears for damage to assess whether they can be reused or need to be replaced. They will likely be using some literature as a guideline that will have some terms in it that are unique to gears. The following section is meant to make you familiar with some common terms related to gears.

GEAR TERMS. ● **SEE FIGURE 11–9** for the different gear terms. These terms will apply to most gears used in heavy equipment.

Involute Curve. If you looked at a gear tooth from the side, you would see its profile or shape. Gear teeth have profiles designed to allow them to mesh properly with a minimum of friction. Most gears used for heavy

force being put out by one tooth is being applied to another tooth but that tooth is 2 ft from the center of the output shaft. ● **SEE FIGURE 11–7** for a gear reduction ratio.

Most heavy equipment drivetrain gear combinations will be gear reductions. This will convert the engine's speed and torque into higher torque and lower speeds.

As you can imagine, if two gears with external teeth are meshed together, then the rotating direction between the two is opposite. If a third gear is put between the drive and driven, then the rotation direction between the drive and driven stays

WEB ACTIVITY

Designing and manufacturing gears is a complicated process. If you want to learn more about gears, there are many good sources of information available by searching gear manufacturers and looking at their resources. The AGMA and JIS are also good sources of information. Search for videos that show how gears are made.

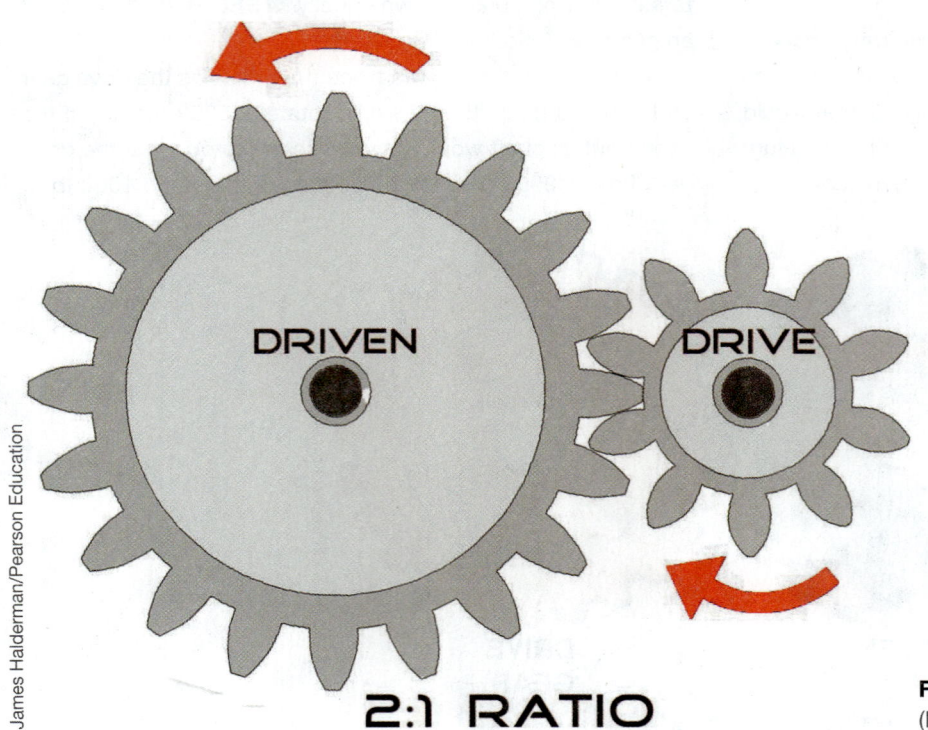

DRIVEN DRIVE

2:1 RATIO

FIGURE 11–7 To see a gear reduction ratio (Max Heard).

330 CHAPTER 11

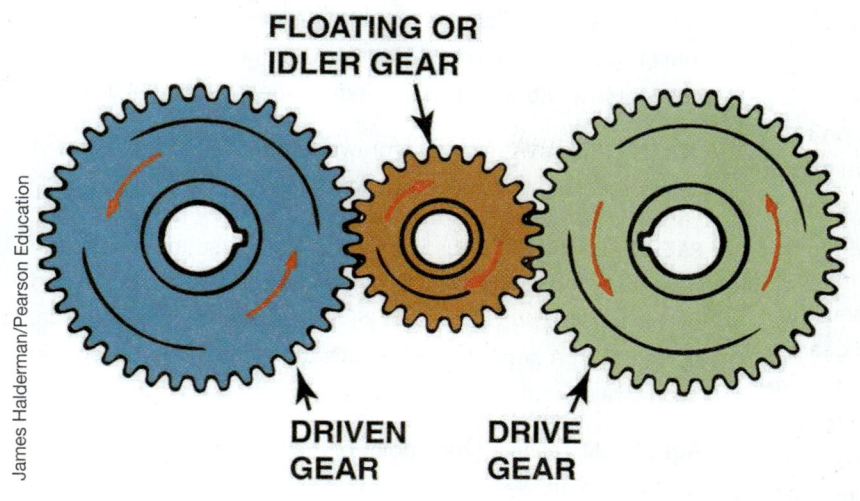

FIGURE 11–8 How an idler gear affects rotation.

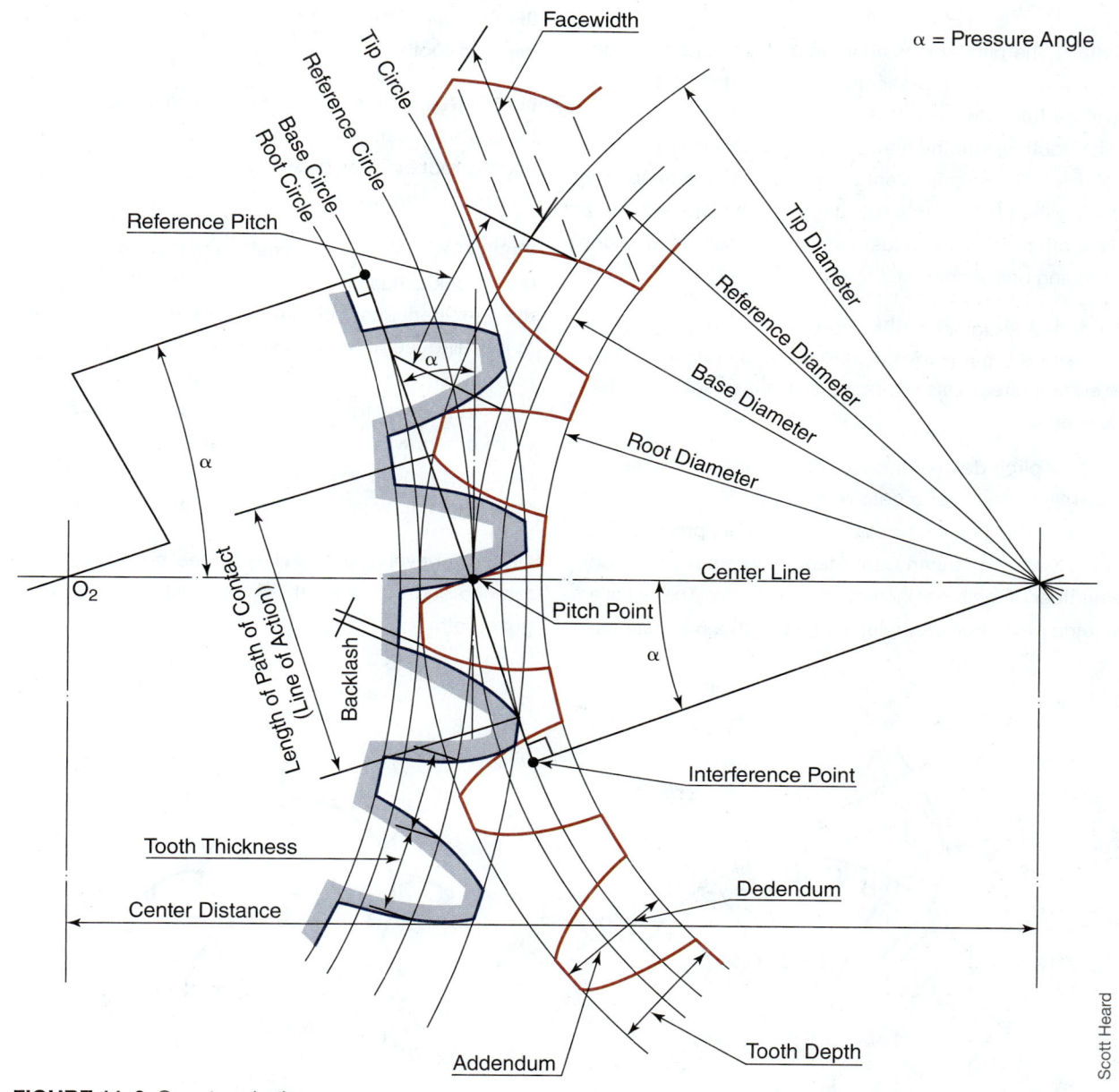

FIGURE 11–9 Gear terminology.

GEARING FUNDAMENTALS AND MANUAL TRANSMISSIONS

equipment will use an involute curve profile. This profile will allow a pair of mating gears to have more than one tooth to be in contact at once to spread the load out. The curve of the gear face is based on a line drawn by a string. If the string is held tight at one point on the circumference of the gear base circle (circle at the bottom of the teeth) and the other end is unrolled away from the base circle shape traced by the end of the string is called an involute curve. This tooth shape makes the meshing teeth of two gears roll together and not slide against each other, thus reducing friction and heat. This profile will also provide a constant speed between the two gears even under load. ● SEE FIGURE 11–10 for how an involute curve is created.

PITCH LINE. The pitch line is an imaginary line drawn across the face of a tooth that represents where two teeth of mating gears will be fully meshed. It is spaced evenly between the points on a tooth where the teeth start to mesh and where they go out of mesh. These points are called the dedendum and the addendum. When two teeth have their pitch lines joined, there will be one other set of teeth just coming into contact and one other set going out of contact.

GEAR FACE WIDTH. This is the width across the face of the gear. A wider face will allow the load to be spread out across a larger surface area. This will make a stronger gear but obviously heavier too.

PITCH. Gear pitch defines the size of a gear tooth in relation to the overall diameter of a gear or how closely the teeth are spaced around the gear circumference. For gears produced according to AGMA (American Gear Manufacturers Association), it is the number of teeth per inch of pitch diameter. This is called the diametric pitch. For example, if a gear with a 6 in. diameter has 24 teeth, then its pitch will be four. Any two gears in mesh must have the same pitch. Therefore, the previously mentioned gear could mesh with an 8 in. diameter gear if it had 32 teeth.

CENTERLINE. A straight line drawn through the center point of a gear reaching both outer circumferences of the gear.

PRESSURE ANGLE. Gear teeth are machined so that they mesh near the pitch line of the tooth. The pressure angle is created by the intersecting pitch lines of two adjacent teeth. A typical angle is 20°. A smaller angle means the teeth engage deeper in each other's root.

ADDENDUM. The highest point of the tooth that will contact a meshing tooth.

DEDENDUM. The lowest point of the tooth that will contact a meshing tooth.

CLEARANCE. When one tooth is meshed between two teeth on another gear, there will be a space between the tooth tip and the root or valley between the opposite teeth. This is called the tooth clearance.

BACKLASH. All gears in mesh need a certain amount of backlash. Backlash is clearance at the pitch line. This will provide the gear lubrication room to get between the contact surfaces of meshing teeth. A specified amount of backlash is needed to prevent binding and premature wear.

Some gear trains will allow the backlash to be adjusted while others will be nonadjustable. Many repair/rebuild procedures will require the setting of gear backlash. An example of adjusting a differential gearset will be covered in Chapter 15.

HEEL. A bevel gear tooth has its heel end at the larger circumference end of the tooth. ● SEE FIGURE 11–11 for a bevel gear tooth.

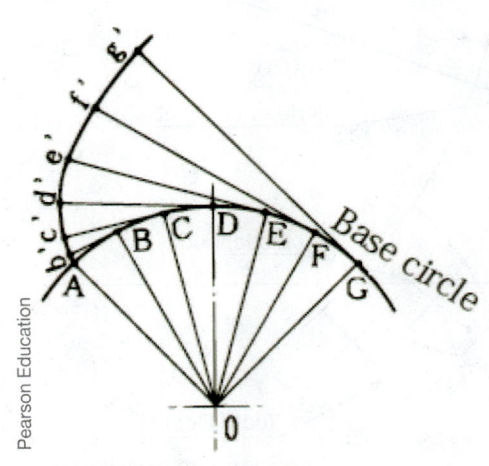

FIGURE 11–10 An involute curve.

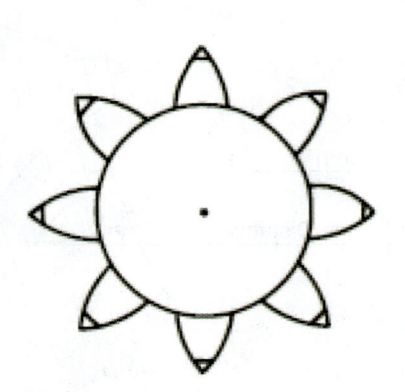

FIGURE 11–11 A bevel tooth gear.

TOE. A bevel gear tooth has its toe end at the smaller circumference of the tooth.

EXTERNAL GEAR. Most gears will have their teeth at the outside diameter of their circumference and are called an external (outside) gear.

INTERNAL GEAR. Some gears such as a planetary ring gear will have their teeth facing in on the inside circumference of its ring.

SPUR OR STRAIGHT CUT GEARS. These gears are the simplest type of gear. Their teeth are cut on the outside circumference of the gear and, as the name implies, are straight as you look at the top of each tooth. They have high load capabilities but are relatively noisy as their teeth mesh together. These gearsets will have parallel axes. **Spur** gears will only create radial loads when they are loaded. A radial load will push the loaded gear away from the drive gear.

When inspecting bearings that support spur gears, look at the area that is opposite the tooth mesh area as it will have the greatest radial load applied and show the greatest wear. ● **SEE FIGURE 11–12** for a spur gear.

HELICAL. This type of gear has its teeth angled as you look down at the top of a tooth. It has the advantage of spreading out the load over more teeth as it meshes with another gear. They are also quieter than spur cut gears. The disadvantage of the helical gear is that it develops thrust loads when under load and must be counteracted with an anti-thrust bearing. Mating helical gears will also have parallel axes. ● **SEE FIGURE 11–13** for a helical gear.

BEVEL GEARS. This type of gearset is used when a mechanical power flow needs to change direction. The meshing gears

FIGURE 11–12 A spur gear.

are said to have intersecting axis. This means the gears' centerlines will intersect. The most common intersect angle is 90°, and this type of gearset will be found in a differential to change power flow direction. The following are variations of bevel gears:

- Straight: The meshing teeth are similar to spur gear teeth but are tapered to allow proper meshing.
- Spiral: This type of bevel gear has spiraled teeth that make it a cross between a helical and a bevel gear.
- Hypoid: This is a variation of the spiral bevel with the drive pinion centerline below the centerline of the driven crown gear. This arrangement allows the input drive shaft to be lowered in the housing.
- Amboid: This is the opposite of the hypoid bevel in that its pinion input centerline is above the centerline of the crown. Some applications that use this arrangement need to have the drive shaft higher in the housing.
● **SEE FIGURE 11–14** for a set of bevel gears.

RACK. This is a flat gear that will create linear movement when reacting with another external tooth gear (pinion).

FIGURE 11–13 A helical gear.

FIGURE 11–14 A set of bevel gears.

TIMED GEARS. Some gears are manufactured as sets and must be timed when assembled to keep their matched surfaces running together. There will be timing marks that must be aligned when the gears are installed.

PINION. This could describe any external gear that is a lot smaller than the one it is meshed with. The best example is a set of differential bevel gears. The smaller drive gear is called the pinion.

CROWN. A common name for the larger gear of a bevel gearset used in a differential.

IDLER. Idler gears do not affect the ratios between the gears they are in between. They only keep the drive and driven gears rotating in the same direction. They are commonly used in manual transmissions for reverse.

PLANETARY GEARSETS. Planetary gearsets consist of an internal outer gear called a ring gear, a center sun gear, a set of planet gears (usually three or four), and a planet carrier. They can be used to create underdrive or overdrive ratios and reverse rotation. Their operation will be explained in Chapter 12.

GEAR CLUSTERS. Some applications for gears require a combination of gears to be driven together. When more than one gear is driven by the same shaft, it is called a gear cluster. They are mostly used in countershaft transmissions.

GEAR LUBRICATION. Proper lubrication of gears is critical to reducing friction while ensuring long life. Always use the recommended gear oil or fluid for any system or component that uses gears. Gears are lubricated to reduce friction at the tooth contact area by promoting sliding between teeth. Lubrication will also take heat away from the contact area. There are three types of lubrication for gears.

Grease lubrication can be used for low-speed gears. If the gears are enclosed in a case, care must be taken not to overgrease or excess heat and friction will build up. The proper type and viscosity of grease is critical for this type of lubrication.

> **WEB ACTIVITY**
>
> Search for and watch any video that demonstrates what happens to high-viscosity oil in freezing temperatures and you will gain a new appreciation for using the correct oil viscosity.

FIGURE 11–15 Pressure or positive lubrication of a gear.

Splash lubrication or oil bath is when the gear teeth pick up oil from the bottom of its housing and carry it around as the gear rotates.

Forced or pressure lubrication occurs when oil is pumped from a reservoir and is dispensed on the gears near the contact point. ● SEE FIGURE 11–15 for how pressure lubrication works.

Gear lubrication could come from oil that is common to other components such as a torque converter or could come from the oil in an axle, final drive, or transmission housing.

Progress Check

1. The ratio of a pair of gears is determined by:
   a. multiplying the total number of teeth and dividing by 10
   b. dividing the number of teeth on the driven by the number of teeth on the drive
   c. dividing the diameter of the drive by the diameter of the driven
   d. dividing the pitch of the drive by the pitch of the driven
2. If the driven gear of a gearset has 40 teeth and the drive gear has 10 teeth, this would make a ratio of:
   a. 4:1
   b. 40:1
   c. 10:1
   d. 0.25:1
3. If an idler gear is smaller than the two gears it is placed between, this will:
   a. speed up the driven gear
   b. slow down the driven gear
   c. increase the torque of the driven gear
   d. not affect the speed of the two gears
4. When compared to a spur gear, a helical gear will:
   a. be noisier
   b. not be able to handle as much torque
   c. spin faster
   d. create a thrust load
5. Gear pitch is determined by:
   a. the amount of torque needed to be transferred
   b. the speed the gear will rotate
   c. the number of teeth on a gear and the material they're made of
   d. the number of teeth on a gear and its diameter

MANUAL TRANSMISSIONS

The next part of a mechanical drivetrain past the clutch or torque converter is the transmission. Transmissions that require the operator to manually move a shift mechanism that will directly change gear ratios or directions inside the unit are called manual transmissions. Manual transmissions could be found in light-duty equipment such as backhoe loaders, off-road forklifts, compactors, and light utility vehicles. Some older dozers, track loaders, and graders used manual shift transmissions but you won't likely see many of these in regular use today. On highway vehicles, use manual transmissions extensively, and these can sometimes be turned into off-road dump trucks, boom trucks, or service trucks. Even though you may not see too many manual transmissions while working as an HDET, understanding how they work will help to reinforce gear function principles. ● SEE FIGURE 11–16 for a highway truck that is used off road.

A transmission is needed to provide different machine travel speeds and directions from a diesel engine that can only rotate in one direction and in a limited rpm range. For example, let's say a backhoe loader has a diesel engine that produces 100 hp at 2200 rpm. It will also create a maximum of 250 ft-lb of torque at 1900 rpm. Its practical rpm range is between 1650 and 2100 rpm where its torque curve varies from 160 ft-lb to its maximum and then levels off to 220 ft-lb. The transmission must take the engine output and transform it to make

FIGURE 11–16 off road dump truck.

FIGURE 11–17 A cutaway view of a twin countershaft transmission.

the machine vary travel speeds from 0 to 25 mph because the gear ratios in the drive axles are fixed. Low speeds are needed when filling its bucket and travelling on a muddy work site. The machine also needs to travel to different work sites and on public roads and without impeding traffic too much. It will do this with four different forward ratios and one reverse because it obviously needs to back up occasionally.

The following is an example of gear ratios from a four-speed transmission:

Gear Range	Reverse	1	2	3	4
Ratio	9.88	9.45	4.90	2.65	1.00

Manual transmissions require the operator to shift speed ranges (or ratios) and directions either when the engine speed is getting close to governed rpm and a faster travel speed is required or when the engine slows down from the load being applied to it. This action is more commonly called shifting gears, which is not always totally accurate as you will learn.

Manual shift transmissions require some degree of operator skill and finesse to prevent gear clash or grinding of gears. This is one reason why they aren't as common in off-highway use. Another reason is highway vehicles have little resistance to movement once they are rolling. This makes it easy to shift gears because there has to be a momentary stoppage of power into the transmission when shifting. Most off-road conditions offer high resistance to machine movement so that any interruption of power to tires or tracks will result in the machine slowing down, and if the shift isn't fast enough, the engine could fall out of its torque range.

Other types of transmissions are more favorable to off-road conditions. Powershift transmissions allow the operator to shift speed ranges by controlling hydraulic oil, which actuates clutch packs. There is a type of hybrid transmission that combines both manual shift and powershift features, and it's called a shuttle shift. Shuttle shift transmissions provide power shifts between forward and reverse, but their speed ranges are actuated manually. Many backhoe loaders use this type of transmission. See Chapter 12 for a complete description of a shuttle shift transmission.

Manual transmissions can be categorized by the number of **countershafts** they have and by the type of shifting they employ. They can have one, two, or three countershafts. All countershafts are parallel with the main shaft. Double and triple countershaft transmissions are mostly used in heavy-duty on highway vehicles. The extra countershafts will spread the gear and bearing loads out more evenly. ● **SEE FIGURE 11–17** for a cutaway view of a twin countershaft transmission.

Manual shift transmissions can use either spur or helical gears or a combination of both.

The simplest manual transmission is the single countershaft transmission that can have anywhere from two to six speed ranges plus one or two reverse speeds.

As with all manual transmissions, it typically bolts to the engine's flywheel housing where it will be driven directly by the flywheel-mounted clutch, but it could also be remotely mounted, meaning there could be a drive shaft between the clutch housing and the transmission.

The simplest way a manual transmission can be shifted is by sliding gears along the main shaft of the transmission. By sliding them sideways, they will get engaged with one of the countershaft gears being driven by the input shaft. The main shaft is now driven at the speed dictated by the ratio of the two gears in mesh. This type of non-constant mesh transmission is primitive in comparison with the constant mesh sliding collar and synchro mesh-type transmissions. By trying to shift the gears together, there will be lots of grinding and mismatching of speeds between the countershaft and main shaft gears. This sliding gear type of transmission is outdated.

Let's take a look at how the gears and shafts are arranged in a single countershaft constant mesh type of transmission. ● SEE FIGURE 11–18 for a cutaway view of a single countershaft transmission.

A single countershaft transmission housing is a cast iron case that holds the input shaft, a main shaft (output shaft), the countershaft, and a reverse idler shaft in place. All shafts are suspended on bearings that are held in machined bores in the case. If the bore for holding a bearing isn't a blind bore—in other words, if there is a hole right through the case, there will have to be a cover over the bearing. The cover will have a gasket to stop any fluid loss. The input and output shaft will have dynamic lip-type seals that will rest in the housing with an interference fit while their lip rides on the shaft.

The case has an opening at its top to allow the internal parts to be installed and has a shifter mechanism as part of its top cover.

The transmission input shaft is driven by the clutch disc and will drive the transmission input gear. This gear then drives the countershaft that is below the main shaft. The countershaft has a series of gears splined to it that will be different in their diameters and pitches. The number of countershaft gears corresponds to the number of gear ranges that the transmission has minus one. The main transmission shaft will have gears that are in constant mesh with the countershaft gears but float on it. For a sliding collar type of transmission, the collars are splined to the main shaft and have external splines on both sides of it that can lock into internal splines on the side of one of the floating gears. The sliding collars are moved sideways on the main shaft by the shift mechanism. The sliding collars will have their splines tapered to ease engagement with the gear.

In neutral, the sliding collars are centered between the floating gears, and when the input shaft is turning the countershaft, the floating gears will rotate with their mating main shaft gears, but no torque transfer happens.

When operators want to move the machine, they will disengage the clutch, the input shaft slows and stops, and then the shift lever can be moved to select first gear. This will move a sliding collar into first gear (the largest diameter) on the main shaft and lock it to the main shaft. When operators engage the clutch, the input shaft starts to turn the input gear that turns the countershaft, which now turns the main shaft because there is a main shaft gear locked to it. First gear will be the highest forward ratio and create the most torque and the biggest speed differential (speed reduction) between the input and output shafts.

Each forward shift to a higher gear (lower ratio) will first disengage the lower gear and then engage the sliding collar for the next higher gear.

Each gear change higher will be a step toward direct drive. This is when the input shaft is locked to the main shaft with a sliding collar. In direct drive, the countershaft is just along for the ride.

If the transmission has an overdrive ratio, then there will be a gear on the countershaft that will mesh with a larger gear on the main shaft to give more than one revolution for each one of its revolutions.

To put the transmission into reverse, a sliding collar will mesh the output shaft with the reverse idler. The **reverse idler** is driven by a gear on the countershaft. By putting an idler gear in between, the output shaft's direction will change.

The rear of the main shaft is the output for the transmission. It will be splined outside the housing to accept a yoke. The **yoke** is fastened to the shaft and a universal joint will send torque into a drive shaft and on to an axle. ● SEE FIGURE 11–19 for the power flow through a five-speed transmission with synchronizers.

A transmission that uses synchronizers will basically work the same as the sliding collar but with the synchronizer taking the place of the sliding collar. The shift fork moves the synchronizer toward a main shaft gear. A **synchronizer** will help match the speeds of the main shaft and the main shaft gears before they are locked together. This will prevent gear grinding and give a much smoother shift. A synchronizer uses friction to ease the speed matching process.

FIGURE 11–18 A cutaway view of a single countershaft transmission.

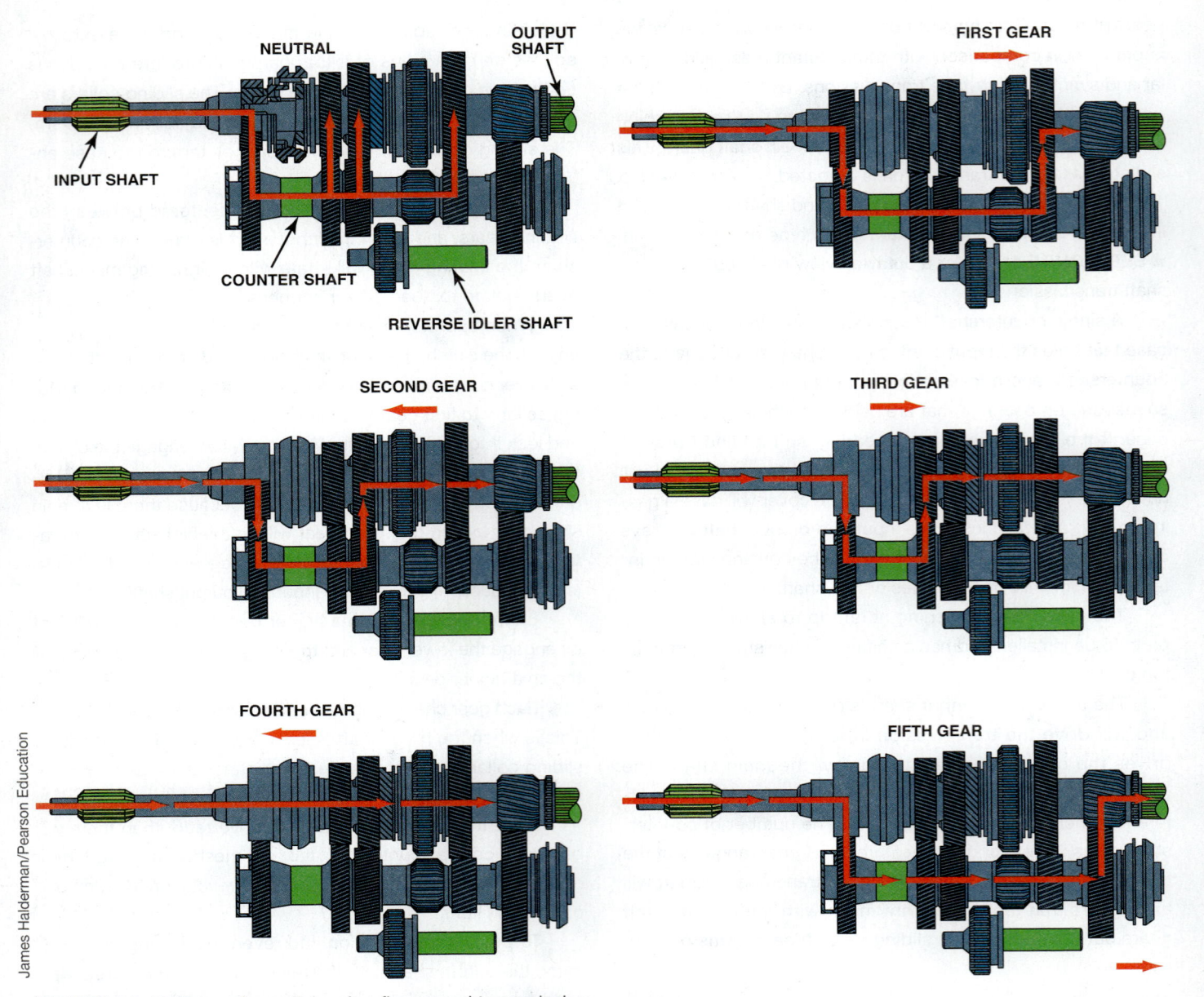

FIGURE 11–19 Power flows through a five-speed transmission.

If a transmission has synchronizers, it will likely use them for only forward speeds from second and up (not first or reverse). To work properly, a synchronizer needs to have the main shaft's gear moving and first or reverse should only be engaged when the machine is stopped.

There are four different variations of synchronizers but their purpose is the same. The four types are block, disc and plate, plain, and pin. ● **SEE FIGURE 11–20** for a synchronizer.

The components of the synchronizer are the hub (F), collar (A), synchronizer ring (D and K), spring (J), and synchronizer detent (pressure piece) (L).

The hub (F) is splined to the intermediate shaft (H). The synchronizer collar (A) is splined to the hub. Three of the splines in the hub are omitted, and a synchronizer detent (L) is inserted between the hub and the collar. Springs (J) push the detents (L) out against the collar.

The synchronizer detents extend into slots in the synchronizer collar. The hub, collar, synchronizer detent, and synchronizer rings always rotate as an assembly with the shaft.

With the synchronizer in neutral, the synchronizer rings do not contact either drive gear. As the gears are not splined to the shaft, they are free to rotate at different speeds.

When a shift is made, the first step is synchronizing the speed of intermediate drive shaft (H) and drive gear (I). As the collar (A) moves across the hub, the synchronizer detents move down and force the synchronizer ring (K) against the tapered

338 CHAPTER 11

SHIFT FORK (B)
SYNCHRONIZER COLLAR (A)
(M) THIRD GEAR ENGAGED
PRESSURE PIECE (L)
(C) FOURTH SPEED DRIVE GEAR
(D) SYNCHRONIZER RING
(K) SYNCHRONIZER RING
(E) NEEDLE BEARING
(J) SPRING
(F) HUB
(I) THIRD SPEED DRIVE GEAR
(G) SNAP RING
(H) INTERMEDIATE SHAFT

(N) SYNCHRONIZER

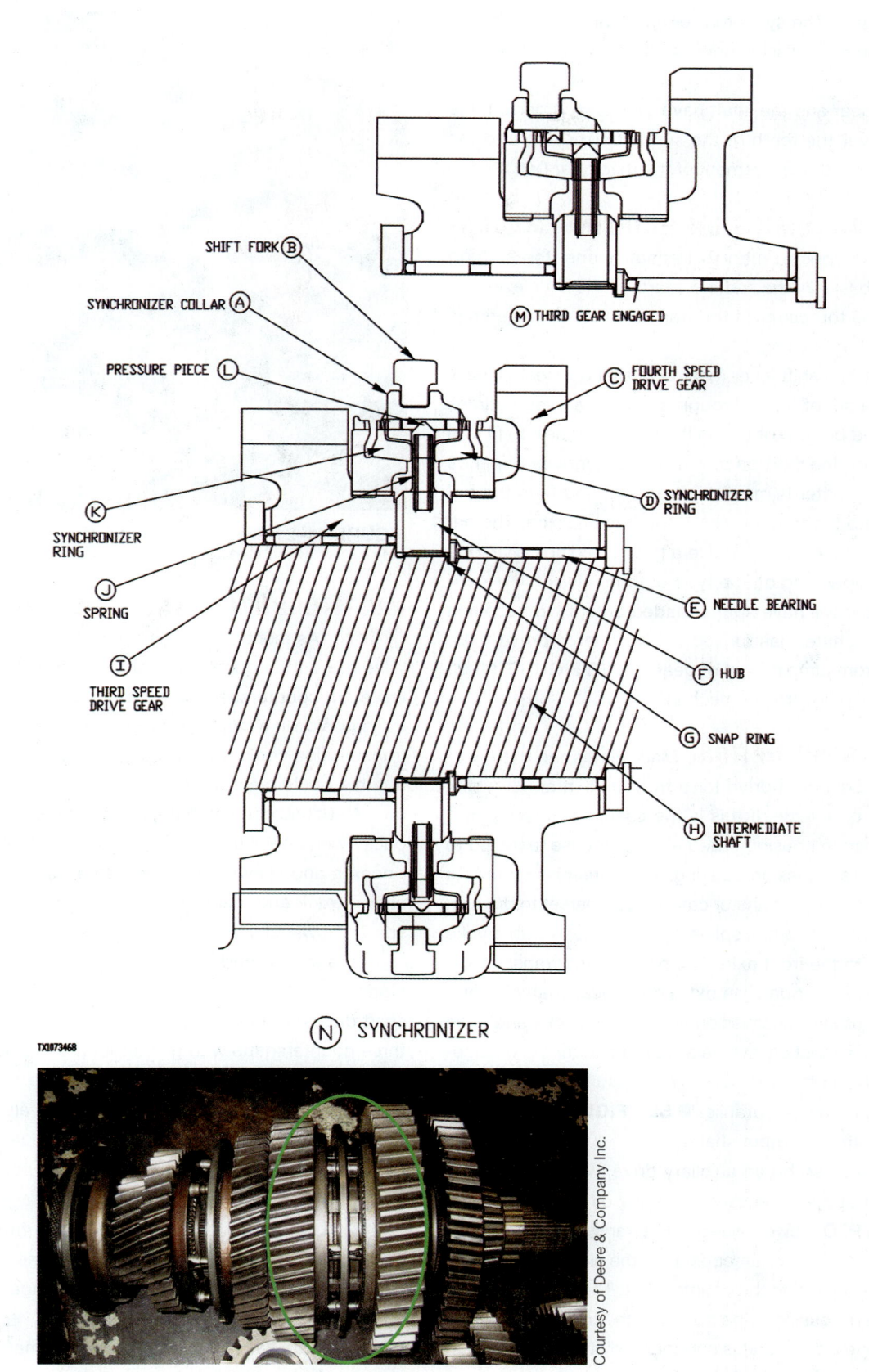

FIGURE 11–20 A synchronizer.

GEARING FUNDAMENTALS AND MANUAL TRANSMISSIONS 339

portion of the gear. The synchronizer rings will accelerate until the speed of the transmission drive shaft matches the speed of the drive gear.

After the gear and the shaft have been synchronized, the collar slides over the teeth of the synchronizer ring and the gear to allow smooth engagement of the third gear (M).

MANUAL TRANSMISSION SHIFTING MECHANISM
For all types of manual transmissions, the shifting mechanism is basically the same. It starts at the shift lever and is housed in the top cover of the transmission or sometimes called the shift bar housing.

The operator's shift lever (gear shift) will be mechanically connected to a set of forks through a set of bars (sometimes called rails). The bottom of the shift lever is notched to fit into different slots on the shift bars. Which bar it moves depends on whether the shifter is moved left or right and then the bar will move toward the front or rear of the transmission. The bar will now move a fork that will move a gear, a sliding collar, or a synchronizer depending on the type of transmission it is controlling. Each bar will have spring-loaded detents to keep it in place once the shifter has moved it. This should prevent the transmission from jumping out of gear. ● SEE FIGURE 11-21 for a transmission cover/shift mechanism.

ALL-WHEEL DRIVE OPTION
Many times the transmission needs to provide a drive for a front axle in an all-wheel drive machine or vehicle. If this is the case, there will sometimes be a separate housing called a transfer case fastened to the rear of the transmission housing. The transmission output shaft will go through the transfer case and a gearset to drive an output that is offset and has splines that will mesh with a yoke to send power to the front axle. Alternatively, the transmission housing will accommodate an extra drive gear that will send drive forward out the transmission housing to a yoke and drive shaft. There will almost always be a way to control the engagement of the drive to the front axle. This could be a mechanical linkage or electric over hydraulic. ● SEE FIGURE 11-22 for a transmission with two output shafts.

There could also be an auxiliary drive available from the transmission that could be used for many different functions. This is called a PTO (power take-off). If a transmission is able to provide a PTO, it will have an opening in the side of its housing that will expose a gear on the countershaft. The opening will allow another gear housing to be bolted to the transmission. The housing will have a shaft that is supported by two bearings and will have a yoke attached to it. A drive shaft can then drive any kind of attachment needed. The auxiliary shaft will always spin at engine speed regardless of the transmission gear selected.

FIGURE 11-21 A transmission cover/shift mechanism.

MULTI-COUNTERSHAFT MANUAL TRANSMISSIONS
As mentioned earlier, manual shift transmissions can have more than one countershaft, and the main purpose for the extra countershafts is to create a higher torque capacity transmission in a relatively small housing. A single countershaft transmission could be made to have higher torque capacities, but this would mean larger gears, shafts, and bearings.

A second countershaft provides a much higher torque capacity with one extra shaft and practically the same size components and likewise for one with a third countershaft. This allows truck and machine designers to not have to make big space allowances to have a much stronger transmission.

The layout and operating principle of these transmissions are the same as a single countershaft type. The input shaft drives an input gear. The input gear will now drive two or three countershafts with gears that are in constant mesh with the floating main shaft gears. There is one main shaft and the shift method and mechanism is the same also as range shifts are still done on the main shaft gears. The reverse idler will be driven by one of the countershafts. The output is out the main shaft as well.

One big difference that you may find with a multi-countershaft transmission is that they usually have an auxiliary section that can provide high and low ranges and sometimes an overdrive range.

A typical twin countershaft arrangement is a five-speed main section with a two-speed auxiliary to make a 10-speed transmission. Some arrangements can provide over 18 speeds.

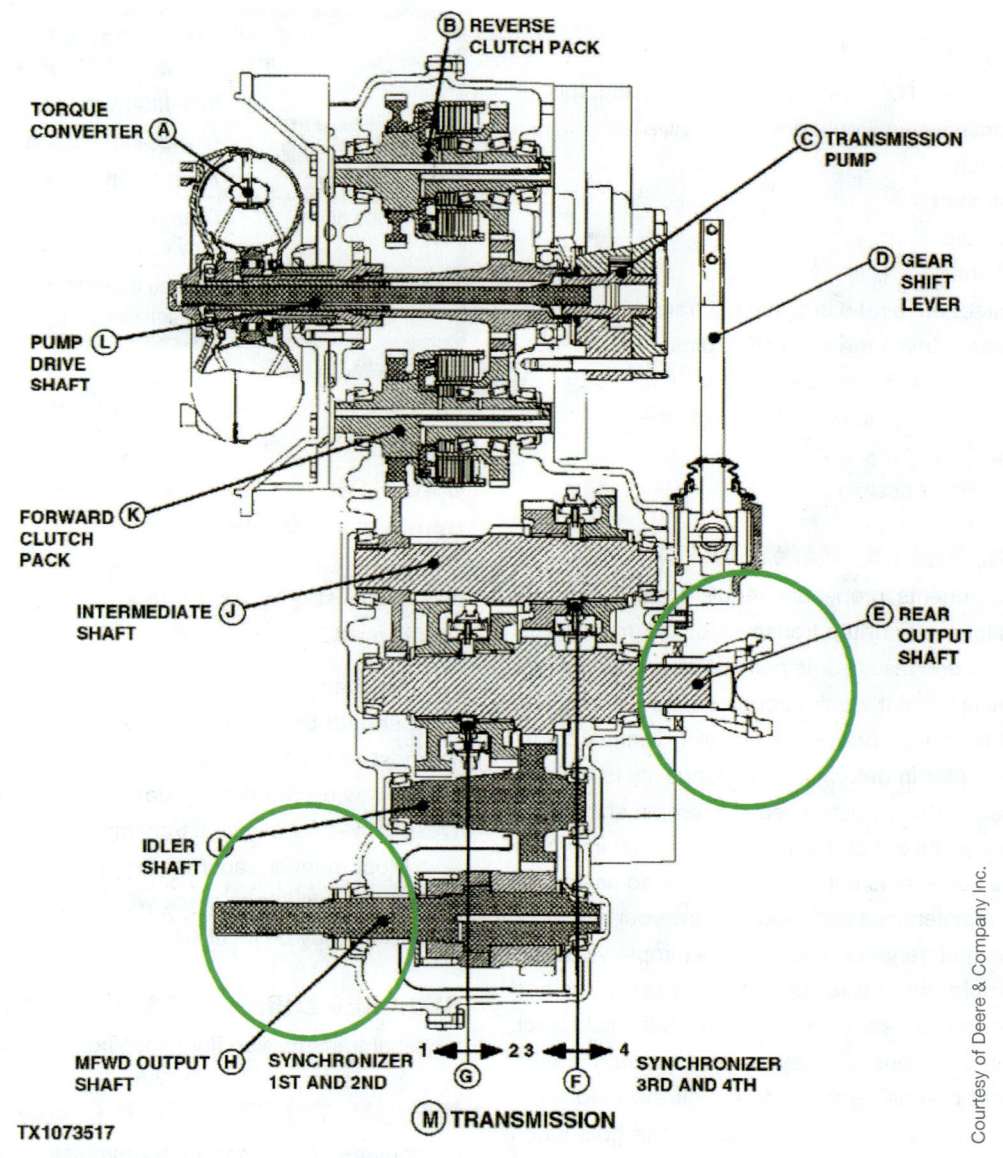

FIGURE 11-22 A transmission with two output shafts.

The auxiliary section is shifted with air pistons that are controlled by a valve attached to the shifter for the mechanically shifted section.

Newer versions of these transmissions are electronically controlled with servo motors doing the job of the shift lever and servo valves controlling the air shifting.

Some auxiliary drives use a planetary gearset and some older configurations had two separate transmissions; the main five-speed unit fed a four-speed auxiliary transmission.

Progress Check

6. The output of a single countershaft transmission is driven by:
   a. the main shaft gear that is locked to the main shaft
   b. the countershaft output gear
   c. the second countershaft idler
   d. the input shaft idler gear

7. A manual transmission may have one or more synchronizers. Their purpose is to:
   a. make sure that the transmission is shifted in order up and down
   b. make sure that the shifts from forward to reverse are smooth
   c. match the speed of the reverse idler to the output shaft
   d. match the speed of the main shaft gear to the main shaft

8. The shift bar housing uses this device to stop the transmission from jumping out of gear.
   a. snap ring
   b. spring loaded detent

GEARING FUNDAMENTALS AND MANUAL TRANSMISSIONS 341

c. magnet
d. solenoid

9. If a machine had a PTO driven by the transmission, this part of the transmission would drive it:
   a. the input shaft
   b. the countershaft
   c. the reverse idler
   d. the output shaft

10. To obtain a direct drive ratio in a manual transmission:
    a. the input shaft drives the main shaft directly
    b. the reverse idler is moved out of the way
    c. the over drive ratio is sent back through the countershaft
    d. direct drive isn't possible

MANUAL TRANSMISSION MAINTENANCE

As with all other machine components, proper and regular maintenance will ensure maximum longevity of the transmission. All mechanical devices will wear out and fail at some point. With proper maintenance, you will ensure that it won't happen before its designed life span unless it is abused or overloaded. With manual transmissions, a major factor in determining its longevity is the person moving the gear shift and clutch pedal. Improper shifting will take as much or more life out of the transmission than improper maintenance. This however is out of your control, so as long as you look after the maintenance part, you've done your part.

Transmission maintenance starts with a simple visual inspection. You should look for leaks and loose fasteners and any other signs of unusual wear. Next is to check the fluid level. Most manual transmissions will have a plug (usually a NPT threaded plug) that can be removed to check the fluid level. The fluid level should be up to the threads of the hole when cold. Even being a little low can easily equal 1 gallon of oil for larger transmissions. Because most manual transmissions rely on splash lube, maintaining the proper oil level is critical.

TECH TIP

Manual transmissions can be abused quite easily. If an operator lets the clutch out too fast when starting the machine moving from a dead stop, there will be a huge shock load through the drivetrain. This could inflict damage on the transmission's gear teeth, shafts, or output yoke. The same type of damage could happen when a tire is spinning and suddenly gains traction. Shifts between ranges that aren't smooth can damage gears.

EATON® ROADRANGER® CD50 TRANSMISSION FLUID

OFF-HIGHWAY USE	
First 30 hours	Change transmission lubricant on new units.
Every 40 hours	Inspect lubricant level. Check for leaks.
Every 500 hours	Change transmission lubricant in severe dirty conditions.
Every 1000 hours	Change transmission lubricant (normal off-highway use).

TABLE 11–1

Typical Transmission Fluid Change Interval.

CAUTION: Be careful when removing the level plug from a warm transmission as there may be pressure in the housing. Use a rag and rubber glove to protect your hand and stay at arm's length when the plug is on its last thread.

Oil changes should be done on a regular basis. Oil sampling should be done regularly also. As with any oil change, it's most effective with warm oil. Remove the fill plug first before removing the drain plug to relieve any pressure. ● SEE TABLE 11–1 for a typical transmission fluid change interval.

Most manual transmission oil is gear oil that is mineral oil based, but this is one place where the added expense of synthetic fluid may be justified.

REQUIRED LUBRICANT CHART

● SEE TABLE 11–2 for a typical transmission fluid specification chart.

MANUAL TRANSMISSION DIAGNOSTICS

Operator complaints about manual transmissions will likely be one of the following: hard shifting, won't shift, excessive vibrations or noise, jumps out of gear, or a fluid leak.

The usual steps are followed when diagnosing a complaint:

Get the information: Gather as much information from the operator as you can like when did it start, under what conditions does it happen, and has anyone worked on the machine lately?

Verify the complaint: This will likely require you to drive the machine or vehicle. See the warnings and cautions below related to that. Try to recreate the conditions that the problem occurs under. Resist the temptation to run the driveline with the machine on jack stands. There are too many things to go wrong when doing this.

Transmission noises and vibrations: Driveline noises and vibrations are sometimes quite hard to pinpoint.

TYPE	GRADE (SAE)	FAHRENHEIT AMBIENT TEMPERATURE
Eaton®-approved synthetic transmission oil	50	All
Heavy-duty engine oil (MIL-L-2104D, API-CD, or Cat TO-4)	50	Above 10°F (−12°C)
	40	Above 10°F (−12°C)
	30	Below 10°F (−12°C)
Automotive Gear Oil API-MT-1	80W–90	Above 10°F (−12°C)
	75W	Below 10°F (−12°C)
Additives and friction modifiers must not be introduced. Never mix engine and gear oils in the same transmission.		

TABLE 11-2

Typical Transmission Fluid Specification Chart.

Do your best to eliminate what is normal and what isn't causing the noise or vibration and you will eventually find out the source of it. Transmission noises and vibrations can be narrowed down to certain speeds, gear ranges, different loads, and whether the machine is accelerating or decelerating.

Hard shifting: Start by seeing if it is one gear, several, or all. If it's one gear, it could be part of the shift mechanism, a synchronizer problem, or a sliding collar problem. If more than one gear is a problem, it could also be a shift mechanism problem, the oil viscosity is wrong, or there is a clutch problem.

Won't shift: Again, see if it is all gears, one, or a combination. If it's all gears, it is likely a clutch problem or a major shift mechanism problem. If it's one or more but not all, it is likely a shift mechanism problem.

Fluid leak: If it's near the rear of the transmission, it may be difficult to find if the oil has got onto the output yoke and has spread out. Cleaning the area may be necessary, but before you clean too much, look for the cleanest dampest spot and this should give an indication of where the leak is originating from. Also check the fluid level to see if the transmission has been overfilled. Leaks near the rear of a transmission are likely from the rear output shaft seal.

Jumps out of gear: This is likely a problem with the shift mechanism's detent or with a synchronizer.

MANUAL TRANSMISSION REPAIR
Once a manual transmission has been diagnosed with a problem, it will likely have to be removed. There are some repairs that can be made without taking the whole unit out of the vehicle, but this is limited to anything related to the shifter mechanism, the rear auxiliary, or replacing the rear seal.

If transmission removal is called for, a proper transmission jack is mandatory. A proper jack should have enough weight capacity for the transmission, should have good wheels for maneuvering, should be fully adjustable to tilt and rotate the transmission, must have adjustable brackets to conform to the transmission shape, and should have a locking mechanism to mechanically keep the jack from drifting down. Transmissions will weigh between 300 and 1000 lb so having a transmission fall off a makeshift jack is not something you want to witness or be close to. ● **SEE FIGURE 11–23** for a proper transmission jack.

General steps involved in removing a manual transmission from a machine are as follows:

1. Park the machine on a level, flat surface, chock the rear wheels, and lock out and tag out the machine.
2. Drain the transmission and take an oil sample.
3. Raise the front or the whole machine and support it with proper stands. It will sometimes be easier to raise the whole machine so that the transmission will come out level. Try to anticipate how high the machine will have to be to get the transmission out once it's lowered on the jack and give yourself a little extra room. You don't want to be lifting the vehicle again just to get the transmission out from under it!

> **SAFETY TIP**
>
> Before starting a vehicle, always make sure that you are seated in the operator's seat. Place the transmission in neutral, set the parking brakes, and disengage the clutch.
>
> Before working on a vehicle, place the transmission in neutral, set the parking brake, and block the wheels.
>
> Before towing the vehicle, place the transmission in neutral, lift the rear wheels off the ground, remove the axle shafts, or disconnect the driveline to avoid damage to the transmission during towing.

FIGURE 11–23 A proper transmission jack.

4. The drive shaft will be removed first and moved out of the way.
5. Remove the speedometer cable and shift lever and clutch release mechanism from the clutch lever, and any other wiring if equipped.
6. Place the jack under the transmission, adjust it to fit the transmission, raise to take a little weight, and secure it to the jack with a chain.
7. Depending on the machine, you may have to support the rear of the engine. If so, this mechanism should be put in place now.
8. Remove the transmission mounts.
9. Remove the clutch housing to flywheel housing bolts.
10. Carefully and slowly back the transmission away from the engine. This may take some creative thinking and actions but avoid using excessive force or you will break something. Make sure that all fasteners are removed and check to see if there are dowels used or threaded pusher holes available.
11. When the transmission is away from the machine, it should be washed. Avoid directing high-pressure water where it may easily enter the transmission because you may be doing some failure analysis on failed parts.

TRANSMISSION RECONDITIONING If a transmission is removed and out of warranty, it is wise to completely disassemble it, inspect all parts, replace all minor parts at a minimum (snap rings, washers, seals, gaskets), and replace any other parts based on manufacturer's guidelines. Most manufacturers will provide a visual guideline on their website.

The following are general guidelines for reconditioning a twin countershaft manual transmission from Eaton:

Cleanliness: Provide a clean place to work. It is important that no dirt or foreign material enters the unit during repairs. Dirt is an abrasive and can damage bearings. It is always a good practice to clean the outside of the unit before starting the planned disassembly.

General practices: Always apply force to shafts, housings, etc. with restraint. Movement of some parts is restricted. Never apply force to driven parts after they stop solidly. The use of soft hammers, soft bars, and mauls for all disassembly work is recommended.

As the cost of a new part is generally a small fraction of the total cost of downtime and labor, avoid reusing a questionable part, which could lead to additional repairs and expense soon after assembly. To aid in determining the reuse

TECH TIP

Now that the transmission is out of the vehicle, what's next? Depending on your skill level, if you have the right tools, the proper service information, and a proper work area, you may or may not want to attempt to repair the transmission. There will be some special tools required to pull and install bearings, but a machine shop can make these if necessary. If you have the right tools and information, rebuilding a transmission is not an insurmountable task even if you have never attempted it. Some patience and no pressure to hurry up will help a lot as well.

If one or all these things aren't in your favor, then buying an exchange transmission is an option or sending it to a transmission shop that is experienced in the type of unit you have is a common way to go. The transmission supplier should stand behind their work and give you a warranty.

WEB ACTIVITY

Search the web for "heavy-duty manual transmission." Find a manufacturer and see if you can find a guideline for reusing parts. There are three or four manufacturers of manual transmissions that provide good information about rebuilding manual transmissions.

344 CHAPTER 11

or replacement of any transmission part, consideration should also be given to the unit's history, mileage, application, and so on.

- **Assemblies:** When disassembling the various assemblies, such as the main shaft, countershafts, and shift bar housing, lay all parts on a clean bench in the same sequence as removed. This procedure will simplify assembly and reduce the possibility of losing parts.
- **Bearings:** Carefully wash and lubricate all usable bearings as removed and protectively wrap until ready for use. Remove bearings planned to be reused with pullers designed for this purpose.
- **Input shaft:** The input shaft can be removed from the transmission without removing the countershafts or main shaft. Special procedures are required and provided in this manual.
- **Snap rings:** Remove snap rings with pliers designed for this purpose. Snap rings removed in this manner can be reused, if they are not sprung or loose.

Recommended inspection procedures are provided in the following checklist.

- **Bearings:**
  - Wash all bearings in clean solvent. Check balls, rollers, and raceways for pitting, discoloration, and spalled areas. Replace bearings that are pitted, discolored, spalled, or damaged during disassembly.
  - Lubricate bearings that are not pitted, discolored, or spalled and check for axial and radial clearances.
  - Replace bearings with excessive clearances.
  - Check bearing fit. Bearing inner races should be tight to shaft; outer races slightly tight to slightly loose in case bore. If the bearing spins freely in the bore, the case should be replaced.

- **Bearing covers:**
  - Check covers for wear from thrust of adjacent bearing. Replace covers damaged from thrust of bearing outer race.
  - Check cover bores for wear. Replace those worn or oversized.

- **Gears:**
  - Check gear teeth for frosting and pitting. Frosting of gear teeth faces presents no threat of transmission failure. Often in continued operation of the unit, frosted gears "heal" and do not progress to the pitting stage. In most cases, gears with light-to-moderate pitted teeth have considerable gear life remaining and can be reused, but gears in the advanced stage of pitting should be replaced.
  - Check for gears with clutching teeth abnormally worn, tapered, or reduced in length from clashing during shifting. Replace gears found in any of these conditions.
  - Check axial clearance of gears.

- **Gear shift lever housing assembly:**
  - Check spring tension on shift lever. Replace tension spring if lever moves too freely.
  - If housing is disassembled, check gear shift lever bottom end and shift finger assembly for wear. Replace both gears if excessively worn.

- **Housings and covers:**
  - Check all gray iron parts for cracks and breaks. Replace parts found to be damaged.

- **O-rings and gaskets:**
  - Replace all O-rings and gaskets.

- **Reverse idler gear assemblies:**
  - Check for excessive wear from action of roller bearings.

- **Shift bar housing assembly:**
  - Check for wear on shift yokes and blocks at pads and lever slot. Replace excessively worn parts.
  - Check yokes for correct alignment. Replace sprung yokes.
  - Check lockscrews in yoke and blocks. Tighten and rewire those found loose.
  - If housing has been disassembled, check neutral notches of shift bars for wear from interlock balls.

- **Splines:**
  - Check splines on all shafts for abnormal wear. If sliding clutch gears, companion flange, or clutch hub has wear marks in the spline sides, replace the specific shaft affected.

- **Synchronizer assembly:**
  - Check synchronizer for burrs, uneven and excessive wear at contact surface, and metal particles.
  - Check blocker pins for excessive wear or looseness.
  - Check synchronizer contact surfaces on the synchronizer cups for wear.

ASSEMBLY PRECAUTIONS Make sure that case interiors and housings are clean. It is important that dirt and other foreign materials are kept out of the transmission during assembly. Dirt is an abrasive and can damage polished surfaces of bearings and washers. Use certain precautions, as follows, during assembly.

- Bearings: Use a flange-end bearing driver for bearing installation. These special drivers apply equal force to both bearing races, preventing damage to balls/rollers and races while maintaining correct bearing alignment with bore and shaft. Avoid using a tubular or sleeve-type driver, whenever possible, as force is applied to only one of the bearing races.
- Cap screws: To prevent oil leakage and loosening, use sealant on all cap screws.
- Initial lubrication: Coat all limit washers, shims, and shaft splines with lubricant during assembly to prevent scoring and galling of such parts.
- O-rings: Lubricate all O-rings with silicon lubricant.
- Universal joint companion flange or yoke: Pull the companion flange or yoke tightly into place with the output shaft nut. Make sure that the speedometer drive gear or a replacement spacer of the same width has been installed. Failure to pull the companion flange or yoke tightly into place can result in damage to the main shaft rear bearing.

Once a new or rebuilt transmission has been installed in a machine, it should be operated by you before it is given back to the operator. A short interval oil change should take place to drain as much of the initial break in material as possible.

Twin countershaft transmissions require the main shaft to be timed to the countershafts upon reassembly. This is sometimes a tricky procedure, but it must be done right or major damage will occur.

Progress Check

11. When checking for proper oil level of a typical manual transmission, you will:
    a. remove the drain plug and measure the oil that drains out
    b. remove the level plug and see if it is at the bottom of the hole
    c. remove the level plug and see if it is within an inch of the bottom of the hole
    d. remove the dipstick and see if it's in the crosshatch area
12. These are all acceptable factors for you to decide to rebuild a transmission except:
    a. your supervisor wants to save money
    b. there is a clean well-lit area with a clear work bench
    c. you have all the special tools needed
    d. you have all the current service information
13. A transmission that has been used in severe conditions should have its oil change interval reduced by:
    a. one half
    b. one quarter
    c. two thirds
    d. 1000 hours
14. If the transmission is being operated in temperatures above 10°F, the proper viscosity of gear oil to be used is:
    a. 10W30
    b. #50
    c. #10
    d. 75W
15. If a transmission jumps out of one gear, this is the likely cause:
    a. loose countershaft bearings
    b. low oil level
    c. defective synchronizer or detent mechanism
    d. input shaft is bent

SHOP ACTIVITY

Find a manual transmission in your shop and:
- Find the oil level and drain plug.
- Remove the level plug and check the oil level.
- Record any leaks and determine where they are coming from.
- Determine how many ratios it has.
- See if there is a location to attach a PTO.

SUMMARY

1. Safety concerns for this chapter include: extra care when working with pullers; when striking gears use a soft hammer or punch; snap rings need to be removed carefully; gears can be very heavy.
2. Many different gear sets are used on any piece of heavy equipment and are used to transfer torque from one drive mechanism to a second driven mechanism.

3. If the meshing gears have a different number of teeth there will be a change in speed and torque from drive to driven.

4. Most gears have teeth equally spaced around their outer circumference (external gear); however, some gears can be a ring with teeth facing inward (internal gear).

5. Gears are usually mounted on shafts that are suspended by bearings. Gears can be connected to a shaft by a key, threaded fasteners, tapered or shrink fit or a combination of these methods.

6. Most gear sets have their center axis parallel to each other; however, some gears can be at 90 degrees to each other.

7. Gear teeth act as levers to transfer torque.

8. Gear ratios are calculated by counting each gears' teeth and then comparing how many revolutions it takes the drive gear to make the driven gear turn one revolution.

9. Underdrive ratios provide a torque increase and a speed decrease when the output is compared to the input. An example would be a ratio of 4:1.

10. Overdrive ratios provide a speed increase and a torque decrease. An example would be a ratio of .75:1.

11. An idler gear will not affect gear ratios but will allow the driven gear to rotate in the same direction as the drive gear.

12. Gear design and manufacturing is complicated and involves elements such as: involute curves, pitch line, gear face width, center line and pressure angle.

13. Gears that run together must have a specific amount of clearance between them. This is called backlash.

14. External gears have teeth on their outer circumference while internal gears have an empty center and their teeth face inward.

15. Gears with straight teeth that are aligned at 90 degrees to the gear centerline are termed spur type gears.

16. Gears that have their teeth on an angle are called helical type gears.

17. Gears that mesh and change the angle of power flow are called bevel gears. The most common use of bevel gears are in differentials where the gear centerlines intersect at 90 degrees.

18. Planetary gearsets have four members: sun gears, planet gears, planet carrier, and ring gear.

19. Gears need lubrication to reduce friction when they mesh and to cool them. Splash lubrication has gears running through oil and using their teeth to pick it up while pressure lube circulates oil and directs it at gear teeth.

20. Manual transmissions are not very common with today's machines. They have at least two main shafts that have a series of meshing gears arranged on them. A shift mechanism allows the operator to select different gear ratios to match engine output to machine travel speed requirements.

21. Simple manual transmissions are called single countershaft and can have up to 6 speeds or ranges while twin and triple countershaft transmissions have 5 speeds in the main section and another 2 or 3 in the auxiliary section. The extra countershafts allow for higher torque capacity.

22. Sliding collars move gears along a splined shaft to change ratios. Some transmissions have synchronizers that help reduce clash when gear ratios are changed.

23. Reverse is provided with an idler gear.

24. Manual transmission service consists of checking oil levels and changing oil at specific intervals.

25. Common problems with manual transmissions are: hard to shift, won't go into gear, noisy, vibrations.

26. Some repairs can be done without removing the transmission from the machine, but most repairs require removal.

27. Extra care should be taken when removing a transmission since they are heavy and usually need to be removed from the bottom of a machine.

28. Repairs can include gear replacement, bearing replacement, shift mechanism repair.

chapter 12
POWERSHIFT TRANSMISSIONS

LEARNING OBJECTIVES

After reading this chapter, the student should be able to:

1. Describe the different types of powershift transmissions.
2. Describe oil flow through a powershift transmission.
3. Explain clutch modulation.
4. Explain clutch engagement.
5. Describe powershift transmission's mechanical, electrical, and electronic controls.
6. Describe manual and electronic automatic shifting.
7. Explain governor operation.
8. Explain autoshift.
9. Describe piston seal types.
10. Describe different clutch materials.
11. Explain the function and types of clutch plates.
12. Explain the function, types, and styles of clutch discs.
13. Describe reusability guidelines for clutch discs and plates.
14. Describe the basic troubleshooting procedure for a transmission operational complaint.

KEY TERMS

Countershaft 356
Friction discs 353
Modulation 355
Planetary 377
Planetary pinion carrier 377
Plates 353
Proportional solenoids 367
PWM 373
Ring gear 377
Seal rings 374
Sun gear 377

INTRODUCTION

SAFETY FIRST A normally operating powershift transmission installed in a machine doesn't pose any real safety threat in itself. However, there are safety concerns you should be aware of when performing repairs or servicing that is related to powershift transmissions.

After working on the control system for the transmission, you must make sure that all safety features incorporated into the control system are working properly. This would include things such as machine anti-start interlocks that would prevent the machine from starting in gear.

Normal safe practices should be exercised when working on transmissions such as properly securing the vehicle before performing work. This would include tag out and lock out procedures if necessary, use of proper hoisting and rigging practices when lifting components over 50 lb, and use of wheel chocks and blocking.

Powershift transmissions use a hydraulic system, so any safety concerns related to it's hydraulic system should be adhered to.

Proper PPE usage should be followed when required. When referencing service information, you will notice many cautions and warnings throughout the manuals. Pay attention to these to keep yourself safe as they are given for a reason.

For example, if you get into transmission rebuilding, you will be cleaning parts in solvent solutions and drying them with compressed air. This will require the use of hand and eye protection. Failure to use these PPE items could result in severe skin irritations and eye damage.

Environmental concerns that you need to be aware of when working with powershift transmissions include fluid spill containment and proper fluid and filter disposal.

WHAT IS A POWERSHIFT TRANSMISSION?

Heavy equipment drivetrains can be grouped into four types of systems that are used to propel the machine or make it travel. Connected to the diesel engine flywheel could be:

1. A hydraulic pump that is used as part of a fluid power system and features a hydraulic circuit for travel (see Chapter 16) or a closed loop hydrostatic system (see Chapter 17). These are typically used for slower travelling (up to 15 mph) small to medium size machines.

2. An electric generator that is connected to the flywheel as part of an electric drive system (see Chapter 13). These are becoming more popular and were mostly used for larger mining trucks and loaders but recently Caterpillar has introduced a track-type tractor with electric drive.

3. A friction clutch that is used as part of a mechanical drive system. They usually drive manual transmission but can also drive power shift transmissions.

4. A torque converter (see Chapter 11), and a hydraulically assisted mechanical transmission or more commonly called a powershift transmission. A powershift transmission could also be directly driven by the engines flywheel although this is less common. This transmission is used to change the torque output from the engine flywheel or torque converter by changing the speed, torque, and direction of rotation through different gear ratios and passing it on to the rest of the machines drivetrain. This could be a differential, a set of axles, or steering clutches, and these components are discussed in Chapters 14, 15, and 20. ● **SEE FIGURE 12–1** for how a typical powershift transmission is arranged as part of the drivetrain.

Many powershift transmissions will also incorporate other components as part of their housing. This could include a torque converter or retarder (see Chapter 10), transfer cases or power take-offs (see Chapter 14), and parking brakes (see Chapter 18).

The powershift-type transmission's purpose is to modify engine torque output to provide the machine with different

FIGURE 12–1 A typical power train arrangement that uses a powershift transmission.

EXTREME EXAMPLES

If you consider a Cat 797 rock truck that has a payload of 380 tons, it will use a 24-cylinder engine that puts out over 3300 hp and roughly 8700 ft-lb of torque. This engine torque will get multiplied through a torque converter and sent to the powershift transmission where it is sent through various gears to give seven forward gear ratios and provide a maximum travel speed of 42 mph! The transmission itself weighs almost 7 tons!

One other point that is quite remarkable about powershift transmissions is how reliable they can be considering the massive torque that is transferred through them and the amount of shifts they may need to complete in an expected life of 12000–15000 hours. For example, an eight-speed scraper will average at least one complete upshift and downshift cycle for each load it hauls. Chances are there will be more than this per load but let's go with that. If an average machine hauls five loads in one hour, this adds up to 1,200,000 shifts over a 15000-hour life span!

MACHINE TYPE	FORWARD SPEEDS	REVERSE SPEEDS
Track-type tractors	3	3
Wheel loaders	3–5	3–4
Motor graders	5–8	5–8
Mining trucks	6–8	1–2
Articulated trucks	5–8	1–2
Track loaders	3	3
Scrapers	6–18	1
Backhoe loader	3–5	2–4
Skidders	4–8	3–4

TABLE 12–1
Different machine types and their transmission speed ranges.

travel speeds and directions to allow the machine to perform different tasks. The term *powershift* comes from its ability to shift speed ranges and directions under load and with the engine producing full power. This is really quite amazing if you imagine a machine that is travelling under full load. There are massive amounts of torque that are multiplied through the drivetrain when the machine is under load and the powershift transmission has to keep that power flowing with just a momentary pause when the operator or ECM requires it to shift to either a different direction or speed, or both. The machine manufacturer will design its own transmission or install one from an OEM (original equipment manufacturer) that is able to handle the maximum flywheel torque from the engine.

Some examples of different extreme drivetrain loads are the following:

1. A track-type tractor pushing up to 80 cubic yards of material.
2. A wheel loader filling a bucket with up to 50 tons of material.
3. A motor grader leveling a rough road pushing a blade up to 24 ft long.
4. A mining truck hauling a load of rock up to 400 tons up out of a mine.
5. A scraper hauling up to 50 cubic yards of material.

Most heavy equipment use diesel engines that have an optimum operating speed at around 1800–2000 rpm. At this speed, engine efficiency is greatest in terms of torque, fuel economy, and exhaust emissions. To keep the engine at this desired speed in relation to different machine travel speeds, there needs to be different gear ratios available, and depending on the type of machine and the kind of work it's trying to do, manufacturers will use transmissions with a variety of different gear ratios. Machines that need to travel relatively long distances and at high speeds can have one or more overdrive ratios.

Powershift transmissions that are used in heavy equipment can have 2–18-speed ranges or "gears" in forward and 1–8 speeds in reverse depending on the type of machine they are used in. ● SEE TABLE 12–1 for some different types of machines that use powershift transmissions and the typical speed ranges they use.

What should be apparent here is that machines that don't need a lot of speed such as bulldozers don't need a lot of gear ratios and machines that need to cover a lot of ground such as trucks and scrapers need a lot of gear ratios to get both high speed and high torque at lower travel speeds.

POWERSHIFT EVOLUTION The need for powershift transmissions became apparent quite early in the evolution of heavy equipment. As heavy equipment evolved from farm machinery just past the turn of the 20th century, it became more specialized. It was soon obvious to equipment owners, operators, and equipment manufacturers that an improved transmission was needed. Mobile heavy equipment transmissions at the time were mechanical standard-type transmissions with flywheel clutches that required clutch disengagement and which meant stopping the travel of the machine to change directions. To improve machine productivity and ease operator effort, the powershift transmission was created. The powershift transmission when coupled with a torque converter allowed the transmission to be shifted to different speed ranges without coming to a full stop and direction changes could be made much faster with very little shock load through the drivetrain.

From an operator's perspective the powershift transmission made their life much easier. If they were operating a machine with a mechanically actuated clutch and standard transmission and they wanted to make a direction change, they would first have to slow the machine down, push the clutch in, wait for the machine to stop, select a different gear, and then ease the clutch out while trying not to slip it too much, which would cause premature wear. This process of changing gears took a lot of extra time and effort, which took away from the machine's productivity. Back in the day, most people were familiar with clutches and standard transmissions because they were popular in automobiles, not so much as these days. If an operator wasn't gentle with the clutch, it could need replacing in a matter of days, and this would be a major repair. The powershift transmission meant the operator could simply select a gear, increase the engine RPM and move the gear shift control to change speeds or direction.

This ability to shift under power is accomplished by a series of hydraulic clutch packs that engage and disengage a series of gears and are controlled by a hydraulic system.
● **SEE FIGURE 12–2** for how the controls for powershift transmissions have changed.

All machines that feature powershift transmissions now will only have one control lever that combines both direction and speed range control. Originally, powershift transmissions were controlled by the operator through mechanical linkage, whereby a series of rods, levers, and ball joints transferred the motion of two gearshift levers from the cab to a series of spool valves in the transmission control valve. This mechanical control system gave way to cables, and now most powershift transmissions are controlled electronically, where the operator is merely actuating electrical switches that control electric solenoids to both speed ranges and direction. Control systems for powershift transmissions have evolved from total operator control into today's machines that sometimes give the operator the option of being in control or letting an ECM control transmission shifting.

FIGURE 12–2 How powershift transmission controls have changed.

POWERSHIFT TRANSMISSIONS 351

Progress Check

1. Powershift transmissions are used in this type of machine propel system:
   a. electric drive
   b. mechanical drive
   c. hydrostatic drive
   d. hydraulically assisted mechanical drive
2. What is the expected life span of a powershift transmission:
   a. two years
   b. 5000 hours
   c. 12000–15000 hours
   d. 20 years
3. Powershift transmissions were mainly developed to:
   a. make the machine quieter
   b. change machine speed and direction under load
   c. increase fuel economy
   d. make transmissions easier to repair
4. The component of a powershift transmission that allows it to make shifts under load is:
   a. the hydraulic clutch
   b. the hydraulic torque converter
   c. the transmission control lever
   d. the transfer case

NEED TO KNOW

There are two main types of powershift transmissions: *planetary* and *countershaft*. These terms refer to the type of gear arrangement that is used to transfer torque through the transmission. Both are constant mesh-type transmissions, meaning that all the gears inside the transmission are always in mesh with another gear. When one or more gears are locked to their shaft by a hydraulic clutch, then torque is transferred through the transmission.

Countershaft powershift transmissions are typically found in small- to medium-sized wheel loaders, graders, trucks, and other machines that are in 100–400 hp range.

Planetary powershift transmissions can be found in any type or size of machine from less than 100 hp to 1000 hp or more.

Depending on the type of machine the transmission used, its output drive will be transferred to a variety of components:

1. If the transmission is used in a two-wheel drive backhoe loader, it will transfer its output torque directly to the axle pinion or through a drive shaft to the axle drive yoke.
2. If the transmission is used with a four-wheel drive backhoe loader, there will be a second output to drive the front axle that can be disengaged when not needed.
3. If the transmission is used in a wheel loader, its output will be to both front and rear axles with no provision for disconnecting the front drive. This part of the transmission is called the transfer case or drop box and is usually part of the main housing, but it could be a separate housing. ● **SEE FIGURE 12–3** for a look at a countershaft transmission used in a wheel loader drivetrain.
4. If the transmission is used in an articulated truck, it will also use a drop box, and if used in a grader, the transmission will have its output go to a differential.
5. If the transmission is used in a track-type tractor, its output will be a pinion gear that directly drives a ring gear to drive either steering clutches or a differential.
6. If the transmission is used in a straight frame rock truck, its output could be a pinion that directly drives the differential for the rear axle, or it could use a drive shaft to drive the truck's axle.

There are some similarities between countershaft and planetary types of transmissions. These are the hydraulic clutches and hydraulic systems that will be discussed next. The main difference between the two types is the mechanical components that transfer torque from the input to the output.

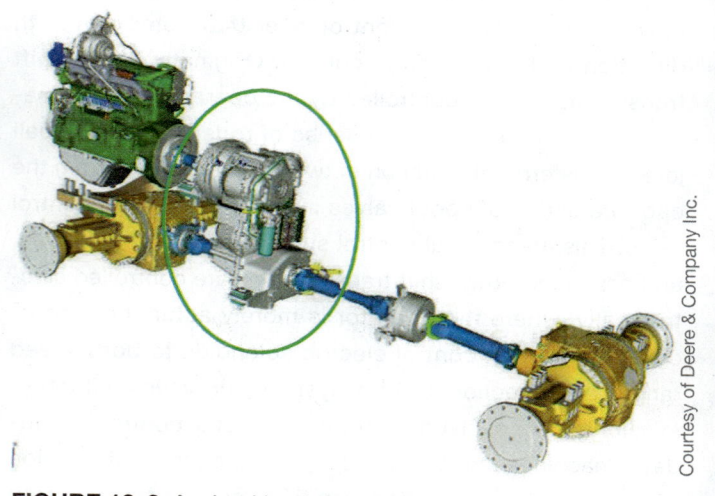

FIGURE 12–3 A wheel loader that uses a countershaft transmission.

HYDRAULICALLY ACTUATED CLUTCHES

The key to making a transmission able to shift speed ranges and directions under full power is the hydraulically applied spring-released clutch. The development of this type of clutch has made the powershift type of transmission a widely used and reliable power train component. ● **SEE FIGURE 12–4** for the components of a hydraulically applied spring-released clutch. Clutches used in a powershift transmission can either lock two rotating components together such as a gear to another gear or stop a rotating component by locking it to a stationary component. Sometimes this clutch will be called a brake when the clutch is used to lock a rotating component to a stationary housing. This may sound a little confusing, but it is really the same component doing a different job. For ease of explanation in this chapter, regardless of their function, the clutches and brakes will be called clutches.

The clutches will use oil pressure that is applied to a piston to squeeze plates and discs together. This means the transmission needs a hydraulic system to supply the oil for clutch engagement. This hydraulic system is also used for one or more functions that will be discussed later.

No matter what type of powershift transmission a machine may use, the clutch operation is basically the same. The purpose of the clutch is to either allow two components to rotate independently or lock them together. The torque transfer components inside a clutch are plates (sometimes called separator plates) and discs (sometimes called **friction discs**) that are alternately stacked together from 2 to 10 each in number. To increase the maximum amount of torque that a clutch can transfer, the transmission manufacturer will either increase the number of plates and discs or make them bigger. Clutch torque capacity relates to surface area. More surface area allows the clutch to handle higher torque loads. Clutch disc and plate diameter for low horsepower machines may be 4 in. in diameter while ones used for the largest machines could be as big as 36 in. The thickness of plates and discs are up to ¼ of an inch each. The plates and discs are designed to run in oil and need this oil for cooling purposes.

The clutch **plates** are thin round slices of steel or cast iron that will have either teeth on the inside or outside diameter or tangs on the outside diameter. ● **SEE FIGURE 12–5** for a variety of clutch plates. The teeth or tangs will lock the plates to a part or component such as a gear hub, ring gear, or housing. The plates should have a smooth surface but not polished so they can retain a film of oil to prevent overheating of the plate during clutch application or release. They are usually ground with a crosshatch pattern to aid in oil retention. Whenever plates are removed during a disassembly procedure, they must be inspected for heat checks, tooth wear, grooving, and discoloration and measured for flatness.

The clutch discs are sandwiched between the clutch plates and could have teeth on their inside or outside diameter, but more typically the teeth will be on the inside diameter. The clutch disc must absorb huge amounts of energy when a shift is made. There is a short period of time between the start of clutch engagement to fully engaged, and in that time these clutch discs need to allow some slippage which, in turn, generates heat. The discs' ability to withstand heat and engage smoothly depends on the type of material they are made of. They will have a base of steel but will have a friction material bonded to both sides. This friction material can be made of various compounds such as combinations of sintered metals like bronze, iron, and copper; paper based or cellulous materials

FIGURE 12–4 A hydraulically applied multi-disc clutch.

FIGURE 12–5 A variety of clutch plates.

POWERSHIFT TRANSMISSIONS 353

> **WEB ACTIVITY**
>
> Visit http://www.frenossauleda.com/friction/?q=frictionmaterial for more details on clutch materials.

FIGURE 12–6 Different types of clutch discs.

(sometimes called organic); and elastomeric materials that contain a mixture of rubber, inorganic fibers, and friction particles. The transmission manufacturer decides on the type of friction material based on the loads and temperatures the clutch will be required to withstand and the shift characteristics desired. When the clutch is engaged under load, the type of material will play a big part in how the shift feels to the operator and how much shock load is transferred through the drivetrain. ● **SEE FIGURE 12–6** for examples of different clutch discs.

Springs will retract the clutch piston and allow the plates and discs to rotate independently and are usually either two types. A series of coil springs spaced around the circumference of the piston could be used or one or more Belleville type of spring. A Belleville type of spring is like a dished washer and can be used singly or stacked up in multiples. The spring or springs will push on the piston to keep it in the released position. ● **SEE FIGURE 12–7** for one type of clutch spring.

To apply a clutch, oil is fed into the bore that houses the piston and oil pressure is applied to the face of the piston, where it is sealed in that area by the piston's seals. Clutch apply pressure will build to its limited maximum of typically 250–350 psi. The piston seals can be many different styles and made of many different materials, with the goal of long life being the main determining factor. The piston does not need to travel very far, usually no more than a quarter of an inch to clamp the plates and discs together; therefore, the quantity of oil needed to engage a clutch is not very large.

The oil to apply the clutch originates from a control valve or solenoid that is controlled by the operator from the machine's operator station. The oil is fed through either a stationary housing or a rotating shaft. If it is fed through a rotating shaft, there must be seals on the shaft so the oil is directed to a passage in the shaft through a drilled hole and comes out to a cavity where the piston is housed. ● **SEE FIGURE 12–8** to understand how oil moves a piston.

POWERSHIFT TRANSMISSION HYDRAULIC SYSTEM

The following description is for a typical powershift hydraulic system. There will be variations to the generic system described here, but this will give you a good idea of the components used to make up the hydraulic system and how the oil flows through the system.

The transmission hydraulic system starts with either an internal- or an external-type gear pump that is driven at engine speed and gets oil from the bottom of the transmission housing usually through a suction strainer. The pump will then supply oil to a filter to remove any contamination from the oil. The filter could be a spin-on type or cartridge type. Oil is then sent to the control valve where its flow is divided and made available to a series of valves to regulate maximum system pressure

FIGURE 12–7 One type of clutch release spring.

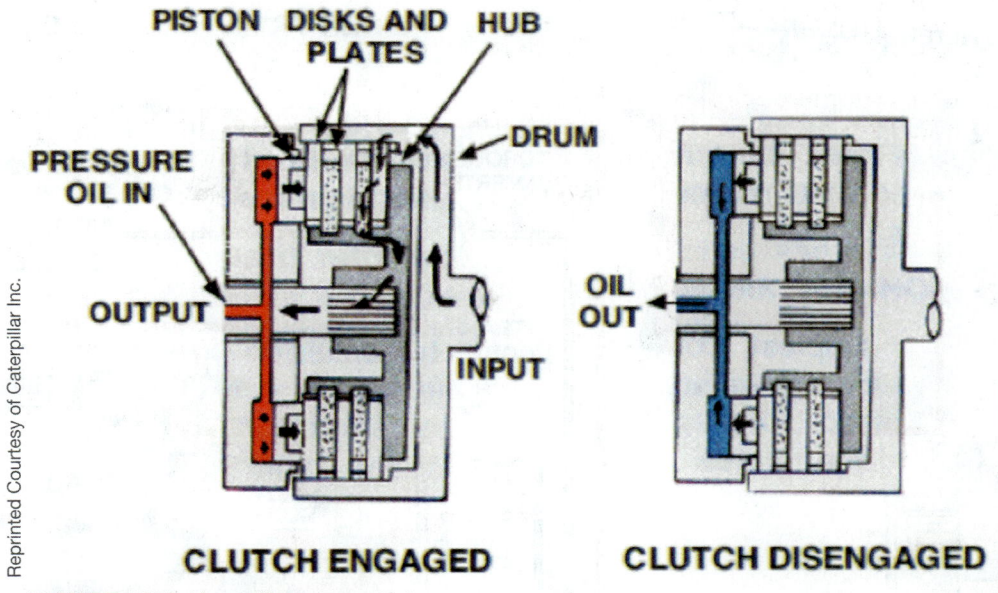

FIGURE 12–8 How oil moves a piston.

to approximately 350 psi and regulate torque converter pressure, cooler pressure, and transmission lubrication pressure. Oil is also made available to one or more clutch modulation valves where the oil can be directed to one or more clutch packs. The oil will move from the transmission control valve to the clutch through sealed passages in the transmission housing and sometimes through rotating shafts. The clutch application pressure needs to be increased at a controlled rate to make a smooth shift and give a cushion effect when a speed or direction shift change is made. This is called **modulation** and is controlled by a valve or orifice in the transmission control valve. The operator has no control over the pressure regulators or modulation valves. These can be adjusted by a technician if required.

Transmission shift control valves for powershift transmissions will be controlled by the operator either mechanically or electrically. If mechanically controlled, there will be spools that are moved by the operator to direct pump oil to the clutch packs, and if electrically controlled, there will be solenoids that receive electrical current based on the operator's input and these solenoids will direct oil to the clutch packs.

Transmission oil will then be sent to supply the torque converter (if the machine uses one) at a lower pressure. The oil will be heated up after running through the torque converter, and to keep the oil from overheating and breaking down, the oil will be circulated through a cooler and then sent back to the transmission to lubricate all rotating parts. This oil could also be used for releasing a parking brake or actuating a differential lock. ● **SEE FIGURE 12–9** for an example of a powershift transmission hydraulic system.

Progress Check

5. This part of the hydraulic clutch will have friction material bonded to it:
   a. piston
   b. disc
   c. plate
   d. spring

6. This part of the hydraulic clutch will have two seals as part of it:
   a. piston
   b. disc
   c. plate
   d. spring

7. The one factor *not* used to determine the type of material formulated when friction material is designed:
   a. how much heat the material will be exposed to
   b. how much torque the clutch will transfer
   c. if it rotates clockwise or counterclockwise
   d. how much slippage occurs at clutch engagement

COUNTERSHAFT POWERSHIFT TRANSMISSIONS

● **SEE FIGURE 12–10** for a cutaway view of a countershaft-type transmission. Internally, countershaft powershift transmissions are arranged similar to manual countershaft-type

POWERSHIFT TRANSMISSIONS 355

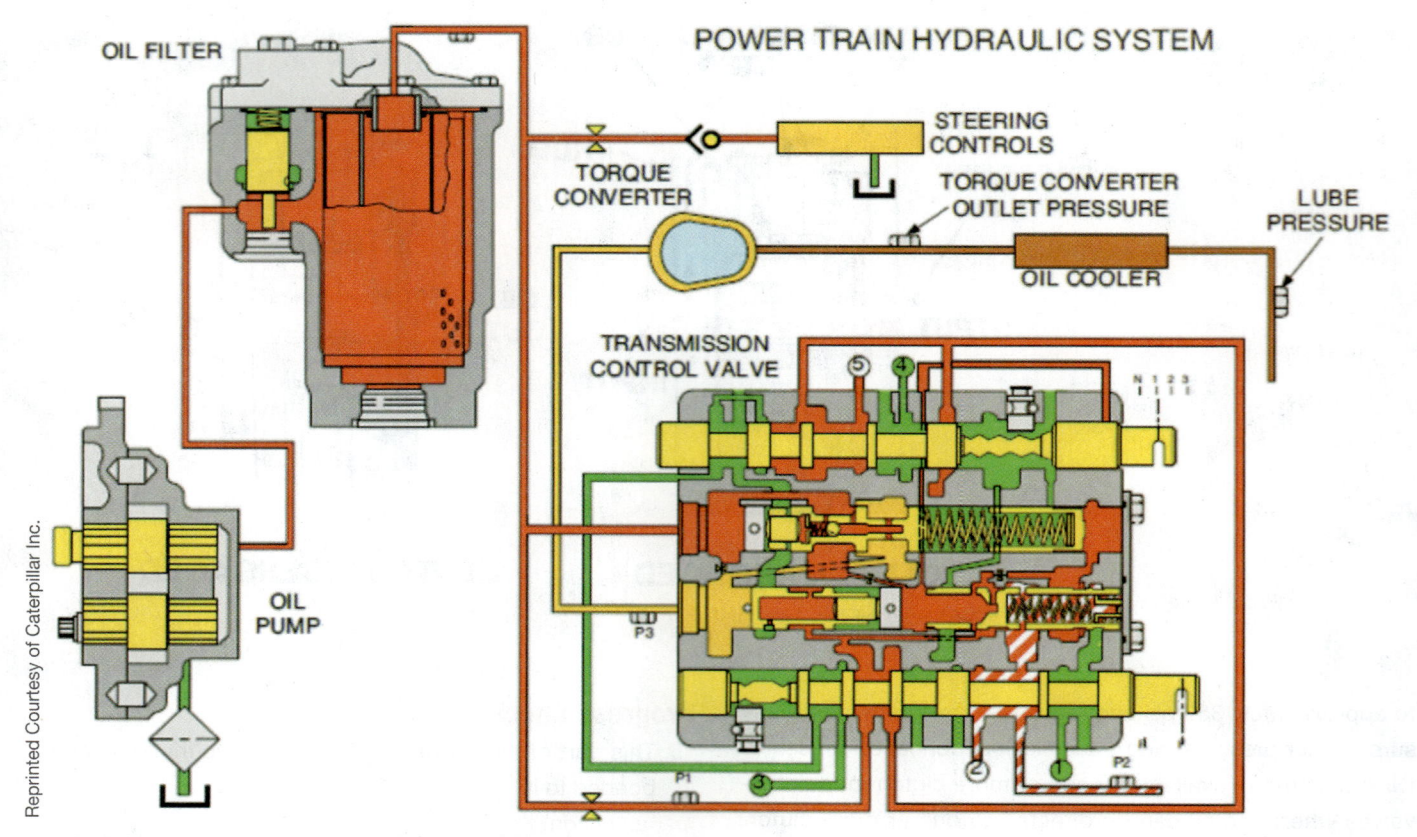

FIGURE 12-9 A powershift transmission hydraulic system.

transmission with the main difference being the method used for changing speed ranges and directions. The powershift transmission will use hydraulically actuated clutch packs to perform gear changes as opposed to having this done mechanically in manual countershaft units.

The term **countershaft** refers to the gear and shaft arrangement that provides the drive flow through the transmission. In a mechanical manual shift countershaft transmission, the input shaft drives a countershaft that has gears that are in constant mesh with gears that float on the output shaft. These two shafts are parallel to each other. The shift mechanism locks one of the gears to the output shaft and then drive is able to be transferred through the transmission.

In the powershift countershaft transmission, the input shaft gets driven by the torque converter output (turbine shaft) or is directly driven from the engine flywheel. The input shaft will have gears that are in constant mesh with gears of one or more other sets of shafts and gears that are parallel to the input shaft. When the transmission is in neutral, all gears will rotate in mesh with each other. Each shaft is driven by a geared hub that has one or two hydraulically actuated clutches that can transfer torque to either one or two output gears that normally float on the shaft. These gears are in constant mesh with gears on another shaft, and if no clutches are engaged, the transmission is in neutral and no torque is transferred through the transmission. Different gear ratios and output shaft rotation direction are determined by which clutches are engaged. As the clutches are engaged, they will transfer torque from the shaft drive hub to one of the shaft output gears. Most transmissions will need two

FIGURE 12-10 A countershaft type of transmission.

clutches, one direction, and one speed engaged to transfer the input drive to the output shaft, but some will need three clutches engaged. The amount of speed ranges the transmission offers will depend on the number of countershafts.

Because of the design of the countershaft transmission, the shafts receive a radial load that increases and decreases in proportion to the load created by the resistance to machine movement. This means as gear load increases, the shafts will be trying to push away from each other, which will put more load on the bearings that the shafts are riding on. The types of gears used are either spur cut or helical cut. Gear types were discussed in Chapter 11. If helical cut gears are used, this will create an additional axial load in the shaft. Shafts will be suspended by various types of roller bearings on each end that will be pressure lubricated from the lubrication circuit of the transmission hydraulic system.

REVERSER OR SHUTTLE SHIFT COUNTERSHAFT TRANSMISSIONS

We'll have a close look at two versions of countershaft-type transmissions. First, let's have a look at a very simple version of this transmission that is sometimes called a reverser or shuttle shift transmission. This type of transmission is a combination of powershift and manual shift transmission and is used in combination with a torque converter. The shuttle shift transmission is popular with many makes of backhoe loaders but could be found in other light-duty machines such as fork lifts. The power shift part of this transmission will only control forward and reverse directional clutches and is sometimes in a separated housing from the speed range section. This simple type of powershift transmission allows the operator to quickly shift from forward to reverse, under load for increased machine productivity. There could be two- to five-speed ranges that will be selected manually with a shifter connected to shift forks that move synchronizers to provide smoother engagement. The speed range section of the shuttle shift is very similar to the mechanical transmissions that were discussed in Chapter 11.

The forward–reverse power shift part of this transmission will use two hydraulically actuated clutches that are controlled by a hydraulic control valve. Older versions of shuttle shift control valves were mechanically controlled, but most versions now are electrically or electronically controlled. The electrical system will incorporate a safety interlock system to prevent the machine from starting or starting in gear unless certain conditions are met such as controls being in neutral and the parking brake being applied. There could also be an

FIGURE 12–11 Shuttle shift transmission controls.

electronic fault code storage system integrated as well as oil temperature monitoring and overheat warning systems provided.

There will usually be a transmission neutralizer function that uses operator-controlled switches to send current to a solenoid that will need to be energized to cut power flow through the transmission so a speed range shift can be made smoothly. The neutralizer simply drains any clutch apply pressure to disengage a clutch and interrupt power flow.

● **SEE FIGURE 12–11** for the transmission controls of a shuttle shift transmission.

Let's see how the power flows through this transmission. We'll start by identifying the main components inside the transmission. ● **SEE FIGURE 12–12** for a cross-sectional look at the John Deere 310E transmission mechanical components.

Its output could be to the rear axle only or to both front and rear axles. The output to the front axle is out the MFWD (mechanical front wheel drive) shaft.

This four-speed transmission's direction and speed range gears are in constant mesh. The torque converter (A) is driven by the engine and its output comes out the turbine shaft, which is coupled to the transmission input shaft. The input shaft drives both direction clutch pack (K and B) floating drive hubs. If neither clutch is actuated, this is as far as the torque is transferred to. When either forward (K) or reverse (B) clutch is filled with oil through drilled passages in the shafts, the discs and plates are squeezed together and torque is sent to the intermediate shaft (J). ● **SEE FIGURE 12–13** for how the power flows through the transmission in first gear forward.

POWERSHIFT TRANSMISSIONS 357

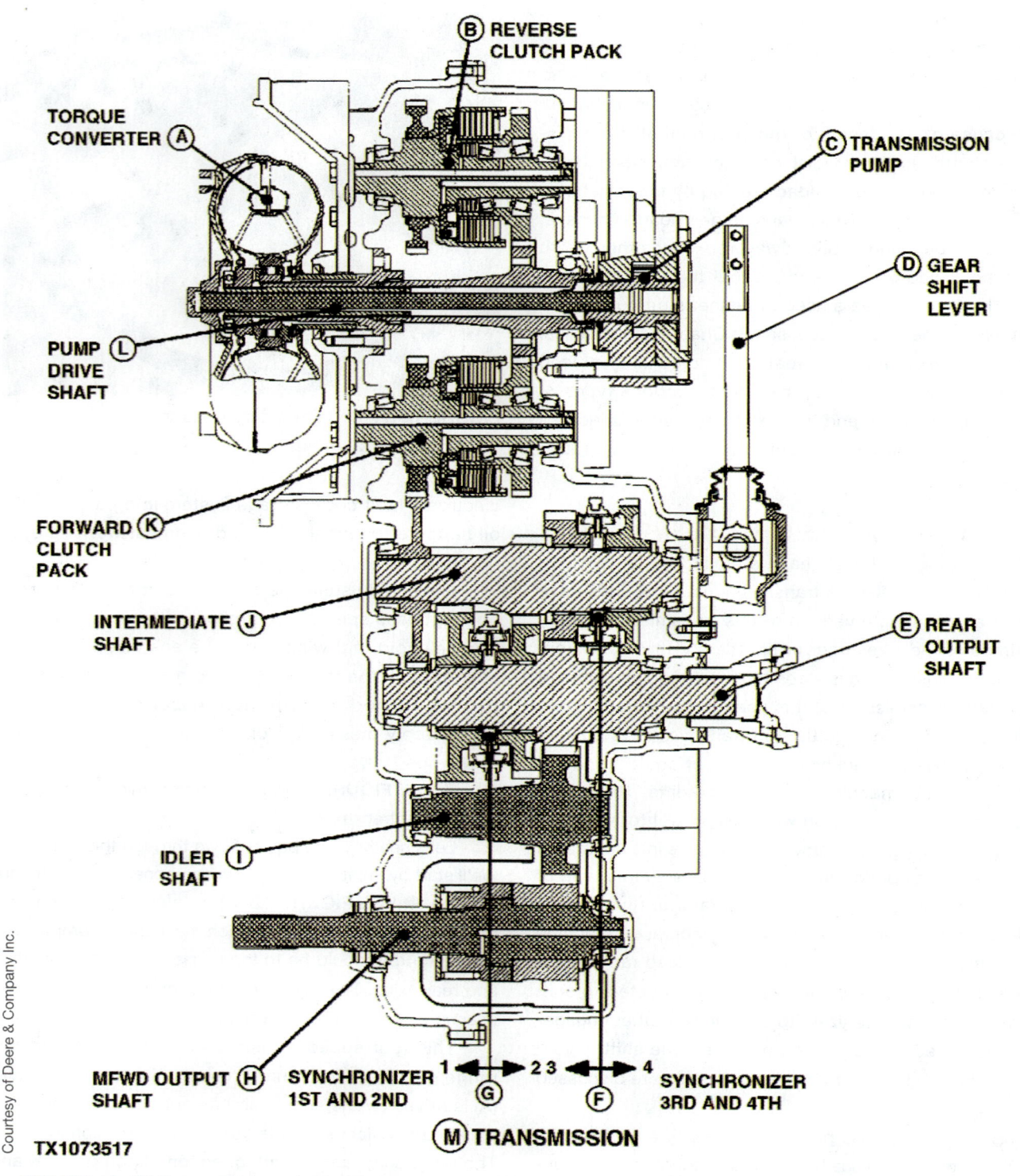

FIGURE 12–12 Cross-section of a 310E transmission.

Following Figure 12–13-Input shaft (B) is always driving the forward clutch hub (D) and the reverse clutch hub. Until one of the direction clutches is pressurized there is no power flow through the transmission. If oil is directed through the drilled passage in clutch shaft (O), this causes the piston (M) to move, compressing the plates and discs (E), causing the forward clutch gear (N) to be driven. This gear is in constant mesh with intermediate gear (L) and shaft (F). If first speed is selected by the operator by moving the gear shift lever for the manually selected part of the transmission, torque will transfer from the intermediate

358 CHAPTER 12

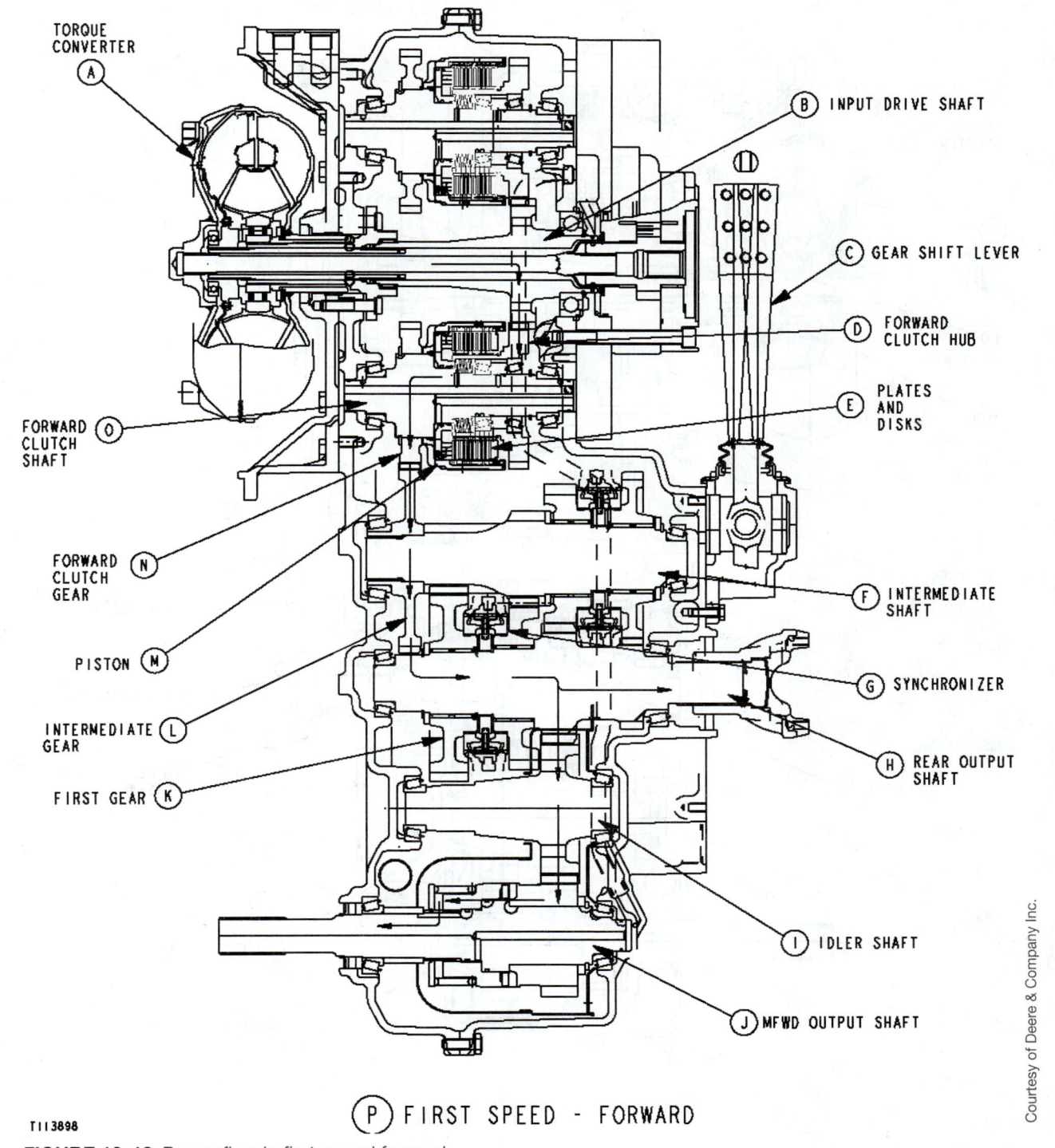

FIGURE 12–13 Power flow in first-speed forward.

shaft to the output shaft where it is sent either to the rear axle or to both front and rear axles. ● **SEE FIGURE 12–14** for how power flows through the transmission in third-speed reverse.

What is not apparent in this diagram is that the reverse clutch output gear (N) and forward clutch output gear (M) are in constant mesh. The gears appear to be not in mesh but in fact they are in the actual transmission. If oil is directed through reverse clutch shaft (P) to engage the reverse clutch, then gear (N) will drive forward output gear (M). This gear acts like an idler gear to reverse rotation to the intermediate shaft

POWERSHIFT TRANSMISSIONS 359

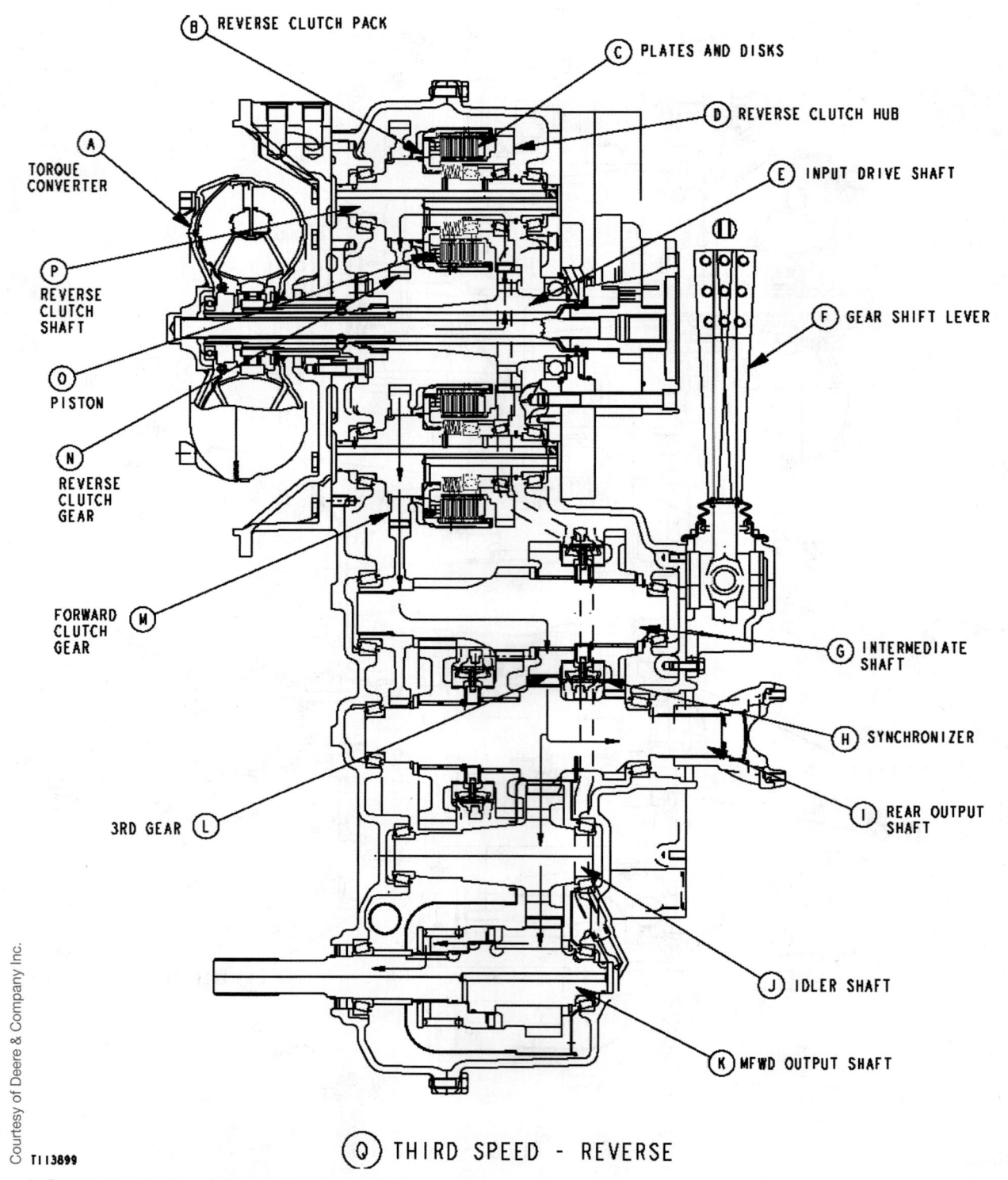

FIGURE 12–14 Power flow in third-speed reverse.

(G). Torque can now be sent through the manually shifted part of the transmission and all gears selected will be in reverse rotation.

Let's see how the hydraulic system works for the 310E. We'll start by looking at the pump.

The transmission pump (C in Figure 12–12) is driven at engine speed by the transmission input shaft and takes oil from the bottom of the transmission housing and supplies flow to the hydraulic control valve. It is an external gear-type pump and provides an oil supply for the hydraulic control section of

360 CHAPTER 12

the transmission for clutch application, cooling, and lubrication.

● SEE FIGURE 12–15 for the disassembled pump.

Next we'll look at the transmission control valve.

● SEE FIGURE 12–16 for a cross-sectional view of the valve.

The components that make up the control valve are the following:

1. The pressure regulating valve (6) controls maximum system pressure at between 230 and 275 psi and is set by adjusting shims behind a spring. This is the pressure that will be applied to the clutch piston when a forward or reverse shift is made.

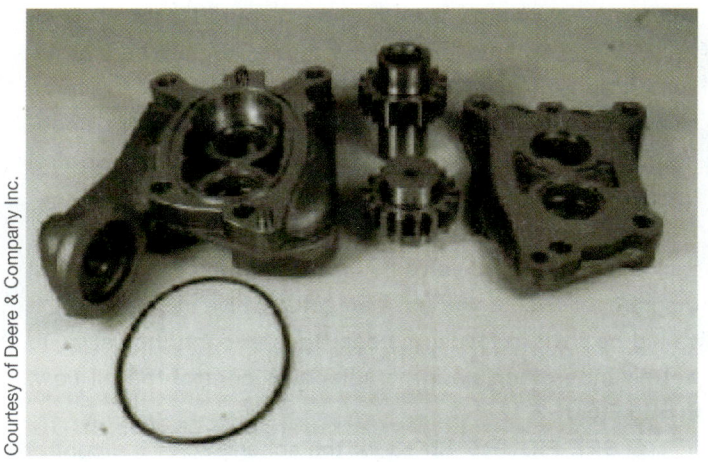

FIGURE 12–15 A disassembled transmission pump.

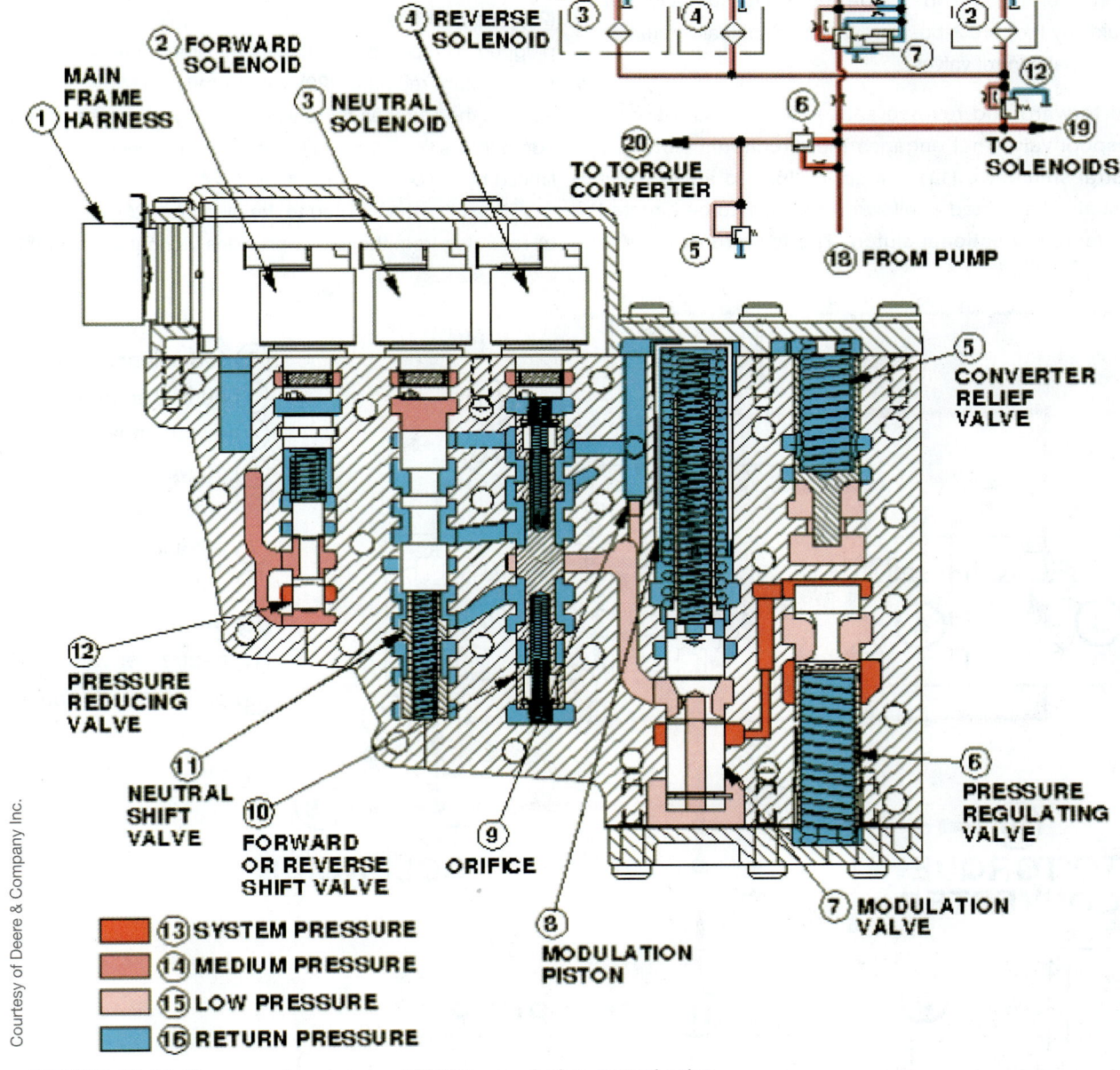

FIGURE 12–16 Cross-sectional view of 310E transmission control valve.

POWERSHIFT TRANSMISSIONS 361

2. The converter relief valve (5) controls maximum pressure to the torque converter, which should be 116–125 psi.

3. Modulation valve (7) controls rate of clutch fill time.

4. Forward and reverse shift valve (10) is a spool-type valve that sends oil to the forward or reverse clutches.

5. Neutral shift valve (11) will dump or drain clutch apply pressure.

6. Pressure-reducing valve (12) reduces pressure oil made available to the solenoid valves and should be 123–152 psi.

7. Forward, neutral, and reverse solenoids (2, 3, and 4) direct reduced pressure to the shift valves. These solenoid valves are on or off valves that allow oil to flow past them when energized. To describe them hydraulically, you would say they are 2-position, 3-way, electrically actuated directional control valves.

The forward and reverse shift valve (10) is a spring loaded spool valve that engages the directional clutches. The neutral shift valve (11) is a spring loaded spool valve that must also be shifted to allow pressure oil to be directed to the selected directional clutch. The forward or reverse shift valve solenoids (2 or 4) are activated by the FNR lever. The neutral solenoid (3) is controlled by the park brake switch and must be energized before forward or reverse clutches can be engaged. The loader and shift lever disconnect switches (sometimes called neutralizers) also control the neutral solenoid.

It may be hard to understand how oil flows through the transmission control valve by looking at the cross-sectional diagram, or even if you have an actual valve in your hand, it would be hard to follow the passages through the valve. To better understand how the solenoids control the oil flow,
● **SEE FIGURE 12-17**.

● **SEE FIGURE 12-18** for the schematic of the control valve when forward is selected. You can see how pump (28) gets oil from the sump (29) and sends the oil through the filter (27). The filter has a bypass valve that will open if the filter gets plugged. The pump flow has its maximum pressure limited by the pressure regulator (22). As not much oil flow is needed to actuate the clutches there will be plenty of excess oil from the pump that is then directed to the torque converter. Its pressure is limited by the converter relief valve (24).

The oil is then routed to the oil cooler (25). A portion of the oil returning from the cooler provides lubrication oil to the shaft

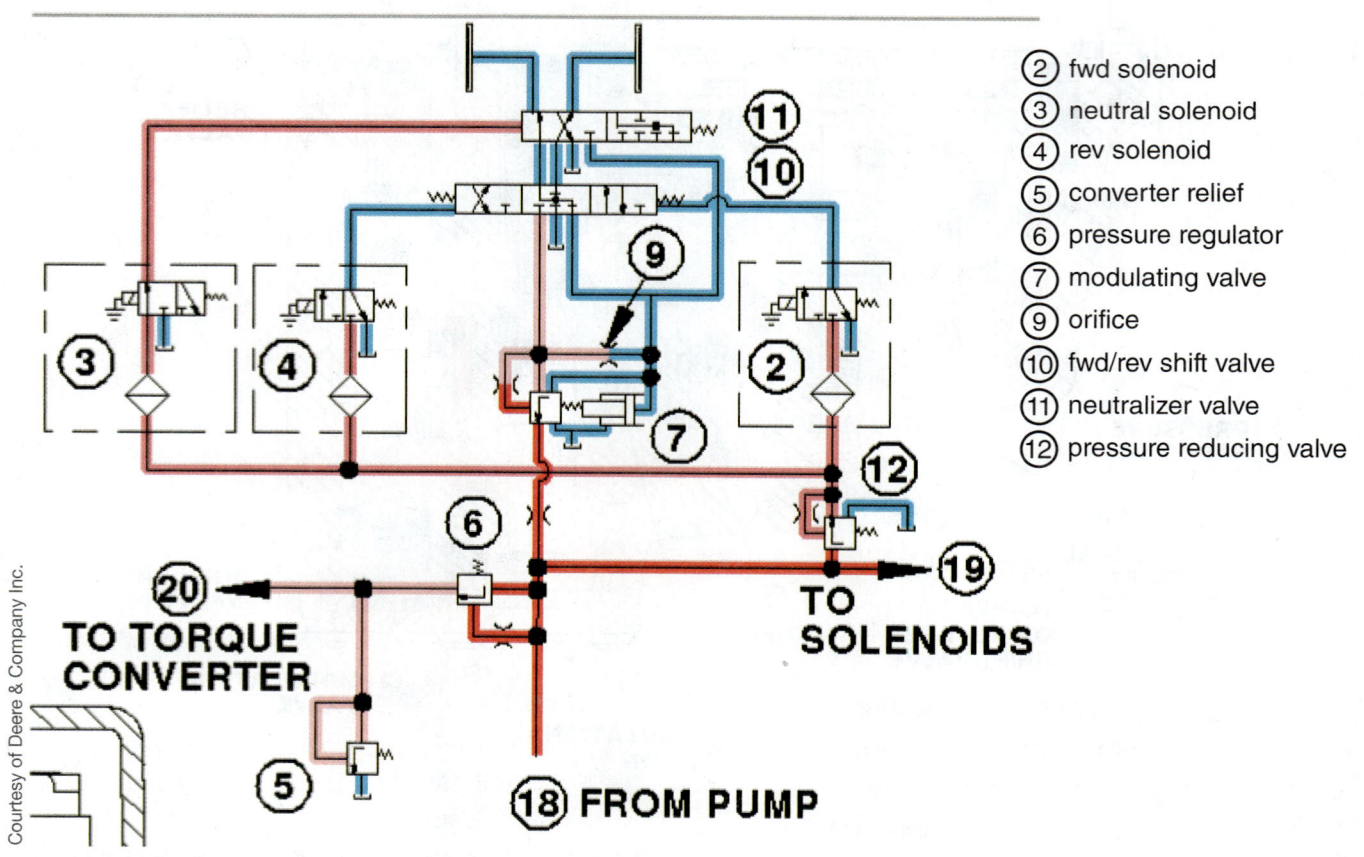

FIGURE 12-17 310E transmission control valve in schematic form.

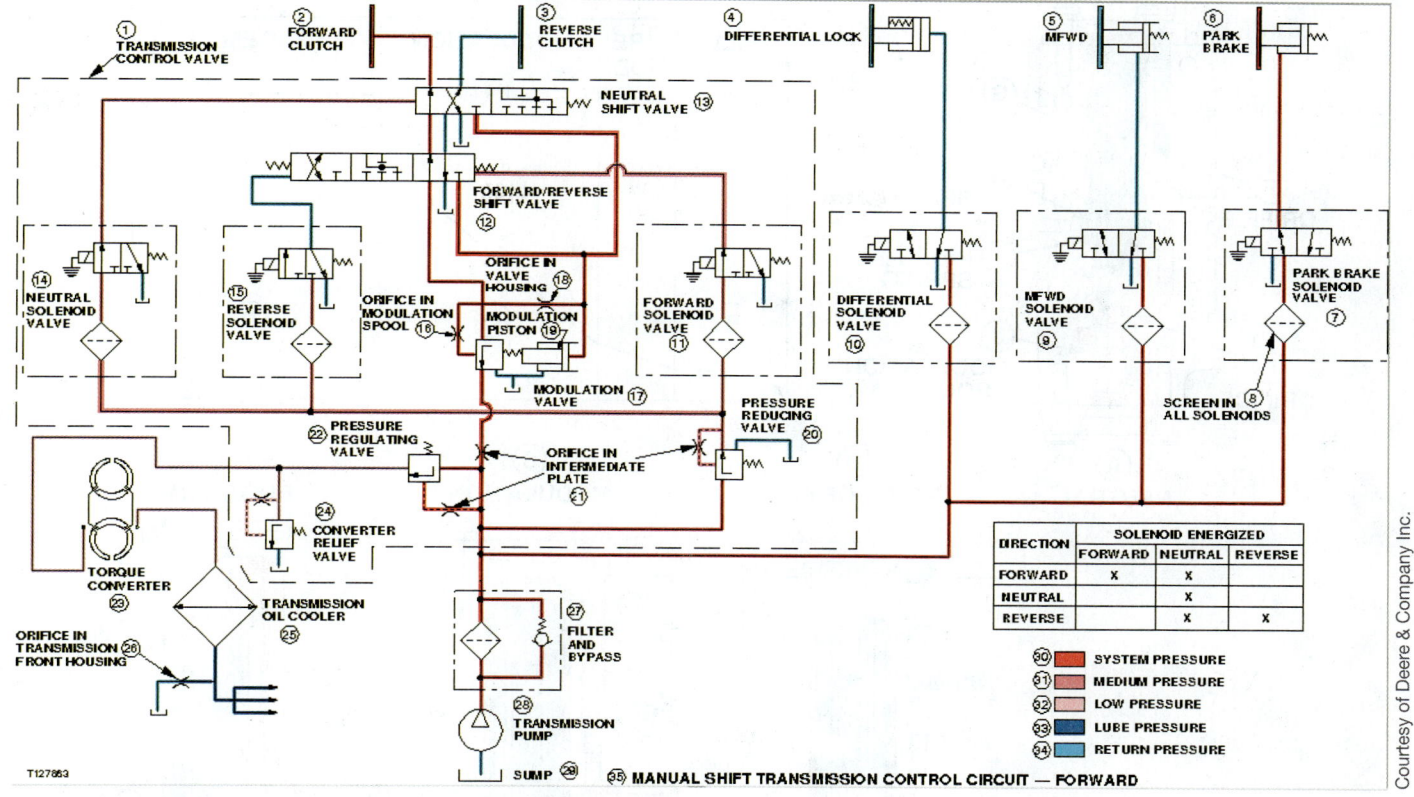

FIGURE 12–18 310E transmission control valve in forward.

bearings and cooling oil to the clutches. The remaining oil goes to sump or to the bottom of the transmission case where it is picked up by the pump again.

The pressure reducing valve (20) supplies medium pressure oil (31) to the solenoid valves (neutral, reverse, and forward). In neutral, with park brake OFF, the forward and reverse solenoids (11 and 15) are de-energized (open), and medium pressure oil is not directed to the ends of the forward/reverse shift valve (12). The neutral solenoid valve (14) is energized (closed) and medium pressure oil (31) is routed to the end of the neutral shift valve. The neutral shift valve in this position allows oil at the forward and reverse clutches to drain through the forward/reverse shift valve.

To control the hydraulic portion of this transmission, the operator will move the FNR lever, which is an electrical switch by the steering wheel or the transmission neutralizer switches on either the speed range lever or the loader control lever.

Three other switches used to control transmission oil flow used for other functions are the park brake switch, the differential lock switch, and the MFWD switch. The switches will send an electrical current to solenoid valves. When energized, the solenoid valves will direct reduced pressure oil to perform various functions.

If the operator moves the FNR lever forward, battery voltage will be given a path to the forward and neutral solenoid valves (11 and 14) to energize them. The neutral solenoid valve (14) sends reduced pressure oil (31) to the neutral shift valve (12) and moves the spool to the right against spring pressure. The energized forward solenoid (11) sends reduced pressure oil to the forward/reverse shift valve (12) moving the spool to the left against spring pressure. Clutch apply oil modulation now starts to provide a smooth transition from disengaged to engaged. Shifting of the forward/reverse spool also blocks the flow of oil downstream of the orifice in valve housing (18) from returning to tank and allows the pressure in the modulation valve to build. As pressure builds, the balance between orifice size and spring load in the modulation valve controls the rate of shift and pushes the modulation piston to the left. The force from the modulation piston pushes on the modulation springs, which move the modulation spool to the fully open position. In the fully open position, high pressure oil (30) from the pressure regulating valve (22) forces the clutch piston against the clutch discs and plates to engage the clutch.

● **SEE FIGURE 12–19** for the cross-sectional view of the modulation valve and a graph to show the different stages of modulation and how pressure rises at a controlled rate.

POWERSHIFT TRANSMISSIONS 363

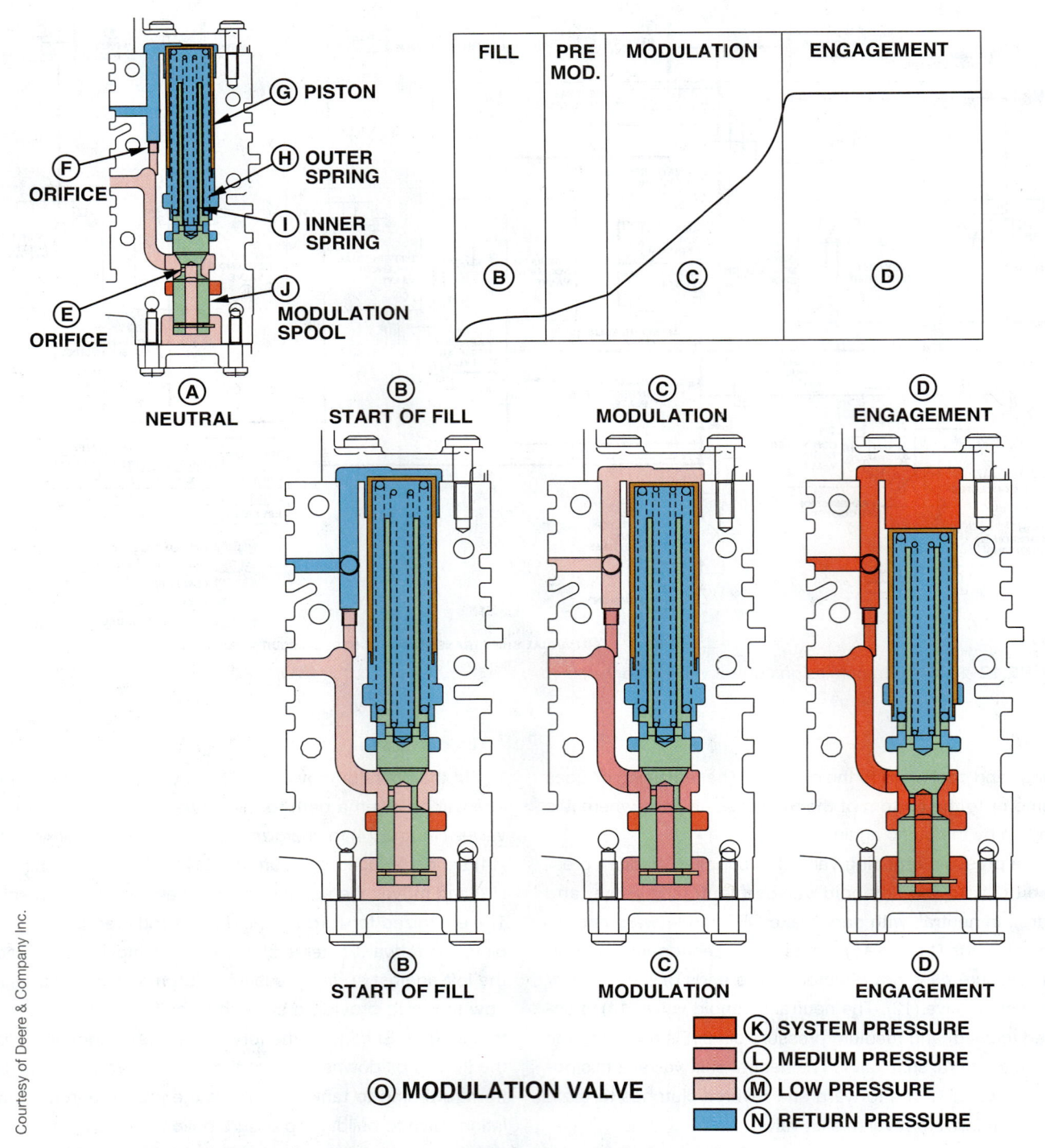

FIGURE 12–19 Modulation valve and how pressure rises at a controlled rate.

The rate of shift from neutral into forward or reverse is controlled by movement of the modulation valve and modulation piston during a shift. By regulating the amount of oil flow being sent to engage the direction pack, the time required is extended to assure a smooth shift. The time for a normal smooth shift is 1.5–2 seconds. This time allows the clutch discs and plates to be squeezed together relatively slowly to provide a cushion effect for the drivetrain. If the modulation was too slow, there would be too much clutch slippage and overheating would occur possibly followed by premature clutch failure. An operator complaint in this case may be lazy, unresponsive shifting. If the clutch engaged

364 CHAPTER 12

too fast because of too little modulation, there would be shock loading of the drivetrain and possibly broken universal joints and other related damage. Some operator complaints would be harsh shifting, banging, clunking, or jerking as the transmission shifts. Some transmission control valves allow for adjustment of clutch modulation by adding or removing shims to change modulation valve spring pressure.

● **SEE FIGURE 12–20** to understand the three modes of modulation.

Neutral (A): System pressure oil (K) bleeds by the edge of the modulation spool (J). This oil (M) (low pressure) flows through orifice (F) and this passage is open to return. Low pressure oil (M) also flows through orifice (E) to the bottom side of modulation spool (J). This oil applies sufficient force on the spring to hold the spool up against spring pressure on the edge of the pressure port opening. Forward and reverse clutch packs are also open to sump when in neutral.

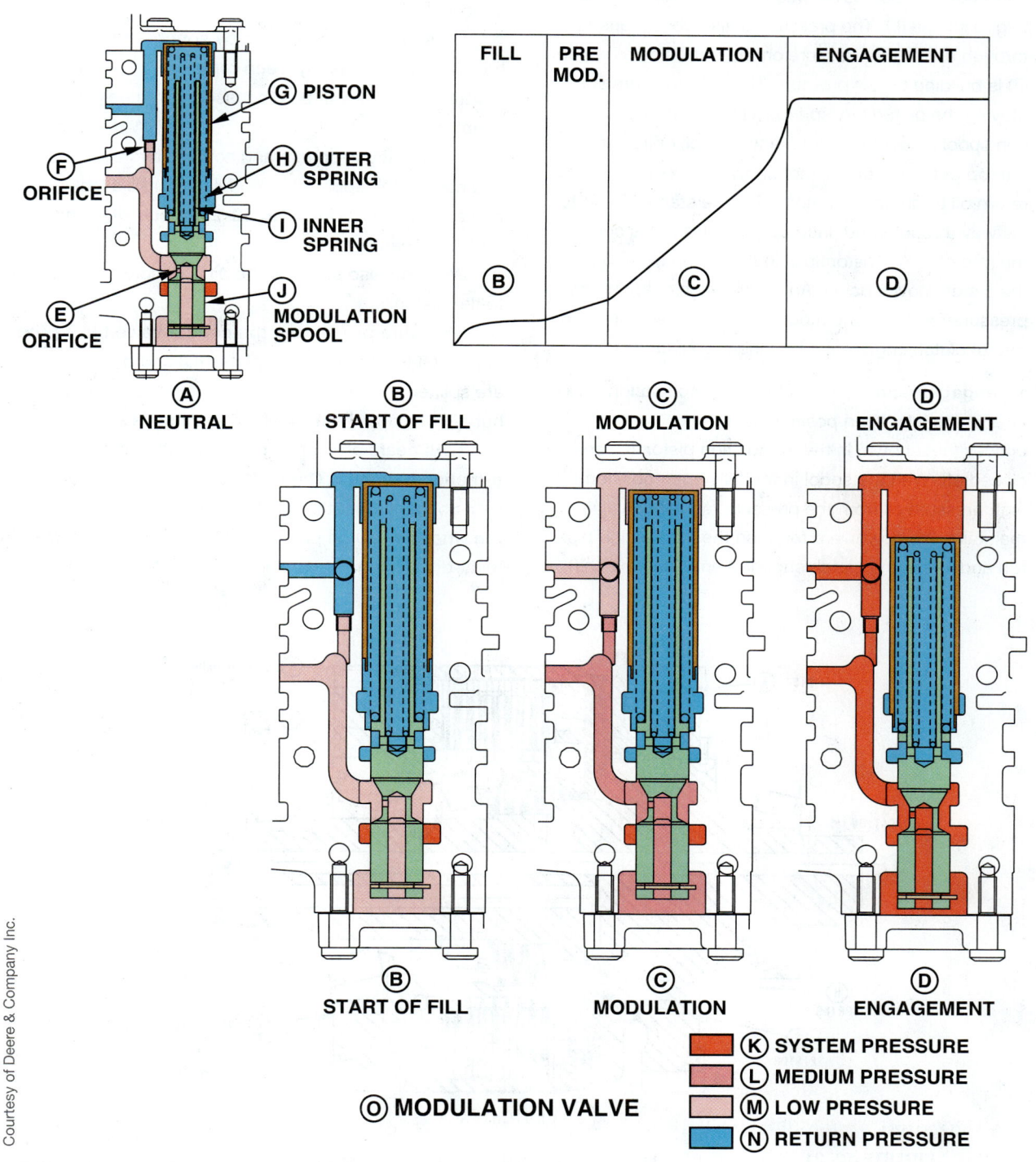

FIGURE 12–20 The three modes of clutch fill oil modulation.

POWERSHIFT TRANSMISSIONS 365

Start of fill (B): Mode begins when a shift to forward or reverse occurs. Shifting of the forward/reverse spool allows oil to flow from the modulation valve to the neutral shift valve and fill the clutch pack. Shifting of the forward/reverse spool also blocks the flow of oil downstream of the orifice (F) from returning to tank and allows the pressure in the modulation valve to build. The clutch pack fills rapidly at the low pressure level.

Modulation: (C) begins as pressure in the clutch pack begins to rise (L). The pressure is also being sensed through orifice (F). Pressure on the back side of orifice (F) is building to low pressure (M). This pressure starts moving the piston (G), springs (H and I), and modulation spool (J) down. There are two small orifices toward the top of the piston (G) (not shown). The orifices allow return oil to dump to sump during the start of fill and halfway through modulation mode. This helps control the rate of shift. The orifices in the piston close off as the piston moves down. An equalization of spring and pressure forces on the modulation valve regulates the rate of clutch engagement during modulation.

Final engagement: (D) is reached, the modulation spool (J) is in the fully open position, and the two small orifices (not shown) in the modulation piston (G) are closed off. With the spool in the fully open position, high pressure oil from the pressure regulating valve maintains the downward force on the modulation piston and keeps the clutch engaged until the next shift.

Cycling the forward/reverse spool or the neutral shift spool allows the high pressure oil in the engaged clutch and between the orifice (F) and the top of the modulation piston to vent to tank. The resulting loss in pressure allows the modulation valve to reset to begin the next shift sequence.

Basically, the pressure rise rate is controlled by the balancing act between the oil pressure on the piston plus spring pressure versus the oil pressure acting on the modulation spool. This oil pressure is also controlled by the effect of the orifices.

This modulation process by springs, spools, and orifices has almost completely been replaced by electric/electronic modulation. This will be discussed in the next section of this chapter.

Now that you understand how the hydraulic system works and how clutch apply oil is modulated, take a look at the clutch cross-section in ● **FIGURE 12–21** to see what the oil does in the clutch pack.

You can also see how the oil is directed through the shaft to the piston.

Pressure oil (D) from the control valve forces the clutch piston (G) to compress the plates (C) and discs (A). The plates are splined to the clutch drum (B) and the discs are splined to hub (E), locking them together. The hub is splined to the drive shaft. All gears are constantly meshed and supported by anti-friction bearings. The bearings and clutches are lubricated with cooled lubrication oil (I). Spring discs push back the piston when the clutch pack (J) is disengaged, thus releasing the multi-disc clutch.

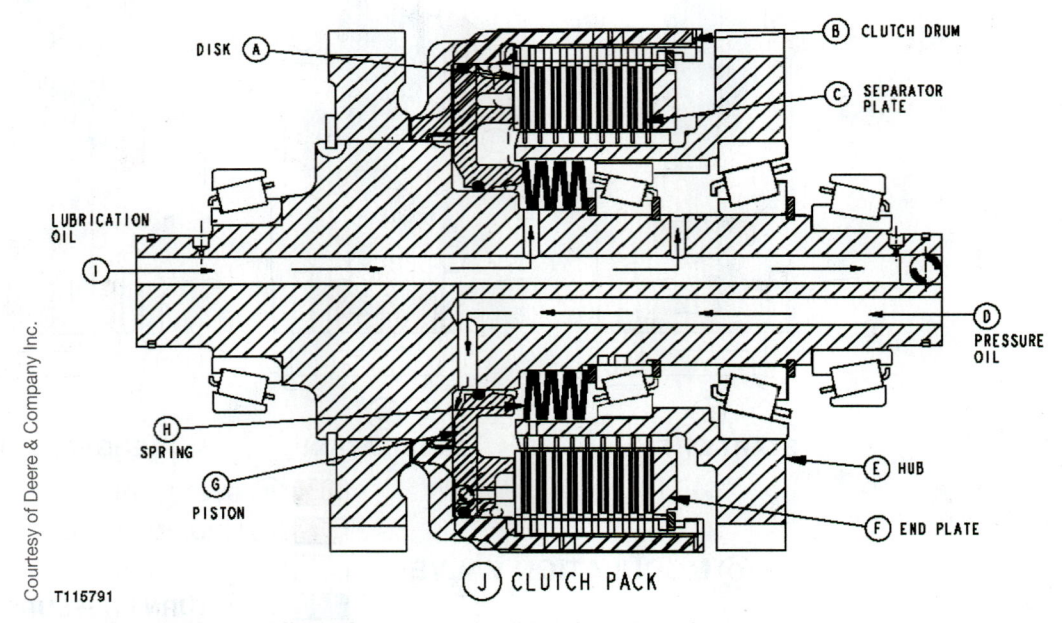

FIGURE 12–21 The effect of clutch apply oil inside a clutch housing.

Progress Check

8. A shuttle shift transmission will provide power shifts in these gears:
   a. first to second
   b. third to fourth
   c. neutral to first
   d. forward to reverse
9. Clutch modulation is:
   a. the time it takes to change speed ranges
   b. the rate of pressure rise for clutch application
   c. the modification of forward and reverse clutches when engaged
   d. the maximum pressure applied to the clutches
10. To get reverse rotation in a shuttle shift transmission:
    a. the torque converter reverser is applied
    b. the reverse idler is rotated
    c. the reverse hub uses the forward hub as a reverse idler
    d. the reverse hub drives the intermediate shaft directly
11. The solenoid valves used in the shuttle shift transmission described earlier are considered to be:
    a. electrically actuated 2-position, 3-way
    b. proportional control valves
    c. mechanically actuated 2-position, 3-way
    d. hydraulically actuated 2-position, 3-way
12. An orifice is used as part of this valve as a:
    a. system pressure regulator
    b. torque converter regulator
    c. lube pressure regulator
    d. modulation valve

FULL POWERSHIFT COUNTERSHAFT TRANSMISSION

The shuttle shift or reverser transmission that was just discussed is a fairly simple design mechanically, but its hydraulic control system is typical of many older style transmissions. The main difference from it to a full powershift transmission control system would be the addition of another spool to control the speed ranges as well as the addition of more clutch packs.

The full powershift countershaft transmission will give the operator the ability to change speed and direction under full power because all gear changes are done with hydraulic clutches. Most powershift transmissions used today are electronically controlled, which may initially seem more complicated but in many ways are simpler. A number of valves are eliminated by using **proportional solenoids** and controlling them electronically. By integrating electronics into the transmission controls, the operator will sometimes have different options or shift modes such as autoshift, which automatically shifts the transmission ranges according to inputs to the transmission ECM. The inputs are usually from sensors such as engine speed, transmission input shaft, transmission output, transmission oil temperature, and engine boost pressure. There will also likely be some kind of fault code logging and diagnostic system and there could be the capability of shift calibration or shift logic where shift rates are tailored to the machine's current operating conditions.

These full powershift countershaft transmissions are commonly used in wheel loaders, graders, smaller trucks, and skidders.

The gear arrangement is a countershaft constant mesh design with two clutches per shaft. To get torque transfer from the input to output shafts, there will need to be two (sometimes three) clutches engaged.

This type of transmission can have up to 18 speeds forward and 6 reverse with the amount of speed ranges only limited by the number of shafts used. Gear ratios can vary from approximately 5:1 to overdrive ratios of 0.75:1. ● **SEE FIGURE 12–22** for a look at the gear arrangement for an 18-speed forward and 6-speed reverse transmission.

For a close look at this type of transmission, we will study the John Deere 872D grader transmission to understand how a full powershift countershaft transmission works. ● **SEE FIGURE 12–23** for the drivetrain minus the engine for the 872D.

This full powershift constant mesh countershaft transmission is driven directly from the engine flywheel, meaning

FIGURE 12–22 Gear arrangement for an 18-speed forward and 6-speed reverse transmission.

POWERSHIFT TRANSMISSIONS 367

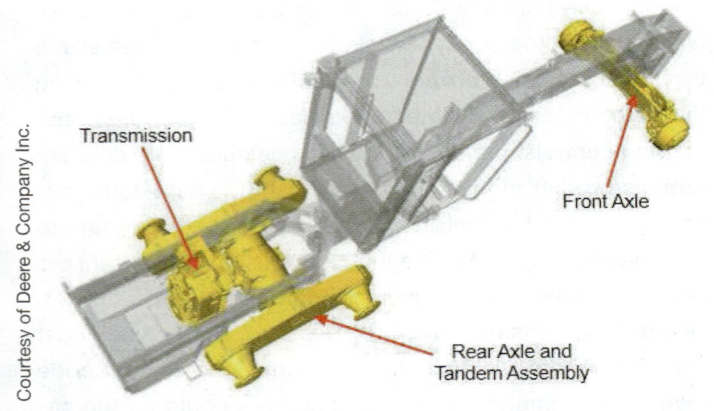

FIGURE 12–23 John Deere 872D drivetrain.

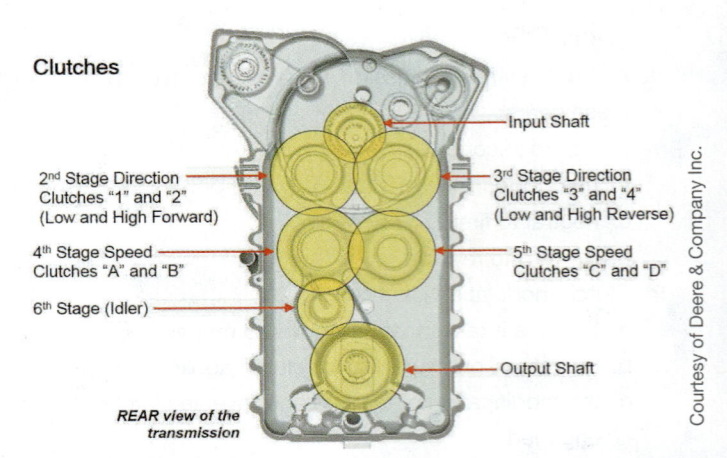

FIGURE 12–24 Clutch and gear arrangement for John Deere 872D transmission.

there is no torque converter and its input shaft is driven at a 1:1 ratio from the engine through a torsional damper. The operator has eight gear ratios in forward and reverse available to choose from. Each gear selection made actuates two clutches: one speed and one direction. Each clutch is engaged hydraulically and controlled by an individual electronically controlled proportional solenoid valve. The rate of pressure rise behind each clutch piston is determined by the transmission control unit (TCU) that provides the solenoids with a 24V PWM signal to each valve.

The transmission's output shaft then drives a differential assembly that contains planetary final drives. The final drive then drives the tandem drive, which in turn drives the grader's wheels and tires.

The operator inputs are the transmission shift control lever, the inching pedal, as well as the autoshift button. The transmission is shifted from neutral when the operator moves the shift lever to any of eight forward or reverse detents in the transmission shift switch. The inching pedal will neutralize the transmission and simulates a mechanical clutch to the operator. If autoshift is selected, the TCU will select upshift and downshift points only from fifth to eighth gear if the selector is in the eighth position. The shift points will depend on engine speed, throttle position, and engine load inputs. These inputs along with several speed and temperature values gathered from sensors are sent to the TCU that will control clutch engagement by energizing proportional solenoid valves with a PWM signal.

GEAR SELECTION

● **SEE FIGURE 12–24** for the actual clutch and gear arrangement of the John Deere 872D powershift transmission.

Two clutches must be engaged to transmit torque from the transmission input to the transmission output shaft. In neutral, all clutches are in the released condition, meaning they are open to sump.

Clutches numbered 1 through 4 are referred to as directional clutches. Clutches 1 and 2 engage forward gears and clutches 3 and 4 engage reverse gears. These clutches have more friction surface area and a higher thermal capacity than the speed clutches. The directional clutches are used for modulation during shifting and when using the inching pedal. The directional clutches connect the input shaft of the transmission through various gear ratios to the output shaft.

The next two shafts below the directional shafts are for speed ranges and also each use two clutches to engage different gear ratios. Clutches lettered A, B, C, and D are referred to as speed clutches. The speed clutches connect the directional shafts to the output shaft of the transmission at various ratios. Working in conjunction with the directional clutches up to eight-speed ranges in both forward and reverse directions is possible. ● **TABLE 12–2** shows which clutches are engaged for each gear.

● **SEE FIGURE 12–25** for how the shafts are arranged in the transmission housing. Note that the illustration is expanded to clearly show each shaft. The input shaft is in constant mesh with the forward and reverse input gears on their shafts. The forward and reverse hub gears are in constant mesh. The forward hub gear is in constant mesh with both the speed clutch input hub gears. Both speed clutch output gears are in constant mesh with the idler gear, which is in constant mesh with the output shaft gear. Take a few moments to learn how the gears and shafts are interconnected by studying both Table 12–2 and Figure 12–24.

	FORWARD		REVERSE	
GEAR RANGE	DIRECTIONAL CLUTCH	SPEED CLUTCH	DIRECTIONAL CLUTCH	SPEED CLUTCH
1st	1 (Low)	A	3 (Low)	A
2nd	2 (High)	A	4 (High)	A
3rd	1 (Low)	C	3 (Low)	C
4th	2 (High)	C	4 (High)	C
5th	1 (Low)	B	3 (Low)	B
6th	2 (High)	B	4 (High)	B
7th	1 (Low)	D	3 (Low)	D
8th	2 (High)	D	4 (High)	D

TABLE 12–2

Clutches engaged for different speeds.

Forward and Reverse Gears

CLUTCH STAGE ASSEMBLY	CLUTCH PACKS
2nd	Direction (1 and 2)
3rd	Direction (3 and 4)
4th	Speed (A and B)
5th	Speed (C and D)

Clutch Packs and Stage Assemblies

Let's see how the power flows through the transmission in first forward speed. ● SEE FIGURE 12–26A to understand how this happens.

When the transmission control lever is in 1st forward, clutches 1 (C) and A (D) are engaged. The solenoids controlling the transmission clutches (solenoids 1 and A) are energized by a signal that produces proportional pressure/flow changes.

When system pressure oil (I) is applied to clutches 1 and A, the piston moves to apply pressure to the clutch plates and, at the same time, moves the lubrication shut-off washer to open the oil passage, which allows lubrication oil (J) to flow through the clutch packs for lubrication. Pressure free oil (K) is present in the clutches that are disengaged.

Engine power comes into the input shaft (A) where it drives the four input floating gears on the directional clutch shafts. If one of the forward clutches is engaged, it will send torque to the two-speed clutch cylinders. The cylinders are the common drum for the two clutches on each shaft. As soon as any one of the four directional clutches is engaged, this will transfer torque to that clutch's output gear. This gear will now transfer torque down to the bottom shafts and out the output shaft.

● SEE FIGURE 12–26B to see power flow in reverse. If one of the reverse range directional clutches is engaged, it will transfer torque from the transmission input shaft to its cylinder, and because this cylinder is in mesh with only the forward cylinder, the torque will flow through this cylinder. This will reverse rotation from the reverse shaft, and when any speed clutch is engaged, its rotation will now be transferred out the transmission in reverse rotation. The forward shaft is now acting as a reverse idler.

When the transmission control lever is in 1st reverse, clutches 3 (C) and A (E) are engaged. The solenoids controlling the transmission clutches (solenoids 3 and A) are driven by a signal that produces proportional pressure/flow changes.

When system pressure oil (J) is applied to clutches 3 and A, the piston moves to apply pressure to the clutch plates and, at the same time, moves the lubrication shut-off washer to open the oil passage, which allows lubrication oil (K) to flow through the clutch packs for lubrication. Pressure free oil (L) is present in the clutches that are disengaged.

Engine power comes into the input shaft (A) to the first-stage gear (back) (B). Power flow is directed to the third-stage direction clutch "3" (C) through the second-stage cylinder gear (D) to the fourth-stage speed clutch "A" (E) and travels through the transmission to the sixth-stage gear (back) (F) and middle (G). Power then flows to the seventh-stage output gear (H) and out the output stage (I).

Let's take a look at the hydraulic system for this transmission. ● SEE FIGURE 12–27 for the schematic of it.

HYDRAULIC SYSTEM OPERATION—POWER SHIFT TRANSMISSION

As mentioned earlier, the transmission hydraulic system for powershift transmissions are basically similar. They all need to supply and modulate oil to engage clutches based on operator requests. The oil also needs to lubricate and cool

FIGURE 12–25 Shaft arrangement in the 872D power shift transmission.

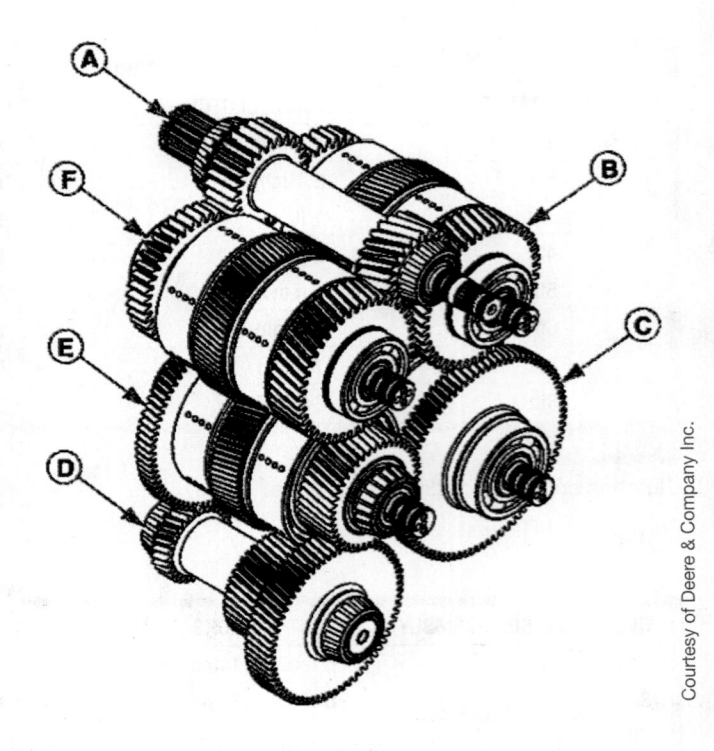

A- Input
B- Hi/low forward
C- A/B speed
D- Idler
E- C/D speed
F- Hi/low reverse

transmission components, and if a torque converter is used, it will need to supply and regulate an oil supply for the torque converter.

The hydraulic system for the 872D grader doesn't need to supply a torque converter, so after the gear pump picks up oil through the suction strainer, the oil is sent to a spin-on filter with a bypass valve in its base. Maximum pressure is controlled by the regulator valve at 255–285 psi and is adjustable with spring pressure. This regulated oil flow is made available to the parking brake valve and the proportional solenoid control valves at this pressure. Any oil that is dumped at the regulator is then sent to the transmission cooler where its pressure is regulated at 120 psi. The cooled oil returning to the valve body from the transmission oil cooler is directed to the lube circuit. Internal passages and manifolds located in the transmission front cover direct the lube oil to the ends of each of

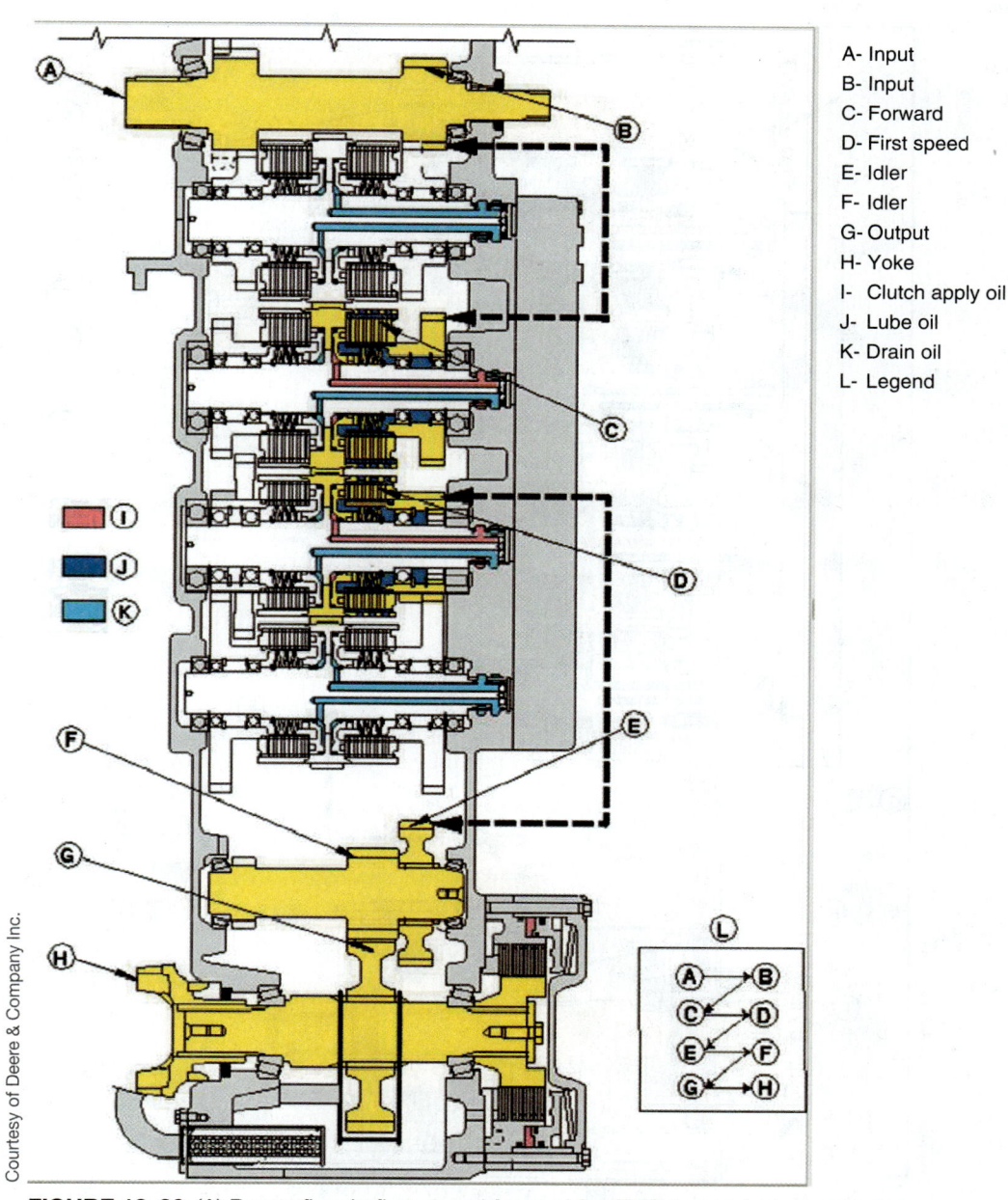

FIGURE 12–26 (A) Power flow in first-speed forward for 872D transmission.

A- Input
B- Input
C- Forward
D- First speed
E- Idler
F- Idler
G- Output
H- Yoke
I- Clutch apply oil
J- Lube oil
K- Drain oil
L- Legend

the clutch shafts. This oil flow passes through the lube shut-off washers on the engaged clutches and provides the cooling for the clutches. There is also leakage flow to the disengaged clutches to lubricate and cool them, and many orifices provide flow to lubricate the bearings on these shafts. After exiting the clutches and bearings on these shafts, the oil returns to the transmission sump where it is again picked up by the charge pump. If the pressure in the lube circuit is excessive, the lube relief valve (6) opens at 45 psi and dumps the excess flow to the transmission sump. This condition is normal when the transmission is in neutral with all lube shut-off washers closed and the engine is at high idle or when the oil is cold.

● **SEE FIGURE 12–28** for a cross-section of the control valve.

Clutch modulation for the shuttle shift transmission was accomplished by using a combination of springs, spools, pistons, and orifices. These parts will eventually wear and clutch modulation will vary greatly with oil viscosity and temperature. By using proportional solenoid valves that are controlled electronically, clutch modulation can change under different conditions and even compensate for any wear past the valve. Each clutch will have its own valve.

POWERSHIFT TRANSMISSIONS 371

Legend

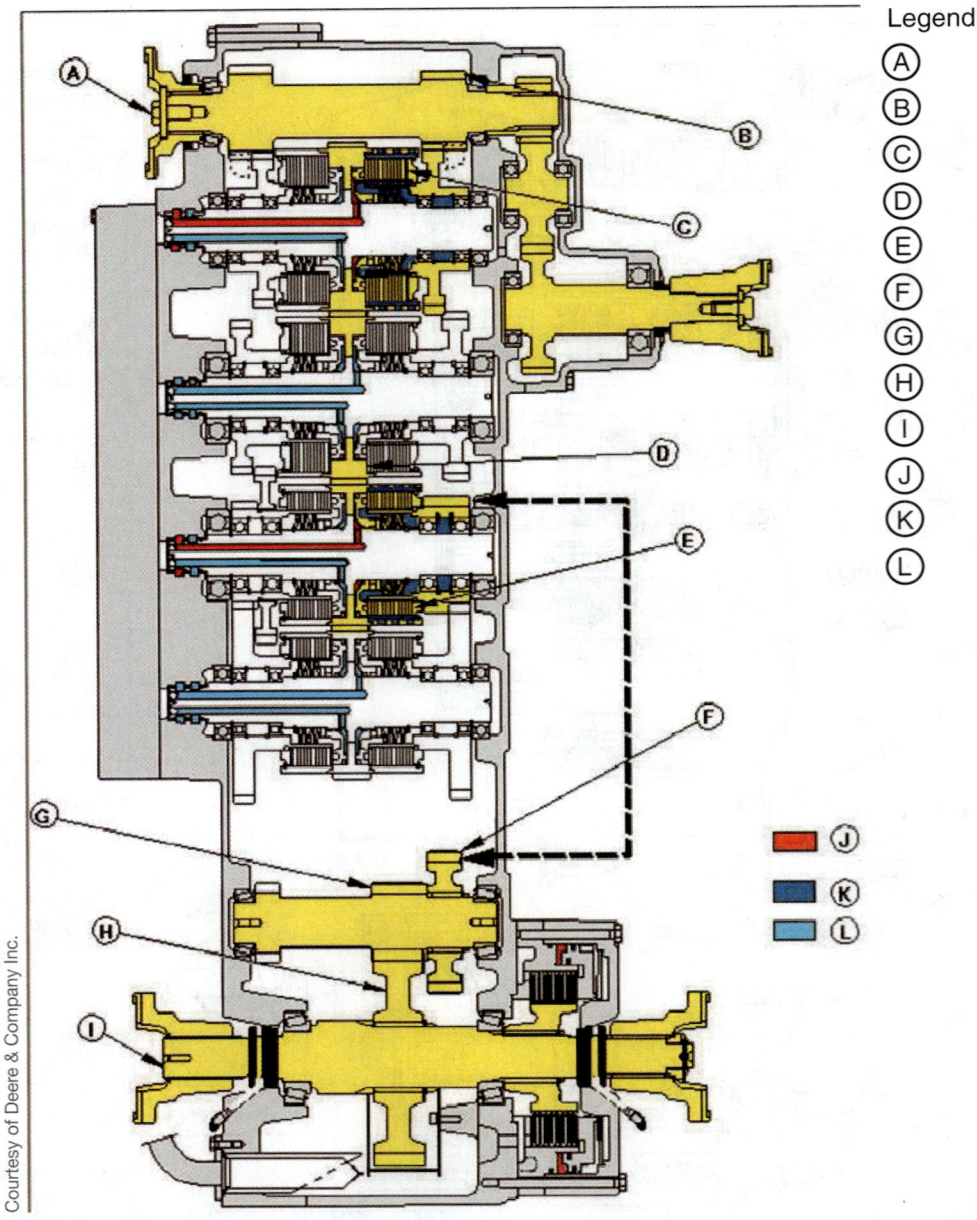

Ⓐ
Ⓑ
Ⓒ
Ⓓ
Ⓔ
Ⓕ
Ⓖ
Ⓗ
Ⓘ
Ⓙ
Ⓚ
Ⓛ

■ Ⓙ
■ Ⓚ
■ Ⓛ

Courtesy of Deere & Company Inc.

FIGURE 12-26 (B) Power flow in reverse

Each proportional solenoid valve is a pilot-operated, pressure-reducing cartridge valve. Its function is to provide a regulated pressure to the clutch port that can be controlled by the TCU to a pressure less than the supply pressure.

In operation, supply pressure enters the inlet supply port where it is blocked by the main spool, which prevents the flow from going to the clutch port. The inlet supply must past through a filter screen before reaching the orifice. The filter screen is used to trap large particles from contaminating the orifice and preventing the valve from shifting. The orifice is used to meter the inlet flow into the pilot cavity. The passage from the inlet supply port is always open to the pilot cavity, regardless of the spool position, to maintain a pressure source to the pilot cavity.

Pilot cavity operation enables the valve to work at a current level that the TCU is capable of controlling. The armature, push pin, and ball configuration controls the pilot cavity pressure, which in turn provides the force required to move the main spool.

With the solenoid valve de-energized, flow is free to travel from the clutch port to the sump port. The orifice is

372 CHAPTER 12

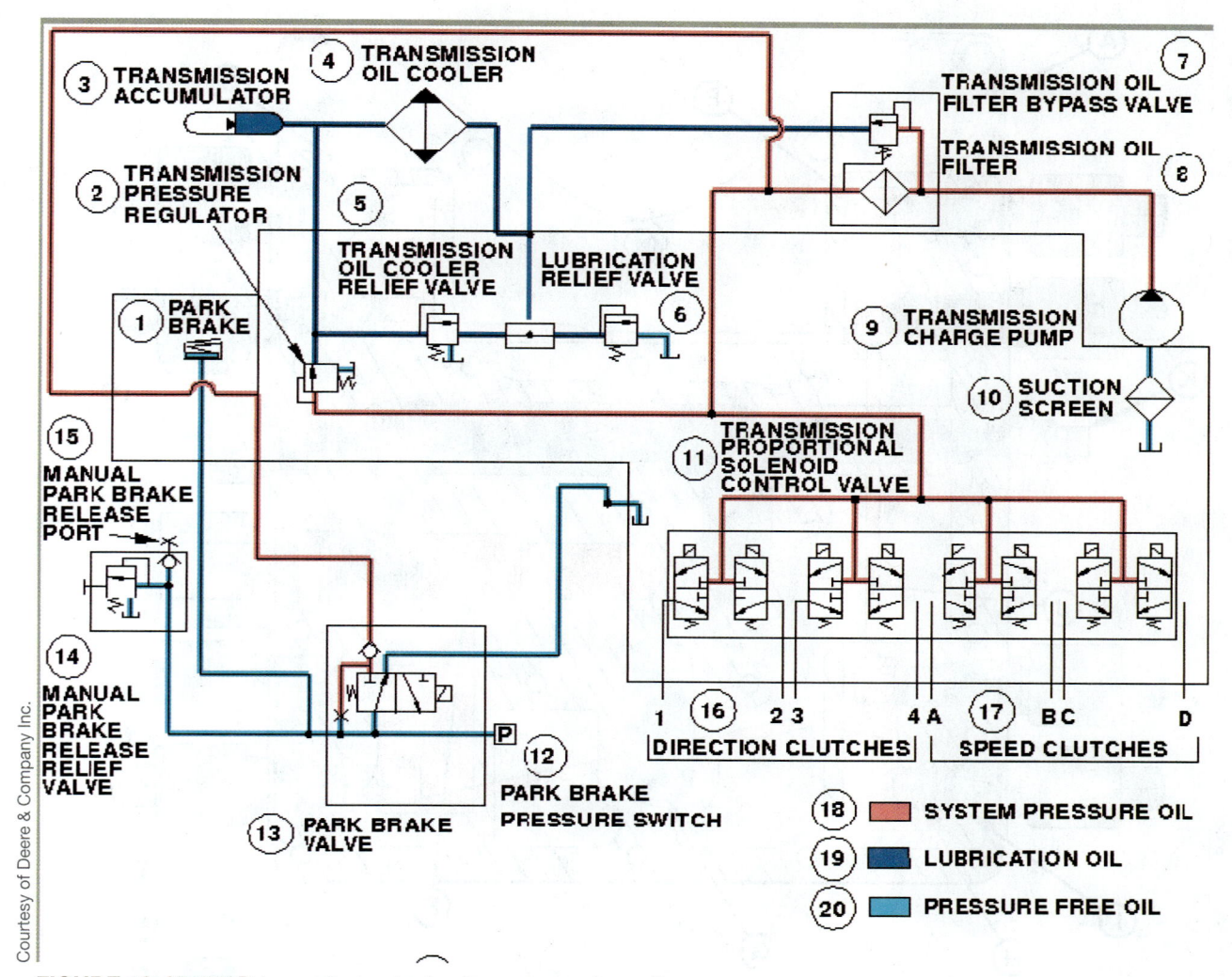

FIGURE 12-27 872D transmission hydraulic system schematic.

used to meter the flow into the pilot cavity. Flow coming into the pilot cavity then passes through the seat and past the ball, which is in the open position when the solenoid coil is de-energized. This flow then travels out through slots on the outside threads of the valve to connect to the sump port.

Because the ball has no restriction to the flow of fluid exiting the pilot cavity, no pressure is built up in the pilot cavity to act against the spool end. A bias spring is located on the opposite end of the spool to maintain spool position and prevents any slight pressure buildup from actuating the clutch. The bias spring requires a nominal pressure of 34.4–55.2 kPa (0.34–0.55 bar or 5.0–8.0 psi) to start spool movement.

To provide proportional control, the solenoid coil is energized by the use of a **PWM** signal by a varying duty cycle of 0–100%. The increased duty cycle acts to increase the effective current in the solenoid coil windings, causing an increase in the magnetic force. As the duty cycle increases, the magnetic force on the armature increases, causing the push pin to push against the ball to restrict flow exiting the pilot cavity. This restriction causes a rise in pressure in the pilot cavity and applies a force on the end of the main spool, causing it to move and compress the bias spring. As the main spool moves, it closes the passage from the clutch port to the sump port and begins to open the inlet supply pressure port slightly. The flow of oil causes the pressure in the clutch to rise.

The clutch port pressure is fed to the spring end of the main spool via the feedback passage, causing the main spool to close as the clutch pressure and spring force become equal to the pilot cavity pressure. The output pressure from the valve is zero at zero current and increases to within 103.4 kPa (1.00 bar or 15.0 psi) of the inlet pressure as the current to the coil is increased. At 100% duty cycle, the ball rests fully against the seat; the flow past the seat is essentially stopped and the pressure in the pilot cavity is the same as the supply pressure at the inlet supply port.

POWERSHIFT TRANSMISSIONS 373

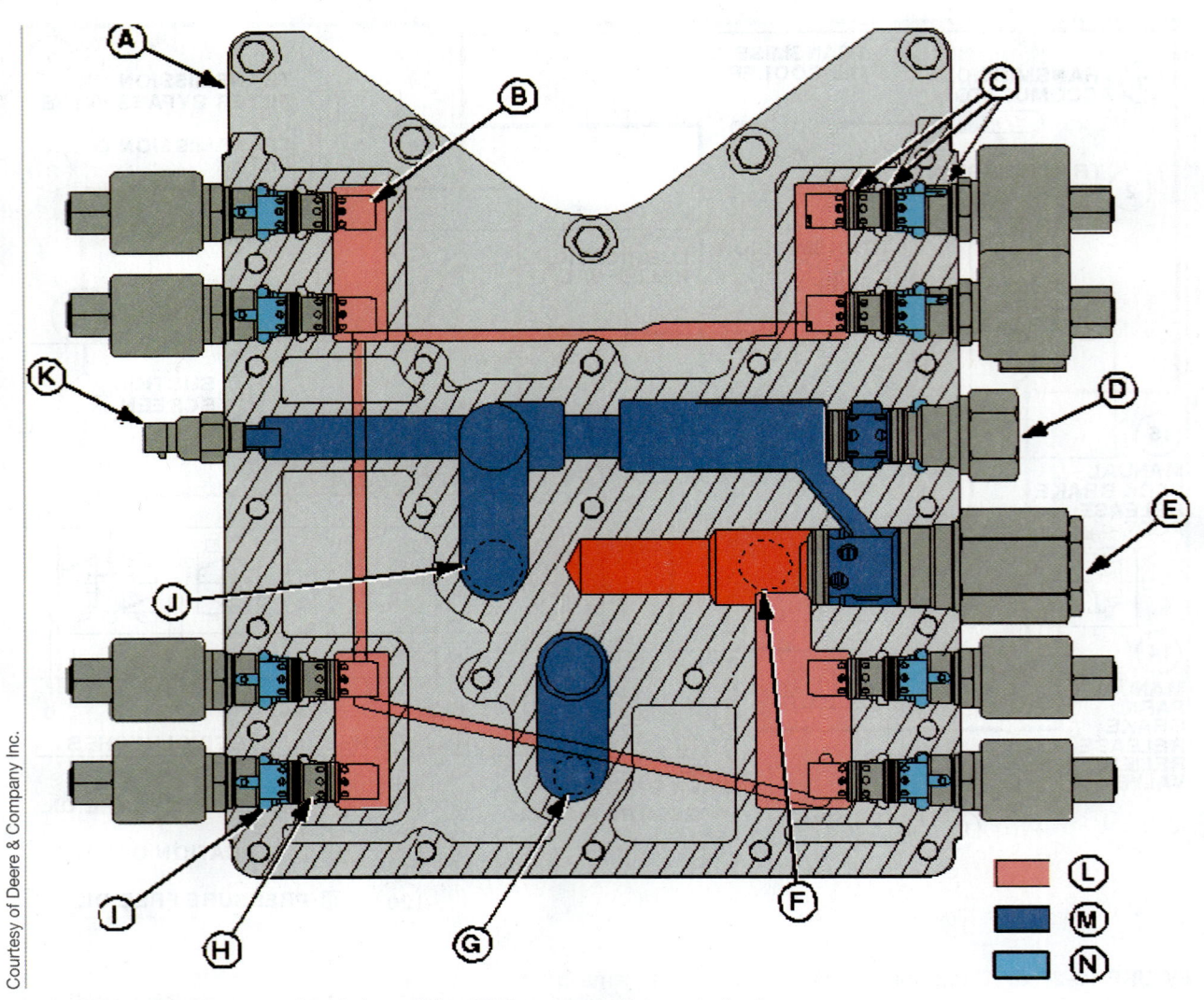

FIGURE 12–28 Cross-sectional view of the 872D transmission control valve.

CLUTCH ACTUATION Clutch actuation for this transmission is the same as the shuttle shift transmission. The oil pressure is transferred from the stationary manifold to the rotating shaft in the area between the shaft **seal rings**. The oil pressure then travels down a hole in the shaft until it enters the area enclosed by clutch cylinder, the clutch apply piston, and the piston seals.
● **SEE FIGURE 12–29** for a directional clutch assembly and how oil applies the clutch. ● **SEE FIGURE 12–30** for the clutch plates, piston, and cylinder from a disassembled clutch.

Now that an ECM can control the transmission shifting based on inputs and stored information, shifting rates and timing can be variable and much more adaptable to conditions such as load, speed, temperature, and operator input and can be manipulated by software. The TCU manages operation of the transmission. These are electronic features that define how the transmission will operate under given conditions.

EVENT-BASED SHIFTING Event-based shifting (EBS) is part of the TCU software that adjusts how a gear is engaged in the transmission. When the operator selects a gear, the TCU follows a standard shift logic (clutch protection, speed matching, downshift inhibit, inching pedal, shuttle shifting, skip shifting, and autoshift—see description of each next). After a gear is determined by the shift logic, the engagement of that gear is governed by the EBS portion of the software. The EBS software may "feather" a clutch to provide a smooth soft shift, such as during transport, or may provide a rapid firm shift to maintain momentum when the machine is under load. EBS is designed to provide optimum shift quality. The TCU uses the sensors on the transmission along with information from other controllers and sensors via the CAN (controller area network) to aid in this operation. Transmission shift duration and timing may vary based on these inputs.

374 CHAPTER 12

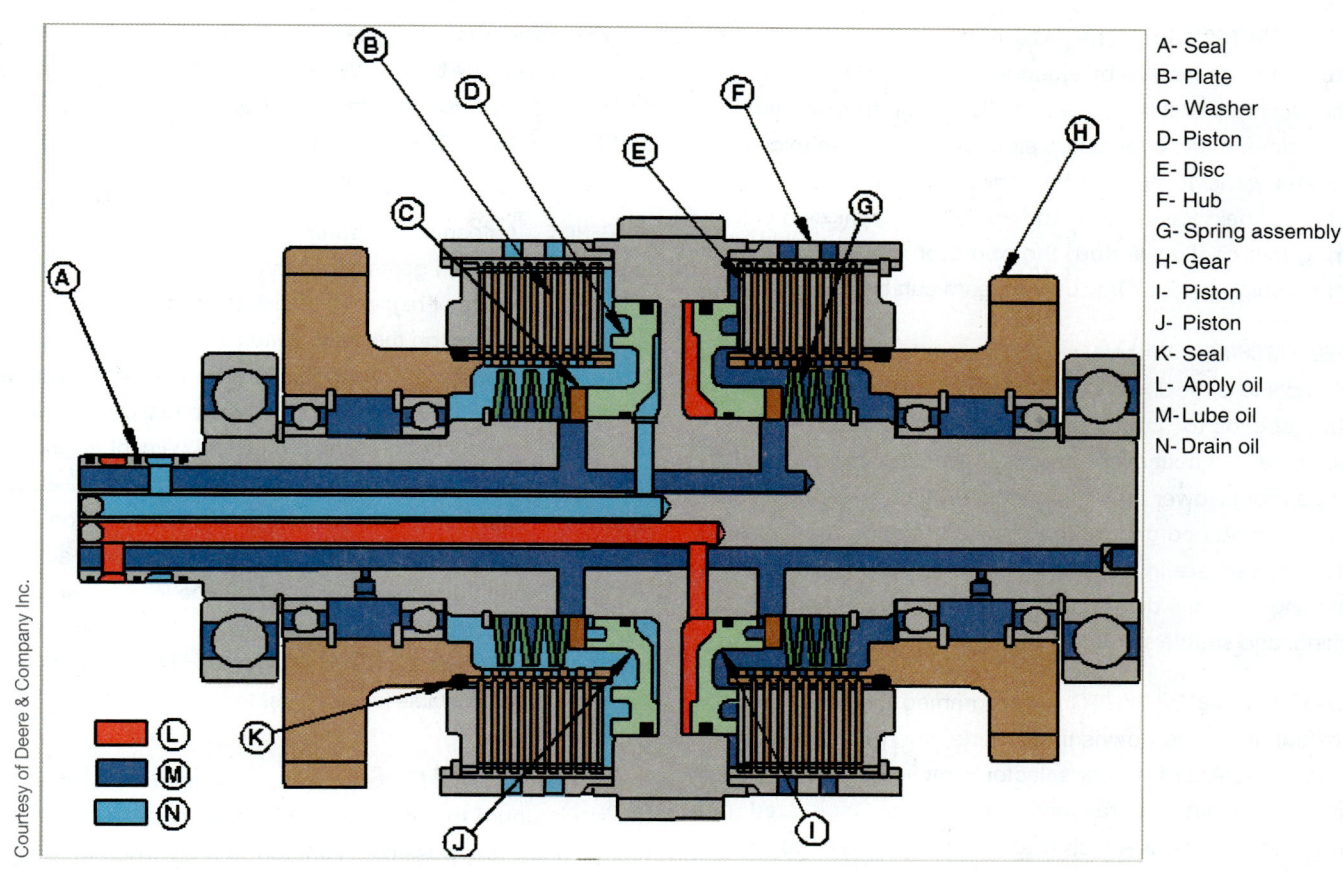

FIGURE 12–29 How oil applies a directional clutch.

A- Seal
B- Plate
C- Washer
D- Piston
E- Disc
F- Hub
G- Spring assembly
H- Gear
I- Piston
J- Piston
K- Seal
L- Apply oil
M- Lube oil
N- Drain oil

FIGURE 12–30 Clutch plates, discs, piston, and cylinder from a disassembled clutch.

POWERSHIFT TRANSMISSIONS 375

CLUTCH PROTECTION. The TCU is programmed to detect clutch slippage caused by extended use of the inching pedal. If the inching pedal is depressed slightly for an extended period of time, directional clutch slippage may cause excessive heat and wear. To prevent this, the TCU will shift the transmission to next lowest gear and cool the over-slipped clutch. During this cooling period, the operator will not be able to shift to a higher gear. Only Lower gears can be selected.

SPEED MATCHING. Whenever a shift is made from neutral while the machine is moving, TCU monitors engine and transmission output speeds to "match" a gear that would have smoothest engagement without over-speeding the engine. If gear selected by operator is lower than the matched gear range, TCU shifts to speed-matched gear until engine speed and transmission output speed are in range to shift to selected gear. Speed matching happens during normal neutral-to-gear shifts, autoshifting, and shuttle shifting.

DOWNSHIFT INHIBIT. The TCU is programed to allow a smooth transition if a large downshift is made, such as from sixth to second gear. After the gear selector is moved, the TCU immediately downshifts the transmission one gear. Once rated rpm is reached, the TCU will shift to the next lowest gear. It will continue to control shifts in this manner until it reaches gear selected by operator.

INCHING PEDAL. If the inching pedal is *not* used and the gear selector is moved from neutral, or from one gear to another in the same or either direction, the transmission will select the appropriate gears to provide the smoothest shift path to reach the selected gear. If the inching pedal is used and the gear selector is moved from neutral to any gear, or from one gear to another gear in the same direction, the transmission will shift directly to the selected gear.

SHUTTLE SHIFTING. A shuttle shift is a shift to a gear in the opposite direction in which the machine is traveling. The TCU only allows a change in direction if the machine is traveling less than 5 mph. If a shuttle shift is made when machine is moving too fast, the TCU will downshift to that gear using downshift inhibit (see "Downshift Inhibit"). Once the machine has slowed to less the 5 mph, the TCU will shift to the currently selected gear based on the use of the inching pedal (see "Inching Pedal").

SKIP SHIFTING. Skip shifting is when the TCU uses one or more intermediate gears to get to a higher gear as commanded by the gear shift lever. When a gear is selected that is higher than what is engaged, the TCU will evaluate the difference in gear selection and determine what gears are necessary to bring the machine up to speed to engage the selected gear. If the machine were in first, and seventh were selected, the TCU may engage third, fifth, sixth, and then seventh. This allows the machine to achieve proper ground speed in the desired gear without stalling the engine.

AUTOSHIFT. Autoshift is a feature that allows the transmission to automatically shift gears fourth through eighth when the gear shift lever is in fifth or higher. Autoshifting is enabled by pressing the autoshift button on the SSM. Autoshift uses operator inputs of engine speed (accelerator pedal or engine speed control set point), gear shift lever position, and inching pedal position. It also uses machine inputs of percent throttle and percent engine load at current speed. It will only shift as high as the gear selected by the operator. If sixth gear is selected, autoshift will only shift through gears fourth to sixth. If eighth gear is selected, autoshift will shift through gears fourth to eighth. The lowest gear available in autoshift mode is fourth, unless otherwise commanded by the gear shift lever (lever manually moved below fourth gear). Autoshift is not available in gears first to third.

MAXIMUM STARTING GEAR

Maximum starting gear is a feature where the maximum neutral-to-gear (NTG) shift or shuttle shift gear possible is limited to third when the engine speed is above 1500 rpm and fourth when the engine speed is below 1500 rpm. If a gear higher than fourth is selected, third or fourth gear will be engaged as determined by the engine speed. Then the current gear will be incremented through open loop, until the selected gear is achieved. This functionality does not apply when the inching pedal is being used.

NEUTRAL-TO-GEAR (SHIFTS WITHOUT INCHING PEDAL)

If shift lever is moved to a gear greater than or equal to the maximum starting gear, the transmission shall shift into the maximum starting gear.

If the shift lever is moved to a gear less than the maximum starting gear, the transmission shall shift into the appropriate vehicle speed matched gear if vehicle speed is greater than the requested gear, otherwise shift to the shifter selected gear.

If the present vehicle speed is greater than or equal to the speed that would be achieved with the maximum starting gear, the transmission shall shift into the appropriate vehicle speed matched gear.

NEUTRAL-TO-GEAR (SHIFTS WITH INCHING PEDAL)

Doing the same NTG (shifts without inching pedal) but done while inching pedal is depressed. The transmission shall shift into the shift lever commanded gear.

Progress Check

13. Oil is sent through shafts to be applied to pistons and engage the clutch. This will direct oil to the piston:
    a. copper washers
    b. steel rings
    c. brass washers
    d. lip-type seals

14. Proportional solenoid valves are used to:
    a. vary the rate of clutch apply pressure
    b. limit system pressure
    c. vary the rate of speed clutch pressure only
    d. increase the clutch apply pressure over system pressure

15. How many clutches need to be applied to turn the output shaft of a typical power shift transmission in an operational machine:
    a. 1
    b. 2
    c. 3
    d. 4

16. The inching pedal is an operator control that will simulate a:
    a. clutch pedal
    b. torque converter
    c. torque divider
    d. mechanical clutch

17. Which one of these is *not* an input to the TCU?
    a. inching pedal
    b. brake pedal
    c. gear selector
    d. autoshift button

PLANETARY POWERSHIFT TRANSMISSIONS

The second main type of powershift transmission that is used in heavy equipment for machine drivetrains is the planetary transmission. These transmissions are used in a wide range of machine types (track-type tractors, graders, wheel loaders, trucks, etc.) and sizes. They can provide from 2- to 12–speed ranges forward and up to 5 speeds in reverse. Planetary powershift transmissions feature a gear arrangement that uses one or more planetary gearsets, a combination of hydraulic clutches, and a control system. Planetary gearsets were briefly covered in Chapter 11, so a closer look is needed before seeing how they are used in a powershift transmission.

Planetary gearsets consist of a combination of three main components that are used to transfer torque. They are the following:

1. **Sun gear**
2. **Planetary pinion carrier** (driven planet pinion gears)
3. **Ring gear**

The advantages of planetary gearsets over conventional gearsets are the following:

1. A balanced load on gears, shafts, and bearings
2. The ability to change ratios and directions easily
3. Compact design

● **SEE FIGURE 12–31** for a planetary gearset.
● **SEE FIGURE 12–32** for how holding one component, driving another will result in the output coming out of the third.

FIGURE 12–31 A planetary gearset.

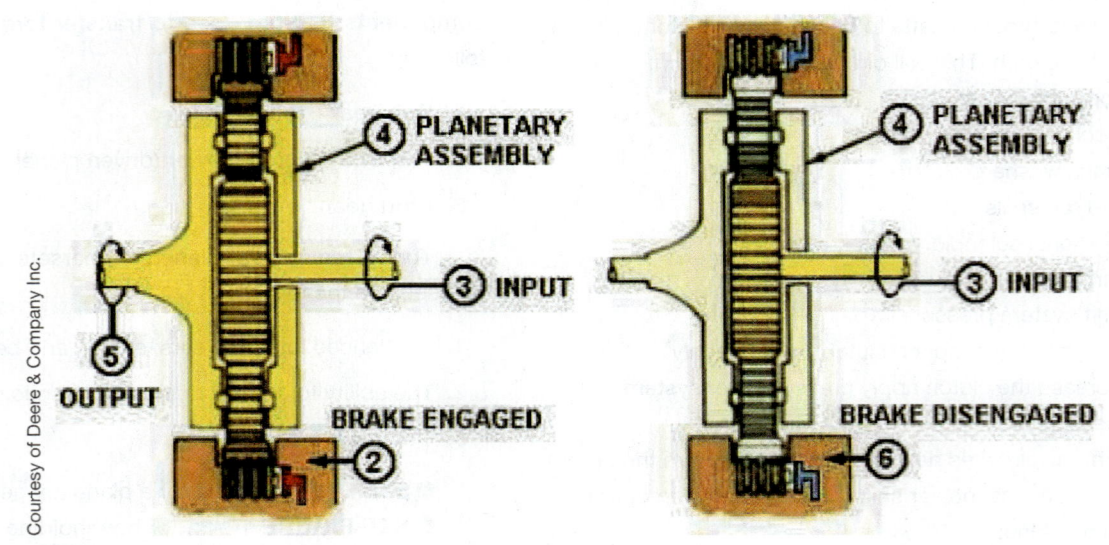

FIGURE 12–32 Power flow through a set of planetary gears.

	SUN	CARRIER	RING	SPEED	TORQUE	DIRECTION
1	Input	Output	Held	Maximum reduction	Increase	Same as input
2	Held	Output	Input	Minimum reduction	Increase	Same as input
3	Output	Input	Held	Maximum increase	Reduction	Same as input
4	Held	Input	Output	Minimum increase	Reduction	Same as input
5	Input	Held	Output	Reduction	Increase	Opposite of input
6	Output	Held	Input	Increase	Reduction	Opposite of input
7	When any two members are held together, speed and direction are same as input and ratio is 1:1.
	1, If the carrier is: the output, under drive results, or speed decrease; 2, the input, overdrive results, or speed increase; and 3, the held member and output direction is reversed.

TABLE 12–3
The seven combinations of power flows through a planetary gearset.

Planetary gearsets can provide an increase in speed/decrease in torque, a decrease in speed/increase in torque, a direct drive of 1:1, or a reverse rotation. ● **SEE TABLE 12–3** for the seven combinations possible when driving one member, holding one, or locking two members together.

To understand how a planetary powershift transmission works, we will now have a look at a planetary powershift transmission that is used in the John Deere 400D articulated rock truck. ● **SEE FIGURE 12–33** for this truck.

This transmission features six forward speeds and one reverse range. It is an Allison model 4500 ORS manufactured by Allison transmission. This transmission is controlled by the operator by a touch keypad and can be shifted manually or automatically depending on the mode selected. ● **SEE FIGURE 12–34** for the transmission control.

The transmission features three sets of planetary gears and five hydraulically actuated clutches. Drive into the transmission starts from the engine flywheel through a torque converter and into the input shaft for the transmission. After two clutches have been engaged, power can flow through the

> **WEB ACTIVITY**
>
> For an excellent visual demonstration of planetary gears, search for a 1953 U.S. army video "Planetary Gears Principles of Operation." It will explain the interaction of the gears of a single set of planetary gears as well as how multiple sets can be used together for different ratios and directions.

FIGURE 12–33 John Deere 400D articulated rock truck.

FIGURE 12–34 Transmission control for the 400D truck.

FIGURE 12–35 Cross-sectional view of the 400D transmission.

transmission and out to the output shaft. The output shaft is connected to a transfer case, which drops the power flow down and makes it available to the front axle and back through to the rear tandem axles. ● **SEE FIGURE 12–35** for a cross-section of the transmission.

The mechanical part of the transmission consists of three sets of planetary gearsets, which provide torque flow through the transmission depending on which combination of planetary gearset members are held and which are allowed to freewheel. This is determined by the five clutches that are attached to various gearsets. As these clutches are engaged at the maximum of two at a time, this will allow torque flow to be transferred through the gearsets.

The five clutches used to change input torque, speed, and direction are called C1, C2 (both rotating clutches), C3, C4, and C5 (stationary clutches).

When any two clutches are engaged, there will either be an under drive, overdrive, or direct drive power flow. ● **SEE TABLE 12–4** for the ratios of each gear.

The torque converter turbine provides input to both C1 and C2 clutches. When the C1 clutch is engaged, it will drive the transmission main shaft at torque converter turbine speed. When C2 clutch is engaged, it will drive the P2 planetary carrier at torque converter turbine speed.

When C3, C4, and C5 clutches are engaged, they will lock the ring gears of P1, P2, or P3 planetary gearsets to the

POWERSHIFT TRANSMISSIONS 379

GEAR	RATIO
1st	4.70:1
2nd	2.21:1
3rd	1.53:1
4th	1:1
5th	0.76:1
6th	0.76:1
Reverse	5.55:1

TABLE 12-4

Gear ratios for the 400D transmission.

TRANSMISSION GEAR	CLUTCHES ENGAGED
Neutral	C5
First forward	C1, C5
Second forward	C1, C4
Third forward	C1, C3
Forth forward	C2, C3
Fifth forward	C2, C4
Reverse	C3, C5

TABLE 12-5

The clutches that are engaged for each speed and direction range.

External components consist of:
Transmission Control Unit or (TCU)
Controller Area Network or CAN bus
Accelerator Position Sensor
Engine Speed Sensor
Turbine Speed Sensor
Output Speed Sensor
Shift Selector
Retarder Request from SSM

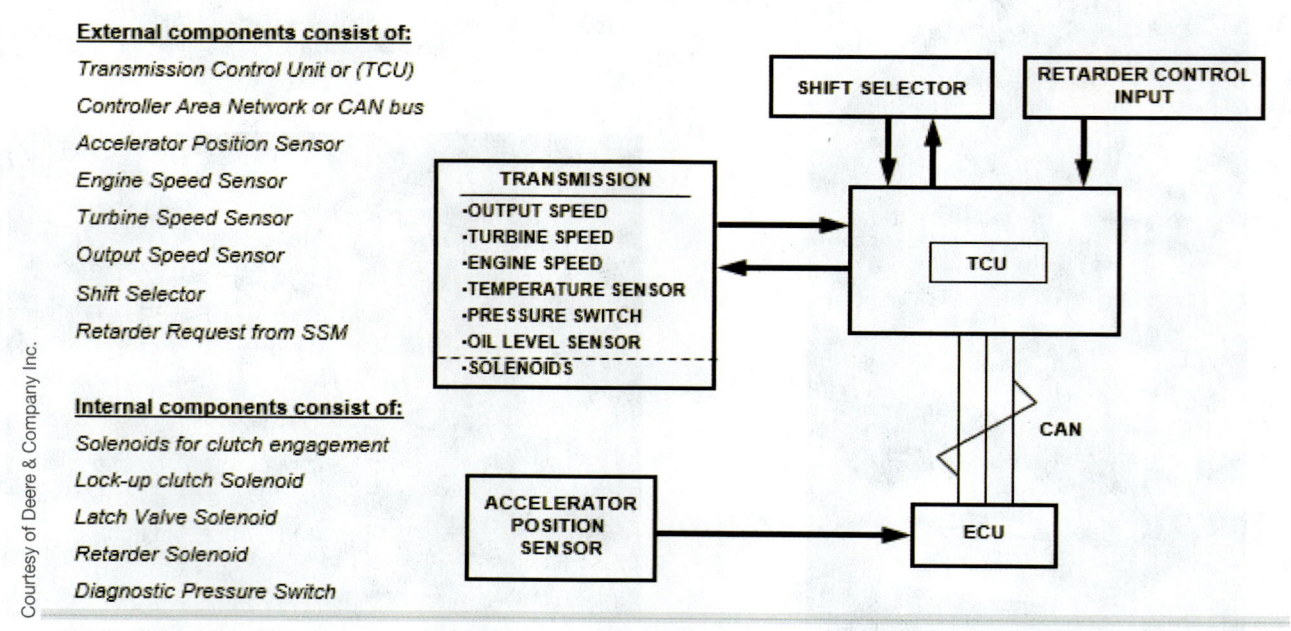

Internal components consist of:
Solenoids for clutch engagement
Lock-up clutch Solenoid
Latch Valve Solenoid
Retarder Solenoid
Diagnostic Pressure Switch

Courtesy of Deere & Company Inc.

FIGURE 12–36 Block diagram of the 400D transmission control system.

transmission housing, respectively. P2 ring gear is also connected to P1 planetary carrier and P3 ring gear is connected to P2 planetary carrier.

Trying to follow power flows through powershift transmissions, especially planetary ones, can be a time-consuming and confusing task. You're probably curious as to how the torque is transferred through the planetary sets, and unless you have a very good illustration, it is almost impossible to do on paper. Even if you have a transmission apart and try to figure out which components are driven, which are held, and what the output is for each gear selection, that can be very difficult also.

Refer to Table 12–5 to see the clutches that are engaged for each speed range. ● **TABLE 12–5**.

Let's see how this transmission is controlled.

● **SEE FIGURE 12–36** for the block diagram of the transmission control system.

The transmission control system consists of the following components:

- Transmission control unit (TCU)
- Chassis control unit (CCU)
- Accelerator position sensor
- Transmission speed sensors
  - Input speed sensor
  - Turbine speed sensor
  - Output speed sensor
- Transmission shift control

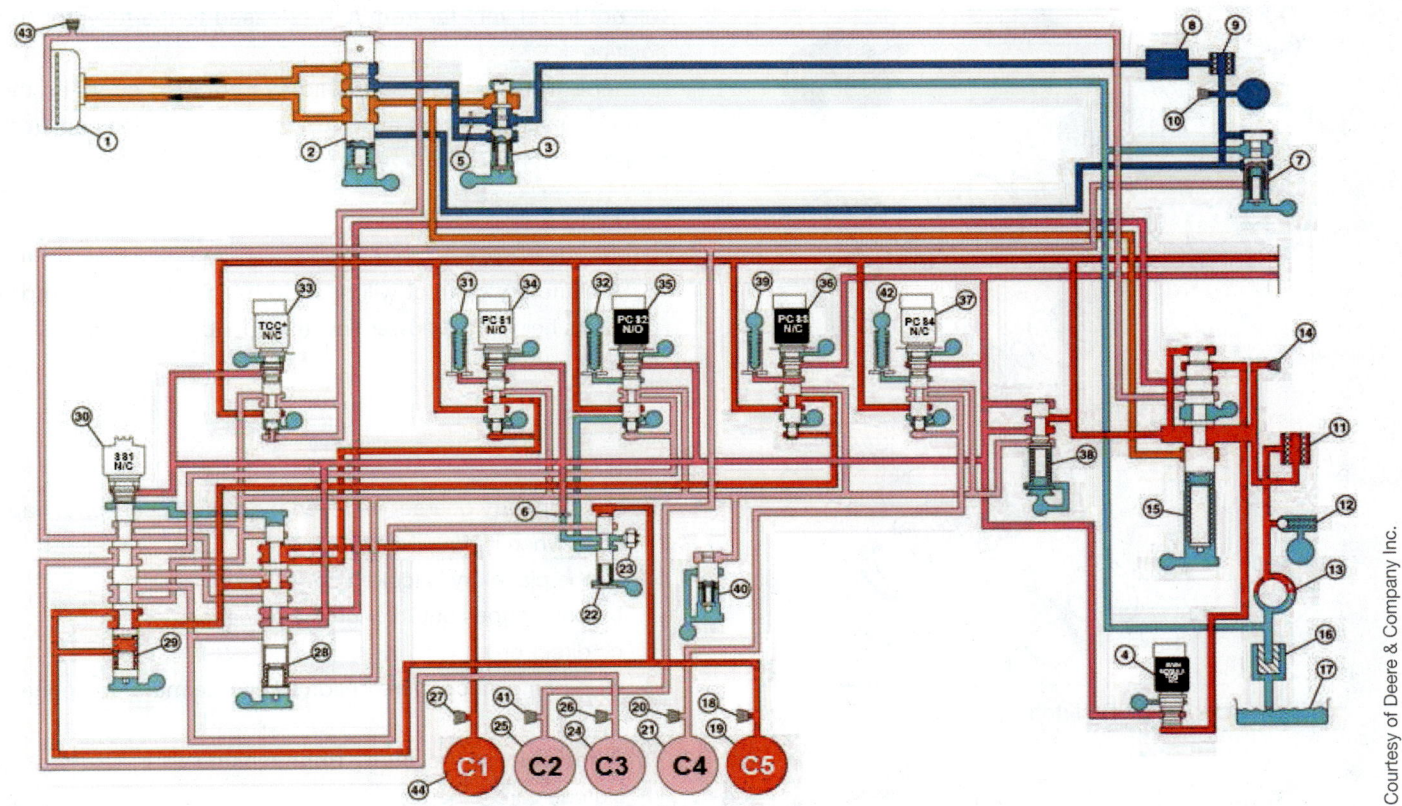

FIGURE 12–37 400D transmission hydraulic system schematic.

- Transmission control module
  - Pressure switch
  - Temperature sensor
  - Oil level sensor
  - Solenoid valves
  - Retarder control
- Diagnostic data connector

The accelerator position sensor, speed sensors, and transmission shift control send information via wiring harnesses to the TCU. The TCU processes this information and sends signals to the transmission control module to actuate specific solenoid valves.

These solenoid valves control increasing and decreasing clutch pressures to provide transmission shifts that match engine speed and operating conditions to protect the transmission from damage.

The TCU has the ability to adapt or "learn" while it operates. Each shift is measured and stored by the TCU so the optimum shift rate is adapted for present operating conditions. The TCU has the ability to "learn" quickly as conditions change and refine shifts at a slower rate as conditions remain constant.

A CAN data line signal is used to communicate between the TCU and CCU. The CCU provides a signal for "body up," which allows the TCU not to upshift the transmission above first gear. The TCU provides the CCU with a signal to activate the neutral start circuit and the backup alarm circuit. ● SEE FIGURE 12–37 for a schematic of the hydraulic control system.

There are five solenoids used to control the clutch engagement. Two of the solenoids are normally open and the rest are normally closed. The normally open solenoids will provide a limp home mode if there is an electrical failure. One other solenoid that is part of the hydraulic system is to engage the torque converter lock up clutch. This will get energized when the torque converter turbine speed is 80% of the impeller speed. Clutch apply pressure will be regulated from 145 to 300 psi depending on the value of the PWM signal from the TCU to the solenoid.

Hydraulic fluid is used to lubricate and cool the transmission, supply the transmission control module to engage the clutches, and supply fluid to the torque converter for transmitting power.

Control pressure oil is sent to each solenoid valve and main pressure oil is sent to each solenoid regulator valve (4). Solenoids and regulator valves work together to control the flow of main pressure oil to the clutch apply circuits.

POWERSHIFT TRANSMISSIONS 381

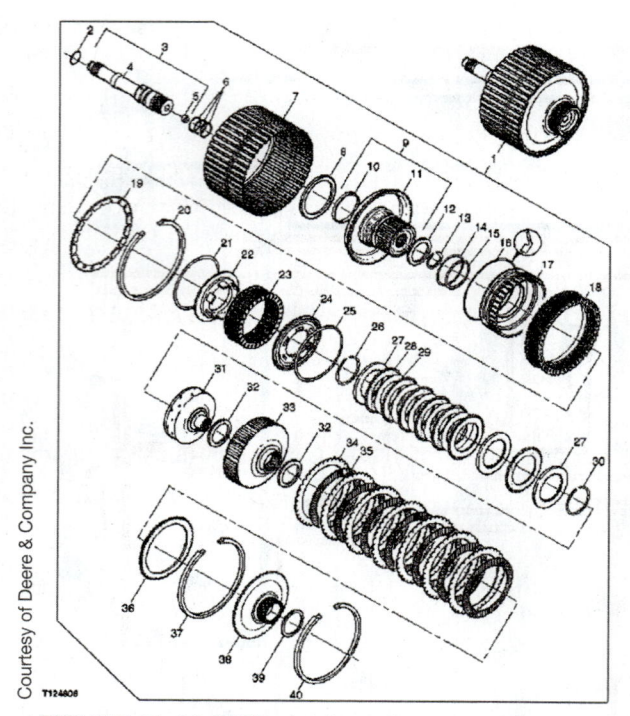

FIGURE 12–38 400D clutch pack.

The transmission uses a series of solenoids and regulator valve assemblies. These solenoids are variable bleed solenoids. The output pressure of these solenoids is proportional to the current supplied by the TCM. As a result, the solenoids are known as pressure control solenoids (PCSs). The solenoids get a signal from the TCM/ECU to actuate a specific solenoid located in the control valve module. Solenoids will then send control pressure oil to the solenoid regulator valve assembly, opening and allowing main pressure oil to the other regulator valves, and activating a specific clutch apply circuit.

Each solenoid and solenoid regulator valve assembly controls a clutch apply circuit.

Energizing and de-energizing specific solenoids applies and releases clutches, allowing the transmission to shift to various gears.

Clutch apply modulation is controlled by the TCU. Clutch operation for this transmission is oil applied and spring released. ● SEE FIGURE 12–38 for a typical clutch pack assembly.

Clutches are held in the release position by springs and are applied when oil pressure is fed underneath the piston, which then squeezes the discs and plates together against spring pressure.

Clutch pistons are moved by transmission pressure oil that is directed by solenoid valves. The pistons will have two seals to make the apply pressure act on their face. They will not travel very far from fully released to the fully applied position.

The transmission hydraulic system also supplies the torque converter with oil and provides oil for cooling and lubricating the transmission.

Progress Check

18. A planetary gearset will need to have this many components held to get power flow through it:
    a. 1
    b. 2
    c. 3
    d. 4

19. If a planetary gearset is driven by its sun gear, what happens when its ring gear is held?
    a. the planetary is locked up
    b. power goes out through the planet carrier
    c. direct drive
    d. nothing since there needs to be one more component held

20. What gear is the 400D transmission in if C3 and C5 clutches are engaged?
    a. neutral
    b. reverse
    c. first
    d. third

21. If none of the components are held for a planetary gearset but the ring gear is driven, what will occur?
    a. reverse under drive
    b. reverse overdrive
    c. neutral
    d. park

22. How many clutches are stationary in this transmission?
    a. all
    b. 2
    c. 3
    d. none

POWERSHIFT TRANSMISSION MAINTENANCE

With proper maintenance, powershift transmissions will usually for last 10000–20000 hours. Some variables that can increase or decrease that timeframe are operator abuse factors, proper fluid usage, including type and viscosity, proper machine warm-up, working conditions, and regular maintenance.

An early initial filter change at the first 100 hours may be required for some transmissions. This is because of the high quantity of break-in and built-in contamination. Typical filter change intervals after this will be 500–1000 hours.

Fluid change intervals will typically be 2000 hours, but this can vary greatly depending on operating conditions.

Fluid sampling is a common practice used to monitor fluid condition to ensure that proper fluid condition is maintained. See Chapter 3 for more information on fluid sampling.

The fluid required for proper operation of a powershift transmission needs to be carefully chosen. The maintenance section of the machine's service manual should have a section to recommend proper transmission fluid. Fluid specifications will be based on clutch material, transmission operating temperature, and loads. Most transmission fluid will be mineral oil based, but some manufacturers could recommend glycol-based or water-based fluids.

Because the transmission incorporates a hydraulic system, the fluid viscosity is an important consideration. The fluid viscosity should be chosen based on ambient temperatures that the machine will be operating in. Severe operating conditions causing high fluid temperature may also be a consideration. For most of North America, a typical transmission fluid viscosity would be number 30 SAE. You may however, see multi-viscosity fluids being recommended.

Fluid filtration typically occurs twice in a powershift transmission. The first location is for the pump inlet when the fluid will pass through a strainer or a combination of magnets and strainer. After the fluid leaves the pump it will be forced through a full-flow spin-on or canister-type filter.

POWERSHIFT TRANSMISSION TROUBLESHOOTING

When troubleshooting a powershift transmission complaint, the first thing to do is verify the complaint. Some typical operator complaints for powershift transmissions would be excessive noise, excessive vibration, oil leaks, overheating, intermittent gear selection, no forward, no reverse, no speed range, and starting in gear.

To verify the complaint, you must understand proper operation of the powershift transmission. If you're unfamiliar with the machine, take a few minutes to read the operator's guide and you may find the operator is not operating the machine properly. For example, the operator could be selecting a gear for the work he is doing, which will cause an overheating condition, or if it's the autoshift type of transmission, the working conditions the machine is in may not be suitable for autoshift mode.

After you've confirmed that the operator has been running the machine properly, at least with the transmission functions, you should now try to simulate the conditions that the machine has to be under to duplicate the complaint. This will be difficult if you're diagnosing a machine at a shop as opposed to being on the job site.

You must bring the transmission fluid to operating temperature, which is typically around 180°F. This is important because of the change in the fluid viscosity as temperature changes. This will play a big part in how transmission controls and clutches respond to operator input. If oil is not warmed up for operating temperatures and viscosity of the oil is too thick, this could introduce a different set of problems or mask the problem that is already there.

✚ SAFETY TIP

When testing a machine to diagnose a powershift transmission problem, you may need to make the machine travel or perform stall testing. These actions will have the potential for accidents and injury if some basic safety precautions are not followed. Make sure that the area around the machine is clear of people and objects that could be damaged by machine movement. If stall testing is to be in or near a shop, make sure that the machine's brake system will hold the machine and place wheel chocks on both sides of the wheels as well.

If travelling around a job site, be aware of all hazards, traffic rules, and company policies regarding machine movement.

You may be distracted by the test procedure you are performing, so it would be helpful to have a spotter keep an eye on you and keep in radio contact.

CASE STUDY

For this section of the chapter, we will go through a case study of a real-world scenario from troubleshooting a powershift transmission operational complaint through to completing the repair procedure.

For this case study, assume that you're a technician working for a medium-size contractor that owns several John Deere 400D articulated dump trucks. The company you work for has invested in laptop computers and subscribes to John Deere's service information system called Service Advisor. Your supervisor has given you a work order that informs you that you need to have a look at a John Deere 400D truck with a transmission shifting problem. You then travel to the job site where the trucks are working to check out the problem. The work site is a highway construction job that has been going on for several months now, where the trucks are hauling material from a cut to a fill for a distance of approximately 1 km and up and down some fairly steep hills. It's late fall, and lately the mornings have been almost down to freezing temperatures.

When you arrive at the job site, you check in with the site supervisor and find that the trucks have been working for about three hours. You get permission to locate the truck that has a shifting problem and ask the foreman if you can look at it for a little while to check it out. You then look for a clear spot with no traffic or dust problems where you can work on it in a safe manner. The first part of any troubleshooting procedure is to find a safe place to work. Sometimes that may be difficult, but to perform your job properly and safely, this is a very important point to follow. Ideally, finding a quiet dust and mud free area is the best scenario.

The next step in troubleshooting is to gather information. For troubleshooting a powershift transmission complaint, a good place to start with is to ask the operator what the problem is and to be specific. If the operator says it's not shifting right, you need to ask questions to get more details. The more details you get, the easier your job becomes. Some examples of questions that you may ask for any operator complaint would be:

a. Is the problem happening now?
b. When did you first notice this problem?
c. Under what condition do you notice the problem (certain ambient temperatures, certain machine temperatures, under load or unloaded, traveling fast or slow)?
d. Is there any other related problems?
e. Has there been any service or repairs done on the machine lately?

Some additional questions you would have asked for this transmission shifting problem are the following:

a. Is there one shift that the problem occurs in or does the problem happen with more than one shift?
b. Does it happen during upshifts or downshifts, or both?

Let's assume from your questioning of the operator that you have determined this problem started one week ago. There has been no service or repair done on the machine for two months, and there is an intermittent problem of soft or slow shifting from third to fourth speed mainly when the machine is cold.

Now you would perform some basic checks. Even though from the operator's reporting of the complaint the problem seems to be isolated to a fairly specific shifting problem under limited conditions, you should always verify the basics. Remember, *simple stuff first*. Basic checks for troubleshooting this problem would include transmission oil level check, visual inspection for leaks, damaged wires, or transmission control damage. Assuming these checks verified to be okay, you could then check for fault codes.

This truck has an electronic control in the cab with a display that can be put into a service mode to check for fault codes and you can hook up a laptop to read codes also. You find a fault code PO 734 logged. You then check the troubleshooting procedure for PO 734.

Incorrect fourth gear ratio: It is probably slipping.

Check ignition voltage: It checks okay.

Check clutch pressures: It requires hooking up gauges to pressure taps on transmission and following a procedure. This also checks okay. This makes sense because at this time the transmission is warm and operating fine. You then decide to pull the transmission oil filter off and cut it open to look for seal or clutch material or metal filings. The filter looks fine.

If this shift problem was occurring now and you found either metal clutch material or seal material, you would then have some serious decisions to make depending on the quantity and type of contamination. You may proceed to send the truck to the shop for transmission removal. You may install gauges to check transmission pressures, or you may have the transmission oil and filter replaced and put the truck back to work.

You would now need to arrange with the foreman and the operator a convenient time as soon as possible to check the machine on a cold morning and make arrangements accordingly.

Let's say you arranged to come back two days to once again troubleshoot the intermittent soft shifting problem.

This time, when simulating all conditions under which the operator said the problem occurs, you in fact discover there is a shifting issue. The PO 734 code is back and the third to fourth shifts under load with cold oil is in fact slow. You put the gauges back on and find that the pressure when shifting is below specification.

You decide to put the truck back to work and over the next two weeks the problem gets progressively worse. Eventually, you have the truck sent to the shop where the transmission is removed, and when disassembled, it's found to have a cracked piston for C4.

TECH TIP

A standard set of gauges for testing transmission pressures would consist of three 500 psi gauges and one 100 psi gauge. These could be traditional gauges or digital gauges. A heat gun is also helpful for troubleshooting overheating problems. A filter cutter can also come in handy when you want to inspect a spin-on-type transmission filter. ● SEE FIGURE 12–39 for these special tools.

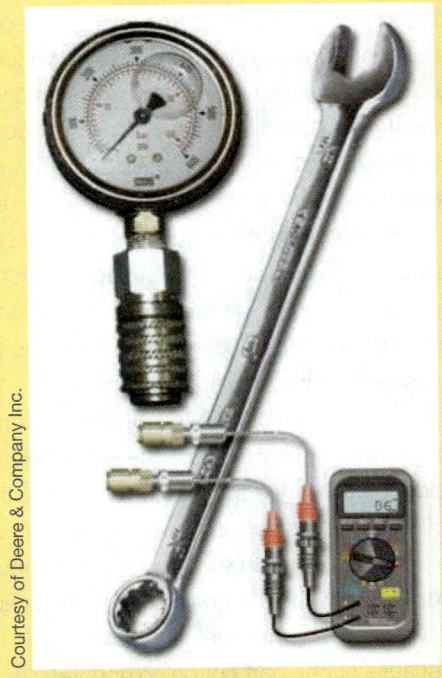

FIGURE 12–39 Some tools needed to diagnose a powershift transmission problem.

TRANSMISSION REMOVAL

Some transmission designs such as the Clark powershift transmissions will allow some internal transmission repairs to be done without removing the transmission from the machine.

Removing a powershift transmission can be a fairly straightforward task or can involve some major dismantling of the machine depending on the different types of machines that use powershift transmissions. For example, Caterpillar track-type tractor machines that use powershift transmissions can have a transmission easily removed at the rear of the machine in a few hours. The transmission is actually rolled out on a set of tracks and rollers that are incorporated into the transmission and machine. One complication to this would be if the tractor had a ripper installed on it, the ripper would need to be removed first. There could be special lifting brackets required to remove the transmission also. ● SEE FIGURE 12–40 for a transmission being removed from a machine.

Wheel loaders, using powershift transmissions, will usually require the cab removed and the transmission can then be removed through the top of the machine. Backhoe loaders will be able to lower the transmission down through the bottom of the machine. This would usually require blocking the machine up securely and lifting the transmission with a heavy-duty jack. Motor graders using powershift transmissions may need to have the engine and transmission removed as a unit or, similarly to the wheel loader, have the cab removed and then the transmission. Articulated rock trucks would have the same options as wheel loaders and straight framed rock trucks could

POWERSHIFT TRANSMISSIONS 385

FIGURE 12–40 A powershift transmission being removed from a machine.

have their transmissions removed with only needing the box locked in the opposition and possibly a special lifting bracket.

The machine should be secured from moving, tagged and locked out. The transmission fluid should be drained, transmission controls disconnected, lifting or lowering devices attached, mounting bolts removed, any drive shafts removed and the transmission can now be removed as a unit.

Transmission installation is usually simply a reverse procedure of transmission removal. However, there could be a need to align driveline part's torque converters and ensure that transmission controls are connected properly. New fluid should be installed and all related power train filters should be replaced, as well as any screens or magnets cleaned. If the transmission that was removed had a major failure and contaminated the transmission hydraulic system, the repair procedure should include transmission cooler cleaning and removals/replacement. There should also be a procedure done at initial start-up to ensure that all contamination has been cleaned out of the system. This could involve an initial fill with a lower viscosity fluid that gets changed soon after it gets warmed up, along with a quick filter change. Some other related repairs to transmission removal and installation would include driveline components such as U-joints, drive shaft's hanging bearings, and any related transmission hoses that may need replacing.

POWERSHIFT TRANSMISSION RECONDITIONING

Once a machine with a powershift transmission has been diagnosed as having a transmission failure, the transmission is removed from the machine and either send to a rebuild

FIGURE 12–41 A transmission rebuild facility.

shop, exchanged with a rebuilt or new unit, or repaired on site.

Most small fleet owners of equipment that is still in warranty will usually opt for an exchange transmission. This will give them some kind of warranty on the installed units.

If a transmission is rebuilt by the machine owner, that owner will assume the risk if there are any repeat failures. A lot of this decision is based on whether the owner has trust in his technician. A failure caused as a result of poor workmanship, overlooked failed parts for installation, and break-in techniques can cost thousands, tens of thousands, or hundreds of thousands of dollars to the machine owner. There is not only the cost of replacement parts to consider but also the time it takes to remove and install and rebuild the transmission and the cost of a replacement machine to take the place of the downed machine. ● SEE FIGURE 12–41 for a transmission rebuild facility.

In order for technicians to properly rebuild powershift transmission, they not only need the required skills and confidence to do the job but they also need to be provided with the proper service information related to the specific transmission that they are rebuilding. There will be a lot of specifications that parts should be measured against to decide if components are reused or replaced. There could also be many updates that have been incorporated into the transmission by the manufacturer. These updates could include things such as improved seal material, improved friction desk material, and possibly hardware updates. So even if technicians perform the transmission rebuild to the best of their ability following the given specifications and procedures that they have been provided, a repeat failure is still possible, simply because of one part not being updated.

Once the powershift transmission has been removed, it can now be disassembled and be evaluated for reusing of internal parts. Specialized transmission rebuild facilities

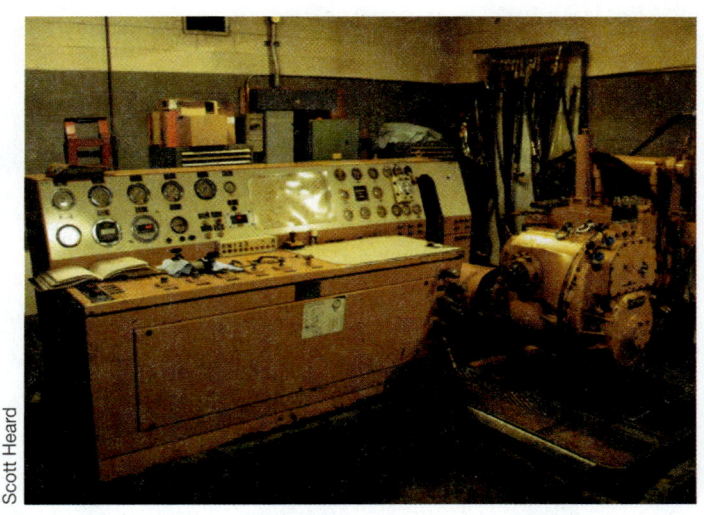

FIGURE 12–42 A transmission test bench.

> ### SHOP ACTIVITY
> - Inspect a countershaft powershift transmission from the outside and draw a sketch of the shafts and gears that you think transfer power flow inside the transmission.
> - Explain how this transmission is controlled.
> - Explain how a complete transmission service would be performed.
> - Explain how transmission pressures are tested on this transmission and what special tools would you need.

will then mount the transmission on a test bench, where it is tested for operation. This process could detect a potential problem, which could save a lot of time as the problem can be rectified before the transmission is installed in the machine. ● SEE FIGURE 12–42 for a transmission test bench.

The rebuild facility ideally would be a climate-controlled environment. Minimum requirements include a clean, well-lit, organized workspace with plenty of properly sized parts' bins for organizing and maintaining transmission parts in sequence. Before starting to disassemble the transmission, it should be cleaned as much as possible. The workspace should be organized and service information should be readily available. Powershift transmissions should be disassembled in a very specific sequence that must be followed in order not to damage parts during the disassembly procedure. Any special tools required for disassembly should also be available.

Transmission rebuilding is sometimes a specialized task reserved for specially trained technicians. However, if required, an experienced technician should be able to perform a routine transmission rebuilt, provided any required specialized tooling is made available.

SUMMARY

1. Some safety concerns related to power shift transmissions are: securing the machine before working on it, using any lifting devices properly, using proper PPE.
2. Power shift transmissions use a combination of gears, shafts, clutches, and oil to transfer torque from a diesel engine or torque converter toward the tires or tracks of a machine.
3. They can have as little as two speed ranges to as many as 18 and be a stand alone unit or have a torque converter, transfer case housing and differential incorporated within the same housing.
4. Different gear ratios will usually provide means to increase engine torque and reduce speed.
5. Power shift transmissions allow the operator to easily shift between speed ranges with only a momentary loss of drive. Most power shift transmission controls allow the operator to change speed and direction with one control and most now use electronic controls.
6. Two main types of power shift transmissions are: planetary and countershaft.
7. Planetary type transmissions are generally more robust while countershaft are mostly used for light to medium duty applications.
8. Both types of transmissions use hydraulically actuated clutches to change speeds and directions by engaging gears to shafts. Hydraulic clutches have a set of friction discs and a set of flat steel plates. A piston is used to squeeze the sets of discs and plates together, which then transfers torque through torque.
9. Friction discs have grooves in them to allow for cooling oil to circulate around them to carry away heat and will have teeth or tangs on their inside or outside circumference.
10. Steels plates are flat and usually have tangs on their outer circumference.
11. Clutch torque capacity is determined by the design engineers and can be increased by adding more plates and discs or increasing their diameter.
12. Clutch apply oil pressure is usually around 250–350 psi and is directed on the surface of a sealed piston.
13. Power shift transmissions need to have a hydraulic system to supply clutch apply pressure and lubrication oil. They usually incorporate the torque converter oil supply system.

14. Countershaft types of power shift transmissions feature gear clusters on shafts that have meshing gears. Oil applied clutches engage gears to shafts and when two clutches are engaged torque can be transferred through the transmission.
15. A simple countershaft transmission is called a shuttle shift or reverser transmission. It provides for power shifts between forward and reverse but manual shifts for speed ranges.
16. A transmission neutralizer will drain the direction clutch to stop drive when the operator needs to have full power for hydraulic functions. It simulates disengaging a mechanical clutch.
17. Transmission control valves will usually house several valves such as: pressure regulating valve, modulating valve, torque converter relief valve, shift selector valve, shift solenoid valves, pressure reducing valves.
18. Modulation valves control how fast a clutch is filled with oil and therefore determines the delay between shifts when the operator commands a speed or direction shift. Older modulation valves used springs, pistons, and orifices while newer ones use proportional solenoids.
19. Transmissions that have electronic modulation can change clutch fill times to compensate for clutch wear.
20. Almost all machines produced today that have powershift transmissions will have an electronic system control them. This system will allow the transmission control to communicate with other machine systems such as engine and hydraulic.
21. Electronic controls can provide features such as: inching pedal, event-based shifting, clutch protection, skip shifting, speed matching, autoshifting, and maximum starting gear. These features will make the machine more efficient and productive.
22. Countershaft powershift transmissions can have up to 18 forward speeds and 6 reverse.
23. Countershaft transmission shafts will have oil passageways drilled through them to direct oil from the transmission housing to the clutch pistons.
24. Planetary type powershift transmissions use combinations of two or more sets of planetary gearsets to provide speed and direction shifts.
25. Machines with planetary powershift transmissions typically have between 3 and 8 speed ranges.
26. Two clutches must be engaged to provide torque transfer through the transmission.
27. Transmission electronic control systems have input sensors such as: operator input switches, speed sensors, oil pressure, and temperature sensors and output sensors such as: proportional solenoids and dash displays.
28. Maintenance required for powershift transmissions consists of regular oil and filter changes. Fluid viscosity and specification requirements must be closely adhered to, else clutch damage will occur.
29. Operator complaints for powershift transmissions could be: no drive, harsh shifting, vibrations, won't upshift, or won't downshift.
30. Troubleshooting procedures related to the above complaints start with visual inspections for fluid levels and leaks, damage, and checking fault codes. Verifying complaints mean operating the machine in a safe manner and trying to replicate the problem.
31. Some tools needed to analyze a powershift transmission are: pressure gauges, temperature gun, and laptop with software.
32. Transmission removal could be fairly simple or a very large job depending on the type of machine. Blocking, rigging, and hoisting equipment should be checked for defects before use.
33. Powershift transmission reconditioning should be performed in a clean well-lit shop with plenty of bench space. Proper tools and service information is critical to ensuring a quality recondition process.
34. Install gauges and record clutch pressures for all speeds.

chapter 13
ELECTRIC DRIVE SYSTEMS

LEARNING OBJECTIVES

After reading this chapter, the student should be able to:

1. Explain what three-phase AC output is.
2. Describe how an inverter works.
3. Explain how DC is switched back to AC to drive propulsion motors.
4. Explain how an induction motor works.
5. Define *switched reluctance*.
6. Describe a typical maintenance procedure for an electric drive system.
7. Explain how an electric drive system can provide machine braking.
8. Describe the operation of electric drive cooling systems.
9. Describe the common safety-related practices for working on electric drive systems.

KEY TERMS

Capacitor 390
CAT III 390
IGBT 403
Inverter 393
Rotor 398
Stator 396
Three phase 397
Windings 397

INTRODUCTION

SAFETY FIRST There are some very serious safety concerns you should be aware of when working with a machine that has any type of high-voltage electric propulsion system on it.

Just like machines with high-pressure hydraulic systems, machines that use high voltage have hidden dangers that can cause serious injury or death. Also similarly, if you use some common sense, follow the manufacturer's related service information including the cautions, warnings, and procedures and anticipate the potential for danger you can safely work on these systems.
● **SEE FIGURE 13–1** for a warning symbol for high voltage.

High-voltage electric propulsion means any system that moves the entire machine whether it has tracks or wheels, or any system that moves part of a machine such as the swing function of an excavator with electrical energy being the power source. Electric propulsion systems need a diesel engine running and driving a generator to create the electrical energy and could have energy storage such as batteries or a **capacitor** to supplement the generator at certain times.

If the machine has capacitors or batteries built into the system, it may store a potential voltage high enough to injure or kill you. Even if a machine doesn't have proper electrical storage devices, there is a possibility for dangerous high voltages to exist in conductors and components for some time after the generator and motors have stopped turning. Many electric propulsion systems normally operate at 480V, and under certain conditions, there could be over 1000V present.

It is highly recommended that before anyone works on a machine with a high-voltage system, that the person is aware of any workplace or government regulation that requires specific training or licensing needed.

You should *never* work on a machine's electrical propulsion system with the generator or motors turning.

If a machine has a high-voltage electrical storage device (batteries or capacitors), it must be isolated from the system you are working on or have its energy level depleted to below 50V to be considered safe.

You *must always* confirm that the voltage of any conductor or component is under a safe limit of 50V before proceeding with service or repairs. This should be done with a **CAT III**–rated multimeter and with proper PPE worn. ● **SEE FIGURE 13–2** for a CAT III multimeter.

You *must always* consult the machine's service information and follow the procedures described to confirm that there is no more than 50V present in any part of the machine's electrical system.

You *must always* wear appropriate PPE as described in the machine's service information until you have confirmed that there is less than 50V present. This includes gloves that meet or exceed ASTM D120-09 specifications and OSHA 29C FR 1910.269 regulations and NFPA 70E–certified gloves that are rated for 1000V AC or 1500V DC. Inspection before use is required and highly recommended. Extra care is needed in the visual examination of the glove and in the avoidance of handling sharp objects. In addition to visual inspection, an air test needs to be performed before each day's use. Perform the test per American Society for Testing and Materials (ASTM) Standard Guide F1236-96.
● **SEE FIGURE 13–3** for a Class "0" glove.

FIGURE 13–1 High-voltage warning symbol.

FIGURE 13–2 CAT III multimeter.

Gloves should be electrically retested within six months of being put into service. Gloves that have not been put into service have a shelf life of 12 months. If gloves have been on the shelf for 12 months and have not been put into service, the gloves must be electrically retested before being put into service. When the old gloves are beyond their service date, they should be retested or shredded and disposed of.

Never remove covers that are part of the machine's electric propulsion system until the generator and motors have stopped turning.

"Orange" insulation on larger conductors indicates high voltage is normally present when the generator or motors are turning or could have high voltage present any time if there is an electrical storage device on the machine.

● **SEE FIGURE 13–4** for orange high-voltage cables.

Any grounding wires that are part of the machine's electrical propulsion system must be connected properly to where they originally were on the machine. All high-voltage connectors *must* be clean and dry when assembled and all threaded fasteners *must* be torque to specification. All original clamps *must* be installed to secure high-voltage cables.

Some electric drive machines will have a hazardous voltage warning lamp to alert anyone working on the machine that there could be dangerous levels of voltage present. There are also safe shutdown procedures that should be followed to ensure that no dangerous voltage is present.

Machines with high-voltage systems may have unique ways to lock out and tag out those systems. Always refer to the manufacturer's service information when performing lock out and tag out procedures. ● **SEE FIGURE 13–5** for a machine's battery disconnect and unsafe voltage light.

FIGURE 13–3 Class "0" glove.

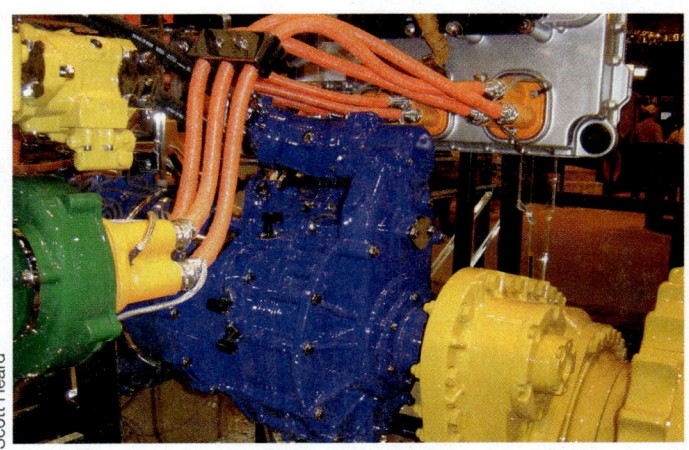

FIGURE 13–4 Orange-colored high-voltage cables.

FIGURE 13–5 Machine's battery disconnect and unsafe voltage light.

> **TECH TIP**
>
> The following is a summary of steps that *must* be followed when servicing the electric drive system:
>
> - De-energize: Eliminate or isolate any electrical system from producing potentially hazardous voltage.
> - Secure (lock out): Ensure that potentially hazardous voltage is not generated without knowledge of service personnel.
> - Verify: Ensure that potentially hazardous voltage is no longer present.
> - Proceed: Perform service to electric drive system. Be aware of hazard and warning decals.
> - Restore: Ensure that all protective equipment is operating properly before resuming operation.

ELECTRIC DRIVE SYSTEMS

Service personnel should be extra careful when pressure-washing machines with high-voltage drive systems to *not* direct high-pressure water at any high-voltage components or cables.

If you are unsure or uncomfortable working on any part of an electric drive machine's electric system, *do not* feel pressured into doing something you are not totally sure about. You need to be confident that you are perfectly safe doing any kind of work on a system that has a potential to cause serious injury or death.

ELECTRIC DRIVE HISTORY

Electric drive systems have been around for a long time. The early 1900s saw a battery-powered car developed and produced for public use. It hasn't been until the last few years that electric drive has started to become accepted in the mainstream automotive world because of the movement to reduce internal combustion engine emissions.

Heavy equipment machines have used diesel electric drive systems for decades. LeTourneau was the first major equipment manufacturer to fully embrace electric drive technology with the first all-wheel drive electric machine produced for the construction industry in the 1940s.

Underground mines utilize full electric drive machines because of the lack of diesel exhaust emissions. The limiting factor is the machine has to have either an umbilical cord or a trolley cable system that are both range limiting.

Diesel electric drive systems have been used in large mining trucks and some wheel loaders and scrapers since the 1950s.

Diesel electric machines have recently become more popular and either are expected to continue to become a drivetrain option or will replace some models of diesel/mechanical drivetrain machines as time moves on.

Some machines will have certain functions or components powered by high-voltage circuits or electrical energy storage devices. A few manufacturers have experimented with true hybrid drive systems. These systems take braking energy that is normally wasted and turn it into stored electrical energy that can be used for acceleration. This is the principle behind the hybrid system used for on-highway vehicles we see now. Several manufacturers have made prototype hybrid drive machines. Energy storage is done with either batteries or capacitors. Battery technology is the limiting factor in making a true electric hybrid machine practical today.

Some electric propulsion systems are used to power the swing mechanism for giant electric shovels. These systems, however, use grid-fed high voltage (they require a massive

TECH TIP

A true hybrid electric drive system has the capability to store energy and release it whenever needed. The energy is created by a braking action and was typically lost energy in the form of heat. Energy can be stored electrically in batteries or capacitors, hydraulically in accumulators, or mechanically in flywheels.

Some "hybrid" systems can recover braking energy but not store it, and therefore, it must be used instantaneously.

FIGURE 13–6 Electric mining shovel.

extension cord!) and should only be serviced by qualified electrical technicians. One massive shovel uses fourteen 700 hp AC motors just to swing the upper structure! ● **SEE FIGURE 13–6** for an electric shovel.

Most electric drive machines can use their drive motors for braking. The motor reverses its purpose and becomes a generator that will put a load on the wheels and create a braking effect. This heat energy is usually sent to a resistor grid where it is dissipated. This is unlike hybrid vehicles that use the drive motors to convert braking energy into electrical energy to recharge the vehicle's batteries.

John Deere recently began manufacturing an electric drive wheel loader that recycles the energy used to slow the machine to assist turning the engine. This in turn will be able to turn the machine's hydraulic pump. For example, if the machine is coasting with a full bucket and a braking action is needed, the propulsion drive motor acts like a generator and the power it produces is fed into the generator, which is now acting like a motor to in turn help drive the diesel engine. If the operator needs to raise the boom when decelerating, this recycled braking energy helps turn the engine that is also turning the hydraulic pumps. The fuel that would normally be needed to perform this function is saved.

This machine is said to show an improvement of 25% in fuel economy when compared to the same machine that has a

FIGURE 13–7 An electric drive loader.

FIGURE 13–8 Komatsu Hybrid energy storage device.

conventional drivetrain (torque converter and powershift transmission). This wheel loader is called a hybrid because it can recycle braking energy for use instantaneously but isn't able to store any energy. It has a generator driven by a diesel engine that drives a single motor. The motor then drives a three-speed powershift transmission that doesn't need directional clutches because the motor can change directions electrically. The transmission then outputs its torque to drive shafts and axles in a conventional manner.

The latest generation of LeTourneau loaders also uses braking energy to drive the engine. LeTourneau currently produces large electric drive wheel loaders and wheel dozers. ● **SEE FIGURE 13–7** for an electric drive loader.

John Deere is also about to introduce a large electric drive wheel loader that will have an electric motor for each wheel.

Komatsu has been selling a production excavator for several years that features an electric drive swing system. It captures the energy used to stop the upper structure when it is swinging and stores it in ultra-capacitors. The ultra-capacitor can release up to 60 hp instantly to drive the upper structure. There is also a generator/motor between the engine flywheel and hydraulic pumps. The generator can charge the ultra-capacitor when necessary and the motor can use stored energy to keep the engine rpm from dropping too far due to sudden load changes. ● **SEE FIGURE 13–8** for the Komatsu Hybrid excavator. This system is said to save 25% in fuel when compared to the same-size machine with a regular hydraulic drive swing function.

Many manufacturers today use electric drive systems for their large haul trucks in mining applications. These trucks are in the 150-ton and higher payload class. These trucks typically feature a diesel engine driven generator that outputs to an **inverter** that converts the AC generator output to DC. Electronics will convert the DC back to AC, which is then sent to a pair of drive motors. These motors send torque to final drives, which then drive the dual wheels at the rear of the truck. The wheel motors can be used to provide braking, and this energy is usually sent to a brake resistor where it is transferred to the atmosphere as heat.

Caterpillar recently came out with an electric drive track-type tractor. This electric drive system uses a diesel engine driven generator that has its output electronically changed from AC to DC and then back to AC where it drives two identical AC motors. These motors drive a common bull gear that drives a set of planetary gears. The planetary gears along with a hydraulic steering motor will provide a left-to-right track speed differential that steers the machine based on operator commands.

ELECTRIC DRIVE SYSTEMS 393

FIGURE 13–9 An electric drive track-type tractor.

WEB ACTIVITY

Search the Internet for manufacturers that produce electric drive heavy equipment. See who makes them and what types of machines are currently being produced. See if there are any new machines and manufacturers coming along.

Torque then leaves the planetary steering section and is sent out to a planetary final drive for each track. ● SEE **FIGURE 13–9** for an electric drive track-type tractor.

WHY ELECTRIC? There are several distinct advantages to electric propulsion systems when compared to conventional drivetrains.

Electric drive machines allow the engine to run at a constant rpm rather than have the operator vary the rpm to suit the varying operating conditions. If an engine can run at a steady rpm, the engine designers are able to make the engine run cleaner by optimizing valve timing and fuel injection for that narrow rpm range. The hydraulic system can also be designed to have its pump work at a steady rpm, which allows the machine designers to better match it to the hydraulic system flow needs.

Fuel economy is also improved if an engine can run at a steady rpm, and if less fuel is consumed, then emissions are reduced. Both the machine owner and the environment benefit here.

Compared to a mechanical or hydrostatic drivetrain, there are less moving parts in an electrical drivetrain. Some manufacturers claim 60% less. The generator, controls, and motors will replace a torque converter, powershift transmission, or pumps and motors. Many electric drive machines eliminate differentials and axles as well. Less moving parts means lower maintenance and repair costs, less downtime for servicing and repairs, and less frictional losses, which equates to higher efficiency.

The integration of electronics to control electric drive systems has increased both their efficiency and their drivability. Electronics have enabled the speed and torque of a motor to be controlled with seamless changes in speed and direction, which makes a much smoother operating machine when compared to a mechanical drive machine. This leads to increased productivity and reduced operator fatigue.

Electric drive systems are claimed to be more efficient than mechanical or hydraulic drives. Larger electric motors are up to 95% efficient. This means they turn 95% of the power they receive into torque.

Electric motors have excellent high-torque characteristics from 0 rpm and up.

Electric components can be standardized to be used for more than one machine. This will keep manufacturing costs down. Some generators and motors are almost identical in construction.

Some machine designs use braking action to assist with engine rpm. This happens when the generator and motor switch roles to convert the energy needed to slow down a machine into assisting with turning the engine crankshaft. Some hydraulic functions may still need to be performed during braking such as steering or lifting the bucket. Fuel consumption during braking will be reduced because of this feature.

To compare an electric drive machine to a mechanical and hydrostatic drive machines, see the following:

	ELECTRIC DRIVE	MECHANICAL DRIVE	HYDROSTATIC DRIVE
Prime mover	Diesel engine	Diesel engine	Diesel engine
Power medium	Electrons/ magnetism	Gears	Fluid
Control type	Electronic	Mechanical/ electronic	Electronic
Final speed reduction/torque multiplication	Gears	Gears	Gears
Power measurement	Watts = $E \times I$	hp = Torque $\times$ rpm/5252	hp = gpm $\times$ psi/1714
Possibility of fluid leaks from drivetrain	Minimal	Greater	Highest

Electric power can be measured in watts and converted to horsepower. The formula to do this is 1 hp = 746 Watts (1000 Watts = 1 kW).

DISADVANTAGES OF ELECTRIC DRIVE
The operating characteristics of an electric drive machine are quite different from other drivetrain systems. Because the engine runs at a steady-state rpm, the operator may not think the machine is working properly, and this may lead to some operator complaints until the operator gets used to the characteristics of the system.

The controls for a machine with an electric drive propulsion system will be different, and the operator will need time to adjust. For example, a wheel loader with electric drive will have a right pedal that controls the travel speed. Direction is selected with a rocker switch on the loader joystick. Conventional drive machines have the right pedal control the engine rpm and a separate control for speed range and direction.

There may be concerns from operators and technicians about working near high voltage. Again this should be put to rest once they become familiar with the machine and knowledgeable about the proper service and repair procedures that are required to stay safe.

There is usually a purchase price premium for electric drive machines versus conventional drive; however, this should be offset by fuel savings.

DC ELECTRIC DRIVE
Electric drive machines originated with a diesel engine driving a generator that produced AC three-phase voltage that was rectified to DC. This DC then went on through drive controllers and on to the DC drive motors. DC drive motors used brushes that transferred current into the motor's rotating armature. These brushes are similar to a large starter motor. The main downfall of this style of motor is that the brushes need regular servicing because they transferred a huge amount of voltage and current through them constantly.

The trend in the last number of years for electric drive machines is the use of AC motors because they are more efficient and require less maintenance because of the elimination of brushes. Many older machines have been converted from DC to AC drive motors. AC technology typically provides a 6–7% higher efficiency than DC drives, averaged over the duty cycle of the motor.

AC ELECTRIC DRIVE
There is a wide variety of configurations for AC electric drive systems that are used for machine propulsion. Most systems will have the generator driven directly by the diesel engine's flywheel, but at least one machine has the generator driven through a gearbox that is between the engine's flywheel and the generator input shaft. ● SEE FIGURE 13–10 for a generator gear drive. For this machine,

FIGURE 13–10 A generator gear drive.

the engine's speed will get multiplied three times to increase the speed of the generator to 5400 rpm when the engine is running at 1800 rpm. The gearbox also drives the machine's hydraulic pump.

The main differences between the different systems is the type and number of generators and the type and number of drive motors used and how they are controlled. Almost all machines employ one generator to feed one or more electric motors. There are, however, machines that use two generators.

The generator will produce three-phase AC that goes to an inverter. The inverter is an electronically controlled switching device that converts the AC to DC and then back to AC. You may wonder why not just send the AC power from the generator to the AC motors directly. To change the speed of a motor, the frequency of the AC voltage needs to change. To do that without converting the generator output to DC first the speed of the generator would have to fluctuate. By using electronic controls, the inverter output to the AC motors can be precisely controlled to match operator demands for speed and direction while having the generator turn at a constant rpm.

From the inverter, the AC flows to the drive motors where it sets up opposing magnetic fields that create an output torque through the motor's shaft. The motor's output torque is then multiplied with gears to drive either tracks or tires.

This briefly explains the electric propulsion system for most electric drive machines. Komatsu produces an excavator that has an electric drive swing system that captures the energy from slowing the swing down, stores it briefly, and reuses it to start the machine swinging again. This energy also assists with keeping the engine rpm steady under varying loads.

AC VERSUS DC After reading Chapter 5, you should have a solid understanding of DC electricity. DC is the movement of electrons through a circuit in one direction starting at a power source (battery) moving through a load and returning to the power source. While this movement of electrons happens at speeds that are hard to comprehend, it's easy to understand because they all flow in the same direction starting at the power source moving through the load and returning to the power source.

AC is still just moving electrons; however, they only move a small distance and in both directions. To compare electron flow between AC and DC, consider this: If in a DC circuit a certain amount (6.28 billion billion) of electrons passes one point in 1 second and in one direction, then 1 ampere of current flow has occurred. If in an AC circuit the same amount of electrons passes the same point in ½ second and then reverses and passes the same point for another ½ second, then there has also been 1 ampere of current flow. It is the "alternating" current flows power source that makes the electrons constantly change direction.

The number of times or frequency the electrons move back and forth in 1 second is measured with a unit called Hertz.

If a 12V DC voltage signal is tracked in real time on an oscilloscope, it will show as a straight line of 12V positive. As there is a steady positive voltage, there is no need to measure a cycle; therefore, the Hertz of DC power doesn't exist.

If a 12V AC voltage signal is tracked on an oscilloscope, it will be a constantly changing sine wave that rises to 6V positive and lowers to 6V negative. This rhythmic wave will keep the same shape as long as the frequency of the sine wave and the voltage value doesn't change. ● SEE FIGURE 13–11 for the comparison of a DC voltage to an AC voltage.

Another example of this is the frequency used for North American household 120V AC voltage that is regulated at 60 Hz.

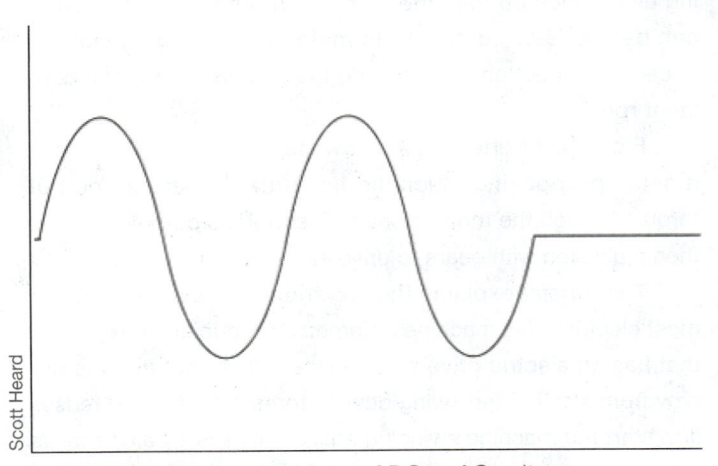

FIGURE 13–11 A comparison of DC to AC voltage.

If a sine wave represents the AC coming out of the wall sockets in North America, it would cycle from positive to negative 60 times each second. If you measured the frequency of European household voltage, you would see a 50 Hz sine wave.

Just like in a DC circuit, it is the movement of electrons that creates work in an AC circuit. For AC electric drive systems, their AC flow will create magnetic fields in the machine's motors, which in turn create work.

The reason why an AC voltage sine wave is constantly moving up and down is because of the power source that is used in an AC system. An AC generator was discussed in Chapter 8. The principle of AC generation is the same process that occurs in a machine's charging system alternator before it is rectified to DC. A rotating series of alternating north and south poles (the rotor) move past stationary windings (**stator**), and this action induces an alternating voltage in the stator. The voltage constantly alternates and changes value from positive to negative because of how the north and south pole magnetic flux lines interact with the stationary winding. ● SEE FIGURE 13–12 for how AC voltage is induced in a stator winding.

The amount of electrical pressure (voltage) generated is dependent on the strength of the magnetic lines of flux and the speed at which they move past the conductor. The stronger the magnetism and the faster they move, the higher the voltage produced. This magnetic field could be produced from a permanent magnet or an electromagnet. To make a stronger electromagnet, there needs to be more current flow or more coils of wire.

The frequency of the AC sine wave output will also change with the speed at which the magnetic fields move past the stator windings.

THREE-PHASE AC VOLTAGE GENERATION If a generator has a rotating magnetic field inside of a housing that has one stator winding formed into two coils, its output is considered to be single-phase AC. ● SEE FIGURE 13–13 for a single-phase voltage generator.

This means that at two points throughout each revolution of the rotor, there is 0 voltage output. This occurs when the magnetic fields are parallel to the stator winding. At two other points, there is maximum voltage created and in alternating polarities. As the magnetic field rotates between the stator coils, the lines of magnetic flux induce current flow in the stator. The polarity changes each time the poles alternate.

This single-phase output may be adequate for a light-duty generator that puts out 120V AC, but it wouldn't be very effective for a heavy-duty application. To make a more stable and consistent power source, generators will have three separate

FIGURE 13–12 How AC voltage is induced in a stator winding.

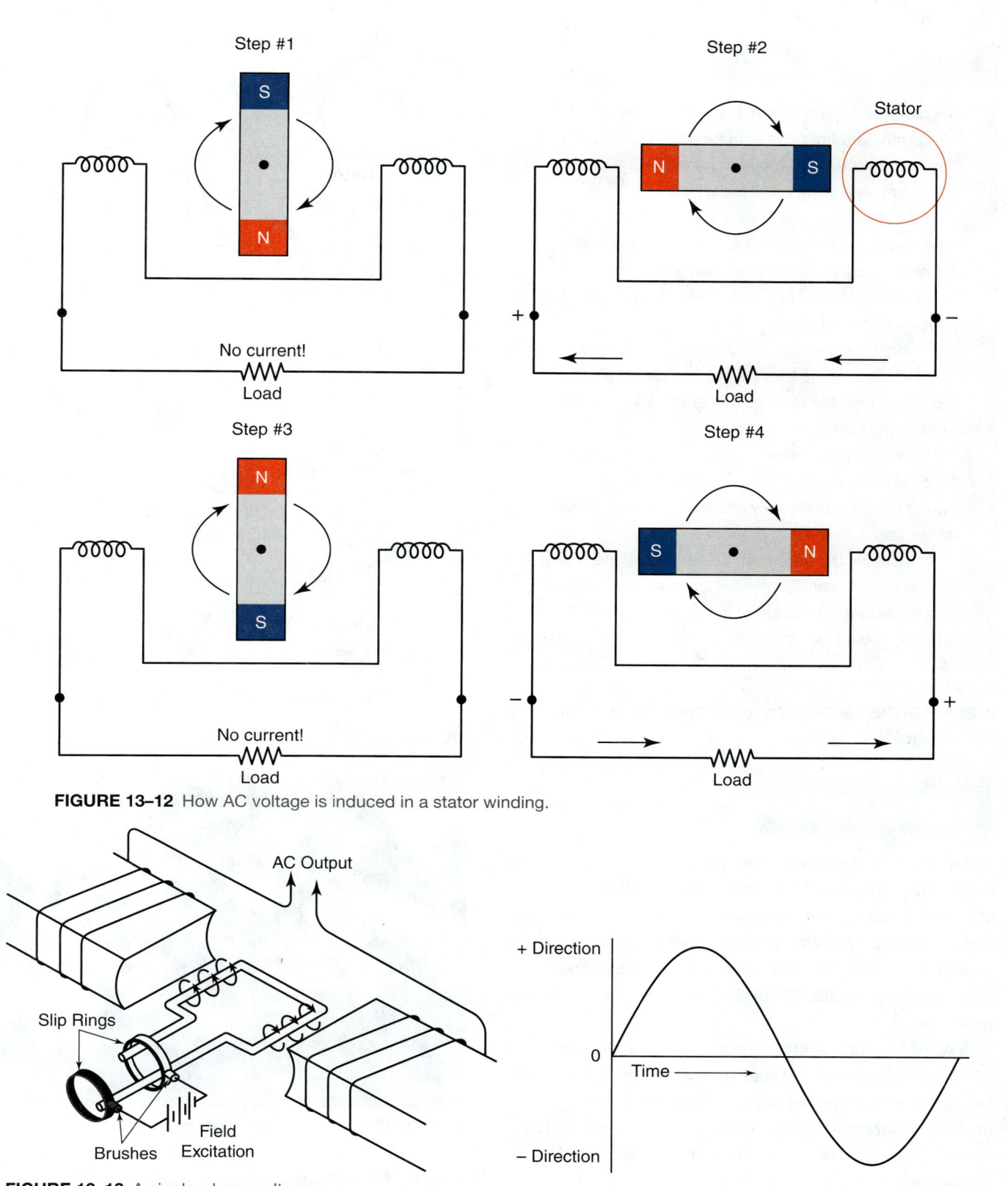

FIGURE 13–13 A single-phase voltage.

stator windings. This means for every revolution of the rotor, there will be three sine waves produced. When three phases are used, there is a much more stable power supply and the generator actually gains a large amount of efficiency when compared to a single-phase generator. This type of generator output is called **three phase**. The word *phase* refers to the timing that occurs between each phase. The **windings** will be spaced evenly at 120° around the generator housing.

ELECTRIC DRIVE SYSTEMS **397**

TECH TIP

The principle behind how a generator works originated with Faraday's law of magnetic induction. If a magnetic field is passed over by a stationary conductor or piece of wire and the wire is connected to a load, then a current flow is induced in the wire. Likewise, if that same piece of wire is moved over a stationary magnet, current will flow in the wire. It is the magnetic lines of flux that move past the wire that induce a current flow.

A more practical and effective way of producing voltage is to arrange the magnetic field on a shaft and spin it past the stationary wire as well as make loops in the wire.

Because a magnet has two opposite poles as the alternating north and south poles move past the stator coil, the current flow changes direction in the stator. This is how AC is produced.

To increase the frequency of the AC voltage, the shaft is spun faster. To increase the amount of current flow, more loops are created in the stator wire or an increase in the magnetic field strength is needed.

● **SEE FIGURE 13–14** for the output of a simple three-phase AC generator.

GENERATORS

The main parts of a typical three-phase AC generator are the rotor, stator, and enclosure. The stator is a stationary series of copper-insulated wires that are wound in place in the generator housing. The rotor is driven by the diesel engine and creates a series of magnetic poles that spin very close to the stator windings. ● **SEE FIGURE 13–15** for a three-phase AC generator.

All generators used for machine electric drive systems will create three-phase voltage. Think of this as one unit making three separate power outputs. This results from the use of three separate windings for the stator assembly. These windings are staggered evenly around the stator frame, which spreads each phase out evenly to create a more even total power output. Each winding will be wound around a series of poles that make up the stator frame. The frame is housed in the enclosure and surrounds the rotor.

The generator's three individual stator windings are heavy insulated copper wire that are looped into coils and placed into

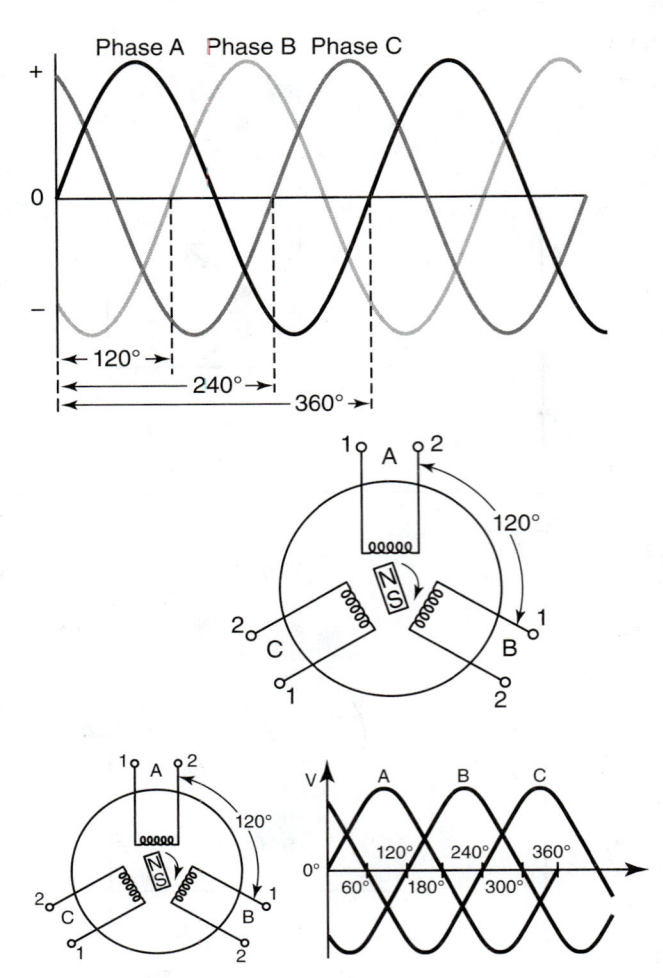

FIGURE 13–14 A simple AC generator.

FIGURE 13–15 A three-phase AC generator.

slots in a laminated soft iron core or pole. The stator leads will go to some type of threaded terminal where heavy conductors can transfer the generator output to an inverter.

Most heavy-duty generators have a **rotor** shaft that spins on a bearing at each end of its shaft. These will likely be running in oil for lubrication and cooling. Some bearings will have temperature sensors that will initiate a fault code if temperatures exceed a specified limit. The rotor needs to be balanced to eliminate any vibration that could make the bearings fail and cause contact with the stator because there is very little clearance between the rotor and stator.

When generators are producing power, they will create heat. If the heat level gets too high, permanent damage can occur to stator windings or rotor bearings. All generators used for heavy equipment electric drive machines will need some type of cooling. Two common cooling mediums are air and oil.

Because of the unfavorable environment, off-road machines usually feature generators that will be sealed from the elements. For air-cooled generators, a pressurized and filtered cooling air system supplied from a central fan will ensure generator cooling.

Other types of generators will have oil circulate through them for cooling purposes.

Because there is really only one moving part in a generator, there is not much maintenance required other than changing the cooling oil when required. The biggest enemy to a generator is heat and dust.

There are three basic types of generators used on electric drive machines: permanent magnet, excited rotor, and switched reluctance.

PERMANENT MAGNET GENERATOR
A permanent magnet generator is the simplest type of generator because the rotor is just a series of spinning magnets. They are arranged to alternate north and south poles past the stator windings as the rotor is driven by the prime mover (diesel engine).

Remember it takes movement of magnetic lines of force (flux lines) to induce current flow in a conductor. The generator's stator is the conductor, and its rotor creates the moving magnetism that will induce current flow in the stator.

In these types of generators, as the strength of the magnet is fixed, the output is variable only by varying the rotor rpm.
● **SEE FIGURE 13–16** for a permanent magnet generator.

A three-phase permanent magnet generator will have three separate windings to create three-phase output as the rotor spins inside the stator.

EXCITED ROTOR GENERATOR
This type of generator most closely resembles the alternator design that is used for the charging system of a machine because the rotor needs to get energized to control the output of the generator. The rotor has a coil of wire that is arranged into a series of loops that are wrapped around a laminated iron core. The energized rotor will create a series of alternating north and south poles that induce

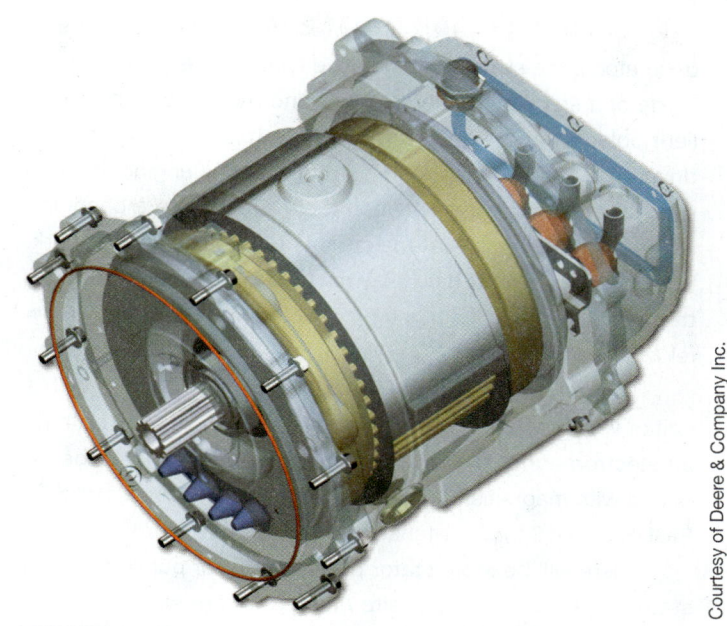

FIGURE 13–16 A permanent magnet generator.

FIGURE 13–17 An excited rotor.

voltage into the main generator stator windings as it spins. The rotor coil, however, is also energized by a rotating exciter coil, and the exciter coil is energized by a stator. When the exciter stator is energized and the exciter coil rotates past it, the coil will have voltage induced in it. This AC voltage is then rectified with rotating diodes that are part of the rotor assembly into DC voltage. This DC voltage is then sent to the main rotor winding to create a strong electromagnet and induce voltage in the main stator. This assembly does the same thing as the permanent magnet rotor except that its strength can be controlled with a signal from an ECM. ● **SEE FIGURE 13–17** for the rotor of an excited rotor generator.

The generator's output can be closely regulated by controlling the current flow in the exciter stator winding. For example, a large mining truck would have the exciter controlled electronically at 144V AC and vary current flow from 0 to 20 amps. This in turn would vary generator output voltage from 0V AC to over 2000V AC and output amperage to over 1000 amps.

ELECTRIC DRIVE SYSTEMS

SWITCHED RELUCTANCE GENERATOR A third type of generator is the switched reluctance type. It uses a rotor that is made of a stack of iron laminations and has projections or salient poles extending out from a base circle. This assembly is driven by the shaft that is driven by the diesel engine. It is the rotor's poles and gaps between them that create a changing magnetic reluctance that makes this type of generator work. Reluctance describes how easy magnetic lines of flux can pass through a material. Air has high reluctance while iron has low reluctance.

The generator's stator has a series of pole pairs with a coil of wire wound around each pair. These poles will both be an electromagnet to start the generation process and be induced with magnetism to have voltage created in them. ● **SEE FIGURE 13–18** for a switched reluctance generator.

There will be more stator poles than rotor poles. A pair of stator poles that are opposite each other are wired in series with a loop of wire. This is called a phase. Pole pairs can be switched on and off quickly with electronic controls. A popular arrangement for a switched reluctance generator is one that has six stator poles (three phases) and four rotor poles.

This type of generator requires sophisticated electronic controls to time the switching of stator poles just right to get the optimum output from the generator.

Switching on a stator pole when it aligns with the rotor puts a magnetic field through the rotor. The mechanical input on the rotor pulls the magnetized poles apart. That action increases the stored energy in the magnetic field. When the electronic switch controlling the phase winding turns off the phase, there is a voltage induced in the phase winding. The outputs of the individual phase combined with the two other phases make the three-phase output of the generator.

LeTourneau machines use switched reluctance generators in combination with switched reluctance drive motors.

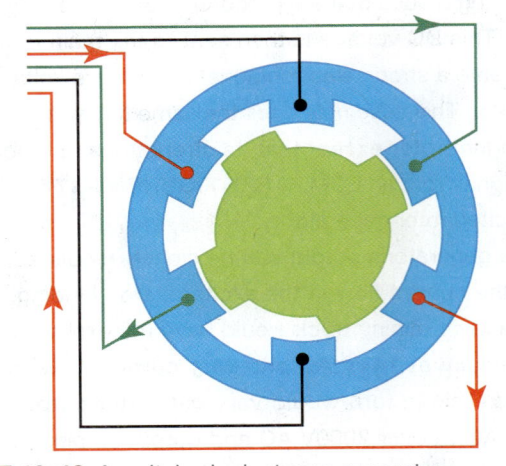

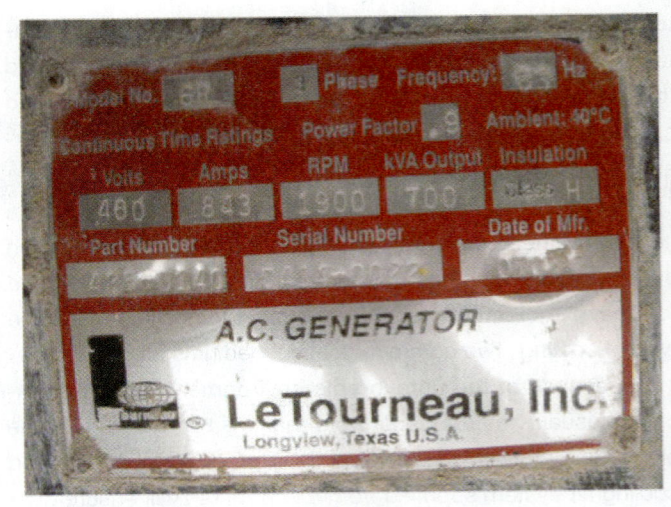

FIGURE 13–19 A generator's rating tag.

GENERATOR RATINGS All generators are rated by their output in kVA at a given rpm. This is a rating given to three-phase generators that represents the product of their voltage and amperage output times a power factor. The power factor relates to the type of load that the generator is feeding power to. ● **SEE FIGURE 13–19** for a generator's rating tag.

To calculate the power output of a generator, multiply its maximum amperage output by its maximum voltage output. This will give the output in watts, which can then be divided by 1000 to get its output in kilowatts. The equivalent horsepower value can be found by dividing kilowatts by 0.746 because 746 watts equals 1 hp.

Progress Check

1. This component would *not* be part of an electric drive system:
   a. generator
   b. torque converter
   c. motor
   d. inverter

2. When working on a high-voltage system, you should use a multimeter that is rated at least:
   a. SAE III
   b. ISO III
   c. CAT III
   d. API III

3. This is *not* one type of generator used on electric drive machines:
   a. switched inductance
   b. permanent magnet
   c. excited rotor
   d. switched reluctance

FIGURE 13–18 A switched reluctance generator.

4. If you were monitoring an AC voltage on an oscilloscope, it would:
   a. be a flat line and negative
   b. be a flat line and positive
   c. be an oscillating line that is positive and negative
   d. be an oscillating line that is positive
5. Three-phase voltage is created by:
   a. having a reversible rotor
   b. having a stator with three sets of windings
   c. having a rotor with three sets of windings
   d. rectifying the single-phase output

CONDUCTORS

All high-voltage conductors will be identified with orange insulation. This should stand out on a machine and equate to potential danger in your mind. It will also help that they will likely be the largest gauge wires on the machine.

They will likely be at least 2/0 gauge or larger. If they need to be larger than 4/0, they will be identified by MCM numbers such as 313 or 777. MCM is an abbreviation for 1000 circular mils. This equates to the cross-sectional diameter in mils. One mil is the equivalent cross-sectional area of a 0.001 in. diameter circle. If a wire is said to be 313 MCM (kcmil), this equates to 313,000 mils.

These high-voltage conductors will be multilayered starting with the current carrying copper core, then rubber insulation, then Mylar tape, then a braided stainless steel grounding shield, and finally an outer orange nylon cover. ● **SEE FIGURE 13–20** for high-voltage conductors.

High-voltage conductors should never be modified and, if damaged, should be replaced as an assembly. They will have heavy-duty connectors at each end that should always be torqued to specification to ensure that they stay tight. If a connector comes loose, it will initially cause a voltage drop then start to arc which could lead to a fire.

Any time a high-voltage connector is about to be assembled, it must be clean and dry. The manufacturer could require a special sealant to be applied after the connection is made.

For the sensors and ECMs that control electric drive systems, the wires are normal gauge and insulation. There could also be fiber optics used for sending digital signals back and forth as part of the CAN.

ELECTRIC DRIVE COOLING SYSTEM

Most large mining trucks and wheel loaders will have a cooling system that provides pressurized and filtered air for the major parts of the electric drive system. This is a critical system to keeping the generator, motors, and inverters cool and clean. There could be a central fan or separate fans providing air flow for this function.

This system starts with one or more electrically or hydraulically driven fans that draw air through a filter and push it through ducting to the major components of the system. The pressurized air ensures that dust and other contaminants are kept out from the sensitive electrical/electronic components. Other electric drive machines will have oil-cooled generators or motors that use drivetrain oil circulating through the components to transfer heat away.

Some inverter assemblies are air cooled by natural convection while others will have their own closed system loop that has coolant flowing through it. ● **SEE FIGURE 13–21** for a cooling system for an inverter assembly.

INVERTER ASSEMBLY Electric drive machines will need to have a way to manage and control the voltage level and frequency to the drive motors. The most accurate way of doing this is to send the three-phase AC voltage from the generator to an inverter. The inverter will change it into DC voltage, and then switch it back to AC voltage to drive the motors. This process is controlled with electronics. ● **SEE FIGURE 13–22** for an inverter assembly.

The three-phase AC voltage that is created by the generator could be anywhere from 480V AC to over 2000V AC depending on the machine with larger machines generating higher voltages. This voltage is then rectified to DC voltage. Again this will vary with machine size and range from 650V DC to over 2000V DC and is maintained at a bus bar at this level (over 1000 amps).

FIGURE 13–20 High-voltage conductors.

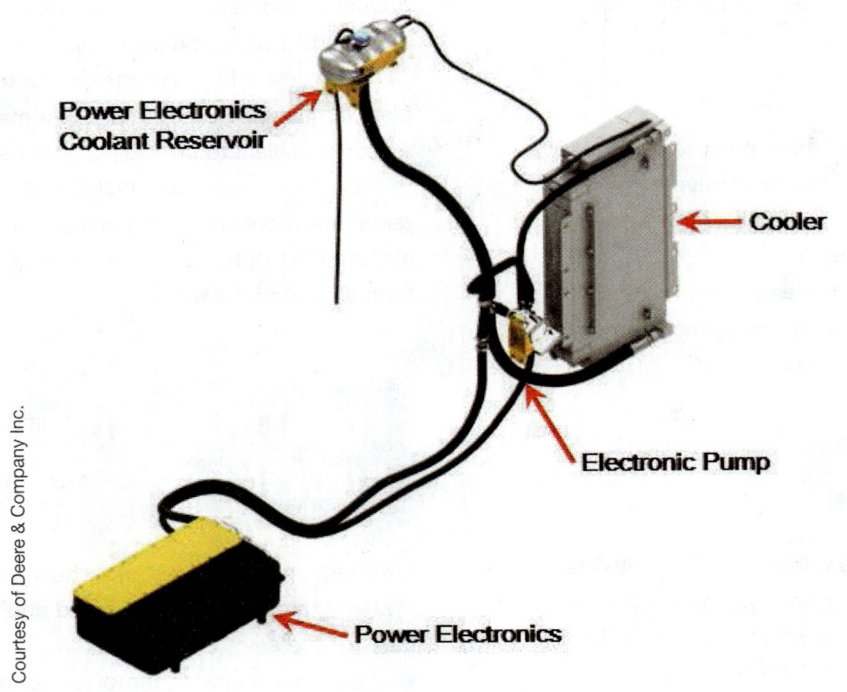

FIGURE 13–21 An inverter cooling system.

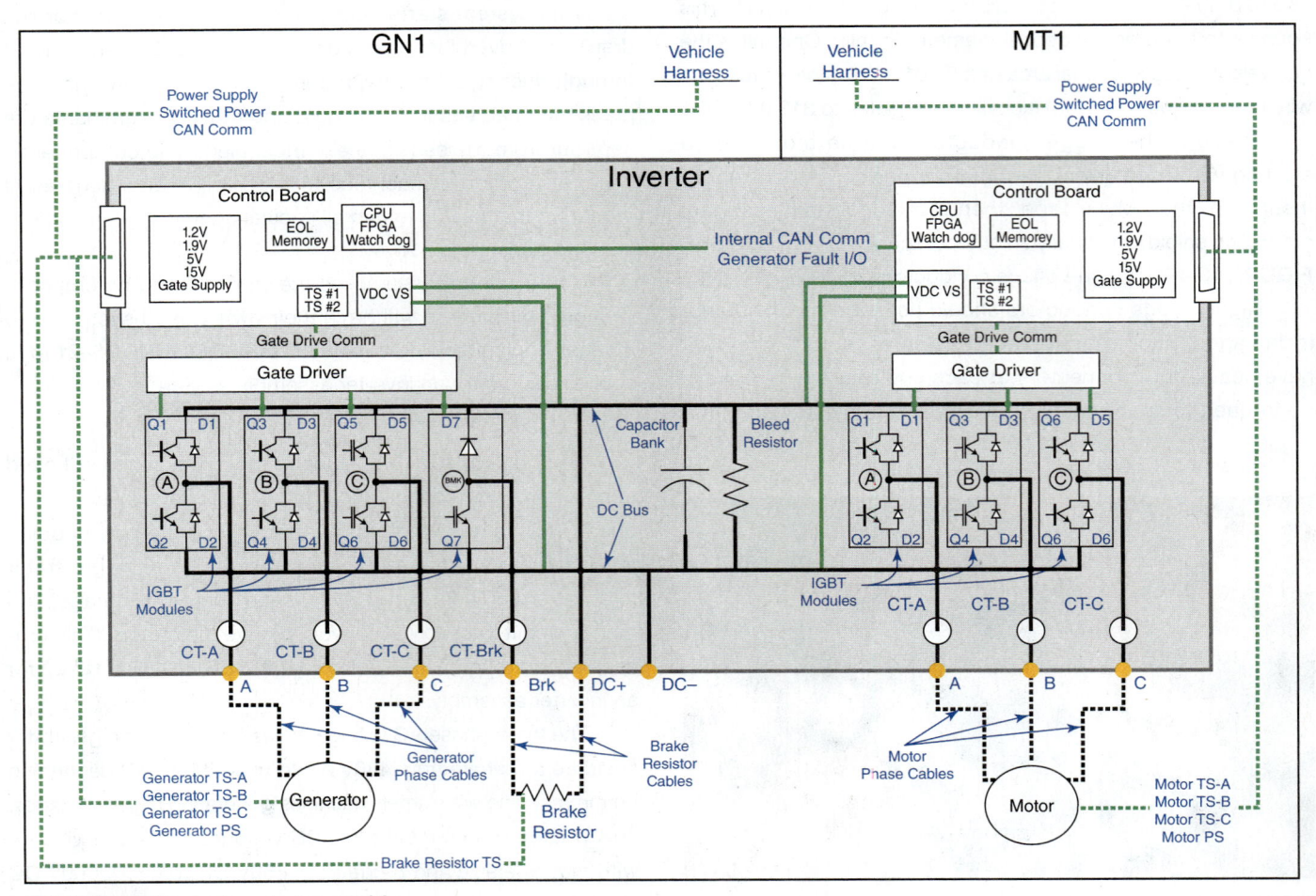

FIGURE 13–22 An inverter assembly.

402 CHAPTER 13

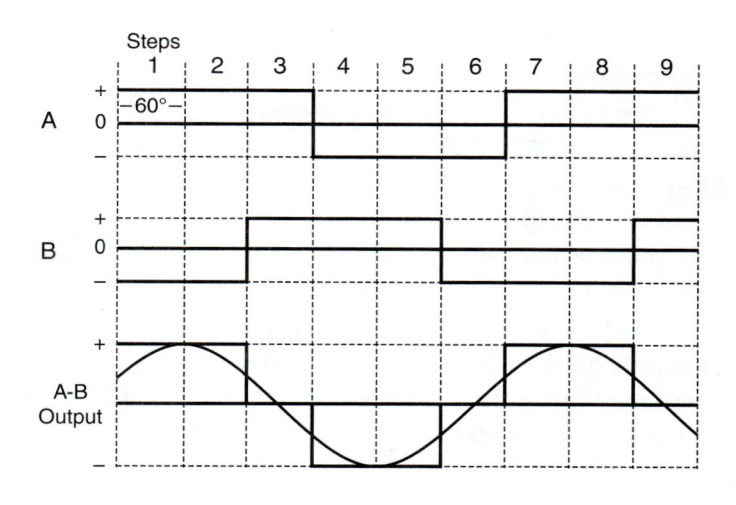

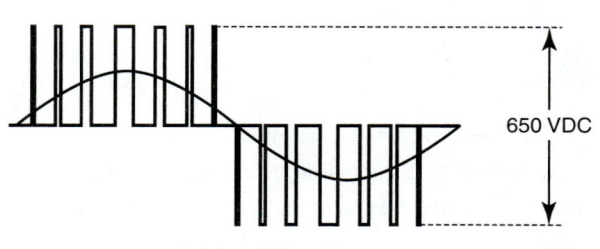

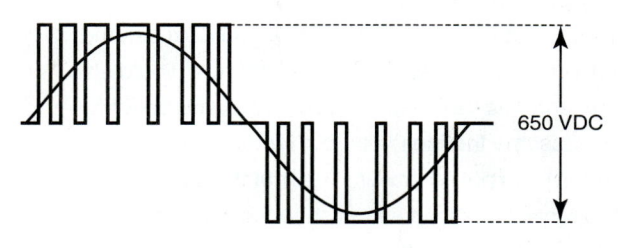

signal from the ECM and close a connection to allow high current flow.

These switching devices are usually controlled with a 15V DC signal that is sent from an ECM. By switching this DC high voltage on and off at high speed, a series of PWM signals that vary in duty cycle can simulate an AC voltage sine wave through a transistor. ● **SEE FIGURE 13–23** for how a PWM signal fed to an IGBT can simulate an AC sine wave.

Generally, a higher frequency AC voltage will result in a "cleaner" sine wave and will also produce a higher motor speed. When voltage and amperage is increased the motors will produce more torque.

ELECTRONIC CONTROLS

As with other machine systems, it's possible and more effective to control the electric drive system with an ECM (Electronic Control Module). The main parts of an electric drive system that can be electronically controlled by an ECM are the generator and the inverter assembly. The electric drive ECM will be part of the machine's multiplex network, meaning it can send and receive information with other control modules on the machine (see Chapter 9).

There needs to be speed, temperature, and position sensors for inputs to the electric drive control module. Operator inputs will include key on, parking brake off, FNR (forward neutral reverse) selector, speed selector, and brake input among others. ● **SEE FIGURE 13–24** for some of the operator inputs for the Cat D7E dozer.

If a speed, position, and direction sensor is used for the generator rotor and motor rotor, it is called a resolver. ● **SEE FIGURE 13–25** for a resolver. As the rotor is directly driven by the engine crankshaft, the ECM could use the engine speed

FIGURE 13–23 How a PWM signal can simulate an AC sine wave.

A DC bus bar is two heavy copper conductors that provide DC voltage to other components. One conductor is positive and the other is negative. The rectification from AC to DC is accomplished with diodes or power transistors.

This DC bus voltage is then manipulated with electronic devices such as thyristors, bipolar transistors, MOSFETs (another type of transistor), and **IGBT**s (insulated gate bipolar transistors) to create an AC three-phase output. Because these transistors are electronic devices, they can be switched at very high speeds. For example, an IGBT can be turned on and off in less than 500 nanoseconds (one nanosecond is one billionth of a second). It will receive a voltage

FIGURE 13–24 D7E tiller.

ELECTRIC DRIVE SYSTEMS 403

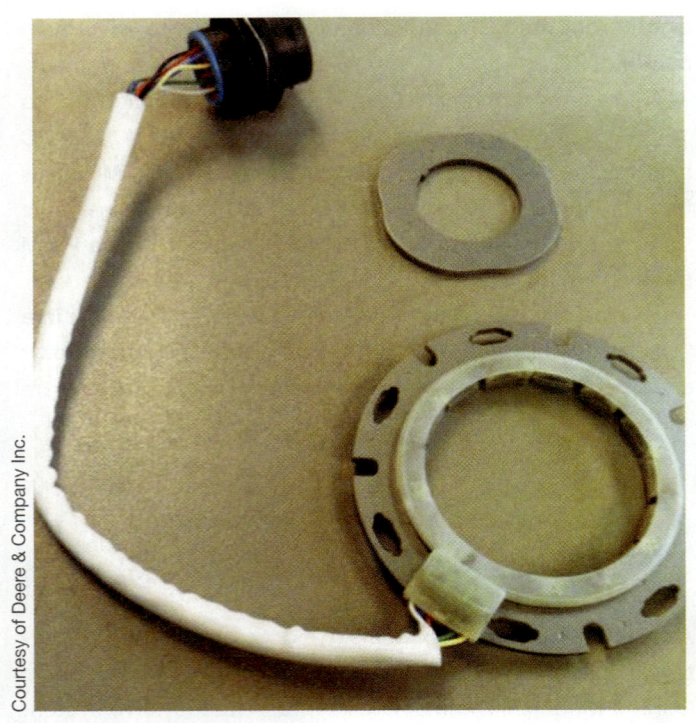

FIGURE 13-25 A resolver.

sensor for monitoring rotor speed. Speed sensors can also be part of a traction control system. If a spinning wheel is detected, the ECM will decrease the output to that wheel.

There can be temperature monitoring for any of the electric drive components. If temperature exceeds acceptable limits, a fault code and operator warning is actuated, and if the high temperature persists, a system derate or limp home process will be started.

For wheeled electric drive machine's position, sensors will be needed to inform the ECM of steering angles from either the articulation joint or the front wheels. This signal is then used to vary the speed of left and right motors to help with turning and to reduce tire wear.

Once the ECM receives these inputs and processes the information, it will send signals to the transistors in the inverter to control their operation. It easy for the ECM to control the transistors with great precision because they are controlled with low-voltage signals that are typically 15V DC. The 15V DC signal is the main output of the ECM.

When a request to change the direction of machine travel is received, the motor ECM will electronically switch two of the three-phase signals. This signal switch will cause the drive motors to reverse the direction of rotation, which will reverse the direction of travel. The ECM will slowly increase or decrease the output power to the drive motors to provide a smooth change of direction to reduce shock loads and increase operator comfort.

The internal electronic driver of the power transistor that receives the ECM pulse signal is expected to respond with an inverted feedback signal pulse. The pulse mirrors the driver signal. The feedback signal indicates to the ECM that the driver circuit and the transistor internal driver circuit are functioning correctly.

Another output could be actuating the backup alarm and controlling the cooling system fan drive. The ECM will also send out fault codes or actuate warning systems if there is a problem with the system.

TECH TIP

Imagine you had two small rectangular-shaped permanent magnets and they were sitting in the bottom of a plastic pipe that had been cut in half and that was several feet long. If you placed the two magnets in the slot with like poles facing each other and slowly pushed one toward the other, there would be a constant distance between them. If you then pushed the first magnet faster, the second magnet would also move faster as the distance stays the same between them (assume friction isn't a big factor here).

Now if you put an eraser in front of the second magnet to represent a load and pushed the first one, the distance between the two magnets would decrease because of the extra load. There would still be a fixed distance between them but would just be shorter.

Now if you doubled the size of the first magnet, the distance between the two magnets would increase even with the extra load on it.

This example is similar to the principle of what happens in a motor. If you move the pushing magnet faster, the second magnet moves faster. This changes the output speed of the motor.

If the load on the second magnet increases, then by increasing the strength of the first magnet the load can be overcome and still moved. This changes the torque of the motor.

These same principles apply to the magnetic forces that keep an electric drive motor turning.

FIGURE 13–26 A wheel motor.

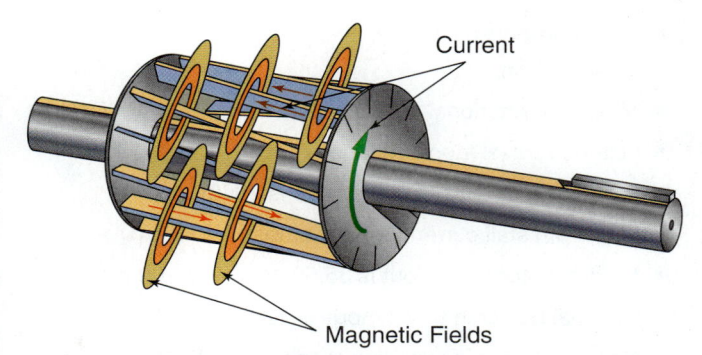

FIGURE 13–27 A squirrel cage motor.

ELECTRIC MOTORS

Electric drive systems need motors to perform work, and they do this by transforming electrical energy into mechanical rotation. They will create the torque needed to propel the machine or perform other work such as the swing function in the Komatsu excavator. Older electric drive systems used DC motors that were more difficult to control and required more maintenance. Almost all high-voltage electric motors used on machines today are AC motors.

All loads moved by electric motors are really moved by magnetism. The purpose of every component in a motor is to help harness, control, and use magnetic force to create torque. When understanding an AC drive system, it helps to remember you are actually creating and using magnetic fields to move a load. To move a load fast requires the magnetic fields to be moved faster. To move a heavier load or acceleration faster, stronger magnetic fields (more torque) are needed. This is the basis for all AC motor applications. ● **SEE FIGURE 13–26** for an electric motor used for driving a wheel loader wheel.

All the previous discussion has focused on the motor as a power consumer that creates torque. Most electric drive machines also use their motors to perform braking. In this role, they turn into generators, and by producing power, their output shaft is now an input shaft that is driven by the moving machine. The output of an electric "motor" can be used to charge a battery or capacitor, can be sent to the generator that now becomes a motor to help drive loads like hydraulic pumps, or could just go to a large resistor where the power gets converted into heat.

Most AC motors used for electric propulsion are either of three main types. Induction-type (squirrel cage) motors and synchronous (permanent magnet) motors are most common while LeTourneau uses switched reluctance–type motors.

INDUCTION MOTORS Induction motors use a stationary stator that creates rotating magnetic fields when it is energized with three-phase AC voltage. The interaction of the three-phase voltage with the three sets of stator windings will naturally produce a constantly rotating set of alternating magnetic poles. The speed and strength of these magnetic poles will change with the changing output of the machine's inverter. As the frequency of the AC voltage going to the motor increases, the speed of the rotating field increases, and therefore, the speed of the rotor increases. As the level of voltage increases, the strength of the magnetic field increases, and therefore, the torque of the motor increases.

The rotor's magnetic fields will get created by the magnetic fields of the stator and the rotor shaft is the output of the motor. The short explanation of the induction motor is that the rotating magnetic field produced by the stator pushes the rotor around ahead of it.

One type of induction motor uses a rotor that is termed a *squirrel cage*. Its laminated frame is machined to allow aluminum conductor bars to be assembled around it to resemble a squirrel cage. These conductor bars will have current induced in them from the AC voltage that is flowing through the stator windings. The induced current in the rotor's individual conductor bars will then create separate magnetic fields around them. These fields will interact with the stator's magnetic fields and will constantly be trying to work toward a balance of magnetic forces without success. This isn't possible because of the changing stator magnetic fields. This constant offset of magnetic fields will keep the rotor turning. ● **SEE FIGURE 13–27** for a cutaway view of a squirrel cage motor.

An easy comparison to think of here is a dog that is constantly chasing its tail. As hard as it tries and no matter what speed it chases its tail, it will never quite catch it.

An example of the specification of an AC induction motor that is used for propelling a 240-ton mining truck is the following:

- The three-phase windings are connected in a wye configuration.
- Maximum rotational speed is 3180 rpm.
- Full-load travel mode voltage is 1960V AC.
- Full-load retarding mode voltage is 2060V AC.
- Maximum stall current is 1300 amps.
- Maximum torque output is 35,523 N-m (26,200 ft-lb).
- Nominal power in travel mode is 1206 kW.
- Nominal power in retarding mode is 2430 kW.
- Weight (each) is 4100 kg (9039 lb).

This type of motor will have slip between the rotor and stator magnetic fields. Slip is the difference in speed between the rotor and the rotating magnetic fields of the stator. An increase in load causes the rotor to slow down, which creates a higher slip and more torque, just what is needed to overcome the higher load. A smaller load means a lower slip value. Slip is necessary in this type of motor to induce current in the rotor windings.

If the operator needs to change the direction of machine travel, a signal is received from the FNR switch and sent to the drivetrain ECM. The drivetrain ECM will send the signal to the motor control ECM, which will then electronically switch two of the three-phase outputs for each traction motor. This phase switch will result in the drive motor reversing the direction of rotation.

PERMANENT MAGNET MOTORS A permanent magnet motor is the simplest design and is used for lower kilowatt applications. Its rotor is made up of a series of magnets that are spaced equally around the rotor shaft. When the stator has three-phase voltage sent to its three windings, there is a rotating set of magnetic fields set up around the perimeter of the rotor. These fields interact with the permanent magnets on the rotor, and because of like poles repelling, the rotor is forced to turn. The rotor shaft is the output of this motor and will then go on to a gear reduction. ● SEE FIGURE 13–28 for a permanent magnet motor.

There is no slip with a permanent magnet because the rotor speed matches the speed of the rotating magnetic fields of the stator. This allows them to be also called synchronous motors.

SWITCHED RELUCTANCE MOTORS In construction, the switched reluctance motor (SRM) is the simplest of all electrical machines. Only the stator has windings.

The rotor contains no conductors or permanent magnets. It consists of steel laminations stacked onto a shaft that are shaped into poles (sometimes called salient poles).

The two previous types of motors (induction and synchronous) relied on two opposing magnetic fields in the stator and rotor to create shaft rotation. The switched reluctance

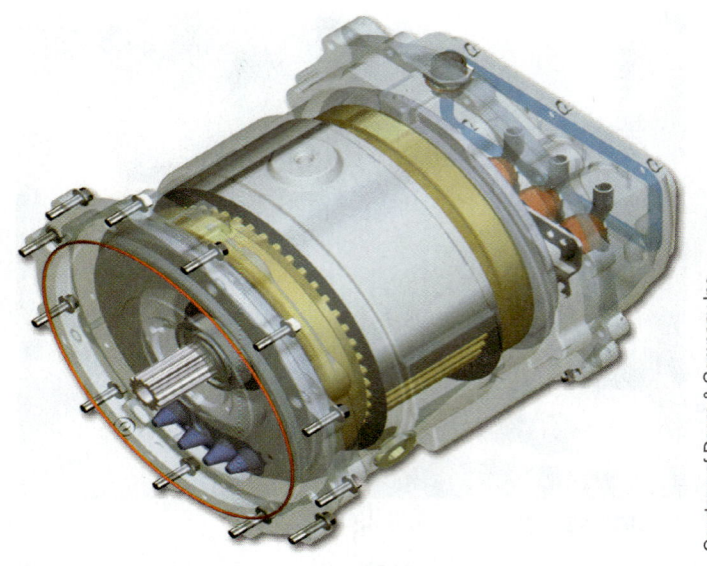

FIGURE 13–28 A permanent magnet motor.

creates shaft rotation as a result of the variable reluctance in the air gap between the rotor and the stator. When a stator winding is energized, producing a single magnetic field, reluctance torque is produced by the tendency of the rotor to move to its minimum reluctance position. Basically, the magnetic field occurring between two opposite coils of the stator wants to align with a set of the rotors poles because that will provide the least reluctance. This is a similar action to the force that attracts iron or steel to permanent magnets. In those cases, reluctance is minimized when the magnet and metal come into physical contact. ● **SEE FIGURE 13–29** for an SRM rotor and stator.

The direction of torque generated is a function of the rotor position with respect to the energized phase and is independent of the direction of current flow through the phase winding. Continuous torque can be produced by intelligently synchronizing each phase's excitation with the rotor position. There will be more stator windings than rotor poles in an SRM.

An SRM achieves rotation by the sequential energizing of stator poles. When the stator pole winding is energized, the nearest rotor pole is attracted into alignment with that stator pole. The rotor will follow this sequence, attempting to align rotor poles with energized stator poles. However, as the rotor and stator poles align, the stator poles switch off and the next group of stator poles switches on, continuing the rotation of the rotor.

The SRM generates continuous movement by consecutively switching the currents on and off, thus ensuring that the poles on the rotor are continually chasing the stator current. The movement achieved is a function of the current flowing through the winding and the characteristics of the iron in the rotor.

SRMs must have their rotor speeds precisely monitored to give feedback to the ECM so it can switch the poles at precise

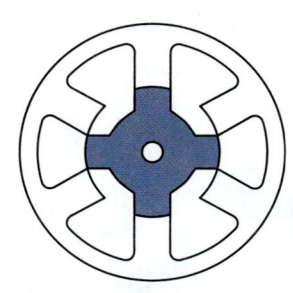

**3-phase,
6 rotor poles/4 stator poles**

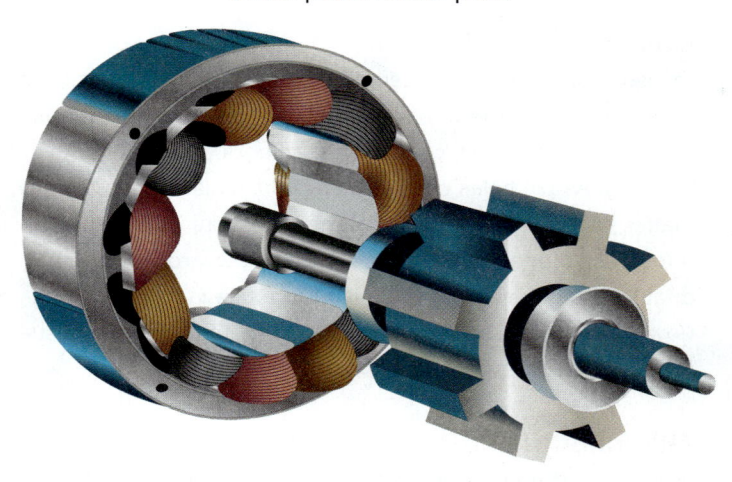

FIGURE 13–29 An SRM motor and stator.

times so just like switched reluctance generators, these motors need sophisticated electronics to precisely control the three-phase power to them.

The simplicity of SRMs is one of the main advantages of using them. The downside is the sophisticated electronics and software that must accompany these motors that stops them from being more popular.

HIGH-VOLTAGE SYSTEM GROUNDING AND BONDING

All electric drivetrain systems operate at high AC and DC voltage levels. The systems are designed to keep these voltages isolated from the machine frames and isolated in protected areas where any accidental human contact cannot occur.

The high-voltage system electrical components are electrically isolated from the machine frame ground. Each of the system component enclosures and housings is grounded to the machine frame ground by a ground strap. The ground straps must remain connected at all times during the operation of the machine.

The electrical system components that are inside of the component housings are electrically isolated from the housing. A ground fault is a "leakage" of electrical current or a short circuit to the machine frame. If a ground fault occurs, the isolation of the system components from frame ground ensures that there is not a path for current flow back to the voltage source.

Most machines will incorporate a ground fault detection system. These systems are in place to monitor the AC and DC high-voltage systems. Very small amounts of current leakage to frame ground can be detected and is normal in a high-voltage power system.

If an amount of current leakage to the frame ground reference is detected that could damage system components, a system control module will activate a ground fault warning. The warning will be activated at a level that will indicate the severity of the problem to the operator and to the technician.

In addition to the component housing mount being grounded to the machine frame, the component housings are connected to each other and to the machine frame by a "bonding" ground wiring system.

The following main components are typically connected to the bonding ground wiring system:

- The generator
- The power inverter
- The electric drive propulsion module
- The accessory power converter

A bonding ground wire is mechanically connected to the housing of each of the components. Some of the bonding ground wires are included as a third or fourth conductor in the DC conductor assemblies. The bond wire is connected internally to the component housing when the DC connector is connected at the component.

If a bonding ground wire or a ground strap must be temporarily disconnected to service a component or a conductor, the bond wire or ground strap must be reconnected before the machine is operated.

BRAKING RESISTOR

Electric drive machines have the benefit of using their drive system to create a braking force. They can turn their motors into generators. The way this happens will differ depending on the type of drive motor used. Once again this is controlled by the ECM based on what the operator is requesting and the stored information in the ECM.

This newly created voltage being generated by the drive motors needs to be dispersed and is sent to a braking resistor.

The most common type is a simple resistor element that is similar to a diesel engine intake heater but only much larger. Think

ELECTRIC DRIVE SYSTEMS **407**

FIGURE 13–30 A resistive-type brake resistor.

of this as a large toaster grid. This retarder grid will usually have a fan mounted to it so that when it heats up, the fan can help dissipate the heat. Sometimes the engine fan is used or there can be a dedicated retarder grid fan that will have its own motor to drive it. This motor will also be controlled by the ECM.

The second type of braking resistor is the one that uses engine coolant as a medium to absorb braking heat when needed. ● **SEE FIGURE 13–30** for a resistive-type brake resistor.

MAINTENANCE

A thorough visual inspection of the electric drive system on a regular basis is very important. You should be looking for any missing or damaged covers or damaged insulation on the high-voltage conductors.

One of the benefits of electric drive components is their low maintenance requirements. Motors and generators really only have one moving part, and as long as the bearings are lubricated and are not overheated, they should last a long time.

As mentioned before, heat and dust are the biggest enemies of electric drive components. The cooling system must be checked for proper operation to ensure that it is providing clean pressurized air to the components. There will likely be filters that need to be changed regularly to keep the cooling system working well.

INSULATION TESTING AND MAINTENANCE

Electrical device and cable failures sometimes can be traced to insulation failure. New electrical insulation deteriorates over time because of the effects of mechanical vibration, excessive heat or cold, dirt, oil, corrosive vapors, and atmospheric humidity. As pin holes or cracks develop, moisture and foreign matter penetrate the surfaces of the insulation, which provides a low resistance path for current to leak to ground, causing a fault. You generally cannot visually notice insulation breakdown, but there is a test that can better determine when insulation is starting to fail.

THE MEGOHMMETER TEST The job of insulation is to keep current flowing along its path in the conductor. Ohm's law will help you better understand how to quantify and measure insulation's value. The law states that the voltage in an electrical circuit must be equal to the current times the resistance, or $V = I \times R$. Resistance represents the insulation value of a device or cable as it is tested with a Megohmmeter, more commonly known as a Megger. A Megger test applies a high voltage to a conductor and its insulation and measures the amount of current that "leaks" through the insulation of the device or cable. Good insulation is insulation that has a high resistance to current flowing through it.

The Megohmmeter insulation tester is a small, portable instrument that directly reads insulation resistance in ohms or mega-ohms. Good insulation readings are normally in the megohm range. This testing method is nondestructive to the insulation. The Megohmmeter develops a high DC voltage that causes a small current to flow through and over the surfaces of the insulation being tested. The Megohmmeter reads and records the current. If Megger testing is done on a regular basis and recorded then components can be replaced before failure and before unsafe limits are reached.

● **SEE FIGURE 13–31** for a Megohmmeter. Megohmmeter tests should be conducted under the same conditions each time a test is performed to provide accurate trending over the life of the cable or device. Temperature, humidity, and other atmospheric conditions can have a huge effect on the results of the Megohmmeter test outcome. The

FIGURE 13–31 A Megohmmeter.

Megohmmeter test also requires the power to be removed from the cable or device during the testing period, so proper scheduling is an important part of the test procedure. The test should be conducted annually unless the cable or device is exposed to the atmosphere, and then testing should be quarterly.

When a cable or device fails a Megger test, it will need to be cleaned, repaired, or replaced.

OPERATOR COMPLAINTS

Operator complaints that could be related to electric drive systems are the following:

- lack of power
- no forward or reverse
- vibrations
- noises
- brakes don't work properly or don't release
- overheating
- uneven steering
- unusual smell

ELECTRIC DRIVE DIAGNOSTICS AND REPAIRS

Manufacturer, employer, and government regulations and policies may prevent you from working on high-voltage systems. Attending a manufacturer-backed and dealer-provided service training course is highly recommended before attempting any repairs on a high-voltage system. This would be a mimimum requirement to make a HDET a "competent" worker that is qualified to service and repair high voltage electric machine systems.

Troubleshooting of electric drive systems should always start with simple checks. This includes visual inspection and checking for fault codes. This will *never* include removing protective covers on a running unit.

You must make yourself familiar with how the system operates, how to check for high voltage, the proper PPE to use when doing this, and the proper test equipment to use. The correct electrical schematic should be available as a reference as well. *Never* proceed with working on a high-voltage electrical system until you are completely familiar with all related safety concerns.

If fault codes are found, there will be a step-by-step procedure to find the root cause of the problem.

A common procedure is to re-flash ECM software if updates are available.

If the power transistors are suspected to be faulty, there are special testing tools to test their operation.

Once an electric drive major component (generator, inverter, motor) is found to be defective, it will likely be replaced as an assembly. This could be a brand-new or exchange unit from a dealer or a reconditioned unit from a third-party vendor.

It is highly unlikely that an HDET will disassemble the main components of an electric drive system unless their employer is set up with specialized tooling and provides the proper training to allow this.

Progress Check

6. An induction motor has its conductor bars turned into electromagnets by:
    a. commutators and brushes
    b. stator phases being energized
    c. a variable reluctance
    d. permanent magnets
7. This color will identify high-voltage conductors:
    a. white
    b. black
    c. red
    d. orange
8. A switched reluctance motor has:
    a. a series of permanent magnets
    b. a series of winding
    c. a series of salient poles
    d. a set of commutator bars

9. You would find an IGBT in this electric drive component:
   a. the motor
   b. the inverter assembly
   c. the generator
   d. the cooling system
10. A Megohmmeter is most commonly used to test:
    a. the generator stator windings
    b. the motor rotor windings
    c. the retarder grid
    d. the insulation on high-voltage conductors

> **SHOP ACTIVITY**
>
> - Go to a vehicle or machine that has a high-voltage electric drive system and record all warnings related to servicing or repairing the system.
> - Refer to the service manual and record the specifications for any recommended PPE.
> - Take note of the orange-colored conductors and identify the major components of the system.

SUMMARY

1. Some safety concerns related to electric drive systems are: lock out and tag out procedures must be followed to prevent high-voltage shock injuries or death; do not expose high-voltage components when machine is operating; heavy lifting injuries, proper PPE usage is critical; CAT III multimeters must be used.
2. Orange wires that are part of the electric drive system indicate they are high-voltage conductors.
3. Five steps to follow when working on an electric drive system are: de-energize, secure, verify, proceed, and restore.
4. There are many types of electric drive machines such as: haul trucks, wheel loaders, dozers, and swing drives for excavators.
5. Electric drive machines allow the prime mover (diesel engine) to run at a steady speed, which translates into greater efficiency.
6. A typical electric drive arrangement is: diesel engine drives a generator; generator output of AC current is rectified to high voltage DC; the operator controls a variable AC current to AC motors that drive final drives.
7. Electric motors have high torque characteristics.
8. The AC generator output is three phase for a smoother total current flow.
9. An inverter will change AC output to DC then back to AC.
10. AC generators have two main parts: the rotating rotor and the stationary stator.
11. Three main types of generators are: permanent magnet, excited rotor, and switched reluctance.
12. Permanent magnet generators use magnets on the rotor to induce voltage into the stator when it is rotated inside the stator.
13. Excited rotor generators have a rotor with coils of wire wrapped around iron cores. When the coils are energized the generator produces current. A rotating exciter coil will send current to the main rotor coil.
14. A switched reluctance generator has a laminated iron rotor with a series of projections called salient poles. Its stator is a series of pairs of coiled wire arranged at equal spacing around the outside of the generator housing. Sophisticated electronics are used to switch on stator windings at the right moment, which creates a magnetic field in the rotor.
15. Generators are rated in kVA units.
16. High-voltage conductors are sized in gauge sizes and MCM and high-voltage connectors must be clean, dry, and secure.
17. Larger electric drive machines will have a cooling system for components. This could be a fan ducting moving air past components, or liquid cooling system that circulates oil, or coolant through components.
18. Inverter assemblies will change generator output from AC current to DC current. Then heavy duty electronic components will change the DC back to three phase AC current. AC voltage and frequency can then be changed to vary the torque output of the motor.
19. Electronic controls allow faster and more precise management of electric drive systems. ECMs can control transistors that will vary the motor output to satisfy operator needs and overcome loads. Machine travel speed and direction can be controlled easily with electronic controls.
20. Electronic control systems need sensor inputs such as speed, temperature, pressure, position, voltage, and current. An ECM will process this data and send signals to output devices such as transistors to control the electric drive system.
21. Electric motors change electric energy into mechanical energy. Electric motors have a rotor that rotates inside its housing. The stator is a series of wire loops arranged around the inside of the housing. Electric voltage and current sent to the stator is used to create magnetic fields that rotate the motors shaft. The shaft will then drive power train components to make the machine travel.
22. An increase in current flow will increase the torque output of the motor. An increase in frequency will increase the speed of the motor.
23. Electric motors can also be used to perform braking since they can act like generators when needed.

24. Three types of electric motors are: induction, synchronous, and switch reluctance.
25. Three phase current sent to an induction motor will create a rotating set of magnetic fields in their stator, which interacts with the rotor to keep it rotating.
26. High-voltage systems need to be properly grounded and bonded to prevent stray voltage and injuries related to it.
27. Braking resistors will dissipate heat from motors when they are used for braking. They resemble large toaster grids, and fans will work to cool them down.
28. Electric drive maintenance consists of: complete thorough visual inspections of all system components and conductors, generator and motor bearing lubrication, and cooling system inspection.
29. Conductor insulation integrity is part of maintenance and consists of using a Megger tester. High voltage is applied to the conductor insulation and excessive leakage is displayed on the test unit.
30. Operator complaints will lead to a diagnostic procedure to resolve the problem. Technicians must be familiar with all safety-related procedures related to diagnosing electric drive systems and components. Strict adherence to service information and successful completion of training courses related to diagnosing electric drive systems will ensure no injuries occur.
31. Most major electric drive components (generators, motors, inverters) will be replaced if found to be defective.

chapter 14
DRIVELINES, TRANSFER CASES, AUXILIARY DRIVES

LEARNING OBJECTIVES

After reading this chapter, the student should be able to:

1. Describe what a typical machine's driveline is composed of.
2. Explain how a transfer case works.
3. Describe what a drive shaft operating angle is.
4. List the different types of yoke styles.
5. Explain what critical speed means.
6. Describe the different features of a transfer case.
7. Explain how to perform driveline inspection.
8. Describe how to maintain common driveline components.
9. List some common driveline repair procedures.

KEY TERMS

Auxiliary drive 413
Cardan 420
Operating angle 415
Phasing 419
Spline 417
Transfer case 413
Trunnion 420
Universal joint 415
Yoke 418

INTRODUCTION

SAFETY FIRST Drivelines that are operational will be rotating with a great amount of torque and sometimes high speed.

Great care must be taken to make sure that you don't get close to an operating driveline (drive shafts and universal joints). There have been too many technicians and agricultural workers killed or severely injured by getting caught up in a working driveline. *Never* work within reach of a live driveline. It will be very easy for any part of a rotating driveline to snag a part of your clothing, hair, skin, or tools and cause serious injury or death.

Always replace any guarding that has been removed that covers a driveline component.

● **SEE FIGURE 14–1** for the symbol that warns of injury from a rotating shaft.

Some driveline components can be too heavy to lift safely. Proper lifting devices and procedures should be followed. If you are unsure about the weight of a component, consult the machine's service information to confirm its weight.

Before a driveline is to be disassembled, the machine needs to be parked on a level surface and secured so it won't move. Wheel chocks should be used because the machine parking brake may be part of the driveline and may be ineffective in holding the machine in place.

MACHINE DRIVELINE

When a machine's engine is running and is part of a mechanical drivetrain, it will need to get its torque output transferred to either tracks or wheels to propel the machine. The mechanical devices it uses to do this make up the machine's driveline. Clutches, torque converters, retarders, transmissions, **transfer cases**, axles, power dividers, and final drives are the major parts of the driveline. ● **SEE FIGURE 14–2** for the components that make up a wheel loader driveline.

If these major parts are separate from each other, there needs to be a way to transfer torque continuously between them when the engine is running and the operator wants to move the machine. There are driveline components such as drive shafts and universal joints to do this.

The driveline can also assist with slowing the machine down. This could be as simple as when the governor control is let off while the machine is in motion and the engine becomes a brake that absorbs energy through the driveline. There could be engine brakes that increase the braking that an engine can create or there could be a hydraulic retarder (see Chapter 10) that is part of the driveline that can absorb much more torque than what an engine can create. This driveline braking will put a great deal of stress on all driveline components and will highlight the need for proper inspection, maintenance, and repair procedures.

There could also need to be mechanical drive sent to an auxiliary system such as a hydraulic pump drive. This auxiliary driveline will have components that are of the same design as the driveline used to propel the machine.

This job of transferring torque from an engine was originally done with chains and sprockets or sometimes cables and pulleys. ● **SEE FIGURE 14–3** for a chain drive rear wheel.

Chain drives were okay for lighter loads and slower speeds, but when used for an external application, they required a lot of maintenance including adjustments and lubrication.

As material technology progressed and as machines got more powerful, the use of transfer cases, **auxiliary drives**, drive shafts, and universal joints became the standard way to transfer mechanical torque.

Drivelines can transfer torque between components on the same plane and inline, or they can transfer torque off center and to a certain degree of height differential. They can also be required to transfer torque between components that move in relation to each other such as a suspended axle.

FIGURE 14–1 The symbol that warns of injury from a rotating shaft.

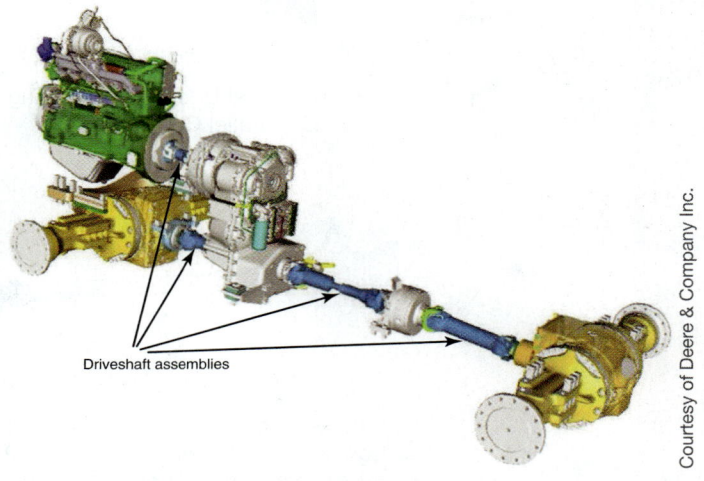

FIGURE 14–2 A wheel loader driveline.

FIGURE 14–3 A chain drive rear wheel.

You can get a good look at an operational driveline if you are in city traffic and get beside a medium-duty delivery truck. You should be able to see a long drive shaft that changes rotational speed in relation to the speed of the truck.

This chapter will cover the drive shafts, universal joints, auxiliary drives, and transfer cases that are used to transfer torque between components in many different machines that use mechanical drivelines.

DRIVELINE ARRANGEMENTS

There are two main arrangements that need to be understood when looking at drivelines. They are called parallel and broken back.
- **SEE FIGURE 14–4** for parallel and broken back drivelines.

TECH TIP

Let's consider the amount of torque created by a diesel engine that's in an articulated truck and that needs to be transferred to its wheels. If the engine creates 2000 ft-lb of torque and a torque converter is attached to the engine's flywheel that could double that torque, there could be 4000 ft-lb of torque going into the transmission. If the first gear ratio of the transmission is 7:1, then 28,000 ft-lb of torque will go to a transfer case that drops the plane of the flow and has a lockable differential. The transfer case will have two outputs to send torque to the front axle and rear axles. These drive shafts then send torque to the axles and will need to handle a maximum of 28,000 ft-lb of torque.

Once the torque is received by the axle, it will once again be multiplied by the differential and final drives. This could be a factor of 20:1 to make the maximum torque applied to the wheel 560,000 ft-lb.

The amount of torque a driveline component can handle is generally related to its heftiness or size of components and wall thickness of tubing. This is called its torsional rigidity.

A. Equal U-joint Working Angles
B. Parallel Centerlines
1. Transmission
2. Driveline
3. Rear Axle

A. Parallel Centerlines
B. Equal U-joint Working Angles
C. Intersecting Centerlines
1. Transmission
2. Driveshaft
3. Forward Rear Axle
4. Driveshaft
5. Rear Rear Axle

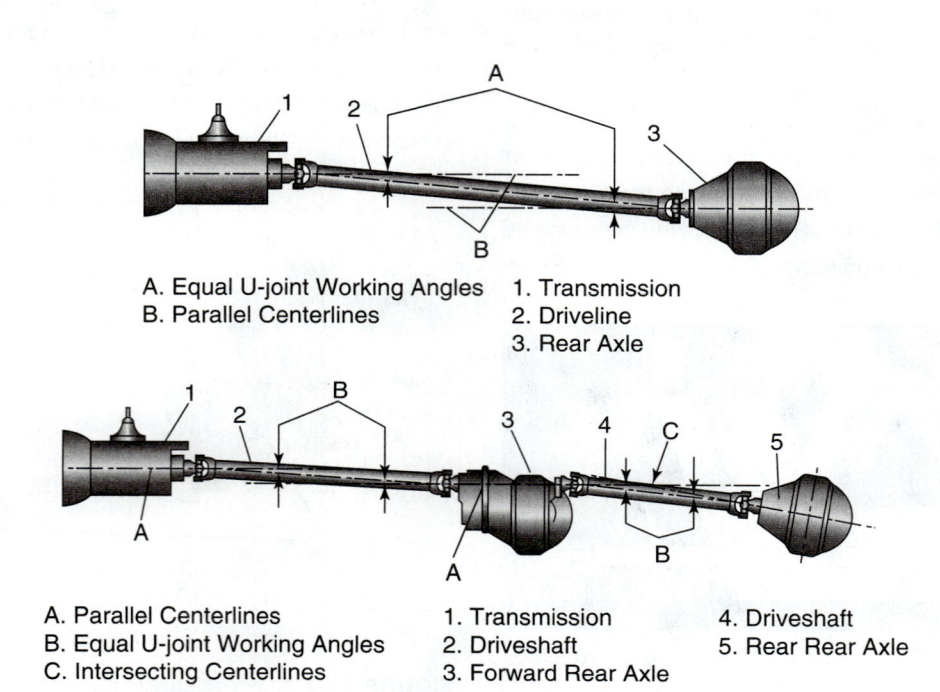

FIGURE 14–4 Parallel (top) and broken back (bottom) drivelines.

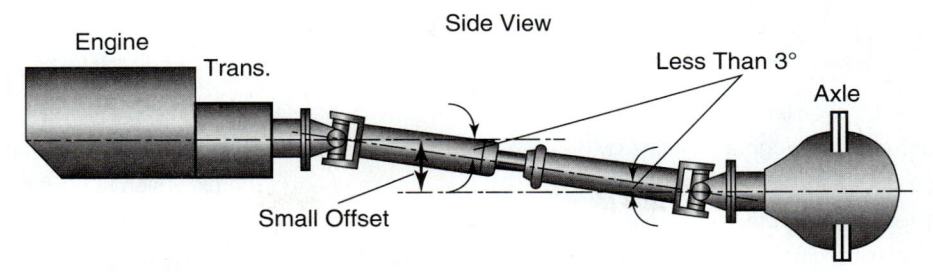

FIGURE 14–5 An operating angle.

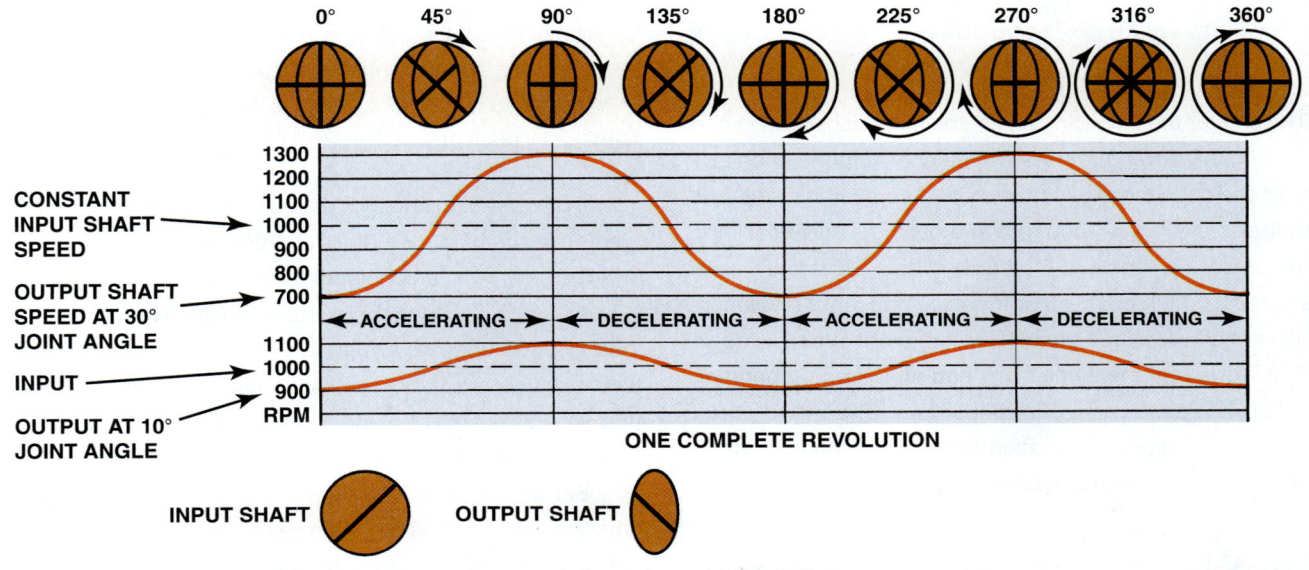

FIGURE 14–6 How a universal joint speeds up and slows down as it rotates.

A parallel driveline has the centerlines or axis of the torque sending or receiving main components parallel to each other but not necessarily on the same plane.

A broken back driveline will have its torque sending or receiving components' centerline not on the same plane and not parallel.

It is more common and advantageous to have two components parallel to each other because this will reduce the **operating angles** of the universal joints connecting the driveline components.

OPERATING ANGLES It is important to understand driveline operating angles because if they are not correct, they can be a source of vibrations and cause early component failure. The operating angle of a machine's driveline should be arranged to give a smooth operation and maximum component life to all related driveline components. However, if repairs require the disassembly of driveline components, then great care needs to be taken to ensure that the original operating angles are maintained.

When major drivetrain components are separated and can't be in perfect alignment, they will need a drive shaft assembly with flexible universal joints at each end to transfer torque between them. The **universal joints** will compensate for any small misalignment between components.

The operating angle of a driveline is measured at the end of a drive shaft that uses a universal joint. It is the maximum angle (compared to a straight line) that is created at the pivot point of a universal joint and corresponds to the intersecting centerlines of the yokes that the universal joint is connected to.

An example is the angle at the universal joint between a drive shaft and an axle pinion shaft yoke. ● **SEE FIGURE 14–5** for an operating angle.

The maximum allowable operating angle for a drive shaft is reduced as running speed increases. Most heavy-duty drive shafts used for heavy equipment will not rotate faster than 2000 rpm. A typical heavy-duty driveline operating angle should be between 1° and 3°.

If a single universal joint angle operates at over 3°, its service life will decrease exponentially as the angle increases. In fact, if the input-to-output angle of a universal joint increases to 6°, its service life will decrease by 50%.

When a universal joint is used on a shaft to transfer torque, there will be a variation in speed between the input and output of the universal joint. This is due to the changing path of travel of the output side of the universal joint. There will be an acceleration and deceleration of the output twice each revolution. ● **SEE FIGURE 14–6** for how a universal joint speeds up and slows down as it rotates.

DRIVELINES, TRANSFER CASES, AUXILIARY DRIVES

> **WEB ACTIVITY**
>
> Search the Internet for an excellent video that illustrates the speed fluctuation through a driveline by using a mockup of a drive shaft that is rotated with an electric motor. There are saw blade–like discs on each shaft and a playing card is held against the saw blade to give an audio indication of speed fluctuations.

> **TECH TIP**
>
> To get an idea of the action of a universal joint, try this. Take a flexible extension (swivel joint) from a ratchet set and attach an extension to each end. Roll one of the extensions across a table with the flex joint driving another extension while hanging the other extension of the side of the table. Guide the output extension in your other hand. You should be able to feel how the speed fluctuation of the output extension changes when the operating angle changes and you should notice how the speed fluctuations increase as the angle increases.

The greater the operating angle, the more accelerating and decelerating the driven member experiences through each revolution of the drive member. An exaggerated example of this is a shaft that is turning at 1000 rpm and driving a universal joint. If the output is at an angle of 8°, there will be a 2% fluctuation in output speed, which means that for every one input revolution, the output shaft will speed up to 1020 rpm twice and slow down to 980 rpm twice. If the same shaft had an extreme operating angle of 20°, a speed differential of just over 12% would occur. This would cause a severe vibration and strain on components.

The following basic rules apply to universal joint operating angles:

1. Universal joint operating angles at each end of a drive shaft should always be at least 1°. This will keep the needle bearings of the universal joints moving slightly so they don't wear into the trunnion.
2. Universal joint operating angles on each end of a drive shaft should always be equal to within 1° of each other.
3. For virtual vibration free performance, universal joint operating angles should not be larger than 3°. If they are, make sure that they do not exceed the maximum recommended angles.

If an operating angle is too great, there will be extra motion created in the universal joint in particular, but the components attached to it will be constantly loading and unloading as well.

As long as the machine operates within its designed payload and nothing has changed as far as the relationship between driveline components such as broken or worn mounts that will increase operating angles, then the operating angle should stay within acceptable limits.

Extra care must be taken when removing and installing driveline components to ensure that the original operating angles are maintained. Some components could have shims or wedges installed under their mounts to provide an adjustment for operating angles.

DRIVE SHAFT ASSEMBLY Drive shaft assemblies are used to transfer torque from an output drive source such as an engine, torque converter, transmission, power divider, transfer case, or power take-off to other components such as transmissions, pump drives, and axles. They are sometimes also called propeller shafts. ● **SEE FIGURE 14–7** for an articulated truck's drive shaft assemblies, including one for a pump drive.

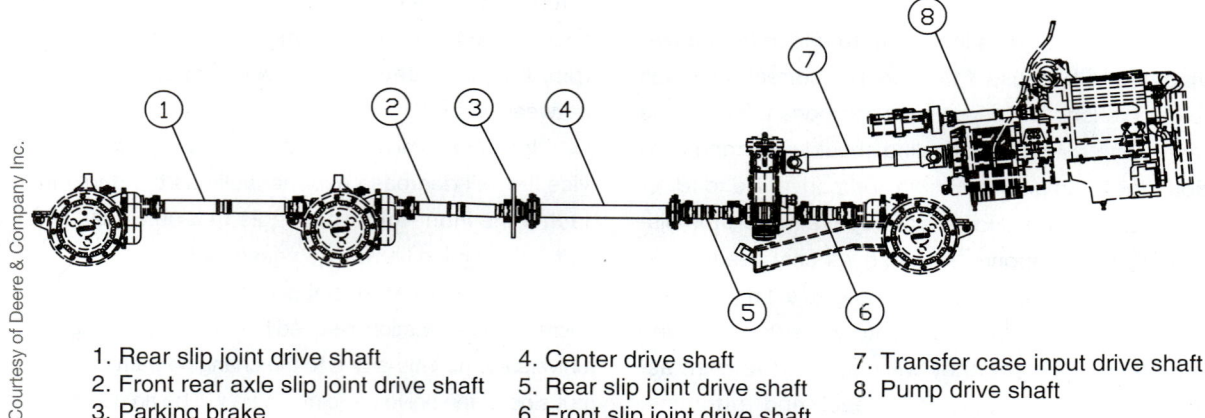

1. Rear slip joint drive shaft
2. Front rear axle slip joint drive shaft
3. Parking brake
4. Center drive shaft
5. Rear slip joint drive shaft
6. Front slip joint drive shaft
7. Transfer case input drive shaft
8. Pump drive shaft

FIGURE 14–7 An articulated truck's drive shaft assemblies, including one for a pump drive.

Drive shaft assemblies used for large machines could weigh several hundred pounds and be required to handle several thousand foot-pounds of torque. They will have a universal joint on each end. Drive shaft assemblies can be one piece for torque transfer between two components that are rigidly mounted to the machine's frame and, therefore, won't move in relationship to each other. A one-piece drive shaft could also be used in combination with a two-piece drive shaft that will have a sliding spline incorporated in it. ● **SEE FIGURE 14–8** for a typical drive shaft assembly.

Sometimes the components that are sending or receiving torque will be allowed to move in relation to the machine's frame because they are mounted with flexible mounts. Many engines and transmissions are mounted to the machine's frame through rubber mounts.

Drive axles are usually allowed to move in relation to the machine's main frame. They could be allowed to move several inches to provide a smoother ride over rough terrain.

If there is relative motion between two components that have torque transferred between them, then the drive shaft needs to be able to change its length. This is accomplished with a two-piece drive shaft assembly with joining male and female **splines** that make up a slip joint. The slip joint allows relative motion lengthwise as the drive shaft assembly rotates.

Some slip joints will have a master spline that will only mate in one position to prevent improper phasing of the two parts of the drive shaft assembly. All other drive shaft assemblies should be marked before disassembly to ensure correct phasing upon reassembly. Phasing of drivelines will be discussed shortly.

One end of the first part of the drive shaft will have a male stub shaft splined section welded to it and the mating end of the second part of the drive shaft will have a female sleeve with a spline section welded to it. The female end will have a lip-type seal assembly that has its lip ride on a smooth-machined

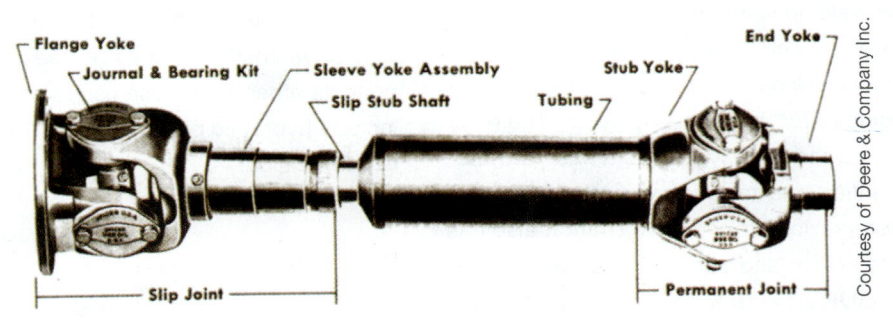

FIGURE 14–8 A drive shaft assembly.

DRIVELINES, TRANSFER CASES, AUXILIARY DRIVES 417

FIGURE 14–9 A slip joint separated.

part of the male section to keep contamination out of the slip joint and grease in. There is usually a nylon type of coating on the male splines to assist with keeping it free to move. The opposite end of the female end of the slip joint will have a light tin plug in it to keep grease inside the tube and dirt out. This is called the Welch plug. ● **SEE FIGURE 14–9** for a slip joint separated.

A drive shaft starts out as a tube that can be either seamless, can be rolled and welded along its length, or can be made from drawn tubing. The tube diameter and wall thickness is determined by the amount of torque that the shaft is required to handle with a larger diameter and thicker wall being stronger. A hollow tube is used because of its lesser mass, which allows it to spin faster, and it allows for longer lengths than a solid shaft. A hollow tube also has a higher torque capacity than a solid shaft.

The tube will have ends called **yokes** that are cast and machined parts that are welded to the tube. The yokes will accept universal joints. The universal joints are necessary to allow for differences in angles between components.

The drive shaft's hollow tube is also designed to twist and flex slightly to allow for shock loads and sudden speed changes.

Once the drive shaft assembly is welded together, it is balanced to eliminate vibrations at speed. This is done by spinning the shaft and welding weights at precise locations to counteract any vibrations caused by an unbalance.

The drive shaft yokes will need to accept the style of universal joint that is designed for use at that location. The most heavy-duty type is called the wing type. It is a cast and machined piece that is drilled and tapped to accept four bolts that will fasten the universal joint to the drive shaft. It also has a machined keyway to locate and secure the universal joint to the yoke. ● **SEE FIGURE 14–10** for a wing-type universal joint.

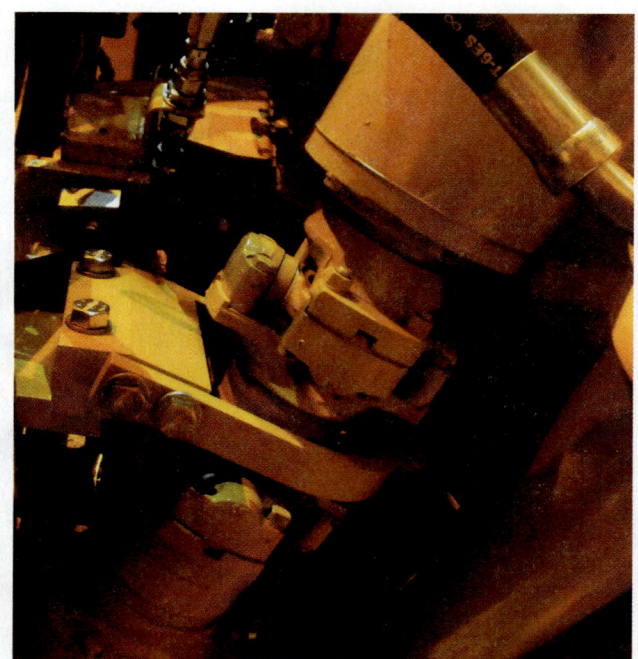

FIGURE 14–10 A wing-type universal joint.

Other types of yokes are full round and half round. Full-round yokes will have the universal joint cap fully surrounded and will have a slight interference fit. Half-round yokes will have an open semicircle seat that will accept a U-bolt or bolt on strap to secure the universal joint caps. ● **SEE FIGURE 14–11** for the full-round and half-round style of drive shaft yokes.

Drive shafts can also be used to transfer torque to components other than those used to propel machines. This could include hydraulic pumps or any other mechanical device that needs rotation.

DRIVELINE PHASING Driveline vibration needs to be minimized to ensure that the longest life can be obtained from all driveline components. Any excessive vibration in a machine

FIGURE 14–11 Full-round and half-round yokes.

driveline can have immediate negative effects or an accumulative long-term effect that will shorten the life of all connected driveline components.

To help cancel the speed fluctuations caused by driveline operating angles, a drive shaft must have its yokes inline. One-piece drive shafts will have their yokes lined up when they are manufactured. This will almost totally cancel out the speed variations between the yokes through each revolution. If the input universal joint drives the shaft at unequal speeds, the output universal joint will be speeding up and slowing down at the opposite time. The total effect is that the driven output yoke will turn at a smooth rpm, which in turn will minimize vibrations.

Drive shaft **phasing** can only be changed if the assembly has a slip joint in it. Care needs to be taken if you are reassembling a slip joint's splines. This makes it important to mark the shaft before disassembly. Splines should always be assembled to give the correct phasing between the opposite yokes. This means that the yokes line up at the opposite end of the shaft. ● SEE FIGURE 14–12 for how a drive shaft should be phased.

CRITICAL SPEED Every drive shaft has a critical speed rating. This is the maximum speed the drive shaft should be run at before it starts to deform due to centrifugal forces. A typical critical speed for a heavy-duty drive shaft is 3000 rpm.

The following factors will decrease a drive shaft's critical speed: increasing length, increasing operating angle, decreasing tube diameter, and imbalance. If a machine is allowed to coast downhill at a higher than normal speed, it could be possible to exceed the critical speed of the drive shaft. For this reason, a machine should never be allowed to coast.

DRIVELINE BRAKES Many machines will use the mechanical driveline as a location to mount secondary or parking brakes. See Chapter 18 for details. These brakes are typically caliper and rotor– or shoe and drum–style brake assemblies. For example, if a machine has a rotor/caliper driveline parking brake, the rotor will be fastened to a drive shaft assembly and the caliper will be mounted stationary to the machine's frame. When the brake pads are squeezed onto the rotor by the caliper, the rotor can't turn and the drive shaft is stopped from rotating.

If the drive shaft is locked stationary, then the machine is stationary. These driveline brakes should only be used for holding or static braking and not for dynamic

A. In Phase

B. Out of Phase

C. Cross-hole centerlines of both yokes must be in alignment

D. Alignment Marks

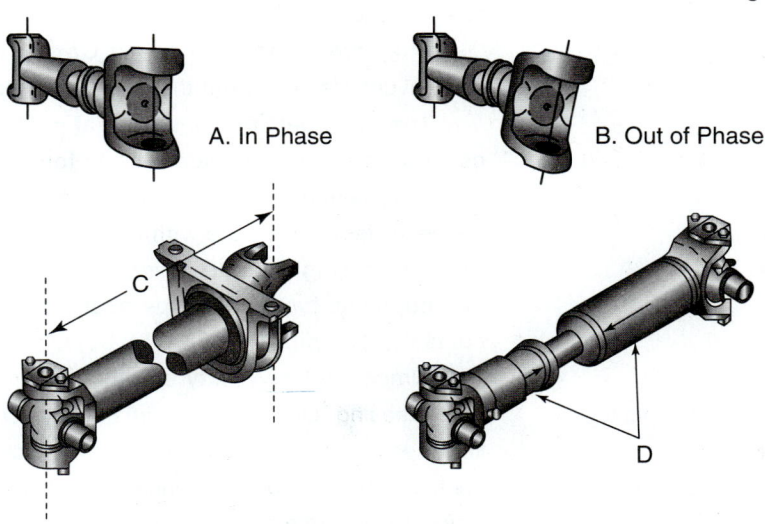

FIGURE 14–12 Drive shaft phasing.

DRIVELINES, TRANSFER CASES, AUXILIARY DRIVES

FIGURE 14–13 A driveline parking brake.

braking. ● **SEE FIGURE 14–13** for a driveline parking brake arrangement.

Progress Check

1. This would *not* be part a machine's driveline:
   a. drive shaft
   b. transmission
   c. fuel tank
   d. torque converter
2. An operating angle should not normally exceed:
   a. 0.5°
   b. 1°
   c. 3°
   d. 6°
3. A drive shaft that uses a slip joint will use this to provide a change of length:
   a. splines
   b. keyway
   c. elastic joint
   d. universal joint
4. A drive shaft assembly that has been properly phased will have its yokes:
   a. inline
   b. 45° apart
   c. 90° apart
   d. at least 3° apart
5. Drive shaft critical speed refers to:
   a. the minimum speed a drive shaft needs to spin to properly lubricate the universal joints
   b. the speed at which the drive shaft makes maximum torque
   c. the maximum speed the drive shaft should turn before it starts to distort
   d. twice the speed that maximum torque occurs at

SUPPORT BEARINGS If a drive shaft assembly is too long, it could start to become distorted and will produce unwanted vibrations as rotational speed and load increase. This will create a whipping effect and will damage driveline components. Generally, a drive shaft assembly longer than 72 in. should include a universal joint connecting two drive shafts and a support bearing holding one section. This support bearing is also sometimes called a center bearing and is usually a sealed ball-bearing assembly. Some machines will have a driveline parking brake assembly that will serve as a support bearing as well. ● **SEE FIGURE 14–14** for a driveline parking brake that incorporates a support bearing.

For lighter-duty applications, the support bearing will attach to a cross-member of the machine's frame and may need shims to set the final location. This is commonly called a hanger bearing. A hanger bearing will be a sealed bearing that is supported in rubber that is bonded to a stamped steel frame. The rubber will allow some movement as the shaft rotates. ● **SEE FIGURE 14–15** for a support bearing of a long drive shaft.

For large heavy-duty drive shafts, some machines will have a bulk head bearing assembly that is used to support the drive shaft. This bearing will be a sealed unit that is lubricated with regular greasing, or it could also be in a separate housing that is oil filled and attached to the machine's main frame.

UNIVERSAL JOINTS Universal joints are used as part of a drive shaft assembly between two components that aren't in perfect alignment or that need to be allowed to move in relation to each other.

They are sometimes called **Cardan** joints based on their inventor's name, but the more common term is *U-joint*.

The main part of the universal joint is the cross that can also be called a spider. It has four lugs or **trunnions** on it that have machined, hardened, and polished surfaces. These surfaces will mate with needle bearings that are held in a cup or wing cap depending on the universal joint style. The cups will have a lip-type seal to keep grease in and contamination out.

Almost all heavy-duty universal joints will require periodic greasing. Light-duty U-joints are typically lubricated for life when installed. For the greasable type of U-joints, the spider has cross-drillings that connect to a tapped and drilled hole, which will accept a grease fitting. The grease passages then allow grease to the ends of the trunnions where the grease fills the cap and surrounds the needle bearings. The caps will also have a thrust surface at their bottom. Some U-joints will have a check

FIGURE 14–14 A machine that uses a driveline parking brake that incorporates a support bearing.

FIGURE 14–15 A support bearing for a long drive shaft.

DRIVELINES, TRANSFER CASES, AUXILIARY DRIVES 421

FIGURE 14–16 Light-duty U-joint.

valve at the end of each trunnion passage to keep warm grease from draining away from a bearing cap after rotation stops.

There are a few different types of universal joints use for heavy equipment drivelines. The style is dictated based on how much torque they have to handle and how they mount to a yoke. The different styles are light- to medium-duty strap and U-bolt type—the yoke that accepts this type of U-joint is an open half-round type. The U-joint itself will have plain bearing caps. ● **SEE FIGURE 14–16** for a light duty U-joint.

Heavy-duty-type U-joints that mate to a full-round yoke can have bearing caps be retained with snap rings or have a plate-type bearing that uses two bolts and a lock to fasten the cap to the yoke. There will be a slight interference fit between the yoke and the bearing cap.

Wing-type U-joints are the most common type for heavy-duty applications. They have cast and machined bearing caps that allow two bolts to fasten each of the four caps to its mating yoke. There will also be a key machined on the mating surface of the cap that will help secure it to the yoke. Most wing-type U-joints will have a light strap welded to each pair of caps to ensure correct assembly and that the caps don't fall off when removed from a drive shaft or yoke. Rubber bands or electrical tape can be used to keep U-joint caps in place when they are not installed on a machine.

Some U-joints will have pairs of different styles of trunnion caps on the same cross.

TRANSFER CASES

Machines that have front and rear drive axles such as wheel loaders, articulated trucks, and backhoe loaders will need a transfer case to send the drive output from the transmission to the front and rear axles. It could be part of the main transmission housing, or it could have its own separate housing. The input to a transfer case typically comes from the output of a powershift transmission, but it could also come from a hydrostatic transmission or an electric motor.

FIGURE 14–17 Transfer case from an all-wheel drive machine.

Transfer cases will usually be arranged to drop the plane of the power flow, and for this reason, they are sometimes called drop boxes. However, some underground mining machines will use a transfer case to raise the level of power flow to reach the axle pinion. ● **SEE FIGURE 14–17** for a transfer case from an all-wheel drive machine.

A stand-alone transfer case can be a simple set of gears that are housed in a two-piece cast iron housing. The input and output gears will be splined to accept input and output yokes. Most transfer cases will have an idler gear as well.

A splined yoke will turn the top gear and receive input torque and then transfer it to an idler gear and to the lower gear (if used for lowering the power flow) where the drive is then sent out through splined yokes toward the front and rear differentials. The gears will be supported with a set of tapered roller bearings and will likely be splash lubricated with gear oil that is contained in the transfer case.

Transfer cases may also incorporate a differential and a parking brake and may be able to disconnect the drive to one

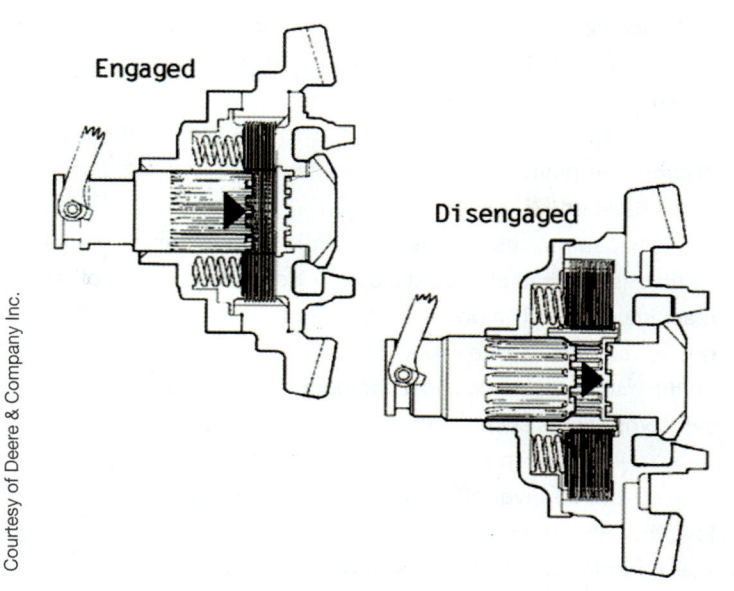

FIGURE 14–18 Differential lock in a transfer case.

of the axles. ●**SEE FIGURE 14–18** for a differential lock used in a transfer case.

Differentials are explained in Chapter 15, but in brief a transfer case that has a differential will allow a speed and torque difference between its two outputs. In most cases, this will be the drive shafts going to the front and rear axles. Most differentials used in transfer cases will be equipped with a differential lock.

Some lighter-duty transfer cases could have a second gear ratio to make it a two-speed unit, but most heavy equipment applications will only provide a 1:1 ratio between the input and output.

There are some articulated trucks that will incorporate a planetary gearset that will reduce the torque ratio to the front wheels of the truck. There could be a 40/60% split between the front and rear output of the transfer case. ●**SEE FIGURE 14–19** for a cutaway of a transfer case with a planetary gearset.

Transfer cases can also be used to drive secondary steering pumps from them because they need to be driven when the machine is moving if the primary steering pump fails.

If a transfer case is a separate remote unit, it could need to be aligned with special tools and adjusted with different thickness shims.

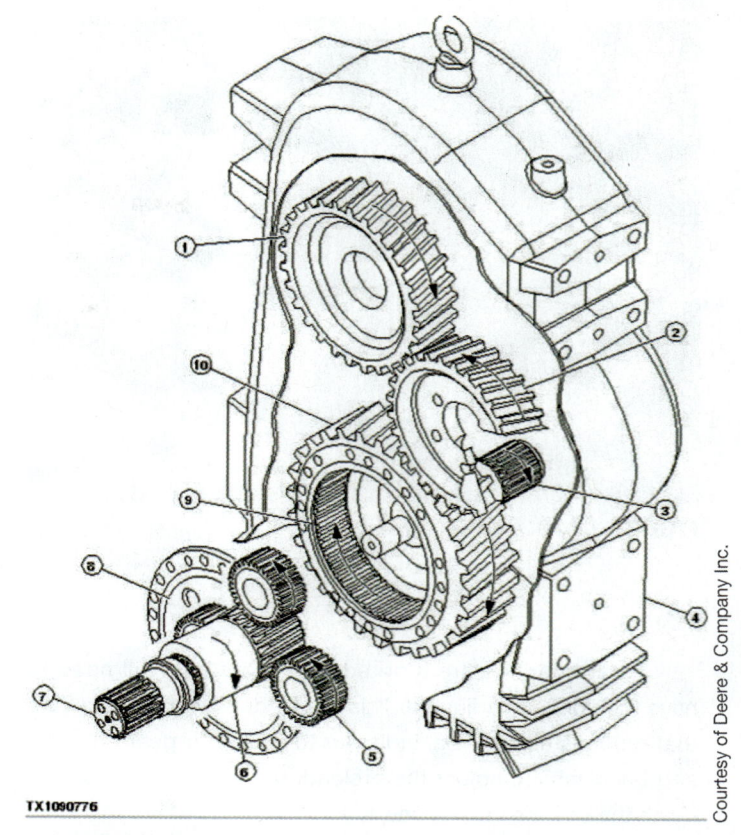

FIGURE 14–19 A cutaway view of a transfer case with a planetary gearset.

AUXILIARY DRIVES

Sometimes mechanical drive is needed for other purposes besides propelling the machine. Most machines will have one or more hydraulic pumps that need to be driven. Hydraulic pumps could be driven from many places such as directly from the engine or transmission. They can also be driven from an auxiliary drive that might be driven directly from the engine's flywheel or a separate unit that is driven by a drive shaft. ●**SEE FIGURE 14–20** for an auxiliary drive.

Whether the drive is for one or multiple pumps, it will be a simple set of gears that are suspended by shafts and bearings and housed in a cast iron two-piece housing. The housing will have its own oil supply and will use this oil for splash lubrication of the gears and bearings. Some auxiliary drives could have their own gear pump to circulate the oil and pump it to the bearings to ensure that they receive a steady supply of oil for cooling and lubricating.

FIGURE 14–20 An auxiliary drive.

Most bearings are tapered roller type and will need to have their preload adjusted. This will be done with a procedure that requires using a dial indicator to check for gear end play and backlash. To adjust the preload, there are shims used to move the bearing cup retainers.

OPERATOR COMPLAINTS RELATED TO DRIVELINES

Operator complaints related to drivelines and transfer cases can include vibrations, noises, no drive, or intermittent drive.

To track down a driveline vibration to its root cause can sometimes be difficult because of the rough conditions most machines work in. There aren't too many smooth-paved roads that most machines can run on and the problem is even trickier if the machine has tracks. The vibration will be more difficult to trace if it only occurs under certain conditions such as heavy loads, going downhill, under braking, or in certain gears or engine rpm.

You should try to identify if the vibration is at the same frequency as engine rpm first. This will eliminate anything that doesn't run at engine rpm. If the vibration isn't engine speed dependent, then try to see at what machine travel speed and direction it occurs. The load on the machine may have an effect on it as well.

It may be helpful to let the machine coast when trying to locate a vibration problem. If this is the case, make sure that the machine's braking system is functioning properly and the machine is not allowed to over-speed or travel in an uncontrolled manner.

Vibration meters can assist with troubleshooting vibration complaints but they are expensive and most shops won't have one.

Unusual noises are identified in much the same way as vibrations and can also be difficult to isolate because of other machine noises. The noise should be created at the same frequency as the driveline speed. If the noise changes as the load on the machine changes, look for loose mounts on major driveline components such as the transmission or pump drive.

You may even get a smell complaint related to drivelines. If the oil level of a pump drive or transfer case is too low, the bearings and gears could start to overheat and create a burning smell. A check of oil levels, looking for discolored paint, or a check with a temperature gun would confirm this problem.

A "no drive" complaint can be identified by visually seeing where the drive stops turning when the machine is in gear with the engine running. In some cases, it may be necessary to place the machine on jack stands and run the machine to see where the drive stops. Use extreme caution whenever you need to do this (see "Safety First").

DRIVELINE MAINTENANCE

Driveline maintenance involves inspection, lubrication, and regular oil changes.

When performing a machine driveline inspection, you would look for loose or missing fasteners; any signs of wearing or rubbing on rotating parts; twisting, warpage, or dents in the drive shaft tube; missing balance weights; broken/missing grease fittings; oil leaks at the transfer case. Transfer case and auxiliary drive oil level should be checked.

Maintenance intervals will be set by the manufacturer and should be adhered to. These intervals could be shortened due to machine working conditions. Many driveline components are lubed for life, and therefore, a regular visual inspection is all that is needed.

If grease fittings are installed in components, then greasing of U-joints and slip joints is an important part of maintaining drive shaft assemblies. The proper type and viscosity of grease should be used according to the machine manufacturer's specification. Always wipe off any dirt on a grease fitting before attaching a grease gun. A U-joint should be greased until you see

FIGURE 14–21 How a U-joint is greased.

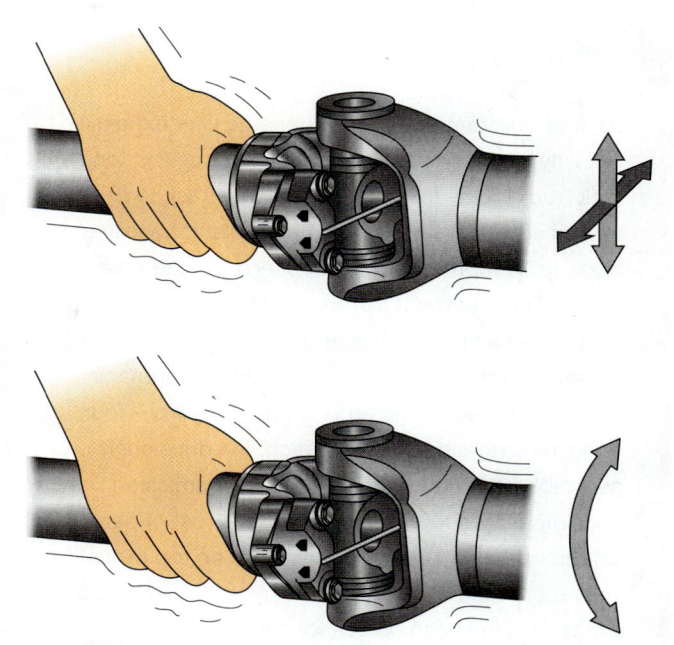

FIGURE 14–22 How to inspect a U-joint.

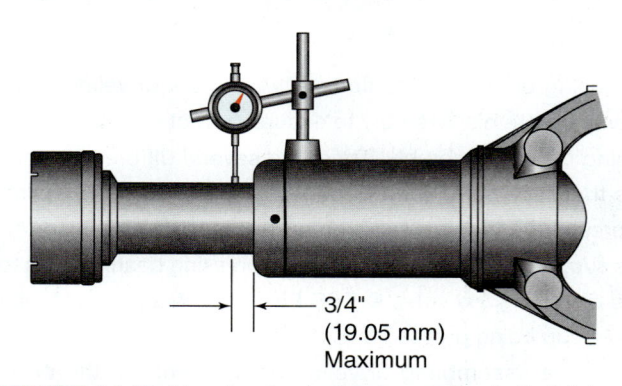

FIGURE 14–23 How to check play in a splined slip joint.

fresh grease coming out around all four trunnion cap seals. Slip joints should be greased as recommended, and care should be taken to not over-grease. Support bearings need to be greased as well if they are serviceable. ● **SEE FIGURE 14–21** for how a U-joint is greased.

U-joints can be measured to see if they should be replaced. This could be as simple as grabbing the yoke of the drive shaft and seeing how much play is there, or it could involve using a dial indicator to measure play while either prying or jacking the yoke. Any play felt should be considered excessive. This should be done before the U-joint is greased so that the grease doesn't mask a worn part. ● **SEE FIGURE 14–22** to see how to inspect a U-joint.

End yokes should be checked for excessive play on their splines as well.

Splined slip joint assemblies should be checked for excessive spline wear or for signs of sticking. To check for spline wear, a dial indicator with a magnetic base is attached to one side of the joint and its pointer is rested on the opposite side. You would then apply up and down pressure to the joint perpendicular to the dial indicator. If total play exceeds allowable specified dimensions, then the drive shaft assembly should be replaced. ● **SEE FIGURE 14–23** for how to check play in a splined slip joint. One manufacturer suggests that any more than 0.017 in. of play measured at a slip joint is excessive. The slip joint seal and Welch plug should also be checked for damage.

Transfer cases could be checked for excessive gear backlash with a dial indicator. They will need to have their oil changed and should have regular oil samples taken.

DRIVE SHAFT/U-JOINT REPAIR

Driveline repairs can occur as a result of failed or worn components or part of a machine rebuild process. It is highly recommended that new fasteners (bolts, locks, snap rings) be used whenever a driveline component is removed or replaced.

Never use excessive force or heat to remove driveline components. Only use soft-faced hammers on driveline components.

Typical driveline repairs could include replacing U-joints. To replace a U-joint that is part of a machine's travel driveline, you must first secure the machine from movement. The drive shaft should also be properly supported to prevent it

TECH TIP

Lack of maintenance quickly became very expensive one day for a company that had a fleet of 100-ton rock trucks. One truck was on its way back to be loaded and was probably in top gear and travelling close to 35 mph when the rear U-joint of the drive shaft going into the transmission failed. The operator said he felt a clunk at first and then the truck started to slow down. The end result was a lot of damage as the drive shaft flung around out of control—four brake cooling tubes at $750 each, the drive shaft assembly at $2500, and a cracked transmission housing at $15,000. This was just the cost of parts with the price of labor and downtime on top of that. During the repair, the failed U-joint was found to be dry and hadn't seen grease for a long time.

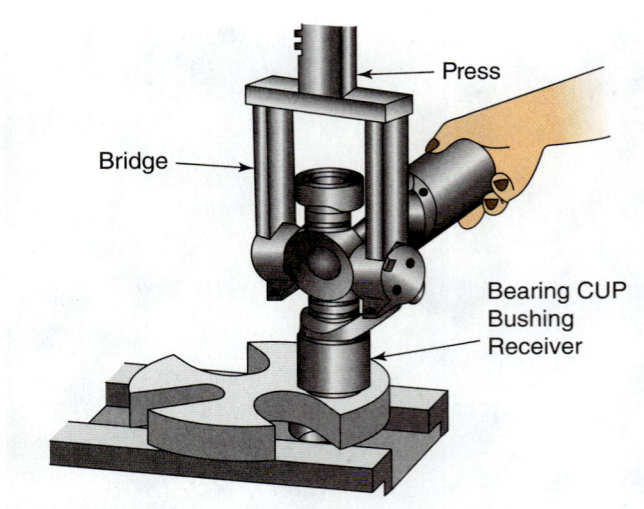

FIGURE 14–24 A bearing cup being pressed out.

from falling on you or the floor. Always mark driveline components before disassembly to ensure correct orientation and phasing upon reassembly. There are several different special tools that may be needed depending on the type of U-joint. For pressed-in bearing caps, an arbor press can be used. Always support the ears of a yoke when pressing bearing caps to avoid distorting the yoke. ● SEE FIGURE 14–24 for a U-joint bearing cup being pressed out.

Before assembling driveline components, a thorough inspection should be done to make sure that there are no cracks, burrs, bent parts, worn splines, or other defects that could lead to failure. All mating surfaces should be cleaned to be free of rust, dirt, oil, and grease before assembly. Degreaser, emery cloth, and round files are common tools/supplies for this task. Snap ring grooves and threaded holes should be cleaned out as well.

When installing new U-joints with a pressing device, make sure the caps pull in straight and take care to not press them in too far. Do not lubricate the outside of caps that are to be pressed in. For wing-type U-joints, be sure to pull the bearing caps in evenly and torque all fasteners to proper specification. New fasteners should always be used, and many times, a thread lock compound will need to be applied to fasteners. Check to be sure the bearing caps are fully seated and try to rotate the U-joint to check for smooth travel.

It's always a good idea to check drive shaft run-out after the assembly has been installed. It can be as simple as holding something stationary against the midpoint of the drive shaft while it is rotated or a dial indicator and magnetic base can be used.

Some drive shaft alignment procedures are required to maximize life span of U-joints and other driveline components. They could require special tools that will align components to minimize operating angles. ● SEE FIGURE 14–25 for the tools needed to align two drive shafts for an articulated truck.

TECH TIP

If a U-joint is removed from a component as part of another repair process and is not going to be replaced, the bearing caps should be reinstalled onto their trunnions and kept in place with tape or a rubber band.

TRANSFER CASE/AUXILIARY DRIVE REPAIR

Transfer cases and auxiliary drives will need repair if they have failed, have been found to have metal contamination from an oil sample taken, or are being reconditioned as part of a machine reconditioning process.

For transfer cases that are part of a transmission, their repair will take place as part of the transmission repair. For stand-alone transfer cases and auxiliary drives, the following general steps should be done to repair them:

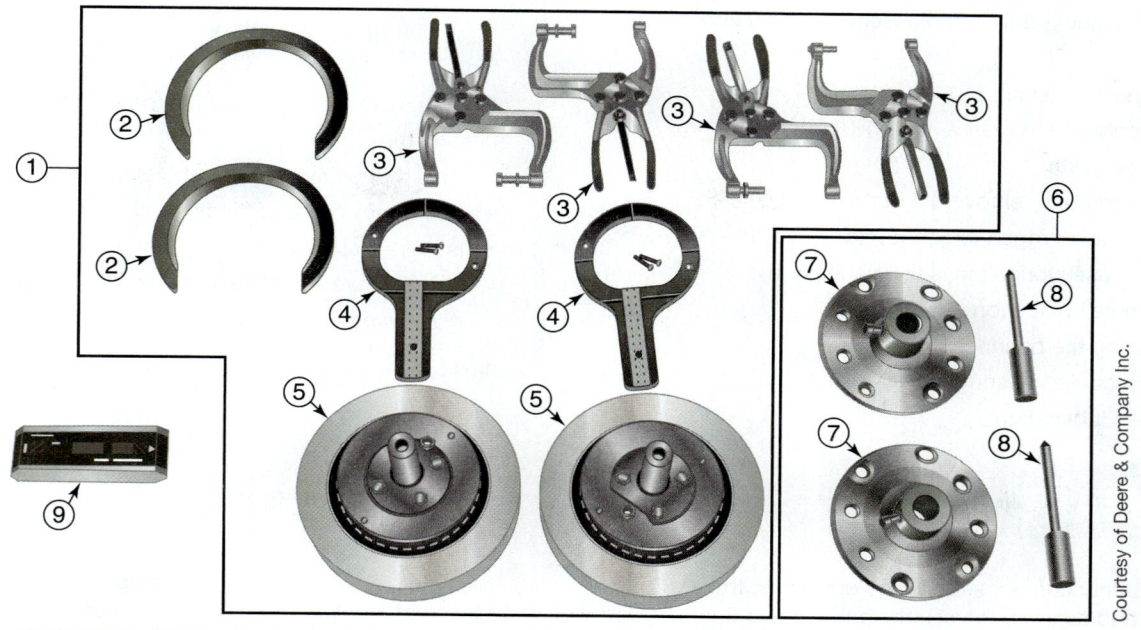

FIGURE 14–25 The tools needed to align two drive shafts for an articulated truck.

1. Remove unit from machine.
2. Clean, perform a visual inspection, and place on a bench or mount to a repair stand.
3. Drain oil and take an oil sample.
4. Remove input and output yokes.
5. Disassemble unit according to manufacturer's procedures. The usual process is to remove bearing retainers and covers and remove gears. Remove bearings from shafts. Disassemble differential and lock up mechanism if equipped.
6. Clean and inspect all parts for defects (cracks, burrs, chips, overheating, scoring, etc.). Replace any parts deemed nonusable.
7. Reassemble unit using new bearings (do not overheat when installing), seals, and gaskets.
8. Perform bearing preload or gear backlash measurement as needed throughout reassembly and torque all fasteners to proper specification.
9. Fill unit with proper type and viscosity oil.
10. Install unit and check operation.
11. May need to align remote mount transfer cases and auxiliary drives. This could include using tools such as electronic levels, protractors, or laser pointers.

WEB ACTIVITY

Search the web for images and information related to driveline/drive shaft failures.

Progress Check

6. This is one feature you won't find in a transfer case that is used for heavy equipment:
   a. an overdrive ratio
   b. a differential
   c. a parking brake
   d. tapered roller bearings
7. What is the maximum length of an unsupported one-piece drive shaft?
   a. 48 in.
   b. 72 in.
   c. 96 in.
   d. 144 in.
8. A wing-type universal joint trunnion cap will be held in place with:
   a. bolts
   b. snap rings
   c. interference fit pressure
   d. U-bolts

9. When rebuilding drive shaft assemblies, you should always:
   a. use new fasteners
   b. rely only on the grease in U-joints that they were shipped with
   c. use a torch to remove old parts
   d. use liquid nitrogen to cool parts before assembly
10. Transfer case lubrication is usually done by:
    a. pressure lubrication
    b. greasing the bearings
    c. teflon coated bearings and gears
    d. splash lubrication

> **SHOP ACTIVITY**
>
> Go to a machine that has a mechanical driveline and answer the following:
> - How many drive shafts can you find?
> - How many grease fittings are there?
> - Is there a slip joint?
> - Does it have a transfer case or auxiliary drive?
> - If yes, where is the oil level checked from?

SUMMARY

1. Some safety concerns related to drivelines, transfer cases, and auxiliary drives are: lock out and tagout before working on machine, never work close to a live exposed driveline, use proper lifting equipment and wheel chocks.
2. Machines with mechanical drivetrains need to have components transfer torque between the drivetrain components such as: drive shafts and universal joints. Auxiliary drive systems also need these driveline components.
3. Driveline components need to handle huge amounts of torque since drivetrains decrease engine speed while increasing torque as the torque is transferred toward tires or tracks.
4. Driveline components are often at different angles and planes to each other and drive shaft assemblies are required to allow for this.
5. Operating angles between components must be within a narrow range. Otherwise the drive shaft assemblies will create vibrations and fail prematurely. This is usually limited to between 1* and 3*.
6. As operating angles increase speed fluctuations that occur twice per revolution increase.
7. Minimum angles are needed to keep the universal joint needle bearings moving.
8. Drive shaft assemblies are sometimes called propeller shaft assemblies. Simple drive shaft assemblies consist of a tube with a flange on each end that have universal joints connected to them.
9. Drivetrain components that need to move in relation to each other will have drive shaft assemblies with slip joints. This will provide a means for the drive shaft to change length.
10. Slip joints have male and female splines that allow the two pieces of a drive shaft to change lengths as needed.
11. Yokes that are welded to the drive shaft tubes provide a means for universal joints to transfer torque to and from them. There are a variety of types of yokes such as: half round, full round, and wing type.
12. Drive shaft assemblies need to be balanced before installation.
13. Two piece drive shafts need to be phased. This will keep both yokes in proper alignment to reduce vibrations.
14. Drive shafts can provide a mounting location for driveline brakes. These can be rotor and caliper–or shoe and drum–style.
15. Long drive shafts will have support bearings reduce the possibility of flexing when they are rotating.
16. Universal joints or U-joints are sometimes called Cardan joints. They consist of a cross, needle bearings, and cups.
17. Fastening U-joints to drive shafts can be done with: snap rings, threaded straps, or threaded fasteners.
18. Transfer cases use gears to transfer drive from an component like a transmission to another like one or more drive shafts. They can also provide a gear reduction and incorporate driveline brakes and differentials.
19. Auxiliary drives can be used to drive pumps.
20. Operator complaints related to drivelines can include: vibrations, noise, or no drive.
21. Driveline maintenance includes visual inspection, greasing, checking oil levels, and checking for excessive play.
22. Driveline repairs include: U-joint replacement, drive shaft replacement, and transfer case recondition.

428 CHAPTER 14

chapter 15
DRIVE AXLES, POWER DIVIDERS, FINAL DRIVES

LEARNING OBJECTIVES

After reading this chapter, the student should be able to:

1. Explain the operation of an open differential during straight travel.
2. Explain the operation of an open differential when cornering.
3. Identify the common components that make a differential work.
4. Describe the mechanism that makes a differential lock.
5. Describe the different components that make a limited slip differential work.
6. Explain how a final drive reduces speed and increases torque.
7. Describe the difference between an inboard and outboard final drive.
8. Describe the steps to service a drive axle safely.

KEY TERMS

Backlash 449
Contact pattern 452
Crown gear 433
Limited slip 441
Locking differential 438
Pinion gear 433
Planet carrier 433
Power divider 430
Ring gear 433
Side gear 441
Sun gear 433

INTRODUCTION

SAFETY FIRST When working on final drives, drive axles, and power dividers, you must be aware of the potential for a machine to roll away. If a machine has inboard or driveline parking brakes, then any disconnection of mechanical drive at the final drive will allow the machine to roll uncontrolled. Wheel chocks must be used to prevent this.

You will be working with components that are heavy. The proper use of lifting devices is critical for staying safe.

You may be required to release or disassemble brake assemblies that have heavy springs as part of them. Take precautions outlined in the machine's service information section to prevent uncontrolled release of spring tension.

As always proper use of appropriate PPE will go a long way in keeping you safe.

ENVIRONMENTAL CONCERNS

Many maintenance and repair procedures will require draining fluids. These fluids should be contained, handled, and disposed of properly to prevent contaminating the environment.

NEED TO KNOW

This chapter will discuss the components of a machine's driveline that receive torque from a transmission, a hydraulic motor, or an electric motor. These components will change the direction of power flow and provide a torque increase and distribute power to the machine's tracks or wheels. These components are called drive axles, power dividers, and final drives. ● SEE FIGURE 15–1 for Caterpillar 826G compactor with mechanical power train components.

Heavy equipment will use these components to create high torque and low speed.

Wheel-type machines will use drive axles to change the direction of power flow and provide a torque increase from their input. The input to a drive axle could be a shaft from a transmission, an electric motor, or a hydrostatic motor. This power flow will almost always go through a final drive gear reduction after it leaves the axle's differential and before it gets to drive the wheels. This final gear reduction is called the final drive.

Most final drives will use a set of gears to provide a gear reduction while some machines will use chains and sprockets to drive their wheels and increase torque. Two types of machines that use chains for final drive are graders and skid steers. The chain drive arrangement used for driving grader rear wheels is called a tandem drive. Most articulated trucks that use a tandem drive axle arrangement as well. ● SEE FIGURE 15–2 for a tandem axle arrangement.

A wheel-type machine that uses a second drive axle directly behind the first one will need a component to transfer torque to the second axle. This component is called a **power divider**, and this axle arrangement is said to be a tandem axle drive. This is similar to a typical highway truck's drive axle arrangement.

FIGURE 15–1 Caterpillar 826G compactor with mechanical power train components.

DRIVE AXLE TYPES

Some axles will just support some of the machine's weight and may or may not steer. If an axle doesn't steer and is just for supporting weight, it will be called a dead axle or tag axle. These are rarely used for heavy equipment because of the adverse travelling conditions machines usually work in; therefore, if an axle doesn't steer, it will likely drive. Steering axles are also most often drive axles when used on machines with straight frames. The main exception is dual rear wheel of highway trucks that use non-driving front wheels. Wheeled machine steering systems will be discussed in Chapter 19.

All drive axles will support part of the weight of the machine because they are attached to the machine's frame. They could be bolted directly to the frame as in the case of the front axle for an articulated wheel loader or could be suspended from the frame to allow the machine's suspension to conform to uneven surfaces as the machine moves over rough terrain and roads such as the rear axles of trucks. Machine suspension systems are relatively simple and are designed to allow a few degrees of axle movement with some

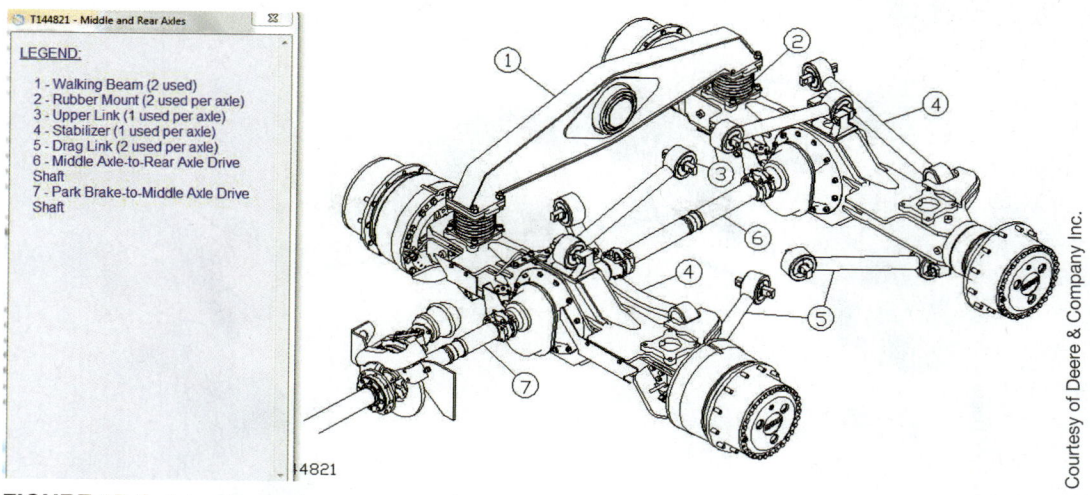

FIGURE 15–2 A tandem axle arrangement.

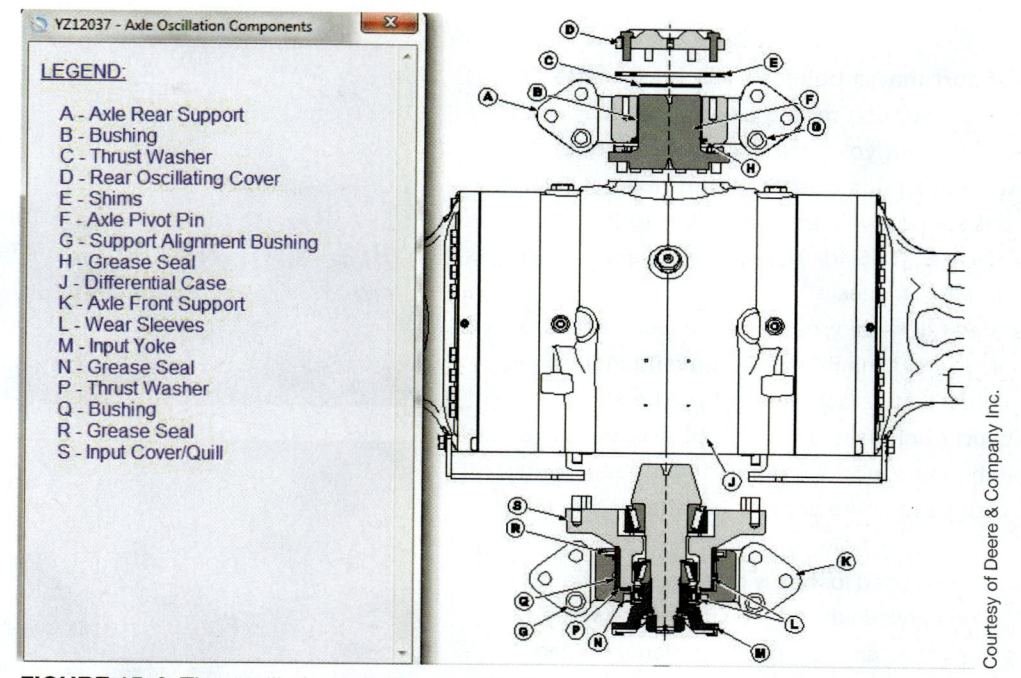

FIGURE 15–3 The oscillation arrangement for the rear axle of a medium wheel loader.

damping or in some cases no damping. Machine suspension systems will be covered in Chapter 21.

The axle could also be mounted to a pivot point that allows the axle to pivot at its center and rotate a few degrees. This is called an oscillating axle and is typical for the rear axle of a rubber-tired wheel loader or the front axle of a backhoe loader. ● **SEE FIGURE 15–3** for the oscillation arrangement of the rear axle of a medium wheel loader.

Oscillation joints are made up of a simple pin that protrudes from the top of the axle housing and is parallel to the centerline of the machine. The axle housing has plain bushings pressed into a bore that the pin can pivot in. The bushings will have grease seals to keep grease in and dirt out. The pin also fits into two bores that are part of one of the machine's main frame cross-members. The pin is held in place with a bolt that goes through the pin ear. This is a typical arrangement for a backhoe loader front axle. ● **SEE FIGURE 15–4** for the front axle oscillation joint of a backhoe loader.

Larger machines such as wheel loaders will have a short trunnion shaft bolted to the rear of the axle housing

DRIVE AXLES, POWER DIVIDERS, FINAL DRIVES 431

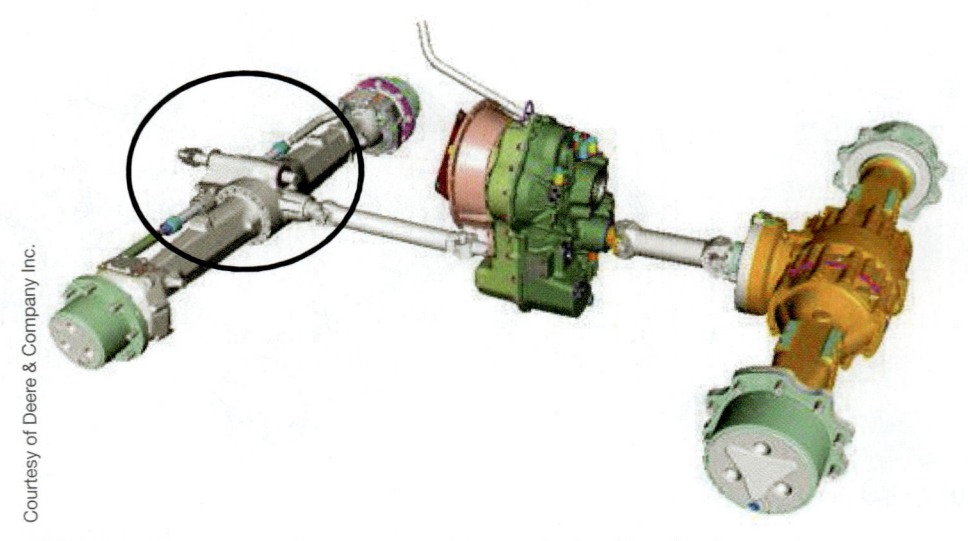

FIGURE 15–4 The front axle oscillation joint of a backhoe loader.

that rides in a support that is bolted to the main frame of the machine. The front support for the axle will have a large opening because the input yoke will be in the center of it. Both supports will have plain bushings that need regular greasing. The front support will also be bolted to the main frame. ● **SEE FIGURE 15–5** for the rear axle oscillating supports for a large wheel loader.

There are several general types of drive axles that are used for heavy equipment machines. A conventional drive axle is used for lighter-duty machines and will be similar to a heavy-duty pickup truck axle. This type of axle will have a one-piece main housing with the differential bolted to an opening in the middle and brake assemblies attached to the outer ends.

Almost all drive axles used for heavy equipment will have a second gear reduction between the differential and the wheels. This is called the final drive and is typically a planetary gearset. Some final drives are incorporated into differential housing. These axles are said to be equipped with inboard final drives. If the final drives are located at the wheel ends of the axle, they are considered outboard final drives.

They could also have brakes near the center of the axle housing, and this is said to be an axle with inboard brakes.

There could be axles that provide two gear ratios that are changeable, and these are called two-speed axles. Some axles may have a second gear reduction incorporated with the differential, and this will be called a double reduction axle assembly.

Depending on the machine type, most drive axles will be driven by a drive shaft, but they could have a hydrostatic

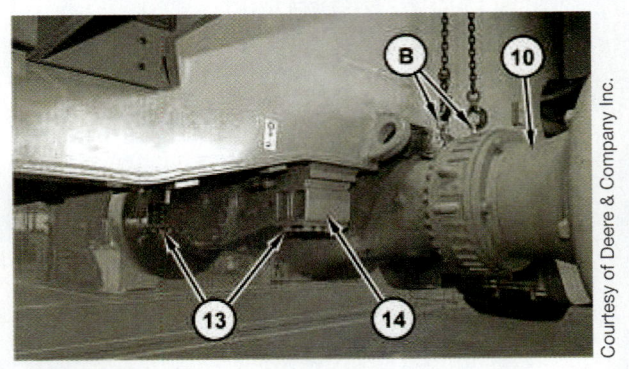

FIGURE 15–5 The rear axle oscillating supports for a large wheel loader.

motor providing input torque directly to them as well. Some telehandlers and small wheel loaders will have this arrangement.

CONVENTIONAL DRIVE AXLE Many heavy equipment machines are four- or six-wheel drive and possibly more. We will start off discussing a simple axle arrangement that could be used for a two- or four-wheel drive machine and call this a conventional axle. A conventional drive axle will only provide gear reduction through the differential and will have brakes mounted at the ends of the axle.

A conventional drive axle will be used for a rubber-tired machine that could be steered by articulation or conventionally steered. With no other gear reduction, this type of axle will only be found on light-duty higher-speed equipment and is directly comparable to a heavy-duty pickup truck axle. Because of their relatively light-duty nature, this type of axle is not common to heavy equipment.

The input to the axle is called the pinion shaft. It will have a yoke splined to it and held in place on the shaft with a large nut or bolt. The yoke is driven by a drive shaft. The pinion shaft will then drive a **pinion gear**. The pinion gear then drives a **crown gear**, which is sometimes called a **ring gear**. This gear will drive the axle's differential. A differential is needed to allow the axle's wheels to rotate at different speeds under certain conditions.

This gearset also changes the direction of power flow. This is accomplished with the type of gears used for the pinion and crown. They are spiral bevel-type gears and their center axis is at 90° to each other.

The differential will send torque out axle shafts that extend past the outer end of the axle housing. The axle shaft flanges will then transfer torque to a wheel and tire assembly. All rotating parts of the axle need to be supported with bearings that are lubricated with oil. There will be many opposing forces acting on the rotating axle components and the bearings will need to counteract these forces.

The axle housing for a conventional-type axle will be a one-piece cast iron housing that has an opening for the differential. The differential is bolted to the axle housing with a ring of bolts or studs and nuts.

The axle shafts are supported at the ends on the axle shaft with bearings and will have a machined surface for the bearings as well as a lip-type seal to ride on. The seal will be pressed into the housing outside of the bearing and will retain axle fluid in the housing. This type of axle is termed a *semi-floating axle* because the axle shaft also supports vehicle weight.

A full-floating axle will have the wheel hub supported by two-tapered roller bearings and the axle shafts will only transfer torque from the differentials to the wheel assemblies. This is a more robust arrangement. The axle shafts of a full-floating axle can be removed to allow towing of a disabled machine.

AXLE WITH INBOARD FINAL DRIVE An axle can have a second gear reduction called a final drive. The first gear reduction comes from the crown and pinion gearset. The name *final drive* derives from the fact that it is the last gear reduction before drive is sent to the wheels. The final drive (second gear reduction) will further decrease speed and increase torque.

Final drives can be located inboard near the center of the axle housing or outboard near the wheels. Most drive axles used for heavy equipment will have outboard final drives although inboard final drive axles seem to becoming more common on small- and medium-sized machines. Outboard final drives will have a housing that is bolted to the ends of the axle housing and house a planetary gearset. They will share oil with the axle and are considered to be full-floating axles because the axle shaft only transfers torque and does not support weight. Final drives will be discussed more in depth later in this chapter.

An inboard final drive type of axle will have a differential that outputs its torque through a short shaft and into a planetary final drive that is in the center section of the housing with the differential. The shaft will drive the **sun gear** of the planetary. As the ring gear is held stationary with the housing, the sun gear will walk the planet pinions around the ring gear. The output therefore is the **planet carrier** that is driven by the planet gear shafts. The planet carrier will then be splined to accept a shaft that will drive the wheel hub. See Chapter 12 for a review of planetary gearsets. ● **SEE FIGURE 15–6** for the differential and final drive arrangement of an inboard final drive axle.

Because of the way the axle shaft both transfers torque and supports some of the weight of the machine, it will be considered to be a semi-floating axle.

This axle type will usually feature inboard brakes as well. ● **SEE FIGURE 15–7** for an exploded view of an axle with inboard final drive.

There could be machines that use axles with inboard final drives and outboard brakes. ● **SEE FIGURE 15–8** for an axle with inboard final drives and outboard brakes.

AXLE WITH INBOARD BRAKES Some axles will have their brakes located in the same housing as the differential. The brakes will usually be located on the high–speed, low-torque part of the axle. In other words, if the axle assembly has a final drive as part of it, the inboard brakes will be between the differential output and the final drive.

The brakes will typically be multi-disc, but for lighter-duty axles, they could be single disc. See Chapter 18 for more details on single- or multi-disc-type brakes. A multi-disc type of brake will use one or more stationary plates and one or

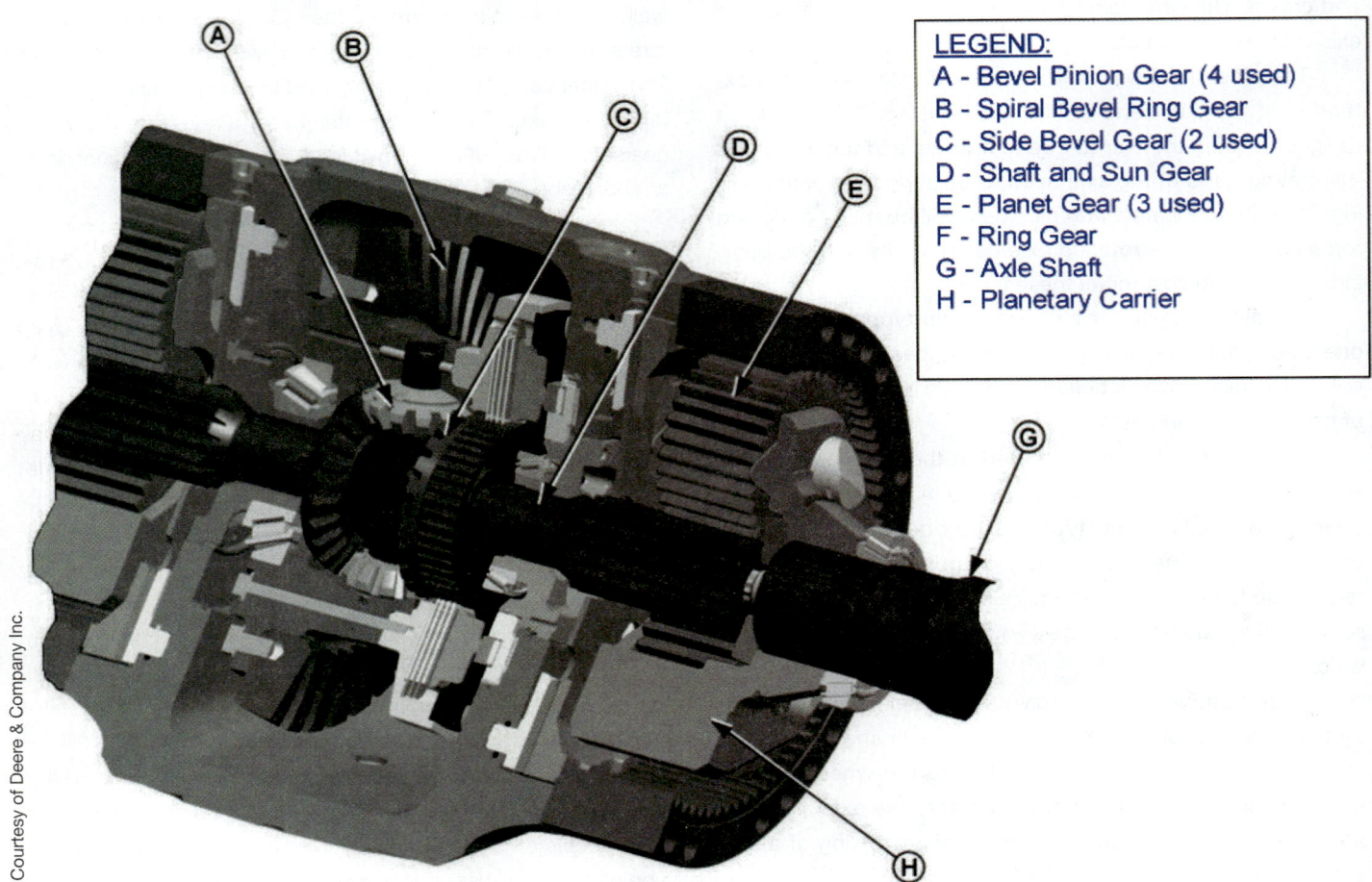

LEGEND:
A - Bevel Pinion Gear (4 used)
B - Spiral Bevel Ring Gear
C - Side Bevel Gear (2 used)
D - Shaft and Sun Gear
E - Planet Gear (3 used)
F - Ring Gear
G - Axle Shaft
H - Planetary Carrier

FIGURE 15–6 The differential and final drive arrangement of an inboard final drive axle.

434   CHAPTER 15

LEGEND:
A - Cap Screws
B - Washer
C - Dipstick or Plug
D - O-ring
E - Axle Housing
F - Plug or Dipstick
G - O-ring
H - Dowel Pin (2 Used)
J - Bearing Cup
K - Bearing Cone
L - Thrust Washer
M - Planetary Carrier Assembly
N - Cap Screw (Special)
P - Lock Plate
Q - Sun Pinion Shaft
R - Ring Gear
S - Shim Pack
T - Bearing Cup
U - Bearing Cone
V - Oil Seal (Face Type)
W - Axle Shaft

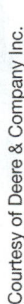

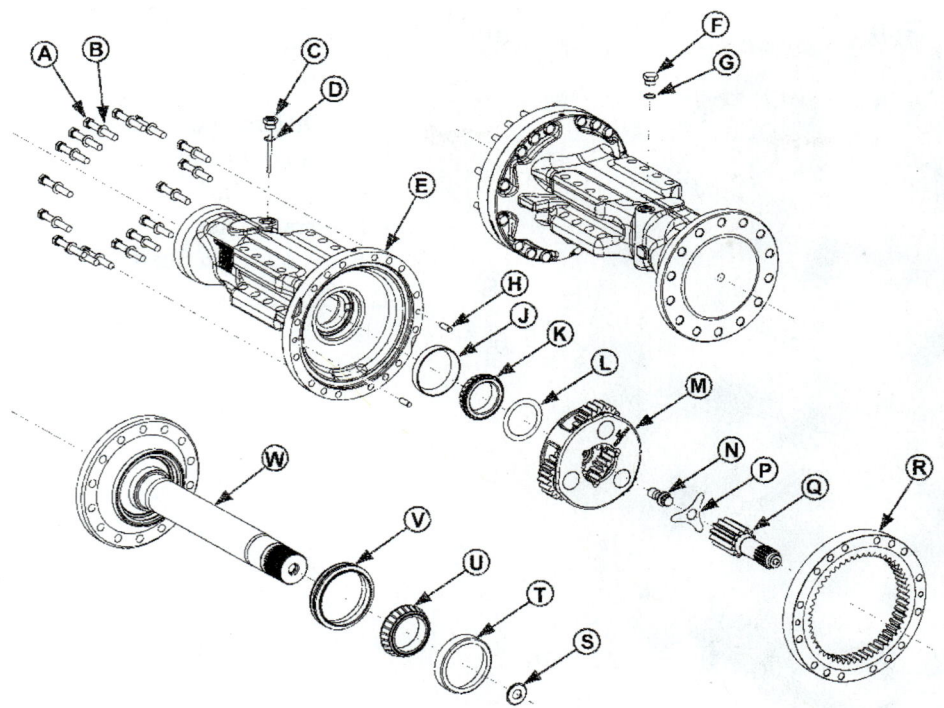

FIGURE 15–7 An exploded view of an axle with inboard final drive.

FIGURE 15–8 An axle with inboard final drives and outboard brakes.

more discs that are splined to a rotating part of the axle. When oil pressure is applied to a piston, it will squeeze the discs and plates together to create a braking action.

As the brakes are oil applied, they will be supplied with pressurized fluid from a different system. There is always the possibility of cross-contamination of the axle fluid when an axle has inboard brakes because a failed seal for the brakes will allow brake fluid into the axle housing.

The brakes will run in axle oil for cooling purposes. This will require the use of oil cooling for the axle for many applications because of the extra heat that will accumulate in the axle oil. The differential oil will also have to be compatible with friction material for use in an axle with inboard brakes.

● **SEE FIGURE 15–9** for an axle with inboard brakes.

DRIVE AXLE WHEEL SEALS Drive axles have stationary housings with a rotating wheel at each end of it. There needs to be a seal to keep the axle oil in and contamination out. Lighter-duty wheel seals will be lip type while most axles used for heavy equipment will use metal face seals that are sometimes

DRIVE AXLES, POWER DIVIDERS, FINAL DRIVES 435

LEGEND:

A - Differential Case
B - Brake Piston
C - Brake Backing Plate
D - Brake Disk
E - Sun Pinion Shaft

FIGURE 15–9 An axle with inboard brakes.

called Duo-cone seals. Lip-type seals will have a metal frame that gets fitted to a bore in the stationary housing. This metal frame may need to have a sealer applied to it but will likely have a soft coating that will partially peel off when the seal is installed. There will be one or more sharp flexible lips that form the inside diameter of the seal and will ride on a smooth hardened surface. The lips will usually have a spring on their backside to hold tension on the lip. The mating surface of the rotating part will be hardened and smooth. There could be a thin wear sleeve pressed onto the shaft that is replaceable if it gets worn.

Extra care should be taken when installing lip-type seals. A proper-sized seal driver will be the best tool for installing the seal, but it should be done slowly and checked to make sure that it goes in evenly.

Metal face seals are a four-part seal (two loads rings and two metal sealing rings) with one-half staying stationary while the other half rotates with the wheel hub. These seals will have a thick rubber O-ring (load ring) that will seal between the housing and the metal part of the seal and create a constant pressure on the metal faces that are in contact. The O-ring will need to be seated in a clean, dry groove to seal properly. Each half of the seal is installed as an assembly and with a special tool that will push the O-ring into its groove.

The metal faces are angled (typically 15°) to give a fine sealing surface when new. As the seal wears, this surface will get wider and the width of the contact area can be measured to see if it can be reused. These seals can withstand up to 50 psi oil pressure and will give long service life under normal conditions (15,000 hours or more). ● SEE FIGURE 15–10 for a metal face–type seal.

STEERABLE DRIVE AXLES Many machines will use a front axle that drives and steers. Non-articulating machines such as backhoes and telehandlers need all wheel drive and a steerable drive axle to turn the machine. Some machines will have steerable front and rear axles.

These axles will have pivoting wheel knuckles at each end of the axle housing that are controlled by a steering system. Wheeled machine steering systems will be covered in Chapter 19. The wheel end will usually have a final drive; therefore, the axle shaft is an input into the sun gear of the planetary final drive.

The knuckle will have an upper and lower pin assembly to allow it to pivot (for steering) on the end of the axle housing. Heavier axles will use tapered roller bearings while light-duty axles will use plain pins and bushings with thrust washers. If the axle uses tapered roller bearings, it will need to be shimmed

FIGURE 15–10 A metal face–type seal.

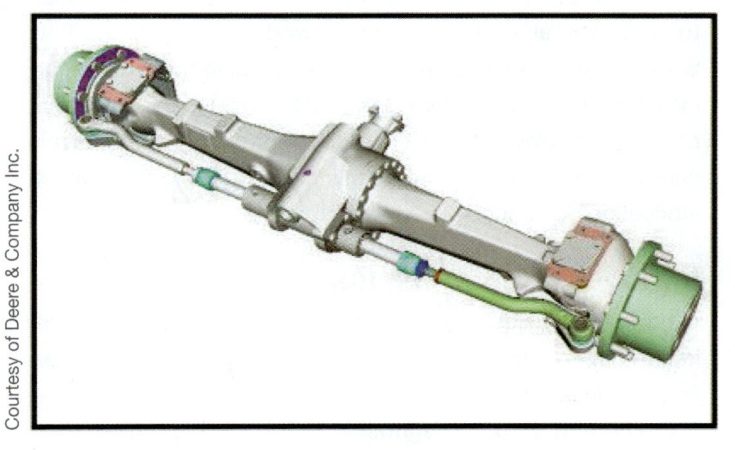

FIGURE 15–11 The steerable axle of a backhoe loader.

to a specific preload that is measured with a spring scale at assembly. There will be seals to keep grease in and dirt out of these pin assemblies. Regular greasing of these pin assemblies is a must. ● **SEE FIGURE 15–11** for a steerable drive axle of a backhoe loader.

The axle's shaft will extend out of the main housing through a seal and be supported by a bearing that is lubricated with axle oil. The axle shaft needs to be flexible at the pivot point of the steering knuckle and needs a universal joint assembly to do this. There are various types of universal joints that will transfer torque, but for this application, it must do so without speed variations, which means it must be a constant velocity joint. A constant velocity joint is a double universal joint assembly. The outer end of the constant velocity joint will drive another shaft that usually drives the sun gear of a final drive. The final drive will be supported by the knuckle and its output drives the wheel.

Some pivot joints are covered by a ball assembly while others are exposed.

AXLE FLUID COOLING AND FILTRATION
Many axles will have a system to cool the lubricating fluid. With the high levels of torque created and the action of meshing gears and rotating bearings constantly shearing and stressing the fluid, there will be a lot of heat generated in the fluid. If the temperature of the fluid is allowed to climb too high, it will start to degrade the lubricating fluid, which could lead to component failure.

Fluid cooling will be even more critical if the axle has inboard brakes. The brakes will transfer heat to the fluid, and this heat needs to be dissipated. Smaller machines and machines that have axles that don't have inboard brakes may rely on the transfer of heat from the fluid to the axle housing where it can transfer to the surrounding air.

Axle cooling circuits are simple systems that use a pump to circulate axle fluid from the axle housing to a cooler where the heat in the fluid is transferred to the air flowing by the cooler.

The cooler will likely be located near the engine cooling fan, but it could also have its own fan. Some machines will have separate cooling systems for each axle. There will be a combination of hoses and steel tubes to connect the axle, pump, and coolers. There could also be fittings that will enable live oil sampling.

Some axle fluid circulation systems will include one or more filters that will clean the fluid as it is pumped around the system. ● **SEE FIGURE 15–12** for an axle oil cooling system.

AXLE LUBRICATION
Axles could use a wide range of fluid for lubrication and cooling. The term *fluid* is sometimes used instead of oil because there could be non-mineral-based fluid required such as synthetic but most axle fluid will be mineral-based oil. For the balance of this chapter, axle fluid will be referred to as axle oil. Axle oil will also lubricate differentials and power dividers.

The axle housing will contain all the required oil for cooling and lubrication unless the axle uses a cooling system. Axles will rarely use a remote reservoir for their oil because the axle housing will contain more than enough volume needed for lubrication.

Some axles will use a circulation pump to move the oil around the inside of the differential housing to provide more

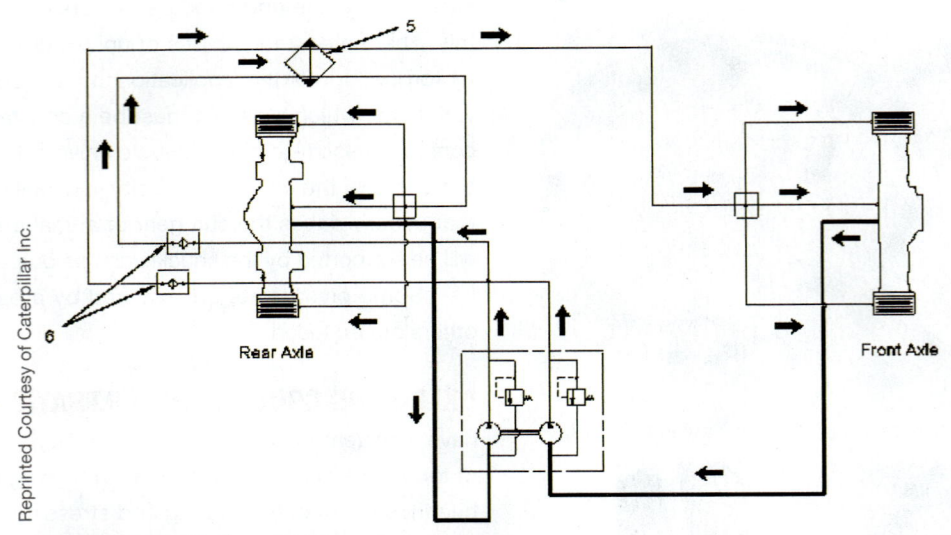

FIGURE 15–12 An axle oil cooling system.

positive lubrication. This is normal for very large axles; however, most axles will not use a pump and will just rely on the crown gear to pick up fluid as it rotates. This action will be enough to provide sufficient lubrication (splash lubrication).

DRIVE AXLE MAINTENANCE
Drive axle maintenance should be done on a regular basis as described in the machine's maintenance manual. Oil levels for the axle and final drive are normally required to be checked every month.

This is usually done by removing a level plug and the oil level should be right at the bottom of the threads. Some axles will have a plug on top with a dipstick attached to it.

It's important for the machine to be parked on a level surface when checking axle oil levels.

There may be a breather for the axle housing that should be cleaned or replaced on a regular basis. This is quite often overlooked and could result in the housing getting over-pressurized and axle fluid leaks could result.

A visual inspection should also be done any time a machine is serviced. Any leaks should be noted, and if they are minor, they can be repaired when convenient. Major leaks should be repaired as soon as possible because oil level will not be properly maintained, and if oil leaks out, then water, dirt, and other contaminants will be able to leak in.

Axle oil changes should normally take place at 2000-hour intervals. A good practice in between oil changes is to take oil samples to monitor oil condition. Axle oil changes should ideally be done after the oil has been circulated and warmed up.

Proper oil refill type and quantity is important to get the maximum longevity of the axle. Cold weather operation can be harmful to differentials because the axle oil is normally thick to start with, and if the oil isn't allowed to warm up and flow before heavy loads are encountered, meshing teeth and bearings can be damaged from lack of lubrication. If the oil is thick enough, the crown gear will cut a path through it, and this is called channeling. With the oil this thick, there won't be much lubrication occurring.

DRIVE AXLE DIAGNOSTICS
Operator complaints related to drive axles could include noises, vibrations, no drive, intermittent drive, unusual smell, pull to one side while travelling, and leaks.

The first step in diagnosing a drive axle problem is to gather information from the operator related to the problem. This includes the conditions that the problem happens under, when it started, how often it occurs, and if there are any other problems that may be related. Next, you should get to know the type of drive axle the machine has; for example, if it has a **locking differential**, then how is it actuated and what types of brakes are on it. If the machine is operational, you would want to operate the machine to verify the complaint.

You would then perform a visual inspection including looking for leaks, checking the fluid level, and perhaps taking an oil

> **SAFETY TIP**
>
> Care should be taken when axle oil level is checked on a warm axle because there could be pressure built up. When the plug is almost ready to be free of its threads, you should place a rag over the plug and stay as far back as possible to avoid having hot fluid sprayed on you.

> ### ⊕ SAFETY TIP
>
> If there was a no drive complaint, you may need to raise the wheels off the ground and see if the wheels for the problem axle turn when the machine is put into gear. If the machine has multi-axle drive, you would want to remove the drive shafts going to the other axles or disengage inter-axle differential locks to safely diagnose one axle. If there is no drive to an axle, you need to see if the drive shaft going into the axle is turning. This *must* be done safely. A small video camera mounted on a magnetic base would work great for this part of diagnosing a no drive complaint.

> ### 🔧 TECH TIP
>
> Many driveline components will use tapered roller bearings to support shafts or other parts. A tapered roller bearing assembly consists of two parts. The outer race (also called the cup) has an outside diameter that is usually pressed into a bore. Its inner surface is tapered and the tapered rollers will ride on it. The inner race is also tapered and has the rollers held onto it with a light gauge framework called a cage. The cage also spaces the rollers evenly around the race. This assembly is called the cone because it is shaped like a traffic cone. The inside surface can be slightly larger than the shaft it rides on, or it could rest against a shoulder.
>
> Tapered roller bearing assemblies will almost always be used in pairs and will usually need to be preloaded. Preloading will make sure that the rollers are seated on the races properly and will align the parts that are being suspended or supported by them. They will control both axial and radial movement once properly preloaded.

sample. If the machine has an axle oil filtration system, you could take the filter off, cut it open, and inspect it for contamination.

For an axle with a no drive problem, it may be fairly easy to remove an axle shaft to see if it has broken or to remove a final drive carrier to see if the problem is with the final drive.

DRIVE AXLE REMOVAL Many times it will be easier to repair an axle if it is removed from the machine. This will require proper lifting and blocking equipment. You may also need another machine to pull the axle out from under the machine. A typical axle removal procedure would involve securing the machine, draining the axle fluid, removing brake lines, undoing axle mounts, and then supporting the axle. The machine would then be supported by its frame and the wheels would then be removed so the axle could be lowered. Depending on the weight of the axle, it could be pulled out from under the machine or the machine could be moved away from the axle.

DRIVE AXLE REPAIR Once a problem has been diagnosed to be an internal drive axle problem, you must determine if the repair can be made with the axle in the machine or whether the machine has to be removed.

Some axle repairs that can be done with the axle left on the machine are differential removal and repair, final drive removal and repair, wheel bearing and seal replacement, axle shaft replacement, and leaks.

Outboard final drive problems can usually be fixed without removing the axle from the machine, but inboard final drive problems will require the axle to be removed and disassembled.

Once an axle is removed, it can be disassembled, inspected, and repaired with new parts. The extent to it being disassembled depends on the type of repair needed. The following procedures will assume the axle will be completely disassembled.

FIGURE 15–13 An axle mounted on a stand.

For smaller axles, they could be mounted to a stand or even put on top of a large bench. ● **SEE FIGURE 15–13** for an axle with inboard brakes and final drives mounted on a stand ready for disassembly.

Larger axles will be supported on stands where they can be disassembled. Axle shafts are removed, and if the axle has outboard final drives, they will be removed next. The differential can then be removed and disassembled if required. There will be more discussion on that in this chapter's differential section.

The axle housing should be inspected for cracks and can be measured for straightness.

DRIVE AXLES, POWER DIVIDERS, FINAL DRIVES 439

For smaller axles, it makes economic sense to replace all bearings, seals, and gaskets. Larger axles will have larger and more expensive bearings. They could be reused if no signs of wear or damage are present, but if there is any doubt, then they must be replaced.

Wheel bearing adjustment is a critical part of axle rebuilding. There are two general ways to set the preload for the tapered roller bearings, and these are shims and lock nuts.

Shim adjustment: If a drive axle wheel bearing adjustment is done with shims, the following generic procedure is likely to be followed. This is a procedure for an axle that has outboard final drives.

a. Install the inner axle bearing cone and metal face wheel seal on the axle spindle. Bearing cone installation may require heating the bearing. *Do not* overheat.
b. Apply axle oil to the bearing and wheel seal face.
c. Install wheel bearing cups and metal face seal in wheel hub and apply axle oil. Bearing cup installation may require cooling the cups.
d. Place wheel hub over spindle. It may be necessary to pull in the hub and support it to have it centered on the bearing.
e. Install the outer bearing cone on the ring gear hub.
f. Install the ring gear hub onto the splines of the spindle.
g. Measure the thickness of the retainer plate.
h. Install the hub retaining plate onto the end of the spindle and tighten bolts in specified pattern and to the specified torque.
i. Rotate the wheel hub at least one revolution to seat bearings.
j. Measure the distance from the outside of the retainer plate to the outside surface of the spindle.
k. Calculate the distance from the bottom of the retainer plate to the outside of the spindle.
l. Use the calculated distance for a reference to install the proper thickness of shims under the retainer plate. This should give the correct amount of preload to the wheel bearings. The specification may require a greater or lesser thickness of shims than the present gap dimension. For example, if the gap was 0.056 in. and the specification required 0.004 in. more, then the correct thickness of shims would be 0.060 in.
m. Adjust the retainer plate fasteners to their final torque.

Lock nut adjustment: If a drive axle wheel bearing adjustment is done with a nut, the following generic procedure is likely to be followed. If the axle has an outboard final drive, the bearing will be behind the final drive ring gear hub and be part of a full-floating axle, whereby the axle shaft merely transfers torque to the sun gear of the final drive. If the axle has an inboard final drive, then the axle shaft supports weight and will have the bearings mounted on the shaft. The bearing nut-type adjustments are similar for both types of axles, but the following is for an inboard final drive semi-floating type of axle. This requires the outer axle housing to be removed.

a. The outer bearing cone and half of the metal face seal is installed on the axle shaft at the wheel end. This may require heating the bearing. The bearing and seal should be lubricated with axle oil.
b. The bearing cups are installed in the axle housing. This may require cooling the cups. The other half of the metal face seal is installed as well.
c. The housing is then installed over the axle shaft onto the axle shaft bearing and seal.
d. The inner bearing is lubricated and installed along with the adjusting nut on the axle shaft.
e. With the nut left slightly loose, a measurement of the torque it takes to turn the axle housing is taken. This needs to be considered when the rolling torque is measured.
f. The adjusting nut is then tightened while the housing is turned and the torque is measured that is required to turn the housing. There will be a specification that will indicate the proper rolling torque. This can be measured with a torque wrench or a string and scale. The adjusting nut can then be tightened or loosened to meet the specified torque. The nut will then be locked in place if it is not a locknut. Sometimes a second check is to see if there is endplay between the axle and the housing.

Whatever the procedure is, it must be followed to a tee and the preload must be confirmed and reset if it is found to be outside of specifications.

One way to check the wheel bearing setting when the axle is assembled on the machine is to check the vertical movement of the wheel in relation to the axle housing. When the weight is taken off the wheel by jacking the axle or using the machine blade or bucket, a dial indicator is mounted to the end of the axle housing and zeroed with its pointer resting against the wheel or axle shaft. The axle is then lowered to put the machine's weight back on it (with brakes applied) and the

FIGURE 15–14 Dial indicator arrangement to measure bearing preload.

dial indicator is read. An allowable maximum reading might be 0.015 in. If the reading is more than this, it would indicate a misadjusted, worn, or damaged bearing. ● **SEE FIGURE 15–14** for how the dial indicator is arranged to measure bearing preload.

Progress Check

1. This type of axle assembly will use the axle shafts to carry the machine's weight as well as transfer torque:
   a. full floating
   b. semi-floating
   c. double reduction
   d. planetary
2. The term used to describe what happens on a very cold morning inside an axle housing with too high of a viscosity oil is:
   a. channeling
   b. tunneling
   c. funneling
   d. barreling
3. The typical arrangement for a wheel loader is:
   a. solid-mounted front axle and oscillating rear axle
   b. front steering axle and rear oscillating axle
   c. front oscillating axle and rear steering axle
   d. oscillating front axle and solid-mounted rear axle
4. What is a drive axle *ring gear* either bolted or riveted to?
   a. axle shafts
   b. differential housing
   c. bearing cage
   d. pinion gear
5. How is the drive torque usually transmitted to the drive axle shafts in a heavy truck application?
   a. internal splines in the crown gear hub
   b. external splines in the crown gear hub
   c. internal splines in the side gears
   d. external splines in the side gears

WEB ACTIVITY

Search the Internet and find an old video that was produced by the U.S. army in the 1937 that explains the operation of a differential with the use of simple props and explanations. By viewing this video before reading on, you will greatly increase your comprehension of how a differential works.

DIFFERENTIAL

All drive axle assemblies will have a differential assembly as part of their arrangement. A differential changes the power flow through a mechanical driveline 90° and splits it to drive two wheels. It also allows each wheel to turn at different speeds under certain conditions. Through the action of a set of gears, a speed "differential" between the axles' wheels is provided. The differential is usually housed in the center of the drive axle housing, but it could be offset to one side for some applications such as the front drive axle of a tractor loader backhoe.

There are several different types of differentials: open, locking, **limited slip**, and double reduction. The basic operating principles of all types of differentials are the same. The common parts of all differentials are the pinion shaft assembly, differential carrier, crown gear, differential assembly, and **side gears**.
● **SEE FIGURE 15–15** for the main parts of a differential.

An open differential is the simplest and will be discussed first.

As the machine goes through a turn, its left and right wheels will turn through different arcs, travel different distances, and travel at different speeds. If they weren't allowed to have this speed differential, the machine would try to keep travelling straight, be difficult to control, and scuff the tires. The differential is necessary to prevent the tires from wearing out prematurely and to reduce the strain on axle components.

The differential gets its input from a drive shaft that is usually driven by a transmission. The drive shaft drives a universal joint that drives the input yoke to the differential. This yoke is splined to the pinion gear shaft and the shaft drives a pinion gear that is part of the shaft.

The pinion gear then drives the crown gear, which is suspended by bearings in the differential housing. The pinion shaft is also supported by a pair of tapered roller bearings that need to be preloaded to provide a specific amount of rolling resistance. This will equate to the correct amount of bearing preload

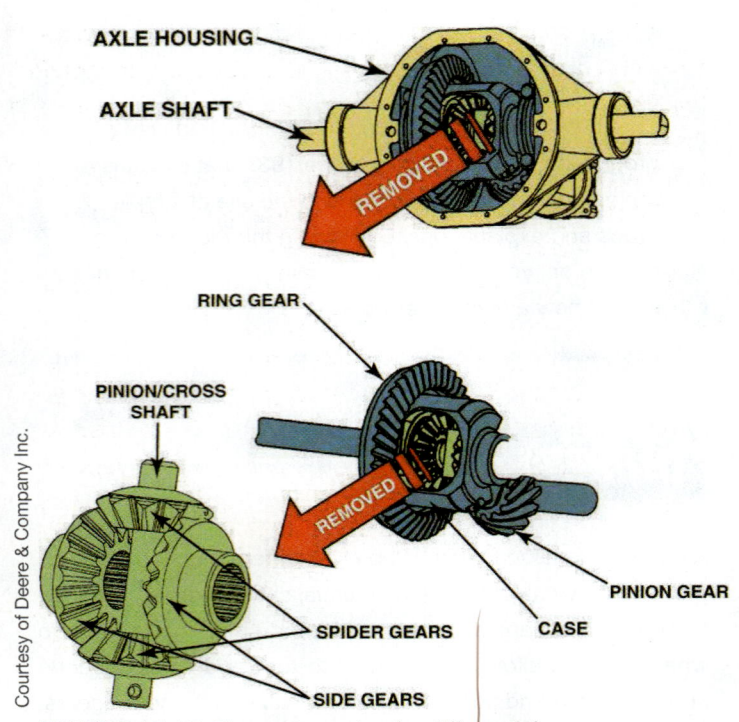

FIGURE 15–15 The main parts of a differential.

to ensure that the bearings will limit endplay and support any load placed on the shaft. The pinion shaft will also have a lip-type seal ride on it to keep dirt out of the axle housing and oil in.

The pinion gear drives the crown gear and the crown gear is bolted to the differential housing. Inside the differential housing is a set of gears that can provide a left wheel to right wheel speed differential. The crown gear and differential housing assembly is supported with a set of tapered roller bearings. These bearings will also need to be preloaded for the same reasons as the pinion shaft bearings.

Inside the differential housing is a set of gears that will send torque out each side of the axle housing through left and right axle shafts. These shafts will then transfer torque to each wheel. The crown and pinion gear action is the first gear interaction that occurs in a differential when the machine is in motion. The number of teeth on each gear will determine the gear ratio that provides a torque increase and speed decrease. For example, if the pinion has 10 teeth and the crown has 41 teeth, the ratio is 4.10:1. This means that the pinion has to rotate just over four revolutions to make the crown gear rotate one revolution. For more information on gear ratios and types of gears, refer to Chapter 11.

Because of the shape of the teeth on the pinion and crown gears, they are considered to be spiral bevel-type gears. Also if the pinion centerline is above the crown gear centerline, it is considered to be an amboid gearset. If the pinion centerline is below the crown gear centerline, it will be considered to be a hypoid gearset.

As the load increases through the drive line, the crown and pinion gears will try to separate as a result. Heavier-duty units will have a thrust screw that will support the bearing cap behind the ring gear when heavy loads are applied to the differential.

The differential housing is machined to drive a spider (sometimes called a cross) at the same speed as the crown gear, and this spider provides an axis for four smaller bevel gears that are held in the differential housing. The bevel gears will have a dished thrust washer between their back and the differential housing. The thrust washer is needed because the force acting on the bevel gear will try to push it into the housing.

The spider is shaped like a cross with four shafts extending from a center hub. The shafts are machined smooth to allow the spider gears to freely rotate on them. The ends of the shafts will be secured in the differential housing to keep it rotating with the crown gear. (Some lighter-duty differentials will use two or three spider gears.)

These bevel gears can also be called pinion gears or spider gears. There are usually four bevel pinion gears that are meshed with two side gears that have internal splines and in turn drive the two axle shafts. The shafts then go out to drive the wheels. The side gears are free to rotate independently of the differential housing so they can allow a change of speed between the crown gear and the side gear. This is the second gear interaction that can occur in a differential and can happen simultaneously while the pinion gear is driving the crown gear.

If the machine is travelling straight ahead, the bevel pinion gears do not rotate on their shaft; therefore, the only gear interaction is the pinion driving the crown gear. When the machine turns a corner or one wheel starts to spin, the bevel pinion gears will start to rotate on their shafts (part of cross) and "walk around" the side gears. It's this second interaction action that provides a speed differential between the left and right wheels. This will change the speed of the axle shafts. For example, if the crown gear is rotating at 100 rpm and the machine is making a gradual left turn, the left side gear and axle shaft will turn 90 rpm and the right side gear and axle shaft will turn 110 rpm. If the machine turns sharper, the left axle shaft may now turn 75 rpm while the right one turns 125 rpm. When the machine then straightens out, all three gears (crown and both side gears) will turn at the same 100 rpm again.

This differential is called an open differential because the input torque is able to transfer wherever it can flow most easily (take the path of least resistance). There is no way of controlling how much torque is sent to each wheel. The main factor in determining how torque is split between the two wheels is the traction the wheels have.

One of the disadvantages of an open differential is that if one wheel was to find poor traction while the other wheel on the same axle had good traction, the majority of torque going into the drive pinion will go to the wheel with the poor traction. This would result in the poor traction wheel spinning uncontrolled and leads to a machine being stuck. This could also lead to differential, axle, or other damage from an over-speeding wheel.

The ultimate speed differential situation would happen if one wheel is sitting on ice and the other is on pavement. An open differential will send all the output torque to the wheel sitting on ice because there is very little resistance compared to the one on pavement. In this case, the speed of the axle shaft coming out of the differential will be twice the speed of the ring gear. This occurs because one side gear is stationary, and as the ring gear drives the cross, the bevel pinions will walk the other side gear around at double the speed of the crown gear. This shouldn't be allowed to happen for long because of the excessive speed of the axle shaft. In this case, if the crown gear rotates at 100 rpm, the wheel that is on pavement turns at 0 rpm while the wheel sitting on ice with poor traction and no resistance will turn at 200 rpm. ● **SEE FIGURE 15–16** for the action of a differential if there is a speed differential between both wheels.

LIMITED SLIP DIFFERENTIAL
Because heavy equipment machines work in adverse conditions where traction is limited, the drivetrain will need to send torque to tires with the best traction to try to keep the machine moving. If a machine has an open differential, is sending torque to two wheels, and has poor or no traction in one wheel, that is the wheel that gets the majority of the torque because it has the least resistance. This means the machine will likely get stuck or at least be spinning the wheel with the poor traction and create unnecessary tire wear.

A limited slip differential is designed to equalize the amount of torque sent to each wheel. This will reduce wheel spin and keep the machine moving longer. As the machine usually isn't trying to turn a corner when torque equalization is needed, it isn't important that a speed differential is allowed between the wheels. In other words, when the limited slip function actuates, there is poor traction conditions and the strain on the axle components is reduced. It is more important to send torque to the wheel with the best traction and to keep the machine moving.

Most limited slip differentials use one or two sets of clutch discs and plates to reduce the differential action. These discs and plates are similar to the discs and plates used in powershift transmissions, as discussed in Chapter 12. ● **SEE FIGURE 15–17** for how a limited slip drive axle works when the tires have equal traction.

When a speed differential action starts to change the speed and torque between the left and right axle shafts, the bevel pinion gears are rotating. As the load increases between the bevel pinion and side gears, they will try to move away from each other. This action in turn creates a side thrust on the side gears of the differential. It is this side thrust action that is

> **TECH TIP**
>
> If you want to see an example of how the bevel pinion gears work, raise the tires off the ground on the axle of a vehicle that has an open differential. Place jack stands under the axle and chock the wheels. With the drive shaft held stationary, you can now turn one wheel and observe the other one. What happened and why?
>
> The pinion shaft and crown gear are being held stationary. When one wheel is turned, it is turning one side gear, which turns the bevel pinion gears. Their shafts are stationary so the only movement that can occur is to have the opposite side gear driven. When that happens, the opposite wheel is driven in the opposite direction.

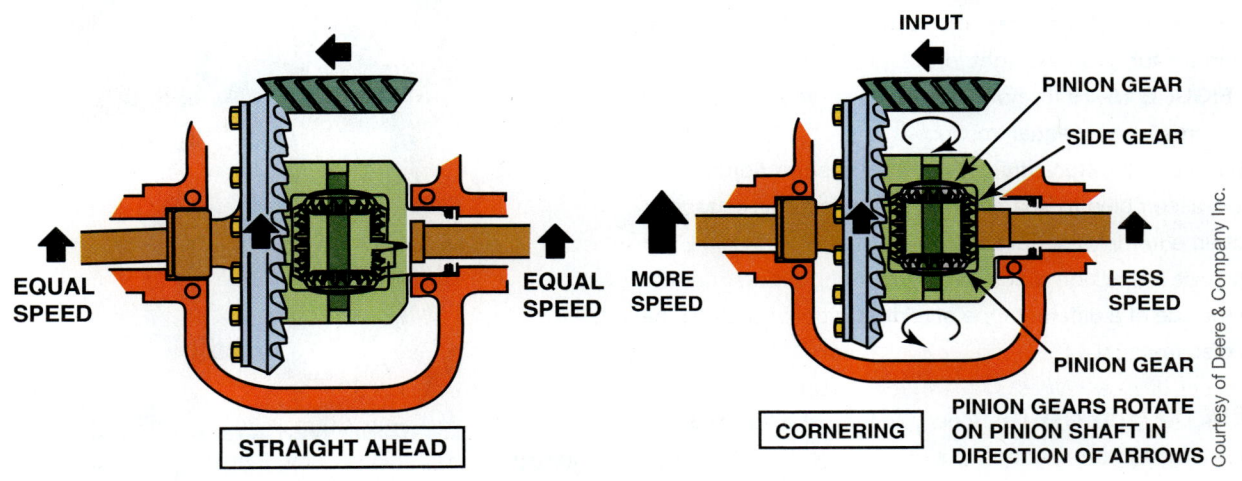

FIGURE 15–16 A differential action when one wheel slows down.

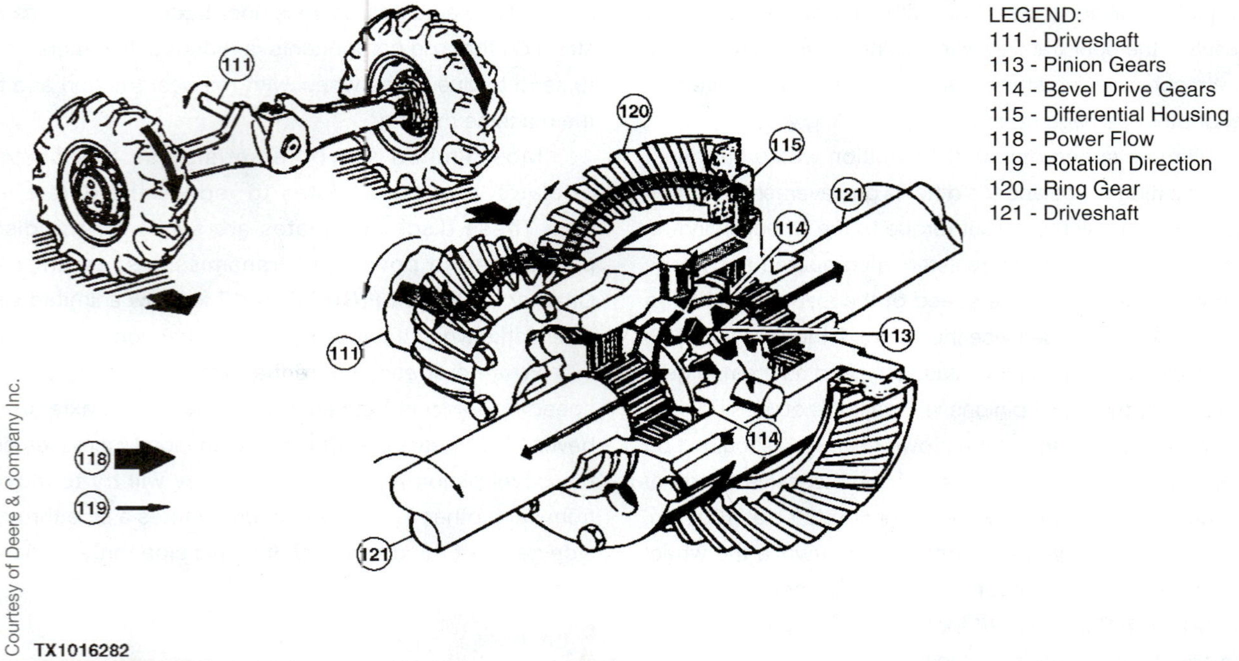

FIGURE 15–17 How a limited slip drive axle works when the tires have equal traction.

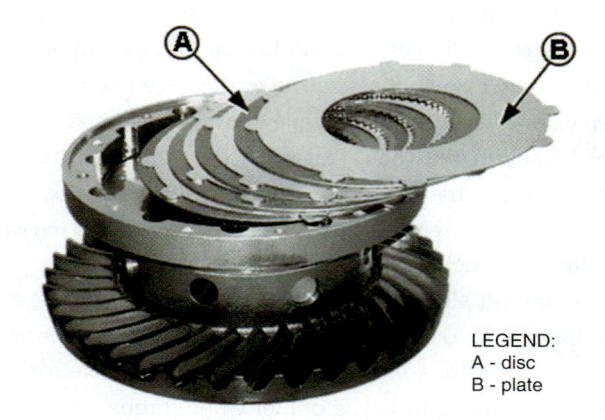

LEGEND:
111 - Driveshaft
113 - Pinion Gears
114 - Bevel Drive Gears
115 - Differential Housing
118 - Power Flow
119 - Rotation Direction
120 - Ring Gear
121 - Driveshaft

used for applying pressure to the clutch packs in a limited slip differential. ● SEE FIGURE 15–18 for how the clutch discs and plates are arranged with the differential case.

The clutch packs would then attempt to match the speed of the side gear to the speed of the differential housing. When this happens, the differential action is reduced, which then equalizes the torque to the left and right axle shafts and ultimately sends torque to the wheel with better traction.

The clutch pack will have friction discs that have internal splines to lock them to the side gears and plates that have lugs that lock them to the differential housing. As the side thrust is applied to the side gear, its thrust face applies force to the plates and discs, which will try to spin the side gear at the same speed as the differential housing. This action limits the slip of the wheel with bad traction but still provides a differential action for turns during light load/high traction conditions.
● SEE FIGURE 15–19 for how a limited slip differential works when the tires have unequal traction.

This type of differential will require a shim adjustment when it is assembled to enable the clutch to start to engage at the right time (under a certain amount of load).

The use of friction discs in a limited slip differential will require the use of a different type of oil that is compatible to the friction disc material.

NO SPIN DIFFERENTIAL A no spin type of differential is a mechanical unit that replaces the side gears, cross, and bevel pinion gears with other mechanical pieces to lock the two axle

LEGEND:
A - disc
B - plate

FIGURE 15–18 How the clutch discs and plates are arranged with the side gear.

444 CHAPTER 15

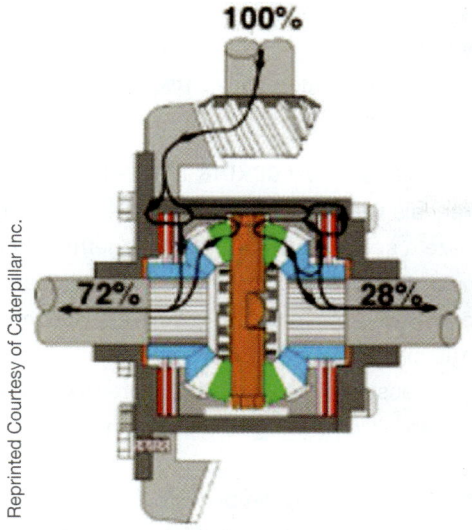

FIGURE 15–19 How a limited slip differential works when the tires have unequal tractions.

shafts together under certain conditions. These differentials can directly replace most standard open differentials and are an option available for most machines.

This type of differential will mechanically keep the left and right axles driving together unless one wheel starts to speed up as in a wheel spin condition. The faster wheel will then be unlocked. This typically only happens when the machine is turning through a corner in high traction conditions.

In simple terms, the no spin differential is really just two spring engaged dog clutches that are normally engaged until one wheel starts to turn faster than the crown gear. In operation, this will keep the differential locked unless the machine turns a corner and then only the slower wheel is driven. In other words, when the machine encounters slippery conditions, both wheels will continue to drive, but when the machine turns on good footing, one wheel will disengage to allow the wheels to travel at different speeds. This will ease turning and reduce tire wear.

There is a spider that is driven by the crown gear, and it has a row of teeth on each side that engage with a driven clutch. The clutch is splined to accept an axle shaft that sends torque out of the differential. There is a cam in the center of the spider and two holdout rings that initiate the disengagement of one of the clutches when a speed difference between the driven clutches happens. ● **SEE FIGURE 15–20** for a no spin differential.

LOCKING DIFFERENTIAL Many machines will be equipped with locking differentials that will perform like an open differential under normal operation until either the operator or an ECM commands a mechanism to lock the side gears to the differential housing. This will make the axle shafts com-

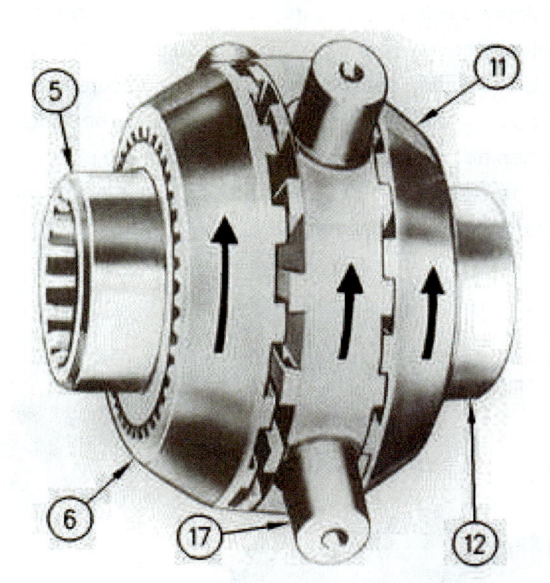

FIGURE 15–20 A no spin differential.

ing out of the differential rotate at the same speed as the crown gear and eliminate any differential action.

These are commonly used for articulated truck axles and inter-axle differentials. If all axles and inter-axle differentials are locked up on a six-wheeled articulated truck, it will be a true six-wheel drive.

DRIVE AXLES, POWER DIVIDERS, FINAL DRIVES   **445**

There are a few different ways of locking a differential: mechanical, air pressure, and hydraulic pressure. Let's explore these different mechanisms.

MECHANICAL. This used to be common for use in backhoe loaders. The operator will step on a pedal in the cab. The pedal is mechanically linked to a lever on the side of the axle housing. The lever will pivot and inside the housing mechanically move a jaw clutch that locks one of the side gears to the differential housing. This eliminates the differential action. When the pedal is released, the side force acting on the side gear is enough to move the clutch away and release the side gear from being locked to the housing.

Because one of the side gears is locked to the differential housing and the bevel pinion gears are carried around with the differential housing, the opposite side gear also has to rotate at the same speed as the differential housing.

This type of clutch could also be actuated by an air cylinder. The operator would step on a valve in the cab that sends air to the cylinder. The cylinder's piston then extends and actuates a lever that moves the clutch. From there the action is the same as the mechanically actuated locking differential. ● SEE FIGURE 15–21 for a locking mechanical differential.

PNEUMATICALLY ACTUATED LOCKING DIFFERENTIAL. A differential can be locked with air pressure only. An air signal goes to a piston that will move out and apply pressure to a jaw clutch. There has to be a rotating seal that seals the air pressure between the rotating part of the differential and the stationary housing.

When air pressure is sent to the piston, the piston will lock one of the side gears to the differential housing to make the axle shafts turn at the same speed.

The air signal will come from a valve in the operator's cab and is usually foot actuated.

This type of locking differential is used for some Caterpillar scraper differentials.

HYDRAULICALLY ACTUATED LOCKING DIFFERENTIAL. Many locking differentials will use a hydraulic signal to move a piston and squeeze a set of plates and discs together. When engaged, the multi-disc clutch locks one of the side gears to the differential housing and prevents any differential action.

As with the pneumatically locking differential, there has to be a rotating seal assembly to allow the oil pressure to transfer from the stationary axle housing to the rotating section of the differential.

Also similar to an axle assembly with inboard brakes, this type of locking differential could create cross-contamination if the seal for the piston were to fail.

This system will likely be controlled electrically by a switch in the cab. It could even be part of the machine's CAN system. A solenoid will get energized and send oil pressure through steel tubes or hoses to the axle housing where it passes through the housing and is fed into the differential through a rotating seal. ● SEE FIGURE 15–22 for a cross-section of a hydraulic locking differential.

When a differential with a hydraulic locking function is reconditioned, the piston should be checked for leaks with air pressure after differential installation.

FIGURE 15–21 A mechanical locking differential.

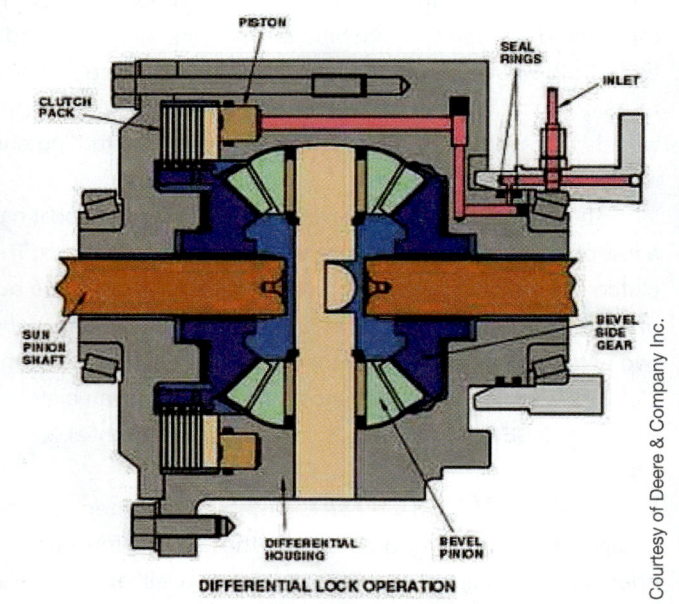

FIGURE 15–22 A cross-section of a hydraulic locking differential.

DOUBLE REDUCTION DIFFERENTIAL Some differentials will feature a second gear reduction to increase torque to the axle shafts even more. There could be a planetary gearset mounted inside the crown gear housing that will provide a gear reduction to the differential or a second gear reduction before the spiral pinion input to the crown. This may be necessary for machines that don't have inboard or outboard final drives. ● **SEE FIGURE 15–23** for a double reduction differential.

TWO-SPEED DIFFERENTIAL Some machines require an extra gear ratio to be provided by the differential. A two-speed unit will have an extra set of gears as described in the preceding section but will also have a way to send torque through the reduction or bypass it. This will give a high and low range feature that can be selected by the operator and engaged with an air cylinder. This type of differential used to be commonly used for medium duty on highway truck applications. ● **SEE FIGURE 15–24** for a two-speed differential.

DIFFERENTIAL DIAGNOSTICS Some differential troubleshooting procedures could stem from an operator complaint such as no drive, machine gets stuck easier than it used to,

FIGURE 15–23 Double reduction differential.

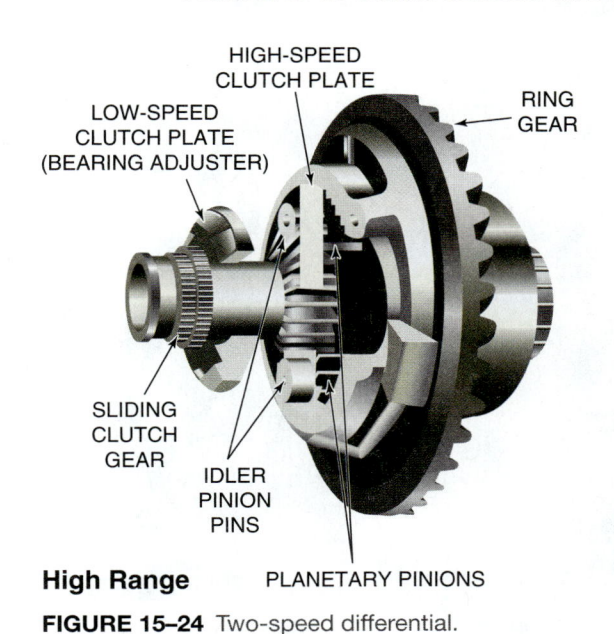

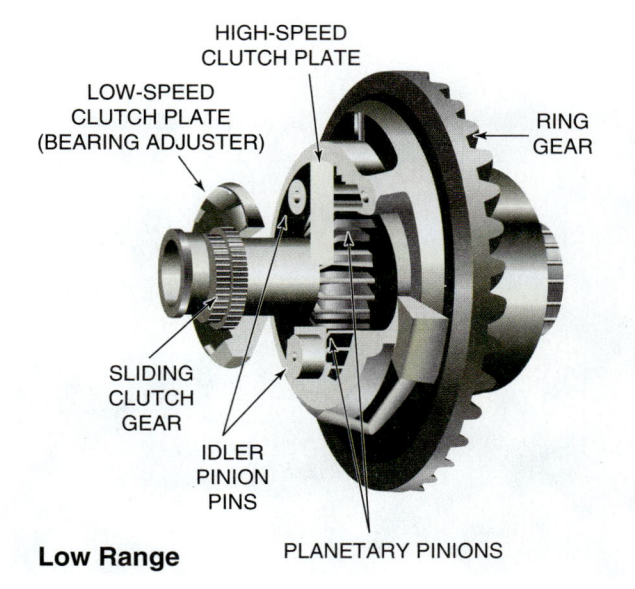

FIGURE 15–24 Two-speed differential.

DRIVE AXLES, POWER DIVIDERS, FINAL DRIVES 447

leaks, noises, vibrations, or machine is hard to turn. These complaints will lead to a technician performing a diagnostic procedure to determine the cause of the problem.

Of course you would need to verify the complaint first. If you are going to operate the machine, make sure that you are familiar with how to operate it safely and there is plenty of room to run the machine.

One of the first things to do is a visual inspection for damage and then check the oil level and take an oil sample while doing this.

No drive means you would need to see if the drive shaft input to the differential was turning. This would eliminate all other preceding driveline components.

If the machine gets stuck easier than it used to, you would need to check and see if the machine has either a limited slip differential or a locking differential and then see if it is working properly.

A leak at the pinion shaft seal should be fairly easy to diagnose. You may need to clean the area first and top up the axle fluid to see if that is where the leak is originating from. If this is the leak location, then you could start a repair process to replace the seal and install a wear sleeve on the pinion yoke. Don't forget to check the axle breather to see if it's plugged and causing the housing to over-pressurize and leak.

Noises and vibrations are going to be hard to narrow down to the differential. It may be necessary to remove the drive shaft and axle shafts for the axle to determine if the differential is the source of the vibration. Oil sample reports that show a high metal content or just a visual inspection of the oil after it's drained could lead you to suspect a failed differential. Be careful not to jump to conclusions because some wear is normal, and if the axle oil hasn't been changed, it could look a lot worse than it is. However, if chunks of teeth come out with the oil, this means without a doubt the differential needs to be removed.

If a machine is hard to turn, it could be a problem with the limited slip or locking differential not releasing. If it is a locking differential, the control for it should be checked for proper operation. If this isn't the problem, then it will require the differential be removed to confirm and repair the problem. This could be a sticking piston, a sticking disc, or a part of the mechanical linkage not returning inside the housing.

You need to be certain the problem lies with the differential because this will require the differential to be removed, and this is usually a big job.

DIFFERENTIAL CARRIER REMOVAL
If a defective differential is suspected from your diagnostic procedure, the only option is to remove the differential and repair it.

Some differentials can be removed without taking the axle assembly out of the machine. This will depend on the clearance available ahead of the differential because it will have to come ahead quite a distance for the crown gear to clear the axle housing. Most often there will not be sufficient room to allow this. Even if there is room to do, this it will likely be more time efficient and safer to remove the entire axle assembly and then remove the differential.

The axle shafts will need to be removed first and the differential could be lifted straight out, or it could be lifted up after the axle housing is turned on its back.

There will be a ring of bolts or nuts to be removed and then likely two or three forcing bolts will be installed to push the differential away from the housing. ● **SEE FIGURE 15–25** for a differential carrier being removed from axle housing.

FIGURE 15–25 A differential carrier being removed from an axle housing (reprinted courtesy of Caterpillar Inc.).

DIFFERENTIAL REPAIR Once the differential is removed, it is easiest to repair it if it is installed on a rotating stand or a fixed stand. The type of stand used will depend on the size of the unit and what is available at the shop you are working in. A rotating stand will enable you to rotate the differential to any position to make the repair easier.

Any time a differential is removed, its components should be thoroughly inspected for wear and damage before it is disassembled. The ring gear should be closely inspected to see if its wear pattern is normal. It's always a good idea to measure **backlash**, measure pinion shaft endplay if it's supposed to be there, and check the tooth wear pattern before disassembly. You should also take several pictures with a digital camera to keep for reference and evidence if there is a warranty issue. ● **SEE FIGURE 15–26** for a differential being adjusted.

As there are many variations of differentials, a rebuild procedure will be highlighted from a medium-duty wheel loader with an open differential. This is a typical process that may be close to the differential that you are rebuilding but in no way should be used as a guideline for any specific unit. *Always* consult the proper service information for the differential you are working on because there could be changes to parts and procedures even within a close serial number range of machines.

This procedure should be carried out in a clean, well-lit area with plenty of bench space. There will be some special tools required so it's a good idea to check that you have the required tools and that they are in proper working condition before starting the procedure.

Mark all mating components of the differential before disassembly to aid in assembly. A paint stick, permanent marker, or carefully placed punch marks can be used for this. To remove bearings, the proper pullers should be used to avoid damaging components that will be reused.

The following is a generic version of a differential assembly process. *Always* use pinion gears and crown gears as a matched set. Use clean oil of the type that will be used in the axle housing to lubricate bearings as they are installed.

1. **Install inner tapered roller bearing onto pinion shaft:** This will likely require heating and pressing the bearing on. The race of the cone will seat against a shoulder on the pinion shaft. ● **SEE FIGURE 15–27** for how a bearing is installed on a pinion shaft.

2. **Adjust pinion preload:** The pinion can rotate in a separate housing that bolts to the differential carrier or in the carrier itself. The pinion shaft is supported by two tapered roller bearing assemblies. The preloading of these bearings can be done with shims or a lock nut. ● **SEE FIGURE 15–28** for a pinion shaft and housing assembly.

   Shim adjustment: Shims will determine the distance between the tapered roller bearings' cups. The inner bearing cone is installed on the pinion shaft. The outer cup is installed in the housing. The inner cup is shimmed

FIGURE 15–26 A differential being adjusted.

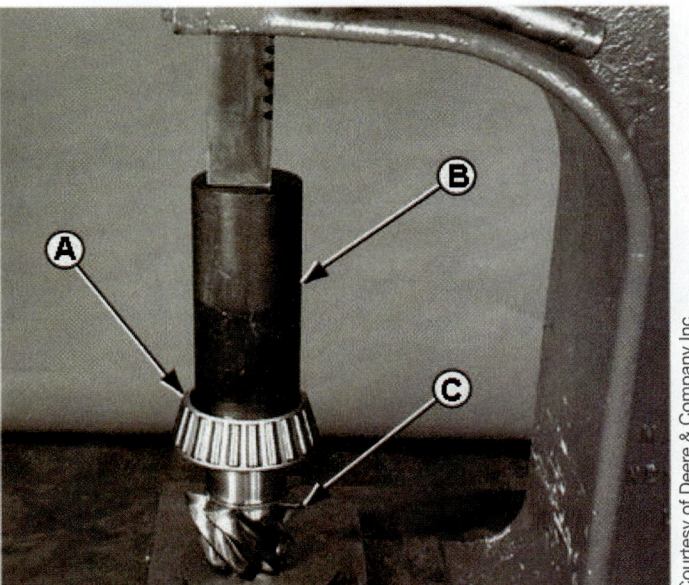

FIGURE 15–27 Installing a bearing onto a pinion shaft.

DRIVE AXLES, POWER DIVIDERS, FINAL DRIVES 449

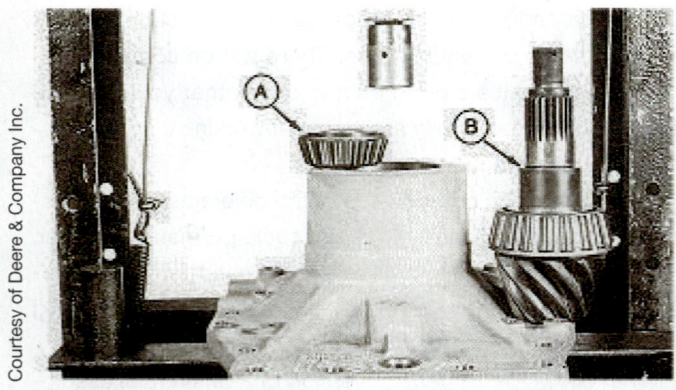

FIGURE 15–28 A pinion shaft and housing assembly.

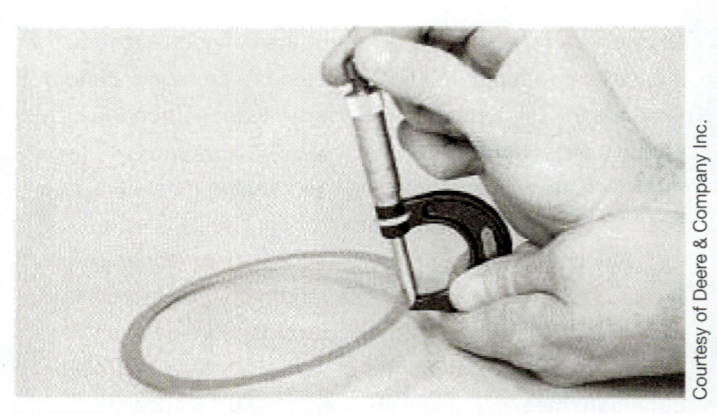

FIGURE 15–29 A shim pack being measured.

to set the pinion depth (see step 4). The pinion shaft is installed in the housing and then a shim pack is install that will go under the outer cone. This will provide either specified preload or endplay to the pinion shaft.
- **SEE FIGURE 15–29** for a shim pack being measured. With the outer cone installed, the endplay can be measured with a dial indicator, or if there is a preload required, it is measured with a torque wrench or string and scale. If the endplay or preload is found to be out of specification, then the shim pack thickness will need to be increased or decreased. **SEE FIGURE 15–30** for pinion shaft endplay being checked.

Sometimes a press is needed to do this, and it will simulate the clamping force created when the yoke is installed on the pinion shaft.

Lock nut adjustment: Other differentials will have pinion preload adjusted with a lock nut on the pinion shaft. The lock nut will put pressure on the outer pinion bearing cone. Sometimes the nut will be under the yoke, or it can be the retaining nut for the yoke.

An example of a proper amount of preload is rolling torque with no seal 9–18 in./lb when turned at 3–5 rpm. This will be done without the pinion seal installed to get a more accurate reading. Once the proper preload is set, then the seal is installed. **SEE FIGURE 15–31** for how rolling torque is measured.

3. **Install yoke:** The drive shaft yoke is installed and its fastener is torqued to specification. Recheck rolling torque. Sometimes a dimension is given for checking shaft endplay. An example is to apply 50 lb of force and measure endplay. A proper dimension might be 0.00–0.001 in.

4. **Pinion depth:** This can be a calculated dimension that relates to how far the pinion protrudes in to the differential carrier and is sometimes called the cone point adjustment (this refers to the shape of the pinion gear). It can also

FIGURE 15–30 A pinion shaft endplay being checked.

FIGURE 15–31 Rolling torque measured.

450 CHAPTER 15

be set according to the amount of backlash between the pinion gear and crown gear.

Cone point adjustment: A measurement will be taken in the differential carrier that relates the centerline of the ring gear to the face of the surface that the inner pinion bearing rests on or the face of the surface that the pinion housing rests on. This dimension will then have the dimension of the inner bearing cup and the dimension marked on the pinion gear subtracted from it. The dimension that is left will be the thickness of the shim pack that will be installed under the inner bearing cup.

Pinion housing adjustment: The pinion depth dimension could also be changed by adjusting the amount of shims under the pinion housing flange when it is installed in the differential carrier. ● **SEE FIGURE 15–32** for shims placed under the pinion housing.

FIGURE 15–32 Shims under pinion housing.

5. **Install pinion assembly:** Install O-ring seal and torque a ring of bolts to specification to hold the pinion assembly to the differential carrier. Remove yoke and install pinion seal in housing. Install yoke and torque nut to specification. This may require special tools to hold the yoke while torqueing the nut. This could be several hundred foot-pounds of torque.

6. **Assemble differential housing:** One side of the housing is placed on a bench with the opening facing up. One side gear is installed into housing. The bevel pinion gears are installed on to the cross and then the thrust washers are installed on the ends of the gears. The assembled cross is placed in the slots in the housing. The second side gear is laid on top of the bevel pinions and then the other half of the housing is installed with ring of fasteners that are torqued to specification. The crown gear is then installed on the housing with a ring of fasteners and torque to specification. If the differential has a limited slip device, the clutch packs would be installed at this stage. If it was a locking differential, the locking mechanism would be installed now.

FIGURE 15–33 An installation of a tapered roller bearing on the differential housing.

7. **Install differential bearings:** The bearing cones will be pressed onto the housing. Sometimes heating the bearings will ease installation. Be sure to not overheat bearings. ● **SEE FIGURE 15–33** for the installation of a tapered roller bearing on the differential housing.

8. **Back out thrust screw:** If the differential has a thrust screw, it should be backed out to avoid interfering with differential installation.

9. **Install differential assembly into carrier:** The proper lifting device needs to be used to carefully lift the assembly into the carrier. ● **SEE FIGURE 15–34** for how a differential assembly is lifted into the carrier.

FIGURE 15–34 Lifting the differential housing into the carrier.

10. **Install outer bearing races and trunnion caps:** With the differential assembly hanging freely from a hoisting device, install bearing cups, caps, and adjusting nuts. The adjusting nuts are specifically designed to push on the bearing cups and usually have a series of lugs protruding from the outside face that allow the nut to be turned with a prybar or a

special socket. The nut will have fine threads that mate with threads in the differential carrier and the bearing caps. The caps need to be left slightly loose when adjusting the nuts.

Care must be taken to not damage the threads in the carrier or caps at this point. Remove lifting device.

11. **Remove axial endplay:** Adjusting nuts should be tightened at this point to remove any endplay the differential assembly may have. Always rotate the crown gear to rotate the bearings when adjusting bearing preload to help seat the rollers. It may be required to turn the adjusting nuts an additional few notches to add some preload to the bearings.
● **SEE FIGURE 15–35** for the adjustment of an adjusting ring while checking backlash. Proper differential carrier bearing preload is usually accomplished when there is zero endplay.

12. **Adjust backlash (clearance between crown and pinion gears):** With the pinion gear locked in place, use the adjusting nuts to adjust backlash according to specification. An example of this is 0.011–0.013 in. This is measured with a dial indicator that has its pointer resting on one tooth of the crown gear. Backlash is necessary to allow the gears to have clearance to allow lubrication between the meshing teeth. It can be adjusted by tightening one adjusting nut and loosening the opposite side same amount. This will move the crown gear closer to or farther away from the pinion gear.

13. **Set rolling drag of differential assembly:** The adjusting nuts will now be turned in slightly to preload the bearings. This will result in a resistance to turning that can be measured with a string and scale or some special tools. An example of the proper amount of rolling drag is 26–35 in./lb.

14. **Check gear contact pattern:** This step will check to see if the crown and pinion gears are meshing properly. With marking grease, Prussian blue or red lead or other paint like liquid applied to the teeth rotate the pinion in both directions while applying a slight load on the ring gear. Compare the mark left on the ring gear to a chart like that shown in ● **FIGURE 15–36**.

If necessary, adjust backlash and pinion depth to obtain correct pattern. This can take a while. Be patient and try to anticipate how much adjustment needs to be made before making one.

15. **Install adjusting nut locks:** There are different ways to lock the adjusting nuts. Roll pins or bolt on locks are two examples. ● **SEE FIGURE 15–37** for how a roll pin is used to lock the adjusting nut.

FIGURE 15–35 An adjustment of an adjusting ring while checking backlash.

A - Ideal Tooth Contact
B - Backlash Must Be Decreased
C - Backlash Must Be Increased
D - Ideal Tooth Contact
E - Backlash Must Be Decreased
F - Backlash Must Be Increased
G - Coast Side (concave)
H - Drive Side (convex

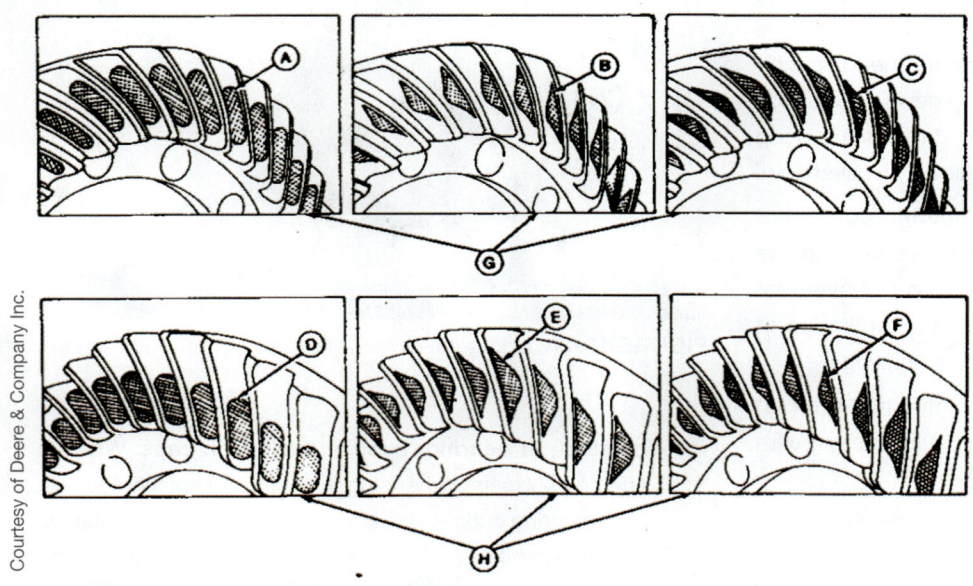

FIGURE 15–36 Gear tooth contact pattern.

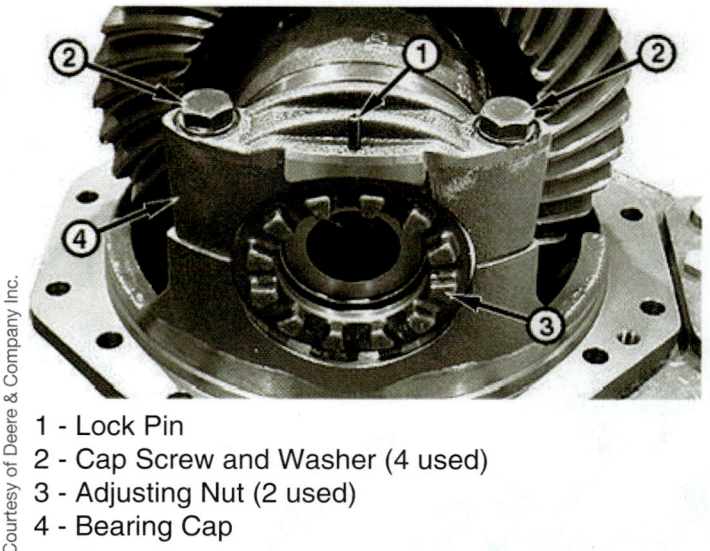

1 - Lock Pin
2 - Cap Screw and Washer (4 used)
3 - Adjusting Nut (2 used)
4 - Bearing Cap

FIGURE 15–37 How a roll pin is used to lock the adjusting nut.

16. **Adjust thrust screw:** The thrust screw needs to be close to the back of the crown gear, and this is done by turning it in until it contacts crown gear and then backing it off slightly. A lock nut is tightened to hold it in place.

The differential is now ready for installation into the axle housing.

POWER DIVIDERS

Power dividers will be used when a machine has a tandem drive axle arrangement. It will transfer drive to the rear axle from the drive shaft that drives the front drive shaft. A power divider is a set of gears that are part of a housing that contains a differential assembly for the forward axle, a second set of gears, and a splined shaft that sends power out the back of the housing to the rear axle. ● **SEE FIGURE 15–38** for the tandem rear axle arrangement for an articulated truck.

Some power dividers will have a differential as part of it that can provide a speed differential between the front and rear

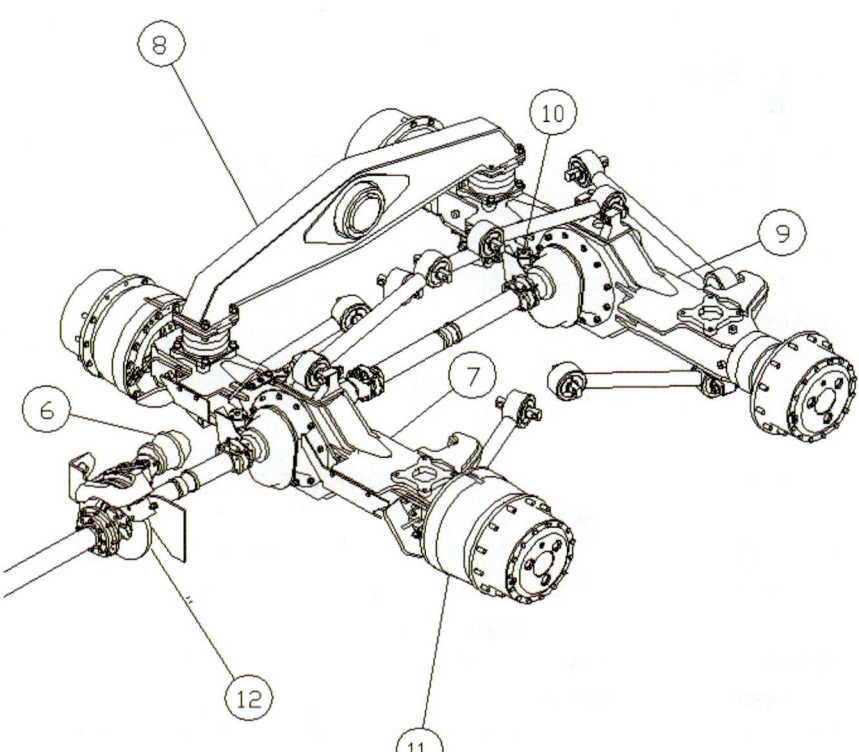

6 - Park Brake Actuator
7 - Middle Axle
8 - Walking Beam
9 - Rear Axle
10 - Differential Lock Actuator
11 - Service Brake Wet Disk
12 - Park Brake Disk
13 - Suspension Strut

FIGURE 15–38 The tandem rear axle arrangement for an articulated truck.

DRIVE AXLES, POWER DIVIDERS, FINAL DRIVES

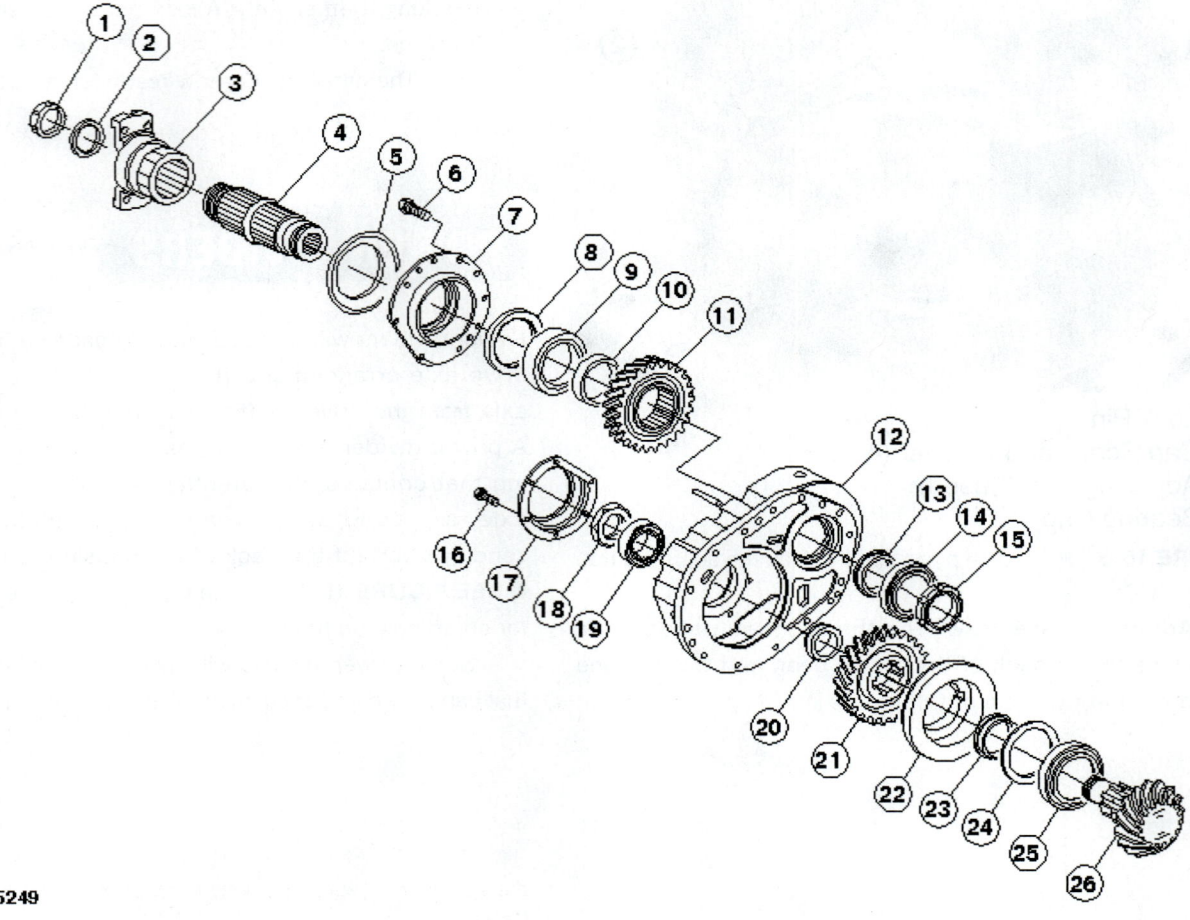

T195249

1 - Spanner Nut	11 - Spur Gear	20 - Shim
3 - Yoke	12 - Adapter Case	21 - Spur Gear
4 - Input Drive Shaft	13 - Shim	22 - Bearing Retainer
5 - Cover	14 - Roller Bearing and Cup	23 - Shim
7 - Bearing Housing	15 - Spanner Nut	24 - Shim
8 - Input Drive Shaft Seal	17 - Cover	25 - Roller Bearing and Cup
9 - Roller Bearing and Cup	18 - Nut	26 - Pinion Shaft
10 - Shim	19 - Roller Bearing and Cup	

Courtesy of Deere & Company Inc.

FIGURE 15–39 An exploded view of a power divider.

tandem axles. This is called an inter-axle differential, and it allows for easier turning and less tire wear when travelling on high traction surfaces. This differential will also have a locking mechanism that is operator controlled for use when the machine encounters soft and muddy conditions.

The input to the power divider is a yoke that is driven by a drive shaft. The yoke is splined to a shaft that is supported by a pair of bearings, and it has a gear splined to it. For non-differential-equipped power dividers, the shaft is splined to accept a second yoke that will send torque to the rear axle through a drive shaft. ● **SEE FIGURE 15–39** for an exploded view of a power divider.

Both types of power dividers will send torque to the front axle differential through the gear that is splined to its input shaft. This gear will mesh with a second gear that has an equal number of teeth. The second gear then drives the pinion for the front axle differential.

For power dividers that use a differential, the input shaft will have splines that drive the inter-axle differential cross. The cross then carries a set of bevel pinions around, and they

454 CHAPTER 15

FIGURE 15–40 A power divider used in an articulated truck.

will drive two side gears. One side gear drives a helical gear that is meshed with a second gear of the same size. This gear is splined to the drive pinion for the front axle differential and the second side gear is splined to the output shaft that will send torque out to the rear axle. The differential lock will simply lock the front side gear to the input shaft, and this will eliminate any differential action between the front and rear axles. When the inter-axle lock is disengaged, the majority of torque will be sent to the axle with the least traction. In an extreme case of a tandem axle arrangement having one wheel on a patch of ice and the other three on pavement a condition called spinout could occur. This can cause severe damage quickly because of the inter-axle differential being driven at drive shaft speed.

Many power dividers will have a gear pump to distribute lubricating oil from the bottom of the axle housing to the inter-axle differential. ● **SEE FIGURE 15–40** for a power divider used in an articulated truck.

FINAL DRIVES

Final drives will be the last gear reduction before either wheels or tracks are driven.

For rubber-tired machines, they will most likely be planetary gearsets, and for track machines, they could be planetary but may also be single or double reduction gearsets.

A typical planetary final drive will have the input shaft rotate the sun gear while the ring gear is held stationary and the planet carrier is the output. The drive axle's output shaft will be splined to the sun gear. There could be two to four planet gears on shafts that drive the planet carrier. ● **SEE FIGURE 15–41** for a planetary final drive.

The planet gears ride on needle bearings that ride on shafts that are fixed to the planet carrier. The shafts will be held in place with dowels or could be press fitted. Each side of the planet gears will have plastic or brass thrust washers.

Some machines will use a double reduction planetary final drive for a higher torque increase and speed reduction. This will allow a smaller gear reduction at the pinion and crown gear combination in the differential. Mining trucks in the ultra-weight class (200 tons plus) will use a double reduction final drive. Both planetary sets' ring gears will be held stationary. The axle shaft will drive the sun gear for the first planetary reduction and its output (planet carrier) will drive the sun gear for the second planetary gearset.

Depending on where the final drive is mounted, it could have its own lubricating oil or could share it with a drive axle. If a final drive is used for a steering axle, it will have its own oil supply.

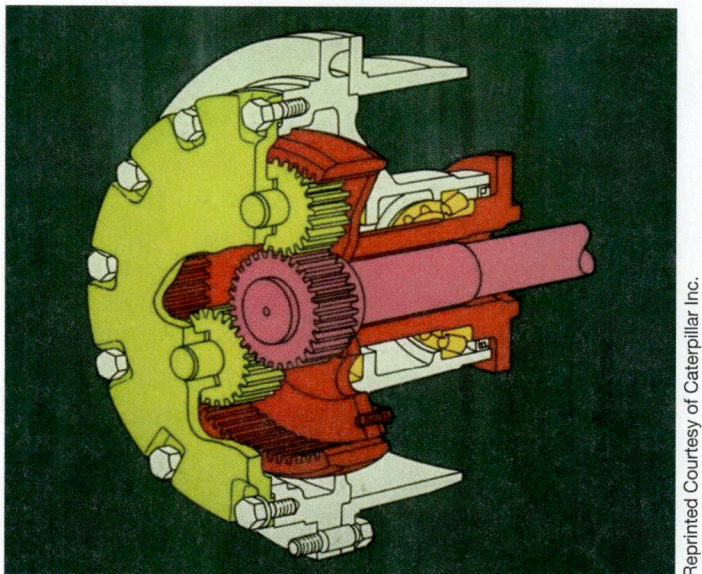

FIGURE 15–41 A planetary final drive.

DRIVE AXLES, POWER DIVIDERS, FINAL DRIVES 455

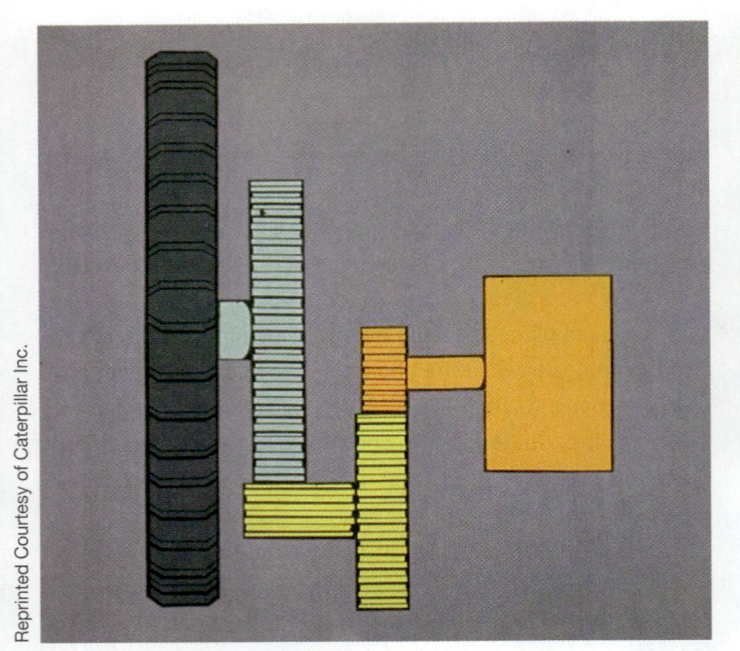

FIGURE 15–42 Double reduction final drive.

Older track machines that featured oval track design had a pinion to bull gear final drive. This was usually driven by a shaft from the steering clutch and the bull gear would drive the sprocket. For a greater gear reduction, an intermediate double gear assembly was used that added a second gear reduction to the final drive. ● **SEE FIGURE 15–42** for a double gear reduction final drive.

CHAIN DRIVES
Chain drives are mostly used for grader tandem drives and skid steer wheel drives.

A grader will have its transmission drive a differential that drives two final drives, which then go on to drive an oscillating tandem drive assembly. The tandem drive assembly has a shaft that drives a double sprocket, which then drives two chains that in turn drive two bigger sprockets. Each sprocket is then connected to a shaft that drives a wheel hub. The chain provides drive to two wheels from one axle. Some articulated rock trucks will have tandem drives for their rear wheels.

There isn't usually a tension adjustment for these chains, but they should be checked for excessive stretching.

The tandem housing is sealed and has a quantity of oil in it to keep the chain lubricated. This oil level needs to checked and changed regularly. ● **SEE FIGURE 15–43** for a grader tandem drive.

A skid steer will have a chain drive arrangement for its wheels as well. There will be one hydrostatic motor for each side of the machine, and it will have a double sprocket that drives two chains. One chain drives the front wheel axle and one drives the rear axle. ● **SEE FIGURE 15–44** for the drivetrain of a skid steer.

FIGURE 15–43 Grader tandem drive.

Progress Check

6. Ring gear run out is measured with a:
   a. micrometer
   b. vernier caliper
   c. dial indicator
   d. feeler gauge

7. Sometimes a ring gear needs to be heated to install it. This is one method of heating that should *not* be used:
   a. torch
   b. induction heater
   c. oil
   d. heat gun

8. Identify one device that is *not* a type of final drive:
   a. hypoid
   b. planetary
   c. single reduction
   d. double reduction

9. The no spin type of differential does this when it senses one wheel turning faster than the other:
   a. it speeds up the faster wheel
   b. it locks both wheels together
   c. it speeds up the slower wheel
   d. it stops driving the faster wheel

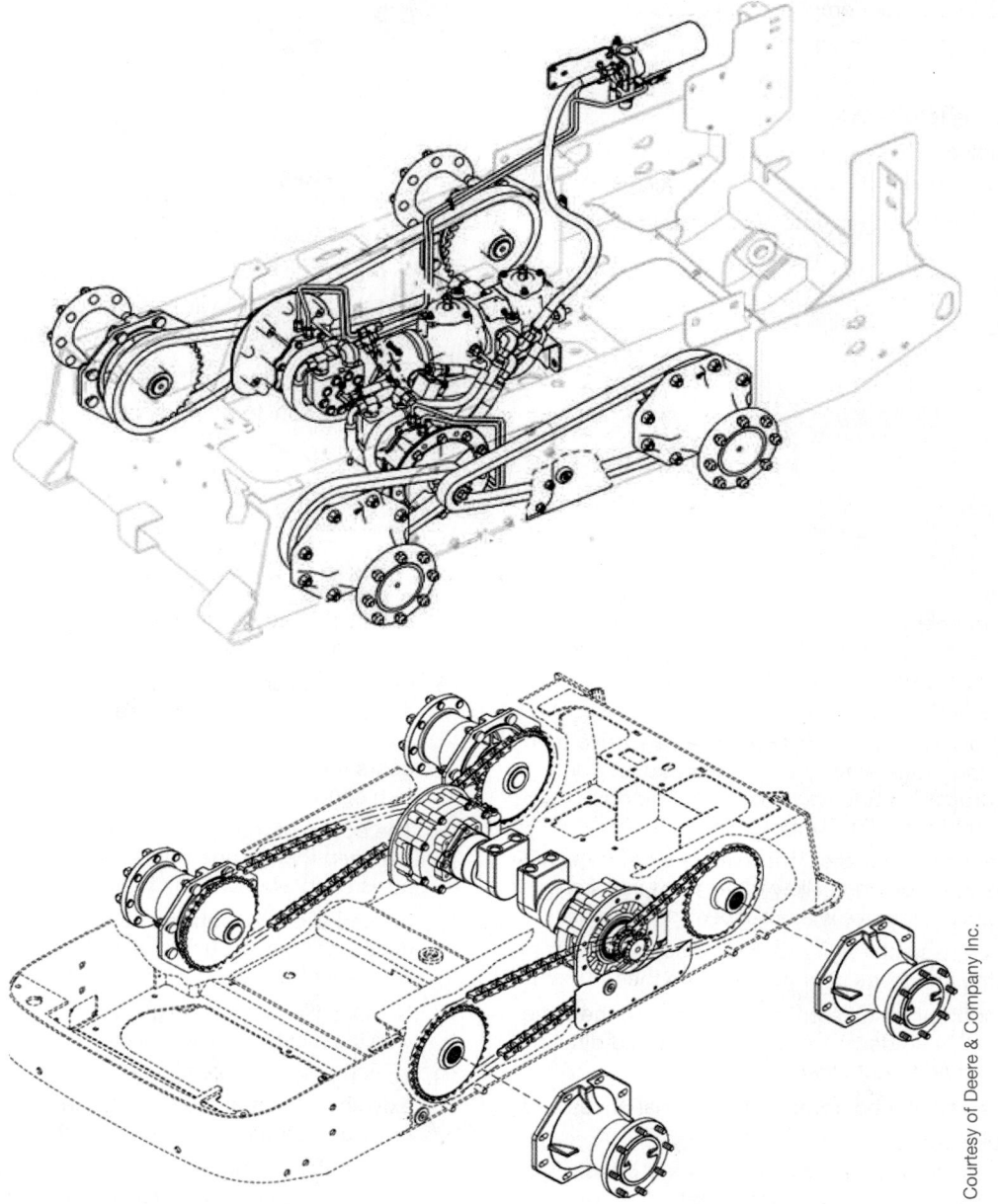

FIGURE 15–44 Drivetrain of a skid steer.

Courtesy of Deere & Company Inc.

10. The term *inboard final drive* refers to this:
    a. a final drive that is used for marine applications
    b. a final drive that is mounted at the wheel end of the axle
    c. a double reduction final drive
    d. a final drive that is next to the differential
11. If a vehicle has an open or standard differential in its drive axle and one wheel has 100% traction while the other is sitting on ice, what happens when the drive shaft turns?
    a. the wheel with 100% traction starts to spin
    b. the wheel sitting on ice turns twice as fast as the crown gear
    c. the drive shaft turns slower than the crown gear
    d. the crown gear reverses
12. The component called a power divider is also referred to as a:
    a. inter-axle differential
    b. inboard planetary
    c. locking differential
    d. torque converter
13. This is the best practice when servicing differentials and final drives:
    a. do it first thing in the morning
    b. get the oil warmed up before draining it
    c. change the filter every other time
    d. only change the oil at the end of the day

DRIVE AXLES, POWER DIVIDERS, FINAL DRIVES

14. This may be an operator complaint that could stem from an excessive backlash setting:
    a. machine wants to travel straight
    b. machine pulls to one side
    c. clunking noise
    d. strong sulfur smell

15. Most planet gears used for final drives would be this type:
    a. helical
    b. spur
    c. sun
    d. moon

> **SHOP ACTIVITY**
>
> - Go to a machine with a drive axle and check to see if it has inboard or outboard final drives.
> - Go to a machine with a steerable drive axle and try to identify the type of universal joints it uses to transfer the drive.
> - Determine the method used to check the axle oil level.
> - Find a machine with a locking differential and describe the type of locking mechanism it has.
> - Find a machine with planetary final drives and describe how to properly check the oil level for it.

SUMMARY

1. Some safety concerns related to drive axles, power dividers and final drives are: lock out tag out of machines when working on them, use wheel chocks and proper lifting and blocking equipment.
2. Drive axles, power dividers and final drives are drive train components that transfer torque from powershift transmissions, hydrostatic transmissions or electric drive motors to other drive train components.
3. Drive axles can be used to steer machines as well as provide drive. They can be mounted solidly to the machines frame or allowed to oscillate or move with a suspension system.
4. Most drive axle assemblies will provide a second gear reduction with final drives incorporated into them.
5. Drive axles incorporate differentials that change the direction of power flow 90 degrees and allow a speed differential between left and right wheels.
6. Drive axles can have inboard or outboard final drives and or service brakes.
7. Drive axles need to have wheel seals that allow the wheel assemblies to rotate and keep contamination out of the housing. They can be lip type or metal face type.
8. Drive axles can be steerable or non steer. Steering axles have turnable knuckles at each end that pivot.
9. The axle shafts extend out from the differential to the final drives. If the axle is steerable the axle shafts will have universal joints to allow pivoting.
10. Some heavy duty axles will have oil cooling and filtration systems to help keep the fluid clean and cool.
11. Drive axle maintenance includes thorough inspections and regular oil changes.
12. Operator complaints related to drive axles could be: no drive, intermittent drive, vibrations, leaks.
13. Drive axle repair usually requires removal of the axle assembly from the machine and once removed special stands and tooling will likely be required.
14. Tapered roller bearing adjustment is usually part of drive axle reconditioning. Adjustment can be either by shim or nut. Proper adjustment will give the specified preload on the bearing to ensure longevity.
15. Differentials are driven by a ring gear that is driven by the pinion gear. The ring gear drives a carrier that drives a cross on which four spider gears can spin. The spider gears then drive side gears that drive axle shafts which in turn send torque to wheels and or final drives.
16. The action of the spider gears allow a speed differential between the left and right axle shafts. When the machine turns each wheel needs to travel at different speeds.
17. Limited slip differentials use clutches to reduce the differential action and transfer some of the torque to the wheel with least traction.
18. No spin differentials use gears, springs and clutches to provide the most torque to the wheel with the best traction.
19. Locking differentials have mechanisms that will ensure the axle shafts turn at the same speed as the crown gear. Oil or air actuated clutches provide the force to do this.
20. Differential diagnostics usually include removing the drain plug and looking for metal or oil contamination.
21. Differential repair requires special tools and specific procedures to ensure pinion and crown gear teeth engage properly. Tapered roller bearings will be adjusted for preload and can move the gears to gain proper backlash and contact pattern.
22. Machines that have tandem drive axles such as articulated rock trucks have a power divider in the forward rear axle. It will send drive to the rear axle and is equipped with a differential that is able to be locked.
23. Final drives provide the last gear reduction before torque is sent to the machines tracks or tires. It can be single reduction with a pinion and bull gear or planetary gearset or double reduction with an idler gear cluster or double planetary gearset.
24. Chain drives can be found on grader tandem drives or skid steer loaders.

chapter 16
ADVANCED HYDRAULIC SYSTEMS

LEARNING OBJECTIVES

After reading this chapter, the student should be able to:

1. Explain how a pressure-compensated hydraulic system works.
2. Describe what is meant by an electrohydraulic system.
3. Explain the benefits of a load-sensing hydraulic system.
4. Describe the function of a main control valve.
5. Explain the meaning of the terms *upstroking* and *destroking*.
6. List the additional valves that may be found in a main control valve.
7. Describe the systematic steps taken when troubleshooting a hydraulic system problem.
8. List the test equipment needed to troubleshoot a hydraulic system problem.
9. List the common root causes of hydraulic component failures.
10. Identify the two main types of variable displacement piston pumps.
11. Describe the operation of a common pump control valve.

KEY TERMS

Axial piston pump 467
Bent axis piston pump 465
Cycle times 500
Destroke 469
Flow meter 500
Heat loss 463
Internal leakage 461
Load-sensing pressure-compensated 470
Margin pressure 471
Pressure gauge 498
Swash plate 467
Upstroke 471

INTRODUCTION

As mentioned in Chapter 4, there are many safety concerns you should be aware of when working near or on hydraulic systems. These same concerns exist with advanced hydraulic systems and are magnified by the higher pressures and flows that are usually found with these systems.

Always make sure that any pressure is released safely in a hydraulic system before removing any lines or components. This could be made more difficult with an electrohydraulic system and extra care needs to be taken to ensure that proper pressure release procedures are followed.

You may be required to work on hydraulic systems that have hot oil and components in them. Be aware of these and take precautions to prevent burns to yourself and coworkers.
● **SEE FIGURE 16–1** for a variety of warnings that you should see on a machine.

HYDRAULIC SYSTEM EVOLUTION

Chapter 4 introduced you to the basic principles of simple hydraulic systems and how they convert mechanical energy to fluid energy and back to mechanical energy to perform work. The components of a simple system were also described and how they work together to transfer energy through fluid. This energy is converted into either motion or heat. It is the energy that is converted into heat that equates to wasted energy or efficiency loss.

This chapter will take the next step and discuss how hydraulic systems have evolved into higher pressure and flow systems while also becoming more energy efficient and easier to operate. These gains in system efficiency and the ease with which they are operated, however, have added some complexity to hydraulic systems.

Currently, there is much research ongoing that is focusing on where to get even more efficiency gains in hydraulic systems. It is estimated that if all working hydraulic systems were able to realize a 10% efficiency gain, this would save $9.8 billion worth of fuel per year. This research is focused on reducing heat losses so that a higher percentage of fluid power energy is used to perform work.

Simple hydraulic systems are controlled by the operator with direct mechanical linkage moving directional control valves, and they generally have pressures no higher than 3000 psi. Their pumps are fixed displacement and have outputs of less than 25 gpm. An example of this would be a small skid steer loader with manual controls and no auxiliary circuits.
● **SEE FIGURE 16–2** for a small skid steer loader with a simple hydraulic system.

Its fixed displacement pump will send oil to an open center directional control valve that can actuate one or two circuits (boom and bucket). The pump output flow can move more than one actuator, but this requires the operator to manually meter pump flow between more than one circuit.

As machine and hydraulic component design has evolved, hydraulic systems have been designed to provide higher pressures and higher flows that are easier and smoother to control by the operator. Several factors have influenced this newer and more advanced hydraulic system such as the need to reduce operator fatigue and increase comfort, equipment manufacturer competition, higher fuel prices, the need for higher fuel efficiency, and tighter regulations on diesel exhaust emissions. Hydraulic system efficiency gains will include pump design changes, control valve changes, and more electronic monitoring and control.

An example of a current benefit of improved hydraulic system efficiency is the ability of machine manufacturers to reduce engine horsepower to get under the 75 hp tier IV emission limit. By making hydraulic systems more efficient,

FIGURE 16–1 Hydraulic safety warnings.

FIGURE 16–2 A small skid steer loader with a simple hydraulic system.

they can be driven by a lower-powered engine. This not only will save fuel and reduce emissions of the engine, for some machines may be able to be downsized enough to be exempt from a higher level of emission regulations, but also will reduce costs of the machine and simplify the design. As with other systems, these efficiency gains are the result of an engine/hydraulic system integration that is monitored and managed with electronics.

Only the most basic small machines today will use a simple hydraulic system as described earlier. Most skid steer machines will feature pilot-operated hydraulic controls at minimum and could include variable displacement pumps with pump control valves, closed center direction control valves with flow compensators, or electronically controlled pump controls. Therefore, the need to understand both basic hydraulic principles and more advanced systems and components by today's HDET is mandatory.

This chapter is meant to give you the knowledge of how common hydraulic system components work individually and together in an advanced hydraulic system and what to do if they aren't working properly. There is no way to cover all the variations of components and systems for advanced hydraulics in this chapter, but if the concepts of how fluid pressure and flow work together to perform work in a system are mastered, then you should be able to understand how almost any system functions. If you can understand how a system functions, then you should be able to diagnose problems with it and repair or adjust system component to restore the system to its intended performance level.

NEED TO KNOW

HYDRAULIC SCHEMATICS If you wanted to find your way to a destination when driving your car, you have a few options. You could just start driving and stop and ask for directions once in a while. You could turn on your GPS and place your faith in the computer-generated voice coming out of it, you could follow a hand-drawn map that may point out some highlights that you need to watch for, or you could get a current paper road map, study it before you leave, and refer to it as you head toward your destination (during a stop of course). Once you have reached your destination, you should be able to find your way back home and you may even find an easier way to get there.

Finding your way through a hydraulic system and learning about it is much like taking a trip in a car. There will be a certain component you need to find and how oil flows to and from it. With the use of a schematic, it will be much easier to find. Hydraulic schematics come in paper and electronic versions. Both versions have their pros and cons.

The ideal schematic is a paper-laminated sheet or sheets. Hydraulic schematics are similar to electrical schematics in that they have symbols that represent the different components that make up the circuit. There are also lines that represent the conductors that connect the different components.

Refer to the chart in Chapter 4 that shows many of the common hydraulic schematic symbols.

You will likely be using a schematic when you are either troubleshooting a hydraulic system or adding an extra hydraulic function to a machine. This means you will be looking at a specific section or part of the schematic. It's always a good idea to take a look at the whole schematic for a few minutes to get you familiar with how the oil flows from the tank to the pump, from the pump to the control valves, and from there to the actuators and back to the tank.

Once you have a general idea of how the whole machine system works, then you can focus on the particular area that you are having a problem with.

OPEN CIRCUIT VERSUS CLOSED CIRCUIT HYDRAULIC SYSTEM The distinction between open and closed circuit systems can be best understood by comparing a hydraulic circuit and a hydrostatic system. A hydrostatic system is typically used to drive a machine and is a closed circuit. This means the pump's output goes directly to the motor's (actuator) inlet and the motor's outlet returns directly to the pump's inlet to create a closed circuit (sometimes called a closed loop). The flow volume and direction in the system is controlled by varying the pump output. A pump used in a closed circuit hydrostatic system will need to be able to send flow out of the pump in either of two directions. This means the pump will be a bidirectional type. There will be some normal **internal leakage** in the system so there must be a charge pump to ensure that there is never a pump starvation condition. A skid steer will use a closed circuit hydrostatic system to drive its wheels but an open circuit system to operate its implement hydraulic system (boom and bucket). Hydrostatic systems will be thoroughly discussed in Chapter 17.

An open circuit hydraulic system is one that has a pump that gets its inlet oil from the tank and then pushes the oil to a directional control valve where it is directed to an actuator. The return oil from the actuator then flows to the control valve where it is then returned to the tank. Oil flow is created by the pressure change between the tank and the pump inlet and then from the pump outlet to the tank.
● **SEE FIGURE 16–3** for a schematic comparison of open and closed circuit systems.

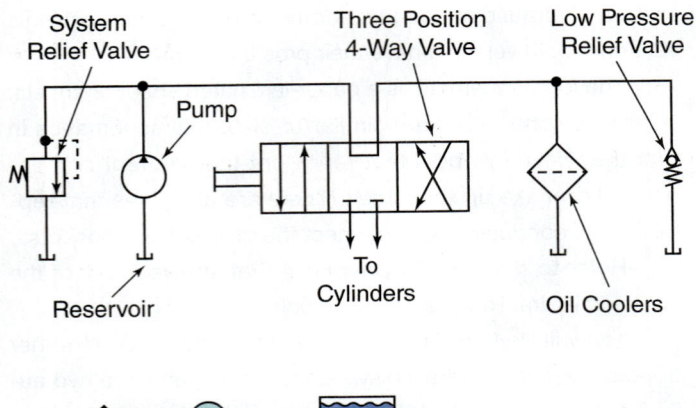

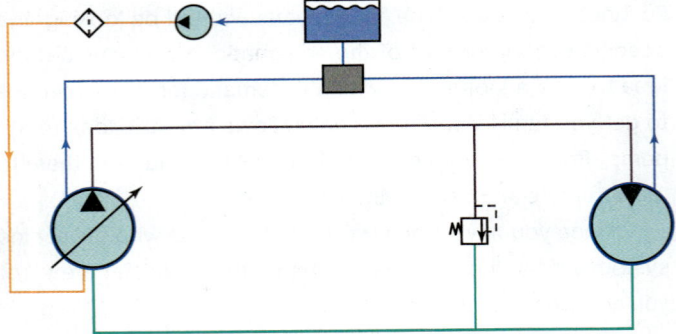

FIGURE 16–3 Open circuit (top) versus closed circuit.

OPEN CENTER VERSUS CLOSED CENTER HYDRAULIC SYSTEMS
Open center systems were briefly mentioned in Chapter 4 when discussing a simple hydraulic system. An open center type of system refers to one that uses a directional control valve having a passage (open center) to allow the flow from the pump to flow through it and on to the tank when the control valve is in neutral. There will likely also be a parallel passage to allow oil flow to other functions when one function is being actuated. ● SEE FIGURE 16–4 for an illustration of an open center hydraulic system.

Most machines will have more than one hydraulic function; therefore, more than one section is included in the main directional control valve. This valve could be a one-piece casting with bores for each directional spool and passages cast or machined into it for all other ports or valves. This would be termed a *monoblock valve*. The directional control valves could also be in individual sections that are stacked and bolted together with end sections, and this style is considered to be a *multi-section valve*. Main control valves will be explained in more detail further on in this chapter. ● SEE FIGURE 16–5 for the two styles of directional control valves—monoblock and multi-section.

The inlet section of the directional control valve will likely have a pressure relief valve to allow pump flow to be routed to the tank if the flow stops or if the system pressure rises too high. The flow will stop either if a cylinder is at the end of its stroke or if the cylinder is stopped in mid travel because of an excessive load. ● SEE FIGURE 16–6 for the schematic of a machine with an open center hydraulic system.

As a fixed displacement pump moves its full displacement of oil as soon as it's turning, it has to be used with an open center control valve. Some systems will use a combination of two or more fixed displacement pumps to provide

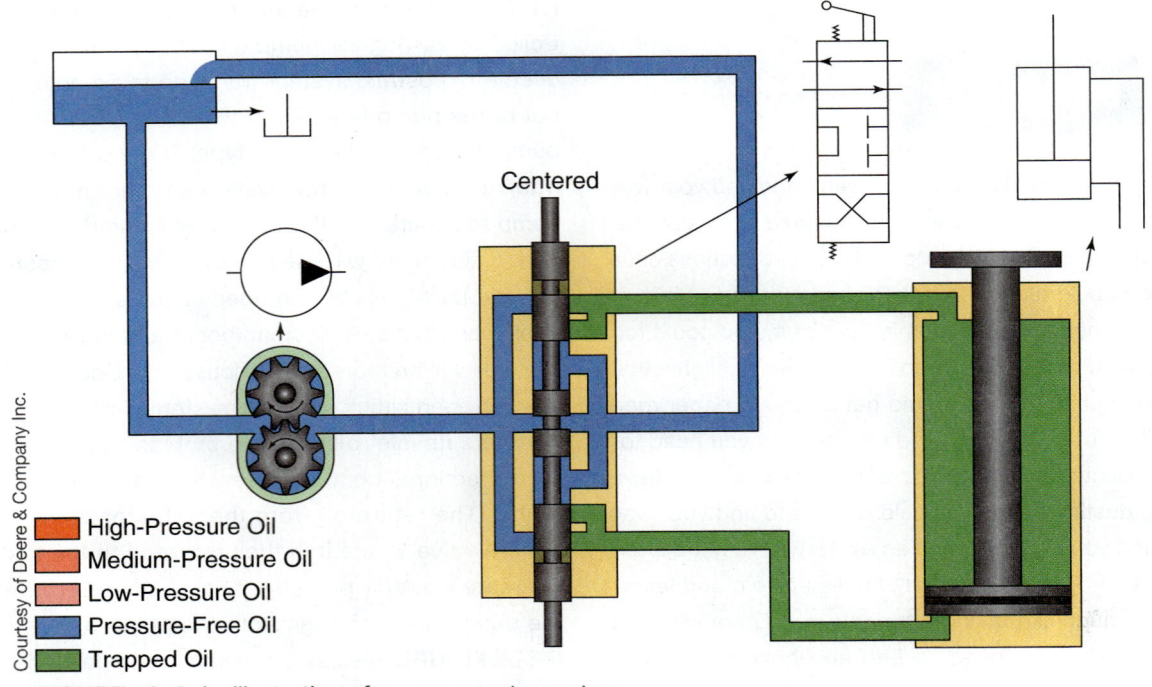

FIGURE 16–4 An illustration of an open center system.

462 CHAPTER 16

system flow. In this case, a pump unloading valve can divert the flow from one pump and direct it to the inlet of the other pumps or back to the tank. Even if an unloading valve is used, there is still some wasted energy as the full flow of oil from the pump is not doing any useful work. A pump unloading valve will sense higher system pressure and divert the flow of the second pump to reduce the load on the engine.

● **SEE FIGURE 16–7** for a double pump with unloader valve arrangement.

Flow losses occur when oil is sent through conductors (hoses, tubes, and fittings), and the friction of the oil on the inside of the conductor creates heat. This heat is a direct loss of energy and leads to inefficiency for a hydraulic system. A machine that has higher flow requirements may use a variable displacement pump to save on flow losses. **Heat losses** can be directly related into wasted horsepower by using the formula: Heat (BTU/hr) = Flow (gpm) × Pressure drop (psi) × 1.5.

Any time there is an internal leak in a hydraulic system, this will equate to a reduced efficiency. This can happen when components get worn or seals fail. Examples of this are pump bodies and pistons that are worn or a cylinder's piston seals fail. These types of failures represent an unwanted pressure drop that creates heat.

For machines that require only low flow and low-pressure hydraulic systems, an open center system is sufficient. However, if the desire of the machine designer is to increase flow and reduce flow losses, then a variable displacement pump and closed center directional control valve must be used. This type of pump will be able to reduce its flow output no matter what speed its drive shaft is turning. By reducing the flow output when it's not needed, the system efficiency raises because of the reduced flow losses.

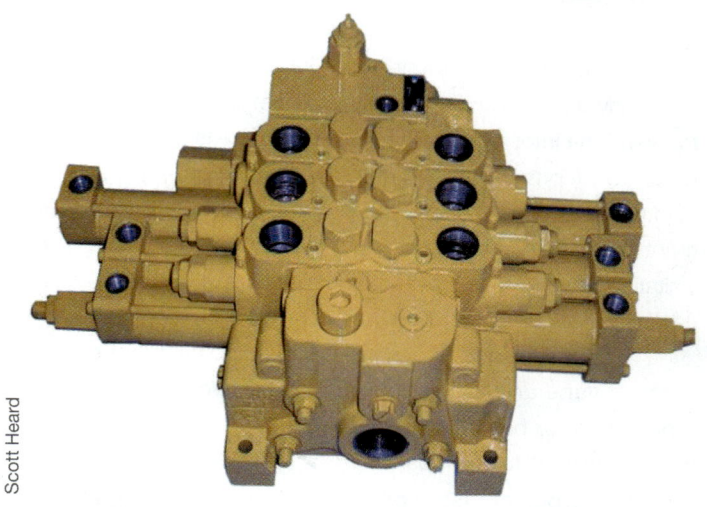

FIGURE 16–5 Two styles of directional control valves.

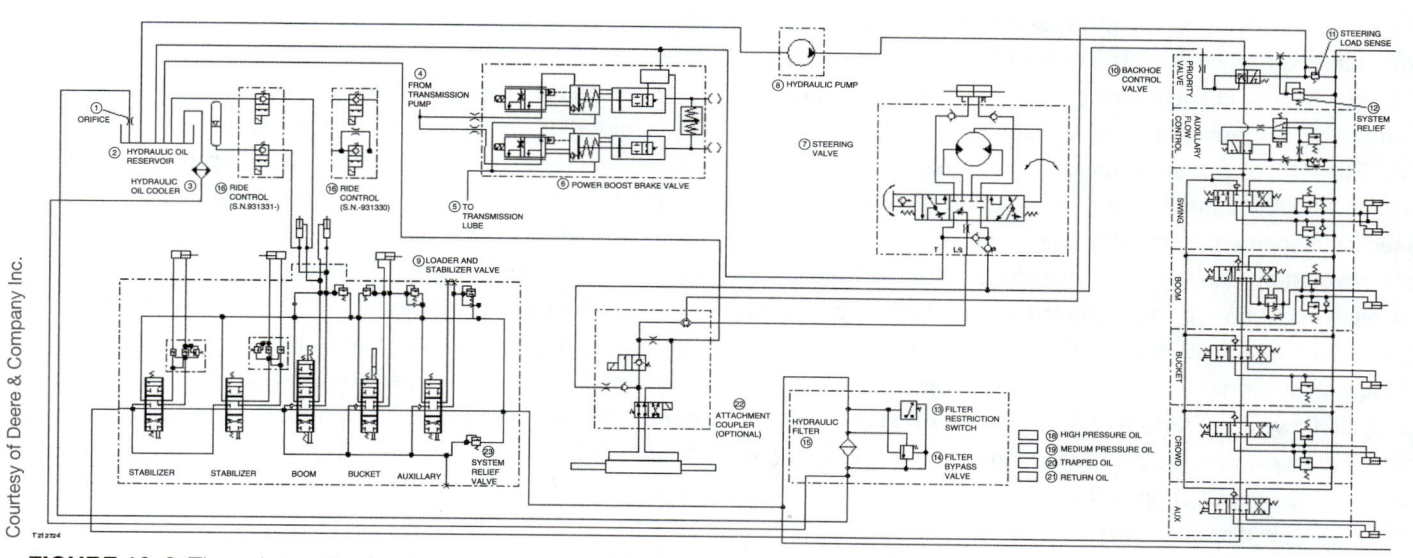

FIGURE 16–6 The schematic of a machine with an open center hydraulic system.

ADVANCED HYDRAULIC SYSTEMS 463

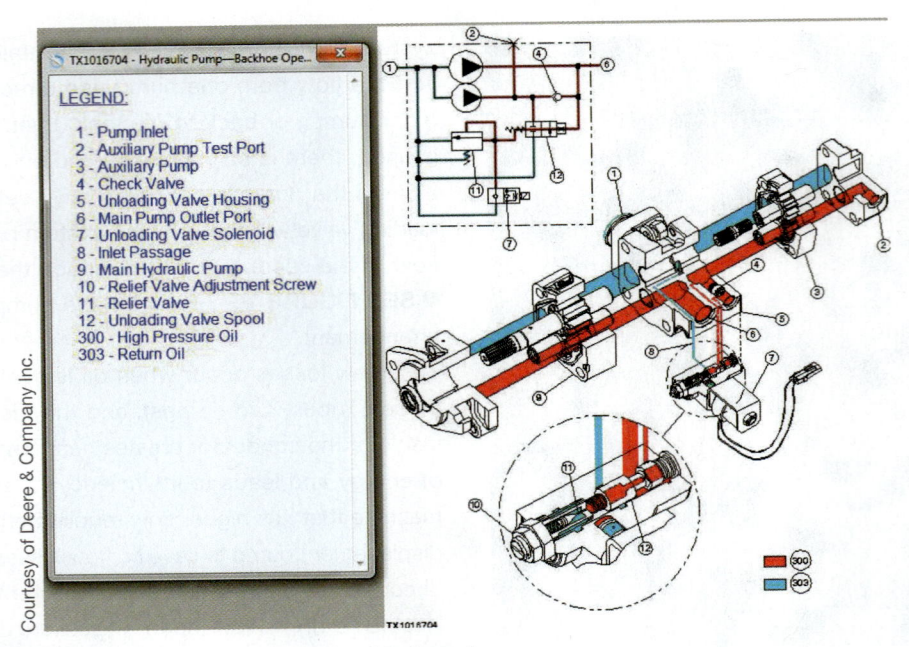

FIGURE 16–7 A pump and unloader valve.

Closed center directional control valves are designed to stop the flow of pump oil at the valve inlet until there is a need to move an actuator. This will reduce flow losses in a system. This type of valve is ideally used with a variable displacement pump because the pump can be "turned off" when there is a low flow demand.

However, there are systems that also use variable displacement pumps with open center valves. Some excavators will have this pump/control valve combination.

VARIABLE DISPLACEMENT PUMPS

Fixed displacement pumps produce the same amount of flow for each shaft revolution. This means its flow output can be changed by only changing the speed it turns. An external gear pump is a good example of this type of pump.

Pump displacement is measured in cubic inches per revolution or cubic centimeters per revolution. An example of a fixed displacement pump that may be used for a small machine would be one that would produce 1 cubic inch of flow for each revolution. If the pump is turning at 2000 rpm, then it would produce 8.65 gpm (2000 cubic inches per minute/231 cubic inches per gallon). If the same pump was slowed to 1000 gpm, its flow would drop to 4.32 gpm. Gear pump displacement will typically range from less than 0.25 cubic inches per revolution to over 61 cubic inches per revolution.

A variable displacement pump can change its output independently of its shaft revolution speed. If this machine example used a variable displacement pump, it could reduce its flow output at any driven speed to almost zero gpm (some flow is needed to overcome internal leakage) whenever there is no requirement for oil flow (control valves in neutral). The output of the pump is controlled by a pump control valve that will change the displacement per revolution to satisfy flow demands and engine power limitations. Piston pump displacements can be well over 1000 cubic centimeter per revolution (61 cubic inches per revolution).

There are several types of variable displacement pumps such as vane and radial piston, but the two most common types used for heavy equipment machines are the bent axis and the axial piston pumps.

Both types of piston pumps use a reciprocating motion created between a cylinder block and a set of pistons to create fluid movement. This action can be compared to the piston motion in an internal combustion engine.

The pistons are fit with very close tolerances to the cylinder block. Usually, any clearance exceeding 0.003 in. is excessive. There may be a metal piston ring-type seal or an O-ring seal at the end of the piston to seal oil between the piston and the cylinder block, or the viscosity of the oil may be relied on to create a seal. If there is excessive clearance between the piston and block or if a seal is damaged, there will be internal leakage and a loss of efficiency. This can also be compared to an internal combustion engine that loses compression because of piston ring wear or cylinder wear. ● **SEE FIGURE 16–8** for a set of pistons and a cylinder block for a variable displacement pump.

These pumps will have two main fluid conductors connected to them, one for tank inlet and the other for pump outlet.

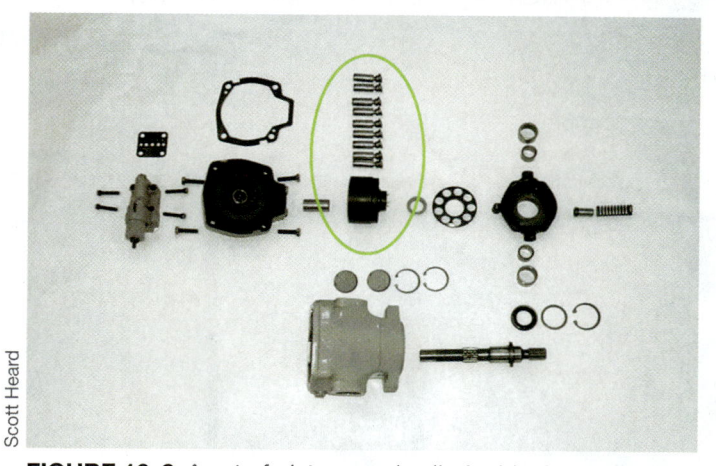

FIGURE 16–8 A set of pistons and cylinder block.

Most piston pumps will also have a case drain line that returns any internal pump leakage back to the tank. All pumps will leak internally a certain amount, and this is normal. This internal leakage equates to pump inefficiency and is usually 15% maximum allowable. In other words, if a pump should produce 100 gpm of flow in theory and is 85% efficient, then no more than

FIGURE 16–9 Piston assembly.

15 gpm should flow out of its case drain port. This inefficiency rises as system pressure increases, increases as the moving parts of the pump wears, and changes with the change in oil viscosity.

BENT AXIS PISTON PUMP
Bent axis piston pumps are used on many small- to medium-size excavators. This type of pump has a moveable cylinder block that when moved changes the effective stoke of the pump's pistons, which in turn changes the pump's displacement. The rotating parts of the pump are driven either directly by the machine's prime mover or from a gear that is part of a pump drive or a gear that is driven by another pump as part of a pump assembly. The pump's shaft transfers this rotation into a piston drive plate that holds the piston slippers in place on it. ● SEE FIGURE 16–9 for a piston assembly.

The pistons are driven by their ball ends that are turned by the drive plate, and the cylinder block is carried around with the pistons. The ball ends are held to the drive plate by a retaining plate. As the cylinder block changes angle in relationship to the pump drive shaft, the stroke of the pistons change. The block has a concave end that moves along a curved end plate opposite of the drive shaft end. The end plate is also called a valve plate or port plate.

The closer the cylinder block and the shaft axis are together, then the pump's relative piston travel is minimal and the pump flow will be at minimum. To increase displacement the pump's control valve piston moves a pin, the pin then moves the cylinder block to a greater angle, and the pistons will travel farther in the block. This will increase the effective stroke of the piston and the pump flow output will increase. The end of the cylinder block is lap finished and mates with the valve plate. The valve plate will have an inlet and outlet port to direct oil in and out of the cylinder block from the pump housing ports. ● SEE FIGURE 16–10 for a cross-section view of a bent axis pump.

ADVANCED HYDRAULIC SYSTEMS

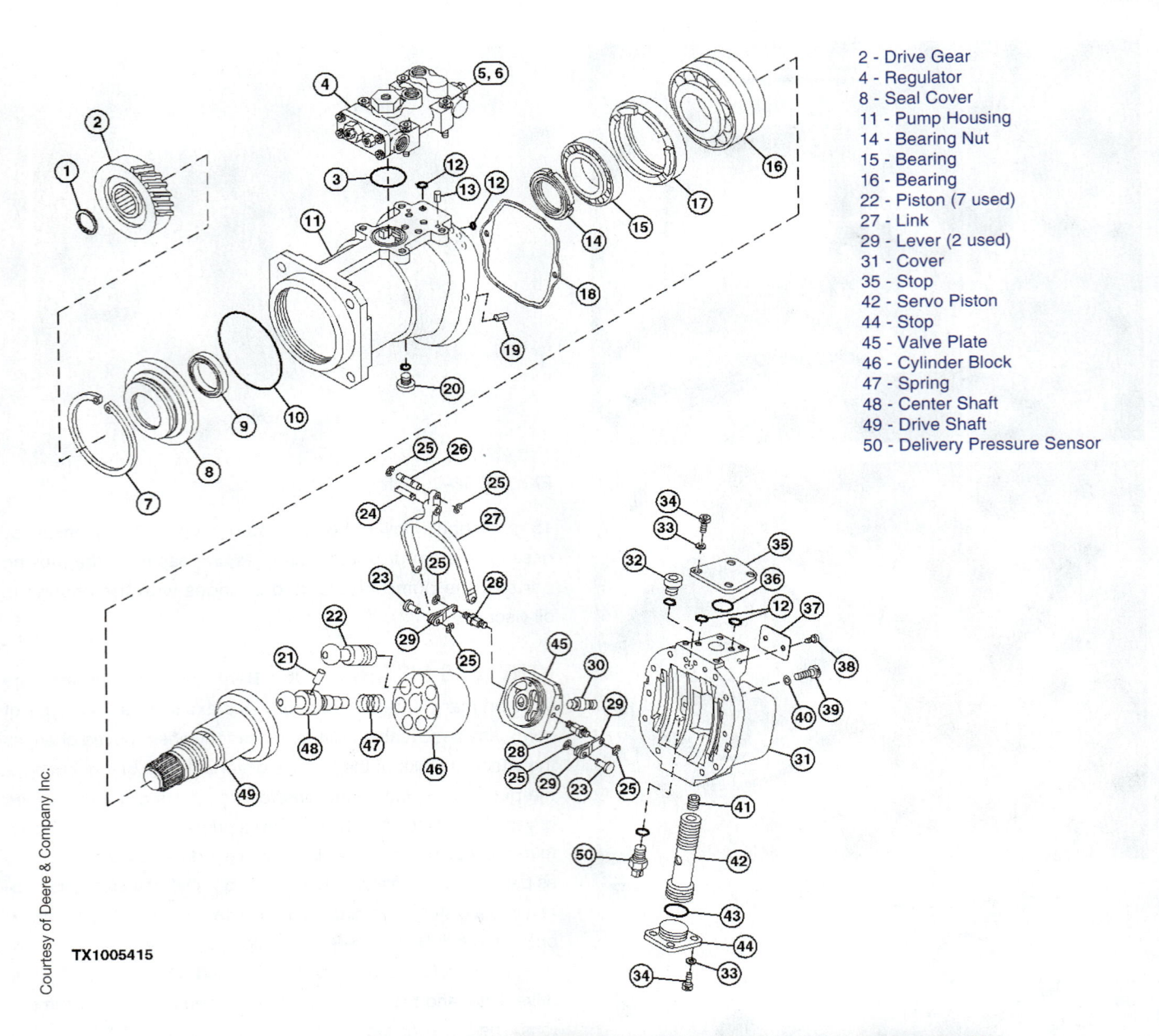

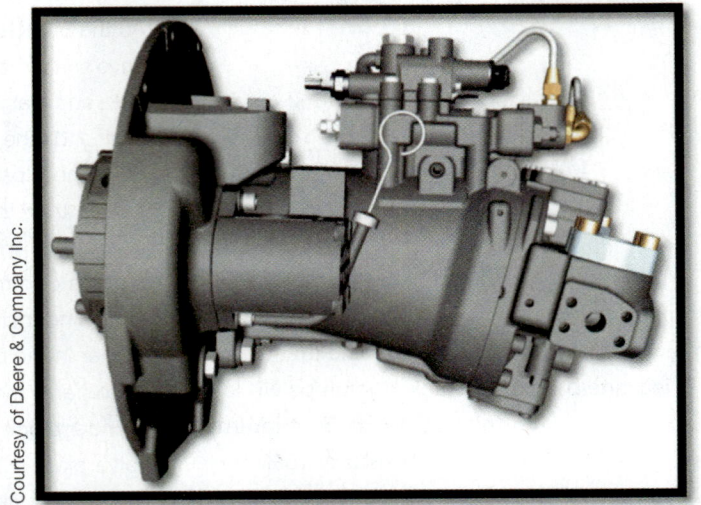

- 2 - Drive Gear
- 4 - Regulator
- 8 - Seal Cover
- 11 - Pump Housing
- 14 - Bearing Nut
- 15 - Bearing
- 16 - Bearing
- 22 - Piston (7 used)
- 27 - Link
- 29 - Lever (2 used)
- 31 - Cover
- 35 - Stop
- 42 - Servo Piston
- 44 - Stop
- 45 - Valve Plate
- 46 - Cylinder Block
- 47 - Spring
- 48 - Center Shaft
- 49 - Drive Shaft
- 50 - Delivery Pressure Sensor

FIGURE 16–10 Bent axis pump.

466 CHAPTER 16

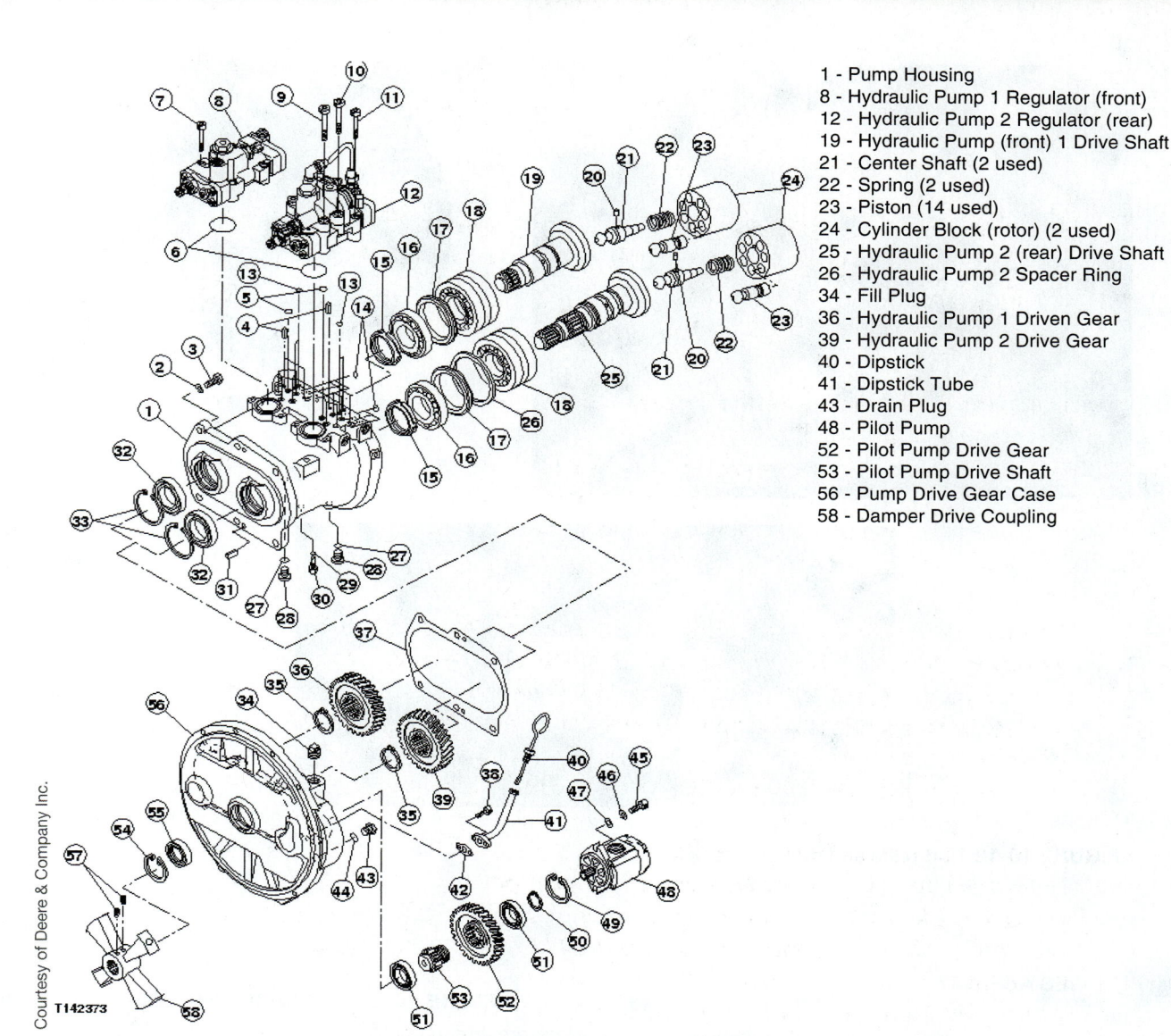

1 - Pump Housing
8 - Hydraulic Pump 1 Regulator (front)
12 - Hydraulic Pump 2 Regulator (rear)
19 - Hydraulic Pump (front) 1 Drive Shaft
21 - Center Shaft (2 used)
22 - Spring (2 used)
23 - Piston (14 used)
24 - Cylinder Block (rotor) (2 used)
25 - Hydraulic Pump 2 (rear) Drive Shaft
26 - Hydraulic Pump 2 Spacer Ring
34 - Fill Plug
36 - Hydraulic Pump 1 Driven Gear
39 - Hydraulic Pump 2 Drive Gear
40 - Dipstick
41 - Dipstick Tube
43 - Drain Plug
48 - Pilot Pump
52 - Pilot Pump Drive Gear
53 - Pilot Pump Drive Shaft
56 - Pump Drive Gear Case
58 - Damper Drive Coupling

FIGURE 16–11 Exploded view of double bent axis piston pump assembly.

Most bent axis pump arrangements that are used for excavators will have two pumping elements or assemblies housed in one common housing. The pump assembly's shaft drives one pump and that drive gets transferred to the second pump by gears. ● **SEE FIGURE 16–11** for an exploded view of a double bent axis piston pump assembly.

AXIAL PISTON PUMP
An **axial piston pump** works on the same principle as a bent axis piston pump as far as the pumping action that occurs between the rotating cylinder and a set of pistons. The pump's drive shaft turns the cylinder block and the block carries around a set of pistons.

The axial piston pump has a **swash plate** that changes the effective stroke of the pistons that are being carried around in a cylinder block. Unlike the bent axis piston pump, the pistons and cylinder block axes are the same as the pump drive shafts. As the pistons and cylinder block rotate together and the swash plate is angled, the pistons will be directed in and out of the cylinder block. If you looked at one of these pumps from the side, you would see the pistons moving up and down inside the block in a continuous motion as the block and pistons rotate.

When the pistons are moving out of the block, they will create a lower pressure than the pump inlet oil. The inlet oil (tank oil) fills the cylinder above the piston. The system's tank oil is available at a port plate that mates with the cylinder block opposite from the drive end. The port plate is lap finished to the end of the cylinder block, is stationary with the pump housing, and will direct oil in and out of the block. ● **SEE FIGURE 16–12** for a port plate and cylinder block.

The outlet port of the port plate gets oil passed through it when the pistons move down into the block. The pistons

ADVANCED HYDRAULIC SYSTEMS 467

FIGURE 16–12 Port plate and cylinder block.

> **WEB ACTIVITY**
>
> Search for and find an animation of a piston pump in operation. Try to find one that shows how the pump can vary its output. This should make variable displacement piston pump operation easier to understand.

have brass slippers that can pivot in a ball joint on the bottom of each piston. These slippers ride on the swash plate as the pistons are carried around by the cylinder block. This ball and slipper arrangement is the same for both styles of piston pumps. ● **SEE FIGURE 16–13** for a cross-section view of an axial piston pump.

Most piston pumps will have an odd number of pistons (usually seven or nine); however, there is one type of axial piston pump that has 10 pistons. This pump will have half of its pistons output to one of two ports on the port plate and the other half output to the other port plate outlet. This will allow the single pump element to act like two pumps.

● **SEE FIGURE 16–14** for how the port plate and cylinder block are arranged.

The swash plate pivots in the pump housing either on stub shafts or a saddle, and its angle can be controlled by several ways. The simplest way is to have a spring trying to keep it at maximum angle and a hydraulically actuated piston (commonly called an actuator piston or control piston) trying to reduce the angle. The piston will get oil supplied to it from one or more valves that are part of a pump control circuit. The simplest pump control circuit will reduce pump flow only when pressure rises to relief valve setting. This valve is called a high-pressure cutoff valve.

The changing of the pump's swash plate from minimum toward maximum angle is called upstroking the pump. An axial piston pump that is at maximum stroke will have its pistons travel in and out of the cylinder blocking the greatest distance possible. This will be limited be a maximum displacement stop that is sometimes adjustable.

When a pump's swash plate is at maximum and starts to move toward minimum angle, it will be said to be destroking. It will destroke when the system is trying to reduce flow. Axial

FIGURE 16–13 An axial piston pump.

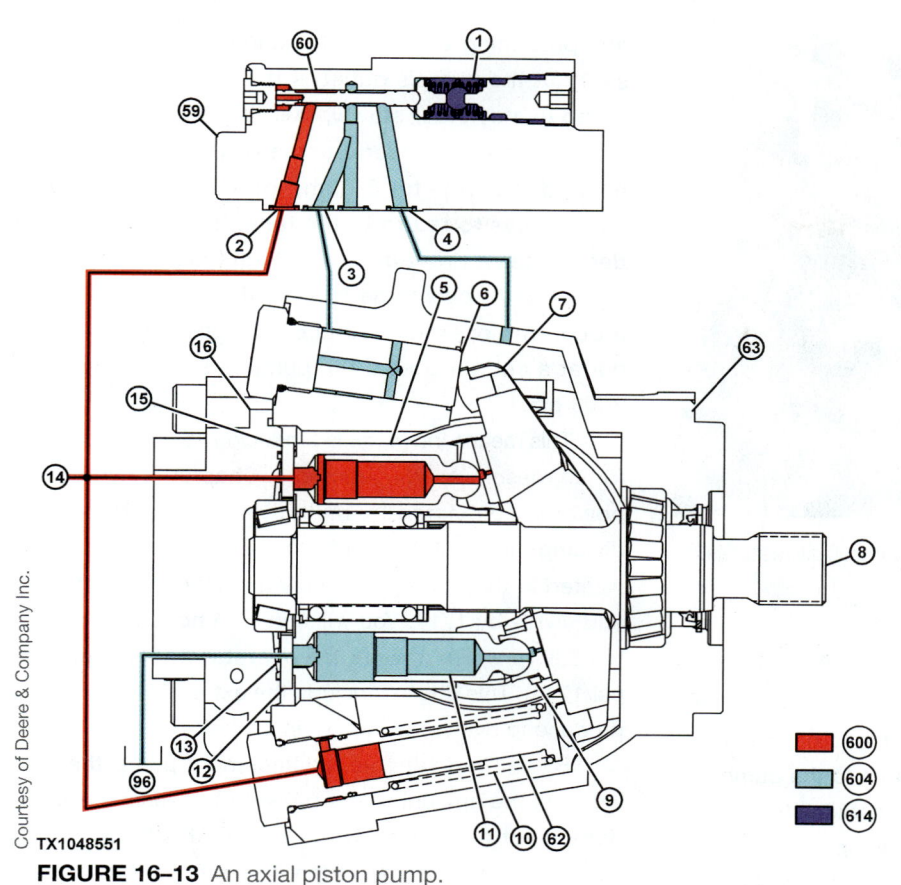

1 - Load Sense System Port
2 - Pump Outlet Port
3 - Displacement Piston Port
4 - Case Return Port
5 - Cylinder Block
6 - Displacement Piston
7 - Swash Plate
8 - Shaft
9 - Piston Retaining Plate
10 - Bias Spring
11 - Piston (9 used)
12 - Valve Plate
13 - Inlet Valve Plate Kidney Slot
14 - Pump Outlet
15 - Outlet Valve Plate Kidney Slot
16 - Valve Block
59 - Load Sense Control Valve
60 - Load Sense Control Valve Spool
62 - Bias Piston
63 - Main Hydraulic Pump
96 - Hydraulic Oil Tank
600 - Supply Oil
604 - Return Oil
614 - Load Sense Oil

FIGURE 16–14 Port plate and cylinder block for dual output pump.

TECH TIP

A simple analogy of a swash plate is the playground ride that gets turned round and round and has spaces for several people to ride on it. If the ride is level, then all the riders stay at the same level. However, if the ride tilts, then all the riders will move up and down in relation to ground level through one revolution as the ride rotates. The greater the tilt of the ride, the greater the distance the riders move up and down through one revolution. The changing level of the riders represents the piston effective stroke that is determined by the swash plate angle.

piston pumps will never completely destroke because there will always be a small internal leakage that has to be overcome.

VARIABLE DISPLACEMENT PUMP CONTROLS

As hydraulic systems get more efficient, they will regulate the pump flow to deliver just enough oil to satisfy system loads and operator demands. There are many variations of systems to do this, but it still comes down to changing the pump flow to meet system's demands. By using a variable displacement pump and a pump control valve, this is easy to do. The pump control valve can have a variety of different inputs, but its main output is used to hydraulically move the swash plate or cylinder block and vary the output of the pump. Oil pressure is applied to a piston that then moves, and its movement will mechanically **destroke** the pump.

ADVANCED HYDRAULIC SYSTEMS **469**

FIGURE 16–15 A variable displacement pump that features a charge pump.

VARIABLE DISPLACEMENT PUMP CHARGE PUMPS Some variable displacement pumps will need a charge pump to ensure a good supply of inlet oil. The location of the tank in terms of length and height to the pump and the pump displacement will determine whether the use of a charge pump is necessary. These charge pumps can be a part of the pump assembly and may be located in the main pump housing.

The pump's driveshaft is typically splined externally and driven by a yoke, shaft, or gear. The gear may be part of a pump drive or may be an internally splined drive output from the rear of another pump. ● SEE FIGURE 16–15 for a variable displacement pump that features a charge pump.

LOAD-SENSING PRESSURE-COMPENSATED SYSTEMS

When trying to understand this system, it is important to remember the basic fact that pumps create flow and not pressure. Pressure is created by the resistance to flow.

What is compensation in terms of hydraulic systems? If a component is pressure compensated it will operate normally no matter what the system pressure is.

An important step in making hydraulic systems more efficient is to make the pump provide only the amount of flow that is necessary to do the work required.

Before we see how this is done, let's first look at a system that uses a fixed displacement pump and an open center control valve. If the pump is turning at 2000 rpm and producing 20 gpm, this quantity of oil is always being sent out the pump's outlet. If the machine operator wants to move a cylinder slowly, then a directional control valve spool is moved to send some of this oil to the cylinder. If the demand is for 5 gpm, for example, then the other 15 gpm has to return to the tank. If the load on the cylinder creates a pressure of 2000 psi, this is the pressure the pump has to overcome to keep the cylinder moving. Even though the operator wants only 5 gpm to move the cylinder at a certain speed, the pump is still moving 20 gpm at 2000 psi.

This means the pump is producing 15 gpm of oil at 2000 psi for no reason. As you learned in Chapter 4, hydraulic horsepower is calculated by multiplying flow and pressure and dividing the result by 1714. To see how much horsepower is wasted in this example, simply multiply 15 gpm by 2000 psi and divide by 1714. You will find 17.5 hp is wasted just to do the 5.8 hp worth of work the operator wants (5 gpm × 2000 psi/1714). This doesn't include the extra 15% the pump has to produce to overcome its own inefficiency.

By using a variable displacement pump that can provide just the amount of flow needed to satisfy the work being done, a lot of horsepower is saved. This translates directly into fuel savings because the prime mover has to work less to drive the pump. Along with fuel savings, a great deal of exhaust emissions is not produced.

In the previous example, the 17.5 hp can be saved by reducing the pump flow to only the quantity necessary to move the cylinder (5 gpm).

There is also much less heat generated in the system, and this leads to reduced cooler sizes and increases the life of the system oil and the hoses.

If a pump is capable of being controlled so that it will produce only the flow that is needed based on the system load, it will be part of a system called this system is called a **load-sensing pressure-compensated** system.

HIGH-PRESSURE COMPENSATION If a system is using a variable displacement pump, it needs to be able to cut the pump flow when a pressure spike occurs; otherwise, a major component failure is likely to occur. To do this, a variable displacement pump is used that will have a high-pressure compensator valve (sometimes called a pressure limiter or high-pressure cutoff valve) to reduce pump flow. This will monitor pump outlet pressure and send oil to the pump control piston to destroke the pump when maximum system pressure is reached. The high-pressure cutoff valve will have an adjustable spring that pushes on a spool. Pump outlet oil works

CHAPTER 16

on the other end of the spool. If the oil pressure overcomes spring pressure, the spool will move and uncover a passage. This passage allows oil to get to the pump control piston and destroke the pump.

This is the simplest type of pressure compensation system, and it is simply compensating or adjusting the pump flow when the system pressure reaches a maximum value set by the spring pressure.

PRESSURE AND FLOW COMPENSATION
A pressure and flow–compensated system is one that will provide oil pump oil flow based on the working pressure of the highest working circuit pressure. For example, if a backhoe is digging a trench and the operator is using the swing, boom, stick, and bucket functions simultaneously, the pump will try to satisfy the circuit that has the highest pressure. This could be any one of the four circuits, and it will be constantly changing depending on what the operator is wanting the machine to do. The point to remember is that the pump control compensator is designed to always make the pump provide enough flow to slightly exceed the pressure and flow demands of the circuit with the greatest needs.

This is usually a few hundred psi higher than what is needed. The difference in working pressure and pump output pressure is sometimes called the differential pressure or **margin pressure** and can range from 200 to 450 psi depending on the manufacturer's specifications.

This margin pressure is added to working pressure and is sent to the pump control piston where it manages pump swash plate or cylinder block angle and, therefore, pump flow output.

The pump control valve will have two spools in it. One is the high-pressure compensator and the other is the pressure/flow compensator. The pump control valve is mounted directly to the pump.

The pressure/flow compensator (sometimes called the load sense valve) has a light spring that acts on one end of a spool valve. The other end of the spool has the pump outlet oil acting on it. When a load-sense signal is sent to the flow compensator spool, it enters the spring chamber where it combines with the spring pressure. This combined pressure moves the spool to allow oil to be metered to the pump control piston.

Most axial piston pumps will have two pistons that control the swash plate: a bias piston and a control piston (sometimes called actuator piston). When the pump is at rest and not turning, the bias piston and spring will keep the swash plate at maximum angle. As the machine is started and the pump turns, it immediately begins to pump oil. If the control valves are in neutral, this flow is blocked and pressure ris This pressure is sensed in the control-valve load-sense passages and sent to the pump control valve where it enters the chamber that is opposite of the spring chamber for the load-sense spool. Once the spring pressure is overcome (200–450 psi), the spool shifts to allow oil flow out of the control valves and into the actuator piston for the pump swash plate. The actuator piston is usually twice the diameter of the bias piston so it can overcome the combination of the bias piston and bias spring.

This will move the pump's swash plate and adjust the pump flow to give just enough flow to overcome the margin spring setting. The result of pump flow in this condition is the standby pressure.

Standby pressure will be the pressure the pump flow creates when the directional control valve is in neutral and the pump is destroked. This is the pressure that the flow compensator spring value is and the result is the pump creates just enough flow to overcome any internal leakage. Standby pressure will make the system more responsive when moving a control lever from neutral. ● **SEE FIGURE 16–16** for an axial piston pump with a pressure compensator valve.

Now that the pump is turning and its flow is creating standby pressure, it is ready to **upstroke**. As soon as the operator actuates a directional control valve to actuate a cylinder or motor, the pump flow will be able to go and do some work. The circuit pressure will now change based on the resistance in the circuit. At this point there needs to be a load-sensing system to control the pump flow based on what the circuit demands are.

An additional pump control that some machines will feature is called a torque-limiting compensator. In operation, most pumps will only be required to produce high flow at lower system pressure or low flow at maximum system pressure. For

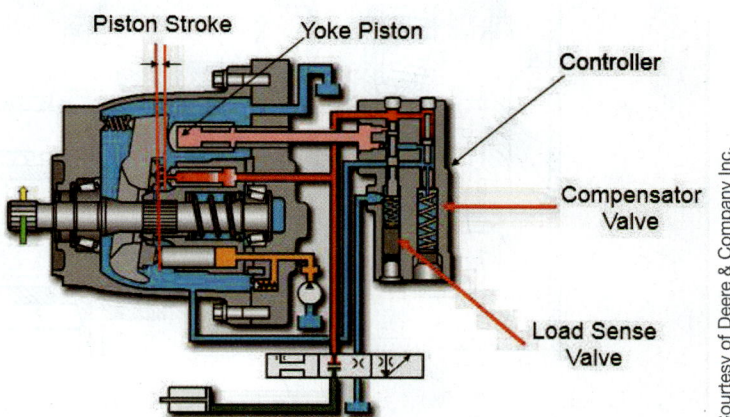

FIGURE 16–16 An axial piston pump with a pressure compensator valve.

you think of how a backhoe loader works when it the highest pressure is needed when the operator to fill the bucket. During this function there isn't flow needed. Then once the bucket is filled, the operator will want to get the bucket out of the trench, swing it to one side, dump it, and return it to the bottom of the trench. All these actions would require high flow and low to medium pressure.

There are instances where high flow and pressure are needed at the same time, but this combination is not very common.

A torque limiter will prevent a high pressure and high flow demand on the pump at the same time. It will sense output pressure and start to destroke the pump as the pressure starts to get close to either high flow or high pressure. As hydraulic horsepower is pressure $\times$ flow/1714, then if a reduction of either of these can be reached, the prime mover size can be reduced to save initial machine cost and operating cost.

To see a variable displacement piston pump with pump control valve including torque control, ● SEE FIGURE 16–17.

LOAD SENSING To be able to change the flow of a pump according to what the system requires will mean that there needs to be a way to monitor the pressure of each circuit in the system. This is the load-sensing function of an advanced hydraulic system.

Once the load is sensed in each circuit and compared to other circuits, the highest load or pressure that is sensed will determine how much flow the pump produces.

If there is no load such as when the controls are in neutral, then the pump will destroke and produce just enough flow to create standby pressure. This standby pressure will likely be 200–300 psi and will make the system more responsive because the pump is producing a small amount of flow at all times. If there was no standby pressure, the system would seem sluggish and lazy.

As soon as a directional control valve is moved and a circuit load is created, the pressure is sensed and a signal is sent to the pump control valve. The pump control valve will now upstroke the pump to provide enough flow to meet the pressure demand in the circuit.

The key to a load-sense system is the ability to monitor the pressures in each circuit. This can be done with passages in the main control valve. These passages will connect all work ports but through the use of a series of check valves will allow only the highest pressure to be sent to the flow/pressure compensator. ● SEE FIGURE 16–18 for a load-sense passage in a control valve.

● SEE FIGURE 16–19 for a control valve section with a load-sense passage represented schematically.

Because of the load-sense system and pump control valve, the pump flow will always be creating more pressure than the system load. For example, if the load on one circuit

1 - Signal line
2 - Flow compensator
3 - Pressure compensator
4 - Pump output
5 - Actuator piston
6 - Cylinder barrel and pistons
7 - Swashplate
8 - Drive shaft
9 - Torque limiter
10 - Spring
11 - Bias spring
12 - Bias piston
13 - Yoke pad
A - Signal oil
B - Pressure oil
C - Return oil
D - Suction Oil and
E - Reduced pressure oil

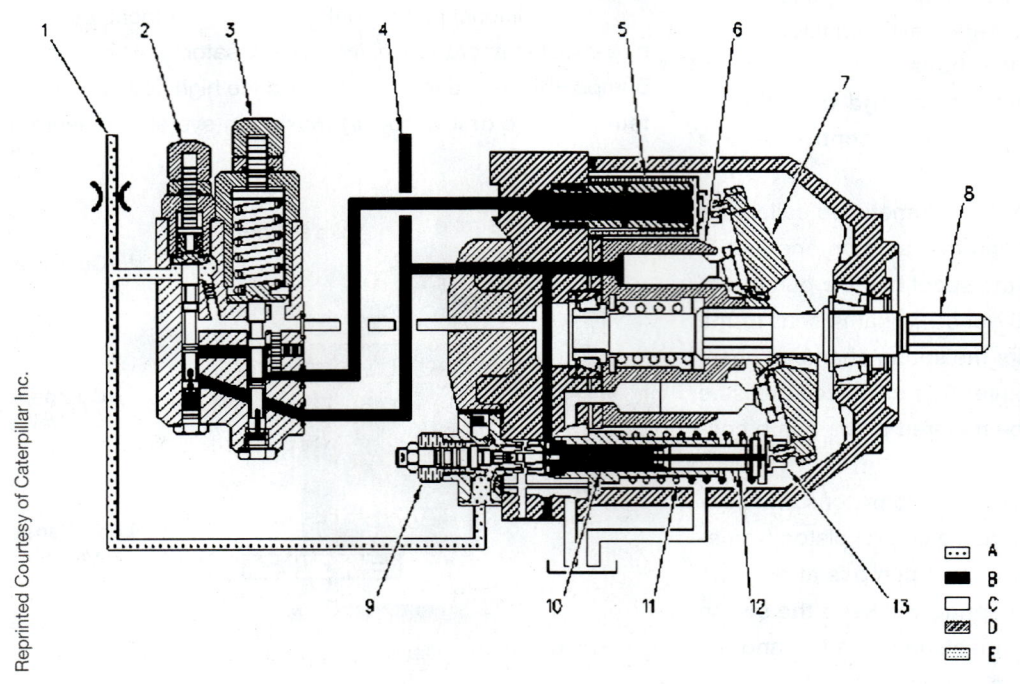

FIGURE 16–17 Pump with torque control.

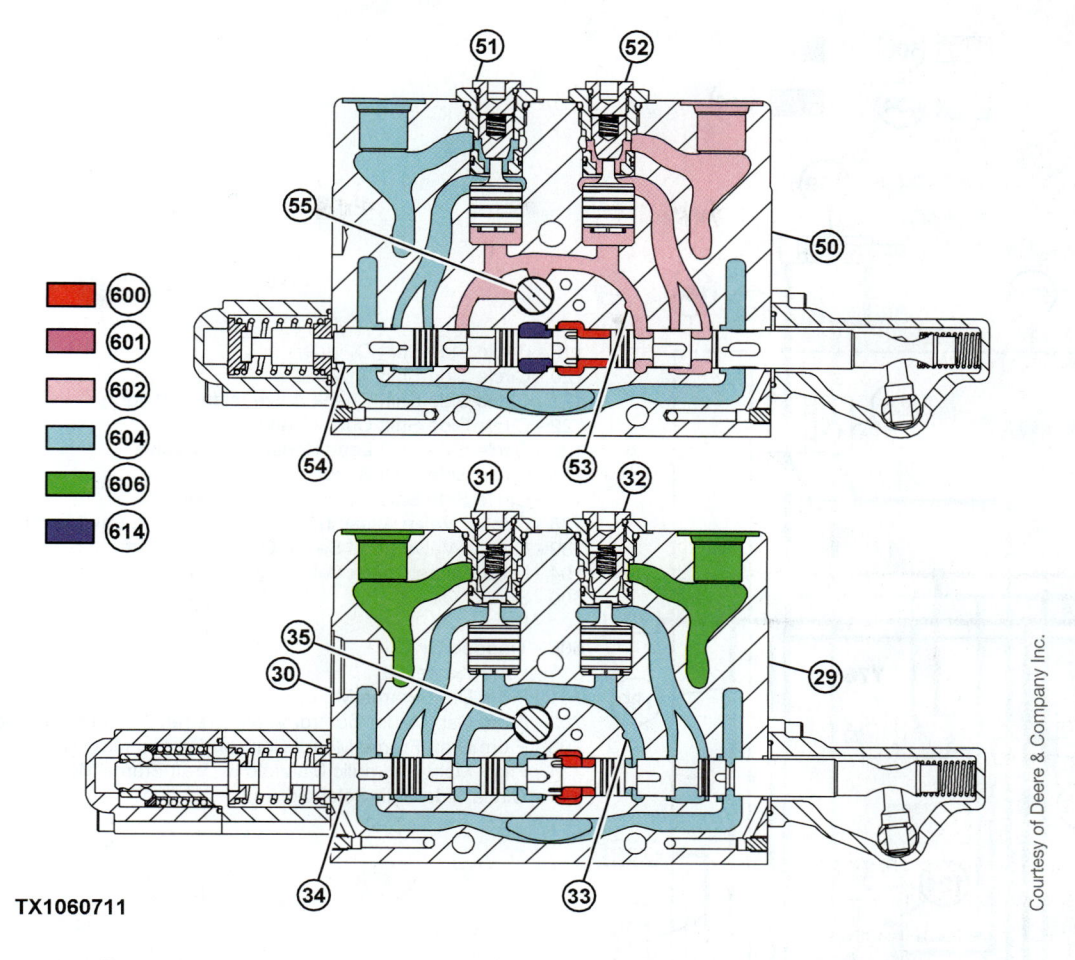

- 29 - Left Blade Lift Valve
- 30 - Left Blade Lift Relief Valve
- 31 - Left Blade Lift Check Valve (left blade lower)
- 32 - Left Blade Lift Check Valve (left blade raise)
- 33 - Left Blade Lift Load Sense Passage
- 34 - Left Blade Lift Spool
- 35 - Left Blade Lift Compensator Spool
- 50 - Blade Pitch Valve
- 51 - Blade Pitch Check Valve (blade pitch forward)
- 52 - Blade Pitch Check Valve (blade pitch back)
- 53 - Blade Pitch Load Sense Passage
- 54 - Blade Pitch Spool
- 55 - Blade Pitch Compensator Spool
- 600 - Supply Oil
- 601 - Supply Oil at a Lower Pressure
- 602 - Supply Oil at Lowest Work Port Pressure
- 604 - Return Oil
- 606 - Trapped Oil
- 614 - Load Sense Oil

FIGURE 16–18 A load-sense passage.

is 2000 psi and the margin spring pressure is 450 psi, then the pump will provide the correct amount of flow to create 2450 psi at its outlet. With a load-sense network if another function is used on this same machine and it creates a load of 3000 psi, then the pump must upstroke to give enough flow to create the additional pressure. The pump outlet pressure would now be 3450 psi.

Another benefit of load-sensing pressure-compensated systems is the fact that there are consistent pressure drops across the control valves. This will make operating the machine much easier because of the consistent feel the operator will have regardless of the amount of oil flow moving past the directional control valve spool.

The pressure drop always equates to the value of the spring pressure in the flow/compensator spool. This is because past the control valve there will be circuit or load pressure, and before the valve, there will be pump pressure. For example, if the load pressure is 2000 psi and the margin spring is 300 psi,

ADVANCED Hydraulic Systems   473

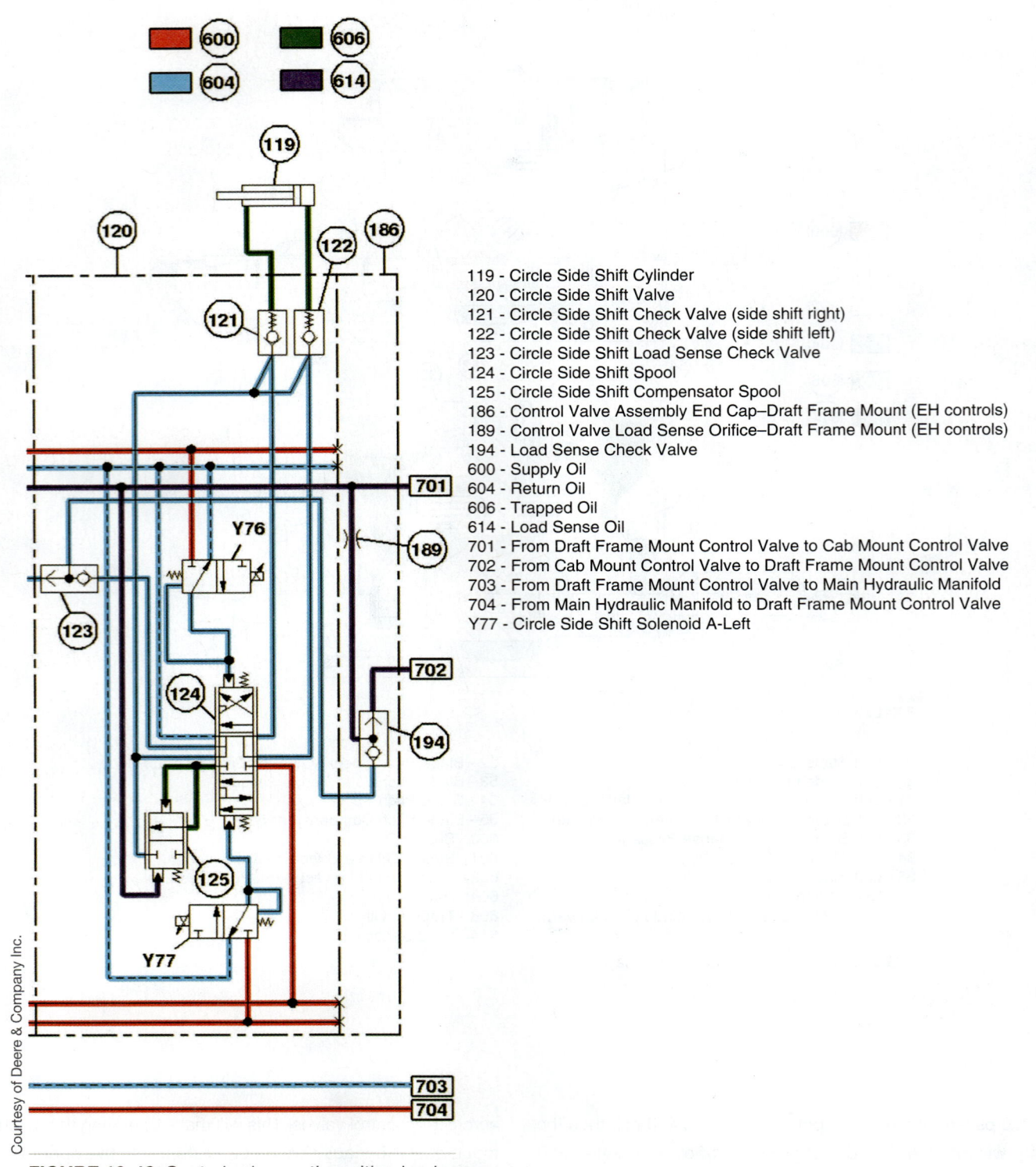

FIGURE 16–19 Control valve section with a load-sense passage.

119 - Circle Side Shift Cylinder
120 - Circle Side Shift Valve
121 - Circle Side Shift Check Valve (side shift right)
122 - Circle Side Shift Check Valve (side shift left)
123 - Circle Side Shift Load Sense Check Valve
124 - Circle Side Shift Spool
125 - Circle Side Shift Compensator Spool
186 - Control Valve Assembly End Cap–Draft Frame Mount (EH controls)
189 - Control Valve Load Sense Orifice–Draft Frame Mount (EH controls)
194 - Load Sense Check Valve
600 - Supply Oil
604 - Return Oil
606 - Trapped Oil
614 - Load Sense Oil
701 - From Draft Frame Mount Control Valve to Cab Mount Control Valve
702 - From Cab Mount Control Valve to Draft Frame Mount Control Valve
703 - From Draft Frame Mount Control Valve to Main Hydraulic Manifold
704 - From Main Hydraulic Manifold to Draft Frame Mount Control Valve
Y77 - Circle Side Shift Solenoid A-Left

then the pump flow will create 2300 psi. This will give a pressure drop across the valve of 300 psi. This pressure difference will always stay the same.

This contrasts to an open center and fixed displacement system that will have varying pressure drops based on how much flow is directed across the spool. This will make the valves harder to actuate smoothly.

A reduced pressure drop equates to less heat loss and higher efficiency as proven by the formula: BTU/hr = Psi drop $\times$ Flow $\times$ 1.5.

474 CHAPTER 16

Progress Check

1. Before performing service or repairs to a hydraulic system, you *must*:
   a. let the oil cool down
   b. remove the pump outlet hose
   c. release all trapped system pressure
   d. warm the oil up to operating temperature
2. An open circuit system differs from a closed circuit system mainly by:
   a. having the pump send its output to a directional control valve
   b. having the pump send its output to the tank
   c. having the pump inlet sometimes be an outlet
   d. having a relief valve on the pump
3. A closed center type of system will most likely feature:
   a. a gear pump
   b. a variable displacement pump
   c. an accumulator
   d. no relief valve
4. A bent axis type of pump will vary its output by:
   a. having its shaft speed change
   b. changing the swash plate angle
   c. rotating its camplate
   d. moving its cylinder block
5. Margin pressure is used to:
   a. ensure that the pump flow output will always be slightly greater than the minimum required
   b. replace the main relief pressure setting
   c. keep system pressure at 0 psi when no flow is required
   d. match pump output to the lowest circuit pressure demand

PILOT CONTROLS

One of the first advances to make hydraulic systems easier to operate was the addition of pilot controls. Manually actuated directional control valves inherently require a higher amount of physical effort to move the valve spools. Pilot controls eliminate much of this manual effort.

Pilot controls use a low pressure and flow system that is controlled by the operator and will in turn control a high pressure and flow system. This is similar to an electrical circuit that uses a relay to control a high-current circuit with a low-current circuit.

As flow volumes increase across a directional control valve, spool flow forces acting on the spool become harder to overcome. The result of these increased forces means that levers have to be longer to gain a mechanical advantage strong enough to move the spool. This resulted in operator fatigue, which in turn led to lost productivity. ● **SEE FIGURE 16–20** for a comparison of manual controls versus pilot controls for the same type of machine.

Pilot oil systems overcame this problem by using hydraulic pressure to directly move the directional control valve spools.

For pilot control systems, the operator joysticks will directly operate directional control poppet valves that get supplied low-pressure (300–400 psi maximum) oil from a dedicated pump or from oil that is diverted from the main system and reduced in pressure. ● **SEE FIGURE 16–21** for a cross-section of a pilot control valve.

FIGURE 16–20 Manual versus pilot controls.

ADVANCED HYDRAULIC SYSTEMS 475

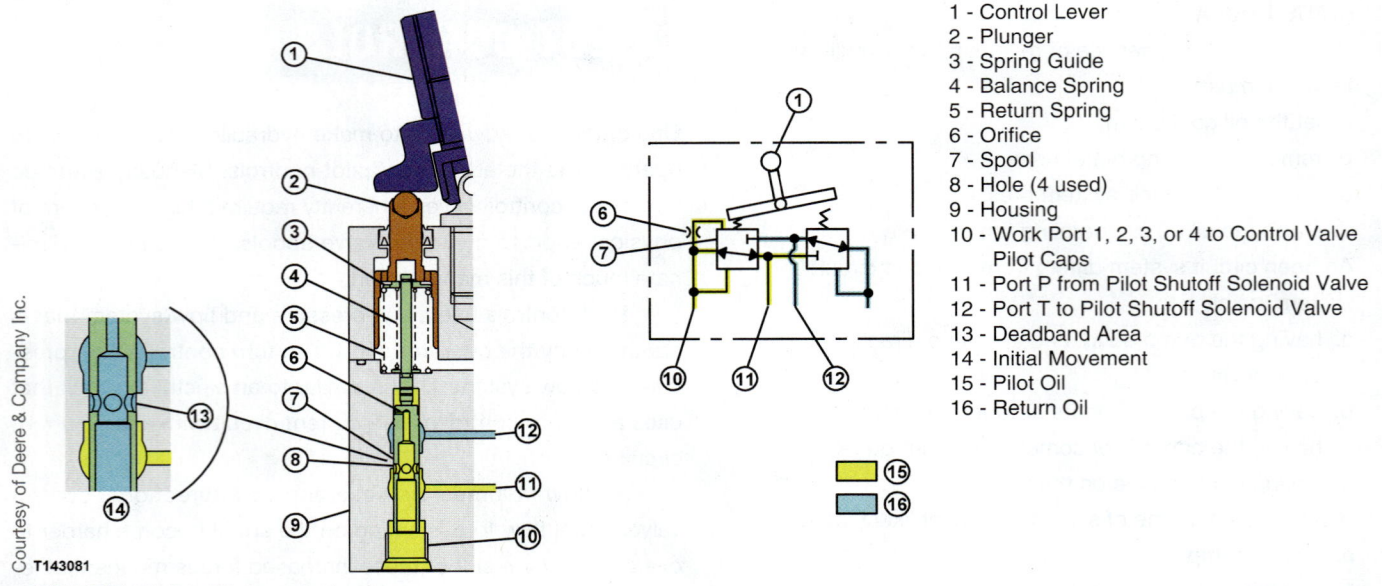

1 - Control Lever
2 - Plunger
3 - Spring Guide
4 - Balance Spring
5 - Return Spring
6 - Orifice
7 - Spool
8 - Hole (4 used)
9 - Housing
10 - Work Port 1, 2, 3, or 4 to Control Valve Pilot Caps
11 - Port P from Pilot Shutoff Solenoid Valve
12 - Port T to Pilot Shutoff Solenoid Valve
13 - Deadband Area
14 - Initial Movement
15 - Pilot Oil
16 - Return Oil

FIGURE 16–21 Pilot control valve.

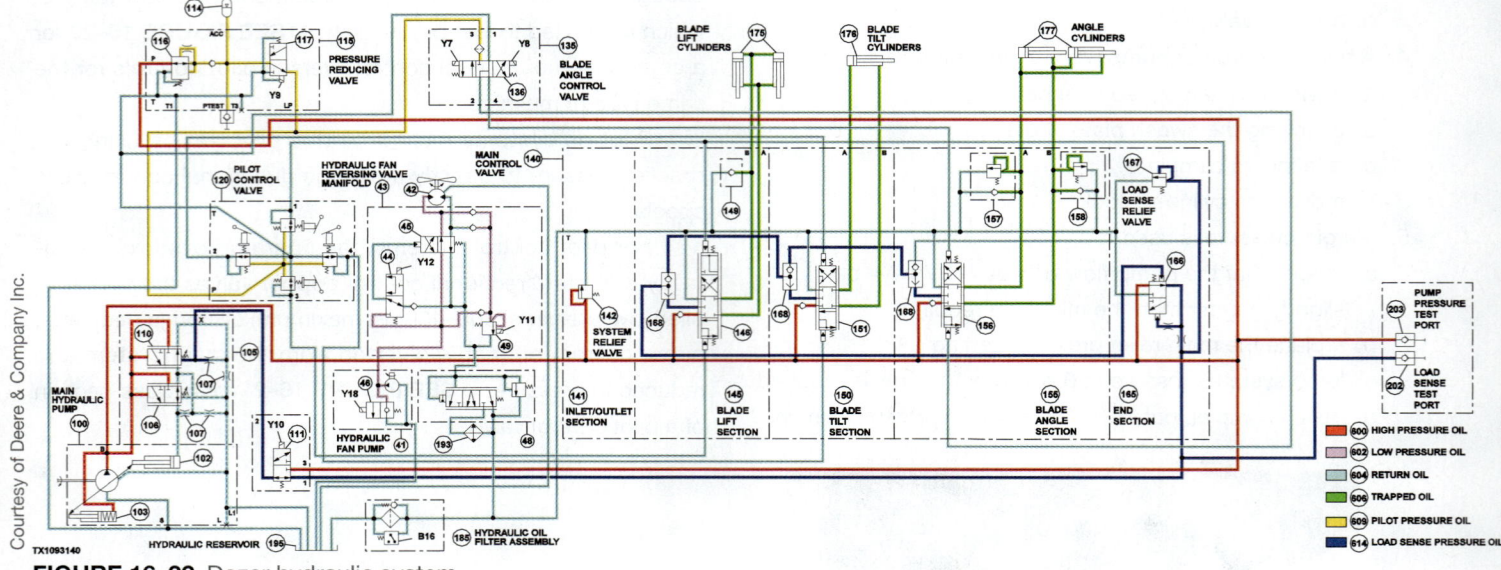

FIGURE 16–22 Dozer hydraulic system.

Pilot system oil pressure is limited by a pilot oil relief valve that will be able to dump excess oil flow to the tank, or it can be controlled by a pressure-reducing valve.

● **SEE FIGURE 16–22** for a schematic of a dozer hydraulic system that uses a pilot oil system.

This pilot oil is then metered by operator-controlled poppet valve movement and sent to the main control valve of the machine where it acts on the end of the main control valve spools to shift them. The distance the spool moves is in proportion to the amount of pressure that is acting on it. The resistance on the spool that creates this pressure is the centering spring on the opposite side of the spool.

When the joystick is returned to its neutral position, spring force will return the poppet valve to neutral. This will drain the oil from the end of the main spool, and the spring force will return the main spool to neutral.

The pilot oil system is usually used for other functions as well such as releasing brakes, running a fan motor, or air-conditioning compressor motor.

MAIN CONTROL VALVES

There are many variations of main control valves and their operation is based on the machine's general hydraulic system function, number of pumps, pump control system, number of circuits, and where they are located. The primary purpose of

476   CHAPTER 16

a main control valve is to provide the operator the means to direct oil flow to individual circuits in either of two directions. This is done with the movement of spool valves that slide in a bore to open and close ports in the valve body. The movement of spool valves can be controlled mechanically through linkage, hydraulically with a pilot oil system or an electro hydraulic system or electrically with electric actuators.

Main control valve housings can be a one-piece cast iron and machined block that houses spool-type directional control valves and other pressure or flow-type valves that are needed for the system. This type of main control valve assembly is called a monoblock valve. ● SEE FIGURE 16–23 for a monoblock type of valve.

This style of main control valve is less likely to leak but is more expensive to manufacture, and if either the housing or one of the spools is defective, the whole assembly must be replaced. Some monoblock-type main control valves can cost over $25,000. Monoblock control valves will have a spool valve for each circuit and possibly other pressure and flow control valves.

Another type of main control valve is called a stacked valve or multi-section valve and consists of individual valve sections that are stacked together between two end housings. The sections are held together with long bolts or threaded rods and nuts. There are seals or gaskets between mating sections to prevent oil leaks. Each section will be dedicated to an individual circuit and will have a directional control valve spool and possibly some other pressure/flow control valves. ● SEE FIGURE 16–24 for a cutaway section of a multi-section valve.

FIGURE 16–23 A monoblock type of valve.

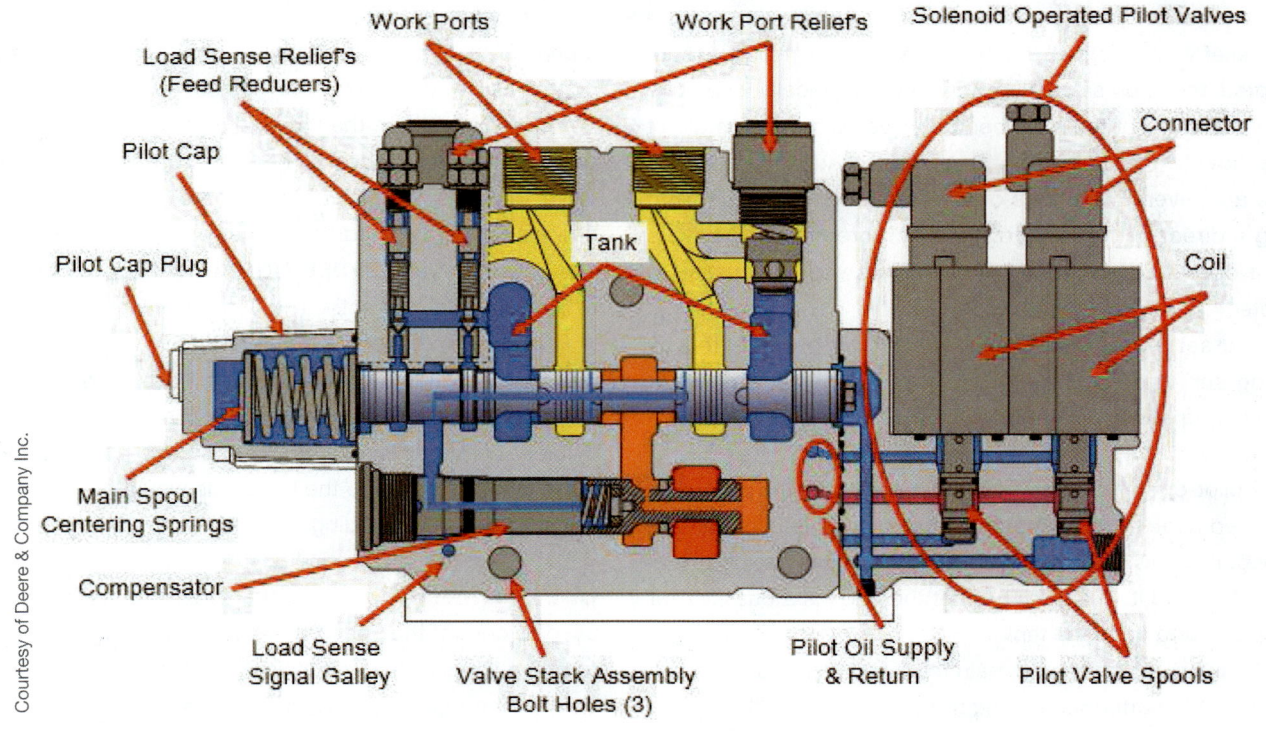

FIGURE 16–24 A cutaway view of one section of a multi-section directional control valve.

ADVANCED HYDRAULIC SYSTEMS **477**

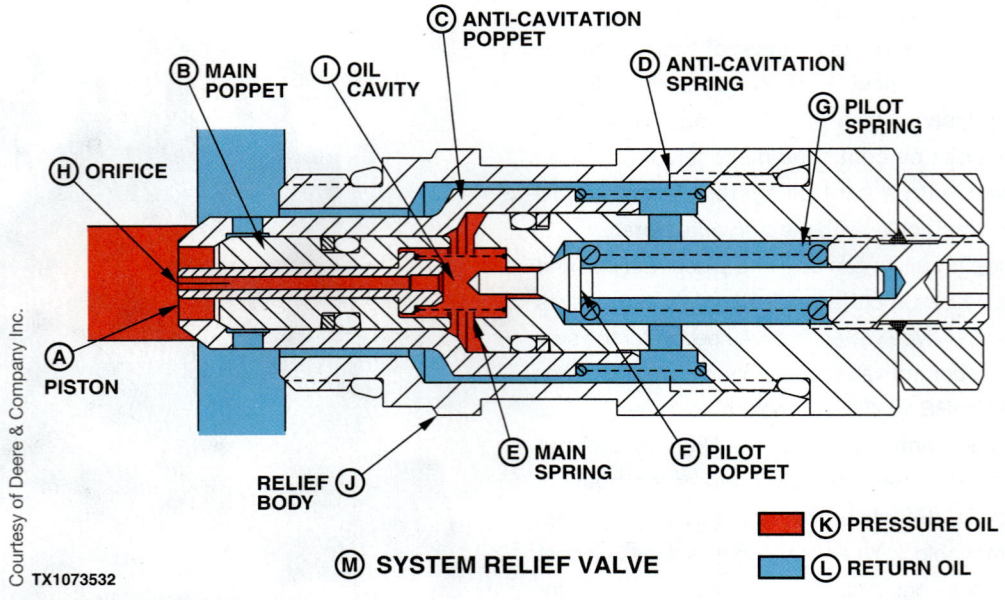

FIGURE 16–25 Main relief valve.

Stacked valves are cheaper to manufacture but can leak between sections and are susceptible to distortion if not assembled properly. The main advantage of this type of valve is that if one section is defective, it can be replaced by removing and disassembling the control valve assembly.

Both types of main control valves will have a combination of one or more of the following valves.

MAIN RELIEF VALVE
Any hydraulic system will need a main relief valve that will limit the maximum system pressure to a predetermined safe value. The valve is typically located in the inlet section of the main control valve so that it can direct pump flow to the tank when its pressure setting is reached. There are several variations of relief valves with the simplest being a direct-acting poppet type. A more common type is a pilot-operated relief valve. Relief valves were discussed in Chapter 4.

The setting of these relief valves will be adjusted by varying spring pressure either with a screw and locknut or with varying shim thicknesses. ● **SEE FIGURE 16–25** for a cross-section of a main relief valve.

A pilot-operated relief valve is opened when the pilot valve is opened against its spring pressure. When the pilot valve is opened, it allows the main poppet to open by releasing the oil trapped behind it. The main poppet then opens a passage for the pump oil to go to the tank.

An example of when a main relief valve would open is if an actuator is bottomed out and the operator holds the control valve in a position to try and keep it moving.

If a main relief valve is part of a system that has a variable displacement pump with a high-pressure cutoff valve, the main relief is only for backup if the high-pressure cutoff fails. In this case, it will typically be set 300–400 psi higher than the pump's high-pressure cutoff valve setting.

LINE OR CIRCUIT RELIEF The circuit pressure between the directional control valve and the cylinder or motor needs to be limited to a safe value. If an external force was to act on the actuator when the control valve is in neutral, this would cause a pressure spike and possibly cause damage to the actuator, control valve, or conductors. These valves are typically set 10–15% higher than the main relief valve pressure setting to prevent them from opening near the setting of the main relief valve. They are not designed to handle high flow volume. They can also be combined with other circuit valves such as an anti-cavitation valve. ● **SEE FIGURE 16–26** for a cross-section of a circuit relief and anti-cavitation valve.

An example of how a line relief would work is if a skid steer was travelling with its bucket in the air and hit an immovable object such as a manhole cover, the line relief in the boom cylinder head end circuit would open and allow the pressure spike created to vent to the tank. This would prevent the boom cylinder rods from bending.

COUNTERBALANCE VALVE
Counterbalance valves are used to control the speed of an actuator if the load on it tries to move it faster than what the operator desires. Loads can be great enough to create uncontrolled movement and could cause serious safety issues.

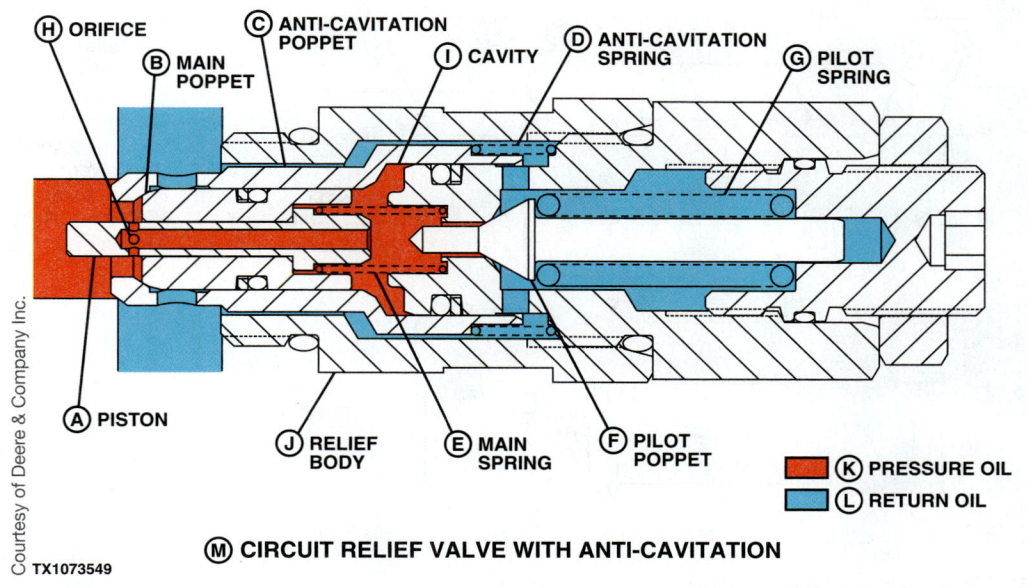

FIGURE 16–26 Combination circuit relief with anti-cavitation valve.

A good example of this is the boom circuit of a man lift. If the operator was in the basket with the boom extended and at a 45 degree angle and moved the control valve to lower the basket, the load would try to force the rod into the boom cylinder at an uncontrolled rate. A counterbalance valve would prevent this.

A counterbalance valve can also prevent actuator movement until pressure is applied to the opposite work port that the load is on. Hydraulic cranes will use counterbalance valves to hold the boom up if there was a failure on the head end return oil boom circuit. These counterbalance valves will be mounted directly on the boom cylinders. In this application, they are called lock valves or anti-drift valves.

These valves are also called pilot-operated check valves and will hold the actuator in place until there is pressure applied to the opposite side of it. They are normally closed valves that open only with pressure applied to their spool. They can be used for rotary actuator circuits as well.

Another common example is a backhoe loader that has lock valves to hold the stabilizer cylinders in the extended position and control the lower of the machine when the operator moves the control valve to do so. The only way to retract the cylinders is to apply pressure to the bottom of the check valve to unseat them. This oil pressure comes from the opposite side of the control valve and needs to be higher than the pressure rod end side of the cylinder. These valves will be part of the directional control valve. ● **SEE FIGURE 16–27** for an illustration of a valve section that uses a lock valve.

LOAD CHECK VALVE A load check valve is needed in most circuits to prevent the load on the actuator from pushing oil back past the main control valve when the directional control valve is moved from neutral. An example of this is if a loader is carrying a bucket full of material while travelling. The bucket would be 1 or 2 ft above ground level. If the operator then wanted to dump the bucket, he or she would move the control valve to direct oil to the bottom of the boom cylinder. If there wasn't a load check valve between the cylinder and the directional control valve, the load would drop until the pump flow could overcome the resistance and create enough pressure to lift it. With a load check, the load stays in place until the pump flow opens it and then lifts the load. The load check won't open until the pump pressure is high enough to overcome the pressure on the other side of load check.

For example, if a pressure of 2000 psi is held in the circuit for the bottom of the boom cylinder, the load check would stay closed until the pump flow created 2001 psi. These valves are also sometimes called lift check valves. ● **SEE FIGURE 16–28** for a lift/load check valve.

MAKEUP/ANTI-CAVITATION VALVE Some situations are created in a hydraulic circuit where the load will try to move an actuator faster than the pump oil can supply to it to keep the circuit from cavitating. Cavitation occurs if there is a pressure drop created that is severe enough to cause the oil to boil. If this happens, the vapor bubbles will burst or implode and cause severe damage to the surrounding components.

An anti-cavitation valve will be placed in a circuit where there is a possibility of this occurring. When circuit pressure

ADVANCED HYDRAULIC SYSTEMS 479

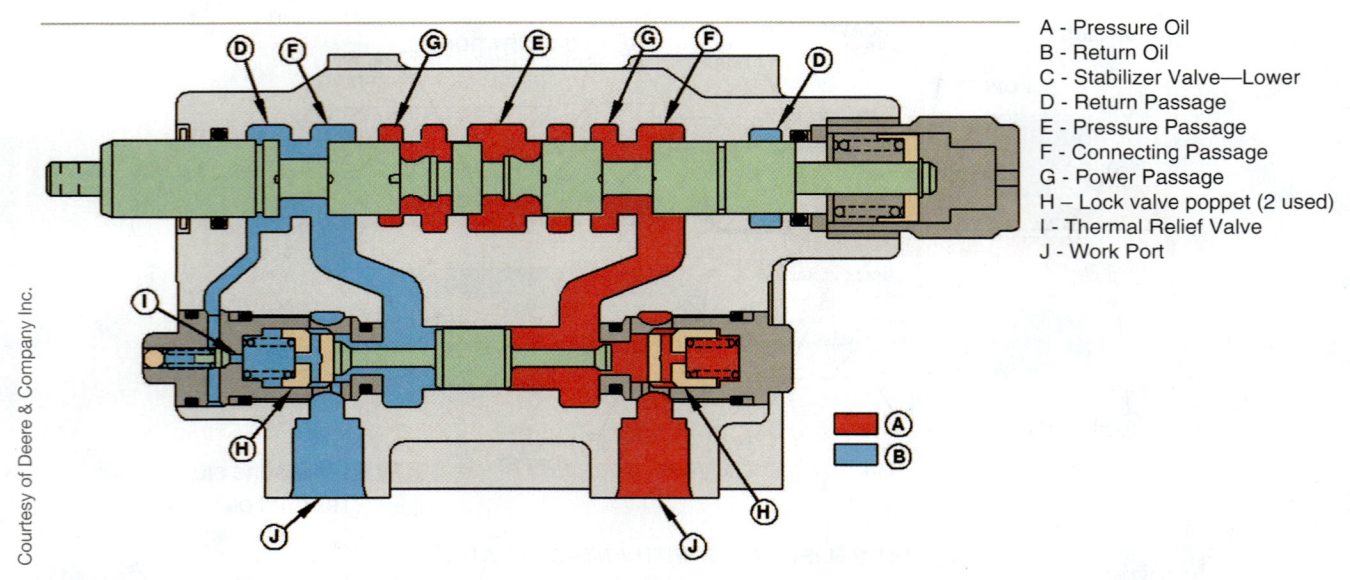

FIGURE 16–27 Valve section that incorporates a lock valve.

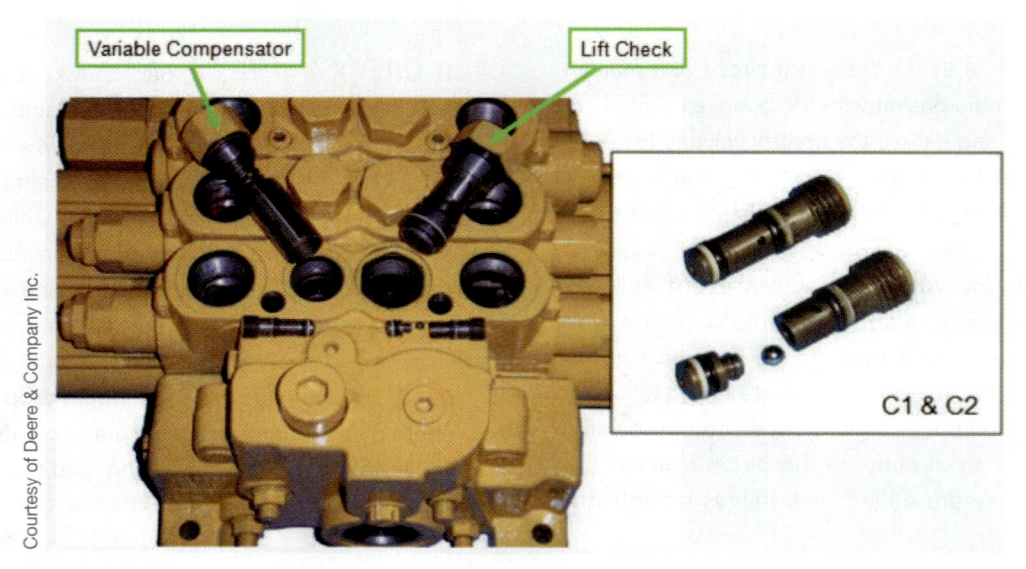

A - Pressure Oil
B - Return Oil
C - Stabilizer Valve—Lower
D - Return Passage
E - Pressure Passage
F - Connecting Passage
G - Power Passage
H – Lock valve poppet (2 used)
I - Thermal Relief Valve
J - Work Port

FIGURE 16–28 Load/lift check valve.

drops, the normally closed valve will open and allow tank oil to make up any oil needed to prevent cavitation. An example of this may be when an excavator is trenching, and as the boom is lowered quickly back into the trench, the pump can't supply enough oil to prevent cavitation on the rod end side of the circuit. The anti-cavitation valve will open and allow tank oil into that part of the circuit. Many times these valves will be part of a multi-function valve such as line relief/anti-cavitation valve. ● **SEE FIGURE 16–29** for how a combination port relief/anti-cavitation valve is part of a directional control valve.

LOAD-SENSE VALVES
Load-sense valves are simple check valves that are exposed to work port pressures and will shift to allow the highest pressure to be sent to the pump control valve. Many manufacturers call these valves shuttles valves or shuttle check valves. ● **SEE FIGURE 16–30** for a section of a directional control valve disassembled that includes a load-sense shuttle valve.

FLOW COMPENSATOR
Some main control valves will include a flow compensator for each section. The individual circuit flow compensator will reduce the pump supply pressure to the section's directional control valve spool. By reducing the pump supply pressure, it will also maintain a constant pressure drop across the directional spool regardless of what the pump supply is. This means the actuator speed is more dependent on spool movement and not just pump flow.

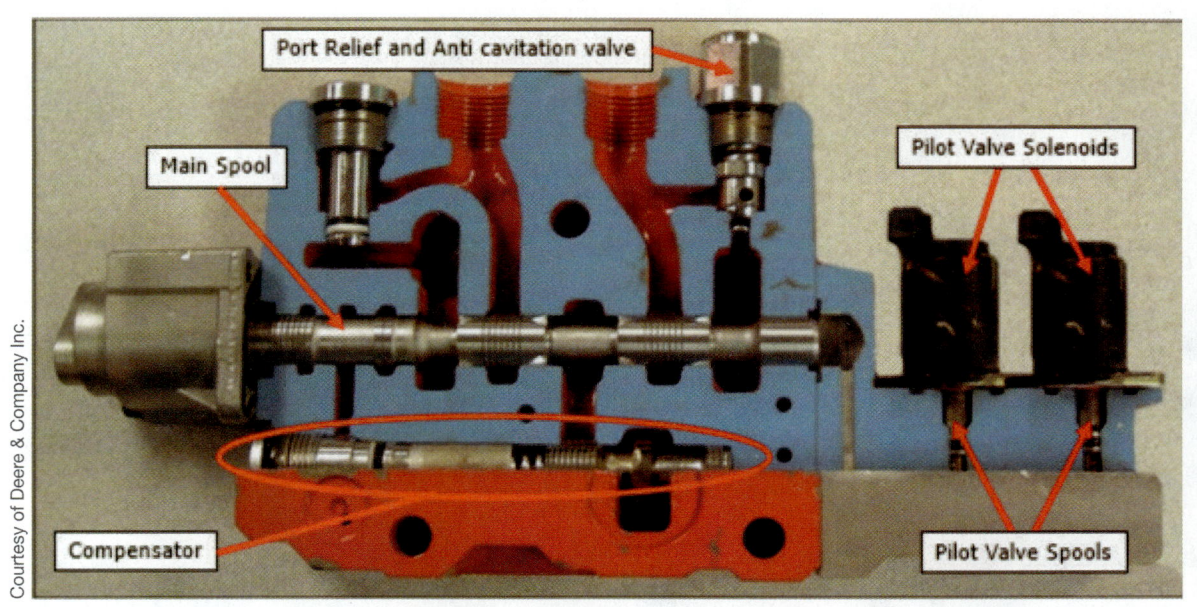

FIGURE 16–29 Combination port relief and anti cavitation valve in a DCV.

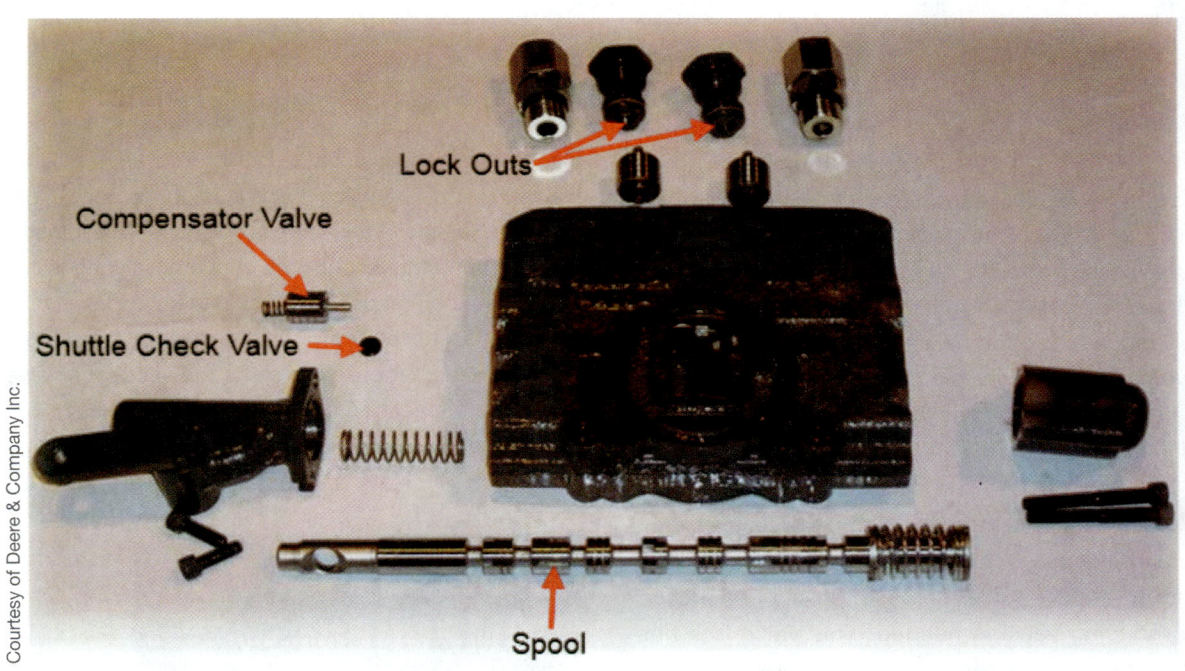

FIGURE 16–30 Valve section with load-sense shuttle valve.

It does this by having load-sense pressure added to a spring pressure on one side of a spool and the pump supply pressure on the opposite end of the spool. Therefore, the pressure drop across the spool is maintained at the value of the spring pressure. An example of this pressure is 50–100 psi. This compensator action will result in an actuator movement that is smooth and predictable.

Some flow compensators will meter flow after it leaves the directional control valve spool. When the pump can supply the total flow demand for all circuits, the circuits will not slow down. When the pump cannot supply the flow demand, the flow is divided proportionally between the activated circuits. The circuits that are active will move at a slower rate of speed because of the reduced flow of oil in each circuit.

This metering of pump flow between circuits in high demand conditions is done with flow compensator valves for each circuit. The compensator valve meters pump flow after it leaves the spool and then directs it back past the spool to the work port for the circuit. ● **SEE FIGURE 16–31** for a directional control valve with a flow compensator.

ADVANCED HYDRAULIC SYSTEMS **481**

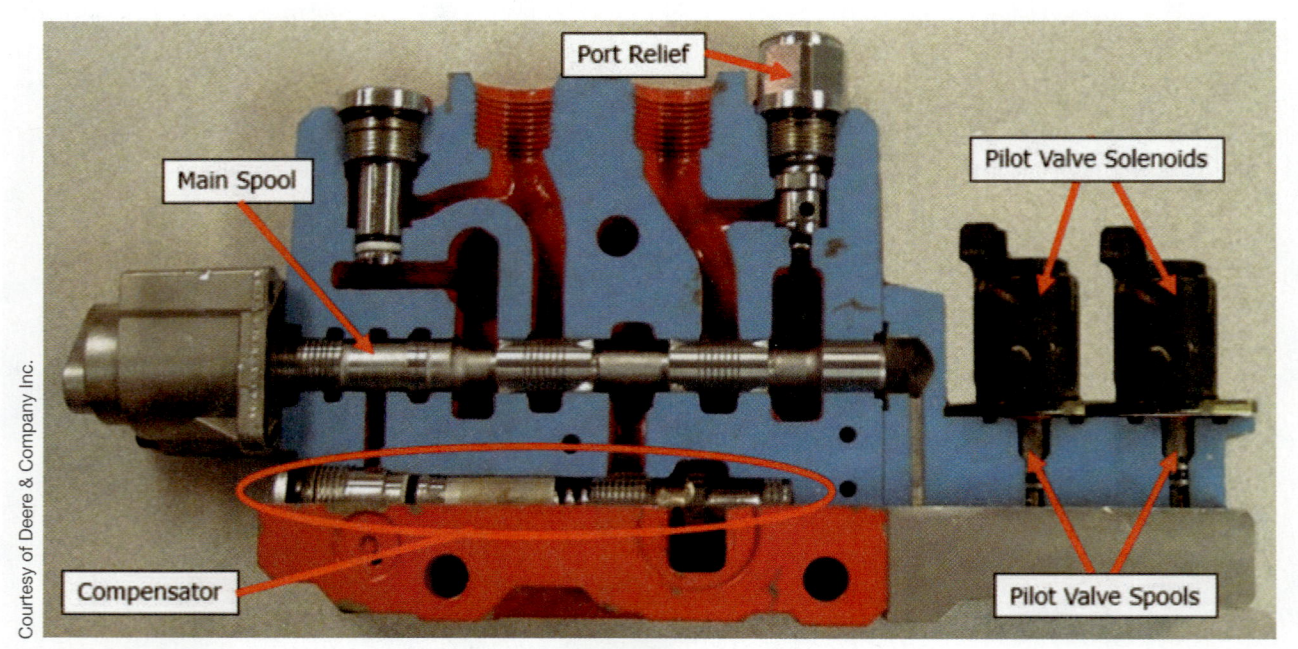

FIGURE 16–31 Directional control valve with a flow compensator.

PRESSURE COMPENSATOR A pressure compensator could be located in each of the main control valve sections, before the control valve spool. The pressure compensator regulates the oil flow rate passing through the spool so that the differential pressure in the circuit before and after the spool is kept constant.

This will reduce the effort needed to move the directional control valve spool and lower the pressure drop, which then lowers the heat generated. Less heat generated equates to higher efficiency.

PRESSURE-REDUCING VALVE Many hydraulic systems that have a pilot control system will reduce main system pressure to a much lower pressure. If this is the case, a pressure reducing valve is needed. This valve is a normally open valve that senses downstream pressure. This pressure is then applied to the valve's spool and is opposed by a spring. Pressure is reduced downstream to the value of the spring. ● **SEE FIGURE 16–32** for a symbol of a pressure-reducing valve.

SEQUENCE VALVE A sequence valve could be included in a main control valve to block flow to one circuit until the complete cycle of another circuit has completed first. If you looked at the symbol for a sequence valve, it would look like a pressure relief valve because it is drawn almost identically and functions almost identically. It is a normally closed valve that will open only when the pressure sensed from the work port of another valve exceeds the spring pressure value that is holding the poppet closed.

An example of this valve is one used for a forestry machine that must perform a sequence of hydraulic actions in order.

Once the first circuit completes its cycle and the cylinder bottoms out or the motor stops turning, the pressure builds up and the higher pressure shifts the sequence valve to allow oil flow to the second circuit. ● **SEE FIGURE 16–33** for a sequence valve in schematic form.

OTHER HYDRAULIC COMPONENTS

FLOW DIVIDERS Some hydraulic functions will have two or more actuators needed for its operation. If this is the case, the flow to each actuator will likely need to be even. This is the purpose of a flow divider. An example of where a flow divider is used is in a hydrostatic propulsion system that uses one motor per track on a four-track machine. If one pump has its flow split to supply two motors, there is the possibility that one track could start to spin under poor traction conditions. This happens because the oil will flow to the motor with the least resistance or lowest load. The flow divider is needed to stop the track from spinning and send flow to the track with good traction.

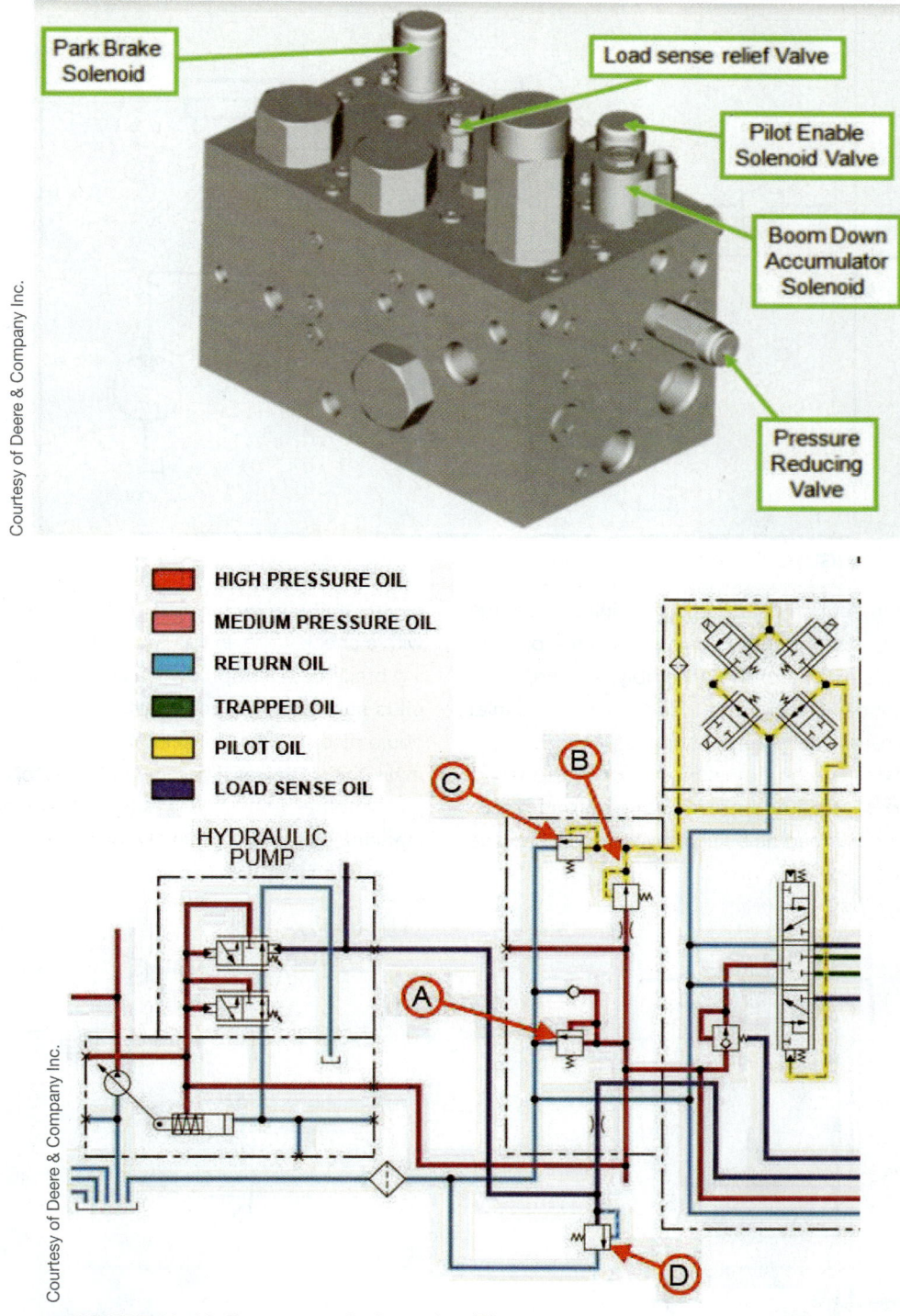

FIGURE 16–32 Pressure-reducing valve (B).

A spool-type flow divider has an inlet from the pump or directional control valve and its outlet goes to the actuators. The inlet oil acts on a free-floating spool that is cross-drilled to allow the inlet oil to go through its center to both ends of the spool. From there the oil gets metered to the outlet ports.

This will give a pressure compensation feature to this type of flow divider. Therefore, if the pressure rises in one of the two outlet ports as the result of a higher resistance to flow, this actuator would slow down. To prevent this, the pressure is sensed in the flow divider valve, and the spool is shifted to block off more flow to the actuator that is moving faster. This will happen quickly to ensure that both actuators move at the same speed.

Another type of flow divider that is used to equalize actuator speed is the gear-type flow divider. This assembly looks

ADVANCED HYDRAULIC SYSTEMS **483**

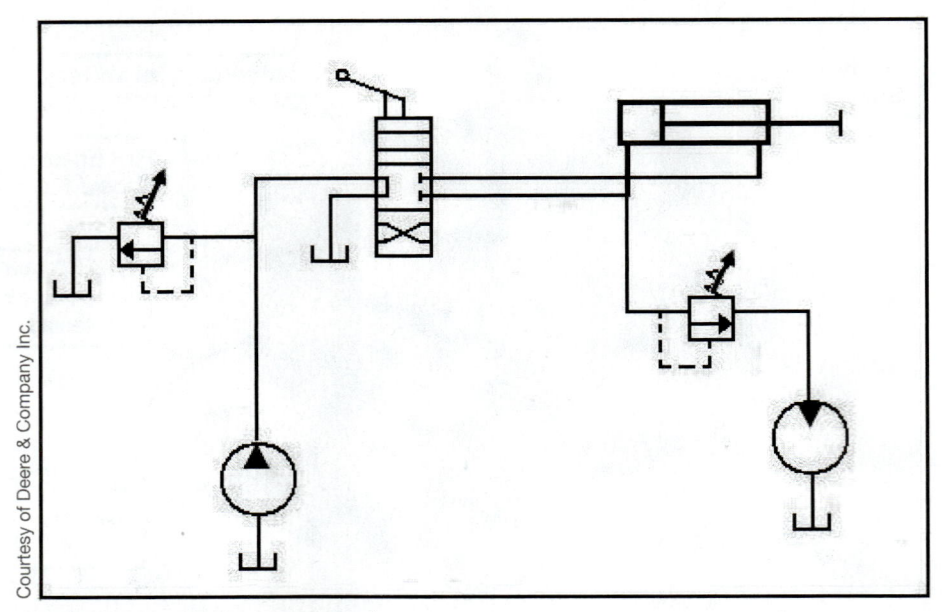

FIGURE 16-33 A sequence valve.

very similar to a multi-section gear pump. The inlet to it is the outlet oil from a directional control valve. Inside the oil gets sent to two or more sections of the assembly. The separate gear sections are connected by a shaft, and this way the inlet oil must be divided evenly between sections.

ACCUMULATORS Accumulators were discussed in Chapter 4 and were described as being a hydraulic/pneumatic device that can be used to store hydraulic energy when used with a check valve. In this application, it could be used for storing brake oil if pump flow was stopped, but brake application oil is still needed (see Chapter 18). Stored hydraulic energy could also be used for pilot oil systems to be able to actuate main control valves if pilot pump flow stops.

Another popular use is for giving a machine or part of a machine suspension (see Chapter 21) or for relieving pressure

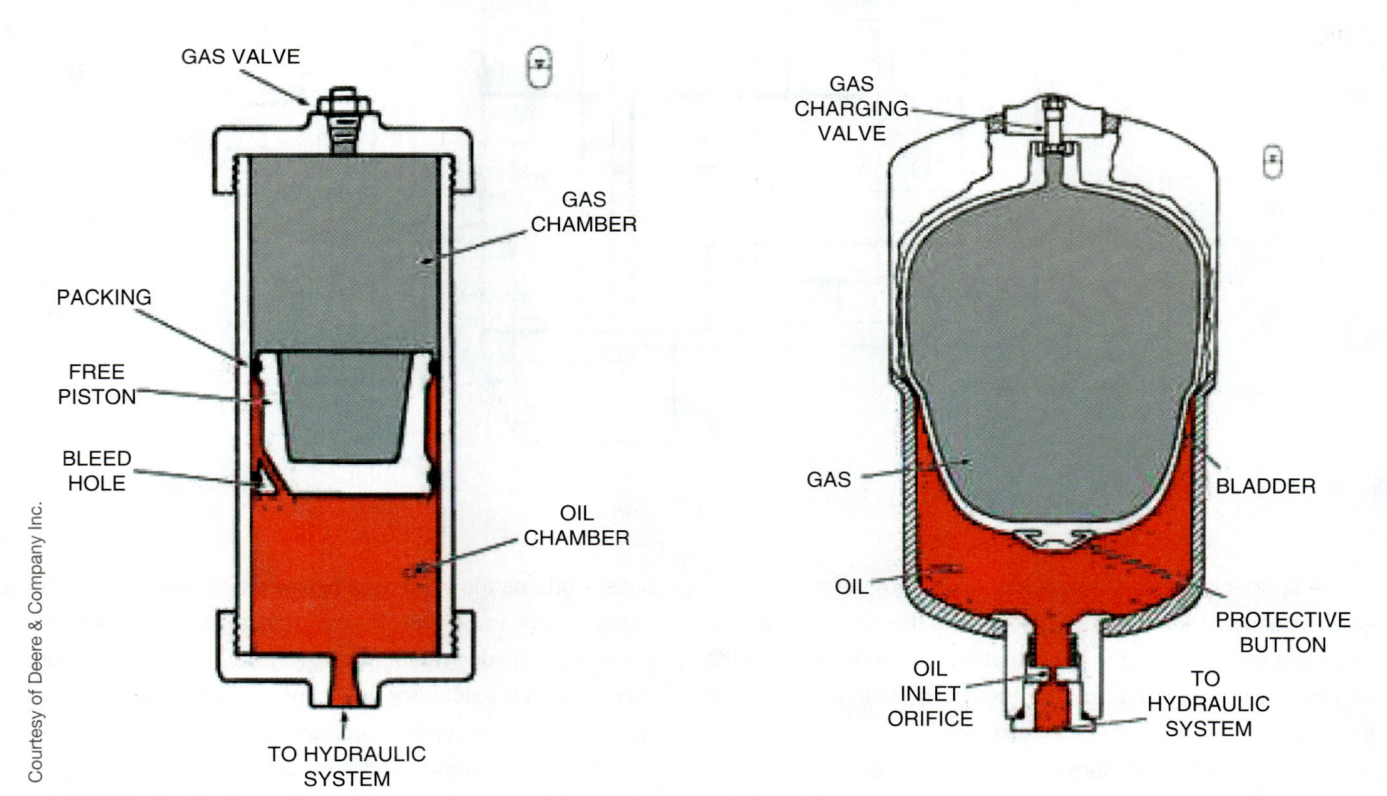

FIGURE 16-34 Two different types of accumulators (piston and bladder).

484 CHAPTER 16

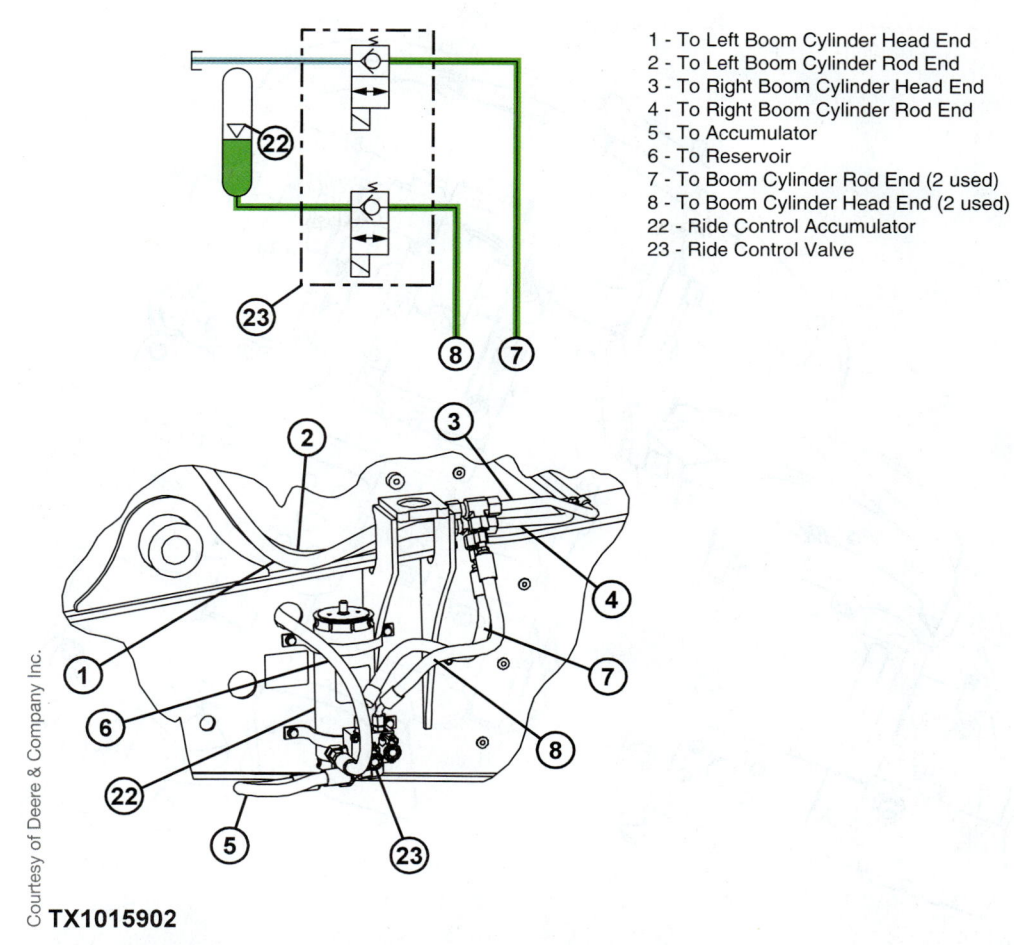

1 - To Left Boom Cylinder Head End
2 - To Left Boom Cylinder Rod End
3 - To Right Boom Cylinder Head End
4 - To Right Boom Cylinder Rod End
5 - To Accumulator
6 - To Reservoir
7 - To Boom Cylinder Rod End (2 used)
8 - To Boom Cylinder Head End (2 used)
22 - Ride Control Accumulator
23 - Ride Control Valve

FIGURE 16–35 An accumulator used for ride control.

spikes in a system much the same way a capacitor does in an electrical system. ● SEE FIGURE 16–34 for the two common types of accumulators.

An accumulator has nitrogen gas at a pre-charged pressure on one side of a piston, bladder, or diaphragm that is inside a sealed container. A charging valve is in the gas end of the accumulator that allows the nitrogen charge to be metered in and adjusted or to be let out for servicing.

As system oil is forced into the opposite end of the accumulator, the piston/bladder/diaphragm compresses the nitrogen. This will give a spring effect to the oil that can perform different functions depending on how the accumulator is connected into the circuit and what valves are used. ● SEE FIGURE 16–35 for an accumulator used of for a ride control function.

EXCAVATOR HYDRAULIC SYSTEMS

Excavators are popular and versatile machines that used to be fairly simple machines hydraulically. However, they have become increasingly complicated in the effort to make them more efficient and easier to operate.

Excavator hydraulic systems have some unique features that should be highlighted. The main functions or circuits of an excavator are boom, stick (sometimes called arm), bucket, swing, and travel. Smaller excavators will typically feature a blade and a boom swing function as well. Most excavators will also have at least one auxiliary circuit for an attachment such as a hammer or swivel bucket.

A medium-to-large excavator will typically have two main pumps that supply the oil flow to the main circuits mentioned earlier and a third pump to supply oil to a pilot system or an auxiliary function. If the third pump instead supplies an auxiliary function, then a forth pump will be needed to supply pilot oil. Smaller excavators with blade, boom swing, or auxiliary functions will need a third pump to supply these circuits and a fourth for pilot oil, or they could use a pressure-reducing valve for pilot oil. ● SEE FIGURE 16–36 for the main pump assembly for a medium-size excavator.

The main pumps will each be dedicated to specific circuits, but there will be control valves that will allow both pump

ADVANCED HYDRAULIC SYSTEMS

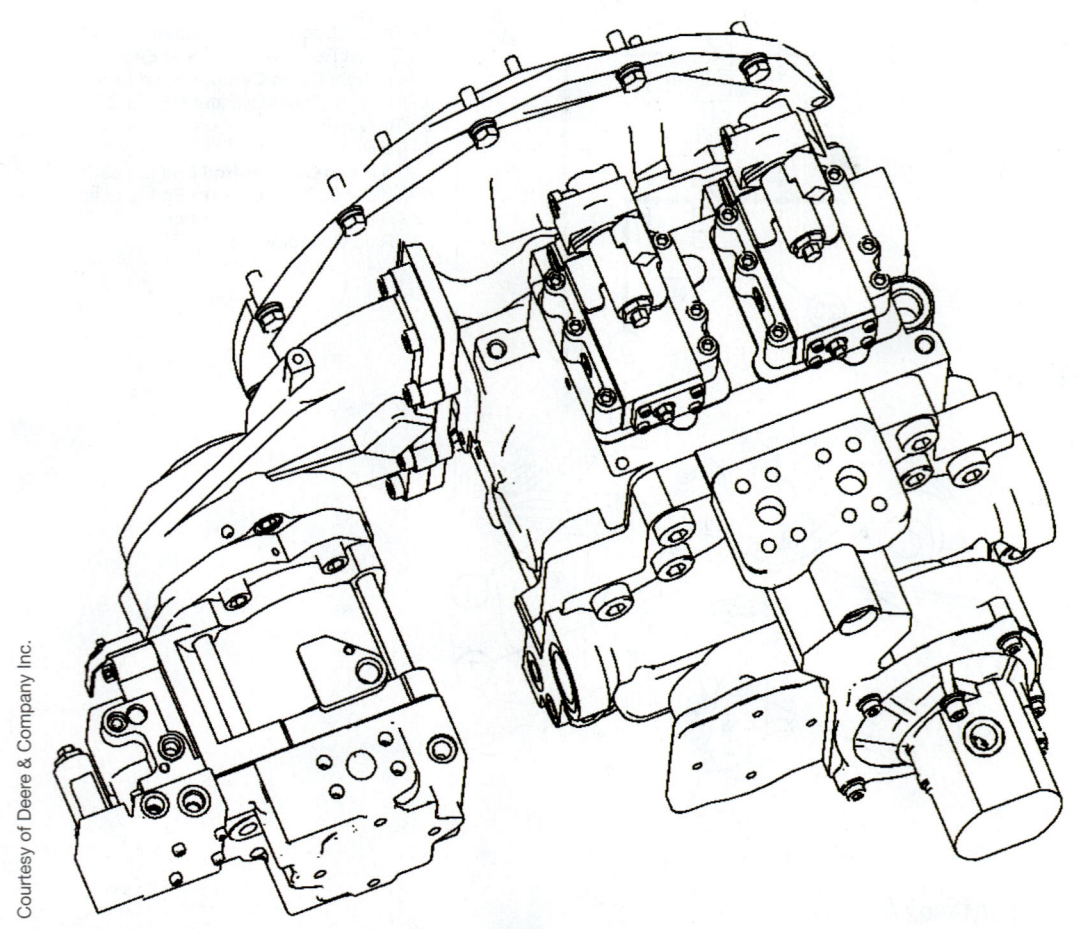

FIGURE 16–36 A pump assembly for a medium-size excavator.

flows to combine for certain circuit conditions. An example of this is for the boom up function when the machine is trenching and two pump flows are desired to get the bucket out of the trench quickly.

Pump control valves can get signals from different sources, and different manufacturers will call these sources different by different names. A typical pump control valve will send oil pressure to an actuator piston that is controlling the displacement of the pump mechanically with a linkage attached to the piston. If the pump is an axial piston type, it will have an actuator piston either directly control swash plate movement or control it through a linkage. If it is a bent axis type of pump, the actuator piston will connect to the cylinder block with a pin. As the actuator piston moves, so does the swash plate or cylinder block, and this will change the output of the pump.

Here's one example of how an excavator pump control valve works: To change the pump output, there will be oil directed to the actuator piston from the pump control valve. There will also be a linkage connected to the actuator piston that will pivot on a pin with the other end of the link moving a spool or sleeve over a spool that is part of the pump control valve. This will give the spool some feedback from the swash plate. ● SEE FIGURE 16–37 for a cross-section view of an excavator pump assembly.

The sleeve is moved by the actuator piston to close off a drain passage to keep oil trapped behind the actuator piston until another part of the regulator valve either adds more oil or drains it.

The goal of the pump control system is to make the pumps supply just enough oil to meet system demands and to limit the amount of oil output of the pumps so the engine doesn't go into a lug condition. This is done by monitoring engine speed, operator input, pump output pressure, and pump displacement with sensors.

Pump control valves can have one or two spools that are moved by oil pressure and direct oil to the pump displacement actuator.

1- pump assembly
4- shaft
17/18- displacement control valve
18-
19- displacement piston
21- cylinder block
22- port plate
25- drive link
26- block link
27- piston
30- pump assembly
31- pump assembly

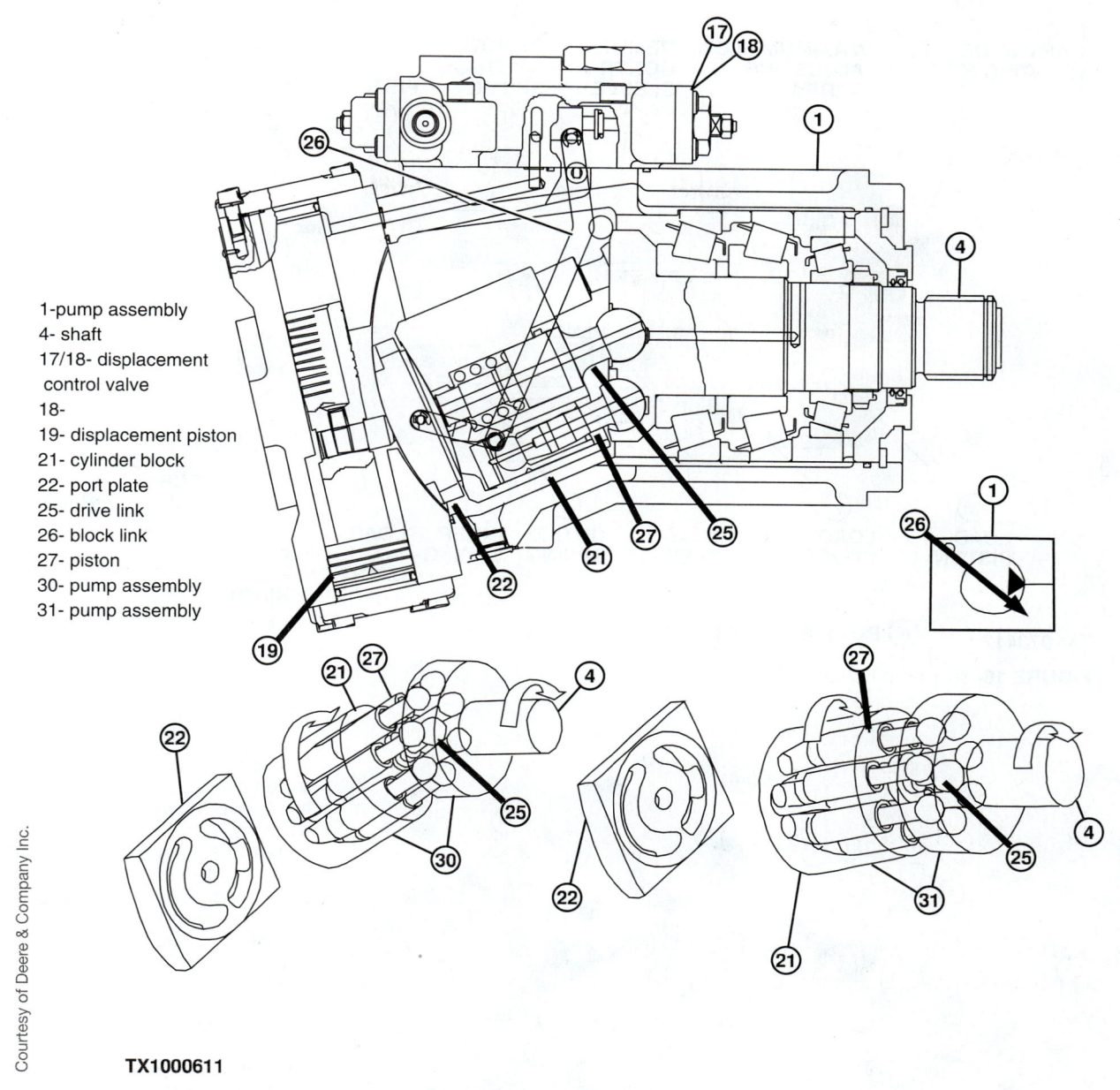

FIGURE 16–37 A cross-section view of excavator pump assembly.

- **SEE FIGURE 16–38** for a cross-section of a pump control valve.
- **SEE FIGURE 16–39** for an exploded view of the actual valve. Any operational pump control valve will be in one of three states:

Upstroking: Changing the pump to a higher displacement. **SEE FIGURE 16–40** for the pump control valve in an upstroking state.

Destroking: Changing the pump to a lower displacement.

Constant flow: Metering oil pressure to the pump displacement actuator to maintain a steady pump flow.

An example of how an excavator's pump control system will keep its prime mover in its most efficient rpm range is as follows: A medium-to-large excavator will typically have an engine that operates most efficiently in a speed range of 1900–2000 rpm. If the hydraulic load starts to drop the engine rpm to close to 1900, then the pump controls will destroke the pumps to lessen the load. The engine rpm will then increase to 2000, and the pumps can then upstroke to supply more oil if operator demands are such. In real time, as the machine is working, this upstroking and destroking action is taking place constantly and quickly to meet flow demands and to keep the engine from lugging.

ADVANCED HYDRAULIC SYSTEMS 487

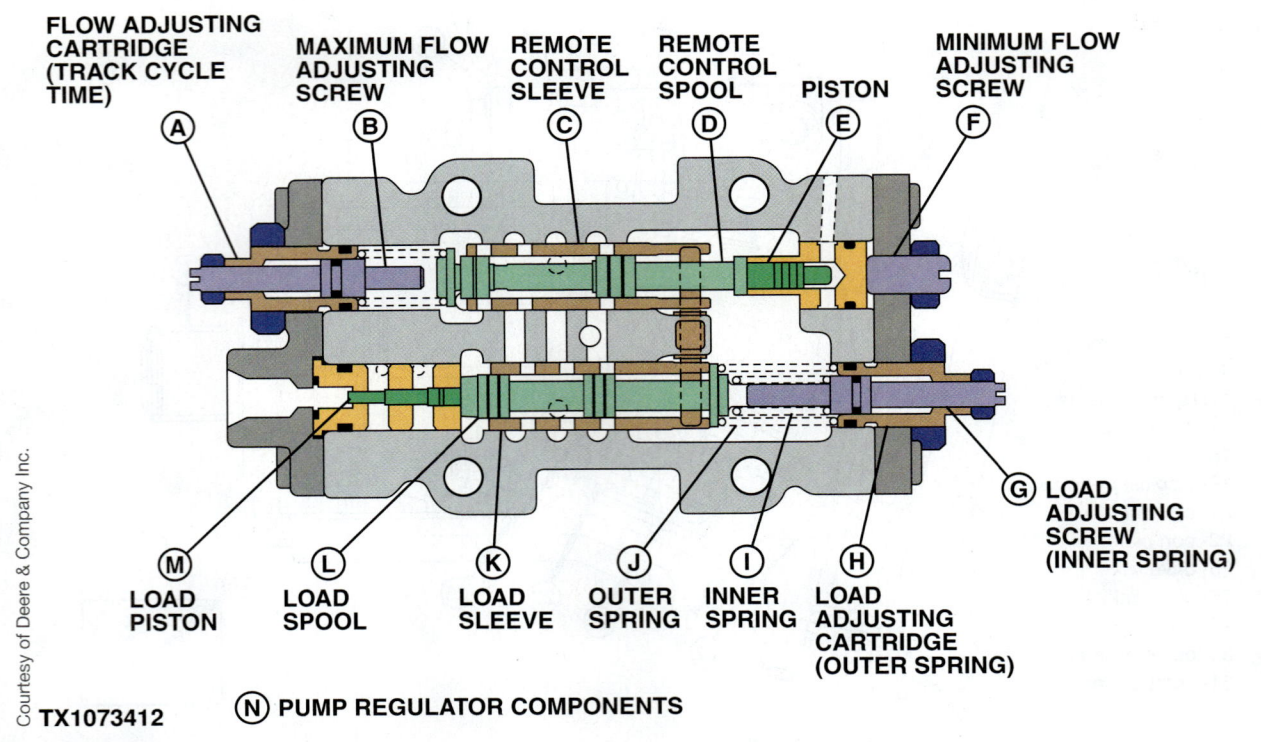

FIGURE 16–38 Pump control valve.

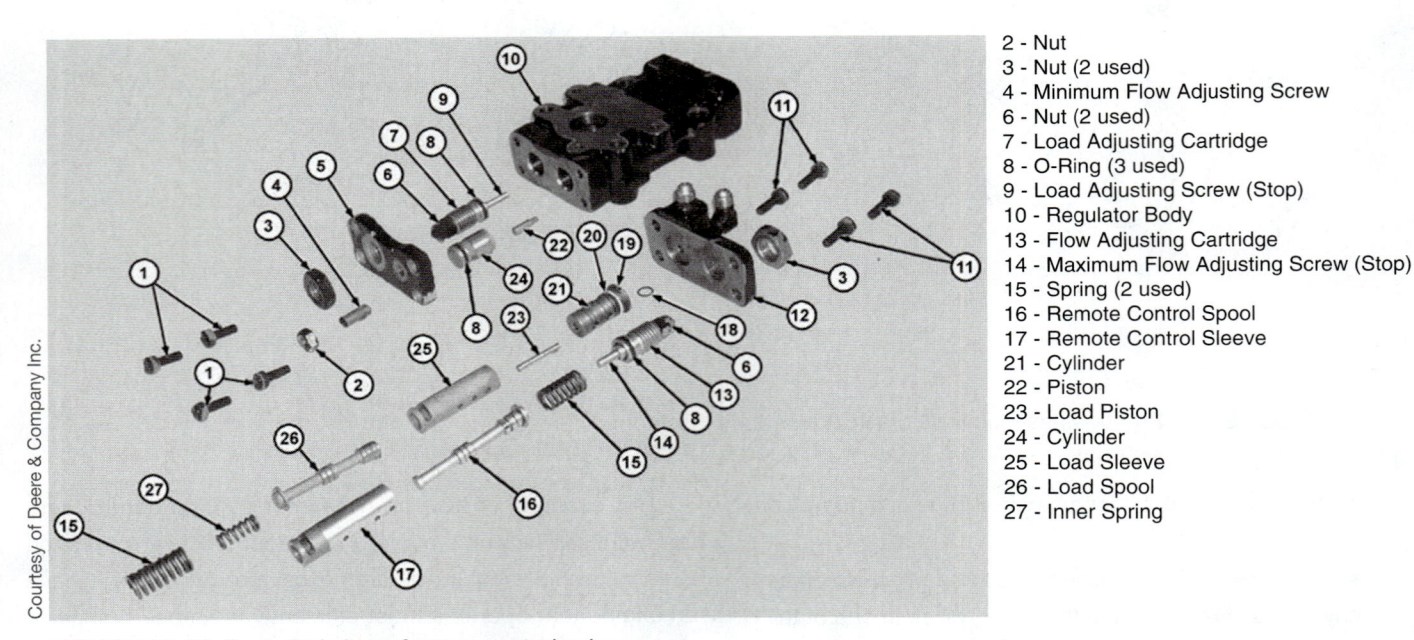

FIGURE 16–39 Exploded view of pump control valve.

The different sources of oil pressure that will act on the pump regulator valve spools are the following:

Pump output oil from the main control valve: For some systems, pump oil will normally flow through a passage in the main control valve. This oil will be sent to the pump control valve through an orifice. If one of the main control valve spools is moved, the oil pressure to the pump control valve is decreased, and this will upstroke the pump. This oil pressure is proportional to the amount of oil sent to a function. The increase or decrease of this oil pressure will shift a spool in the

488 CHAPTER 16

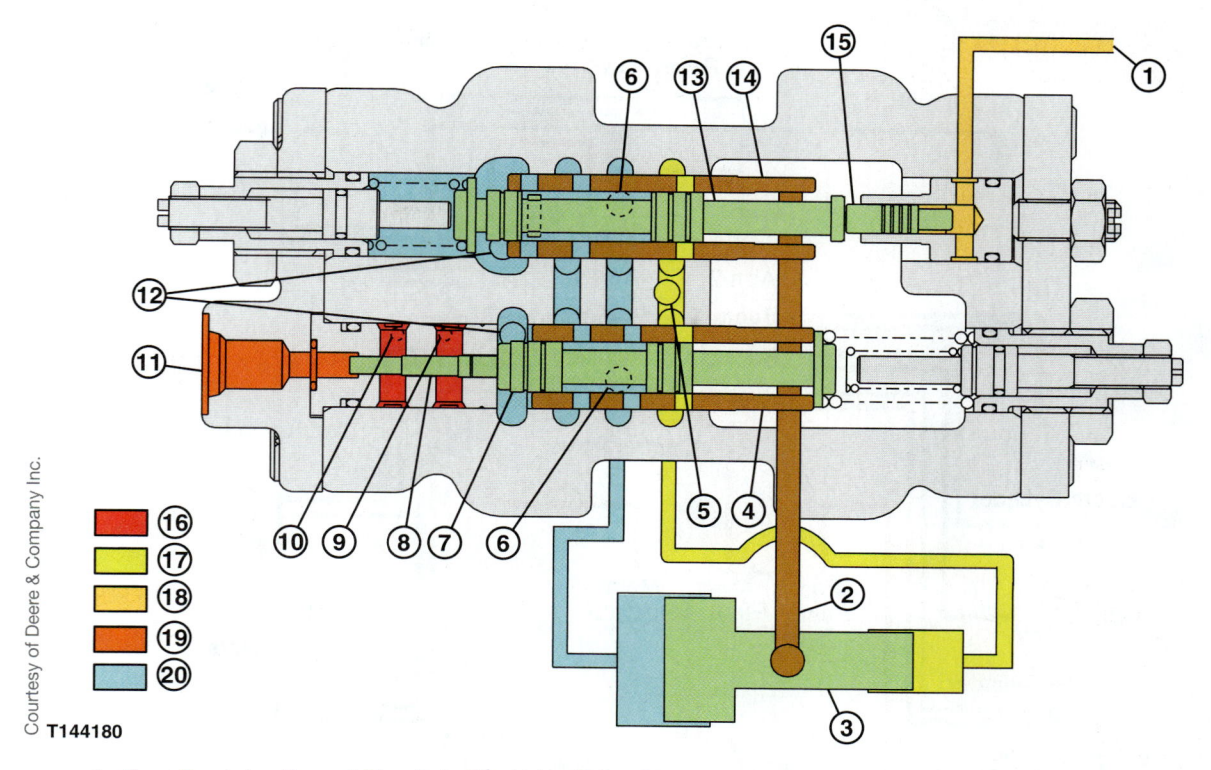

1 - From Pump 1 or Pump 2 Flow Rate Pilot Valve (SA or SB)
2 - Feedback Link
3 - Servo Piston
4 - Load Sleeve
5 - Pilot Oil Inlet
6 - To Large End of Servo Piston
7 - Load Spool
8 - Load Piston
9 - Pump 1 Pressure Inlet
10 - Pump 2 Pressure Inlet
11 - Torque Sensing Port
12 - Return to Pump Housing
13 - Remote Control Spool
14 - Remote Control Sleeve
15 - Piston
16 - Supply Oil
17 - Pilot Oil
18 - Flow Rate Valve Pilot Oil
19 - Torque Sensing Pilot Oil
20 - Return or Pressure-Free Oil

FIGURE 16–40 Control valve in an upstroking state.

TECH TIP

A simpler version of pump controls for excavators used the flow from a fixed displacement pump that was routed through an orifice to determine engine speed. The pump was driven from the same pump drive as the two main variable displacement pumps. Because the pressure drop across the orifice changed as pump flow changed, this pressure drop could be interpreted as an engine speed change. A signal was sent to the pump control based on this pressure drop, and this is how the pump displacement was changed to prevent the pumps from taking too much horsepower out of the engine. This system used to be used on older medium-to-large excavators and is still used on some smaller excavators.

pump control valve. The spool shifts to direct oil to the actuator piston or drain it away. This will move the actuator piston and change the pump output.

Average pump outlet pressure: For an excavator that has two main pumps, it is critical the pump control system considers what the output of both pumps is. This is done with a set of orifices and check valves. This pressure also moves the pump control valve spool to direct oil to the actuator piston or drain it away.

Solenoid controlled engine speed sense oil: The hydraulic/engine controller will have an engine speed sensor as an input. One of its outputs based on engine speed and programmed memory will be a voltage signal to a proportional solenoid. This solenoid will allow oil to be sent to the pump regulator valve or drain it away. This another way of regulating the pump output is regulated based on engine speed.

ADVANCED HYDRAULIC SYSTEMS

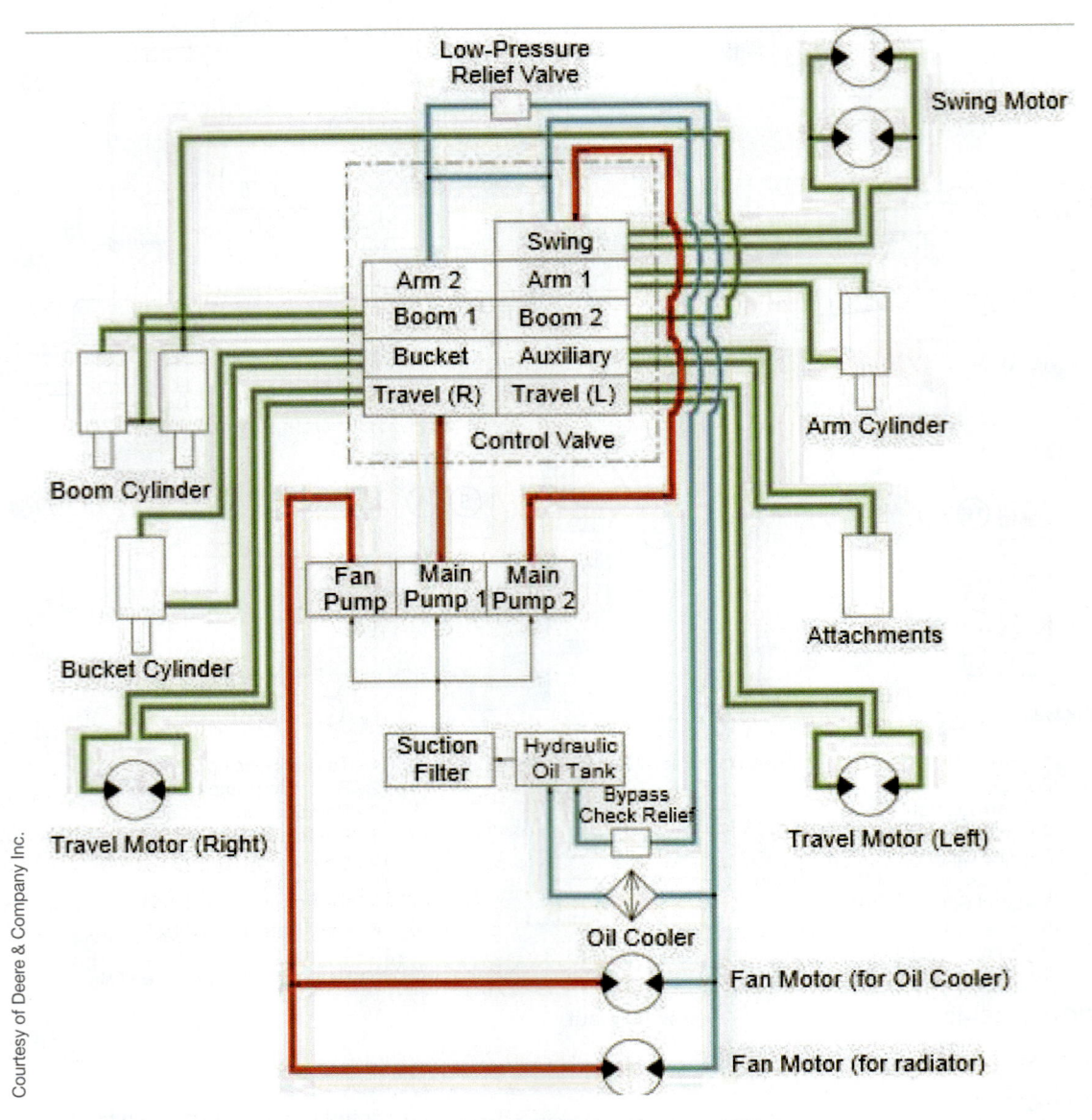

FIGURE 16–41 A typical main control valve arrangement for an excavator.

Even though both pumps operate independently, there is a limit to how much horsepower one pump can take. This is usually 85% of the available engine horsepower.

EXCAVATOR MAIN CONTROL VALVES

A typical arrangement for excavator main control valves is to have one pump supply a five-spool main control valve and have the other pump supply a four-spool main control valve.

The five-valve spool will supply one travel motor, auxiliary circuit, stick cylinder, boom combiner valve, and swing motor.

The four-valve spool will supply the other travel motor, bucket cylinder, boom cylinders, and stick combiner valve.

- **SEE FIGURE 16–41** for a typical main control valve arrangement for an excavator.

There are many other features/circuits that a hydraulic excavator could have as part of its hydraulic system. The following are a few examples:

COMBINER VALVES The combiner circuits are activated to enable both pumps to supply oil for high-flow circuits such as the stick and boom. They will be an extra spool that is activated based on extra movement of the joystick, which creates a higher pilot oil pressure.

REGENERATION CIRCUITS Some excavator functions will move with the force of gravity faster than the pump can

490 CHAPTER 16

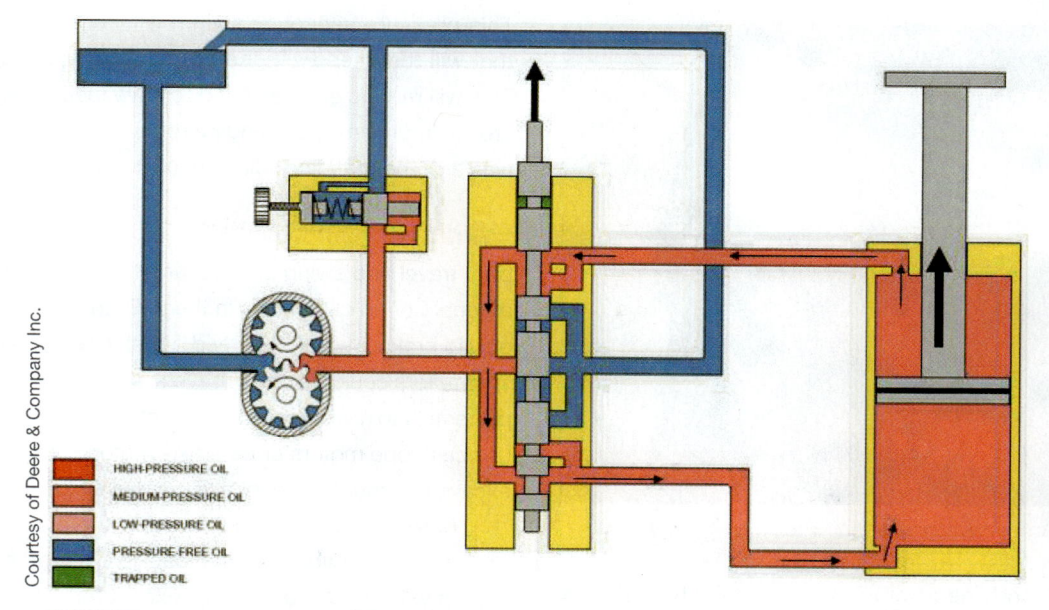

FIGURE 16–42 Regeneration circuit schematic.

supply oil to them. When this happens, there is a possibility of cylinder cavitation, which is a destructive force that can damage components. The cylinder is also not in control, which is a safety concern.

A regeneration circuit allows the return oil from a cylinder or motor that would normally be directed to the tank to be sent to the opposite side of the cylinder piston.

An example of this is when an excavator has its stick extended with the bucket in the air and the operator wants to bring the stick in. The cylinder needs to extend, and this will require a large volume of oil. As the pump can't supply the required oil fast enough, gravity will try to pull the stick and bucket down, and this could create a void in the cylinder and cause a cavitation condition. The regeneration valve will combine the oil that is leaving the rod end of the cylinder with pump supply oil to give the required volume that is needed to prevent cavitation.

A regeneration circuit could be part of the main control valve spool or the spool could be used in combination with an electrically controlled solenoid valve.

Regeneration circuits can also be used to supply oil from one circuit to another under certain conditions. An example of this is when an excavator is booming up and moving the stick in the rod end boom oil can go to the stick cylinder head end oil in combination with pump oil. ● **SEE FIGURE 16–42** for a schematic of a regeneration circuit.

AUXILIARY CIRCUITS
Excavators can have many different attachments put in place or in addition to their bucket. Some examples are hydraulic hammer, swivel bucket, tamper, thumb, and wood chipper.

There will be an extra spool valve in the main control valve that will likely be shifted with pilot oil and controlled electrically. The operator will start this function operating with a button, trigger, or foot pedal. Some machines will have an electronic auxiliary flow controller that can be used to match pump flow to the attachment's flow requirements.

There will be hoses and lines from the main control valve to shutoff valves and quick couplers that are used to connect to the attachment.

STRAIGHT TRAVEL FUNCTION
Many excavators will have a feature that will allow the operator to easily make the machine travel in a straight line if another circuit is activated. Normally, each travel motor is fed oil from one of the main pumps. If another function is activated while the machine is travelling, there would be a decrease of flow to one travel motor, which could make the machine turn. The straight travel function will dedicate one pump to drive both motors to prevent an unwanted turn if there is another circuit activated.

WORK/POWER MODES
Most excavators produced today will feature operator selectable hydraulic power modes. There are usually three or four different settings that the machine can be chosen to work in. If the machine has selectable power modes, the operator could choose a power mode to match the type of job being done. If the machine is working in light soil conditions, it won't need full hydraulic power. If the machine is being used for heavy lifts, then it will likely have a dual stage main relief valve that will get shifted to increase the maximum

ADVANCED HYDRAULIC SYSTEMS **491**

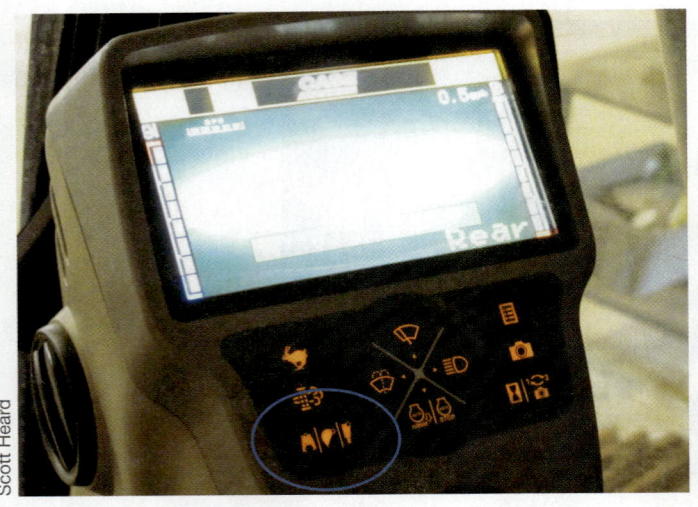

FIGURE 16–43 An excavator working mode control.

system pressure and the pump flow will be limited. ● SEE FIGURE 16–43 for an excavator working mode control.

These modes could also prioritize different circuits to get more pump flow than others. If there is a need to swing the machine a lot, then a swing priority mode could be chosen to give more flow to the swing circuit.

BACKUP MODES If the machine has a major problem with its hydraulic/electronic control system, there needs to be a way to keep the machine running to enable the operator to get the machine off the job site or just out of the way.

Most excavators will have a backup or limp home mode that bypasses the normal operation of the electronic control system to allow the engine to just provide enough power to drive the hydraulic system so that the machine can be moved. There will likely be one or more switches that the operator needs to move to change into backup mode.

OVERLOAD WARNING An excavator is used for lifting heavy object sometimes. For example, if it is being used on a sewer and water main job, it will typically dig a trench and then place a length of pipe as it is going. There may be pressure sensors that will alert the operator to overload conditions.

AUTO IDLE Most of today's excavators feature a function called auto idle or a similarly named system that is designed to save fuel, reduce noise, and increase engine life when there is no hydraulic system demand. This is a feature that will detect when the hydraulic system is not being used for a few seconds and automatically reduce the engine speed.

Most of these systems will operate by having a pressure sensor in the pump outlet, the main control valve, or the pilot oil system to tell when there is a minimum system pressure.

This pressure sensor or switch will send a signal to an ECM that will then send a signal to the engine fuel system control. This will reduce engine speed to save fuel. When there is an increase in pressure, the engine rpm will rise to where it originally was and the machine will work as it was.

SWING AND TRAVEL CIRCUITS Excavators use motors for both travel and swing circuits, and these circuits will likely have a couple of unique valves in them. Both motors will likely incorporate brakes, which will need a brake release valve and circuit. The motors could be one of several designs and may be dual displacement to give a two-speed function. Smaller excavators could use cam lobe motors or bent axis motors while medium-to-large excavators could use bent axis or axial piston motors.

Makeup valves are required in travel and swing circuits to eliminate the ability of the load to overcome the pump flow capacity. In other words, if an excavator is travelling downhill, there would be a tendency for the machine to freewheel out of control. The pump flow could not keep up to the travel motor flow requirement and a cavitation condition would occur. This would mean the operator could not control the machine travel speed, which would be a huge safety concern. In this valve, a spring will hold a poppet closed and be assisted by the pressure that builds on the outlet of the loaded actuator. The poppet stays closed until enough pressure is built up to overcome the spring pressure. This will mean the actuator will move only if there is pressure applied to the inlet side of the actuator.

Swing circuits will also need crossover relief valves. These normally closed valves will open when the upperstructure is in motion and the control lever is returned to neutral. The oil will be trapped in the circuit between the main control spool and the motor. The inertia of the upperstructure will want to keep it turning, and this is why the crossover relief is needed.

It will open when pressure builds during a swing stop condition and let oil go to the opposite side of the motor to prevent cavitation. This forces oil through an orifice in the relief valve and provides a braking action. This valve will also provide a cushion when starting and stopping a swing function.

Progress Check

6. A hydraulic system that uses a pilot oil system will:
   a. use lower pressure pilot oil to move the actuators slower
   b. always have a pilot oil pump
   c. use lower pressure oil to move the main control valve spools
   d. always have its relief valve set 1000 psi lower than main system pressure

7. What is the least likely way you would see a directional control valve spool moved on a current machine?
   a. pneumatic
   b. pilot oil
   c. mechanically
   d. electrically

8. The purpose of a line relief or port relief valve is to:
   a. allow some oil to escape from a line going to an actuator
   b. limit the load-sense pressure going to the pump control
   c. increase the pressure in a circuit over main relief pressure
   d. relieve pressure spikes in a circuit when the directional control valve is in neutral

9. If a load on an actuator causes it to move faster than the pump flow is able to supply oil to it, there should be a _____ valve to prevent cavitation in the actuator:
   a. line relief
   b. load-sense check
   c. load check
   d. makeup valve

10. The purpose of a combiner valve that is used for an excavator hydraulic system it to:
    a. increase the main relief pressure for heavy lifting
    b. combine the two main pump flows for certain high-flow functions
    c. combine the pilot oil pump with one of the main pump flows
    d. combine the bucket and stick circuits for faster digging

CARTRIDGE VALVES

Cartridge-type valve assemblies are used as a way to reduce the amount of hoses, tubes, and connecting seals that are needed to connect separate valves when compared to conventional custom-made cast valve manifolds. ● SEE FIGURE 16–44 for a cartridge-type valve block.

They are also used as a way to save space that makes their use on smaller machines popular.

They are typically used between a pump and main control valve or between the main control valves and can even take the place of a conventional main control valve assembly. They can house a single valve or many different valves and will perform the same function of any of the previously mentioned valves such as main relief, port relief, anti-cavitation, lock valve, load check, load-sense relief, and pressure reducing. They can have high pressure and flow ratings as well.

Cartridge valves are housed in a block of aluminum or steel that has been machined with passageways to allow oil to flow into and out of the block as well as between the different valves that are installed in the block. This is also a more economical way to produce an arrangement of valves that are needed to work together within a circuit or system. Each cartridge is installed in the block to perform a specific task and to work with other cartridge valves if needed. They will have seals on their outer circumference to keep the oil from leaking internally and externally. These valve blocks are usually fitted with test ports for checking pressures.

FIGURE 16–44 A cartridge-type valve block.

There are two main types of cartridge valves and these are slip in and screw in. ● SEE FIGURE 16–45 for the different types of cartridge valves.

Slip in valves are pushed into their bores and held in place with a hold down clamp. They will have a series of O-rings on them to seal the different sections of the valve with appropriate passageways of the valve block.

Cartridge valves can be used to control pressure or flow and can be actuated manually with pilot oil but are usually solenoid actuated. There are some cartridge valves that are considered to be smart valves because they incorporate microprocessors that are multiplexed with other valves and machine ECMs.

They will have a moveable spool, ball, or poppet in them to open or close one or more paths between passageways in the block they are installed in. Cartridge valves can be on/off valves (sometimes called bang/bang valves) or can be proportional valves. This means their spool or poppet can be controlled in infinite steps from one end of travel to the other to give complete variability. They can also give a feedback signal

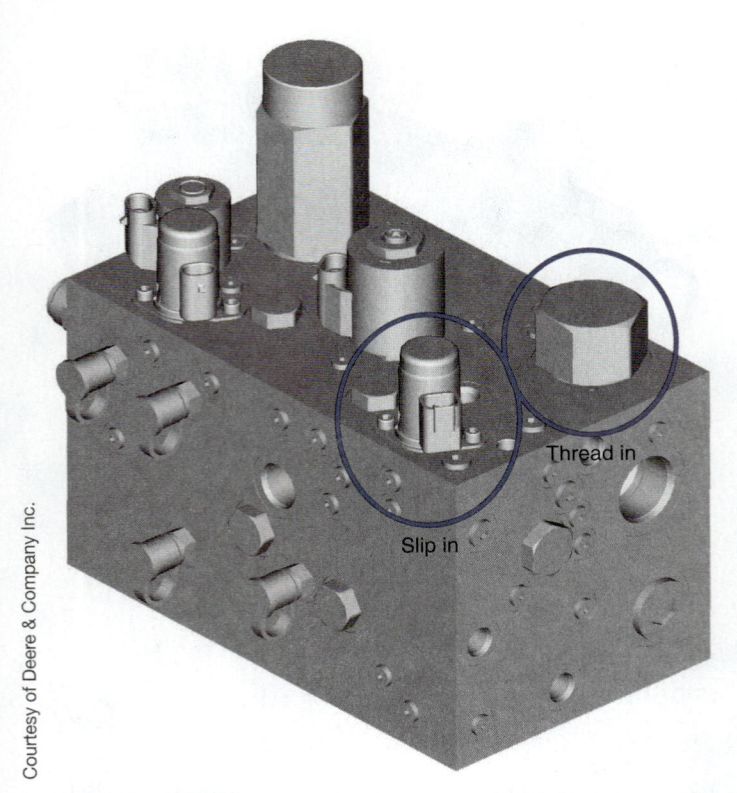

FIGURE 16–45 Two different types of cartridge valves.

to other ECMs to provide information about where they are in terms of travel.

They can be easy to troubleshoot because if there is more than one valve of the same kind in the block, you could swap valves to see if there is a fault with the cartridge assembly.

They could be two-, three-, or four-way valves and could be used for many functions such as directional control valves, relief valves, check valves, sequence valves, and pressure-reducing valves. Some cartridge valves will be combination valves that will save even more space.

They will be rated in terms of maximum flow across them, maximum pressure capability, and the size of bore they fit into. This is commonly a dash size from –4 to –16 with pressure ratings of up to 6500 psi and flow ratings up to several hundred gpm. ● **SEE FIGURE 16–46** for an example of a cartridge valve used for a grader and its schematic representation.

ELECTROHYDRAULIC SYSTEMS

Electrohydraulic is a general term that could include any hydraulic circuit that uses electrical components. However, a more specific way to define the term is to say that it is a hydraulic system that includes electronic components such as electronic control modules and system sensors as well as operator inputs such as electronic joysticks that are inputs to the system controller. The electronic controller of an electrohydraulic system will then control the pump output with an electronic signal and control directional control valves by energizing solenoid valves.

This is a natural progression for hydraulic systems because it provides a way to include the hydraulic system with the rest of the machine's multiplexed electronically controlled systems. There are many benefits of this integration such as allowing the hydraulic system to create fault codes; allowing technicians to monitor live pressures and flows for diagnostic and calibration procedures; allowing the engine to react quickly to changing hydraulic loads; and allowing the system to learn how the machine is typically operated and tailor the system to most efficiently work within the typical operating conditions. Some more advanced systems will allow the electronic joystick's controllability to be changed. This will change the electrical output response of the joysticks in relation to their movement.

The main parts of an electrohydraulic system are: the controller (ECM), inputs such as: pressure, flow, temperature, and position sensors, operator inputs such as electronic joysticks, speed control and work mode switch, solenoid operated pump control, electrically actuated control valves, operator display, and variable relief valves. Input sensors were covered in Chapter 9.
● **SEE FIGURE 16–47** for an electrohydraulic main control valve.

An example of an electronically controlled hydraulic circuit is a modern machine's engine cooling fan system.

CONTROLLER The ECM electronic control module will be the brains of the system. It performs the same duties as any other ECM in that it will gather information from sensors, process this information, and compare it to stored programming. It will then send out electrical signals to actuators. If you read Chapter 9, you should recognize this arrangement as the ECM inputs and outputs.

INPUTS

SENSORS. Some examples of sensors the ECM will need information from are engine speed sensor—much of the purpose of an electrohydraulic system is to keep the engine working in its ideal rpm and power range. Therefore, the ECM will need to monitor the engine speed.

Pump outlet pressure will also need to be measured and reported to the ECM. There will be oil temperature sensors in one or more areas. These may be needed to allow

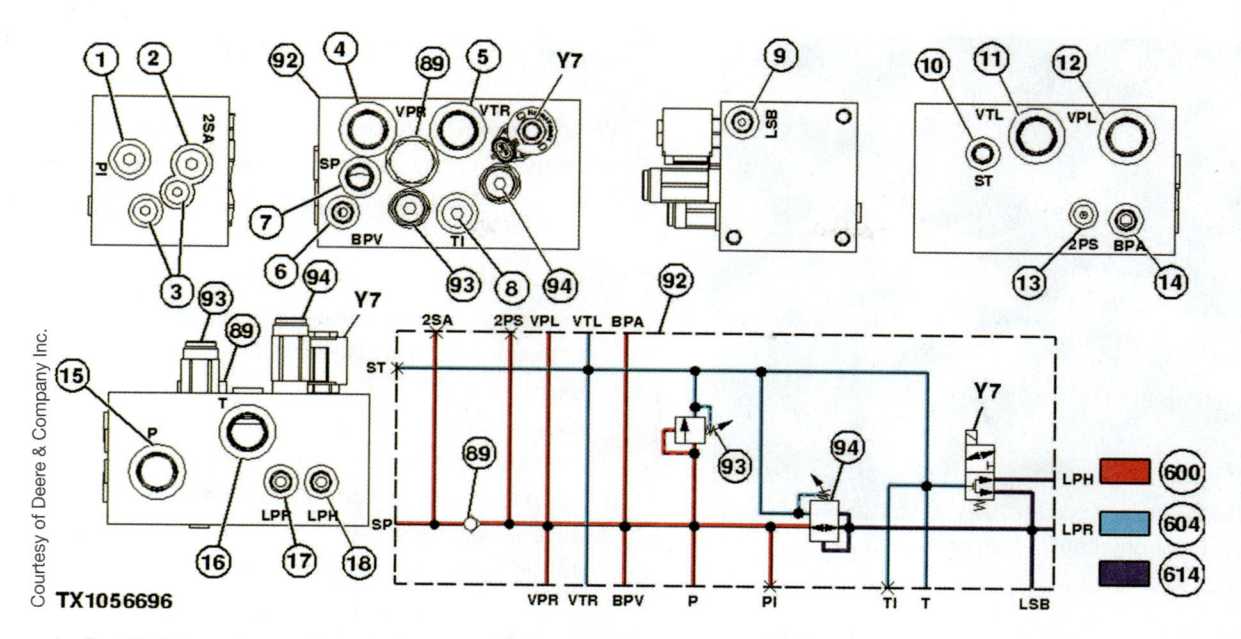

1 - Port P1 (plugged—provisions for integrated grader control system)
2 - Port 2SA (to secondary steering accumulator—if equipped)
3 - Plug (2 used)
4 - Port VPR (to right control valve assembly—standard controls) (to cab mount control valve assembly—EH controls)
5 - Port VTR (from right control valve assembly—standard controls) (to cab mount control valve assembly—EH controls)
6 - Port BPV (to brake accumulator shuttle valve)
7 - Port SP (to steering valve)
9 - Port LSB (to mid-inlet valve of right control valve assembly—standard controls) (to cab mount control valve assembly—EH controls)
10 - Port ST (from steering valve)
11 - Port VTL (from left control valve assembly—standard controls) (to draft frame mount control valve assembly—EH controls)
12 - Port VPL (to left control valve assembly—standard controls) (to draft frame mount control valve assembly—EH controls)
13 - Port 2PS (to hydraulic pump pressure sensor)
14 - Port BPA (to service brake accumulator)
15 - Port P (from main hydraulic pump)
16 - Port T (to hydraulic oil tank)
17 - Port LPR (to saddle locking pin—rod end)
18 - Port LPH (to saddle locking pin—head end)
89 - Secondary Steering Check Valve (if equipped)
92 - Hydraulic System Manifold
93 - System Relief Valve
94 - Pressure Reducing Valve
600 - Supply Oil
604 - Return Oil
614 - Load Sense Oil
Y7 - Saddle Locking Pin Solenoid

FIGURE 16–46 A cartridge valve used on a grader.

the ECM to decide whether it should allow full pressure and flow if the system oil is too hot or too cold. Hydraulic tank oil level sensors are needed to alert the operator of low oil levels, and if the level is critically low, it could derate or even put the machine into a shutdown condition. Position sensors could be needed to give the operator inputs an electronic value to the controller. They can also be used to tell the controller where the actuators are in their travel. For example, a loader boom sensor will tell the controller when the boom is close to the ground during a boom lower function so it can slow the boom down. Some cylinder position sensors can be inside the cylinders to give an accurate indication of rod position. Speed sensors could be used to monitor motor speeds of rotary actuators.

SWITCHES. There will likely be several switches that are controlled by the operator to enable auxiliary flow, change work modes, set engine speeds, pilot oil disable, and others.

OUTPUTS Electrohydraulic system outputs are used to control system flow and pressure. They can also be electronic signals to other ECMs and or machine displays.

Main control valve spools are actuated with pilot oil that is controlled by proportional solenoid valves. These

FIGURE 16-47 Electrohydraulic main control valve.

valves get an electrical signal from the ECM based on operator inputs, engine operating conditions, and stored programing.

Many main control valves will have pilot oil pressure applied to the ends of the spools at all times. This pressure along with spring pressure keeps the spool centered. To move a spool, the ECM will energize a solenoid on the opposite end of the spool, and this will open a passage to the tank.

Pump controls are electrically controlled in electrohydraulic systems. This is done with proportional solenoid valves.

ON/OFF SOLENOID VALVES

Simple low-flow and low-pressure circuits could use on/off solenoids to control pump oil flow to an actuator and its return oil back to the tank.

They will have a coil of wire that gets energized from an ECM, and this will create a magnetic field that pulls a spool plunger into the solenoid body. When this happens, the spool opens passages in the valve body that it is part of. The passages allow oil to flow as described earlier. When the coil is de-energized, there will be a spring to return the spool to its normal position.

These valves could also be used to send oil to one end of a main control valve spool. In this case, there will be a pair of solenoids used. This is an example of where on/off solenoids are used to control higher flows and pressures.

On/off solenoids could also be used for many other functions such as allowing pilot oil to flow to pilot control valves. This would be a safety feature that would not allow any oil to be available to pilot oil joysticks until certain conditions are met to provide an electrical signal to the on/off solenoid (seat occupied, door closed, seat belt done up, etc.).

> **TECH TIP**
>
> When it comes to electrohydraulics, there are three terms that you may hear and wonder what they mean:
>
> **Hysteresis:** In electrohydraulic control systems, a certain result is expected from a certain input, and the same input may produce two different results, depending on whether the input level was increasing or decreasing. This difference is the hysteresis. For example, a 500 mA signal to a proportional solenoid valve may produce 100 psi. If it changes to 1000 mA to give 200 psi and then is reduced to 500 mA, again the result may be 105 psi. The 5 psi difference is hysteresis. Electrohydraulic controls have to consider the hysteresis factor of components.
>
> **Stiction:** When a spool valve or plunger is stationary, it wants to stay stationary. This is called stiction. Many electrohydraulic controls will slightly pulse an electronic signal to the valve actuator to slightly oscillate the valve to prevent stiction.
>
> **Dither:** The term *dither* refers to how a solenoid valve is made to oscillate slightly to prevent the spool it is controlling from sticking or reduce stiction.

PROPORTIONAL SOLENOIDS

Proportional solenoids are used to control oil pressure and flow based on a pulsed electrical signal that is sent to them from an ECM. They can be found as part of main control valves, pump control valves, and part of distributed control valves. The signal is sent from a transistor in the ECM that is called a driver. Transistors were covered in Chapter 9 and are an electronic switching device. They are controlled by very small signals from the ECM. When the signal is pulsed on and off quickly, the output from the driver is called a PWM signal.

A proportional solenoid will have a plunger that is held in either a normally closed or a normally open state by a spring. When the solenoid's coil of wire is energized with a PWM signal, it will become a magnet that will move the plunger. The amount of movement of the plunger will be determined by the amount of current that is sent to the solenoid; therefore, these valves are infinitely variable.

A PWM signal is a series of on and off voltage pulses that are varied by a transistor. The frequency of the signal can range from 400 to 5000 Hz. The amount of on time versus off time will determine the amount of current sent to the coil and the strength of the magnetic field, which will determine how far the coil's plunger travels. ● **SEE FIGURE 16-48** for a stacked valve with proportional solenoids.

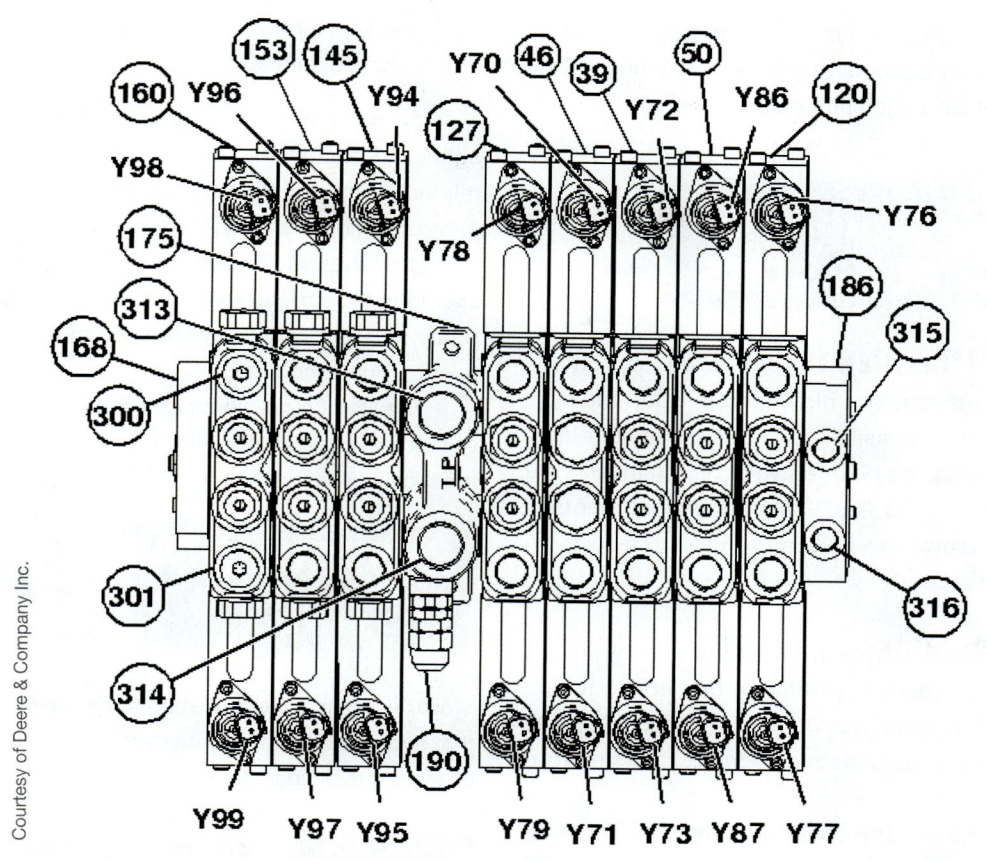

FIGURE 16–48 Stacked valve with proportional solenoids.

DEERE/HUSCO-DISTRIBUTED VALVE SYSTEM
An example of a true complete multiplexed electrohydraulic system is the latest generation of John Deere backhoes that feature the distributed valve system.

The operator inputs for the hydraulic system consist of a pair of joysticks that are electronic inputs to an ECM that is part of the joystick assembly.

Each backhoe cylinder has a valve assembly mounted directly to the cylinder. The valve assembly has four proportional poppet valves that are ECM controlled to give an almost infinite control of oil flow to the cylinder. These valve assemblies have a pump supply and tank return line to route oil to and from them.

This system relies on the software that is programmed into the ECMs to make the cylinders function as requested by the operator's joystick inputs.

CALIBRATION OF FUNCTIONS
Some machine electrohydraulic systems will require a calibration procedure to set function limitations, to adjust the operation of the input for operator preference, or to compensate for component wear.

An example would be the swing function calibration for the previously described John Deere backhoe. If the operator didn't like how the swing function reacted to joystick movement—in other words, the operator thought it was too slow/lazy or too fast/jerky, a calibration of the joystick would change the reaction time.

The calibration of a function typically requires the hydraulic oil temperature to be at least 125°F. Then the service mode on the machine display is entered, and a set of steps is followed that will match the joystick movement to when the operator sees the swing function work. This can then be fine-tuned to give the swing function the exact desired relationship between joystick movement and swing cylinder movement.

Another example would be for a machine that incorporates a weight scale into its electrohydraulic system. A wheel loader that loads highway trucks needs to be as close as possible to the truck's maximum payload without going over. To keep this weighing function accurate, the system should be calibrated regularly. This will require the machine to lift a known weight and the operator to go through a series of instructions that are shown on the machine's display. The first step to this is to get an empty bucket weight that allows the operator to zero the scale. There will be pressure transducers that translate hydraulic cylinder pressure into an electrical signal, which is then interpreted by an ECM. At this point the value of the known weight is entered by the operator into the machine keypad.

This step tells the ECM that the electrical signal coming out of the pressure transducer equates to the weight in the bucket. This is the end of the calibration procedure.

ADVANCED HYDRAULIC SYSTEM DIAGNOSTICS

ENVIRONMENTAL CONCERNS When working on advanced hydraulic systems, you may need to open the system that could lead to a possibility of a hydraulic fluid leak. You must do everything possible to avoid contaminating the environment with hydraulic fluid. There is environmentally friendly hydraulic fluid available that can be used to reduce the impact of hydraulic leaks.

GENERAL PROCEDURES Troubleshooting hydraulic system problems isn't really any different than trying to find problems with other systems. A systematic approach needs to be taken to efficiently troubleshoot a hydraulic system problem.

You should first verify the operator complaint. This may require you to get familiar with how the machine is supposed to work unless you are already knowledgeable about the machine. This could mean reading the operator's manual or asking someone who is familiar with the machine how it should normally work. The best information source will be the most current service information for the exact machine you are working on. Referring to machine service information from similar machines could possibly lead to misdiagnosing the problem. The key is that you won't be able to know if there is a problem if you don't know how the system is supposed to work.

At this point you should also determine there are no other hydraulic system problems and try to define whether the problem is just with one part of one circuit, one complete circuit, or the complete system.

Many hydraulic system problems will occur only under certain conditions such as when the oil is a certain temperature or when there is a load on the circuit. This can be difficult to replicate at times especially if the machine is at the shop and not on the job site.

Once you understand the system and can verify that there is a problem, you should then perform a visual inspection. For any hydraulic system problem, this will include:

- Checking for external leaks
- Checking hydraulic fluid level, visual appearance, and temperature (should be clear, no foam)
- Checking for damaged hydraulic components (bent cylinder rod, dented tank, crushed tube, hose damage)

If a machine has an operator display that will show diagnostic information, you should access this and record any related fault codes or any other related information such as pressures, flows, and ECM input/output values.

Now that you know the problem and the system, you should be able to put together a list of possible causes for the problem. Most service information manuals will have a diagnostic troubleshooting process that should be followed.

The list of possible causes can then be prioritized in order of the easiest and quickest to test first with the hardest and most difficult last.

HYDRAULIC SYSTEM TESTING

Most hydraulic system testing tools are precision instruments that should be handled with care and stored properly. Hydraulic testing tools include the following:

PRESSURE GAUGES Pressure gauges can be either analog or digital. ● SEE FIGURE 16–49 for a digital type of gauge.

> **SAFETY TIP**
>
> You will be tapping into hot hydraulic fluid that will have the potential to seriously injure or kill you. You need to be aware of this potential and take steps to protect yourself and others from this hazard. Use only recommended pressure gauges, fittings, hoses, and flow meters to perform testing. Any adapter fittings that are needed for hooking up test equipment must be certified to handle the system pressure applied to them. If you are handling any test equipment that has high pressure and flow in them, be aware of the potential for serious injury if it failed.
>
> If testing requires you to operate a machine in a shop, make sure that the machine's engine is adequately ventilated and all other personal are kept out of the danger area surrounding the machine.
>
> The testing procedures for hydraulic system troubleshooting will likely require some special tools. Proper diagnosis of problems cannot be accomplished if the proper tools aren't available and aren't used properly.

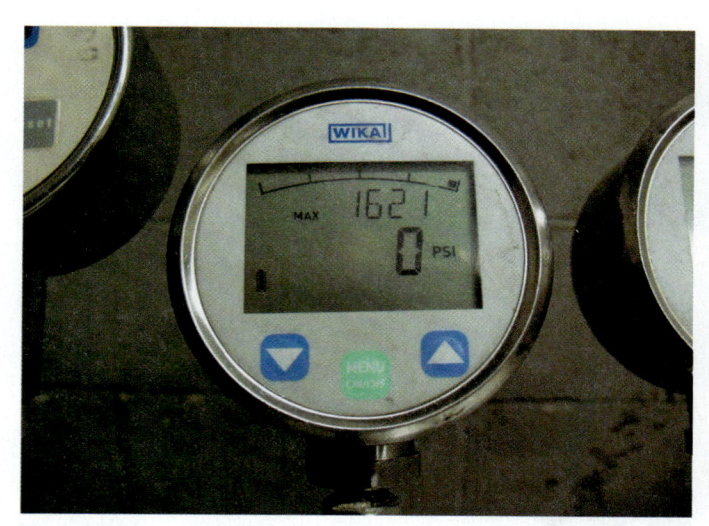

FIGURE 16–49 A digital gauge.

You should be aware of the normal operating pressure and maximum system pressure of the system you are about to test and use the appropriate gauge so you don't damage gauges or connecting hoses/fittings. The maximum working pressure of the system should be 75% of the maximum pressure of the gauge. You will be measuring pressures from a negative pressure to close to 10,000 psi.

Unless a gauge indicates it is showing pressure in PSIA, you should assume the pressure you read is adding atmospheric pressure, and this is called PSIG or psi "gauge." This will matter only for lower pressure readings. In other words, if the gauge reads 75 psi and it's an absolute gauge, the pressure is 75 psi minus the atmospheric pressure at that location and elevation. If there were no markings on it, then the pressure would be considered to be a PSIG gauge and the reading would actually mean 75 psi plus 14.7 psi.

The easiest way to think of this is if you have two gauges, one that reads absolute and the other PSIG, and you were standing on a beach beside the ocean on a summer day, the absolute gauge would read 14.7 psi and the other one would read 0 psi.

Analog gauges will have a curved tube (Bourdon tube) inside that will try to straighten out as the system pressure is applied to it. As the tube moves, it will turn a gear rack that will move a pinion gear that will mechanically move the gauge's needle. The needle will move according to how much pressure is applied to it.

Ideally, an analog gauge should be selected so that the needle will be in the 10 o'clock to 2 o'clock position for the anticipated pressure range being tested.

Digital pressure gauges consist of pressure transducers and a display unit. The transducers will have test pressure applied to them and convert this pressure into a variable electrical signal.

FIGURE 16–50 A digital gauge and transducer.

The transducer sends this signal to the display unit through a wire. The display will then give a readout in psi or kPa on the display. There are three different pressure ranges of transducers that can be used, and they are identified with colors. They are blue—500 psi; red—5,000 psi; and orange—10,000 psi. Always use a transducer that is rated for a pressure that is higher than system pressure. Digital pressure meters can be used to easily give pressure differentials between two pressure sources. This is a common test for measuring margin pressure on a variable displacement pump. ● **SEE FIGURE 16–50** for a digital pressure gauge transducer and display.

Digital pressure meters are safer to use because the display unit has no pressurized oil in it. One downfall of them is it if there is a fluctuating pressure, it may not be able to give an accurate reading. An analog gauge would be better suited for this.

Calibrating pressure gauges on a regular basis will give you confidence that the pressure you are reading is accurate. Most gauges are accurate within 2%.

FLOW METERS **Flow meters** are used to measure the rate of oil movement through a hose or tube. This is usually measured close to the pump outlet but they could be used anywhere in a circuit. Depending on the unit they can measure from 1 to 1000 gpm and can withstand pressures of several thousand psi.

There are a few variations of flow meters. Some are simple mechanical inline meters used to measure flow through the meter. These are relatively inexpensive and robust units that don't require any additional hardware. This type of simple flow meter uses magnetism to show flow by having a disc move according to the amount of flow, and the disc then magnetically moves a follower that has a line on it. Where the line matches with a scale on the outside of the flow meter, it will indicate the amount of oil flowing through the flow meter.

Some are equipped with an adjustable restriction that can be used to put a load on the pump. This will increase system pressure and is a true test of how the pump will operate in a working system. The flow meter reading with this type of unit will be compared to specifications that will give a minimum amount of flow at a certain pressure. As the pressure increases, the flow will decrease. This type of flow meter will have a burst disc to protect the unit and connecting hoses from rupturing if the flow is restricted too much. ● **SEE FIGURE 16–51** for a flow meter with an adjustable restriction.

Other flow meters will measure flow with a turbine wheel that generates an electrical signal that is sent to a digital readout.

TEST FITTINGS AND HOSES You may need a variety of test fittings and hoses to connect gauges and flow meters. You must be sure that the threads match and that any type of seal used is compatible with system fluid type and pressure levels. Always check for leaks on any test fittings before pressure, flow, and temperature levels rise.

STOPWATCH A typical test for a hydraulic function is to measure how fast a cylinder rod cycles. A digital stopwatch is necessary to measure these speeds, and this is called measuring **cycle times**. You can also test the speed of a rotary actuator with a stopwatch.

FIGURE 16–51 A flow meter with an adjustable restriction.

TACHOMETER Hydraulic motor speed may need to be measured and a digital tachometer can be used to do this. A good digital tachometer can also measure length and velocity, which may be useful for checking cylinder rod speed.

PORTABLE HYDRAULIC HAND PUMP Some problems can be diagnosed with a "porta-power" by applying pressure to a component and seeing if it will hold pressure. This is called leakage testing. This type of pump can also be used to release parking brakes occasionally.

TEMPERATURE PROBE AND HEAT GUN A critical condition of hydraulic fluid is its temperature. A thermo couple adapter that is connected to a multimeter is a valuable tool to measure this. The probe needs to be placed in the hydraulic fluid so caution needs to be taken if the tank is pressurized and the oil can get hot enough to cause severe burns.

Most hydraulic systems will have oil coolers that are needed to regulate oil temperature. To check temperature drop across the oil cooler, an infrared heat gun can be used for this.

Some problems such as failed cylinder seals can be detected with a heat gun by checking along a cylinder barrel and seeing where a temperature change is. A failed seal is similar to an orifice, and if oil flows through an orifice, it will generate heat.

TESTING AND ADJUSTING

There are some common testing and adjusting procedures that may need to be performed occasionally to keep hydraulic systems running at their most efficient level, producing the necessary flow and pressure and being responsive to the operator's input. These procedures are sometimes part of a hydraulic tune-up. With the increasing popularity of electronic controls, hydraulic tune-ups may be required less frequently because the control system can compensate for wear and flow losses to a certain extent. Hydraulic tune-ups will require the technician to measure flows, pressures, temperatures, and cycle times while comparing them to specifications and then making adjustments if needed.

One test is to set the minimum flow for a variable displacement pump. This requires a flow meter to be hooked up to the pump outlet and a way to measure engine speed. There will be a minimum displacement screw that can be adjusted to match the pump flow with the specification.

Other tests are for pressure relief valve settings that control pressures such as main system, pilot system, or line relief settings.

HYDRAULIC SYSTEM PROBLEMS

There are five common types of problems that are possible when dealing with hydraulic systems. The root cause of many of these problems is often system contamination.

PRESSURE
If an operator complaint is related to the lack of force a hydraulic circuit has, this directly relates to the pressure or lack thereof in the circuit. An example of a typical operator complaint would be the machine's bucket won't fill as easy as it used to.

FLOW
If an operator complaint is related to the speed of an actuator, this is a flow problem. A simple saying that summarizes this is "Flow makes it go."

HEAT
Hydraulic system transfers energy to either heat or motion. If too much heat is being created, it is likely a result of a loss of motion, and if the hydraulic oil cooling system can't manage this excessive heat, the system will overheat. Problems with the cooling system will also cause an overheating condition.

LEAKAGE
External leaks are usually obvious but what isn't obvious is if fluid can leak out, then contamination can enter the system. Internal leaks aren't so obvious but keep in mind that fluid takes the path of least resistance. Many components will seal oil internally, and if this seal is compromised, then an internal leak will occur. This could show up as a cylinder drift, excessive pump or motor case drain flow, or an overheated system.

NOISE AND VIBRATION
Hydraulic systems will create a certain amount of noise and vibration because of oil flow and pressure pulses. Some hydraulic systems will use an attenuator on the pump outlet to reduce pump pulsations. As components wear, there may be an increase in noise and vibrations.

COMPONENT REPLACEMENT PROCEDURES (SAFE START-UP)

Once a problem has been found with one or more hydraulic system components that can't be corrected with an adjustment, you may need to remove the components for repair or replacement.

This procedure will start with depressurizing the system and may include draining or blocking off oil flow to the component. Try to minimize any oil spills during system repairs. Always refer to the machine's service information for proper removal and repair procedures.

Once the new or reconditioned component is installed, proper start-up procedures *must* be followed or a repeat failure is likely to occur. For example, if a pump or motor is installed, it will likely need to be filled with oil first to ensure that it doesn't fail at start-up. There may need to be an air-bleeding procedure take place to eliminate air from the system.

Once the system is started, you should slowly work all functions to see that they all operate normally and to work any air out of the system before putting a load on the new components.

It is always a good idea to change the hydraulic filter soon after changing any component. You can then cut it open and inspect it for any unusual wear particles.

If there was a major component failure that contaminated the system, you will need to perform a cleanout of the hydraulic tank at minimum, and this may include removing and dismantling the entire hydraulic system. A good piece of insurance after this is to perform a kidney circuit cleaning on the system with a filter cart that has a particle counter built in.
● **SEE FIGURE 16–52** for a filter cart.

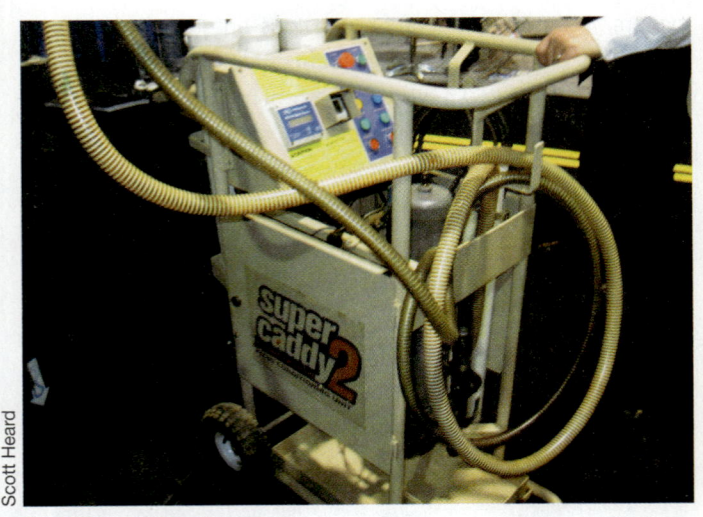

FIGURE 16–52 Filter cart.

CAUSES OF COMPONENT FAILURES

CONTAMINATION Contamination is the leading cause of component failures. There are many sources of contamination, and this was discussed in Chapter 4.

AERATION Aeration occurs when air enters the inlet of the pump along with hydraulic fluid. The air enters either from a leak in the inlet lines or because of low fluid levels in the reservoir. Air can also be mixed with the fluid returning from the hydraulic system or from turbulence in the reservoir. The low pressure at the inlet causes the air bubbles to expand. When the fluid is pressurized, the bubbles collapse. This causes an implosion and will release large amounts of energy as each bubble forms a micro jet. This will cause rapid erosion of the internal pump components.

CAVITATION The difference between cavitation and aeration is that the vapor bubbles form in the fluid at the inlet, rather than the fluid already having air in it as it approaches the inlet. The bubbles form at the inlet because of excessively low inlet pressure for the fluid being used. Once in the pump, the bubbles cause the same damage as aeration. A common cause for cavitation is a restriction in the flow of the hydraulic fluid to the inlet of the pump.

OVERHEATING Hydraulic systems are usually not meant to operate over 150°F. The most important part of a system is the oil circulating through it, and if it gets overheated, it will start to degrade quickly. When this happens, it loses its ability to protect the system components that it comes into contact with.

OVERLOADING Most hydraulic system components are fairly forgiving when it comes to overloading them. However, it is possible to abuse a system and push it past its design limitation. This is how pump shafts get twisted, cylinder rods get bent, and valve bodies crack.

> **TECH TIP**
>
> Don't rely on a machine's electronic system to troubleshoot problems for you. I was called to a wheel loader with an electrohydraulic system that had an intermittent hydraulic problem. When trying to verify the complaint, I ran this large wheel loader, and in every fifth or sixth bucket load, the boom would reset to an unusual level. It normally would lower to a preset level when the boom lever was set to float after dumping. There were no active or logged fault codes related to this problem displayed.
>
> Relying on previous experience, I checked the boom location sensor and found the return spring on its lever broken. There was nothing wrong electrically, but as the sensor's arm couldn't follow the ramp properly because of the broken spring, it was sending a false signal to the ECM.
>
> In other words, the ECM didn't know where the boom was in its travel. The only way to know if there was a problem with this was to visually inspect the sensor.

> **CASE STUDY**
>
> I once had a machine (100 ton rock truck) that had a hydraulic pump starting to fail and should have been taken out of service. However, it was put back into service and had a major failure that sent metal contamination through the entire hydraulic system. This was bad enough, but the hydraulic oil was also used to cool the brakes and was used for the torque converter oil. This meant the truck was out of service for over six weeks while all these systems were cleaned out. After the truck was repaired and before it was put back into service, the hydraulic oil was filtered with a mobile filtration system (kidney circuit) to ensure that the oil was clean.

Progress Check

11. Cartridge valves differ from conventional cast valve assemblies mainly because they:
    a. can withstand higher pressures
    b. are easier to troubleshoot and cheaper to repair
    c. can accept only low-flow amounts
    d. increase the amount of hoses, tubes, and fittings on a machine

12. An example of an input to an ECM that is part of an electrohydraulic system is:
    a. proportional solenoid
    b. pump control solenoid
    c. variable relief valve sensor
    d. pump speed sensor

13. To prevent stiction from occurring to electro-hydraulic valve, this is done:
    a. pulse an electrical signal to the valve
    b. use a lower viscosity hydraulic fluid
    c. use a higher voltage signal to the valve
    d. reduce the main relief pressure setting

14. One advantage of using a digital pressure gauge to test hydraulic systems is:
    a. they can withstand higher pressures
    b. there is no high-pressure oil in the handheld display
    c. they can read quickly fluctuating pressures
    d. they are cheaper than analog gauges

15. If an operator complains that the hydraulic system is slow, you would be looking for this type of problem:
    a. overheating
    b. excessive pressure
    c. reduction of flow
    d. vibration and noise

SHOP ACTIVITY

- Go to a machine with a variable displacement type of pump and determine how the pump's displacement is controlled.
- Go to a machine with electric joysticks and try to find the ECM that receives the signal from them. Now find the main control valve and see how the spools are moved.
- Go to a machine with an operator display and try to see how much hydraulic system data you can view while the machine is running.

SUMMARY

1. Some safety concerns related to advanced hydraulic systems include: all safety concerns that were outlined in the Basic Hydraulics chapter (Chapter 4). System pressure must be released before service or repairs are made to the system. Hydraulic oil and system components can get hot and cause burns

2. Advanced hydraulic systems incorporate components and design features that make the system more efficient and easier to operate

3. Hydraulic schematics provide the technician with a way to understand how oil flows through the system and are a valuable tool that can be used to assist with troubleshooting system problems. They are similar to electrical schematics and use symbols to represent different components. They can come in paper and electronic formats

4. Open circuit systems are most common and have oil reservoirs that supply oil to the pump/s and the pumps will send oil to directional control valves. The valves can then direct the oil to actuators, return oil from the actuators goes through the directional control valves and returns to the tank. Closed circuit systems have their bi-directional pumps send oil directly to motors, the oil then returns to the pump inlet. They are mostly used for hydrostatic systems

5. Open center and closed center type hydraulic systems refer to the type of directional control valves and whether oil flows freely through them when the spools are in neutral

6. Most systems with fixed displacement pumps use open center control valves

7. Monoblock directional control valves (DCVs) have bodies that are one piece cast iron while multisection DCVs have individual sections for each spool and are stacked together

8. Closed center systems need to be used with variable displacement control valves since the pumps can reduce their flow to almost zero when the valve is in neutral

9. Variable displacement pumps will be able to vary their flow output according to systems needs and operator commands.

10. Axial piston and bent axis piston type pumps are the most common types of variable displacement pumps. These pumps rely on reciprocating piston action to move oil from the pump inlet to the outlet and through the system

11. A pressure differential is created between the pumps inlet and the hydraulic reservoir to get oil into the pump

12. Bent axis pumps vary the angle of their cylinder block to the pumps shaft centerline while axial piston pumps have a movable swashplate that can change angles. These actions change the pistons effective stroke which changes the pumps displacement
13. Variable displacement pumps need to have control mechanisms that change the displacement of the pump. The control is usually a small spool that has system oil pressure working on one end against an adjustable spring on the other. As the piston moves it will direct oil to a pump control piston that mechanically change the pumps displacement. These controls can also be electronically controlled and monitored
14. Load sensing systems will monitor system pressure and adjust pump output flow to compensate for pressure changes. Pump control mechanisms can be used for this
15. Margin pressure is the pressure set above system pressure that controls the pump control and is usually 300–400 psi higher than system pressure
16. When a pump control increases pump displacement its called upstroking and downstroking is the opposite action
17. Pilot control systems use low pressure oil to move directional control valve spools. They reduce operator effort. Typical pilot system pressure is 300-400 psi and operator controlled joysticks meter oil to either end of the DCV spools to move them against spring pressure
18. Pilot operated main relief valves are needed in all systems to limit system pressure to safe levels. They can be adjusted with springs or shims
19. Line or circuit relief valves limit pressures in one section of a circuit to safe levels when the spool valves is in neutral
20. Counter balance valves will stop actuators from moving in an uncontrolled manner if the load tries to overcome them
21. Load check valves will keep a load in place until the pump flow builds enough pressure to overcome the circuit pressure
22. Anti-cavitation valves will allow tank oil to flow into a circuit if the circuit pressure drops below tank pressure to prevent cavitation (can be called make up valves)
23. Flow compensators can be part of DCVs and for each circuit will be able to vary pump flow to an actuator and maintain a fixed pressure drop across the spool
24. Pressure reducing valves can be used to supply pilot oil systems from main system oil
25. Sequence valves can be used to only send oil to a circuit after complete movement of a first actuator
26. Flow dividers are sometimes used to split oil flow evenly between two or more actuators
27. Accumulators have an inert gas acting on one side of a bladder or piston while oil is sent to the other side. They can be used to store oil pressure or to reduce pressure spikes. They are precharged with a specific amount of gas pressure
28. Excavator hydraulic systems can be complicated and are usually arranged to have two main pumps supply two main DCVs. The DCVs will then direct oil to the machines actuators
29. A regeneration circuit will redirect return oil from an actuator to the opposite side of it to prevent cavitation
30. Excavators with electronically controlled pumps can have the capability to change modes for different working situations such as heavy lifting or truck loading
31. Cartridge type valves are made with machine steel or aluminum bodies and have cartridge type valves threaded in or held in place with fasteners. The cartridges can be DCVs, pressure relief valves or other types of valves
32. Electro hydraulic systems incorporate electronic and electrical components to allow the system to be precisely controlled and be more efficient. This also allows the hydraulic system to be part of the machines multi plexed network and communicate with other machine systems
33. Inputs for electro hydraulic systems include: pressure, temperature and position sensors
34. Outputs for electro hydraulic systems include: pump control valves, DCVs, relief valves
35. Some electronic components will need to be calibrated upon installation or as part of a repair
36. Contamination of the environment should be avoided when servicing or repairing hydraulic systems
37. Hydraulic system testing involves measuring pressures and flows with gauges and flowmeters. Reading are compared to specifications and adjustments can be made if needed to bring the system back within designed parameters
38. Gauges can be analog or digital
39. Other tools needed can include: heat gun, stopwatch, tachometer, laptop with software
40. When hydraulic components are replaced because of a failure it is good practice to: change all filters, fill component with clean oil, bleed air out before starting, work the system slowly with light loads initially, ensure proper oil level, take an oil sample before sending machine to work
41. Hydraulic systems problems can be related to: pressure, flow, heat, leakage, noise/vibration
42. Hydraulic system components can fail as a result of: contamination, aeration, cavitation. overheating, overloading,

chapter 17
HYDROSTATIC DRIVE SYSTEMS

LEARNING OBJECTIVES

After reading this chapter, the student should be able to:

1. Describe how a simple hydrostatic system works.
2. Describe the different control systems used to control hydrostatic systems.
3. Explain how to test a hydrostatic drive system for several common problems.
4. Explain the purpose of the charge pump circuit in a hydrostatic system.
5. Describe how to make basic adjustments to a typical hydrostatic system.
6. Describe the operation of the flushing valve.
7. Identify the main parts of simple hydrostatic systems.
8. Describe the operation of the different types of pumps used in hydrostatic systems.

KEY TERMS

Case drain filter 532
Charge pump 508
Closed loop 508
Combination valve 512
Dual path system 510
Flushing valve 512
Joystick 511
Position sensor 530
Spot turn 512

SAFETY FIRST

Hydrostatic drive systems use hydraulic fluid flow and pressure to transfer energy from the flywheel of a diesel engine to drive a machine's tracks or tires. When working on hydrostatic systems, you need to be aware of high oil pressures and flows as well as hot oil. The hazards related to working on hydrostatic systems are much the same as the ones on hydraulic systems such as oil injection and burns. Refer to Chapter 6 for more details.

As the hydrostatic system is part of the drive system for a machine, you must be sure that the machine is secured from rolling in an uncontrolled manner if you are working on it.

Some machines will have hydrostatic components underneath the machine's cab. To work on these components, the cab will likely need to be raised. Ensure that the cab is secured in the raised position before getting underneath it.

INTRODUCTION

Hydrostatic drive systems have been used to propel heavy equipment machines since the 1950s.

Small machines such as skid steers first used hydrostatic drive systems after the original friction clutch drive system proved to be inadequate. The first hydrostatic drive had simple controls and its smooth operation provided seamless tight turning with plenty of power, which worked perfectly for this type of machine. Also because of the machine's small size, the compact components of a hydrostatic system were a good match for this drivetrain application.

In simple terms, a hydrostatic system is one that uses a hydraulic pump driven by a prime mover (usually a diesel engine) to convert mechanical energy into fluid energy. The fluid flow from the pump is sent directly to a hydraulic motor. The motor's internal components receive the oil and the interaction between the oil flow and the motor's internal elements makes the motor's output shaft rotate. The motor changes the fluid energy back to mechanical energy. Once the oil has done its job in the motor, it leaves the motor and is then sent back to the pump inlet. This oil flow continues in a loop as long as the pump keeps moving oil. ● **SEE FIGURE 17–1** for how oil flows between a variable displacement pump and fixed displacement motor.

The direction and amount of flow output of the pump determines the direction and speed output of the motor's shaft. The flow amount and direction of output of the pump is controlled by the machine's operator. The pump or pumps used are described as bidirectional variable displacement. This means the pump can vary its output flow volume and change the direction that its output flow is independent of its input shaft speed and direction of rotation.

The amount of torque output created in a hydrostatic drive system is dependent upon the load that the drive system is trying to overcome. As the load increases, then the system fluid pressure must increase to overcome the load. For example, a track-type dozer with a hydrostatic drive that is travelling with its blade in the air will not need a high amount of torque to drive its tracks. However, when the blade is dropped and the operator wants to cut through some material, there needs to be a high amount of torque produced to keep the machine moving. The hydrostatic system pressure increases to produce more driving torque to keep the machine moving. If a gauge

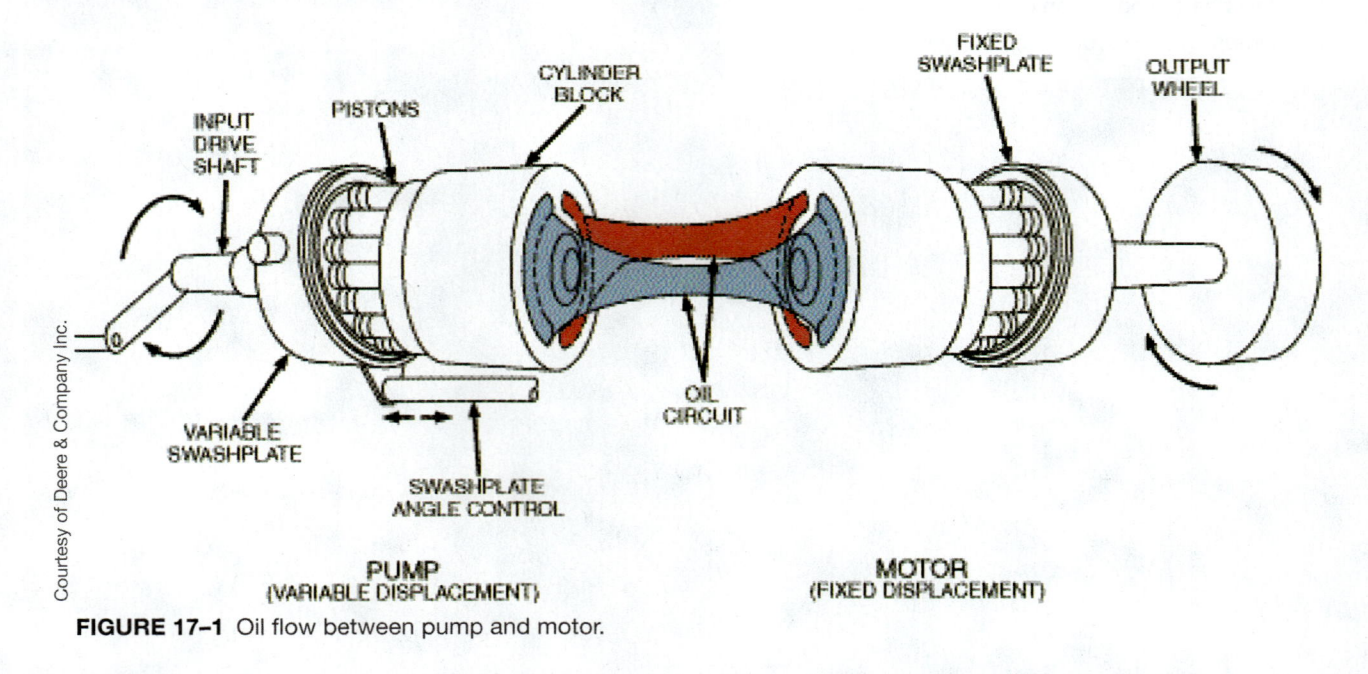

FIGURE 17–1 Oil flow between pump and motor.

was reading system pressure, it may change from 500 to 5000 psi between the two different conditions.

If the same machine's operator wants to change travel speed from slow to fast, then the hydrostatic pumps will be controlled to increase the flow rate. If a flow meter was measuring the output of the pumps, it may read a change from 5 to 40 gpm under these two different conditions (slow to fast). If the pump flow is reduced to 0 gpm, then the motor shaft will stop turning. This is considered to be neutral.

HYDROSTATIC DRIVE ARRANGEMENTS

There are many different configurations of hydrostatic drive systems and a few examples are discussed here. These drive systems could be used to propel machines to well over 20 mph as well as generate high torque at low speed. Some hydrostatic systems can be used to steer a machine and some motors are also variable displacement to give an extra high speed/low torque or low speed/high torque feature. ● SEE FIGURE 17–2 for a basic layout of a hydrostatic drive for a track machine.

Small machines (fewer than 50 hp) that use a hydrostatic system may use an arrangement that combines the pump and motor into one housing. This could even be housed together with a differential. One example of this type of machine is the agricultural-based utility tractor. This is commonly called a hydrostatic transmission although some multi-pump assemblies are also called hydrostatic transmissions. This compact design works well for small equipment, and its seamless operation makes it very user friendly. ● SEE FIGURE 17–3 for a hydrostatic drive utility tractor.

This type of unit will use passageways in the transmission housing to transfer oil flow. All other hydrostatic systems use pumps and motors that are separate and the oil flows between them through hoses, tubes, and fittings. ● SEE FIGURE 17–4 for a hydrostatic pump and motor with connecting hoses.

Although the compact design of hydrostatic systems are ideal for smaller machines where space can be limited, many medium and large machines will also use hydrostatic systems to drive their tracks or wheels. Because of the excellent torque multiplication capabilities of hydrostatic drives, some track-type tractors with over 400 hp engines have their tracks driven with hydrostatic drive systems. One manufacturer makes a wheel loader that has a 340 hp engine driving two pumps that drive two motors. These motors then drive a transfer case that sends torque to the front and rear axles via drive shafts. ● SEE FIGURE 17–5 for hydrostatic drive arrangement for a wheel loader.

Hydrostatic systems can be used to drive a variety of wheel-type machines such as skid steers, forestry machines, wheel loaders, the front wheels of a grader, telehandlers, and

FIGURE 17–3 Utility tractor.

FIGURE 17–2 Basic hydrostatic drive system for a track machine.

FIGURE 17–4 Hydrostatic pump and motor.

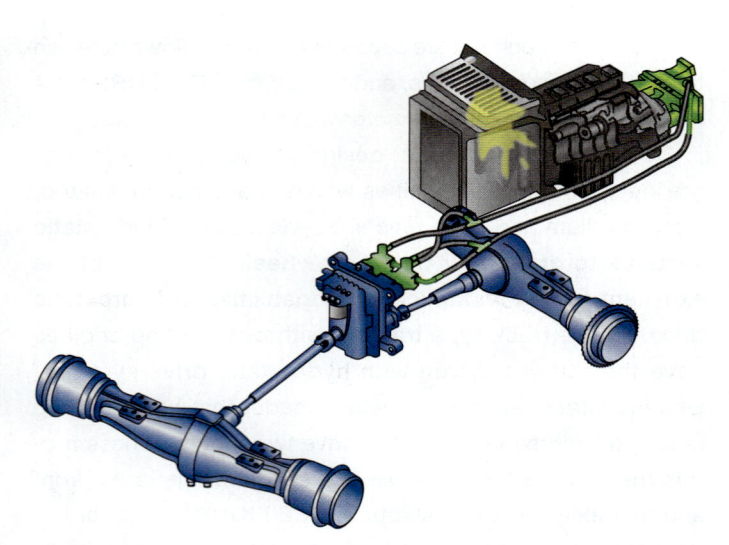

FIGURE 17-5 Wheel loader with hydrostatic drive system.

pavement grinders. Machines with drums (compactors) will also use hydrostatic drive systems to rotate the drum.

Hydrostatic drives can also be used to drive track-type machines such as bulldozers, track loaders, paving machines, drilling machines, and track-type skid steer machines. Almost all small- and medium-size track machines have had their drivetrains evolve from mechanical to hydrostatic over the last 20–30 years.

While the majority of hydrostatic systems are used for machine travel, they can be used for other functions that need rotation. An example of this is the differential steering system used on some larger track-type tractors. They use the motor of a hydrostatic system to give the needed rotational torque to steer the machine by rotating a set of gears (see Chapter 20 for more information).

For machines that need to drive the left- and right-side wheels or tracks independently of each other (skid steer machines, track type tractors, and track loaders), they will use an arrangement that features two pumps and two motors. One pump will send oil to the left-side drive motor and the other pump supplies the right-side motor. The motors will then typically drive an input to a gear reduction whose output then drives a sprocket to turn a track.

These machines will use a speed differential between the left- and right-side track drives to steer the machine. The tracks or tires can even be driven in opposite directions to provide spot turning or counter-rotation.

The motors for small skid steer machines drive a sprocket that in turn drives a chain that transfers drive to the wheels.

Dual pump and dual motor arrangements are called dual path hydrostatic systems. Some large track machines will have each pump split its flow and send oil to two motors for each track. The motors will drive a gear reduction to further slow down rotation and increase torque.

Hydrostatic machines that feature a driven steering axle, have a driven solid (non-steering) axle, or use an articulated steering can have a variety of pump and motor arrangements. For example, a small articulating wheel loader may have one motor directly drive the front axle and one motor directly drive the rear axle. These motors will be driven by the same pump and rotate at the same speed.

Another articulated machine that is driven hydrostatically is a drum-type compactor. This machine will use one pump to drive one motor that drives the drum and another pump to drive one motor to drive the rear axle or two motors to drive the rear wheels.

A telehandler that is driven with a hydrostatic system will have steering axles front and rear and use one pump to drive one motor. The motor will drive one axle through a set of gears, and there will be a drive shaft that is driven by the first axle that will send torque to the other drive axle.

As you can tell, there may be a wide variety of hydrostatic arrangements that all use the **closed loop** pump/motor system (see the following Tech Tip) for propelling machines. However, the same basic components are used for the hydrostatic portion of all hydrostatic drive arrangements: One or more bidirectional variable displacement pumps sending oil to one or more fixed or variable displacement motors.

If a hydrostatic machine has more than one pump, it may have a pump drive. This is a set of gears that divides the engine input to it into two splined outputs. Pump drives will have their own oil to lubricate the gears and bearings. ● **SEE FIGURE 17-6** for a pump drive.

Larger pump drives will have a dampening device between the flywheel and the pump drive input gear.

Some smaller hydrostatic pump assemblies can be belt driven. The pumps could also be piggybacked (one pump drives the other) on each other and driven from the flywheel. The pumps' shaft could also provide drive to other machine pumps or to the system **charge pump**. ● **SEE FIGURE 17-7** for two hydrostatic pumps that are piggybacked.

FIGURE 17-6 Pump drive.

When compared to mechanical-style drivetrains, hydrostatic drives offer many advantages. One that is similar to why electric drives are gaining popularity is that the prime mover (diesel engine) can be operated in a narrow rpm range and, therefore, designed for optimum efficiency with minimum emissions.

Here is a list of advantages and disadvantages of hydrostatic drivetrains:

HYDROSTATIC ADVANTAGES	HYDROSTATIC DISADVANTAGES
Seamless transition of speeds and directions	Components are sensitive to contamination
Allows engine to run at a steady rpm for improved efficiency and lower emissions	Components are expensive and require special tooling to repair and adjust
Track machines can power both tracks on turns	Can require specialized diagnostic tooling to troubleshoot problems
Excellent power to weight ratio	susceptible to overheating if overloaded
Locations of main components are very flexible	
Provides dynamic braking	
Can be stalled without damaging components	
Allows for counter-rotation of tires or tracks	

FIGURE 17–7 Hydrostatic dual pump arrangement.

TECH TIP

This chapter will focus on closed loop hydrostatic systems that propel machines. Closed loop systems feature a pump that sends oil directly to a motor and the return oil from the motor goes directly back to the pump. Some oil will leak out of the loop (internally in pumps, motors, and valves) along the way and is replaced by a charge pump. Some oil is intentionally bled out of the loop as well for cleaning and cooling purposes.

Some machines such as wheel-type or track-type excavators will use open loop hydrostatic systems to propel them. An open loop hydrostatic system is one that has a pump, gets its oil from the tank, and sends it to a directional control valve that is one section of a main control valve assembly. The directional control valve is actuated by the operator to send the pump flow to either of two ports of a travel motor. The travel motor will then rotate and its shaft will drive a gear reduction (final drive) that will drive the machine's track or axle. This is similar to any other hydraulic circuit except that it is moving a rotary actuator (motor) and not a linear actuator (cylinder). ● **SEE FIGURE 17–8** for an open loop travel circuit for an excavator.

FIGURE 17–8 Open loop travel circuit for an excavator.

HYDROSTATIC DRIVE SYSTEMS 509

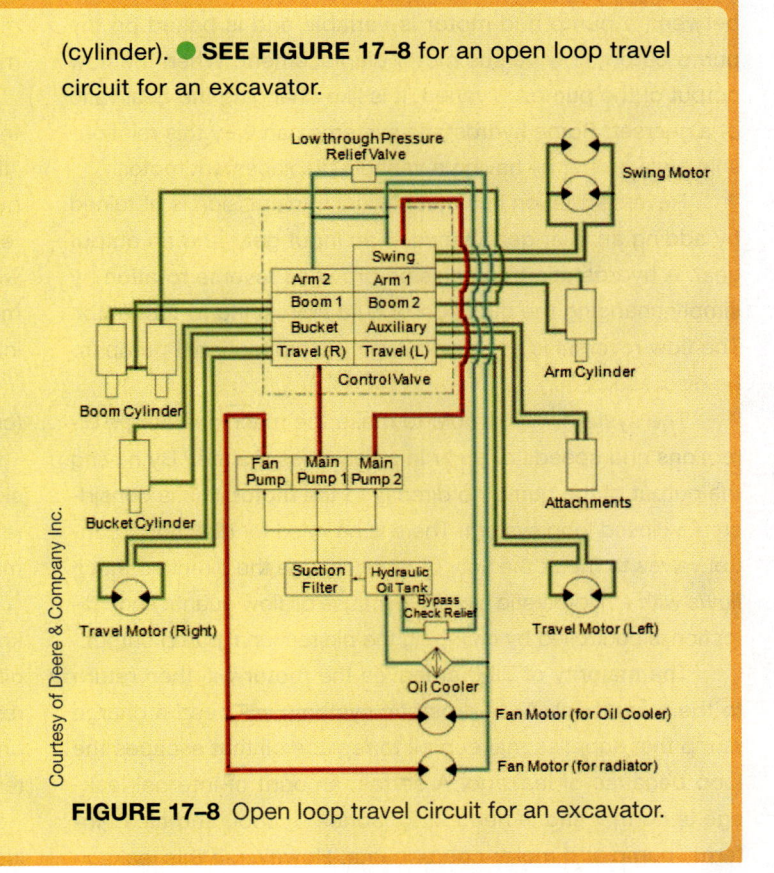

This chapter will discuss the most common arrangements of hydrostatic systems and look at the major components of these systems. Mechanical, hydraulic, and electronic control operation will be covered as well as common servicing, diagnostic, and repair procedures.

HYDROSTATIC SYSTEM OVERVIEW

Hydrostatic systems are sometimes also called hydrostatic transmissions because they transmit torque from the machine's prime mover (diesel engine) to tracks or wheels. A mechanical drivetrain is able to change the speed and torque of the machine's diesel engine output to increase torque and reduce speed of the drivetrain's output. It does this with combinations of gear ratios in different components that work to slow down rotational speed and increase rotational torque.

The hydrostatic transmission does this with one or more pumps, one or more motors, and hydraulic fluid. If you were to compare the interaction of a pump and motor to a mechanical gearset, the pump would be the drive gear and the motor the driven gear. The ratio of a gearset is fixed and is based on the number of teeth on the drive and driven gears. The relationship between a pump and motor is variable and is based on the pump output flow versus motor displacement. When the flow output of the pumps is varied, it is like changing the gear ratio of a gearset. Some hydrostatic systems can vary this relationship even farther by having a variable displacement motor.

Reverse rotation in a mechanical transmission is obtained by adding an idler gear between an input gear and an output gear. A hydrostatic transmission provides reverse rotation by simply changing the direction of fluid flow going to the motor. The flow reversal is enabled by using a bidirectional pump (to be discussed later).

The system will be able to make the machine change directions and speeds or steer in a seamless manner. By having the output of the pump go directly to the motor, this is considered a closed loop system. There is no need for directional control valves to meter the flow of oil or change the direction that it flows with a hydrostatic system because oil flow quantity and direction is controlled by changing the bidirectional pump output.

The majority of oil that leaves the motor will then return to the pump inlet. All hydrostatic systems will need a charge pump that supplies makeup oil to replace oil that escapes the loop because of leakage. A certain amount of internal leakage is normal and is needed to lubricate, cool, and clean internal pump and motor components. However, if this leakage becomes excessive, then there will be a loss of efficiency, which equates to lower torque and speed output. The charge oil circuit will have an oil cooler and filter as part of it to clean and cool system oil.

DUAL PATH HYDROSTATIC SYSTEM

Many hydrostatic machines will need two motors to drive them and in turn two pumps to supply oil to the two motors. This can mean the two motors drive separate axles, final drives, or chain drives. This is considered to be a dual path hydrostatic system.

The main components of a typical **dual path system** are two pumps and two motors. The output of the motors will transfer torque into a gear reduction to further increase torque and decrease speed. The output of the gear reduction will then drive the machine's tracks or wheels. The pumps will be able to discharge oil in either of two directions, which will make the motors turn their output shaft in either direction and will then change the direction of the machine's travel. A pump used for a hydrostatic system is considered to be bidirectional. Its shaft will be turned in only one direction by the machine's engine, but because of its design, it will be able to send oil out either of two ports. This means the pump's inlet and outlet can trade places. Likewise, the motors will have two ports that can be either an inlet or an outlet.

The speed of the machine's travel is changed by changing the flow volume output of the pumps (remember the phrase—"flow makes it go"). By varying the flow output volume of the pump, the speed output of the motors is changed. Simple systems will use fixed displacement motors, but many other systems will use dual displacement or variable displacement motors. The motor speed can be calculated by knowing the flow volume going into the motor and the motor displacement and using the formula:

Motor rpm = Flow (gpm) × 231/Motor displacement (cubic inches per revolution)

An example for this is a motor that drives the wheels of a skid steer machine has a displacement of 25 cubic inches per revolution and has a flow of 30 gpm going to it. Using this formula, you would calculate that its shaft would turn at 277 rpm.

The torque of the motor's output shaft is determined by knowing its displacement per revolution and the pressure of the oil going into the motor. The oil pressure will be created by the resistance on the shaft and the resistance is created by the load on the wheels or tracks. Remember, pressure is created by the resistance to flow. The formula for motor torque output is:

Torque (inch pounds) = PSI × Displacement (cubic inches per revolution)/2π

> **TECH TIP**
>
> An important rating for pumps and motors that are used in hydraulic and hydrostatic systems is displacement. This can be measured in cubic inches per revolution (CIR) or cubic centimeters per revolution (CCR). For a pump, its displacement rating is the volume of fluid it can move during one revolution of its shaft. For a variable displacement pump, this will range from zero and increase to its maximum rating. All pumps for hydrostatic systems are axial piston style, and their displacement is based on swash plate angle.
>
> Motor displacement refers to the amount of oil required to rotate its output shaft one revolution. Motors can be fixed or variable displacement.

FIGURE 17–9 Motor and brake cutaway.

> **TECH TIP**
>
> One manufacturer's crawler dozer machine will also use one of its implement hydraulic pumps to provide an extra load on the engine when extra braking effort is needed. This system will run the pump's output through a restriction, thereby putting a load on the engine and, therefore, making the engine resist being driven by the hydrostatic system even more.

An example for this would be a motor that is used to drive one set of wheels for a skid steer. If it has a displacement of 25 CIR and the oil pressure is 5000 psi, then using this formula to calculate the torque output of this motor you would find it produces just fewer than 20,000 in.-lb or approximately 1,600 ft-lb of torque.

The maximum pressure will be limited by pressure relief valves in each half of the closed loop.

When calculating hydrostatic system values, there needs to be a consideration of the amount of internal leakage within the pumps and motors of the system to get an actual value that can be expected. Like hydraulic systems, this leakage is the inefficiency of the system and is usually 15–20%. This inefficiency will decrease the speed of the motors and when factored into calculations will give a more realistic value of the actual motor speed.

This leakage from the pumps and motors needs to be sent to the tank, and this is done through ports in the housing. This oil flow is commonly called case drain oil and can sometimes be measured during a troubleshooting procedure that checks for pump and motor wear. As pumps and motors wear, their case drain flow and pressure increase.

Hydrostatic systems can also naturally provide dynamic machine braking (driveline braking), which means the need for a service brake system is eliminated on most hydrostatic machines. Because of the closed loop arrangement, there is nowhere for the oil on the high-pressure side of the loop to go as long as the relief valve doesn't open. If the weight of the machine starts to try to push the machine, then the motor becomes a pump. The pump will provide a braking action because its shaft will resist turning faster because of the resistance of the engine.

There will still be a need for a spring-applied parking brake system because hydraulic pressure will leak off over time. Hydrostatic systems will usually have the parking brake on the motor's output shaft to give a braking effect on the final drive input. The brake release oil circuit is usually part of the hydrostatic system's charge oil system.
● **SEE FIGURE 17–9** for a cutaway view of a hydrostatic motor and brake assembly.

Hydrostatic systems can be controlled by the operator by controlling the pump, and this can be done in many different ways including mechanical linkages or cables that are hand or foot controlled (used for lower flow and pressure systems), pilot oil-actuated pumps that are **joystick** controlled, and electrical/electronic controls that are joystick controlled. These different operator controls will make the machine move from neutral towards either direction at infinitely variable speeds (to a maximum limited speed) and could also steer the machine depending on the type of steering system used for the machine.

HYDROSTATIC DRIVE SYSTEMS 511

HYDROSTATIC CIRCUIT OPERATION

To understand how fluid flows in a simple manually controlled hydrostatic system, we will look at a skid steer loader's system. The operation of this machine's hydrostatic system is typical of how a simple hydrostatic system works. It is considered to be a dual path system because there are two pumps and two motors. It is really two separate hydrostatic systems that use a common fluid, charge oil system, hydraulic tank, filtration, and oil cooler. ● SEE FIGURE 17–10 for the dual path hydrostatic arrangement of a typical skid steer.

This machine has two flywheel-mounted bidirectional variable displacement axial piston pumps that supply oil to two geroller-type motors. Pump and motor types will be discussed later in the chapter. The motors then send torque to sprockets and chains to drive the machine's wheels. The speed and direction of each drive sprocket is changed by the operator moving joysticks back and forth that are connected to the pump swash plates. As the operator adjusts the swash plate angles, the pump flow output to the motors change. This will change the speed and direction of the left and right wheels independently to speed up, slow down, steer the machine, or perform a **spot turn**.

There is another fixed displacement pump on this machine that is used for the machine hydraulic functions (boom and bucket). The pump is also used to supply charge oil for the hydrostatic system. This machine uses a common reservoir, cooler, and filter for both the hydraulic and hydrostatic systems. This is a common arrangement for many small skid steer machines. ● SEE FIGURE 17–11 for the schematic of the Deere CT 315 skid steer hydrostatic system.

Refer to the schematic and follow the path of oil flow for the high-pressure closed loop circuit.

If you look at the right pump (4), imagine it is sending oil out its top line to the right motor (7). Let's say the machine direction that this oil flow creates is forward. The oil would enter the motor and cause its shaft to rotate, leave the motor, and then head back to the pump. If there was a high enough resistance to stop the shaft, then pump outlet pressure would rise quickly and the relief valve (57) near the pump would open. This would route the pump output oil back to the low-pressure side of the pump through the inlet side makeup valve (57). These valves (57) are actually **combination valves** that act as both a relief valve and a check valve depending on whether they are exposed to oil pressure on the high-pressure side or the low pressure-side of the loop.

When pressure builds in the high-pressure side of the loop, it is sensed at the end of the shuttle valve (61), which is part of each motor. The shuttle valve then shifts to allow some of the return oil to pass by the shuttle valve where it is sensed at the flushing relief valve (79). This is a low-pressure relief valve that will open to allow some oil to flow through an orifice and into the motor housing. The oil flow will then go through the pump housing to help cool it and then go to the tank. This valve is sometimes called a **flushing valve** and provides a constant flow of oil going to the tank to cool and lubricate the motor and pump when the pump is moved from neutral. This valve will also keep a minimum amount of oil pressure in the low-pressure side of the closed loop. The flushing relief valve is usually set 20–50 psi lower than charge pump relief, which is typically in the 200–350 psi range.

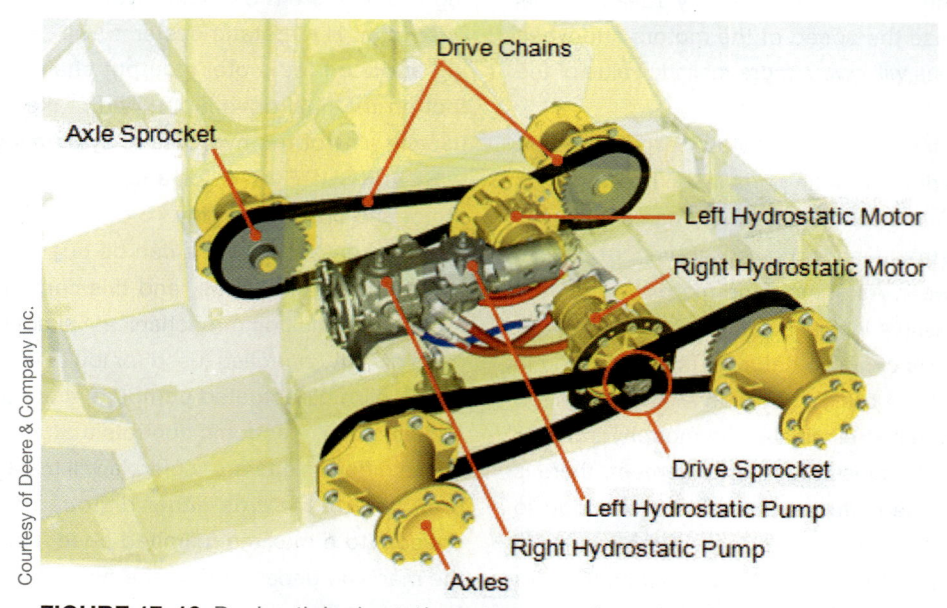

FIGURE 17–10 Dual path hydrostatic arrangement for a skid steer.

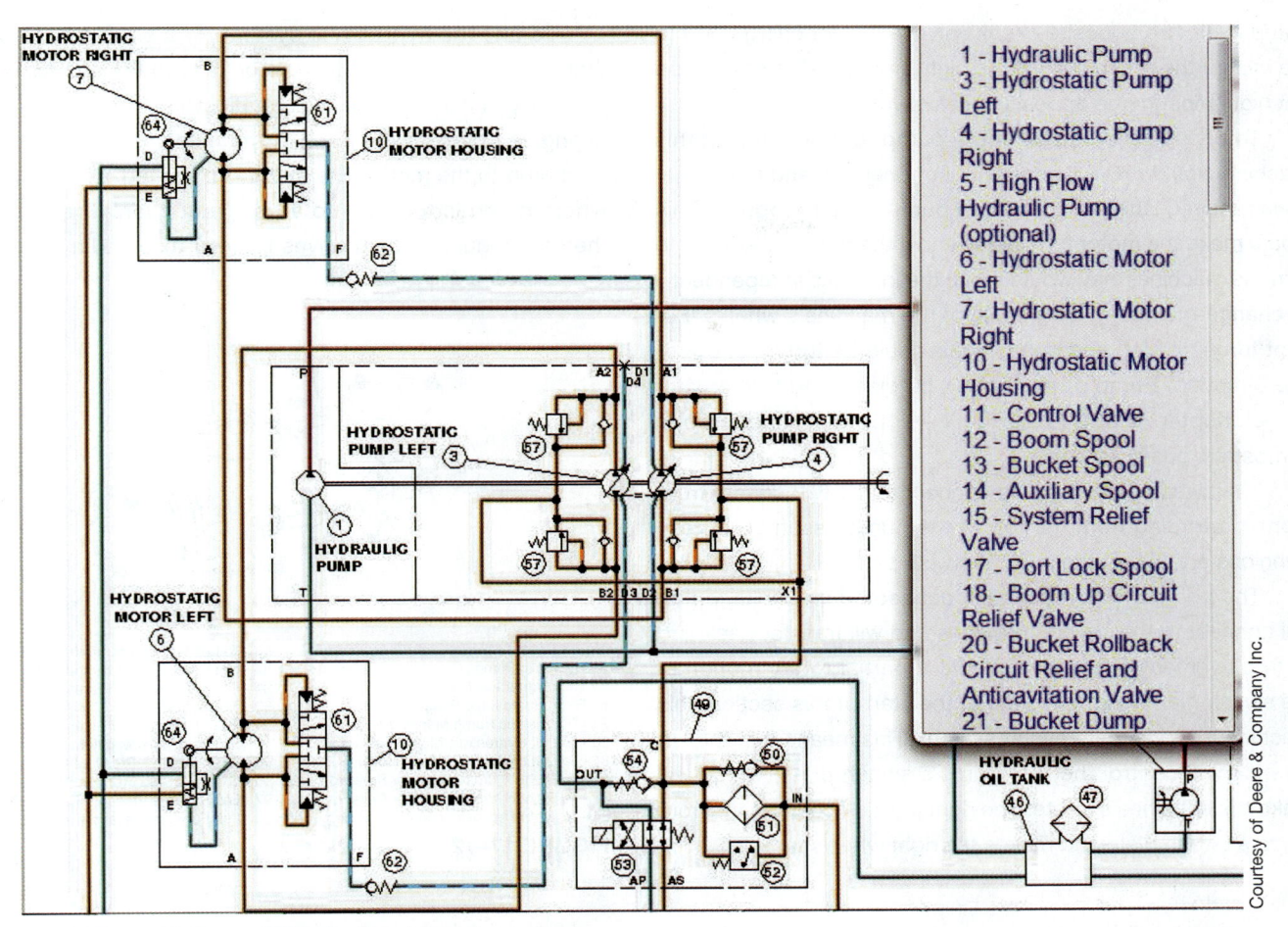

FIGURE 17-11 Deere CT 315 skid steer schematic.

1 - Hydraulic Pump
3 - Hydrostatic Pump Left
4 - Hydrostatic Pump Right
5 - High Flow Hydraulic Pump (optional)
6 - Hydrostatic Motor Left
7 - Hydrostatic Motor Right
10 - Hydrostatic Motor Housing
11 - Control Valve
12 - Boom Spool
13 - Bucket Spool
14 - Auxiliary Spool
15 - System Relief Valve
17 - Port Lock Spool
18 - Boom Up Circuit Relief Valve
20 - Bucket Rollback Circuit Relief and Anticavitation Valve
21 - Bucket Dump

Courtesy of Deere & Company Inc.

As there is also a loss of oil in the loop from internal leakage past the pump and motor rotating elements as well as the controlled loss through the flushing valve, there must be a constant supply of oil to replenish this oil at the low-pressure side of the loop. This oil is supplied by the fixed displacement gear pump (that is also the implement pump for this machine) through the makeup valves (57) that are in the hydrostatic pump housing. These are simple check valves that open to allow oil into the pump inlet side if the return oil pressure from the motor falls below charge pressure. The valves close when the pump port is a pump outlet and the pressure rises. This part of the system is needed to prevent pump cavitation and is common to any hydrostatic system. The charge oil supply circuit is called as such because the oil charges the inlet side of the pump. It also keeps all parts of the system full of oil, circulates oil to help cool and lubricate components, and keeps back pressure on the pump's pistons when the pump is in neutral. Charge pump pressure is typically limited to between 200 and 350 psi depending on the manufacturer and is limited by a charge relief valve. When the charge relief valve opens, it will dump the pump flow back to tank.

The gear pump flow for this system first goes to supply the implement functions (boom, bucket, and auxiliary) of this machine and then flows through the cooler and filter before it becomes available for charge oil.

Larger hydrostatic systems will have a dedicated charge pump that is used for supplying charge oil and one or more other pumps to supply the hydraulic system flow requirements. These machines will also likely have a dedicated hydrostatic oil tank so that if there is a catastrophic failure, then there won't be cross-contamination between the two systems. Charge oil can also be used for brake release oil, pump servo controls, pilot controls, fan drive, and any other low pressure hydraulic functions.

As you can see by the pump symbols on the schematic, the pumps are variable displacement bidirectional, which means they can send oil out either one of two ports. If the pump sent oil out the bottom line, this would make the motor rotate in the opposite direction from what was just discussed. This would also shift the shuttle spool in the other direction but would still send some of the motor exhaust oil out past the flushing valve to keep the motor cool and clean. The pump main port that was the pump's outlet for forward is now the

HYDROSTATIC DRIVE SYSTEMS **513**

pump inlet. The opposite check valve would let charge oil into the inlet of the pump. Now if the high-pressure side of the loop got high enough, the opposite relief valve would open.

To change the output of the pump, the operator of this machine would move the joystick from neutral, and the farther it was moved, the more flow the pump would produce. This would make the motor turn faster, and if the operator wanted to turn the machine, they would move the joysticks independently to change the speed differential of the machine's wheels. For spot turns, the left and right wheels could be turned in opposite direction. Pump operation will be discussed further later in the chapter. ● SEE FIGURE 17–12 for an illustration of a hydrostatic pump assembly.

The swash plate controls connect to the two square trunnions protruding from the top of the pump assembly. These trunnions are part of the pump's swash plates.

The operator's joysticks are connected to a linkage that will connect to the square trunnions and will transfer back and forth movement of the joysticks to a partial rotary motion of the swash plate. As mentioned at the start of this section, this machine is a dual path hystat system. This means that the left joystick will control the rear pump's swash plate, which will make the left wheels rotate. The right joystick controls the front pump's swash plate and makes the right wheels turn.

To see the hydrostatic system for a small wheel loader that uses one pump and one motor, refer to ● FIGURE 17–13.

As you can see, the hydrostatic pump and charge pump is engine driven at the back of the machine. Hoses connect the pump to the motor. The motor is mounted on the rear axle where it sends torque into a gear reduction. The output of the gear reduction then drives the rear axle and a drive shaft

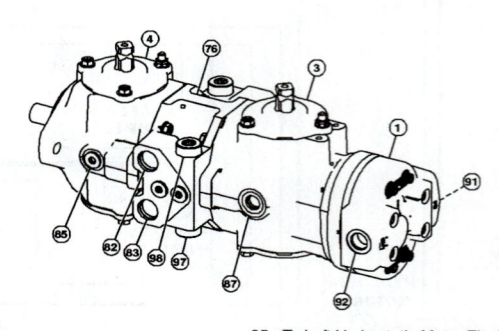

1 - Hydraulic Pump
3 - Left Hydrostatic Pump
4 - Right Hydrostatic Pump
76 - Hydrostatic Pump Manifold
80 - To Left Hydrostatic Motor Forward Port
81 - To Left Hydrostatic Motor Reverse Port
82 - To Right Hydrostatic Motor Forward Port
83 - To Right Hydrostatic Motor Reverse Port
84 - To Right Hydrostatic Motor Flushing Port
85 - To Left Hydrostatic Motor Flushing Port
87 - Not Used, 88 - To Hydraulic Oil Tank
91 - From Hydraulic Oil Tank
92 - To Control Valve
94 - From Hydraulic Oil Filter and Park Brake Solenoid Valve Manifold
95 - Right Reverse Pressure Relief Valve
96 - Right Forward Pressure Relief Valve
97 - Left Reverse Pressure Relief Valve
98 - Left Forward Pressure Relief Valve

FIGURE 17–12 Hydrostatic pump assembly.

1- cooler
2- filter
8- pump assembly
11- rear axle
14- motor
17- front axle

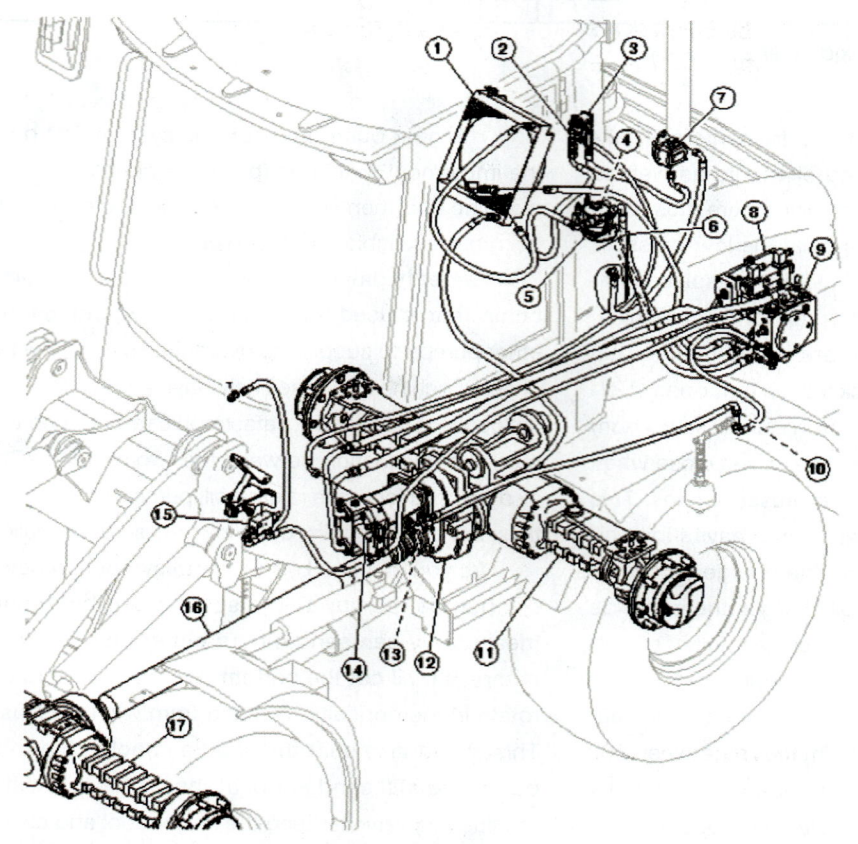

FIGURE 17–13 Power train for a small wheel loader.

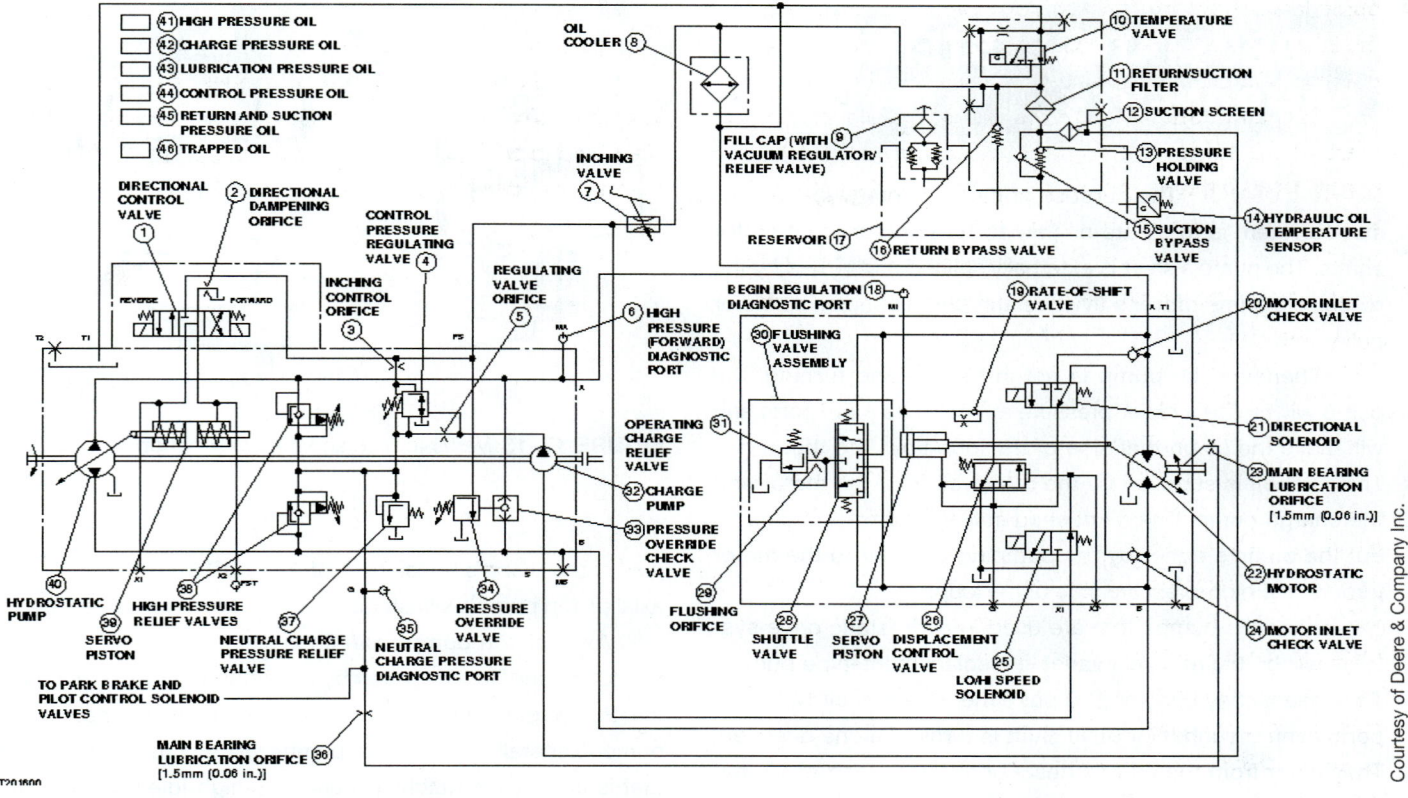

FIGURE 17–14 Hydrostatic schematic for a single path system.

transfer's torque to the front axle. This would be considered a single-path hydrostatic system. ● SEE FIGURE 17–14 for a schematic of a single pump and single motor hydrostatic system for a small wheel loader.

This is a simple hydrostatic system that features an electronically controlled pump and a two-speed motor. There is a charge pump that will provide a steady supply of makeup oil.

Progress Check

1. This is *not* one advantage of a hydrostatic system:
   a. gives smooth transitions between forward and reverse
   b. is cheap and simple to repair
   c. has a compact and flexible component arrangement
   d. is easy to operate
2. This controls the direction and amount of oil flow in a hydrostatic system:
   a. the main directional control valve
   b. the speed of the engine
   c. the motor swash plate
   d. the pump swash plate
3. The output drive of a hydrostatic motor will most likely go to:
   a. the track sprocket
   b. a final drive gear or chain sprocket
   c. steering clutches
   d. an axle shaft
4. A closed loop hydrostatic system is one that:
   a. has a sealed reservoir
   b. has a closed loop pilot system
   c. has a directional control valve
   d. has its pump and motor flow recirculate between each other
5. If a motor has a displacement of 7.2 cubic inches and there is 3000 psi oil pressure applied to, it will develop _____ in.-lb of torque:
   a. 6879
   b. 3438
   c. 2697
   d. 1633
6. Tech A says hydrostatic systems are less expensive to repair than all other types of power transmission systems. Tech B says hydrostatic systems can easily withstand fluid contamination. Who is correct?
   a. Tech A
   b. Tech B
   c. both A and B
   d. neither A nor B

HYDROSTATIC DRIVE SYSTEMS 515

HYDROSTATIC SYSTEM COMPONENTS

MAIN PUMPS The purpose of the main pump in a hydrostatic system is to create oil flow to turn one or more motor shafts. The pump's shaft is externally splined and turned by the machine's prime mover's flywheel through a coupler, spline or pulley and belt.

Whenever the pump is not in neutral and turning, the pump will create a lower pressure at one of its main ports and will make the oil on that side of the loop flow into the pump. This inlet oil is supplied by the return oil from the motor and the charge pump. The oil pushed into the pump is then sent out the pump's outlet by the pumping element to the motor through the high-pressure side of the loop.

Hydraulic pumps that are used in hydrostatic drive systems will be bidirectional variable displacement–type pumps. This means they can send oil out either one of their two main ports even though their drive shaft is turning in one direction. They differ from the pumps described in the Chapter 16 that are used for hydraulic systems because those pumps had ports that would only be inlet or outlet and, therefore, could only send oil out one port. ● **SEE FIGURE 17–15** for a variable displacement bidirectional pump.

The main style of pump used for hydrostatic systems is the axial piston. This type of pump was described in detail in Chapter 16 but for a quick review:

- The pump shaft is suspended on bearings in the pump housing and drives a cylinder block.
- The cylinder block is rotated by the shaft and has seven to nine bores that each carries a piston around inside of it.
- The bottom of the steel pistons is shaped into a round ball, and over the ball, a brass slipper is fitted that pivots with the tilting swash plate.
- The brass slippers move around the face of the swash plate and are guided by a retaining plate so they stay at the same angle as the swash plate. Sometimes swash plates can also be called camplates.
  ● **SEE FIGURE 17–16** for a disassembled piston pump.

As the swash plate is angled, it will make the pistons move in and out of the cylinder block as the block is rotated. A typical maximum inclination of 15° means the piston travels the greatest distance when the swash plate is tilted 15° away from being perpendicular to the centerline of the pump shaft.

The end of the cylinder block opposite the piston slippers is lap finished to a stationary port plate that allows oil in and out of the block. The port plate lines up with the two ports that are part of the pump's housing. ● **SEE FIGURE 17–17** for how an axial piston pump's swash plate is controlled to go over center.

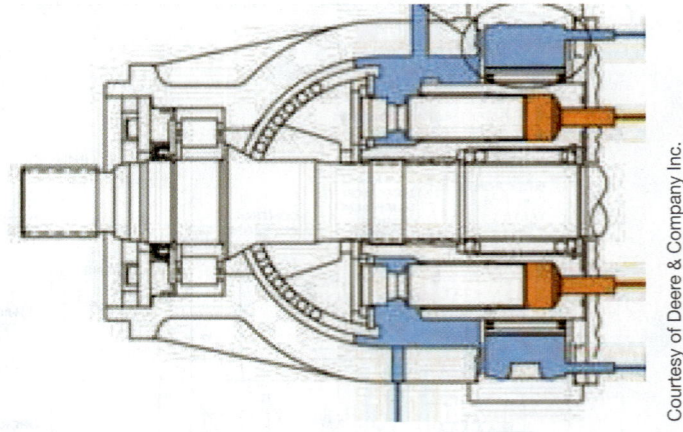

FIGURE 17–15 Variable displacement bidirectional pump.

Axial piston pumps can change their displacement per revolution and flow direction by changing the swash plate angle. Pump rotation speed will also affect the flow output of the pump. Generally, hydrostatic pumps will be run at a fixed rpm that is usually the machine's engine's high-idle rpm. This will allow engineers to design the engine to provide maximum performance and low emissions at this speed.

Changing the pump swash plate angle changes the piston effective stroke, which changes the pump's displacement and ultimately changes the machine's travel speed and direction. When a pump is in neutral, the swash plate is at a 90° angle to the pump drive shaft. This means there is no axial movement of the pistons as they rotate around the drive shaft in the cylinder block, and therefore, there is no oil flow from the pump. As the swash plate angle is changed, the pistons will start to move up and down inside the cylinder block. The easiest way to picture piston movement is to think of children's manually powered merry-go-round that you might see in a city park.
● **SEE FIGURE 17–18** for a merry-go-round.

If there was one person on the ride spinning around and the ride was level they wouldn't be moving up and down. If there was a way to tilt the ride, then people would be moving up and down as they spin around the ride's axis.

To relate this to the axial pump operation, focus on one piston through one revolution with the swash plate at its maximum angle. The pump's two ports are half-moon shaped and are separated. As the piston moves from one port to another, it will be moving from its lowest to highest position in the cylinder block. This will be when the piston is pushing oil out of the pump through almost half of one revolution. As the piston continues through its second half of the revolution, it will move

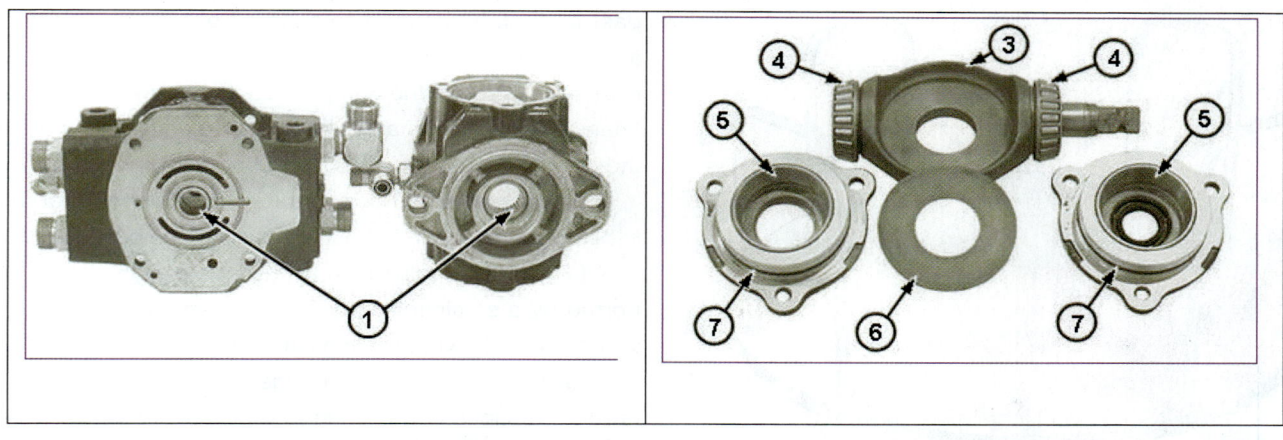

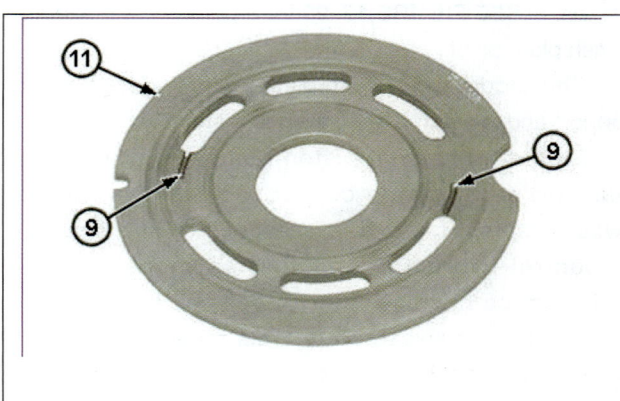

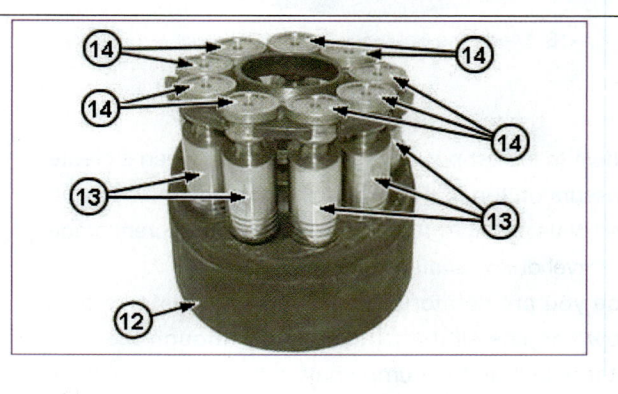

1- Drive shaft bearings; 3- swashplate; 4- tapered roller bearings; 5- bearing cups; 6- washer; 7- O rings
9- lubrication notch; 11- port plate; 12- barrel; 13- pistons; 14- slippers

FIGURE 17-16 Piston pump.

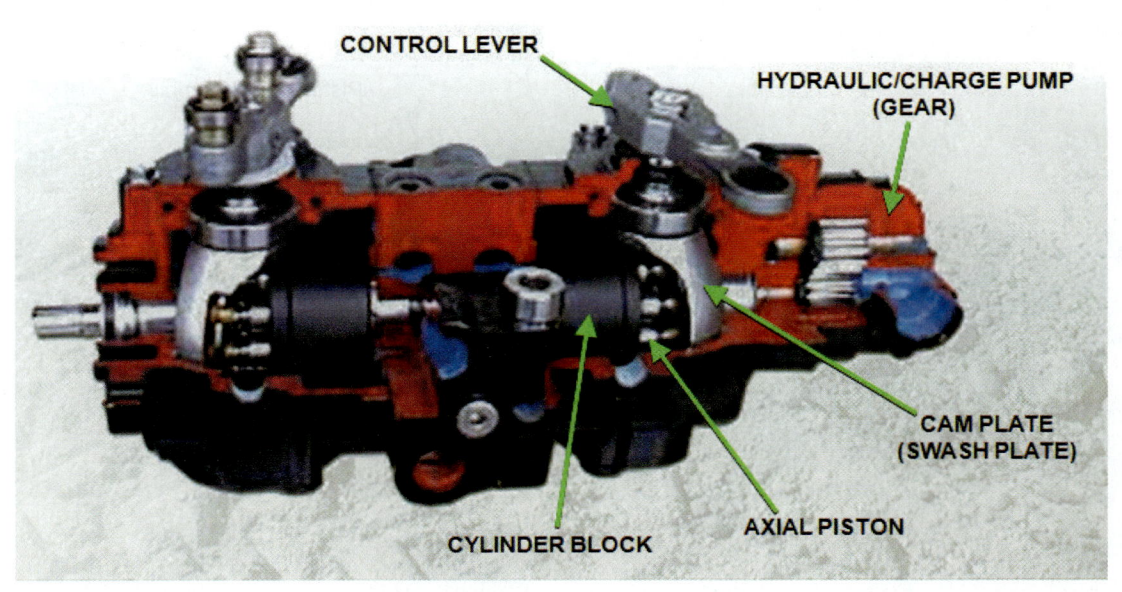

FIGURE 17-17 Axial piston pump swash plate control.

HYDROSTATIC DRIVE SYSTEMS **517**

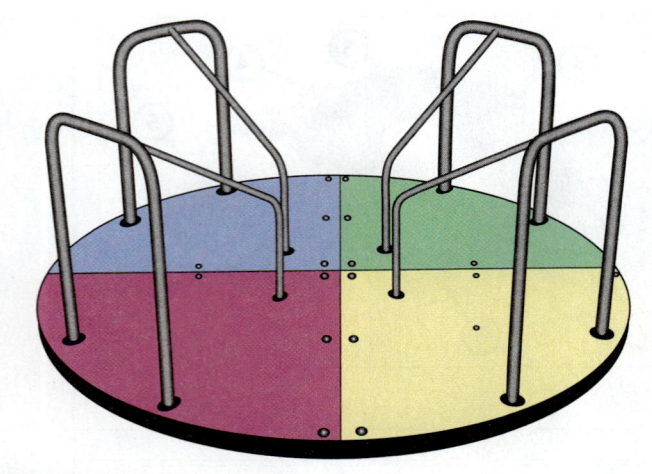

FIGURE 17–18 Merry-go-round.

from highest to lowest position, and this will be when it creates a low pressure on top of it to bring oil in through the other half of the port plate. ● SEE FIGURE 17–19 for the different stages of piston travel during shaft rotation.

Once you are comfortable with a single piston's operation, now imagine all the pistons going through the same constant motions as the pump shaft is turned. This will create a constant flow of oil from the pump inlet and out the outlet. As mentioned, this was when the pump swash plate was at a maximum angle. As the angle decreases, the piston travel decreases and, therefore, pump's displacement per revolution decreases. When the swash plate then moves past neutral, it will start to send oil out the opposite port as the inlet and outlet ports trade places. This will in effect reverse the flow of the pump. Then as the swash plate continues to increase its angle, the flow volume will increase in the opposite direction.

Swash plates can be supported by trunnions that rotate in bearings that are in the pump housing. The bearings can be needle or tapered roller type. Swash plates can also be supported by a semicircle saddle arrangement that allows the swash plate to pivot as it is pushed around the saddle. This is also called a cradle-bearing arrangement sometimes, and it will have a U-shaped roller bearing for the curved swash plate to ride on. ● SEE FIGURE 17–20 for a pump with a saddle-type swash plate pivot.

The saddle-type swash plate will provide a more robust support and is controlled by a servo piston.

Lighter-duty axial piston pumps can be controlled manually with linkage connecting the joysticks to the swash plates. This is a less-expensive and simpler way to control hydrostatic pumps. As pumps get bigger and produce higher flows, it becomes harder to manually control the swash plate. This means there needs to be hydraulic assistance to change the swash plate angle. This is done with one or more servo pistons that are connected to the swash plate that will move the swash plate when oil pressure is applied to the servos. ● SEE FIGURE 17–21 for an axial piston pump that has a servo piston to move its swash plate.

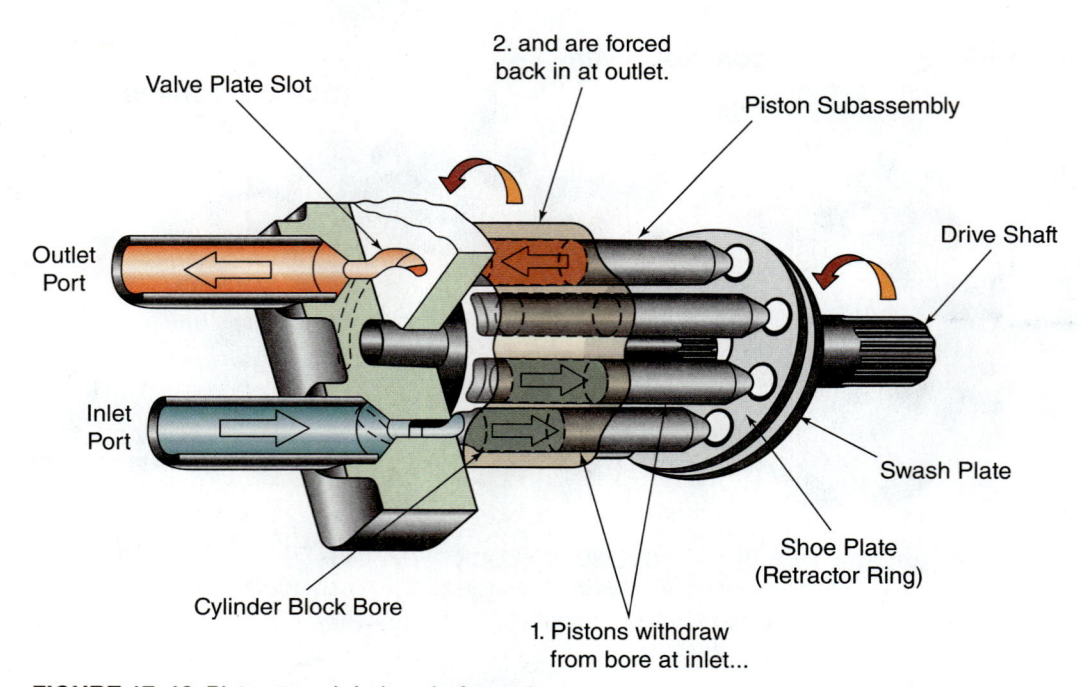

FIGURE 17–19 Piston travel during shaft rotation.

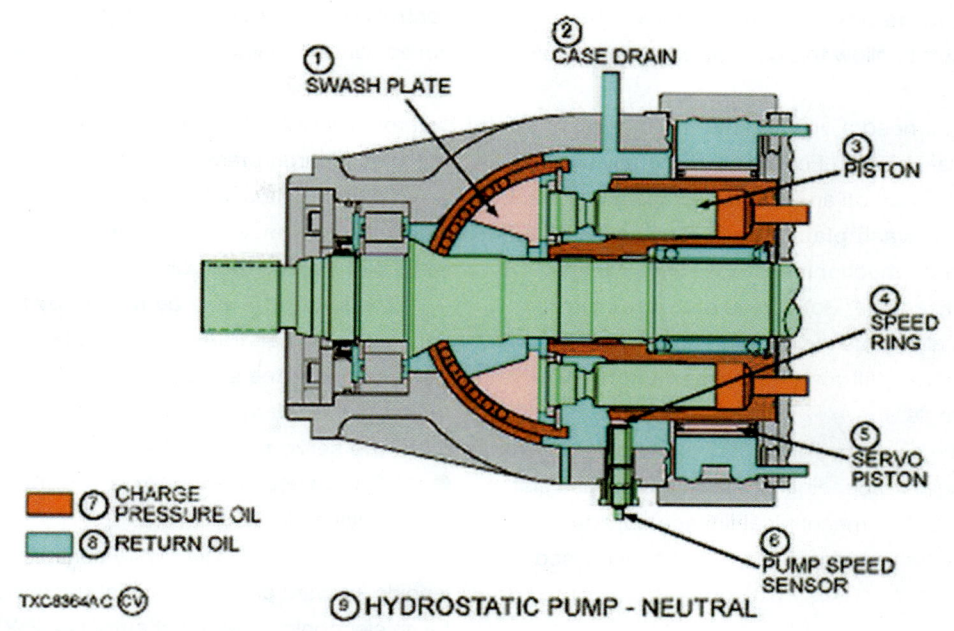

FIGURE 17–20 Saddle-type swash plate.

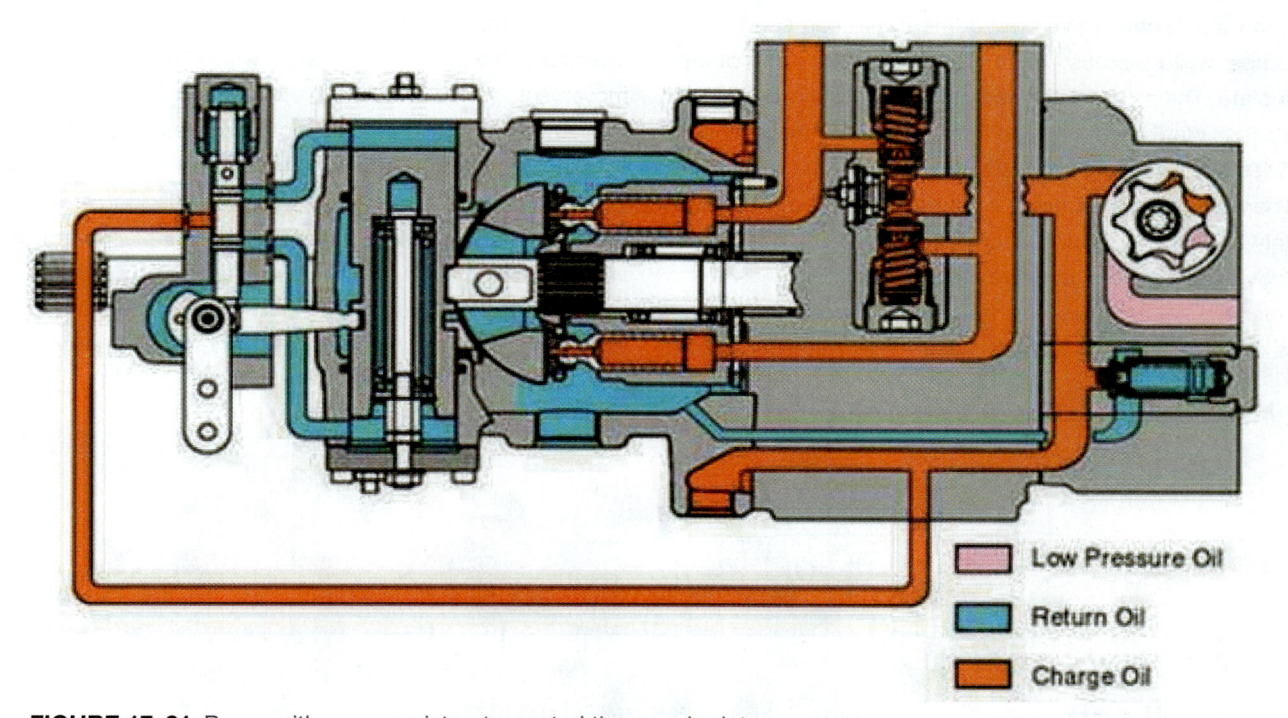

FIGURE 17–21 Pump with a servo piston to control the swash plate.

The pump's servo pistons get oil sent to them from a displacement control valve. This valve can be manually shifted or moved with solenoid-controlled oil pressure. As it is shifted, it will send oil to one of the servo pistons to change the swash plate angle, or if there is one servo piston, it will send oil to one end of the piston. Most pumps will have this valve mounted on the pump housing.

As hydrostatic system pressures can reach over 6000 psi, the pump's pistons and barrels need to have very close tolerances to seal this pressure. They will be lap finished to provide between 5 and 40 microns clearance. Some pistons will have seals that ride in grooves on their outer circumference to seal the pressure on top of the piston. Even though new pumps have small clearances, there will still be a certain amount of

HYDROSTATIC DRIVE SYSTEMS **519**

leakage that needs to be drained back to tank. The pump housing will have a port to allow this oil to return to tank. This is called the case drain port.

These pumps will need a way to limit the maximum angle that the swash plate can tilt. For the mechanically actuated swash plate pump, this can be an external linkage adjustment. For servo-controlled swash plate pumps, the servo piston movement is limited by mechanical stops. This will limit the movement of the swash plate. Both types of swash plate limiters are adjustable to allow maximum pump flow (motor speed) to be set. Some pumps will also have internal mechanical stops that are adjustable.

There will also be a way to adjust the neutral point for the swash plate angle. Again depending on the type of swash plate control, this will be either a mechanical linkage adjustment or a servo piston adjustment. More details will be discussed on this later.

SERVO CONTROL VALVES To manage pump flow from a hydrostatic pump that is not mechanically controlled, there needs to be a control valve arrangement that can send oil to the pumps servo pistons, which in turn will move the pump swash plate. The servo piston will be mechanically connected to the swash plate either through a linkage or directly with a pin that slides in a slot in the servo piston.

There are many variations of pump servo control valves and only a few will be discussed here. Pumps can use either one or two servo pistons to move the swash plate.

Servo control valves can be manually actuated spool valves or spool valves that are moved with pilot oil that is controlled with electrically actuated solenoids. Manually controlled valves will likely be spring centered or cable controlled, and as the operator moves a FNR (forward-neutral-reverse) control, this will move the spool from neutral. The spool valve will allow charge pressure oil to go to one end of the pump's servo piston. The spool will also allow the oil at the other end of the servo to be returned to drain. This happens with a single servo piston pump control.

The spool will also be moved by a feedback link that is connected to the swash plate. The feedback link will re-center the spool after the swash plate has moved, and if the operator control is left in one position, the spool will slightly meter oil to the servo to keep the swash plate at a desired angle.

● **SEE FIGURE 17–22** for a manually controlled servo control valve.

If the pump is electronically controlled, there will be two solenoids (forward and reverse) that get energized individually and by an electronic system that outputs a PWM signal or by potentiometers that send a variable voltage signal. The solenoid armature will move proportionately to the value of the signal sent to it. The armature movement will shift a valve that directs oil to one end of the pump servo piston, which will cause swash plate movement.

Another variation of electrical/electronic pump servo controls is one that uses a linear torque motor that uses two solenoids to move a flapper valve. When the flapper valve moves, it will send charge oil to either end of a displacement control valve spool. This will shift the spool and send charge oil to either end of the swash plate servo piston to move the swash plate. There is a feedback linkage connecting the swash plate

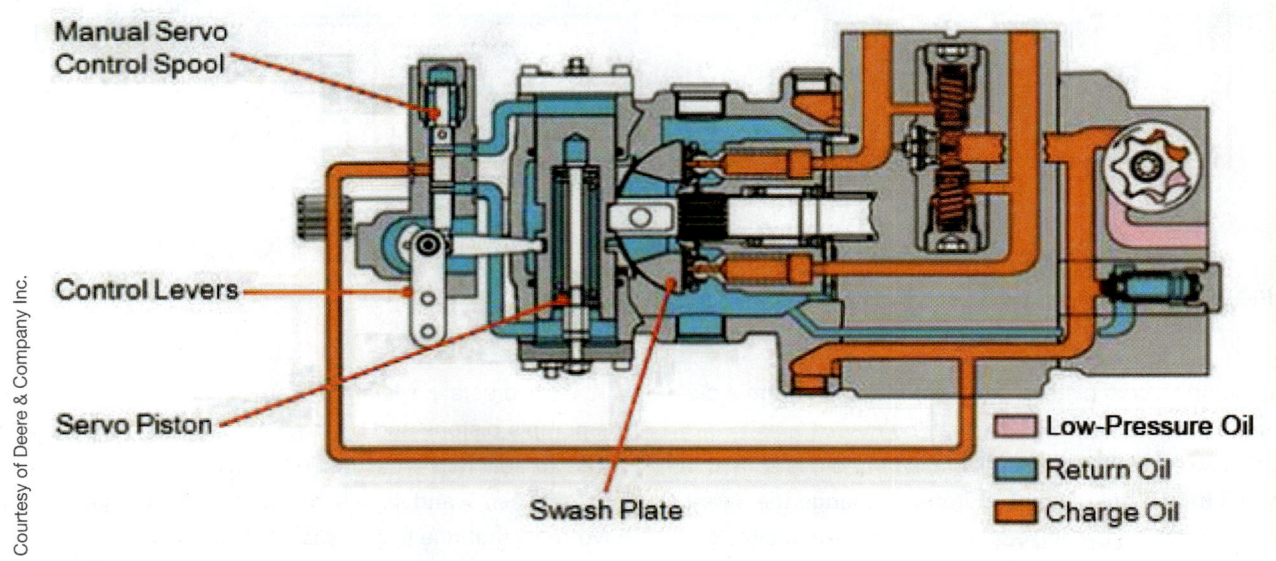

FIGURE 17–22 Manual servo control valve.

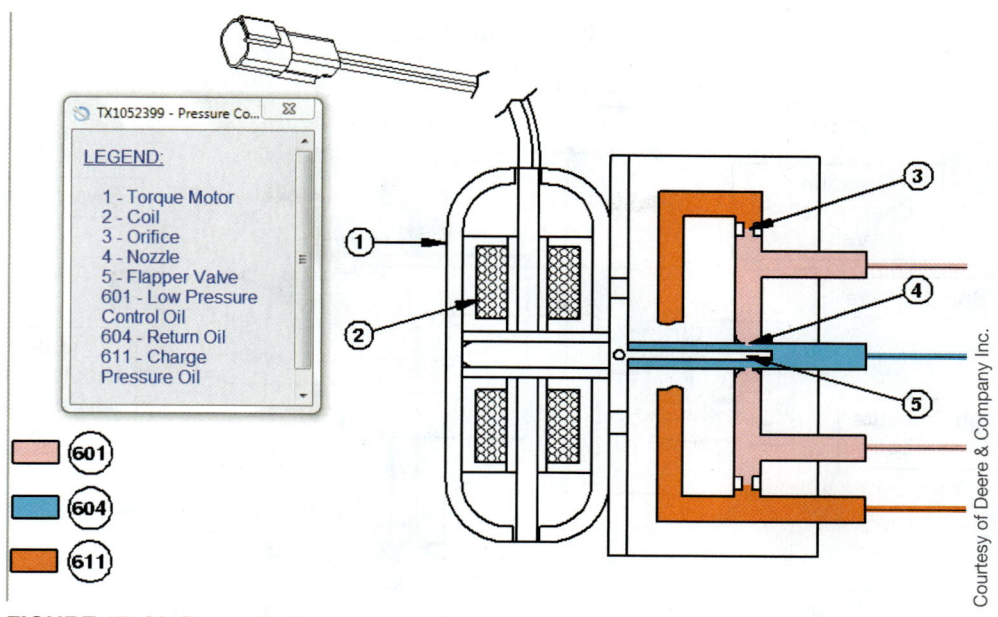

FIGURE 17–23 Pump displacement control valve.

and spool valve. This will center the spool to stop and hold the swash plate in place if the operator's demand requires this.
- **SEE FIGURE 17–23** for a pump control valve that uses a flapper valve for the displacement control valve.

MOTORS Hydrostatic system motors receive oil flow from the pump and convert this flow into mechanical torque that is transferred out the motor's shaft and sent to the machine's final drive mechanism.

Hydrostatic system motors can be one of several types: axial piston, bent axis piston, cam lobe–type motors, and smaller machines can also use geroller motors.

Motors that are fixed displacement will only give a low speed/high torque output.

Many hydrostatic systems will feature variable displacement motors that will provide multiple travel modes from high torque/low speed to high speed/low torque. To change to high speed, the motor displacement will need to decrease. This increases the output shaft revolutions per unit of oil flow to the motor. As a result, it decreases output torque and increases motor shaft speed.

All motors will incorporate a flushing valve or case drain port to allow a constant flow of cooling and cleaning oil flow out of the loop on the low-pressure return side after the oil leaves the motor element. There will also be a normal amount of internal leakage that needs to be returned to tank as well.

Most motors will incorporate a parking brake into their housing that will lock the output shaft to the housing. These are typically spring-applied oil-released multidisc brakes.

Motors, like pumps, are rated as to their displacement per revolution as well as their maximum pressure capability. For example, a variable displacement motor for a compact wheel loader can have a maximum displacement of 4.8 CIR (80 CCR), can have a minimum displacement of 2.0 CIR (34 CCR), and is able to withstand a working pressure of 7000 psi.

Some machines that have a dual path system will use a flow divider to equalize flow to the left and right motors. This will give the same effect as a differential lock that is used in a mechanical drivetrain. The flow divider ensures that the left and right motors turn at the same speed.

AXIAL PISTON MOTOR Some hydrostatic systems will use axial piston motors to convert fluid flow to mechanical rotation. This type of motor is very similar to the axial piston pumps that are used to create the flow for the system in that they use a swash plate, pistons, and cylinder block to transfer piston movement created by fluid flow to shaft rotation for the motor output.

Judging by physical appearance, it would be hard to tell the difference between an axial piston pump and motor. This is true for most pumps and motors that operate on the same principle.

Oil is sent into one of the two main ports of the motor housing and goes through a port plate that is stationary.

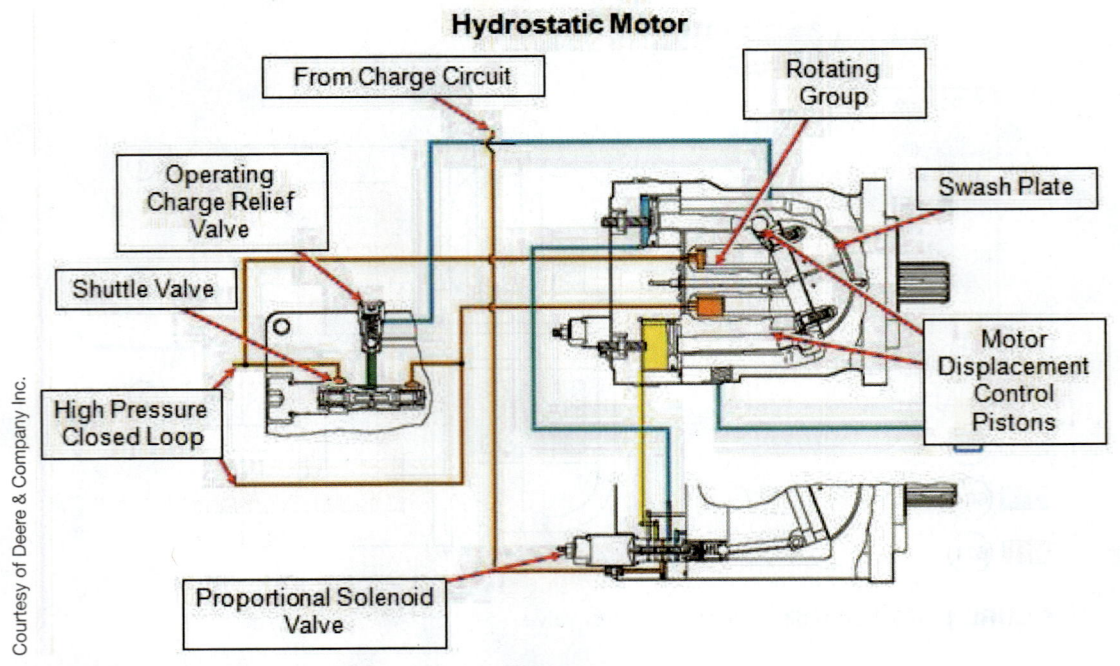

FIGURE 17–24 Variable displacement axial piston motor.

The port plate has a lap finished surface that mates with the cylinder block. As the oil flows into the cylinder block, it will push on half of the pistons down the swash plate while the other half of the pistons are moving up the swash plate and pushing oil out the other main port. As the pistons are pushed down the swash plate, they carry the cylinder block with it. The block then drives the output shaft of the motor. The oil that leaves the motor returns to the pump inlet.

This style of motor can be fixed or variable displacement.

To vary the displacement of an axial piston motor, its swash plate angle will need to be changed. If the motor is a two-speed type, it will likely default to the low speed/high torque mode, which means its swash plate angle will be at maximum, and therefore, it will be at maximum displacement. An oil signal is sent to a servo piston in the motor that will move the swash plate to a smaller angle. An example of this is a motor used for a track loader that has a displacement of 3.5 cubic inches for low speed (maximum displacement) and can change to a displacement of 2.0 cubic inches for high speed (minimum displacement).

To change displacement, this type of motor will have two control pistons to move the swash plate to either minimum or maximum displacement. The oil pressure to do this can come from a solenoid valve on the motor or one that is mounted remotely. ● **SEE FIGURE 17–24** for a cross-section illustration of a variable displacement axial piston motor.

BENT AXIS MOTOR Bent axis motors have their cylinder block centerline at an angle to the output shaft centerline. The motors' shaft assembly is supported by two tapered roller bearings in the housing and is driven by the cylinder block and pistons.

Oil is sent into the motor through the inlet port of the housing and goes through the port, past a bearing plate, and into the valve segment. The valve segment has two slots to allow oil to transfer between the bearing plate and pump housing end cap. The bearing plate provides a bearing surface between the rotating cylinder block and the nonrotating valve segment. The valve segment will be able to move a few degrees if the motor is a variable displacement type; otherwise it is fixed. The valve segment has a stub shaft protruding from it that fits into a bearing in the end of the cylinder block. This will support the cylinder block in the housing. ● **SEE FIGURE 17–25** for a cross-sectional illustration of a bent axis motor.

WEB ACTIVITY

There are many animations of hydraulic motor operation available for viewing on the Internet.

Search for an animation of an axial piston motor and see if it helps you visualize how a hydraulic motor works.

Hydrostatic Motor

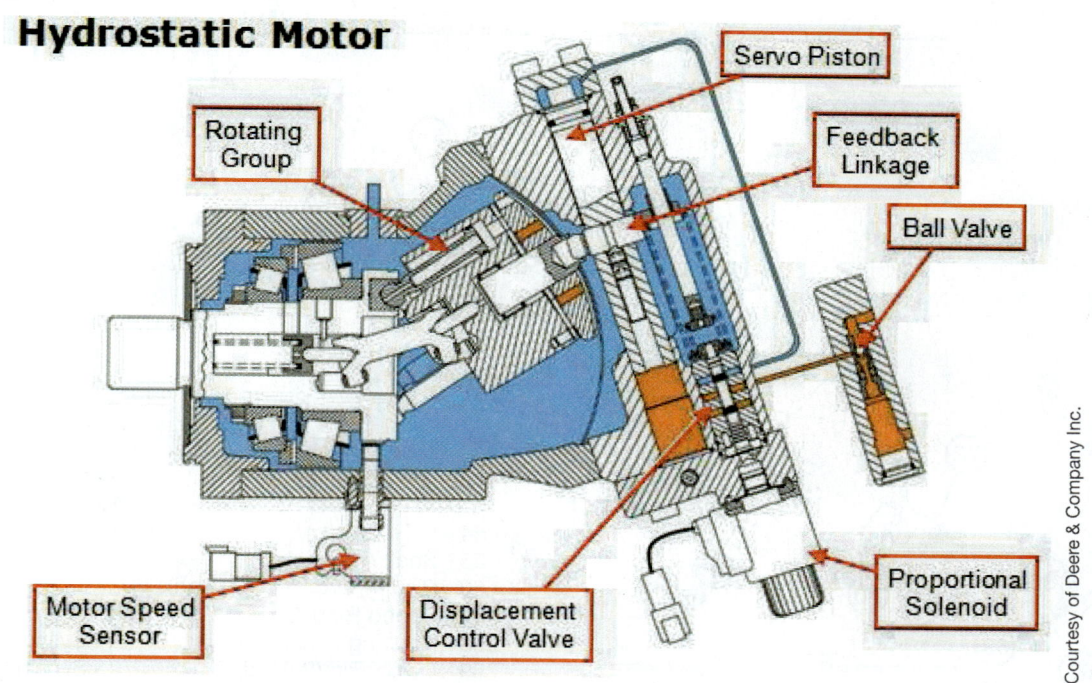

FIGURE 17–25 Variable displacement Bent axis motor.

A variable displacement bent axis motor has the opposite side of the valve segment featuring an outward curve shape that is matched to a bearing surface on the bottom of the motor end housing. When oil is fed into the cylinder block where the pistons are at the top of their stroke, the oil will force the pistons down into the barrel for half of the motor's rotation. The pistons have a head with a groove that locates a seal to keep the oil from leaking between the cylinder block and piston. When the pistons are forced down by oil pressure, this will make the cylinder block rotate.

The cylinder block is connected to the motor's shaft with a synchronizing shaft with two sets of three evenly spaced rollers protruding out from it. The rollers will fit into grooves in the shaft assembly and the cylinder block. This is similar to a type of constant velocity joint that is used for a front wheel drive car. There is a spring on the inside of the shaft assembly that will keep the cylinder block seated against the bearing plate and valve segment.

The bottoms of the pistons have a ball-shaped end and are supported in the shaft assembly in sockets.
● **SEE FIGURE 17–26** for an exploded illustration of a bent axis piston motor.

A variable bent axis motor will have minimum and maximum displacement stops to physically stop the movement of the barrel between two points. There will be a servo piston with a recess in it where the pin from the back of the valve segment is engaged. When the motor's displacement control (electric solenoid controlled) sends oil to one end of the servo, it will move the cylinder block with it to move toward a lower displacement. This will be the high speed/low torque mode for the motor. A spring will return the piston when the displacement control drains the oil.

CAM LOBE MOTOR This type of motor can also be called a radial piston motor.

A cam lobe motor uses a series of eight pistons that are oriented in a radial arrangement in a carrier and have rollers attached to their bottom end. The carrier turns the output shaft for the motor.

The pistons will move in and out of the carrier as their rollers ride around the inside of an internal cam. The internal cam has six equally spaced lobes on the inside diameter of the cam ring. Oil pressure is fed into the motor housing through one of two ports. The other port will direct oil out of the motor. The ports feed oil to a manifold that has passageways machined into it. These manifold passageways connect to ports on the outer circumference of it. There are six ports connected to one of the main motor ports and six ports connected to the other main port. When oil is supplied by the system pump to one of the motor's main ports, the manifold will be able to send oil to four of the pistons at any time. The other four pistons will be aligned with the return port. The remaining four ports are blocked off temporarily. To start the motor turning, the oil pressure pushes four pistons and rollers down cam lobes, and because the piston are in bores in the carrier, the carrier

HYDROSTATIC DRIVE SYSTEMS 523

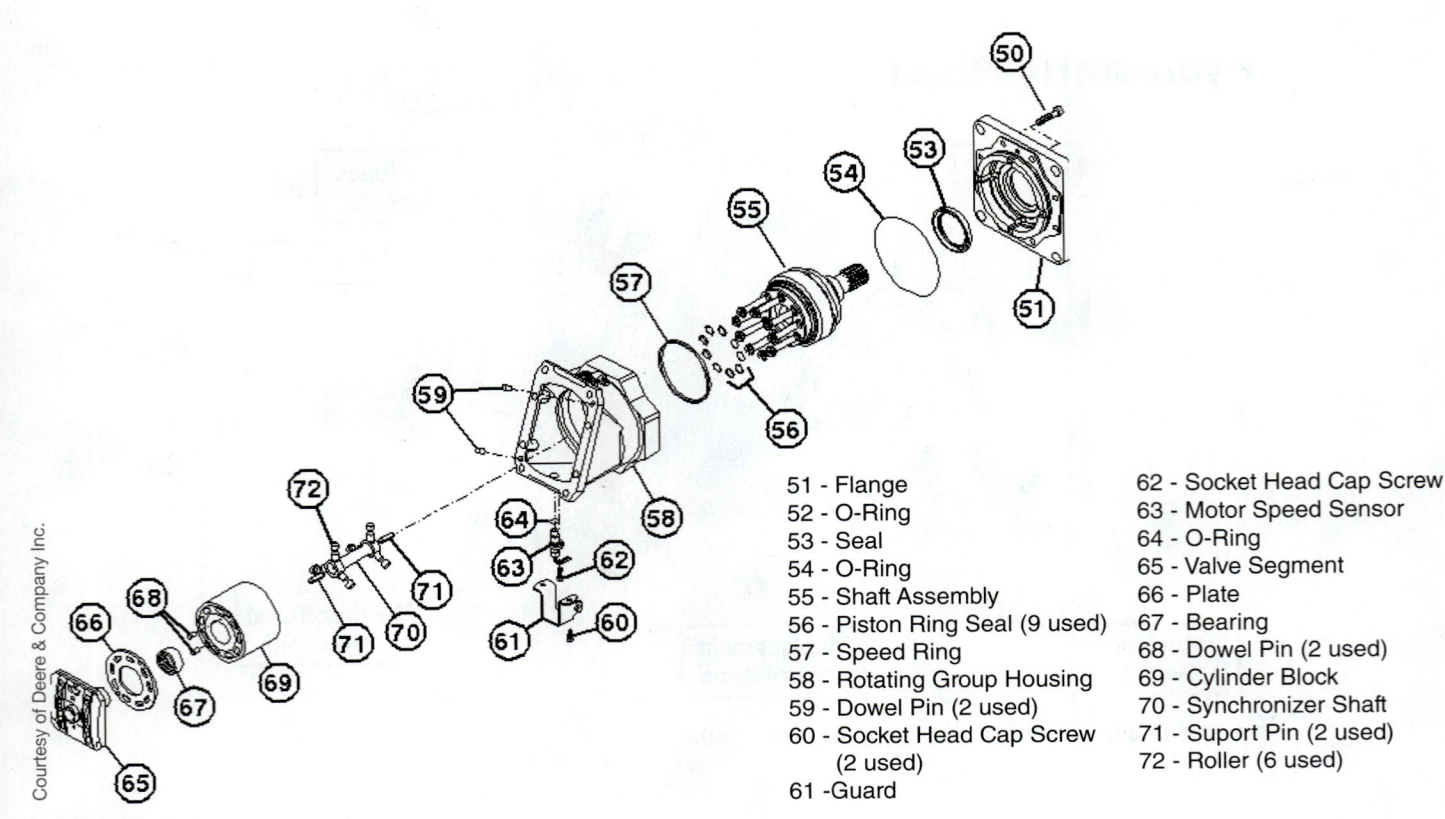

51 - Flange
52 - O-Ring
53 - Seal
54 - O-Ring
55 - Shaft Assembly
56 - Piston Ring Seal (9 used)
57 - Speed Ring
58 - Rotating Group Housing
59 - Dowel Pin (2 used)
60 - Socket Head Cap Screw (2 used)
61 - Guard
62 - Socket Head Cap Screw
63 - Motor Speed Sensor
64 - O-Ring
65 - Valve Segment
66 - Plate
67 - Bearing
68 - Dowel Pin (2 used)
69 - Cylinder Block
70 - Synchronizer Shaft
71 - Suport Pin (2 used)
72 - Roller (6 used)

FIGURE 17-26 Bent axis piston motor.

will rotate. The other four pistons are also carried around and will be riding up cam lobes. This will push oil out of the carrier through the manifold and out of the motor into the low-pressure side of the loop. Once the motor starts turning, all pistons are constantly moving in and out of the carrier and riding up and down the cam profiles.

As you can imagine, port timing is critical to the operation of this style of motor because there needs to be a seamless transition between the oil flow in and out of the motor to the pistons that either are in being pushed down cam lobes or are returning back into the carrier. ● **SEE FIGURE 17-27** for a cross-section illustration of a cam lobe motor.

Cam lobe motors can also be dual displacement motors. If the same amount of oil flows to the motor but is only supplying half the pistons, then the motor will turn at twice the rpm. This is done by supplying pressure oil to only half of the pistons. The other half receives only charge pressure oil. This is done by moving a displacement control valve, which will switch half of the discharge ports from high-pressure pump flow to charge oil pressure.

ORBITAL MOTOR Lighter-duty hydrostatic systems will sometimes use an orbital motor. There are several variations of these motors that can also be called gerotor and geroller motors, but they fall into three main groups called disc valve, spool valve, and valve in star. These different groups of orbital motors relate to how the oil is distributed within the motor as it moves to and from the main ports.

They all work on the same principle of having an inner and outer rotor with coarse "teeth" or lobes. The inner rotor will have one less tooth than the outer rotor and their centers will be offset. The outer rotors center is on the same axis as the output shaft and the inner rotor's axis will be offset from the outer rotor's axis.

As oil is sent into the motor, it will enter a small chamber that is created between inner and outer rotor teeth. The applied pressure will make the rotors turn as the chamber expands. The expanding chamber will reach a maximum volume, then as rotation continues, chamber volume decreases, forcing fluid out of the chamber and out the return port. The process occurs constantly for each chamber, providing a smooth pumping action. If oil is sent to the opposite port, the motor will turn in the opposite direction.

The inner rotor has internal splines that will drive an externally splined drive shaft. The splines will be machined to allow the offset between the inner rotor and the output shaft as rotation occurs. This shaft is sometimes called a cardan shaft.

A geroller motor will have rollers on the ends of the outer rotor teeth to reduce friction and improve longevity. The inner rotor has one less lobe than the rollers so that one lobe will

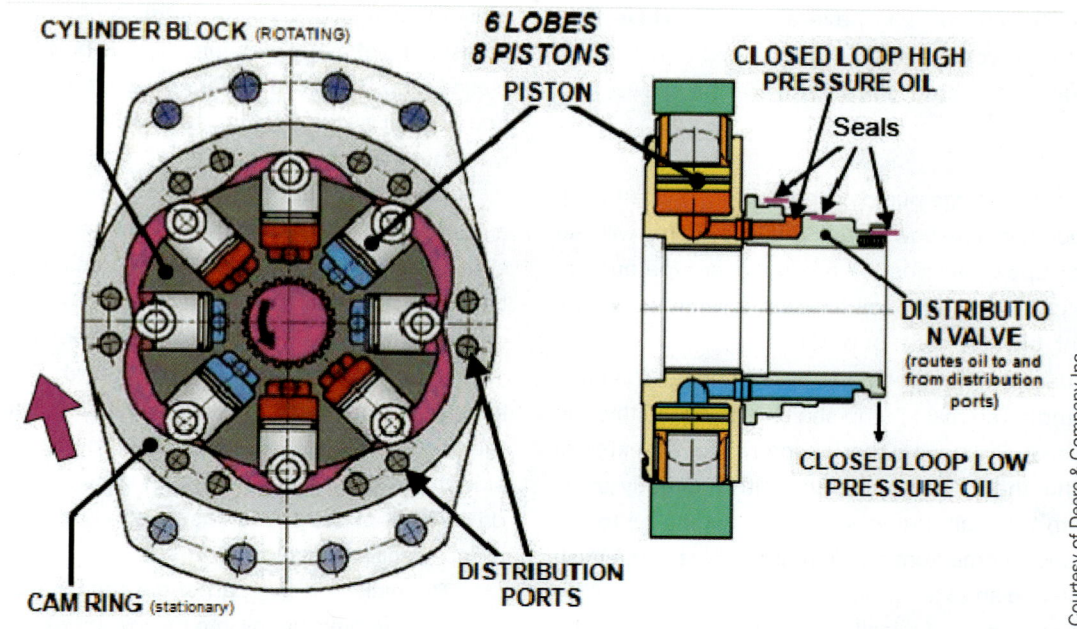

FIGURE 17–27 Cam lobe motor.

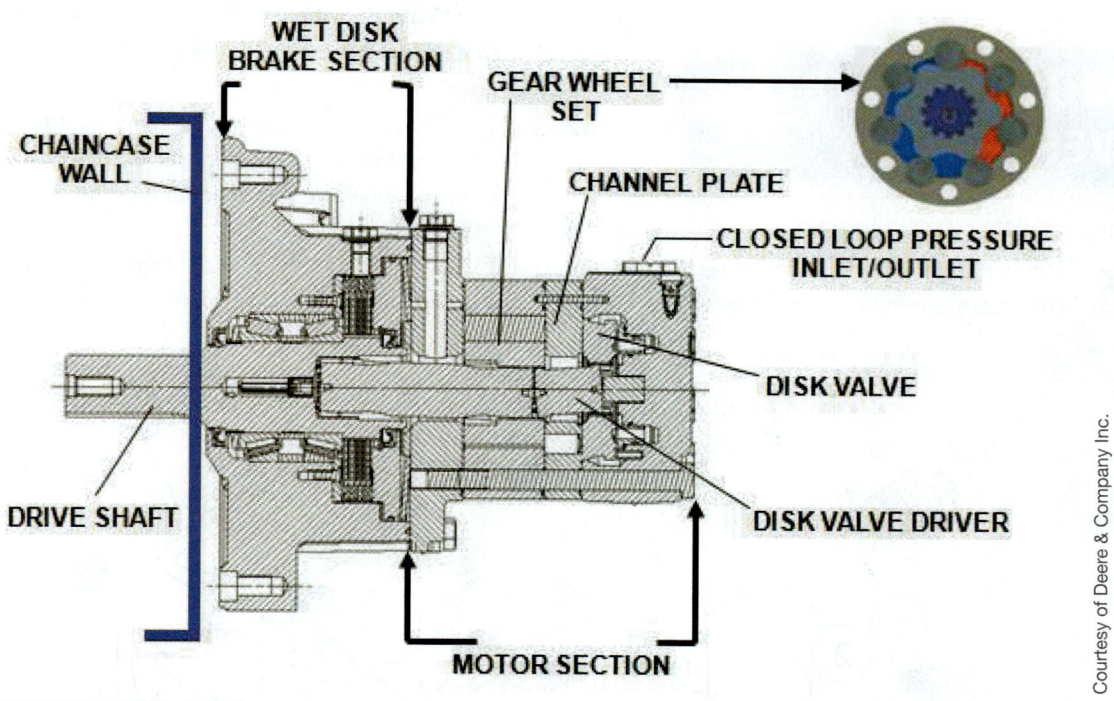

FIGURE 17–28 Geroller motor.

always be in full engagement with the rollers at any one time. This allows the rotor lobes to slide over the rollers creating a seal to prevent pressure oil from returning to the inlet side of the motor.

This change causes the orbiting gerotor to make as many power strokes as it has teeth for every revolution of the output shaft. The six-tooth gear shown makes six power strokes while the output shaft turns once. ● **SEE FIGURE 17–28** for a geroller-type motor.

These motors can also be equipped with a flushing port to allow oil to circulate within the closed loop circuit to aid in cooling and cleaning.

Gerotor motors give at or near full torque from about 25 rpm and normally do not go higher than 250–300 rpm. Maximum output torque is directly related to the width of the gerotor element, which may be as narrow as ¼–2 in. Pressure ratings as high as 4000 psi are common from most manufacturers.

HYDROSTATIC DRIVE SYSTEMS 525

Some gerotor motors can have a selector valve that changes the internal rotary valve output to feed only half the chambers, causing the motor to run at twice the speed, and half the torque.

CHARGE PUMP Charge pumps used in hystat systems will be fixed displacement low flow units. Smaller systems will use gerotor-type pumps while most others use gear-type pumps. These pumps can also supply oil for other systems such as main implement, brake release, or pilot oil.

They are usually mounted on the back of one of the main hydrostatic pumps. The main pump shaft will continue through the pump housing and have either an internal or an external spline on it. The charge pump is driven either directly or by a coupling from the main pump shaft. The supply oil to the charge pump could come from an internal passage in the main pump, or it could be an external line.

Charge pump flow is typically 20% of the volume of the main pump's flow.

HYDROSTATIC SYSTEM VALVES

COMBINATION VALVES Almost all hydrostatic systems use combination or multifunction valves on each side of the closed loop. These are a combination of relief and makeup (check) valves. Every hydrostatic system will need a high-pressure relief and a makeup valve as part of each side of the closed loop. These valves are usually threaded into the pump housing but quite often appear on a schematic as being separate from the pump.

If a pump is upstroked to provide oil flow to one port of a motor, then one side of the loop becomes the high-pressure side. This side needs to have its pressure limited to a certain level to prevent component and conductor failure. This could be from 4000 to 7500 psi or higher. The high pressure is sensed at the normally closed poppet, which is held closed by spring pressure. If the pressure exceeds the spring value, then the poppet opens and allows the oil to go to the low-pressure side of the loop. This will happen if the load on the machine's drivetrain is so great, it stops the motor's output shaft and the pump is still producing flow.

The makeup valve part of the combination valve opens to allow charge pressure oil into the low-pressure side of the loop. It is a simple check valve that opens to prevent cavitation and overheating of the oil. The makeup oil is to replenish oil lost to drain from component leakage and from flow through the flushing valve.

These valves are usually nonadjustable and must be replaced if found to be defective. ● **SEE FIGURE 17–29** for the combination valves (1) used for a pump assembly.

FLUSHING VALVES Flushing valves are designed to give a constant controlled flow of fresh oil into the closed loop for

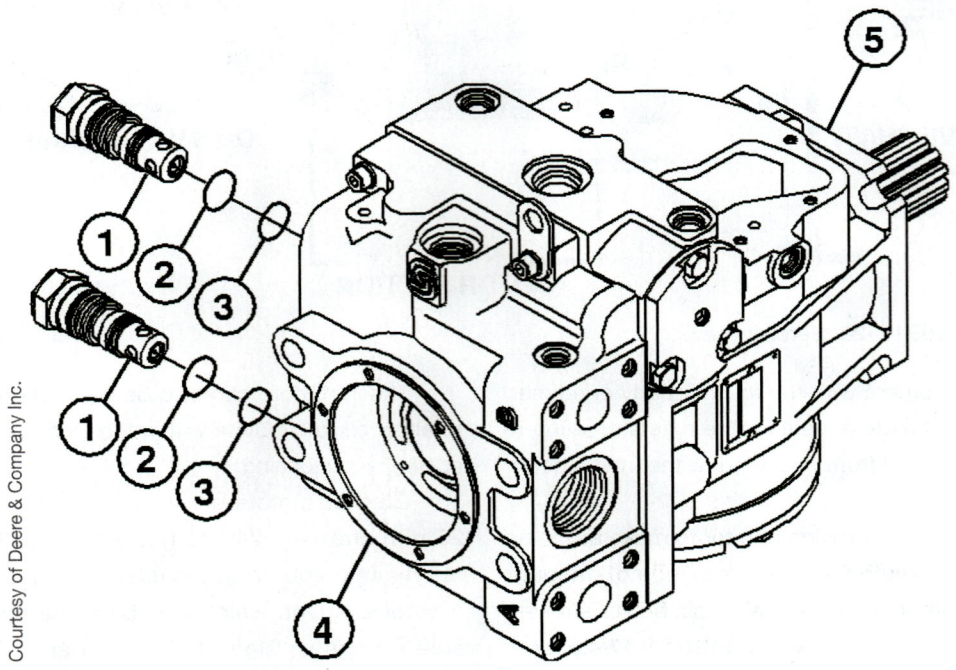

FIGURE 17–29 Combination valve (#1).

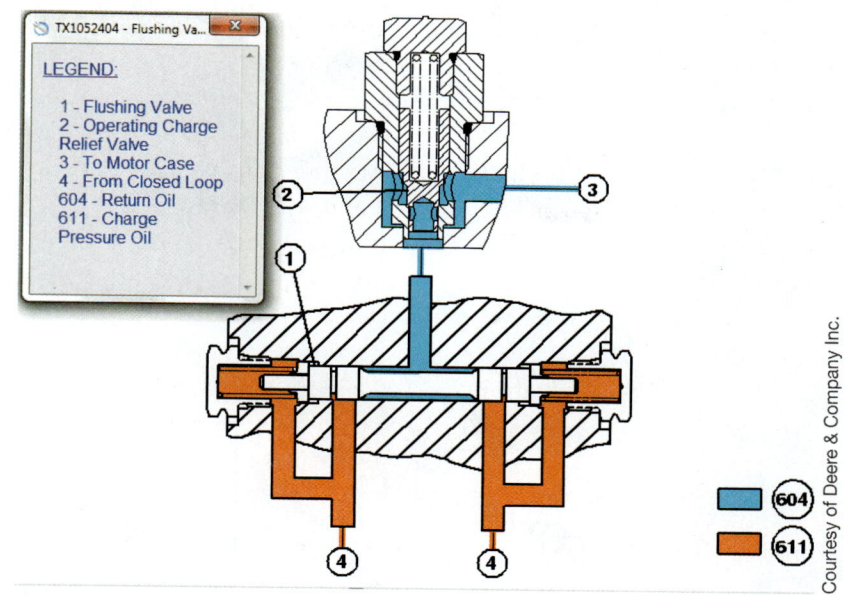

FIGURE 17–30 Flushing valve.

cooling and cleaning purposes. Because a hydrostatic system operates with a closed loop, there is only a small amount of leakage oil that needs to be replenished. If there were no flushing valve, then the oil in the closed loop would eventually overheat and break down. By draining a constant flow of oil out of the loop, the charge pump will be continuously renewing this oil with clean cool oil from the tank. This constant drain flow is created by the flushing valve that is located on the motor housing. The flushing valve is sometimes called a hot oil purge valve and must be set at a lower pressure than the charge relief valve.

● **SEE FIGURE 17–30** for a flushing valve. You can how see the valve will get shifted when the high-pressure side of the loop is sensed at the end of the spool and shifts it.

It is important that the flushing valve pressure is set below the charge relief setting; otherwise there won't be a steady flow of oil. The flushing valve is usually set around 30 psi lower than the charge relief, and this is not normally adjustable.

For example, if there is a leakage drain of 3 gpm from pump and 3 gpm from the motor and the charge pump is producing a flow of 12 gpm, then there will be a constant flow of 6 gpm through the motor housing when the flushing valve is shifted. This flow will carry away heat from the motor and pump to the oil cooler or tank.

CHARGE RELIEF
The charge oil system needs to have a relief valve to limit its pressure. This will be a simple direct-acting poppet-type valve that will be part of the pump assembly if the charge pump is part of the pump assembly. The charge pressure relief setting is adjustable and is usually easily checked through a pressure tap.

FLOW DIVIDER
Some machines that use hydrostatic systems with individual left and right travel motors may use a flow divider between the pump and motors. It will ensure either wheels or tracks get equal oil flow when needed. This will be necessary when the machine is travelling in muddy or slippery conditions. If one pump is supplying two motors, the majority of the flow will go to the motor with the least resistance. If the machine is turning, this will allow a speed difference between the left and right drives much like a differential. However, if the machine starts to spin one drive, then all the oil flow will go to that drive and the machine will get stuck. The flow divider will act like a differential lock to give equal flow to both drives.

BRAKE RELEASE VALVES
Parking brake release valves are usually solenoid-actuated on/off directional control-type valves.

The brake release circuit will usually be supplied oil from the charge pump flow. When the operator wants to release the parking brake, a rocker or toggle switch is moved to send an electrical signal to the solenoid. This will move a plunger that allows oil to flow to the spring-applied brake piston to compress the spring and release the brakes. ● **SEE FIGURE 17–31** for a brake release valve.

OTHER HYDROSTATIC SYSTEM COMPONENTS

CASE DRAIN CIRCUIT
Hystat systems will usually direct case drain oil to the oil cooler because this oil is usually the hottest oil in the loop. This pressure should

HYDROSTATIC DRIVE SYSTEMS 527

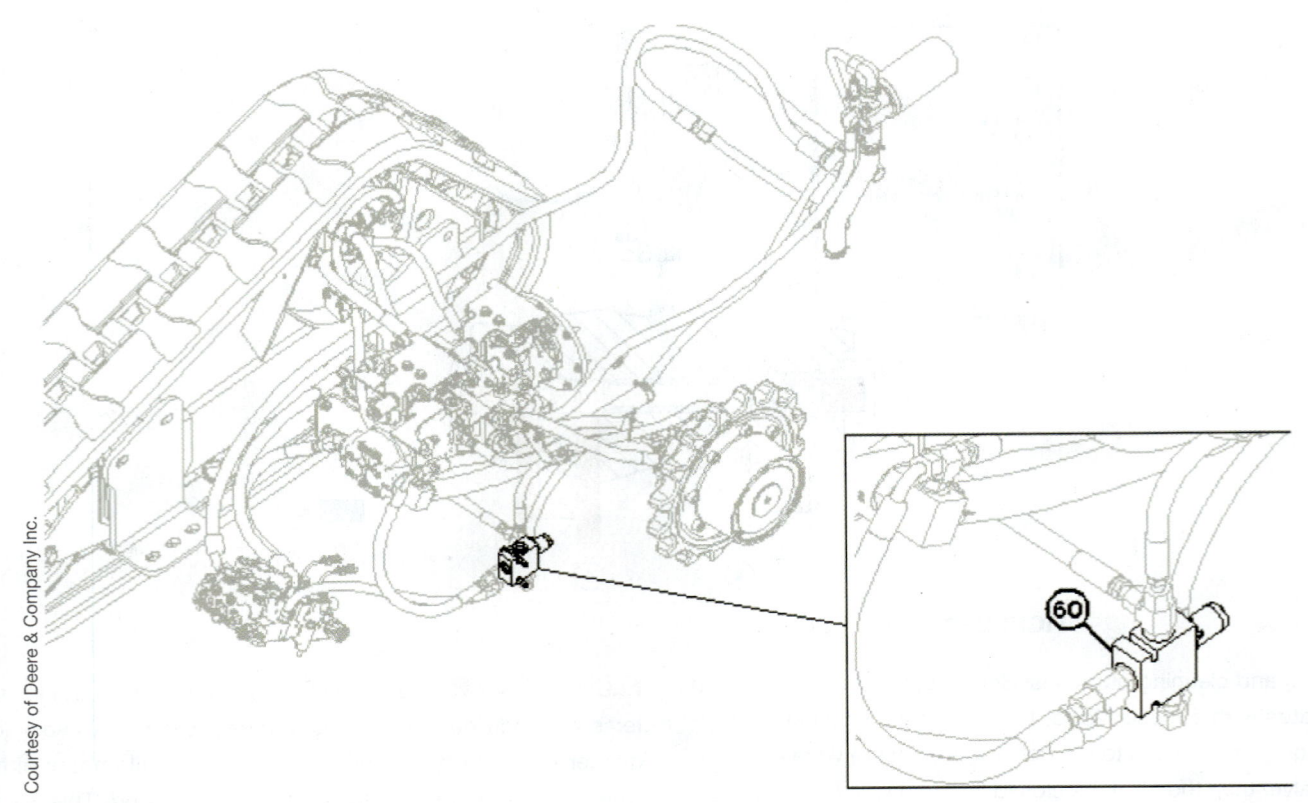

FIGURE 17–31 Brake release valve.

never exceed 60–70 psi for most systems and will quite often have a much lower maximum pressure. Otherwise shaft seals and housing gaskets on the pump or motor could be damaged.

FILTRATION As is the case with hydraulic systems, hystat systems use the mantra "clean, clean, clean." Hydrostatic systems must have the oil flowing through their circuits cleaned to a minimum specific level of cleanliness. This is usually set to an ISO 4406 standard of 22/18/13 minimum (see Chapter 4).

There will be a main filter for all systems, and it is almost always located directly downstream from the charge pump. Some systems will also use a charge pump inlet filter sometimes called a suction filter and have their case drain flow filtered.

All filters should be equipped with filter bypass valves to ensure oil flow even if the filter becomes plugged due to contamination or oil viscosity that is too high. ● SEE FIGURE 17–32 for an oil filter arrangement of a hydrostatic system.

COOLING As hystat systems work with very high pressures, there will be a lot of heat created that needs to be managed. The oil will need to be kept well below 240°F to ensure that the oil doesn't start to break down. Oil will be

FIGURE 17–32 Oil filter.

directed to an oil cooler that has air pushed past it by a fan. The cooler will have a series of tubes that divide the oil flow from one side to the other and the tubes will have small fins on them that dissipate the heat from the oil to the surrounding air. ● SEE FIGURE 17–33 for a schematic with oil cooler.

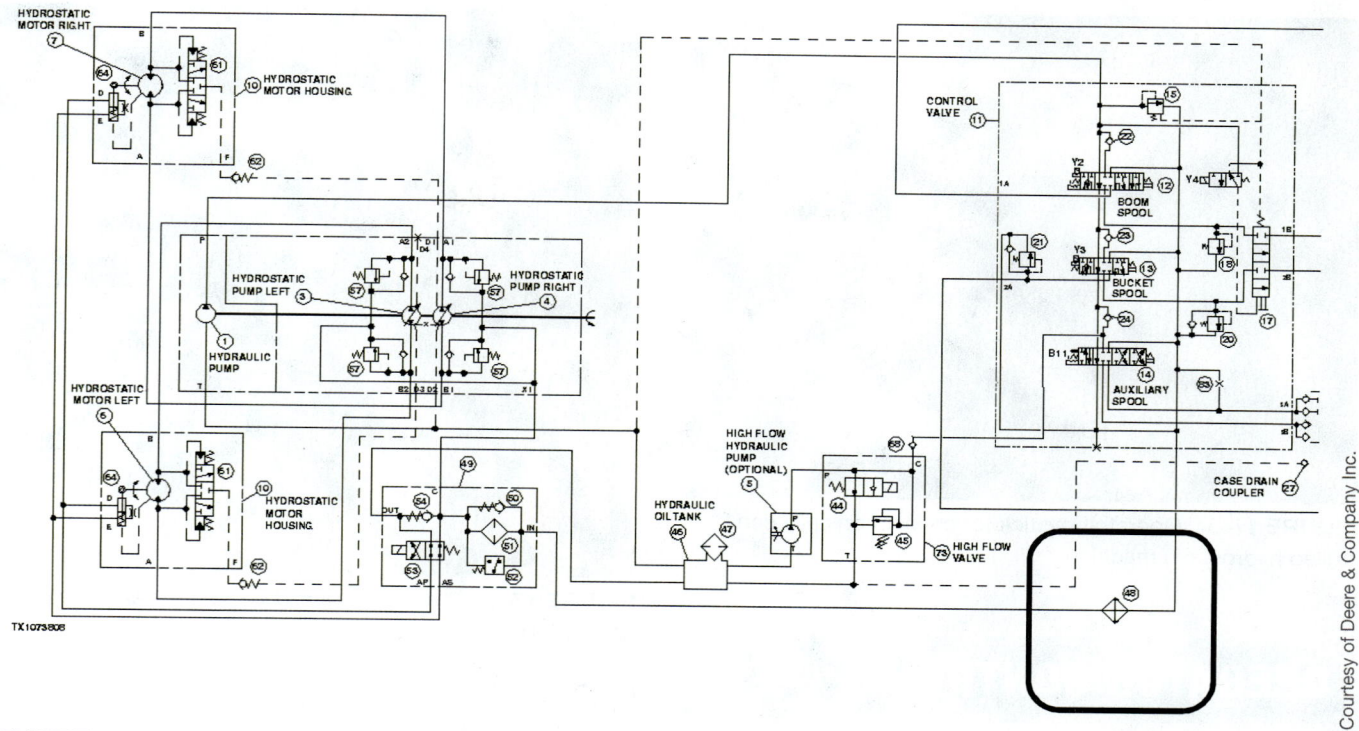

FIGURE 17-33 Schematic with oil cooler.

The fan could be engine driven by belts but will likely be driven by a hydraulic motor that is part of the charge oil circuit. For simple systems, this will be a fixed displacement motor, but for many systems, there will be a variable displacement motor and the circuit could provide a means to reverse the rotation of the motor. This is done to clean out any accumulated debris on the cooler fins.

Cooling circuits may have a bypass valve to allow cold oil to bypass the cooler. The valve would open when the pressure climbs at the cooler inlet to prevent cooler damage. The valve is likely set to open at around 30 psi.

Some systems will have a cooler bypass valve that is electronically managed to make the return oil bypass the cooler before it reaches the tank. This will help to bring the oil up to operating temperature quicker.

HYDROSTATIC FLUID TANK
Hydrostatic system reservoirs are sized to provide the proper volume of oil to keep the pumps and motors supplied with clean and cool oil.

Most small- to medium-sized machines will share the same oil for both the hydraulic and the hydrostatic systems while larger machines will have a dedicated hydrostatic fluid tank. The tank will have a level check device (sight glass, dipstick) for the operator to monitor fluid level. The dipstick or sight glass level marks may be temperature dependent to allow for fluid expansion when it warms up.

Some tanks will have a sensor to warn the operator of low level.

The tank should have a breather to prevent a vacuum being created that could starve the charge pump inlet.

FLUID
It is sometimes said the most important component in a hydrostatic system is the fluid. When engineers design a system, they will recommend a specific type and viscosity of fluid.

The fluid will need to have oxidation, rust, and foam inhibitors in it to protect the hydrostatic system components and ensure their longevity.

Only the specified fluid or one that is directly compatible should be used. Never mix different types of fluids.

Fire-resistant and biodegradable fluids may be allowed for some systems but always check with the manufacturer to be sure that it's acceptable.

Hydrostatic fluid will virtually be the same as that used in hydraulic systems, and for many machines, it will be the same fluid.

A typical viscosity for hydrostatic fluid at an ambient operating temperature of −20°C–40°C is SAE 10W.

HYDROSTATIC DRIVE SYSTEMS

FIGURE 17–34 Operator joystick for an electronically controlled hydrostatic machine.

FIGURE 17–35 Diagnostic port and communication adapter connected to a hydrostatic machine.

OPERATOR CONTROLS

Operator controls for hydrostatic machines vary greatly from mechanical foot pedal controls to electronic joysticks. Only the simplest and smallest machines will still use mechanical controls today. For example, some skid steer loaders will use hand levers that convert forward and backward movement directly into swash plate angle changes. If the machine's hydrostatic system is used for steering, then the operator controls will need to control left and right machine movement as well as forward and reverse.

Pilot oil controls will use low-pressure oil that is typically around 300–400 psi maximum to feed oil to pilot joystick control valves. As the operator moves the joystick, this oil pressure is metered at various lower pressures according to how far the joystick is moved and sent to the pump control piston. The joysticks have two or four poppet valves depending on whether there are one or two pumps.

Electronic joystick controls are simply **position sensors** and will be an input to an ECM. ● **SEE FIGURE 17–34** for a joystick that is part of an electronic control system for a hydrostatic dozer.

HYDROSTATIC ELECTRONIC CONTROL SYSTEMS

There has been a trend over the last several years that see more machines with hydrostatic systems being managed by an electronic system. Hydrostatic electronic control systems will be multiplexed with other machine ECMs. Multiplexing enables sharing of information from sensors by all ECMs on the machine, and this will allow the machine to be more efficient by being able to manage the engine output most effectively. There will also be the capability to create operator warnings if the system operates outside of set parameters, and it will allow a technician to diagnose problems and make adjustments.

To manage the hystat system, there will need to be a number of input sensors and switches to the ECM that may include joystick position, FNR switch, swash plate position sensor, oil temperature sensor, engine speed sensor, motor speed sensor, oil level sensor, filter bypass switch, park brake switch, charge pressure sensor, throttle position sensor, and seat switch. These inputs send electrical signals to the ECM that get processed and compared to stored information in the ECM memory.

The ECM then sends a signal out to a variety of actuators. The main outputs are the solenoids that will control the pump swash plate angle and the motor displacement. Some other outputs include park brake solenoid, pilot enable solenoid, fan drive solenoid, CAN signal, and operator indicator/warning lights/display.

The electronic control system will also provide a way to connect an electronic diagnostic tool through a diagnostic connector port. ● **SEE FIGURE 17–35** for a communication adapter connected to a hydrostatic machine.

Progress Check

7. Tech A says when a hydrostatic motor is in maximum displacement, its output shaft can turn the fastest. Tech B says when a hydrostatic motor is at maximum displacement, its output shaft will make the highest torque. Who is correct?
   a. Tech A
   b. Tech B
   c. both A and B
   d. neither A nor B

8. The most common type of pump used for a hydrostatic system is called the:
   a. bent axis
   b. geroller
   c. radial piston
   d. axial piston

9. Pump flow for a hydrostatic system is reversed by:
   a. moving the pump swash plate past neutral
   b. changing the pump drive gear selector
   c. changing the motor swash plate angle
   d. shifting the directional control valve

10. You would find a flapper valve doing this for a hydrostatic system:
    a. providing makeup oil to the low-pressure side of the loop
    b. allowing oil to bypass the cooler
    c. directing oil to the pump's servo piston
    d. directing oil to the motor's servo piston

11. If a motor is switched from maximum to minimum displacement, this would result:
    a. the motor would produce more speed and less torque
    b. the motor would produce more torque and less speed
    c. the motor would return oil to the pump outlet
    d. the motor's shaft would reverse directions

12. Sometimes a hydraulically driven fan will reverse directions to:
    a. heat the oil
    b. cool the oil faster
    c. reverse the oil flow
    d. clean the cooler fins

13. Tech A says that an electronically controlled hydrostatic system is simpler than a mechanically controlled system. Tech B says any problem can be diagnosed with a laptop for an electronically controlled hydrostatic system. Who is correct?
    a. Tech A
    b. Tech B
    c. both A and B
    d. neither A nor B

14. When diagnosing a problem with a hydrostatic system and a technician checks the case drain pressure for a motor an excessive value would indicate:
    a. the flushing valve is working good
    b. there is excessive internal leakage
    c. the case drain relief valve is stuck open
    d. the oil viscosity is too high

15. A cam lobe–type motor could be:
    a. only one speed
    b. completely variable speed
    c. used for only high-speed applications
    d. high or low speed

DIAGNOSTICS

Hydrostatic system troubleshooting is similar to hydraulic system diagnostics because the same general principles of fluid power apply to both.

1. If there is a low speed problem, there is likely a lack of flow.
2. If there is a weak drive problem, there is likely a lack of pressure.

Operational complaints for hydrostatic machines can be grouped into a few categories:

Machine won't drive

Machine is slow

Drive is weak

No reverse or forward

Machine won't travel straight

Hydrostatic fluid overheating

For any of these or other complaints, the technician would start with verifying the complaint. This may mean that you need to read the operator's manual to understand how the machine should work properly. Many problems will be condition specific, and this will need to be reproduced to perform a proper diagnosis. For example, if a problem is related to the system oil temperature being at a certain level, then you will need to match this such as getting the oil up to 150°F.

Similar to all troubleshooting procedures, don't overlook the simple checks. For any type of complaint, simple checks include:

Check oil level and condition.

Check for fault codes if applicable.

TECH TIP

I was called to a hydrostatic drive track loader once that was said to have a bad fluid leak. An initial inspection didn't reveal anything obvious, but it was clear the hydrostatic fluid was low. After topping it up with 5 gallons of fluid, I ran the machine and checked for leaks again. Once again there was nothing leaking. I told the operator to keep running the machine but keep a close watch on the fluid level and look for any signs of leakage.

I returned the next day to find the level low, added another 5 gallons, checked for leaks, and repeated my instructions to the operator.

This went on for another six days before a leak finally appeared. It was coming from a door hinge fastener on the machine's main frame! This finally made me realize what was happening and was confirmed when the machine was sent to the shop for repairs.

One of the track motor's brake housings developed a crack and the oil was leaking into the cavity of the main frame of the machine. There was no way to confirm this until the oil started leaking out of the frame!

Needless to say, the inside of the frame will never rust out.

FIGURE 17–36 Test ports.

FIGURE 17–37 Case drain filter.

- Check for external leaks.
- Check for physical damage to components.
- Check track tension for track drive machines.
- Check tire pressure for wheeled machines.
- Check operator controls and interlock system for proper operation.

If there is a speed problem, most manufacturers will provide a test procedure to confirm whether there is a problem. In general, this involves getting the oil to operating temperature first, ensuring the engine rpm is as specified, and making sure that the controls are functioning properly. The machine's brake system must also be confirmed to be released completely. Track machines also must have proper track tension because tracks that are too tight will rob horsepower and cause weak or slow travel.

Wheel or track speed can be measured with different methods and mainly depends on the size of the machine. For skid steer loaders with wheels, they can be put on stands and the wheel speed is measured with a photo tachometer.

System pressure is usually tested by keeping the brakes applied and trying to drive the machine while monitoring the high-pressure side values. ● **SEE FIGURE 17–36** for a group of test ports that can be used to check hydrostatic system pressures.

Case drain pressure or flow could be tested and compared to specifications. If the results exceed specifications, this would indicate internal wear and most often leads to pump or motor replacement. If there is a **case drain filter**, the filter can be removed, cut open, and inspected. ● **SEE FIGURE 17–37** for a case drain filter with excessive debris accumulated.

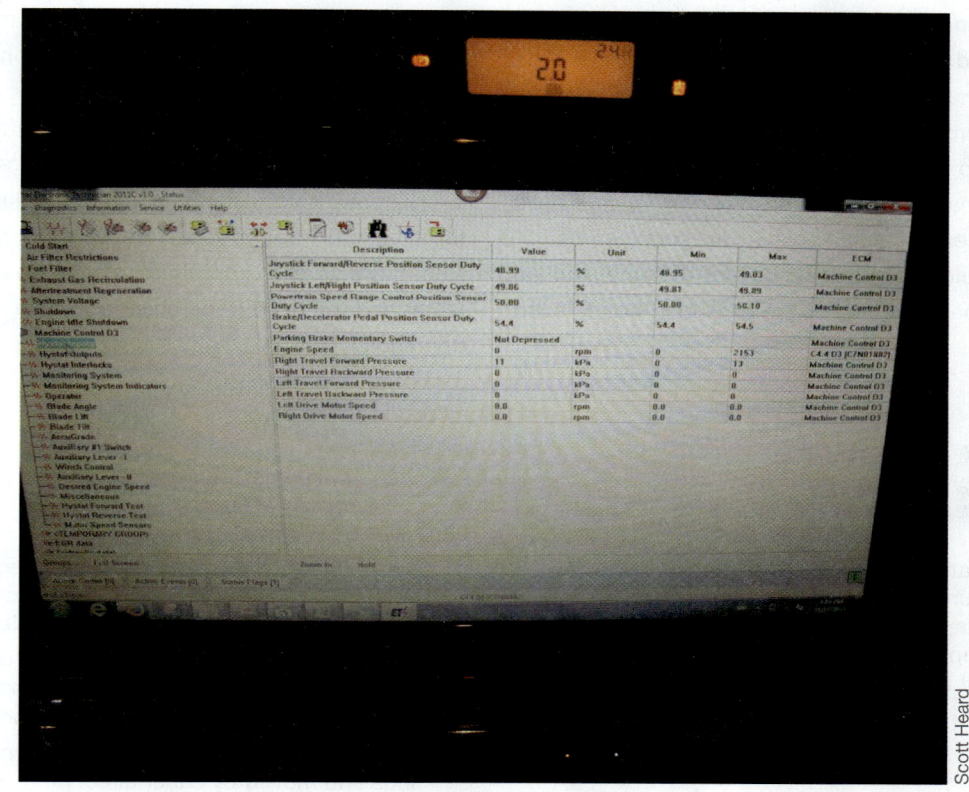

FIGURE 17–38 Hydrostatic drive system being calibrated.

HYDROSTATIC SYSTEM ADJUSTMENTS AND REPAIRS

There could be adjustments made to hystat systems that will restore proper operation.

One common adjustment that is made to hydrostatic pumps is neutral adjustment. This will ensure that the machine doesn't creep when the travel controls are in neutral.

If a machine has left and right drives and isn't driving straight when requested, then a tracking adjustment may be needed. This will match the speed of both sides to make the machine travel straight. This could be a linkage adjustment for machines with simple controls or could be an electronic calibration procedure performed with a laptop computer.
● **SEE FIGURE 17–38** for a laptop being used to calibrate a hydrostatic drive system.

COMPONENT REPAIR Once a hydrostatic system component is found to be defective, it will need to be repaired or replaced. Pump and motor repairs should be performed only if there is a clean, well-lit shop that has the necessary tools to complete the repair. The technician should also be comfortable with these types of repairs and with using measuring tools because the decision to reuse many parts will be based on precise dimensions.

HYDROSTATIC SYSTEM POST REPAIR START-UP Once a hydrostatic system has been repaired, you will need to

> **SHOP ACTIVITY**
>
> - Go to a machine with hydraulic motors, follow the main lines that go to it, and determine if the machine uses an open or closed circuit system to drive the motor.
> - Find a machine that uses a closed loop hydrostatic system and describe how it is controlled, how the oil is cooled, and whether the motors are fixed or variable displacement. For the same machine, describe three test procedures and list three tools used to diagnose a travel problem.

HYDROSTATIC DRIVE SYSTEMS 533

follow a specific procedure when first starting and running the machine to avoid a repeat failure or causing another type of failure. Always check the machine manufacturer's service information for this procedure. Generally speaking, you need to make sure that pump or motor housings are filled with clean oil before starting the machine. The pump inlet must always have all air bled from it to ensure that it gets a solid supply of oil.

All filters should be replaced before starting and oil cleanliness should be confirmed or the oil should be replaced to be sure.

If there was a catastrophic failure, you may need to dismantle the entire system and clean all components. Sometimes a kidney loop procedure will need to be done as well. This means filtering the system oil with an external filter and monitoring contamination levels until they fall within an acceptable range.

There may be cleanout filters available to be installed for initial start-up. These filters will have finer micron ratings to capture smaller contaminants. Theses filters will plug up faster and will need to be changed within the first 10 hours of operation.

SUMMARY

1. Some safety related concerns related to hydrostatic systems are: pressure must be released before servicing or repair work is started, heavy components must be lifted with proper equipment, raised cabs must be secured. Lock out/tag out and wheel chocks must be used.
2. Hydrostatic systems use a prime mover (diesel engine) to drive a pump. The pump sends oil directly to a motor and the motor shaft can be used to turn a variety of components. Hydrostatic travel systems have the motor drive final drive components that then drive tires or tracks. Return oil from the motor is sent back to the pump inlet and this is called a closed loop system.
3. Motor speed is changed by sending more pump flow to the motor. Motor shaft direction is changed by reversing pump flow. Hydrostatic pumps therefore are considered to be bi-directional.
4. There are many different hydrostatic drive configurations used for machine drive trains such as: single pump/motor; dual pump/dual motor; motors driving drive axles; motors driving chain drives.
5. Dual path hydrostatic systems have two pumps and two motors. Track type tractors will use this system to provide drive and steering.
6. Motor rpm can be calculated with the formula: Flow (gpm) $\times$ 231/ motor displacement (cubic inches/revolution).
7. The formula for motor torque output is: Torque (inch pounds) = PSI $\times$ Displacement (cubic inches/revolution) / 2π.
8. System pressure is created by resistance on the motor's shaft and limited by relief valves. System inefficiency is a result of internal leakage and is typically 10–15%. Internal leakage is returned to tank through case drain conductors. It is sometimes filtered as well.
9. Hydrostatic systems can give dynamic braking since the motor is in effect turned into a pump and the pump resists being turned faster by the prime mover.
10. Parking brakes are quite often part of the motor assembly and are spring applied oil released.
11. Hydrostatic pumps are bidirectional axial piston type. They have a swashplate that can stay at 0 degrees which gives the pistons a zero effective stroke. The operator controlled swashplate can then angle in either direction which will send oil out either port of the pump to the motor. This will provide forward or reverse oil flow. Greater angle = greater flow.
12. Pump pistons are carried around in a cylinder block and oil is sealed by a port plate that is lap finished. Port plates openings align with pump housing ports.
13. Swashplates are supported in the pump housing by bearings and moved by either direct mechanical linkage, a hydraulic servo piston or an electrohydraulic servo system.
14. There are many variations of pump control systems. One way is to have a cable actuated spool valve direct oil to a servo piston that then moves the swashplate.
15. Hydrostatic motors can be different types such as: axial piston, bent axis, cam lobe or orbital type.
16. Motors can be fixed displacement or variable displacement. To increase speed a motors displacement will be reduced.
17. Motors incorporate a flushing valve that maintains a constant flow of oil through the housing to cool and clean the motor. This is called case drain oil and can be filtered before it returns to tank.
18. All hydrostatic systems need a charge pump to ensure the closed loop is kept full of oil since there will always be some internal leakage. The charge pump is usually a gear type pump and provides roughly 20% of the flow volume.
19. Combination valves are used on each side of the loop to provide high pressure relief and allow charge oil in depending on the state of pressure at the time.
20. Flushing valves ensure a minimum amount of oil leaks internally from the closed loop. They are set at a pressure slightly below charge relief pressure.
21. Charge relief valves limit the charge supply pressure and are usually in the 300– 500 psi range.
22. Flow dividers can be used in some systems to ensure oil flow is equal to two motors that are fed from one pump.
23. Brake release valves send oil to spring applied parking brakes.
24. Oil filtration is critical to component longevity and usually is done on the charge pump outlet. Case drain oil can also be filtered.

25. Oil cooling circuits will ensure oil is normally kept below maximum operating temperatures.
26. Hydrostatic systems can have their own oil reservoirs or may share the machines hydraulic system's.
27. Hydrostatic fluid has properties that are the same as most hydraulic systems and typical viscosity is 10 W.
28. Operator controls vary from handles that are directly linked to pump swashplates to pilot oil controls to electric joysticks.
29. Electronic control systems can have inputs to monitor position, pressure and temperature and outputs that control swashplate position and fan speed.
30. A few examples of operator complaints related to hydrostatic systems are: no drive, drifting when travelling, overheating, noises, vibrations.
31. Diagnostic procedures follow the standard practice of: knowing the system, gathering information, verifying the complaint, list possible causes, find the root cause, repair the problem, verify the repair.
32. Simple checks include checking oil level, look for leaks, check tire pressure or track tension, check control movement, check for fault codes.
33. If a speed related problem is being diagnosed pump flow may need to be measured with a flowmeter.
34. If a low power problem occurs a pressure check is needed and parking brakes may hold the machine.
35. Case drain pressure or flow checks could indicate excessive component wear.
36. Adjustments can be made to hydrostatic systems to make the machine travel straight.
37. Component repair/replacement is the same as hydraulic systems'. Cleanliness is critical and startup procedures must be followed.

HEAVY-DUTY HYDRAULIC FRICTIONAL BRAKE SYSTEMS

LEARNING OBJECTIVES

After reading this chapter, the student should be able to:

1. Explain braking principles as applied to heavy equipment machines.
2. Identify safety concerns related to hydraulic brake systems.
3. Describe how hydraulic system fundamentals relate to hydraulic brake systems.
4. Describe the advantages of a hydraulic brake system over a mechanical brake system.
5. Identify and explain the operation of non-boosted brake system components.
6. Identify and explain the operation of boosted brake system components.
7. Describe basic hydraulic brake system troubleshooting procedures.
8. Explain how kinetic energy is converted to heat energy.
9. Identify the different types of hydraulic foundation brakes.
10. Describe the operating principles of foundation brakes.

KEY TERMS

Accumulator 537
Actuator 544
Caliper 538
Charging valve 551
Dynamic braking 538
Foundation brakes 538
Levers 543
Master cylinder 544
Multi-disc 544
Non-servo 545
Pump 551
Self-energizing 546
Servo 545
Static braking 538
Torque 540
Volume 552
Wheel cylinder 545

INTRODUCTION

SAFETY FIRST Hydraulic brake systems sometimes use hot, high-pressure fluids; high-energy springs; and harmful friction materials to transfer large amounts of energy. There is a high potential for injury or death if a technician isn't proactive in preventing accidents.

Incidents can happen as a result of improper testing methods or use of improper test equipment such as using underrated hoses or gauges. Use of inadequate replacement parts such as hoses, seals, or tubes can result in the rupture of these components and not only allow the escape of hot, high-pressure fluid but the sudden uncontrolled movement of a machine could have disastrous results. A little common sense and proper use of PPE will go a long way to keep you safe when working on hydraulic brake systems. You should also always refer to the machine's specific manufacturer's service information before performing any servicing or repairs to hydraulic brake systems. There will be lots of warnings and cautions on machines and in service information to heed so don't overlook them and become another statistic. Here are a few types of specific safety concerns that you should be aware of when working with hydraulic brake systems and some tips to avoid hurting yourself or others and staying healthy:

FRICTION MATERIAL DUST Caution should be observed when servicing/repairing brake components that use dry friction materials. Although most friction material used these days do not use asbestos, there can still be harmful materials used that become airborne when disturbed and if ingested could cause serious health problems. The safest way to work around these materials is to assume they contain asbestos and use the appropriate shop equipment and PPE.

OIL INJECTION Oil injection injuries can be deadly. Refer to Chapter 4 for information regarding these injuries.

TRAPPED POTENTIAL ENERGY Many braking systems use **accumulators** that store high-pressure oil. Make sure that you are aware of any accumulator in the system before servicing or repairing it. Ensure that all pressure is bled off before opening the system. Many braking systems will use spring pressure. Use extreme caution when working with springs that are not relaxed.

CRUSHING HAZARDS Because the brake system is designed to slow, stop, and hold a machine, it is very important that the machine be secured from uncontrolled movement when you are working on the brake system. This means using wheel chocks, proper lifting, and blocking techniques and tooling. Proper lock out tag out procedures must also be used when working on braking systems.

BURNS Brake components will get very hot because of what they do and as such will need to be handled with care if the machine has been moving recently. Appropriate PPE should be used.

SLIPS AND FALLS When working on a brake system, you could be climbing up and down from the machine's cab with tools and gauges. Take care not to fall by keeping three-point contact and use fall-restraint harnesses when necessary. Any oil spill should be cleaned up immediately to avoid a slip hazard.

FIRE HAZARDS Most brake systems use hydraulic fluid that is mineral-based oil, which is flammable. There have been more than one machine catching fire and burning up because of a hydraulic leak that has either been sprayed onto a turbocharger or been ignited by welding sparks or a torch. You need to be careful when welding or using a torch on a machine that has a brake hydraulic leak. Brakes can also become hot enough to ignite components so be aware of excessive brake temperatures.

ENVIRONMENTAL CONCERNS A hydraulic brake system should be a sealed system, and if it is intact, there are minimal environmental concerns. However, if your actions as a technician cause a fluid leak by opening a system, the fluid leak *must* be contained. Drain buckets or trays, fluid adsorbent cloth, and floor dry should be used to contain leaks. The result of not containing fluid leaks can be very harmful to the environment and very costly, with cases of fines of $20,000 per square foot reported.

BRAKE SYSTEM INTEGRITY The braking systems on any piece of heavy equipment machinery could be considered the most important safety feature of that machine. If a technician fails to repair a brake system fault properly, there are huge potential negative consequences. From monetary damage to machines and surrounding equipment to injury and death of operators and workers, a machine's brake system must be as functional as it is designed to be. From simple things such as properly checking brake fluid level to brake system air bleeding, there are many crucial checks and procedures that need to be carefully followed to ensure that maximum brake performance is available. There are several videos that are available on the Internet that demonstrate the serious results of brake system failure. Machines will usually have a fail-safe brake system. This means that if there is a major failure of the service brake system, there should be a secondary brake system that will bring the machine to a controlled safe stop. This should happen automatically, and the secondary system could also have a backup system. This would be called a triple redundant system.

NEED TO KNOW

BRAKES FOR HEAVY EQUIPMENT The automotive industry has had a mostly consistent configuration for frictional/hydraulic braking systems. Generally, hydraulic **calipers** and rotors are used for the front wheel braking action, and the same arrangement or drum brakes are used for the rear wheel brakes. The brake components that actually perform the braking action are called the **foundation brakes**. These systems are usually vacuum boosted to create a more powerful brake system. Vehicles larger than a standard half-ton pickup truck could use deviations from this standard and highway trucks that are considered to be 5-ton and larger will mostly use an air brake system that generally use air chamber actuated S-cam drum brakes for the foundation brakes (see Chapter 22).

These are general statements that overlook the influence lately of electronics being incorporated into brake systems to give anti-lock, stability, and traction control.

To be able to make general statements like this in relation to heavy equipment brake systems is impossible. If a method of slowing or stopping a machine can be imagined, it is probably used somehow in the braking system on a piece of heavy equipment. One machine can have as many as four different types of brake systems used to slow it down, stop it, and hold it in place. This chapter will focus mainly on hydraulically actuated, spring released and spring actuated, hydraulically released types of brakes that are used on wheeled machines. Brakes can be used for different functions on heavy equipment such as steering the machine and for controlling winches, but this chapter will deal with brakes that are used for slowing and stopping moving machines (**dynamic braking**) and holding machine's stationary (**static braking**). ● SEE FIGURE 18–1 for a rock truck traveling downhill.

Brakes that are used to slow a moving machine under normal conditions are called service brakes while brakes used to stop a machine in an emergency or to hold a parked machine stationary are called secondary brakes.

FIGURE 18–1 Rock truck going downhill.

Brakes systems can use different energy sources such as air, hydrodynamic, electromagnetic, and engine compression, but this chapter will discuss only brake systems that use hydraulic and spring pressure.

The kinds of machines that use these types of brake systems are the following:

Track-type machines: for steering and static braking
Wheel loaders: for service and parking brakes
Mining trucks: for service and parking brakes
Articulated trucks: for service and parking brakes
Graders: for service and parking brakes
Forestry machines: for service and parking brakes
Backhoe loaders: for steering, service, and parking brakes
Fork lifts: for service and parking brakes

BRAKING FUNDAMENTALS Any brake system used to slow down (dynamic braking) and stop (static braking) a vehicle in motion is merely an energy conversion machine. The law of conservation of energy states that energy can't be destroyed, it can only change states. For a heavy equipment machine, an energy source (usually diesel fuel) gets converted into heat energy by the machine's drivetrain to create motion, and once the machine is moving, its momentum or inertia wants to keep it moving. This is called kinetic energy. If the machine was to continue to coast on a level surface, then frictional losses in the drivetrain would overcome the kinetic energy and the machine would eventually stop moving. To allow the operator to bring the machine to a controlled stop or decelerate on command, there has to be a frictional brake system to convert the kinetic energy back into heat energy. This heat energy is ultimately dissipated to the atmosphere on most machines.

FIGURE 18–2 Hybrid wheel loader.

Hybrid drivetrain systems try to recycle the braking energy that is normally wasted. ● **SEE FIGURE 18–2** for a hybrid wheel loader.

The kinetic energy of a moving machine is converted into heat energy by friction. A simple example of how friction is turned into heat is by imagining what you do if your hands are cold. By rubbing your hands together, you are creating friction and heat. The faster you rub and the harder you press them together, the more heat you create. This is the same principle used for the dynamic braking system of a machine. As mentioned, brake systems are designed to create heat; however, if this heat isn't dissipated properly, it becomes excessive and will then transfer into the brake system and other related components. Premature failure of these components is likely to occur soon after.

Smaller machines will rely on the surrounding air to cool brakes while larger machines will use oil to carry heat away from brake components.

The system and component that creates braking friction for heavy equipment is the topic for this chapter.

The actual braking effect of a wheeled machine takes place between the surface it is traveling over and the tire. If the amount of friction applied to by the foundation brake is higher than the friction available between the tire and the traveling surface, the tire will skid. If the amount of friction applied by the foundation brake is slightly less than the amount needed to make the tire skid, this will give the maximum amount of deceleration. An operator should try to avoid skidding the machine's tires as this is actually reducing braking effect and also reducing the life of some very expensive tires. Some machines will now feature an anti-lock system that will release the brake momentarily if it sensed to be skidding the tire.

BRAKING EFFORT Brake effort is usually measured in inch pounds and can be thought of as the opposite force that is created by an engine's crankshaft, which is measured in foot pounds. If a heavy-duty diesel engine can produce 2000 ft-lb of torque at the flywheel, the opposite but equal value of brake effect would be 24,000 in.-lb. While an engine's crankshaft is trying to twist or rotate a load at the start of the machine's drivetrain to move the machine from a stop or accelerate it to a higher speed, the brake system is trying to slow or stop rotation at the opposite end of the drivetrain to slow or stop the machine's motion.

The brake effect that will be discussed in this chapter is created by frictional forces that vary depending on a force

TECH TIP

Here's an example of what can happen when excessive heat builds up from a malfunctioning brake system:

A limestone quarry that I worked at had a large wheel loader that had a leak at the front axle. I was asked to see where the leak was coming from and determine how to fix it. When the machine was closely inspected, it was found to be leaking at both front wheel seals. When the axle was filled up, the oil would leak out fairly steadily so I reported that both wheel seals had to be replaced. I also noticed that the axle oil smelled burnt. The foreman wanted the machine to go back to work, so the axle was topped up and the machine was given back to the operator. I noticed when the operator pulled the machine out of the shop and went from reverse to forward that it rocked more than usual. I asked the operator to stop and then got in the cab myself. Upon further investigation, I found that the brake pedal was sticking down slightly because of a mud build up on the floor of the cab.

The sticking pedal had caused the brakes to always be slightly applied, and the longer the machine was run like this, the more heat built up. This excessive heat eventually caused the wheel seals to fail, which resulted in a several thousand dollar repair and the machine to be down for an unplanned repair.

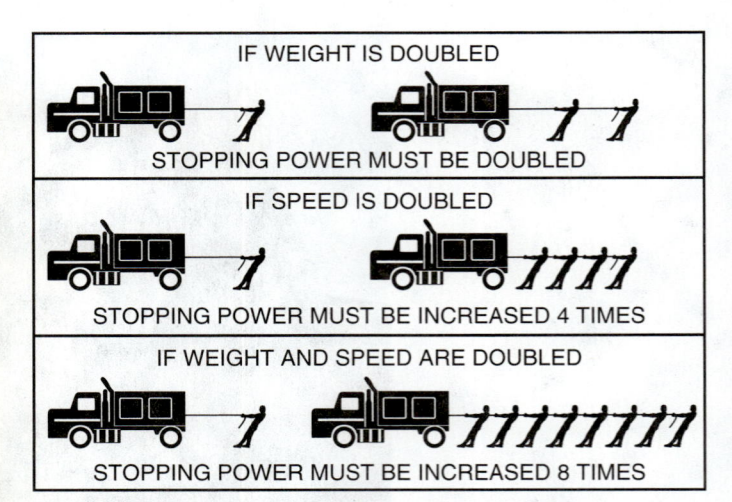

FIGURE 18–3 How speed and weight affect stopping distances.

FIGURE 18–4 A primitive friction brake system.

applied to a friction material against another material (usually steel). One of the two components will need to be stationary, and as pressure is applied, friction increases, and the speed differential between the two components will get reduced.

Braking effect can also be measured in horsepower. The amount of horsepower consumed by trying to slow a moving machine is massive when compared to its engine's horsepower. Horsepower is calculated based on the amount of rotational force (**torque**) created and the speed (rpm) or rate of time that force is spinning at.

To stop a machine from full speed in a very short distance requires a huge braking effort. For example, consider a rock truck loaded with 400 tons of material and traveling at a top speed of 42 mph. The total weight of the vehicle is 1,375,000 lb, and it uses a 4000 hp engine to propel the machine. Its total brake surface area is just over 52,000 in.² The 4000 hp engine's torque is sent through the machine's drivetrain and is able to take the loaded truck from 0 to 42 mph in 60 seconds. If the truck is required to stop from this speed in less than 6 seconds, the brake system must be capable of converting 40,000 hp worth of kinetic energy back into heat. Although the amount of heat energy created is massive and hard to fathom, it is still just created by the friction of the rotating brake material against stationary steel.

If a machine's speed is doubled, it must have four times the braking power to stop it, and if its weight is doubled, it must have twice the braking power to stop in the same distance. ● **SEE FIGURE 18–3** for how speed and weight affect stopping distances.

In older, very primitive brake systems, the effect was created by forcing a stationary piece of wood against a rotating steel rim. ● **SEE FIGURE 18–4** for a primitive braking system.

The amount of braking effect created by these two materials is determined by their coefficient of friction and the clamping force applied to the block of wood. Coefficient of friction is a major factor in determining the rate of deceleration a braking

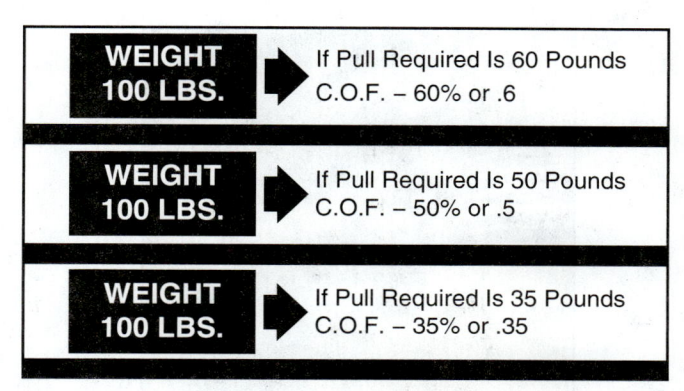

FIGURE 18–5 Different coefficient of frictions.

FIGURE 18–6 Different friction materials (reprinted courtesy of Caterpillar Inc.).

system can create. The amount of force applied is determined by mechanical advantage (leverage), spring pressure, hydraulic pressure, or air pressure.

Coefficient of friction is defined as the force required to move a material of a certain weight across the surface of another material. ● SEE FIGURE 18–5 for an example of this.

If it takes 28 lb of force to move the 100-lb block of hardwood across a steel plate, then the hardwood is said to have a coefficient of friction of 0.28. If another type of material (that also weighs 100 lb) takes 45 lb of force to move it across the same steel plate, then its coefficient of friction is 0.45.

If the friction material is intended to be used in a dry state and a lubricant is introduced, this will lower the coefficient of friction. For example, this will happen if there is a leak from the brake hydraulic system onto the friction material.

Some friction material is designed to be operated wet or have oil circulate around it. The oil circulating around the brake material will carry heat away from the friction creating components.

Brake friction material manufacturers test their materials and will list their products with a coefficient of friction. The higher the value, the higher the force required to move the material across a steel surface and, therefore, the better braking effect you would get for that material. The trade-off for this is usually a shorter friction material life span and more heat generated.

Brake friction materials consist mainly of non-asbestos organic or metallic materials (fiberglass, Kevlar, ceramic, etc.), friction modifiers (alumina or silica), and a resin binding agent that keep these compounds together. The materials are combined and cured in a baking process and formed into different shapes. These shaped friction materials are then attached to a metal backing that makes a disc, shoe, or pad assembly. They are chemically bonded (glued), riveted, or both to the metal backing. These assemblies are then attached to the machine as part of a stationary or moving component of the brake system. Friction material can withstand intense high temperatures, but if the operating temperature gets close to the material's manufacturing temperature, the material will start to break down. If you have ever watched a car race at night and are observing the action at a hard braking zone, you may have noticed the brake rotors glowing red soon after the driver applies the brakes. This is a good example of the heat energy that has been created by the brake system to slow down the speeding race car. The friction material for racing applications must be formulated to withstand these extreme temperatures.

In Europe, there is a racing series that features large highway trucks that are highly modified and reach high speeds. To keep the friction material from melting, the teams will install a water spray system to help cool the brakes after a heavy application. You can view several videos of these trucks in action and see the steam coming off the brakes as the truck is braked hard into a corner.

Generally, softer friction material will have a higher coefficient of friction but will not last as long as harder material that has a lower coefficient of friction. ● SEE FIGURE 18–6 for different friction materials.

FACTORS THAT AFFECT BRAKING To increase braking effort, there are four factors that can each be increased to accomplish this.

Braking leverage: The larger the radius that the friction material works on away from the center line of the component being braked means higher brake torque or more effective braking. This factor is part of the design of the machine and will be determined by engineers to give the brake system enough torque to properly slow and stop the machine. An automotive example would be the larger diameter brake rotors that are installed on high-performance cars that increase braking leverage and, therefore, braking performance.

Total swept area: The brake swept area is the total surface area that the friction material has in contact with its opposing brake surface. An increase in the swept area means more friction surface area that is creating the brake effect and, therefore, better brakes. This is also determined by engineers and is not an item to be modified. Some articulated rock trucks will use as many as three sets of calipers per rotor to increase the braking effort by increasing the swept area of friction material exposed to the rotor. ● SEE FIGURE 18–7 for a rotor and dual caliper brake arrangement used on an articulated truck.

Coefficient of friction: If the friction material has a higher value, this means more grab or more brake effect. This is another item that is not typically changed from what the engineers originally designed to be used for a particular machine. If there were instances of less than expected life of friction material, it may be possible to buy different types of material, but this isn't very common for heavy equipment machines.

Higher clamping force: More force applied to the friction material means more brake effect. Whether this force comes from springs or a hydraulic system, it must be as high as originally designed or the machine's brakes will not be performing as they should. This part of the brake system is the one area that a technician will have the most influence on.

For the service brake system, the clamping force must be infinitely variable from zero pressure to maximum system pressure as requested by the operator. If operators can't control the clamping force as they should, there will complaints like the brakes lock up or won't slow the machine as much as desired.

For the parking or secondary brake system, the clamping force needs to be as high as originally designed. Anything less than this will mean the parking or secondary braking system is not working properly and the machine may roll away uncontrolled after the parking brake is applied.

The clamping force applied to the friction material can come from three sources:

1. Mechanical linkage (lever applied, spring released)
2. Hydraulic fluid pressure (hydraulic applied, spring released)
3. Spring force (spring applied, hydraulic released)

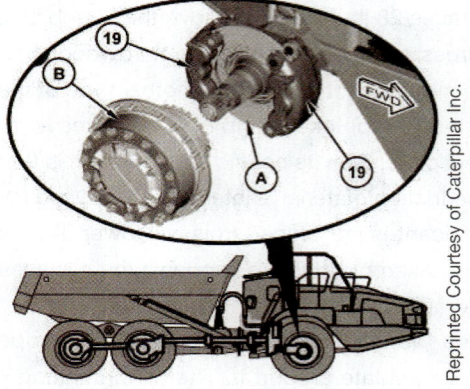

FIGURE 18–7 Rotor and dual caliper (reprinted courtesy of Caterpillar Inc.).

The key to keeping the braking effect as it was designed is to keep the brake system components relatively cool, use the recommended fluid, keep components adjusted properly, and keep them in good condition.

Progress Check

1. Ideally, secondary or parking brakes will always be:
   a. oil applied
   b. spring applied
   c. air applied
   d. air released
2. When vehicle speed doubles, its braking force must increase ___ times to stop in the same distance.
   a. 4
   b. 8
   c. 10
   d. 12

3. If it takes 33 lb of force to move a 100-lb block of friction material across a surface, it is said to:
   a. need 33 psi applied to it to stop it
   b. have a surface coefficient of friction of 0.33
   c. need more force to keep it going than to stop it
   d. have a surface coefficient of friction of 3.3
4. If a machine has a faulty service brake system, this will be the result:
   a. the machine's brakes are overheating
   b. the machine will roll on a hill when the parking brakes are applied
   c. the machine will slow down normally
   d. the machine will not slow down normally
5. This is one factor that will *not* increase a machine brake system braking effect:
   a. increase the clamping force on the friction material
   b. increase the friction material coefficient of friction
   c. increase the brake fluid viscosity
   d. increase the swept area of friction material

FRICTIONAL BRAKE HYDRAULIC SYSTEMS

To apply or release a machine's dynamic hydraulic brake system, the operator controls the flow of hydraulic fluid to do one of two things. The fluid will either force a friction material against a steel component to slow down and stop a machine or release (drain) fluid pressure to allow spring pressure to force a friction material against a steel component. If the operator is sending fluid to move a component against spring pressure, this fluid is getting modulated to differing pressures by the operator. If the operator is releasing fluid pressure or draining it, this is called a reverse-modulating system.

A reverse-modulating system will have the brake's spring applied and fluid pressure is then applied to compress the spring and release the brakes. As this pressure is then released, this will allow the spring to apply the brakes. This is one way to provide a fail-safe type of brake because if the hydraulic system fails, then the brakes are automatically applied by spring pressure.

For a machine that uses a hydraulic system for its parking brake system, the operator will direct fluid to release a spring-applied brake.

These brake hydraulic systems can be very simple or very complicated, and many machines will now incorporate electronics into their brake systems.

WHY USE HYDRAULIC SYSTEMS FOR BRAKING?

A hydraulic system as it applies to heavy equipment brakes can be loosely defined as any system that transfers mechanical energy into a confined liquid that then is used to transfer that energy back into mechanical energy to perform work. As mentioned in Chapter 5, a hydraulic system is like a fluid lever that can be used to multiply input force to get an increased output force. This increased output force is needed to apply machine brakes with the force necessary to slow down and stop a moving machine.

Service brake application usually starts with pedal movement. ● **SEE FIGURE 18–8** for a typical hydraulic system brake pedal and valve.

Machine brake systems will usually use mineral-based oil but sometimes automotive-type brake fluid is specified. These fluids are virtually non-compressible, and this will give the brake system an immediate and powerful response to the operator's input. Air, however, is compressible and is not something you want in a brake hydraulic system. If a hydraulic brake system is suspected to have air in it, then a proper bleeding procedure should be followed to remove all traces of air from inside the system. If air does get into a brake hydraulic system, the brake pedal will feel spongy and the brake performance will not be as responsive or effective as it should. This can lead to a dangerous situation and should be corrected as soon as possible.

USE OF HYDRAULIC SYSTEMS FOR MACHINE BRAKING
As mentioned in the introduction, early primitive friction-type brake systems relied on strictly mechanical linkage to provide the force needed to create enough friction to slow and stop the machine. The limiting factor to the required brake effect was how much force the operator could apply to the linkage and how much this could be multiplied by the mechanical advantage that was created through the use of **levers**.

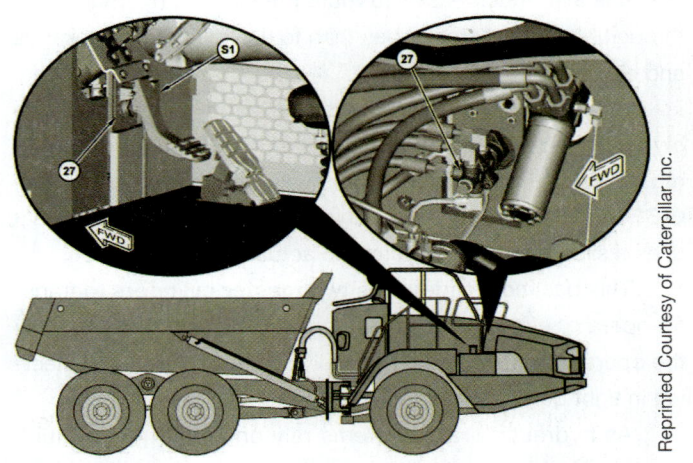

FIGURE 18–8 Brake pedal and valve (reprinted courtesy of Caterpillar Inc.).

HEAVY-DUTY HYDRAULIC FRICTIONAL BRAKE SYSTEMS

FIGURE 18–9 Mechanical park brake.

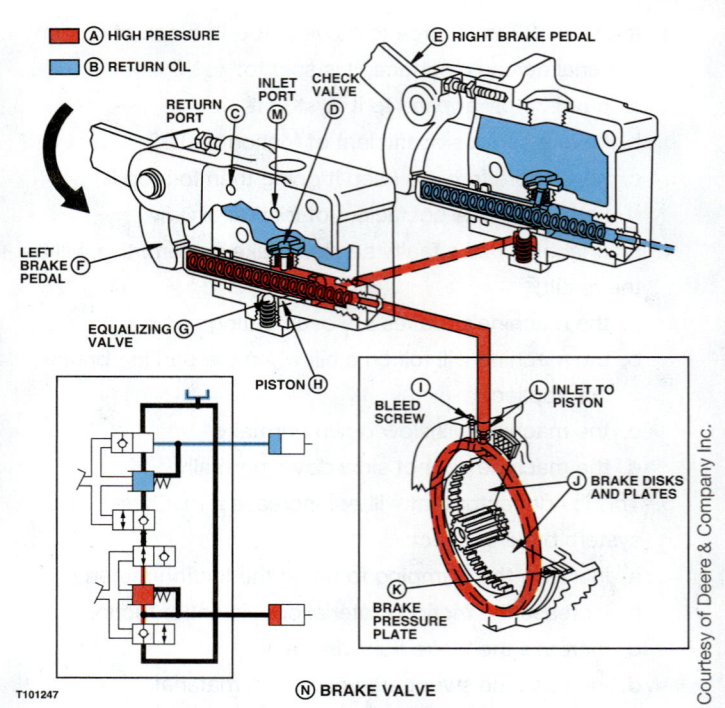

FIGURE 18–10 A simple hydraulic brake system.

Today, only very small machines will use strictly mechanically actuated service brakes and some will still use mechanically applied parking brakes. ● **SEE FIGURE 18–9** for a mechanically applied parking brake.

Machine braking systems evolved many years ago to include a hydraulic system to create more braking effect and eliminate the maintenance involved with the mechanical linkages. As you learned in Chapter 5, hydraulic pressure is created by the resistance to flow. As a hydraulic brake circuit ends at the foundation brake **actuator** (caliper, wheel cylinder, piston), the moment the friction material comes into contact with its opposing member fluid flow virtually stops and pressure rises quickly. There is very little flow needed to move the actuators, maybe a few cubic inches, which is way less than a gallon (231 cubic inches/U.S. gallon). This makes brake applications quick and powerful if needed because the control system has to move so little fluid.

It is also much easier to route pressurized hydraulic fluid through hoses and steel tubes than to use mechanical linkages and cables to transfer force.

Besides smaller equipment and some graders that use glycol-based automotive-type brake fluid (DOT Type 3 or 4), most heavy-duty brake systems will use mineral-based oil. ● **SEE FIGURE 18–10** for a backhoe loader brake system that uses a simple hydraulic system to actuate the brakes.

This backhoe loader uses two **master cylinders** to transfer operator effort to the friction brakes. This system will allow the operator to use the brakes individually to assist with steering in tight quarters.

As hydraulic brake systems rely on some type of fluid, these systems require fluid maintenance and must stay sealed to prevent contamination.

FRICTION-TYPE FOUNDATION BRAKES

The term *foundation brake* refers to the business part of the brake system or the part that provides the actual brake effect. A foundation brake assembly will have one or more rotating members and one or more stationary members. As brake application pressure is applied, the rotating member will be slowed down to try and match the state of the stationary member. This is done by creating friction between the two members.

The foundation brakes will have an actuator component (piston) to convert the hydraulic pressure that is created by the brake system supply components into mechanical movement. This mechanical movement is used to squeeze two or more components together to create friction. All actuators will have a means to bleed air from them so that only brake fluid is in the system.

There are some different types of foundation brakes:

1. Expanding shoe and drum
2. Caliper and rotor: fixed, sliding, multi-caliper, **multi-disc**
3. Single and multi-wet disc
4. Bladder type

EXPANDING SHOE AND DRUM

More commonly called drum brake, this type of brake is becoming less popular for heavy equipment machines. A drum brake assembly consists of two half-moon-shaped "shoes" that match the internal radius of the rotating drum they fit inside of as well as one or more hydraulic actuators (pistons), an adjustment mechanism, and related hardware to keep everything in place. The shoes are stationary (they don't rotate) and are lined on their external surface with a friction material. They are anchored to the axle housing, and as they are expanded inside the rotating drum, their friction material will grab the drum and provide a braking action. As the drum is attached to the rotating wheel, this will in turn slow the machine down providing there is good traction between the tire and the roadway surface. These brakes work best when they are dry and, therefore, are not as favorable as other types of brakes for most types of heavy equipment because of the muddy and wet conditions that most machines work in. To prevent the drum brakes from being affected by moisture and mud, the brakes are usually covered as much as possible. This will then create another problem from the covers not allowing air circulation to cool the brakes down. When excessive heat is created in drum-type brakes, the drum will expand, which means the shoes have to travel farther to maintain the same amount of force against the drum. The friction material can also become overheated and lose some of its coefficient of friction. This leads to a scenario called brake fade, where the braking effect decreases as the apply pressure stays the same.
● SEE FIGURE 18–11 for a typical drum/expanding shoe-type brake.

Drum-type brakes can also be used for static braking (parking brake) on a machine's driveline where they are typically spring applied but sometimes mechanically applied with a cable and ratchet mechanism.

They will be applied with a lever that turns a cam to expand the brakes against the drum. The lever could also be moved to apply the brake with spring pressure but could be released with air or hydraulic pressure.

Hydraulic drum brakes can be broken down into two types: **servo** and **non-servo** and both can be actuated by one or two hydraulic **wheel cylinders** that are mounted to the stationary axle housing. These cylinders receive hydraulic fluid from a master cylinder that is controlled by the operator and usually started with moving a foot pedal. As fluid enters the cylinder, it acts on a piston or pistons that move out of the cylinder housing and will in turn act on the brake shoes to move them into the brake drum. The amount of piston movement and force it creates depends on the amount of fluid flow and pressure that is sent to the wheel cylinder. This will be determined by how far and how hard the operator pushes the brake pedal or moves a lever. Maximum brake pressure can range from 750 to 2500 psi depending on the mechanical advantage of the pedal/lever and whether the master cylinder is boosted or not.

The wheel cylinder piston will have at least one seal on its outside diameter to keep the brake fluid from leaking out past the piston. This is usually a lip-type seal that has the lip facing toward the fluid. There will also be a seal or rubber boot to keep dirt away from the piston seal. ● SEE FIGURE 18–12 for a typical wheel cylinder.

To return the piston and brake shoes to the released position after the brake pedal is released, there is usually return springs that are attached to the shoe and axle hub. As the shoe is pulled back to the released position, it will also bottom the piston back in the housing.

Drum-type brakes will need to be adjusted to compensate for shoe and drum wear and will typically have some type of threaded adjuster that may be automatically or manually adjusted.

FIGURE 18–11 A typical expanding shoe and drum-type brake.

FIGURE 18–12 A typical wheel cylinder.

NON-SERVO TYPE. If a drum brake is called a non-servo type, this means that the wheel cylinder acts on the toe of each shoe and the heel of each shoe pivots on a common anchor pin. This will provide a sturdy platform for the shoes and is a very simple arrangement. There will be one return spring to bring the toes of the shoes back away from the drum and an adjuster mechanism to provide a means of keeping the shoes fairly close to the drum. As the friction material wears off the shoes, the gap between the shoes and drum gets bigger, and this will delay the start of braking and will also decrease the effective braking. The adjuster can be used to minimize the gap between the shoes and drum, and the clearance between them should be checked to ensure that it is within specification.

As there is one leading and one trailing shoe with this type of drum brake for each direction of machine travel, there is only one shoe that is being self-energized. ● **SEE FIGURE 18–13** to understand how a brake shoe gets self-energized.

Self-energizing happens when the toe of a shoe gets pushed into the direction of rotation that the drum is turning, and this in turn pulls the shoe into the drum. The trailing shoe will not be nearly as effective because it relies strictly on the wheel cylinder to push the shoe into the drum against the direction of rotation.

SERVO TYPE. Lighter-duty drum brakes will likely be servo type, which means they will use two shoes that are joined together with an adjuster and will work together as one shoe. This will allow both shoes to be self-energized.

WEDGE TYPE. A less popular hydraulically actuated drum brake is the wedge type of brake. A hydraulic cylinder is used to push a wedge out that moves a roller out, which in turn moves the toe end of the shoes out to the inside diameter of the drum.

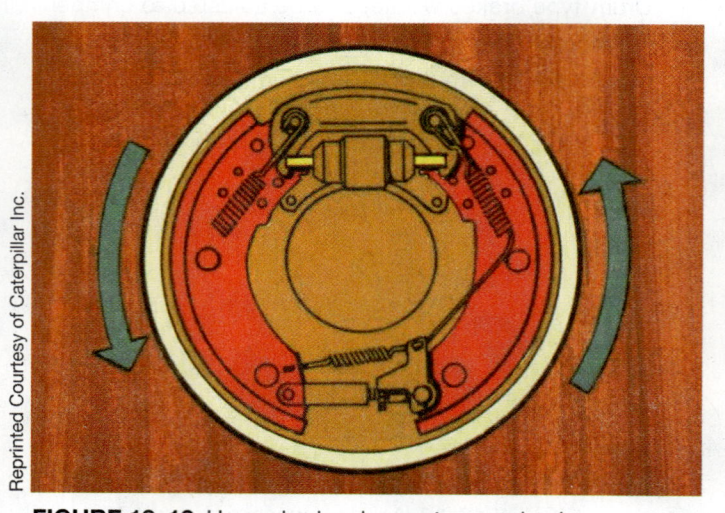

FIGURE 18–13 How a brake shoe gets energized.

CALIPER AND ROTOR–TYPE BRAKES

This type of brake is very common in the automotive world and is also the most common type of hydraulically actuated "dry" brake used on heavy equipment. The term *dry* relates to the friction material that is not running in oil.

Caliper and rotor–type brakes create friction by clamping a stationary friction material (brake pad) against a rotating steel rotor. The rotor is attached to a wheel hub that when slowed down will transfer the braking action to the machine's tires, which then slows the machine down.

For heavy equipment applications, it is not uncommon that they can be used as multi-caliper per wheel arrangements or even multi-rotor per wheel to improve the overall effective braking results. Once again the multiple caliper or multiple rotor arrangement provides more swept area to create a higher braking effect.

Caliper brakes are mostly oil applied, but they can also be spring applied and oil or air released. Spring-applied caliper brakes will likely be found as a driveline brake for static or parking brake purposes. The SAHR (spring applied, hydraulic released) type of caliper will have springs behind the caliper pistons and oil will move the piston back to release the brake.

When used for service braking, this type of brake is less susceptible to brake fade because as the rotor heats up, it will expand into the brake pad and actually improve the brake effect. The brake components are also much more exposed and, therefore, will cool down faster than drum brakes. This of course depends on the application, and if the machine is normally working in a lot of wet, muddy conditions, there will likely be covers to try and protect the brakes, which will also reduce the cooling effect.

Rotor and caliper brakes can be broken into two types: sliding and fixed caliper.

Sliding caliper brakes are not very common on heavy equipment as they are usually single or double piston and, therefore, are used for light-duty applications. The term *sliding caliper* refers to how the caliper is mounted to the machine's axle or spindle. Because the caliper has its pistons only on one side, the housing must float or slide to allow for even brake pad wear.

The fixed caliper brake is more popular because it will use pistons on both sides of the rotor to apply force to the brake pads. Because of this the pads should wear even and there is no need to allow the caliper to slide. Fixed calipers can have up to eight pistons in them. ● **SEE FIGURE 18–14** for a fixed caliper type of brake.

Some hydraulically applied calipers will have springs to assist in retracting the pistons after a brake application but most will rely on the piston seals. These square seals will bend to allow the pistons out and seal the fluid, but after the fluid pressure is released, the seals will straighten up and retract

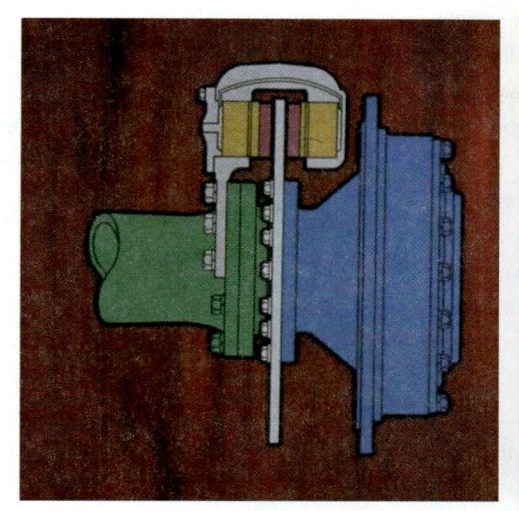

FIGURE 18–14 A fixed caliper-type of brake (reprinted courtesy of Caterpillar Inc.).

the pistons back into the caliper. This will also give this type of brake a self-adjusting effect because as the friction material or rotor wears, the piston will keep moving out past the seal slightly to compensate for the wear. A slight movement of the rotor and the cooling down of the friction material will provide the necessary clearance between the brake pads and rotor.

The rotors will have a relatively rough surface when new, which will burnish (to make smooth or polish) the friction material after the first few brake applications. Until this burnishing process is complete, the friction material will have about half its designed coefficient of friction. Caution needs to be taken when new brake material is installed, and a machine's brakes are applied the first few times. The rotor will then have a polished appearance after as well.

Most rotors will be ventilated to allow them to cool faster. There will be ribbed vents between the braking surfaces, which gives the rotor more exposed surface area to improve cooling. However, once the vents get filled with mud, which easily happens if the machine works in less than ideal conditions, this cooling effect is negated.

SINGLE AND MULTI-DISC FOUNDATION BRAKES

Single and multi-disc–type foundation brakes are likely the most common type of brake you will find on medium- to large-size heavy equipment. They can be used for both spring-applied, oil-released and oil-applied, spring-released types of brakes. You will find them used for service and parking brakes as well. Some machine applications will use the same multi-disc foundation brakes for both service and parking brakes. They can be used for steering track machines as well as for braking other

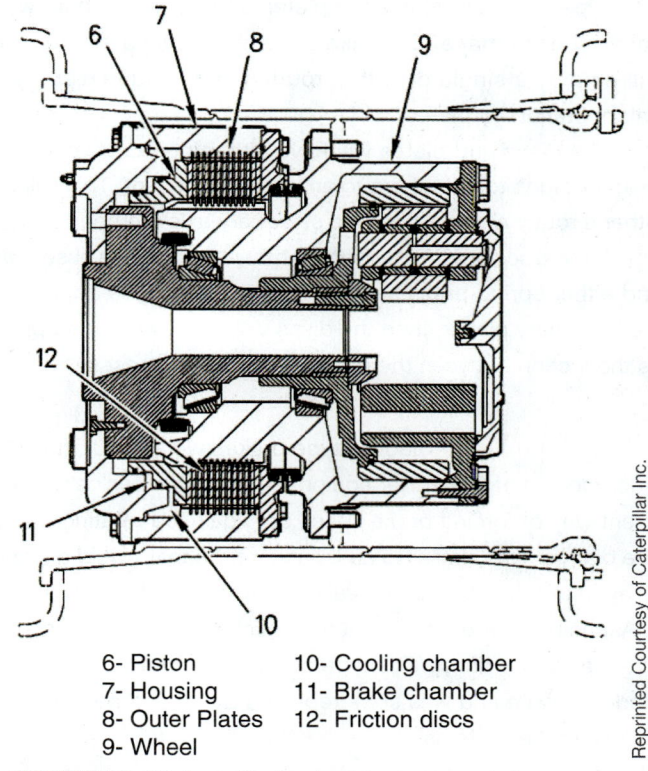

6- Piston
7- Housing
8- Outer Plates
9- Wheel
10- Cooling chamber
11- Brake chamber
12- Friction discs

FIGURE 18–15 A multi-disc brake arrangement.

functions such as winches and the swing function of an excavator. ● **SEE FIGURE 18–15** for a multi-disc brake arrangement.

If these disc-type brakes are used for a wheeled machine, they could be mounted out at the end of the axle near the wheels, in which case they are called outboard brakes or they could be mounted close to the center of the differential where they are called inboard brakes.

Disc-type brakes will almost always be wet type; this means they will be running in oil and the oil will carry away heat from the components to be dissipated naturally or to a heat exchanger.

The friction material used for this type of brake is designed to be run in oil and still provide its designed coefficient of friction. The discs will have a series of grooves formed into the friction material to allow the oil to remove heat from them. Another big advantage of this type of brake is that they will run in a sealed compartment and will not be exposed to the environment. These two factors mean this type of brake will last a long time. It is not uncommon for a set of multidisc brakes to last well over 15,000 hours if properly maintained and used.

The main components of these disc brakes are one or more discs (lined with friction material), one or more plates that are sometimes called reaction plates (some type of steel material), a spring or springs to apply or release the brake, and a piston to transfer hydraulic pressure to apply or release the brake. The piston will apply even pressure as it is a similar shape and size to the discs and plates. Multi-disc-type brakes will have the plates and discs alternate with each other to make the brake assembly while a single disc will employ a single disc that rotates and is squeezed by a pair of stationary plates.

The discs and plates will have either internal or external teeth or tangs to hold them to another component. This will be either a rotating wheel hub or a stationary axle member.

If the discs are the rotating member of the brake assembly and either spring pressure or hydraulic pressure is applied to the stationary plates, then the discs will be slowed or stopped as the friction between the plates and discs increases.

BLADDER TYPE Bladder-type brakes were used on some older models of Caterpillar equipment and were basically a different way of forcing brake shoes out against a rotating drum. The bladder was similar to an tire inner tube that would expand as hydraulic fluid flow and pressure was applied to the inside of it. As the bladder expanded, it then pushed the friction material out against the drum. This type of brake system was prone to bladder failure and was soon replaced by other types of more reliable brake systems.

Progress Check

6. This would *not* normally be found in a hydraulic brake system:
   a. mineral oil
   b. brake fluid
   c. air
   d. hydraulic oil

7. Brake fade is a condition that happens to this type of brake:
   a. shoe and drum
   b. rotor and caliper
   c. multi-disc
   d. bladder

8. This component will return the brake pads away from the rotor:
   a. piston seal
   b. air pressure
   c. adjuster spring
   d. hydraulic pressure

9. This type of brake will never be spring applied:
   a. drum
   b. rotor caliper
   c. multi-disc
   d. bladder

10. One main advantage that multi-disc-type brakes have over other types of hydraulically applied foundation brakes is:
    a. they will apply faster
    b. they are easier to repair
    c. they are air cooled
    d. they are sealed from the environment

HYDRAULIC BRAKE APPLICATION SYSTEMS

Hydraulic brake application systems can be broken into three categories: non-boosted, boosted, and full power. Machine designers will choose one of the three types for a machine based on the weight and travel speed of the machine and the braking energy required for slowing and stopping the machine.

A non-boosted hydraulic brake system consists of a master cylinder, brake lines (hoses or steel lines), wheel cylinders (actuators), foundation brakes, and brake fluid. The master cylinder receives an input force from the operator. This is usually through a foot pedal, but in some cases a hand lever could be used.

Boosted brake systems will have a boost mechanism incorporated into the master cylinder assembly between the brake pedal and the master cylinder. The foot pedal pushes on a lever that uses the boost system to multiply the force input to the master cylinder.

The master cylinder moves fluid out through the brake lines to the wheel cylinders and their pistons move as a result. The wheel cylinder piston movement is transferred to the foundation brakes and a brake application is made. When the brake pedal is released, the wheel cylinders are withdrawn and the master cylinder and foot pedal return to their starting position and are ready for the next application.

A full-power system has brake lines and wheel cylinders as well but has a hydraulic system and modulating valve that are capable of creating higher pressures. The modulating valve is actuated by the operator to make brake applications with fluid movement.

BRAKE CONTROLS
Some brake controls are combined with power train controls. Most medium- to large-size wheel loaders will have two brake pedals. The right pedal is usually just for brake application while the left one is a combination of drivetrain neutralizer and brake. For wheel loaders that use a powershift transmission, the first part of the left brake control movement will quite often neutralize the transmission. This is sometimes called an inching or declutch function.
● SEE FIGURE 18–16 for a typical three-pedal arrangement.

NON-BOOSTED HYDRAULIC BRAKE SYSTEMS
This type of system will be found only on very light-duty equipment because the brake application pressure is limited to the maximum force the operator can apply to the brake pedal plus any gain through a mechanical lever between the pedal and the master cylinder. Maximum pressure that is sent the brake actuators will be around 1000 psi.

The operator's input is transferred to a master cylinder that will force brake fluid out to the foundation brake actuators. A master cylinder will have a rod that is pushed into its bore by the operator and the rod will push on a piston that has brake fluid sealed in front of it with a cup or seal. The size of the bore will determine how much pressure is created in the system.

> **TECH TIP**
>
> As with any hydraulic system, all hydraulic brake systems transfer energy through a fluid to create work. Brake fluid is moved through lines to actuators and the actuators convert the fluid movement into mechanical movement that actuates the foundation brakes. Pressure is created in the system by resistance to flow, which is created by the foundation brakes. For example, if a brake caliper piston pushes a brake pad onto a brake rotor, the application pressure rises as soon as contact between the pad and rotor is made. As more force is applied to the fluid at the master cylinder end, the pressure increases, which will increase the braking action.

For example, if 500 lb of force is created by the operator pushing the brake pedal through a mechanical advantage and this force is then used to push a cup that has a 1 in.^{2} diameter, then there will be 500 psi of oil pressure in the system. If the same force acts on a 2 in.^{2} cup, then 250 psi of pressure will force the foundation brakes to work (remember $F = P/A$).

The rod pushes on a piston that has two (for a dual circuit system) cups or seals seated on it. These cups then move the brake fluid out of the master cylinder through hoses or tubes on to the foundation brake actuators to create braking action. The brake fluid is always available to the piston from the reservoir. After the fluid has been pushed out of the master cylinder and the operator releases the brake pedal, a return spring will push the piston back into the master cylinder where it will be

FIGURE 18–16 A typical three-pedal arrangement.

HEAVY-DUTY HYDRAULIC FRICTIONAL BRAKE SYSTEMS   549

ready for the next application. The returning actuator piston will also return the brake fluid back the reservoir.

If the master cylinder has its own reservoir, it will have a vented cap to allow air on top of a diaphragm. This will keep atmospheric moisture from getting into the fluid but still allow the fluid level to lower slightly without creating a vacuum on top of it.

There may be a sensor in the reservoir that will turn a warning light on if the fluid level gets too low. ● SEE FIGURE 18–17 for a non-boosted master cylinder.

The master cylinder may have its own fluid reservoir, may have a remote reservoir, or may use fluid from another machine hydraulic system like steering or implement. The master cylinder will also likely incorporate one or more additional valves. One may be a residual check valve that will keep a small amount of pressure in the brake lines to ensure responsive brake action. Another may be a pressure differential valve that will sense any pressure loss in one half of the dual circuit system and close that circuit off. It will also turn on a warning light in the cab to warn the operator of a brake system problem. These additional valves could be housed separately in a combination valve that is between the master cylinder and the wheel cylinders.

BOOSTED HYDRAULIC BRAKE SYSTEMS To increase the brake fluid apply pressure, many brake systems are boosted. This term refers to a system that has an additional force besides the operator's foot or hand pressure applying input force to the master cylinder. This is most common with small- to medium-size machine brake systems and some older haul trucks. If more force is exerted on the master cylinder, more pressure can be applied to the foundation brakes. This in turn creates a higher braking effect.

The boost could be provided from an air system or a hydraulic system. An air-boosted system uses an air over oil master cylinder. Air systems will be covered more in depth in Chapter 22. For example, an air-boosted master cylinder gets regulated air from a treadle valve that the operator pushes on with foot pressure or a hand-operated lever (spike) that regulates air. This air pressure acts on a diaphragm in the air section of the master cylinder that in turn pushes a plunger into the master cylinder. The master cylinder operation after that point is basically the same as a non-boosted master cylinder. If the diaphragm of the air chamber has a surface area of 10 in.^{2} and there is 100 psi applied to it, there will be 1000 lb of force applied to the master cylinder. The master cylinder can multiply this force to create up to 2000 psi brake application pressure.
● SEE FIGURE 18–18 for an air over oil master cylinder.

A hydraulic boost system works in a similar manner except the hydraulic oil is at a higher pressure and, therefore, doesn't need as big of an area to work on to increase force to the master cylinder.

A hydraulically boosted master cylinder uses operator input to push a plunger that will seat a valve. When this valve closes, hydraulic boost pressure acts on a larger piston that in turn pushes on the master cylinder piston. From there the master cylinder works like any other master cylinder to send oil pressure to the wheel brakes. There will be a pressure relief valve in the boost section to limit boost pressure from getting too high. When the operator stops pushing the brake pedal farther, the valve comes off the seat and the master cylinder doesn't get moved any farther. When the brake pedal is

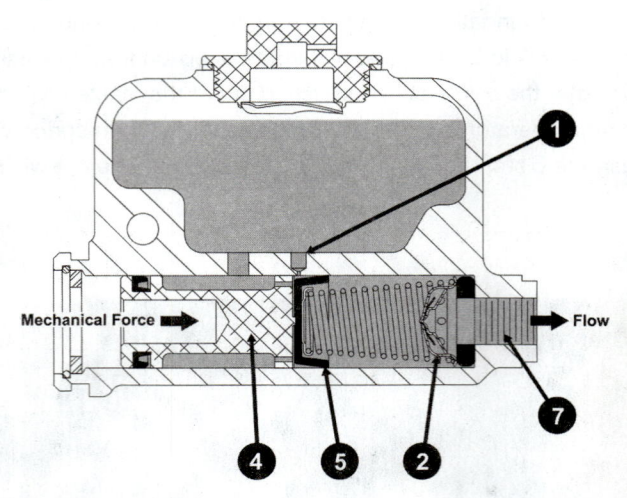

1- Equalizing port
2- Residual check valve
3- Check valve seat
4- Piston
5- Piston cup
6- Air outlet
7- Outlet port
8- Spring

FIGURE 18–17 A non-boosted master cylinder.

FIGURE 18–18 Air over oil master cylinder.

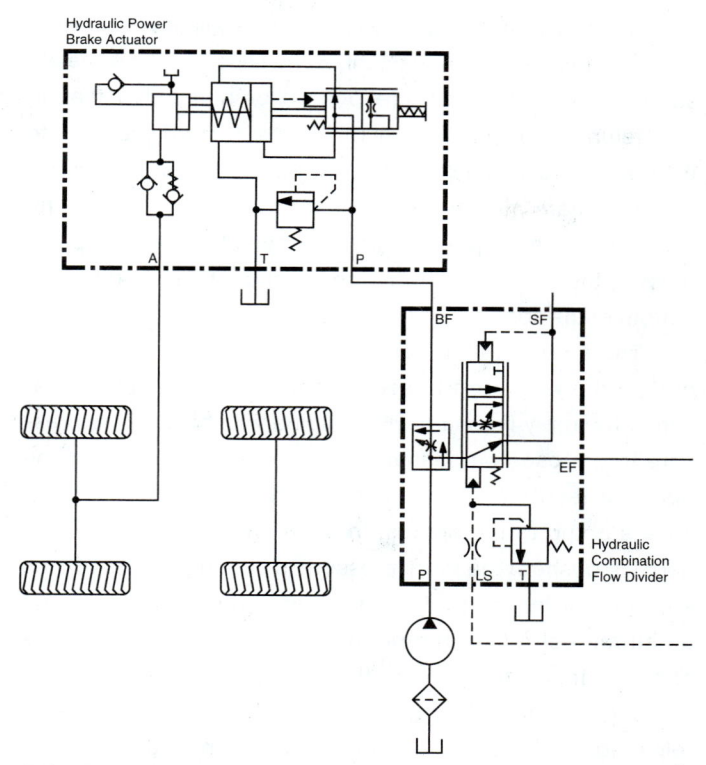

FIGURE 18–19 A simple hydraulic brake system used on a small forklift.

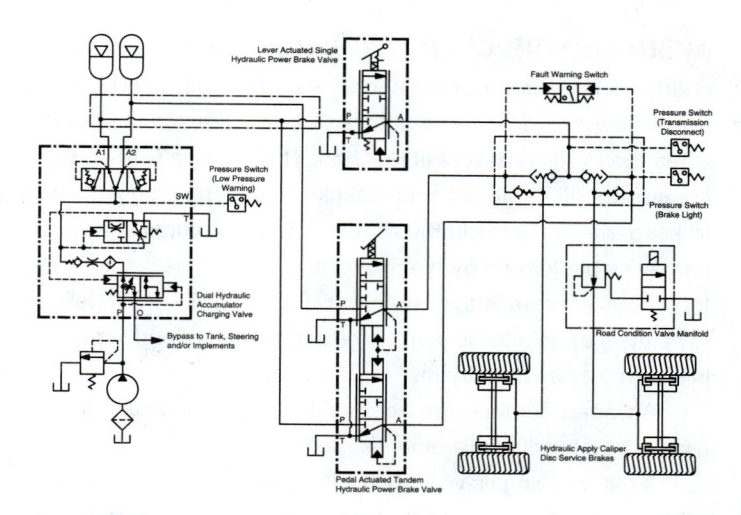

FIGURE 18–20 A full-power brake system.

released, return springs return the spool and piston to their starting positions. ● **SEE FIGURE 18–19** for a schematic of a simple hydraulic brake system for a forklift that uses drum-type foundation brakes with a hydraulically boosted master cylinder.

In this case, the forklift brake boost system uses common oil that is shared with the steering and implement systems.

All hydraulically boosted system master cylinders can provide brake application without boost pressure if the boost system fails or the engine is off. In this case, brake pedal movement is transferred directly to the master cylinder pistons without boost assist and will provide some brake pressure to slow the machine.

Supplementary boost systems will provide boost pressure if the normal boost system fails. This is a safety backup feature and is usually an electrically driven pump that is powered from the machine's 12V or 24V DC electrical system. There will be flow and pressure sensors that will turn on the supplementary system if the sensors detect a pressure or flow loss. This is an important safety feature that should be checked on a regular basis.

FULL-POWER HYDRAULIC BRAKE SYSTEMS Full-power hydraulic brake systems will require the operator to create only a small amount of force at the operator input valve. This will reduce operator fatigue. Full-power systems will usually create higher apply pressures and, therefore, more effective braking.

The apply pressure is created by a hydraulic system that could be dedicated to brakes or part of another machine hydraulic system. These systems will use an accumulator to store hydraulic pressure to supplement pump supply or supply brake apply pressure if the supply pump stops producing flow. There will be a specific minimum number of full brake applications that the system should be capable of when there is a power off (dead engine) situation, and this should be verified whenever a brake system is serviced or repaired. ● **SEE FIGURE 18–20** for a schematic of a full-power brake system.

This hydraulic brake supply system is usually used only with rotor and caliper or disc-type brakes.

The main components of a typical full-power hydraulic brake supply and control system are **pump** (fixed or variable displacement), accumulator **charging valve**, accumulators, modulating or reverse-modulating valve, lines, and foundation brakes.

The system shown in ● **FIGURE 18–20** is a dual circuit arrangement that provides brake application to two separate circuits. If this were used on a wheel loader, it could be one circuit for each axle or a diagonal front left/right rear and right front/left rear arrangement. This system uses a dual charging valve to keep two accumulators charged up and keep them charged within a high- and low-pressure range. These pressures are called the cut-in and cut-out pressures. Sometimes these pressures are adjustable, and a typical setting is 2100 psi cut out and 1700 cut in. In other words, the accumulator pressure should stay within this range if the accumulator charge valve is working properly. If there are enough brake applications made to drop the accumulator charge pressure to 1700 psi, then the charge valve should direct pump oil flow to the accumulators and stop it when it reaches 2100 psi.

HEAVY-DUTY HYDRAULIC FRICTIONAL BRAKE SYSTEMS 551

ACCUMULATOR CHARGING VALVE The accumulator charging valve is a combination valve that will divert some pump flow from another system's pump to the accumulators when the cut-in pressure is reached. This is done with an unloading spool. There is also a check valve in the charge valve to keep the oil charge in the accumulators from draining away and available for use by the modulating valve. The last part of the charging valve is the cut-in/cut-out spool. This will determine at what point oil is sent to the accumulators and when oil is diverted away from them.

An open center–type accumulator charge valve will be used with a fixed displacement–type pump.

If the charging valve is using a variable-type pump, then it will have a load-sense section that will send a signal to the pump to upstroke the pump if flow to the accumulators is needed. A dual charging valve will also have poppet valves that will always charge the lowest charged accumulator first. Some charging valves will also incorporate a system relief pressure valve.

Full-power brake systems need modulating valves to allow the operator to use the available built-up accumulator pressure for applying the brakes with as much or as little force as required. They will also give the operator a feedback feel to give him or her a sense of more or less application pressure they are sending to the wheel ends. ● SEE FIGURE 18–21 for a modulating valve.

MODULATING VALVES Modulating valves are usually floor mounted for foot operation but could also be lever operated or remotely operated to allow operation from a different location.

A typical dual circuit modulating valve will have four main moving parts and four ports connecting lines to it. One line will supply accumulator pressure, one will be a drain line that allows return oil to go to tank, and the other two will supply oil to the brake circuits to actuate the wheel end brakes.

The main moving parts inside are the compensating spring that receives pressure from the operator through a pedal or lever, the upper and lower spools, and a bias spring at the bottom of the spools.

The spools will meter application oil from the supply port to the work port based on compensator spring pressure. The apply pressure will combine with bias spring pressure to oppose compensator spring pressure and give the operator feedback that directly relates to how much pressure is going to the brakes. In other words, the harder the pedal is pushed, the more pressure is sent out to the brakes and the harder the pedal is to push. This gives the pedal a natural feel. When the pedal is held part way down, the spool is balanced between the two springs and the apply pressure is trapped in the line to the brake. If the pedal is released, then the spool moves up and the trapped oil is allowed to go to tank through the return port. This allows the brakes to release.

REVERSE-MODULATING VALVES Reverse-modulating valves are used with spring-applied, oil-released brakes. This type of brake is termed a *fail-safe brake* as it will apply whenever pressure is lost. The most common type of foundation brake used with this system is the multi-disc type. The reverse-modulating valve works the exact opposite way of the modulating valve that was just explained. When the machine is first started, oil is sent through the valve to the spring-applied brakes to release them. Then as the pedal pressure is applied, this oil is drained away to apply the brakes. A full brake pedal application will completely drain the brake and fully apply it.

OTHER HYDRAULIC BRAKE COMPONENTS

SLACK ADJUSTERS. Slack adjusters are used for larger foundation brakes that require larger **volumes** of application oil. A slack adjuster is used to hydraulically compensate for wear in the foundation brakes. It will always ensure that the quantity of brake application oil will need to be only the minimum amount required to move the foundation brakes to make an application. This is done in the slack adjuster with the use of two different diameter pistons and a check valve. If the friction material wears enough that the large piston is allowed to travel to the end of the slack adjuster housing, the check

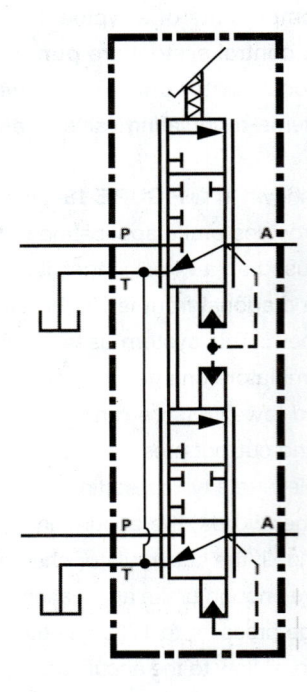

FIGURE 18–21 A modulating valve.

valve will open, and when the oil pressure is released, the large piston will reset or move back to the bottom of the slack adjuster.

RELAY VALVE. A relay valve is sometimes used on larger machines so the operator's control command will be more responsive. The relay valve will be located close to the foundation brake that it is actuating. The relay valve is similar in operation to an air relay valve. The oil from the modulating valve is used as a signal that then sends oil from the relay valve on to apply the brakes.

BRAKE RELEASE PUMP. Some larger machines will be equipped with a brake release system to be used in the event the machine loses hydraulic pressure. If the machine needs to be towed, then an electrically driven pump can supply hydraulic pressure to release the spring-applied parking brake.

BRAKE COOLING SYSTEMS

Many medium- to large-size machines that travel fast and use multi-disc foundation brakes will have a brake oil cooling system that will circulate oil around the brake friction components, absorb the heat generated by them into the oil, and send the oil to a cooler. These systems can be simple with a minimum of hoses, a pump, and an oil to air heat exchanger or complex with diverting valves being controlled by an ECM that uses temperature sensors as inputs. The ECM can turn on a warning light and alarm if the brake temperature gets too high and a fault code will be set.

BRAKE OPERATION TESTING

Like all other machine systems, you need to know how a machine's brake system should work properly before you can determine if there is a problem with it. Because of the importance of the safe operation of the brake system and the negative consequences that could occur if the brakes were not operating properly, you should also be testing a machine's braking performance and operation as part of any routine maintenance check.

Always check the machine's manual for the proper procedure to test the braking system. Some examples of machine brake system tests are the following:

- For a machine with a powershift transmission, put the machine in its highest speed range forward or reverse apply the service brakes and set the engine to high idle. The machine should not move.
- For a machine with a hydrostatic drivetrain, disable the parking brake release and try to drive the machine through the brakes. Some machines should be left at low idle and others at high idle.
- For a machine with a drive line parking brake, apply the parking brake and try to drive over the brake in high-speed range.
- For machines with reverse-modulated, spring-applied brakes, try to drive the machine with the brake pedal depressed.

Some test procedures will ask that you park the machine on a certain percentage of slope, apply the parking brake, and see if the machine stays stationary. Another example of a test for a grader's brakes is to put the machine into second gear and get the machine moving at high idle. While the machine is moving, put it in neutral and apply the parking brake. If the parking brakes are working properly, the wheels should skid.

Again these are general test procedures and you should always consult the machine's manual for the exact test method.

If you are checking brake performance on any machine, you should make sure that the machine stops straight with no drifting to one side and it shouldn't demonstrate any unusual noises or vibrations when the brakes are applied.

Service and parking brakes should also be checked for releasing fully because a dragging brake will waste fuel and can cause major drivetrain component damage by overheating.

BRAKE SERVICING

The importance of servicing machine brakes regularly and thoroughly cannot be overstated. As mentioned, it can easily be argued that the brake system is the most important safety-related system on the machine and proper servicing will lead to ensuring maximum brake performance.

Part of many machine service procedures will focus on servicing the brake system. Depending on the type of brake system, servicing can be a simple process or a complex procedure.

If the machine is newer and has an electronic fault code logging system, a good place to start a brake service is to check for any brake-related fault codes that have been logged or are still active.

The hydraulic supply and control section of any brake system will have some common service procedures that should be performed. They are the following:

- Check the fluid level. This could involve looking at a sight glass, pulling a dipstick, or looking at a transparent reservoir. ● **SEE FIGURE 18–22** for a brake fluid reservoir.
- Check the fluid condition. There should be no air visible and the fluid should appear clean with no burnt smell.
- Check the brake system malfunction warning system. This could involve turning the key to a certain position and looking for a warning light or pushing a button or toggle switch.

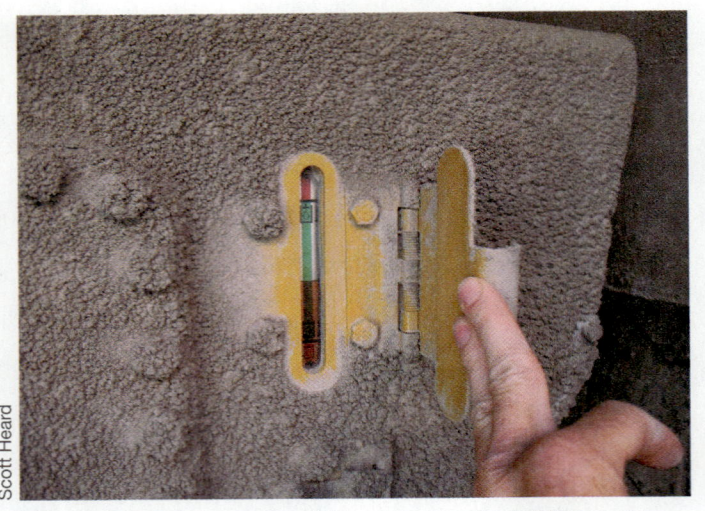

FIGURE 18–22 Checking the level in a brake fluid reservoir.

FIGURE 18–23 A drum and shoe brake with an automatic adjuster.

Check for fault codes related to the brake system.

Check for leaks.

Check for any damage to seals or boots at the calipers or wheel cylinders.

Check for proper operation of the service brake system.

Check for proper operation of the parking brake system.

Perform a visual inspection of the controls for the brake system.

Check any brake lights the machine may have for proper operation and damage.

Check the brake cooling system for proper operation.

Check for any unusual or excessive wear of friction materials or drums/rotors.

Check for loose or missing covers around foundation brakes.

The following are some examples of brake servicing procedures for the different foundation brakes:

1. **Drum and shoe:** This type of brake could require an adjustment to ensure proper operation. This adjustment will keep the shoes close to the drum with a minimum clearance to provide quick positive brake action and ensure that there is no brake drag.

    If there is an automatic adjuster mechanism for a drum and shoe brake, it should be checked to ensure that it isn't seized and it could also require lubrication. ● SEE FIGURE 18–23 for a shoe and drum brake with an automatic adjuster.

    Drum and shoe brakes could require manual adjustment and could require friction material measurements to ensure that the drums don't get damaged if there is metal-to-metal contact. Shoe adjustments could include special tools that are needed to turn the adjusting mechanism to set the proper shoe to drum clearance.

    A visual inspection is performed to check for damaged components and unusual wear patterns.

    Sometimes a service will require the wheel to be rotated with the axle raised to see if the brakes drag.

2. **Rotor and caliper:** Servicing rotor and caliper brakes is fairly simple in that it usually requires only inspection. Some machines that use this style of brake will require a measurement of the pads to ensure that they get replaced before too much friction material is gone and there is a possibility of the metal backing contacting the rotors. The measuring involves using a steel rod that is inserted through a hole in the caliper casting that measures the thickness of the pad.

    Rotors should be inspected for discoloring (indicates overheating), cracks, warping, grooves, and glazing.
    ● SEE FIGURE 18–24 for an example of a large brake rotor.

3. **Multi-disc:** Because these brakes are sealed, there is little that can be done for a service procedure. Some machines will have an access hole that allows the technician to perform a measuring procedure to warn when friction material is getting worn enough to need replacement.

BRAKE FLUID SERVICE If automotive-type brake fluid is used, it will occasionally have to be replaced. This type of fluid is hygroscopic, which means that it absorbs moisture. Over time the fluid's water content will increase and will start to allow rust to form inside the system. A higher water content also means a lower boiling point for the brake fluid. If the foundation brakes get too hot and transfer enough heat into the fluid, it will

FIGURE 18–24 A large brake rotor.

FIGURE 18–25 Different types of brake fluid.

boil or vaporize. This will result in poor brake pressure because the fluid now has gas in it.

Make sure that if the system requires glycol-based automotive fluid (DOT 3 or DOT 4) or silicone-based fluid (DOT 5), no mineral-based fluid (hydraulic oil) gets mixed with it and vice versa.

New automotive-type brake fluid has a boiling point much higher than water. For example, a typical DOT 3 fluid will have a minimum dry (new) BP of 400°F, DOT 4 a BP of 450°F, and DOT 5 a BP of 500°F. ● **SEE FIGURE 18–25** for different types of automotive brake fluid.

BRAKE SYSTEM TROUBLESHOOTING There are some general troubleshooting tips that can be applied to all hydraulic brake systems, but because of the diversity of differing systems, there is no way to cover even a fraction of the procedures you may use to find the root cause of a brake system malfunction.

The following are some general tips:

1. Verify the complaint. This will involve performing brake performance tests so make sure that you are comfortable with running the machine and know the proper test procedure.
2. Know the system. Because of the great variety of systems, don't assume you know how the system works. Read the operator's manual or service manual to familiarize yourself with it.
3. Check the simple stuff first (fluid level/condition, leaks, damage, control operation, fault codes, and check any recent repair history). One simple test would be to use a heat gun to compare the heat build up between individual foundation brakes (both wheels on the same axle should be close to the same temperature).
4. Determine if it's a complete system problem or an individual or multi-foundation brake problem. If a machine is pulling one way when the brakes are applied, it is likely the brake on the side the machine is pulling too is working too soon or the opposite side brake is not working as it should.
5. Perform instrument testing. You may need to install pressure gauges to assess the hydraulic brake system. Some checks could include application pressures, boost system pressure, accumulator pre charge pressure, and accumulator cut-in/cut-out pressure.

BRAKE REPAIRS Brake repairs are performed when either an operational problem is found or the friction material is found at or near its wear limit.

Some examples of typical brake repairs are the following:

Leak repairs: Any hydraulic brake system leaks needs to be repaired as soon as possible. This would include seals, hoses, tubes, valves, actuators, and accumulators. These components could be repaired or replaced.

Friction material replacement: For rotor and caliper–type brakes, this will be a relatively easy repair. After the wheel is removed, some calipers will allow the brake pads to be replaced without removing the caliper. There may be some retainer clips or bolts to remove and the pads can be replaced after the piston is pushed back into the caliper bore. Other calipers may have to be removed to enable the pads to be removed. Care needs to be taken whenever a caliper is removed to not damage the flex lines or hoses that connect the caliper to the brake hydraulic system.

Shoe and drum–type brakes that need friction material replacement will need the wheel and drum to be removed to allow access to the shoes. Depending on the axle configuration, this could range from sliding the drum over the wheel studs to removing the final drive assembly and wheel bearings. Some larger brake shoes that are riveted to only their backing will allow the technician to reline the shoes. This process requires the technician to drill all rivets out, remove the old friction material (sometimes called blocks), clean the shoe-mounting surface, and rivet the new material on.

Single and multi-disc brakes will require the most time and skill to replace the friction material. This is mainly because they will be inside a sealed compartment and the process will likely mean removing a wheel, removing a final drive, removing a hub assembly for a wheeled machine, and possibly dismantling the axle assembly if the brakes are the inboard type. Once the discs and plates are exposed, they will be measured and inspected and can be reused if they meet certain criteria. Generally, the friction discs should not show signs of discoloration, have any teeth/tangs missing, be warped or cracked, and be a certain minimum thickness. The same goes for the plates, and there should be no grooves worn into them.

Wheel cylinder resealing or replacement: If a wheel seal is found to show signs of leakage, it must be resealed or replaced. After the wheel cylinder is removed, it is disassembled and check for damage to its bore. If there is light scoring, these marks can be removed with a brake cylinder hone, and the cylinder is reassembled with new seals. If there is heavy scoring or rust, the cylinder should be replaced. Pistons are also inspected and replaced if found to be damaged. Care must be taken when installing new piston seals to ensure that the lip is pointing in the right direction.

Caliper resealing or rebuilding: This process is much the same as for rebuilding wheel cylinders.

Master cylinder rebuilding or replacement: Master cylinders may need to be resealed, and this would involve disassembly, inspection, cleaning, possible honing, resealing, and reassembly. Some valves in the master cylinder may need to be replaced as well. The master cylinder is then bench bled to remove all air before it is installed.

Modulating valve repair/replacement: Most modulating valves are mounted on the floor of a machine where they are subject to mud, dirt, and moisture. There is usually a protective boot that will keep this contamination away from the valve spool, but if this boot fails, valve replacement/resealing will soon follow.

Accumulator repair or replacement: Most larger accumulators can be resealed and put back into service if there is no damage to the bore or piston. Care must be taken to release all pressure before removing and disassembling any accumulator.

Other brake valves: Many brake valves can be resealed, but if the valve body or seat is damaged, they must be replaced. If there is any doubt about the integrity of a brake valve's condition, it is better to err on the side of caution and replace it. This point also applies to any brake component.

Brake bleeding: One of the last steps of any brake system repair is to remove all air from the system. Always refer to the machine's service manual for the proper procedure to bleed the brake system.

In general, you will always start with the wheel cylinder/caliper/disc that is farthest away from master cylinder. Pressure is built in the brake's lines either by an assistant working the brake control or by a vacuum pump. When pressure is built, the bleeder valve is opened and any air in the system will be purged out into a container through a hose attached to the bleeder. The procedure is repeated until nothing but clean air–free fluid comes out of the bleeder. This is then repeated for the rest of the brake actuators. The brake reservoir must be monitored and maintained to ensure that no new air is introduced into the system.

Progress Check

11. One source of boost for brake systems that would NOT be typically found on heavy equipment is:
    a. vacuum
    b. air pressure
    c. oil pressure
    d. electric
12. This type of machine will use individual master cylinders to brake individual wheels to assist with steering:
    a. backhoe loader
    b. wheel loader
    c. grader
    d. skid steer
13. The purpose of a residual check valve that is used in a master cylinder for hydraulic brakes is:
    a. to allow fluid in behind the piston from the reservoir
    b. to maintain boost pressure after the engine is shut down

c. to allow air into the reservoir

d. to keep a small amount of pressure in the brake lines at all times

14. If you adjusting cut-out pressure on an accumulator charge valve, this would:
    a. change the maximum pressure the accumulator would be charged to
    b. change the minimum pressure the accumulator would be charged to
    c. change the maximum pressure that the boost pump could produce
    d. change the minimum pressure the boost pump could produce

15. A reverse modulating valve is used to:
    a. improve braking when machine is reversing
    b. modulate park release oil pressure
    c. boost brake apply pressure
    d. drain pressure and apply service brakes

NICE TO KNOW

ELECTROHYDRAULIC BRAKES With advancements in electronics, the addition of sensors, ECMs, and proportional control valves has made brake systems more effective, efficient, and smarter. Brake bias (front to rear proportioning) can be adjusted by the ECM to give more effective braking. Brake wear can also be compensated for by electronic control systems.

Many machines will now offer anti-lock, traction control stability control systems, or a combination of all that work with the machine's hydraulic brake system. Electronic sensors monitor wheel speed and gravitational forces and send information to an ECM that will apply individual brakes with proportional solenoid valves to prevent wheel lock up, spinning tires, and loss of control.

HANDS-ON ACTIVITY

Go to a piece of heavy equipment, and by visual inspection, try to determine:
- What type of foundation brakes does it use?
- What type of fluid does the system use?
- How is the brake fluid level checked?
- Is there any brake system malfunction operator warning system in the cab?
- What type of parking brake does the machine use?
- Are there any adjustments that can be made to the braking system?

SUMMARY

1. Some safety concerns related to hydraulic frictional brake systems include: hot pressurized oil, strong compressed springs, defective test equipment, friction material dust. Brake systems should be tested for proper operation before the machine is put back to work.

2. There is a wide variety of braking systems used for heavy equipment machines. Hydraulic fluid pressure and its application to friction material is the most common method to slow down and or hold in place a machine.

3. Brakes that are used to slow down a machine are usually called service brakes.

4. Brakes that are used to hold a machine in place are usually called parking brakes but are sometimes called secondary brakes. This is called dynamic braking.

5. Brake systems that slow down machines do so by converting kinetic energy into heat energy. This is called static braking.

6. Braking effort is the resistance to rotating torque and can be measured in ft/lbs or Nm.

7. A heavily loaded machine that is travelling fast will need a massive amount of braking effort to slow it down.

8. Friction brake effort is influenced by the coefficient of friction of the friction material, the surface area of the friction material and the amount of pressure applied to the friction material.

9. Coefficient of friction of a material is determined by measuring the force it takes to move a certain weight of material across the surface of a second material. A higher coefficient of friction for a material means it grabs better.

10. Manufacturers of friction material use different chemical formulas to arrive at the composition of materials that can be used for brake components.

11. Hydraulic systems needed for braking will need to create a varying amount of pressure and not much flow. The pressure is used to either apply or release brake friction components.

12. The type of fluid used in the hydraulic system can be the same as that used in any hydraulic system or can be automotive type glycol based brake fluid.

13. Foundation brake assemblies are found near wheel assemblies and consist of rotating components and frictional components. As brake application pressure is applied to the rotating component its speed will start to match the stationary component.
14. Some examples of types of foundation types of brakes are: expanding shoe and drum; caliper and rotor: fixed, sliding, multi-caliper, multi disc; single and multi wet disc; bladder type.
15. Expanding shoe and drum foundation type brakes feature a rotating drum and non rotating shoes. When actuated the shoes move out against the drum. Shoes are moved with hydraulic pistons at one end and can pivot on an anchor on the other.
16. Drums will expand when heated which decreases the brake effort.
17. Drum brakes can be used for static or dynamic braking.
18. Return springs will pull the shoes back away from the drum and there will be a adjusting mechanism to allow the shoe to drum clearance to be kept within a specified tolerance.
19. Caliper and rotor type brakes consist of a rotating steel disc (rotor) and a stationary caliper. The caliper has one or more a movable pistons that squeeze friction material (pads) against the rotor.
20. Caliper and rotor brakes can be hydraulically applied or spring applied and can be used for static or dynamic braking.
21. Piston seal design provides the means to return the pistons after an application to keep a slight clearance between the pad and rotor.
22. Single and multi disc brakes can be used for static or dynamic braking. They are sealed in a housing and oil pressure applied to a piston moves one rotating component towards a stationary component. They are almost always "wet" brakes which means they have oil circulating around them to dissipate heat away.
23. The two main components are discs (friction material) and plates (steel discs). They can have either external or internal teeth or tangs to lock them to either a stationary or rotating component.
24. Hydraulic brake application systems can be one of three types: non boosted, boosted and full power.
25. Non boosted would only be found on very light duty machines and have the operator pedal input go directly to a master cylinder that transfers oil pressure to a piston or caliper. They will usually only develop 1000 psi maximum.
26. Boosted systems have a master cylinder that combines a second energy source with the operator input to create a higher apply pressure. It could use vacuum, air pressure or low pressure oil and will likely have an electric back up pump to provide boost in case of engine failure.
27. Full power systems use a dedicated hydraulic circuit with an operator actuated modulating valve to meter apply pressure. These systems will use an accumulator to store a quantity of pressurized oil in case of engine failure. An electric or ground driven pump will supply oil flow in emergencies.
28. Some multi disc brake systems will incorporate a cooling system that circulates oil past the brake friction components. The oil pulls heat away and then transfers the heat to the engine cooling system.
29. Testing brake operation should be part of regular maintenance checks and any deficiencies should be corrected. Check the machines service information to ensure the proper procedure is used.
30. Brake system maintenance may include: thorough visual inspection, adjustments to shoes, adjustments to linkages or cables, checking fluid levels, changing oil and filters.
31. Brake system complaints can include: no brakes, weak brakes, brakes pulling, brakes grabbing, overheating, won't release.
32. Brake system troubleshooting involves: knowing the system, verifying the complaint, list possible causes, repair the root cause, verify the repair, test the brake system operation.
33. Brake system repairs include: friction material replacement, piston resealing, drum reconditioning, rotor replacement, valve resealing, valve replacement, accumulator repair/recharging, brake line repair, brake system bleeding.

chapter 19
WHEELED MACHINE STEERING SYSTEMS

LEARNING OBJECTIVES

After reading this chapter, the student should be able to:
1. Describe the different types of steering systems used for wheeled machines.
2. Identify different types of steering controls for wheeled machines.
3. Explain how to safely work on an articulated machine's steering system.
4. Identify different personal protective equipment.
5. Describe how an orbital steering valve works.
6. Explain how a neutralizer valve works.
7. Describe why a secondary steering system is needed.
8. Describe the different modes of an all-wheel steering system.

KEY TERMS

Ackerman angle 568
Crab steering 571
Double rod cylinder 568
Ground driven pump 577
Kingpin 563
Neutralizer valve 578
Orbital valve 568
Secondary steering 577
Toe-in 567

INTRODUCTION

SAFETY FIRST Some machines can travel up to 40 mph with enormous payloads and will be near other machines and people. They may even travel on public roads to go between job sites. Safe machine operation will rely on a properly operating steering system, and a steering system failure could have deadly results for the operator and many others.

The technician working on a machine's steering systems must be familiar with the system before performing any service or repairs and must always confirm it operates properly before the machine is put back into service.

Whenever working on an articulated machine and you are near the pivot point, the steering lock must be in place to prevent the machine from turning. Wheel chocks should also be used to prevent the machine from rolling. ● SEE FIGURE 19–1 for a machine steering lock being installed.

FIGURE 19–1 Machine steering lock installed.

NEED TO KNOW

Heavy equipment machines can be grouped into two general categories by how they are propelled: tracks and tires. Between the two types of machines, there are many different ways to steer the machine.

Machines that travel on wheels can change direction or steer in one of four different ways: skid steer (wheels don't pivot), articulated steering, conventional steering, and all-wheel steering. This chapter will discuss how articulated conventionally and all-wheel steered machines. Skid steer machines were covered in Chapter 17.

Most machines that are steered with conventional steering systems will have a one-piece frame (straight frame) with articulated graders being the exception to this. Some examples are backhoe loaders, wheel-type excavators, and off-highway trucks.

Conventional steering systems will have many similarities to heavy-duty automotive steering system principles. A few light-duty machines, backhoe loaders, and straight frame haul trucks may use similar components and principles to heavy-duty automotive steering systems.

Conventional steering systems have both front wheels pivot either left or right to direct the front of the machine in the direction that the operator desires. As the steering angle increases to provide a smaller turning radius, the left and right wheels need to turn at different angles because the inside wheel will turn through a smaller radius than the outside wheel. ● SEE FIGURE 19–2 for conventional steering during cornering.

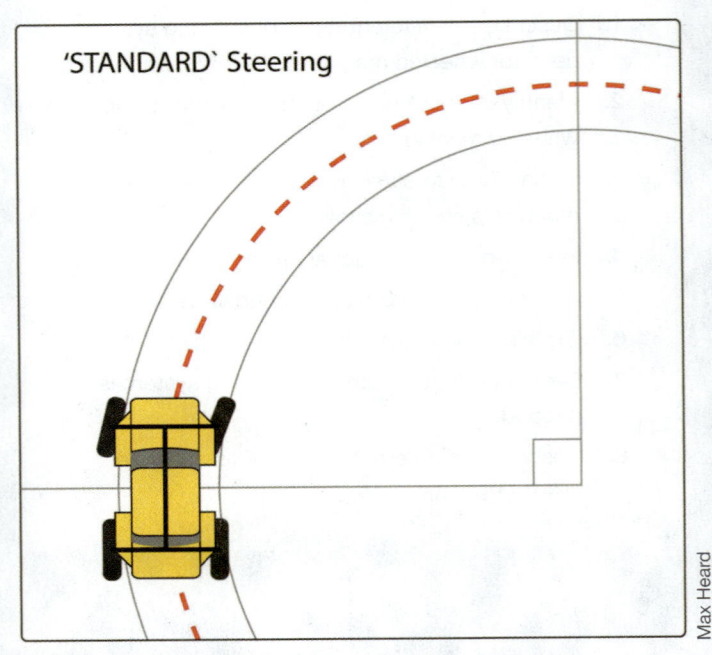

FIGURE 19–2 Conventional steering during cornering.

Some older straight frame wheel loaders and many current forklifts use a non-steering drive axle at the front and a steering axle at their rear to make a rear wheel conventional steer system. ● SEE FIGURE 19–3 for a machine with a rear steering axle.

Some other types of straight frame machines will use all-wheel steer systems. This means they have front and rear steerable axles. A couple of examples are telehandlers, forklifts, and rough terrain cranes. Some large cranes that have

FIGURE 19–3 Rear steering machine.

FIGURE 19–4 Crane with multiple steer axles.

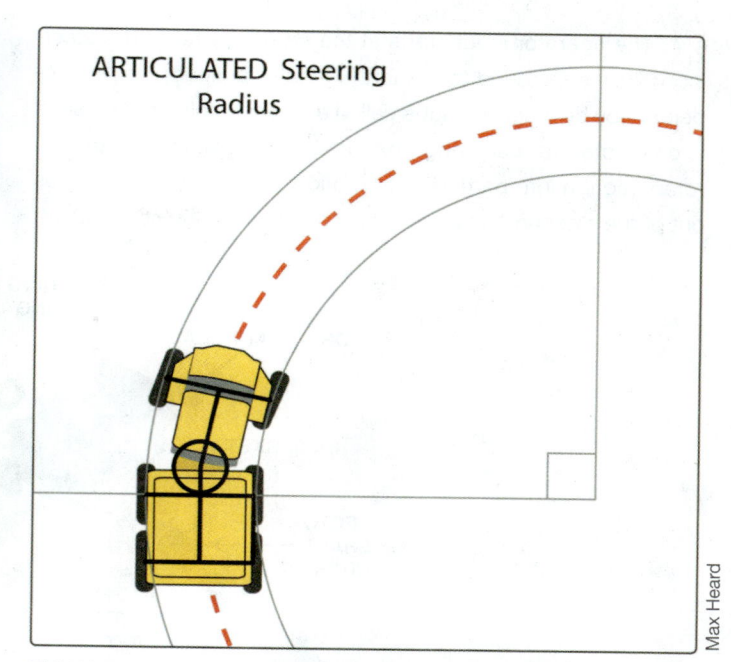

FIGURE 19–5 Articulated machine's turning radius.

multi-axle arrangements will have two or more axles at each end of the machine that are steer axles. ● **SEE FIGURE 19–4** for a crane with many steering axles.

Articulated steering machines have a two-piece frame and a steering system to make the two-piece frame pivot on a center hinge pin arrangement. This pivoting action makes the machine turn in the direction that it pivots, and the greater its pivot angle, the smaller turning radius it will travel through. ● **SEE FIGURE 19–5** for an articulated machine's turning radius.

Both types of steering systems (conventional and articulated) will have some similar components for their control section, and except for very small machines that use all mechanical systems or boosted mechanical systems, they will use hydraulic systems to operate them.

The greatest load on a machine's steering system is when the machine is loaded, is stationary, has its brakes applied, and is on soft terrain. Once the machine starts to move, is on a hard surface, or is unloaded, the effort needed to steer the machine is greatly reduced. Many steering complaints will result from problems associated to the system being under load.

Tires for heavy equipment can be very expensive, and their life will be shortened by improper steering geometry. This can be caused by worn or improperly adjusted components, which cause the tires to scrub and grind rubber off their tread. It is important for an HDET to have a good understanding of machine steering systems. The rest of the chapter will discuss different types of steering systems for wheel-type machines in more detail.

MECHANICAL STEERING SYSTEMS

As mentioned, only very small or very old small machines will use mechanical steering systems, but a brief look at their operation and components is a good starting point for the other more common steering systems seen today. A machine with this type of steering system will use mechanical components to transfer operator input to pivoting the wheels that turn the machine.

WHEELED MACHINE STEERING SYSTEMS

The operator input starts at the steering wheel. The steering wheel is fastened to a shaft (keyed or splined) that is suspended by bearings in a tube called a steering column. The shaft is connected to a steering box (or steering gear) that transfers rotary motion into partial rotary motion, and this motion comes out of the steering box on a splined shaft called the sector shaft.

The direction of power flow is also changed at 90° through the steering box. There are two main types of steering boxes called worm and roller gear and recirculating ball type. These steering boxes are usually mounted to the machine's frame rail and are filled with oil to keep all internal moving parts lubricated. ● SEE FIGURE 19–6 for a manual steering box.

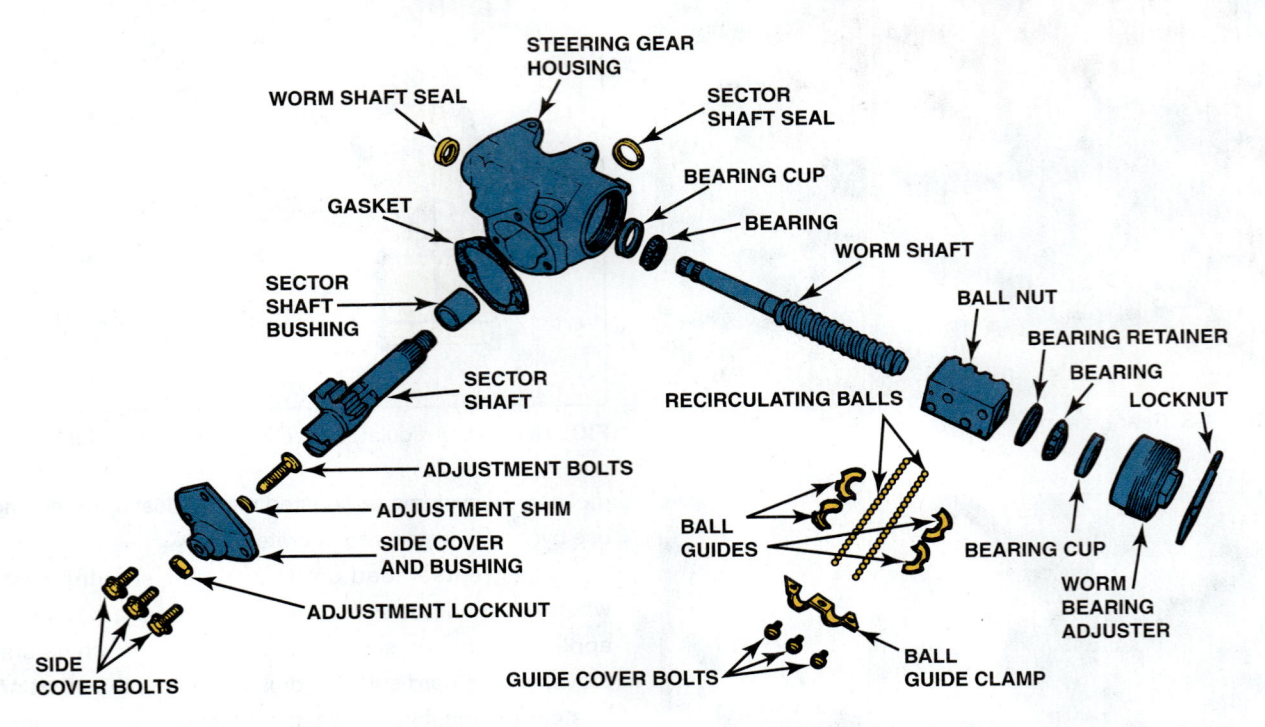

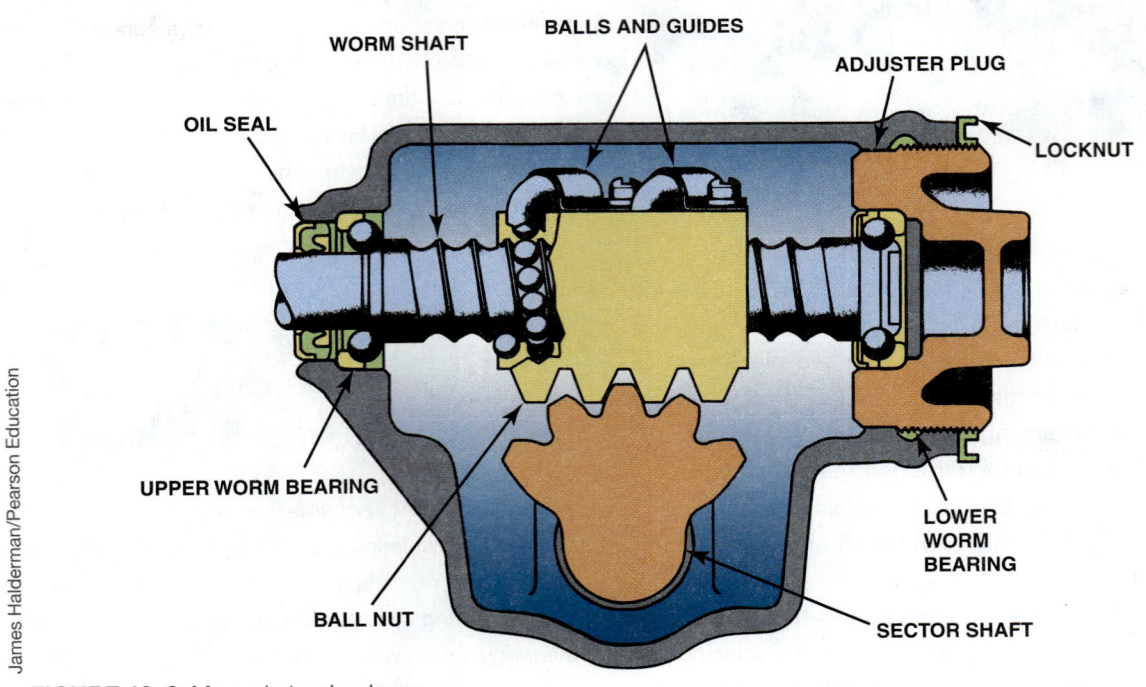

FIGURE 19–6 Manual steering box.

562 CHAPTER 19

Only very light-duty or older machines will use a manual steering box.

From the steering box output, there will then be a lever called a pitman arm attached to the steering box's splined output shaft that will transfer motion to a drag link.

The drag link may be adjustable in length and will have a ball joint on each end of it. The opposite end of the drag link is connected to a steering arm. The steering arm is connected to a steering knuckle that has a spindle as part of it. The spindle is able to swivel on a pivot point at the end of the axle housing. The pivot is either a one-piece **kingpin** or upper and lower kingpins. Attached to the bottom of one of the spindles is a tie rod arm and attached to that through a ball joint is a tie rod end. The tie rod is threaded to the left and right tie rod ends and provides a connection between left and right spindles as well as a means of adjustment for toe angle.

The spindle has two tapered roller bearings on it that allow the hub, wheel, and tire to rotate on it. The hub can be oil or grease filled. The grease or oil is retained by a dynamic seal on the inside of the hub and a cover with a gasket or seal on the outside. ● **SEE FIGURE 19–7** for the components of a typical heavy-duty steering axle arrangement.

Steering axles can just steer and carry some of the machine's weight, or they can also provide drive. Most steering axles on machines today will drive and steer. Non-drive steer axles will have a plain-steering knuckle at each end

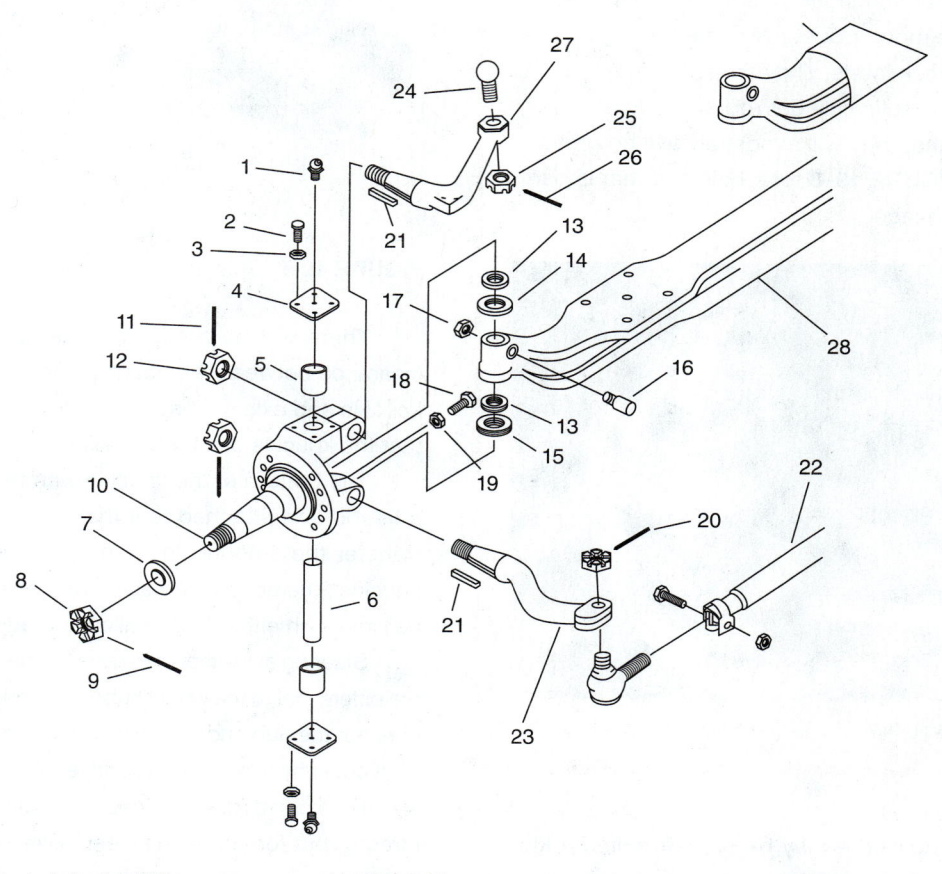

1 - Lubrication Fitting
2 - Hex Bolt
3 - Washer
4 - Knuckle Cap
5 - Bushing
6 - King Pin
7 - Washer, Spindle
8 - Hex Nut, Spindle
9 - Cotter Pin
10 - Steering Knuckle

11 - Cotter Pin
12 - Hex Nut, Steer Arm
13 - Seal Knuckle
14 - Shim
15 - Thrust Bearing
16 - Draw Key
17 - Nut, Draw Key
18 - Stop Screw
19 - Nut, Jam
20 - Cotter Pin

21 - Woodruff Key
22 - Tie Rod Assembly
23 - Tie Rod Arm
24 - Steer Arm Ball Stud
25 - Ball Stud Nut
26 - Ball Stud Cotter Pin
27 - Steering Arm
28 - I-Beam
29 - Tubular Beam

FIGURE 19–7 The components of a typical heavy-duty steering axle arrangement.

TECH TIP

Ball joints can be used for many purposes on heavy equipment. Any time a part of linkage has to transfer motion to a component or another linkage and there needs to be an allowance for some rotary motion, a ball joint is used. For control linkages that move spool valves, they are a simple swivel ball that is allowed to move inside a housing. The swivel ball will either have a stud protruding from it or have a hole drilled in it to allow a bolt through. Small ball joints are typically dry but can also have a grease fitting to allow lubrication to be applied to the swivel.

Ball joints that are part of steering linkages will provide the same functions as smaller ball joints but need to be more robust to handle the high turning and shock loads. They are sometimes threaded to hydraulic cylinder rod ends. When used for steering, they will almost always be greasable. ● SEE FIGURE 19–8 for a ball joint that is part of a steering linkage.

FIGURE 19–8 Ball joint.

that is able to pivot. There may be a shaft called a kingpin that is held stationary in the steering axle, and the steering knuckle has pressed in upper and lower bushings that allow movement between the knuckle and axle. Another arrangement will have the knuckle pivot on upper and lower kingpins that are held in the knuckle and protrude into bushings that are pressed into the axle housing. Lighter-duty axles will use plain bushings while most other axles will use tapered roller bearings. ● SEE FIGURE 19–9 for a steering knuckle with upper and lower kingpins.

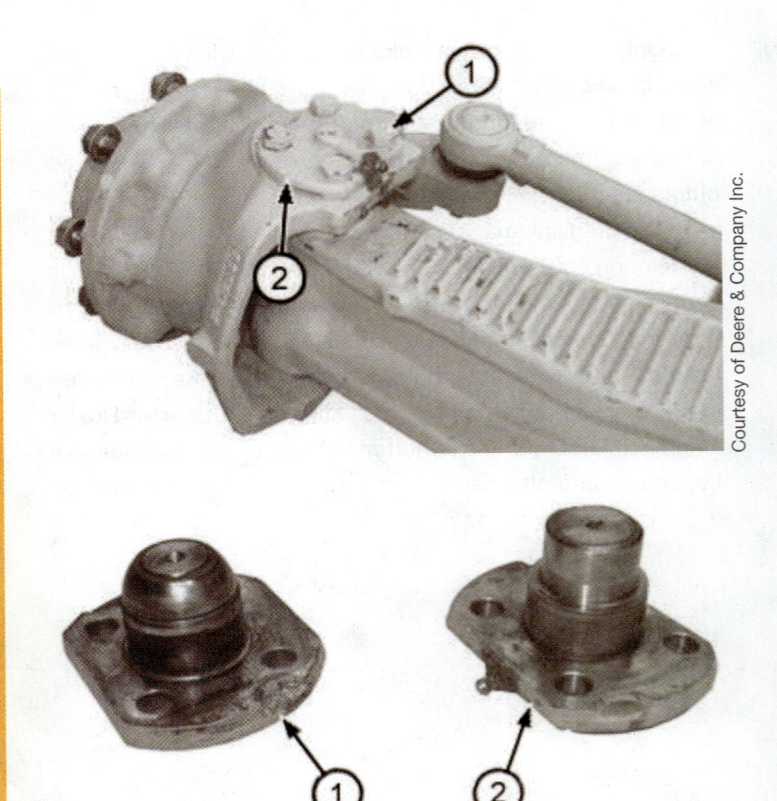

FIGURE 19–9 Steering knuckle and upper and lower kingpins.

There will be grease fittings to allow grease to fill the clearance between the bushings and the kingpin. The steering knuckle and axle will have stops to limit the amount of degrees that the knuckle can turn.

The bottom of the knuckle with the steering arm will have a tie rod arm attached to it that allows a connected tie rod to transfer the steering movement to the other side's steering knuckle. The tie rod will have threaded tie rod ends that include ball joints on either end to allow toe adjustment and pivoting.

Steering axles that also drive will need to transfer torque from the differential to the wheel hub. There needs to be an axle shaft to connect the hub and differential side gear. ● SEE FIGURE 19–10 for a cutaway view of a steer/drive axle.

For light-duty machines, the axle shaft will drive the hub directly, but for almost all steer/drive axles, the shaft will drive another gear reduction that is part of the steering knuckle called a final drive. This will be a planetary gearset that is filled with oil. See Chapter 15 for a complete explanation of drive/steer axles. ● SEE FIGURE 19–11 for a drive/steer axle steering knuckle.

HYDRAULIC ASSIST STEERING SYSTEMS As machine size and weight increase, there needs to be a system to assist the operator's steering effort to pivot the steering wheels.

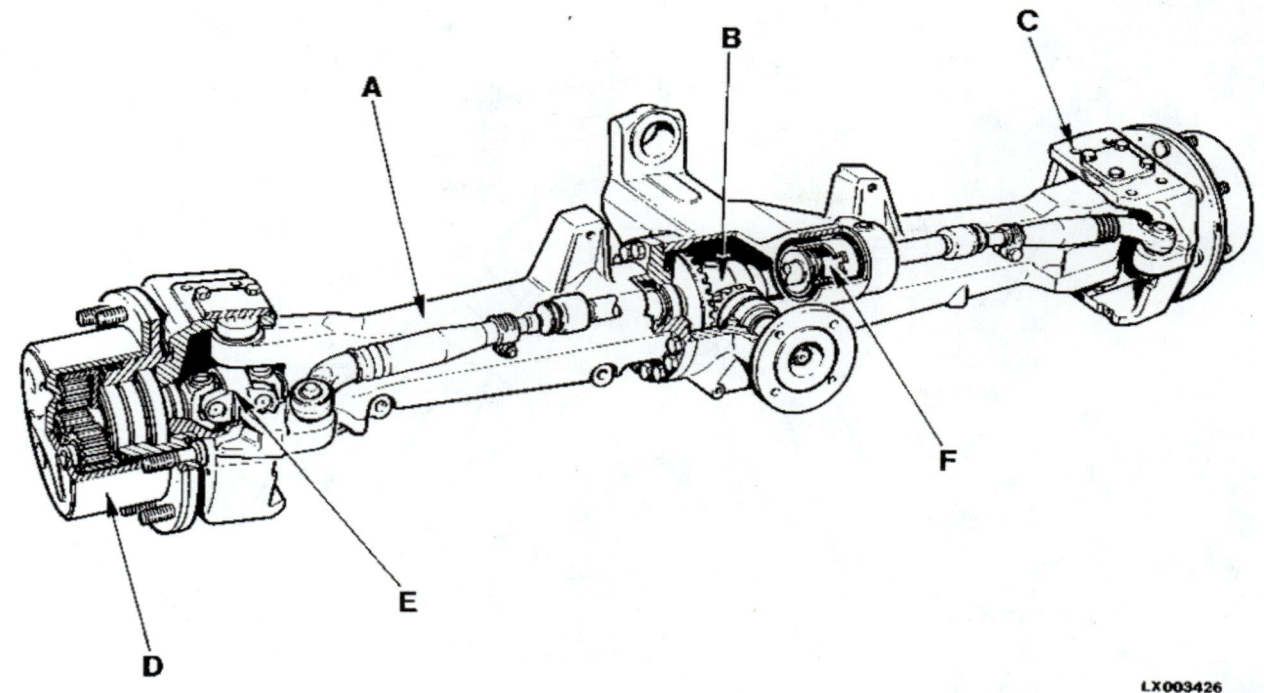

A- Axle housing
B- Differential
C- Right steering knuckle
D- Left final drive
E- Constant velocity joint
F- Right axle shaft

FIGURE 19–10 Steer/drive axle.

This requires a hydraulic assist system to be integrated into the steering box that would then move all the normal steering components past it with greater force. ● SEE FIGURE 19–12 for a steering box with hydraulic assist.

This could also be accomplished by adding a hydraulic cylinder to the steering linkage.

This steering arrangement first evolved into a power-assisted steering system with the addition of a low-pressure hydraulic pump, reservoir, and the replacement of the manual steering box with a power steering unit. The hydraulic steering box will use pump oil flow to apply pressure to a piston inside the unit. The oil flow is controlled by a rotary valve that gets turned by the operator. The piston then moves up or down depending on which way the steering wheel is turned. The piston has a rack of teeth machined on the outside of it that engage with a sector gear and will turn the sector gear as it moves. The sector gear turns a splined shaft that extends outside of the steering box and the pitman arm is turned by the sector shaft. The oil pressure for this system is usually limited to 1500 psi, but this provides plenty of power assist to allow the operator to easily turn the steering wheel and make a heavily loaded machine steering wheels pivot. ● SEE FIGURE 19–13 for a cross-section of a power steering box.

Manual steering units can be adjusted to reduce steering wheel play. Power steering units can be tested with a flow meter. Manual and power steering boxes can be rebuilt by a technician, but if found to be defective, they are usually replaced with a rebuilt or new unit.

Most rubber-tired machines won't provide for much as far as adjustments to the geometry of steering axle components. There are a few terms you should understand when it comes to steering geometry, however:

Caster: If you look at a machine from the side and draw a straight line through the center of the kingpins, then compare this to a line that is 90* to the flat surface that the machine is sitting on you would find a difference of a few degrees. When the top of

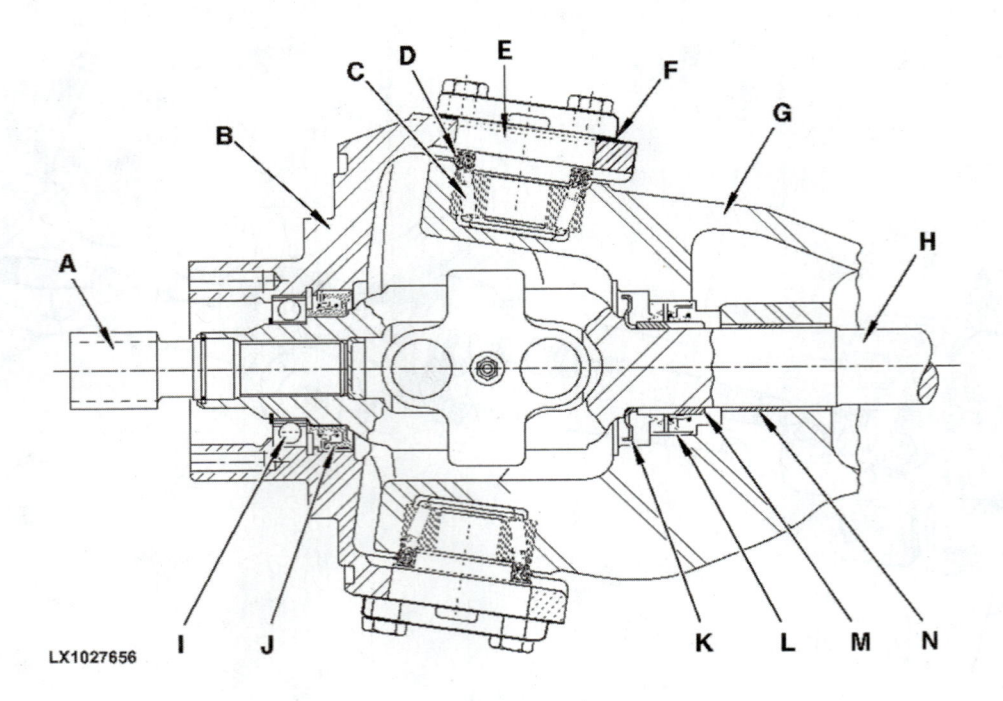

A- Shaft to wheel
B- Steering knuckle
C- Tapered roller bearing
D- Seal
E- Upper kingpin
F- Shim
G- Axle housing
H- Axle shaft
I- Roller bearing
J- Seal

FIGURE 19–11 Drive steer axle steering knuckle.

FIGURE 19–12 Power steering box.

the kingpin is tilted backwards it is considered to be positive caster. Positive caster help to self-center the wheels after a turn and make the machine more stable as speed increases. The only way to adjust caster angle on a heavy-duty steering axle is if it has provision to put tapered shims between the axle and its leaf springs. It's this part of steering geometry that makes one side of a vehicle lift as the wheels are turned. ● SEE FIGURE 19–14 for an automotive example of negative caster.

Camber: If you look at one of the machine's steering axle wheels from the front of the machine, you would see that it may be leaning slightly at the top one way or the other. For an extreme example of camber, think of oval track racing cars that only turn left when

566   CHAPTER 19

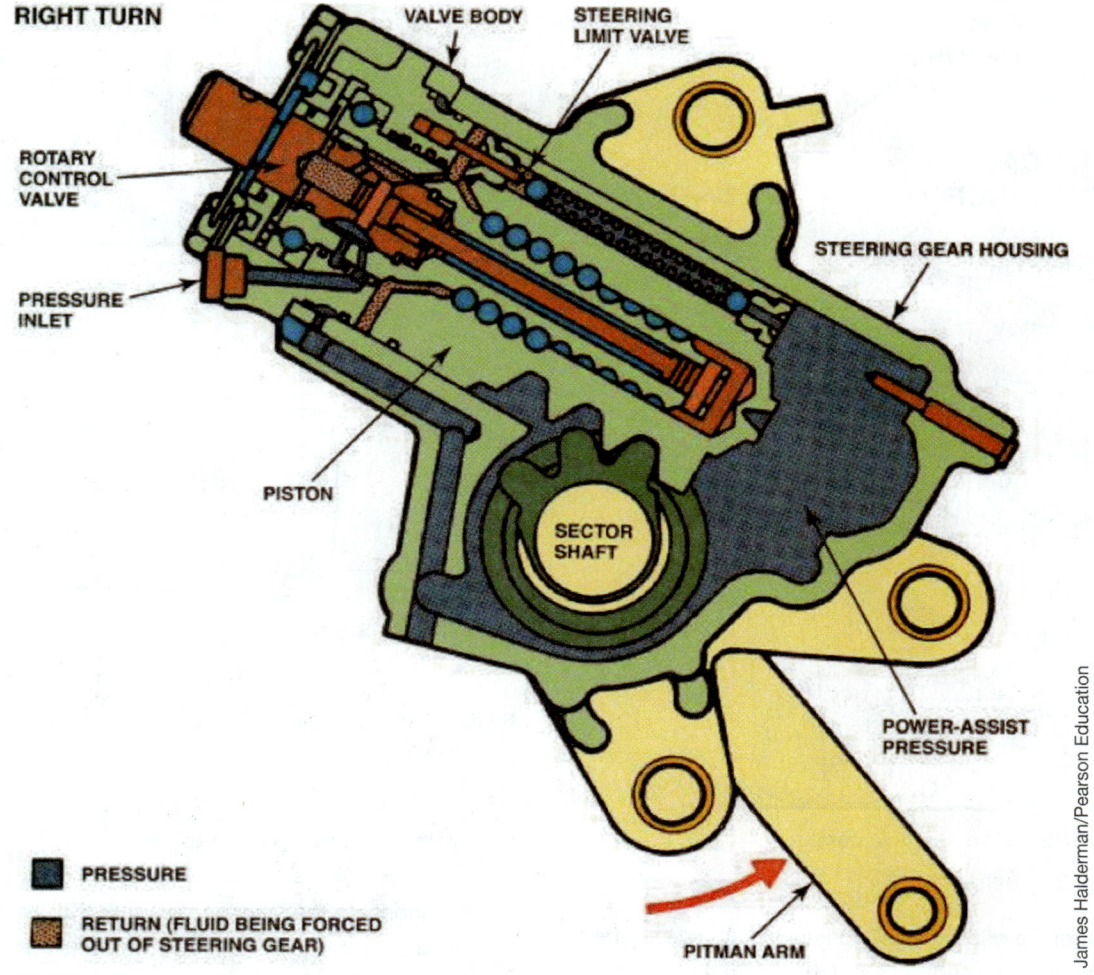

FIGURE 19–13 Power steering box cross-section.

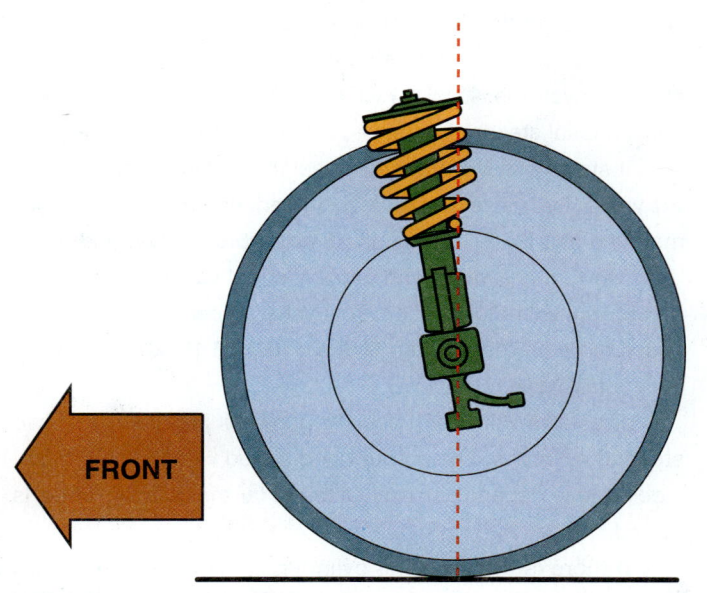

FIGURE 19–14 Negative caster.

racing. Their right front wheel will have a very noticeable camber angle to counteract the weight transfer when turning left hard. Most machines will have a 0–1° camber angle, and this is non-adjustable.
● **SEE FIGURE 19–15** for a positive camber.

Toe angle: If you were able to look down on the steering axle from high up, you may notice that the forward inside edge of the tires are closer together than the rear inside edge. This is considered to be a **toe-in** dimension. As the machine's speed increases, there is a tendency for the wheels to toe out. A slight toe in is the most common setting. This is one angle that can be adjusted on heavy-duty steering axles by changing the length of the tie rods. ● **SEE FIGURE 19–16** for what toe angle looks like.

Ackerman angle: This is the angle created between an imaginary point that extends from the centerline of the rear axle to one side of the machine and intersects

WHEELED MACHINE STEERING SYSTEMS **567**

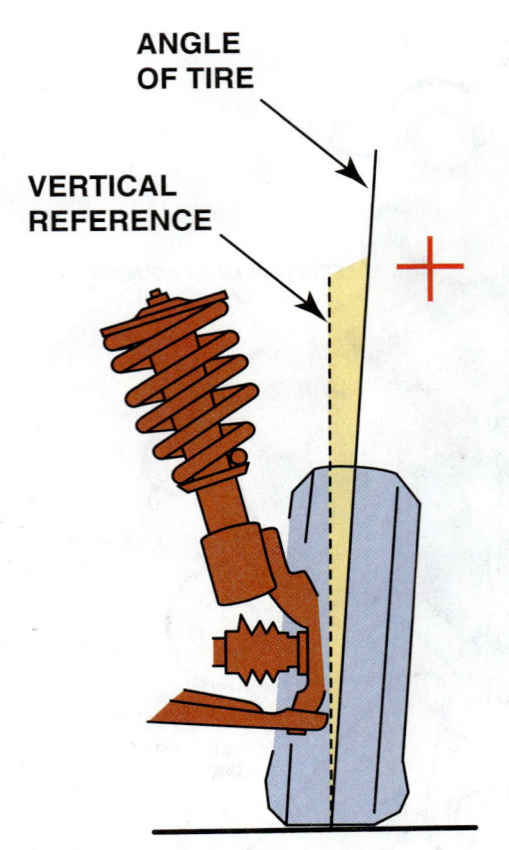

FIGURE 19–15 Positive camber.

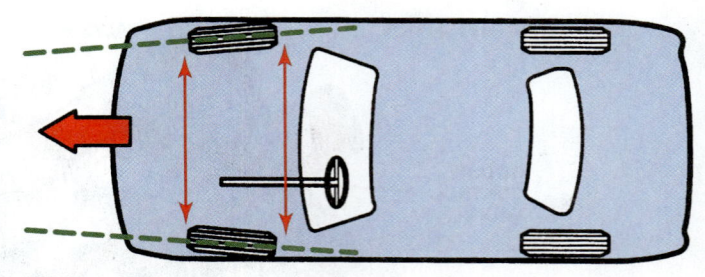

FIGURE 19–16 Toe angle.

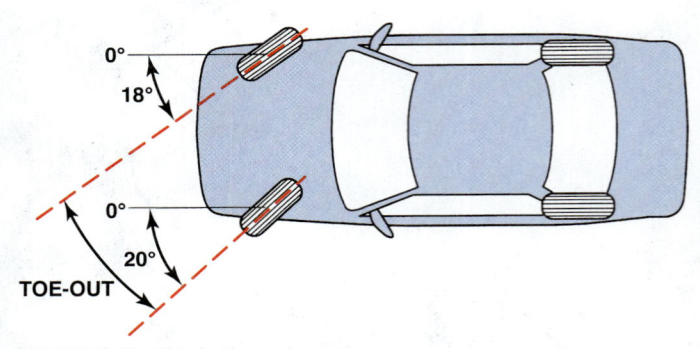

FIGURE 19–17 Ackerman angle.

the centerline of both steering knuckle spindle shafts when they are turned. As a machine with a steering axle is turned, the steering arms that are used to pivot the wheels will naturally turn each wheel at differing individual angles. The angles differ proportionally to the amount of steering input that is put into the tie rods.
● **SEE FIGURE 19–17** for an **Ackerman angle**.

As the steering wheels are turned farther, the geometry between the tie rod and tie rod arms will naturally provide a differing of steering angles between the left and right steering wheels. This will make the wheels travel through different arcs as the machine moves through a turn, and this will decrease tire scrub to lengthen tire life. If the steering components are designed, adjusted, and functioning properly, the Ackerman angle intersection points should meet during all points of left and right turns.

CONVENTIONAL STEERING SYSTEMS—FULL HYDRAULIC
Almost all steering systems on machines today that have a conventional steering system (straight frame with front wheels pivoting) will use a hydraulic system to either control it or actuate it. The steering system will likely be part of the machine's main hydraulic system, and the system will be designed to ensure that the steering circuit always has sufficient oil flow to allow the operator to safely steer the machine under any condition. This is sometimes called a priority valve because it gives priority to the steering circuit. ● **SEE FIGURE 19–18** for a hydraulic schematic of a forklift that has a flow divider valve to ensure that the steering circuit always has oil.

Larger machines such as mining haul trucks will have a completely separate hydraulic system for steering (tank, pump, valve, filter, lines).

The full hydraulic steering system will replace the steering box, pitman arm, and steering arm of older conventional steering systems. ● **SEE FIGURE 19–19** for a machine with conventional steering system—full hydraulic.

It starts with the steering wheel that turns a steering control valve. This valve has several different names such as hand-metering unit (HMU), metering pump, **orbital valve**, and Char-Lynn valve. It is a valve with a rotary spool valve section and a gerotor pump/motor section. This valve's operation will be discussed in more detail later. ● **SEE FIGURE 19–20** for a steering control valve.

The valve will direct steering pump oil flow to one or two steering cylinders. The amount and speed with which the cylinders move depends on how fast and far the steering wheel is turned.

If there is one steering cylinder, it will be a **double rod cylinder** that has each of its rods attached to a tie rod. The tie rods are attached to the steering knuckles that pivot on the

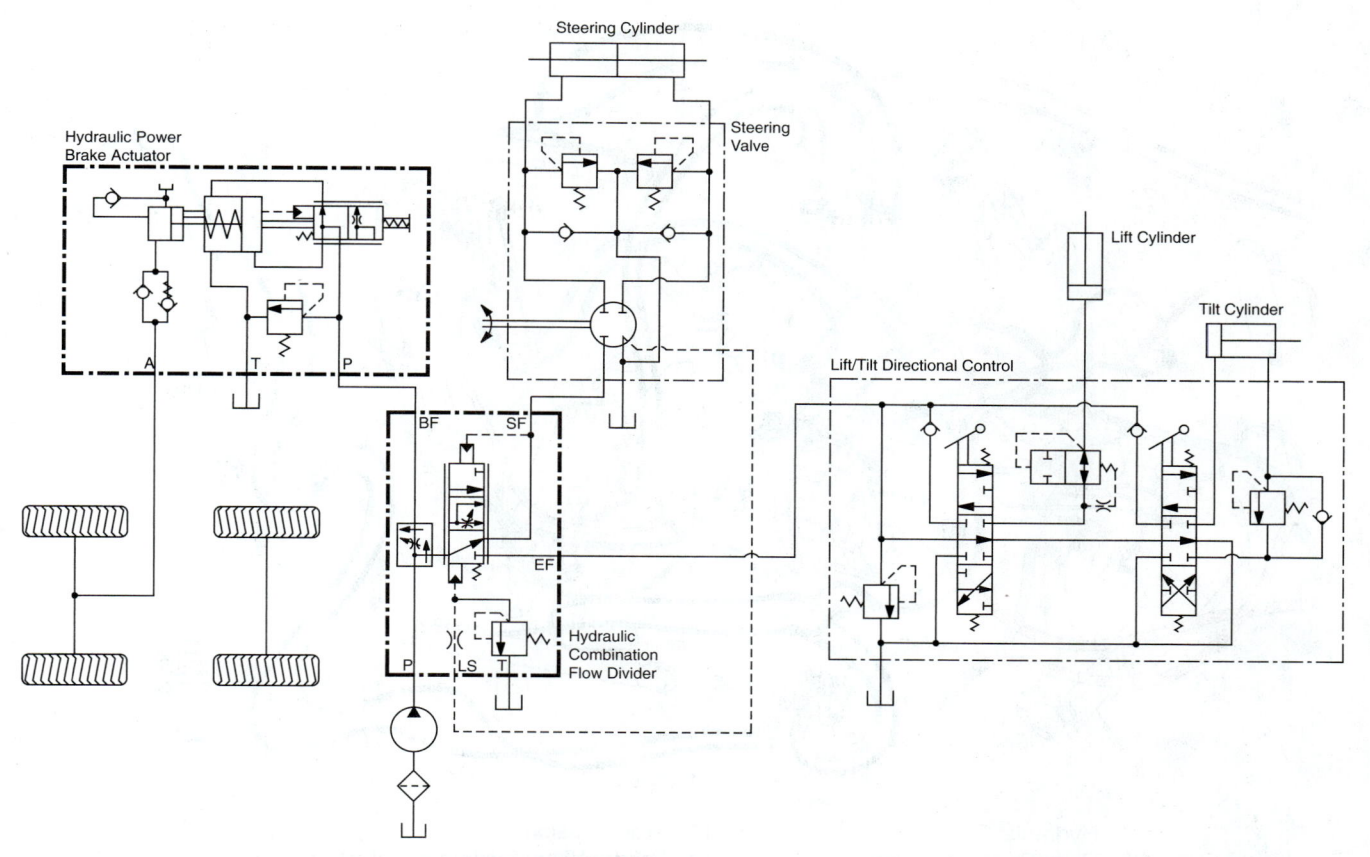

FIGURE 19–18 Forklift hydraulic system with flow divider.

outer ends of the axle housing. As the cylinder has oil sent to it from the steering valve, the rods will move left or right, and this causes the wheels to steer left or right. Oil must also leave the cylinder, and it returns to tank through the steering valve. This is a common arrangement for backhoe loaders, rough terrain cranes, and telehandlers. ● SEE FIGURE 19–21 for a double rod steering cylinder.

If there are two steering cylinders, they will be fed oil from the machine's steering control valve. It will send oil to one cylinder's head end and one cylinder's rod end. This will move one rod in and one rod out to move each wheel. There will also be a mechanical link between the wheels to ensure that they move together and the Ackerman angle stays correct. This can be a one-piece tie rod or a two-piece tie rod that is attached to a center pivoting link. ● SEE FIGURE 19–22 for the steering cylinder and tie rod arrangement for a 100-ton rock truck.

This is a common arrangement for a large straight frame haul truck or some larger rough terrain cranes.

ARTICULATED STEERING SYSTEMS
There are many machines that use two-piece main frames such as wheel loaders, articulated truck, graders, skidders, forwarders, feller/bunchers, and compactors.

The two-piece frame for these machines will have a pivot point that can be as simple as a top and bottom stationary pin that is supported by two plain bushings or top and bottom pins that are supported by tapered roller bearings. These pins require regular lubrication.

The steering systems for today's articulated machines will turn the machine by pivoting the front and rear frames to the left or right. Some very small machines will use a single cylinder to do this, but the vast majority of machines will use two steering cylinders. They typically have their head end fastened to the rear frame and their rods attached to the front frame. A left steer would be made by sending oil to one left cylinder's rod end and one right cylinder's head end. Reversing oil flow to the cylinders will make the machine pivot the opposite direction. ● SEE FIGURE 19–23 for an articulated machine frame and steering cylinders.

Machines with a steering wheel will use a steering control valve that is identical to the one used for conventional steering systems. It will meter oil to the steering cylinders.

Some machines will feature joystick-controlled steering. The joystick control can be a valve that is supplied pilot oil from the machine's low-pressure system and then sends metered oil to a main control valve. The amount of oil metered is determined by how far the joystick is moved.

WHEELED MACHINE STEERING SYSTEMS

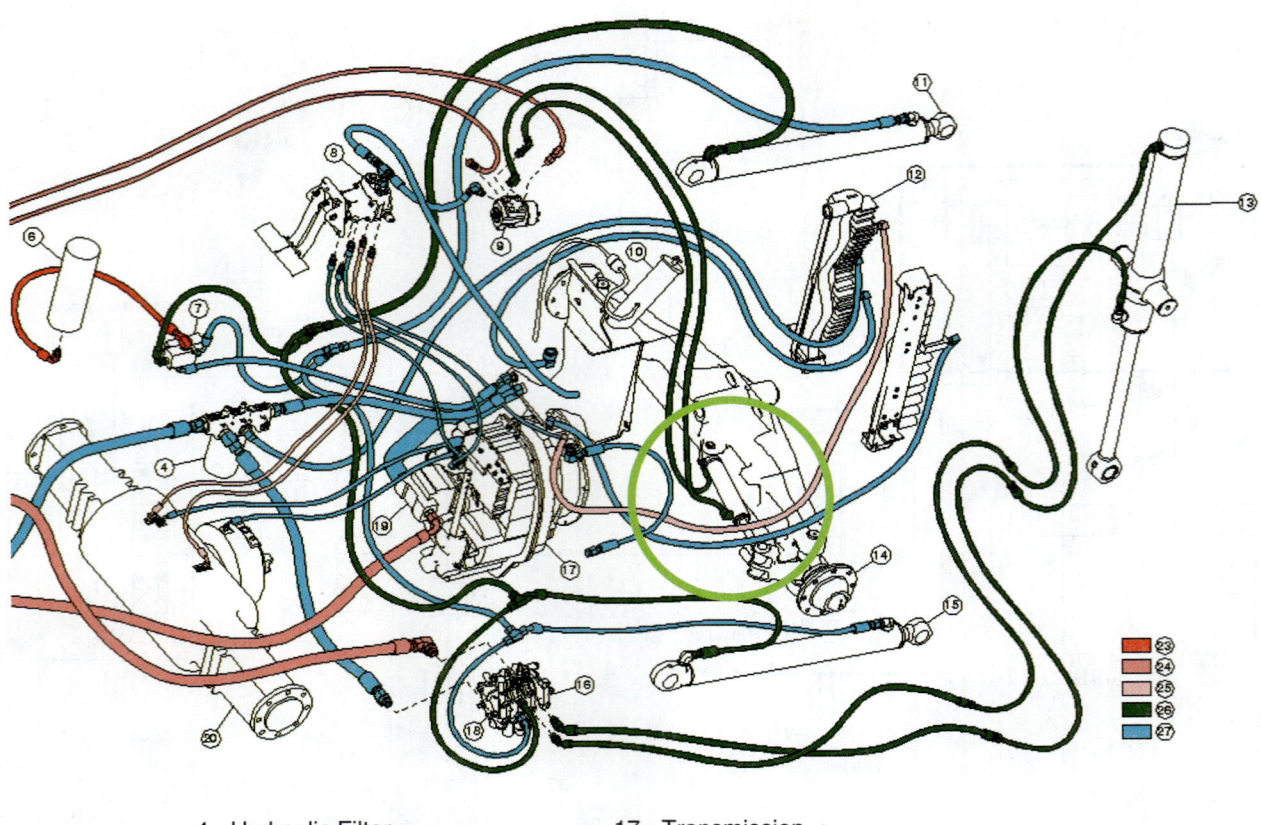

4 - Hydraulic Filter
5 - Right Stabilizer Cylinder
8 - Service Brake
9 - Steering Valve
10 - Hydraulic Reservoir
11 - Left Loader Cylinder
12 - Hydraulic Cooler
13 - Bucket Cylinder
14 - Front Axle
15 - Right Loader Cylinder
16 - Loader Valve
17 - Transmission
18 - Loader Valve System Relief Valve
19 - Hydraulic Pump
20 - Rear Axle
23 - High Pressure Oil
24 - Medium Pressure Oil
25 - Low Pressure Oil
26 - Trapped Oil
27 - Return Oil
Steering cylinder in circle

FIGURE 19–19 A machine with conventional steering system—full hydraulic.

Courtesy of Deere & Company Inc.

The main control valve then sends oil to the steering cylinders. Another version of joystick steering is one that uses a joystick to send proportional electrical signals to an ECM. The ECM then sends a proportional signal to an electrohydraulic control valve's solenoids. When the solenoid is energized, it will shift a spool that sends oil to the cylinders. There are machines that can be steered by either a steering wheel or a joystick. The two modes can be changed by an electrical solenoid-actuated valve.

Some older systems used a mechanical follow-up linkage that is moved by the pivoting frames to re-center the spool.

ALL-WHEEL STEERING SYSTEMS
Some types of machines with straight frames such as telehandlers, backhoe loaders, and rough terrain cranes will feature all-wheel steering systems. This means that both the front and rear axles can steer. These machines are usually able to switch to one of four following different modes:

Conventional steering: Only the front axle wheels steer.

All-wheel steering: Both axles steer at the same time in the opposite direction to make the machine turn in a smaller radius.

Crab steering: Both axles steer at the same time in the same direction to make the machine move diagonally.

Rear axle steering: Some forklifts will have the ability to steer only the rear axle wheels.

● SEE FIGURE 19–24 for the different modes of all-wheel steering system.

570   CHAPTER 19

When the other solenoid is energized, the spool is shifted to the farthest opposite position, and this will make all wheels turn in the same directions for **crab steering**.

The operator control for this could be as simple as a three-position toggle switch or a switch that is an input to an ECM. There could also be wheel position sensors that the ECM needs to see information from to see if the wheels are straight ahead when they are supposed to be.

Progress Check

1. All these machines feature conventional steering but one can also have articulated steering:
    a. tractor loader backhoe
    b. mining haul truck
    c. fork lift
    d. grader
2. An articulated machine will feature this:
    a. a two-piece main frame
    b. a steering box
    c. a pitman arm
    d. four steering knuckles
3. The greatest load is placed on steering components when:
    a. the machine is stationary and there is a steering action
    b. when the machine is loaded going uphill
    c. when the machine is loaded going downhill
    d. when the machine is reversing with a full load
4. If you were determining a machine's toe angle, you would:
    a. measure the lean the steering wheels have
    b. calculate the difference in dimension between the inside edges of the front and rear of the tires
    c. measure the tilt of the kingpin
    d. measure the pitman arm angle at full turn
5. This is *not* one mode of steering for an all-wheel steer machine:
    a. articulated
    b. crab
    c. conventional
    d. rear axle

FIGURE 19–20 A steering control valve.

To switch between steering modes, a three- or four-position, four-way, solenoid-operated directional control valve is installed between the front and rear steering cylinders. When the valve's two solenoids are not energized, the valve spool is centered by springs. This position blocks oil flow to the rear axle steering and the system performs steering in the conventional mode.

When one of the valve's solenoids is energized, it will shift the spool and allow oil flow to go to both axles for steering. This will steer both axles in the opposite direction for all-wheel steer.

FIGURE 19–21 Double rod cylinder.

WHEELED MACHINE STEERING SYSTEMS 571

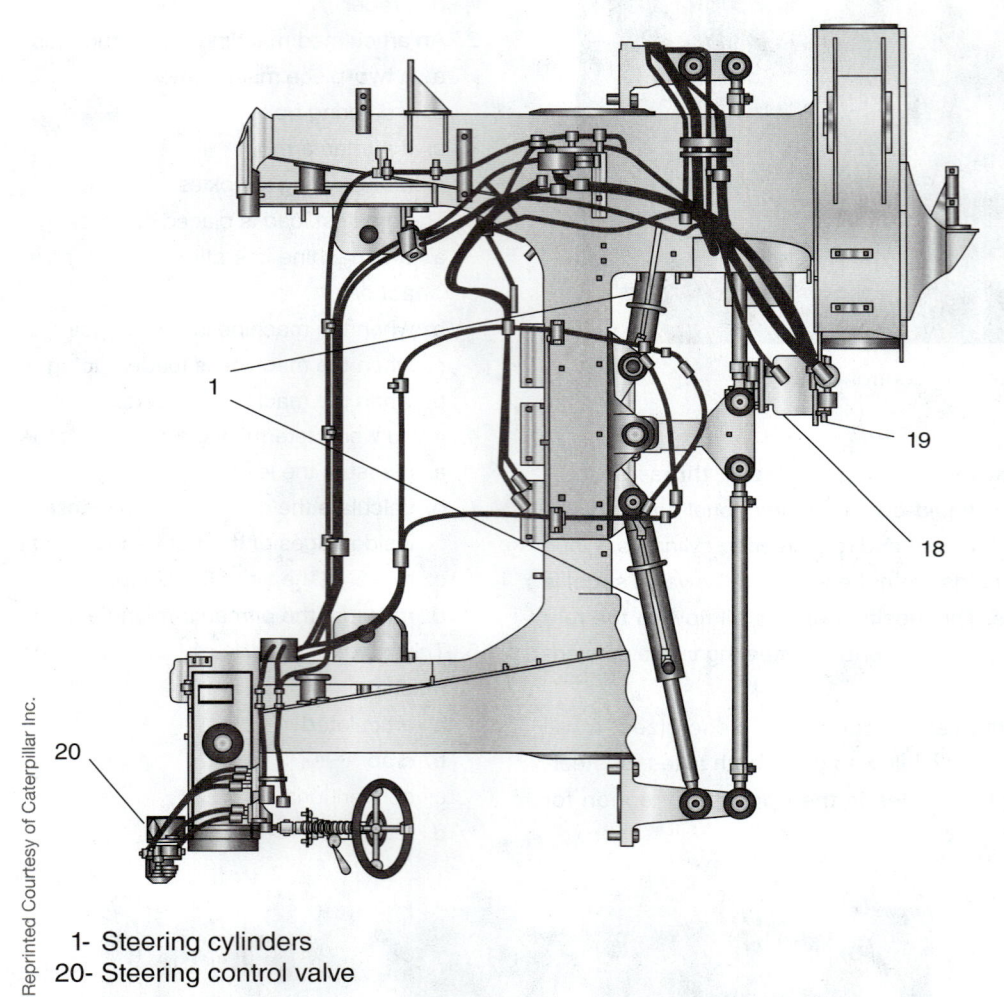

1- Steering cylinders
20- Steering control valve

FIGURE 19–22 The steering cylinder and tie rod arrangement for a 100-ton rock truck.

572   CHAPTER 19

FIGURE 19–23 An articulated machine frame and steering cylinders.

'CRAB' Steering

'ALL-WHEEL' Steering

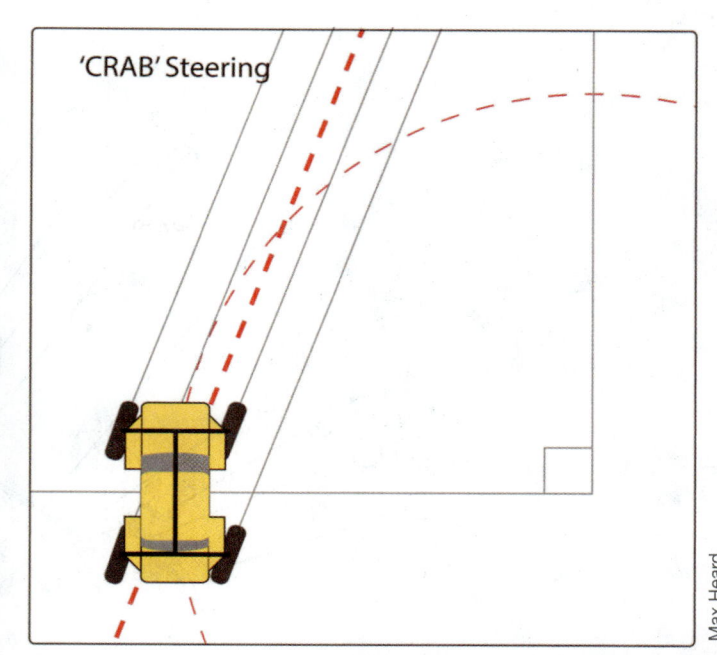

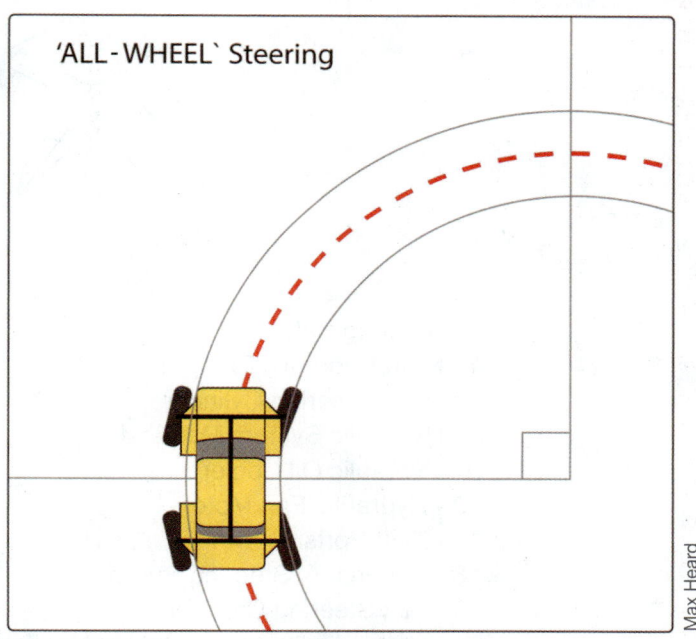

FIGURE 19–24 All-wheel steer mode.

HYDRAULIC STEERING SYSTEM OPERATION

Hydraulic steering system components and operation will vary between different machines, but in general the following is true for all machines that have a hydraulic steering system:

There needs to be a pump to create oil flow. The pump can be a dedicated steering pump and this will be the case for most large machines, or steering oil flow can be shared with main hydraulic system pump flow. If the steering system uses pump flow from the main pump, there will be a priority valve that makes sure that the steering circuit gets its flow requirements satisfied before any other circuits.

There needs to be a way to direct oil flow to the steering cylinders. This can be done with a rotary steering control valve that is turned by a steering wheel or with a joystick that sends either pilot oil to a main control valve or an electrical signal to an ECM. In the latter case, the ECM then sends oil to a solenoid-operated directional control valve. The valve will also direct return oil back to tank.

There needs to be one or more linear actuators (cylinders) to convert pump oil flow into movement.

There also needs to be all the other components that are normally needed to make a complete hydraulic system such as

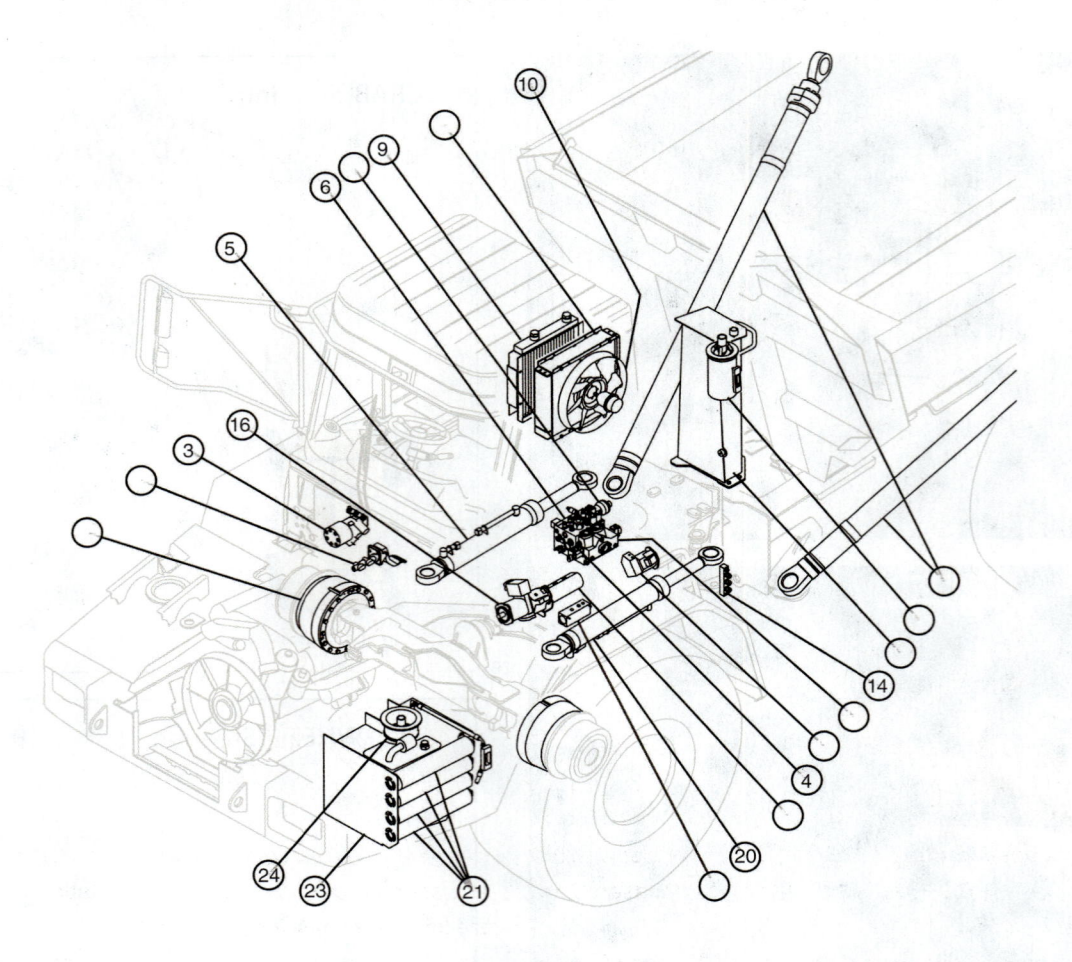

3 - Steering Valve
4 - Main Hydraulic Pump
5 - Right Steering Cylinder
6 - Hydraulic System Manifold
9 - Hydraulic Oil Cooler
10 - Hydraulic Fan Motor
14 - Test Ports
16 - Secondary Steering Pump
20 - Left Steering Cylinder
21 - Hydraulic Reservoir Return Manifold
23 - Hydraulic Reservoir
24 - Return Filter

FIGURE 19–25 Articulated truck hydraulic system.

fluid, one or more filters, a tank and lines, fittings, gaskets, and seals. ● **SEE FIGURE 19–25** for a look at an articulated truck's hydraulic steering system.

HYDRAULIC STEERING PUMP
For simple systems, the steering system could share oil flow with another machine hydraulic circuit because steering flow is usually not very high even in high-demand steering situations. The pump could be a fixed displacement gear or vane pump, but most systems will use a variable displacement pump that gets a load-sense signal from the steering valve. For steering systems with their own pump the pump could be external gear, vane or piston type.

ROTARY STEERING VALVE OPERATION
As mentioned earlier, this valve has several different names such as HMU, steering control unit, orbital valve, and Char-Lynn valve. It is used in many types of steering systems, and in general, it

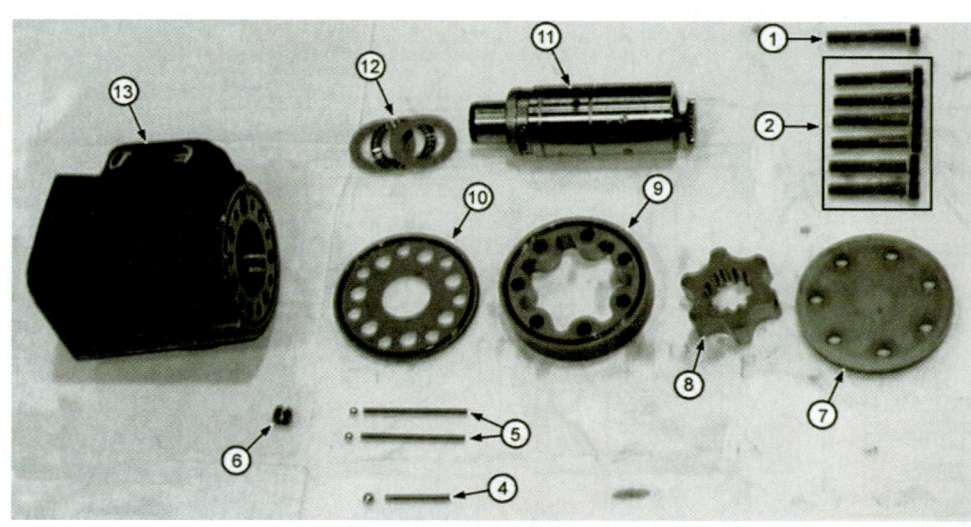

FIGURE 19–26 A disassembled steering valve.

1,2- fasteners
4,5- springs
6- ball seat
7- cover
8- rotor
9- housing
10- plate
11- rotary valve
12- washers
13- housing

Courtesy of Deere & Company Inc.

sends steering pump oil to one or two steering cylinders and returns oil back to the tank. It is controlled by the operator turning its input shaft and is a rotary directional control valve. The input shaft is splined to a shaft that is turned by the machine's steering wheel such as in a steering column or the steering wheel could mount directly to the valve depending on the machine. ● **SEE FIGURE 19–26** for a disassembled steering valve.

There are several different variations of this type of valve depending on the system it is used in, and they can be used with fixed displacement pumps or in a load-sensing pressure-compensated system. Most variations of these valves will allow the operator to steer the machine manually if there is no pump flow. This will happen if the engine quits running or the pump fails.

The input to the valve is through the inner section of the rotary valve. The inner section is allowed to turn a few degrees inside of an outer sleeve. These two pieces will always try to stay centered with each other due to a set of leaf springs that link them. The inner and outer valve sections of this unit are part of the rotary valve and can stop or allow oil flow between them based on their rotational relationship to each other. This is similar to the action between a spool and body in a spool-type directional control valve.

Inside the inner sleeve is a shaft that transfers drive from the sleeve down to the inner rotor of the gerotor pump/motor. When the steering wheel is turned continuously, the inner sleeve, outer sleeve, and gerotor all turn together inside the valve's outer housing. The outer housing has four ports that are threaded to accept hydraulic lines for: P, pump; T, tank; A, left; B, right. If the valve is part of a load-sensing system, it will have a fifth port to send a load-sense signal to the pump.

To get an idea of how oil flows through one of these valves, (● **SEE FIGURE 19–27**).

The steering valve (O) consists of a spool (I) inside a sleeve (H) within a housing. When the steering wheel is not moving, the valve is in the neutral (A) position. In neutral, the spool and sleeve are aligned so that oil flow through the valve is blocked. The steering cylinders (B) are held stationary by trapped oil in the left and right work ports.

When the steering valve is turned to the right, the inner sleeve rotates relative to the outer sleeve and opens passages that allow pump flow through the spool and sleeve assembly. Oil flows to the gerotor (E), and from the gerotor flows back into the valve where it is directed out the right work port to the respective ends of the steering cylinders.

A bypass orifice is machined into the spool and sleeve assembly. It is a variable orifice that introduces a small leak into the pressure side of the steering valve. Its purpose is to dampen the initial pressure surge when the steering wheel is partially turned. When the steering wheel is fully turned, the leak is closed off.

Return oil flows back in through the left work port through the spool and sleeve assembly to return. The load-sensing orifice is located between the sleeve and the gerotor. This orifice feeds the load-sensing pressure to the priority valve.

When the rotation of the steering wheel stops, the gerotor gear continues to move, turning the sleeve, until the sleeve stops the flow to the gerotor. At this point, the valve is back in the neutral position and will remain there until the steering wheel is moved again.

The valve provides a variable steering rate that is proportional to the speed the steering wheel is rotated. A variable

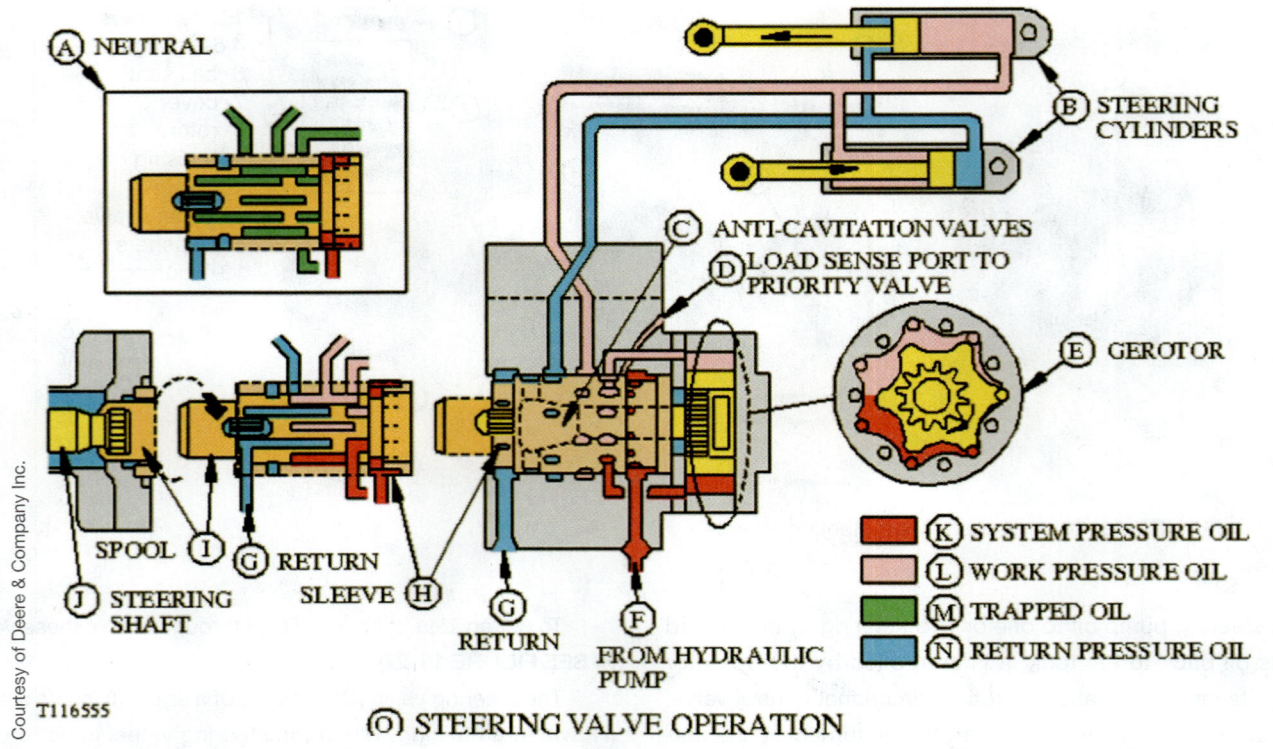

FIGURE 19–27 HMU oil flow.

orifice bypasses oil around the gerotor. Turning the steering wheel full lock to lock takes approximately 1.5–4.5 turns, depending on machine model. Four anti-cavitation balls are located inside the spool/sleeve to prevent cavitation in the system.

Crossover relief valves in the steering valve protect the steering circuit from spikes caused by sudden excessive load. For example, if an external force tries to make the machine move and the steering valve is in neutral, the crossover relief valve opens and lets the oil pressure relieve to tank. This may happen if a stuck machine is being towed out.

JOYSTICK STEERING CONTROLS Some articulated machines will provide an option to steer the machine with a joystick in addition to a steering wheel or by joystick only. ● **SEE FIGURE 19–28** for a large wheel loader that uses only a joystick for steering.

If it is used on a machine equipped with a powershift transmission, the joystick could also have a switch for FNR and two switches for upshifting and downshifting. There could be other switches on it as well.

Joystick steering could also be used to actuate the steering wheels on a conventional steering axle for a slow moving machine like a man lift or for slow maneuvering with a wheel-type excavator.

Joysticks used for steering could actuate poppet-type pilot oil control valves that send a proportional oil pressure signal to a main control valve, or they could be electrical position sensors that send an electronic proportional signal to an ECM.

This signal gets processed by the ECM and then sends an output signal out to one of the main steering control valve solenoids ● **SEE FIGURE 19–29** for an electronically controlled joystick steering control valve.

> **TECH TIP**
>
> Rotary steering valves can be rebuilt. However, unless you have experience with reconditioning these valves, it is recommended that you replace a faulty unit with either a rebuilt or a new valve assembly. To be properly reconditioned, they require a lot of patience, some special tools, and a clean, well-lit work area.
>
> Some tasks are better left to technicians who are specialized in one area and work on only a limited number of components on a regular basis. Reconditioning a HMU is a good example of a repair procedure that should be left to someone else to perform.

The main control valve could have the capability to give a feedback signal to the ECM so that fault codes can be generated if the expected spool movement hasn't been met.

STEERING CYLINDERS Steering cylinders used on articulated steer machines are no different than those used for other purposes on a machine. One variation of steering cylinder that isn't very common though is the double rod type that may be used for a steering axle's wheels. Double rod cylinders have one piston with a rod attached to each end. ● SEE FIGURE 19–30 for a disassembled double rod cylinder.

Steering cylinders used for articulated machines will likely have snubbers on the end of the rod to prevent the cylinders from bottoming out.

SECONDARY STEERING Many medium to larger systems will have a secondary steering system to serve as a backup oil supply when the machine's regular steering system oil supply system fails. It will use a different pump that can be either mechanically ground driven or electrically driven. **Secondary steering** is a safety backup system to ensure that the machine can be steered to a parked position.

A **ground driven pump** is one that has its shaft turned whenever the machine is moving. It will be driven by part of the machine's driveline that is driven by the wheels and is usually mounted on the transfer case. ● SEE FIGURE 19–31 for a ground driven pump.

FIGURE 19–28 Joystick steering.

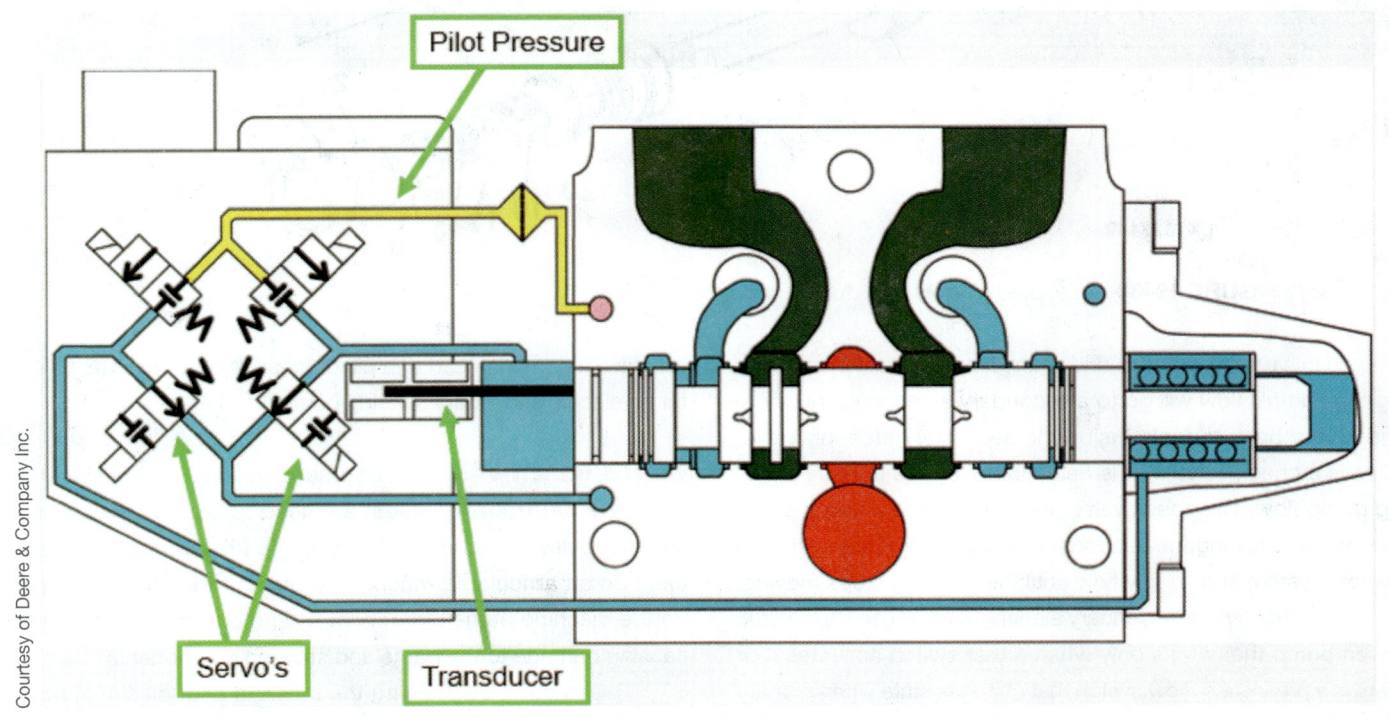

FIGURE 19–29 Electronically controlled joystick steering control valve.

WHEELED MACHINE STEERING SYSTEMS

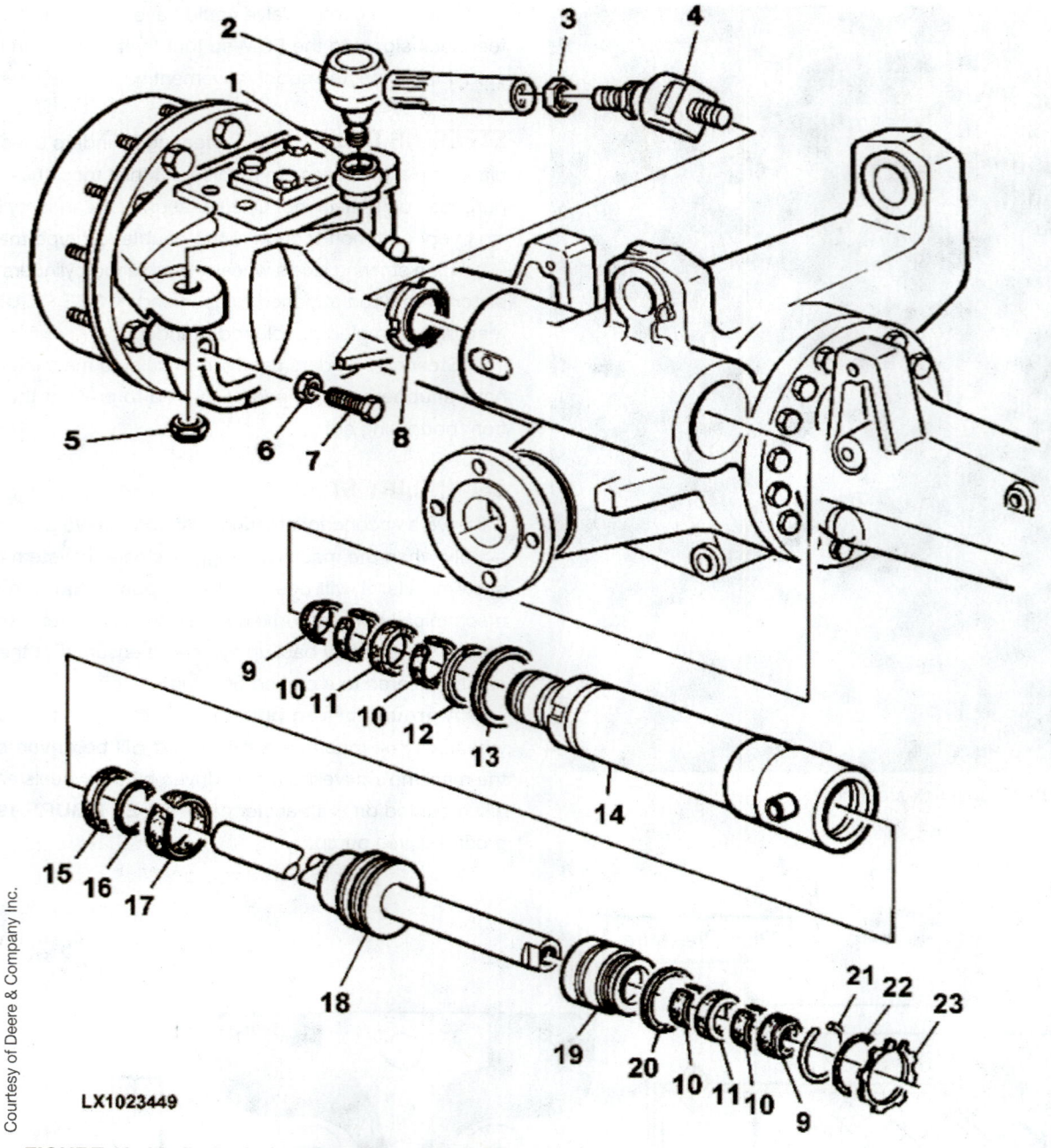

FIGURE 19–30 An exploded view of a double rod cylinder.

As this type of pump is driven any time the machine is moving the pump's flow will go to a secondary steering valve where it is routed back to tank. Inside the secondary steering valve, there is a check valve that is held closed by the primary steering pump flow. The check valve opens when the primary pump flow stops, allowing the secondary pump flow to move into the primary system and supply flow until the machine stops moving.

The other type of secondary steering system uses an electric driven pump that works only when a test switch activates it or there is a pressure or flow switch that changes state when primary pump flow stops. The motor is similar to a starter motor; only it will drive a pump instead of a pinion gear. ● SEE FIGURE 19–32 for an electric secondary steering pump.

NEUTRALIZER VALVES Hydraulic steering control systems that have **neutralizer valves**, are used to stop oil flow to the steering cylinders when the machine is fully turned. They are used on an articulated machine to prevent the steering stops on the machine frames from coming into contact and possibly causing damage to the frame and steering components.

This is done by draining the steering pilot oil signal from the pilot control valve that is going to the main control valve.

CHAPTER 19

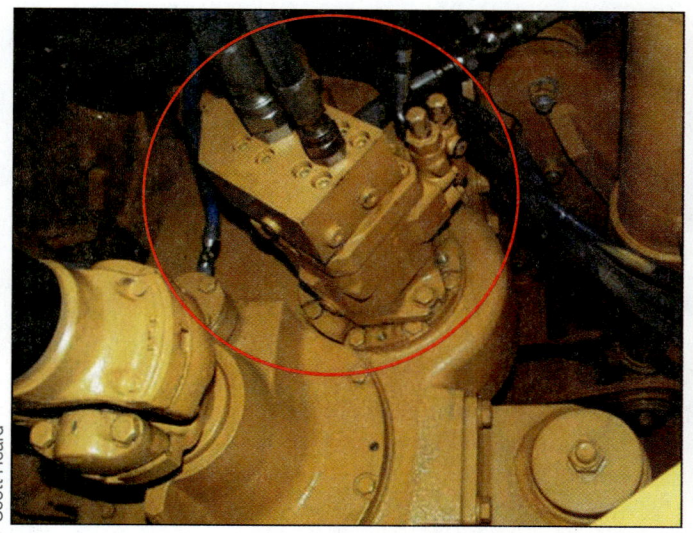

FIGURE 19–31 A ground driven pump.

FIGURE 19–32 Electric drive pump.

This allows the main control valve to return to neutral. There will be two neutralizer valves that are normally closed spool-type valves that open to tank when their spool is pushed by an adjustable stop. The stops are near the top center pin under the cab for most articulated machines. They should be adjusted to stop the steering motion just before the main frames contact each other. ● **SEE FIGURE 19–33** for a neutralizer valve.

LOAD-SENSING HYDRAULIC STEERING SYSTEMS
Most hydraulic steering systems on machines today will be part of a load-sensing hydraulic system. This type of system was covered in detail in Chapter 16, but a brief overview is as follows: The pump will be a pressure-compensated variable displacement type and usually an axial piston pump. There will be a load-sense line going to the pump displacement control to change the pump flow based on steering demands. This will reduce wasted energy when there is little or no need for oil flow because the pump will destroke

FIGURE 19–33 Neutralizer valve.

to near zero, which will minimize pump output. When a steering function is sensed at the steering control valve, a signal is sent to the pump control valve, and it will upstroke to supply slightly more oil flow than is needed for the steering function.

ELECTRONIC STEERING CONTROLS As mentioned earlier, some steering systems will be controlled by electronic components that are part of an electrohydraulic system. These systems are not only used for articulated and all-wheel steering machines, but they could be part of conventional steering systems as well.

For example, one machine manufacturer makes a skid steer that can switch between normal skid steering and all-wheel steering. Controlling these steering functions is an electronic system that has input sensors, an ECM, and output actuators. ● **SEE FIGURE 19–34** for an all-wheel skid steer loader.

The main inputs are the operator controls that include a steering mode switch and electronic joystick and wheel position sensors that tell the ECM what angle the wheels are turned at.

> **WEB ACTIVITY**
>
> Search the web to find a video of a machine that features different steering modes such as all-wheel steering, crab steering, and rear wheel steering. Does it show how the modes are selected? Do you think there are any sensors on the machine to determine wheel position?

WHEELED MACHINE STEERING SYSTEMS **579**

FIGURE 19–34 AWS skid steer loader.

The ECM processes this information and sends one or more signals to a set of solenoid valves. These valves will then direct oil to the four steering cylinders individually. This system can steer the four wheels totally independent of each other; therefore, it relies on the wheel position sensors for feedback so the machine can turn smoothly and as desired by the operator.

The system may need calibrating, and it will provide a port to hook up a laptop for this purpose and for diagnosing problems.

STEERING SERVICE Servicing steering systems is an important part of any maintenance procedure. This ranges from simple inspections such as looking at oil levels, checking for leaks, checking for excessive wear and damage, changing oil and filters, and taking an oil sample. Greasing all steering-related grease points is critical to ensuring safe operation of the machine and longevity of the components. Some machines will have an automatic greaser that will need to have its level checked. All steering-related grease points need to be checked to see that grease is reaching them because many steering grease points are hard to reach.

Steering systems should be checked for proper operation when the machine is serviced to ensure that it works smoothly and evenly. If the steering system has different modes, they should also be checked for proper operation.

Secondary steering systems should also be checked for proper operation because they likely don't get actuated too often.

CHECKING FOR WEAR Mechanical steering components can be checked for excessive wear to see if they are within acceptable specification. Some examples of these components are kingpins and bushings, tie rod ends, and steering columns.

This could be a simple process such as working the steering wheel back and forth slightly and watching for excessive wear at all moving joints.

Also a machine's steering wheel should have a specific amount of resistance, and this could be measured with a fish weighing scale to see if the resistance is within specification.

STEERING DIAGNOSTICS See Table 19- for a list of steering system symptoms, problems and possible solutions.

Some tools that may be required for diagnosing steering system complaints are pressure gauges, flow meter, heat gun, stopwatch, fish scale, filter cutter, and tape measure.

STEERING SYSTEM ADJUSTMENTS Toe-in adjustments may be required on occasion if excessive tire wear is noticed. This is easily done if the machine has a double rod steering cylinder because the rod can be turned in the cylinder to change the toe dimension. For other types of steering arrangements, the tie rod tube will need to be rotated. Its female-threaded ends mate with tie rod ends. ● **SEE FIGURE 19–35** for how conventional steering axle toe in is measured.

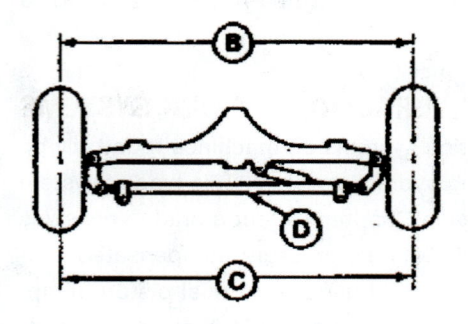

A - Center of Hub
B - Front of Tire
C - Rear of Tire
D - Tie Rod Tube

FIGURE 19–35 Measuring toe in.

STEERING CYLINDER LEAKAGE CHECK If a steering problem relates to a possibility of the steering cylinder leaking, a procedure to check for cylinder leakage or bypass can be performed. This involves turning the wheel all the way one way, turning the machine off, removing the hose on the cylinder that would return oil to tank, and installing a hose on the cylinder that will direct any bypassing oil into a measuring container. The machine is then started and the steering wheel turned to send oil into the hose that is still attached to the cylinder for a specific amount of time. There should be very little oil going into the container. For example, the specification for a backhoe loader is no more than 5ml/min.

STEERING SYSTEM REPAIR Steering system repairs can vary from fairly simple mechanical repairs to hydraulic component repairs to electrical repairs.

An example of each follows:

SYMPTOM	PROBLEM	SOLUTION
Slow or no steering function	Articulation locking bar installed	Disconnect articulation locking bar and put in storage position.
	Oil level low	Check oil level in hydraulic reservoir.
	Steering load-sense relief valve pressure setting too low or malfunctioning	Check steering load-sense relief valve pressure. Inspect steering load-sense relief valve.
	Steering lines damaged	Inspect and replace lines.
	Priority valve in hydraulic system manifold	Check priority valve operation. Remove priority valve and inspect.
	Steering valve	Disassemble steering valve and inspect.
	Steering cylinder piston seals	Check steering cylinders for leakage. Replace piston seals as needed.
	Secondary steering system not functioning properly	
Constant steering to maintain straight travel	Air in steering system	Check for foamy oil. Tighten loose fitting. Replace damaged lines.
	Steering cylinder piston seals	Check steering cylinders for leakage. Replace piston seals as needed.
	Steering valve	Disassemble steering valve and inspect.
Erratic steering	Air in steering system	Check for foamy oil. Tighten loose fitting. Replace damaged lines.
	Oil level low	Check oil level in hydraulic reservoir. Add hydraulic oil.
	Cylinder piston loose	Disassemble cylinder and inspect.
	Steering valve	Disassemble steering valve and inspect.
Spongy or soft steering	Air in steering system	Check for foamy oil. Tighten loose fitting. Replace damaged lines.
	Oil level low	Check oil level in hydraulic reservoir. Add hydraulic oil.
Free play at steering wheel	Steering wheel-to-shaft nut loose	Tighten nut.
	Splines on steering shaft or valve worn or damaged	Inspect and replace worn or damaged parts.
Steering locks up	Large particles of contamination in steering valve	Inspect return filters for contamination. Repair cause of contamination. Flush hydraulic system. Disassemble steering valve and inspect.
Abrupt steering wheel oscillation	Steering valve gerotor not timed correctly	Time gerotor gear.
Steering wheel turns by itself	Lines connected to wrong ports	Connect lines to correct ports.
Machine turns in opposite direction	Lines to steering cylinders connected to wrong ports at steering valve	Connect lines to correct ports.
Machine turns when steering valve is in neutral	Steering valve leakage	Disassemble steering valve and inspect.

STEERING SYSTEM MECHANICAL REPAIR
One component that takes a lot of stress is a tie rod. If a machine is moving along a rough road, there is a lot of shock loads transferred into the tie rod, and it will eventually wear out. Most tie rods will need to be greased regularly, and if they aren't, they will wear out quickly. Tie rods are taper fitted to the steering knuckle and held in place with a lock nut or castellated nut and cotter pin. The other end of the tie rod is threaded into a tie rod tube or a steering cylinder depending on the machine. There will be either a jam nut to hold it in place in the rod or a clamp to lock it in the tie rod tube.

The tapered end of the tied rod can be separated from the steering knuckle several ways. For smaller tie rods, a simple tool called a pickle fork can be used. It's always a good idea to mark the old tie rod's threads or count the ones showing so you know where to locate the new one. For large tie rods, the tapered end may need to be pressed out. When they have been in place for many hours, they are going to be rusted into place. Although it may be tempting to apply heat to the steering knuckle to aid in removing it, *do not* use any form of heat. There are proper pressing tools that can be used to remove any tie rod.

Once the tapered end is loose, the threaded end of the tie rod can be loosened and removed. ● SEE FIGURE 19–36 for how to remove a tie rod end.

Install the new tie rod with the same amount of threads exposed as the old one and then install the tapered end into the steering knuckle and torque the nut to specified torque. Before locking the tie rod tube or rod end, check the toe-in dimension according to the service manual.

STEERING SYSTEM HYDRAULIC REPAIR
One common repair for a steering system's hydraulic system is to replace the HMU. These valves are very susceptible to contamination and will easily get damaged if any contamination gets in between its moving parts. A thorough diagnosis to ensure that this valve is faulty is mandatory because many of these valves have been replaced for no good reason.

Once you have determined that the valve is faulty, proceed with the usual steps taken for any hydraulic repair. This includes releasing pressure and having oil drain buckets and floor dry handy.

It's very important to mark the lines going to the valve before removing because they could be easily mixed up. Taking a couple pictures at this time will be a good idea as well. Don't leave the lines open for any longer than necessary to avoid contaminating the replacement valve. Replacing the valve itself can sometimes be a challenge depending on its location, but patience and the right tools will get the job done right.

A filter replacement should be a standard part of this procedure. Make sure that the hydraulic oil level is right before starting the machine. Start the engine, leave it at low idle, and slowly work the steering wheel back and forth just short of a full turn a few times to remove all the air from the system. Do this until the steering system seems to react normally to steering wheel input.

After the oil has warmed up a little, you can operate the machine to see if the steering works as it should. When the oil is up to operating temperature, you can perform cycle time checks to see if the system is working within specified cycle times.

STEERING SYSTEM ELECTRICAL REPAIR
Some steering systems will include electrohydraulic components. Most of these systems include solenoid valves that occasionally fail. To diagnose a solenoid you suspect if faulty, it is helpful if there is more than one solenoid that is the same. This way it is fairly easy to swap solenoids when troubleshooting to isolate the problem. They are usually held onto the solenoid valve assembly loosely with a light-duty nut. Replacing a faulty solenoid is easy to do because of this.

Progress Check
6. This is *not* one other name for a rotary steering valve:
   a. hand-metering unit
   b. electrohydraulic valve
   c. Char-Lynn valve
   d. orbital valve

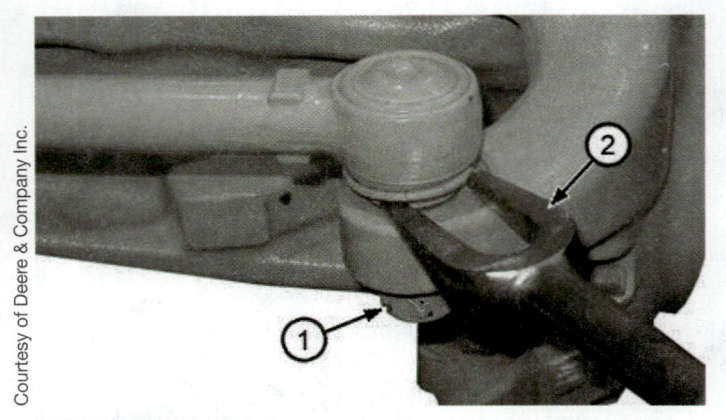

FIGURE 19–36 How to remove a tie rod.

7. This is what tries to keep the two moving parts of a rotary steering valve centered:
   a. coil spring
   b. leaf spring
   c. heavy-duty O-ring
   d. proportional solenoid
8. Joystick steering could use these *two* different ways to control the main steering control valve:
   a. mechanical linkage and pilot oil
   b. position sensors and poppet valves
   c. mechanical linkage and poppet valves
   d. poppet valves and rotary valve
9. Neutralizer valves are used as part of hydraulic steering systems to:
   a. destroke the pump when no steering occurs
   b. stop the flow of power through the drivetrain when the machine is turning
   c. drain the pilot signal to the main control valve to stop the machine from turning further
   d. upstroke the pump during a steer function to improve response
10. If a machine needs to have tie rod ends replaced, this is one method that should *not* be used:
    a. heat with acetylene torch
    b. mechanical puller
    c. hydraulic puller
    d. soft hammer

SHOP ACTIVITY

Go to a machine with a conventional steering system and see if there is a toe adjustment.
- Go to a machine with a double rod steering cylinder and describe how the cylinder is mounted to the machine.
- Got to a machine that has articulated steering and draw a schematic of the steering cylinders and HMU valve.

SUMMARY

1. Some safety concerns related to wheeled machine steering systems include: proper operation of the steering system must be confirmed after service or repair work, steering lock must be in place for articulated machines when service or repair work is done.
2. Wheeled steering machine can be steered four different ways- skid steer (wheels don't pivot), articulated steering, conventional steering and all wheel steering.
3. Articulated machines have two piece frames that pivot to allow steering.
4. Straight frame or one piece frame machines can have one or more steering axles.
5. Some light duty machines may have mechanical steering boxes that transfer operator input from a steering wheel to a steering linkage that pivots the wheels on a steering axle.
6. Steering box output is sent to a Pitman arm then on to a drag link that moves a steering arm. The steering arm will turn one steering knuckle.
7. Both steering knuckles are connected together with tie rods that have ball joints on each end.
8. Steering knuckles have spindles on them that allow wheel bearings and wheels to turn on them. Steering knuckles pivot on kingins or tapered roller bearings.
9. Steering axles can be non drive or provide drive for all wheel drive arrangements.
10. Hydraulic assist steering features a power assist steering box that uses hydraulic pressure from a pump to multiply operator steering input effort.
11. Steering geometry terms include: caster (lean of kingpin when seen from the side), camber (lean of kingpin when viewed from front), toe angle (turn in between two wheels).
12. Full hydraulic steering systems feature a steering valve turned by a steering wheel, hydraulic pump, steering cylinder/s and lines. The steering cylinders will move the wheels through mechanical linkage.
13. Articulated steering systems feature a steering valve turned by a steering wheel, oil is sent to a pair of steering cylinders that pivot the two piece frame.
14. Machines with joystick steering have a pilot valve that sends oil to a main control valve and the pump oil goes to the steering cylinders.
15. All wheel steering machines can have up to four steering modes: conventional steer, crab steer, all wheel steer. The operator selects the mode with a switch that shifts a spool valve. The spool valve redirects oil to the steering cylinders.
16. Hydraulic steering system DCVs can be called: orbital valves, Char Lynn valves, rotary valves, hand metering unit. They are a combination of rotary DCV, gerotor pump/motor and pressure relief valve.

17. Steering cylinders can be conventional differential type when used in pairs for articulated steering or double rod type when used for a steering axle.
18. Secondary steering systems provide a backup oil flow in case the primary pump fails or stops turning. This can be a ground driven or electrically driven pump.
19. Neutralizer valves are used on articulated machines to stop the turning action when the machine is fully turned.
20. Steering system service includes: thorough visual inspection, greasing ball joints and steering linkages, checking fluid levels for hydraulic system, changing oil and filters, check operation of system.
21. Some steering system related complaints include: slow or no steering, abnormal tire wear, no steer one way, won't change steering mode, drifting, steering wheel stiff.
22. Steering linkage can be adjusted to correct some problems such as toe-in adjustments with tie rod adjustments.
23. Steering system repairs can include: worn component replacement, hydraulic system leaks and component replacement, electrical circuit repairs.

chapter 20
TRACK-TYPE MACHINE STEERING SYSTEMS

LEARNING OBJECTIVES

After reading this chapter, the student should be able to:

1. Describe the different types of steering systems used for track-type machines.
2. Identify the main components used for a differential steering system.
3. Describe how a spring-applied/oil-released clutch works.
4. Explain how a band brake stops track motion.
5. Describe how steering brakes can be used as parking brakes.
6. Describe what counter-rotation means.
7. Describe the initials checks performed when diagnosing a steering problem.
8. Explain the safety concerns related to working on a steering system for a track-type machine.
9. Describe some basic troubleshooting procedures for track machine steering systems.

KEY TERMS

Band brake 591
Counter-rotate 586
Equalizer planetary 598
Oil applied 588
Spring applied 588
Steering lever 594
Steering motor 597
Steering pedal 600
Steering planetary 597

INTRODUCTION

SAFETY FIRST You may be required to lift heavy components when repairing a steering system. Make sure that you are competent with proper lifting methods and use only approved and inspected lifting devices. Machine cabs may need to be tilted to allow access to steering components. *Do not* work under a raised cab unless it is securely held in place.

Some friction material may contain asbestos fibers. Treat any dust that has been created from friction material wear as if it contains asbestos. This will include wearing appropriate breathing protection and hand protection.

Some steering systems will use extremely high hydraulic pressures. Take caution and wear appropriate PPE before working on high-pressure steering systems. Always refer to the manufacturer's service information before working on steering systems. See the safety information in Chapters 4 and 16 for more cautions related to hydraulic systems.

A malfunctioning steering system has the potential to make the machine a safety hazard. If the operator is unable to control the machine when it is traveling, then it could put the operator or any workers close by at risk. For this reason, the technician must always confirm the steering system operates properly before the machine is put back into service.

NEED TO KNOW

Heavy equipment machines can be grouped into two general categories by how they are propelled—tracks or tires. Between these two main types of machines, there are many ways to steer them. As the tracks on almost all track-type machines stay parallel to each other, the only way to steer most track-type machines is to change the speed of the tracks in relation to each other.

This chapter will cover track machines that have one-piece main frames (straight frame) and are steered with mostly mechanical components. There are three different systems to do this.

One way to steer a straight frame track-type machine is by stopping the drive to one track and driving the opposite track around it with a mechanical drive arrangement. This is done by disconnecting the drive to one track with a clutch and applying a brake to it while the other track drive stays connected and continues driving. This is sometimes called skid steering because of the stopped track skidding on the ground.
● **SEE FIGURE 20–1** for a simple illustration of how a track machine turns (one track driving).

> **TECH TIP**
>
> There are some track machines that have two-piece main frames that are hinged together and are steered by pivoting the frames. These machines have four smaller track assemblies. This is called articulated steering and was covered in Chapter 19.
>
> Some track machines used for paving and curb forming can be steered by pivoting from two to four of their tracks. The tracks are located at the bottom of legs that allow hydraulically actuated height adjustments. An electrohydraulic system allows the operator to turn the inside of the leg, which then turns the machine. Electrohydraulic systems are covered in Chapter 16.

The second way to do it is to use a differential steering arrangement to adjust the speeds between the two tracks to make the machine turn. With differential steering, both tracks are driven when the machine is turning through a gradual turn. Differential steering uses a combination of gears and a hydraulic drive to drive the tracks at different speeds relative to each other. It is also possible to make the tracks turn in opposite directions or **counter-rotate** with differential steering systems. This will enable the machine to perform spot turns for turning in tight quarters. ● **SEE FIGURE 20–2** for an illustration of a differential steer machine turning with both tracks driving.

A third way to steer a track-type machine is to incorporate a two-speed mechanical drive system for the left and right tracks. This will give the operator a way to turn the machine without stopping the drive to one track. The downfall is that the steering will be a fixed arc because of the fixed speed differential between high and low. With this system when the operator wants to turn while driving with both tracks, then one track is put in low range and the other is put in high range. A machine with this steering feature will also have steering clutches and brakes to provide one track drive for sharp turns (skid steering turn).

Two other drive systems for track machines are hydrostatic and electric. Hydrostatic drive provides infinitely variable speed changes independently between the two tracks. It also allows the tracks to be turned in opposite directions (counter-rotation) for spot turns. Hydrostatic drives were covered in Chapter 17. Electric drive can also be used to drive and steer track machines. Caterpillar uses a combination of electric drive and differential steering to steer one of its dozers. This drive system also provides infinitely variable speeds and counter-rotation. Electric drive was covered in Chapter 13.

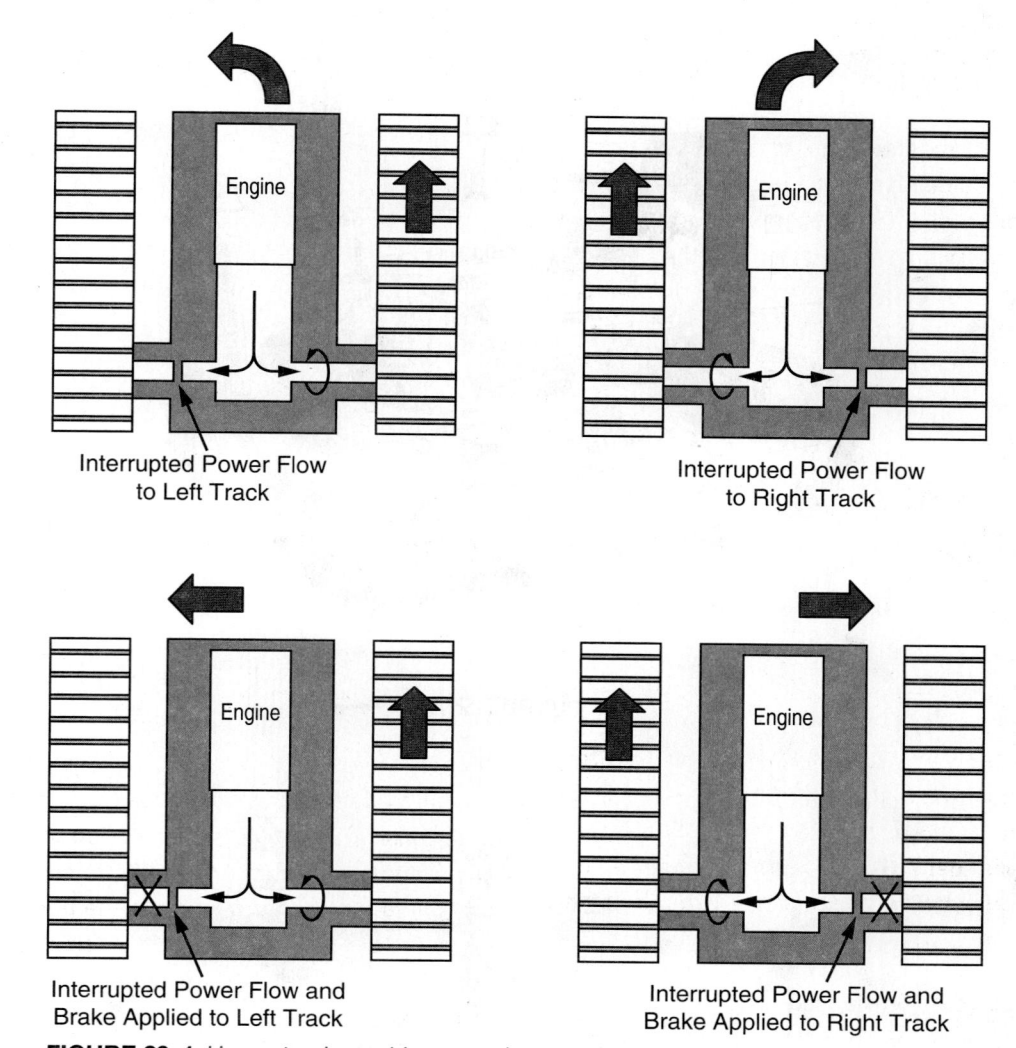

FIGURE 20–1 How a track machine turns (one track driving).

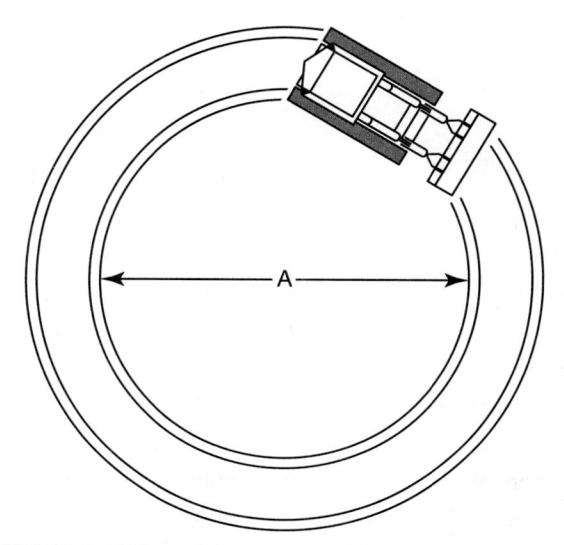

FIGURE 20–2 Differential steer machine turning (both tracks driving).

TRACK MACHINE MECHANICAL STEERING SYSTEMS

Track-type machines that may use mechanical steering systems are track-type tractors (crawler dozers/bulldozers) and track loaders.

The power flow for older track-type machines that use mechanical methods to steer the machine is as follows: diesel engine flywheel to clutch or torque converter to transmission (manual shift or powershift) to pinion gear to bevel gear to steering clutches to final drives to sprockets to tracks. Between the steering clutches and final drives, there are also

TRACK-TYPE MACHINE STEERING SYSTEMS **587**

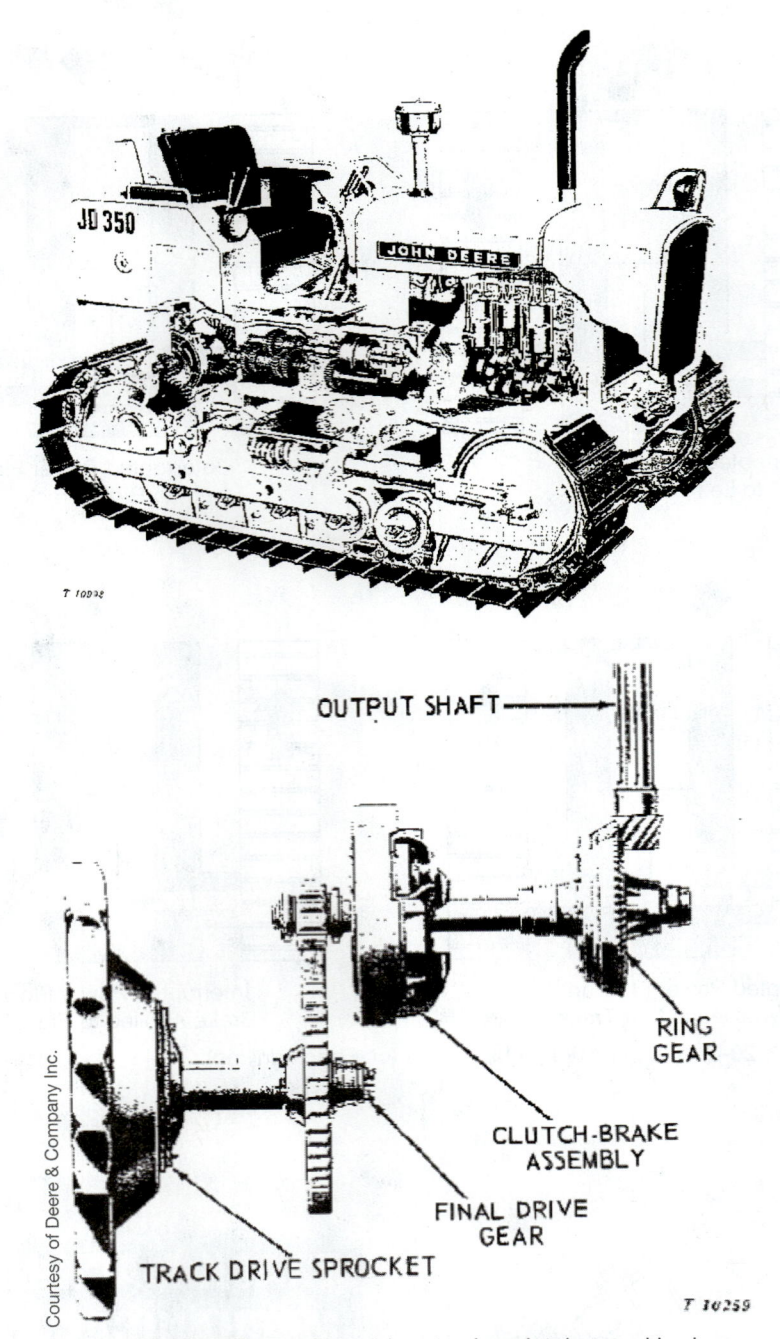

FIGURE 20–3 Power flow older steering clutches and brakes.

brakes to stop the tracks from turning. ● SEE FIGURE 20–3 for the power flow of a machine that uses linkage-controlled steering clutches and brakes.

Steering clutches allow the operator to stop and start drive to each track individually. They use friction material that is squeezed against a smooth metal surface to transfer torque through them. Springs or oil pressure can provide the squeezing force. Some slippage is acceptable as the clutch is engaged or disengaged, but when the clutch is fully engaged, there shouldn't be any slipping between the input and output of the clutch.

While there are several variations of steering clutches currently, they are mostly based on a multi-disc-type clutch that can be either a **spring applied**/oil released or an **oil applied** type of clutch.

Steering brakes will also use friction to stop one track from turning so that the machine can pivot on the stopped track while the other track continues to drive. They can either be drum and external band or multi-disc-type brakes.

As machine horsepower increases for larger machines, steering clutch and brake torque capacity must also increase.

For multi-disc-type clutches and brakes, this is done by increasing either the size or the number of discs and plates in the clutches or both. To increase torque capacity of brakes, the manufacturers increase either the diameter or number of discs or the width and diameter of the brake drum and the corresponding friction material. Clutch and brake torque capacity directly relates to surface area because they transfer torque through friction, and more surface area means more torque can be transferred. Clamping force and the friction material's coefficient of friction could also be increased to increase clutch capacity. Clamping force is increased by increasing the amount or tension of the springs for a spring-applied clutch or brake. For an oil-applied clutch or brake to have its torque capacity increased, either the oil pressure or the piston surface area is increased.

When operators want to steer a machine with steering clutches and brakes, they will release one steering clutch to stop driving one track. This could give a partial turn depending on the ground conditions and the load on the machine's blade, bucket, or ripper. For a more positive turn, they would then apply the brake to the track that has no power flow to it and continue to drive the opposite track.

The original way for the operators to control this was to move mechanical levers or pedals in the operator station that would be connected by linkages to the steering clutches and brakes. This evolved into mechanically controlled hydraulically actuated clutches and brakes, and finally electronic controls replaced all mechanical controls for hydraulically actuated clutches and brakes. This allows the operator of a very large track-type machine, Caterpillar's 850 HP D11, for example, to steer it with the pull of one finger.

SPRING-APPLIED STEERING CLUTCHES
Spring-applied steering clutches can be found in small to large older dozers and loaders and are still used to steer some small current dozers. They rely on spring pressure to squeeze a stack of alternating friction discs and steel plates together to transfer torque through them. The discs are squeezed between a smooth-faced hub and a smooth-faced movable pressure plate. Their operating principles are similar to the spring-applied flywheel clutches discussed in Chapter 10. ● **SEE FIGURE 20–5** for a spring-applied clutch.

The clutch is driven by a flanged hub (input) that is driven by a bevel gear that is driven by the transmission output pinion gear. The bevel gear and left and right clutch drive hubs are fastened to a shaft that is supported by bearings that are supported in the machine's frame. This assembly is located roughly under the operator's seat inside the machine's main frame. There are usually three compartments that are part of the machine's frame that house the bevel gear and the left and right steering clutches. Each clutch assembly drives a

> **TECH TIP**
>
> Steering brakes and clutches that use multiple discs will use different terms for the wear parts that transfer torque through the clutch or brake. One set of discs will have friction material bonded to their face, and they will be squeezed against another set of smooth steel plates. The following is an example of some different terms you may see used.
>
> **Discs:** friction linings, fiber plates, clutch facing discs, clutch discs
>
> **Plates:** steel plates, reaction plates, steels
>
> Discs and plates can have internal or external (one or the other) teeth or tangs to hold them to one part of the clutch or brake assembly. ● **SEE FIGURE 20–4** for a set of brake and clutch discs and plates.

FIGURE 20–4 Plates and discs.

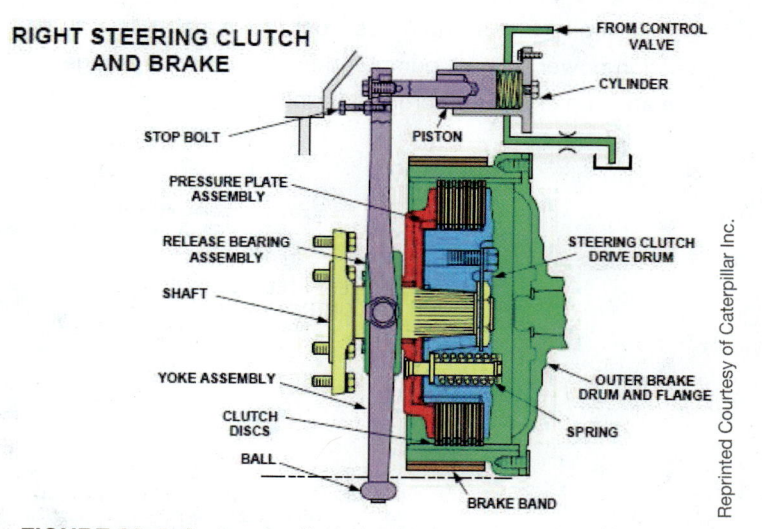

FIGURE 20–5 Spring-applied clutch.

brake drum (output) that is bolted to a flange that drives a final drive pinion gear. The brake drum will be surrounded by a band-type brake.

TRACK-TYPE MACHINE STEERING SYSTEMS 589

The flanges (clutch input and final drive input) are usually pressed onto tapered and splined shafts (bevel gear and final drive input) with 5–50 tons of force before a large nut and locking device is installed. This is necessitated because of the extreme torque forces and shock loads that are transmitted through them as the machine is stopped, started, and steered. The flanges are fastened to the steering clutch with a ring of threaded fasteners.

For a spring-applied clutch, there could be one of two types of springs used to create the necessary clamping force. One is a Belleville type of spring (dished washer shape) that could be singular or stacked back to back. The other is a coil spring type that will be arranged in multiples around the outside circumference of the clutch assembly. Both types of springs will apply pressure to a series of alternating friction discs and plates by squeezing the drive hub and a pressure plate together. The plates have internal splines or tangs that are driven by the drive hub. The discs have friction material bonded to a metal plate that is externally splined. The external splines will drive the brake drum that then drives the input to the final drive.

The friction material on the discs will be grooved to allow heat to dissipate from them. Most steering clutches and brakes will have oil circulating past them for cooling and lubrication. If oil circulates through a steering clutch housing and around the clutches and brakes, they are considered to be "wet" steering clutches and brakes. For a dry steering clutch arrangement, the bevel clutch housing will have its own oil that is sealed from the steering clutch compartments. For machines that have wet steering clutches, all compartments will likely use the same oil.

The oil for this originates from one section of a multi-section power train oil pump and flows through passages in the steering clutch and brake housing. Once it flows past the clutches and brakes, it will drain into the machine's steering clutch case. Another section of the multisection pump will move the oil through an oil cooler to transfer heat into the engine's coolant.

Overheated clutch and brake components will quickly deteriorate to being worn out or damaged and nonfunctional. Dry-type clutches and brakes are used on low horsepower machines and not subject to high temperatures. Wet-type clutches will handle higher horsepower applications and almost always have the oil cooled to keep the clutch temperatures below damaging levels to ensure longevity.

For mechanically controlled spring-applied steering clutches, there is a mechanism called the release bearing that will pull the pressure plate away from the drive hub against spring pressure when moved by the clutch release linkage. When the pressure plate is moved, this allows the plates and discs to turn independently and torque transfer through the clutch is stopped. Linkage connects the operator control (lever or pedal) to a pivoting yoke that moves the release bearing cage with the release bearing inside it. The release bearing outer race does not rotate and is moved sideways while the inner race rotates with the clutch and acts on the pressure plate.

The next step in design for this type of clutch was to use hydraulic assist to release it to lessen the operator effort required to control it. For this system, a mechanical linkage actuates a spool valve that sends oil to a cylinder. The cylinder pushes on the release bearing yoke to release the clutch. ● **SEE FIGURE 20–6** for an oil-assisted clutch release mechanism.

The oil pressure for this type of assist is sourced from the transmission control system and is fairly low at around 300 psi. This oil is most often prioritized in the transmission system to be available to the clutch or brake system first.

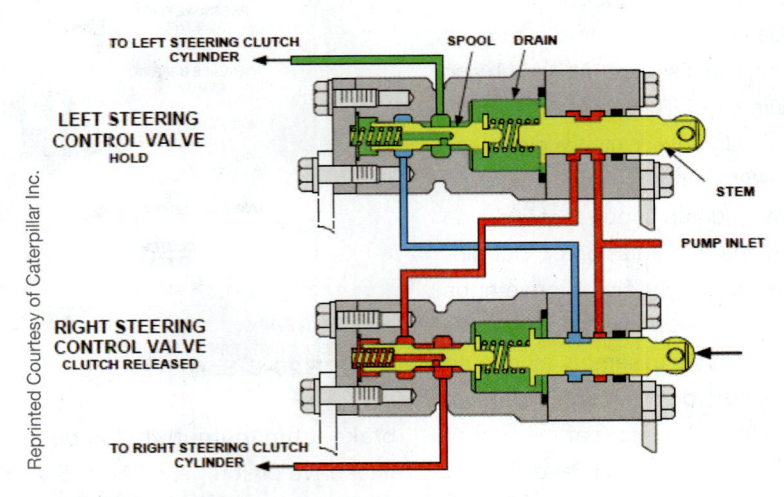

FIGURE 20–6 Oil-assisted clutch release valve.

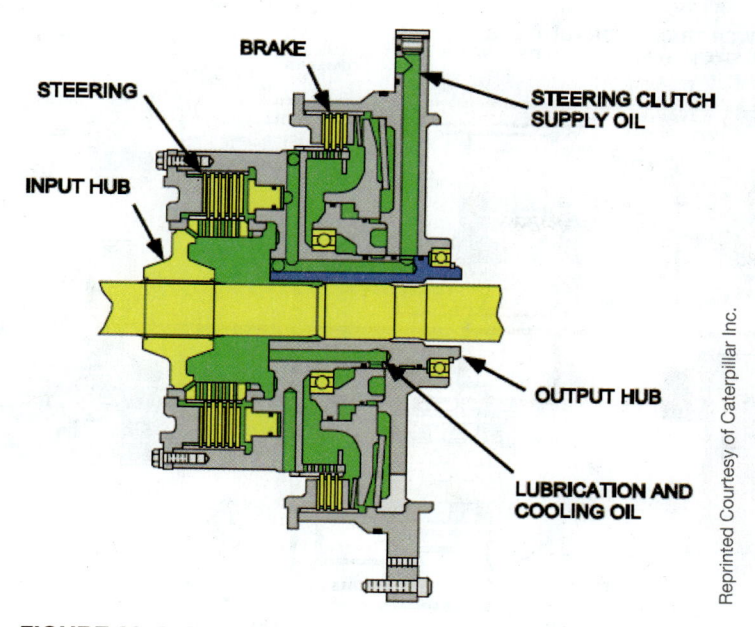

FIGURE 20–7 Oil-applied steering clutch.

OIL-APPLIED STEERING CLUTCHES
Spring-applied steering clutches are always engaged until released by oil pressure or mechanical movement. Oil-applied clutches are disengaged until there is oil pressure applied to a piston. They are similar in design but don't have any springs. As oil pressure builds behind the piston, it pushes on the clutch plates and discs and torque starts to get transferred between the clutches' drive hub (input) and the brake drum (output).
● **SEE FIGURE 20–7** for an illustration of an oil-applied steering clutch.

This type of clutch can be found on all sizes of track-type machines. The typical power flow for a machine that uses this style of steering clutch is the machine's transmission output pinion drives a ring gear that drives the clutch shaft. The clutch shaft drives the left and right clutch hubs that are splined externally to drive the clutch discs. When the clutch is engaged, the discs drive the steel plates through friction and the plates have tangs on their outer circumference. These tangs then transfer drive into the brake drum and the brake drum sends torque to the final drive. This arrangement would use a band-style brake to stop the drum.

Oil-applied clutches can also be used with multi-disc-type brakes, and if this is the case, then the discs will drive a housing that is the output of the clutch and is splined to accept the splines of the brake discs.

When there is no oil pressure applied to the steering clutch piston, the discs and plates are able to rotate independently. Oil is sent to the piston through a rotating seal from a spool-type valve that is moved by operator-actuated linkage. Oil pressure is normally present at the clutch to provide drive torque to the final drive, but when the operator wants to steer the machine, the spool valve is moved and oil is drained from the clutch. With the clutch released, the machine makes a gradual turn until the brake is applied on the same side to lock the track and then the machine makes a sharp turn.

Like the oil-assisted clutches, oil-applied clutches receive their oil from the transmission control system.
● **SEE FIGURE 20–8** for the hydraulic system of a power train for a track type machine.

STEERING BRAKES—BAND TYPE
Older machines with steering clutches and brakes used band-type brakes to slow down or lock one track up when steering the machine. These brakes also often double as parking brakes. Smaller and older machines used mechanical linkage from the operator input (levers or pedals) to apply the left and right brake bands. The brake bands surround the brake drums and are lined with friction material. The friction material is riveted and bonded to a flexible 7/8 circle metal band. The friction material is likely in multiple pieces and may or may not have grooves in its contact face. The grooves will allow oil to circulate around the friction material to cool and clean it. ● **SEE FIGURE 20–9** for a **band brake**.

When the open part of the band is pulled together, the friction material grabs the smooth exterior surface of the brake drum and slows it or stops it from turning. After the steering clutch is released, the brake is applied to stop the final drive

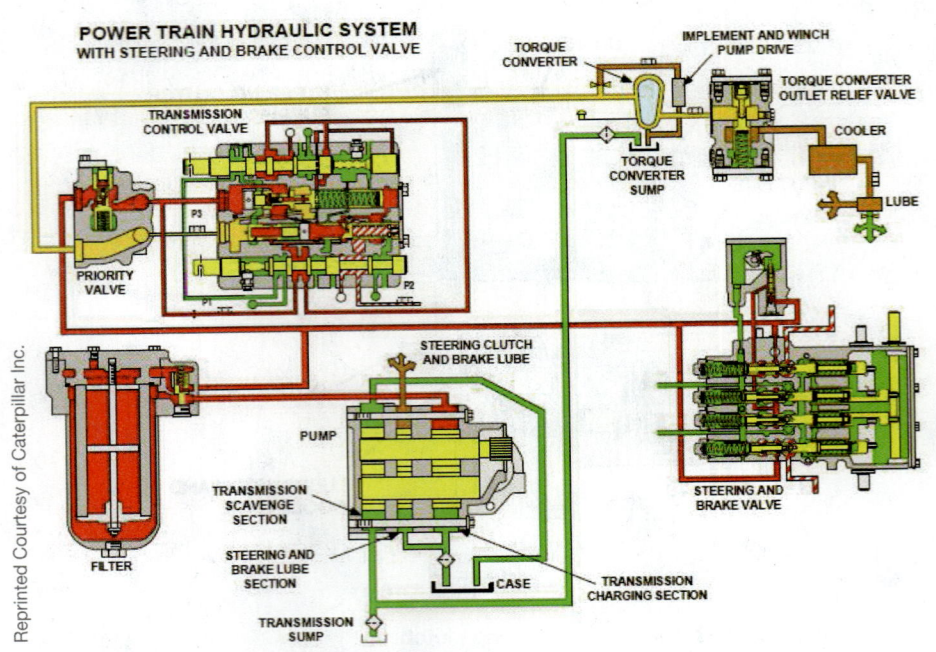

FIGURE 20–8 Hydraulic system for a track type machine.

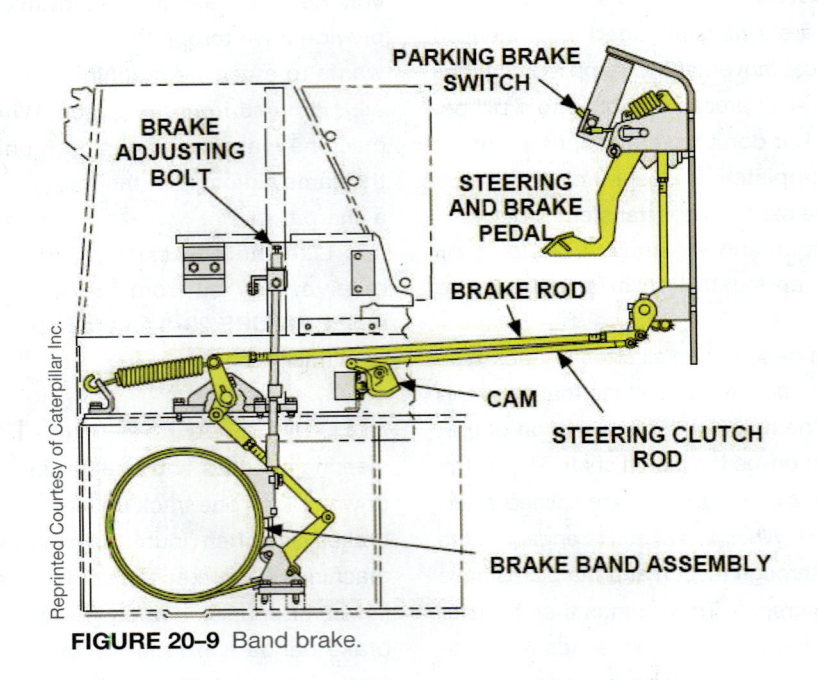

FIGURE 20–9 Band brake.

pinion shaft, which stops the track. When the brake is released, there will be a support screw or a couple of light springs to hold the brake away from the drum so it doesn't drag.

For larger and newer machines, the band brake actuation was updated to either a spring-applied/oil-released or oil-applied brake that uses a booster piston or cylinder to apply the brake. ● **SEE FIGURE 20–10** for an oil-boosted band brake mechanism.

This arrangement provides a way to use the band brakes for parking brakes when oil is drained away from the brake cylinder.

MULTI-DISC STEERING BRAKES Most new medium to large track-type machines that use steering clutches and brakes will use multi-disc-type brakes. Just like the band-type brakes, they will also be between the steering clutch

1. Link
2. Shaft
3. Piston
4. Bellcrank
5. Spring
6. Rod
7. Pawl
8. Pin
9. Pin
10. Lever
11. Lever
12. Shaft
13. Strut
14. Pin
15. Strut
16. Shaft

Reprinted Courtesy of Caterpillar Inc.

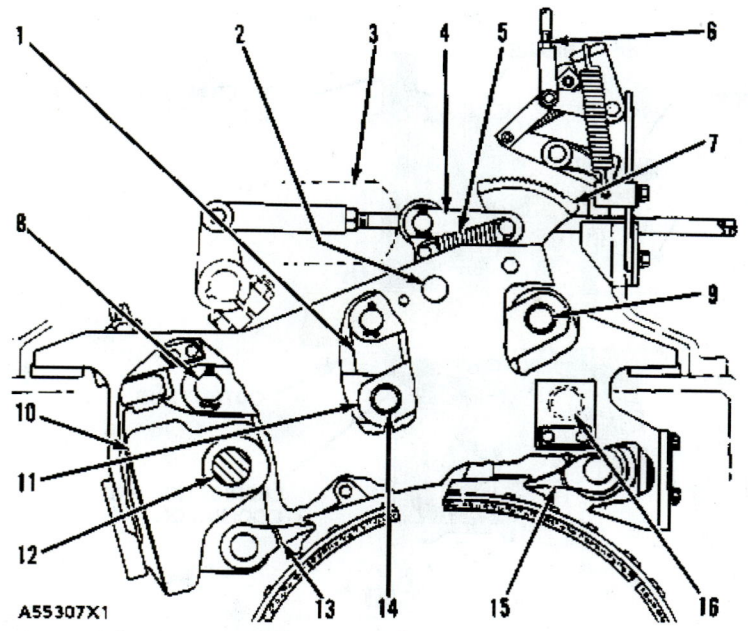

FIGURE 20–10 Oil-boosted band brake mechanism.

output and the final drive input in the driveline torque transfer sequence. They are spring applied and oil released with a Belleville washer type of spring used to apply them. The spring will push directly on the piston, and the opposite side of the piston pushes on the discs and plates. The appearance of these brakes could be mistaken for a clutch unless you were able to see where they are located in the drivetrain.
● **SEE FIGURE 20–11** for a multi-disc steering brake.

The machine will have a steering and brake control valve that sends oil to the clutches and brakes to steer the machine. These brakes will double as parking brakes because they are spring applied.

When machine operators want to turn a machine with this type of steering brake, they will first disengage one steering clutch and then oil is drained from behind the same side's brake piston. Once the oil pressure is drained, the brake applies and the track on that side of the machine is locked. The opposite side track continues driving and the machine is turned.

TWO-SPEED STEERING Currently, the heavy equipment manufacturer Dressta produces medium (160 hp) to large (515 hp) dozers that feature a two-speed steering system. It consists of a two-speed planetary-geared steering module that provides gradual turns while maintaining full power to both tracks with a conventional clutch–brake mechanism for tight or pivot turns.

Coupled to a three-speed transmission, the two-speed steering module provides six speeds forward and six reverse. The steering module contains a planetary gearset, wet multi-disc low, high, and brake clutches. ● **SEE FIGURE 20–12** for a two-speed steering system.

When one track is driven in high range and the other at low range, there will be a 30% speed difference between the

> **TECH TIP**
>
> Dry steering clutches and brakes are only used for the smallest- and lightest-duty applications. Clutches and brakes that are dry means they don't run in oil.
>
> All steering clutches and brakes used today will use friction material that allows it to be run in oil. The friction material is just a different formulation that allows it to be compatible with oil.
>
> The oil will also carry away contamination and wear particles where it can be left in the power train oil filter. The main purpose of having clutches run in oil is to have the oil absorb heat so they won't break down and fail prematurely.
>
> The heat in the oil is then transferred into an oil cooler where the oil cooler then transfers the heat to engine coolant.
>
> A variety of friction material can be used for steering clutches and brakes. Generally, this includes paper-based, elastomer-based, or sintered metal-based materials. Added to these base materials are binders, glass fibers, friction modifiers, fillers, and curatives.

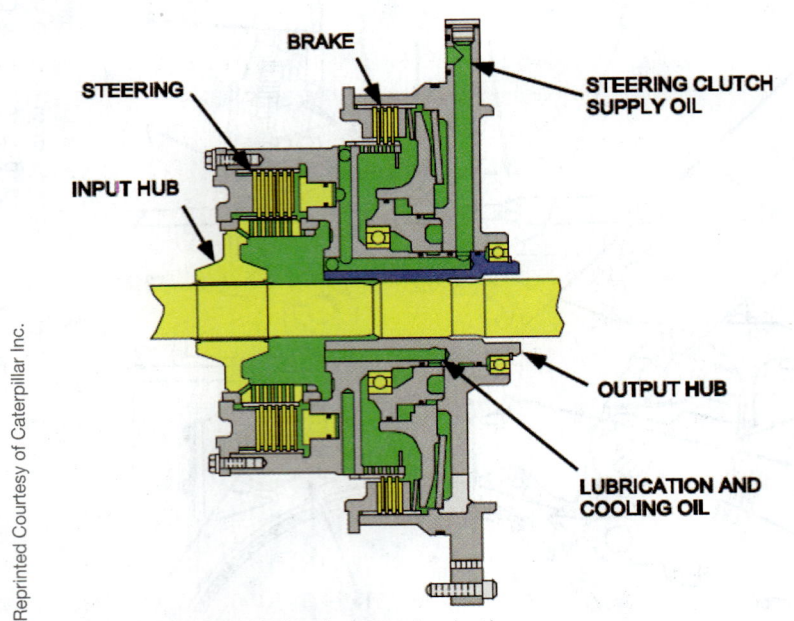

FIGURE 20–11 Multi-disc steering brake.

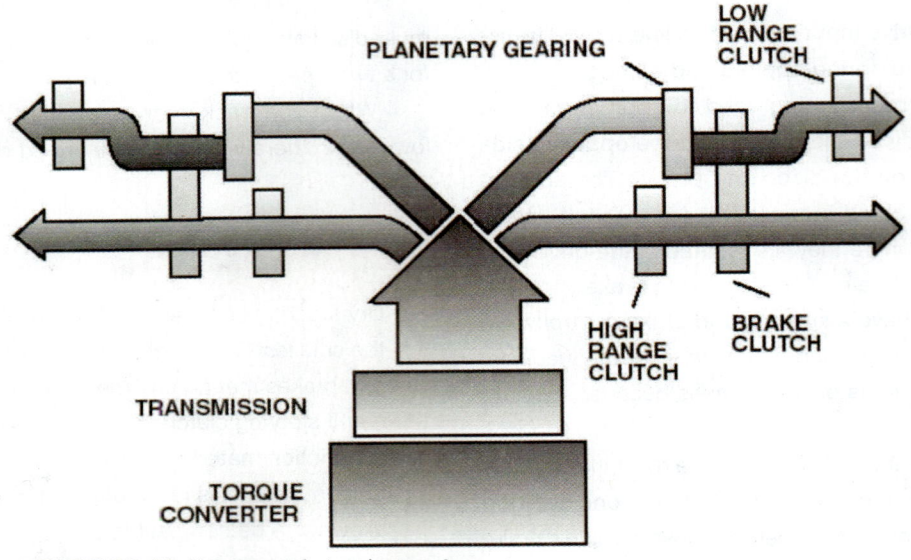

FIGURE 20–12 Two-speed steering system.

two tracks. For example, if the machine is traveling straight ahead in second-gear high range, it will travel at 4.2 mph. If the operator wants a gradual turn to the left, then the right track is switched to low range, and it slows down 30% to 3.2 mph. ● SEE FIGURE 20–13 for a cross-section illustration of one side of the steering module.

If there needs to be a sharper turn than 30%, then the operator disengages the steering clutch and applies the brake to one track while the opposite track drives. The two-speed steering system gives only a fixed amount of turning radius for power turns.

The left-hand joystick controls transmission and steering drive for up and down shifting, steering, Hi/Lo selection, and LH/RH gradual geared turn. Foot pedals apply both brakes for parking and downhill control. Brakes are spring applied and hydraulically released. ● SEE FIGURE 20–14 for the operator controls of a two-speed steering system.

STEERING CONTROLS FOR STEERING CLUTCHES AND BRAKES
Older machines originally used a combination of levers and pedals to steer a machine that had steering clutches and brakes. The levers would disengage the clutches and the pedals apply the brakes. ● SEE FIGURE 20–15 for an older machine with **steering levers** and foot pedals.

594 CHAPTER 20

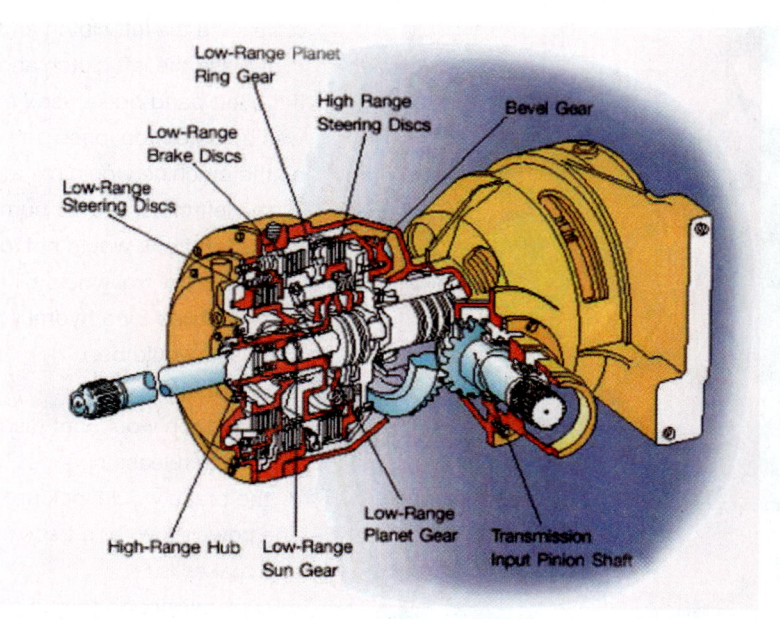

FIGURE 20–13 A cross-section of steering module.

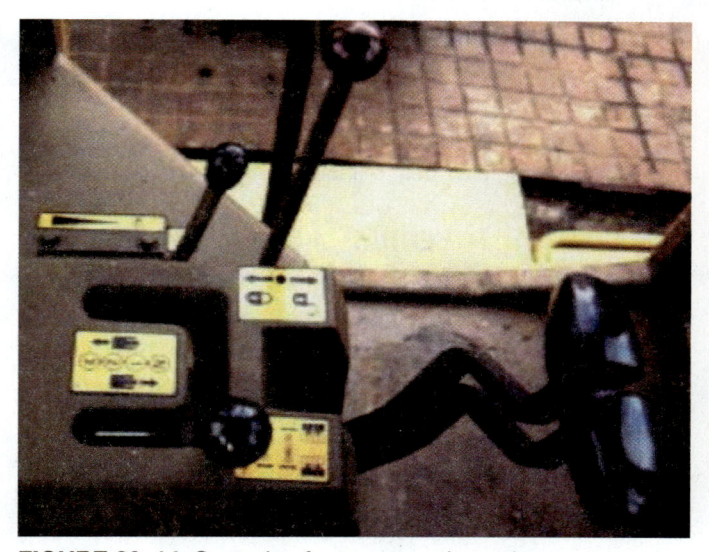

FIGURE 20–14 Controls of a two-speed steering system.

FIGURE 20–15 Older steering controls.

FIGURE 20–16 Pedal steer machine.

This evolved into either two pedals or two levers to actuate the clutches and brakes. This would have incorporated the use of hydraulic pressure to assist the operator. ● **SEE FIGURE 20–16** for a machine that is steered with pedals.

When differential steering machines (discussed in next section) were introduced, a single lever called a tiller could be used to steer the machine by pushing or pulling it. It moved a pilot control valve spool that directed oil to a main control valve spool. This changed to the tiller moving a position sensor that sends a signal to an ECM. The ECM then sends an electrical signal to the steering pump displacement control. ● **SEE FIGURE 20–17** for a tiller steering control.

The newest machines that use steering clutches and brakes use fingertip levers that move electronic position sensors. The position sensors then send an electrical signal to an ECM, and the ECM will generate an output signal

TRACK-TYPE MACHINE STEERING SYSTEMS 595

FIGURE 20–17 Tiller steering.

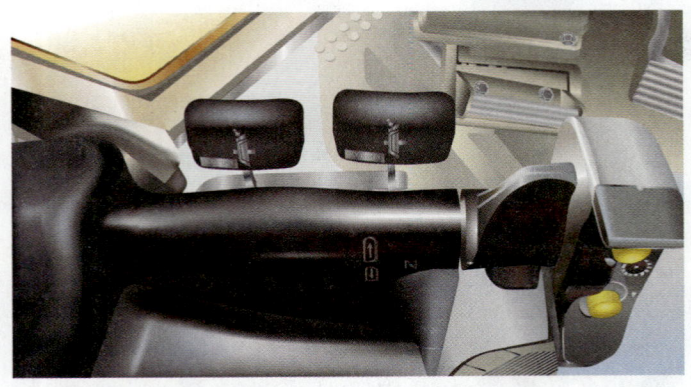

FIGURE 20–18 Fingertip controls.

to two or more solenoids. The solenoids will then direct oil to the steering clutches and brakes to steer the machine.
● **SEE FIGURE 20–18** for fingertip controls.

A 200,000 lb machine could be steered with one finger if it has electronically controlled steering clutches and brakes.

Two-speed turning systems are controlled with two levers that are pulled back to achieve low range and then pulled back farther to release the clutch and apply the brake.

Progress Check

1. A track machine that uses steering clutches and brakes to steer will *never* use:
   a. a differential
   b. final drives
   c. foot controls
   d. hand controls
2. If a track machine turns sharp left, the operator must actuate controls that will:
   a. release the right clutch and apply the left brake
   b. release the left clutch and apply the left brake
   c. release the left clutch and apply the right brake
   d. release the left clutch and release the left brake
3. If the left band brake used on a track machine for steering was adjusted too loose, the result would be:
   a. the machine would roll away if parked on a slope
   b. the left clutch would burn out
   c. the left track would not lock up
   d. the left control would bind up
4. If a seal failed for a hydraulically released steering clutch, the result would be:
   a. a slipping clutch
   b. the clutch would not disengage
   c. broken release springs
   d. the brake would lock up
5. The power flow for a track machine with steering clutches and brakes is:
   a. transmission to torque converter to steering clutch to final drive
   b. hydrostatic motor to steering clutch to final drive
   c. transmission to differential to steering clutches to final drive
   d. torque converter to transmission to pinion to ring gear to steering clutches to final drive

DIFFERENTIAL STEERING As mentioned before, track-type machines that use only steering clutches and brakes can drive only one track when they are turning. This is a big disadvantage when the machine is under load, and it doesn't provide a means to make a smooth turn. A machine that uses a differential steering system will always drive both tracks, can easily make smooth turns, and can even turn the tracks in opposite directions for tight maneuvering.

A track-type machine with differential steering operates in a similar principle to a wheeled machine's drive axle with a differential. When a machine with wheels turns a corner, the wheels must follow two different arcs. If the machine turns left, the left wheel must slow down and the right wheel must speed up to follow the different arcs. The average speed between the two wheels is the speed that the machine is traveling.

Similarly for a track-type machine, the differential steering action slows down one track and speeds up the opposite one to make the machine turn. In other words, it creates a speed differential between the two tracks. Again the average speed of the two tracks is the speed of the machine.

Differential steering has been around for a long time but not widely used for heavy equipment until around the late 1980s. Currently, Caterpillar medium to large track-type tractors use differential steering (D6–D9). ● **SEE FIGURE 20–19** for a track-type dozer with differential steering.

FIGURE 20–19 Differential steer machine.

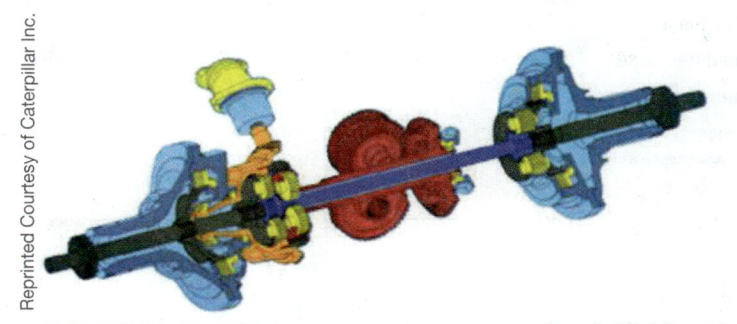

FIGURE 20–20 Differential assembly.

> **WEB ACTIVITY**
>
> For an excellent visual demonstration of planetary gears, search for a 1953 U.S. army video "Planetary Gears Principles of Operation." It will explain the interaction of the gears of a single set of planetary gears as well as how multiple sets can be used to steer a tank by using them as a differential.

They also incorporate differential steering with their electric drive dozer where a generator and two electric motors take the place of a powershift transmission in the drivetrain (see Chapter 13 for electric drive information). Differential steering uses a combination of gears and a hydraulic system to make the machine turn. For a review of planetary gearsets, see Chapter 12.

The system's main components are the transmission output pinion to give forward, reverse, and three-speed ranges (or a gear driven by two electric motors to give infinitely variable speed and direction changes with an electric drive); the planetary differential assembly, a hydraulic steering pump, hydraulic steering motor; and the steering controls. The **steering motor** is controlled by the operator, and as it rotates, it will make the machine turn left or right. The speed of the motor will determine how sharp the machine will turn, and the direction it turns will determine the direction the machine turns. In other words, there are two possible inputs to a differential steering system: the input from a powershift transmission or an electric drive and the steering motor.

The planetary differential assembly consists of three sets of planetary gearsets. ● **SEE FIGURE 20–20** for the differential assembly.

One set (furthest left) has its ring gear driven by the steering motor and is called the steer planetary. Another set (furthest right) is called the equalizing planetary and its ring gear is held stationary at all times. The third set is called the drive planetary. Its planet carrier is driven by the transmission through the pinion and ring gear because they are side by side. The outputs from the differential assembly are the steer planetary whose planet carrier drives an axle shaft that drives the left final drive and the equalizing planetary whose planet carrier drives an axle shaft that drives the right final drive. It's the interaction of these three planetary gearsets and the two inputs that make the machine's final drives, sprockets, and tracks turn.

When traveling straight, the steering motor doesn't turn and both final drives are driven at the same speed with the same torque. ● **SEE FIGURE 20–21** for how the system works in a straight line.

The only input for straight travel is the transmission pinion, and whether it turns clockwise or counterclockwise for forward or reverse, the torque and speed output is divided evenly between the left and right tracks. The transmission pinion drives the drive planetary planet carrier and its planet gears will rotate the sun and ring gear. The sun gear is the input for the equalizing planetary. The equalizing planetary compensates for the difference in diameters of the sun and ring gears of the drive planetary. As the ring gear of the equalizing planetary is stationary, its planet carrier is driven, and it drives the right final drive's input axle shaft. The ring gear of the drive planetary drives the planet carrier of the **steering planetary**, and it in turn drives the left axle shaft.

The extreme opposite of straight travel is a spot turn (tracks will counter-rotate) when there is no transmission input, but the operator moves the steering control lever, which makes the steering motor turn. This will cause the tracks to turn in opposite directions. The steering motor turns the steering planetary ring gear. As the drive planetary's carrier is stationary, the drive starts with the steering planetary ring gear that is turned by the steering motor. It turns the planet gears and sun gear. The planet

TRACK-TYPE MACHINE STEERING SYSTEMS 597

TECH TIP

A planetary gearset consists of a sun gear, planet gears, planet gear carrier, and ring gear. The planet gear shafts are held in the planet gear carrier. To transfer drive through this combination, one member is the input (drive), one member is held stationary, and the third member is the output (driven). By holding and driving different members, a combination of seven different gear ratios and rotation directions result. ● SEE TABLE 20–1 for the different combinations.

	SUN	CARRIER	RING	SPEED	TORQUE	DIRECTION
1	Input	Output	Held	Maximum Reduction	Increase	Same as Input
2	Held	Output	Input	Minimum Reduction	Increase	Same as Input
3	Output	Input	Held	Maximum Increase	Reduction	Same as Input
4	Held	Input	Output	Minimum Increase	Reduction	Same as Input
5	Input	Held	Output	Reduction	Increase	Opposite of Input
6	Output	Held	Input	Increase	Reduction	Opposite of Input
7	When any two members are held together, speed and direction are same as Input; ratio is 1:1						

If the carrier is: 1- The output, under drive results, or speed decrease
2- The input, overdrive results or speed increase
3- The held member, output direction is reversed

TABLE 20–1

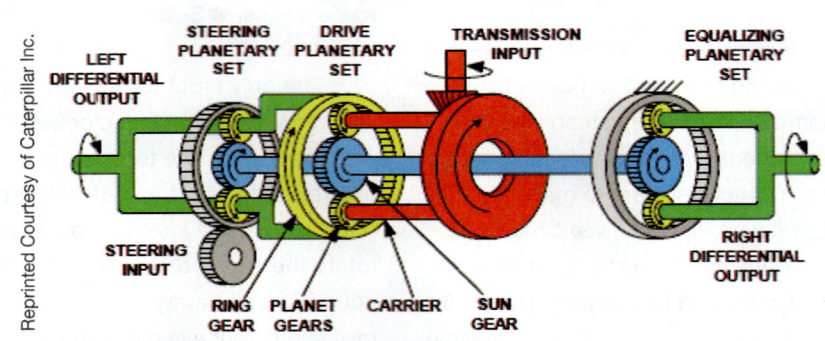

FIGURE 20–21 Differential steer, straight.

carrier is driven by the planet gears and turns the left axle shaft. The sun gear drives the **equalizer planetary** gears, which turn the planet carrier that turns the right axle shaft. If the only input to the steering arrangement is the steering motor whenever the motor turns the ring gear of the steering planetary, it causes the left and right axle shafts to turn in the opposite direction.

The faster the motor turns with no transmission input, the faster the tracks counter-rotate (a counter-rotation will only turn the tracks slowly) and the direction the motor turns will determine the direction the tracks turn. ● SEE FIGURE 20–22 for an illustration for a counter-rotation maneuver.

In between the two extremes of straight travel and a spot turn is a gradual turn. This is when there is input from both the transmission or electric drive and the steering motor. How fast the transmission pinion or electric drive output turns and how fast the motor turns will determine how sharp the machine turns. ● SEE FIGURE 20–23 for an illustration of a gradual turn.

An example of a gradual turn is if the machine is driving straight ahead at 3 mph and the operator moves the steering control to turn the machine to the left, then the left track will slow to 2 mph and the right track will speed up to 4 mph.

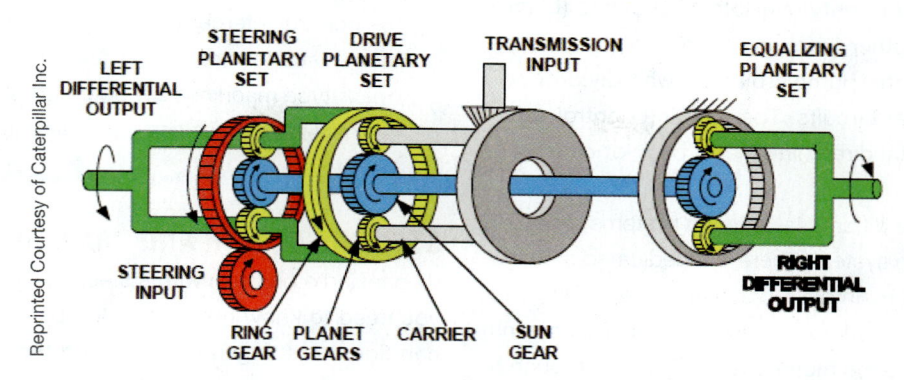

FIGURE 20–22 Differential steer, counter-rotation.

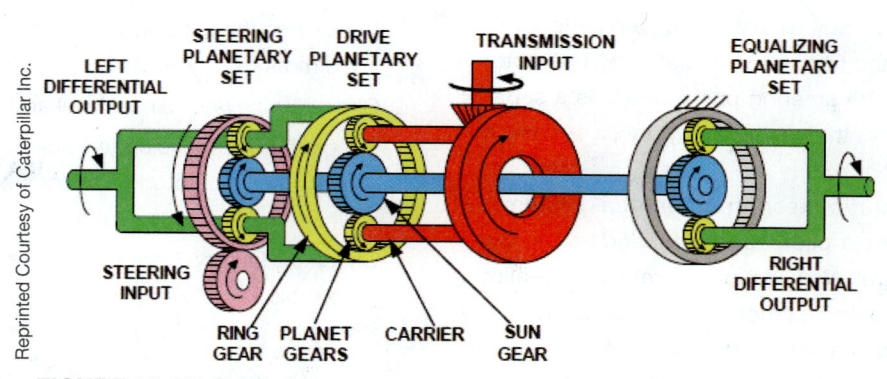

FIGURE 20–23 Gradual turn.

	LEFT-TURN FORWARD	LEFT-TURN REVERSE	RIGHT-TURN FORWARD	RIGHT-TURN REVERSE
Rotation of steering motor input	Clockwise	Counterclockwise	Counterclockwise	Clockwise
Rotation of transmission pinion	Clockwise	Counterclockwise	Clockwise	Counterclockwise
Position of steering control lever	Pushed forward	Pulled back	Pulled back	Pushed forward

TABLE 20–2

Direction of rotation during the various operations.

Differential steering in summary: The power transferred to the final drives is provided by the transmission or electric drive in all cases except a spot turn. Also the direction of rotation of the axle shafts is controlled by the transmission when its output pinion is turning to make the machine travel forward or backward.

The amount of speed difference between the axle shafts and the direction of the machine's turn are controlled by the steering motor. The speed of the motor shaft determines the tightness of the turn. A faster motor speed causes a sharper turn. The direction of rotation of the steering motor controls the direction of the turn.

● **SEE TABLE 20–2** for the direction of rotation during the various operations.

Differential steering controls for the operator are simple. A lever to the left of the seat is called the tiller and pivots at the left end of it. It can be pushed ahead and pulled back. The tiller handle also has controls to shift the transmission direction and speed range.

Some differential steer machines have their differential steering motors getting their oil supply from a steering control valve that is part of the implement hydraulic system. This system uses one pump as part of a load-sensing

TRACK-TYPE MACHINE STEERING SYSTEMS 599

pressure-compensated system for all hydraulic functions. As the steering section is first in the main control valve assembly, the steering system had priority over other functions (blade, ripper). However, when other functions are activated, steering slowed down because the pump's oil flow was divided between steering and other circuits. The steering control valve spool could be shifted by direct linkage or by pilot oil from a pilot control valve.

The motor for newer differential steering systems is part of a hydrostatic closed loop system. There is a dedicated steering pump for this system to maintain a consistent steering speed whether the implement functions are used or not. This means the pump supplies flow to the motor directly and not through a directional control valve. The return flow from the motor goes directly to the pump inlet. The direction and speed that the motor turns is determined by the output of the pump. There will be a charge pump to replenish any oil lost in the loop because of normal internal leakage and from losses created by the motor flushing valve. Although the steering pump/motor is a separate system from the rest of the machine's hydraulic system, it shares a common reservoir.

For original hydrostatic steering differential systems, the operator control shifts a pilot valve that sends oil to the pump's swashplate control. Newer systems replace the pilot oil system with an electronic/electrical system to control the pump. The system pump is a bidirectional variable displacement pump with a swashplate control piston. The direction the operator moves the control lever will make the swashplate move, which will make the pump oil flow out of either of two pump ports and to one of two motor ports. If one port receives oil, the motor will turn clockwise, and if the opposite port receives oil, it turns counterclockwise. This in turn makes the machine turn left or right. The motor is a fixed displacement bent axis type, and its output shaft drives a pinion gear. The pinion gear is the input for the steering planetary and drives the ring gear.

For more details on hydrostatic systems, refer to Chapter 17.

STEERING SYSTEM DIAGNOSTICS
Track machine steering systems that are working properly should allow the operator to steer the machine left and right in either forward or reverse with ease and consistency. Depending on the type of steering system the machine uses, this could be a seamless power turn with a differential steer system, a fixed arc power turn as with the two-speed system, or a turn that is made in two steps to drive one track around the other one that is either not driving or locked up as with the clutch and brake system. Because these systems operate quite differently, we will look at diagnostic procedures for each separately.

> **WEB ACTIVITY**
>
> Search the Internet to see if you can find video demonstrations on how different steering systems for track-type machines are operated. What other types of track-type machines use differential steering?

STEERING CLUTCH AND BRAKE DIAGNOSTICS
When diagnosing a problem with a steering clutch and brake system, you need to know how it should operate normally before you can figure out why it isn't working the way it should. Unless you are familiar with how the system should work, you need to read the operator's manual. After that there are some basic checks to perform initially:

- Check that the controls are operating smoothly and with full motion.
- Check the power train oil level and condition.
- Check the track tension. A track that is too tight will be hard to drive and could affect steering operation. See Chapter 23 for an explanation of track tension.
- Check for fault codes if the machine has electronic controls.
- Check for oil leaks and obvious damage.

If these initial checks don't reveal any problems, then proceed to confirm the complaint by operating the machine. Make sure that you are familiar with all safety features of the machine and find a flat and open area with a consistent surface material to check its operation. The machine steering controls should turn the machine in a smooth and consistent manner equally left and right. You should be able to tell when the steering clutch is disengaged and when the brake is applied. If the brake comes on too soon after the clutch is released, the machine will jerk. There should be a noticeable space in the control travel between clutch release and brake application.

For older machines without hydraulic assist clutches or brakes, there will be specifications for dimensions that should be checked and measured for the control pedal and lever linkage. You may find that an adjustment is needed to either a stop screw or a linkage length to restore proper operation. Be aware that you may need to perform a series of adjustments in a specific order. For example, the following is a list of adjustments that must be done in this specified order to reset the **steering pedal** linkage on a John Deere 450G dozer:

1. Loosen Brake Bands
2. Adjust Linkage Inside Transverse Housing

3. Adjust Stops
4. Adjust Linkage
5. Adjust Steering Valve
6. Adjust Brakes

Some operational problems you may have to troubleshoot are when the machine steers only one way, doesn't steer either way, steers erratically, steers slowly one way, and won't travel.

For machines with hydraulic steering systems, you may need to install pressure gauges to check apply or release pressures. These pressures will likely come from the transmission oil control system and be limited to around 250 psi. You may also need to check that pressure drains to zero when it is supposed to.

If the oil is supplied from the transmission oil circuit and you suspect there is a problem, you should check to see if there are any operational problems with the transmission as well. This could lead you to an oil pump or oil supply problem. The oil system filter and suction screen should be removed and inspected for excessive wear particles and contamination. An oil sample should be taken as well if there are doubts about the state of the oil in the system.

TWO-SPEED STEERING DIAGNOSTICS The same initial checks should be performed for this steering system as for the steering clutch and brake checks. The complaint should also be verified by the technician operating the machine. As this system relies on oil pressure to be applied to its clutches and brakes, the hydraulic control system pressures may need to be checked. There will be pressure taps available to install gauges for these pressure checks. This system gets its oil supplied from the power train hydraulic system and the entire system should be checked to see if it is operating properly. There may be steering system problems that are common to another power train hydraulic problem.

If a mechanical problem with the system's planetary gears is suspected, you should drain the steering case oil and look for metal flakes or pieces.

Oil and filter condition could also be checked.

DIFFERENTIAL STEERING DIAGNOSTICS The same initial instructions apply to diagnosing a differential steering machine problem as with the steering clutch and brake machine. Get yourself familiar with how it should operate.

Some operational problems you may have to troubleshoot are when the machine steers only one way, doesn't steer either way, steers erratically, won't counter-rotate, and oil overheats.

As this steering system uses a hydraulic system to steer the machine, there are some additional initial checks to perform when troubleshooting: check oil level for hydraulics, check for leaks, check for fault codes, check to see if there are other hydraulic system problems.

If the machine is older and shares pump flow to the steering motor with other implements, make sure that the other functions are working properly if there is a steering problem. This should eliminate the pump and its control system as the source of the problem.

STEERING SYSTEM ADJUSTMENTS/CALIBRATIONS

Machines with steering clutches and brakes will sometimes need regular adjustments. This could be adjustments to the band brakes to compensate for friction material wear. This will likely mean removing a small cover and turning an adjuster that will move the ends of the band brake closer together. The proper adjustment is determined by pedal travel and should be referenced to a specification in the machine's service information. If the pedals travel too far and don't stop the tracks, then an adjustment or repair is needed.

A calibration procedure may need to be done for a steering system that has electronic controls. See Chapter 9 for more details on calibration procedures. This will keep the operator input for controls matched to what is happening with the actuators or pistons and cylinders.

STEERING SYSTEM REPAIR Steering system repairs can vary from fairly simple mechanical repairs to hydraulic component repairs to electrical repairs.

An example of each follows:

A simple mechanical repair would be one that involves repairing the control linkage. This could be as easy as applying penetrating fluid to a rod end or greasing a pivot shaft because any binding of linkages will affect steering operation.

If excessive wear is found in this linkage, the worn parts should be replaced. The manufacturer's linkage dimensions will need to be reset whenever any linkage components are replaced.

If a steering system has clutches and brakes that are hydraulically actuated, their seals can occasionally fail and cause pressure loss. This would mean removing the clutch or brake assembly and reconditioning the assembly including replacing all seals and bearings at a minimum.

A hydraulic component repair could be a simple leak repair or a total pump or motor recondition. Whenever the steering hydraulic system is opened up, extreme cleanliness is critical on the technician's part. Cleanout filters are sometimes used after a hydraulic system repair for an extra bit of insurance against a repeat failure. A cleanout filter will have a smaller micron rating and is only designed for short-term use because it will plug up faster than a regular filter.

An electrical repair related to a steering system could be the replacement of a wiring harness, solenoid, or position sensor.

STEERING CLUTCH AND BRAKE RECONDITIONING

If a steering clutch or brake is found to be defective, it can be reconditioned. The main reason for reconditioning a steering clutch is the friction material has been reduced to less than specified minimum dimension. There will need to be a certain thickness of friction material on the steering and brakes discs and the brake band. The teeth or tangs on the discs should be checked for excessive wear.

Friction discs that are worn too thin can only be replaced while brake bands can have friction material replaced. Discs and plates must be replaced if there are any cracks or excessive warpage found. ● SEE FIGURE 20–24 for an example of damaged plates and discs.

The steel plates can be reconditioned if there aren't deep grooves in them or aren't warped or dished. They can be ground slightly to renew the surface finish as long as the minimum thickness isn't exceeded, and they should be ground to a specified roughness so they can retain oil. They are checked for straightness with a straightedge.

FIGURE 20–24 Non-reusable plates and discs.

Progress Check

6. A track machine that can counter-rotate will probably:
    a. use an HMU
    b. use a lock up torque converter
    c. have a differential steering system
    d. have double reduction final drives

7. A differential steer machine that is performing a fast counter-rotation turn will have:
    a. its left and right steering clutches turning opposite
    b. its transmission output gear turning at maximum speed
    c. its steering motor turning at maximum speed and transmission output stationary
    d. its left and right brakes turning opposite

8. If a track machine counter-rotates, it must:
    a. have planetary final drives
    b. have two-speed track motors
    c. have hydraulically actuated steering clutches
    d. be either hydrostatic drive or differential steer

9. An operator running a track machine is complaining that the machine is hard to steer to the left. Which of the following would be a good first troubleshooting step?
    a. run the machine to verify the complaint
    b. check the air filter
    c. check left track tension
    d. check the power train oil filter

10. A differential steering system uses this many planetary gearsets to steer the machine:
    a. 1
    b. 2
    c. 3
    d. 4

SHOP ACTIVITY

- Go to a machine with steering clutches and brakes and see how the operator controls them to steer the machine. Is the steering brakes part of the parking brake system?
- What adjustments can be made to the steering clutch and brake system?

SUMMARY

1. Safety concerns related to track machine steering systems include: proper lifting methods and equipment should be used for heavy components; friction material can contain hazardous material and should be treated with care; high pressure oil may be present and appropriate precautions need to be taken when working near it; proper steering operation must be confirmed before a machine is put back into service.

2. Track-type machines with one piece frames can only steer by having their two tracks drive at different speeds. Three ways to do this are: steering clutches and brakes, differential steering, two speed gearing.

3. Track-type machines with steering clutches and brakes have them located past the ring gear shaft and before the final drives. Disengaging one steering clutch stops drive then applying the steering brake locks the track, the other track can still drive and this turns the machine.

4. Most steering clutches are multi-disc spring applied oil released. Steering brakes can be external band and drum type or multi-disc.

5. Steering clutches and brakes can be controlled with: mechanical linkage, oil-assisted mechanical linkage, electro hydraulic systems.

6. A steering clutch is driven by a flange that is driven by a bevel gear shaft that is driven by the transmission output pinion gear.

7. The steering clutch output is its friction discs that have external teeth which mesh with a brake drum. The brake drum has a smooth outer surface and it drives an input flange to the final drive.

8. Clutch apply pressure can come from springs (Belleville or coil type) or oil pressure.

9. Most clutches are wet type that run in oil to transfer heat away from friction material.

10. Clutch release devices will remove clamping pressure from the friction discs to stop torque transfer. They will compress springs or drain oil pressure.

11. Newer machines that use steering clutches will use multi disc steering brakes that are spring applied and oil released. These brakes also function as parking brakes.

12. One machine manufacturer uses a two speed steering system that can provide drive to both tracks while the machine steers. Two planetary gearsets along with steering clutches and brakes is used.

13. If one track is driven in high range and the other in low there is a 30% speed difference which makes the machine turn.

14. Differential steering systems for track machines use the same principle as drive axle differentials to provide a speed differential for the tracks which in turn makes the machine turn.

15. Larger track type dozers can feature differential steering and have the transmission output pinion drive through three planetary gearsets which sends drive toward each tracks final drive. A hydraulic motor is also used in conjunction with the steering planetary to provide a speed difference between the left and right tracks.

16. For straight travel the motor doesn't turn. To steer the machine the motor has oil sent to it and when it rotates its output gear turns the ring gear of the steering planetary which changes the speed between the two tracks.

17. Differential steering systems can provide track counter rotation and steering is controlled with movement of a tiller lever.

18. Track-type machine steering system operator concerns can include: no steering, do drive one side, rough engagement, slipping, overheating, strong odor.

19. Diagnostics start with knowing the system, thorough visual inspection, verifying the complaint, performing tests.

20. Inspections include: check of oil level and condition, leaks, control operation, track tension.

21. Adjustments may be needed to restore proper clutch and brake operation since wear of friction material changes clearances.

22. Oil pressure checks are common and can detect seal failure or valve issues.

23. Differential steering system diagnostics can include hydraulic pressure and flow testing.

24. Steering system repairs include friction disc replacement, plate replacement, brake band relining, hydraulic motor replacement.

chapter 21
MACHINE FRAMES AND SUSPENSION SYSTEMS

LEARNING OBJECTIVES

After reading this chapter, the student should be able to:

1. Explain the purpose of the main frame for a machine.
2. Describe why main frames may fail.
3. Identify likely locations for high stress areas on machine frames.
4. Explain why suspension systems are needed for some heavy equipment machines.
5. Describe how a suspension cylinder works.
6. Explain how a suspension cylinder gets charged.
7. Identify the different fluids and gases that a suspension cylinder could use.
8. Explain how a walking beam suspension system works.
9. Identify different types of springs used for suspension systems.
10. Describe how a ride control system works.
11. Identify the different kinds of springs used for heavy equipment suspension systems.
12. Explain how an electronic payload measuring system works.
13. Describe some different troubleshooting procedures for identifying suspension system faults.

KEY TERMS

Gooseneck 614
Leveling valve 614
Load cylinder 614
Locked down 615
Neo-con 612
Nitrogen 612
Regulator 620
Spring rate 607
Spring shackle 607
Strut 610
Torque rod 616

INTRODUCTION

SAFETY FIRST When working on a machine's main frame and suspension system you need to pay attention to safety related concerns such as:

- Be aware of all hazards related to cutting and welding on machine frames such as the build-up of pressures in closed vessels.
- Be aware of the potential for uncontrolled movement of the machine or component when working on frame and suspension systems. Use proper lifting and blocking procedures.
- Be sure not to mix different types of gases when filling suspension cylinders or accumulators.
- Make sure that any stored energy in accumulators is released before opening the system it is part of.
- Make sure that spring tension is released slowly and in a controlled manner.
- Use proper lifting and blocking equipment and procedures.
- Some suspension systems use high pressure gas or fluid. Be aware of the safety concerns related to working with these high pressure systems.

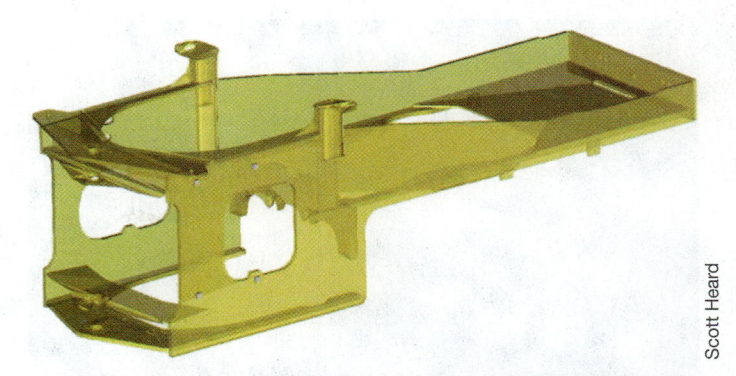

FIGURE 21–1 Machine main frame examples.

NEED TO KNOW

MACHINE FRAMES All heavy equipment machines will have some type of main frame. The frame of the machine is its backbone and will serve as the mounting point for all other systems and components that the machine needs to function as it is designed to. Main frames will vary greatly in their construction depending on the type of machine it is part of. Main frames are a welded and machined assembly that start out as plate steel or cast sections. The plate steel is cut and formed into box sections with large presses and the separate pieces are welded together to form frame sections. These individual box and cast (if used) sections are then welded together to form a strong structure.

Once the assembled frame is completed it is painted to prevent rust from weakening it. ● **SEE FIGURE 21–1** for a machine's main frame.

Machine manufacturers will design the main frame to withstand the most severe forces that may normally be applied to it and it should be able to last the expected life of the machine without any type of failure. If the design limits are exceeded a variety of failures could occur. A frame will get overstressed from overloading, rollover, traveling over extremely rough ground, or operator abuse. Any of these events will lead to a frame failure.

Frame failures usually start out as small cracks or slight bending and will likely progress to larger cracks and possibly complete failures where the frame completely breaks. Most frames can be repaired with straightening or welding procedures that may include adding reinforcing.

Machine frames can be one piece assemblies but can also be multi piece as well. A wheel loader will use a two piece frame while a wheeled tractor-scraper uses a three piece frame. There are pivot points between the main frame sections for a multi-piece frame that use pins and bearings to allow movement and hold the sections together. The pins and bearings must be lubricated according to the machine's maintenance manual. An articulated truck must also allow rotation between the front and rear sections of the truck in case the box flops over while it is dumping. A pair of large bearings allows this rotation.

An excavator has upper and lower frames that are held together with one large bearing. This bearing is usually called the swing bearing. Part of the bearing has teeth that allow the machine's swing motor to turn the upper structure. ● **SEE FIGURE 21–2** for a swing bearing.

Machine frame design flaws may appear at some point during the machine's life. If this happens the manufacturer should

FIGURE 21–2 Swing bearing.

create an update to the frame that could include a welding procedure or adding reinforcements to repair and improve the frame.

Whenever technicians are working on a machine, they should consider where the high stress areas of the frame are and perform visual inspections on these areas. Typical high stress areas are where:

- Main component mounts are located such as transmission, engine, or track frames.
- A cast section to box section weld is located.
- Threaded holes have been cut into the frame.
- Main boxed sections are welded together.
- Suspension components are mounted.
- Stresses are concentrated such as cylinder mounting points.

To perform a thorough main frame inspection it will be necessary to remove all covers and wash the frame. Part of a complete machine reconditioning will include the machine being stripped down to the main frame and a crack checking procedure will be performed on the main frame. Some types of nondestructive test methods for frame defects are magnetic particle testing; dye penetrant testing; fluorescent liquid testing; and X-ray imaging. Any defects found will be repaired with industry-accepted welding and fabricating procedures and equipment. ● SEE FIGURE 21–3 for a main frame repair.

SUSPENSION SYSTEMS

Some heavy equipment machines need to have suspension systems to provide a smoother ride for the operator and to allow the machine to travel faster over rough roads and terrain. This makes the machine more productive; it will last longer because there is less jarring and vibration to its frame-mounted machine systems. There will also be less stress on the machine's frame if some independent movement is allowed between the

FIGURE 21–3 Main frame repair.

machine's wheels or tracks and its frame. A wheel type machine with no suspension must rely on the cushion its tires give as the machine travels over rough ground.

As the name implies, suspension systems are intended to suspend some of the machine and allow relative movement between two or more main parts of the machine. This motion is mainly between the machine's axles or tracks and its frame but

> **TECH TIP**
>
> I have witnessed several machine frame failures during my time as a HDET. These were failures of varying degrees, from small cracks that can be noticed by rust marks occurring where there shouldn't be to major failures where machines were literally ready to break apart. One large excavator I was sent to had a strange noise that would occur when the bucket was prying in a deep trench and it turned out to have a major crack in its upper frame. The operator was lucky the frame didn't completely fail because the boom would have likely taken the machine's cab off! This machine had to be sent to the repair shop where the frame was stripped down and a major weld repair had to be completed to allow the machine to return to work.

can also be between two other parts of the machine such as the tractor and scraper frames of a wheel tractor-scraper. Movement will occur when the machine travels over anything other than flat hard ground and when the machine accelerates and brakes.

Suspension systems will vary greatly depending on the type of machine they are on and what they are intended to do. They can be simple mechanical devices like springs or complex electro hydraulic systems that include ECMs and sensors.

A machine with a suspension system that isn't working properly will not likely be as productive as it should be because the operator will likely slow down over rough terrain. It may also cause damage to other parts of the machine because of the extra shock loads placed on machine components. For this reason it is important to understand how different types of suspension systems work, why they need to be maintained, and what may be required to diagnose problems and repair them.

SPRINGS The simplest type of suspension system is one that uses springs. Older rock trucks and some small utility equipment manufactured today use springs to allow movement between the machine's frame and its drive axles or steering axles. The two types of springs most commonly found on heavy equipment are leaf and coil and they are usually combined with shock absorbers to allow axle and wheel movement. The springs will allow movement by flexing under compression while the shock absorbers will control the movement hydraulically.

Springs are made from a specific high alloy of spring steel that should meet the SAE 5160 specification that refers to the metals composition. This alloy is formed into flat bars or wires (round bars) that then become leaf or coil springs.

The load-carrying characteristics of coil springs are related to their **spring rate**. Spring rate is a measure of the ability of a spring to resist deflection. It is usually measured as the amount of force required to change the length of the spring by 1 in. In other words, a spring that has a rate of 500 lb means it takes 500 lb of force to change the spring's length by 1 in.

Two terms related to how springs perform and how they are designed are *linear spring rate* and *variable spring rate*. A spring with a linear spring rate will compress at the same rate throughout its normal deflection. A variable rate spring will become increasingly hard to deflect as it is compressed.

Individual leaf springs are stacked in an assembly that's size relates to the machine's weight and its payload capacity. A multi-leaf spring or leaf spring assembly will have a main leaf that is formed into an eye at one or both ends. The single eye main leaf is then anchored to the machine's frame at the front end while the opposite end is allowed to float on a pad. Lighter-duty spring assemblies will have eyes on both ends with one end being anchored and the opposite end being allowed to move because it is connected to the machine's frame through a **spring shackle**. This is necessary because as the spring deflects it will change lengths. ● **SEE FIGURE 21–4** for a machine that uses leaf springs.

The spring eye (or eyes) gives the spring a second function of locating the attached axle in relation to the machine's fame.

The main leaf will have one or more shorter leafs stacked beneath it and they are all held together with one long bolt. Each leaf is progressively shorter as they are stacked below the main leaf. Newer styles of leaf springs tend to use less leafs that are thicker so that there doesn't need to be as many. Less leafs will mean less friction created between them, which can give a better ride. Some leafs are tapered to give a more progressive spring rate as the spring deflects more.

There will likely be band straps or U-bolts and brackets that will keep the multi-leaves aligned with the main leaf.

FIGURE 21–4 Machine that uses leaf springs.

TECH TIP

Some heavy equipment machines that use leaf springs will have a system to eliminate spring action when the machine is digging. In the case of rubber-tired excavators that travel at fairly high road speeds the springs are needed to allow the axles to conform to road imperfections as the machine travels. However, when the machine is stationary and digging, spring action would be counterproductive. This machine has hydraulic cylinders that lock the axle in place and eliminate the springs from deflecting. ● SEE FIGURE 21–5 for axle lock cylinders.

Coil springs can be used to provide for movement between the machine's frame and axles and as with leaf springs these aren't used for suspension much anymore except for small equipment. Coil springs only provide tension to suspend the machine and its payload weight and don't help to locate the axle to the machine's frame, unlike leaf springs. Some older small straight frame haul trucks used large coil springs in the front end. ● SEE FIGURE 21–6 for a heavy-duty coil spring.

In this case the coil spring is located between the machine's frame and a pivoting trailing arm. In cases where they are used on machines for suspending drive axles, the axles will need links to keep the axle located relative to the frame but still allow it to move up and down.

Coil springs are formed from spring steel wire into spirals. As they are compressed the wire twists and the resistance it overcomes during the twisting gives the spring its strength.

Coil spring capacity increases with an increase in the diameter of the coil, the diameter of the wire, the length of the coil, and the number of coils.

FIGURE 21–5 Axle lock cylinders.

FIGURE 21–6 Large coil spring.

They will wrap around the spring assembly at the mid points between the axle and the ends of the assembly.

Leaf springs are made with an arch that counteracts the weight of the machine and its payload. The arch will get straightened as the load is increased and there will be stops to limit the amount of deflection in the spring assembly caused by the load or by the machine traveling over rough terrain. The amount of arch in a spring assembly is determined when the spring is manufactured and will stay in the spring as a "memory." The spring may lose this memory over time, reducing the spring arch. The spring will then start to sag and the axle stops will be contacted easier.

The top of the axle housing will contact stops if the spring is fully compressed. They shouldn't normally make contact but as the spring starts to sag then it becomes easier to contact the stops.

All the leaves of the assembly are bolted together with a long bolt going through their center. This bolt will then serve to locate the axle to the spring assembly. There will be U-bolts that fasten the axle to the spring assembly through brackets.

SHOCK ABSORBERS
Some machines will use shock absorbers to reduce the suspension oscillations created when they travel over rough ground. Shock absorbers are usually used in combination with springs to provide a dampening effect and they are sometimes called dampers. If there weren't shock absorbers the machine would move up and down in a series of slightly decreasing rebounds and slowly come back to rest.

They are a sealed and oil-filled device that has a movable piston inside. The piston is connected to a rod that extends out of the barrel past seals. The rod will be fastened to the frame of the machine while the barrel is fastened to the axle or steering arm. The two parts of the machine are allowed to move in relation to each other and are limited to the maximum amount of travel by axle stops.

The shock absorbers piston has one or more valves or orifices that allow the transfer of oil from one side of the piston to the other. This happens as the rod is pulled out or pushed into the barrel. As the oil moves through a restriction in the piston it is slowed and this action makes the suspension travel slow down.

● **SEE FIGURE 21–7** for a heavy-duty shock absorber.

OSCILLATING AXLES
All recently made wheel loaders will have an oscillating rear axle, and all backhoe loaders will have an oscillating front axle. The axle is mounted to the machine's main frame through a pivot at its center. This allows the axle to oscillate or tilt a few degrees either way from being parallel to the machine's frame. This occurs when the machine travels over uneven terrain to keep the machine level longer and to help keep all the tires on the ground where they can get traction and help move the machine.

The front axle of a backhoe will have a raised portion of its housing machine bored to provide for the installation of two plain bushings. The machine's frame has two cross members that have holes machined into them. The axle is held to the frame by a pin that is held stationary to the cross members with a bolt. The bushings are drilled to accept grease

> **TECH TIP**
>
> If you have ever driven on a multi-lane highway and noticed a vehicle with one wheel that moves up and down rapidly and uncontrolled, then you have seen a vehicle with a failed shock absorber. This is not only unsafe, because the wheel spends as much time in the air as it does on the road, it is very hard on the suspension and drive axle components.
>
> Heavy equipment machines that have one or more failed shock absorbers will experience the same problems as an automobile with a failed shock absorber would.

fittings and as grease is applied to them it will form a lubricating barrier between the pin and bushings. There are lip-type seals on both sides of each bushing that will try to keep grease in while keeping dirt out. The relative rotational movement between the bushings and pin is what allows the axle to pivot. The amount that it can pivot is limited by stops on the axle housing that will contact the bottom of the machine's frame. ● **SEE FIGURE 21–8** for how much the front axle of a tractor loader backhoe will oscillate.

Wheel loader rear axles pivot on trunnions that are mounted to the front and rear of the center axle housing. The front trunnion is a smooth machined surface that is on the outside of the pinion shaft housing. The rear trunnion is bolted to

FIGURE 21–7 Heavy-duty shock absorber.

MACHINE FRAMES AND SUSPENSION SYSTEMS **609**

FIGURE 21–8 TLB front axle oscillating.

FIGURE 21–9 Rear axle trunnion mounting.

> **TECH TIP**
>
> An important part of any service interval on a machine that has an oscillating axle is to ensure that the pin and bushing for the axle pivot is getting greased regularly. Moist grease should be visible either around the pivot bushing or coming out of a grease relief valve. Another check on these machines is to look at the axle stops to see if they are getting contacted frequently. This could be an indication that the machine is being abused.

the main axle housing and also has a machined surface on its outer surface. There are two trunnion supports (front and rear) that are bolted to the machine's main frame. Each trunnion support will be fastened to the bottom of each side of the main frame. ● SEE FIGURE 21–9 for how the trunnion is mounted to the frame.

The trunnion supports have plain bushings or sleeves that allow movement between the axle trunnion and the machine's frame. There are grease fittings on each trunnion support to allow grease to create a lubricating barrier between the trunnions and trunnion supports. The trunnion supports will also have grease seals on each side of the bushing. There may also be a thrust plate that limits the fore and aft travel of the trunnions in the supports.

Rear axle oscillation is limited to a few degrees because of stops on the axle housing that will contact the main frame. Some rubber-tired excavators feature an oscillating axle that can be locked in place to make the machine more stable when it was in digging mode. The grease fittings for these oscillating axles will likely have hoses or tubes going to them to allow the operator to grease the bushings easily from a convenient location. Because of the high forces applied to them the pins and trunnions need to be lubricated on a regular basis. Always check with the manufacturer's service information to see when the correct greasing interval is.

SUSPENSION CYLINDERS
There are several types of machines that use suspension cylinders to provide both a spring action and a damping function for the machine. One machine that uses them is the straight frame haul truck. It will have one suspension cylinder at each corner of the machine to allow movement between the frame and the machine's axles, wheels, and tires. Another type of machine that can use suspension cylinders is the articulated truck and they can be used for the front wheels only or for front and rear suspension of the truck. Another common term for this suspension component is *suspension strut* or just **strut**.

There are two main types of suspension strut systems: active and passive. An active strut will be part of the machine's main hydraulic system and have a remote accumulator. Passive struts are sealed units that contain gas and oil.

PASSIVE SUSPENSION CYLINDER
Each passive suspension cylinder uses a combination of oil and gas to act as a spring and to provide a damping or shock absorber effect. Struts have three main parts: a barrel, a rod, and a piston. They are similar to a hydraulic cylinder or a shock absorber because the rod moves in and out of the barrel.

The gas inside the strut is precharged to a certain pressure and when compressed under load by either suspension action or payload it acts like a spring. The precharge pressure value will usually extend the strut's rod a little more than halfway (in terms of total rod travel) out of the barrel when the machine is empty. When the machine is loaded the gas is compressed and the rod is pushed into the barrel. Damping is needed to reduce the oscillations caused by the compression and expansion of the gas as the machine moves over rough ground.

Some machines will have the front suspension cylinders use their rod as a kingpin similar to how conventional steering systems use kingpins as pivots for the steering knuckle. The cylinder's barrel is fixed to the machine's frame and its rod is allowed to move in and out. These haul trucks will have a steering knuckle attached to the rod of the suspension cylinder. The rod and knuckle can turn in the barrel of the cylinder for steering purposes and the rod can also move in and out of the barrel to follow the contours of the haul road. ● **SEE FIGURE 21-10** for a strut that is part of the steering system.

The steering knuckle has a wheel spindle attached to it as well as a steering arm that has a tie-rod and steering cylinder move it to rotate the rod. Wheeled machine steering systems were covered in Chapter 19.

For most other applications the strut simply provides a suspension function. The rod and barrel are attached to the machine with pins. The pins go through self-aligning bearings that are held in the strut and the ends of the pins are held to the machine's frame and axle or suspension component. These bearings will need to be greased regularly because they take the weight of the machine and its payload.

Another manufacturer of haul trucks has the front suspension cylinder's rods attached to trailing arms. Trailing arms on these trucks pivot on large tubes that are fastened to the bottom of the machine's frame just behind the front bumper. ● **SEE FIGURE 21-11** for a trailing arm suspension system.

The struts that are used with trailing arms suspend machine weight between the trailing arm and the frame.

For all types of suspension cylinders the barrel is sealed at the rod to keep oil and gas in and dirt and other contamination out.

The rod will have a piston fastened to it and like a hydraulic cylinder it will have seals around its outer diameter to create a seal between the piston and the inside of the barrel. There could be as many as four seals and they will likely have back up rings to support them.

FIGURE 21-10 Strut with steering knuckle.

FIGURE 21-11 Trailing arm suspension.

MACHINE FRAMES AND SUSPENSION SYSTEMS

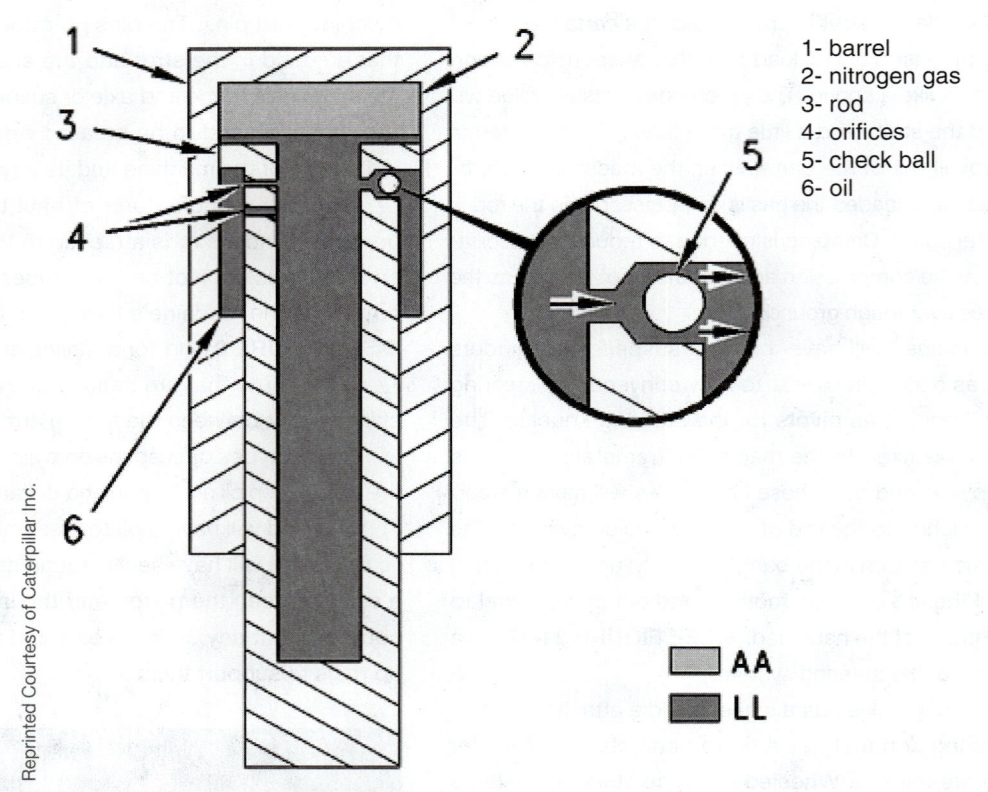

FIGURE 21–12 Oil movement in a strut.

1- barrel
2- nitrogen gas
3- rod
4- orifices
5- check ball
6- oil

The rod has a cavity inside of it that is open to the head end of the cylinder. There is a check valve and orifices that allow oil to move from the rod cavity to the opposite side of the piston.

The **nitrogen** at the top of the piston gives a spring effect to the strut while the oil lubricates the moving rod and gives a damping effect. This will slow the movement of the rod and reduce the oscillations created by the gas spring effect as the machine travels over rough terrain.

The damping effect is accomplished with the check valve and orifices. Oil is forced to move from inside the rod to the top of the rod either past the check valve or through the orifices. The way the rod is moving (in or out) will determine which way the oil has to move and will open and close the check valve. If the oil opens that check valve, it will flow freely past it to allow the rod to move quickly. This is when the rod is moving into the barrel. If the oil reverses flow, it will close the check valve and this forces the oil to travel through the orifices in the rod, which slows the travel of the rod back out of the barrel. ● **SEE FIGURE 21–12** for how the oil moves in a strut.

When the oil movement is slowed down it will perform a damping effect to reduce suspension oscillations.

The majority of suspension struts will use nitrogen gas as a precharge and SAE transmission oil. Some manufacturers however will use helium as a precharge gas. These struts will also use a different type of fluid that is called **neo-con**. It is a silicone-based fluid that is exclusive to Hitachi truck suspension cylinders.

Strut gas precharge is sometimes given as a fixed pressure and other times it is based on the proper amount of extension of the strut rod. Total strut travel is usually only a few inches.

ACTIVE STRUTS Active struts have the capability to use oil from an external source to extend or retract to rod. This is done to adjust the machine's ride height based on the load it is carrying. A typical application is an articulated truck that has suspension cylinders for only the front suspension. They sometimes feature a separate accumulator to give the strut a spring effect; otherwise, there is part of the strut that has a nitrogen charge in it.

WEB ACTIVITY

Check the Internet to see if you can find video of a working machine that has suspension cylinders to get an idea of how they operate. Search for "articulated truck or mining truck suspension."

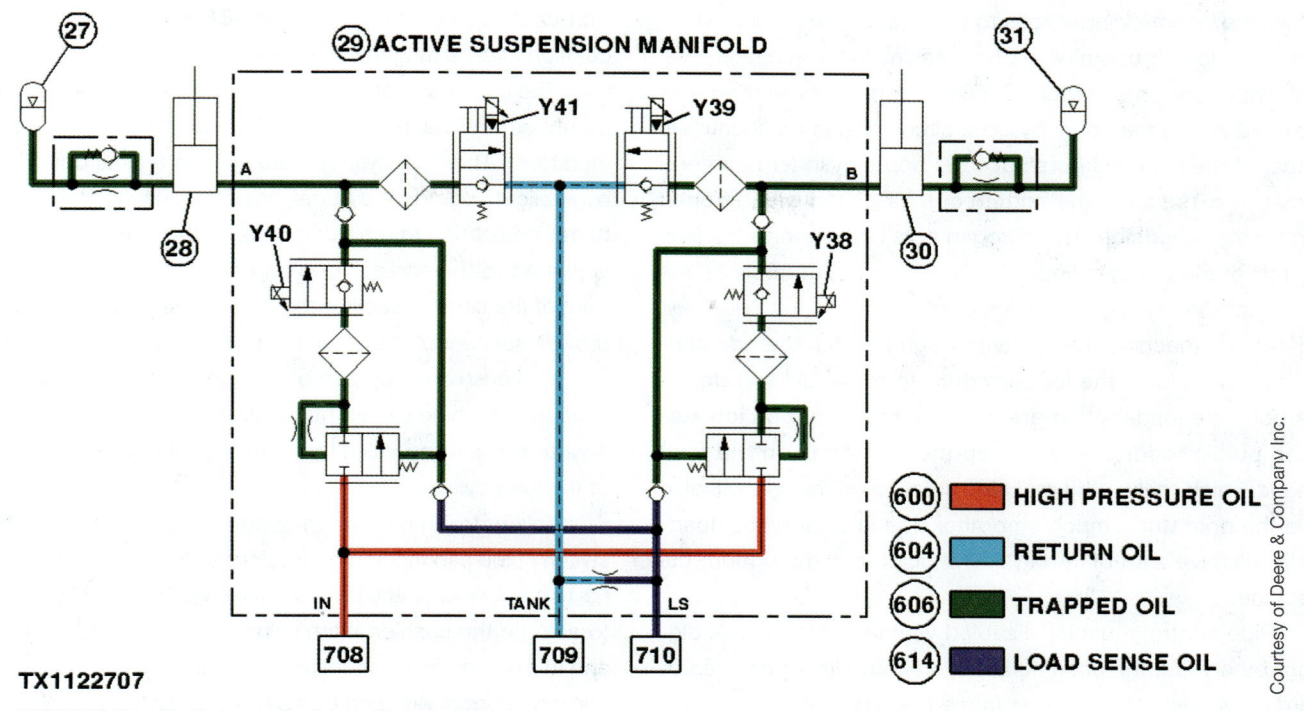

FIGURE 21–13 Schematic for active suspension.

There is a manifold with electronically actuated solenoids to direct main pump oil into the bottom of the strut and ride height sensors will send a signal to the chassis ECM. The ECM sends electrical signals to the solenoids based on preprogrammed information and attempts to keep the struts extended within a certain range. ● SEE FIGURE 21–13 for the schematic of an active suspension system.

Progress Check

1. A machine with leaf springs will also feature:
   a. suspension cylinders
   b. coil springs
   c. rubber blocks
   d. spring shackles
2. The term *spring rate* refers to:
   a. the price per pound that a spring costs
   b. the amount of stretch a spring provides
   c. the amount of force needed to deflect a spring a certain distance
   d. the amount of distance a spring will deflect when a certain weight is applied
3. Oscillating axles are most commonly found on these types of machines:
   a. wheel loaders and backhoe loaders
   b. scrapers and track loaders
   c. haul trucks and drills
   d. excavators and dozers
4. Suspension cylinders could use either of these two kinds of gas:
   a. helium or oxygen
   b. helium or nitrogen
   c. nitrogen or acetylene
   d. butane or propane
5. This will provide a damping effect in a suspension cylinder:
   a. nitrogen gas
   b. helium gas
   c. orifice
   d. seals

LEVELING SYSTEMS Some machines such as telehandlers will need to stay as close to level as possible. This is necessary if the machine is used for lifting materials in a safe manner. These machines will have hydraulic suspension cylinders at each corner.

There will be an electronic control system that adjusts the height of the struts to keep the machine level when lifting a load. The control system can change the rod extension for the suspension cylinders by adding or draining oil to the suspension cylinder as required.

RIDE CONTROL A variation of a suspension system that has recently become popular is one called ride control, which is found on many machines with loader booms. This system allows the boom cylinder rods to move and give a cushion effect to the loader boom when the machine is traveling.

There are one or more accumulators that are teed into the head end of the boom cylinder circuit. When ride control is actuated the accumulators have the boom cylinder head end circuit oil exposed to the bottom of their piston with nitrogen gas on the other side. The nitrogen acts like a spring to allow the rod of the boom cylinder to move in and out freely as the gas is compressed.

As the machine travels over rough ground, the ride control system allows the loader boom to move up and down. For backhoe loaders that are traveling fast either on jobsites or on public roads this will keep the front tires on the road longer, which makes the machine safer to operate. It will also give the operator a much smoother ride and for wheel loaders it will save a lot of jarring and shock loads throughout the machine.

Ride control must be disabled when the machine is digging; otherwise, the accumulator/s will not allow positive upward movement of the loader frame because the nitrogen gas will compress.

Some systems can have electronics incorporated into them to only allow ride control to be enabled if the machine is traveling over a certain speed. ● SEE FIGURE 21–14 for a schematic if a ride control system.

CUSHION HITCH Wheel type tractor/scraper machines (scrapers) can travel fast (up to 35 mph) with heavy loads and usually work on uneven ground.

Scrapers have a suspension system to allow the two main parts of the machine to move independently of each other to reduce operator fatigue and machine damage. There is a pair of H-bar linkages that allow vertical movement between the two main parts of the machine and a hydro/pneumatic system that controls this movement. ● SEE FIGURE 21–15 for the cushion hitch arrangement for a scraper.

The main components of the system include: cushion hitch pump, control valve (**leveling valve**), **load cylinder**, and accumulator/s. The load cylinder looks like a typical double acting hydraulic cylinder with a barrel, head, rod, piston, and seals. The barrel is fastened to the tractor part of the machine and the rod is pinned to the scraper **gooseneck**. The rod will support the front of the scraper section and move in and out of the barrel to provide suspension between the two main parts of the machine.

The cushion hitch pump is driven by the transmission and uses the machine's main hydraulic system oil. Pump output is sent to the leveling valve that controls oil flow to the head end of the load cylinder.

To control the cushion system, newer scrapers have a switch in the cab that the operator uses to send an electrical signal to a solenoid on the leveling valve. When the switch is moved to actuate the cushion hitch system, the solenoid is energized and this will make the spool move in the leveling valve. This position of the spool will send pump oil to the bottom of the load cylinder and the bottom of the accumulator. Older scrapers used a pilot oil valve that was moved by the operator and sent oil to the leveling valve. The leveling valve is mounted on the cushion hitch assembly and is a single circuit directional control valve.

When the machine is first started, the load cylinder piston is bottomed out in the load cylinder. When the machine is running and the operator wants the cushion hitch to provide suspension, the operator will shift the leveling valve spool either by electric control or by a remote pilot valve in the cab. When the spool shifts, pump oil is directed to the head end of the load cylinder and it will push the rod out the load cylinder. There is a leveling system that will shift the valve back to neutral when the load cylinder rod is extended approximately half way through its travel. At this point pump oil flow is diverted to the tank, and the

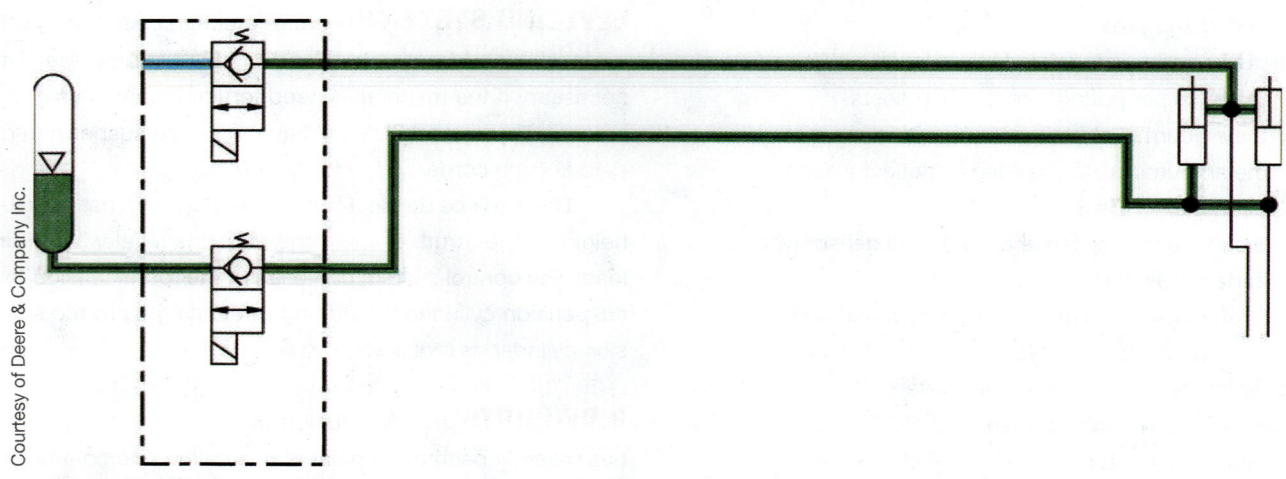

FIGURE 21–14 Ride control schematic.

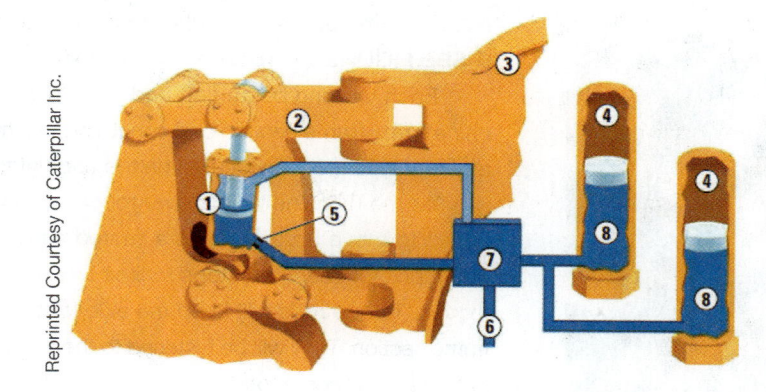

1- Load cylinder
2- Hitch frame
3- Draft tube frame
4- Nitrogen charge
5- Orifice
6- Pump supply
7- Levelling valve
8- Hydraulic oil

FIGURE 21–15 Cushion hitch for a scraper (reprinted courtesy of Caterpillar Inc.).

head end of the cylinder is opened to the oil end of the cushion hitch accumulator. Now when the machine is traveling and runs over some rough ground the accumulator acts like a spring because its precharged gas will compress to allow its piston to move, which allows the load cylinder rod to move in and out. The oil is free to flow between the accumulator and load cylinder but is slowed down by an orifice. The orifice gives a damping effect to slow down the cushion hitch oscillations.

The leveling valve has a leaf spring assembly that is attached to the leveling valve spool at one end and is attached to the scraper part of the cushion hitch frame at the other end. As the scraper part of the machine is raised because of the load cylinder rod extending the leaf springs tension will change. There is a point where the change of tension will shift the valve spool and this is when the pump oil flow stops raising the load cylinder and opens the accumulator to the cylinder. Now as the machine travels over rough ground the leveling valve will try to maintain a specific ride height between the tractor and scraper. After the machine is loaded, the extra weight of the load will be automatically compensated for by the leveling valve because it will allow more oil into the head end of the circuit before the valve is shifted back to neutral. As the load is dumped the load cylinder rod will extend because of the reduced weight but the leaf spring will shift the leveling valve and drain some oil from under the load cylinder piston. This will maintain a rod extension that allows the load cylinder to provide a cushion between the main frame sections because of the accumulator action.

● **SEE FIGURE 21–16** for a leveling valve.

When the machine is getting loaded, the load cylinder must be **locked down** to prevent movement between the tractor and scraper. The leveling valve will create a hydraulic lock on the rod side of the cylinder's piston to keep the rod retracted for this part of machine operation.

Although loading only occurs at low speed, the forces involved in loading the bowl means the rod would get full extended if it wasn't locked down. This could damage the load cylinder. The control valve creates a hydraulic lock on top of the load cylinder piston.

There could be one or two accumulators used for these cushion hitch systems and they are precharged with nitrogen

MACHINE FRAMES AND SUSPENSION SYSTEMS **615**

FIGURE 21–16 Leveling valve.

FIGURE 21–17 Cushion hitch accumulator.

gas to a specific pressure. ● SEE FIGURE 21–17 for the accumulator for a cushion hitch.

WALKING BEAM SUSPENSION
This type of suspension is used for rear tandem axle arrangements on articulated trucks.

Articulated trucks will have a pair of walking beams and a series of linkages that hold the rear axles in place but also allows them to move in relation to the trucks frame. These linkages are sometimes called **torque rods** or panhard rods.
● SEE FIGURE 21–18 for a walking beam suspension.

Each walking beam (left and right) is mounted to the machine's rear frame section and can pivot on a trunnion. The front and rear axle are then located on the ends of the walking beam.

As the machine travels over rough terrain, the axles are allowed to move up and down a limited amount as the walking beam pivots. There are stops to limit the amount that each axle moves. Torque rods will keep the axle located in relation to the frame section. They will be fastened to the frame and axle with ball joints to allow some rotational movement. ● SEE FIGURE 21–19 for a torque rod.

Walking beam suspensions will use rubber blocks to absorb suspension movement. Many highway trucks have rubber block suspension.

The rubber blocks are made of a dense rubber material that will compress and expand as the load changes on them.
● SEE FIGURE 21–20 for a rubber block.

AIR BAGS
Some lighter-duty machines and over-the-road cranes may use air bags for their suspension. They are a reinforced rubber bag that gets filled with air from the machine's pneumatic system to a certain pressure. The pressure is determined by a leveling valve that is mounted to the machine's frame. The valve will have an arm with linkage going to the axle they are attached to. As the air bags fill, the frame rises from the axle and the leveling valve arm moves to stop the air supply to the bags, which in turn limits the air pressure to the bags. Some machines may have their cab suspended on air bags. ● SEE FIGURE 21–21 for an air bag.

ELECTRONIC CONTROLS FOR SUSPENSION SYSTEMS
Some machine suspension systems could incorporate an electronic control system. This could allow the operator to adjust or change modes of suspension. It could also allow some diagnostic functions to be carried out by a technician. The system could also be part of the machine's CAN bus network and integrate suspension operation with other machine systems.

PAYLOAD SCALE
An added benefit of some machines with hydro pneumatic suspension systems is that an electronic scale can be incorporated into it. The most common use for this is with haul trucks used for mining operations.

All that is needed is a pressure sensor (transducer) to be added to each suspension strut and the signal from these sensors is then sent to an ECM. Once the system has been calibrated to zero with an empty box and a known weight it can then accurately measure the payload the truck has been loaded

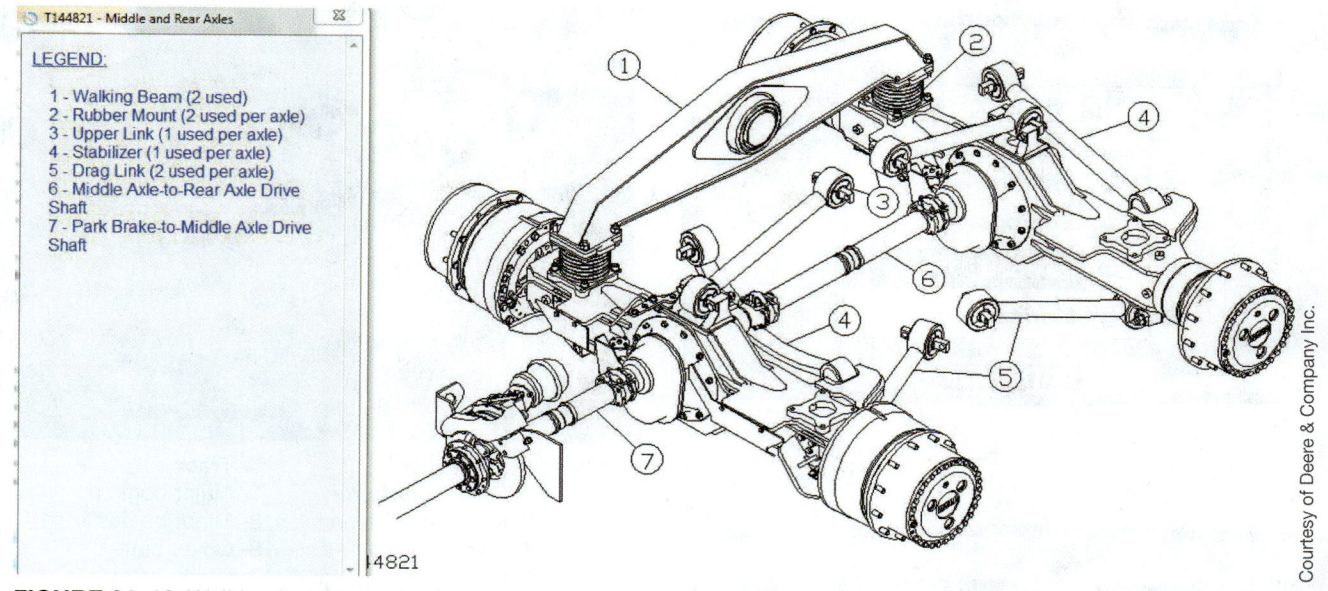

FIGURE 21–18 Walking beam suspension.

FIGURE 21–19 Torque rod.

FIGURE 21–21 Air bag.

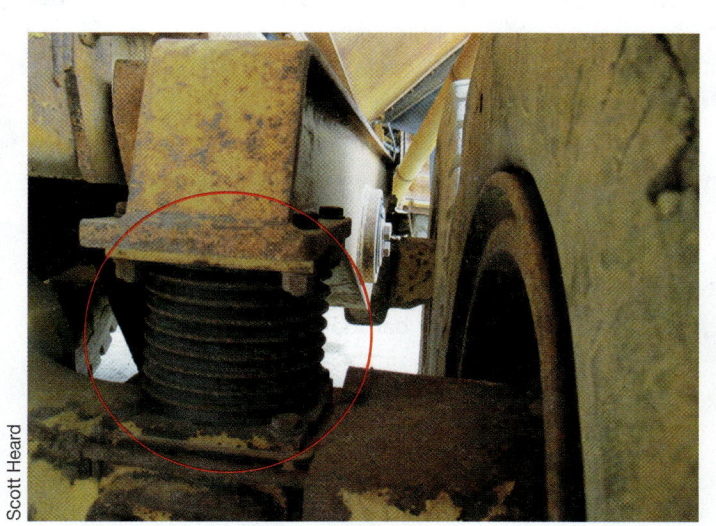

FIGURE 21–20 Rubber block.

with. The weight that accumulates in the truck as the truck is loaded is then displayed to the loader operator by an LED readout. This should be monitored to prevent overloading the truck.

The ECM can also record all loads and total weights for each shift. ● **SEE FIGURE 21–22** for a payload scale display.

TRACK MACHINE SUSPENSION SYSTEMS

TRACK ROLLER BOGIE SYSTEMS Some track type dozers will feature a bogie suspension system that uses rubber blocks and pivoting track roller assemblies to allow the ma-

MACHINE FRAMES AND SUSPENSION SYSTEMS **617**

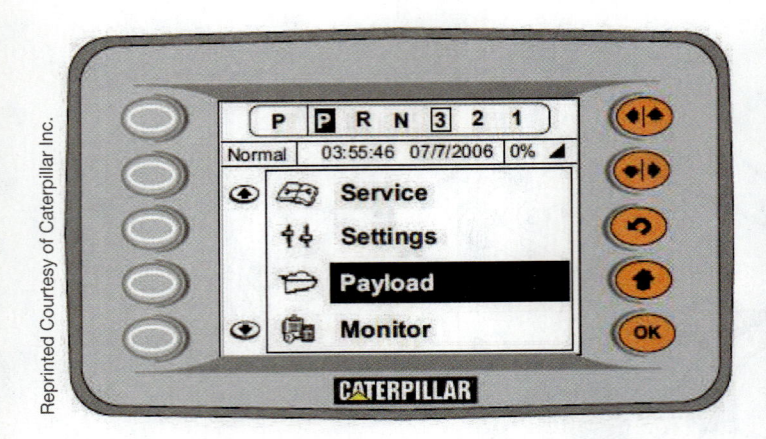

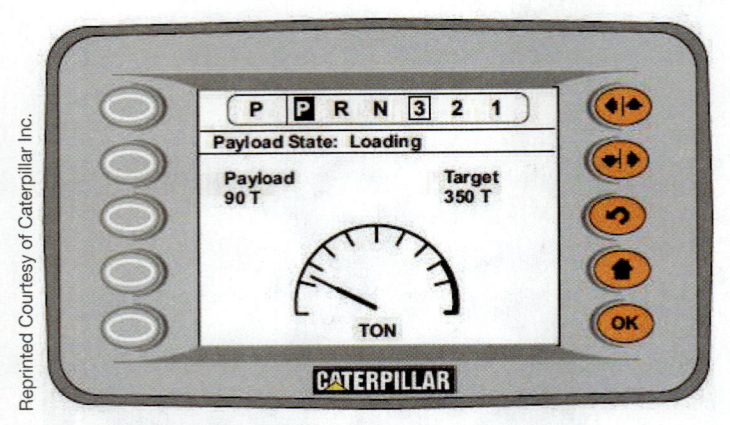

FIGURE 21-22 Payload scale.

1- Front idler
2- Front track frame
3- Rear track frame
4- Pivot shaft
5- Rear idler
6- Track
7- Major bogies
8- Minor bogies
9- Cover plate
10- Guide cover

FIGURE 21-23 Bogie suspension system.

chine's undercarriage to conform to uneven terrain. This not only provides a smoother ride for the operator but also reduces the shock load on the track rollers and the rest of the machine. Most other large track-drive machines have their track rollers mounted directly to the track frame so that any shock caused by uneven ground the machine travels over is transferred directly to the machine's frame. ● SEE FIGURE 21-23 for a track machine with a bogie suspension system.

Larger high drive track dozers use this system and they have a pair of bottom rollers mounted to a bracket that can swivel on pins at one end of a large bracket. The large bracket pivots on a pin at one end of it that is held in the track frame. The top of the opposite end of the large bracket rests on a rubber block that is held to the bottom side of the track-roller frame. The compression of the rubber block as well as the pivoting of the roller assembly small bracket will provide some suspension as the machine is traveling.

Some smaller (100 hp and under) track-loader machines feature rollers that are suspended on springs to give a smoother ride.

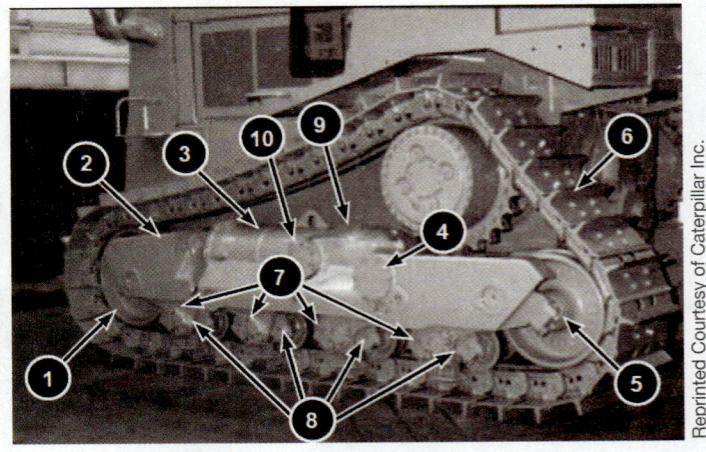

EQUALIZER BAR Most track machines have a basic form of suspension that allows the track frames to pivot a few degrees. If you were looking at the side of a track machine the front idler could move up and down slightly to keep the track engaged with the ground longer. This system can be used for track loaders and dozers.

Some machines that don't feature high-drive track arrangements (see Chapter 23) will have their track frames pivot on a shaft that extends out from the main frame. Older low-drive machines could have that pivot at two points at the rear of the machine. One is a large bearing that extends from the final drive and the other is a shaft at the bottom center rear of the machine.

These machines will also use an equalizer bar that extends across the frame under the engine and pivots in the center on a pin. Each end of the equalizer bar rests on a pocket in each track frame. This is where the weight of part of the machine is transferred to the track frames.

The equalizer bar will limit the maximum up-and-down movement of the track frames and when one side moves up it makes the other side tilt down the equal amount. An example is for a large track loader that may allow for a maximum of 6 in. of up and down travel at the idler. The travel is limited by stops on the equalizer bar that contact the bottom of the main frame.

The center pivot for equalizer bars is a plain or self-aligning bearing that needs to be greased regularly. ● SEE FIGURE 21-24 for an equalizer bar.

Track frame pivot shafts will support the track frame on large plain bushings that are usually lubricated with gear oil.

FIGURE 21–24 Equalizer bar.

This oil could have its own separate reservoir and needs to be checked regularly. There are seals to keep the oil in place and if these seals leak causing a loss of oil an expensive and time intensive repair will follow.

DIAGNOSING SUSPENSION SYSTEMS
Operator complaints that stem from suspension system faults include the following: the machine is rough riding; the machine is not level; the machine continues to bounce after it hits a bump; and there are noises when the machine travels over rough ground.

A thorough visual inspection of all suspension components will reveal the cause of many problems. All components should be fastened securely to the machine's frame or other components. Any loose suspension component should be repaired immediately.

As suspension system operation requires moving parts, there must be lubrication applied regularly to allow free movement and to keep wear to a minimum. Look for any signs of dry joints that should have grease.

If a machine with a suspension system is not sitting level on level ground when unloaded, you can make a few simple checks based on the type of system it has. A rubber-tired machine should first be checked to see if it has the proper size of tires on it and then look for broken springs or leaking suspension cylinders. Look for dry pivot joints that should have grease in them.

An electronically controlled suspension may have fault codes set that are related to its malfunction. This is a good place to start if there is a complaint related to the suspension system.

SERVICING SUSPENSION SYSTEMS
Regular lubrication of many suspension components is necessary because of the moving parts that need to be separated by a lubricant. This could include torque rod ends, strut mounts, and leaf spring shackle bushings. Always check with the machine manufacturers' service information for the correct type of lubrication to use and the proper frequency of greasing intervals. Many machines

FIGURE 21–25 Grease lines on for a suspension cylinder.

will have automatic greasing systems installed and these must be checked to ensure grease is getting to where it should.
● **SEE FIGURE 21–25** for the grease lines of a suspension cylinder.

Track machine pivot shafts may have oil reservoirs that are often overlooked. Be sure to check the oil level in the reservoir.

STRUT RECHARGE PROCEDURE
A common service or repair procedure that is performed on suspension systems is a strut recharge. This is necessary when the nitrogen precharge inside the strut has leaked down. If a strut has a low charge it will cause the machine to ride uneven and could lead to the strut bottoming out when traveling over rough ground. If there are signs of oil leakage from a strut it will have to be removed and resealed at a minimum and may have more damage such as a scored barrel or piston. This would require machining or parts replacement.

To measure a struts rod extension the unloaded machine must be allowed to coast to a stop on firm level ground. This ensures an accurate reading. Some machines require special tooling to properly measure the rod extension and the proper dimension can be found in the machine's service information. If the measured dimensions are correct and there are no visual leaks or damage, nothing else needs to be done unless there is an operational complaint. If so, this will require further troubleshooting.

Some struts will also allow a technician to purge the old oil from them. This could be part of a regular service procedure because the oil is under a great deal of stress inside the strut and will eventually break down.

The nitrogen gas charge in a strut needs to be at a specified pressure for the strut to work properly and for the machine to have the proper ride height. Always refer to the machine's service information to find the proper precharge pressure.

An accumulator charging kit will be the basis for the special tooling needed for a strut recharge. If the oil level needs to be adjusted or the oil needs to be purged, then additional oil pumping tooling is also needed. Some procedures will suggest using the machine's brake system to purge the oil. Some extra fittings and hoses are needed for this.

For most machines this procedure requires both struts on one end of the machine to be charged at the same time providing the struts on the other end of the machine are charged properly. The tooling to do this will Tee together both struts into one valve assembly that is fed the gas precharge from a **regulator** that is fastened to a gas bottle. Pressure from the bottle is regulated to around 400 psi and a gate valve can be opened to allow gas to flow into the struts.

The first part of this process is to drain the struts of their precharge gas. The strut charging tooling is fastened to the struts charging valves and the precharge pressure is bled out through the Tee before the gas line is attached. This should fully retract the rods of the struts. A measurement from the top and bottom mounting pin of both cylinders will confirm this.

The next step is to charge the struts. The gate valve is opened and when both struts reach the specified extension the valve is closed and the tooling is removed. The machine should then be sent to work, loaded through a normal work cycle, and then returned to have the strut extension measured. Another charge pressure adjustment may be required at this point.
● **SEE FIGURE 21–26** for a charging hose hookup.

Oil viscosity is an important consideration when servicing struts because the damping effect is dependent on oil flowing through orifices. Having the wrong viscosity oil in a strut will make the strut rod to move in and out either too fast or too slow. Make sure that the specified viscosity and oil type is used for the strut oil. This is usually 10 W.

REPAIRING SUSPENSIONS SYSTEMS
Suspension systems will need repairs as their components wear past acceptable limits. Components that need oil and gas sealed in them may need to be resealed on occasion if they start to leak either internally or externally. ● **SEE FIGURE 21–27** for a strut change repair.

Caution needs to be taken when disassembling struts because there may still be pressurized oil and gas in them.

FIGURE 21–26 Charging hose hookup.

FIGURE 21–27 Strut change repair.

SHOP ACTIVITY

- Find a track machine that has a suspension system that was covered in this chapter and describe its main components and how they work.
- Find a rubber-tired machine that has a suspension system that was covered in this chapter and describe its main components and how they work.

Suspension system components that need to be greased regularly will need to be repaired if they run dry and have normal wear accelerated. Most repairs to greasable suspension components are fairly straightforward procedures.

Progress Check

6. Why should you use extreme caution when servicing suspension cylinders?
   a. because they get extremely hot
   b. because they can contain high pressures
   c. because they contain extreme vacuum
   d. because they are extremely expensive

7. A payload system uses this type of information obtained from suspension cylinders to provide a payload reading to the operator:
   a. height
   b. weight
   c. temperature
   d. pressure

8. A cushion hitch suspension system will have a leveling valve that does this:
   a. keeps the suspension cylinders level
   b. diverts pump flow away from the load cylinder when the proper rod extension is reached
   c. diverts pump flow to the nitrogen side of the accumulator when the proper precharge is reached
   d. charges the accumulator to the proper pressure to achieve the proper load cylinder rod extension

9. A machine that is equipped with ride control will allow this component to move up and down to provide a suspension effect:
   a. loader boom
   b. ripper
   c. bucket
   d. cab

10. Ride control should only be activated on a machine when:
    a. it is reversing
    b. it is digging from a bank
    c. it is leveling rough ground
    d. it is traveling fast

SUMMARY

1. Safety-related concerns when working with machine frames and suspension systems include: torch cutting on sealed vessels; proper lifting and blocking equipment and procedures must be used; accumulator pressures need to be bled before disassembly; high pressure oil could be present and must be handled with care.
2. Machine frames are the backbone of the machine and all machine systems will be mounted to it. There will be extreme forces applied to the frame and it needs to be strong enough to withstand these forces.
3. Frames are constructed from welded together plate steel and cast steel sections.
4. Frames will likely fail at one of the welds if the frame is pushed past design limits or the weld is defective.
5. Articulated machines have two piece frames and excavators have an upper and lower frame held together with a rotex bearing.
6. Frame inspection should be part of regular maintenance and focus should be concentrated on welds and threaded holes.
7. Frames can be repaired and reinforced to prevent recurring failures. Frames should always be covered with paint to prevent rust.
8. Suspension systems allow movement between a machines frame and its drivetrain. This gives the operator a smoother ride and keeps the tires or tracks engaged with the ground longer.
9. Some older and or small machines will use springs for suspension. Springs can be leaf type or coil type.
10. Leaf-type springs are stacked up in shortening multiples and held stationary at one end. The other end can pivot in a shackle and a drive axle is held to the spring assemblies' centre with U-bolts.
11. Coil springs can be for axle suspension but the axle needs to have links to keep it in place.
12. A sprung axle will likely have a set of two shock absorbers mounted to it. The other end is mounted to the frame. Shock absorbers use a hydraulic piston, oil and, an orifice to reduce spring oscillations.
13. Front axles on backhoe loaders and rear axles on wheel loaders can oscillate. This is a simple type of suspension that allows the axle to pivot a few degrees on a center pin or bolt on trunnions. These moving parts need regular greasing.
14. Suspension cylinders act like springs and shock absorbers and are mostly found on trucks to give a smooth ride and allow the machine to travel fast over rough ground.
15. Most suspension cylinders use nitrogen gas for the spring effect and hydraulic oil for the damping action. They must be precharged to a certain pressure or to make the rod extend a certain length.
16. Active suspension cylinders can have oil fed into them or drained out to change ride height or level the machine.
17. Ride control is a form of suspension for backhoe loaders and wheel loaders that introduces an accumulator to the head end of the boom circuit. This gives a spring action to the boom when the machine is travelling fast and or over rough ground.

18. Scrapers have cushion hitches that allow some movement between the tractor section and the scraper section of the machine. A load cylinder and linkage allows movement between the frame sections and an accumulator provides spring action.

19. Walking beam suspension is found on the rear tandem axles of articulated rock trucks and allows a few degrees of oscillation between all four wheels. Travel is cushioned with rubber blocks and axles are located with torque rods.

20. Payload scale systems can use suspension components to translate pressures into payload weights.

21. Track-type machines can have suspension systems to increase operator comfort. Equalizer bars will allow track frames to pivot a few degrees and track rollers can have a bogie system to allow them to follow the underfoot profile.

22. Operator complaints related to machine frames and suspension systems can include: rough ride, noises, excessive oscillation, machine not level.

23. Diagnostic procedures start with knowing the system and a thorough visual inspection.

24. Most suspension components that allow movement will require regular greasing.

25. Adjustments could be made to accumulator or strut pre-charge pressures. This requires a charging kit that includes special tools.

WHEELS, TIRES, HUBS, AND PNEUMATIC SYSTEMS

LEARNING OBJECTIVES

After reading this chapter, the student should be able to:

1. Describe the different types of wheels that could be found on machines.
2. Explain why excessive heat should never be applied to wheels.
3. Describe the correct procedure to tighten wheel fasteners.
4. Explain what the numbers that are used for tire sizing mean.
5. Describe the procedure that should be followed to check tire pressure and to inflate a tire.
6. Explain how to adjust a wheel hub bearing.
7. Identify the main components of a compressed air system.
8. Describe how to adjust an air system governor.
9. Describe how an air dryer works.
10. Explain how to cage a spring-applied air brake.
11. Describe how to adjust S-cam brakes.

KEY TERMS

Air compressors 647
Aspect ratio 639
Bead 633
Bearing adjusting 645
Brake chamber 651
Governor 649
Lock ring 635
Plies 637
Ply rated 638

Pneumatic 637
Radial tires 638
Rock ejectors 641
Service brakes 651
Slack adjuster 652
Spindle 644
Tubeless 635
Valve stems 635

INTRODUCTION

SAFETY FIRST There are several safety concerns when working with wheels, tires, hubs, and pneumatic systems.

Rubber-tired heavy equipment machines need their tires to support the weight of the machine, and tires therefore contain enormous potential energy in the form of air pressure. Tire pressures can vary from as low as 15 psi to well over 100 psi. Many deaths and serious injuries have resulted from the mishandling of tires during heavy equipment servicing or performing repairs. Tires should *always* be deflated when any type of servicing or repairs are performed by a technician on wheel assemblies. If this is done by removing the valve core, there is a possibility of ice forming in the valve as air rushes out. To be sure that all air pressure is drained down in the tire, carefully run a piece of wire or small screwdriver into the valve stem.

Tire explosion is a *very violent* event and can easily cause death if anyone is close-by when it happens. This happens when the temperature of the air inside the tire or the tire itself reaches 250°C. The resulting reaction is called pyrolysis. The use of nitrogen to inflate a tire should prevent the explosion from happening.

A tire explosion is different than a blowout. Blowouts occur because of a tire failure when the machine is traveling or because of overinflation. Tire explosions can occur when a tire starts to burn on the inside and pressure eventually rises high enough to make the tire explode. This can be caused by using excessive heat on a rim (welding or cutting), a brake that is dragging, or a machine that has caught fire. If you notice a burning rubber smell or see smoke from a tire, do not approach the machine and evacuate the area immediately. Exploding tires can propel debris more than 1500 ft away. Wait at least 8 hours before approaching the machine to let the tire cool down.

Care must be taken by technicians to *never* apply excessive heat to wheels, tires, or pneumatic systems because heat will increase pressure and increased pressure can cause component failure. Never use torches, welding equipment, or any other type of heat source near tires or wheels.

Large tires and wheels must be handled properly to avoid injury caused by crushing or pinching. In most cases, a heavy equipment technician is wiser to call a tire technician to deal with tire issues. Tire technicians will have the proper tools and equipment as well as the necessary training to safely handle machine tires.

Machines should *never* be left on hydraulic jacks alone when they are being serviced. Hydraulic jacks are meant only for lifting and lowering and not for holding the machine in place. The machine should be lowered onto jack stands or wooden cribbing after it has been lifted.

CAUTION A haul truck was traveling on a quarry site once with the box left up when it came into contact with a high-voltage power line. The operator continued driving for a distance and then decided to stop and get out of the truck. Luckily, he was far enough away from the truck when its tires exploded that he didn't receive any serious injury. The force of the blast was so severe that it completely destroyed the rear axle and final drives of the 100-ton rock truck. Search the Internet to see if you can find a video that demonstrates the force of an exploding tire.

Some machines may have the main part of the rim bolted together unlike most rims that are welded together when they are manufactured. If this is the case, special care must be taken to not loosen the fasteners that hold the two pieces (inner and outer) together. If this happens and the tire is pressurized, the rim halves could separate and the pieces could cause serious injury or death.

Pneumatic systems can also pose serious injury threats because they contain pressurized air. Even though most pneumatic systems will have a maximum pressure of 120 psi, which is relatively low, there is still the possibility of component failure that can cause serious injury.

NEED TO KNOW

WHEELS Rubber-tired heavy equipment machines can have several different types of wheels on them. Wheels can also be called rims. Small equipment such as skid steers and backhoe loaders will have one-piece rims. This simply means they are an automotive style of rim made from two pieces of stamped, pressed, and rolled steel welded together with a center hole to fit over the hub and several holes drilled around the hub hole for wheel studs or bolts. Their outer circumference will have rolled lips for the tire **bead** to be seated against to seal air pressure in the tire. Toward the inside of the wheel on the surface that is exposed to air pressure, the wheel will step down. This is to allow the tire bead to drop into a smaller circumference area when the tire is installed or removed.
- **SEE FIGURE 22–1** for a cross-section view of a one-piece rim.

One-piece wheels are sometimes called disc-type wheels.

Some road-worthy rubber-tired cranes will use aluminum rims that are forged as one piece and then finished to make an attractive and functional wheel. These machines can travel at close to highway speeds and need a wheel similar to a highway truck's wheel.

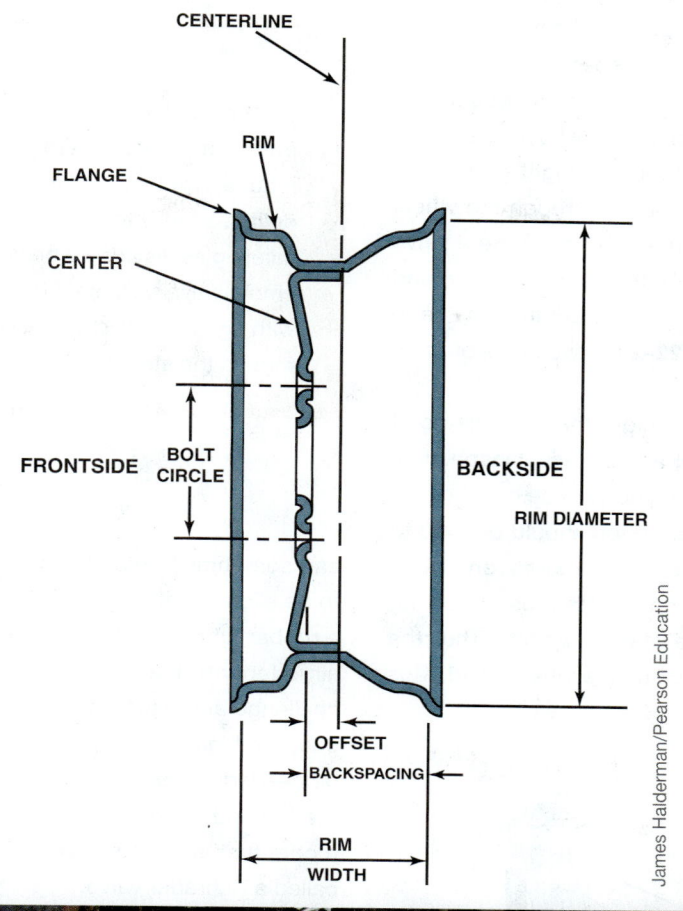

FIGURE 22–1 Cross-section of a one-piece rim.

WHEELS, TIRES, HUBS, AND PNEUMATIC SYSTEMS 625

Most wheel and tire assemblies used for heavy equipment are tubeless today. In the past, most tires needed an inner tube to hold air, and this meant the tube's valve stem would protrude through a hole in the rim. **Tubeless** tire/wheel assemblies need a valve stem that is sealed to the rim. Light-duty tubeless tires have a valve stem that is pulled through the wheel from the inside while heavy-duty tubeless tires have a valve stem assembly that has a seal on both sides of the rim and then a washer and nut is tightened onto it from the outside to complete the seal. ● SEE FIGURE 22–2 for two types of valve stems for tubeless tires.

As these rims are simple by design, they usually don't require any maintenance outside of a good visual inspection. When inspecting one-piece rims, they should be checked for rust, cracks, and bending. Close attention should be paid to the hub-mounting area, wheel fastener hole area, and bead seat area. Tires are installed onto one-piece rims by forcing the inside tire bead over the outside wheel bead first. Then the outside tire bead is forced over the outside wheel bead. This can sometimes be a challenge because the tire bead needs to be slightly deformed to do this. Once the tire is on the rim, a rubber lubricant is spread around the wheel bead area and air is fed into the tire valve. Seating the bead can also be a challenge and extra caution should be taken while doing this because higher than specified operating air pressure will likely be needed. To remove a tire from a one-piece rim, first deflate the tire and remove the valve core. All **valve stems** will have a core that is threaded into the stem. The core is sometimes called a Schrader valve.

As tires get larger, they become more difficult to remove and install from a rim, and this is why multi-piece rims are used for larger tire sizes. Multi-piece rims provide a way to disassemble the rim, which allows the tire to be removed without having to deform the tire bead to get it over top of the rim bead seat area.

Medium-size rubber-tired machines will likely use a three-piece rim. These rims will have a main rim base with inner bead seat flange as one piece and the outer bead seat and locking ring making up the other two pieces. To install a tire on a three-piece rim, the tire is fitted over the rim base and then the outer bead and lock ring go on. When the tire is inflated, the outer bead is pushed out, and this will push out the outer rim bead against the lock ring. The **lock ring** then seats in a groove in the rim base. ● SEE FIGURE 22–3 for a cross-section view of a three-piece rim.

Five-piece rims are used for large wheel and tire assemblies. They are similar to three-piece rims but have an additional separate inner bead seat and a bead seat band for the outer flange. Multi-piece rims will need to have O-rings to seal the areas between their base and removable bead seats. ● SEE FIGURE 22–4 for a cross-section view of a five piece rim.

Multi-piece rims could also feature bead locks or keys that are used to prevent rim bead seats from spinning on the

> **TECH TIP**
>
> Any time a wheel assembly needs to be removed from a machine, the tire should be deflated first to reduce the chance of an accident. The easiest way to deflate a tire is to remove the valve core. Gloves and safety glasses should be worn when doing this. The empty valve stem should be checked for blockage with a piece of wire because it is common for ice to form in the stem from the air rushing out.

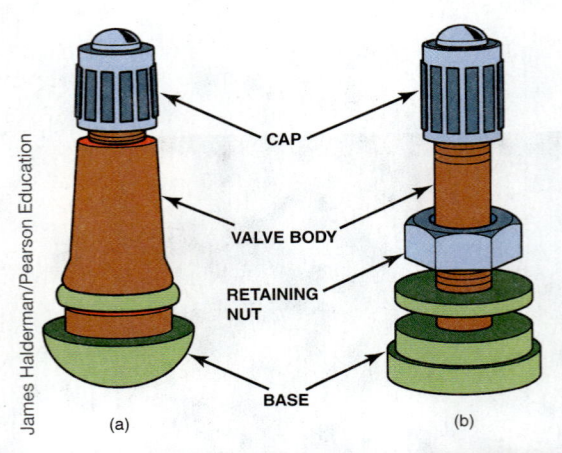

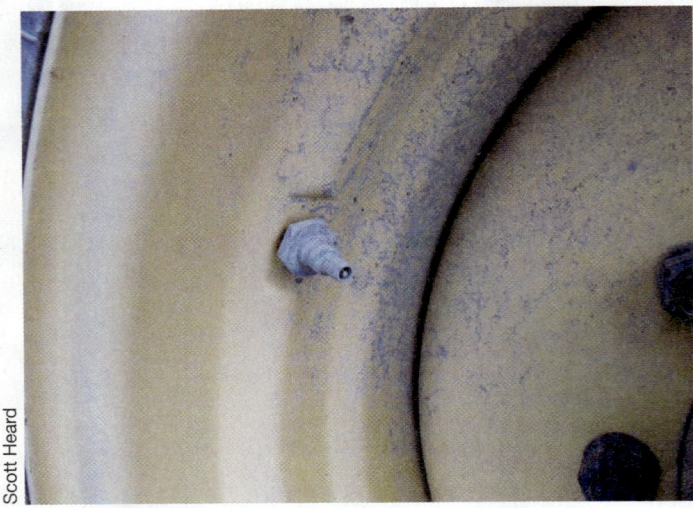

FIGURE 22–2 Tubeless valve stems.

626 CHAPTER 22

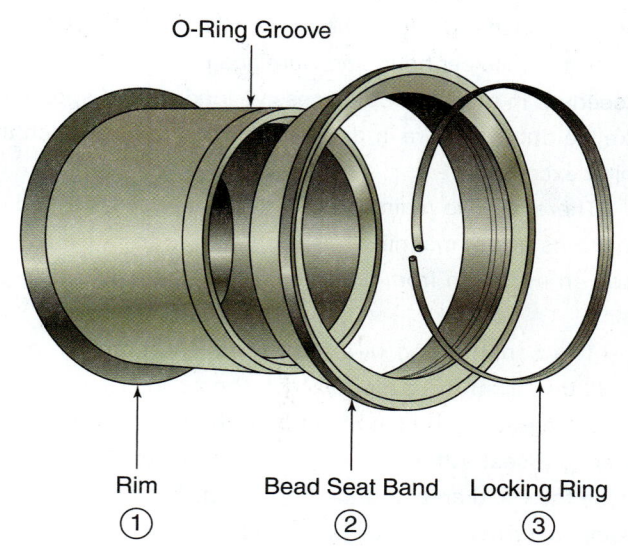

FIGURE 22-3 Three-piece rim.

Labels: O-Ring Groove; Rim ①; Bead Seat Band ②; Locking Ring ③

capped and be inspected for damage on a regular basis. Dual wheel arrangements will usually have valve extensions on the inside wheel to allow for easy pressure checks.

When installing rims onto a hub, the mating surfaces should always be clean and free of dust, dirt, and grease. Fasteners should also be installed properly and torqued to specification with a calibrated torque wrench.

Tightening and torqueing of wheel fasteners should be done with a star pattern method. In other words, always move to the opposite side of the rim to tighten the next fastener. If the rim needs to be drawn in over the hub, make sure it gets drawn in evenly so it doesn't get distorted. The rim and hub should be cleaned before installing.

TIRES

Tires that are used on heavy equipment machines represent a large portion of machine maintenance and running costs. If they are abused or not maintained properly, the replacement cost and downtime can be very expensive. Tires for the biggest haul trucks can cost over $100,000 each.

For wheel-type machines, tires are the contact point between the machine and the ground the machine travels on. The tires' tread pattern and sidewall construction will play a huge part in how the machine performs and feels to the operator.

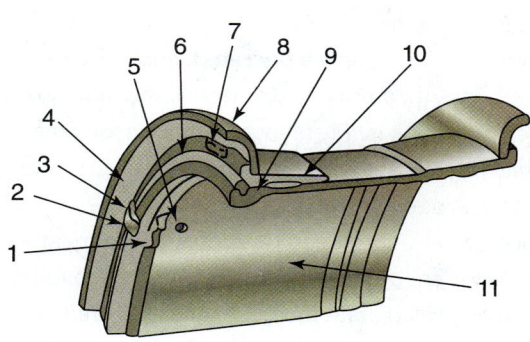

rim base. This could happen when the machine is under high-load, low-speed working conditions.

Multi-piece rim inspections include basic visual inspections that are found on one-piece rims as well as inspecting lock rings and rim locks for cracks, bends, and damage.

Heavy equipment wheels are a tubeless design. This means the valve stem is not part of a tube and is fitted to the rim base. Valve stems are simply one-way check valves that feature a removable core. The valve stems should always be

1. Locator-Demountable Rim Only
2. Lock Ring Driver
3. Gutter Notch for Lock Ring Driver
4. 5° Bead Seat Band, Tapered Ring
5. Tubeless Valve Hole
6. Lock Ring
7. Pry Bar Slot
8. Flanges (Side Ring)
9. O-Ring Gasket
10. Bead Seat Band Toe
11. Rim Base

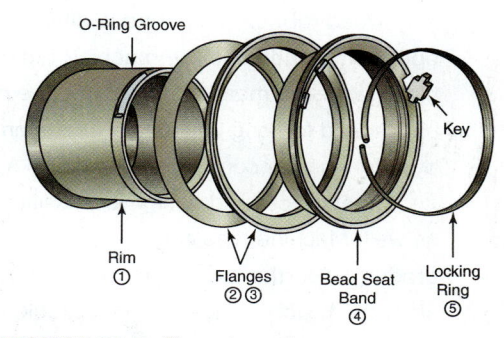

FIGURE 22-4 Five-piece rim.

Labels: O-Ring Groove; Key; Rim ①; Flanges ②③; Bead Seat Band ④; Locking Ring ⑤

WHEELS, TIRES, HUBS, AND PNEUMATIC SYSTEMS 627

Tire size, type, air pressure, and tire condition will greatly affect the operating characteristics of a machine.

Tires are mostly made of rubber-based compounds and are reinforced with steel cords and layers (also called belts or **plies**) of materials such as nylon, fiberglass, polyester, and rayon. Most tires used on machines are **pneumatic**, meaning they are hollow and must be filled with pressurized air to support the load of the machine and its payload or bucket full of material. You can think of a pneumatic tire as a cushioning container for the air pressure needed to keep the machine from riding on its rims.

There are certain machine applications, however, that require the use of solid rubber tires. Work sites where there is a constant threat of tire puncture such as a scrap yard are a place where solid tires may be used. The downfall of solid rubber tires is that there is very little flex to them, which gives a rough ride for the operator. They will also heat up if the machine travels fast for long periods. The air inside an air-filled tire provides a cooling effect to tires.

The main factors that affect tire life are operating conditions, operator abuse, proper tire selection for work site conditions, and proper inflation. Obviously, most of these factors are out of the technicians' control with the exception being proper tire inflation. Proper tire inflation is based on the weight placed on the tires and the speed at which the machine travels. Tire manufacturers will provide inflation information in cases where the machine manufacturer doesn't make this information available.

One of the biggest enemies to tires is heat, and if heat is allowed to build up in a tire from overloading or under inflation, the structure of the tires rubber will start to break down.

If a tire is underinflated, its sidewalls will flex excessively, which causes heat buildup. Some tire manufacturers set limits as to how far a tire should travel in one hour at a certain load. If the tire exceeds this distance, it will likely start to run hot.

TECH TIP

Some tire manufacturers suggest that if a tire is run with less than 80% of its proper inflation pressure at any time, it has likely sustained permanent damage and should be removed for inspection or replacement.

The tire's sidewalls have likely flexed excessively, which causes cord damage, and an overheating condition has likely occurred, which causes rubber degradation.

There are charts to inform the tire user of the tire's maximum TMPH (ton mile per hour) and warnings that this shouldn't be exceeded. In other words, excessive loading and speed will take life out of the tire in direct proportion to how much the limit is exceeded.

There are two main types of tires that can be found on machines. Most machines use radial tires today. This has been an evolution from an older tire design called bias ply tires.

The construction of a tire starts with loops of metal cords that form the base for each tire bead. The tire bead is the area that makes contact with the rim and must form an airtight seal with the rim. Tire sidewalls are the areas between the tread area and the bead area. The tire carcass or casing is formed with multiple layers or plies of material that form a base for the tire from bead to bead. With **radial tires**, these plies run at 90° to the tread. This gives the tire more flexible sidewalls, which results in a smoother ride and bigger contact patch.

Bias ply tires have their plies run at angles to the tread and the sidewall. Tire plies are made of many different types of materials to give the tires different characteristics such as higher strength, smoother ride, or less rolling resistance. The inside of the tire will be a smooth rubber layer to seal the air on inside of the tire and eliminate the need for tubes. This type of tire design is called tubeless. Before tubeless tires became the norm, an inner tube was used to seal the air and gave the effect of having a separate bladder on the inside of the tire.

Tires are **ply rated**, and this relates to the amount of plies that a tire has. Generally, the more plies a tire has, the stronger it is. Ply ratings for heavy equipment tires range from 12 to 78. Some tire manufacturers use a star rating to indicate the strength of a radial tires carcass.

Tire tread designs are an almost endless variety of different patterns. Machine working conditions and traction needs will dictate the best suited tire tread design for its most common working application. ● **SEE FIGURE 22–5** for some different tread designs.

Tread rubber compound is different based on the type of operating conditions the machine typically works in such as highly abrasive material or whether the machine is traveling at high speed for long distances. Tread compounds are identified by letter number combinations such as A, A4, B, B4, C, and C4.

Tread depth and design is classified by letter and number as well. Machine type can be designated as C, compactor; G, grader; E, earthmover; L, loader and tread style/depth as S, smooth; 1, ribbed; 2, traction; 3, rock; 4, deep rock; 5, extra deep rock; 7, floatation.

FIGURE 22–5 Different tread designs.

TIRE SIZING Tire sizing is based on three dimensions of the tire. The first dimension is the bead diameter and is typically measured in inches. The second dimension is the section width, which for smaller tires could be measured in millimeters but for most heavy equipment tires is measured in inches. The section width is measured at the widest part of the tire if you were looking directly at the tire tread face. The third dimension that may be part of a tire size is the **aspect ratio**. It is not measured in inches, but it is a ratio of the height of the sidewall in comparison to the section width. If these two dimensions are the same, this means the ratio is one to one or 100%, meaning the tire has a normal aspect ratio. If a tire has a one-to-one aspect ratio, there is no indication of this in the tire sizing.

However, if the sidewall height is shorter than the tread face width, then this ratio will be included with the tire size and indicates a low-profile tire. The ratio will be identified as a percentage that the sidewall dimension is in comparison to the tread face.

Also, if the tire size includes an R, this indicates the tire is a radial design.

For example, a tire size of 27.00-49 would indicate a bead diameter of 49 in. and a tire section width of 27 in. It would also tell you that the section width and the sidewall height are equal. If the tire size was 33.25 R 29, this indicates the section width is 33.25 in., it is a radial tire, and the rim diameter that it fits on is 29 in.

WHEELS, TIRES, HUBS, AND PNEUMATIC SYSTEMS **629**

TECH TIP

The value of tires is no more apparent when replacement tires aren't available. Several years ago when the world economy was running at full capacity, there was a shortage of large tires for heavy equipment. This not only drove the price of tires through the roof, there were many stories of machines sitting idle because of there were no tires to put on them. Some brand-new machines were being shipped to customers with no tires. Many companies were forced to find any tire that could hold air to keep machines running. They would also limit top speeds to make the tires last as long as possible. This also spurred on many companies to install tire pressure–monitoring systems.

Not only was this a major frustration to the end user, it put a major emphasis on maintaining tires that were on machines that were working.

To illustrate the importance of making tires last as long as possible, there are some companies that will send an operator home if spinning tires are seen. Many machines come with traction control systems to prevent tire spin as well.

FIGURE 22–6 Two different tire sizes.

Another example would be 45/65 R 45. This indicates a bead diameter of 45 in., a section width of 45 in., and a sidewall dimension that is 65% of the section width. This example can be related to automotive tires that are sometimes termed low-profile tires. ● **SEE FIGURE 22–6** for tire sizing information from the side of two different tires.

An extreme example of low-profile tires would be 290/35 R 17. This would be a low-profile tire found on a high-performance car.

Some tire section widths are indicated in metric dimensions. An example of this is a tire size 800/65 R 29. This means the tire section width is 800 mm. It is also a radial and has an aspect ratio of 65.

PROPER INFLATION As mentioned previously, proper tire inflation is one of the major contributors to determining tire life. An underinflated tire will allow the sidewall to flex excessively as the machine travels while loaded. This causes excessive heat buildup in the tire casing and will start to weaken the rubber. To avoid starting to break down the tire, the critical temperature of the air inside the tire that shouldn't be exceeded is 80°C or 176°F.

An overinflated tire will cause the tread face to bulge and cause excessive wear on the center of the tire's tread. A slight tire overinflation may be acceptable in certain circumstances to compensate for an overloaded machine, but normally tires should be inflated as close as possible to the recommended *cold* psi specification. Ideally, tire pressure should be checked when tires are cold to provide a more consistent reading. Cold means that the machine has not been working for at least six hours and should be close to room temperature (20°C or 70°F). Once the machine begins to operate, its tires will warm up, and this causes an increase in tire pressure.

A rule of thumb is that a 10°F temperature change will change the tire pressure 1 psi.

Underinflation is much more damaging to tires than overinflation because excessive sidewall flex will dramatically increase tire temperature.

Ideally, tires should be inflated with nitrogen, which is an inert gas. This will help to maintain a more stable pressure, keep moisture outside of the tire, and reduce the risk of tire explosion because there is no oxygen to support combustion. As nitrogen comes in bottles that are charged with over 2000 psi, the technician should be properly trained before putting nitrogen into a tire.

In reality, most tires will get inflated with air from shop or mobile air compressors, and this is fine for most tire applications.

As mentioned earlier, proper tire inflation is critical to getting maximum life from the machine's tires. Small machine tires may cost as little as $250 while large mining machines such as haul trucks and rubber-tired loaders could use tires that cost close to $100,000 each. If an improperly inflated tire reduces tire life by 20%, the cost of having a technician monitoring tire pressure is a wise investment.

Most machines can be equipped with tire pressure–monitoring systems to provide a constant watch on tire pressures. These are very cost-effective systems when tire prices get into the several thousand dollar range. These systems use tire pressure monitors mounted to the machine's rims that send pressure information wirelessly to a display unit in the machine's cab. ● **SEE FIGURE 22–7** for a tire pressure monitor.

DUAL TIRES Single drive axle haul trucks used in mining operations will have dual tire arrangements on their rear drive axle. Dual tire arrangements can be seen on most highway trucks as well. The payload capacity of the truck can be increased by running dual tires on its rear axle. It is important that tires be identical in size, tires be the same type, and model of tires must be closely matched for wear.

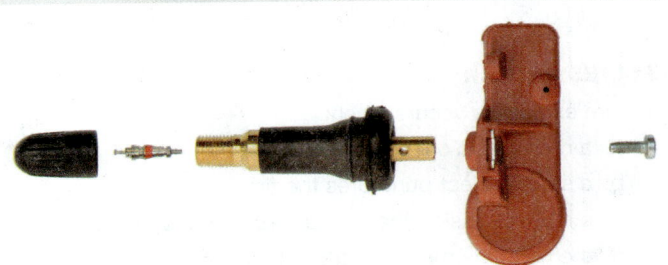

FIGURE 22–7 Tire pressure monitor.

TECH TIP

Technicians must always exercise caution when both checking and inflating tires. Only stand beside the sidewall of a tire when absolutely necessary such as checking pressures or connecting an air check. The ideal tire inflator tool is one that clips onto the valve stem and has a long enough hose extension to allow the technician to get away from the sidewall of the tire. It also has a gauge and a hand-operated valve.
● **SEE FIGURE 22–8** for a tire inflator.

If a tire is to be inflated off the machine, it should be placed in a tire cage or have chains wrapped around the tire to create containment if the tire or rim was to come apart while it was being inflated.

FIGURE 22–8 Tire inflator.

Most trucks with dual tires will use rock ejectors between the dual tires in case a rock gets stuck between dual tires while operating. If not removed, the rock becomes a serious safety hazard and will severely damage the tire sidewalls. **Rock ejectors** are metal bars that hang down from the truck's box. ● **SEE FIGURE 22–9** for a rock ejector.

TIRE HANDLING There will be times when a heavy equipment technician will need to remove and install tire assemblies as a part of other repair procedures. There are several different ways of safely handling tires and many ways to handle them unsafely. Proper tire-handling methods will be based on tire size, machine type, and available equipment and tooling. Medium to

WHEELS, TIRES, HUBS, AND PNEUMATIC SYSTEMS 631

FIGURE 22–10 Tire-handling machine (reprinted courtesy of Caterpillar Inc.).

FIGURE 22–9 Rock ejector.

large tires should be handled with an appropriate-sized forklift and rigging or a truck with a tire clamp arm.

Extremely large tires should be handled only with the proper equipment. These specialized machines are sometimes converted wheel loaders or forklifts. ● **SEE FIGURE 22–10** for a large tire-handling machine.

If a machine is in the shop and needs one or more of its tires removed, the shop crane with tire slings or chains can sometimes be used if overhead access is available.

It is important to *always* deflate tires before starting to remove them from a machine by removing the valve core.

TIRE CHAINS Some work sites can be very damaging to rubber tires. Machines that work in quarries where sharp rocks are present will require extra protection to prolong tire life. Tire chains are one way to protect tires from damage caused by sharp rocks. They will completely surround the tire and become a protective barrier for the tire. ● **SEE FIGURE 22–11** for a machine with tire chains.

Tire chains can also be used for traction. Some forestry machines will use tire chains to improve the machine's maneuverability. They are a more open design and can have a variety of different patterns.

As you can imagine, tire chains will wear as they are used for a longer period of time and, therefore, will require some extra maintenance. This mostly involves tightening the tire chains to keep a proper tension on them. If the chains become too loose, they can cause damage to the machine. They also require extra work if a tire needs to be removed from the rim because they have to be removed first.

TIRE REPAIR As tires can be very expensive to replace, many damaged tires can be repaired to give them a second life. The repair process will be done at a specialized tire repair facility. Even if a tire has severe damage, it may be possible to repair it by putting a new section in the tire's casing. Tires with treads worn down can be retreaded to give the tire a second life, providing their casing is in good shape.

Progress Check

1. Tire explosion occurs when:
    a. a machine is overloaded for too long
    b. a sharp object punctures the tire
    c. a reaction inside the tire causes extreme pressures
    d. the machine runs over a gas line

FIGURE 22–11 Machine with tire chains.

2. A tire has the numbers 750/65 R 25 on its sidewall, meaning:
   a. the rim is 750 mm in diameter
   b. the tire section is 750 mm wide
   c. the rim is 65 in. tall
   d. the tire is 25 in. wide

3. When is the safest time to weld a rim:
   a. in the dark
   b. on the weekend
   c. any time as long as the tire is deflated
   d. never

4. A locking key or driving lug is used with multi-piece rims to:
   a. prevent the tire slipping on the flange
   b. prevent the flange from slipping on the rim base
   c. protect the valve stem
   d. keep the locking ring from coming out

5. Most machines that feature dual wheels will use this to save the tires' sidewalls:
   a. sidewall cleaners
   b. sidewall covers
   c. rock ejectors
   d. mud flaps

WHEEL HUBS

Rubber-tired machines have hub assemblies that the wheel assembly is fastened to. The hub allows smooth wheel rotation while it supports part of the weight of the machine.

Wheel hubs can vary a great deal between different types of machines. Many wheel hubs will transfer torque from the drivetrain to the wheel and will have a final drive planetary gearset incorporated into it. Refer to Chapter 16 to learn about final drives and how torque is transferred through them.

Non-drive axles have plain wheel hubs that are supported on the axles' spindle shaft by a set of tapered roller bearings. When the machine's wheels are fastened to plain wheel hubs, the hubs allow wheel rotation on a stationary axle spindle.
● **SEE FIGURE 22–12** for a plain wheel hub on the front axle of a two-wheel drive backhoe.

Some plain wheel hubs could also be connected to the machine's brake system to transfer braking torque from the rotating wheel to the stationary axle spindle.

Plain wheel hubs have inner and outer tapered roller bearings. The roller bearing cones (inner race and roller) will rest on

FIGURE 22–12 Plain wheel hub.

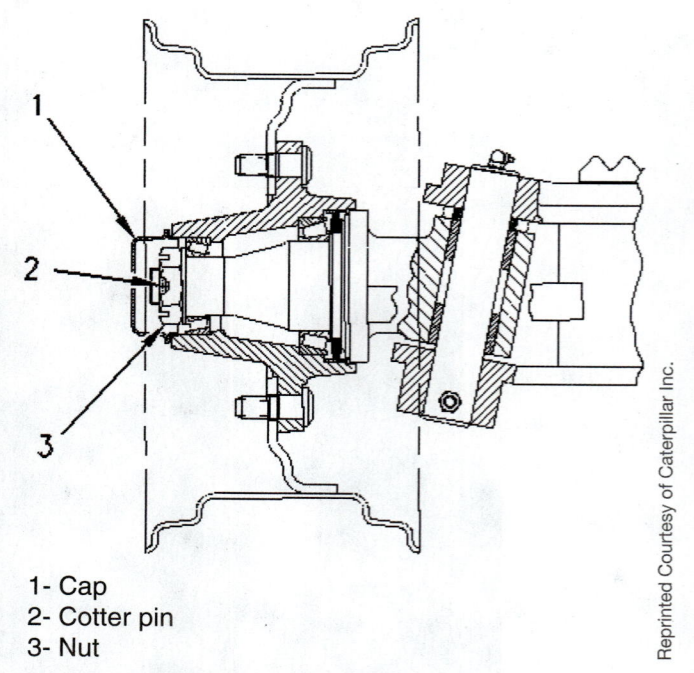

1- Cap
2- Cotter pin
3- Nut

FIGURE 22–13 Plain wheel hub cutaway.

SHOP ACTIVITY

Go to a machine that has rubber tires and record the tire size information from the tire sidewall.

Look up the tire inflation information in the machine's manual and check the tire inflation pressures to see if they are within recommended ranges.

Describe the type of rim used and record any defects found after thoroughly inspecting the tire and rim visually.

the axles' **spindle** and their cups (outer race) will be pressed into the rotating wheel hub. The inside of the wheel hub will also have a lip-type seal pressed into it to retain lubricating oil or grease inside the wheel hub and keep contamination out. The lip of the seal will rotate on a smooth metal sleeve or surface on the spindle shaft. When the hub is installed, the inner bearing cone slides over the shaft and stops against a step on the shaft. The outer bearing is then installed and held in place with a washer and nut.
● SEE FIGURE 22–13 for an illustration of a plain wheel hub.

Lighter-duty plain wheel hub adjustment is typically as follows: Tighten the nut while rotating the wheel assembly until the wheel starts to rotate smoothly and then tighten a little more until the wheel starts to bind slightly. This indicates that all the rollers are in contact with the races. Then back off the nut to allow a measured movement between the hub and the spindle when the hub is moved in and out. This can be specified from .001 in. to .025 in. and should always be set to manufacturers' current service information. This movement is called freeplay.

The **bearing adjusting** nut must be secured in place because there is no clamping force to hold it. There are several ways to do this including a castellated nut and a cotter pin or lock wire that goes through a hole in the spindle; a locking washer that has an internal tab to hold it to the spindle and then a tab is bent over one flat of the nut; or a double nut arrangement with a locking washer between them. Some plain wheel hubs won't require any adjustment procedure and will rely on machined steps on the spindle and the accurately manufactured specifications of the tapered roller bearings to give the correct bearing freeplay.

Light-duty wheel hubs on machines that don't travel very fast will use grease to lubricate their bearings. The grease will be applied to the bearings upon assembly, and this is called packing the bearings. There will also be some extra space in the wheel hub for grease. The hub will have a grease fitting so that some grease can be applied during servicing. Care must be taken to not overgrease wheel hubs as this will blow out the grease seals.

Wheel bearing adjustment for larger wheel hubs is usually based on measuring the rolling resistance of the wheel.

If a machine has an axle with an inboard final drive arrangement, the wheel hub is part of the axle shaft and the shaft is supported by a set of bearings. The bearing cones are on the axle shaft and the bearing cups are pressed into the hub assembly. Two types of seals could be used for this type of axle.

634 CHAPTER 22

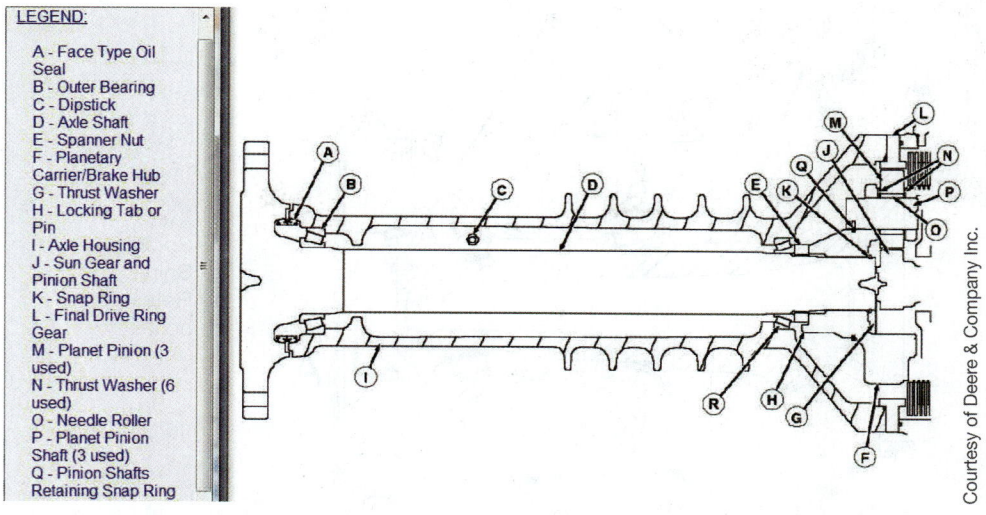

LEGEND:
A - Face Type Oil Seal
B - Outer Bearing
C - Dipstick
D - Axle Shaft
E - Spanner Nut
F - Planetary Carrier/Brake Hub
G - Thrust Washer
H - Locking Tab or Pin
I - Axle Housing
J - Sun Gear and Pinion Shaft
K - Snap Ring
L - Final Drive Ring Gear
M - Planet Pinion (3 used)
N - Thrust Washer (6 used)
O - Needle Roller
P - Planet Pinion Shaft (3 used)
Q - Pinion Shafts Retaining Snap Ring

FIGURE 22–14 Inboard axle hub.

Lighter-duty axles will use a lip seal that is installed on the inside of the hub assembly and its lip rides on the axle shaft. This seal keeps the axle lubricating fluid from leaking out. Higher capacity axles will use metal face–type seals (duo-cone) to seal between the axle housing and hub assembly. ● SEE FIGURE 22–14 for an axle hub that is part of an inboard final drive axle assembly.

If a wheel hub incorporates a final drive, there may be a procedure that uses shims to set the bearing preload to get the correct rolling torque of the hub. This could require some special tools such as hub-turning tool, depth micrometer, dial-type torque wrench, fish scale, and outside micrometer. After assembling the hub onto the spindle with new bearings and with or without the wheel seal (check manufacturer's procedure), the rolling resistance is measured. If it is within specification, final assembly can be completed.

If the rolling torque is out of specification, there is a shim adjustment that must be made to adjust the bearing preload. ● SEE FIGURE 22–15 for how the shim pack thickness is measured.

Wheel assemblies will be fastened to wheel hubs with some type of threaded fasteners. Some hubs may be tapped so the wheels can be fastened with bolts, but usually there are wheel studs that provide a way for wheel nuts and washers to secure the wheel to the hub. The studs could have a serrated shoulder that grabs the hole they get driven into when they are installed, or they could have a clipped head that rests against a step in the back of the hub. Both styles are designed to prevent the stud from turning once they are installed. ● SEE FIGURE 22–16 for the two styles of wheel studs.

WHEEL HUB SERVICING
Hubs for larger machines and ones that travel at higher speeds will use gear oil to lubricate their bearings. This oil will need to be changed on a regular basis and will need to be a proper viscosity for the machine's operating ambient temperatures. The oil level will likely be about half way up the hub housing. It should not be overfilled because this can also lead to seal failure because of the oil expanding after it warms up.

A typical oil viscosity for wheel hubs is SAE 30 or GL 80W90, and a typical oil change interval would be 1000 hours.

Care needs to be taken when removing the plug for checking the level. If the machine has been running and the oil is warm, pressure will build up in the hub cavity. This pressure can force oil out the plug hole when it is removed. Use gloves or a rag to protect yourself when checking hub oil levels. Some hubs may have a magnetic drain plug to collect metal wear particles. It should always be cleaned off at service time. ● SEE FIGURE 22–17 for the oil level check/fill plug of a wheel hub.

Wheel hub repairs usually start as a result of a fluid leak. This could be caused by a seal failure only or from a seal that failed as a result of a wheel bearing failure. If a wheel bearing fails, major damage can occur to the hub, axle shaft, axle housing, and final drive components.

SHOP ACTIVITY

Go to a machine that has oil-filled wheel hubs and describe the proper way to check the oil level for this machine. Check the oil level and inspect the hub for defects (leaks, damage). Record the recommended type and viscosity of oil called for as refill oil in the hub.

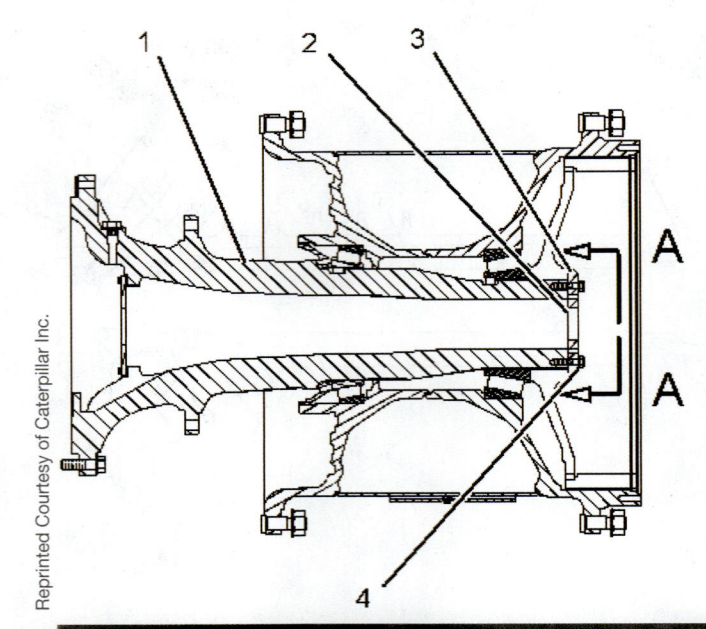

1 - Spindle
2 - Shims
3 - Retainer plate
4 - bolts

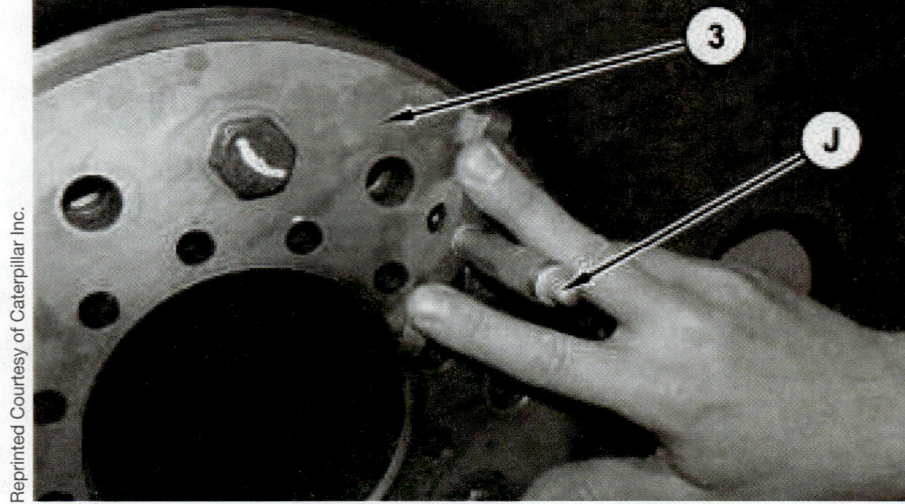

FIGURE 22–15 Shim pack thickness being measured for rolling torque wheel bearing adjustment.

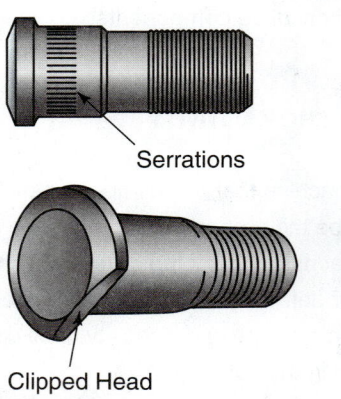

FIGURE 22–16 Two styles of wheel studs.

Serrations

Clipped Head

FIGURE 22–17 Level plug.

PNEUMATIC SYSTEMS

Heavy equipment machines can feature pneumatic systems to actuate air brake systems, air starter systems, differential locks, air seats, air suspension systems, and air horns or to control hydraulic systems. Some examples are machines that may still use air-actuated service brakes such as wheel loaders and scrapers, any diesel engine could be equipped with an air starter, and some older cranes had pneumatically controlled hydraulic systems.

Rock drills use high-pressure and high-volume pneumatic systems to operate the hammer part of the drill and to flush out material from the drill hole so the drill doesn't get stuck.

The main components of the pneumatic system are the air compressor, governor, air dryer, one or more air tanks, airlines, and a variety of different valves to control the pressure and flow of air. ● SEE FIGURE 22–18 for how the main components of a pneumatic brake system are arranged.

COMPRESSED AIR SUPPLY SYSTEM Air compressors can be belt driven, but most are gear driven directly from the engine timing gears.

Most machines with pneumatic systems will use reciprocating piston-type compressors. They usually have one or two cylinders; however, if there is a large airflow needed, V-four or inline four-cylinder compressors are available. A gear or pulley drives the compressor crankshaft, which in turn moves one or more pistons up and down via a connecting rod inside a bore. The compressor head has check valves in it to control the flow of air in and out of the compressor.

They are usually cooled by engine coolant and are engine oil lubricated. Inlet air to the air compressor is usually sourced

> **WEB ACTIVITY**
>
> For an excellent visual demonstration of a pneumatic brake system, search for a 1967 U.S. army video "Air Brakes Principles of Operation." It will explain the interaction of the main components of an air brake system, and even though it was made over 40 years ago, not much has changed as far as the main components and their fundamental operation.

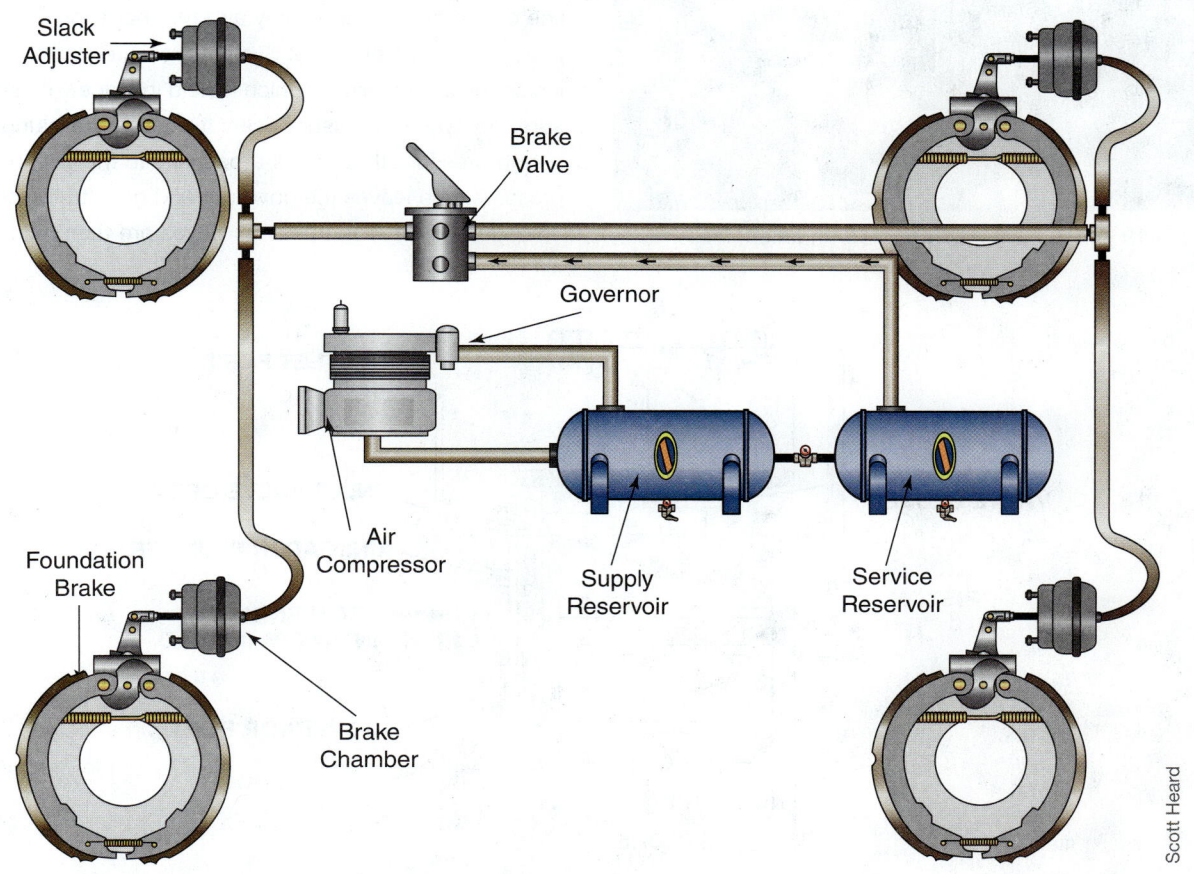

FIGURE 22–18 Pneumatic brake system components.

WHEELS, TIRES, HUBS, AND PNEUMATIC SYSTEMS

from the engine's inlet air once it has been cleaned by the engine air filter. ● **SEE FIGURE 22–19** for an air compressor mounted to a diesel engine.

Coolant from the engine flows through the head to keep it cool because compressing air will generate heat. Pressurized engine oil is fed to the bearings for the crankshaft to create a film of oil between the crankshaft and the bearings. This oil also goes through the crankshaft to journals where the connecting rods are attached to it. The oil keeps the connecting rod bearings from making direct contact with the crankshaft journals. The oil then drains back to the engine sump either through a drain line or through an opening in the compressor body and into the oil sump of the engine.

The connecting rods transfer rotary torque from the crankshaft to the pistons to drive them up and down. The pistons are similar to combustion engine pistons in that they have metal piston rings to create a seal between the piston and its cylinder bore.

As the air compressor piston is drawn down by the crankshaft, air moves in past the unloader valve (check valve) and fills the void on top of the piston. The discharge valve (check valve) is held closed by the pressure differential on it during the piston downstroke.

The unloader valve is a simple check valve that is controlled by the system's governor. ● **SEE FIGURE 22–20** for an illustration of the compressor components when the piston is moving down.

As the piston then travels up in the bore, the unloader valve is closed and air is pushed past the discharge valve that is now open.

As the air compressor is constantly driven by the engine timing gears when the engine is running, there must be a way to regulate system air pressure. The governor will have an airline connected to the supply air tank that allows the governor to sense system pressure. System pressure works on a plunger inside the air **governor**, which is held in place with spring pressure. As system pressure rises, it pushes the plunger against spring pressure that opens a passage in the governor. The air pressure then leaves the governor and goes to the compressor unloader valves. The unloader valves are then held open, and

FIGURE 22–19 Air compressor mounted on an engine.

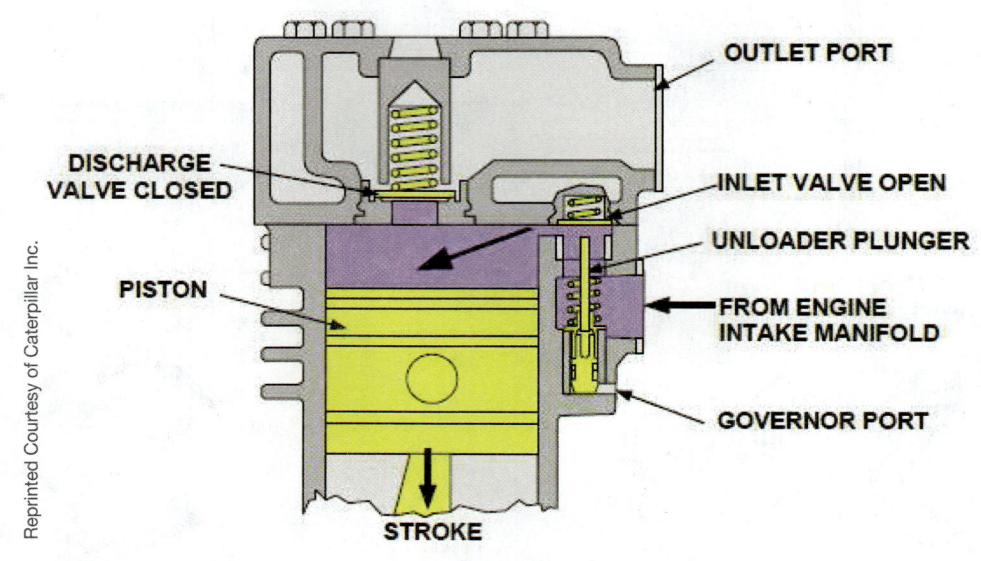

FIGURE 22–20 Compressor components on inlet stroke.

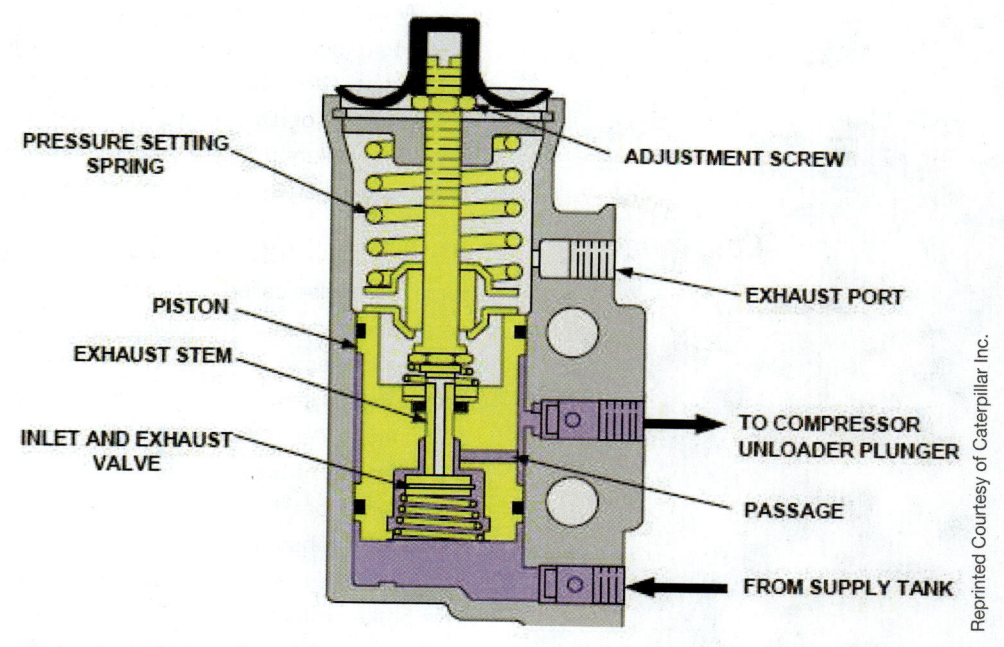

FIGURE 22–21 Governor in cut-out mode.

the piston just pushes air in and out the inlet line or back and forth between the two compressor pistons, depending on the type of compressor. This pressure setting is called cut-out pressure. ● SEE FIGURE 22–21 for a governor in cut-out mode.

There is a check valve in the outlet line of the compressor that stops pressure from returning into the compressor when it is in the unloaded mode.

Spring tension in the governor is adjustable, and therefore, by varying spring tension, the technician can set maximum system pressure. As system pressure drops, the governor plunger will raise and block off the outlet to the unloader valve and the compressor starts pushing air out the discharge line. This pressure is called cut-in pressure and is not adjustable but is a fixed difference below the cut-out pressure of 20–30 psi.

A typical value for maximum pressure is 120 psi, which would make cut-in pressure 100 psi. There will likely be a dash gauge to display the supply air pressure.

Compressors are sized by the volume of air they can pump at a certain pressure level. An example of a compressor used to supply air for an air brake system is 15 CFM @ 100 psi.

They are usually sized so they move air only 25% of the time that the compressor crankshaft is turning. This would be a 25% duty cycle and the other 75% of run time gives the compressor time to cool down. The amount of time that they are pumping is controlled by the governor and the governor determines this by monitoring system pressure.

If there is an air leak downstream and the duty cycle increases to compensate for the air loss, the compressor will likely overheat.

Some machines, such as rock drills, need high air volumes (sometimes over 500 CFM) and high air pressures (sometimes over 300 psi) to perform their job. These machines use screw-type compressors that have their own lubrication system.

AIR DRYER Air flow from the compressor outlet will go to an air dryer next to get most of the moisture removed from the air. An air dryer is a canister that contains desiccant, which is a substance that will absorb moisture. It will also feature a purge valve at the bottom of it. The purge valve is opened when cut-out pressure is reached and will exhaust any accumulated moisture from inside the air dryer. There is an airline going from the governor to the bottom of the air dryer that will unseat the purge valve. ● SEE FIGURE 22–22 for an air dryer.

Once the mostly dry air leaves the dryer, it will go to the supply tank, which is sometimes called the wet tank. When the air is compressed in the compressor, it is heated up, and when it cools down in the primary air tank, any moisture left in the air will condense and collect in the bottom of the tank. This makes it necessary to drain the air tank on a daily basis to purge any collected moisture or contamination from the tank. Some systems will have automatic drain valves on the supply tank.

There should be safety valves in all components of the supply system. In case the governor doesn't stop the compressor or there is a blockage somewhere downstream from the compressor, a safety valve will open and release air pressure. They are usually set for 250 psi and should open only if system pressure reaches dangerous levels.

WHEELS, TIRES, HUBS, AND PNEUMATIC SYSTEMS

FIGURE 22–22 Air dryer.

Airlines that leave the compressor must be heat resistant. This could mean they are copper- or stainless steel–reinforced nylon hose. Other airlines could be rubber or plastic and could have a variety of different types of connectors on their ends.

Air pressure that is stored in the supply tank can now be used for a variety of uses as mentioned earlier.

PNEUMATIC BRAKE SYSTEMS Pneumatic brake systems, otherwise called air brake systems, use compressed air as an energy source to actuate service brakes and to release parking brakes. Again there are many variations on these systems such as service brakes only; service and parking brakes; and primary, secondary, and parking brakes.

Service brakes are used to slow down and stop the machine while parking brakes are used to hold the machine in place after it has been stopped. Primary brakes are the rear service brake system and secondary brakes are the front service brake system. They are split into two systems to minimize the chance of complete brake failure. ● **SEE FIGURE 22–23** for a schematic of this system.

For an explanation of braking fundamentals, refer to the start of Chapter 18.

Compressed air that has been created by the air supply system is available in the supply tank. From there it goes

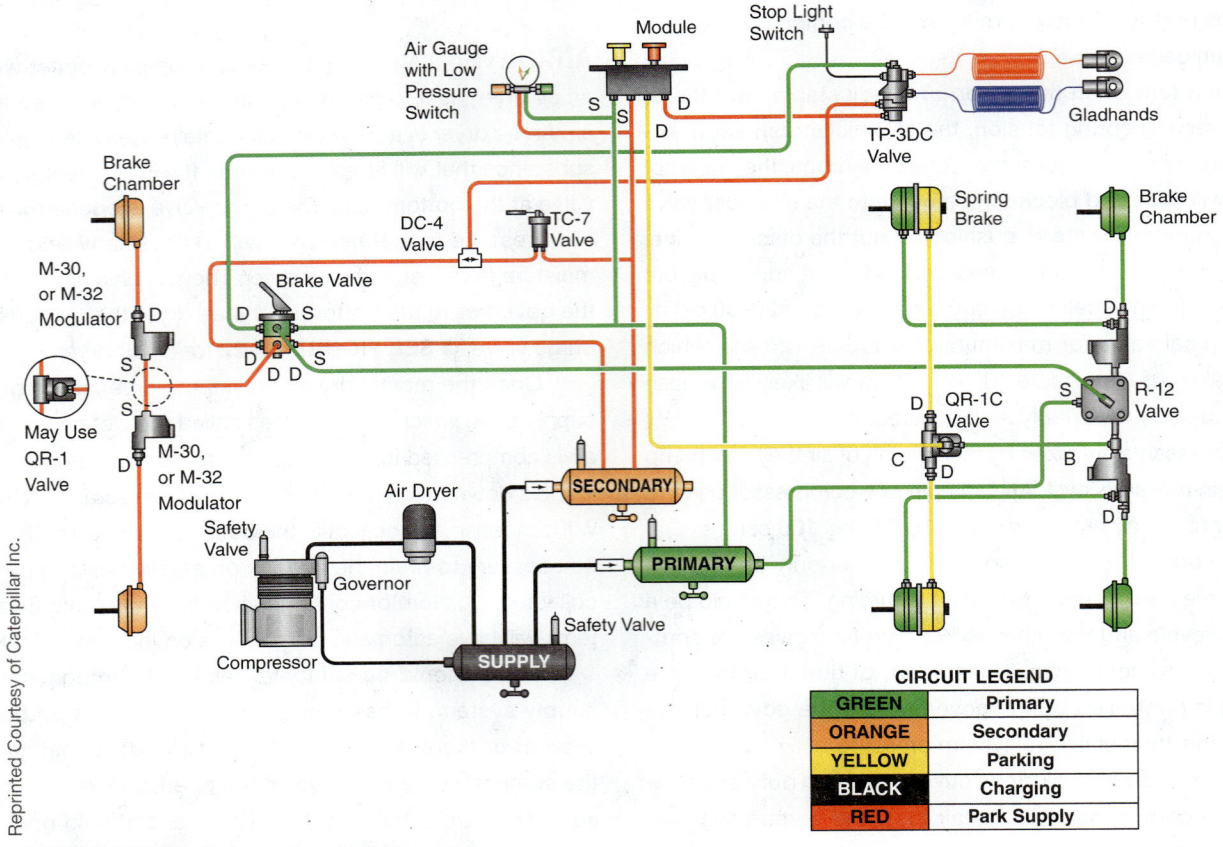

FIGURE 22–23 Schematic of brake system.

CIRCUIT LEGEND	
GREEN	Primary
ORANGE	Secondary
YELLOW	Parking
BLACK	Charging
RED	Park Supply

640 CHAPTER 22

through check valves and on to primary and secondary air tanks. These tanks supply one or more control valves that could be foot pedal (treadle) operated or hand (wand) operated.

The secondary tank supplies air to one section of the treadle valve and the primary tank supplies the other. If the brake pedal is pushed down, air is metered out of both valve sections to the primary and secondary service brake chambers.

Inside the **brake chamber** is a rubber diaphragm. Air pressure pushes on the diaphragm against spring pressure on the opposite side of the diaphragm. The diaphragm then pushes on a plate that pushes the brake rods out of the brake chamber, which in turn mechanically actuates the foundation brakes.

Air is also supplied to a parking brake valve from both tanks. When the valve is pushed in, air is sent to the parking brake chambers to release the parking brakes, which are typically only on the rear brakes. The rear brake chambers have two chambers that are stacked together and actuate the same rod coming out of it.

The spring brake chamber (parking) part of it has a very strong spring in it that keeps the rod pushed out and the brake applied until the parking brake release valve sends air to it. This will act on a diaphragm and compress the spring, which in turn pulls the rod in to release the brake. The service brake chamber is then able to push the rod out to apply the brakes when there is air pressure modulated to it. ● SEE FIGURE 22–24 for an illustration of a service/parking brake chamber.

On-highway vehicles with air brakes have more valves in their systems to assist with removable trailers, anti-lock brakes, and tighter regulations for brake systems.

S-CAM FOUNDATION BRAKES There are a couple of variations of air-actuated foundation brakes such as wedge-

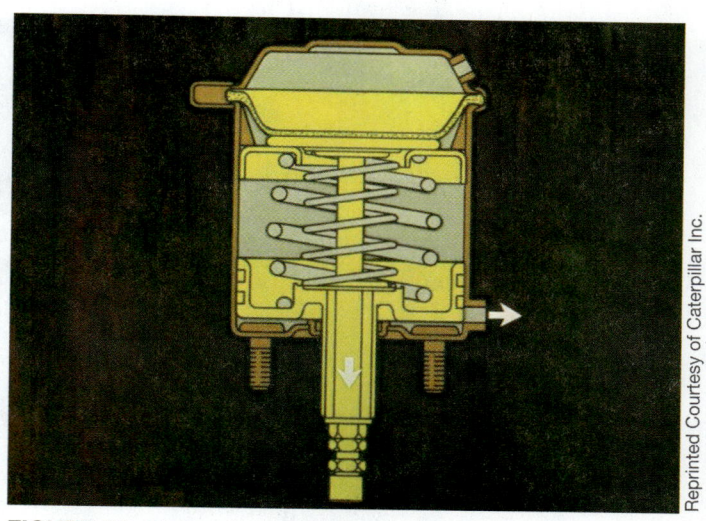

FIGURE 22–24 Service and parking brake chamber.

type brakes and disc and caliper brakes, but the most common is the S-cam type.

For this type of brake, the chamber's rods will push on the end of a slack adjuster, which causes the adjuster to rotate. The opposite end of the **slack adjuster** has a large female spline that mates with one end of a shaft. Therefore, as the slack adjuster is moved by the brake chamber, it rotates the shaft.

The shaft has an S-cam on its opposite end that will push on a pair of rollers. The rollers are hooked onto one end of each brake shoe. As the rollers ride up the S-cam profile, they spread the brake shoes. Braking action takes place when the brake shoes are pushed out against the inner surface of the brake drum. As the brake shoes are anchored to the axle and the brake drum rotates at wheel speed, the friction material on the brake drums will try to make the brake drum match the speed of the axle housing (0 rpm).

> ### TECH TIP
>
> Some service procedures will require spring brakes to be caged. This means that they are held in a released position mechanically. To do this, a caging bolt is inserted through the back of the brake chamber. It has two small pins protruding from the end of it that when turned 180° will lock into the spring brake's plate. Next, a nut and washer is tightened against the housing, and as the nut is turned, the tool will pull the plate back into the chamber housing as if air pressure was being applied to the opposite end of the diaphragm.
>
> This is done whenever there is a need to release the parking brakes mechanically. ● SEE FIGURE 22–25 for a brake release tool.
>
> **FIGURE 22–25** Brake release tool.
>
> Be extra cautious whenever a brake is caged because the spring-applied feature is now defeated.

WHEELS, TIRES, HUBS, AND PNEUMATIC SYSTEMS 641

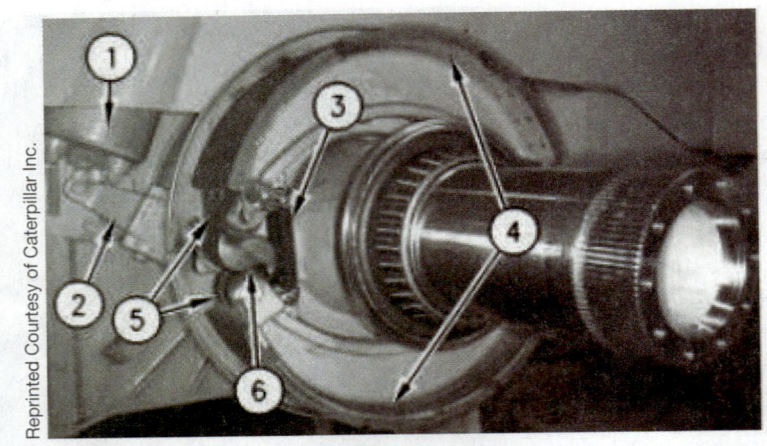

FIGURE 22–26 S-cam brakes.

(1) Brake actuator
(2) Slack adjuster
(3) Return springs
(4) Brake shoes
(5) Rollers
(6) Brake camshaft

The amount of braking action is directly proportional to the force applied by the brake shoes to the brake drum. This force is increased as air pressure applied to the brake chamber is increased, which is metered by the operator. Two other factors could also affect the force: the size of the brake chamber diaphragm increases and the length of the slack adjuster are increased. Normally, the only variable is the air pressure that gets metered to the chamber because the size of the brake chambers and the length of the slack adjusters are determined by the machine design engineers.
● SEE FIGURE 22–26 for an S-cam brake configuration.

Brake shoe friction material will eventually wear down, and this loss of material must be compensated for by the slack adjuster. Slack adjusters will rotate the S-cam shaft with a worm gear while the brake chamber rod is stationary.

Manually adjusted slack adjusters allow the technician to keep the brake shoes close to the drum. There are also automatic slack adjusters that will keep shoe to drum clearance to a minimum, but these are not usually found on heavy equipment.

Another type of brake chamber used for S-cam type foundation brakes is called the roto-chamber. It is similar to the previously mentioned brake chambers except that it will use a different style of diaphragm that allows it to produce a longer stroke and maintain a constant force throughout the entire stroke. They also have a different parking brake apply and release mechanism. There is a series of balls that work on the rod of the chamber to hold the rod in place. The balls are held in place by air pressure, and when released, they get wedged between the housing and rod to lock the rod in the applied position.

AIR OVER OIL BRAKE SYSTEMS Some machine brake systems will need to have their hydraulic brake pressure boosted by pneumatic pressure. This type of brake system is commonly called an air over oil system. The pneumatic part of the system is very similar to the control system for pneumatic brakes. System air pressure is available to a treadle valve that meters air pressure according to how hard the operator steps on the brake pedal. This metered air pressure is applied to a piston in the booster cylinder. The piston then applies pressure to a rod, which in turn pushes the master cylinder piston. The opposite side of the master cylinder piston is the brake oil system, and this was covered in Chapter 18. As there is an area differential between the pneumatic piston and the brake oil piston in the booster cylinder, there will be pressure amplification from the air system to the oil system. For example, if 50 psi is supplied to the air piston, then 250 psi may be created in the brake oil system.

Some older mining haul trucks and medium-size wheel loaders use air over oil brake systems. ● SEE FIGURE 22–27 for an air over oil master cylinder.

FIGURE 22–27 Air over oil master cylinder.

OTHER PNEUMATIC SYSTEMS

Air pressure and flow can also be used for a power source for rock drilling equipment. Drill equipment quite often use rotary hammers to penetrate rock. The holes left behind are then filled with explosives, and the rock is blasted to be hauled away for processing.

Drilling equipment may use air pressure at up to 350 psi and several hundred CFM. To provide this kind of pressure and flow, a screw-type compressor is most likely used. They are similar to a Roots-type blower that was used on two-stroke engines. The rock drill's diesel engine drives one of the shafts and a gear from this shaft drives the gear on the second shaft. There may be a clutch driving the first gear, or it could be directly driven from the flywheel.

Some rock drills could be powered completely by a remotely located compressor. In this case, the machine's tracks are driven by air motors.

DIAGNOSING PNEUMATIC SYSTEM PROBLEMS

Some operator complaints for machines that use pneumatic systems could be no air pressure, low air pressure, air compressor cycling too fast, too much moisture in the air, and air leaks.

As with any other diagnostic procedure, the technician should start by verifying the complaint. Then move on to performing a visual inspection such as looking and listening for air leaks, checking the wet tank for oil or excessive moisture, checking system pressure, checking the air dryer for regular purging, and timing the compressor cut-in to cut-out cycle.

SERVICING PNEUMATIC SYSTEMS

To keep a pneumatic system operating properly, a few simple service steps should be performed. This mainly involves removing moisture from the system air tanks. This should be done daily; otherwise, air systems will accumulate moisture, which will lead to operational problems. This is mainly a problem in winter time when the moisture in the air will freeze and block passageways in the air system. However, excessive moisture and pneumatic systems can also lead to rust and corrosion on internal components.

Sometimes it may be necessary to install an alcohol injector into a pneumatic system. The injector will meter alcohol into the system, which will eliminate moisture to prevent freezing.

PNEUMATIC SYSTEM REPAIRS

Airline leaks can be repaired by replacing the entire line assembly or by using a line repair kit.

Pneumatic system's valves can be rebuilt, but it may be cheaper to replace the entire valve assembly. The deciding factor with valve replacements may be the condition that the airlines connected to the valve body are in. If the lines are corroded or they are steel lines that can't be moved easily, it may be farther ahead to rebuild the valve in place, if that's possible.

> **TECH TIP**
>
> One of the main production loaders for a large limestone quarry had an air over oil brake system on it. The air supply system was a fairly standard arrangement and just needed to provide air for the brakes and an air horn. Over time it became common practice to drain a significant amount of water from the machine's air tanks on a daily basis. In the wintertime this machine would also suffer a lot of downtime because of ice forming in the machine's airlines. One technician's solution to this was to install a 24 V floodlight near the air tanks with the hope that the heat from the light would thaw the frozen airlines so the machine could continue to work. This was a temporary fix that worked with limited success. Another technician wondered why so much moisture continuously accumulated in the air system. This technician went on to replace the machine's factory installed air dryer with an aftermarket unit that was a larger capacity air dryer. This air dryer was also a completely different style unit, and soon after it was installed, the excessive moisture problems went away.

An air compressor that needs to be repaired will usually be replaced as an assembly. If the air compressor has had a major failure, then a system cleanout may be required. Some machines will have air compressors that drive other components such as fuel pumps or small hydraulic pumps. In this case, the air compressor replacement will get more complicated.

Progress Check

6. Non-drive wheel hubs will have this type and quantity of wheel bearings:
   a. two plain bearing
   b. two self-aligning bearings
   c. one tapered roller bearing
   d. two tapered roller bearings
7. Wheel hubs that transfer drive from inboard final drives will:
   a. be part of the axle shaft
   b. be driven by the ring gear
   c. only be lubricated by grease
   d. only be used to drive dual wheels
8. Air compressors used for machine pneumatic systems will:
   a. always be air cooled
   b. always be coolant cooled
   c. have their own oil supply
   d. always be belt driven

9. A pneumatic system governor will send air pressure to:
   a. the wet tank
   b. the primary piston in the compressor
   c. the unloader valve
   d. the treadle valve
10. The parking brake chamber used with S-cam brakes is:
    a. air actuated spring released
    b. oil actuated spring released
    c. spring actuated air released
    d. spring actuated oil released

SHOP ACTIVITY

- Go to a machine that has a pneumatic system on it and describe the location of the following components: air compressor, governor, air dryer, supply tank, service tank.
- Describe the warnings that alert the operator to a low air pressure condition for this machine.

SUMMARY

1. Extreme caution must be used when handling and working near tires and rims that are found on heavy equipment. Tire explosion can occur if a tire is overheated and will suddenly release tremendous energy.
2. One-piece steel rims used on light-duty machines are similar to one that may be found on commercial heavy-duty pickup trucks.
3. *Always* deflate tires before starting to remove them from a machine.
4. Lighter-duty machines will use one-piece rims similar to automotive-style rims.
5. Multi-piece rims allow for tires to be changed without deforming the tire to get its bead over the bead flange.
6. Medium-duty machines will use three-piece rims with a lock ring to hold one removable flange in place.
7. Large heavy-duty rims are five-piece assemblies with lock lugs to prevent the bead flange from slipping on the rim base.
8. Tires are a major cost of operating heavy equipment.
9. They can be considered as a container for pressurized air that supports the weight of the machine, and they also assist with providing a smoother ride as the tire sidewalls flex.
10. Some tires will not use air pressure and either are solid rubber or have voids in the sidewall to allow some flex.
11. An HDET will be responsible for checking and maintaining air pressure in tires, and proper inflation pressure is a big factor in getting maximum life from tires.
12. Most tires found on machines today are radial design. Older design types were called bias ply. This refers to the arrangement of the supporting plies of the tire carcass.
13. Tire ply rating refers to how many layers of supporting plies are used for the carcass, and generally, more plies create a stronger tire. Some tires will have over 70 plies.
14. Tire tread designs vary greatly, and machine application will dictate which design is best.
15. There are many different tire composition and design ratings such as rubber compound, tread depth, and tread design.
16. Tire sizing is based on rim diameter, tread face width, and aspect ratio.
17. Aspect ratio is a comparison of tire sidewall height to tread face width. Low-profile tires will have an aspect ratio 80, 70, or 65, for example.
18. Tire inflation should be checked when the tire is cold (not at operating temperature).
19. Nitrogen can be used to inflate tires and is preferable to air because it is a pure inert gas and, therefore, is more stable and will not support combustion.
20. Tire pressure–monitoring systems are becoming popular and can alert equipment operators and owners if a tire is losing pressure.
21. Proper inflating equipment should be used to prevent an HDET from being in a dangerous area when inflating tires.
22. Most haul trucks will have dual rear wheels and will have rock ejectors to keep rocks from getting stuck between the tires.
23. Large tires will need to be handled with special machines or trucks for tire installation and removal.
24. Some machine applications will require machines to have tire chains installed to preserve the tires (e.g., rock quarry loaders).
25. Wheel hubs allow rotation of wheel assemblies on axles. They usually use two sets of tapered roller bearings and can incorporate planetary final drives.
26. Wheel bearing adjustment can involve shims or nut adjustment procedures.
27. Pneumatic systems can be found on machines for brake systems, starting systems, or other systems controls.
28. Air compressors use reciprocating pistons to compress air and charge the supply circuit where pressure is maintained between 105 and 125 psi (generally). This pressure is controlled by the governor that "turns on and off" the compressor.
29. Air dryers are needed to remove moisture from the air after it has been compressed and cools down.
30. Service brakes slow down machines while parking brakes hold the machine stationary after it is stopped.
31. Air brake systems will use S-cam shoe-type foundation brakes.
32. Dual function brake chambers will have spring-applied, air-released parking brake and air-applied service brake functions.

chapter 23
UNDERCARRIAGE SYSTEMS

LEARNING OBJECTIVES

After reading this chapter, the student should be able to:

1. Identify the main components that make up an undercarriage system.
2. Describe the common methods used to check and adjust track tension.
3. Explain the different factors that create wear in undercarriage.
4. Identify the different types of track chain.
5. Identify the three different styles of track pads.
6. Describe the different types of track chain master links.
7. Describe the steps taken to remove and install most tracks.
8. Explain how to measure undercarriage for wear.
9. Describe the common undercarriage component reconditioning procedures.
10. Describe the dangers associated with working on or near undercarriage systems.

KEY TERMS

Bottom roller 650
Carrier roller 648
Equalizer bar 648
Master link 653
Pivot shaft 649
Sealed and lubricated 654
Track frame 647
Track pitch 653
Track tension 646

INTRODUCTION

SAFETY FIRST Because of the extreme forces applied to undercarriage components, they need to be made from heavy gauge steel and cast iron and can be very heavy. If attempting to move undercarriage components without the assistance of cranes, proper personal lifting techniques should be used to avoid strain injuries. If lifting devices such as shop or service truck cranes are used, then proper hoisting and rigging equipment and techniques should be used to reduce the chance of serious injury or death. Poor techniques and equipment usage for lifting and moving heavy undercarriage components can result in them being dropped or tipped over resulting in a risk of major injury or death.

Track adjusters and recoil springs will be under a great amount of pressure and must be treated with respect. Recoil springs *must* be disassembled with proper tooling.
● **SEE FIGURE 23–1** for a track recoil spring of a small dozer.

Do not remove the **track tension** grease fitting or the release valve. You risk high-pressure fluid injection injuries if these warnings aren't followed.

If both tracks are removed from a machine, be aware that it could roll in an uncontrolled manner because the sprocket may not engage with the track chain. Proper blocking should be used to prevent this.

Some track chain master pins will need to be pressed out with a great deal of force (possibly over 100 tons). Make sure that all press tooling is in good condition, and you stay back from the pin as far as possible when hydraulic pressure starts to rise on the press pump gauge.

Heat may need to be applied to fasteners or components during undercarriage repairs so extra caution should be taken to avoid burns, fires, and explosions.

FIGURE 23–1 Recoil spring from small dozer.

NEED TO KNOW

Track-type machines have been around for many years and were developed as a way to farm wetlands that would leave rubber-tired tractors stuck. Track machines have better flotation characteristics because the track chain and pads that are engaged to the ground are able to spread the weight of the machine out over a larger area. This reduces the ground pressure of the machine. Traction is also increased because there is a greater contact area.

A wide variety of track-type machines will use undercarriage components such as track-type tractors (dozers), track loaders, excavators, some skid steer loaders, crawler cranes, and other specialized types of equipment that need good flotation.

Track-type machines have a variety of different components that combine to make up their undercarriage system. The machines' drivetrain will send torque to the sprockets that transfer torque to a chain assembly, which in turn has track pads fastened to it. The pads have grouser bars that engage with the ground, and as the chain is pulled toward the sprocket, the machine moves. ● **SEE FIGURE 23–2** for the main components of a typical undercarriage.

The undercarriage of a track-type machine accounts for approximately 20% of the purchase price of the machine and a significant portion of the operating cost. It is important that a technician understands how an undercarriage system works,

646 CHAPTER 23

FIGURE 23–2 Undercarriage components.

TECH TIP

There are many factors that influence how the moving undercarriage components wear (the moving components are sprocket, idlers, rollers, and chain). Most of these factors have very little to do with an HDET. The conditions the machine works in will greatly affect how long the moving undercarriage components last. Dry light soil will cause very little wear in comparison to wet coarse sand. Heavy wet soil will cause packing of the track chain, which in turn increases track tension and causes accelerated wear. Some machines will work in harsh chemicals or extreme temperature conditions, and this also has a negative effect on undercarriage life. If a track machine is working in rocky conditions, the undercarriage components will suffer from impact shocks.

The operator has a large influence on how fast the undercarriage wears as well. Operating on side slopes, excessive turning, and excessive high-speed reverse will significantly decrease the potential life of the moving undercarriage components.

While an HDET has nothing to do with the working conditions the machine works in, he or she can influence the operator to practice more undercarriage friendly operating techniques.

how to maintain it, how to diagnose problems with it, and how to repair it. This will help the machine owner to reduce the operating costs of the machine.

FIGURE 23–3 Machine with oval track arrangement.

UNDERCARRIAGE CONFIGURATIONS

There are two main arrangements for undercarriage: one is called oval tracks and the other is called high drive. The term *oval track* comes from the shape of the track if you looked at it from the side of the machine. This design is also sometimes called conventional because the configuration has been around since track machines were invented. ● **SEE FIGURE 23–3** for a machine with an oval track arrangement.

High-drive machines have their drive sprocket positioned above the **track frame**. ● **SEE FIGURE 23–4** for a high-drive undercarriage.

The track shape would be a triangle in comparison to the oval track design.

UNDERCARRIAGE SYSTEMS **647**

FIGURE 23–4 High-drive undercarriage.

High-drive machines have been around for many years. One of the big advantages of a high-drive machine is having the final drive and sprockets up and away from the ground. This reduces the chance of the sprocket and final drive receiving shock loads from running over rocks. The operator also sits higher in a high-drive machine, which increases visibility of the blade. The final drive seal is also much less susceptible to damage from mud and water because it is up and away from the majority of it.

This arrangement can also eliminate the need for a roller to support the track on the top side of the track frame. The disadvantages to high-drive machines are that the chain will need to be slightly longer and there needs to be an extra idler on the track frame. This will add some expense to the initial cost of the machine as well as add expense every time the undercarriage is replaced.

TRACK FRAME

To support track machines and most of the undercarriage components, there needs to be a left and right track frame. These track frames will support the weight of the machine and be the mounting point for track rollers, **carrier rollers**, track recoil spring, track adjuster, and one or two idlers. The track frame must be a very strong part of the machine because it will have several extreme forces acting on it. Many track-type dozers will use the track frame as a mounting point for the blade frame.

Like the machine's main frame, the track frame is constructed from formed plate steel and possibly cast iron sections that are welded together. Some additional machining processes to form threaded holes and smooth surfaces will complete the track frame. It will then get a coat of paint to prevent rust from occurring as much as possible. It may have sheet metal covers to keep material off of the recoil spring and track adjuster or some larger machines will have these components enclosed in

FIGURE 23–5 Wishbone type of track frame.

the track frame housing. This style of track frame will also likely have a quantity of oil in it to keep the recoil spring from rusting.

The frame should be inspected for cracks in areas where stresses are concentrated. The track frame could be mounted solidly to the machine's main frame or could be allowed to pivot. This depends on the type of machine the undercarriage is attached to. Excavators will have their track frames rigidly mounted to their main frame while track-type tractors and most track loaders will have pivoting track frames.

The pivoting action allows the front of the track frame to move up and down slightly from being parallel to the machine's main frame. This will help to keep the track pads engaged with the ground longer if the machine is working in rough or uneven terrain.

Some older machines with pivoting track frames were in a wishbone or "Y" shape. The rear Y part of the track frame was supported in two places. One was at the outside of the final drive and the other was a smaller support at the rear center of the machine. ● **SEE FIGURE 23–5** for how an older machine's track frame is mounted to the main frame.

The front of the track frames are sometimes tied together by an **equalizer bar** that pivots on a pin in the bottom center of the machine main frame. ● **SEE FIGURE 23–6** for an equalizer bar.

The weight of the front of the machine rests on the ends of the equalizer bar, which in turn are supported by the track frame.

The equalizer bar is also sometimes called a crossbar, and it can pivot on the pin but is limited by stops on the top of the crossbar and the bottom of the main frame. There are bushings to support the crossbar on the pin, and these bushings need to be greased regularly. There will likely be a grease hose going to the pin, and it should be inspected regularly to ensure that grease is getting to the pin because the hose is exposed to

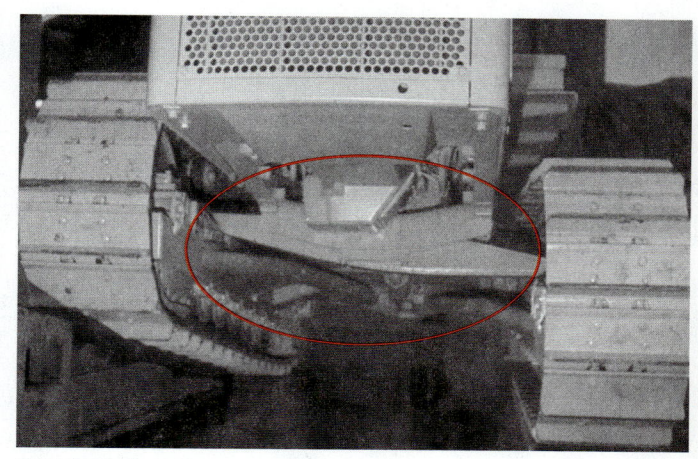

FIGURE 23–6 Equalizer bar.

FIGURE 23–7 Location of the pivot shaft.

unfriendly conditions and prone to failure. If the pin seizes up, a large labor-intensive repair will be needed.

The outer ends of the equalizer bar can either simply rest on pads on the top of the track frames or be fastened to the track frame with a self-aligning bearing and cap. The self-aligning bearing will also need to be greased regularly.

Newer machines will have track frames that mount on to **pivot shafts**. The pivot shafts extend from the side of the main frame. They have two smooth surfaces that align with plain bushings in the track frame. The track frame is machined to accept a seal and two plain bushings, and when the frame is installed onto the pivot shaft, a plate is fastened to the end of the pivot shaft. A cover with a seal then creates a sealed compartment where oil is added to keep the shaft and bushings lubricated. ● SEE FIGURE 23–7 for the location of the pivot shaft.

TRACK ROLLERS
Mounted to the bottom of the track frame is a series of heavy-duty rollers. These rollers are designed to support the weight of the machine on top of the track chain assembly as the revolving chain is laid on the ground. There are single- and double-flange rollers, and they are usually arranged on the track frame to have single-flange rollers at both ends of the track frame and alternate with single- and double-flange rollers in between.
● SEE FIGURE 23–8 for single- and double-flange rollers.

The rollers have a small amount of oil in them to keep their shaft and bushings lubricated. Metal-faced seals keep oil in and contamination out. If any sign of oil leakage is noticed from a track roller, plans should be made to replace it before it fails. Larger rollers can be "re-shelled" to keep repair costs down, but smaller rollers will likely be replaced as an assembly when they are worn out. Re-shelling a roller involves replacing the outer shell or wear surface of the roller with a new shell. This will enable the shaft to be reused and is only done at a specialized undercarriage shop.

FIGURE 23–8 Single- and double-flange rollers.

Some manufacturers recommend testing track rollers for seal leakage before installing them on a machine. This is done by removing the oil fill plug and installing an air fitting and then regulating low-pressure air to the inside of the roller and looking for leaks.

Some larger machines will have a track roller suspension system that allows the rollers to move up and down slightly to conform to uneven ground. See Chapter 21 for an explanation of the track roller bogie system.

Once a track roller's seal leaks the oil out, it will likely seize, and when this happens, major wear occurs to the track chain links. An indication of a seized track roller will be a high-pitch squeal when the machine is traveling.

Regular undercarriage inspections by the operator and technician should include looking for signs of oil leaks and broken, missing, or loose fasteners.

FIGURE 23–9 Track frame with carrier roller and idler.

FIGURE 23–10 Full roller guard.

CARRIER ROLLER
Mounted on top of the machine's track frame is one, two, or three carrier rollers, and their purpose is to support the chain assembly as it moves between the sprockets and the idler on the top side of the track frame.
● **SEE FIGURE 23–9** for a track frame with carrier roller and idler.

They are usually mounted on a pedestal that is bolted to the track frame. Some carrier rollers may be supported on both sides, but most are supported on only one side.

Some high-drive machines won't need carrier rollers because the elevated sprocket will keep the track suspended and away from the track frame.

Carrier rollers can be adjusted side to side to ensure that the track stays in alignment with the idler and sprocket. A straightedge or string can be used to align the carrier rollers with the idler and sprocket.

Carrier rollers are similar in design to **bottom rollers** and should be checked for signs of leakage. Their shell should be inspected and measured for excessive wear.

TRACK ROLLER GUARDS
Many machines will have track roller guards that are mounted to the bottom of the track frame. There are partial guards that protect only one or two rollers or full guards that protect all rollers. Their purpose is to keep rocks and large debris from getting between the rollers and track frame and to also help guide the track along the track frame. These track guards will usually have hardened wear surfaces on them that can be replaced when needed.

Track guards can be a problem in freezing conditions if mud gets in behind them and freezes. This may prevent the rollers from turning and cause accelerated track link wear.
● **SEE FIGURE 23–10** for a track frame with a full roller guard.

IDLERS
At the front (and sometimes rear) of the track frame is a large steel wheel that is called an idler, and its purpose is to guide the track assembly around the front and rear of the track frame. Machines with conventional undercarriage will have only a front idler while high-drive machines will have a front and rear idler.

A rear idler has its shaft held into saddles on the frame by bolt-on caps. Its only purpose is to guide the track around the rear of the track frame and up to the sprocket. Its construction is identical to front idlers.

At times the idler will take some of the machine's weight, but its main purpose is to guide the track chain on to and off of the track frame. The track rollers are meant to support the majority of the weight of the machine. The idler assembly has bushings that ride on a shaft and the shaft supports the idler. There is a small amount of oil sealed into the center of the idler to keep the shaft lubricated, which is similar to the track rollers. Front idlers will also be used to keep proper track tension. They have their shaft mounted to a C-shaped yoke, which is pushed out by a track-adjusting cylinder. Idlers should also be inspected for oil leakage and measured for excessive wear.

Idler wheels can have their outer wear surface built up after they become worn down over time. This is similar to retreading a tire. ● **SEE FIGURE 23–11** for a front idler.

TRACK ADJUSTER
All track-type machines need a way to have their track tension adjusted. Proper track tension is critical to ensuring proper machine operation, reducing fuel consumption, and getting the maximum life out of undercarriage components. Different working conditions can cause the tracks to become too tight and normal wear will make the tracks loose over time. These conditions will make checking and adjusting track tension a regular operational requirement for any track-type machine.

Track tension is set and maintained by the front track idler position. The idler position can be changed by pumping

FIGURE 23–11 Front idler.

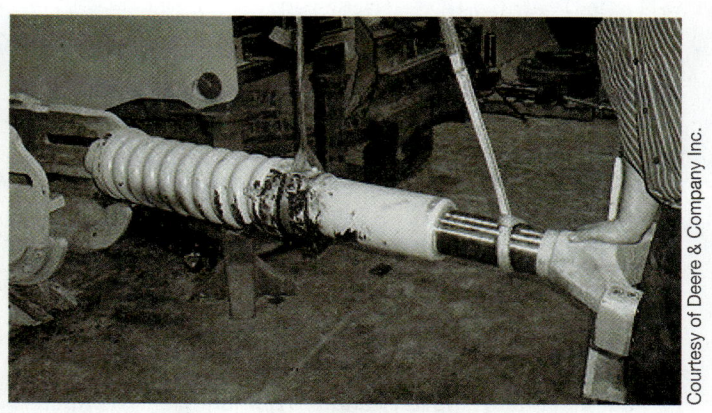

FIGURE 23–12 Track adjuster and recoil spring.

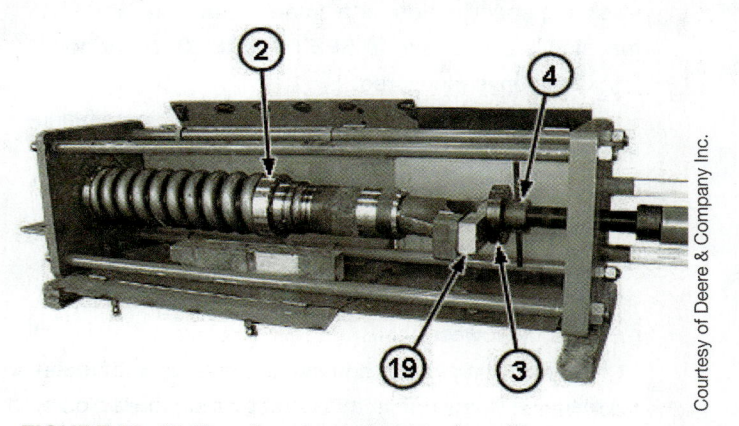

FIGURE 23–13 Recoil spring being tensioned.

grease to a track adjuster cylinder that is inside the track frame. The grease will push a piston out of a barrel and the rod attached to the piston moves the idler out to increase track tension. The track adjuster cylinder is really a simple version of a single-acting hydraulic cylinder and the grease acts like oil to move the piston and then keep it in place. The piston has seals on its outer diameter to create a seal between the piston and barrel and the barrel has seals on its inside diameter to keep dirt out.

To decrease track tension, a valve on the track adjuster barrel is loosened. This allows grease to escape from behind the piston and the idler will move back into the track frame from the weight of the track chain. The tensioner will have to be fully retracted whenever the tracks are removed or installed. This may require pushing the idler in with either a pry bar or another machine.

Some cranes will have a hydraulic cylinder to keep track tension. It is supplied a fixed amount of low-pressure oil to keep a fixed amount of force on the track.

Older machines used a large threaded adjuster to move the idler and adjust track tension. ● **SEE FIGURE 23–12** for a track adjuster and recoil spring.

RECOIL SPRING

Behind the track adjuster is a large recoil spring that will allow the idler to retract into the track frame if something like a rock gets caught in the track or the track becomes too tight because of changing working conditions. The spring will not normally get compressed, but when it does, it prevents damage to other machine components. This spring will be preloaded so use *extreme caution* when working with or near recoil springs because they can release enormous amounts of energy very quickly if they are mishandled.
● **SEE FIGURE 23–13** for a recoil spring being tensioned with special tooling.

It may be necessary to use special tools to cage or compress the recoil spring to make sure that it can't release its energy in an uncontrolled manner.

Some larger machines will have the inside of the track frame hold a quantity of oil to keep the recoil spring and track adjuster from rusting.

There is a dipstick to check the oil level attached to the cover that allows access to the track adjuster.

Some machines will use an accumulator instead of a recoil spring to allow the idler to move under extreme conditions.

SPROCKET

To drive the chain assembly around the track frame, there needs to be a sprocket. The sprocket is the output of a track-type machine's power train and transfers rotary torque

TECH TIP

Track tension can be measured in a couple of different ways depending on the type of machine. For most track dozers and loaders, the most common method is to check the sag across the top of the grouser bars along the top section of track. The machine should be allowed to coast to a stop on firm level ground before measuring track tension. A straightedge or string line is placed across the top of the track between the sprocket and the carrier roller or the idler and the carrier roller. The track sag is measured at the lowest grouser bar. A typical track sag specification for a small- to medium-size track machine is 2 in. ± ½ in. ● SEE FIGURE 23–14 for how track tension is measured.

To check track tension on an excavator, some manufacturers will recommend lifting the track frame in the air by pushing down with the bucket. The track should then be rotated at least two complete revolutions and slowly brought to a stop. Track sag is then measured from the bottom of the track frame to the bottom of the track pad that is at the lowest point.

Tension can change depending on the type of material machine is working in and must be monitored closely by the operator.

If track tension is too great, the chain bushing to pin wear will be increased dramatically because a lot of extra strain is placed on all pin to bushing joints. One example is for a small dozer that has track tension force measured at the idler as 800 lb when the chain is correctly adjusted versus 5600 lb if the chain is too tight. Tracks that are too tight will also place a greater load on the final drives and drivetrain in general, which will also increase fuel consumption.

If track tension is too loose, there will be excessive movements in the chain assembly as the tracks move between the carrier rollers and the sprocket and idler. This will also cause accelerated wear. If track tension gets very loose, there is the added chance the tracks may fall off.

There will be warnings related to adjusting tracks because there are extreme pressures and forces involved. *Never* remove the track adjuster valve unless the track is removed from the track frame.

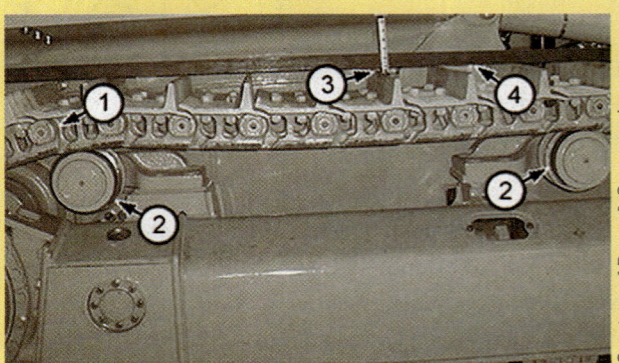

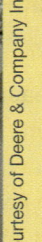

FIGURE 23–14 Measuring track tension.

---

into linear motion to the chain and pads to move the machine. The track-type machine's sprocket can be compared to the front bicycle chain sprocket, which transfers torque into the bicycle's chain. Some sprockets are one-piece assemblies that bolt on to the final drive output shaft or hub. These are usually found on smaller machines. Other sprockets use a one-piece ring that bolts onto a drive hub. The sprocket ring can be replaced when its teeth are worn down. Larger track machines will use multi-piece sprocket segments with four to five teeth on each segment. Each segment is replaceable when they wear down. Visual inspection of bolt-on sprocket fasteners is critical. These fasteners are subject to high forces, and if they loosen or fall out, it will damage the output hub of the final drive. This will result in an expensive repair.

FIGURE 23–15 Loose/missing sprocket bolts.

- **SEE FIGURE 23–15** for some loose and missing sprocket segment bolts.

Monitoring sprocket wear is a good way to monitor the wear rate of the machine's undercarriage.

CHAIN ASSEMBLY

The track chain assembly is similar to the chain on a bicycle. The big difference being that a track machine's chain has track pads bolted to it that will transfer chain movement into machine movement as the sprockets pulls the chain into it and the track pads dig into the work surface. ● **SEE FIGURE 23–16** for a track chain with no pads.

The chain assembly is a series of link assemblies that are joined together and are able to pivot at their joints.

FIGURE 23–16 Track chain.

Track chains will sustain internal and external wear as the machine travels. Internal wear occurs between the pin and bushing while external wear occurs in several places. The bushings wear on the sprocket as the chain is driven by it while the chain link rails wear on the idlers and rollers

There are three main types of chain assemblies:

NON-LUBRICATED CHAIN The simplest style is called the sealed (non-lubricated) track assembly. Each chain link assembly consists of two links (left and right), a pin, a bushing, and a set of seals. The left and right links are pressed onto a pin and bushing. The links' bushing will go over the next links pin and so on until the chain is one continuous length. The chain links are assembled in a continuous sequence to make a length of chain long enough to form a loop around the idlers and sprocket. The pin is a loose fit inside the bushing to allow the chain to follow the radius of the idlers and sprocket. You could think of the pin and bushing as a hinge. There are seals between the pins and bushings to keep dirt out from the wear areas of the pin and bushing. The seals are metal rings that have pressure applied to one side of them with Belleville washers. ● **SEE FIGURE 23–17** for a sealed track.

At each end of the chain, there will be one half of a **master link** that joins together to complete the loop. Master links are described later in this chapter.

SEALED AND LUBRICATED CHAIN The most common chain style is called seal and lubricated, and it has the same basic components of the sealed style. The big difference is that each pin and bushing joint is lubricated with a small amount of oil to reduce wear. The pins are cross-drilled through their center, and this serves as a reservoir for the lubricating oil for each joint. The links have polyurethane seals at each joint that are used to keep the oil where it needs to be. This style of chain will last approximately 50% longer than non-lubricated chains

> ### TECH TIP
>
> **Track pitch** is the measurement taken between the center points of two track pins. This is how internal wear is measured between the outside of the pin and the inside of the bushing. As this internal wear increases, track pitch increases. This will also increase sprocket wear because the bushings will be farther apart.

UNDERCARRIAGE SYSTEMS 653

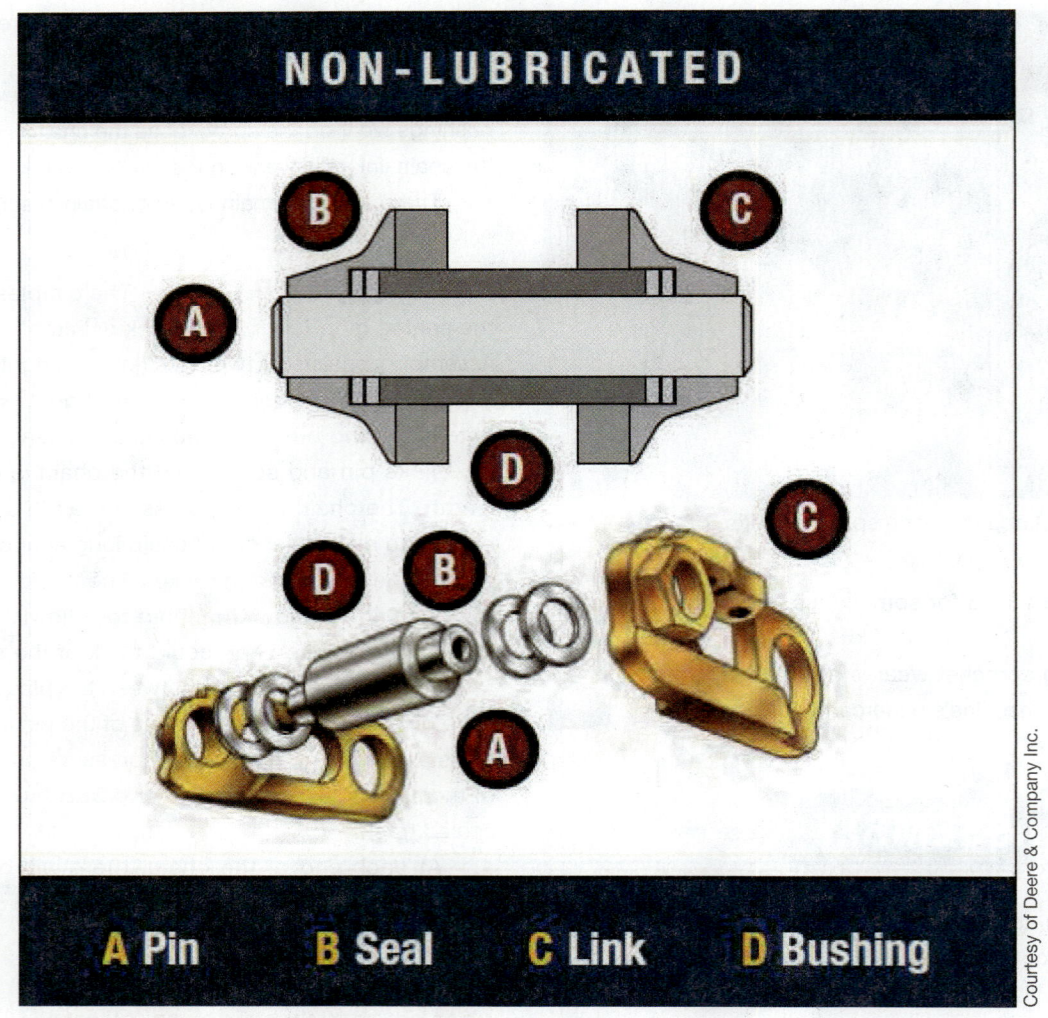

FIGURE 23–17 Sealed track.

and will reduce sprocket wear because the track pitch won't change because of the reduced wear between the pins and bushings. ● **SEE FIGURE 23–18** for a **sealed and lubricated track link**.

ROTATING BUSHING CHAIN A new type of chain assembly has recently been put on machines that have rotating bushings. There is a sealed and lubricated pin and bushing assembly that holds both chain rails together and will greatly reduce external bushing wear. Special tools are required to service and repair this type of chain.

MASTER LINK

To join a track assembly and make it a continuous loop, there needs to be a master link and again there are several styles of master links. When removing and installing a track, the master link needs to be disassembled. This is called splitting a track.

The simplest master link has one pin that is slightly undersized to fit loosely through the master link and goes through the bushing. It is held in place by a cotter pin or snap ring. This type of master link is mainly found on excavators. ● **SEE FIGURE 23–19** for a master pin that is retained with a cotter pin.

The second type of master link is one that has a master pin that is slightly smaller than all other pins but must be pressed in with a special tool called a track press. It will be pressed into the rails of the joining link. It is identified by a small dimple in the end of the pin. ● **SEE FIGURE 23–20** for a master pin.

The third type of master link is a two-piece split link assembly. The two pieces of this link have a matching pattern that is joined with fasteners to keep the chain in one continuous loop.

To join this master link together, there are four bolts that go down through the track pad and clamp the master link together. There are a few variations of this split link, and one style is a serrated master link. ● **SEE FIGURE 23–21** for a serrated master link.

LUBRICATED — SEALED

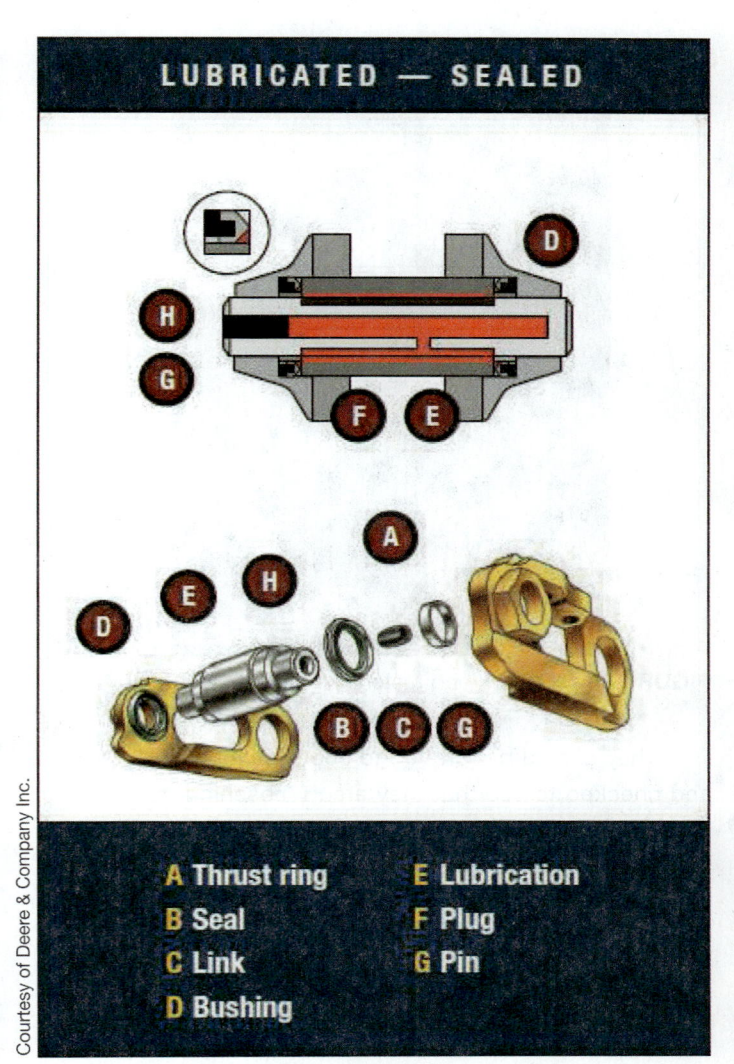

- **A** Thrust ring
- **B** Seal
- **C** Link
- **D** Bushing
- **E** Lubrication
- **F** Plug
- **G** Pin

FIGURE 23–18 Sealed and lubricated track link.

FIGURE 23–19 Master pin with a cotter pin.

FIGURE 23–20 Master pin with dimple.

FIGURE 23–21 Serrated master link.

UNDERCARRIAGE SYSTEMS **655**

TECH TIP

One of the least popular repairs a field service technician could be required to perform is to install a track that has come off while a machine is working. The cause of this is usually a combination of excessively worn undercarriage, loose track tension, and an operator that isn't paying attention to the tracks as the machine travels.

The first part of this job is to try and get the machine to a solid and level area. This may be the hardest part of the job because chances are the track has come off in the worst area of the job site. The next part is to recover the track, separate it at the master link, and lay it out so the machine can get back on it.

If the machine is an excavator, the bucket can be used to assist with the job of laying out the track; otherwise you will likely need a second machine or use the crane on the field service truck if possible. The machine's idler should be retracted as far as possible next and then the track can be fed onto the machine. Once the two ends of the master link are joined, the track tension can be adjusted and the machine can go back to work providing the reason why the track came off has been resolved.

Extra care must be taken when joining a two-piece master link to be sure that the mating surfaces are clean and free of rust. The threads in the bottom half of the master link should be cleaned out as well.

TRACK PADS

There are many different types and sizes of track pads that can be found on track-type machines. The ideal width of the track pad used is one that is as narrow as possible but will still give the machine the necessary flotation required. Track pads can also be called track shoes.

Each pad is fastened to the track chain with four fasteners. Special fastening hardware is required for fastening track pads to chains. The nuts are square and fit into a slot in the track chain. The bolts have a long shoulder on them to go through the pad and chain. ● **SEE FIGURE 23–22** for an example of track pad fasteners.

FIGURE 23–22 Track pad fastener.

They should be torqued to specification when installed and checked to see that they aren't loosening on a regular basis. If a track pad fastener becomes loose and stays undetected, the holes in the track chains can become elongated and damage to the chain could be irreversible.

The two main determining factors on the style of track pad used are the type of machine and the most common conditions the machine works in. Machines with blades (dozers) will have more aggressive pads because they need to dig into the terrain the machine is working on to keep pushing the load in front of the blade with minimum slippage. ● **SEE FIGURE 23–23** for a variety of track pads.

These pads will likely have only one grouser bar per pad. The pads can be light–duty pads, which are made from thinner gauge metal and only meant for soft rock free conditions, or extreme heavy-duty pads, which are made from very thick material and made to withstand broken rock conditions and anything in between. Low ground pressure machines need to have wide track pads to spread the machine's weight across a wider surface area. The downfall of this is the wider pads will place more twisting stress on the track chains and reduce their life compared to the same chain with a standard width pad on the chain. ● **SEE FIGURE 23–24** for a chain with wide pads.

Track machines that don't have blades such as track loaders and excavators will usually have track pads that have more than one grouser bar. There are exceptions to this, but generally these machines don't need to be as aggressive because a track loader can use its hydraulics to fill its bucket

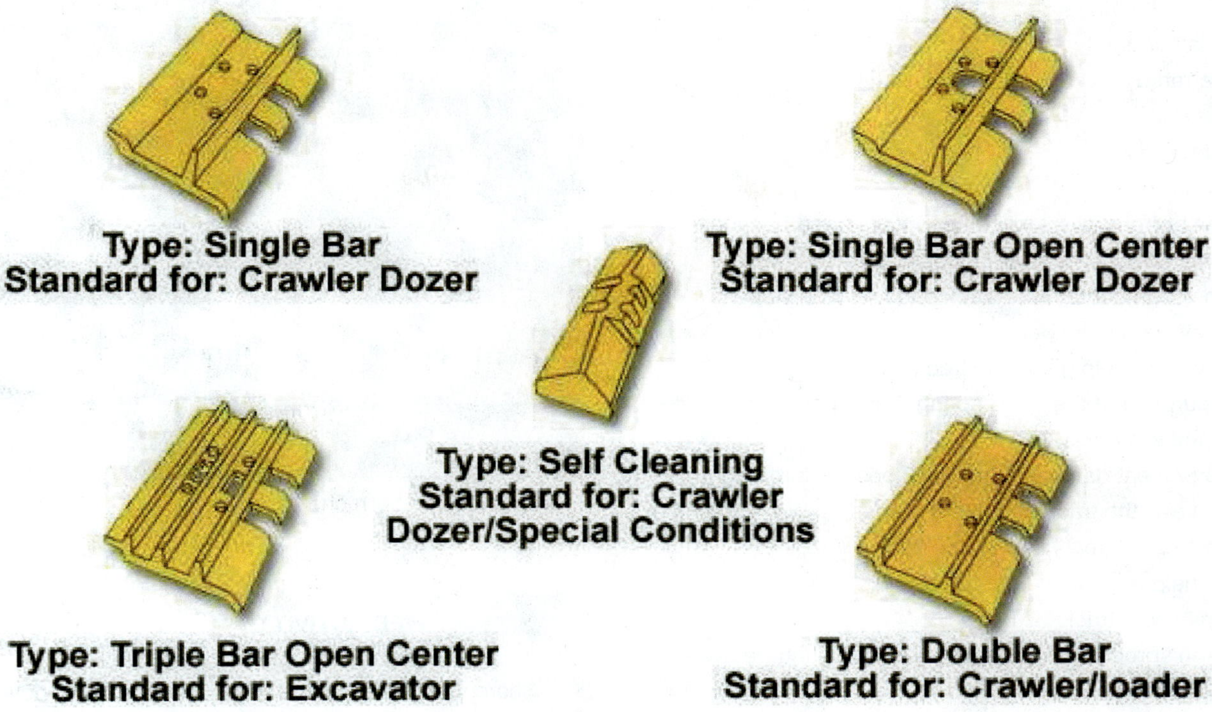

Type: Single Bar
Standard for: Crawler Dozer

Type: Single Bar Open Center
Standard for: Crawler Dozer

Type: Self Cleaning
Standard for: Crawler Dozer/Special Conditions

Type: Triple Bar Open Center
Standard for: Excavator

Type: Double Bar
Standard for: Crawler/loader

Courtesy of Deere & Company Inc.

FIGURE 23–23 Variety of track pads.

FIGURE 23–24 Chain with wide pads.

and an excavator is mostly stationary when it's working. Track loaders will also be turning a lot so lower grouser bars will allow easier turning.

For larger machines that have worn down grouser bars, it is possible to get new grouser bars welded on to the pad to avoid scrapping the pad.

There are special application track pads that can be installed to any track chain. Some examples are chopper pads that are used on machines working in landfills; swamp pads to help keep the machine afloat; pads with rubber inserts for working on pavement.

Whenever new track pads or old pads are installed on new chains, the paint must be removed at the mating surfaces. If not, the pads will become loose eventually and damage the pads, chains, fasteners, or a combination of all these components.

UNDERCARRIAGE SYSTEMS 657

Progress Check

1. This is *not* one type of master link:
   a. serrated
   b. weld in
   c. pressed in
   d. cotter pin
2. Some bottom rollers will be double flanged because:
   a. they will have less rolling resistance
   b. they can contact the track pads on side slopes
   c. they will last longer
   d. they can help guide the track better
3. The purpose of the recoil spring that is part of an undercarriage system is:
   a. to prevent damage if the track becomes too tight
   b. to keep the track tensioned properly
   c. to help the machine ride better
   d. to help push the idle out when splitting a track
4. Bolt-on sprocket segments have this advantage over-pressed on sprockets:
   a. they are much easier to replace
   b. they allow quick changing for changing work conditions
   c. fastener torque isn't important
   d. they can be reversed when worn 50%
5. If a machine has an equalizer bar, it will:
   a. ensure both tracks turn at the same speed during straight travel
   b. make removing the track frame easier
   c. allow the track frames to pivot slightly
   d. keep the track frames rigid at all times

UNDERCARRIAGE WEAR

As mentioned earlier, undercarriage is an expensive part of running costs for track-type machines. The initial purchase price of a track-type machine includes the cost of undercarriage at approximately 20% of the total. Some very large track-type machines can have undercarriage replacement costs of well over $250,000. Many large mining shovels will travel on their tracks only when absolutely necessary to reduce wear and costs. ●SEE FIGURE 23–25 for an extremely worn set of chains.

All moving undercarriage components will wear, and by minimizing this wear, a lot of money can be saved. The main three factors that contribute to undercarriage wear are contact, load, and motion. Any combination of at least two of these factors will create wear, and as the level of each increases, so too does the rate of wear. The cause of these factors is mostly in the hands of the operator and a result of the working conditions the machine is in. Educating an operator to try to reduce these factors will go a long way to increasing undercarriage life and saving money.

An example of undercarriage contact occurs when a track link bushing external surface contacts the sprocket as the sprocket turns.

The load between moving undercarriage parts is determined by the resistance of the load the machine is pushing and the type of material the machine is working in. The load between pins and bushings and other components is also affected by the track tension.

There will be relative motion between the bushing and the sprocket when the bushing initially contacts the sprocket, but the motion stops until the bushing is leaving the sprocket.

External wear surfaces of moving undercarriage components are hardened to make them last longer and components such as roller shells, idler wear surfaces, track links, track bushings, and sprockets will be hardened. Ideally, two different components that run on each other will be hardened to the same value so that they wear evenly.

FIGURE 23–25 Worn chains.

WEB ACTIVITY

Search the Internet for a video explanation of how contact load and motion work together to create wear.

MEASURING UNDERCARRIAGE

There is no escaping the fact that every time you drive your vehicle, you are wearing down the tires. Several factors such as how you drive your car, the road conditions, how far you drive, how much weight is in it, and so on will determine how much wear accumulates on the tires. Similarly, undercarriage components that move and contact other components or contact the ground will wear. The factors that affect the amount of wear have been discussed previously. Measuring tire wear is a fairly simple procedure unlike measuring the wear of undercarriage components.

Measuring undercarriage is an important service procedure and requires some special tooling. Ideally, undercarriage will get replaced when it is 100% worn. This means the components are worn to the point of the last bit of hardness remaining on the wear surfaces. If it is run past this point, then the components will wear rapidly and there is no chance of reconditioning it. The undercarriage is said to be on run out then because the components are being run to destruction.

At 100% wear, the chains can be sent to an undercarriage shop where each link is disassembled and the pins and bushings are turned 180°. They are then reassembled and installed back onto the machine. This procedure is called a pin and bushing turn, and it will give the undercarriage a second life. Once a set of chains has had a pin and bushing turn, they are left on the machine as long as possible.

UNDERCARRIAGE MEASURING TOOLS For many years, the standard tooling to measure undercarriage wear was a toolkit that contained a set of calipers, a steel ruler, a depth gauge, and a tape measure. These tools are still effective and don't need battery power. ● SEE FIGURE 23–26 for an example of nonelectronic undercarriage measuring tools.

A few years ago, machine manufacturers started making ultrasonic measuring equipment available for their dealers to assess undercarriage wear. These instruments require some calibration and practice using but can be connected to a laptop to make entering data to an electronic spreadsheet easy.

The instrument has a transducer probe and a handheld keypad with display. It will accurately measure the thickness of material that the probe is applied to. A new file is started in the handheld unit for the undercarriage to be measured. To measure a component accurately, all dirt and rust must first be removed to expose the metal surface. A couplant fluid is then applied to where the transducer probe is placed. The probe is then firmly held in one spot and the thickness of the material it is resting on is shown on the display. With the push of a button, the value is entered into the file for this machine. Undercarriage components such as rollers, idlers, track chain, track bushings, and grouser bars can be measured this way.

Records can be maintained to create a history of undercarriage wear for any machine. This can become part of the machine's service and repair log and is part of a valuable cost tracking system for the machine owner.

UNDERCARRIAGE COMPONENTS MEASURED The following is a list of undercarriage components that can be measured with an ultrasonic tool when an undercarriage evaluation is performed on a track machine.

TRACK LINK. The link is measured to see how much the rail is worn down. Three links are measured on each track, and the one with the highest wear is recorded. ● SEE FIGURE 23–27 for a track link being measured.

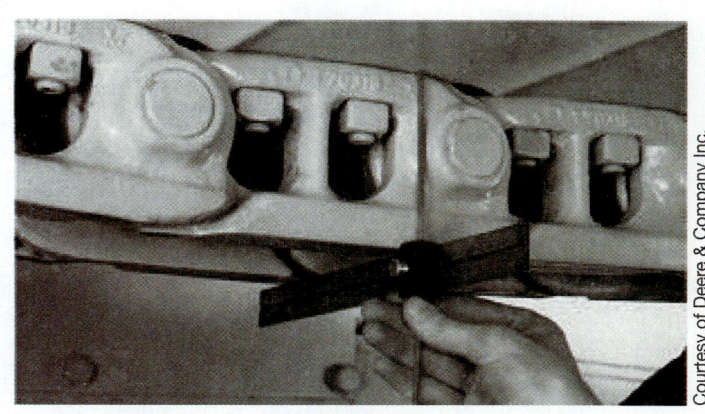

FIGURE 23–26 Undercarriage measuring tools.

FIGURE 23–27 Track link measured.

UNDERCARRIAGE SYSTEMS 659

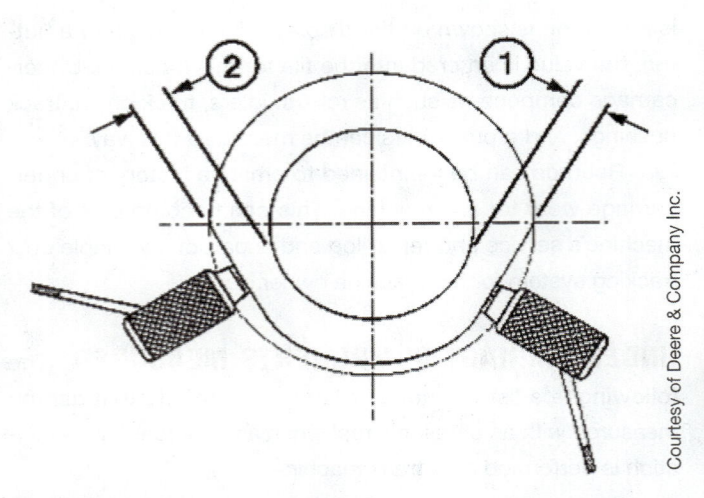

FIGURE 23–28 Track bushing being measured.

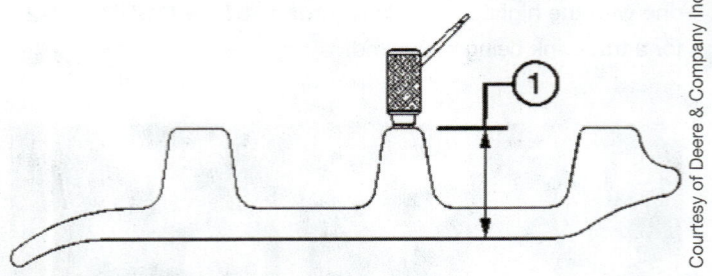

FIGURE 23–29 Grouser bar measurement.

TRACK CHAIN BUSHING. The ultrasonic tool is able to measure the track bushings and measure the internal and external wear that has occurred. ● SEE FIGURE 23–28 for how a track bushing is measure.

GROUSER BAR HEIGHT. Grouser bar height is easily measured with an ultrasonic tool. Most times track pads can be reused, or for larger pads new grouser bars can be welded on to recondition the pads. ● SEE FIGURE 23–29 for how grouser bar height is measured.

TRACK AND CARRIER ROLLERS. Roller shell thickness can be measured with an ultrasonic tool, but an accurate reading may be hard to obtain because it has to be measured in the highest wear point. The transducer should be placed on the most worn part of the roller. ● SEE FIGURE 23–30 for how a roller is measured.

All the previously mentioned components can also be measured with mechanical measuring tools and the following must be measured that way:

IDLER FLANGE HEIGHT. To measure the wear on idler flanges, a depth gauge is used. The flanges contact the link rails and are subject to wear.

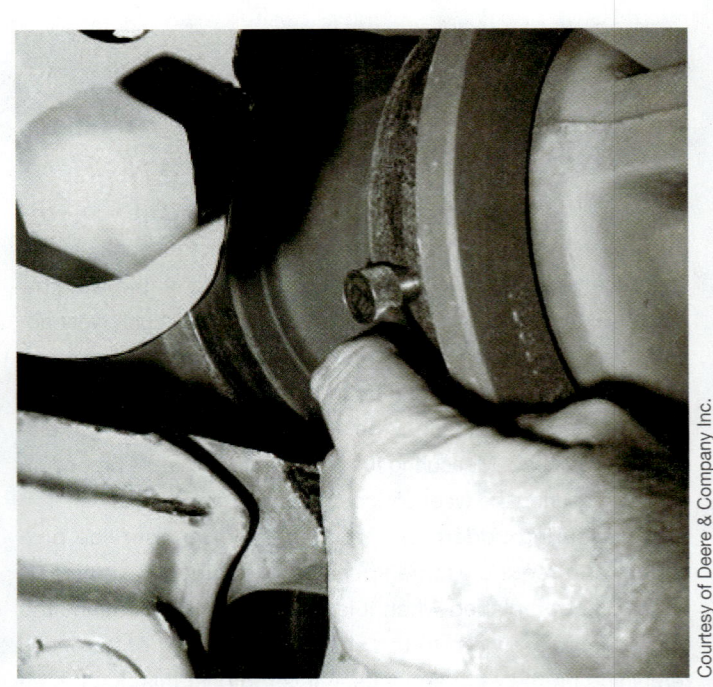

FIGURE 23–30 Measuring a roller.

TRACK CHAIN PITCH. Track pitch is the distance between the center of track pins and is usually measured over a set of three or four link assemblies. As track chains rotate, there is wear created between the pin and bushing. The only way to measure this wear is to measure the track pitch. To measure chain pitch, the track must be tightened temporarily. An easy way to do this is to put an old master pin in between two sprocket teeth and slowly rotate the sprocket to take all slack out of the chain. A measurement can then be taken across a set of three or four track pins. This should be done for at least three different sections of the track to get an accurate picture of the total chain wear. The longest track pitch measurement should be recorded.

SPROCKET. Sprockets can be measured but not very accurately. Their replacement is based on track pitch measurement because sprocket wear mirrors track pitch growth.

When all undercarriage components are new, the sprocket teeth spacing will match the track pitch. As the pitch changes, the sprocket teeth will wear accordingly. The sprocket wear is a visual indication of a change in track pitch. ● SEE FIGURE 23–31 for how track pitch is measured.

Once all components are measured, an estimate of the percentage that the components are worn is calculated and can be converted into hours. This will be part of a predictive maintenance program and allow the machine owner to schedule undercarriage repairs as close to 100% wear as possible.

RUBBER TRACKS

Some track-type machines are originally equipped with rubber tracks. You will most likely see rubber tracks on mini-excavators, skid steer loaders, curb forming machines, and agricultural tractors.

Rubber tracks are a one-piece seamless design and are constructed in a similar manner to heavy-duty tires. Heavy steel wire makes the foundation of the track, and layers of cords are applied on top. Metal drive lugs are incorporated

Machines with rubber tracks will need to keep the tracks tensioned differently than steel track machines. Small machines will use a threaded rod to adjust tension with a recoil spring to allow idler movement under shock loads.

For larger machines, they typically use a hydraulic cylinder with an accumulator to move an idler out and keep the track tensioned that way.

Rubber track removal can be more difficult than steel tracks because the track can't be separated. ● SEE FIGURE 23–32 for a machine with rubber tracks.

REMOVING AND INSTALLING TRACKS Part of being an HDET includes removing and installing tracks. Some field service technicians will be required to install a track that has come off of a machine. There are several different ways to install tracks on machines, and some variations come down to personal preference. The type of machine will also determine the method chosen to remove and install tracks.

DIAGNOSING UNDERCARRIAGE PROBLEMS

The following is a list of operator complaints and possible causes that could be related to undercarriage components:

1. Excessive squealing (track chains are dry, seized roller, dry roller shaft, dry idler shaft)
2. Machine won't turn properly (one track too tight, one track too loose)
3. Heavy clunking noise (track too loose and sprocket is skipping in chain)
4. Track won't stay tight (track adjuster leaking)
5. Track is snapping (tension too tight for the working conditions)

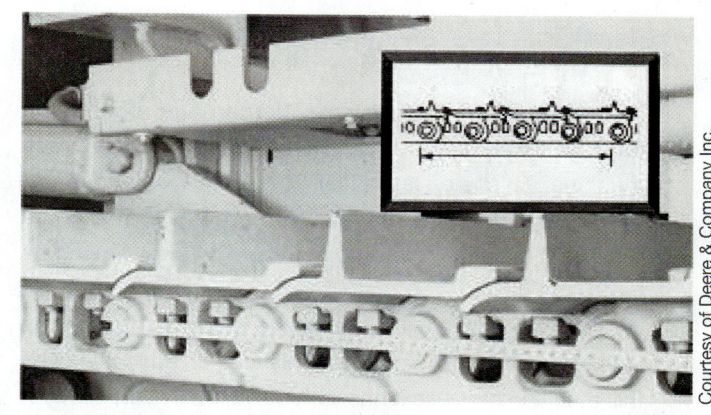

FIGURE 23–31 Track pitch measure.

FIGURE 23–32 Machine with rubber tracks.

6. Track spins too easy (grouser bars worn)
7. Track falls off (track too loose, excessive chain wear)
8. Can't loosen track tension (seized idler yoke assembly)
9. Can't tighten track (recoil spring broken, track adjuster leak)

Many undercarriage problems can be diagnosed with a thorough visual inspection and any signs of unusual wear, loose fasteners, oil leaks, or discoloration related to any undercarriage component will indicate a problem. If the machine is still operating, any signs of excessive heat would indicate either a lack of oil for rotating parts or misalignment.

Progress Check

6. This is *not* one of the three major factors related to undercarriage component wear:
   a. load
   b. motion
   c. contact
   d. age

CASE STUDY

We'll now take a look at an undercarriage recondition job that takes place on a conventional oval type-track machine. This is a Caterpillar 307 excavator whose undercarriage was left to wear to the point that it could no longer work in muddy conditions without having its tracks fall off.

The job starts with getting the machine blocked up with its tracks off the ground. This was an easy task with this machine because it has a blade. Its own hydraulic system lifted the machine up and hardwood cribbing placed under the main frame will keep it secure.

As is the case with most excavators, the track pads can be reused. At this point, the track pads are removed from the chains.

The next task is to release the track tension to get the track as loose as possible. On this and many other machines, the release valve is loosened slightly and the weight of the tracks will move the idler back into the track frame. If this doesn't work, a piece of wood can be placed between the sprocket and chain, and with the valve loose if the sprocket is reversed slowly, the idler will be pulled back.

The master pin should get positioned just past the top of the idler toward the front of the machine. This machine has a simple master pin that is loosely held in place by a cotter pin. When the master pin is removed, the track will separate, and it can then be run off the top roller and sprocket with by turning the sprocket.

● SEE FIGURE 23–33 for one track being removed from an excavator.

The rollers, idler, and sprocket were then removed.

All mating surfaces should be cleaned up to remove any rust or paint and the threaded holes in the track frame should have a tap run in and out of them.

It is usually good practice to replace all fasteners when installing new undercarriage.

All rollers, sprockets, and idlers are installed and torque to specification.

The new chain is laid out on the floor and slid under the track frame. One end is fed onto the sprocket to go up, around, and over the top of the track frame. Care must be taken to feed the proper end on first. The wide end of the link usually goes on first. As the sprocket is slowly turned, the chain is lifted and guided over the carrier rollers and eventually onto the top of the idler. When the other end of the chain is close to going under the idler, the sprocket is stopped. The chain on the floor is lifted up and blocked at about the 4 o'clock position (if you looked at it from the outside of the track frame). With the idler back as far as it will go, the master link end of top of the chain should reach the master link end of the bottom chain. The master link is now joined with the master pin.

The same process is repeated for the opposite track.

The track pads can now be installed. Once again all paint and rust is removed from mating surfaces first and new fasteners should be used.

The machine can now be let down and the track can be adjusted. Whenever new undercarriage is installed on a machine, it's always a good idea to check it visually after a few hours of operation.

FIGURE 23–33 One track being removed from an excavator.

---

7. This is *not* one common measurement taken when assessing undercarriage component wear:
   a. idler shaft
   b. bushing outside diameter
   c. grouser bar height
   d. track pitch

8. If you are measuring undercarriage with an electronic device, you would be using:
   a. a transducer probe
   b. a laptop
   c. a pressure transducer
   d. a thermal image camera

9. A set of rubber tracks will have this material as its foundation:
   a. rubber
   b. heavy steel wire
   c. Kevlar
   d. carbon fiber

10. If a set of undercarriage is _____ % worn, it will be ready for replacement:
    a. 10
    b. 50
    c. 75
    d. 100

WEB ACTIVITY

Search the Web for videos of tracks being removed and installed. You should notice there are different methods. Also look for the proper PPE being worn. You will likely find some backyard mechanics and questionable techniques.

SHOP ACTIVITY

- Go to a track-type machine and inspect the undercarriage. Are there any signs of visual wear? Check the track tension. Is it correct? Loosen the tension and adjust to specification following the machine's service information.
- What type of master link does the machine have?
- How many bottom rollers are single flange and how many are double?

SUMMARY

1. Undercarriage components are heavy and proper lifting techniques and equipment must be used to prevent injury.
2. Undercarriage accounts for 20% of a track machine's purchase price and a large portion of its operating cost.
3. The two main arrangements of undercarriage are oval track and high drive.
4. Track frames are the backbone of the machine's undercarriage because most of the undercarriage components will be mounted to it.
5. Track frames will be able to pivot on most track-type machines and some will have an equalizer bar that connects the front of the track frames and allow a few degrees of movement.
6. Equalizer bar pivot pins need regular greasing, and if they become dry, an expensive repair bill will result.
7. Some track frames are mounted to the machine's main frame on pivot shafts. These are usually lubricated with oil and need to be checked regularly.
8. Track rollers are mounted to the bottom of the track frame and support the weight of the machine. They can be single or double flange.
9. Carrier rollers support the track as it moves between the idler and sprocket on to top side of the track frame.
10. Idlers are large steel wheels that guide the track around the front of an oval undercarriage or front and rear of a high-drive undercarriage.
11. Track adjusters push on the front idler to maintain track tension. They use grease as a medium, and as it is pumped into the adjuster, a rod moves out.
12. Track tension can be measured in a few different ways. By measuring track sag, the correct tension can be checked and adjusted if necessary.
13. Recoil springs are very heavy coil springs that allow idler movement if something gets caught in the track chain. These are extremely strong springs that get pretensioned and must be handled with extreme caution.
14. Sprockets are the output of the machine's drivetrain that engage with the track chain. There are several types of sprockets, but most have bolt-on replaceable wear segments.
15. Track chain is similar to a bicycle chain but only much heavier. Three types of chain are sealed, sealed and lubricated, and rotating bushing.
16. Chains need to have master links to enable HDETs to remove and install the tracks. Three types are loose pin with cotter pin, pressed in master pin, and serrated master link.
17. Track pads come in many variations to suit both the type of machine they are on and the type of work conditions they will be used for.
18. Undercarriage wear needs to be monitored closely so some parts don't become nonserviceable.
19. To measure undercarriage components, there are a variety of tools/instruments available. These can be electronic with the capability to directly enter data to a software program.
20. Some machines will have rubber tracks that are one-piece assemblies.
21. Undercarriage repairs are a common task that most HDETs will become familiar with at some point in their career.

chapter 24
OPERATOR'S STATION

LEARNING OBJECTIVES

After reading this chapter, the student should be able to:

1. Describe the inspection procedure for ROPS/FOPS components.
2. Describe how a mechanical gauge works.
3. Describe how an electric gauge works.
4. Explain the operating principles of a mobile HVAC system.
5. Identify the main components of most air-conditioning systems.
6. List the inputs and outputs to the ECM for a climate control HVAC system.
7. Explain the procedure to identify a cab heater problem.
8. Describe the steps taken to identify the location of an air-conditioning system leak.

KEY TERMS

Accumulator 677
Compressor 677
Electronic display 665
Evacuate 687
FOPS 672
R-134 678
Receiver drier 677
Recharge 688
Recover 688
ROPS 672

INTRODUCTION

SAFETY FIRST Some safety concerns related to operator stations are as follows:

- Operator safety features need to be checked for integrity, such as the rollover protective structure needs to be structurally sound; seat belts need to be checked on a regular basis; machine interlock systems; windows and mirrors need to be in good condition; if a machine is equipped with supplemental vision systems (cameras) or object warning systems (radar), these systems must be verified to be working correctly before the machine is released back to the operator.
- When working on the outside of a machine's operator station, you should be careful to not fall off the machine. For larger machines, you may need to wear a fall arrest apparatus.
- Air-conditioning systems can produce extreme temperatures. There is always a risk of frostbite or burns when working near an operating air-conditioning system.

NEED TO KNOW

Operator stations, also more commonly called cabs, have changed a great deal in the last 50 years of machine development. Older machines didn't have any operator protection from machine rollover, bad weather, or falling objects. The machine controls started as big and heavy levers and pedals that were connected to mechanical linkages that transferred operator effort into power train and hydraulic system control movement. Operator seats were lightly padded or not padded and not very comfortable.

Machine system monitoring consisted of the operator having to keep an eye on mechanical gauges if there were any. All these factors meant long days in the operator seat were hot or cold; dusty, windy, wet, and generally tiring, unhealthy, and unsafe. ● **SEE FIGURE 24–1** for an operator's station of an older machine.

The need to make operators safer and more comfortable to keep them being more productive for longer periods of time has changed the design and complexity of operator stations over the years. Creature comforts such as heating and air conditioning, sealed cabs, seats with suspension, music-playing devices, and smaller, lighter fingertip controls have made operator cabs much more user friendly.

FIGURE 24–1 Older machine's operator station.

Increasing visibility out of the cab is always a desire and challenge for designers. The better an operator is able to see what the machine is doing, the more productive the machine will be. Awareness of other machines and people near the machine will also be improved, which increases the safety factor of the work site. To improve visibility, cab glass area has increased, which makes the efficiency of the machine's HVAC (heating, ventilation and air conditioning) system even more critical. A common feature of new machines is to have one or more cameras mounted around the perimeter of the machine with a display in the cab. This allows the operator to view hidden blind spots to reduce the chance of the machine contacting other machines, people, or objects. One roadblock to better visibility lately is the extra hardware that has been added to the low-emission diesel engines, which has added a lot of bulk to machines. Compare the profile of any engine compartment of two similar-size machines that are 10 years apart, and you'll see how machine design has changed.

Machine system's monitoring has also vastly improved with **electronic displays** and gauges being the norm now. Most machines can be equipped with a remote-monitoring system that can communicate a wide variety of information to owners and service personnel anywhere in the world.

MACHINE CONTROLS

The operator station is the interface between the operator and the machine. Older machine controls were mainly mechanical linkages that started with levers and pedals that connected to rod ends, rods, bell cranks, cables, and shafts. Motion was transferred through these mechanisms to engine, drivetrain, brake, steering, and hydraulic components. Many of these mechanisms required lubrication and adjustment regularly. ● **SEE FIGURE 24–2** for an older machine with mechanical controls.

FIGURE 24–2 Mechanical controls.

FIGURE 24–3 Hydraulic leak in a cab.

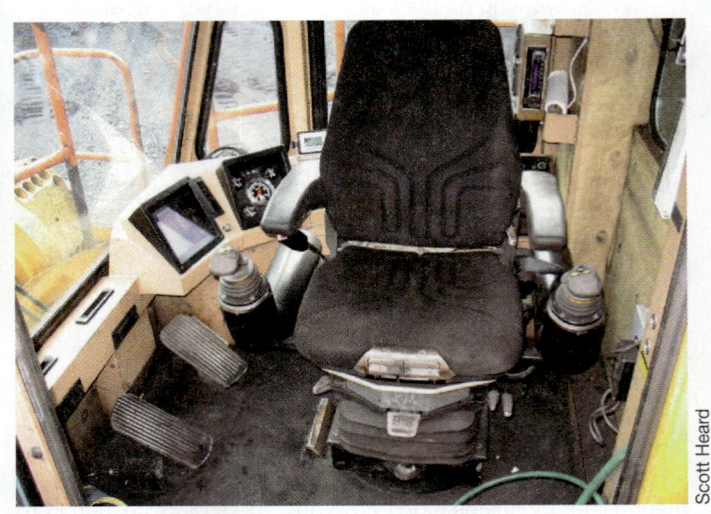

FIGURE 24–4 Electronic controls.

These mechanisms were replaced by maintenance-free components that would last longer. This included components that were made of higher-quality metals and other materials that would resist wear and corrosion better.

The next step was to replace mechanical controls for hydraulic systems with hydraulic or pilot controls. See Chapter 16 for more details on pilot controls. These controls required less effort for the operator to move and eliminated a lot of mechanical wear points. However, the possibility of hydraulic leaks in or near the cab was introduced. ● **SEE FIGURE 24–3** for a hydraulic leak in a cab.

Brake controls also upgraded from mechanical to hydraulic. See Chapter 18 for more details on hydraulic brake controls.

To get away from hydraulic leaks in cabs and decrease the effort required to manipulate a machine, the next step was to feature electric or electronic controls in cabs. Many machines now have controls for drivetrain, engine, and hydraulics that are electrical devices such as switches or dials that either directly control machine systems or are inputs to ECMs, which in turn send outputs out to the various machine systems. This is where the term *fly by wire* is borrowed from the airline industry and applies to newer heavy equipment. ● **SEE FIGURE 24–4** for a set of electronic operator controls.

See Chapter 9 for more details on ECM input devices. Many of these operator inputs require very little effort. For example, an operator could control the hydraulic system for a wheel loader that is capable of lifting 50 tons of material with one finger!

OPERATOR DISPLAYS/ MACHINE GAUGES

Throughout the day when a machine is working, its various systems on the machine need to be monitored to ensure that they stay within predetermined safe temperatures and pressures and there are no faults. There are different devices for doing this.

TECH TIP

Did you know some machines don't need an operator on the machine? Bobcat has a remote control system that allows an operator to run the machine from a safe distance away from the machine (up to 1500 ft). This may be necessary if the machine needs to work in hazardous environments that aren't safe for a human being.

There are other machines that can run on their own with the help of GPS or cell phone technology that links the machine to a remote command center. Machines such as haul truck, drills, track-type dozers, and loaders can also be operated without an operator.

The term *autonomous machine* is often used when referring to a machine that can function without an operator and refers to a smart machine that can learn the work site and function with very limited assistance. Search the Internet to see the latest developments on autonomous machines.

FIGURE 24–5 An example of mechanical gauges.

MECHANICAL GAUGES Older machines used mechanical gauges to monitor things such as engine coolant temperature and oil pressure. This type of gauge had a needle that moved mechanically across a range of colors or numbers to inform the operator as to what a system's fluid temperature, pressure, or level was. ● SEE FIGURE 24–5 for an example of mechanical gauges.

Mechanical pressure gauges need to have a line connected to them that is connected to the component that needs its pressure monitored. An engine oil pressure gauge, for example, would be fed oil pressure through a small hose that is connected to the main oil gallery of the engine. The pressure is then able to be sensed at the gauge. The gauge's needle moves as a result of the pressure acting on a thin tube that is normally curved. As more pressure is applied to the tube, it tries to straighten out and an attached linkage moves the needle. This is called a Bourdon Tube type of gauge. The numbers on the gauge will be calibrated to match the needle movement and give the operator an accurate reading of engine oil pressure.
● SEE FIGURE 24–6 for a Bourdon Tube–type pressure gauge.

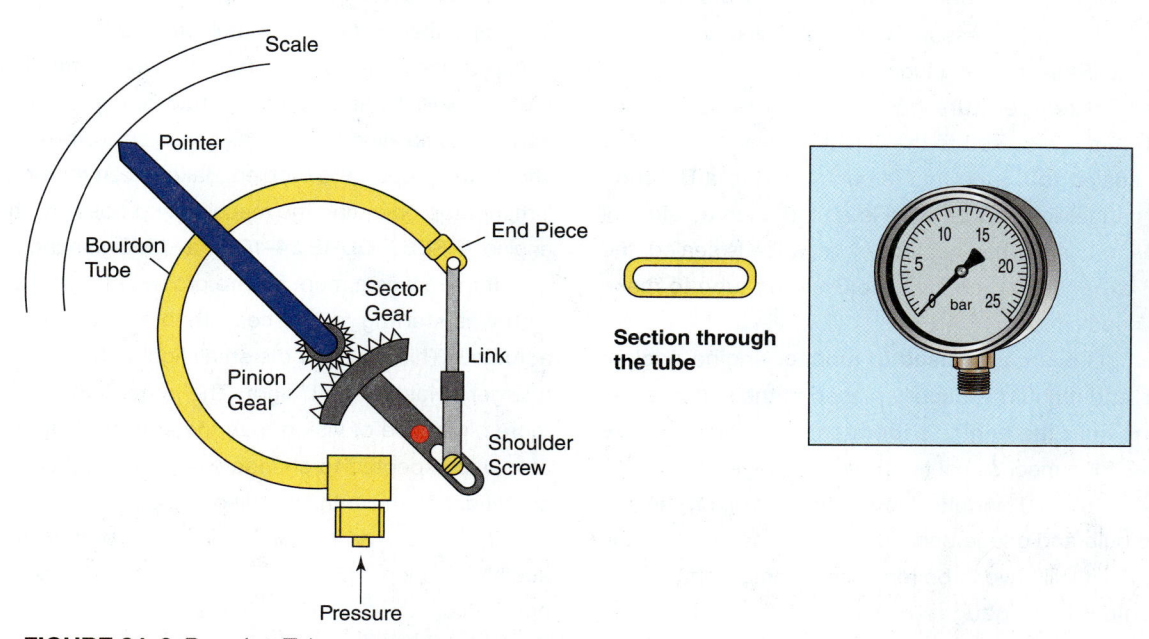

FIGURE 24–6 Bourdon Tube pressure gauge mechanism.

OPERATOR'S STATION **667**

FIGURE 24–7 Mechanical low-pressure fuel gauge.

FIGURE 24–8 Mechanical temperature gauge.

The line between the engine and the gauge needs to be protected and secured because a failure will cause a pressurized engine oil leak that could be disastrous.

Some low-pressure fuel systems have a mechanical gauge that displays fuel pressure. ● **SEE FIGURE 24–7** for a mechanical low-pressure fuel gauge.

Mechanical temperature gauges use a temperature-sensing bulb that is exposed to the fluid it is meant to monitor. The bulb is sealed to a tube that is connected to a Bourdon tube at the gauge. Inside that tube is a fluid such as alcohol that will expand easily when heated. As the tube is sealed, the heated fluid expands and acts on the Bourdon tube to make the gauge's needle move.

An example is a gauge used to monitor engine-coolant temperature and its bulb is usually placed in the water jacket of the engine near the engine's thermostat housing. ● **SEE FIGURE 24–8** for a mechanical temperature gauge.

Great care should be taken to not kink or pinch the line between the bulb and gauge because this will make the gauge inoperable, and it will have to be replaced as an assembly.

Mechanical level gauges use a float that raises and lowers with the fluid level, and this movement is transferred mechanically into needle movement. An example is a fuel tank that has a gauge on the side of it to allow the person filling the tank to be aware of how full the tank is getting.

OPERATOR-WARNING SYSTEMS It soon became obvious to machine manufacturers that an operator couldn't be expected to constantly monitor the gauges for a machine, and this led to adding a warning system that alerted the operator if operating parameters exceeded normal ranges. An alerting system would turn on a light or buzzer to grab the operator's attention and make them aware of what was happening with a machine system. Some machines have different levels of warning systems that change depending on the severity of the problem.

An example of this is Caterpillar's EMS (electronic-monitoring system). If there was a minor problem such as the engine air filter being plugged, then a small red light would start flashing on the EMS display. It was then the operator's responsibility to find out why it was flashing as soon as it was convenient. The operator should then report the problem, take action to correct the problem, or stop operating the machine and possibly shut down the engine. ● **SEE FIGURE 24–10** for an EMS warning display.

If there was a more severe problem such as the transmission was starting to overheat, then a higher level warning is actuated. This includes the small light in the EMS display and a larger "Master Fault" light. The larger light was placed in a more direct line of vision than the smaller light, and the operator was expected to change operating methods or work site conditions to make the condition stop.

The third level of warning included both lights and a loud audible warning. This buzzer or horn should indicate to the operator that immediate action is needed or severe machine damage will occur. An example of this is low engine oil pressure.

TECH TIP

Machines can be equipped with a different type of gauge that is sometimes called a Murphy Gauge. Although Murphy is a manufacturer of many different types of gauges and operator displays, their gauges are also part of a warning/shutdown system that are seen as a standard for this type of gauge.

These are normal-looking gauges, in that there is a moving needle but feature integrated switches that are used either to turn on a warning system or to shut the engine down if an extreme condition exists.

The switch closes when the gauge needle moves too far, and this will trigger a warning or disable the engine's fuel system, thereby causing it to shut down.

They are found on many stationary engine applications because these engines are usually only monitored briefly when someone starts them.

● SEE FIGURE 24–9 for a Murphy Gauge.

FIGURE 24–9 Murphy Gauge.

FIGURE 24–10 EMS display.

ELECTRIC GAUGES Electrically actuated gauges look similar to mechanically actuated versions, but the needle movement is created by changing electrical signals. These variable signals start at the sender unit (sensing device) as a varying resistance. The sender unit is grounded to the machine and the wire connecting the sender to the gauge provides a variable resistance to ground for the gauge assembly. The variable resistance will change current flow to the gauge, which in turn creates needle movement. Needle movement can be created by varying the current flow through two coils of wire in the gauge or by changing current flow through a bimetal strip that bends as it heats up or cools. The amount of current flowing through the strip will change its temperature.

An electric gauge has two balancing coils that are placed on either side of the gauge's needle. As the current flow through these coils changes, their magnetic strength changes, which in turn will make the needle move. System voltage is fed to one side of both coils. One coil's other end goes directly to ground while the second coil has the sender's variable resistance to ground placed in series with it. The variable ground value creates a variable current flow through one coil, which creates a variable strength magnetic field, and this is what will change the gauge needle's position. ● SEE FIGURE 24–11 for an illustration of a balancing coil type of electric gauge.

Diagnosing and repairing faulty electrical gauges will require a good understanding of basic Ohm's law principles and electrical troubleshooting procedures.

OPERATOR'S STATION

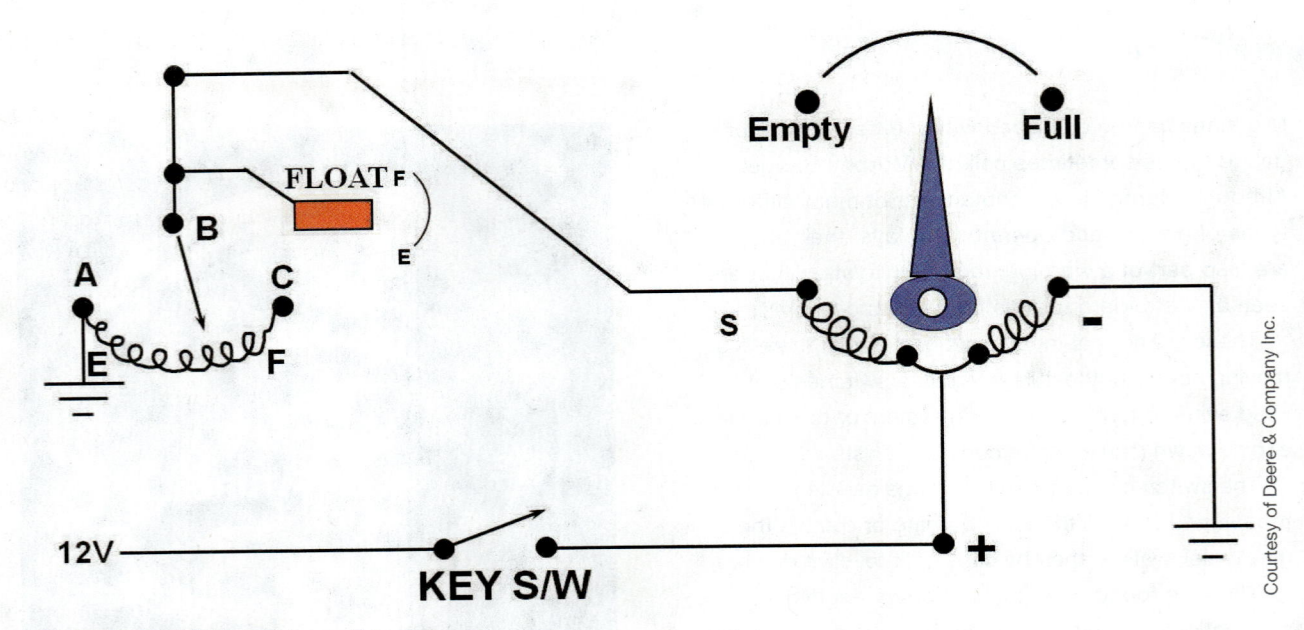

FIGURE 24–11 Balancing coil-type gauge.

The gauge circuit's sender is exposed to system fluid for pressure, temperature, level monitoring, or a sender that is close to a moving component for speed monitoring.

An electric pressure gauge sender has a diaphragm that is exposed to the system's fluid pressure. As the diaphragm is deflected by system pressure, there is a contact that moves across a variable resistance that goes to ground. As the sender's resistance value changes, a wire from the sender sends this changing resistance to ground value to the gauge assembly. A common type of electrical pressure gauge is one used to measure engine oil pressure. ● SEE FIGURE 24–12 for a diaphragm-type pressure sender.

An electric temperature gauge sender will change its resistance when the fluid it is exposed to changes in temperature. The most common type of temperature sender is a thermistor that contains a type of material that changes its resistance based on its temperature. There are two types of thermistors: NTC (negative temperature coefficient) that decrease resistance as temperature increases and PTC (positive temperature coefficient) that increases resistance as temperature increases.

Another less common type of temperature sender has a bi-metal strip that bends with temperature changes. As the strip bends, a contact moves across a variable resistance to ground, and this changing value is sent to the gauge assembly through a wire. This changing value creates gauge needle movement in the same manner as the electric pressure gauge.

An electric level gauge has its needle move as a reaction to the level sender. The sender has a float that raises and lowers with fluid level, and this action is changed into a variable resistance

FIGURE 24–12 Diaphragm-type sender.

to ground value through a contact that moves across a variable resistor. There is a single wire leaving the sender to carry the resistance value to the gauge. The movement of the gauge is the same as the previously explained pressure and temperature gauges.

ELECTRONIC GAUGES Electronic gauges are ones that use the output of an ECM to create needle movement. There are different types of senders that could send inputs to the ECM such as variable resistance, variable capacitance, hall effect, and piezo resistive. These senders were explained in detail in Chapter 9. Once the ECM receives input values, it compares this information to data that is stored in the ECM memory and an output is sent to an electronic gauge.

FIGURE 24–13 LCD gauge.

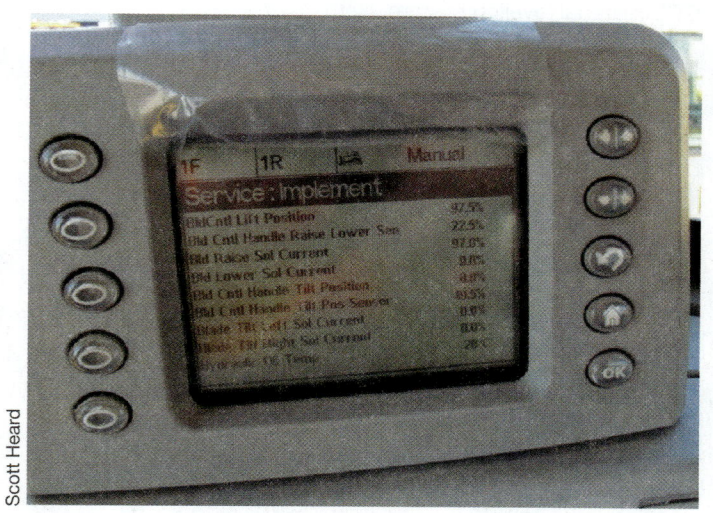

FIGURE 24–14 Electronic display.

There are different types of electronic gauges. Some are similar to electric gauges and have real moving needles while others are part of an LCD that could show one or more separate gauges. ● SEE FIGURE 24–13 for an LCD gauge arrangement.

Many machines today will feature an interactive display that allows the operator to scroll through different menus as well as view virtual gauges.

The menus could include pressure and temperature readings in numerical form, payload information, or be used as a display for blind spot cameras. ● SEE FIGURE 24–14 for an electronic display.

TELEMATIC SYSTEMS
The word *telematic* comes from telecommunication, which refers to wireless communication, and informatics, which refers to the exchange of information. Telematics refers directly to mobile equipment and the wireless exchange of information between the machine and a remote device.

Many of today's machines are equipped with electronic communication systems that allow the machine to be monitored from remote locations, allow partial control of the machine, or allow full control of the machine. These systems are sometimes called telematic systems.

These systems are based on cell phone or satellite communication systems that allow either one- or two-way transfer of data to and from a machine to anywhere in the world that has access to the Internet or is within range of a cell phone tower.

The machine needs certain hardware to be telematics capable such as a system controller that contains an ECM, modem and GPS chipset, an SIM plug to enable cellular communication, and a GSM/GPS antenna. To communicate with a machine that is equipped with the proper hardware, the machine owner needs the compatible software, a subscription, and an Internet connection. Subscriptions are usually for three- to five-year terms and are either sold outright or hidden in the cost of a machine.

There are typically different levels of communication that are possible with telematic systems. A base level would allow the machine owner to monitor where the machine is, if it is running, and some basic data such as fuel consumption and hours. There could also be a "geo fence" constructed that allows the owner to disallow the machine to start if the machine is outside of a selected area. If the machine is moved outside of the selected area, a text message can be sent to the owner. If the machine owner has access to the machine manufacturer's electronic service information system, then a remote link to that machine's CAN is available through the telematics system. This would allow viewing of live data, checking fault codes, and monitoring machine systems.

A higher level of service will send alerts to the machine owner for events such as active fault codes, new fault codes, or aborted DPF regenerations. Sometimes customers can even download and install flash files into the machine ECM when updates become available. Machine production reports can be created and downloaded on a regular basis. This can be a very valuable tool for machine management if used properly. ● SEE FIGURE 24–15 for one of the screens of John Deere's JD Link system.

Dealer access will allow all the previous functions plus they are able to change parameters that are stored in the machine's ECM such as increasing or decreasing engine horsepower.

OPERATOR'S STATION

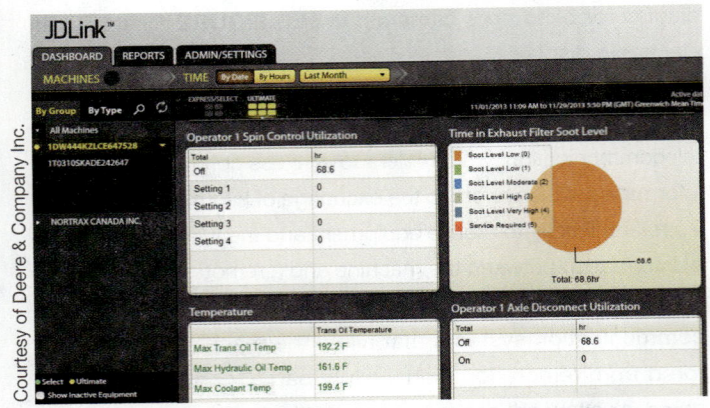

FIGURE 24–15 JD Link, a telematics information system.

FIGURE 24–16 GPS material leveling system.

Telematic systems are a huge benefit to machine owners with medium to large fleets because of the ability to track machines easily.

This was made even easier recently when most major manufacturers got together and standardized the format for the wireless transfer of commonly requested machine data. For example, machine location, machine hours, fuel consumption, whether the machine is running can all be monitored and downloaded to the same spreadsheet even though the machine owner might own several different makes of machines.

Close range wireless communication is also possible with Bluetooth technology.

GPS MATERIAL LEVELING SYSTEMS Many machines will have a GPS (global positioning system) system added onto their hydraulic control system to provide automatic material leveling capability. These systems can be installed at the factory or added on at the dealer.

Some examples of machines using this technology are graders, excavators, and dozers. ● **SEE FIGURE 24–16** for a GPS material leveling system.

OPERATOR PROTECTION— ROPS/FOPS/CANOPIES

Rollover protection systems (**ROPS**) are designed to keep the cab intact when the machine turns either on its side or completely upside down. The ROPS is usually an external frame that surrounds the cab at four corners. It can also be an integral part of the machines cab structure. It can be compared to a roll cage found in a racing vehicle. It is designed to support several times the weight of the machine.

It is fastened to the machine's main frame with large high-grade fasteners.

The ROPS must *never* have holes drilled in it, be welded on, or have any excessive heat applied to it. It should also be rust free. ● **SEE FIGURE 24–17** for a machine with a ROPS canopy.

A falling object protection system (**FOPS**) is one that is designed to protect the operator from heavy objects that may fall on the machine's cab.

ROPS and FOPS structures must always be inspected at *all* service intervals. Their fasteners' torque should also be checked at intervals as recommended in the maintenance guide for the machine.

FIGURE 24–17 Machine with a ROPS canopy.

RADAR-WARNING/VISION SUPPLEMENT CAMERAS

Radar-warning systems use a transponder device to detect objects that are close to the machine and in its path of travel. This system is capable of warning the operator of a collision that is about to happen with enough notice that an avoidance maneuver should be possible.

Another common feature on newer machines is to have one or more video cameras mounted on the machine that will feed a signal to a display in the cab. The operator is then able to see blind spots around the machine to see if any object or person is in the path of travel for the machine.

These camera systems could be standard equipment on many large machines now as well as an option that could be requested and installed by the dealer before machine delivery. ● SEE FIGURE 24–18 for an operator display of a blind spot camera.

OPERATOR SEATING
Older machines had very basic operator seats that may not even have had padded cushions. Today's machine seats are meant to keep the operator as comfortable as possible and support their spine and lower torso completely.

They could be mounted to the cab frame work or mounted to a suspension mechanism to reduce operator fatigue and the occurrence of health problems such as sore backs. Fast traveling machines that tend to travel over rough terrain such as scrapers and rock trucks will have suspended seats, but they can also be found on any type of machine.

The suspension systems could be simple linkages with spring suspension and a damping mechanism; linkages with an air bag suspension and a damping mechanism or some

FIGURE 24–18 Blind spot camera and display.

older scrapers had a linkage system with hydropneumatic suspension and damping.

Most suspended seats are height adjustable, and the ones with dampened suspension can have adjustable damping.

SEAT BELTS
One of the most critical safety features that a machine has is its operator restraint or seat belt. All operator manuals will recommend that the operator wear a seat belt any time the machine is being operated.

OPERATOR'S STATION **673**

FIGURE 24–19 Three-point belt.

FIGURE 24–20 Tilted cab.

Most machine seat belts are lap belts that are adjustable, but some machines are featuring three-point belts with shoulder restraints just like those found in automobiles.
● SEE FIGURE 24–19 for a three-point seat belt.

Seat belts *must* be inspected on a regular basis for tears, rips, and others defects to the material and *must* be replaced in accordance with the machine's operating manual. Their retracting and adjusting mechanisms must be inspected as well.

The seat belt anchor and latching mechanism *must* also be inspected for proper torque and proper operation.

CAB NOISE PREVENTION
Recently, manufacturers have started to put more focus on noise level reduction inside cabs. Part of the motivation for this has been tighter noise restriction regulations imposed by governing bodies. This will also reduce operator fatigue and increase safety.

This is accomplished by installing more and thicker insulating material on the interior surfaces of the cab including floor mats.

What this means to technicians is that if panels are removed in a cab or on a machine, they must be replaced; otherwise sound levels inside the cab could exceed safe levels.

CAB MOUNTS
Many cabs are mounted on isolating mounts that separate the cab from machine-generated vibrations and noise. Some cabs are suspended on spring devices and have shock absorbers to reduce oscillations.

If a cab is removed, make sure that all isolating mounts/cab suspension components are installed or replaced. Many times a faulty cab mount can be misdiagnosed as a drivetrain issue.

TILTING AND REMOVING CABS
Many machine repair procedures require the cab to be tilted or removed. Some machines allow the cab to be tilted for access to drivetrain components. Some track loaders, track dozers, and articulated trucks are examples of machines with tilting cabs. There will likely be a hydraulic cylinder on the machine to hoist one side of the cab as it pivots on the other. *Never* get under a tilted cab unless it is mechanically locked in place.

Cab removal can be necessary for larger repairs such as transmission and pump removal. It can also be part of a complete machine-reconditioning process. Make sure that *all* wiring, hoses, tubing, and linkages are disconnected before starting to hoist the cab away from the machine. ● SEE FIGURE 24–20 for a machine with its cab tilted.

Progress Check

1. If a cab is equipped with pilot controls, this means:
    a. its hydraulic controls use low-pressure hydraulic oil
    b. it has aircraft like controls
    c. it has electronic sensors
    d. it has two sets of controls
2. If a machine is called "autonomous," it will:
    a. have an automatic transmission
    b. not have an operator
    c. have satellite communication
    d. have an automatic hydraulic system

3. A mechanical gauge is one that:
   a. is harder to move than other types
   b. is less accurate than other types
   c. has oil pressure fed right to it
   d. is faster acting than other types

4. If a gauge has "balancing coils," it will be considered to be this type of gauge:
   a. electric
   b. electronic
   c. mechanical
   d. virtual

5. This is one function that *is not* possible when a telematics system is functioning on a machine:
   a. receiving notice of fault codes
   b. remote control of the machine
   c. viewing live data
   d. setting up a geo fence

HVAC SYSTEMS

Heating, ventilation, and air-conditioning (HVAC) systems can be found in varying forms and are standard equipment on most machines that have an enclosed cab sold today. They are a key part in keeping the operator comfortable and, therefore, more productive. Both cold winter mornings and hot sunny days present challenges to HVAC systems, but if they are designed and maintained properly, they can overcome any heat, cold, or moisture influences from weather or from heat created by the machine itself.

CAB HEATER AND PRESSURIZER The most basic version of an HVAC system is one that pressurizes a sealed cab to keep the dust out and provides heat to keep the inside temperature at a comfortable temperature during cold weather. Pressurizing the cab prevents outside air that may be dirty, dust filled, and possibly hazardous from entering the cab.

Air is drawn in past a cab filter by a fan that is controlled by the operator. The fan usually has three or four speeds to allow the operator to vary the amount of air flow through the cab. Refer to Chapter 7 to review the operation of a DC electric motor that drives the fan. ● **SEE FIGURE 24–21** for a cab filter.

A simple electrical circuit to operate a fan motor will start with a protected power supply (fused or circuit breaker). This lead then goes to a switch that sends voltage out one of two to four terminals on the back of the switch. Wires will connect these leads to a stepped resistor. The stepped resistor has two to four input terminals and one output terminal and the resistance changes between each input and output terminal.

FIGURE 24–21 Cab filter.

As the switch is turned through various positions, different outputs of the switch leads connect to corresponding different input terminals on the stepped resistor. The output of the stepped resistor is connected to the feed terminal on the fan motor. The differing amount of resistance between the different terminals in the stepped resistor will determine how much current goes to the motor and, therefore, how fast the fan turns that is fastened to the motor shaft.

Many larger machines will use more than one fan and motor. This will increase the air flow through the cab to ensure that it stays pressurized. Most fans will be the squirrel cage style.
● **SEE FIGURE 24–22** for a squirrel cage fan motor.

FIGURE 24–22 Squirrel cage fan motor.

Some newer machines have electronically controlled fan motors. The fan switch is an input to an ECM and the output from the ECM is likely a voltage signal that goes either directly to the fan motor or to a relay coil that once energized will flow current to the fan motor.

All simple HVAC systems will have a means to vary the amount of airflow in the cab, but some may not have the means to reroute the airflow direction from the fan motor. These systems may just have one outlet from the fan through a heater core and out into the cab interior. Other machines will have ventilation systems that are similar to cars and pickup trucks that will allow the operator to change the airflow to different parts of the cab. The use of ducts and doors makes this possible. Most HVAC vent doors are moved by sliding levers that are connected to cables. The inner wire of the cable gets moved to make the door pivot. Some machines with electronically controlled HVAC systems use stepper motors to move the vent doors. These systems are most often controlled by a sealed touch pad, will likely have their own ECM that may or may not be multiplexed with the other machine ECMs, and will be able to generate fault codes to help with diagnosing problems.

Because all windows and doors are sealed with weather stripping, the air in the cab is pressurized as it tries to escape through a small ventilation opening that also has a filter over it.

The cab air filter is an often overlooked maintenance item that needs to be cleaned or replaced on a regular basis. If it becomes plugged with dust, the pressurizing effect won't work and the cab air becomes stale. All windows and doors must be closed, and their weather stripping must be intact for this pressurization to take place. If a machine is operated with the doors and windows open, it won't be long before the cab is coated in dust. Besides being messy, this could also become a health concern.

A cab heater will use warm engine coolant to add heat to the cab air. Coolant is routed to a heat exchanger (heater core) where cab air is blown across it to warm the cab up. The coolant flows through a pair of heater hoses that are typically ½ in. or ⅝ in. in diameter. The inlet to the heater core starts from the high-pressure side of the cooling system (water pump outlet) and the outlet of the heater core returns to the low-pressure side of the cooling system (engine block usually).

The amount of coolant flowing through the heater core is controlled by a valve that is controlled by the operator. A rotary knob or slide that is mounted in the cab can be moved, and this will increase or decrease the amount of coolant flow to increase or decrease the heat coming out of the heater core. Some controls will move a cable that turns a ball-type valve, some move a plunger in and out, and others will send an electrical signal to an ECM. The ECM then sends an electrical

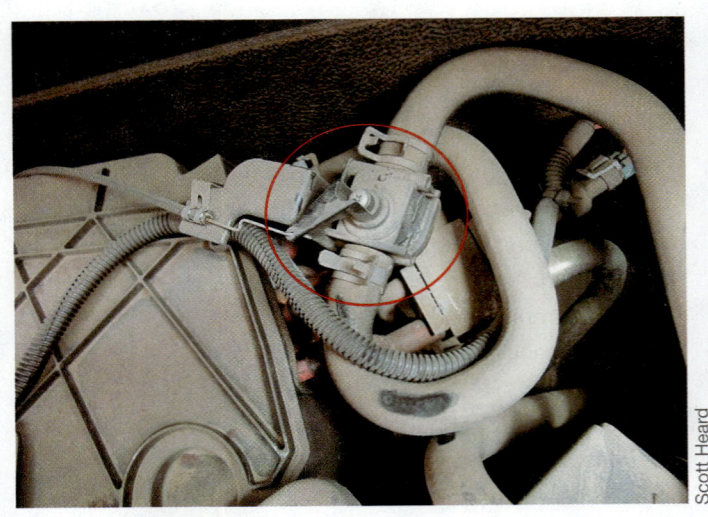

FIGURE 24–23 Heater control valve.

signal to a motor that turns a rotary valve through a set of gears. ● SEE FIGURE 24–23 for a heater control valve.

Some machines will use a separate set of shutoff valves to allow the system flow to be disabled, no matter the position of the control valve. These valves are typically used in the summer to stop any heat from entering the cab or when servicing the system to prevent excessive loss of coolant. They are usually located where the hoses are connected to the engine.
● SEE FIGURE 24–24 for a heater shutoff valve.

AIR CONDITIONING Most machines with cabs will have large areas of glass. On sunny days, the cab may seem like a greenhouse because the heat load from the sun will create enormous amounts of heat inside the cab. Even underground mining machines will eventually heat the cab up just from the heat transferring from other machine systems into the cab.

The heat can accumulate to make unbearable operating conditions for the operator. If the windows and doors are opened to allow some cooler air into the cab, then the cab will likely become dusty and unhealthy. Unlike automobiles with heavy equipment machines, you can't roll down the windows and get cooled by the air rushing by at highway speeds. It is not uncommon for a machine operator to refuse to operate a machine on a hot sunny day because the air-conditioning system isn't working properly.

For these reasons, understanding air-conditioning system operation and diagnostic procedures to correct air-conditioning problems are vital parts of what a technician needs to know.

AIR-CONDITIONING OVERVIEW To understand how an air-conditioning system works, you must understand how the cab air is cooled. To do this, you need to understand how heat is transferred out of the cab air and accept that the

FIGURE 24–24 Heater shutoff valve.

air-conditioning system doesn't add cold air to the cab but it *removes heat* from it. To help understand this, you should know that heat energy will always travel from hot to cold.

An air-conditioning system is a closed loop fluid system that uses a fixed quantity of refrigerant sealed inside hoses, lines, and other main components. Like any fluid system, it needs a pump to make the fluid flow and the pump for an air-conditioning system is called a **compressor**. When the operator turns on the system's compressor, the refrigerant starts

FIGURE 24–25 Three states of matter (Halderman, *Automotive Technology*, 4th ed., Fig. 62-1).

to flow through the system and will do so in a continuous loop as long as the compressor is running.

As well as having a pump, there needs to be a pressure differential to make the fluid flow and all air-conditioning systems will have a restriction to provide this.

Any substance can be in three forms: solid, liquid, and vapor. ● SEE FIGURE 24–25 for the three different states water can be in.

The refrigerant in an air-conditioning system is in either liquid or vapor state, depending on where it is in the system and whether the compressor is pumping. After the compressor stops pumping for a few minutes, all the refrigerant goes back to a liquid state and stays at a low pressure of approximately 35–50 psi. This will vary depending on ambient air temperature.

The amount of refrigerant that is installed in the system is determined by the volume inside the system's hoses, lines, and components. The refrigerant is measured by weight, and its liquid volume will leave room for expansion inside the system as its pressure and temperature rises.

When the compressor starts pumping the refrigerant through the system, the refrigerant changes temperature, pressure, and state (liquid to gas and gas to liquid), depending on where it is in the system and what it is being influenced by. It is the change of temperature, the change of pressure, and the

WEB ACTIVITY

For an excellent visual demonstration of how a refrigeration system works, search for a U.S. Department of Education 1944 video "Principles of Refrigeration." It will explain the fundamentals of the change of states, different pressures, and temperatures involved in a refrigeration process. There are many similar principles involved in a mobile air-conditioning system.

TECH TIP

There are two main types of air-conditioning systems that you may find on a heavy equipment machine. They are defined as to the type of restriction that is used in the system. An air-conditioning system is a sealed fluid system, and as mentioned previously, all fluid systems need a pump and a restriction to make fluid flow.

Both types of restrictions are located downstream from the compressor outlet and past the condenser.

One type of system is called an orifice tube because of the type of restriction it has. It also has an accumulator that is unique to this type of system. The accumulator is located at the evaporator outlet. ● SEE FIGURE 24–26 for the main components of an orifice tube system.

You'll notice the compressor and the orifice tube/expansion valve are the dividing lines between the high- and low-pressure sides of the system.

The orifice tube is a fixed orifice that causes the pressure to rise between it and the compressor outlet.

The other type is a thermal expansion valve (TXV) system. It has a receiver drier that is unique to it, and it is located between the TXV and the condenser. ● SEE FIGURE 24–27 for a TXV system.

The expansion valve is a variable orifice that will open and close slightly based on the system needs.

Some TXVs will have a thermal bulb to vary the orifice while others are called an H block and will regulate the orifice internally.

■ HIGH TEMPERATURE AND HIGH PRESSURE
■ LOW TEMPERATURE AND LOW PRESSURE

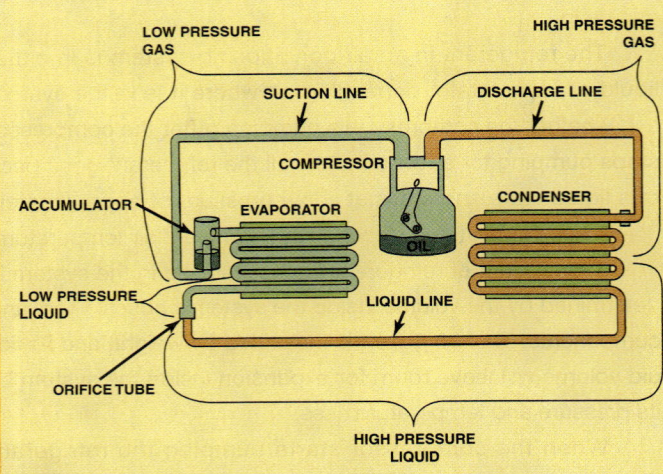

FIGURE 24–26 Orifice tube system (Halderman, *Automotive Technology*, 4th ed., Fig. 62-11).

■ HIGH TEMPERATURE AND HIGH PRESSURE
■ LOW TEMPERATURE AND LOW PRESSURE

FIGURE 24–27 TXV system (Halderman, *Automotive Technology*, 4th ed., Fig. 62-10).

---

change of state of the system's refrigerant that provides for the removal of heat from the inside of a cab.

An air-conditioning system uses two heat exchangers: one to absorb heat from inside the cab (evaporator) and one to dissipate (get rid of) that heat away from the cab (condenser). Proper air flow past these two heat exchangers is critical to proper air-conditioning system operation.

An engine's heat exchanger is the radiator and it uses air flowing past it to dissipate heat from the engine coolant to the atmosphere.

The other main parts of the system are the compressor, an orifice, hoses, tubes, fittings, controls, either an **accumulator** or a **receiver drier** (depending on the type of system), and of course the refrigerant and some oil. There also needs to be two fans to move air past both the evaporator and the condenser. The cab fan is part of the cab pressurization system and moves air past the evaporator, and the condenser may have its own fans or may use the engine cooling fan to move air past it. These two fans are critical to the proper operation of any AC system.

REFRIGERANT AND OIL There are many different substances that are used as refrigerants. Household air conditioners, commercial air conditioners, and huge cold food storage facilities will all use different refrigerants in their systems. They will have slightly different properties that make them ideal for their application. An ideal refrigerant is one that can mix with lubricating oil, has a boiling point of less than 32°F at atmospheric pressure, and can be noncorrosive, nonpoisonous, and nonexplosive.

Approximately 20 years ago, the refrigerant used in all vehicles and mobile equipment was R-12. However, it was found to be creating holes in the earth's ozone layer and contributing to harming our environment. A replacement was found that is much less harmful to our environment and is called **R-134**. ● SEE FIGURE 24–28 for the effects of harmful refrigerant on the ozone layer.

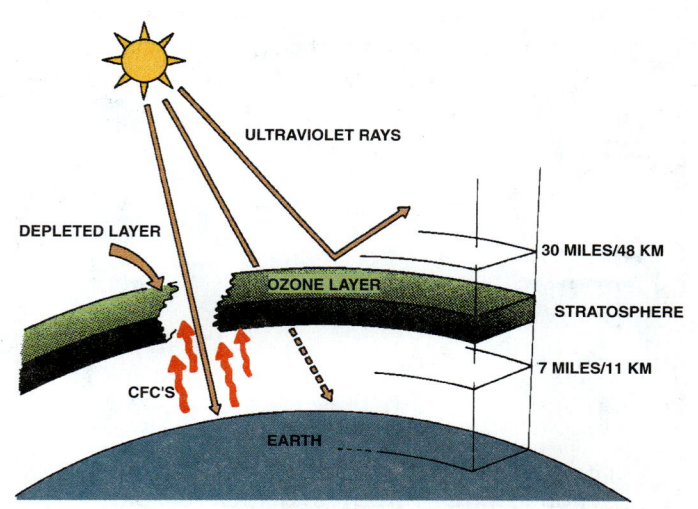

FIGURE 24–28 Effects of R12 (Halderman, *Automotive Technology*, 4th ed., Fig. 62-15).

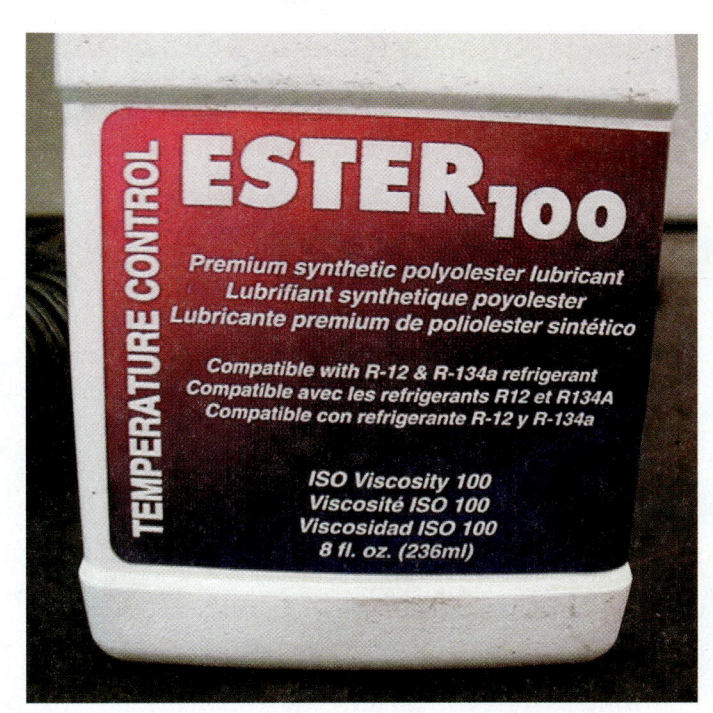

FIGURE 24–29 Air-conditioning oil (Halderman, *Automotive Technology*, 4th ed., Fig. 62-19).

TECH TIP

In 1987, a conference was held in Montreal where 22 countries had representatives attend and decide on how to move forward with better refrigerants and established better refrigerant handling practices. The Montreal Protocol had several significant outcomes. The main ones affecting all HDETs were the following:

R-12 production would cease and its replacement (R-134) would start to be used in 1996.
All technicians who service or repair mobile air-conditioning systems *must* be properly trained and certified by an accredited organization.
Refrigerant recovery and recycling equipment must be properly approved.

Each jurisdiction in North America has their own certification processes and standards that must be adhered to in relation to working on mobile air-conditioning systems and handling refrigerant. Make sure that you are compliant with these standards before working on any air-conditioning system. There could be severe fines or jail time involved otherwise.

The refrigerant used in machines produced today is called R-134. Its properties give it a low boiling point of −14.9°F. Even when it is pressurized to 20 psi in the evaporator, it will still easily change states when the cab air temperature is transferred into it.

All air-conditioning systems will have a specific type and amount of oil that is mixed in with the refrigerant and gets circulated throughout the system. Unless there is a leak or the system is connected to test equipment, there is no way that oil can escape. If service or repairs are needed to the system and oil has to be added, it is *very* important that the correct amount and type of oil is added. Many types of oil will have dye in it to assist with leak detection. ● SEE FIGURE 24–29 for one type of air-conditioning oil.

EVAPORATOR The evaporator is in the cab and is usually next to the heater core. It performs the opposite function of a heater core in that it absorbs heat in the cab. This is done by blowing the warm cab air through the external fins of the evaporator and the refrigerant that flows through it absorbs heat from the air. ● SEE FIGURE 24–30 for the thermal effects of an evaporator.

OPERATOR'S STATION **679**

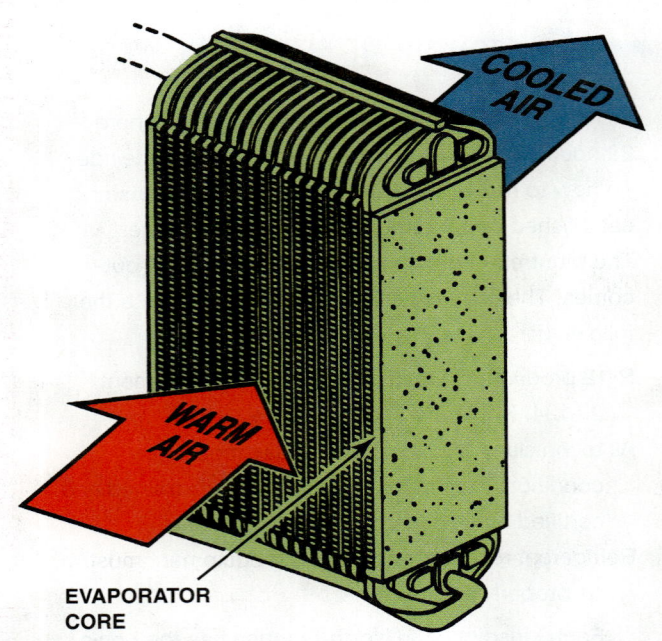

FIGURE 24–30 Evaporator (Halderman, *Automotive Technology*, 4th ed., Fig. 62-7).

The evaporator is a finned tube that weaves back and forth to allow enough time for the refrigerant flowing through it to absorb cab heat.

Because of the different properties of the system refrigerant, it boils at a low temperature. This is made easier because it is also under low pressure because it just flowed past the restriction (orifice tube or TXV valve). It enters the evaporator as a low-pressure and low-temperature liquid. The low temperature of the refrigerant cools down the evaporator.

As the warm air flows past the cool evaporator, its heat gets transferred through the evaporator to the refrigerant. The refrigerant absorbs the heat and changes state from a liquid to a vapor. In other words, it boils or evaporates. Remember that heat always travels from hot to cold, and this is what happens with the warm cab air. It transfers its heat into the cold refrigerant.

About $2/3$ of the way through the evaporator, the refrigerant will have changed from a liquid to a gas.

The low pressure in the evaporator is usually between 15 and 30 psi. Most refrigerants will boil at the equivalent temperature value as their pressure value. For example, if the pressure is 20 psi, the refrigerant will boil at 20°F.

When any liquid reaches its boiling point temperature, it changes state to a gas. Think of what you see coming out of a kettle when the water starts to boil. The steam seen rising from the kettle is vaporized water that was liquid water just moments before. This change of state takes a great deal of heat energy, and this massive transfer of heat is what can remove heat from the cab air with great efficiency.

TECH TIP

BTU is a measure of the amount of heat required to raise the temperature of 1 lb of water 1°F at sea level. For example, if you wanted to heat 1 lb of water up from 50°F to 150°F, it would take 100 BTUs. However, once that water is heated to its boiling point of 212°F, it then takes 970 BTUs to change its state. The same amount of heat is needed to be removed to change the steam back into water. ● **SEE FIGURE 24–31** for how water changes to vapor.

These two processes are called vaporization (liquid to vapor) and condensation (vapor to liquid).

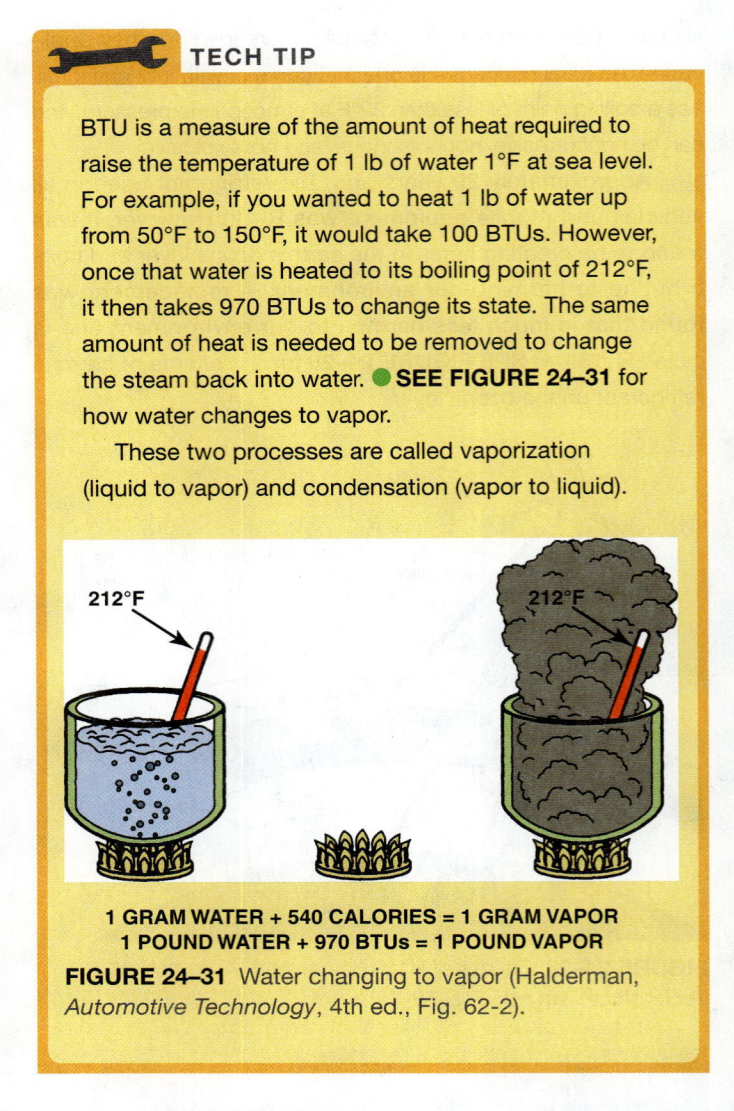

1 GRAM WATER + 540 CALORIES = 1 GRAM VAPOR
1 POUND WATER + 970 BTUs = 1 POUND VAPOR

FIGURE 24–31 Water changing to vapor (Halderman, *Automotive Technology*, 4th ed., Fig. 62-2).

The low-pressure and low-temperature liquid that entered the evaporator is now leaving as a low-pressure and high-temperature vapor.

COMPRESSOR The pump for the system is called the compressor because the refrigerant is in a gaseous state when it reaches the compressor and is therefore compressible. Most compressors are reciprocating piston type and use an angled swash plate to have the pistons travel up and down in their bores as they rotate inside the compressor housing. This is a similar action to a hydraulic piston pump with a swash plate. ● **SEE FIGURE 24–32** for an air conditioner compressor.

There are check valves in the compressor to ensure refrigerant flow continues in the proper direction. The low-pressure and high-temperature vapor is drawn into the compressor inlet as it leaves the evaporator outlet. This gas is then compressed in the compressor when the piston pushes it out through the outlet of the compressor. The resistance to flow that creates the higher pressure is the restriction downstream of the condenser.

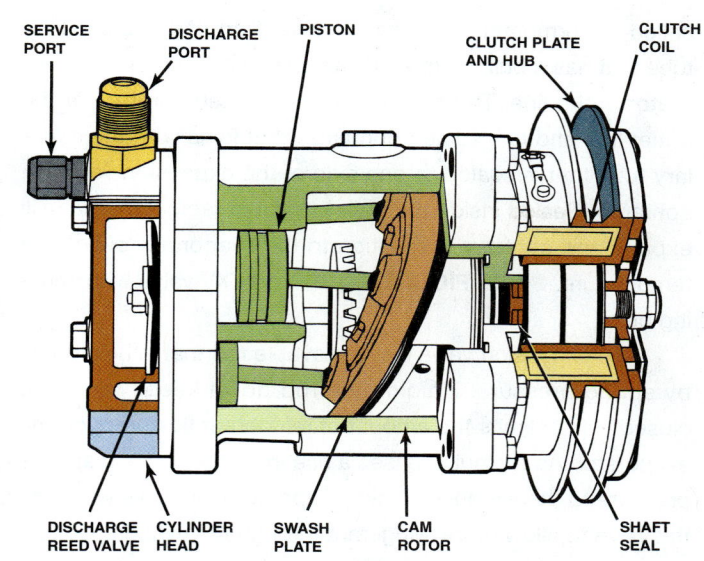

FIGURE 24-32 Air conditioner compressor (Halderman, *Automotive Technology*, 4th ed., Fig. 62-8).

The oil that is mixed in with the refrigerant will keep all the moving parts in the compressor lubricated.

COMPRESSOR DRIVE The compressor is belt driven and the belt could be a V-belt or multi-rib belt. Belt tension is critical to proper operation of the air conditioner system because the compressor must turn at the proper speed to get the proper refrigerant flow.

The compressor pulley is driven by the belt, and between the pulley and the compressor shaft is an electromagnetic clutch.

This clutch consists of one or more coils of wire that turn into an electromagnet when energized. One end of the coil is connected to frame ground and the other gets a voltage signal from the air-conditioning controls. When energized, the clutch's magnetism pulls a drive disc in and locks the pulley to the shaft. This action allows the pulley to drive the compressor shaft. The clutch can be cycled on and off to start and stop refrigerant flow in the system. This cycling can be a noticeable sound, and for lower horsepower, machines will likely change the sound of the engine because of the extra load on it.

CONDENSER When a gas is compressed, it heats up. This is the same principle that applies to diesel engines to make their intake air hot enough to ignite diesel fuel.

> **WEB ACTIVITY**
>
> Find a video of an air-conditioning compressor that is cycling and see if you can hear a change in the sound of the engine that is driving it!

FIGURE 24-33 Condenser.

The refrigerant pressure usually reaches 250–300 psi as it is pushed out from the compressor against the system restriction to another heat exchanger that is called the condenser. As its pressure rises, its temperature will rise in proportion.

To get rid of the heat in the refrigerant that was collected at the evaporator, it is passed through the heat exchanger that is located downstream from the compressor that is called the condenser.

The condenser is a tube that contains the refrigerant and has light fins attached to it to allow easy heat transfer. The tube weaves its way back and forth across the condenser assembly to allow enough time for the refrigerant's heat to be exposed to the condenser and be able to shed heat. Once again it is the principle of heat traveling from hot to cold that transfers heat in the refrigerant through the condenser to the outside air because the outside air is much colder than the high pressure/temperature refrigerant.

About $2/3$ of the way through the condenser, the refrigerant should be back to a liquid state.

The condenser is usually located near the engine cooling fan because it needs air flowing past it to get rid of the heat in the refrigerant. ● **SEE FIGURE 24-33** for a typical air-conditioning condenser.

Some machines will have the condenser located somewhere else on the machine and have one or more fans dedicated to moving air past it. These fans will be driven with electric motors. ● **SEE FIGURE 24-34** for a condenser with its own fans.

As the heat leaves the refrigerant, it cools down, changes state, and condenses back to a liquid. It then flows to the system restriction where its pressure is dropped.

When the refrigerant changes state, it is able to absorb and dissipate great amounts of heat.

FIGURE 24–34 Condenser with electric fans.

THERMOSTATIC EXPANSION VALVE As mentioned previously, there can be two different devices used to make a restriction in the system. One of these is a variable restriction called a thermostatic expansion valve. Part of it is a capillary tube that has a bulb at its end. The bulb is placed on the evaporator outlet line. The bulb will sense the temperature of the outlet line and is a sealed container that joins with the capillary tube that is sealed to one side of the expansion valve diaphragm. Sealed inside the bulb is a liquid refrigerant that will expand and contract in reaction to the evaporator outlet line temperature. ● **SEE FIGURE 24–35** for a TXV valve with sensing bulb.

The expansion valve has a movable rod that is held down by spring pressure. Holding the rod down keeps the valve closed, which limits the amount of refrigerant flow through the evaporator. As the bulb senses a rise in temperature, it applies pressure to the expansion valve diaphragm, and this will open the valve to allow more refrigerant through the evaporator. This is how a variable orifice is created.

There will be a metering of refrigerant through the evaporator based on its outlet temperature.

The other type of thermal expansion valve is called an H-block valve. ● **SEE FIGURE 24–36** for an H-block valve mounted on an evaporator.

It works on the same principle as the valve previously described except there is no remote-sensing bulb. Both the lines going into and leaving the evaporator are attached to the valve and the temperature-sensing element is in the top of the valve where it senses the outlet temperature directly. It also works with a spring and diaphragm to move a pushrod, which in turn varies the orifice in the line leading into the evaporator. ● **SEE FIGURE 24–37** for an H-block valve.

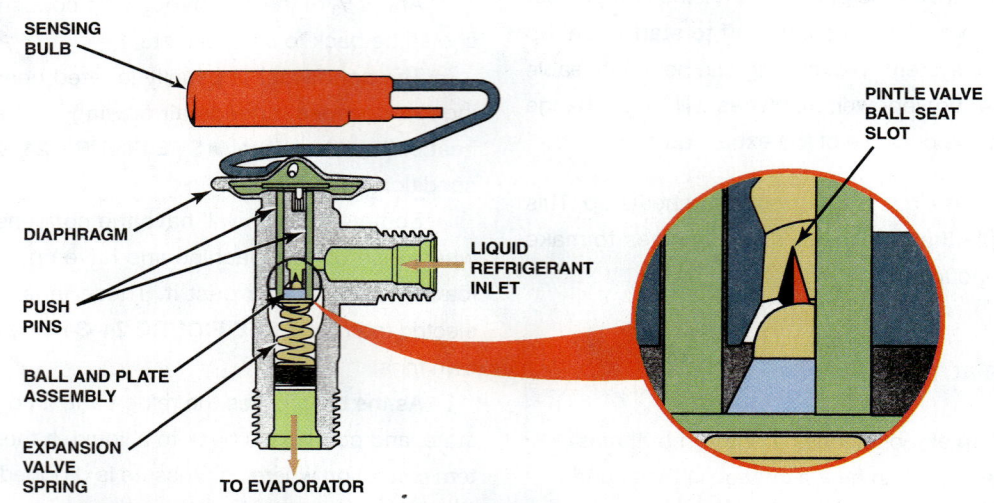

FIGURE 24–35 TXV valve with sensing bulb (Halderman, *Automotive Technology*, 4th ed., Fig. 62-27).

682 CHAPTER 24

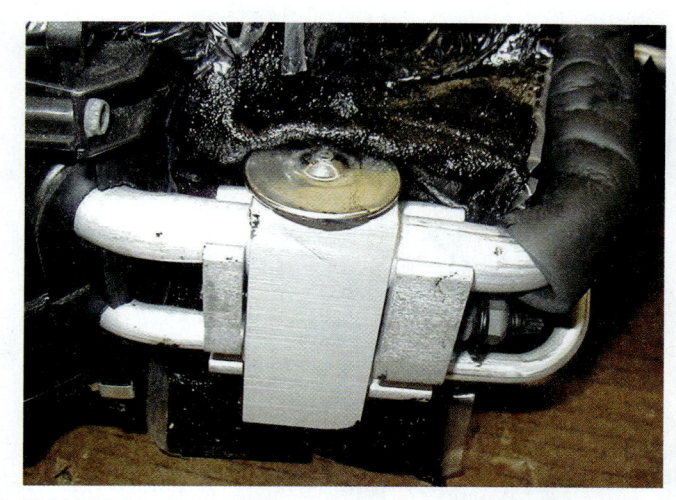

FIGURE 24–36 H-block on an evaporator (Halderman, *Automotive Technology*, 4th ed., Fig. 62-31).

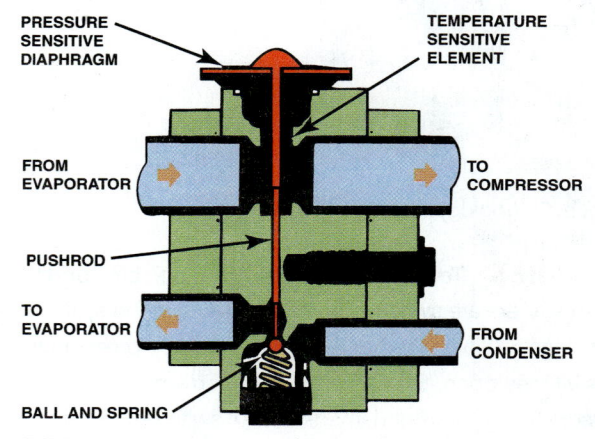

FIGURE 24–37 H-block valve (Halderman, *Automotive Technology*, 4th ed., Fig. 62-30).

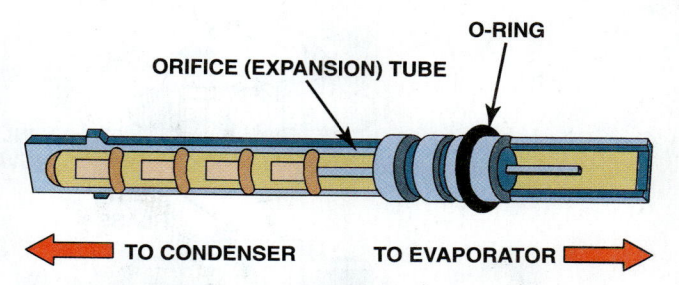

FIGURE 24–38 Orifice tube (Halderman, *Automotive Technology*, 4th ed., Fig. 62-12).

ORIFICE TUBE
The second main type of mobile air-conditioning system is called an orifice tube system. It uses a fixed restriction that is placed in the line between the condenser outlet and the evaporator inlet. It is usually incorporated with a screen to stop any contamination from entering the evaporator. ● SEE FIGURE 24–38 for an orifice tube.

ACCUMULATOR
Orifice tube systems use accumulators, and they can be found between the outlet of the evaporator and the inlet of the compressor. The accumulator prevents any liquid refrigerant from getting into the compressor. Because the compressor will only pump vapor, any liquid reaching it would cause it to lock and possibly cause permanent damage.

It also serves as a reservoir of refrigerant and contains desiccant that will remove any moisture (water) in the system. ● SEE FIGURE 24–39 for an accumulator.

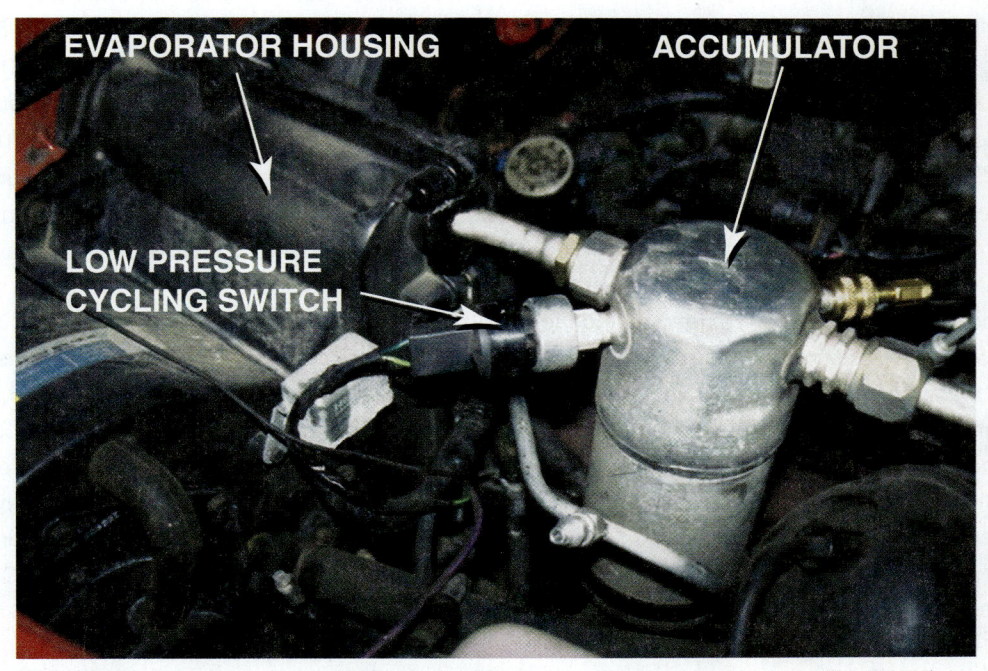

FIGURE 24–39 Accumulator (Halderman, *Automotive Technology*, 4th ed., Fig. 62-24).

OPERATOR'S STATION

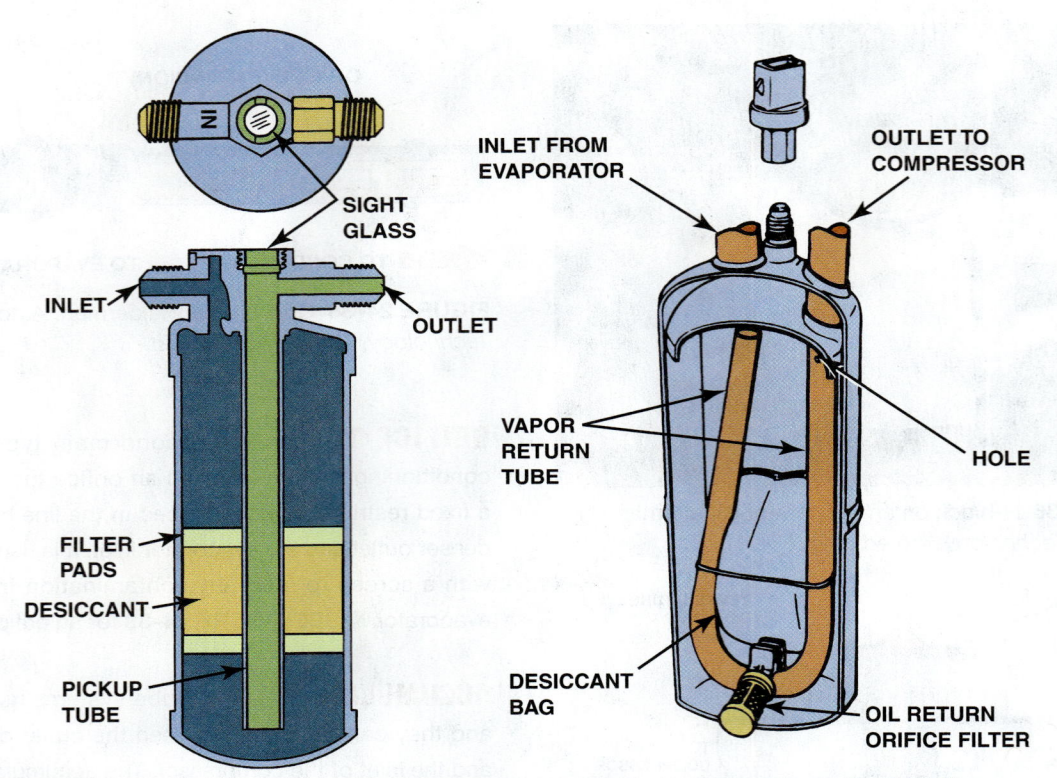

FIGURE 24–40 Receiver drier (Halderman, *Automotive Technology*, 4th ed., Fig. 62-23).

RECEIVER DRIER
Air-conditioning systems that use expansion valves use a receiver drier to serve as storage for liquid refrigerant and to remove any water moisture from the system. It is located in the line past the condenser before the expansion valve. Like the accumulator, it has desiccant in it to remove moisture. ● SEE FIGURE 24–40 for a cutaway view of a receiver drier.

Many receiver driers will have a sight glass on them to allow a technician to view refrigerant flow when the system is operating.

HOSES, LINES, AND FITTINGS
As the air-conditioning system is a sealed pressurized system, the refrigerant inside must be positively contained in the system. To connect all system components, there will be a combination of hoses, lines, and fittings.

Hoses allow some movement between components and are constructed of layers of rubber and fabric. The rubber compound must be compatible with R-134 refrigerant and the oil used for the system. Lines are made of aluminum tubing and formed to fit the machine contours. There should be several clamps to hold all lines and hoses secure from vibration.

When hoses, lines, and fittings join each other or other components, there will be O-ring seals to ensure that no leakage occurs. These seals must also be compatible with the refrigerant and oil.

Some components like receiver driers will have quick couplers to allow for easier servicing. ● SEE FIGURE 24–41 for system hoses and lines.

SWITCHES
To help control and monitor air-conditioning systems, there are switches that are installed in the system refrigerant lines and are wired in series with the compressor clutch. If the switches open, the compressor clutch won't drive the compressor shaft to prevent damage. These switches are as follows:

Low-pressure switch: If the refrigerant pressure is lower than 25 psi, this switch will open and not allow the compressor clutch to energize. A pressure lower than 25 psi would indicate there has been a leak. This switch prevents the compressor from working if some of the refrigerant has leaked out and with it some lubricating oil.

High-pressure switch: If system pressure exceeds safe limits, a high-pressure switch will open. A typical limit is 375 psi. ● SEE FIGURE 24–42 for an example of a pressure switch.

HUMIDITY REDUCTION
Cab humidity is also controlled with an AC system. This occurs when the cab air is moved past the cold evaporator, the moisture in the air condenses on the outside of the evaporator. It will drain off the evaporator and should be channeled outside of the cab where it drains on the ground. You'll notice this action on a hot humid summer day if your vehicle is sitting stationary with the AC system running. There will be a puddle of water accumulating on the pavement under the vehicle, and this is the moisture that has been removed from the air inside your vehicle.

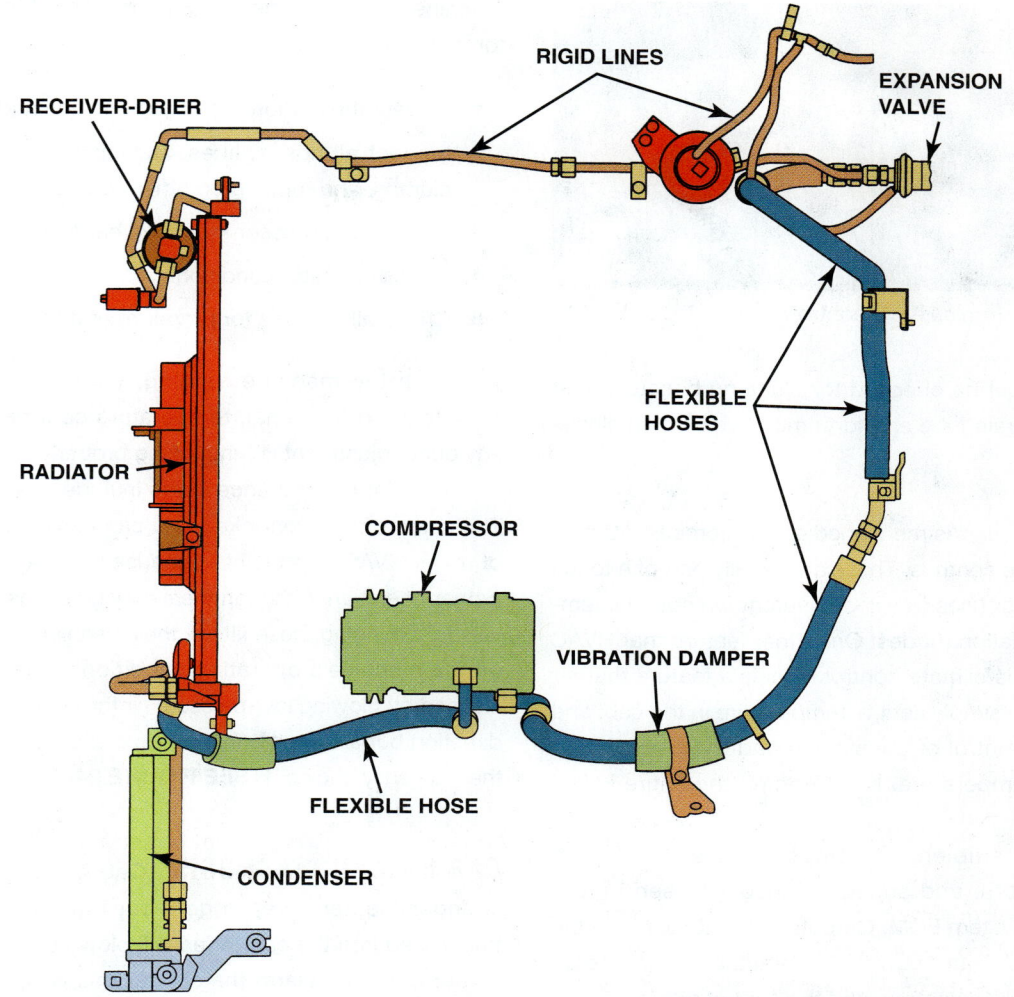

FIGURE 24–41 System hoses and tubes (Halderman, *Automotive Technology*, 4th ed., Fig. 62-25).

FIGURE 24–42 Pressure switch (Halderman, *Automotive Technology*, 4th ed., Fig. 62–38).

OPERATOR'S STATION **685**

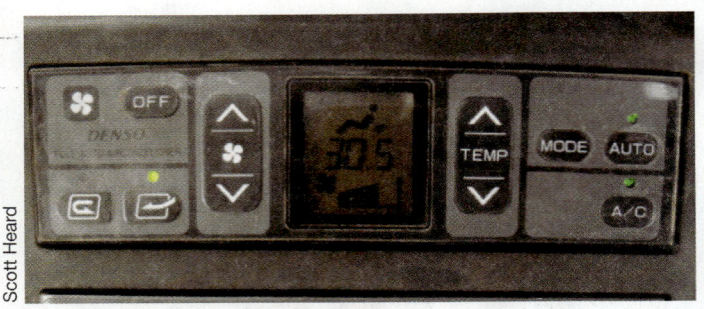

FIGURE 24–43 Electronic HVAC control.

Cab drains must be checked for plugging because dust will collect on the drain tube and form mud, which then closes off the drain.

CLIMATE CONTROL As mentioned earlier, some HVAC systems have electronic controls. These are usually part of a touch pad that the operator uses to select different fan speeds, temperature, and ventilation modes. One other feature that HVAC systems may have is climate control. This is a feature that allows the operator to simply set the temperature in the cab, and an ECM will take control of the system and try to reach and maintain the set temperature. It will control the entire HVAC system.

There will be ambient air temperature sensors, cab temperature sensors, and sunlight sensors to send input information to the system ECM. Outputs from the ECM are the compressor clutch, heater control valve actuator, fan speed, ventilation duct doors actuators, and the control display.

This system is usually capable of displaying fault codes. ● SEE FIGURE 24–43 for an electronic control of an HVAC system.

SERVICING AND REPAIRING HVAC SYSTEMS

Although HVAC systems are not a critical operating system of the machine as far as it being productive or not, a great deal of an HDET's time will be spent looking at HVAC system problems. This is because many operators place their comfort level ahead of machine production. It is understandable that if the HVAC system isn't keeping the cab warm in the winter and cool in the summer, then the operator won't be happy but some complaints can be exaggerated.

Regular maintenance of HVAC systems is fairly straightforward. It includes thorough visual inspections of all HVAC components that are easy to see when performing other machine servicing. The following is a list of inspection points for an HVAC system:

1. Inspect the compressor belt condition and tension.
2. Inspect all hoses, lines, and fittings for leaks, missing clamps, and rubbing on other components.
3. Inspect compressor-mounting hardware.
4. Check cab filter condition.
5. Check all controls for proper operation.

With the machine running, you should check all fan speeds are operational, temperature control is working, and any duct adjustment is functioning properly.

Regular maintenance items include changing cab filters. These are often overlooked and can easily be the root cause of many HVAC complaints because good airflow is critical to proper operation. Although there may be suggested hourly intervals for changing these filters, their cleanliness mainly depends on the machine's operating conditions. These filters can be cleaned by blowing low-pressure air through them in the reverse direction but are usually inexpensive enough to make changing them an easy choice. ● SEE FIGURE 24–44 for a dirty cab filter.

CAB HEATER DIAGNOSIS Cab heaters use a heat exchanger (heater core) to get heat from hot engine coolant transferred into the cab. A fan will blow cold air across the hot heater core and warm the cab air this way. If there is a lack of heat in the cab complaint, the following is a generic list of steps to take:

1. Verify the complaint. Make sure that all the controls are in the right place.
2. Make sure that the engine is reaching proper operating temperature.
3. Make sure that the fan is working.
4. Check to see that the air-conditioning system isn't working as well.
5. Check for related fault codes if the system is part of a climate control system.
6. Check the condition of any air filters in the cab.
7. Carefully feel the heater hoses going in and coming out of the heater core if all previous checks don't reveal any problems. They should be too hot to touch. If they are hot, the heater should be hot as well and there is an air circulation problem. If the hoses aren't hot, there is a coolant circulation problem that needs to be addressed. Look for any shutoff valves that may be off or

686 CHAPTER 24

FIGURE 24–44 Dirty cab filter.

FIGURE 24–45 Thermometer in air vent.

FIGURE 24–46 Refrigerant-identifying machine (Halderman, *Automotive Technology*, 4th ed., Fig. 64-5).

restricting flow. The machine's service information will likely guide you through a troubleshooting exercise to find the cause.

8. Find the root cause of the problem and perform repair.
9. Verify the repair has corrected the problem.

AIR-CONDITIONING SYSTEM DIAGNOSIS Air-conditioning problems will almost always be related to the cab being too hot. To verify this complaint, perform the following steps:

1. Perform a visual inspection to verify compressor belt condition and tension, heater controls are off (heater hose taps should also be turned off if equipped), condenser or heat exchangers stacked near it are not plugged, check condenser fan operation, check cab filters, look for any signs of leakage from hoses or lines.
2. Start the engine and turn the air-conditioning controls to maximum. Check to see if the compressor pulley is driving the compressor (center of the clutch should be turning). This should confirm that there is enough refrigerant in the system to keep the low-pressure switch closed.
3. Confirm the cab fan is running and there is plenty of air moving out the vents. Plugged filters will restrict air flow and cause low cooling.
4. Raise the engine rpm to about 1200 and hold there for at least 10 minutes to stabilize the system.
5. Place a thermometer in one of the vents, and after waiting another five minutes, it should read between 35°F and 45°F (2°C and 7°C). If you find this reading, the system is working as it should.

If it reads higher than 45°F, then a pressure check will need to be performed. ● **SEE FIGURE 24–45** for a thermometer in an air vent.

6. Follow the instructions of the refrigerant identifying machine and identify the type of refrigerant in the system. If refrigerant other than R-134 has been in the system, it has been contaminated and will need to be **evacuated**. ● **SEE FIGURE 24–46** for a refrigerant identifier.

OPERATOR'S STATION **687**

TECH TIP

Most automotive parts' stores will sell you an air-conditioning system top-up kit. Read the label to see what you are about to put into your vehicle. Many of these kits contain hydrocarbon-based refrigerant that may work okay temporarily. First, the leak source should be detected and repaired. If any of this non-R-134 is put into your system and you take it to a garage with an air-conditioning problem, they may refuse to work on it because of what is in the system. Always be leery of "repairs in a can."

7. If it has R-134 in it, connect a set of gauges to the service valves. Run the system again as before and observe the pressures.

8. At 70°F ambient temperature, normal low-side pressure should be 25–30 psi and high-side pressure should be 150–200 psi. Variations will occur because of different ambient temperatures, humidity levels, and different system configurations. Always refer to the manufacturer's specifications for the machine you are working on.

9. If pressures read outside these ranges, then a troubleshooting procedure needs to be performed to find the root cause. Refer to the machine's service information for specific information.

REFRIGERANT LEAK REPAIR Any time it is suspected refrigerant has leaked from a system the leak source must be found and a repair made before the refrigerant can be topped up. In order to get the proper amount of refrigerant in the system the system will need to be emptied and recharged with the exact amount as specified by the machine manufacturer.

Be aware of all legal requirements related to recovering and recharging refrigerant procedures and equipment standards. Many regulating bodies require proper documentation be filled out and kept on file to record all air-conditioning system repairs as well as tracking usage of all new refrigerant.

A system leak can be an obvious visual defect like a hose rubbed through, or it can be indicated by pressure gauge findings that point to low refrigerant levels. Leak-detecting equipment can also be used to look for leaks if the system is still functional. There are electronic sniffers, black lights used with dyed oil, and soapy water in a spray bottle to check for leaks.

The following steps should be taken to repair a refrigerant leak:

1. Identify the type of refrigerant in the system
2a. If the system is still operational find the leak
2b. If the system is not operational use nitrogen to find the leak
3. **Recover** the refrigerant and oil
4. Repair the leak
5. Evacuate the system
6. Refrigerant recycling
7. Recharge the system
8. Run system and check for leaks

1. Identifying the type of refrigerant is important to keep any non-R-134 substances from contaminating your equipment. Use an electronic tool to identify the type of refrigerant and if any contaminants are present.

If non-R-134 substances are detected, they must be removed into a separate container that is clearly marked as contaminated. This can be done with a vacuum pump and container.

2a. If the system is still operational, run the system at its coldest setting and use leak-detecting equipment to find the leak source. ● **SEE FIGURE 24–47** to see a black light being used to find a leak.

Start at all connection locations and any locations where hoses or tubes may be rubbing on something else. Then go through the rest of the system slowly and methodically.

FIGURE 24–47 Black light in use (Halderman, *Automotive Technology*, 4th ed., Fig. 64-18).

Service valves allow testing and refilling equipment to be connected to the high and low sides of the system. R-134 systems are identified with red (high) and blue (low) caps that are threaded on and sealed with an O-ring. They are a type of quick coupler and are different sizes from each other to prevent mixing up low and high sides.

R-12 systems also had service valves but had threads on their outside to allow test equipment connection. There is no way of interchanging R-12 and R-134 test or refill equipment.

2b. If the system is not operational, in other words the compressor won't pump, check to see if there is any refrigerant in it by hooking up gauges to the service ports. There should be 35–50 psi in the system when the compressor isn't running. This pressure will vary with ambient temperature. If it is less than 25 psi, the low-pressure switch won't let the compressor clutch engage to start pumping, and this should tell you there is a leak.

If this is the case and you have confirmed there are no contaminants in the system, then recover any R-134 left in it (see step 3). Then the system can be filled with nitrogen gas and pressurized to 50 psi. If there is dye in the system, a black light can be used to find the leak or use soapy water. Again start looking at the most likely locations for leaks.

3. Refrigerant recovery allows good R-134 to be reused. A refrigerant recovery machine is connected to the service fittings and a slight vacuum is applied (5 in Hg). The machine will also remove oil from the system and it is separated into a bottle that allows the amount to be recorded.

4. Once the exact leak location is found, determine what parts need to be replaced to repair the leak. Replace faulty components and always install new O-rings with them.

It is also common practice to replace the receiver drier whenever the system is opened to the atmosphere because moisture will enter the system.

5. Evacuating the system ensures that there is no moisture or contamination in the system. A strong vacuum is applied to the system, and this should lower the boiling point of water enough that any water moisture in the system with vaporize and then be drawn out by the vacuum

FIGURE 24–48 An automatic air-conditioning service machine.

pump. A general rule of thumb is at least 26 in Hg for at least 45 minutes to ensure complete evacuation of any non-R-134 substances. This is a critical step in repairing air-conditioning systems and can't be rushed. In fact, stronger vacuum for a longer period of time is better.

6. Refrigerant recycling is possible if good R-134 was recovered from the system. Most air-conditioning service machines will filter the recovered R-134 to remove any water, air, and oil. ● **SEE FIGURE 24–48** to see an automatic air-conditioning service machine.

7. Recharging the system with the proper type and amount of refrigerant and oil is a critical step in ensuring that the air-conditioning system works properly. There are different methods for doing this, and the one used depends on the type of system that is on the machine being serviced and the type of service equipment used. Always follow recommended procedures to get this right.

Progress Check

6. If a cab is pressurized, it means:
   a. the cab has pilot controls
   b. the machine has power brakes
   c. there is negative pressure inside the cab
   d. there is positive pressure inside the cab

7. The stepped resistor that is used in some fan motor circuits will:
   a. turn the motor on and off
   b. provide reverse rotation
   c. provide different fan speeds
   d. make the motor last longer
8. An air-conditioning system is a _____ system:
   a. sealed fluid
   b. open liquid
   c. sealed gas
   d. open gas
9. The heat exchanger that gets rid of heat from the refrigerant is called the:
   a. evaporator
   b. condenser
   c. receiver drier
   d. accumulator
10. The refrigerant that was formerly used in mobile air-conditioning systems and had its production stopped because of its ozone destroying properties is called:
    a. H-12
    b. R-12
    c. H-134
    d. R-134
11. At an ambient temperature of 70°F, normal pressure range on the high side of an air-conditioning system is:
    a. 15–30 psi
    b. 30–50 psi
    c. 100–150 psi
    d. 150–200 psi
12. An air-conditioning system is called a _____ system if it has a fixed restriction as part of it:
    a. TXV
    b. H-block
    e. accumulator
    f. orifice tube
13. This would NOT be one method/tool of finding a refrigerant leak in the system:
    a. black light
    b. smell
    c. electronic sniffer
    d. soapy water

TECH TIP

Whenever you are around a machine that is running and the air conditioning is on, pay attention to how the compressor cycles and if the engine changes tone. Also carefully feel for different temperatures with your hands at various locations such as hoses going in and out of the evaporator, hoses going in and out of the condenser, and air vents in the cab. A check around with an infrared temperature gun is also a good idea.

14. This air-conditioning component ensures no liquid refrigerant reaches the compressor inlet:
    a. thermal expansion valve
    b. evaporator
    c. H-block valve
    d. accumulator
15. What is the stage of an air-conditioning repair process where a strong vacuum is applied:
    a. evacuation
    b. leak detection
    c. refrigerant identification
    d. recharge

SHOP ACTIVITY

- Go to a machine with air conditioning and identify its main components. Determine which type of system it is (orifice tube or TXV).
- Locate the service ports for it. Are they near the compressor?
- Is there a tag that tells you how much refrigerant the system contains? Does the tag also tell you what type of oil is in the system?

SUMMARY

1. Safety concerns related to operator stations include: checking all safety related features that are part of the operator station to ensure their proper operation (such as: seat belt, mirrors, steps, handrails, machine controls); fall arrest apparatus may be required, air-conditioning systems can produce extreme temperatures so burn and frostbite is a possibility.
2. Operator stations or cabs have changed a great deal in the past decades to ensure operators can be comfortable to reduce fatigue and increase production.
3. Machine controls have evolved from heavy levers and pedals to small electronic joysticks.
4. Operator displays provide information to the operator related to how the machine systems are reacting to the operating conditions the machine is under.
5. Displays can be simple mechanical gauges that move their needle with a Bourdon tube to show temperatures, levels, and pressures.
6. Dash display warning systems alert the operator to machine system conditions that are outside of normal parameters. This should prompt the operator to change how the machine is being operated or component damage is likely to occur.
7. Electrical gauges use sending units to send variable electrical signals that make their needle move.
8. Electronic gauges have their needle move in reaction to an ECM output. The ECM receives an input from a sending unit.
9. Machines can be equipped with telematic systems that can transmit live and stored data from the machine to a remote location. They can be Bluetooth, cell based or satellite based systems.
10. Machines used to grade material can have GPS integrated into the machine hydraulic system.
11. Operator stations need to protect the operator from falling objects (FOPS) or rollover (ROPS). These structures need to be inspected regularly for damage and CANNOT be welded on or drilled into or modified in any way.
12. Many machine displays now incorporate blind spot display systems that use one or more cameras to show the operator what they normally can't see around the machine.
13. Operator seats can have suspension systems with air compressors.
14. Seat belts are a critical safety feature and require regular inspection and replacement.
15. If a cab must be tilted the proper lock must be in place before anyone goes under the cab.
16. HVAC (heating, ventilation, air conditioning) systems should pressurize the cab and keep heat and humidity levels to a comfortable level.
17. Cab fans will draw outside air past a filter to give the inside a slightly higher than atmospheric pressure to keep out dust. Cab fans usually have three speeds and may use a stepped resistor to give it different speeds.
18. Cab heaters circulate engine coolant through a heat exchanger and a fan pushes air past it to heat the cab up on cold days. Heater control valves can regulate the amount of coolant running through the heater.
19. Air-conditioning systems remove heat from cab interiors on hot days. They will also remove humidity.
20. An A/C compressor moves a refrigerant through a sealed system that has two heat exchangers. The properties of the refrigerant allow it to boil when exposed to warm cab air in the evaporator. The refrigerant then moves to the condenser where it condenses as it cools down. The cycle continues as long as the compressor moves the refrigerant and air moves past the heat exchangers.
21. Two main types of HVAC systems are: orifice tube and thermal expansion valve (TXV). These components create a pressure drop in the system.
22. Presently machines use R-134 for refrigerant but this will change to a more environmentally friendly refrigerant in the future. Technicians MUST be aware of ALL government regulations related to handling refrigerants and servicing A/C systems.
23. A/C systems require a specific type and amount of oil in them.
24. HVAC evaporators are located in the cab usually with the cab heater and fan and will have a filter to prevent their fins from getting dirty.
25. HVAC compressors have an electromagnetic clutch that is belt driven and engaged when the HVAC controls send an electric current to it.
26. HVAC condensers are usually located near the engines cooling fan to use the air flow from the fan to shed heat. They will sometimes have their own fans that use electric motors.
27. TXV valves are variable orifices with a temperature sensing bulb, an orifice tube is a fixed restriction with a screen.
28. Orifice tube systems use accumulators to trap moisture.
29. TXV systems use receiver driers to trap moisture.
30. HVAC hoses, lines, and fittings need to be compatible with refrigerant and withstand heat and temperature changes.
31. A pressure switch will be in the clutch circuit and open if system pressure gets too high.
32. HVAC systems with climate control allow the operator to set a temperature and an ECM will control the system to achieve it.

33. HVAC systems need regular maintenance to operate properly. This includes: filter cleaning/changing, belt tension adjustment.
34. Cab heater problems can include: no heat, too much heat, no air movement, leaks.
35. Heater diagnostics includes: a thorough visual inspection, coolant level check, control valve check, filter check, fan check.
36. A/C problems include: cab too hot, cab too cold, no air movement.
37. A/C diagnostics include: thorough visual check, set A/C to maximum and check clutch engagement, vent air temp, air flow at evaporator and condenser.
38. Refrigerant leaks must be found and repaired before recharging. System can be pressurized with nitrogen and dye injected to assist finding leak.
39. Recovery and recycling of refrigerant is mandatory and easy to perform with proper equipment and training
40. System must be recharged with correct amount and type of refrigerant and oil.

Glossary

Accumulator (hydraulic) A device to store high pressure oil.
Accumulator (air conditioning) Prevents any liquid refrigerant from getting into the compressor, serves as a reservoir of refrigerant, and contains desiccant that will remove any moisture (water) in the system.
Ackerman angle The angle created between an imaginary point that extends from the centerline of the rear axle to one side of the machine and intersects the centerline of both steering knuckle spindle shafts when they are turned.
Actuator (hydraulic) A system that converts the hydraulic energy created by the pump into mechanical energy.
Actuator (hydraulic brakes) Wheel cylinders in hydraulic brake system.
Aeration A process by which air is circulated or mixed in a liquid.
Air compressors A belt driven or gear driven component that drives other components such as fuel pumps or small hydraulic pumps.
ampere A unit measuring electron flow.
Area A section or a portion or a part.
Armature Rotating part of the motor that has several windings that have each of their ends loop to a commutator bar.
Aspect ratio A ratio of the height of the sidewall in comparison to the section width.
Atmospheric Relating to atmosphere; the pressure is measured in psi and is the weight of a 1 square inch column of air
Auxiliary drive Drives hydraulic pumps that need to be driven.
Axial piston pump Works on the same principle as a bent axis piston pump as far as the pumping action that occurs between the rotating cylinder and a set of pistons.
Backlash Allows gears to have lubrication between the meshing teeth.
Band brake Slows down or locks one track up when steering the machine.
Bar Metric unit for pressure; kilograms per centimeter squared (kg/cm^2)
Bead A term for the edge of a tire that sits on a wheel.
Bearing adjusting A washer and nut assembly that allows smooth wheel rotation while supporting part of the weight of the machine.
Bent axis piston pump Pumps used on small to medium sized excavators with a moveable cylinder block that when moved changes the effective stoke of the pump's pistons, which in turn changes the pump's displacement.
Biased Occurs when the anode is connected to the negative side of the power source.
Bipolar Two diodes that are combined back to back with the two similar parts of the diodes joined in the center as one.
Bottom roller An assembly at the bottom of a track frame to support the weight of the machine.
Brake chamber A machine component that mechanically actuates the foundation brakes.
Brushes A graphite material that rides on the rotor slip rings and transfers current into the rotor.
Brushes Made of a hard conductive combination of carbon and copper material and constantly transfer current to and from the armature commutator.
Brushless A stationary coil with pole pieces rotating over top of it.
Caliper One or more movable piston that squeezes friction material (pads) against the rotor.
Camshaft An internal combustion engine that makes it possible for the engine's valves to open and close.
Capacitor Stores energy to supplement the generator at certain times.

Car body The upper structu... indefinitely with a counterweig...
Cardan Universal joints used between two components that a... need to be allowed to move in re...
Carrier roller Mounted on top o... supports the chain assembly as it... the idler on the top side of the trac...
Cartridge filter A filter designed ... material from fluids.
Case drain filter A filter that can re... any internal wear.
CAT III Multimeter used to confirm th... conductor or component.
Cavitation Occurs when the vibrations pressures inside the cylinder cause small... outside of the cylinder liner.
Charge pump A gear type of pump that ... the volume.
Charging valve Diverts pump flow from an... the accumulators when the cut-in pressure is...
Closed loop A system that sends the return... pump.
Clutch disc Driven component of the clutch th... riveted or bonded or both on to a metal disc.
Combination valve A valve that acts both as a ... a check valve depending on exposed to oil pressu... pressure side or the low pressure-side of the loop.
Commutator Moving part of a rotary electrical sw...
Compressor The pump for an air-conditioning syst...
connector A device that joins wires to each other or t...
Contact pattern A step that checks if the crown and pi... are meshing properly.
Cooler Dissipates the heat from the clutch using air or engi... coolant.
Cooling fan A device for creating a current of air for cooling.
Corrosion A condition where sulphate deposited on the plates during discharge is left too long.
Counter-rotate Tracks turning in opposite directions while enabling the machine to perform spot turns for turn in tight quarters.
Countershaft (single) Gear and shaft arrangement that provides the drive flow through the transmission.
Countershaft (multiple) Run parallel with the main shaft and extra countershafts spread the gear and bearing loads out more evenly.
Crab steering Both axles steer at the same time in the same direction to make the machine move diagonally.
Crankshaft Helps to smooth out the pulsing action occurring between the various cylinders; the heaviest moving component in the diesel engine.
Crown gear A gear which drives the axle's differential; also changes the direction of power flow.
Cycle times The total time from the beginning to the end of the process.
Dash size One dash size equals 1/16 of an inch.
Destroke A state in which a pump control valve changes the pump to a lower displacement.
Diode Eliminates induced current in coils or rectifies a C output from alternators.

GLOSSARY 693

Directional control valve Allows the operator to direct pump flow to one or more cylinders.

Disconnect switch Occurs on the ground side but is occasionally on the positive side of the system.

Double rod cylinder If there is one cylinder, it will be attached to a tie rod. The double rod cylinder that has each end attached to the steering tie rods are attached to the steering knuckles that pivot on the outer ends of the axle housing.

Drivetrain Makes machines move; starts at the prime mover and ends at the machine's tires.

Dual path system A system that transfers torque into a gear reduction to further increase torque but decrease speed.

Dynamic braking A system used to hold a machine in place; are usually called retarders but are sometimes called secondary brakes. This is dynamic braking.

Ejector Lifts the material out of the bowl after the apron is lifted.

electroly... LCD gauge arrangement that featuring an ... allows the operator to scroll through different termin... virtual gauges.

elect... produces electron flow when the battery's ... a complete circuit.

El... part of a very small particle called an atom.

...at the front of the track frames together, it pivots from center of the machine main frame. It is limited by top of the crossbar and the bottom of the main frame.

...ary Gear which turns the planet carrier that turns ... shaft.

...move or empty the contents.

...eavy copper windings that create a strong magnetic field ... flows through them.

...t transistor A device that enables to use one electrical ...ontrol another.

...splacement Displaces a fixed volume of fluid for every ...on of the input shaft.

...meter A device to measure the rate of oil movement through a ...or tube.

...w To move freely.

...id sampling A sample taken using a sampling valve when the machine is running and the fluid is warm.

Flushing valve A valve that provides a constant flow of oil to the tank to cool and lubricate the motor and pump when the pump is moved from neutral.

FOPS Falling object protection systems are designed to protect the operator from heavy objects that may fall on the machine's cab.

Foundation brakes A device found near wheel assemblies consisting of rotating components and frictional components.

Friction discs Torque transfer components (plates and discs) inside a clutch that are alternately stacked together from 2 to 10 each in number.

Gear pump The two rotating shafts inside the pump having gears with teeth on their outside diameter that rotate inside the pump housing.

Gear reduction Ratio of the teeth on the drive gear versus teeth on the drive where the drive speed is under the output speed.

Gooseneck Used in Cushion Hitch where it facilitates the movement of a rod in and out of the barrel to provide suspension between the two main parts of the machine.

Governor An air line connected to the supply air tank that allows to sense system pressure.

Ground driven pump A type of steering system that has its shaft turned whenever the machine is moving.

ground The negative side of the circuit.

Heat loss Wasted horsepower; Heat (BTU/hr) = Flow (gpm) × Pressure drop (psi) × 1.5.

Helical Type of gear with angled teeth, spreads out the load over more teeth as it meshes with another gear.

Hold in One of the two windings in Solenoids.

Horsepower The rate of work done by an engine.

Hose Made to withstand very high pressure or moderate vacuum; mostly used when two components need to move in relation to each other.

Hydrostatic A type of drivetrain that uses hydraulic pumps driven by the prime mover to send fluid to hydraulic motors.

IGBT Electronic device that can be turned on and off in less than 500 nanoseconds, manipulates DC bus voltage.

Impeller Component of a torque converter which spins at engine speed and throws oil with centrifugal force at the turbine.

Inputs Receives electrical information (signals) from sensors and switches as inputs.

Internal leakage The unintentional movement of fluid from one component to another.

Inverter Receives signals from the ECM in its transistors to control the operations of the transistors.

ISO viscosity rating ISO viscosity numbers go from ISO VG (viscosity grade) 22 to ISO VG 100.

Joystick A part of an electronic control system for a hydrostatic dozer.

Kingpin A shaft that is held stationary in the steering axle.

Leveling valve A single circuit directional control valve that controls oil flow to the head end of the load cylinder in Cushion Hitch.

Levers A device that can be used to multiply input force to get an increased output force.

Limited slip A gear arrangement that equalizes the amount of torque sent to each wheel.

Lip-type seal Creates a seal between the seal lip and the mating circumferential surface of a component or shaft.

Load cylinder An assembly of barrel, head, rod, piston, and seals that provides a cushion between the main frame sections. It must be locked down to prevent movement between the tractor and scraper.

load The device that the circuit powers when current flows through it.

Load-sensing pressure compensated A system that controls a pump to produce only the flow that is needed based on the system load.

Lock out Isolates energy from the machine or system which physically locks the system in a safe mode.

Lock ring Seated in a grove in the rim base it secures the bead in the wheel hub.

Lock up clutch Locks the impeller to the stator and eliminates the problem of internal leakage due to necessary clearances between the moving parts of a torque converter.

Locked down Prevents movement between the tractor and scraper.

Locking differential Mechanisms that ensure that the axle shafts turn at the same speed as the crown gear.

Long block Refers to the main mechanical components that are needed to complete the combustion process.

Margin pressure The pressure set above the system pressure that controls the pump control and is usually 300–400 psi higher than system pressure.

Master cylinder Transfers operator effort to the friction brakes.

Master link Joins a track assembly and makes it a continuous loop.

Micron A very small measurement based on a meter.

Modulation Increases the clutch application pressure at a controlled rate to make a smooth shift or give a cushion effect when a speed or direction shift change is made.

Motor A machine that produces movement.

Multi-disc Brakes that could be mounted out at the end of the axle near the wheels or close to the center of the differential.

Neo-con A silicone-based fluid that is exclusive to Hitachi truck suspension cylinders.

Neutralizer valve A valve that stops oil flow to the steering cylinders when the machine is fully turned; used in articulated machines to stop the turning action when the machine is fully turned.

Nitrogen Gas used by the majority of suspension struts as a precharge and SAE transmission oil.

Non-servo A type of drum brake in which the wheel cylinder acts on the toe of each shoe and the heel of each shoe pivots on a common anchor pin.

Ohm Electrical resistance measured in units called ohms and represented by the Greek letter O.

Oil applied Type of steering clutches that stay disengaged until oil pressure is applied to a piston.

open An open circuit fault simply means a part of the circuit that is normally complete has opened. This is just like opening a switch and results in the load not working (light stays off, wipers don't move, etc.).

Operating angle Angles between parts of a machine's driveline that, when properly aligned, give a smooth operation and maximum component life to all related driveline components.

Orbital valve A valve with a rotary spool valve section and a gerotor pump/motor section.

Outputs Stores information and sends out electrical signals to control machine systems or display information to the operator.

Overdrive Ratio of the teeth on the drive gear versus teeth on the driven where the drive speed is over the driven speed.

Oxy-acetylene Torches used for metal cutting, gas welding or heating.

Packer A machine that compacts material as it moves over it.

Pascal's law Pressure created by an external force acting on a fluid inside a sealed container applies the same amount of pressure to any and all surfaces equally and at right angles to the inside of that container. F = P * A (F is force, P is pressure, and A is area).

Payload The carrying capacity of a vehicle.

Phasing Positioning of drive shafts to ensure minimal driveline vibrations.

Pinion gear A gear with small number of teeth.

Piston A solid cylinder or disk that fits snugly into a larger cylinder and moves under fluid pressure.

Pivot shaft Extends from the side of the main frame and has two smooth surfaces that align with plain bushings in the track frame.

Planet carrier A gear that contains the shaft upon which the planet pinion rotates.

Planetary Gear arrangement that uses one or more planetary gear sets, a combination of hydraulic clutches, and a control system.

Planetary pinion carrier Component of a planetary gear set used to transfer torque.

Plates Thin round slices of steel or cast iron that will have either teeth on the inside or outside diameter or tangs on the outside diameter.

Plies Multiple layers of material in the tire carcass or casing that form a base for the tire from bead to bead.

Ply rated Relates to the amount of plies that a tire has.

Pneumatic State of the tires on a machine being hollow.

Position sensor Electronic joystick controls are position sensors; an input to an ECM.

Power divider A set of gears that are part of a housing that contains a differential assembly for the forward axle, a second set of gears, and a splined shaft that sends power out the back of the housing to the rear axle.

Power shift A method that sends power mechanically through other driveline components to drive the tires or tracks of the machines.

Pressure gauge A device that measures the pressure of fluid.

Pressure plate Movable spring loaded and machined disc that applies pressure to the clutch disc when released toward the flywheel.

Pressure Amount of force applied to a specific area.

Prime mover A group of machines that transforms energy from thermal, electrical, or pressure into mechanical form in different sources.

Proportional solenoids Output sensor in a transmission electronic control system.

Pull in One of the two windings in Solenoids.

Pump A component that creates fluid flow.

PWM Signal provided to each solenoid valve.

R-134 A replacement for the refrigerant R-12 that is much less harmful to the environment.

Radial tires A type of tire construction that gives the tire more flexible sidewalls, which results in a smoother ride and bigger contact patch.

Ratio Comparison between two things that helps to understand their relationship.

Receiver drier Serves as storage for liquid refrigerant and removes any water moisture from the system.

Recharge The process of emptying refilling with the exact amount of refrigerant as specified by the machine manufacturer.

Reciprocating Converts chemical energy in a fuel into work by combusting the fuel in air to produce heat.

Recover A process that when done right allows the reuse of a refrigerant.

Rectification The conversion of AC to DC.

Rectifier bridge A set of diodes that converts the AC to DC.

Regulator (alternator) A device that controls alternator output voltage levels and ultimately the system voltage level.

Regulator (gas pressure) An assembly that feeds the gas precharge to regulate the pressure to around 400 psi.

relay An electromagnetic switching device that allows a low-current circuit to control a high-current circuit.

Release bearing Commonly known as throw out bearing, indirectly moves the pressure plate away from the flywheel.

Reservoir Holds a supply of hydraulic fluid for the system to use whenever needed.

resistance Resistance in an electrical circuit opposes current flow and is measured in ohms (Ω) with an ohmmeter.

Reverse idler Changes the output shafts direction and is driven by a gear on the countershaft.

Ring gear Component of a planetary gear set used to transfer torque.

Ring gear Drives a carrier that drives a cross on which four spider gears can spin.

Rock ejectors Metal bars that hang down from the truck's box and help to remove rocks stuck between dual tires while operating.

ROPS Rollover protection systems keep the cab intact when the machine turns either on its side or completely upside down.

Rotary flow Occurs when the oil is mostly moving in the same direction as the impeller, and is hard to achieve if there is any load on the turbine.

Rotor (alternator) A rotating part of an electrical device that creates several sets of alternating magnetic fields.

Rotor (retarder) Moving part of the retarder and has vanes or fins on one or both sides.

Rotor (electric drive) Shaft that spins on a bearing at each end, runs in oil for lubrication and cooling.

SAE viscosity rating SAE viscosity numbers for machine fluids go from SAE 0W to SAE 60.

Scraper An articulated four-wheeled machine that pulls a bowl that collects material (earth) when the cutting edge "scrapes" the ground.

Seal Joins hoses and tubes or connect components together.

Seal rings Contains the rotating shaft which gets the transferred oil pressure from the stationary manifold during clutch actuation.

Sealed and lubricated The most common chain style that has the same basic components of the sealed style and has each pin and bushing joint lubricated with a small amount of oil to reduce wear.

Secondary steering A safety backup system to ensure that the machine can be steered to a parked position.

Self-energizing Happens when the toe of a shoe gets pushed into the direction of rotation that the drum is turning, and this in turn pulls the shoe into the drum.

Semiconductor Unique structures that can generate voltage when they are squeezed and move when they have voltage applied to them.

Service brakes Slows down and stops the machine.

Service intervals Time specified for undertaking maintenance procedures based on the working hours of the machine or fuel consumption.

Servo A type of lighter-duty drum brakes that use two shoes that are joined together with an adjuster and will work together as one shoe. This will allow both shoes to be self-energized.

GLOSSARY **695**

Set point Voltage regulation for 12 volt systems establishes a maximum charging voltage known as the set point.
short Occurs when the normal path for current flow is diverted to another load.
Side gear Allows a change of speed between the crown gear and the side gear.
Silicon Chemical component available in sand and glass; silicon is a metal-like material.
Slack adjuster A part of the brake system used to control the brakes as needed.
Spindle A part of suspension system that carries the hub for the wheel and attaches to the upper and lower control arms.
Spin-on filter A metal housing with a threaded base for mounting the filter media, filter media end plates, and a spring to keep the media assembly in place inside the housing.
Spline Teeth on a shaft assembly that transfer torque to components.
Spot turn Enables tracks or tires to drive in the opposite direction.
Spring applied Type of steering clutches that rely on spring pressure to squeeze a stack of friction discs and steel plates together to transfer torque through them.
Spring rate A measure of the ability of a spring to resist deflection.
Spring shackle A spring assembly that has eyes on both ends with one end being anchored and the opposite end being allowed to move.
Spur Simplest type of gears with cut teeth on the outside circumference of the gear, creates radial loads when they are loaded.
Starter interlock A system that verifies conditions with switches, sensors, or a keypad before the engine starts.
Starter relay A system that switches larger current to start the motor.
Static braking Brake systems that slow down machines do so by converting kinetic energy into heat energy.
Stator (torque converter) Component of a torque convertor, redirects the oil that leaves turbine back to the impeller.
Stator (electric drive) Stationary series of copper-insulated wires that are wound in place in the generator housing.
Stator Three lengths of wire arranged in many loops; induces AC voltage and sends the current to the rectifier for conversion to DC.
Steering lever Lever used in old steering controls.
Steering motor Controlled by the operator. As it rotates, it makes the machine turn left and right.
Steering pedal Linkage that is reset by making adjustments such as loosening brake bands, adjusting linkage inside transverse housing, adjusting stops, and adjusting steering valve etc.
Steering planetary Provides a speed difference between the left and right tracks when used in conjunction with a hydraulic motor.
Stick A part attached to the end of the boom of an excavator; has a bucket attached to it.
Strut A suspension component that provides both a spring action and a damping function for the machine.
Sun gear Moves the planet pinions around the ring gear.
Swash plate A device that changes the effective stroke of the pistons that are carried around in a cylinder block.
switch A simple switch would be a single pole, single throw, where the switch has two terminals, and when actuated it either opens or closes contacts to provide a path for the electrons to flow through or complete the circuit.
Synchronizer Helps match the speeds of the main shaft and the main shaft gears before they are locked together.
Tag out A safety procedure that helps to ensure that the machine does not get started without prior approval from the person who tagged the machine.
Tapered roller bearing Cone shaped rollers used in wheel bearings, pinion gear shafts and crown gear assemblies.

Tensile strength Relates to the maximum stretch that a fastener will withstand before becoming deformed.
Thread crest The surface of the thread which connects the crest with the root.
Thread pitch The distance between threads.
Three phase Stable electricity generated by three separate stator windings which produce three sine waves for every revolution of the rotor.
Toe-in Angle that can be adjusted on heavy-duty steering axles by changing the length of the tie rods.
Torque multiplier A set of gears that increases the torque input and transfers it to the outer drive.
Torque rod A series of linkages that hold the rear axles in place but also allow them to move in relation to the trucks frame.
Torque Rotational force
Track frame A very strong part of the machine that supports the weight of the machine. It is the mounting point for track rollers, carrier rollers, track recoil spring, track adjuster, and one or two idlers.
Track pitch The measurement taken between the center points of two track pins.
Track tension A hub component to regulate the amount of grease in the track frame.
Transfer case Sends the drive output from the transmission to the front and rear axles.
Transistor An electronic device that has three terminals; base terminal controls the current flow from the collector to the emitter.
Trunnion Lugs on the cross part of a universal joint with hardened surfaces that mate with needle bearings that are held in a cup or wing cap depending on the universal joint style.
Tube Moves hydraulic fluid from one component to another.
Tubeless An aspect of tires that does not mandate an inner tube to hold air.
Turbine Torque convertor output that drives the transmission's input shaft.
Underdrive Ratio of the teeth on the drive gear versus teeth on the drive where the drive speed is under the output speed.
Universal joint Part of a drive shaft assembly between two components that are not in perfect alignment or that need to be allowed to move in relation to each other.
Upstroke A state in which a pump control valve changes the pump to a higher displacement.
Valve stems An assembly of base, retaining nut, body, and cap that extends from the tire in the wheel hub.
Viscosity Measures the flow of the hydraulic fluid.
Viscosity Refers to the fluid's resistance to flow.
voltage An expression of the potential difference in charge between two points; referred as electromotive force (EMF).
Voltage drop The voltage loss that occurs in a circuit.
Volume Space enclosed within a container
Vortex flow State where the load on a turbine stops it completely or stalls it while the impeller oil flow is trying to turn it.
Waste oil container A device that collects used machine oil drawn out by a vacuum source through evacuation valves.
Wheel cylinder Converts hydraulic pressure coming from the master cylinder into force.
Windings Stagger evenly around the stator frame.
Yoke Ends of a drive shaft that are cast and machined parts and are welded to the tube.
Zener diode A special diode that breaks down and permits a reverse flow of current when the voltage reaches a certain value, without damaging the semiconductor material.

Index

Page numbers in *italics* indicate figures:

A

Absorbers, shock, 609
  heavy-duty, *609*
Accumulators, 126–127, 484–485, 537
  air-conditioning system, 683, *683*
  charging valves, 552, *552*
  cushion hitch, *616*
  piston and bladder, *484*
  piston-type, *127*
  used for ride control, *485*
Ackerman angle, 567–568, *568*
Active struts, 612–613
Actuators, 116–119, 544
  rotary (motors), 118–119
Addendum, 332
Additives, 86
Advanced hydraulic systems, 459–502
  component replacement procedures, 501
  diagnostics
    environmental concerns, 498
  problems, 501
    flow, 501
    heat, 501
    leakage, 501
    noise and vibration, 501
    pressure, 501
  testing, 498–501
    and adjusting, 501
    fittings and hoses, 500
    flow meters, 500
    portable hydraulic hand pump, 500
    pressure gauges, 498–500
    tachometer, 500
    temperature probe and heat gun, 500
Aeration, 112
Air bags, 616, *617*
Air compressors, 647
Air-conditioning system, 676–678
  accumulators, 683, *683*
  closed loop fluid system, 677
  compressor, 677, 680–681, *681*
  condenser, 681, *681*, *682*
  diagnosis, 687–688
  evaporator, 679–680, *680*
  fittings, 684
  heat exchangers, 678
  hoses, 684, *685*
  humidity reduction, 684, 686
  lines, 684
  oil, 679, *679*
  orifice tube system, 678, *678*, 683, *683*
  receiver drier, 684, *684*
  refrigerant, 677, 678–679
  switches, 684, *685*
  TXV, 678, *678*, 682, *682*, *683*
  types, 678, *678*
Air cooled engine, 196, *197*
Air dryer, 639–640, *640*
Air filters, 95–96
  restriction gauge, 191, *193*
Air over oil brake systems, 550, *550*, 642
Air starting systems
  components, 230
  procedures, 229–231
Air vent, thermometer in, 687, *687*
Allen keys. *See* Torx wrenches
All-wheel drive machine, *422*
All-wheel drive option, 340
All-wheel steering systems, 570–571
  mode, 570, *573*
Alternative current (AC) electric drive, 395
Alternators
  arrangement on machine, *240*
  bench testing, 256
  brushless rotor, 244
  brush-type rotor, 243–244

cable voltage drop test, 257
  components, 243
  cooling, 243
  cutaway of, *243*
  disadvantages, 257
  drive, 241–242
  drive pulleys, 242, *242*
  housing, 243
  operating principles, 241–246
  self-exciting, 244
  stator, 244–246
  24V battery pack, *241*
  wiring connections, 252–253
American Wire Gauge (AWG), 158
  vs. metric wire sizes, 158
Ammeter, 143, *144*, 165–166
  clamp-on, *166*
Ampere, 139
Analog gauges, 499
AND gate, 275
Anti cavitation valve
  and combination port relief, *481*
Anti-friction bearings, 41–44
  ball bearings, 41–42, *43*
  roller bearings, 42, *43*
Arcair cutting equipment, 62, *62*
Area, 107, 108–109
Armature, 218, *221–222*
Articulated steering systems, 561, *561*, 569–570, *573*
Articulated truck
  hydraulic system, *574*
  inspection checklist of, 76, *77*
Aspect ratio, 639
Atmospheric pressure, 135
Atoms, 140
  copper, *141*
  with full valence shell, *141*
Attachments, 17, *17*
Auto idle, 492
Automatic greaser pump unit, *84*
Autonomous machine, 667
Autoshifting, 376
Auxiliary circuits, 491
Auxiliary drives, 413, 423–424, *424*
Axial piston motor
  bent, 523, *524*
  hydrostatic drive systems, 521–522
  variable displacement, *522*
Axial piston pumps, 467–469, 516
  cross-section view, 468, *469*
  lighter-duty, 518
  with pressure compensator valve, 471, *471*
  and swash plate, 467
  swash plate control, *517*
Axle filters, 83
Axle/final drive fluid, 92
  change, 83
Axle lock cylinders, *608*
Axle toe measurement, *580*

B

Backhoe, tractor loader, 13–14, *14*
Backlash, 332, 449
Back support, 8
Backup cameras and vision cameras, 76
Backup modes, 492
Backup/travel alarm, 74
Ball bearings, 41–42, *43*
Ball joint, *564*
Band brake, 591, *592*
Bar, 110
Batteries
  boosting, 156
  charging, 156
  cutaway view of, *154*
  lead-acid, 153–156
  maintenance, 84

maintenance-free, 155
  ratings, 155
  safety warning, *138*
  secured and maintained, *155*
  testing, 155–156
  two different battery connections, *153*
Battery cables, 227–228, *228*
Battery positive cable, 253
Bead, 633
Bearings, 40–44, 345
  adjusting, 645
  anti-friction, 41–44
    ball bearings, 41–42, *43*
    roller bearings, 42, *43*
  failures, 44–45, *45*
  friction type, 41
    plain, 41
  and heavy equipment, 40
Belt driven fan *vs.* hydraulically driven fan, *198*
Belts, seat, 75, 673–674, *674*
Bench testing
  alternators, 256
  and electric starter, 227
Bent axis motor
  hydrostatic drive systems, 522–523
  piston motor, 465–467, *466*, 523, *524*
  variable displacement, *523*
Bevel gears, 333
Biased, 263
Biodegradable hydraulic fluid, 92, *92*, *122*
Biodiesel identification, *90*
Bipolar transistors
  base, 267
  collector, 267
  definition, 266
  emitter, 267
Bladder type brakes, 548
Blind spot camera, 673, *673*
Blocking equipment, 59, *59*
Blowby, excessive, 199
Bobcat, 12
Bogie suspension system, *618*
Bolt grades and tensile strength, 34–35, *35*
Boosted hydraulic brake systems, 550–551
Boosting batteries, 156
Bottom roller, 650
Bourdon Tube
  pressure gauge, 667, *667*
  temperature gauges, 668
Brake controls, 549, 666
Brake cutaway and motor, *511*
Brake pedal and valve, *543*
Brake(s)
  air over oil, 642
  bladder type, 548
  caliper and rotor, 546–547
  chamber, 651
  components, 552–553
  cooling systems, 553
  effort, 539–541
  factors that affect
    brake swept area, 542
    braking leverage, 541–542
    coefficient of friction, 542
    higher clamping force, 542
  fixed caliper-type, *547*
  fluid, 92, 555, *555*
    reservoir, 554
    service, 554–555
  foundation, 538
  friction-type foundation, 544–548
  fundamentals, 538–539
  for heavy equipment, 538
  hydraulic brake system, *544*
  and machines, 538
  mechanical park, *544*

multi-disc steering, 592–593, *594*
  operation testing, 553–554
  power flow older steering clutches and, *588*
  primitive braking system, *540*
  reconditioning and steering clutch, 602
  release pump, 553
  release tool, *641*
  release valves, 527, *528*
  repairs, 555–556
  rotor, 554, *555*
  S-cam foundation, 641–642, *642*
  schematic, *640*
  service and parking chamber of, *641*
  shoe, 546
  single and multi-disc foundation, *547*, 547–548
  steering, 591–592
  troubleshooting, 555
Brake system integrity, and hydraulic brake systems, 537
Braking leverage, 541–542
Braking resistor, 407–408
Broken fastener removers, 49
Brushes, 218, *222*, 244
Brush holders, 218
Brushless alternator rotor, 244, *245*
Brush-type rotor, 243–244, *244*
Bull dozer, 10
  with simple electrical system, 139, *139*
Burns, 101
  and hydraulic brake systems, 537
Bypass orifice, 575

C

Cab(s). *See also* HVAC (heating, ventilation, and air-conditioning) systems
  air-conditioning system. *See* Air-conditioning system
  brake controls, 666
  controls, 665–666
  electric gauges, 669–670
  electronic gauges, 670–671
  GPS (global positioning system) system, 672, *672*
  heaters, 675–676, 686–687
  humidity, 684, 686
  mechanical gauges, *667*, 667–668, *668*
  mounts, 674
  noise prevention, 674
  pressurizer, 675–676
  protections, 672–674
  removal, 674
  safety concerns, 665
  telematic systems, 671–672, *672*
  tilting, 674, *674*
Cab air filter, 676
Cab escape tools, 75
Cab filter, 675, *675*
Cab filter change, 84
Cab heater, 200
Cable assemblies, 58, *58*
Cable excavator, *104*
Cab lights, 75
Calibrating procedures, 296
Calibration
  of functions, 497–498
  pressure gauges, 500
Caliper, 538
  brakes, 546–547
Camber, 566–567
Cam-in-block design, *188*
Cam lobe motor, 523–524, *525*
Camshaft, 188
  bearing, 41, *41*
Capacitors, 248, 271–272, *272*, 390

INDEX 697

Carbon resistors, 270
Cardan joints, 420
Carrier roller, 648, 650
Cartridge filters, 94–95, *95*
Cartridge-type valve block, *493*
Cartridge valves, 493–494
    types of, *494*
    used on grader, *495*
Case drain circuit, 527–528
Case drain filter, 532, *532*
Case drain oil, 511
Caster, 565–566
    negative, *567*
CAT III multimeter, 390, *390*
Cavitation, 199
Centerline, 332
Central processing units (CPUs), 274–275
Chain assembly, 653–654
Chain drive rear wheel, *414*
Chain drives, 456
Chains, 57–58, *58*
Charge pump, 508, 526
Charge relief, 527
Charging systems
    components, 240
    diagnosis, 254–256
    fault conditions and possible causes, 256
    machine *vs.* hydraulic, 240
    operating procedures, 240–241
    preventive maintenance practices, 254–257
    safety concerns, 240
Charging valve, 551
Char-Lynn valve, 574
Check valves, 125, *125*
Chisels, and punches, 48–49, *49*
Circle checks, 76
Circle dimensions, of cylinder, 108
Circuit breakers, 157–158, *158*
Clamping forces, 542
    created by different sizes of fasteners, *36*
Clamping tools, 50
Cleaning equipment, 50, *50*, 63, *65*
Clearance, 332
Climate control, HVAC system, 686
Closed center directional control valves, 464
Closed center hydraulic system *vs.* open center hydraulic systems, 462–464
Closed circuit *vs.* open circuit hydraulic system, 461, *462*
Closed loop, 508
    pump, 508
    systems
        hydrostatic dual pump, 509
Closed loop fluid system, 677. *See also* Air-conditioning system
Clutch
    actuation, 373
    brake, *308*
    cooling system, 309
    diagnosing problems, 309–310
    dog, 311
    dry-type, 309
    flywheel, 307–308, *308*
    friction-type flywheel clutches, 302–307
    maintenance, 309
    oil-applied steering, 591, *591*
    oil-cooled, 309
    one-way, 311
    over-center-type, 306
    overheated, 590
    pilot bearing, 308–309
    power flow older brakes, *588*
    protection, 376
    release mechanism, 307, *307*
    repair/replacement, 310–311
    safety concerns, 302
    shoe, 311
    sprag, 311
    spring-applied steering, *589*, 589–590
    wet, 309
Clutch cover, 304
Clutch disc, 304–305
Coefficient of friction, 540–541, *541*, 542
Combination port relief and anti cavitation valve, *481*
Combination valves, 479, 512, 526, *526*
Combiner valves, 490
Commutator, 221
Compactors, 13, *14*
Compensator pressure, 482

Component replacement procedures
    causes of, 501, 502
    and hydraulic systems, 501
Compressed air supply system, 637–639
Compression ignition, 177
Compression ratio, 178
Compressor
    air-conditioning system, 677, 680–681, *681*
    components on inlet stroke, *638*
    on engine, *638*
Compressor pulley, 681
Computer memory
    electronically erasable programmable read-only memory, 277
    programmable read-only memory, 276
    random-access memory, 277
    read-only memory, 275–276
Condenser, air-conditioning system, 681, *681*, *682*
Conductors, 158–159
    high-voltage, 401, *401*
Connecting rods, 184, *185*
Connectors, 159, 160
    fittings, hydraulic, 120, *120*
    pin extractor tools, *160*
    wiring, *159*
Constant flow, 487
Contact pattern, 452
Container handlers
    and heavy-duty forklifts, 16–17, *17*
Controller, 494
Controller area network (CAN) bus system
    diagnostics, 292–293
    operating procedures, 289–291
    usage of, 293
Controls, cab, 665–666
Control valve
    cab heater, 676, *676*
    load sensing system, 472, *473*
    section with load-sense passage, *474*
Conventional current flow, *142*
Conventional steering systems, 560, 568–569
    axle toe, *580*
    during cornering, *560*
    full hydraulic, 568, *570*
Conventional switches, 160
Conventional theory. *See* Electron
Coolant, 676
    change, 83
    engine, 91, 198–199
    heaters, 199–200
    temperature out of normal range, 207
Coolers, 125, 199
    fluid, 199
Cooling equipment, 62–63, 195–200
    large air tools, 63, *64*
    undercarriage tooling, 63, *64*
Cooling fan, 197–198
    belt driven fan *vs.* hydraulically driven fan, *198*
Copper atom, 141
Copper wires, 158
Corded electric tools, 52
Corroded fasteners, 35–36
Costs
    fuel, 19, *19*
    maintenance, 19
    operating and replacement parts, 19
    rental replacement, 20, *20*
    repair, 19
Cotter pin, and master pin, *655*
Counterbalance valves, 478–479
Counter-rotate, 586
Countershaft, 336, 356
Countershaft powershift transmission
    definition, 352, 356
    description, 355–357
    reverser/shuttle shift, 357–366
Coveralls/safety vest, 8
CPUs. *See* Central processing units
Crab steering, 571
Cranes, 11–12, *12*
    with multiple steer axles, *561*
Crankshaft, 187
    four-cylinder, *187*
    throw, 178
    throw equals more torque, 180
Crawler tractor. *See* Dozer
Critical speed, 419
Crossover relief valves, 576
Cross-section
    pilot control valve, *476*

power steering box, 565, *567*
of steering module, 594, *595*
Crown gears, 334, 433
Crushing hazards, 102–103
    and hydraulic brake systems, 537
Current flow
    hold-in and pull-in windings, 224
    magnetic lines of force inducing, *241*
    opposing magnetic fields, 220
    series of wound motor, 222
Current rectification
    alternator wiring connections, 252–253
    description, 246–247
    diode trio, 248
    rectifier bridge, 247, 247–248
    smoothing capacitors, 248
    voltage regulators, 248–252
Cushion hitch, 614–616
    accumulator, 616
    for scraper, 615
Cutaway view of batteries, 154
Cutting and filing tools, 49, *49*
Cycle times, *131*, 500
    chart, *131*
Cylinder
    circle dimensions, 108
    cutaway view of, 117
    diameter, 134
    liners, 184–185, *185*
    log splitter, 134, *134*
    misfires, 207
Cylinder block, 189
    and pistons, 464–465
    and port plate, 467, *468*
    and port plate for dual output pump, *469*

D

D7E tiller, *403*
Darlington pair, 268
Dash size, 119
DC electrical circuits, 146–147
    negative side, 147
    positive side, 147
Dealership, heavy equipment, 18, *19*
Dedendum, 332
Deere CT 315
    skid steer, 512, *513*
Deere/husco-distributed valve system, 497
Destroking state, 487, 499
Diagnose no-load test, 227
Diagnostic port
    and communication adapter connected to hydrostatic machine, *530*
Diaphragm-type pressure sender, 670, *670*. *See also* Electric gauges
Diesel, Rudolph, 177
Diesel engines, 177–180
    accessory systems, 191–204
        intake and exhaust systems, 191–195
    complex air inlet system, 192, *193*
    diagnosing, testing, adjusting, and repairing, 205
    dynamometer testing, *209*
    exhaust system with EGR, *194*
    fluid leaks, 205, 205–206
    liquid cooling system, *196*
    lubrication systems, 201
    main internal components, 178–179
    major failure, 205
    noises, 206
    overhead cam arrangement, *189*
    reconditioning, 208–209
    repairs and maintenance, 204–205
    tune-ups, 207–208, *208*
    vibrations, 206
Diesel engine starting systems
    importance of, 215–217
    safety concerns, 215
    troubleshooting, 233–236
    types of, 217
Diesel fuel, 90
    coolant, *91*
    label on a container of, *91*
Differential lock transfer case, *423*
Differential pressure. *See* Margin pressure
Differentials
    carrier removal, 448
    components, 441–442, *442*
    diagnostics, 447–448

double reduction, 447, *447*
limited slip, 443–444, *445*
locking, 445–446
no spin, 444–445, *445*
repair, 449–453
two-speed, 447, *447*
types of, 441
working procedures, 441–443
Differential steer machine, 596–600, *597*
    counter-rotation, 598, *599*
    diagnostics, 601
    straight, 597, *598*
    turning, 586, *587*
Digging cycle, 19–20
Digital gauge, *499*
    and transducer, *499*
Digital information outputs, 287
Digital multimeter, 165
Digital pressure meters, 500
Digital *vs.* analog devices, 273
Dimple, and master pin, *655*
Diodes
    light-emitting diode, 265–266
    operating process, 262–264
    photodiodes, 266
    rectifying, 264–265
    symbol for, *263*
    Zener, 265
Diode trio, 248
Direct current (DC) electric drive, 395
Directd current electrical systems, 137–173
    introduction to basic, 139
    with three equal series resistances, *148*
Direct drive electric starter
    components, 217–219
    definition, 217
Directional control valve (DCV), 115–116
    cutaway view of, 116
    with flow compensator, *481*
    in schematic form, *117*
    two styles of, 462, *463*
Dirty cab filter, 686, *687*
Disassembled transmission pump, *361*
Disconnect switch, 235
Discs, 589
Disc-type wheels, 624
Dither, 496
Dog clutches, 311
Doors/window, latches and handles, 75
Double reduction differentials, 447, *447*
Double reduction final drives, 456
Double-rod cylinder, *118*, 568, 569, *571*
Dowel pullers, 52
Downtime, 19–20, 32
Dozer, 10
    hydraulic system, *476*
DPF (diesel particulate filter) cleaning, 84
Drilling machines, 15–16, *16*, 49
Drill presses and grinders, 63–65
    cleaning equipment, 63, *65*
    fluid evacuation, filtering, and refilling equipment, 63–64
    welding machines, 65
D-ring seals, 37–38, *38*
Drive axles
    conventional, 432–433
    diagnostics, 438–439
    fluid cooling and filtration, 437
    inboard brakes, 433–435, *436*
    inboard final drive, 433
    lubrication, 437–438
    maintenance, 438
    removal, 439
    repair, 439–441
    safety concerns, 430
    steerable, 436–437
    types, 430–432
    wheel seals, 435–436
Drivelines
    arrangements, 414–415
    brakes, 419
    critical speed, 419
    maintenance, 424–425
    operating angles, 415–416
    operator complaints, 424
    overview of, 413–414
    parallel and broken back, *414*
    phasing, 418–419
    safety concerns, 413
    shaft assemblies, 416–418
    shaft/U-joint repair, 425–426
    support bearings, 420, *421*
    universal joints, 420–422
    wheel loader, *413*

Drive pulleys, 242, 249, 50
Drivers, and pullers, ing knuckle,
Drive steer axle st
 564, 566, panding shoe and
Drum brake. S
 drum
Dry-type c motor, 542
Dual calip 510
Dual pa
 co. See Metal-to-metal
Duals
 Du ing, 538

 Event-based shifting
 e Electronic control modules
 M. See Electronically erasable
 rogrammable read-only
 memory
 ical circuit
 uxiliary, 491
 with common ground, 147
 components, 153–157
   battery system, 153
 direct current, 146–147
 effect of resistance, 143, 144
 faults, 167–168
 with open circuit, 169, 169
 protection, 157–163
   circuit breakers, 157–158
   conductors, 158–159
   fuses, 157
   fusible link, 158
 regeneration, 490–491, 491
 with short-to-ground fault, 149
 simple, 147
 types, 147–150
   parallel circuit, 149, 149
   series circuit, 147–148, 148
   series-parallel circuits, 150, 150
 voltage drop, 168
Electrical components vs. hydraulic components, 140
Electrical loads, 162–163
   electric DC motors, 162–163, 163
   engine preheating devices, 163, 163
   horns/backup alarms, 163
   lights, 162, 162
Electrical resistance, 143–144
Electrical system fault diagnostics, 166–167
Electrical test
   equipment, 164–171
   multimeter, 164
   and repair tools, 50, 51
Electric DC motors, 162–163, 163
Electric drive loader, 393
Electric drive systems
   AC vs. DC, 396, 396
   advantages of, 394
   alternative current, 395
   braking resistor, 407–408
   diagnostics and repairs, 409
   direct current, 395
   disadvantages of, 395
   history of, 392–394
   inverter cooling systems, 401–403, 402
   maintenance, 408
   operator complaints, 409
   reasons for, 394
   servicing of, 391
   three-phase AC voltage generation, 396–397
   wheel loader, 138
Electric drive track-type tractor, 394
Electric gauges, 669–670
   balancing circuit, 669, 670
   diaphragm-type pressure sender, 670, 670
   needle, 670
   pressure gauge, 670, 670
   temperature gauge, 670
Electric mining shovel, 392
Electric power tools, and pneumatic power tool, 52
Electric propulsion, high-voltage, 390
Electric starter motor assembly
   battery cables, 227–228
   bench testing, 227
   control circuit, 228–229, 229
   cutaway, 218
   direct drive electric starter, 217

double starter assembly arrangement, 216
flywheel ring gear, 215
gear reduction starter, 217
operation procedures, 220–223
replacing starters, 226
soft start starters, 224
solenoid assembly, 223–224
starter drives, 225
thermal over-crank protection, 224
Electrohydraulic main control valve, 496
Electrohydraulic systems, 494–498
   controller, 494
   defined, 494
   inputs, 494–495
     sensors, 494–495
     switches, 495
   outputs, 495–496
   parts of, 494
Electrolyte (acid) solution, 153
Electromagnetism, and magnetism, 145–146
Electromotive force (EMF). See Voltage
Electron, 140–144
   flow measurement, 143, 144
   flow vs. oil flow, 142, 142
   free, 141
   movement in conductor by crossing flux lines, 146
   reasons for movement of, 141–142
   theory, 141
   work performance, 142–143
Electronically erasable programmable read-only memory (EEPROM), 277
Electronic control modules (ECMs), 139
   calibrating procedure, 296
   controller area network bus system, 289–293
   data link, 286
   multiplexing, 289
   replacement, 296–297
   software, 289
Electronic control modules (ECMs) inputs
   sensors, 283–286
   switches, 281–283
Electronic control modules (ECMs) outputs
   description, 286–287
   digital information outputs, 287
   high-side PWM proportional driver outputs, 287
   low-side/sinking driver outputs, 287
   on/off high-side driver outputs, 287
   sensor power supply outputs, 287
Electronic controls, 403–404
   for suspension systems, 616
Electronic devices
   CAB displays, 288
   capacitors, 271–272
   central processing unit, 274–275
   computer memory, 275–277
   digital vs. analog, 273
   diodes, 262–266
   ECM inputs and outputs, 280–287
   vs. electrical, 261
   electronic control modules, 277–278
   failure reasons, 278–279
   fault codes, 293–294
   information cycle, 279–280
   integrated circuits, 274
   logic gates, 275
   resistors, 269–271
   safety concerns, 260
   semiconductors, 262
   signal conditioning, 273–274
   transistor, 266–269
   usage reasons, 260–261
   wiring system, 280
Electronic gauges, 670–671
Electronic service information, 296
Electronic service tool (EST), 295
Electronic steering controls, 579
Electrostatic discharge (ESD), 268
Emission regulations, 18, 190
EMS (electronic-monitoring system) display, 668, 669
End cap, 218
Engine air filters/filtration devices, 80
Engine coolant, 91, 198–199
Engine governor control, 76
Engine inlet heater, 163, 163
Engine oil, 90–91
   filters, 81
   pressure gauge, 667

Engine starter switch, 160
Environment protection
   and heavy-duty equipment technician (HDET), 27
   and hydraulic brake systems, 537
   and hydraulic system, 103
Equalizer bar, 618–619, 619, 648, 649
Equalizer planetary, 598
ESD. See Electrostatic discharge
EST. See Electronic service tool
Evaporator, air-conditioning system, 679–680, 680
Event-based shifting (EBS), 373
Excavators, 11, 11
   control valve arrangement, 490
   cross-section view of, 487
   hydraulic systems, 485–490
   main control valves, 490, 490–494
   one track removed, 662
   open loop travel circuit for, 509
   pump assembly for a medium-size, 486
   working, 132
   mode control, 492
Excited rotor generators, 399, 399
Exhaust filters, 96
Exhaust gas recirculation (EGR) technology
   Tier 3 emission regulations, 194
Exhaust system, and intake systems, 191–195, 193
Expanding shoe and drum brakes, 545, 545
   with automatic adjuster, 554
   nonservo type, 546
   servo type, 546
Extensions, 47, 47–48
External gears, 333
External gear-type pump, 114
Extreme weather protection, 8

F

Fail-safe brake. See Reverse-modulating valves
Falling object protection system (FOPS), 672
Fan motors, 675, 675–676
Fasteners, 31–41
   anti-loosening devices/methods, 36
   clamping forces created by different sizes of, 36
   normal fastener rotation, 32, 32
   repair tools, 50, 51
   and rivets, 32, 32
   threaded fastener, typical, 33, 33
   inspection, 36
   repair tool, 51
   torqueing of, 36
   track pads, 657
Fault codes, 167, 167, 293–294
FET. See Field-effect transistor
Field coil, 218, 220
Field-effect transistor (FET), 266
Field service diagnostic call, 130
Field service HDET, 67, 68
Filing tools, and cutting tools, 49, 49
Fill plugs, and drain plugs, 83
Filter cart, 502
Filter change, 53, 53
Filters
   air, 95–96
   cab, 675, 675, 676
   cart and particle counter, 124
   cartridge, 94–95, 95
   exhaust, 95–96
   liquid, 93–94
   locations, 94
   machine, 93–96
   spin-on, 94, 95
   spin-on hydraulic, 123
Final drives
   chain drives, 456
   double reduction, 456
   planetary, 455
   safety concerns, 430
Fingertip controls, 596, 596
Fire hazards, 101
   and hydraulic brake systems, 537
First track-type machines, 9
Five-piece rim, 626, 627
Fixed caliper-type of brake, 547
Fixed displacement, 113
Flaring tools, and tube cutting, 51
Flat flywheel, 308

Flow, 110
   with adjustable restriction, 500
   compensation and pressure, 471–472
   compensator and directional control valve, 481
   divider, 527
   meter, 500
Fluid
   axle and final drive, 92
   brake, 92
   cleanliness, 86–87
   cleanliness ISO code, 123, 124
   conditioners, 123–125
     coolers, 125
     filters, 123–124
     heaters, 125, 125
   conductors and fittings, 119–121
   coolers, 199
   disposal/recycling, 84–85
   evacuation filtering, and refilling equipment, 63–64
   hydraulic, 92, 121–123
   leaks, 205, 205–206
   powershift transmission, 92
   sampling, 87–89
   through external gear-type pump, 113
   transfers, 207
   viscosity, 86
   volume calculation and relation to movement, 111
Flushing valves, 512, 526–527, 527
Flux lines movement, 145, 146
Flywheel
   in clutches, 307–308
   flat, 308
   pot-type, 308
Flywheel damper
   and vibration damper, 189–190
Flywheel ring gear, 215
FOPS. See Falling object protection system (FOPS)
Fops, and rops, 75
Force, 108–109
   vs. pressure, 109–110
Forklift hydraulic system
   flow divider, 568, 569
Forwarders, 14, 14
Foundation brakes, 538
   defined, 544
   types of, 545–548
     bladder, 548
     caliper and rotor, 546–547
     expanding shoe and drum, 545, 545
     single and multi-disc foundation brakes, 547–548
Four-stroke cycle, 180–190, 181
Free electron, 141
Frictional brake
   hydraulic systems, 543–544
   use for braking, 543
   use for machine braking, 543
Friction bearings. See Plain bearings
Friction discs, 353
Friction materials, 541
   dust, 537
Friction-type flywheel clutches
   clutch cover, 304
   clutch disc, 304–305
   components, 303–304
   intermediate plate, 307
   mechanical drivetrain, 302–303, 303
   pressure plate, 305, 305, 307
Friction-type foundation brakes, 544–548
Fuel efficiency, 18
Fuel supply systems, 202–204
   low-pressure, 204
Fuel system filters/water separators, 82
Fuel tank drain, 82–83
Fuel water separator, 82
Full-floating axle, 433
Full-power brake system, 551, 551
Full powershift countershaft transmission, 367–368
Fully equipped lube truck, 96
Fuses, 157, 157
   color identification, 158
Fusible links, 158

G

Gaskets, 39–40, 40
Gasoline-powered internal combustion engine, 133

INDEX **699**

Gauges
- analog, 499
- digital, *499*
- electric, 669–670
- electronic, 670–671
- mechanical, *667*, 667–668, *668*
- pressure, 498–500. See also Pressure gauges

Gear pump, 113
Gear reduction starter, *219*
- components, 219
- definition, 217

Gears
- clusters, 334
- face width, 332
- heavy equipment, 326–327
- as levers, 327–328
- lubrication, 334–335
- ratios, 329
- reduction, 329
- Rotex, *327*
- safety concerns, 326
- set of wooden teeth, *326*
- terminologies, 330–335, *331*
- torque *vs.* speed, 328–329

Generator gear drive, *395*

Generators
- components, 398–399
- excited rotor, 399, *399*
- permanent magnet, 399, *399*
- ratings, 400, *400*
- simple AC, *398*
- switched reluctance, 400, *400*
- three-phase AC, *398*

Geroller motor, 524–525, *525*
GMAW (Gas Metal Arc Welding). See MIG (Metal Inert Gas) welding
Gold collar, 21
Gooseneck, 614
Governor, 649
- governor-in, cut-out mode, *639*
GPS (global positioning system), 672, *672*
Graders, 11, *12*
Gradual turn, 598, *599*
Grease, 92–93, *93*
- lines on suspension cylinder, *619*
Grinders, and drill presses, 63–65
Groomers, 16
Ground, 170
- driven pump, 577
Grounded circuit, 169
Ground speed radar sensor, 285
Grouser bar, measurement, *660*

H
Hall-effect sensors, 286
Hall-effect switch, 282
Hammers, 48, *48*
Handholds, and steps, 75
Hand-metering unit, 568, *571*
Hand protection, 7
Hand tools
- proper tool usage, 46
Harvesters, 14, *14*
H-block valve, 682, *683*
HDET checking oil level, *80*
Head protection, 7
Heater, cab, 675–676
- control valve, 676, *676*
- coolant, 676
- diagnosis, 686–687
- shutoff valve, 676, *677*
Heat exchanging system. See Cooling equipment
Heat gun, temperature probe, 500
Heating tools, 62, *63*
Heat loss, 463
Heavy-duty diesel engine block, *189*
Heavy-duty electric starter
- diesel engine, *215*
- starter relay, *233*
Heavy-duty equipment technician (HDET), 21
- benefits of being, 26
- employer looking for in, 25–26
- and environment protection, 27
- factors relating to minimum selection of hand and power tools, 45–46, 52, 53
- field service, 67–68
- future of, 26–27
  - field service supervisor, 26
  - sales representative, 26–27
  - shop foreman, 26
- technical advisor, 26
- technical trainer/educator, 26
- instructor teaching class of future, 26
- machine servicing, 24, *24*
- personality traits of, 25
- personal protective equipment to protect an, 6–8
- physical and mental characteristics of, 24–25
- responsibilities of, 23–24
- role of, 23
- skills upgrading, 18
- value of a safe and efficient, 28
- working area of, 26

Heavy-duty forklifts
- and container handlers, 16–17, *17*
Heavy-duty hydraulic frictional brake systems, 536–556
Heavy-duty shock absorber, 609, *609*
Heavy-duty steering axle arrangement components, *563*
Heavy-duty-type U-joints, 422
Heavy equipment, 8–18
- and bearings, 40
- becoming safe, 1–5
- brakes for, 538
- connecting rod bearings, *188*
- dealership, 18, *19*
- demage, *2*
- and fasteners, 31, *32*
- industry groups, 10
- intake and exhaust system, *192*
- introduction, 1–5
- machine types, 10–17
- machine with tracks, *9*
- minimize risk of injury, 3–5
- reconditioned, 21–23
- repair and safety, 1–17
- repair shop, 25, *25*
- safe work practice, 2

Heel gears, 332
Helical gears, 333
Hex wrenches, 47, *47*
High-drive undercarriage, *648*
High-pressure compensation, 470–471
High-pressure cutoff valve, 468. See also Pressure limiter
High-pressure switches, 684
High-side PWM proportional driver outputs, 287
High stress areas, 606
High-voltage conductors, 401, *401*
High-voltage electric propulsion, 390
High-voltage grounding and bonding, 407
High voltage safety warning, *138*
High-voltage warning symbol, *390*
HMU oil flow, 575, *576*
Hoisting equipment, 57, *57*
Hold-in windings, 218, *224*
Horns/Backup alarms, 163
Horsepower, 180
Hoses, 119, 198, *199*
- clamp, 76
- hookup
  - charging, 620
- hydraulic, 120, *120*
- making equipment, 63
- securely fastened, 120, *120*
- and test fittings, 500
Hoses, air-conditioning system, 684, *685*
Humidity, cab, 684, 686
HVAC (heating, ventilation, and air-conditioning) systems, 675–689. See also Air-conditioning system
- climate control system, 686
- electronic control, 686, *686*
- heaters, 675–676, 686–687
- inspection points, 686
- pressurizer, 675–676
- regular maintenance, 686
- servicing and repairing, 686–689
Hybrid wheel loader, 539
Hydraulically actuated locking differential, 446
Hydraulically driven fan *vs.* belt driven fan, 198
Hydraulic assist steering systems, 564–568
Hydraulic brake systems, 544
- application systems, 548–557
  - accumulator charging valve, 552, *552*
- boosted hydraulic brake systems, 550–551
- brake controls, 549
- full-power hydraulic brake systems, 551, *551*
- non-boosted, 549–550, *550*
- reverse-modulating valves, 552
- brake system integrity, 537–538
- burns, 537
- components, 552–553
- crushing hazards, 537
- environmental concerns, 537
- fire hazards, 537
- friction material dust, 537
- introduction, 537–538
- oil injection, 537
- safety concern, 537
- slips and falls, 537
- trapped potential energy, 537
- used on small forklift, 551

Hydraulic charging systems, 240
Hydraulic components *vs.* electrical components, 140
Hydraulic connectors/fittings, 120, *120*
Hydraulic cylinder, 118
Hydraulic energy, 484
Hydraulic excavator
- powerflow components, 105
Hydraulic filter change, 82
Hydraulic flow meters, 53
Hydraulic fluids, 92, 121–123
- biodegradable, *122*
- change, 82
- requirement of, 121
Hydraulic motor, 118
Hydraulic powershift transmission system, 354–355, *356*
- autoshifting, 376
- clutch actuation, 373
- clutch protection, 376
- components, 369–373
- downshift inhibit, 376
- event-based shifting, 373
- inching pedal, 376–377
- maximum starting gear, 376
- neutral-to-gear, 376
- shuttle shifting, 376
- skip shifting, 376
- speed matching, 376

Hydraulic pullers, 53, *54*
Hydraulic repair, of steering system, 582
Hydraulic reservoirs, 114, *114*
Hydraulic retarder
- definition, 321
- problems, 321
- repair/replacement, 321–322
Hydraulic safety warnings, *460*
Hydraulic schematics, 461
Hydraulic schematic symbols, 127, *128*
- backhoe loader's front lift and tilt circuits, *129*
Hydraulic seals, 120, *121*
Hydraulic shop press, 57
Hydraulic starting systems, 231, *231*
Hydraulic steering pump, 574
Hydraulic systems
- basics, 101–135
- components, 112–125
  - hydraulic reservoirs, 114
  - prime mover, 112
  - pump, 112–114
- efficiency, 460
- environmental concerns, 103
- and environment protection, 103
- evolution, 105–106
- flow, 110–111
- fluid movement, 106
- frictional brake, 543–544
- heat generation, 101
- maintenance, 130
- *vs.* mechanical systems, 104
- need of using, 103–106
- open center *vs.* closed center, 462–464
- open circuit *vs.* closed circuit, 461, *462*
- repairs, 132–133
- slippery, 101
- testing
  - pressure gauges, 498–500
- for track type machine, 591, *592*
- troubleshooting, 130
Hydraulic wrenches, 47, *47*

Hydrostatic drive systems, 12, 505–534
- adjustments and repairs, 533–534
- component repairs, 533–534
- post repair start, 533
- advantages and disadvantages of, 509
- arrangements, 507–508
- axial piston motor, 521
- bent axis motor, 522–523
- calibrated, 533
- circuit operation, 512–515
- components, 516–526
  - axial piston motor, 521
  - bent axis motor, 522
  - cam lobe motor, 523–524
  - case drain circuit, 527–528
  - charge pump, 526
  - filtration, 528, *528*
  - hydrostatic fluid tank, *529*
  - main pumps, 516–520
  - oil cooler, 528–529, *529*
  - orbital motor, 523
  - servo control valves, 520–521
- dual path, 510–511
- hazards, 506
- introduction, 506–507
- motors, 521
- operator controls, 530
- overview, 510
- for track machine, *507*
- troubleshooting, 531–532
- values calculation, 511
- valves, 526–527
  - brake release, 527
  - charge relief, 527
  - combination, 526, *526*
  - flow divider, 527
  - flushing, 526–527
- wheel loader and, *508*
Hydrostatic dual pump arrangement, *509*
- closed loop systems, 509
Hydrostatic electronic control systems, 530
Hydrostatic fluid, 529
- tank, 529
Hydrostatic motor
- and pump, *507*
Hydrostatic system component repair, 533
Hydrostatic transmissions. See Hydrostatic drive systems
Hysteresis, 496

I
Idler gears, 334
Idlers, 650, *651*
IGBT, 403
Ignition excite, 253
Inboard axle hub, *635*
Induction motors, 405–406
Inflating equipment, and tire handling, 56, *56*
Inflation, proper, 630–631
Infrared temperature gun, 50, *52*
Inputs to ECM, 281
Insulation testing and maintenance, 408–409
Intake and exhaust valves, 187
Intake systems, and exhaust system, 191–195
Integrated circuits, 274
Intermediate plate, 307
Intermittent problem, 172–173
Internal gears, 333
Internal leakage, 461
Inverter, 393
Inverter cooling systems, 401–403, *402*
Involute curve, 330, 332
ISO 4406 contamination chart, 87

J
Jacking equipment, 56, *56*
John Deere 310E transmission control valve
- clutch fill oil modulation, 365
- cross-section of, 358, 361
- in forward, 363
- modulation valve, 364
- power flow in first-speed forward, 359
- power flow in third-speed forward, 360
- schematic form, 362
John Deere 400D transmission control system
- block diagram, 380

clutches, *380*
clutch pack assembly, *382*
cross-sectional view, *379*
gear ratios, *380*
schematic hydraulic system, *381*
John Deere backhoes, 497
Joystick, 511
   operator, *530*
   steering controls, 576–577, *577*
   valves, 577
Jumbo drill, 16

K

KAM. *See* Keep alive memory
Keep alive memory (KAM), 277
Keypads, *229*, 283
Kinetic energy, 538–539
Kingpin, 563
Komatsu hybrid excavator, *393*

L

Laptop computer, 166
Large coil spring, *608*
LCD gauge, 671, *671*
LCDs. *See* Liquid crystal displays
Lead-acid battery, 153–156
   components, 154
      electrolyte, 154
      negative plates, 154
      positive plates, 154
Leaf springs
   machine, *607*
Leaking seals, *37*
LED. *See* Light-emitting diode
Leveling systems, 613
Leveling valve, 614, 615, *616*
Level plug, *636*
Levers, 104, 543
Life expectancy
   of machine, 20
Lifting equipment, 57
Light bulb, 143
Light-duty U-joint, 420, *422*
Light-emitting diode (LED), 265–266
Lighter-duty axial, *635*
   piston pumps, 518
Lights, 162, *162*
Limited slip, 441
Limited slip differentials, 443–444, *445*
Line/circuit relief, 478
Lines, air-conditioning system, 684
Lip-type seals, 38–39, *39*
Liquid cooling system, and diesel
   engines, *196*
Liquid crystal displays (LCDs) 288
Liquid filters, 93–94
Load, electrical, 163
Load adjusters, 58–59, *59*
Load check valve, 479, *480*
Load cylinder, 614
Load-sense passage
   control valve section with, *474*
Load-sense valves, 471, 480, *481*
Load-sensing hydraulic steering
   systems, 579
Load-sensing orifice, 575
Load-sensing pressure- compensated
   systems, 470–475, *473*
   high-pressure compensation,
      470–471
   pressure and flow compensation,
      471–472
Load sensing system, 472–474
   in control valve, 472, *473*
Locked down, 615
Locking differentials, 438, 445–446
Lock ring, *635*
Lock up torque converters, 316–318, *317*
Logic gates, 275
Log splitter, 133, *133*
   cylinder, 134, *134*
Long block, 182
   components, 182–190
   connecting rods, 184, *185*
   crankshaft and bearings, 187–188
   cylinder head, 185–186, *186*
   cylinder liners, 184–185, *185*
   defined, 182
   intake and exhaust valves, 187
   piston rings, 184, *184*
   pistons, 182–183, *183*
Low-pressure fuel gauge, mechanical
   gauges, 668, *668*

Low-pressure switches, 684
Low-side/sinking driver outputs, 287
Lubricated chain, and sealed chain,
   653–654
Lubricated track, and sealed
   track, 654
Lubrication systems
   diesel engine, 201
   fuel supply systems, 202–204
   oil filtration, 200
   oil pan, 201
   oil pump, 201, *203*
   pre-lubrication systems, 202
   scavenge pump, 201

M

Machine
   and brakes, 538
   cost to purchase, 18–19
   cycle times chart, *131*
   diagnostic systems, 18
   downtime, 19–20, 32
   efficiency, 18
   emission regulation, 18
   evolution, 18
   filters, 93–96
   fluids, 86–87
   frames and suspension systems,
      604–620
   fuel efficiency, 18
   implement controls, 75
   leaf springs, *607*
   life expectancy of, 20
   main frame, 605, 605–606
   maintenance costs, 19
   operating costs, 19
   operator comfort, 18
   out of service, 22–23, *23*
   repair costs, 19
   in repair shop, 20, *20*
   replacement parts costs, 19
   rubber tracks, *661*
   safety system checks, 74–76
   servicing, 24, *24*
   steering lock installed, *560*
   steering system, 75
   types, 10–17
      articulated trucks, 12, *12*
      attachments, 17, *17*
      compactors, 13, *14*
      container handlers and heavy-duty
         forklifts, 16–17, *17*
      cranes, 11–12, *12*
      drilling machines, 15–16, *16*
      excavators, 11
      forwarders, 13–14, *14*
      graders, 11, *12*
      groomers, 16
      harvesters, 14, *14*
      mining shovels, 17, *17*
      off-highway trucks, 12, *12*
      paving equipment, 13
      pipelayers, 15, *15*
      rubber-tired excavator, 17, *17*
      scooptrams, 15, *16*
      scrapers, 11, *11*
      skidder, 14, *14*
      skid steer loaders, 12, *13*
      telehandlers, 12–13
      track loaders, 15, *16*
      track-type tractor, 10
      tractor loader backhoe,
         13–14, *14*
      trencher, 16
      wheel loaders, 15, *15*
      wheel/track dozers, 15, *15*
   warning, *103*
   wiring harness bulkhead, *159*
Machine charging systems, 240
Machine drivelines
   arrangements, 414–415
   brakes, 419
   critical speed, 419
   operating angles, 415–416
   overview of, 413–414
   phasing, 418–419
   shaft assemblies, 416–418
   support bearings, 420, *421*
   universal joints, 420–422
Machine fluid properties, 90–93
Machine hour meter, 78
Machine lubrication, 84
Machine model designation, 9–10
Machine servicing
   documentation, 85

Machine-specific diagnostic
   reconditioning, and adjusting tools,
   54, *55*
Machine *vs.* hydraulic charging systems,
   240
Magnetism, and electromagnetism,
   145–146
Main control valves, 476–482
Main frame, 605–606
Main relief valve, 478, *478*
Maintenance costs, of machines, 19
Maintenance-free battery, 155
Maintenance manual, and operation
   manual, 73–74, *74*
Makeup/anti-cavitation valve, 479–480
Manual control *vs.* pilot controls, 475
Manual steering box, 562
Manual transmissions
   all-wheel drive option, 340
   asembly precautions, 345–346
   description, 335–340
   diagnostics, 342–343
   maintenance, 342
   multi-countershaft, 340–341
   reconditioning, 344–345
   repairs, 343–344
   safety concerns, 343
   shift mechanism, 340, *340*
Margin pressure, 471
Master cylinder, 544
Master link, 648, 654–656
Master pin
   with cotter pin, *655*
   with dimple, *655*
Master switch, 235
Material safety data sheet (MSDS), 5–6
Maximum starting gear, 376
Measuring tools, 51, *52*, 54, *55*
Mechanical excavator
   powerflow components of, 105
Mechanical gauges, *667*, 667–668
   low-pressure fuel gauge, 668, *668*
   pressure gauges, 667, *667*
   temperature gauges, 668, *668*
Mechanical locking differential, 446, *446*
Mechanical park brake, 544
Mechanical steering systems, 561–573
   checking for wear, 580
Mechanical switches, 161
Mechanical systems
   vs. hydraulic systems, 104
Mechanics, 21
Medium-size rubber-tired machines
   and three-piece rim, 626
Megohmmeter test, 408–409, *409*
Merry-go-round, 516, *518*
Metal oxide semiconductor field-effect
   transistor (MOSFET), 268–269
Metal-to-metal face seals, 39, *40*
Metric wire sizes
   vs. American Wire Gauge (AWG), 158
Micron, 123
MIG (Metal Inert Gas) welding, 67, *67*
Mining shovels, 17, *17*
Mirrors, 76
Modulating valves, 552
Modulation, 355
Monoblock valves, 462, *477*
MOSFET. *See* Metal oxide semiconductor
   field-effect transistor
Motors, 118
   and brake cutaway, *511*
   fixed /variable displacement, 522
   geroller, 524–525, *525*
   gerotor, 525–526
   hydrostatic drive systems, 521
   induction, 405–406
   oil flow between pump and, *506*
   permanent magnet, 406, *406*
   squirrel cage, 405
   switched reluctance, 406–407
   wheel, 405
Mounts, cabs, 674
Multi-countershaft manual transmissions,
   340–341
Multi-disc, 544
   steering brakes, 592–593, *594*
   type clutches, 589
Multi-disc clutches, 303
Multimeter, 164, *165*
   and ammeter, 165–166
   digital, *165*
   and ohmmeter, 166, *166*
   and voltmeter, 164–165
Multi-piece rims, 627

Multiplexing, 289
Multi-position switch, 281–282
Multi-section directional control valves
   cutaway view of, *477*
Multi-section valves, 462. *See also*
   Stacked valves
Murphy Gauge, 669, *669*

N

NAND gate, 275
Needle bearings, 42
Negative cable, 253
Negative plates
   lead-acid battery, 154
Neo-con, 612
Neutralizer valves, 578–579, *579*
Neutral-to-gear (NTG), 376
Neutrons, 141
Nitrogen, 612
Noise prevention, cabs, 674
Noises
   diesel engines, 206
Non-boosted hydraulic brake systems,
   549–550, *550*
Non-lubricated chain, 653
Non-reusable plates and discs, *602*
Nonservo brakes, 545
   expanding shoe and drum brakes,
      546
NOR gate, 275
Normal fastener rotation, 32, *32*
Nose piece, 218
No spin differentials, 444–445, *445*
NOT gate, 275
NTG. *See* Neutral-to-gear
N-type material, 262
Nucleus, 140
Nuts, threaded, 35
Nylon slings/straps, 58–60, *59*
   blocking equipment, 59, *59*
   heating, cutting, and cooling
      equipment, 60
   load adjusters, 58–59 *59*

O

Off-highway trucks, 12, *12*
Ohm, Georg, 144
Ohmmeter, 166, *166*
Ohms, 144, 145
Ohm's law, 145, 408
   for parallel circuit, 149–150
   practical uses, 151
   for series circuits, 148–149, *149*
   for series parallel circuits, 150, *150*
Oil. *See also* Refrigerant
   air-conditioning system, 679, *679*
Oil applied, 588
Oil-applied steering clutches, 591, *591*
Oil-assisted clutch release valve, *590*
Oil-boosted band
   brake mechanism, 592, *593*
Oil-cooled clutch, 309
Oil filter, 200, *528*
Oil flow *vs.* electron flow, 142, *142*
Oil injection, 102, 537
   injury, *102*
Oil movement in strut, *612*
Oil pan, 201
Oil pressure
   abnormal, 207
Oil pump, 201, *203*
Oil sample lab test machine, *88*
Oil sample ports, 88
Oil sample report, *89*
Onboard diagnostics, 166
One-piece rim cross-section, *625*
One-piece wheels. *See* Disc-type wheels
One-way clutches, 311
On/Off high-side driver outputs, 287
On/off solenoid valves, 496
On/Of toggle switch, 281
Open center hydraulic systems, 128–130
   vs. closed center hydraulic system,
      462–464
   schematic machine with, *463*
Open circuit
   vs. closed circuit hydraulic system,
      461, *462*
   and electrical circuit, 169, *169*
   fault, 168
Open circuit voltage (OCV), 156
   vs. state of charge, *156*
Open series circuit, 148
Operating angle, 415

INDEX **701**

Operational steam shovel, 177
Operational water-powered grist mill, 105
Operation manual, 73–74, 74
　and maintenance manual, 73–74, 74
Operator comfort, 18
Operator complaints
　drivelines, 424
　electric drive systems, 409
Operator controls, 530
Operator displays, 666–672. See also Gauges
Operator joystick, 530
Operator keypads, 283
Operator protections, 672–674
Operator seats, 673
Operator stations. See Cab(s)
Orbital motor, 524–526
Orbital valve, 568
OR gate, 275
Orifices, 125–126, 126
　purposes of, 125–126
　variable, 125
Orifice tube system, 678, 678, 683, 683. See also Air-conditioning system
O-ring seals, 37–38, 38
Oscillating axles, 609–610
　TLB front, 610
Outputs from ECM, 286
Over-center-type clutches, 306
Over-crank protection (OCP) switches, 224, 225
Overdrive, 329
Overhead cam design engine, 189
Overhead shop crane, 57
Overheated clutches, 590
Overload warning, 492
Overrunning clutch, 219, 225
Oxy-acetylene setup, 60, 60
Oxy-fuel
　torches, 60, 60–61
　welding, 65

P

Packers. See Compactors
Parallel circuit, 149, 149
　Ohm's law for, 149–150
Parking brake operation, 74, 74
Pascal's law, 106
Passive suspension cylinder, 610–612
Pavement grinder, 13, 13
Payload capacity, 9
Payload scale, 616–617, 618
Permanent magnet generators, 399, 399
Permanent magnet motors, 406, 406
Personality traits of heavy-duty equipment technician (HDET), 25
Personal protective equipment, 6–8
　examples and usage of, 6–8
　　back support, 8
　　extreme weather protection, 8
　　hand protection, 7
　　head protection, 7
　　hearing protection, 7–8, 8
　　respiratory protection, 8, 8
　　safety boots, 7, 7
　　safety glasses, 6–7, 7
Personal tools, 45–51
Pessure vs. force, 109–110
Phasing, 419
Photodiodes, 266
Pictorial schematic, 128, 129
Piezoelectric pressure sensor, 285
Piezo-resistive pressure sensors, 285
Pilot controls, 475–476
　vs. manual control, 475
Pilot control valve
　cross-section, 476
Pilot oil systems, 475
Pinion gears, 218, 334, 433
Pipelayers, 15, 15
Piston, 177
　accumulator, 127, 127
　assembly, 465
　and cylinder block, 464–465, 465
　and long block, 182–183, 183
　pump, 516, 517
　rings, 184, 184
　type seals, 39
　stroke, 179
　travel during shaft rotation, 518, 518
Pitch, 332
Pitch line, 332

Pivot shaft, 648
　location of, 649
Plain bearings, 41, 41
Plain wheel hub cutaway, 634
Planetary final drive, 455
Planetary gearsets, 334
　advantages of, 377
　components, 377, 377
　power flow of, 378
　transmission control system, 378–382
Planetary pinion carrier, 377
Planetary powershift transmission
　definition, 352
　operating procedures, 377–382
Planet carrier, 433
Plasma cutter, 61–62
Plates, 353, 589
Plies, 637
Plugged fuel tank
　breather, 211
Ply rated, 638
Pneumatically actuated locking differential, 446
Pneumatic brake systems, 640–641
　diagnosing, 643
　repairs, 643
　servicing, 643
Pneumatic power tool and electric power tools, 52
Pneumatic systems, 637–643
　components, 637
　and safety measures, 624
　wheels, tires, hubs, and, 623–643
Pole shoes, 218, 224
Portable hydraulic hand pump, 500
Portable particle counter, 89
Port plate
　and cylinder block, 467, 468
　cylinder block for dual output pump, 469
Position sensor, 530
Positive plates
　lead-acid battery, 154
Potentiometer, 284
Pot-type flywheel, 308
Power dividers, 430
　articulated truck, 455
　exploded view, 454
　safety concerns, 430
　working procedures, 453–455
Power flow
　first-speed forward, 359
　five-speed transmission, 338
　older steering clutches and brakes, 588
　planetary gearset, 378
　third-speed reverse, 360
Powershift transmission
　countershaft, 352, 355–366
　description, 349–350
　drivetrain loads, 350
　evolution, 350–351
　fluid, 92
　full countershaft transmission, 367–368
　gear selection, 368–369
　hydraulic system, 354–355, 356, 369–377
　maintenance, 382–383
　planetary, 352
　reconditioning, 386–387
　removal of, 385–386, 386
　safety concerns, 349, 383
　troubleshooting, 383–385
　types of, 352
Power steering box
　cross-section, 565, 567
Power take-off (PTO), 340
Pre-lubrication systems, 202
Pressing equipment, 56–57, 57
Pressure, 108–109
　abnormal oil, 207
　and advanced hydraulic systems, 501
　in bottle jack, 109
　compensator, 482
　control valve, 115, 116
　differential/ margin, 471
　and flow compensation, 471–472
　pump outlet, 489
　reducing valve, 482, 483
　standby, 471
　transferred through liquid, 106, 107
Pressure, trapped, 102
Pressure angle, 332
Pressure compensator valve
　and axial piston pump, 471, 471

Pressure/flow compensator, 471
Pressure gauges, 53, 110, 498–500
　electric, 670, 670
　mechanical, 667, 667
Pressure limiter, 47
Pressure plate, 305, 305, 307
Pressure switches, 684, 685
Prime movers, 112, 112, 175–211, 176
　defined, 176
　warning signs, 176
Priming pumps, and low-pressure fuel system, 204
Primitive braking system, 540
Priority valve, 568
Programmable read-only memory (PROM), 276
PROM. See Programmable read-only memory
Proportional solenoids, 367, 496–497
　stacked valve with, 497
Protons, 141
Pry bars, 48, 48
PSIG. See PSI gauge
PSI gauge, 499
PTO. See Power take-off
Pullers, and drivers, 49, 50
Pull-in windings, 218, 224
Pulse width modulation (PWM)
　definition, 279–280
　position sensor, 285
Pump, 106, 112–114, 551
　assembly for a medium-size excavator, 586
　axial piston, 467–469, 516, 517
　control valves, 469, 486, 488
　　three states of, 487
　　constant flow, 487
　　destroking, 487
　　upstroking, 487
　destroke, 469
　displacement control valve, 521
　displacement measurement, 464
　drive, 508
　exploded view of assembly, 467
　oil flow between motor and, 506
　outlet pressure, 489
　output oil from main control valve, 488–489
　piston, 516, 517
　portable hydraulic hand, 500
　with torque control, 472
　variable displacement bidirectional, 516
　variable displacement pump charge, 470
Punches and chisels, 48–49, 49
Pup engine starter motor, 215
Push button switch, 281
PWM. See Pulse width modulation

R

Rack gears, 333
Radar-warning systems, 673
Radial tires, 638
Radiator, 197
　clean, 198
RAM. See Random-access memory
Random-access memory (RAM), 277
Ratchets, 47, 47–48
Ratio, 329
Read-only memory (ROM), 275–276
Rear axle
　oscillation, 610
　trunnion mounting, 610
Rear steering machine, 560, 561
Receiver drier, 684, 684
Reciprocating pistons, 212
Recoil spring, 651
　being tensioned, 651
　from small dozer, 646
Rectification
　definition, 246
　vs. inversion, 248
Rectifier bridge, 247, 247–248
Reed switch, 282
Reference voltage, 280
Reflective devices, 75
Refrigerant, 677, 678–679. See also Air-conditioning system
　identifier, 687, 687
　leak repair, 688–689
Regeneration circuits, 490–491, 491
Regulator, 241, 612

Relays, 161, 162, 162
　testing, 170–171, 172
　valves, 553
Relay (R) terminal, 253
Release bearing, 307, 307
Remote voltage sense, 253
Rental replacement cost, 20, 20
Repair costs machine, 19
Repair shop
　machine in, 20, 20, 25
Repair tools
　and electrical test, 50, 51
　fastener, 50, 51
Replacement parts costs machine, 19
Reservoir, 114
Resistance, 144
　to current flow, 168
　and electrical circuit, 143, 144
Resistance, electrical, 143–144
Resistive-type brake resistor, 408
Resistors
　carbon, 270
　for circuit board, 270
　color code chart, 270
　definition, 269
　de-spike a coil, 271
　simple, 270
　stepped, 271
Resolver, 403, 404
Respiratory protection, 8, 8
Retainers, 36–37, 37
Retarder, 321
Reverse idler, 337
Reverse-modulating valves, 552
Reverser countershaft powershift transmission, 357–366
Ride control, 613–614
　accumulators used for, 485
　schematic, 614
Rigging equipment, 57
Ring gear, 377, 433
Rivets, 32, 32
Rock ejectors, 631, 632
Rocker switch, 281
Rock truck
　going downhill, 538
　hour meter, 22
Roller
　measuring, 660
　track and carrier, 660
Roller bearings, 42, 43
　tapered, 42–44, 43
　adjustment procedure, 43, 44
Rolling torque, 635
　shim pack thickness measurement, 636
Rollover protection systems (ROPS), 672, 673
ROM. See Read-only memory
ROPS. See Rollover protection systems (ROPS)
Rops and fops, 75
Rosebud heating tips, 61, 62
Rotary actuators (Motors), 118–119, 119
Rotary flow, 312, 313
Rotary steering valve operation, 574–576
Rotary switch, 281
Rotating bushing chain, 654
Rotex gears, 327
Rotor, 243, 398
　brake, 554, 555
　brakes, 546–547
　and dual caliper, 542
Rubber tracks, 661
　machine, 661
　removing and installing, 661

S

Saddle-type
　swash plates, 518, 519
Safety boots, 7, 7
Safety concerns
　cabs, 665
　charging systems, 240
　clutches, 302
　diesel engine starting systems, 215
　drive axles, 430
　drivelines, 413
　electronic devices, 260
　final drives, 430
　gears, 326
　high-voltage electric propulsion, 390
　manual transmissions, 343
　power dividers, 430

powershift transmission, 349, 383
torque converters, 302
torque dividers, 302
torque retarders, 302
Safety glasses, 6–7, 7
Safety signs/decals/plates, 76, 76
Safety system checks, machine, 74–76
  backup and vision cameras, 76
  backup/travel alarm, 74
  cab escape tools or mechanisms, 75
  door and window latches and handles, 75
  engine governor control, 76
  fire suppression equipment, 75
  guarding to covers moving parts or hydraulic lines, 75
  implement controls, 75
  machine steering system, 75
  mirrors, 76
  parking brake operation, 74, 74
  rops and fops, 75
  safety signs/decals/plates, 76, 76
  seat belt, 75
  seat condition and adjustments, 75
  service brake operation, 74
  starting system interlock, 74
  steering or boom locks, 75, 75
  steps and handholds, 75
  windshields/windows, 74
  work lights, reflective devices, warning lights, and cab lights, 75
Safety warning
  for batteries, 138
  for high voltage, 138
Sales representative, 26–27
S-cam foundation brakes, 641–642, 642
Scavenge pump, 201
Schematic machine
  for active suspension, 613
  with an open center hydraulic system, 463
  of brake system, 640
Schrader valve, 626
Scooptrams, 15, 16
Scrapers, 11, 11
  cushion hitch for, 615
Screwdrivers, 47, 48
Sealed chain and lubricated chain, 653–654
Sealed track, 654
  lubricated track, 654
Sealed track link
  lubricated track link, 655
Seal rings, 374
Seals, 37–39, 120
  hydraulic, 120
  leaking, 37
  lip-type, 38–39, 39
  metal-to-metal face, 39
  O-ring, D-ring, square-ring, 37–38, 38
  picks, 51, 52
  piston ring-type, 39
Seat belts, 75, 673–674, 674
Seat condition, and adjustments, 75
Seats/seating, operator, 673
Secondary steering, 577, 579
Securely fastened hose, 120, 120
Seering valve disassembled, 575
Self-energizing, 546
Self-exciting alternator, 244
Semiconductors, 262
Semi-floating axle, 433
Sensor power supply outputs, 287
Sensors
  and electrohydraulic systems, 494–495
  ground speed radar, 285
  Hall-effect, 286
  piezoelectric pressure, 285
  piezo-resistive pressure, 285
  PWM position, 285
  thermistor temperature, 283–284
  thermocouple, 286
  variable capacitance pressure, 285
  variable reluctance, 283
  variable resistance, 283–284
Sequence valve, 482, 484
Series circuits, 147–149, 148
  Ohm's law for, 148–149, 149
  open, 148
Series-parallel circuits, 150, 150
  Ohm's law, 150, 150
Serrated master link, 655

Service and parking chamber of brakes, 641
Service brakes, 651
  operation, 74
Service intervals, 78–79
  chart, 79
  determination, 76–78
Servicing machine, 71–99
  belly pans and heavy guards, 72
  climbing/descending, 73
  disable, 72, 72
  machine, 72, 72
  fall arrest, 72
  in the field, 96
  hazardous materials and, 73
  hot fluids and components, 72
  ladders/steps to, 73
  machine component mechanical locks, 72, 73
  moving parts, 73
  operation and maintenance manual, 73–74, 74
  procedures, 80–84
    axle filters, 83
    axle or final drive oil change, 83
    battery maintenance, 84
    cab filter change, 84
    coolant change, 83
    DPF (diesel particulate filter) cleaning, 84
    engine air filters/filtration devices, 80
    engine oil change, 80
    engine oil filters, 81
    filter change, 83
    fuel system filters/water separators, 82
    fuel tank drain, 82–83
    hydraulic filter change, 82
    hydraulic fluid change, 82
    machine lubrication, 84
    transmission filter change, 82
    transmission oil change, 81–82
  slips and falls, 72
  stored fluid power energy, 72
Servo, 545
  control valves, 520–521
Servo type
  expanding shoe and drum brakes, 546
Set point, 252
Shielded metal arc welding (SMAW). See Stick welding
Shift lever, 219
Shock absorbers, 609
  heavy-duty, 609
Shoe clutches, 311
Shop equipment, 56–58
  cable assemblies, 58, 58
  chains, 57–58, 58
  hoisting/lifting equipment, 57, 57
  jacking equipment, 56, 56
  pressing equipment, 56–57, 57
  rigging equipment, 57
  tire handling and inflating equipment, 56, 56
Shop foreman, 26
Short circuit, 169, 170, 170
Short to ground circuit. See Grounded circuit
Shutoff valve, cab heater, 676, 677
Shuttle shift countershaft powershift transmission, 357–366
Shuttle shifting, 376
Side gear, 441
Signal conditioning, 273–274
Silicon, 262
Simple AC generators, 398
Simple homemade electric motor, 221
Simple resistors, 270
Single and multi-disc foundation, brakes, 547, 547–548
Single-phase AC voltage generation, 397
Single-stage single-phase torque converter, 315
Skidder, 14, 14
Skid steer
  Deere CT 315, 512, 513
  dual path hydrostatic arrangement, 512
  loader, 12, 13, 139, 139
  loader with simple hydraulic system, 460
Skip shifting, 376

Slack adjusters, 552–553, 652
Slipping clutch, 303
Slips and falls, and hydraulic brake systems, 537
Small dozer, recoil spring, 646
Smoke exhaust emissions
  blue exhaution, 207
  excessive visual, 206–207, 207
  white exhaution, 207
Smoothing capacitors, 248
Snap ring pliers, 49, 49
Sockets, 47, 47–48
Soft start starters, 224
Solenoid
  assembly, 223–224
  proportional, 367
Solenoid controlled engine speed sense oil, 489–490
Solenoids, 161–162, 162
  testing, 171–172
Special tools, 52–55
  dowel pullers/stud removers, 52
  hydraulic flow meters, 53
  hydraulic pullers, 53, 54
  machine-specific diagnostic, reconditioning, and adjusting tools, 54, 55
  measuring tools, 54, 55
  pressure gauges, 53
  surface reconditioning tools, 52–53
  tachometers, 53
  torque multiplier, 53–54, 54
Speed and weight affect stopping distance, 540
Spindle, 644
Spin-on filters, 94, 95
Spin-on hydraulic filter, 123
Spin-on oil filter, 81
Spline, 417
Splined slip joint, 425, 425
Spongy, 107
Spool-type flow divider, 483
Spool valve, 521
Spot turn, 512
Sprag clutches, 311
Spring applied, 588
Spring-applied steering clutches, 589, 589–590
Spring rate, 612
Spring shackle, 607
Springs, 607–609
  leaf, 607
Sprocket, 651–652, 660
  loose/missing bolts, 653
Spur gears, 333
Square-ring seals, 37–38, 38
Squirrel cage fan motor, 675, 675
Squirrel cage motor, 405
SRMs. See Switched reluctance motors
Stacked valves, 477, 478
Standby pressure, 471
Starter drives, 225
Starter housing, 217
Starter interlock, 232
Starter relay, 223
Starting circuit relays, 232–233
Starting control circuit
  electric motor assembly, 228–229, 229
  interlock systems, 232
  voltage drops, 234
Starting system interlock, 74
State of charge vs. open circuit voltage (OCV), 156
Static braking, 538
Stator, 244–246, 396
Steel cords, 628
Steerable drive axles, 436–437
Steer/drive axle, 564
Steering box, with hydraulic assist, 565, 566
Steering brakes
  band types, 591–592
  diagnostics, 600–601
Steering clutches
  brake diagnostics, 600–601
  and brake reconditioning, 602
  for steering control, 594–596
Steering controls, for steering clutches and brakes, 594–596
Steering cylinders, 577
  leakage check, 581
  sand tie rod arrangement, 569, 572
Steering knuckle, 564, 564
  strut, 611

Steering lever, 594
Steering motor, 597
Steering or boom locks, 75, 75
Steering pedal, 600
Steering planetary, 597
Steering system
  adjustments/calibrations, 580, 601
  all-wheel, 570–571
    mode, 570, 573
  articulated, 569–570, 573
  diagnostics, 580, 600
  electrical repair, 582
  hydraulic assist, 564–568
  hydraulic repair, 582
  load-sensing hydraulic, 579
  mechanical, 561–573
    repair, 582
  oeration, 573–582
  repair, 581, 601–602
  track-type machine, 585–602
  two-speed, 593–594, 594
Stepped resistor, 271
Steps and handholds, 75
Stick welding, 65–66, 66
Stiction, 496
Stopwatch, 500
Straight cut gears, 333
Straight travel function, 491
Strut, 610
  active, 612–613
  change repair, 620
  oil movement in, 612
  recharge procedure, 619–620
  with steering knuckle, 611
Stud removers. See Dowel pullers
Sun gear, 377, 433
Surface reconditioning tools, 52–53
Suspension cylinders, 610
  passive, 610–612
Suspension systems, 606–607
  diagnosing, 619
  electronic controls for, 616
  repairing, 620
  servicing, 619
  track machine, 617–620
Swash plates, 467
  saddle-type, 518, 519
  trunnions, 518
Swept area, brake, 542
Swing bearing, 605, 606
Swing circuits, and travel circuits, 492
Switched reluctance generators, 400, 400
Switched reluctance motors (SRMs), 406–407
Switches, 159–160, 161, 684, 685
  conventional, 160
  and electrohydraulic systems, 494–495
  engine starter, 160
  Hall-effect, 282
  high-pressure, 684
  inputs, 281
  low-pressure, 684
  mechanical, 161
  multi-position, 281–282
  other uses, 282–283
  reed, 282
  testing, 170, 171
  types, 281–283
Synchronizer, 337, 339

T

Tachometers, 53, 500
Tapered roller bearings, 4
  adjustment procedure
Technical advisor, 26
Technical trainer/educat
Telehandlers, 12–13, 13
Telematic systems, 671
Temperature, probe an
Temperature gauges
  electric, 670
  mechanical, 668
Tensile strength, 33
  and bolt grades,
Terminating resisto
Test fittings, and h
Test ports, 532
Thermal expansio
  678, 678, 68
  See also Air-
  system
Thermal over-crank

thermistor temperature sensors, 283–284
Thermocouple, 286
Thermometer in air vent, 687, 687
Thread crest, 33
Threaded fastener, typical, 33, 33
   classification, 33–36
      corroded fasteners, 35–36
      tensile strength, 33–35
      tensil straingth, 35
   inspection, 36
   making threads, 36
   retaining/locating devices, 36–37
   sizing, 34
   thread repair, 36
   washers, 36
Threaded fastener sizing gauge, 35
Threaded nuts, 35, 35
Thread pitch, 33
Thread repair, 36
Three-pedal arrangement, 549
Three phase, 397
Three-phase AC generators, 398
Three-phase AC voltage generation, 396–397
Three-piece rim, 627
   and medium-size rubber-tired machines, 626
Tier 3 emission regulations
   Exhaust gas recirculation (EGR) technology, 194
Tie rod, removal of, 582
Tilting, cabs, 674, 674
Timed gears, 334
Tires, 627–633
   chains, 632, 633
   dual, 631
   handling, 631–632, 632
   inflating equipment, 56, 56
   inflation, 630
   inflator, 631
   pressure monitor, 631
   repair, 632
   sizing, 629–630, 630
   underinflation, 630
TLB front axle oscillating, 610
Toe angle, 567, 568
Toe gears, 333
Toe-in, 567
Torque, 118, 178, 540
   control and pump, 472
   crankshaft throw equals more, 180
   motor's output shaft, 510
   multiplier, 33, 53–54, 54
   output, 118
   rod, 616, 617
   rolling, 635
   turn, 36
   wrenches, 36, 49–50
Torque converters
   components, 313
   configurations of, 315–316
   freewheeling stator, 316
   lock up, 316–318, 317
   machine's drivetrain, 314
   oil cooling system, 318, 319
   oil system, 318
   operating mechanism, 311–315
   problems, 318–319
   safety concerns, 302
   two-piece turbines, 318
   variable capacity, 317
   variable impellers, 316
   variable vane stator, 318
Torque ..., 302, 319–321, 320
Torq...
... 47, 47
... 650–651, 651
... 60
...
... 648–652
... and idler, 650
... 648

Track machine
   mechanical steering systems, 587–602
   suspension systems, 617–620
   turns, 586, 587
Track pads, 656–657
   chain with, 657
   fastener, 656
   variety, 657
Track pitch, 653
   measure, 661
Track rollers
   bogie systems, 617–618
   guards, 650, 650
   single- and double-flange rollers, 649
Track tension, 646
   measurement, 652
Track-type machine
   introduction, 586
   steering systems, 585–602
Tractor loader backhoe, 13–14, 14
Trailing arm suspension, 611
Transducer, and digital gauge, 499
Transfer case, 413
   all-wheel drive machine, 422
   auxiliary drives repair, 426–427
   cutaway view of, 423
   differential lock, 423
   operating procedures, 422–423
Transistor
   bipolar, 266–267
   Darlington pair, 268
   definition, 266
   drivers, 268
   field-effect, 266
   MOSFET, 268–269
   operations, 267–268
   symbols, 267
   terminology, 268
   testing, 269
Transmission filter change, 82
Transmission oil change, 81–82
Trapped pressure, 102
Travel circuits
   and swing circuits, 492
Tread designs, 628, 629
Trencher, 16
Troubleshooting, starting systems
   battery systems, 233–235
   simple stuff first, 233
   slow cranking, 235–236
   verifying complaint, 233
Trunnion, 420
   swash plates, 518
Tube, 120
   cutting and flaring tools, 51
Tubeless tire/wheel, 635
   valve stems, 626
Turbocharger technology, 193
   with variable geometry technology, 194
24V battery pack, 241
Two-speed differentials, 447, 447
Two-speed steering, 593–594, 594
   diagnostics, 601
TXV. See Thermal expansion valve (TXV)

U

Ultracapacitors, 272
Undercarriage systems, 645–661
   components, 647
      measured, 659–660
   configurations, 647–648
   diagnosing problems of, 661
   high-drive, 648
   introduction, 646
   measuring, 659–661
   tools, 659, 659
   wear, 658
Underdrive, 329
Underinflation, of tires, 630
Universal joints, 415
   heavy-duty-type, 420, 422
   light-duty-type, 420, 422
   operating angles, 416
   wing-type, 422

Unloader valves
   pump, 464
Upstroking state, 471, 487
Used oil tank, 85
Utility tractor, 507

V

Valence shells, 141
   atom with full, 141
Valves
   accumulator charging, 552, 552
   brake pedal, 543
   cartridge, 493–494
   charging, 551
   clearance, 208
   combination, 479, 512, 526, 526
   combiner, 490
   counterbalance, 478–479
   crossover relief, 576
   deere/husco-distributed, 497
   drivetrain, 188–189
   exploded view of actual, 488
   flushing, 512
   H-block valve, 682, 683
   heater control valve, 676, 676
   heater shutoff valve, 676, 677
   high-pressure cutoff, 468
   hydrostatic drive systems, 526–527
   joystick steering control, 577
   leveling, 614
   load check, 479, 480
   load-sense, 480
   main relief, 478, 478
   makeup/anti-cavitation, 479–480
   manually controlled servo control, 520
   modulating, 552
   monoblock, 462
   monoblock type of, 477
   multi-section, 462
   neutralizer, 578–579, 579
   oil-assisted clutch release, 590
   on/off solenoid, 496
   orbital, 568
   pressure reducing, 482
   priority, 568
   pump and unloader, 464
   pump control, 469, 486
   pump displacement control, 521
   reverse-modulating, 552
   schrader, 626
   section with load-sense shuttle valve, 481
   sequence, 482, 484
   servo control, 520, 520–521
   spool, 521
   stacked, 477, 478
Valve stems, 635
Variable capacitance pressure sensors, 285
Variable capacity torque converter. See Variable impellers
Variable displacement bidirectional pump, 516
Variable displacement pumps, 464–470
   charge pumps, 470
   controls, 469
Variable frequency, 279
Variable impellers, 316
Variable orifice, 126, 126
Variable reluctance sensors, 283
Variable resistance, 279
Variable resistance sensors
   potentiometer, 284
   thermistor temperature sensors, 283–284
Variable vane stator, 318
Variable voltage, 279
Vehicle's electrical system, 252
Vibrations
   damper and flywheel damper, 189–190
   diesel engines, 206
Viscosity, fluid, 86, 121
   chart, 86, 122
   test device, 122

Vision camaras, and backup cameras, 76
Voltage, 142
   drop in circuit, 168
   measurement, 148
   open circuit, 156
   and voltmeter, 143
Voltage regulator
   classification, 248
   engine cranking, 250
   generating electricity, 250–251
   operation, 249–250
   set point, 252
   shutoff, 251
   types, 249
Voltmeter, 164–165
   and voltage, 143
Volume, 110, 552
Vortex flow, 312

W

Walk-around inspections. See Circle checks
Walking beam suspension, 617
Warning
   fire, 103
   hydraulic safety, 460
   labels, 4, 4
   lights, 75
   machine, 103
   overload, 492
   signs near prime movers, 176
Washers, 36
Waste oil storage container, 27, 85
Water pump, 197, 197
Watt's law, 145
Weight and speed affect stopping distance, 540
Welding machines, 65
   oxy-fuel, 65
Wet clutch, 309
Wheeled machine steering systems, 559–582
Wheel loader drivelines, 413
Wheel motor, 405
Wheels, 624–627
   cylinder, 545, 545
   dozers, 15, 15
   hubs, 633–636, 634
      servicing, 635–636
   loaders, 15, 15
   and hydrostatic drive systems, 508
      servicing, 97
   studs, 636
Windings, 397
Windshields/windows, 74
Wing-type U-joints, 422
Wireless communication, 295–296. See also Telematic systems
Wiring
   machine wiring harness bulkhead, 159
   repair, 170, 170
Wiring diagnostics, 297
Wiring harness, and connectors, 159, 159
Wiring schematics, and symbols, 151, 152
Wiring system, 280
Wishbone track frame, 648
Work lights, 75
Work/power modes, 491–492
Worn chains, 658
Wrenches, 46, 46–47
   hex, 47, 47
   hydraulic, 47, 47
   torque, 36, 49–50

Y

Yoke, 337, 418

Z

Zener diode, 249, 265